T0392456

Advanced Drug Delivery Systems in the Management of Cancer

Advanced Drug Delivery Systems in the Management of Cancer

Edited by

Kamal Dua
Discipline of Pharmacy, Graduate School of Health, University of Technology Sydney, Sydney, NSW, Australia

Meenu Mehta
Discipline of Pharmacy, Graduate School of Health, University of Technology Sydney, Sydney, NSW, Australia

Terezinha de Jesus Andreoli Pinto
University of Sao Paulo, Sao Paulo, Brazil

Lisa G. Pont
Discipline of Pharmacy, Graduate School of Health, University of Technology Sydney, Sydney, NSW, Australia

Kylie A. Williams
Discipline of Pharmacy, Graduate School of Health, University of Technology Sydney, Sydney, NSW, Australia

Michael J. Rathbone
ULTI Pharmaceuticals, Turangi, New Zealand

Academic Press is an imprint of Elsevier
125 London Wall, London EC2Y 5AS, United Kingdom
525 B Street, Suite 1650, San Diego, CA 92101, United States
50 Hampshire Street, 5th Floor, Cambridge, MA 02139, United States
The Boulevard, Langford Lane, Kidlington, Oxford OX5 1GB, United Kingdom

Notices

Knowledge and best practice in this field are constantly changing. As new research and experience broaden our understanding, changes in research methods, professional practices, or medical treatment may become necessary.

Practitioners and researchers must always rely on their own experience and knowledge in evaluating and using any information, methods, compounds, or experiments described herein. In using such information or methods they should be mindful of their own safety and the safety of others, including parties for whom they have a professional responsibility.

To the fullest extent of the law, neither the Publisher nor the authors, contributors, or editors, assume any liability for any injury and/or damage to persons or property as a matter of products liability, negligence or otherwise, or from any use or operation of any methods, products, instructions, or ideas contained in the material herein.

Library of Congress Cataloging-in-Publication Data
A catalog record for this book is available from the Library of Congress

British Library Cataloguing-in-Publication Data
A catalogue record for this book is available from the British Library

ISBN 978-0-323-85503-7

Publisher: Andre Gerhard Wolff
Acquisitions Editor: Erin Hill-Parks
Editorial Project Manager: Mona Zahir
Production Project Manager: Niranjan Bhaskaran
Cover Designer: Greg Harris

Typeset by SPi Global, India

Dedication

The publication of this book was finalized during the coronavirus (COVID-19) pandemic. We would like to dedicate this book to all those who were affected by the pandemic and, in particular, to our health workforce around the world for their dedication and care during this difficult time.

Contents

5. Target drug delivery in cancer

Sangita Saini

6. Material and strategies used in oncology drug delivery

Nitin Verma, Komal Thapa, and Kamal Dua

7. Hydrogel-based drug delivery systems for cancer therapy

Brahmeshwar Mishra and Juhi Singh

8. Recent advances in drug formulation development for targeting lung cancer

Charles Gnanaraj, Ching-Yee Loo, Faizan Naeem Razali, and Wing-Hin Lee

9. Advanced drug delivery systems in lung cancer

Anil Philip and Betty Annie Samuel

10. Advanced drug delivery systems in breast cancer

Samipta Singh, Priya Singh, Nidhi Mishra, Priyanka Maurya, Neelu Singh, Raquibun Nisha, and Shubhini A. Saraf

11. Advanced drug delivery systems in the treatment of ovarian cancer

Santwana Padhi and Anindita Behera

12. Advanced drug delivery systems in blood cancer

Ashish Garg, Sweta Garg, Neeraj Mishra, Sreenivas Enaganti, and Ajay Shukla

Contributors

Numbers in parentheses indicate the pages on which the authors' contributions begin.

Muzaffar Abbas (207), Faculty of Pharmacy, Capital University of Science & Technology, Islamabad, Pakistan

Nasir Abbas (207), Punjab University College of Pharmacy, University of Punjab, Allama Iqbal Campus, Lahore, Pakistan

Alaa A.A. Aljabali (351,447), Faculty of Pharmacy, Department of Pharmaceutics and Pharmaceutical Technology, Yarmouk University, Irbid, Jordan; School of Pharmacy, International Medical University (IMU), Kuala Lumpur, Malaysia

Walhan Alshaer (447), Cell Therapy Center, The University of Jordan, Amman, Jordan

Nizar A. Al-Shar'i (267), Department of Medicinal Chemistry and Pharmacognosy, Faculty of Pharmacy, Jordan University of Science and Technology, Irbid, Jordan

Krishnan Anand (243,373), Department of Chemical Pathology, School of Pathology, Faculty of Health Sciences and National Health Laboratory Service, University of the Free State, Bloemfontein, South Africa

Roy Anitha (217), Department of Pharmacology, Biomedical Research Unit and Laboratory Animal Centre, Saveetha Dental College and Hospitals, Saveetha Institute of Medical and Technical Sciences, Chennai, TN, India

Alice Ao (9), Pharmacy Department, Royal North Shore Hospital, Sydney, NSW, Australia

Vimal Arora (243), University Institute of Pharma Sciences, Chandigarh University, Mohali, Punjab, India

Vijaya Anand Arumugam (373), Department of Human Genetics and Molecular Biology, Bharathiar University, Coimbatore, TN, India

Balamuralikrishnan Balasubramanian (373), Department of Food Science and Biotechnology, College of Life Science, Sejong University, Seoul, South Korea

Ratnali Bania (267), Pratiksha Institute of Pharmaceutical Sciences, Guwahati, Assam, India

Seema Bansal (17), Department of Pharmacology, PGIMER, Chandigarh, India

Anindita Behera (127), School of Pharmaceutical Sciences, Siksha 'O' Anusandhan Deemed to be University, Bhubaneswar, Odisha, India

Helen Benson (9), Discipline of Pharmacy, Graduate School of Health, University of Technology Sydney, Ultimo, NSW, Australia

Sanjay Kumar Bharti (409), Institute of Pharmaceutical Sciences, Guru Ghasidas Vishwavidyalaya (A Central University), Bilaspur, CG, India

Amit Bhatia (351), Department of Pharmaceutical Sciences and Technology, Maharaja Ranjit Singh Punjab Technical University, Punjab, India

Shvetank Bhatt (225), Amity Institute of Pharmacy, Amity University Madhya Pradesh (AUMP), Gwalior, Madhya Pradesh, India

Pobitra Borah (267), Pratiksha Institute of Pharmaceutical Sciences, Guwahati, Assam; School of Pharmacy, Graphic Era Hill University, Dehradun, Uttarakhand, India

Victoria Garcia Cardenas (9), Discipline of Pharmacy, Graduate School of Health, University of Technology Sydney, Ultimo, NSW, Australia

Natalia Neto Pereira Cerize (331), Bionanomanufacturing Center, Institute for Technological Research, São Paulo, Brazil

Yinghan Chan (281,431), School of Pharmacy, International Medical University (IMU), Bukit Jalil, Kuala Lumpur; Department of Pharmacology, Faculty of Medicine, University of Malaya, Kuala Lumpur, Malaysia

Balakumar Chandrasekaran (373), Faculty of Pharmacy, Philadelphia University, Amman, Jordan; Faculty of Pharmaceutical Sciences, Block-G, UCSI University, Kuala Lumpur, Malaysia

Nitin Bharat Charbe (447), Departamento de Quimica Orgánica, Facultad de Química y de Farmacia, Pontificia Universidad Católica de Chile, Santiago, Chile

Shashank Chaturvedi (299), Institute of Pharmaceutical Research, GLA University, Mathura, Uttar Pradesh, India

Pooja A. Chawla (343), Department of Pharmaceutical Chemistry and Analysis, ISF College of Pharmacy, Moga, Punjab, India

Viney Chawla (343), University Institute of Pharmaceutical Sciences and Research, Baba Farid University of Health Sciences, Faridkot, Punjab, India

Dinesh Kumar Chellappan (235,243,281,351,373,447), Department of Life Sciences, School of Pharmacy, International Medical University (IMU), Kuala Lumpur, Malaysia

Kiran Kumar Chereddy (431), Cell and Gene Therapy Development and Manufacturing, Lonza Pharma and Biotech, Houston, TX, United States

Shruti Chopra (351), Amity Institute of Pharmacy, Amity University, Noida, Uttar Pradesh, India

Rachelle L. Cutler (9), Discipline of Pharmacy, Graduate School of Health, University of Technology Sydney, Ultimo, NSW, Australia

Lina A. Dahabiyeh (267), Department of Pharmaceutical Sciences, School of Pharmacy, The University of Jordan, Amman, Jordan

Rajiv Dahiya (467), School of Pharmacy, Faculty of Medical Sciences, The University of the West Indies, St. Augustine, Trinidad and Tobago

Sunita Dahiya (467), Department of Pharmaceutical Sciences, School of Pharmacy, Medical Sciences Campus, University of Puerto Rico, San Juan, PR, United States

Joydeep Das (499), Advance School of Chemical Sciences, Shoolini University, Solan, HP, India

Terezinha de Jesus Andreoli Pinto (331), School of Pharmaceutical Sciences, University of São Paulo, São Paulo, Brazil

Pran Kishore Deb (267), Department of Pharmaceutical Sciences, Faculty of Pharmacy, Philadelphia University, Amman, Jordan

Satyendra Deka (267), Pratiksha Institute of Pharmaceutical Sciences, Guwahati, Assam, India

Anju Dhiman (343), Department of Pharmaceutical Sciences, M.D. University, Rohtak, Haryana, India

Kamal Dua (1,9,27,47,235,281,331,351,373,447), Discipline of Pharmacy, Graduate School of Health, University of Technology Sydney, Ultimo; Priority Research Centre for Healthy Lungs, Hunter Medical Research Institute (HMRI) & School of Biomedical Sciences and Pharmacy, University of Newcastle, Callaghan, NSW, Australia; School of Pharmaceutical Sciences, Shoolini University of Biotechnology and Management Sciences, Solan, Himachal Pradesh, India

Harish Dureja (243,373), Department of Pharmaceutical Sciences, Maharshi Dayanand University, Rohtak, Haryana, India

Sreenivas Enaganti (141), Bioinformatics Division, Averin Biotech Pvt. Ltd., Nallakunta, Hyderabad, TG, India

Devaraj Ezhilarasan (217), Department of Pharmacology, Biomedical Research Unit and Laboratory Animal Centre, Saveetha Dental College and Hospitals, Saveetha Institute of Medical and Technical Sciences, Chennai, TN, India

Valker Araujo Feitosa (331), Bionanomanufacturing Center, Institute for Technological Research; School of Pharmaceutical Sciences, University of São Paulo, São Paulo, Brazil

Ireen Femeela (373), Department of Human Genetics and Molecular Biology, Bharathiar University, Coimbatore, TN, India

Gasper Fernandes (415), Department of Pharmaceutics, Manipal College of Pharmaceutical Sciences, Manipal Academy of Higher Education, Manipal, KA, India

Lakshmi Pallavi Ganipineni (431), Advanced Drug Delivery and Biomaterials, Louvain Drug Research Institute (LDRI), Université catholique de Louvain, Brussels, Belgium

Ashish Garg (141), Department of P.G. Studies and Research in Chemistry and Pharmacy, Rani Durgavati University, Jabalpur, MP, India

Sweta Garg (141), Shri Ram Institute of Pharmacy, Jabalpur, MP, India

Rupesh K. Gautam (387), Department of Pharmacology, MM School of Pharmacy, Maharishi Markandeshwar University, Sadopur-Ambala, Haryana, Indiia

Ritu Gilhotra (1), School of Pharmacy, Suresh Gyan Vihar University, Jagatpura, Jaipur, India

Charles Gnanaraj (75), Faculty of Pharmacy and Health Sciences, Royal College of Medicine Perak, Universiti Kuala Lumpur (RCMP UniKL), Ipoh, Perak, Malaysia

Manoj Goyal (267), Department of Anesthesia Technology, College of Applied Medical Sciences in Jubail, Imam Abdulrahman bin Faisal University, Dammam, Saudi Arabia

Monica Gulati (373), School of Pharmaceutical Sciences, Lovely Professional University, Phagwara, Punjab, India,

Gaurav Gupta (1,351,373), School of Pharmacy, Suresh Gyan Vihar University, Jagatpura, Jaipur, India

Mehra Hagi (9), Discipline of Pharmacy, Graduate School of Health, University of Technology Sydney, Ultimo, NSW, Australia

Khalid Hussain (207), Punjab University College of Pharmacy, University of Punjab, Allama Iqbal Campus, Lahore, Pakistan

Salman Hussain (1), Department of Pharmaceutical Medicine, School of Pharmaceutical Education and Research, Jamia Hamdard, New Delhi, India

Nadeem Irfan Bukhari (207), Punjab University College of Pharmacy, University of Punjab, Allama Iqbal Campus, Lahore, Pakistan

Gaurav Kumar Jain (259), Delhi Pharmaceutical Science and Research University, New Delhi, India

Rupa Joshi (521), Department of Pharmacology, PGIMER, Chandigarh, India

Saikrishna Kandalam (431), Advanced Drug Delivery and Biomaterials, Louvain Drug Research Institute (LDRI), Université catholique de Louvain, Brussels, Belgium

Ummarah Kanwal (207), Punjab University College of Pharmacy, University of Punjab, Allama Iqbal Campus, Lahore, Pakistan

Deepak N. Kapoor (351), School of Pharmaceutical Sciences, Shoolini University of Biotechnology and Management Sciences, Solan, Himachal Pradesh, India

Ritu Karwasra (259), Department of Pathology, ICMR-National Institute of Pathology, New Delhi, India

Harpinder Kaur (521), Department of Pharmacology, PGIMER, Chandigarh, India

Sanjay Kulkarni (415), Department of Pharmaceutics, Manipal College of Pharmaceutical Sciences, Manipal Academy of Higher Education, Manipal, KA, India

Deepak Kumar (499), Department of Pharmaceutical Chemistry, School of Pharmaceutical Sciences, Shoolini University, Solan, HP, India

Dileep Kumar (409), Department of Pharmaceutical Chemistry, Poona College of Pharmacy, Bharati Vidyapeeth (Deemed to be University), Pune, MH, India

Nitesh Kumar (359,509), Amity Institute for Advanced Research and Studies (Materials & Devices), Amity University, Noida, India

Subodh Kumar (17), Department of Pharmacology, PGIMER, Chandigarh, India

Varun Kumar (359,509), Amity Institute for Advanced Research and Studies (Materials & Devices), Amity University, Noida, India

H. Lalhlenmawia (499), Department of Pharmacy, Regional Institute of Paramedical and Nursing Sciences, Aizawl, MZ, India

Wing-Hin Lee (75), Faculty of Pharmacy and Health Sciences, Royal College of Medicine Perak, Universiti Kuala Lumpur (RCMP UniKL), Ipoh, Perak, Malaysia

Ching-Yee Loo (75), Faculty of Pharmacy and Health Sciences, Royal College of Medicine Perak, Universiti Kuala Lumpur (RCMP UniKL), Ipoh, Perak, Malaysia

Debarshi Kar Mahapatra (409), Department of Pharmaceutical Chemistry, Dadasaheb Balpande College of Pharmacy, Nagpur, MH, India

Saniya Mahendiratta (17), Department of Pharmacology, PGIMER, Chandigarh, India

Deepti Malik (521), Department of Pharmacology, PGIMER, Chandigarh, India

Subha Manoharan (235), Department of Oral Medicine and Radiology, Saveetha Dental College & Hospital, Saveetha Institute of Medical and Technical Science, Chennai, India

Pawan Kumar Maurya (359,509), Department of Biochemistry, Central University of Haryana, Mahendergarh, Haryana, India

Priyanka Maurya (107), Department of Pharmaceutical Sciences, Babasaheb Bhimrao Ambedkar University, Lucknow, Uttar Pradesh, India

Bikash Medhi (17,521), Department of Pharmacology, PGIMER, Chandigarh, India

Akansha Mehra (359,509), Amity Institute for Advanced Research and Studies (Materials & Devices), Amity University, Noida, India

Meenu Mehta (1), Discipline of Pharmacy, Graduate School of Health, University of Technology Sydney, Ultimo, NSW, Australia; School of Pharmaceutical Sciences, Lovely Professional University, Phagwara, Punjab, India

Brahmeshwar Mishra (63,319), Department of Pharmaceutical Engineering & Technology, Indian Institute of Technology (BHU), Varanasi, Uttar Pradesh, India

Neeraj Mishra (141), Amity Institute of Pharmacy, Amity University Madhya Pradesh, Gwalior, MP, India

Nidhi Mishra (107), Department of Pharmaceutical Sciences, Babasaheb Bhimrao Ambedkar University, Lucknow, Uttar Pradesh, India

Dhrubojyoti Mukherjee (225), Department of Pharmaceutics, Faculty of Pharmacy, Ramaiah University of Applied Sciences, Bengaluru, Karnataka, India

Srinivas Mutalik (415), Department of Pharmaceutics, Manipal College of Pharmaceutical Sciences, Manipal Academy of Higher Education, Manipal, KA, India

Anroop B. Nair (267), Department of Pharmaceutical Sciences, College of Clinical Pharmacy, King Faisal University, Al-Ahsa, Saudi Arabia

Mukesh Nandave (387), Department of Pharmacology, Delhi Pharmaceutical Sciences and Research University (DPSRU), New Delhi, India

Sin Wi Ng (281,431), School of Pharmacy, International Medical University (IMU), Bukit Jalil, Kuala Lumpur; Head and Neck Cancer Research Team, Cancer Research Malaysia, Subang Jaya, Selangor, Malaysia

Xin Yi Ng (281), Program in Neuroscience and Behavioural Disorders, Duke-NUS Medical School, National University of Singapore, Singapore, Singapore

Ajinkya NIkam (Nitin) (415), Department of Pharmaceutics, Manipal College of Pharmaceutical Sciences, Manipal Academy of Higher Education, Manipal, KA, India

Nimisha (155), Amity Institute of Pharmacy Lucknow, Amity University, Uttar Pradesh, Noida, India

Raquibun Nisha (107), Department of Pharmaceutical Sciences, Babasaheb Bhimrao Ambedkar University, Lucknow, Uttar Pradesh, India

Mohammad A. Obeid (447), Faculty of Pharmacy, Department of Pharmaceutics and Pharmaceutical Technology, Yarmouk University, Irbid, Jordan

Santwana Padhi (127), KIIT Technology Business Incubator, KIIT Deemed to be University, Bhubaneswar, Odisha, India

Bharath Singh Padya (415), Department of Pharmaceutics, Manipal College of Pharmaceutical Sciences, Manipal Academy of Higher Education, Manipal, KA, India

Ajay Kumar Pal (387), Department of Pharmacology, Delhi Pharmaceutical Sciences and Research University (DPSRU), New Delhi, India

Abhijeet Pandey (415), Department of Pharmaceutics, Manipal College of Pharmaceutical Sciences, Manipal Academy of Higher Education, Manipal, KA, India

Kalpana Pandey (155), Amity Institute of Pharmacy Lucknow, Amity University, Uttar Pradesh, Noida, India

Chandrakantsing V. Pardeshi (183), Department of Pharmaceutics and Pharmaceutical Technology, R. C. Patel Institute of Pharmaceutical Education and Research, Shirpur, India

Kamla Pathak (299), Pharmacy College Saifai, Uttar Pradesh University of Medical Sciences, Etawah, Uttar Pradesh, India

Tania Patwal (509), Department of Biotechnology, Thapar Institute of Engineering and Technology, Patiala, India

Atmaram Pawar (409), Department of Pharmaceutical Chemistry, Poona College of Pharmacy, Bharati Vidyapeeth (Deemed to be University), Pune, MH, India

Anil Philip (101), School of Pharmacy, College of Pharmacy and Nursing, University of Nizwa, Nizwa, Oman

Lisa G. Pont (9), Discipline of Pharmacy, Graduate School of Health, University of Technology Sydney, Ultimo, NSW, Australia

Ajay Prakash (17,521), Department of Pharmacology, PGIMER, Chandigarh, India

Parteek Prasher (27,351), Department of Chemistry, University of Petroleum & Energy Studies, Dehradun, India

Gopal Kumar Rai (409), S. P. College of Pharmaceutical Sciences and Research, Bijnor, UP, India

Alan Raj (197), College of Pharmaceutical Sciences, Government Medical College Kannur, Kerala, India

Sushma Rawat (1), School of Pharmacy, Suresh Gyan Vihar University, Jagatpura, Jaipur, India

Abida Raza (207), NILOP Nanomedicine Research Laboratories, National Institute of Lasers and Optronics College, PIEAS, Islamabad, Pakistan

Faizan Naeem Razali (75), Faculty of Pharmacy and Health Sciences, Royal College of Medicine Perak, Universiti Kuala Lumpur (RCMP UniKL), Ipoh, Perak, Malaysia

Sangita Saini (37), BabaMast Nath University, Asthal Bohar, Rohtak, India

Betty Annie Samuel (101), School of Pharmacy, College of Pharmacy and Nursing, University of Nizwa, Nizwa, Oman

Shubhini A. Saraf (107), Department of Pharmaceutical Sciences, Babasaheb Bhimrao Ambedkar University, Lucknow, Uttar Pradesh, India

Mohammad Arshad Shaikh (1), R P College of Pharmacy, Osmanabad; School of Pharmacy, Suresh Gyan Vihar University, Jagatpura, Jaipur, India

C. Sarath Chandran (197), College of Pharmaceutical Sciences, Government Medical College Kannur, Kerala, India

Phulen Sarma (17), Department of Pharmacology, PGIMER, Chandigarh, India

Saurabh Satija (1,447), Discipline of Pharmacy, Graduate School of Health, University of Technology Sydney, Ultimo, NSW, Australia; School of Pharmaceutical Sciences, Lovely Professional University, Phagwara, Punjab, India

T.K. Shahin Muhammed (197), College of Pharmaceutical Sciences, Government Medical College Kannur, Kerala, India

Mousmee Sharma (27,351), Department of Chemistry, Uttaranchal University, Dehradun, India

Nitin Sharma (259), Department of Pharmaceutical Technology, Meerut Institute of Engineering and Technology, Meerut, Uttar Pradesh, India

Ganesh B. Shevalkar (183), Department of Pharmaceutics and Pharmaceutical Technology, R. C. Patel Institute of Pharmaceutical Education and Research, Shirpur, India

Priya Shrivastava (359,509), Amity Institute for Advanced Research and Studies (Materials & Devices), Amity University, Noida, India

Ajay Shukla (141), Department of Pharmacy, Institute of Technology and Management (ITM), Gida, Gorakhpur, UP, India

Apoorva Singh (155), Amity Institute of Pharmacy Lucknow, Amity University, Uttar Pradesh, Noida, India

Juhi Singh (63), NTU Institute for Health Technologies, Nanyang Technological University, Singapore, Singapore

Lubhan Singh (499), Kharvel Subharti College of Pharmacy, Swami Vivekanand Subharti University, Meerut, India

Neelu Singh (107), Department of Pharmaceutical Sciences, Babasaheb Bhimrao Ambedkar University, Lucknow, Uttar Pradesh, India

Priya Singh (107), Department of Pharmaceutical Sciences, Babasaheb Bhimrao Ambedkar University, Lucknow, Uttar Pradesh, India

Sachin Kumar Singh (373), School of Pharmaceutical Sciences, Lovely Professional University, Phagwara, Punjab, India

Samipta Singh (107), Department of Pharmaceutical Sciences, Babasaheb Bhimrao Ambedkar University, Lucknow, Uttar Pradesh, India

Santosh Kumar Singh (1), School of Pharmacy, Suresh Gyan Vihar University, Jagatpura, Jaipur, India

Vinayak Singh (267), Drug Discovery and Development Centre (H3D); South African Medical Research Council Drug Discovery and Development Research Unit, Department of Chemistry and Institute of Infectious Disease and Molecular Medicine, University of Cape Town, Rondebosch, South Africa

Yogendra Singh (1), Department of Pharmacology, Maharishi Arvind College of Pharmacy, Ambabari Circle, Ambabari, Jaipur, India

Chloe C.H. Smit (9), Discipline of Pharmacy, Graduate School of Health, University of Technology Sydney, Ultimo, NSW, Australia

Sanjay J. Surana (183), Department of Pharmaceutics and Pharmaceutical Technology, R. C. Patel Institute of Pharmaceutical Education and Research, Shirpur, India

Murtaza M. Tambuwala (351,447), SAAD Centre for Pharmacy and Diabetes, School of Pharmacy and Pharmaceutical Sciences, Ulster University, Coleraine, Northern Ireland, United Kingdom

Rakesh Kumar Tekade (267), National Institute of Pharmaceutical Education and Research–Ahmedabad (NIPER-A), Gandhinagar, Gujarat, India

Lakshmi Thangavelu (235), Department of Pharmacology, Saveetha Dental College & Hospital, Saveetha Institute of Medical and Technical Science, Chennai, India

Komal Thapa (47), Chitkara School of Pharmacy, Chitkara University, Solan, Himachal Pradesh, India

Rajiv K. Tonk (499), Department of Pharmaceutical Chemistry, Delhi Pharmaceutical Sciences & Research University, New Delhi, India

Mansi Upadhyay (319), Department of Pharmaceutical Engineering & Technology, Indian Institute of Technology (BHU), Varanasi, Uttar Pradesh, India

Katharigatta N. Venugopala (267), Department of Pharmaceutical Sciences, College of Clinical Pharmacy, King Faisal University, Al-Ahsa, Saudi Arabia; Department of Biotechnology and Food Technology, Durban University of Technology, Durban, South Africa

Dhriti Verma (351), School of Pharmaceutical Sciences, Shoolini University of Biotechnology and Management Sciences, Solan, Himachal Pradesh, India

Nitin Verma (47), Chitkara School of Pharmacy, Chitkara University, Solan, Himachal Pradesh, India

Kylie A. Williams (9), Discipline of Pharmacy, Graduate School of Health, University of Technology Sydney, Ultimo, NSW, Australia

Nisha R. Yadav (183), Edhaa Innovations Pvt. Ltd., Mumbai, India

Rati Yadav (499), Department of Pharmaceutical Chemistry, School of Pharmaceutical Sciences, Shoolini University, Solan, HP, India

Farrukh Zeeshan (461), Department of Pharmaceutical Technology, School of Pharmacy, International Medical University (IMU), Kuala Lumpur, Malaysia

Editor biographies

Kamal Dua is a Senior Lecturer, Discipline of Pharmacy, Graduate School of Health, University of Technology Sydney (UTS), has a research experience of over 12 years working in the field of drug delivery targeting inflammatory diseases and cancer. Dr. Dua is also a node leader of Drug Delivery Research in the Centre for Inflammation at Centenary Institute/UTS, where the targets identified from the research projects are pursued to develop novel formulations as the first step toward translation into clinics. Dr. Dua researches in two complementary areas: drug delivery and immunology, specifically addressing how these disciplines can advance one another helping the community to live longer and healthier. This is evidenced by his extensive publication record in reputed journals. Dr. Dua's research interests focus on harnessing the pharmaceutical potential of modulating critical regulators such as interleukins and microRNAs and developing new and effective drug delivery formulations for the management of chronic airway diseases. Dr. Dua is also an adjunct assistant professor in the Faculty of Pharmacy and Pharmaceutical Sciences, University of Alberta, Canada; Adjunct Fellow with the NICM Health Research Institute at Western Sydney University, Australia and Honorary Research Associate with the Australia-China Relations Institute (UTS:ACRI) and Consultant at ULTI Pharmaceuticals. He also held a Conjoint Lecturer position with the School of Biomedical Sciences and Pharmacy, The University of Newcastle, Australia. He has also published more than 80 research articles in the peer-reviewed international journals and authored or coauthored four books and various book chapters.

Meenu Mehta is a research scholar at the Discipline of Pharmacy, Graduate School of Health, the University of Technology Sydney. Her research is focused in the area of pharmaceutical technologies developing novel drug delivery systems for chronic respiratory diseases such as asthma, chronic obstructive lung disease (COPD), and lung cancer. She has obtained her M. Pharm degree in Pharmaceutical Sciences in India, where she has gained working experience in several areas such as drug development, analytical methods, pharmacognosy studies, microwave techniques, and nanotechnology-based novel drug delivery systems. Mehta has received various awards and carries an impressive bibliography of scientific papers published in journals of international repute.

Terezinha de Jesus Andreoli Pinto is a Professor at the School of Pharmaceutical Sciences, University of Sao Paulo. Professor Terezinha holds more than 40 years of sound experience in academia and researches on parenterals (formulation, analytical, microbiological, and performance methods) and medical devices. Professor Terezinha authored 180 articles in scientific journals, more than 12 book chapters and also holds two patents. Alongside her career, Professor Terezinha took on management roles in the University, such as being dean of School of Pharmacy for two mandates (2004–2008 and 2012–2016) and also being chair of Deliberative Board of FURP—a pharmaceutical firm that manufactures products from the WHO Essential Medicines list, run by Sao Paulo State Government. Strong scientific skills allied to leadership and management allowed her to establish agreements with internationally prestigious institutions including the University of Alberta, Lisbon, and Bath. Under her supervision, the CONFAR Laboratory was set up, the only Brazilian state university laboratory accredited by the National Institute of Metrology, Quality and Technology (INMETRO) (ISO/IEC 17025) and, authorized by the National Agency of Sanitary Surveillance (ANVISA) and the Ministry of Agriculture, Livestock, and Supplies (MAPA), and is considered a reference both in Brazil and overseas, in the analytical area. Professor Terezinha is also involved with standard-setting activities in agencies such as Brazilian Pharmacopeia (as coordinator of the Brazilian Pharmacopeia Drug Product and Medical Devices technical committee, and currently member of Biological and Biotechnological Products Committee), International Standard Organization (ISO) as Brazilian Association of Technical Norms (ABNT) representative and also the United States Pharmacopeia stakeholder with participation in convention current cycle.

Lisa G. Pont is a Professor in the Discipline of Pharmacy, Graduate School of Health at the University of Technology Sydney. Pont is a pharmacoepidemiologist with expertise in medicine utilization and health service research to evaluate, understand, and improve the quality and safety of medicine use in practice. Her research explores current practices and patterns of medicines use across populations with a focus on the safety and quality of prescribing and medicines use. Throughout her career, Professor Pont has received a number of awards including two national NPS MedicineWise awards in 2014 and in 2016 for her research. She is a registered pharmacist and was awarded the Society of Hospital Pharmacist of Australia Medal of Merit in 2018.

Kylie A. Williams is a pharmacist and academic leader. She is a founding member of the Graduate School of Health at University of Technology Sydney (UTS), where she is currently the Head of Pharmacy and the Deputy Head of School Learning & Teaching. She has over 25 years of academic experience in learning and teaching, curricula development, research, and academic leadership. Professor Williams leads a research team focused on development, implementation, and evaluation of new models of pharmacy and health service delivery. Alongside her research expertise, Professor Williams is internationally recognized for her curriculum development, and has received a number of teaching grants and awards and has coauthored numerous teaching-related peer-reviewed journal articles, professional books, and educational articles for pharmacists.

Michael J. Rathbone is founder and managing director of his own company, ULTI Pharmaceuticals, New Zealand. He is also an associate of the School, Discipline of Pharmacy, Graduate School of Health, University of Technology Sydney, NSW, Australia. He was formerly professor of Pharmaceutical Technology and dean, School of Pharmacy, International Medical University, Kuala Lumpur, Malaysia. Prior to this he was associate professor of Pharmaceutics, School of Pharmacy, Griffith University, Australia where his responsibilities included Acting Head of School. Previous to his appointment at Griffith University, he was the director of Research and General Manager InterAg, New Zealand, where he spearheaded the companies veterinary controlled drug delivery research and directed their national and global collaborative research activities. Dr. Rathbone obtained his undergraduate degree in Pharmacy at Leicester Polytechnic (De Montfort University), UK (1980), and PhD in Pharmaceutics from the University of Aston, Birmingham, UK (1986). Dr. Rathbone has innovated many novel veterinary drug delivery systems, several of which have been registered on the New Zealand, Australian and United States markets. He is a fellow of the Controlled Release Society, has served on the Board of Directors of the CRS and has received several prestigious awards for his contribution to the science and technology of controlled release. He has edited eight books in the area of modified release drug delivery and 10 special theme issues of journals such as Advanced Drug Delivery Reviews and Journal of Controlled Release. His knowledge of the entire spectrum of innovation, product research & development, cGMP analysis, manufacturing scale-up, QC analysis, stability testing, and registration provides him with an extensive overview, and unparalleled experience of the veterinary pharmaceutical industry.

Chapter 1

Introduction to cancer cell biology

Mohammad Arshad Shaikh[a,c], Salman Hussain[b], Ritu Gilhotra[c], Santosh Kumar Singh[c], Sushma Rawat[c], Yogendra Singh[d], Saurabh Satija[e,f], Meenu Mehta[e,f], Kamal Dua[e], and Gaurav Gupta[c]

[a]*R P College of Pharmacy, Osmanabad, India,* [b]*Department of Pharmaceutical Medicine, School of Pharmaceutical Education and Research, Jamia Hamdard, New Delhi, India,* [c]*School of Pharmacy, Suresh Gyan Vihar University, Jagatpura, Jaipur, India,* [d]*Department of Pharmacology, Maharishi Arvind College of Pharmacy, Ambabari Circle, Ambabari, Jaipur, India,* [e]*Discipline of Pharmacy, Graduate School of Health, University of Technology Sydney, Ultimo, NSW, Australia,* [f]*School of Pharmaceutical Sciences, Lovely Professional University, Phagwara, Punjab, India*

1 Introduction

Cancer includes more than 200 various diseases related to genetics in which information stored in cell DNA is altered, leading to change in expression pattern of genes that causes cells to develop, multiply, and metastasize abnormally [1, 2]. Conversion of a normal functioning cell to a genetically modified cell, known as a neoplastic cell, is referred to as somatic mutation [3]. The number of patients suffering from cancer is increasing daily. According to a US survey, 1,660,290 new cases of cancer occurred in 2014 among both males and females. Of this cohort of cancer patients, there were 580,350 deaths and 339,810 cases of cancer of the genital system. The data suggest that prostate cancer is the most common cancer among men, while breast cancer is the most common among women. Children aged 1–14 years mostly suffer from leukemia as well as brain cancer, sarcoma of soft tissue, neuroblastoma, and renal tumors, but these occur less frequently [4, 5]. Factors such as genetics, environment, lifestyle, occupation, and some biological factors play a vital role in causing different types of cancer [6, 7] (Table 1).

2 Types of cancer

Cancer is classified into the following types according to type of cells or tissues involved:

- carcinoma: neoplastic cells in the skin and tissues in the linings of internal organs
- sarcoma: cancer in connective tissues like bone, cartilage, blood vessels, and muscles
- leukemia: cancerous cells in bone marrow, leading to formation of abnormal blood cells
- lymphoma: cancer of immune cells

3 Pathophysiology of cancer

Hanahan and Weinberg mentioned in their review that there are eight distinguishing and harmonizing features or hallmarks responsible for tumor growth and metastasis, as shown in Fig. 1.

- sustained multiplication (proliferative) signaling
- avoidance of immune annihilation (destruction)
- deregulation of cell energetics
- avoidance of apoptosis
- shunning of anti-growth signals
- metastasis and tissue invasion
- endless multiplication (replication) potential
- extended (sustained) tumor angiogenesis

Further, they suggest that there are two characteristics that enable in cancer cell survival, growth, and proliferation [8, 9].

- genetic instability and mutability
- inflammation

3.1 Sustained multiplication (proliferative) signaling

Normal cells maintain their homeostasis of multiplication, proliferation, and function by regulating signals of growth promotion. In cancerous cells, these signals are enabled in a sustained manner so that multiplication and proliferation processes continue for prolonged periods of time [10]. Van Roosbroeck suggests that protein kinases of the cyclin-dependent kinase (CDK) family such as CDK4 and CKD6 control the cell cycle and its proliferation in the G1 phase. In cancer genes and kinase proteins, abnormal expression leads to dysregulation of the CDK signaling pathway [11]. The PI3K-Akt-mTOR or MAPK pathway reregulation

Advanced Drug Delivery Systems in the Management of Cancer. https://doi.org/10.1016/B978-0-323-85503-7.00013-4

TABLE 1 Common causes of cancer.

Lifestyle factors	Tobacco chewing
	Cigarette smoking
	Alcohol consumption
	Cosmetics
Environmental factors	Vehicle emissions
	Pesticides
	Industrial effluents
	Radiation (X-rays, soil radon, UV radiations)
Occupational factors	Exposure to radioactive chemicals, asbestos, petrochemicals, some metals, plastic, etc.
Bacterial viruses	*Helicobacter pylori*,
	Hepatitis B and hepatitis C,
	Human papilloma virus,
	Epstein-Barr
Drugs	Hydrochlorothiazide (skin cancer)
	Oxazepam (liver cancer)

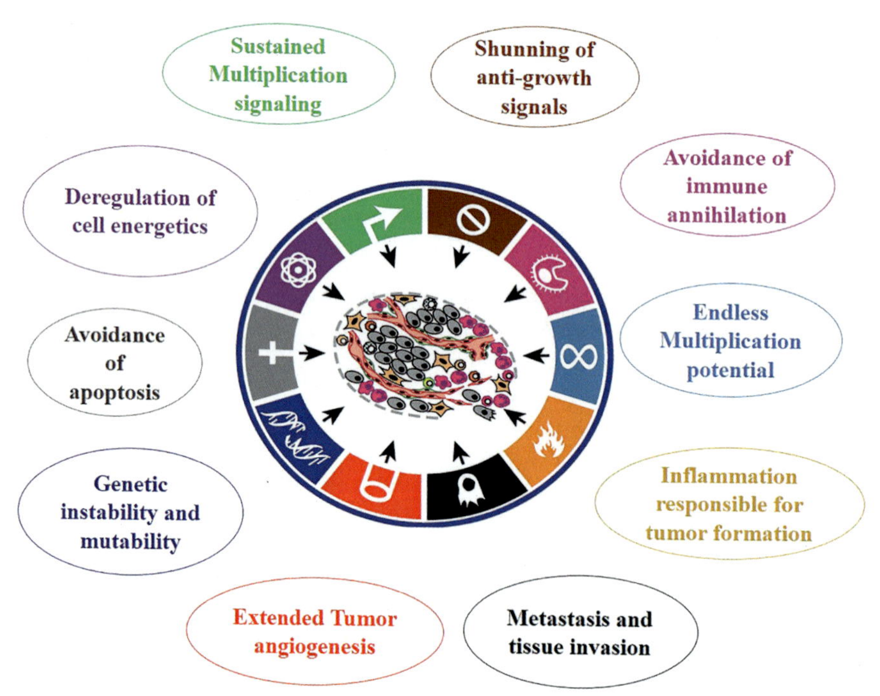

FIG. 1 Hallmarks and characteristics of cancer and metastasis.

leads to sustained proliferative signaling, as these pathways are part of mitogenesis [12].

3.2 Shunning of anti-growth signals

In normal cells there are some checkpoint signals to stop cell growth. These are known as antigrowth signals. In tumor cells these anti-growth signals become insensitive, leading to loss of control of growth mechanisms that results in unlimited replication of tumor cells and inability to inhibit tumors and prevent their shedding, stagnation, and aging [13].

Disruption of anti-growth control can be accomplished by avoiding p53, retinoblastoma protein, AT-rich interaction domain 1A, hippo, growth differentiation factor 15, Notch, insulin-like growth factor, phosphatase and tensin homologs, and Krüppel-like factor 5 pathways [14].

3.3 Avoidance of immune annihilation (destruction)

Immune surveillance theory suggests that all tissues and cells work under the control of the immune system, which resists tumor formation and propagation by early detection and removal of cancerous cells [10]. Cancer alters immunity, hence immune cells are unable to detect cancerous cells and thus these cells remain for prolonged time. Tumor cells also modify the immune-suppressive mediator and antigen detection, as well as disturb T-cell function and alter the immune system [15]. In 2012, Kobayashi determined that receptor tyrosine kinases, JAK-STAT pathway, MAPK pathway, Hippo, PI3K-AKT-mTORpathway, and Wnt immunity pathway are mainly disturbed in cancer as well as deregulation of immune checkpoints [16]. The study demonstrated that alteration of phagocytosis of tumor cells occurs via the CD47–SIRPa pathway, resulting in an increase in the number of tumor cells and inhibition of their elimination [17].

3.4 Avoidance of apoptosis

Apoptosis is vital process for normal cell which is govern by pro and antiapoptotic factor as well as endogenous and exogenous factor. DNA damage causes proliferation blocked and the possibility of DNA repair pay attention. When the damage exceeds the repair capacity, the balance between positive and negative signals changes, causing the cell to suffer some form of regulatory cell death, which prevents DNA damage and avoids the risk of the mutation spreading to future generations of cell division. The imbalance between apoptosis and antiapoptotic factors leads to the production of cancer cells [18]. Naturally, apoptosis is carried out by intrinsic pathway, extrinsic, p53, BCL2, endoplasmic reticulum stress-induced apoptotic pathways and necroptosis cell death pathway. Cancer alters the expression pattern of microRNA, resulting in deregulation of these pathways and evasion of programmed cell death [19, 20].

3.5 Tissue invasion and metastasis

Cancer cells can invade adjacent cells and spread to other cells as well as tissue through blood circulation or the lymphatic system. Distal metastasis causes 90% of cancer deaths. Tissue invasion and. metastasis consist of a sequence of complex biological processes. First, cancer cells leave their immediate state local neighbors and stroma. Then, there is enzymatic breakdown of the matrix present outside the cell, which then moves in a particular direction. Next, infiltration (penetration) of blood or lymph vessels and tumors embolism; survival of tumor cell in blood circulation till reaching to place where metastasis occurs and where convenience in supply of growth factor; attachment to the endothelium of the blood vessel at its target and extravasation of the blood vessel; and new place for diffusion and self-invasion as well as adaptation of the new blood supply [2].

Research by Zhao suggests that decreased expression of calpain small subunit 1/Capn4 leads to reduced expression of matrix metalloproteinase (MMP9) through the Wnt/β-catenin signaling pathway whose activation leads to suppression of tissue invasion and metastasis in gastric cancer [21]. Activated leukocyte cell adhesion molecule (ALCAM)/CD166, participates in tumor invasion and metastasis producing hemophilia and xenogeneic interactions [22].

3.6 Extended (sustained) tumor angiogenesis

The development of blood vessels in cancerous cells or at a cancerous site is called tumor angiogenesis. Tumor cells use the body's blood supply to grow. They appear as clumps of vascular cells that undergo malignant transformation. Growing cancer cells need to recruit blood vessels to promote the release of oxygen and nutrients and the removal of waste products. The rapid proliferation of cancer cells creates a hypoxic environment, represented by glucose. These environmental pressures cause changes in angiogenesis and secrete growth factors that promote angiogenesis, such as vascular. endothelial growth factor, platelet-derived growth factor, basic. fibroblast growth factor, fibroblast growth. factor, transforming. growth factor α and β, and tumor necrosis factor α [2]. The secretion of tumor-induced growth factors shows an imbalance between pro-angiogenic factors and anti-angiogenic factors, which stimulate and germinate potential endothelial cells from nearby blood vessels and transfer them to cancer cells [23].

3.7 Endless multiplication (replication) potential

According to the Hayflick phenomenon, somatic noncancerous cells have limited ability to replicate before entering the G_0 phase in which cell growth completely stops. As replication occurs, telomeres appear over time and they become shorter and shorter, acting as a molecular clock that shows cell life. However, tumor cells become immortal by maintaining their telomere length via overexpression of the telomerase enzyme, which results in infinite replication of tumor cells.

3.8 Deregulation of cell energetics

Deregulation of cell energetics is a hallmark of cancer in which change in normal metabolic patterns or ways of energy production are changed in order for cancerous cells to meet their needs for energy expenditure, stress tolerance, growth, proliferation, and change of the microenvironment [24]. To meet the high energy requirements for proliferation, cancer cells must coordinate metabolism to reprogram the metabolic pathway through glycolysis, oxidative phosphorylation, amino acid, nucleotide, lipid synthesis, the pentose phosphate pathway, the tricarboxylic acid cycle, β-oxidation and degradation of glutamine [25]. Cancer cells function through the Warburg effect, which occurs under oxygen-containing conditions. Cancer cells tend to switch their metabolism to the preferred use of glycogenolysis to produce lactic acid. Metabolic reprogramming can be combined with changes in the hypoxic regulatory phenotype that results in tumor and stromal cell plasticity and intratumoral heterogeneity. In some cancer cells, the hypoxic microenvironment can stimulate neuroendocrine differentiation and epithelial-to-mesenchymal migration, as well as induce cancer stem cells. These processes are associated with an increased tendency for cancer cells to metastasize [2, 26].

3.9 Oncogenes and tumor suppressor genes

Nenclares and Harrington suggest that oncogenes are activated and dominate in cancer, whereas tumor suppressor genes (TSGs) are inactivated and phenotypically recessive in pathogenesis of cancer. Activation of oncogenes occurs by the following three different ways:

- DNA code is mutated, leading to alteration in amino acid sequence and increase in its activity
- Excess of amplification process leads to increased protein production
- Transfer of a segment of DNA to a new position on a chromosome leads to formation of highly biologically active fusion protein (Fig. 2)

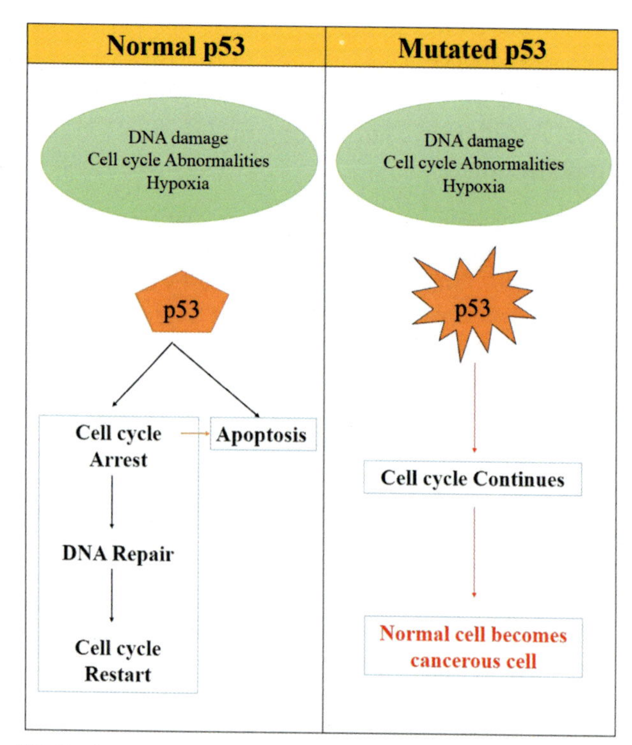

FIG. 2 Oncogenes and tumor suppressor genes.

TSGs are cellular genes that prevent cell proliferation and help in the apoptosis process. In pathogenesis of many cancers, TSGs are inactivated, leading to abnormal proliferation and development of tumor cells. *Rb* is the first TSG discovered and its suppression leads to sarcoma, retinoblastoma, and breast, lung, and bladder cancer. Another TSG, *p53*, is associated with brain, colon, breast, liver, esophageal, and lung cancers as well as sarcoma, leukemia, and lymphoma.

3.10 Breast cancer biology

Breast cancer is the most common cancer in women. It is caused by development of malignant tumor cells. Identifying whether the gland unit or duct unit of the breast is affected enables pathologists to classify breast cancer. First the tumor starts in the breast, but once it infiltrates it progresses into the lymph (armpit/internal breast) node; then it has transfer capability. Among all types of breast cancer, 70%–80% are the ductal type (either papillary, medullary, mucinous, or tubular), while lobular breast cancer makes up the remaining percentage [27]. Breast cancer is categorized into three different types on the basis of histological and molecular basis: (1) breast cancer due to expression of female sex hormone of receptor (estrogen ER^+ or progesterone PR^+), (2) breast cancer due to expression of human epidermal receptor 2 ($HER2^+$), and (3) breast cancer due to negative expression of progesterone, estrogen, and human epidermal receptor 2 (ER^-, PR^-, $HER2^-$) [28].

3.11 Gastric cancer biology

Stomach cancer is typically diagnosed as malignant because diagnosis usually occurs at an advanced stage. Gastric cancer has a high incidence of mortality, making it the third leading cause of cancer-related deaths. In 2018, there were 784,000 deaths due to gastric cancer worldwide. Gastric cancer is classified into two types based on location: (1) cardiac stomach cancer occurring in the area of the stomach near the junction of the esophagus and stomach, and (2) non-cardiogenic gastric cancer occurring in the distal area of the stomach [29].

Gastric cancer is attributed to *Helicobacter pylori* infection elated to bacterial virulence and host genetic polymorphism and environmental factors. Most *H. pylori* consists of gene A related to cytotoxin (CagA) encode 120–140 kDa. CagA protein which is an oncoprotein that affects cell expression signal protein. About 10% of all gastric cancer cases are associated with heredity and 1%–3% of gastric cancers occur due to genetic mutations. Hereditary forms of stomach cancer can be divided into three categories: (1) diffuse inheritance gastric cancer (2) familial gastric cancer, and (3) gastric adenocarcinoma with proximal gastric polyps [30].

3.12 Pancreatic cancer biology

Pancreatic cancer is a deadly disease with a high mortality rate. It is the fourth leading cause of cancer-related death. Risk factors for pancreatic cancer include diabetes mellitus, genetics, pancreatitis, cigarette smoking, obesity, and age. About 50%–60% of pancreatic cancer causes are due to family history, and 5%–10% develop due to gene mutations, for example, Kirsten rat sarcoma (KRAS). From an evolutionary perspective, pancreatic cancer is divided into three stages controlled by genetic mutations, including normal cell activation caused by environmental exposure. However, clone proliferation is the continuous division of mutated cells. A population of clones is introduced and ultimately a foreign body microenvironment is established in which tumor cells penetrate the basement membrane and launch into the surrounding matrix [31]. Epigenctic abnormalities in tumor genes and the presence of closed TSGs (such as p16, TP53, and cyclin-dependent kinase inhibitor 2A) are risk factors for pancreatic cancer [32].

3.13 Lung cancer biology

Lung cancer is often associated with smoking. It is a heterogeneous disorder with multiple genetic and epigenetic changes and a high mortality rate. About 15% of lung cancers are subtypes of small cell lung cancer (SCLC) and the remaining 85% are non-small cell lung cancer (NSCLC). Lung adenocarcinoma and squamous cell lung cancer are subtypes of NSCLC [33]. Lung adenocarcinoma generally originates in the epithelial cells of the last airways, exhibits various morphological features such as acinar, squamous, and papillary, and generally expresses thyroid transcription factor 1 and cytokeratin 7. Oncogenes associated with adenocarcinoma include epidermal growth factor receptor mutations and anaplastic lymphoma kinase translocations, reactive oxygen species 1 gene (chromosome 6q), hepatocyte growth factor mutations, and human epidermal growth factor receptor 2 (HER2). Non-smoking adenocarcinoma has a greater frequency of epidermal growth factor receptor mutations, reactive oxygen species, and undifferentiated lymphoma kinase, while smokers have a greater frequency of KRAS mutations, indicating that smokers and nonsmokers may have different mechanisms of tumorigenesis. Squamous cell carcinoma is usually a more central lesion caused by basal bronchial epithelial cells and is typically part of a squamous metaplasia. Morphologically, these lesions produce keratinization and intercellular bridging, and express both p40 and p63 [34].

SCLC has unique clinical and pathological features. Despite its initial sensitivity, it has aggressiveness and early metastasis, as well as a high incidence of disease recurrence [35].

3.14 Leukemia biology

Leukemia refers to the overproduction of white blood cells (WBCs) in the body. Types of leukemia include acute lymphocytic leukemia (ALL), chronic lymphocytic leukemia (CLL), acute myeloid leukemia (AML), and chronic myeloid leukemia (CML) [36]. Leukemia can occur in all age groups, from infants to the elderly, but different morphologies have different age distributions. ALL usually occurs in early childhood and is rare in adults. AML occurs less frequently in children but is more common in the elderly. CMK is infrequent in young children, but CLL is the most common leukemia in the Western world, almost unique to people older than 40 years, with an average age of detection of 70 years or older [37].

Acute myeloid leukemia (AML) is a malignant disease characterized by the proliferation of cloned embryonic cells that are abnormally differentiated in the myeloid system. This results in immature bone marrow cell proliferation including the accumulation of immature precursors (embryonic cells) with reduced normal hematopoietic function, leading to severe infections, anemia, and bleeding [38].

Chronic myeloid leukemia (CML) is characterized by translocation of the Philadelphia (Ph) chromosome to between chromosome 9 and chromosome 22, which forms the oncogenic fusion gene BCR-ABL1 on chromosome 22 [39]. The presence of the Philadelphia chromosome leads to the fusion of the BCR-ABL oncogene due to the juxtaposition of the 30 portion of the Abelson gene (ABL) from the long arm of chromosome 9 with the 50 portion of the breakpoint cluster region (BCR) on the long arm of the chromosome 22. CML can be divided into three stages: chronic, accelerated, and explosive. Nearly 85 percent of CML patients are diagnosed in the chronic phase, while 15 percent of patients are diagnosed in the aggressive or explosive phase [40].

Acute lymphocytic leukemia (ALL) is a blood disease caused by the proliferation of immature lymphoid cells in the bone marrow, blood, or other organs [41].

Chronic lymphocytic leukemia (CLL) refers to a proliferative lymphocytic disease consisting of monomorphic B lymphocytes involving peripheral blood, bone marrow, and the lymphatic system. CLL lymphocytes are cloned CD19-positive B cells with specific immunophenotypic characteristics and can co-express CD5, CD23, and low-density CD20 (weak CD20) [42]. The risk factors for CLL include genetic factors, exposure to insecticides or herbicides, hepatitis C, decreased sun exposure, and atopic health history [43].

References

[1] Cheon DJ, Orsulic S. Mouse models of cancer. Annu Rev Pathol Mech Dis 2011;6:95–119. https://doi.org/10.1146/annurev.pathol.3.121806.154244.

[2] Nenclares P, Harrington KJ. The biology of cancer. Medicine (UK) 2020;48(2):67–72. https://doi.org/10.1016/j.mpmed.2019.11.001.

[3] Alexandrov LB, Diego S, States U. Mutational signatures and the etiology of human cancers. In: Encyclopedia of cancer. 3rd ed; 2018. https://doi.org/10.1016/B978-0-12-801238-3.65046-8.

[4] Hassanpour SH, Dehghani M. Review of cancer from perspective of molecular. J Cancer Res Pract 2017. https://doi.org/10.1016/j.jcrpr.2017.07.001.

[5] Siegel R, Naishadham D, Jemal A. Cancer Statistics, 2013. Cancer J Clin 2013;63(1):11–30. https://doi.org/10.3322/caac.21166.

[6] Friedman GD, Udaltsova N, Chan J, Quesenberry CP, Habel LA. Screening pharmaceuticals for possible carcinogenic effects: initial positive results for drugs not previously screened. Cancer Causes Control 2009;20(10):1821–35. https://doi.org/10.1007/s10552-009-9375-2.

[7] Verma M, Ph D. Biomarkers for risk assessment in molecular epidemiology of cancer. Technol Cancer Res Treat 2004;3(5):505–14. https://doi.org/10.1177/153303460400300512.

[8] Hanahan D, Weinberg RA. Hallmarks of cancer: the next generation. Cell 2011;144(5):646–74. https://doi.org/10.1016/j.cell.2011.02.013.

[9] McNamara J. The etiology of cancer. Brit Med J 1923;2(3274):586–7. https://doi.org/10.1136/bmj.2.3274.586-a.

[10] Caon I, Bartolini B, Parnigoni A, Caravà E, Moretto P, Viola M, Passi A. Revisiting the hallmarks of cancer: the role of hyaluronan. Semin Cancer Biol 2020;62(June):9–19. https://doi.org/10.1016/j.semcancer.2019.07.007.

[11] Van Roosbroeck K, Calin GA. Cancer hallmarks and MicroRNAs: the therapeutic connection. In: Advances in cancer research. 1st ed, vol. 135; 2017. https://doi.org/10.1016/bs.acr.2017.06.002.

[12] Frost FG, Cherukuri PF, Milanovich S, Boerkoel CF. Pan-cancer RNA-seq data stratifies tumours by some hallmarks of cancer. J Cell Mol Med 2020;24(1):418–30. https://doi.org/10.1111/jcmm.14746.

[13] Nishijima David L, Wisner DH, Holmes JF, D. K. S. Evasion of anti-growth signaling: a key step in tumorigenesis and potential target for treatment and prophylaxis by natural compounds. Physiol Behav 2016;176. https://doi.org/10.1016/j.physbeh.2017.03.040.

[14] Amin ARMR, Karpowicz PA, Carey TE, Arbiser J, Nahta R, Chen ZG, Shin DM. Evasion of anti-growth signaling: a key step in tumorigenesis and potential target for treatment and prophylaxis by natural compounds. Semin Cancer Biol 2015;35. https://doi.org/10.1016/j.semcancer.2015.02.005.

[15] Vinay DS, Ryan EP, Pawelec G, Talib WH, Stagg J, Elkord E, Kwon BS. Immune evasion in cancer: mechanistic basis and therapeutic strategies. Semin Cancer Biol 2015;35:S185–98. https://doi.org/10.1016/j.semcancer.2015.03.004.

[16] Kobayashi Y, Lim SO, Yamaguchi H. Oncogenic signaling pathways associated with immune evasion and resistance to immune checkpoint inhibitors in cancer. Semin Cancer Biol 2020;65(September):51–64. https://doi.org/10.1016/j.semcancer.2019.11.011.

[17] Chao MP, Weissman IL, Majeti R. The CD47-SIRPα pathway in cancer immune evasion and potential therapeutic implications. Curr Opin Immunol 2012;24(2):225–32. https://doi.org/10.1016/j.coi.2012.01.010.

[18] Cree IA. Cancer biology. Methods Mol Biol 2011;731(1):79–91. https://doi.org/10.1007/978-1-61779-080-5.

[19] Fernald K, Kurokawa M. Evading apoptosis in cancer. Trends Cell Biol 2013;23(12):620–33. https://doi.org/10.1016/j.tcb.2013.07.006.

[20] Shirjang S, Mansoori B, Asghari S, Duijf PHG, Mohammadi A, Gjerstorff M, Baradaran B. MicroRNAs in cancer cell death pathways: apoptosis and necroptosis. Free Rad Biol Med 2019;139(February):1–15. https://doi.org/10.1016/j.freeradbiomed.2019.05.017.

[21] Zhao C, Yuan G, Jiang Y, Xu J, Ye L, Zhan W, Wang J. Capn4 contributes to tumor invasion and metastasis in gastric cancer via activation of the Wnt/β-catenin/MMP9 signalling pathways. Exp Cell Res 2020;395(2):112220. https://doi.org/10.1016/j.yexcr.2020.112220.

[22] Darvishi B, Boroumandieh S, Majidzadeh-A K, Salehi M, Jafari F, Farahmand L. The role of activated leukocyte cell adhesion molecule (ALCAM) in cancer progression, invasion, metastasis and recurrence: a novel cancer stem cell marker and tumor-specific prognostic marker. Exp Mol Pathol 2020;115(146):104443. https://doi.org/10.1016/j.yexmp.2020.104443.

[23] Roudsari LC, West JL. Studying the influence of angiogenesis in in vitro cancer model systems. Adv Drug Deliv Rev 2016;97:250–9. https://doi.org/10.1016/j.addr.2015.11.004.

[24] Wettersten HI. Reprogramming of metabolism in kidney cancer. Semin Nephrol 2020;40(1):2–13. https://doi.org/10.1016/j.semnephrol.2019.12.002.

[25] Torresano L, Nuevo-Tapioles C, Santacatterina F, Cuezva JM. Metabolic reprogramming and disease progression in cancer patients. Biochim Biophys Acta Mol Basis Dis 1866;2020(5):165721. https://doi.org/10.1016/j.bbadis.2020.165721.

[26] Li L, Li W. Epithelial-mesenchymal transition in human cancer: Comprehensive reprogramming of metabolism, epigenetics, and differentiation. Pharmacol Therapeutics 2015;150:33–46. https://doi.org/10.1016/j.pharmthera.2015.01.004.

[27] Chopra S, Davies EL. Breast cancer. Medicine (UK) 2020;48(2):113–8. https://doi.org/10.1016/j.mpmed.2019.11.009.

[28] Barzaman K, Karami J, Zarei Z, Hosseinzadeh A, Kazemi MH, Moradi-Kalbolandi S, Farahmand L. Breast cancer: biology, biomarkers, and treatments. Int Immunopharmacol 2020;84(February). https://doi.org/10.1016/j.intimp.2020.106535.

[29] Thrift AP, El-Serag HB. Burden of gastric cancer. Clin Gastroenterol Hepatol 2020;18(3):534–42. https://doi.org/10.1016/j.cgh.2019.07.045.

[30] Smyth EC, Nilsson M, Grabsch HI, van Grieken NC, Lordick F. Gastric cancer. Lancet 2020;396(10251):635–48. https://doi.org/10.1016/S0140-6736(20)31288-5.

[31] Dariya B, Alam A, Nagaraju GP. Biology, pathophysiology, and epidemiology of pancreatic cancer. Theranos Approach Pancreat Cancer 2019. https://doi.org/10.1016/b978-0-12-819457-7.00001-3.

[32] Jain AG, Saleem T, Kumar R, Khetpal N, Zafar H, Rashid MU, Ahmad S. Overview of pancreatic cancer biology; 2019. p. 1–11. https://doi.org/10.1016/B978-0-12-817661-0.00001-9.

[33] Salehi-Rad R, Li R, Paul MK, Dubinett SM, Liu B. The biology of lung cancer: development of more effective methods for prevention, diagnosis, and treatment. Clin Chest Med 2020;41(1):25–38. https://doi.org/10.1016/j.ccm.2019.10.003.

[34] Dorward DA, Walsh K, Oniscu A, Wallace WA. Molecular pathology of non-small cell lung cancer. Diagn Histopathol 2017;23(10):450–7. https://doi.org/10.1016/j.mpdhp.2017.08.002.

[35] Tay RY, Heigener D, Reck M, Califano R. Immune checkpoint blockade in small cell lg cancer. Lung Cancer 2019;137(June):31–7. https://doi.org/10.1016/j.lungcan.2019.08.024.

[36] Blackburn LM, Bender S, Brown S. Acute leukemia: diagnosis and treatment. Semin Oncol Nurs 2019;35(6). https://doi.org/10.1016/j.soncn.2019.150950.

[37] Juliusson G, Hough R. Leukemia. Progr Tumor Res 2016;43:87–100. https://doi.org/10.1159/000447076.

[38] Short NJ, Rytting ME, Cortes JE. Acute myeloid leukaemia. Lancet 2018;392(10147):593–606. https://doi.org/10.1016/S0140-6736(18)31041-9.

[39] Yurttaş NÖ, Eşkazan AE. Novel therapeutic approaches in chronic myeloid leukemia. Leukemia Res 2020;91(December 2019):106337. https://doi.org/10.1016/j.leukres.2020.106337.

[40] Hanlon K. Chronic myeloid leukaemia key points. Medicine 2017;1–5. https://doi.org/10.1016/j.mpmed.2017.02.004.

[41] Marinescu C, Vladareanu A, Mihai F. Acute lymphocytic leukemia in adults. Pathol Feat Prog 2015;1–6. https://doi.org/10.1515/rjim-2015-0004.

[42] Scarfò L, Ferreri AJM, Ghia P. Chronic lymphocytic leukaemia. Crit Rev Oncol Hematol 2016;104:169–82. https://doi.org/10.1016/j.critrevonc.2016.06.003.

[43] Hallek M, Shanafelt TD, Eichhorst B. Chronic lymphocytic leukaemia. The Lancet 2018;391(10129):1524–37. https://doi.org/10.1016/S0140-6736(18)30422-7.

Chapter 2

Current practice in cancer pharmacotherapy

Lisa G. Pont[a], Kamal Dua[a], Rachelle L. Cutler[a], Helen Benson[a], Mehra Hagi[a], Victoria Garcia Cardenas[a], Chloe C.H. Smit[a], Alice Ao[b], and Kylie A. Williams[a]

[a]*Discipline of Pharmacy, Graduate School of Health, University of Technology Sydney, Ultimo, NSW, Australia,* [b]*Pharmacy Department, Royal North Shore Hospital, Sydney, NSW, Australia*

1 Introduction

Cancer is one of the leading causes of death throughout the world. In 2018, over 18 million individuals developed cancer, and almost 10 million deaths worldwide were due to cancer [1]. Both cancer incidence and mortality continue to grow globally, and it is estimated that by 2050 the global cancer mortality rate will double [2]. There are hundreds of different cancer types. Lung cancer is the most common form of cancer among both men and women. In terms of total burden, that is, the number of cases and number of cancer deaths, the leading 10 cancers in 2018 were lung, breast (female), prostate, colon, skin (nonmelanoma), stomach, liver, rectum, esophagus, and cervix [1].

A range of treatment modalities are used in the management of cancer, with choice of treatment modality primarily dependent on the location, type, and extent of the cancer being treated. The three main approaches in the management of most cancers are surgery, radiation, and pharmacotherapy, and each of these modalities may be used either alone or in combination. While surgery is the first-line treatment option for many cancers, oncological pharmacotherapy may be used for primary treatment of cancer, where the goal of treatment is to completely remove the cancer. Adjuvant pharmacotherapy, or pharmacotherapy following surgery is used where the goal is to target any remaining cancer cells following primary treatment and neoadjuvant pharmacotherapy, or pharmacotherapy prior to surgery, to shrink tumors and reduce cancer, prior to surgery. Pharmacotherapy may also be used in palliative care, where the goal of treatment is symptom management through tumor reduction, rather than removal of the cancer [3].

Traditionally, pharmacotherapy in oncology has consisted of a range of cytotoxic agents which could be primarily classified as alkylating agents, antimetabolites, cytotoxic antibiotics, and plant derivatives based on their origin or mechanism of action [4]. While many of these pharmacotherapies are effective in the treatment of different cancers, the adverse effects associated with their use can be severe, often limiting their use and thus their effectiveness. In recent times, much attention has been placed on the development of effective pharmacotherapies for the treatment of cancer which are more targeted, better tolerated, and have less adverse impacts on the individuals being treated [5].

Over the past decades, developments in our understanding of the physiology of cancer have identified a number of new oncological drug targets and resulted in a number of novel pharmacotherapies including hormones, immunotherapies, monoclonal antibodies, protein kinase inhibitors, retinoid X receptor agonists, proteasome inhibitors, enzymes, and photoactivated cytotoxic complexes, further extending the pharmacological treatment options for many cancers [4, 5]. With the choice of oncological pharmacotherapy largely dependent on cancer location and type, the current practices in cancer pharmacotherapy should be considered by cancer location and type. In the following sections, we consider the current practice regarding pharmacotherapy treatment options for the five most common cancers.

1.1 Lung

Lung cancer is the most common cancer worldwide, both in terms of incidence and mortality. The health burden due to lung cancer remains substantial, with lung cancer accounting for 11.6% of all new cancer cases and 18.4% of all cancer deaths worldwide in 2018 [1]. Globally, exposure to tobacco smoking accounts for approximately three-quarters of all lung cancer cases, with the remaining quarter attributed to occupational and environmental carcinogen exposure [6]. A diagnosis of lung cancer relies on computer tomography (CT) of the thorax followed by tumor biopsy to allow staging and subtyping [6].

Traditionally, lung cancer has been classified as small cell or nonsmall cell carcinoma, however with advances in

Advanced Drug Delivery Systems in the Management of Cancer. https://doi.org/10.1016/B978-0-323-85503-7.00040-7

targeted pharmacotherapeutic options, subtyping small cell tumors as squamous or nonsquamous small cell carcinoma has become increasingly important. Management options for lung cancer differ depending on the stage, histology, genetic alterations, and patient characteristics and preferences. Surgery remains the treatment of choice for patients with early stage lung cancer, with radiotherapy second line for patients not considered suitable for surgical lobectomy [7]. For individuals with locally advanced disease, surgery with adjuvant chemotherapy or combination chemotherapy and radiotherapy is recommended, while for metastatic disease pharmacotherapy guided by tumor type, including the presence of genetic mutation, is used [8].

While surgery remains first line for primary treatment of nonsmall cell lung cancer (NSCLC), pharmacological options include the use of platinum-based chemoradiotherapy followed by immunotherapy with a programmed cell death 1/programmed death ligand (PD-1/PD-L1) checkpoint inhibitor or tyrosine kinase inhibitors (Table 1) [8, 9, 11]. For treatment of advanced NSSCLC, the current best practice recommends the use of a checkpoint inhibitor, with the choice of checkpoint inhibitor dependent on the specific tumor genetic mutation involved in the following molecular testing [12]. The addition of adjuvant or neoadjuvant chemotherapy with a platinum-based regimen either prior to or following surgery may also be considered. [8, 9, 11]. Other potential targets currently under investigation include neurotrophic tyrosine kinase (NTRK) fusion positive gene inhibitors such as larotrectinib or entrectinib [8]. Management of metastatic NSCLC may include mono- or combination therapy consisting of combination platinum-based and plant-based regimens, tyrosine kinase inhibitors (TKIs), BRAF inhibitors, or (PD-1/PD-L1) checkpoint inhibitors [12].

Like NSSCLC, surgery is first line for primary treatment of squamous nonsmall cell lung cancer (SNSCLC). Pharmacotherapies used in the management of SNSCLC are generally similar to those used in local nonsquamous small cell lung cancer, that is, platinum-based chemoradiation followed by a PD-L1 checkpoint inhibitor [6]. For advanced SNSCLC treatment, options differ depending on the extent of PD-L1 expression. Tyrosine kinase inhibitors (pembrolizumab) either as monotherapy or combined with platinum-based and taxane chemotherapy is first line for PD-L1 negative and -positive tumors [6].

Surgery is also first line for primary treatment of early stage small cell lung cancer (SCLC) [8]. For more advanced SCLC, platinum-based regimens with or without radiotherapy may be used [8, 11], and SCLC relapse may be managed with either platinum-based or anthracycline-based regimens. However, among individuals who did not respond to first-line chemotherapy, there is little evidence that second-line chemotherapy will be beneficial [8]. For more extensive SCLC, chemotherapy with platinum-based regimens or plant-based agents may be used [8, 11].

1.2 Breast (female)

Cancer of the breast is common with 11.6% of all cancer cases and 6.6% of all cancer deaths attributed to breast

TABLE 1 Pharmacotherapy used in the management of nonsmall cell carcinoma (NSCLC) lung cancer [7–10].

Type	Drug class	Local disease	Advanced or metastatic disease
Chemotherapy	Platinum derivatives	Cisplatin	Cisplatin
		Carboplatin	Carboplatin
	Antifolate	Pemetrexed	Pemetrexed
	Vinca alkaloid	Vinorelbine	
	Taxane		Docetaxel
Biologicals	Checkpoint inhibitor		Durvalumab
			(PD-L1)
			Atezolizumab
			(EGFR-TK mutation)
	Tyrosine kinase inhibitor		Entrectinib, Larotrectinib (NTRK fusion–positive, ROS1-positive)
			Osimertinib, Dacomitinib (EGFR-TK mutation)
			Alectinib, Brigatinib, Ceritinib, Crizotinib, Lorlatinib (ALK-positive)
			Nintedanib, Atezolizumab, Nivolumab, Nintedanib, Pembrolizumab (ROS1-positive)

cancer [1]. Breast cancer is the most commonly occurring female cancer throughout the world, accounting for 24.2% of all female cancers and 15.0% of all female cancer deaths [1]. It is estimated that one in every eight women living in the United States will develop breast cancer at some point throughout their lifetime [13].

Management of breast cancer is dependent on a range of factors, namely general health, tumor location, involvement of nodes and tumor stage, and molecular target [14]. Two main molecular targets have been identified in breast cancers, the estrogen receptor (ER) and epidermal growth factor 2 (ERBB2) which is also known as the human epithelial growth factor receptor 2 (HER2) [15]. Breast tumors can be classified into three distinct cancer subtypes, HR+, HER2+, and triple negative, and the choice of systemic therapy is primarily dependent on the molecular biomarker subtype [15]. Oncological pharmacotherapies used for the treatment of breast cancer can be classified into three primary drug classes: chemotherapy, biological HER-targeted treatments, and endocrine therapy. These classes may be used as either neoadjuvant or adjuvant therapy depending on tumor stage and biomarker subtype (Table 2).

Chemotherapy is recommended for the majority of HER2+ or triple negative tumors, as well as in some HR-negative tumors [14, 16, 17]. The most commonly used adjuvant regimens for treatment of all breast cancers include taxanes and/or anthracyclines, although regimens containing cyclophosphamide, generally in combination with methotrexate and 5-fluorouracil, may also be used [14, 16]. Alongside adjuvant regimens, neoadjuvant platinum-based chemotherapy regimens may also be used in the treatment of triple negative tumors [16].

First-line systemic therapy for HER2+ tumors includes adjuvant immunotherapy with trastuzumab, an antiepithelial growth factor (EGF) agent. In HER2+ tumors, trastuzumab is generally used in combination with a second immunological agent, pertuzumab, which directly targets the HER2 receptor and prevents HER2 signaling, and docetaxel, a taxane chemotherapeutic agent [14, 16, 17]. Second-line systemic treatment for HER2+ tumors includes neratinib, a tyrosine kinase inhibitor [16, 17].

Endocrine therapy to counter estrogen-driven tumor growth is the mainstay of management for HR+ tumors. Hormone antagonists, such as tamoxifen, inhibit estrogen binding to the estrogen receptor, thus blocking estrogen-driven tumor growth [15]. Aromatase inhibitors may also be used in the treatment of HR+ tumors. Aromatase inhibitors (anastrozole, exemestane, and letrozole) inhibit the conversion of androgens to estrogen, thus decreasing circulating estrogen levels [15]. Both hormone antagonists and aromatase inhibitors may be used initially and then long term for up to 5 years post primary treatment.

1.3 Prostate

Prostate cancer is the third most common cancer overall globally, accounting for 7.1% of new cancer cases and 3.8% of all cancer deaths [1]. However, among men, prostate cancer is the second most common cancer both in terms of

TABLE 2 Pharmacotherapy options for the treatment of breast cancer by molecular biomarker status [14, 16, 17].

		Hormone receptor positive (HR+)	Human epithelial growth factor receptor 2 positive (HER2+)	Triple negative
Chemotherapy	Anthracyclines (doxorubicin, epirubicin)	Yes (some)	Yes	Yes
	Taxanes (docetaxel, paclitaxel)			
	Platinums			
	Cyclophosphamide			
	Antimetabolites (methotrexate, capecitabine, 5-fluorouracil)			
Endocrine therapy	Hormone antagonists (tamoxifen)	Yes		
	Aromatase inhibitors (anastrazole, exemestane, letrozole)			
HER-targeted therapy	Anti-HER2 monoclonal antibodies (trastuzumab, pertuzumab)		Yes	
	Tyrosine kinase inhibitors (neratinib)			

incidence and mortality [1]. Management of prostate cancer is dependent on tumor location and risk stratification, with risk category generally based on a combination of prostate-specific antigen (PSA), Gleason score [18], and clinical stage [19, 20].

Radiotherapy or prostatectomy remains first-line treatment for the treatment of localized prostate cancer with adjuvant and/neoadjuvant androgen deprivation pharmacotherapy (ADT) being added dependent on risk category (Table 3) [19–21]. Adjuvant or neoadjuvant ADT along with taxane-based chemotherapy (docetaxel) in combination with radiotherapy is used for the management of high-risk localized and locally advanced prostate cancer [19–21]. ADT may be achieved surgically via bilateral orchidectomy, or more commonly, pharmacologically via the use of gonadotropin-releasing hormone agonists or antagonists. Androgen inhibitors used in the management of prostate cancer include bicalutamide, flutamide, nilutamide, and cyproterone acetate [22]. These agents inhibit the binding of dihydrotestosterone but do not reduce serum testosterone levels and are less effective than bilateral orchidectomy or lutenizing hormone–releasing hormone (LHRH) agonists (also known as gonadotropin-releasing hormone (GnRH) agonists) or antagonists. LHRH agonists include leuprolide acetate, triptorelin embonate, goserelin actetate, and histrelin acetate, and these agents act by reducing androgen signaling [22]. Darolutamide is a competitive androgen receptor inhibitor, used in combination with other ADT agents in individuals with nonmetastatic disease [8].

In hormone-relapsed metastatic cancer, current management strategies recommend the use of CYP17 inhibitors or androgen receptor signaling inhibitors followed by taxane-based chemotherapy (docetaxel) [19–21]. In patients with metastatic disease who fail docetaxel chemotherapy, the use of alternate taxane derivatives such as cabazitaxel [19] or the use of androgen inhibitors such as abiraterone or enzalutamide may be considered [19, 20]. Abiraterone is a structural analog of pregnenolone which acts by inhibiting

TABLE 3 Pharmacotherapy used in the treatment of prostate cancer by risk and tumor location [19–23].

	Drug class	Commonly used agents	Tumor risk and location
Androgen deprivation Therapy (ADT)	Antiandrogen	Bicalutamide	Adjuvant or neoadjuvant therapy for:
		Flutamide	Intermediate or high-risk localized
		Nilutamide	Locally advanced
		Cyproterone acetate	
	LHRH agonists	Leuprolide acetate	Adjuvant or neoadjuvant therapy for:
		Triptorelin embonate	Intermediate or high-risk localized
		Goserelin acetate	Locally advanced
		Histrelin acetate	
	LHRH antagonists	Degarelix	Adjuvant or neoadjuvant therapy for:
			Intermediate or high-risk localized
			Locally advanced
	Androgen receptor inhibitors	Darolutamide	Nonmetastatic
	Tyrosine kinase inhibitors	Enzalutamide	Metastatic
	CYP17 inhibitors	Abiraterone	Metastatic
Chemotherapy	Taxanes	Docetaxel	Adjuvant or neoadjuvant therapy for:
			Intermediate or high-risk localized
			Locally advanced
			Metastatic
		Cabazitaxel	Metastatic disease relapse post docetaxel
Genomic biomarker–based treatments	Tyrosine kinase inhibitor	Entrectinib	Unknown

the CYP17 enzyme necessary for androgen synthesis [22]. Enzalutamide is an androgen receptor signaling inhibitor which acts via three pathways: direct inhibition of the androgen receptor, prevention of translocation of the androgen receptor from the cytoplasm to the nucleus, and inhibition of binding of the androgen receptor to chromosomal DNA preventing tumor gene transcription [24]. The role of genomic biomarker–based treatments, such as the tyrosine kinase inhibitor entrectinib, in the treatment of prostate cancer is yet to be determined [19].

1.4 Colon

Alongside lung, female breast, and prostate cancer, colon cancer has one of the highest cancer burdens among both men and women worldwide. It accounts for 6.1% of all new cancer cases worldwide and 5.8% of all cancer deaths [1]. Colon cancer is more common in countries with a high human development index including Europe, Australia, New Zealand, and North America [1]. Like lung, breast, and prostate cancer, surgery is the primary option for treatment of colon cancer with pharmacotherapy reserved for later stages [25–28]. In nonmetastatic colon cancer, the choice of treatment strategy depends on both tumor and patient characteristics. Chemotherapy is used for tumors classified as stage III or above using the tumor, nodes, metastases (TNM) classification of malignant tumors. The TNM classification uses tumor size and extent (T), the number of nearby lymph nodes involved (N), and if the cancer has metastasized (M), to classify tumors [29].

A wide range of chemotherapeutic agents are used in the management of colon cancer (Table 4). Treatment regimens include both monotherapy and combination regimens. Management of nonmetastatic colon cancer generally involves chemotherapy, with agents being used either as monotherapy or as combination regimens. FOLFOX (oxaliplatin in combination with 5-fluorouracil (5-FU) and leucovorin (folinic acid)), CAPOX (capecitabine and oxaliplatin), and FOLFIRI (fluorouracil, leucovorin, and irinotecan) are the regimens currently most commonly used in the management of nonmetastatic stage III or higher tumors [25–27]. Capecitabine may be used both in combination regimens and as monotherapy [25–27].

Biological pharmacotherapeutics, generally in combination with chemotherapy regimens, are commonly used in the management of metastatic disease [25–27]. A number of molecular targets have been identified, including mutations in tumor BRAF, RAS, or MMR genes as well as

TABLE 4 Pharmacotherapies used in the treatment of colon cancer [25–28].

Pharmacotherapy	Drug class	Commonly used agents	Tumor type
Chemotherapy	Platinums	Oxaliplatin	Nonmetastatic
			Metastatic
	Antimetabolites	5-Fluorouracil ± leucovorin	Nonmetastatic
			Metastatic
	Antimetabolites	Capecitabine	Nonmetastatic
			Metastatic
	Antimetabolite	Tegafur + uracil	Metastatic
	Antimetabolite	Trifluridine + tipiracil	Metastatic (not first line)
	Campothecins	Irinotecan	Metastatic
Biologicals	Anti-VEGF monoclonal antibody	Bevacizumab	Metastatic
		ziv-aflibercept	
		ramucirumab	
	Anti-EGFR monoclonal antibodies	Cetuximab	Metastatic with KRAS/NRAS WT or BRAF V600E mutations
		Panitumumab	
	BRAF inhibitor	Vemurafenib	Metastatic with BRAF V600E mutation
	PD-1 receptor check point inhibitor	Nivolumab	cMMR/MSI-H only
	Multityrosine kinase inhibitor	Regorafenib	Metastatic (not first line)

other biomarkers such as vascular endothelial growth factor (VEGF) involved in tumor growth [28]. Choice of biological agent in the treatment of metastatic disease depends on tumor biomarker and previous treatments (Table 3).

1.5 Skin

Nonmelanoma skin cancer (NMSC) includes basal cell carcinoma, squamous cell carcinoma alongside a number of rare skin tumors. While the incidence of nonmelanoma skin cancer is high, with nonmelanoma skin cancer the fifth most common cancer globally comprising 5.8% of all new cancer cases, mortality is considerably lower at 0.7% of all cancer deaths [1]. Australia and New Zealand have the highest age standardized rate of nonmelanoma skin cancer at 229.2 per 100,000 men and 66.7 per 100,000 women, followed by North America with 76.9 per 100,000 men and 36.8 per 100,000 women [1]. Management of nonmelanoma skin cancer is dependent upon the type of malignancy.

Basal cell carcinoma (BCC) is the most common type of nonmelanoma skin cancer, accounting for more than half of all NMSC cases [30]. BCC removal, generally surgical, remains the gold standard for the management of BCC [30, 31]. Surgical removal may occur via a range of modalities including standard excision, curettage and electrodessication, and micrographic surgery [30]. Topical pharmacotherapeutic removal may also be used, with the primary agents used being topical imiquimod or topical 5-fluorouracil [30, 32]. For locally advanced lesions or metastatic disease, the current best practice includes the use of vismodegib, a smoothened inhibitor targeting the hedgehog pathway (Table 4) [6].

Squamous cell carcinoma (SCC) is the second most common skin cancer following BCC [33]. Like BCC, the majority of local SCCs are removed surgically using surgery, curettage and electrodessication, or micrographic surgery [33]. There are no data to support the use of adjuvant pharmacotherapy in the treatment of SCC [34]. While there have been a number of studies exploring the use of topical pharmacotherapies for SCC removal, the current research does not support their routine use, unless the SCCs exist among multiple in situ actinic keratoses or tumors. In these circumstances, topical imiquimod, 5-fluorouracil, diclofenac, or ingenol mebutate may be used to remove the surrounding tumors and keratoses [25, 33]. In the management of SCC, pharmacotherapy is primarily used for the management of locally advanced or metastatic disease. A wide range of chemotherapeutic agents are used in the management of advanced SCC, either as monotherapy or as combination regimens (Table 5). Unlike other cancers such as colon cancer, where well-established chemotherapy regimens exist, there is no standard regimen for the management of advanced SCC [34]. Current biological therapy for SCC has focused on EGFR signaling. To date, EGRF inhibitors such as monoclonal antibodies cetuximab or panitumumab and small molecule kinase inhibitors erlotinib, gefitinib, and dasatinib have all been used in the management of advanced disease.

However, these remain second line in advanced disease for disease progression following chemotherapy [33, 34].

TABLE 5 Pharmacotherapy used in the management of nonmelanoma skin cancer [33, 34].

		Local disease	Locally advanced or metastatic disease
Basal cell carcinoma	Topical agents	Imiquimod	
		5-Fluorouracil	
	Chemotherapy		Platinum-based
	Biologicals		Vismodegib
Squamous cell carcinoma	Chemotherapy		Platinum-based
			5-Fluorouracil, bleomycin, methotrexate, adriamycin, taxanes, gemcitabine
			Ifosfamide
	Biologicals		Cemiplimab
			Cetuximab
			Panitumumab
			Erlotinib
			Gefitinib
			Dasatinib

2 Conclusions

A wide range of pharmacotherapies are used in the management of the most commonly occurring cancers. With increased understanding of the molecular pathways and biomarkers involved in tumor growth and replication, and the development of biological agents targeting these, pharmacological options for the management of many cancers are changing rapidly. As our knowledge and understanding of the molecular processes underlying cancer continue to grow, future treatment options allowing targeted drug delivery will enable new treatments maximizing efficacy while minimizing adverse effects.

References

[1] Bray F, Ferlay J, Soerjomataram I, Siegel RL, Torre LA, Jemal A. Global cancer statistics 2018: GLOBOCAN estimates of incidence and mortality worldwide for 36 cancers in 185 countries. CA Cancer J Clin 2018;68(6):394–424.

[2] Sonnenschein E, Brody JA. Effect of population aging on proportionate mortality from heart disease and cancer, U.S. 2000-2050. J Gerontol B Psychol Sci Soc Sci 2005;60(2):S110–2.

[3] Chemocare. Chemotherapy terms: Cleveland Clinic. Available from http://chemocare.com/chemotherapy/what-is-chemotherapy/chemotherapy-terms.aspx; 2020.

[4] Ritter J, Flower R, Henderson G, Loke YK, MacEwan D, Rang H. Rang and Dale's pharmacology. 9th ed. Edinburgh: Elsevier; 2019.

[5] Schirrmacher V. From chemotherapy to biological therapy: a review of novel concepts to reduce the side effects of systemic cancer treatment (review). Int J Oncol 2019;54(2):407–19.

[6] Neal RD, Sun F, Emery JD, Callister ME. Lung cancer. BMJ 2019;365:l1725.

[7] Alexander M, Kim SY, Cheng H. Update 2020: management of non-small cell lung cancer. Lung 2020;198(6):897–907.

[8] National Institute for Health and Care Excellence. Lung cancer: diagnosis and management (NICE guideline 122). Updated; 2019.

[9] Postmus PE, Kerr KM, Oudkerk M, Senan S, Waller DA, Vansteenkiste J, et al. Early and locally advanced non-small-cell lung cancer (NSCLC): ESMO Clinical Practice Guidelines for diagnosis, treatment and follow-up. Ann Oncol 2017;28(suppl_4):iv1–iv21.

[10] ESMO Guidelines Committee. eUpdate – Early and Locally Advanced Non-Small-Cell Lung Cancer (NSCLC) Treatment Recommendations; 2020.

[11] Cancer Council Australia Lung Cancer Guidelines Working Party. Clinical practice guidelines for the treatment of lung cancer. Sydney: Cancer Council Australia; 2021. Available from: https://wiki.cancer.org.au/australia/Guidelines:Lung_cancer.

[12] Arbour KC, Riely GJ. Systemic therapy for locally advanced and metastatic non-small cell lung cancer: a review. JAMA 2019;322(8):764–74.

[13] Howlader N, Noone AM, Krapcho M, Miller D, Brest A, Yu M, Cronin KA, editors. SEER Cancer Statistics Review, 1975–2017. Bethesda, MD: National Cancer Institute; 2021. [Available from https://seer.cancer.gov/csr/1975_2017/.

[14] Cardoso F, Kyriakides S, Ohno S, Penault-Llorca F, Poortmans P, Rubio IT, et al. Early breast cancer: ESMO clinical practice guidelines for diagnosis, treatment and follow-up. Ann Oncol 2019;30(10):1674.

[15] Waks AG, Winer EP. Breast cancer treatment: a review. JAMA 2019;321(3):288–300.

[16] National Institute for Health and Care Excellence. Early and locally advanced breast cancer: diagnosis and management. NICE guideline [NG101]; 2018.

[17] Australia C. Guidance for the management of early breast cancer: Recommendations and practice points. Cancer Australia: Surry Hills, NSW, Australia; 2020.

[18] Gleason DF and the Veterans Administration Cooperative Urological Research Group. Histologic grading and staging of prostatic carcinoma. In: Tannenbaum M, editor. Urologic Pathology. Philadelphia, PA: Lea & Febiger; 1977. p. 171–87.

[19] National Institute for Health and Care Excellence. Prostate cancer: diagnosis and management. NICE guideline [NG131]; 2019.

[20] Parker C, Castro E, Fizazi K, Heidenreich A, Ost P, Procopio G, et al. Prostate cancer: ESMO clinical practice guidelines for diagnosis, treatment and follow-up. Ann Oncol 2020;31(9):1119–34.

[21] Sanda MG, Cadeddu JA, Kirkby E, Chen RC, Crispino T, Fontanarosa J, et al. Clinically localized prostate Cancer: AUA/ASTRO/SUO guideline. Part II: recommended approaches and details of specific care options. J Urol 2018;199(4):990–7.

[22] Crawford ED, Heidenreich A, Lawrentschuk N, Tombal B, Pompeo ACL, Mendoza-Valdes A, et al. Androgen-targeted therapy in men with prostate cancer: evolving practice and future considerations. Prostate Cancer Prostatic Dis 2019;22(1):24–38.

[23] Dellis A, Zagouri F, Liontos M, Mitropoulos D, Bamias A, Papatsoris AG, et al. Management of advanced prostate cancer: a systematic review of existing guidelines and recommendations. Cancer Treat Rev 2019;73:54–61.

[24] Saad F. Evidence for the efficacy of enzalutamide in postchemotherapy metastatic castrate-resistant prostate cancer. Ther Adv Urol 2013;5(4):201–10.

[25] Argiles G, Tabernero J, Labianca R, Hochhauser D, Salazar R, Iveson T, et al. Localised colon cancer: ESMO clinical practice guidelines for diagnosis, treatment and follow-up. Ann Oncol 2020;31(10):1291–305.

[26] Benson AB, Venook AP, Al-Hawary MM, Cederquist L, Chen YJ, Ciombor KK, et al. NCCN guidelines insights: colon cancer, version 2.2018. J Natl Compr Canc Netw 2018;16(4):359–69.

[27] National Institute for Health and Care Excellence. Colorectal cancer. NICE guideline [NG151]; 2020.

[28] Van Cutsem E, Cervantes A, Adam R, Sobrero A, Van Krieken JH, Aderka D, et al. ESMO consensus guidelines for the management of patients with metastatic colorectal cancer. Ann Oncol 2016;27(8):1386–422.

[29] Brierley JD, Gospodarowicz MK, Wittekind C, editors. TNM classification of malignant tumours. 8th ed. Oxford: Wiley-Blackwell; 2017.

[30] Bichakjian CK, Olencki T, Aasi SZ, Alam M, Andersen JS, Berg D, et al. Basal cell skin Cancer, version 1.2016, NCCN clinical practice guidelines in oncology. J Natl Compr Canc Netw 2016;14(5):574–97.

[31] Griffin LL, Ali FR, Lear JT. Non-melanoma skin cancer. Clin Med (Lond) 2016;16(1):62–5.

[32] Peris K, Fargnoli MC, Garbe C, Kaufmann R, Bastholt L, Seguin NB, et al. Diagnosis and treatment of basal cell carcinoma: European consensus-based interdisciplinary guidelines. Eur J Cancer 2019;118:10–34.

[33] Work G, Invited R, Kim JYS, Kozlow JH, Mittal B, Moyer J, et al. Guidelines of care for the management of cutaneous squamous cell carcinoma. J Am Acad Dermatol 2018;78(3):560–78.

[34] Stratigos AJ, Garbe C, Dessinioti C, Lebbe C, Bataille V, Bastholt L, et al. European interdisciplinary guideline on invasive squamous cell carcinoma of the skin: Part 2. Treatment. Eur J Cancer 2020;128:83–102.

Chapter 3

Current practices in oncology drug delivery

Saniya Mahendiratta*, Seema Bansal*, Subodh Kumar, Phulen Sarma, Ajay Prakash, and Bikash Medhi
Department of Pharmacology, PGIMER, Chandigarh, India

1 Introduction

1.1 Historical background

Worldwide, cancer is the one of the leading causes of mortality and is still the biggest challenge for scientists due to lack of complete cure. However with increase of our knowledge, early detection and treatment is a game changer where millions of cancer patients extend their survival. Cancer terminology originated from the Greek word karkinos and was given by a physician Hippocrates (460–370 BC) but cancer is not a new disease for our world and its evidences are present in ancient manuscript dates and in ancient Egypt in about 1600 BC where the presence of human bone cancer was found in mummies. Similarly, the oldest recorded breast cancer case is evident from ancient Egypt in 1500 BC [1–3].

1.2 Ancient theories about cancer

There are various ancient theories about cancer. One of them is the Humoral theory, in which Hippocrates believed that the human body is made up of basically four types of body fluids which include phlegm, blood, yellow bile, and black bile. Any imbalance in the proportion of these fluids leads to disease and it was thought that cancer is the outcome of black bile in a particular organ site. Over more than 1300 years, this theory of cancer was thought to be standard, as autopsies were not allowed due to religious reasons and very limited knowledge. In the 17th century, another supported ancient theory in rule was Lymph theory, which proposed that the cancer formation takes place due to a fluid called lymph. Life was supposed to be functional due to a continuous circulation of lymph fluid and cancer formation was supposed to be due to continuous throwing out of lymph from blood. In a theory called the Parasite theory till 18th century, cancer development was thought to be due to parasites. In another theory starting from 1800 to 1920s, cancer development was considered to be due to trauma. In 1838, Muller was the scientist who suggested that instead of lymph, cancer is made of cells and called it as the Blastema theory. Further his student Virchow (1821–1902) observed that cancer cells originate from other normal cells and due to chronic irritation and proposed the Chronic irritation theory [4]. Thereafter, Theirsch speculated that instead of any unidentified fluid, cancers metastasize due to the spread of malignant cells [5]. In the earlier and middle decades of the 20th century, scientists started to decode the complexity behind the cancer and later Watson and Crick discovered the DNA helical structure and received the Nobel Prize in 1962. Later, scientists learned how mutations damage the workflow of genes and how these DNA damages could be inherited or may be caused through radiations, carcinogens, and viruses. In this regard, scientists discovered two important gene families which are oncogenes and tumor suppressor genes. Both the gene families play an important role in cell cycle regulation, and mutations in these cells cause deregulation of cell division and DNA repair resulting in an uncontrolled cell growth which ultimately leads to cancer. With increasing knowledge about the molecular basis of tumorigenesis, several genes have been discovered which are associated with different types of cancers, just as BRCA1 and BRCA2 genes are associated with breast cancer.

1.3 Description and epidemiology of cancer

There are different types of tumors but one thing is common that all tumors originate when normal cells of a particular body organ instead of dying start growing in an uncontrolled manner as a result of changes at the genetic level. Our cellular system has DNA damage/mutation-specific repair mechanisms and cancer cells are the outcome of escaped repair mechanism. Cancers are also inherited when a person carries the damaged DNA from their parents. Types of cancers are named for the area where they originate and

* Equal contribution, so designated as combined first author.

Advanced Drug Delivery Systems in the Management of Cancer. https://doi.org/10.1016/B978-0-323-85503-7.00006-7

the type of cells involved, even if they are circulated to other organs of the body through circulation (metastasis) like when a breast cancer cell spreads to liver, the cancer is still called as breast cancer. There are some tumors which are not cancerous and called benign tumors which do not grow and are not life threatening. Different clinical terms are designated for the type of cancer on the basis of their origin such as Carcinoma (cancer originated from skin or the tissues that line other organs); Sarcoma (involvement of connective tissues); Leukemia (bone marrow); Lymphoma and myeloma (involvement of the immune system). There are several risk factors, which are responsible for cancer, such as exposure to DNA damaging chemicals (carcinogens), radiation exposure, smoke exposure, environmental, and lifestyle choices which also play a major role such as choices of diet and level of physical activities. As with a healthy lifestyle, chances of cancer can be drastically reduced, while with early stage of detection, better drug delivery, and treatment options the survival rate of a person can be increased for many years.

2 Past therapeutic approaches in cancer (1900 and earlier)

In ancient times, there was no known treatment for cancer and only palliative treatment was available. According to available inscriptions in the ancient times, surgeons knew about the reoccurrence of cancer even after surgery and only surface tumors were surgically removed in way similar to that in recent times. The invention of anesthesia took place in 1846 and for the first time, surgeons Billroth, Handley, and Halsted directed cancer surgeries by eradicating the entire tumor along with lymph nodes. After that a surgeon, Paget reported that the phenomena of cancer spread through metastasis and reported the challenges of cancer surgery. Later in 1878, Thomas Beatson revealed that the removal of rabbit ovaries resulted in the lack of milk production. Later, regression of metastatic prostate cancer was observed through removal of testes. In 1896, the discovery of "X-ray" by Roentgen came as revolution, as 3 years later the uses of radiations were applied for cancer diagnosis and in treatment [5, 6].

3 Current approaches of drug delivery

Novel therapeutic approaches in oncology are required, as they improve the effective delivery of chemotherapeutic agents to the cancer cells. The conventional anticancer agents pose a problem of getting accumulated in cancerous as well as normal cells, owing to their nonspecific nature. Therefore, it becomes a prerequisite to go for novel targeted agents that can reduce the systemic toxicity as well as improve the quality of life. Various categories of the targeted therapies are discussed.

3.1 Targeted therapy in cancer

3.1.1 *Molecular targeted therapy*

The aim of targeted therapy is to overcome the nonspecific toxicity corresponding to the conventional chemotherapy. The use of biological agents can be done individually but most commonly it is given in combination with the standard chemotherapeutic agents. This way, malignant cells are killed either through targeting the expression of specific molecules or via activating different molecular pathways that lead to the transformation of the tumor [1].

Most commonly targeted therapies are: (1) Small molecule inhibitors, (2) monoclonal antibodies, and (3) immunotoxins.

Small molecule inhibitors: The function of small molecule drugs is to inhibit certain growth factors (GFRs) and enzymes, for instance, epidermal growth factor receptor (EGFR) which leads to the proliferation of cancer cells. Cancer cells evade the normal cell cycle constraints, owing to an inappropriate kinase activity functioning as the major pathway. These molecules bind competitively to the active and inactive sites of tyrosine kinase targeting those proteins which are either upregulated or downregulated during the progression of cancer (e.g., Akt or mTOR, BCR-ABL) [7].

Targeting BCR-ABL: Targeting BCR-ABL is one of the successful drug-targeted therapies and is the most common targeted fusion protein in chronic myeloid leukemia (CML) patients. Currently, United States is marketing two drugs created by Novartis and Bristol-Myers Squibb and which are marketed under the names Gleevec and Sprycel (second-generation small molecule inhibitor) [3]. Gleevec (imatinib mesylate, Novartis) has its application in various cancers which also includes chronic myeloid leukemia positive for Philadelphia chromosome positive (Ph + CML) and gastrointestinal stromal tumors (GISTs) positive for c-kit [4]. It has specificity toward tyrosine kinase BCR-ABL with minimal effects on other tyrosine kinases.

Sprycel (dasatinib, a product of Bristol-Myers Squibb) is a second-generation inhibitor and has the same activity as that of Gleevec, i.e., interfering with the binding at ATP sites of BCR-ABL and its use is limited to Ph + CML patients who are resistant to Gleevec [3, 5].

Targeting Vascular Endothelial Growth Factor (VEGF): Angiogenesis is the most important process during disease development and progression, and VEGF plays quite an important role as a proangiogenic molecule. It has three individual receptors: VEGFR-1 or Flt-1, VEGFR-2 or Flk-1, and VEGFR-3 or Flt-4 [8]. These receptors are present on the surface of the cells of different kinds which include endothelial cells leading to its dimerization and activation on transphosphorylation after binding to the ligand. VEGFR-2 activation leads to an increased vascular permeability and migration with increased proliferation [9]. This makes VEGF an ideal therapeutic target. Small molecules targeting VEGF have been discussed.

Nexavar (sorafenib, Bayer HealthCare): It not only targets VEGFR but also has its effects on c-kit and PDGFRA. Among VEGFRs, it has activity on both VEGFR-1 and VEGFR-3. FDA has approved if for the treatment of patients with advanced renal cell carcinoma (RCC) and unresectable hepatocellular carcinoma and [7].

Targeting Epidermal growth factor receptor (EGFR) and Human epidermal growth factor 2 (HER-2): EGFR being a transmembrane receptor plays a role in cell processes such as proliferation, migration, cell growth, tissue invasion, and survival. There is an overexpression of HER-2 in breast carcinomas and is associated with more aggressive forms of cancer.

Iressa (gefitinib, AstraZeneca) has its specific action on EGFR ATP-binding site leading to the inhibition of EGFR activation of Ras signaling cascade and hence suppressing the cell growth. It has been approved to be used for the management of metastatic or advanced nonsmall cell lung cancer (NSCLC) by pairing it with conventional chemotherapy [10].

A small molecular inhibitor Tykerb (lapatinib, GlaxoSmithKline) has its specificity against both HER-2 and EGFR and possesses a greater affinity toward both the receptors in comparison to Iressa [11]. This drug is certainly given to those patients who have been unresponsive to Herceptin. It has received approval by FDA to be administered with Xeloda (capecitabine) which is an oral anticancer agent having it utility in breast cancer tumors with overexpressed HER-2 [12].

Targeting mTOR: The original discovery of mammalian target of rapamycin (mTOR) involves the use of rapamycin where it acts as a protein kinase regulating growth and proliferation of the cells. It also includes several other signaling pathways with a possible role of tumor suppressor phosphatase and tensin homolog (PTEN) [8].

Temsirolimus (Torisel, Wyeth Research) is one such small molecule inhibitor given approval for RCC. Median survival time has been increased (10.9–7.3 months) in Phase III clinical trials, compared to the patients given interferon alfa-2a [13].

A similar drug known as Certican (Everolimus, Novartis) has effects on the immune system by suppressing it and preventing transplanted organs from rejection. Therefore, it was first used in combination with transplantation of the organs (heart and kidney) and is yet another powerful inhibitor of RCC progression [14].

3.1.2 Monoclonal antibodies

Monoclonal antibodies can specifically target a protein aiding in promoting tumorigenesis, owing to its dysregulation. It was first made in laboratory with the fusion of cancer cells from mouse (myeloma) with its B-cells and leading to the formation of a hybridoma. Earlier antibodies were all produced with the use of mouse cells but now-a-days they are either fully human or humanized which marks their safety and efficacy for their treatment in humans.

Bevacizumab (Avastin, Genentech): It is a monoclonal antibody targeting VEGF. It showed increased survival in Phase II clinical trials when it was given in combination with 5-fluorouracil (5-FU) [15].

Trastuzumab (Herceptin, Genentech): It has been synthesized against the ectodomain region of HER-2 where it binds with HER-2 interfering with receptor dimerization and disrupting the intracellular signaling cascade leading to cell growth inhibition [16]. It has its utility in breast cancer positive for HER-2 who have recurrence to previous chemotherapy. There have been studies demonstrating the use of Herceptin in combination with doxorubicin, which is an anthracycline and increasing the survival by 20% in comparison to anthracycline alone [12].

Rituxan: Studies have demonstrated dysregulation in the growth of B-cells in B-cell non-Hodgkin's lymphoma and CML displaying protein CD20 on the surface of the cell [17]. Rituxan has the capability of targeting CD20 resulting in cytotoxicity of both normal B-cells as well as cancerous cells [13]. It has also been approved by FDA as a combination with other anticancer agents such as doxorubicin, cyclophosphamide, prednisone, and vincristine (CHOP) [18, 19].

Bexxar and Zevalin: Tositumomab (Bexxar, GlaxoSmithKline) has similar actions as that of Rituxan of binding to CD20, with the exception that it contains Iodine-131 (I-131) bound to the antibody [16]. Its mechanism of action mostly relies on radiations with the decay of I-131 and also effects mediated by antibody [20]. Zevalin (Cell Therapeutics), which is a similar drug, has the capability of binding Yttrium-90 (Y-90) to anti-CD20 antibody [21].

Alemtuzumab (Campath, Genzyme): It is a monoclonal antibody produced against CD52 which on binding makes prone the cancer cells toward killing by the immune system. It has been given approval for the treatment of B-CLL as a single agent [22].

3.1.3 Immunotoxins

This category of drugs involves monoclonal antibodies which are modified or growth factors where their production was originally done by their chemical binding to a toxin or a protein [18]. Now, they have been produced with the creation of recombinant DNA generating a fusion protein containing an additional domain on the antibody or growth factor in the form of toxin [19].

Denileukin diftitox (Ontak, Ligand Pharmaceuticals): It is a fusion protein of interleukin-2/DT category which has its utility in the treatment of patients with cutaneous T-cell lymphomas (CTCLs). Thirty percent of response rate has been shown in CTCL patients in Phase III clinical trials including side effects such as hypotension, edema, and nausea [23, 24].

BL22: A B-cell related cancer, Hairy cell leukemia (HCL) can be treated with immunotoxins against CD22 which is a specific marker for B-cells which will lead to a reduction of B-cells [21]. Among this category, BL22 is a fusion protein which aims at CD22 as its target and produces rDNA which encodes an Fv fragment of an anti-CD22 antibody. Phase I clinical trials evaluated the remission rates which were 61% for complete remission and 19% for partial remission in patients with HCL who did not respond to conventional chemotherapy [25].

3.2 Gene therapy

During the last two decades, there have been advances in human genomics which reveal host genome somatic aberrations as a very common cause of cancer. It was Rogers et al. who gave the first proof-of-concept for the transfer of the gene via viral vectors, where he observed the utility of virus to transfer foreign genetic material into the cells of interest [26] which he further tested in humans. These experiments made Roger the first person for performing the gene therapy trials in humans. In his study, a wild-type Shope papilloma virus was used to transfer the arginase gene into two subjects who were suffering from disorder of urea cycle (i.e., hyperargininemias) [26, 27]. According to his hypothesis, this gene would have naturally encoded the gene for arginase activity, but the results were negative showing no change in arginine levels as well as no variations in clinical consequences.

Finally, the US FDA approved first gene therapy protocol conducted in 1989. This protocol mentioned the use of tumor-infiltrating lymphocytes (TILs) extracted from patients of advanced melanoma conjugated with a marker gene ex vivo and further expanded in vitro, and final reinfusion into the patients [23]. Similarly, Cline et al. made another breakthrough in gene therapy by treating thalassemia patients, by extracting their bone marrow cells followed by transfection with human globulin gene–containing plasmids ex vivo and delivering back to the patients [4]. This study was considered a milestone as it was carried out without the consent of the University of California, Los Angeles (UCLA) Institutional Review Board and also demonstrated the technicality and complexity of gene therapy.

3.2.1 Gene therapy in clinics

Despite several advances in the field of molecular biology and genetic engineering, the approach of using gene therapy has not shown much promise. Nonetheless, it can hold a potential of providing success in future, owing to its promising results of long-term survival in a couple of studies. Some gene therapies are being successfully translated clinically:

p53, one of the most important genes getting affected in many cancer types, has been a subject for gene therapy. In 2001, a genetically modified Ad gene in E1B region known as ONYX-015 (*dl*1520), with an absence of a 55 kd gene (for restricting replication in p53 mutant cells), was assessed in NSCLC patients [28]. It was observed in Phase I clinical trials that the virus was well tolerated and was capable of shrinking the tumor with a partial response rate not up to the expectations, as demonstrated in preclinical studies (response only in 3 patients out of 15). During a follow-up in Phase II trial, intratumoral injections were administered in two cycles giving a response rate of 28% in the treatment group in contrast to 12% in controls. When ONYX-50 was given alone or along with a chemotherapeutic agent in Phase III clinical trial, a response rate of 72.7% was seen, compared to 40.4% in the control group. Apart from this, another clinical study was conducted utilizing CG7870 which is a prostate-specific therapeutic adenovirus with the replication of the virus under the regulation of a promoter specific for prostate cancer. However, the results were disappointing, as there were no signs of tumor regression except a reduction of 25%–49% in PSA levels in five patients [29].

Gendicine was the first gene therapy, which was commercially available for head and neck squamous cell cancer treatment, synthesized from an engineered adenovirus expressing *p53* of the wild type regulated by *Rous sarcoma virus* [30]. Phase III trial was done to access the efficacy of this therapy by recruiting 132 patients where a complete regression was observed in 64% of patients when used along with radiotherapy. Similarly, Advexin® (INGN 201, which was originally a product of Introgen Therapeutics, Inc., Austin, TX) was also reviewed for its clinical benefit in several cancers [31] (Gabrilovich 2006).

Additionally, clinical trials conducted with other tumor suppressor genes such as TNF-alpha, TRAIL, IL-2, and *mda-7*/IL-24 genes exhibited promising results. In 2005, several investigators in their study published efficacy and well tolerability of replication-incompetent adenovirus expressing *mda-7*/IL-24, Ad.*mda-7* (INGN-241) with advanced cancer patients [32]. Induction of apoptosis was observed with multiple intratumoral injections of Ad.*mda-7*, owing to MDA-7/IL-24 expression resulting in the shrinkage of more than 50% of the lesion, with no signs of toxicity in 44% of the patients suggesting it to be safe and efficacious. Interestingly, it was demonstrated that MDA-7/IL-24 administration to patients induced a significantly greater apoptosis compared to the ones who received Ad-p53 in NSCLC. A 64-year-old patient of metastatic melanoma showed a considerable response compared to a patient who had squamous cell carcinoma of the penis that showed a less dramatic response [33].

Multiple researches conducted over the past few years have shown a clear correlation between the altered expression of microRNAs (miRNAs) leading to the development of cancer [29]. Typical structures of microRNAs have RNA short and single-stranded sequences (19–23 nucleotides) which can bind with 3′-UTR sequences of genes resulting in a gradual degradation of mRNA and reduced gene levels [34]. This leads to the postulation that mutation in miRNAs results in the overexpression of an oncogene or inversely reduction in the tumor suppressor gene [35]. Certain mRNAs like miR-21, miR-155, miR-23, miR-191, and miR-196b are typically overexpressed in breast cancer [36]. Other miRNAs like miR-191, miR-199a, and miR-155 are overexpressed in AML. miR-2909 has a very high potential to control immunomodulation [36,37] and its altered expression can be the specific cause of multiple diseases including rheumatoid arthritis [38].

Another form of gene therapy is the enzyme prodrug therapy that converts substances of nontoxic origin (prodrug) into agents which are active physiologically. The most frequent virus used is HSV-TK which converts ganciclovir, a prodrug, into cytotoxic triphosphate ganciclovir. On the basis of this approach, Cerepro was developed by the Anglo Finnish Company Ark Therapeutics to treat malignant glioma under a promoter of cytomegalovirus origin. The survival benefit of Cerepro was quite significant in comparison to retroviral or standard therapy, with fever being the only adverse effect.

3.3 Bacterial-mediated therapy

Bacterial-mediated therapy is one of the important alternative approaches of conventional cancer therapies. Commonly used bacteria in cancer therapy are *Salmonella*, *Bifidobacterium*, *Escherichia*, *Pseudomonas*, *Clostridium*, etc. Its most important feature is that it can be engineered by genetic methods to alter their ability to synthesize and release specific compounds. Most of the time, these bacteria target the hypoxic area of tumors and penetrate into the tissue resulting in the release of toxins or enzymes which is to be tested such as proteases and lipases. Sometimes, tumoricidal or immunotherapeutic agents can be transferred in the targeted tissues in the form of bacterial vectors [39]. Mode of action of bacteria-mediated therapy is as per the following:

Genetically modified bacteria: These are the bacteria which are genetically modified for the expression of reporter genes, proteins which are toxic, and tumor-specific antigens. Studies have reported that genetically modified bacteria play a promising role as anticancer therapy via lowering pathogenicity to the host and via increasing the antitumor efficacy [40,41]. Study by Ref. [42] reported that to selectively colonize tumors, *Salmonella typhimurium* and *Clostridium butyricum* are used as delivery vectors in mouse tumor cells without toxic adverse effects and immunogenic reactions. Clinical trial studies have reported that the administration of genetically engineered *Salmonella typhimurium* and *Clostridium novyi* activates the host immune system via expressing cytokines (IL-2, IL-4, IL-18 and chemokines-21) which results in regression and necrosis of tumors [43].

Bacteria as immunotherapeutic agents: Study by Ref. [40] reported that CD8+ lymphocytes isolated from *Clostridium novyi* stimulate acquired immunity in a tumor-specific model. Further study by Avogadri et al. reported that *Salmonella typhimurium* releases a type 3 secretion system in tumorous cells. Study by Furomoto et al. reported that compounds derived from bacteria, like CpG oligonucleotides, activated dendritic cells and result in a complete regression of B16F10 melanoma tumors.

Bacterial toxins/enzymes in cancer therapy: Bacterial toxins which block tumor cell cycles are known as cyclomodulins. Various microorganisms release protein toxins which suppress immune response of the infected host. Cytolysin A is an enzyme toxin released by *S. typhimurium/E. coli* which acts as anticancer therapy via building pores in the tumor cell membrane and triggers caspase-mediated cell death. The authors of the study cited here [44] have developed a bacterial metabolite therapy by combining carbon nitride with a bacterial strain that produced nitric oxide.

Bacterial spores/vectors as tumoricidal agents: Development of live bacterial vectors is the most valuable tool as anticancer therapy. Studies have reported that anerobic bacteria specifically target solid tumors via triggering inflammatory cytokines within a tumor cell. Bacterial spores are also used as agents for the delivery of cytotoxic peptides, therapeutic protein, and anticancer agents for gene therapy [39].

3.4 Nanomedicine

Nanomedicine is a valuable method to treat cancer. The first clinically approved nanomedicine was a PEGylated liposomal formulation of doxorubicin. Till date, 15 anticancer nanodrugs have been approved and used clinically. Properties of nanomedicines, i.e., nanoscale sizes, high surface/volume ratio, release of drug in a controlled manner at the targeted site, protection from enzymatic and mechanical degradation of drugs, and addition of additives that enable drug solubilization make them unique to treat cancer. Common nanoparticles that have been used in cancer therapy are liposomes, polymer, lipid nanocapsules and nanoparticles, polymer drug conjugates, and pluronic micelle [40].

Nanomedicines are safe alternative to injurious formulations. For example, albumin-based paclitaxel nanoparticles, abraxane, are used in different types of cancers. Nanoparticles protect drugs from mechanical and enzymatic degradation. Nanoparticles protect drugs that are sensitive to degradation by various gene silencing agents, small interfering RNAs, and microRNAs. Currently, several oligonucleotide-containing nanoparticles are clinically approved; however, the development of several SiRNA nanodrugs for cancer therapy has been discontinued due to insufficient patient outcomes. Another form of nanomedicine is the use of components that assist in immunoevasion. Most common strategies to avoid immunological recognition and clearance of nanoparticles include the use of antifouling polymers, cell coatings, and self-peptides for surface modifications. Liposomal formulations of anticancer drugs, Doxil and Onivyde, are pegylated nanoparticles [45,46].

Nanomedicine is advantageous because of its ability to deliver combination of drugs in one nanoparticle. Vyxeos is the first combination nanomedicine of daunorubicin and cytarabine used in acute myeloid leukemia. Targeted nanomedicines are based on the principle on surface conjugation of molecular targeting ligands. Many of the targeted nanomedicines are under clinical trials (MM310 is an ephrin receptor 2 targeted docetaxel-containing liposomes which are in Phase 1 clinical trials), however, none of them is clinically approved [46].

Nanomedicines also trigger the release of therapeutic agents in the tumor. An important example of this is CX-072 which is a probody, antiprogrammed death ligand-1 in Phase I/II of clinical trials used in patients with lymphoma and solid tumor. Thus, site-specific drug delivery as compared to free drug administration is the major property of nanomedicines which make it unique. However, despite advancement in the biodistribution of drugs, less than 1% of the systemically administered nanomedicines reach the targeted tumor. As much as 90% of the injected nanoparticles are absorbed in liver and spleen. Moreover, failure during clinical trials is the major limitation of nanomedicines as anticancer therapies [46].

3.5 Immune checkpoint inhibitors

Immunotherapy also plays a promising role for the treatment of cancer patients. Studies have reported that T lymphocytes/effector molecules released by tumor cells play a significant role to control spontaneous or transplanted tumors. Study by Yuan et al. [47] reported that CTLA-4 negatively regulates the activation of T-cells and gets upregulated during the course of activation of T-cells. In response to this, the first fully human monoclonal antibody "IgG1k anti CTLA-4 known as ipilimumab was approved to treat the patients with cutaneous melanoma." The same antibody is also under investigation to treat a variety of cancer types, including renal carcinoma, prostate cancer, etc. Apart from ipilimumab, the other IgG2 anti-CTLA-4 monoclonal antibody under clinical trials is tremelimumab. Although as a monoclonal therapy tremelimumab has not shown any positive effect on cancer, it is also investigated in combination with duravalumab (immune checkpoint inhibitor targeting PD-L1) [43].

After the discovery of immune checkpoint CTLA-4, FDA gave approval to a fully human IgG4k monoclonal antibody known as nivolumab for metastatic melanoma. After that, another anti-PD-1 immune checkpoint inhibitor was investigated known as pembrolizumab, human IgG4k monoclonal antibody which was a marketed product of Merck for the treatment of solid tumors and identified as the first anticancer drug for receiving tissue/site-agonistic approval on the basis of biomarker analyses [48].

After the disruption of the pathway related to PD-1 immune checkpoint, the targeting of PDL1 with immune checkpoint inhibitors was proven useful as anticancer approach. Atezolizumab was identified as the first human monoclonal antibody released by Roche Genentech approved for checkpoint blockade therapy for locally advanced or metastatic urothelial carcinoma. After this, two other fully human IgG1k monoclonal antibody PD-L1 inhibitors avelumab and duravalumab were prepared. Apart from this, there was an exceptional response by the combination of ipilimumab and nivolumab in metastatic melanoma patients and hence was given approval for treating patients with BRAF mutations. There are certain other combinatorial checkpoint blockade therapies such as pembrolizumab and ipilimumab and duravalumab and tremelimumab which are in clinical trials for various kinds of tumors [44].

3.6 Radiomics and pathomics

Radiomics, a novel concept, had its beginning in 2012 when it was introduced in the form of advanced level of imaging analysis. It obtains the data from other image analyses such as magnetic resonance imaging (MRI), computed tomography (CT), and positron emission tomography (PET) [49]. This idea of radiomics can be useful to the clinicians for efficient treatment, predicting the location of tumor metastasis, correlating the results with histological examination. Radiomics combined with other testing techniques makes available to the patients a personalized treatment plan for advanced examination and treatment. On the other hand, *Pathomics* can be defined as the process of generating, interrogating, and characterizing large volumes of quantitative features from high-resolution tissue images [50].

3.6.1 Combining radiomics and pathomics with genomics

Radiogenomics can also play an important role in precision medicine where it provides voxel-by-voxel genetic information as well as information on the primary tumor to guide in the clinical treatment [51]. Additionally, it specifically defines lesions and differentiates appropriately between benign and malignant tumors giving a clearer picture of the screening of tumors [27]. Mu Zhou et al. 2015 from a study of 113 patients of NSCLC demonstrated certain features which include nodular shape, texture, margin, environment of the tumor, and overall characteristics of the lungs after collecting the CT data preoperatively and from tumor specimens postoperatively. The study concluded with the involvement of 10 meta-genes in various molecular pathways which included EGF and also irregular nodules, frosted glass opaque shadows, and undefined knots on the margins.

3.6.2 *Combining radiomics and pathomics with artificial intelligence*

Artificial Intelligence (AI) can be defined as the intelligence demonstrated by machines, but correctly it can be defined as interpretation and extracting information from external data in order to achieve specific goals and tasks [52]. Recently, the field of radiomics has provided additional improvements when combined with AI, such as the operational workflow being improved with the improvement in financial management and quality. For instance, the application of AI was done prior to MRI studies for differentiating between low- and high-level tumors with the use of image texture features for achieving World Health Organization (WHO) tumor grading levels. Further, diagnosis of different molecular subtypes of glial cells such as mutations present in isocitrate dehydrogenase (IDH) can be made easy with intensive learning by training machines with the use of convolutional neural network (CNN) method [53]. Even the response of treatment toward breast cancer has improved with the advances in radiomics and computer analysis capabilities. Although AI has been successful in cancer imaging, there are certain constraints which need to be addressed before it can be used extensively in the clinics. Briefly, it can be said that AI has wide applications in medical imaging which can be clearly seen from early research.

3.6.3 *Combining radiomics and pathomics with tumor markers*

"Biomarkers" has been defined as a crucial indicator to measure and assess any changes in normal physiological process, pathological process, or any treatment intervention. If radiomics is combined with biomarkers, which is a complex developmental process, it is capable of improving the accuracy in diagnosing cancer with reducing the cost and trauma to the patients [54]. This can be illustrated with the study of Professor Wang Hongbin and his team who retrospectively studied 177 patients with rectal adenocarcinoma and collected 385 radiomic groups. Results of the experiment showed that the combination of clinical radiology with CA199 and CEA models can be useful as noninvasive biomarkers for the identification of high-risk patients, thus customizing the strategies of treatment [25].

Immunohistochemistry (IHC) being a developed method is used by histopathologists for the improvement in diagnosis related to surgical pathology [26]. Therefore, in future if integrated radiology and IHC expression could be correlated it can provide the physicians with more information on diagnosis and evidence for predicting the clinical outcomes [55]. This will aid them to formulate an appropriate treatment which can ultimately be helpful for improving the quality as well as efficiency of medical services [56].

3.6.4 *Combining radiomics with pathomics immunological detection*

Immunotherapy for cancer has changed the scenario of its treatment by significantly improving the outcomes in patients with melanoma, lymphoma, and lung cancer, with the exception being the solid tumors where only 20%–50% of the patients showed response to treatment. Therefore, making it is a prerequisite to develop a test which can be frequently used for the determination of efficacy of immunotherapy [57,58].

Radiomics has the capability of assessing the immune infiltration in the tumor cells making itself as the novel predictor of efficacy. This has been demonstrated by Roger Sun and his colleagues who reported cell expression profiles of CD8 on the basis of eight feature radiology. To confirm the consistency between the radiology histological features with the gene expression for markers, CD8 cells were used. Hence, this study has evaluated the potential of radiomics for personalized immunotherapy for simulation analysis [59].

With the advancement of precision medicine, the innovative approach of radiomics provides an opportunity of cost-effective, noninvasiveness technique which avoids the unnecessary risk of toxicity related to treatment. Further, radiomics should be combined with other extra sources of information (e.g., genomics, biology, and immunology) for improving the diagnosis and anticipating personalized medicine treatments [60].

3.7 Theranostics

Theranostics combines two words "therapeutics" and "diagnostics" which refers to agents or techniques coupled with diagnostic imaging for targeted therapy [32]. The field of theranostics has provided with multiple benefits by inserting nanoparticles into the patients with the use of photodynamic therapy. Nanoparticles are actually the basis of theranostics, owing to their properties of precise imaging and diagnostic capability and simultaneously treating the disease [61]. Photosensitizers were first used for treating cancer in early 20th century in Munich, Germany but clinical approval was first given to Photofrin in 1993 by the Canadian Medical Association for the treatment of bladder cancer and other types of tumors. The use of Photofrin enabled both diagnosis and treatment resulting in what is known as "Theranostics" [62].

Theranostics provides the basic objective of treating the tumors or cancers, especially breast cancer efficiently and ethically providing a better treatment approach. Clinicians are of the notion that Theranostic nuclear oncology should be adopted into routine clinical practice for managing neuroendocrine tumors (NETs) [63]. This concept was initialized with the publication of results from Phase III randomized clinical trial, NETTER-1. This trial was the first one to provide evidence of level 1b regarding the safety and efficacy of "68-gallium/177-lutetium-DOTA-octreotate peptide receptor radionuclide therapy of mid-gut neuroendocrine tumors." Even the results were quite encouraging of "multicentric Phase II studies of 68-gallium/177-lutetium-prostate specific membrane antigen theranostic approach for managing end-stage metastatic castrate-resistant prostate cancer." The primary outcome of the study was progression-free survival (PFS), which was measured at month 20 and it was estimated to be 65.2% compared to 10.8% in controls when treated with standard somatostatin receptor analogs [64].

Recently, advancement has been seen for the management of metastatic castrate–resistant prostate cancer (mCRPC) where the dramatic effects of 68-gallium-prostate specific membrane antigen (PSMA) PET/CT have been demonstrated. The results have been documented in a prospective, multicenter Australian study of 431 patients for whom the imaging was done for primary staging (25%) and for restaging/biochemical recurrence (75%) [65].

The field of theranostic nuclear oncology is quite challenging to the clinicians for personalized treatment of cancer. The nuclear physicians require to incorporate in them the knowledge of molecular oncology with background concepts of genomics and proteomics. This will assist the physicians in choosing the therapy according to the requirements of the patient for an effective control of the tumor and to prolong their survival [66].

4 Future directions

Although there has been advancement in cancer research in many prospects, still its efficacy has been limited by a number of challenges posing problems in clinical translation for the treatment of various kinds of cancers. Cancer is unstable genomically, therefore next-generation sequencing data can be a novel technology for showing the inside molecular machinery in cancer cells. Currently, very few nanodrugs are available to treat cancer, the main reason behind this being the unknown reason of toxicity of nanoformulations. Thus, advancement in nanomedicines via improvement in materials and smart designing of nanomedicines can offer promising anticancer therapeutics.

References

[1] History of medicine | History & Facts | Britannica [Internet]. [cited 2020 Sep 30]. Available from: https://www.britannica.com/science/history-of-medicine.

[2] Cg K, Jw Y. A conceptual history of cancer. Semin Oncol 1979;6(4):396–408.

[3] Diamandopoulos GT. Cancer: an historical perspective. Anticancer Res 1996;16(4A):1595–602.

[4] The Cline affair. Abstract - Europe PMC [Internet]. [cited 2020 Sep 29]. Available from: https://europepmc.org/article/med/11708875.

[5] Timeline: Milestones in Cancer Treatment [Internet]. Cure Today. [cited 2020 Sep 30]. Available from: https://www.curetoday.com/view/timeline-milestones-in-cancer-treatment.

[6] Harvey AM. Early contributions to the surgery of cancer: William S. Halsted, Hugh H. Young and John G. Clark. Johns Hopkins Med J 1974;135(6):399–417.

[7] Gerber DE. Targeted therapies: a new generation of cancer treatments. Am Fam Physician 2008;77(3):311–9.

[8] Aprile G, Ongaro E, Del Re M, Lutrino SE, Bonotto M, Ferrari L, et al. Angiogenic inhibitors in gastric cancers and gastroesophageal junction carcinomas: a critical insight. Crit Rev Oncol Hematol 2015;95(2):165–78.

[9] Overview of Current Targeted Anti-Cancer Drugs for Therapy in Onco-Hematology [Internet]. [cited 2020 Sep 28]. Available from: https://www.ncbi.nlm.nih.gov/pmc/articles/PMC6723645/#B1-medicina-55-00414.

[10] Targeting cancer with small molecule kinase inhibitors | Nature Reviews Cancer [Internet]. [cited 2020 Sep 28]. Available from: https://www.nature.com/articles/nrc2559.

[11] bcr-abl, the hallmark of chronic myeloid leukaemia in man, induces multiple haemopoietic neoplasms in mice. [Internet]. [cited 2020 Sep 28]. Available from: https://www.ncbi.nlm.nih.gov/pmc/articles/PMC551781/.

[12] Le QA, Hay JW. Cost-effectiveness analysis of lapatinib in HER-2-positive advanced breast cancer. Cancer 2009;115(3):489–98.

[13] Temsirolimus, Interferon Alfa, or Both for Advanced Renal-Cell Carcinoma | NEJM [Internet]. [cited 2020 Sep 28]. available from: https://www.nejm.org/doi/full/10.1056/nejmoa066838.

[14] Dancey JE. Therapeutic targets: MTOR and related pathways. Cancer Biol Ther 2006;5(9):1065–73.

[15] Ma L, Wang DD, Huang Y, Yan H, Wong MP, Lee VH. EGFR mutant structural database: computationally predicted 3D structures and the corresponding binding free energies with gefitinib and erlotinib. BMC Bioinformatics 2015;16(1):85.

[16] Regulation of phosphorylation of the c-erbB-2/HER2 gene product by a monoclonal antibody and serum growth factor(s) in human mammary carcinoma cells. [Internet]. 2020 [cited 2020 Sep 28]. Available from: https://www.ncbi.nlm.nih.gov/pmc/articles/PMC359762/.

[17] Plosker GL, Figgitt DP. Rituximab: a review of its use in non-Hodgkin's lymphoma and chronic lymphocytic leukaemia. Drugs 2003;63(8):803–43.

[18] Jazirehi AR, Bonavida B. Cellular and molecular signal transduction pathways modulated by rituximab (rituxan, anti-CD20 mAb) in non-Hodgkin's lymphoma: implications in chemosensitization and therapeutic intervention. Oncogene 2005;24(13):2121–43.

[19] Reff ME, Carner K, Chambers KS, Chinn PC, Leonard JE, Raab R, et al. Depletion of B cells in vivo by a chimeric mouse human monoclonal antibody to CD20. Blood 1994;83(2):435–45.

[20] Cobleigh MA, Vogel CL, Tripathy D, Robert NJ, Scholl S, Fehrenbacher L, et al. Multinational study of the efficacy and safety of humanized anti-HER2 monoclonal antibody in women who have HER2-overexpressing metastatic breast cancer that has progressed after chemotherapy for metastatic disease. J Clin Oncol Off J Am Soc Clin Oncol 1999;17(9):2639–48.

[21] Sharkey RM, Press OW, Goldenberg DM. A re-examination of radioimmunotherapy in the treatment of non-Hodgkin lymphoma: prospects for dual-targeted antibody/radioantibody therapy. Blood 2009;113(17):3891–5.
[22] Harjunpää A, Junnikkala S, Meri S. Rituximab (anti-CD20) therapy of B-cell lymphomas: direct complement killing is superior to cellular effector mechanisms. Scand J Immunol 2000;51(6):634–41.
[23] A phase-1 trial of bexarotene and denileukin diftitox in patients with relapsed or refractory cutaneous T-cell lymphoma. - Abstract - Europe PMC [Internet]. [cited 2020 Sep 28]. Available from: https://europepmc.org/article/med/15811959.
[24] Kaminski MS, Tuck M, Estes J, Kolstad A, Ross CW, Zasadny K, et al. 131I-tositumomab therapy as initial treatment for follicular lymphoma. N Engl J Med 2005;352(5):441–9.
[25] Differential Cellular Internalization of Anti-CD19 and -CD22 Immunotoxins Results in Different Cytotoxic Activity | Cancer Research [Internet]. [cited 2020 Sep 28]. Available from: https://cancerres.aacrjournals.org/content/68/15/6300.
[26] Rogers S, Lowenthal A, Terheggen HG, Columbo JP. Induction of arginase activity with the shope papilloma virus in tissue culture cells from an argininemic patient. J Exp Med 1973;137(4):1091–6.
[27] Alewine C, Hassan R, Pastan I. Advances in anticancer immunotoxin therapy. Oncologist 2015;20(2):176–85.
[28] Ahrendt SA, Hu Y, Buta M, McDermott MP, Benoit N, Yang SC, et al. p53 mutations and survival in stage I non-small-cell lung cancer: results of a prospective study. J Natl Cancer Inst 2003;95(13):961–70.
[29] Reid T, Warren R, Kirn D. Intravascular adenoviral agents in cancer patients: lessons from clinical trials. Cancer Gene Ther 2002;9(12):979–86.
[30] Räty JK, Pikkarainen JT, Wirth T, Ylä-Herttuala S. Gene therapy: the first approved gene-based medicines, molecular mechanisms and clinical indications. Curr Mol Pharmacol 2008;1(1):13–23.
[31] Kratz F, Senter P, Steinhagen H. Drug Delivery in Oncology, 3 Volume Set: From Basic Research to Cancer Therapy. John Wiley & Sons; 2011. p. 1823.
[32] Telomerase Inhibition in Cancer Therapeutics: Molecular-Based Approaches [Internet]. [cited 2020 Sep 29]. Available from: https://www.ncbi.nlm.nih.gov/pmc/articles/PMC2423208/.
[33] Fisher PB. Is mda-7/IL-24 a "magic bullet" for cancer? Cancer Res 2005;65(22):10128–38.
[34] Deregulated Blood Cellular miR-2909 RNomics Observed in Rheumatoid Arthritis Subjects | Insight Medical Publishing [Internet]. [cited 2020 Sep 29]. Available from: https://www.archivesofmedicine.com/medicine/deregulated-blood-cellular-mir2909-rnomics-observed-inrheumatoid-arthritis-subjects.php?aid=3699.
[35] Wahid F, Shehzad A, Khan T, Kim YY. MicroRNAs: synthesis, mechanism, function, and recent clinical trials. Biochim Biophys Acta 2010 Nov;1803(11):1231–43.
[36] Human cellular mitochondrial remodelling is governed by miR-2909 RNomics [Internet]. [cited 2020 Sep 29]. Available from: https://journals.plos.org/plosone/article?id=10.1371/journal.pone.0203614.
[37] Kaushik H, Malik D, Parsad D, Kaul D. Mitochondrial respiration is restricted by miR-2909 within human melanocytes. Pigment Cell Melanoma Res 2019;32(4):584–7.
[38] Malik D, Kaul D. KLF4 genome: a double edged sword. J Solid Tumors 2015 Mar 26;5(1):49.
[39] Therapeutic bacteria to combat cancer; current advances, challenges, and opportunities [Internet]. [cited 2020 Sep 30]. Available from: https://www.ncbi.nlm.nih.gov/pmc/articles/PMC6558487/.
[40] Cancer immunotherapy based on killing of Salmonella-infected tumor cells - PubMed [Internet]. [cited 2020 Sep 30]. Available from: https://pubmed.ncbi.nlm.nih.gov/15867392/.
[41] Tumor-targeting bacteria engineered to fight cancer [Internet]. [cited 2020 Sep 30]. Available from: https://www.ncbi.nlm.nih.gov/pmc/articles/PMC6902869/.
[42] Staedtke V, Roberts NJ, Bai R-Y, Zhou S. Clostridium novyi-NT in cancer therapy. Genes Dis 2016 Jun 1;3(2):144–52.
[43] Oncolytic Bacteria and their potential role in bacterium-mediated tumour therapy: a conceptual analysis [Internet]. [cited 2020 Sep 30]. Available from: https://www.jcancer.org/v10p4442.htm.
[44] Zheng D-W, Chen Y, Li Z-H, Xu L, Li C-X, Li B, et al. Optically-controlled bacterial metabolite for cancer therapy. Nat Commun 2018;9(1):1680.
[45] Tran S, DeGiovanni P-J, Piel B, Rai P. Cancer nanomedicine: a review of recent success in drug delivery. Clin Transl Med 2017;6(1):44.
[46] Wolfram J, Ferrari M. Clinical Cancer Nanomedicine. Nano Today 2019;25:85–98.
[47] Yuan J, Hegde PS, Clynes R, Foukas PG, Harari A, Kleen TO, et al. Novel technologies and emerging biomarkers for personalized cancer immunotherapy. J Immunother Cancer 2016;4:3.
[48] Ribas A, Wolchok JD. Cancer immunotherapy using checkpoint blockade. Science 2018;359(6382):1350–5.
[49] Buckler AJ, Bresolin L, Dunnick NR, Sullivan DC, Aerts HJWL, Bendriem B, et al. Quantitative imaging test approval and biomarker qualification: interrelated but distinct activities. Radiology 2011;259(3):875–84.
[50] Saltz J, Almeida J, Gao Y, Sharma A, Bremer E, DiPrima T, et al. Towards generation, management, and exploration of combined Radiomics and Pathomics datasets for Cancer research. AMIA Summits Transl Sci Proc 2017;2017:85–94.
[51] Precision medicine and molecular imaging: new targeted approaches toward cancer therapeutic and diagnosis [Internet]. [cited 2020 Sep 28]. Available from: https://www.ncbi.nlm.nih.gov/pmc/articles/PMC5218860/.
[52] Visvikis D, Cheze Le Rest C, Jaouen V, Hatt M. Artificial intelligence, machine (deep) learning and radio(geno)mics: definitions and nuclear medicine imaging applications. Eur J Nucl Med Mol Imaging 2019;46(13):2630–7.
[53] Immunomodulatory effects of RXR rexinoids: modulation of high-affinity IL-2R expression enhances susceptibility to denileukin diftitox | Blood | American Society of Hematology [Internet]. [cited 2020 Sep 28]. Available from: https://ashpublications.org/blood/article/100/4/1399/106210/Immunomodulatory-effects-of-RXR-rexinoids.
[54] Mey U, Strehl J, Gorschlüter M, Ziske C, Glasmacher A, Pralle H, et al. Advances in the treatment of hairy-cell leukaemia. Lancet Oncol 2003;4(2):86–94.
[55] Prescott JW. Quantitative imaging biomarkers: the application of advanced image processing and analysis to clinical and preclinical decision making. J Digit Imaging 2013;26(1):97–108.
[56] Overview of established and emerging immunohistochemical biomarkers and their role in correlative studies in MRI - Alvi - 2020 - Journal of Magnetic Resonance Imaging - Wiley Online Library [Internet]. [cited 2020 Sep 29]. Available from: https://onlinelibrary.wiley.com/doi/abs/10.1002/jmri.26763.
[57] Immuno-Oncology: Emerging Targets and Combination Therapies - PubMed [Internet]. [cited 2020 Sep 29]. Available from: https://pubmed.ncbi.nlm.nih.gov/30191140/.

[58] New strategies in immunotherapy for lung cancer: beyond PD-1/PD-L1 - PubMed [Internet]. [cited 2020 Sep 29]. Available from: https://pubmed.ncbi.nlm.nih.gov/30215300/.
[59] Sun R, Limkin EJ, Vakalopoulou M, Dercle L, Champiat S, Han SR, et al. A radiomics approach to assess tumour-infiltrating CD8 cells and response to anti-PD-1 or anti-PD-L1 immunotherapy: an imaging biomarker, retrospective multicohort study. Lancet Oncol 2018;19(9):1180–91.
[60] Chen S, Feng S, Wei J, Liu F, Li B, Li X, et al. Pretreatment prediction of immunoscore in hepatocellular cancer: a radiomics-based clinical model based on Gd-EOB-DTPA-enhanced MRI imaging. Eur Radiol 2019 Aug;29(8):4177–87.
[61] Madamsetty VS, Mukherjee A, Mukherjee S. Recent trends of the bio-inspired nanoparticles in cancer theranostics. Front Pharmacol [Internet] 2019. Oct 25 [cited 2020 Sep 29];10. Available from https://www.ncbi.nlm.nih.gov/pmc/articles/PMC6823240/.
[62] Schaffer M, Schaffer PM, Corti L, Gardiman M, Sotti G, Hofstetter A, et al. Photofrin as a specific radiosensitizing agent for tumors: studies in comparison to other porphyrins, in an experimental in vivo model. J Photochem Photobiol B 2002 Apr;66(3):157–64.
[63] Czernin J. Molecular imaging and therapy with a purpose: a renaissance of nuclear medicine. J Nucl Med 2017 Jan 1;58(1):21A–2A.
[64] Strosberg J, El-Haddad G, Wolin E, Hendifar A, Yao J, Chasen B, et al. Phase 3 trial of 177Lu-Dotatate for midgut neuroendocrine tumors. N Engl J Med 2017 Jan 12;376(2):125–35.
[65] Roach PJ, Francis R, Emmett L, Hsiao E, Kneebone A, Hruby G, et al. The impact of 68Ga-PSMA PET/CT on management intent in prostate Cancer: results of an Australian prospective multicenter study. J Nucl Med Off Publ Soc Nucl Med 2018;59(1):82–8.
[66] Theranostics in nuclear medicine practice [Internet]. [cited 2020 Sep 29]. Available from: https://www.ncbi.nlm.nih.gov/pmc/articles/PMC5633297/.

Chapter 4

Emerging need of advanced drug delivery systems in cancer

Parteek Prasher[a], Mousmee Sharma[b], and Kamal Dua[c,d,e]

[a]Department of Chemistry, University of Petroleum & Energy Studies, Dehradun, India, [b]Department of Chemistry, Uttaranchal University, Dehradun, India, [c]School of Pharmaceutical Sciences, Shoolini University of Biotechnology and Management Sciences, Solan, Himachal Pradesh, India, [d]Discipline of Pharmacy, Graduate School of Health, University of Technology Sydney, Ultimo, NSW, Australia, [e]Priority Research Centre for Healthy Lungs, Hunter Medical Research Institute (HMRI) & School of Biomedical Sciences and Pharmacy, University of Newcastle, Callaghan, NSW, Australia

1 Introduction

Cancer continues to claim a considerable mortality rate worldwide, with treatment regime relying on chemotherapy, which suffers certain limitations such as adversely affecting the healthy tissues [1]. The involvement of multiple cellular, biochemical, and metabolic pathways in cancer progression further complicates the situation thereby deterring the therapeutic regimes focused on managing limited targets [2]. The development of hybrid molecules and practice of polypharmacology added prominence to the chemotherapeutic programs [3]. However, the current exigencies associated with anticancer therapy necessitated the advertent of advanced drug delivery systems [4] with site-specific action, superior biotolerance and bioavailability, better pharmacokinetic, rapid clearance rates, biodegradability [5], and sustaining a controlled release of the payload drug molecules [6]. The desire for a pluralistic approach to target the multiple contributing pathways in cancer pathogenesis realized the development of state-of-the-art therapeutics and delivery vectors that demonstrate high tolerance toward biological systems [7], while restraining from switching the innate immune response, which may further potentiate the complications [8]. The exploration of highly efficient and safer drug delivery systems led to the identification of smart drug delivery systems including liposomes, polymersomes, dendrimers, nanoclusters [9], and micelles enabling the sustained and target-specific release of the carrier drug molecules [10]. The smart drug delivery systems comprise stimuli-responsive drug release, prompted by the local physicochemical properties of cancer cells [11], such as anomalous pH, temperature, free radical species such as ROS and RNS, inducible enzymes, and erroneous metabolism [12]. Similarly, the functionalized metal nanoparticle-based drug delivery vehicles offer theranostic functions by enabling the in vivo monitoring of the biological event [13], while annihilating the metastatic cells. In this chapter, we discuss about the advanced anticancer drug delivery strategies and their precedence over the customary approaches for delivering anticancer drugs [14].

2 Multidrug resistance in cancer cells

Majority of the chemotherapeutic efforts overlaying cancer treatment suffer cellular efflux thereby rendering the deliberated drugs as ineffective [15]. The multidrug resistance, which offers resistance to several structurally and mechanically diverse pharmaceuticals, offers a significant challenge in anticancer chemotherapy [16]. Mainly, the onset of multidrug resistance triggers via ATP-binding cassette containing efflux pumps, which pumps out the deliberated therapeutics [17]. In addition, the adaptations of cancer cells to microenvironment, such as oncogenic mutations, prompt the development of resistance against previous treatments [18]. Similarly, the activation of cell growth factors and cellular defense pathways in response to the pharmaceutical alleviate the cancer cell susceptibility to chemotherapy [19]. The rapidly proliferating cancer cells require higher supply of oxygen and signal transmission for their rapid metabolism [20]. This results in an inadequate oxygen delivery thereby causing tissue hypoxia, which causes cancer cells to adapt less oxygen cellular environment by upregulating the key pathways including hypoxia-inducible factor-1α (HIF-1α) [21]. Tissue hypoxia potentially instigates multidrug resistance by inducing the expression of ABCB1 and ABCG2 for mediating the intracellular pumping out of chemotherapeutic agents [22]. The self-immolating autophagy mechanism for the degradation of intracellular components serves as an energy regulator, in response to nutrient stress in cells [23].

Advanced Drug Delivery Systems in the Management of Cancer. https://doi.org/10.1016/B978-0-323-85503-7.00032-8

Autophagy supports survival and resistance mechanisms of cells during chemotherapy, and its inhibition potentially re-sensitizes multidrug resistance cells while enhancing the effect of chemotherapeutics [24]. Several factors including signal transducer and the activator of the transcription (STAT3), TP53 [25], and nuclear factor kappa-light-chain-enhancer of activated B cells (NF-κB) regulate the cell proliferation, hematopoiesis, and apoptosis, including tumor growth and metastasis of resistant cancers [26]. In addition to these factors, the alterations in lipid metabolism of drug-resistant cells trigger biophysical changes in lipid bilayer, which adversely affects the drug uptake by these cells [27].

3 Toxic effects of chemotherapy

The chemotherapeutics deliberated to annihilate the fast-growing cancer cells damage healthy tissues and cells in every part of the body. However, the targeted anticancer therapy owing to its specificity causes less harm; nevertheless, this approach often dysregulates the signaling pathways critical to the optimal functioning of the healthy cells [28]. Importantly, a limited information regarding the participation and mechanism of the composite, intertwined metabolic events responsible for cancer pathogenesis deter the optimization and quantization of toxicity effects of chemotherapy [29]. Similarly, the dearth of preclinical models and assays, and limited funding for the toxicity and symptom research adversely influence the efforts to identify the perspective therapeutic approaches deliberated at toxicity mitigation [30]. The antineoplastic drugs cisplatin, paclitaxel, and vincristine enhance the events of neurotoxicity whereas the drugs cytarabine, carmustine, and 5-fluorouracil cause encephalopathy when used as a high-dose chemotherapy [31]. Similarly, the drugs procarbazine and carboplatin cause peripheral neuropathy. The chemotherapeutic program with drugs etoposide, methotrexate, and ifosamide result in alopecia [32]. Moreover, the drugs irinotican, mitomycin, bleomycin, and metoclopramide reportedly cause less severe side effects such as diarrhea, constipation, and stomatitis [33]. Apart from this, the contemporary anticancer drugs manifest significant renal, pulmonary, and hepatic toxicity, which necessitates the obligation for alternative drugs or advanced delivery systems to overcome these side effects by ensuring targeted delivery, optimal drug distribution at the target site, and rapid clearance from systemic circulation [32]. The controlled release biodegradable systems that undergo a gradual disintegration in vivo after performing the sustained drug release prove highly effective in alleviating the limitations of contemporary chemotherapy. As such, the natural polymers serve as a fine alternative to design the advanced drug delivery systems for anticancer drugs.

4 Hazardous effects of radiation oncology

Radiation therapy in cancer selectively annihilates the cancer cells by using high-energy radiation that cause irreparable damage to the tumor cell DNA thereby causing an onset to cancer cell death [34]. The techniques such as external beam radiation therapy selectively target the tumor localized cellular environment without inflecting damage to the surrounding healthy tissues in most cases. Similarly, the internal radiation therapy or brachytherapy exhibits selectivity to the tumor microenvironment eventually causing its annihilation [35]. However, the radiation causes potentially hazardous effects to the healthy tissues, which necessitates a high precision and skill while administrating the radiation dosage. The healthy cells damaged by radiation typically recover within months after receiving the therapy; however, the ensuing late effects take extended time to alleviate [36]. Likewise, the patients undergoing stereotactic radiosurgery for brain tumors face high risk and the side effects such as brain swelling, memory loss, stroke, and poor brain functioning. The radiation therapy targeting breast cancer causes side effects such as lymphedema, which takes longer time to go away [37]. Reportedly, the radiation therapy directed at chest causes lung inflammation or radiation pneumonitis, and brachial plexopathy. The persons suffering from diseases such as emphysema become more vulnerable to these side effects, which further deteriorate to pulmonary fibrosis [38]. Similarly, the radiation therapy directed at prostate cancer or bladder cancer causes cystitis that persists as a long-term ailment in the therapy recipients. The development of fistula presents another grave consequence of radiation therapy, which requires surgical treatment to achieve recovery [39]. Besides, the radiation therapy may cause permanent damage to the essential genes causing an anomalous expression in vital life processes. The perilous effects of radiation therapy obligate improved approaches for cancer therapy that cause minimal damage to the healthy cells and ensure marginal side effects [40]. Apparently, the search for safer alternatives becomes a cornerstone for perspective anticancer therapies.

5 Polypharmacology in anticancer therapy

The recent decades witnessed tremendous challenges in developing anticancer therapy owing to the tumor heterogeneity, evolution, and complicated interactions between administered drugs and human proteome [41]. In addition to inducing cytotoxicity, the present integrative approaches focus on regulating the tumor microenvironment for producing holistic results [42]. Polypharmacology provides stimulating possibilities for manipulating off-target effects by drug repurposing that assures a simultaneous obstruction

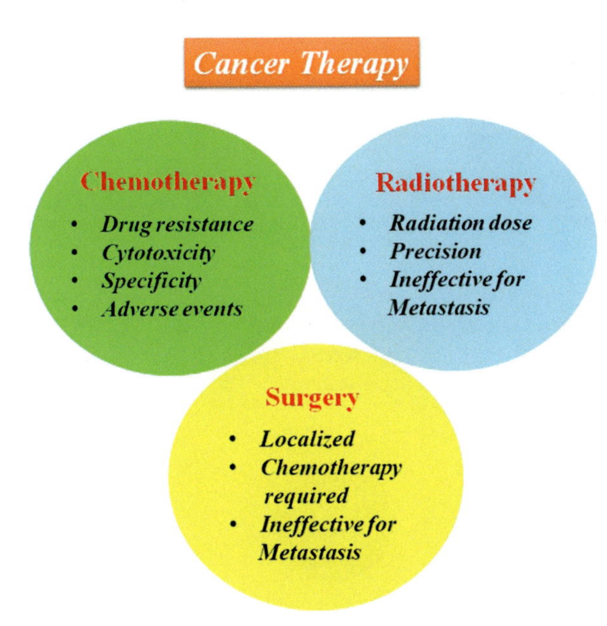

FIG. 1 Conventional cancer therapies and their limitations.

of multiple pathways thereby decelerating or evading drug resistance [43]. The administration of a cocktail of drugs as polypharmacology provided encouraging results in managing polygenic cancer drug resistance; however, the exponential possibilities of drug combinations make it hard to identify and analyze the most feasible drug blends [44]. In addition, most of the cancer-causing genes mutate at a low frequency due to which the rational development of drugs for targeting these genes proves futile [45]. Limited conclusive evidence is available for the beneficial effects of the practice of polypharmacology and drug repurposing intended to produce synergistic effect in cancer therapy [46]. A major limitation of polypharmacology approaches is a partial understanding of essential pathways associated with cancer and the mechanism via which they operate [47]. The incomplete data make it difficult to assert a comprehensive pharmacological profile for an intended drug deliberated for targeting complex physiological processes associated with cancer development [48].

Nevertheless, the patient compliance and perception toward the polypharmacology approach, especially in aging population presents a major hindrance in the further progression of this approach [49]. Considering the above intricacies associated with the traditional and contemporary anticancer therapies, it further creates an urgency for the development of advanced drug delivery regime for combating the cancer pathophysiology [50]. Fig. 1 presents the contemporary cancer therapy strategies and their limitations.

6 Advanced drug delivery systems

The advanced drug delivery vehicles successfully overcome the drawbacks posed by conventional chemotherapy such as low bioavailability, poor therapeutic index, non-specific targeting, and requirement of high drug-dose. As such, the rapidly dissolving drug delivery systems present high prominence in pediatrics and geriatrics, compared to the conventional rapid dissolving tablets. Similarly, the self-emulsifying drug delivery vehicles present a high potency for oral administration, typically by improving the oral bioavailability of hydrophobic drugs. The stimuli-responsive drug delivery systems present the most advanced drug delivery vectors that prompt a controlled drug release in response to the physical stimuli created by target site, such as pH, temperature, and the presence of oxidative stress. These drug delivery vehicles present applications in several diseases associated with considerable changes in the physicochemical characteristics of cellular microenvironment at the morbid site. The osmotic devices present another robust drug delivery system for controlled release of cargo drug molecules especially in the gastrointestinal tract, while using osmosis as a driving force. Similarly, the advanced vesicular drug delivery systems protect the encapsulated therapeutic molecules from oxidative, enzymatic, and proteolytic damage and ensure a safe delivery to the target site. The drug delivery systems based on natural polymers such as polysaccharides offer a superior safety profile and high physiological tolerance due to biodegradability, while ensuring a sustained release of the carrier drug molecules. Similarly, the drug delivery vehicles based on metal-nanoparticles represent the most advanced systems with dual functions. These systems provide theranostic applications owing to the unique physicochemical properties of the metal nanoparticles that enable in vivo imaging and diagnostics.

Similarly, the aggregation effect of the drug-loaded metal nanoparticles promotes the evading of efflux pumps by maintaining high concentration of drug molecules difficult to efflux-out by ATP-dependent protein pumps, thereby ensuring an optimal concentration to kill the diseased cells. Fig. 2 highlights the physicochemical conditions at tumor microenvironment that act as a stimuli for the advanced drug delivery systems in anticancer therapy.

7 Stimuli-responsive drug delivery systems

7.1 pH-responsive

Several intrinsic and extrinsic stimuli cause the drug release from carriers. The internal stimuli constitute pH, enzymes, redox imbalance, and temperature, whereas the external stimuli include exposure to radiation, and electromagnetic field. Tumor cells display low pH caused by hypoxia, poor perfusion, and metabolic anomalies owing to unregulated cell growth. The pH-sensitive drug delivery systems facilitate the drug release in response to acidic pH in tumor

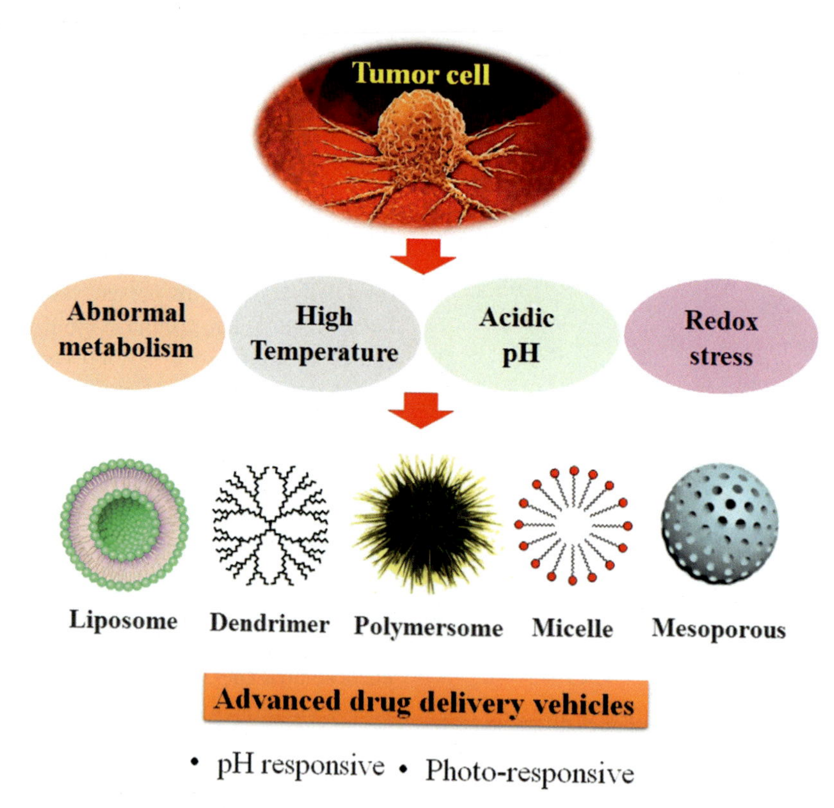

FIG. 2 Advanced drug delivery vehicles in cancer therapy.

microenvironment or intracellular compartments in target tumor cells including endosomes, cytoplasm, and lysosomes. The immense drug release in this situation results in amassing of cytotoxic drug molecules in tumor tissues, which may also exceed the efflux pump capacity, thereby resulting in evading the multidrug resistance. Similarly, the enhanced permeability and retention effect resulting in drug accumulation in tumor tissue enables an optimal efficiency of the deliberated drug molecules at the target site. Primarily, the designing of pH-triggered drug delivery systems consist of incorporating carboxyl, imidazolyl, and amino functionalities, which protonate at acidic pH at tumor microenvironment thereby resulting in destabilization of the carrier system hence prompting the drug release. The pH-sensitive polyacidic and polybasic materials containing morpholine, piperazine, pyridines, and boronic acid, sulfonic acid, and phosphonic acid provide the best alternative for the generation of drug delivery vectors. Similarly, the peptides with predictable conformations, presence of diverse functionalities, and extraordinary stability provide a suitable material for building pH-sensitive drug delivery vehicles. Furthermore, the pH-responsive polymeric micelle presents an appealing profile that exploits the pH gradient between the tumor tissues and the surrounding healthy tissues. The design of pH-sensitive micelles constitutes hydrophobic/hydrophilic polymers joined via acid-liable linkers such as oxime, acetal, ester, and hydrazone that hydrolyze in response to pH stimuli thereby releasing the encapsulated drug molecules at the target site. Jin et al. [51] reported peptide conjugates of anticancer drug doxorubicin bridged via a pH-sensitive hydrazone spacer. The peptide-guided drug delivery system displayed substantial stability and exhibited a minimal drug release at neutral pH. The drug release happened at mildly acidic conditions present in endolysosomes. The system selectively identified AP2H peptide and LAPTM4B protein thereby showing a specific binding to cancer cells by internalizing to endolysosomes, which further provides minimal cytotoxicity and side effects. Yang et al. [52] reported a smart drug delivery system based on poly (ethylene glycol)-phenylhydrazone-dilaurate micelles loaded with anticancer drug paclitaxel. The micelles displayed commendable colloidal stability and drug release properties, ensuring a rapid release of the encapsulated drug on lowering pH from 7.4 to 5.5. The intracellular delivery of drug-carrying micelle system displayed endosomal escape and an enhanced toxicity to cancer cells hence indicating a high therapeutic efficacy. Men et al. [53] composed pH-sensitive drug delivery system from hyaluronic acid and poly (β-amino ester) nanoparticles for the delivery of doxorubicin. The pH-sensitive nano-drug delivery system displayed a higher drug loading capacity, appropriate particle size, and superior serum stability for delivering the cargo drug molecule. The nanosystem responded to low pH and facilitated a sustained release of cargo drug to the human non-small lung carcinoma (A549) cells by efficiently inducing apoptosis. The negative surface charge present on nanosystem and serum stability indicated a prolonged circulation time and increased accumulation at the target site for optimal anticancer effect. Furthermore, the nanosystem displayed a minimal cytotoxicity toward healthy cells, a prerequisite for accounting for the high therapeutic of drug delivery vehicles. Yilmaz et al. [54] reported glycopolymer-coated metal nanoparticles as pH-sensitive drug delivery vehicles for the delivery of doxorubicin. The drug conjugated to the glycopolymer via hydrazone bonding. Fluorescence cell imaging of drug-glycopolymer conjugates indicated cellular internalization to show a marked cytotoxicity toward cervical cancer cells, lung cancer cells, and neuroblastoma cells. The presence of higher mannose residues in drug-glycopolymer conjugates presented the most potent anticancer activity toward these cell lines. Palanikumar et al. [55] reported highly stable, pH-triggered polymeric nanoparticles for successful delivery of anticancer drugs. The nanosystem comprised of polylactic-*co*-glycolic acid core conjugated to a shell of bovine serum albumin. The transported drug was loaded to the core, while the presence of serum proteins minimized the interactions with macrophages that prompted target selectivity. The presence of acidity-triggered rational membrane peptide facilitated specific internalization to cancer cells in acidic microenvironment in tumor cells. Gao et al. [56] developed an effective, pH-triggered co-delivery platform for chemotherapy and photothermal therapy. The system consisted of loading of doxorubicin to nanosheets of black phosphorus, further conjugated with poly (2-ethyl-2-oxazoline) polydopamine ligands. The nanosystem displayed a long circulation time and higher cellular uptake and assisted in the co-delivery of bortezomib along with doxorubicin. Notably, the surface charge present on nanosystem changed from negative to positive, in response to the extracellular pH at tumor microenvironment thereby ionizing tertiary amide groups present on poly (2-ethyl-2-oxazoline) chains. This event facilitated the cellular internalization of the carrier system, with cytotoxic therapeutic effect further improved in the presence of near-infrared laser irradiation. Nogueira et al. [57] further developed magnetic-assisted nanocarriers for pH-triggered release of doxorubicin in cancer therapy. The nanocarriers constituted of magnetic iron oxide core wrapped with siliceous shells and carrageenan. The drug loading occurred on hybrid shells, whereas the iron oxide core facilitated maneuvering via an external magnetic field. The system provided the most advanced platform for targeted drug release in anticancer therapy.

7.2 Light-responsive

The irradiation of tumor cells with near-infrared light triggers the drug release by responsive drug delivery systems thereby facilitating targeted therapy without causing damage to surrounding healthy tissues. Wang et al. [58] reported lipid nanoparticles consisting of eutectic mixture of lauric acid and stearic acid for near-infrared sensitive release of doxorubicin. The coencapsulation of near-infrared absorbing dye methylenebisacrylamide promoted light-sensitivity thereby promoting a quick drug release by the nanosystem. This novel, light-sensitive drug delivery nanosystem displayed a superior cancer targeting efficiency with noticeable photothermal effect. Notably, the nanoparticles accumulate at the tumor site thereby ensuring a targeted drug release for an optimal anticancer effect. Chen et al. [59, 60] reported near-infrared light-responsive drug co-delivery system for cancer treatment. The system architecture consisted of hollow, mesoporous Prussian blue nanoparticles filled with phase change materials such as 1-tetradecanol that functioned as thermosensor and as a medium for loading of hydrophilic/hydrophobic drugs. The thermosensing effect in the presence of infrared radiation results in the loss of phase change materials from the hollow Prussian blue nanoparticles thereby releasing the encapsulated drug molecules doxorubicin and camptothesin. This drug co-delivery system displayed significant photothermal-chemo synergistic effect on cancer cell lines with high efficiency. Gao et al. [61] reported RBCs decorated with photosensitizer as ultrasensitive light-responsive drug delivery system for the delivery of doxorubicin. The system is comprised of RBCs membrane decorated with chlorin e6 without altering the membrane integrity and stability. The light irradiation triggers the release of singlet oxygen by chlorin e6 resulting in the RBC membrane disruption thereby releasing the doxorubicin enclosed within the cells. The light-controlled chemotherapy by photosensitizer-loaded RBCs presents high biocompatibility in cancer drug delivery. Feng et al. [62] developed near-infrared light-responsive programmed drug delivery systems for tumor-targeting magnetic resonance imaging and chemo-phototherapy. The system is comprised of doxorubicin-coated CuS nanoparticles, further decorated with superparamagnetic iron oxide nanoparticles. The functionalized iron oxide nanoparticles prevented the premature drug release and minimized the harmful side effects during in vivo drug delivery. The nanosystem displayed enhanced photothermal effect arising mainly due to the coupling of plasmonic resonances resulting in enhanced heat-generation capacity. Notably, the performance of superparamagnetic iron oxide nanoparticles in T_2-weighted MRI enabled the identification of cancerous lesions in vivo. Yang et al. [63] developed a novel drug delivery system based on mesoporous silica nanorods decorated with photosensitizer chlorin e6 that promotes on-demand drug release in the presence of near-infrared light. Further coating of this nanosystem with bovine serum albumin via singlet oxygen-sensitive bis-(alkylthio) alkene spacer followed by modification with polyethylene glycol resulted in the effective loading of small drug molecules including doxorubicin, and large-sized *cis*-platin inside the mesoporous structure. The irradiation of nanosystem with infrared radiation generated singlet oxygen, which cleaved bis-(alkylthio) alkene spacer thereby detaching BSA-PEG from the nanorods surface, eventually releasing doxorubicin and cis-platin molecules. This nanosystem offered a commendable synergistic therapeutic efficiency in the anticancer therapy. Yang et al. [64] developed light-sensitive drug delivery systems based on mesoporous silica-coated gold nanorods functionalized with aptamer DNA. Upon irradiation, the photothermal effect of gold nanorods resulted in a temperature rise that caused dehybridization of linkage DNA duplex anchored to quadruplex DNA thereby releasing the entrapped drug molecule. The nanosystem displayed multifunctional properties with an effective integration of photothermotherapy, chemotherapy, and imaging in a single system. Notably, the superior physiological compatibility, cancer-cell identification potency, and controlled release characteristics of the nanosystem played a critical role in their efficacy as drug delivery vehicles. Cheng et al. [65] developed photocontrollable drug delivery system based on light-responsive self-assembled polymeric micelles. The drug delivery system consisted of light-sensitive block copolymer made of water-soluble poly (ethylene glycol)-b-poly (caprolactone) polymer, and photosensitive maleimide-anthracene spacers. The system displayed a low critical micelle concentration with high stability, controlled light-responsivity, maneuverable drug-loading, highly sensitive light-triggered drug release, and enhanced safety profile. In vitro investigations indicated that the drug-carrying micelles demonstrated high serum stability and extremely low toxicity under physiological conditions. The treatment of cancer cells with these drug-loaded micelles and irradiation with UV light for 10s caused an extremely rapid intracellular release of cargo drug molecules eventually inducing a robust antiproliferative activity on the target cancer cells. Notably, the cytosol and cellular nuclei rapidly uptake the drug-loaded micelles whereby the rapid drug release by UV irradiation caused cancer cell death by apoptosis. The micelle-based drug delivery system demonstrated an exceptional reliability and high performance owing to an ultra-sensitivity to UV radiation for releasing the payload drug molecules. The radiation responsive drug delivery systems therefore present a high potency in anticancer therapy for a targeted annihilation of the tumor tissues without harming the surrounding healthy tissues.

7.3 Redox-responsive

Tumor microenvironment displays a heightened redox stress owing to the erroneous functioning of several critical pathways that lead to the production of reactive oxygen/nitrogen species. Li et al. [66] developed redox-sensitive drug delivery system obtained from RGD peptide-functionalized mesoporous silica nanoparticles. The anticancer drug doxorubicin is enclosed in mesoporous silica nanoparticles. The redox stress caused by intracellular glutathione in tumor microenvironment triggered the removal of surface RGD peptide appended to mesoporous silica nanoparticles via disulfide bonds. Redox-responsive detachment of this layer resulted in drug release in tumor cells. Zhang et al. [67] reported redox-responsive drug delivery system based on cytochrome *c* end-capped mesoporous silica nanoparticles. The fabrication of nanocarrier occurred with immobilized cytochrome *c* (Cyt C) on the surface of mesoporous silica nanoparticles via disulfide linkers. The intracellular glutathione in tumor microenvironment triggered the drug release resulting in apoptosis in vitro as reflected by change in tumor size. Nanocarriers displayed biodistribution among various organs in test animals. Notably, a great deal of nanoparticles were lost in lung after 30 days of administration owing to higher homoperfusion rates in lungs, whereas the heart and kidney displayed a moderate accumulation in the liver and kidney of test animals, which gradually lowered after 30 days. Similarly, an enhanced permeability and retention effect caused a significant accumulation of nanocarriers in tumor xenograft causing tumor annihilation. Chen et al. [59, 60] developed multifunctional drug delivery systems triggered by intracellular reduction potential and displaying a high stability under physiological conditions. The presence of reductants such as glutathione resulted in the swelling and subsequent disintegration of drug delivery system caused by protonation of amino groups and reductive cleavage of disulfide bonds. The system proved highly effective in the delivery of docetaxel even in acidic pH of endosomes thereby pausing the tumor growth. These drug delivery vehicles exhibited higher potency in inhibiting tumor growth compared to the free docetaxel, hence suggesting a synergistic effect. Laskar et al. [68] developed prodrug dendrimer by tethering PEGylated and 3-diaminobutyric polypropylenimine dendrimer to anticancer drug camptothecin via redox-responsive disulfide bond. The self-assembly of prodrug dendrimer into cationic vesicles caused an increased dendrimer concentration and enhanced stability for a week. These dendrimers effectively released around 70% of encapsulated camptothesin in the presence of 50 mM glutathione. Notably, the DNA condensation occurred instantly and sustained for 24 h. These events resulted in enhanced DNA uptake by cells eventually amplifying gene transfection in cancer cells. Jiao et al. [69] developed smart, redox-responsive nanocarrier by conjugating fluorescent carbon dots to mesoporous silica nanoparticles via disulfide bonds. The carbon dots developed from polyanion poly (acrylic acid) polymer by hydrothermal polymerization enabled in vivo imaging. These carbon dots tethered to the mesoporous silica nanoparticles by amidation that enable the trapping of drug molecules in the porous structure. The nanocarrier system displayed significant responsiveness to redox conditions for the in vitro release of doxorubicin at pH 7.4 and 5.0 in PBS solution. The delivery system displayed an extraordinary biocompatibility and fluorescence properties. The nanocarrier system exhibited enhanced cellular uptake by cancer cells that further potentiated their therapeutic efficiency. These investigations validated the application of nanocarriers for simultaneous real-time imaging and controlled drug delivery for probing the performance of drug-loaded nanocarriers in tumor therapy. Lei et al. [70] designed redox-sensitive metal–organic framework drug carriers based on Fe, Zn, and Zr as metal nodes and 4,4′-dithiobisbenzoic acid (4,4′-DTBA) as organic ligand joined by disulfide bond, which cleaves in the presence of intracellular glutathione in cancer cells. The MOFs displayed rapid drug release behavior by releasing curcumin in vitro. The in vivo anticancer investigations revealed higher antitumor efficacy of MOF-based drug carrier compared to the free curcumin. Liu et al. [71] developed redox-sensitive nanosized drug delivery vehicles based on amphiphilic hyperbranched multiarm copolyphosphates with superior biocompatibility and biodegradability. In aqueous medium, the nanosystem displayed self-assembly to spherical micellar nanoparticles with size in the range 70–100 nm. The reductive conditions disturbed micellar structure of nanoconstructs thereby releasing the encapsulated drug molecules of doxorubicin that exhibited marked cell proliferation against glutathione monoester impregnated HeLa cells, as compared to the untreated cells. The nanosystem reportedly exhibited lower inhibition of HeLa cells treated with buthionine sulfoximine. Lv et al. [72] reported amphiphilic triblock copolymers with enhanced biodegradability as redox-responsive drug delivery vehicles for intracellular delivery and controlled release of doxorubicin. The polymeric nanoparticles self-assembled to form micelle structures in aqueous media while encapsulating doxorubicin, hence forming a lower mobility, rigid core. Notably, the cross-linked copolymers improved the micellar stability and improved the drug-loading capacity in addition to alleviating entrapment efficiency. The reductive environment in tumor cells accelerated the in vitro sustained release of doxorubicin via the nanosystem. The in vivo investigations indicated highly efficient cellular uptake of doxorubicin-loaded micelles, which further improved the therapeutic efficacy of the encapsulated drug. Therefore, the redox-responsive core-crosslinked micelle-based drug delivery system maintained a highly efficient delivery of hydrophobic drug molecules, while demonstrating extraordinary physiological stability, and high drug-loading capacity. The redox-responsive drug delivery systems therefore boosted the advanced drug delivery paradigm in anticancer therapy by utilizing the physical conditions in the tumor microenvironment, thereby ensuring targeted delivery and minimal damage to the vicinal healthy cells.

7.4 Enzyme-responsive

The cancer cells exhibit abnormal expression of enzymes such as matrix metalloproteinases and hyaluronidase. The former executes degradation of extracellular matrix, whereas the latter mediates the degradation of hyaluronic acid in acidic environment of cancer cells. Zhou et al. [73] reported enzyme-responsive drug delivery systems based on extracellular matrix components shelled mesoporous silica nanoparticles with minimal side effects for cancer therapy. The covalently conjugated biomacromolecules in ECM, collagen, and hyaluronic acid improved the physiological compatibility of mesoporous silica nanoparticles and prevented the premature drug leakage. This feature further promotes selective biodegradation instigated by hyaluronidase and matrix metalloproteinase overexpressed in cancer tissues. The nanosystem displayed controlled release efficiency in response to these enzymes in cancer microenvironment, while displaying a commendable antitumor activity. Guan et al. [74] developed enzyme-responsive supramolecular drug delivery system based on biocompatible sulfato-β-cyclodextrin conjugated to anticancer drug chlorambucil. The drug delivery system release the conjugated drug in response to butyrylcholinesterase enzyme, which cleaves ester bond in nanosystem. The drug delivery system contains hydrophilic choline moiety and hydrophobic chlorambucil member, and its amphiphilic framework aggregates by sulfato-β-cyclodextrin into large assemblies mainly via host-guest interactions. The negatively charged macrocyclic receptor sulfato-β-cyclodextrin presents water solubility, lower toxicity, and cost-effectiveness effectively lowers the critical aggregation concentration of encapsulated prodrug mainly via electrostatic interactions. Jia et al. [75] reported enzyme-responsive novel drug delivery system containing organic–inorganic nanovehicle carrying anticancer drug, which releases in response to overexpressed glutathione S-transferase enzyme in the tumor microenvironment. In response to the enzyme, the degradation of protective organosilica shell occurs leading to drug release followed by the release of nitric oxide particularly in the tumor cells. This event protects healthy cells from the cytotoxicity effect of the anticancer drug. The nanosystem displays a passive targeting of tumor cells by enzyme-responsive drug release. Li et al. [76] reported multifunctional nanovehicles for enzyme-responsive in vivo release of doxorubicin. The system also displayed magnetic imaging properties due to the presence of iron oxide nanoparticles coated with mesoporous silica. The matrix metalloproteinase-responsive peptide anchored to the nanoparticle surface contains the drug molecules in its porous cavities that influenced controlled drug release with high specificity. The external magnetic field reportedly guiding the targeted delivery of drug delivery vehicle at the tumor site by utilizing magnetic properties of iron oxide nanocore formed a unique feature of this system. This property enabled MRI application for real-time probing of the tumor treatment. Li et al. [77] developed enzyme-sensitive PEGylated amphiphilic drug delivery system for selectively delivering doxorubicin to cancer cells. The drug delivery system reportedly self-assembled into negatively charged nanoparticles that exhibited an extraordinary accumulation and retention in tumor cells for an extended time. The nanoparticles displayed a superior anticancer potency that potentiated over the activity of doxorubicin alone with a marked antiproliferative effect and apoptosis induction in 4T1 murine breast cancer cells. Importantly, the drug delivery system lowered the doxorubicin-induced cytotoxicity while minimizing the hazardous effect on healthy tissue vicinal to the tumor site. The fast drug release assay suggested a rapid drug release from the carrier system upon its internalization in the tumor cells. Zha et al. [78] developed enzyme-responsive CuS nanoparticles for multifunctional purposes. The drug delivery system enabled photoacoustic imaging, photothermal therapy, and chemotherapy in cancer treatment. The drug delivery vehicle enabled the selective release of encapsulated drug due to enzymatic degradation of nanosystem, in addition to exhibiting synergistic annihilation of cancer cells. Near-infrared laser irradiation of drug delivery vehicle caused photothermal effect, which effectively killed cancer cells. Importantly, the hyperthermia caused by laser irradiation improved the cell metabolism and membrane permeability of cancer cells for an enhanced drug penetration. The hyperthermal effect effectively reduced the dosage of chemotherapy pharmaceuticals thereby mitigating the side effects ensued by the latter. Wang et al. [79] reported endogenous enzyme-sensitive cisplatin drug nanoplatforms for cascade-promoted photo/chemotherapy aimed at annihilating drug-resistant cancers. The nanoplatforms were prepared from photosensitizers as indocyanine green and cisplatin polydrug amphiphiles constituting enzyme-degradable units. The cellular uptake of drug carrier in lysosomes caused its degradation to cisplatin prodrug, followed by excitation of the photosensitizer by laser irradiation to generate reactive oxygen species and hyperthermia thereby causing the onset of photothermal effect. Importantly, the simultaneous photothermal and photodynamic processes effectively damaged lysosomes, followed by the enzymatic reduction into active drug cisplatin thereby triggering cascade chemotherapy. In addition to these features, the polydrug nanoplatforms afforded dual-mode fluorescence imaging and photoacoustic effect to monitor the chemotherapeutic treatment. The nanoplatforms proved highly beneficial in targeting cisplatin-resistant A_{549}/DDP cancers via imaging-guided cascade treatment strategy. The cascade photo-chemotherapy further improved the therapeutic efficacy of cisplatin thereby displaying superior anticancer properties compared to the drug alone.

8 Conclusion and future perspectives

The limitations countered by representative anticancer drugs prompted the development of advanced drug delivery system and strategies to overcome the existing constraints in chemotherapy. The necessity for targeted delivery led to the development of smart strategies such as photothermal therapy, magnetothermal approach, and the utility of smart drug delivery vehicles for targeted cancer therapy without harming the normal cells. The presently developed smart delivery vehicles comprise the effective exploitation of tumor microenvironment, which provides stimuli such as abnormal temperature and pH, erroneously expressed enzymes, redox stress, and metabolic anomalies for prompting drug release. The incorporation of metallic nanoparticles to these drug delivery systems further improves their applications. As such, the magnetic nanoparticles provide external field-guided drug delivery to the target site, while enabling the magnetic resonance imaging of the affected area for better understanding of dynamics and drug pharmacokinetics. Similarly, the photothermal effect posed by metal nanoparticles prompts selective annihilation or ablation of cancer cells by irradiating the nanometallic drug carriers while they reach the target site. The most recent developments in the smart nanotechnology for cancer drug delivery present hybrid carriers that combine all the features pertaining to a typical anticancer therapy regime, including chemotherapy, magnetic imaging, phototherapy, and photoacoustic applications in a single formulation. Another important aspect of advanced drug delivery vehicles obligates higher biological tolerance and minimal toxicity toward the healthy cells. The drug delivery vehicles based on natural polymers or physiological benevolent entities fulfill this obligation while displaying stimuli-sensitive degradation in vivo, eventually causing a controlled drug release. The advanced drug-nanocarriers display a superior stability in systemic circulation, protect the encapsulated drug molecules from degrading enzymes in the circulatory system, and release the drug load in a controlled manner on sensing the physical or chemical stimulus presented by the tumor microenvironment. Nevertheless, the smart nanotechnology for improving the contemporary cancer treatment regime has righteously fulfilled the emerging need for advanced drug delivery systems in anticancer therapy.

References

[1] Moorthi C, Manavalan R, Kathiresan K. Nanotherapeutics to overcome conventional cancer chemotherapy limitations. J Pharm Pharmaceut Sci 2011;14:67–77.
[2] Boroughs LK, DeBerardinis RJ. Metabolic pathways promoting cancer cell survival and growth. Nature Cell Biol 2015;17:351–9.
[3] Decker M. Hybrid molecules incorporating natural products: applications in Cancer therapy, neurodegenerative disorders and beyond. Curr Med Chem 2011;18:1464–75.
[4] Senapati S, Mahanta AK, Kumar S, Maiti P. Controlled drug delivery vehicles for cancer treatment and their performance. Signal Transduct Target Ther 2018;3. Article 7.
[5] Hossen S, Hossain MK, Basher MK, Mia MNH, Rahman MT, Uddin MJ. Smart nanocarrier-based drug delivery systems for cancer therapy and toxicity studies: a review. J Adv Res 2019;15:1–18.
[6] Kalaydina R-V, Bajwa K, Qorri B, Decarlo A, Szewczuk MR. Recent advances in "smart" delivery systems for extended drug release in cancer therapy. Int J Nanomedicine 2018;13:4727–45.
[7] McCarty MF. Targeting multiple signaling pathways as a strategy for managing prostate cancer: multifocal signal modulation therapy. Integr Cancer Ther 2004;2004:349–80.
[8] Gonzalez H, Hagerling C, Werb Z. Roles of the immune system in cancer: from tumor initiation to metastatic progression. Genes Dev 2018;32:1267–84.
[9] Jahangirian H, Lemraski EG, Webster TJ, Moghaddam RR, Abdollahi Y. A review of drug delivery systems based on nanotechnology and green chemistry: green nanomedicine. Int J Nanomedicine 2017;12:2957–78.
[10] Safari J, Zarnegar Z. Advanced drug delivery systems: nanotechnology of health design a review. J Saudi Chem Soc 2014;18:85–99.
[11] Liu D, Yang F, Xiong F, Gu N. The smart drug delivery system and its clinical potential. Theranostics 2016;6:1306–23.
[12] Lorenzo CA, Concheiro A. Smart drug delivery systems: from fundamentals to the clinic. Chem Commun 2014;50:7743–65.
[13] Sharma A, Goyal AK, Rath G. Recent advances in metal nanoparticles in cancer therapy. J Drug Target 2018;26:617–32.
[14] Evans ER, Bugga P, Asthana V, Drezek R. Metallic nanoparticles for cancer immunotherapy. Mater Today 2018;21:673–85.
[15] Housman G, Byler S, Heerboth S, Lapinska K, Longacre M, Snyder N. Sarkar, S. Drug resistance in cancer: An overview 2014;6:1769–92.
[16] Vasan N, Baselga J, Hyman DM. A view on drug resistance in cancer. Nature 2019;575:299–309.
[17] Haider T, Pandey V, Banjare N, Gupta PN, Soni V. Drug resistance in cancer: mechanisms and tackling strategies. Pharmacol Rep doi 2020. https://doi.org/10.1007/s43440-020-00138-7.
[18] Ward RA, Fawell S, Floch N, Flemington V, McKerrecher D, Smith PD. Challenges and opportunities in cancer drug resistance. Chem Rev 2020. https://doi.org/10.1021/acs.chemrev.0c00383.
[19] Gottesman MM. Mechanisms of cancer drug resistance. Ann Rev Med 2002;53:615–27.
[20] Hawkes N. Drug resistance: the next target for cancer treatment. BMJ 2019;365. Article 12228.
[21] Ozben T. Mechanisms and strategies to overcome multiple drug resistance in cancer. FEBS Lett 2006;580:2903–9.
[22] Zahreddine H, Borden KLB. Mechanisms and insights into drug resistance in cancer. Front Pharmacol 2013;4. Article 28.
[23] Donnenberg VS, Donnenberg AD. Multiple drug resistance in cancer revisited: the cancer stem cell hypothesis. J Clin Pharmacol 2005;45:872–7.
[24] Nooter K. Stoter, G. Molecular mechanisms of multidrug resistance in cancer chemotherapy 1996;192:768–80.
[25] Asghari MH, Ghobadi E, Moloudizargari M, Fallah M, Abdollahi M. Does the use of melatonin overcome drug resistance in cancer chemotherapy? Life Sci 2018;196:143–55.
[26] Frenkel GD, Caffrey PB. A prevention strategy for circumventing drug resistance in cancer chemotherapy. Curr Pharm Des 2001;7:1595–614.
[27] Wu CP, Hsieh C-H, Wu Y-S. The emergence of drug-transporter-mediated multidrug resistance to cancer chemotherapy. Mol Pharm 2011;8:1996–2011.
[28] Chabner BA, Roberts TG. Chemotherapy and the war on cancer. Nat Rev Cancer 2005;5:65–72.
[29] Liu H, Lv L, Yang K. Chemotherapy targeting cancer stem cells. Am J Cancer Res 2015;5:880–93.
[30] Batlle JF, Arranz EE, Carpeno JC, Saez EC, Aunon PC, Sanchez AR, Baron MG. Oral chemotherapy: potential benefits and limitations. Revista de Oncologia 2004;6:335–40.
[31] Monsuez J-J, Charniot JC, Vignat N, Artigou J-Y. Cardiac side-effects of cancer chemotherapy. Int J Cardiol 2010;144:3–15.
[32] Zimmermann S, Dziadziuszko R, Peters S. Indications and limitations of chemotherapy and targeted agents in non-small cell lung cancer brain metastases. Cancer Treat Rev 2014;40:716–22.
[33] Zugazagoitia J, Guedes C, Ponce S, Ferrer I, Pinelo SM, Ares LP. Current challenges in cancer treatment. Clin Ther 2016;38:1551–66.
[34] Gamez ME, Blakaj A, Zoller W, Bonomi M, Blakaj DM. Emerging concepts and novel strategies in radiation therapy for laryngeal Cancer management. Cancers 2020;12. Article 1651.
[35] Nina M, Kai B, Alan K, Landon W, Psy CAL, Stephanie C, Waylene W, Ning C, Simon L, George S, Juergen M. Reducing cardiac radiation dose frombreast Cancerradiation therapywith breath hold training and-cognitive behavioral therapy. Topics Mag Res Imag 2020;29:135–48.
[36] Luo D, Johnson A, Wang X, Li H, Erokwu BO, Springer S, Lou J, Ramamurthy G, Flask CA, Burda C, Meade TJ, Basilion JP. Targeted Radiosensitizers for MR-guided radiation therapy of prostate Cancer. Nano Lett 2020. https://doi.org/10.1021/acs.nanolett.0c02487.
[37] Wang CC, Blitzer PH, Suit HD. Twice-a-day radiation therapy for cancer of the head and neck. Cancer 1985;55:2100–4.
[38] Shepard DM, Ferris MC, Olivera GH, Mackie TR. Optimizing the delivery of radiation therapy to cancer patients. SIAM Rev 2006;41:721–44.
[39] Demaria S, Golden EB, Formenti SC. Role of local radiation therapy in cancer immunotherapy. JAMA Oncol 2015;1:1325–32.
[40] Reyngold M, Parikh P, Crane CH. Ablative radiation therapy for locally advanced pancreatic cancer: techniques and results. Radiation Oncol 2019;14. Article 95.
[41] Gujral TS, Peshkin L, Kirschner MW. Exploiting polypharmacology for drug target deconvolution. Proc Natl Acad Sci 2014;111:5048–53.
[42] Ravikumar B, Aittokallio T. Improving the efficacy-safety balance of polypharmacology in multi-target drug discovery. Expert Opin Drug Disc 2018;13:179–92.
[43] Cheng F. In silico oncology drug repositioning and Polypharmacology. In: Krasnitz A, editor. Cancer bioinformatics. Methods in molecular biology, vol 1878. New York, NY: Humana Press; 2019.
[44] Lopez MH, Echeberria ML, Catalan EB. The multitarget activity of natural extracts on cancer: surgery and xenohormesis. Medicines 2019;6. Article 6.
[45] Ma X, Lv X, Zhang J. Exploiting polypharmacology for improving therapeutic outcome of kinase inhibitors (KIs): an update of recent medicinal chemistry efforts. Eur J Med Chem 2017;143:449–63.

[46] Blanc TW, Neil MJ, Kamal AH, Currow DC, Abernethy AP. Polypharmacy in patients with advanced cancer and the role of medication discontinuation. The Lancet Oncol 2015;16:E333–41.
[47] Reddy AS, Zhang S. Polypharmacology: drug discovery for the future. Expert rev Clin Pharmacol 2013;6:41–7.
[48] Amelio I, Lisitsa A, Knight RA, Melino G, Antonov AV. Polypharmacology of approved anticancer drugs. Curr Drug Target 2017;18:534–43.
[49] Apaya MK, Chang M-T, Shyur L-F. Phytomedicine polypharmacology: Cancer therapy through modulating the tumor microenvironment and oxylipin dynamics. Pharmacol Ther 2016;162:58–68.
[50] Knight ZA, Lin H, Shokat KM. Targeting the cancer kinome through pharmacology. Nat Rev Cancer 2010;10:130–7.
[51] Jin Y, Huang Y, Yang H, Liu G, Zhao R. A peptide-based pH-sensitive drug delivery system for targeted ablation of cancer cells. Chem Commun 2015;51:14454–7.
[52] Yang Y, Wang Z, Peng Y, Ding J, Zhou W. A smart pH-sensitive delivery system for enhanced anticancer efficacy via paclitaxel endosomal escape. Front Pharmacol 2019;10. Article 10.
[53] Men W, Zhu P, Dong S, Liu W, Zhou K, Bai Y, Liu X, Gong S, Zhang S. Layer-by-layer pH-sensitive nanoparticles for drug delivery and controlled release with improved therapeutic efficacy *in vivo*. Drug Deliv 2020;27:180–90.
[54] Yilmaz G, Guler E, Geyik C, Demir B, Ozkan M, Demirkol OD, Ozcelik S, Timur S, Becer CR. pH responsive glycopolymer nanoparticles for targeted delivery of anti-cancer drugs. Mol Syst Des Eng 2018;3:150–8.
[55] Palanikumar L, Al-Hosani S, Kalmouni M, Nguyen VP, Ali L, Pasricha R, Barrera FN, Magzoub M. pH-responsive high stability polymeric nanoparticles for targeted delivery of anticancer therapeutics. Commun Biol 2020;3. Article 95.
[56] Gao N, Xing C, Wang H, Feng L, Zeng X, Mei L, Peng Z. pH-responsive dual drug-loaded Nanocarriers based on poly (2-Ethyl-2-Oxazoline) modified black phosphorus Nanosheets for Cancer chemo/Photothermal therapy. Front Pharmacol 2019;10. Article 270.
[57] Nogueira J, Soares SF, Amorim CO, Amaral JS, Silva C, Martel F, Trindade T, Silva ALD. Magnetic driven nanocarriers for pH-responsive doxorubicin release in cancer therapy. Molecules 2020;25. Article 333.
[58] Wang F, Yuan Z, McMullen P, Li R, Zheng J, Xu Y, Xu M, He Q, Li B, Chen H. Near-infrared-light-responsive lipid nanoparticles as an intelligent drug release system for Cancer therapy. Chem Mater 2019;31:3948–56.
[59] Chen H, Ma Y, Wang X, Zha Z. Multifunctional phase-change hollow mesoporous Prussian blue nanoparticles as a NIR light responsive drug co-delivery system to overcome cancer therapeutic resistance. J Mater Chem B 2017;5:7051–8.
[60] Chen Y, Su M, Li Y, Gao J, Zhang C, Zhou J, Liu J, Jiang Z. Enzymatic PEG-Poly (amine-co-disulfide ester) Nanoparticles as pH- and Redox-Responsive Drug Nanocarriers for Efficient Antitumor Treatment. ACS Appl Mater Interface 2017;9:30519–35.
[61] Gao M, Hu A, Sun X, Wang C, Dong Z, Feng L, Liu Z. Photosensitizer decorated red blood cells as an ultrasensitive light-responsive drug delivery system. ACS Appl Mater Interface 2017;9:5855–63.
[62] Feng Q, Zhang Y, Zhang W, Hao Y, Wang Y, Zhang H, Hou L, Zhang Z. Programmed near-infrared light-responsive drug delivery system for combined magnetic tumor-targeting magnetic resonance imaging and chemo-phototherapy. Acta Biomater 2017;49:402–13.
[63] Yang G, Sun X, Liu J, Feng L, Liu Z. Light-responsive, singlet-oxygen-triggered on-demand drug release from photosensitizer-doped mesoporous silica nanorods for cancer combination therapy. Adv Funct Mater 2016;26:4722–32.
[64] Yang X, Liu X, Liu Z, Pu F, Ren J, Qu X. Near-infrared light-triggered, targeted drug delivery to cancer cells by aptamer gated nanovehicles. Adv Mater 2012;24:2890–5.
[65] Cheng C-C, Huang J-J, Lee A-W, Huang S-Y, Huang C-Y, Lai J-Y. Highly effective Photocontrollable drug delivery systems based on ultrasensitive light-responsive self-assembled polymeric micelles: anin vitrotherapeutic evaluation. ACS Appl Bio Mater 2019;2:2162–70.
[66] Li Z-Y, Hu J-J, Xu Q, Chen S, Jia H-Z, Sun Y-X, Zhuo R-X, Zhang X-Z. A redox-responsive drug delivery system based on RGD containing peptide-capped mesoporous silica nanoparticles. J Mater Chem B 2015;3:39–44.
[67] Zhang B, Luo Z, Liu J, Ding X, Li J, Cai K. Cytochrome c end-capped mesoporous silica nanoparticles as redox-responsive drug delivery vehicles for liver tumor-targeted triplex therapyin vitro and in vivo. J Control Release 2014;192:192–201.
[68] Laskar P, Somani S, Campbell SJ, Mullin M, Keating P, Tate RJ, Irving C, Leung HY, Dufes C. Camptothecin-based dendrimersomes for gene delivery and redox-responsive drug delivery to cancer cells. Nanoscale 2019;11:20058–71.
[69] Jiao J, Liu C, Li X, Liu J, Di D, Zhang Y, Zhao Q, Wang S. Fluorescent carbon dot modified mesoporous silica nanocarriers for redox-responsive controlled drug delivery and bioimaging. J Colloid Interface Sci 2016;483:343–52.
[70] Lei B, Wang M, Jiang Z, Qi W, Su R, He Z. Constructing redox-responsive metal–organic framework Nanocarriers for anticancer drug delivery. ACS Appl Mater Interface 2018;10:16698–706.
[71] Liu J, Pang Y, Huang W, Zhu Z, Zhu X, Zhou Y, Yan D. Redox-responsive polyphosphate Nanosized assemblies: a smart drug delivery platform for Cancer therapy. Biomacromolecules 2011;12:2407–15.
[72] Lv Y, Yang B, Li Y-M, Wu Y, He F, Zhou R-X. Crosslinked triblock copolymeric micelle for redox-responsive drug delivery. Colloid Surf B 2014;122:223–30.
[73] Zhou J, Wang M, Ying H, Su D, Zhang H, Lu G, Chen J. Extracellular matrix component shelled nanoparticles as dual enzyme-responsive drug delivery vehicles for cancer therapy. ACS Biomater Sci Eng 2018;4:2404–11.
[74] Guan X, Chen Y, Wu X, Li P, Liu Y. Enzyme-responsive sulfatocyclodextrin/prodrug supramolecular assembly for controlled release of anti-cancer drug chlorambucil. Chem Commun 2019;55:953–6.
[75] Jia X, Zhang Y, Zou Y, Wang Y, Niu D, He Q, Huang Z, Zhu W, Tian H, Shi J, Li Y. Dual Intratumoral redox/enzyme-responsive NO-releasing nanomedicine for the specific, high-efficacy, and low-toxic cancer therapy. Adv Mater 2018;30. Article 1704490.
[76] Li E, Yang Y, Hao G, Yi X, Zhang S, Pan Y, Xing B, Gao M. Multifunctional magnetic mesoporous silica Nanoagents for in vivo enzyme-responsive drug delivery and MR imaging. Nanotheranostic 2018;2:233–42.
[77] Li N, Li N, Yi Q, Luo K, Guo C, Pan D, Gu Z. Amphiphilic peptide dendritic copolymer-doxorubicin nanoscale conjugate self-assembled to enzyme-responsive anti-cancer agent. Biomaterials 2014;35:9529–45.
[78] Zha Z, Zhang S, Deng Z, Li Y, Li C, Dai Z. Enzyme-responsive copper sulphide nanoparticles for combined photoacoustic imaging, tumor-selective chemotherapy and photothermal therapy. Chem Commun 2013;49:3455–7.
[79] Wang W, Liang G, Zhang W, Xing D, Hu X. Cascade-promoted photo-chemotherapy against resistant cancers by enzyme-responsive Polyprodrug Nanoplatforms. Chem Mater 2018;30:3486–98.

Chapter 5

Target drug delivery in cancer

Sangita Saini
BabaMast Nath University, Asthal Bohar, Rohtak, India

1 Introduction

Cancer is a major health issue and the most common cause of mortality worldwide. It is abnormal growth of cells in the body that can invade any organ, cells, or other body parts. Most but not all tumors are cancerous. Malignant tumors are due to the continuous dividing and spreading of cancerous cells to other tissues, whereas benign tumors are noncancerous and do not spread to other body tissues or organs. According to the American Cancer Society (ACS), the cancer death rates are decreasing due to early detection and treatment; there was approximately a 29% decrease in cancer deaths in 2017. In addition, the 5-year relative survival rate for all cancers combined has increased substantially since the early 1960s. The decline in death rate is due to improved treatments for the four leading cancers: breast, lung, colorectal, and prostate. Specifically, breast cancer and colorectal cancer have a high recovery rate, whereas lung cancer causes the most deaths. Early detection and advanced treatment options are required for the fight against cancer [1].

Chemotherapeutic drugs can eliminate the progression of rapidly dividing cancerous cells, but, unfortunately, they also interfere with the normal cell division process, which causes serious side effects in the body. Some common side effects with this therapy include nausea, vomiting, diarrhea, weight loss, fatigue, and permanent alopecia. Chemotherapeutic agents can even damage the nervous system [2, 3]. These drug-induced side effects can be ameliorated by using a targeted drug delivery system. The clinical significance of targeted drug delivery is that it allows a drug or drug carrier to accumulate in a targeted zone or to a specific site, thus avoiding healthy tissues, organs, or cells [4]. Therefore, the main goal of drug targeting is to improve and extend patient life by reducing systemic toxic effects and increasing the therapeutic efficacy of the drug. In the last few years, cancer targeting has become a multistrategy approach that can be achieved by using drug carriers [5]. Several nanoparticle drug delivery systems are used to overcome the limitations of traditional chemotherapy. Nanoparticle size varies from a few to a hundred nanometers. These nanosized particles have the ability to encapsulate various drug particles and have highly selective accumulation in tumors via enhanced permeability effects and cellular uptake. The most used nanoscale drug carriers are liposomes, polymeric nanoparticles, magnetic nanoparticles, dendrimers, micelles, and lipid nanoparticles, all of which have great potential to combat cancer. A few examples of approved nano-chemotherapeutics include Doxil (doxorubicin pegylated liposome), Abraxane (albumin-bound paclitaxel), and Daunoxome (liposomal daunorubicin). Despite having advantages for targeted drug delivery, only a few drug carriers have obtained approval from the US Food and Drug Administration (FDA) [6]. This discrepancy is due to high vascular permeability and interstitial pressure, which obstruct the path of novel drug delivery with an effective concentration. Therefore, there is an urgent need to design new controlled drug delivery systems with enhanced ability to target selective sites with improved therapeutic index and reduce drug-resistance problems in chemotherapy. There are two strategies available for designing targeted drug delivery: active drug delivery or surface modification and passive targeting. Active drug delivery is also known as ligand-mediated drug delivery. In this approach, a ligand is attached to the polymer used in the formulation or ligands are conjugated directly with drugs. Ligands interact with specific receptors present on tumor cells or overexpressed receptors (Fig. 1). Passive targeting exploits leaky vasculature and enhanced permeability retention effects.

2 Ligand-mediated targeting (surface modification)

In ligand-mediated targeting or active targeting, the ligand is conjugated on the surface of drug carriers for strapping of the receptor structure. Ligand-based drug delivery utilizes the overexpressed receptors on tumor cells and increases the drug accumulation via an increase in cellular uptake of drugs by receptor-mediated endocytosis. For receptor–ligand interaction, there should be a complete and homogenous distribution of receptors or antigens on the surface of cancer cells rather than healthy cells [7].

Advanced Drug Delivery Systems in the Management of Cancer. https://doi.org/10.1016/B978-0-323-85503-7.00009-2

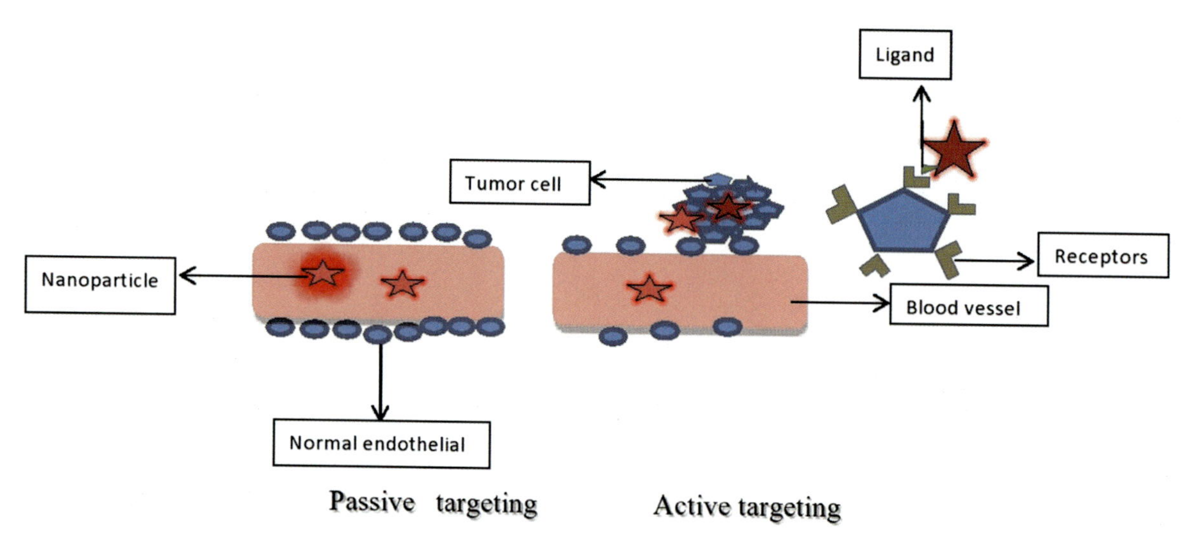

FIG. 1 Passive and active targeting of drug.

There are several approaches to functionalizing ligands on nanoparticles (Table 1).

Lignads can either be coupled on the surface of a nanocarrier, or the polymer can be modified with ligands. The synthetic route is selected based on the size of the ligands to which polymers attaches. Compared to single ligation, multi-ligand-decorated nanoparticles show more activity. An advantage to coupling ligands prior to nanocarrier formation is that the density of the ligands can be tailored for their effective binding to the receptors [8,16]. A broad variety of ligands have been studied and evaluated for targeted drug delivery. Table 2 lists some of these ligands along with their advantages and disadvantages. These ligands include peptides, aptamers, sugars, folic acid, antibody proteins, and small molecules.

2.1 Peptide as targeting ligand

Among the ligands in targeted drug delivery, peptides have unique properties like moderate size, high specificity and affinity, biocompatibility, simple structure, and easy modification [17]. A peptide is a molecule originating from a natural or synthetic source that contains more than one amino acid linked by a covalent bond. Peptides, excluding cyclic peptides, also comprise a carboxy at one terminus and an amine at the other. Peptide ligands can be selected by screening the peptide library produced by phage display or chemical synthesis. The phage display is most widely employed to identify a target molecule that is already known [18]. Moreover, phage display can be used to study molecular imaging and cancer monitoring. Some peptides have anti-cancer properties, while others are conjugated with nanoparticles to target tumor-specific antigens or receptors [19]. Cell-penetrating peptides are short peptides (5–30 residues) that transport the drug into the cell via direct penetration and exhibit endocytosis-mediated translocation that delivers many drugs and small molecules. A few examples of cell-penetrating anticancer peptides are transactivating transcriptional activator (TAT), arginylglycylaspartic acid (RGD), La peptide (LA), and epidermal growth factor receptor (EGFR). RGD is a tripeptide motif (Arg-Gly-Asp). Integrin is a heterodimeric glycoprotein that contains alpha and beta subunits. Mammals have 24 subunits, 18 of which are alpha subunits and 8 of which are beta subunits. Integrin shows a strong affinity for fibronectin, vitronectin, collagen, and laminin extracellular protein. After binding, integrin activates signal transduction pathways that mediate cellular growth, survival, division, and migration. Overexpression of integrin extracellular protein occurs in some particular types of cancers, for example, fibronectin appears in head and neck cancer, whereas vitronectin appears in lung cancer [20]. Integrin plays a vital role in the progression of tumors, and some subtypes are overexpressed on many cancer cells. RGD binds with integrin and blocks the proliferation of tumor cells [21]. Doxorubicin is a drug used for treating breast cancer and lymphoma. Modification of Dox nanoparticles with RGD peptide improved safety, biocompatibility, and targeting ability [22]. RGD peptide-modified dual drugs (paclitaxel and cisplatin) loaded with lipid polymer nanoparticles showed high antitumor activity for lung cancer [23]. Yu et al. synthesized cyclic RGD (cRGD)-conjugated, gemcitabine-loaded human serum albumin (HSA) nanoparticles and evaluated them in vitro on cell lines. The cRGD–HSA nanoparticles effectively inhibited the growth of pancreatic cell lines with moderate toxicity [24]. The gH625 peptide is a 19-residue peptide identified in the domain of glycoprotein H of herpes simplex. This is a peptide from a cell-penetrating peptide family that transverses the biological cell membrane and delivers the cargo in the cytoplasm and blood–brain barrier. This peptide can

TABLE 1 Examples of ligand-mediated drug delivery.

	Ligand	Drug carrier	Active drug	Tumor target	Condition	References
1.	Folate	Nanoemulsion	Docetaxel	Overexpressed folate receptors	Ovarian cancer	[8]
2.	Transferrin	Nanoparticle	Doxorubicin Sorafinib	Iron deficient transferrin receptors	Liver cancer	[9]
3.	Aptamers	Superparamagnetic iron oxide nanoparticle	Docetaxel	Prostate-specific membrane antigen	Prostate cancer	[10]
4.	Lectin	Nanoliposomes	Ova antigen	C-type lectin receptors	Cancer immunotherapy	[11]
5.	Peptide (RGD)	Chitosan micelles	Doxorubicin	Overexpressed Integrin	Solid tumor	[12]
6.	Antibody (Farletuzufab)	Iron oxide nanoparticle	–	Folate receptor alpha	Ovarian cancer	[13]
7.	Anisamide	Micelles	Paclitaxel	Sigma 1 receptors	Prostate cancer	[14]
8.	Hyaluronic acid	Nanomedicine	Doxorubicin	CD44 receptors	Diagnostic and cancer therapy	[15]

TABLE 2 Advantage and disadvantage of ligands in targeted drug delivery.

	Ligand	Advantages	Disadvantages
1.	Aptamer	Good solubility and permeability Biocompatible	Difficult to escape from the reticuloendothelial system
2.	Antibody	High affinity and specificity	Large size, poor permeability, immunogenic, and expensive
3.	Peptide	Moderate size and high specificity and immunity Easy synthesis and chemical modification	Prone to proteolysis
4.	Folate	Water soluble, low antigenicity	Undesirable accumulation in cancerous cells
5.	Galactose	High specificity, affinity, and selectivity	Utilizes glucose metabolism

transport quantum dots, liposomes, nanoparticles, and dendrimers through a translocation mechanism rather than endocytosis [25]. In one study, the gH625 peptide, which is a cell-penetrating peptide, was coupled with superparamagnetic iron oxide (SPION) nanoparticles. The gH625 peptide contains cysteine residues with the carboxy (C)-terminus. The C-terminus was conjugated via polyethylene coating (PEG 5000) of nanoparticles. The conjugation of CPP–SPION nanoparticles increased the cellular uptake threefold as compared to SPION nanoparticles alone [26]. Therefore, peptide-conjugated nanoparticles showed better anticancer activity.

2.2 Aptamer-mediated drug targeting

Aptamers are a class of oligonucleotides that are small (20–60 nucleotides), single-stranded DNA or RNA nucleic acids capable of binding various molecules, including proteins and cells, with high affinity and specificity. An aptamer has a unique three-dimensional structure [27]. Nucleic acids play a crucial role in the storage, processing, and expression of genetic information. Aptamers can be selected from Systemic Evolution of Ligands by Exponential Enrichment (SELEX), a combinatorial library developed in 1990 that contains a large number of oligonucleotides. Using this technique, aptamers can be screened in vitro in a few cycles (5–10 rounds) of polymerase chain reaction (PCR), transcribed to develop an RNA pool, and partitioned on the base of affinity of binding to their specific ligands [28, 29]. Paclitaxel is a common anticancer drug used for treating ovarian, breast, lung, and head and neck cancers, as well as Kaposi sarcoma. To reduce the toxic effects of the drug, paclitaxel-loaded nanoparticles were conjugated by the mucin1 (MUC1) aptamer through a DNA spacer. The MUC1 aptamer is a 25-base long variable region that binds specifically to the target MUC1, which is a transmembrane glycoprotein found abundantly in malignant cells of the breast, prostate, ovaries, and pancreas (i.e., the most commonly diagnosed cancers) [30, 31]. SN-38 is a new anticancer drug against colon cancer, but it cannot be used clinically due to poor solubility and high toxicity. MUC1 DNA aptamers conjugated with self-assembled chitosan nanoparticles increased efficacy of drug SN-38 through an increase in solubility and specific delivery to target tissue [32]. MUC1 aptamer-conjugated nanoparticles showed improved efficacy in MCF cell lines in vitro compared to paclitaxel-loaded nanoparticles [33]. RNA-aptamer conjugated micelles are inserted into liposomes called aptamosomes, which are prepared by postinsertion methods. RNA aptamers are specific for prostate-specific membrane antigens (PSMA) that are overexpressed on the surface of prostate cancer cells. These aptamosomes can be used to target the anticancer drug doxorubicin to cancerous cells rather than non-target tissues. This targeted drug delivery showed very few side effects specific to prostate cancer only [34]. In a study by Gray et al., monomethyl auristatin drugs (MMAE and MAMF) inhibited microtubules that are too toxic in the free state. When drugs are conjugated with the E3 aptamer, which is specific to prostate cancer, the drugs targeted and killed cancerous cells not normal prostate cells. Animal studies indicate that highly toxic drug and aptamer conjugations represent a targeted approach to treat cancer patients. This conjugation can further reduce patient morbidity [35].

2.3 Small molecule-mediated targeted drug delivery

2.3.1 *Folate receptor targeting*

Folate receptors such as glycophosphatidylinositol are expressed in normal cells for the synthesis of DNA and cell proliferation. The alpha isoform (FRα) overexpression in different cancer cells can contribute to cell growth regulation and signaling functions [36]. Therefore folate receptors are attractive targets for drug delivery and imaging. Folic acid-conjugated liposomes of celastrol and irinotecan target drugs to breast cancer cells. In vitro studies showed greater uptake of both drugs in folate receptor cells. In a mouse tumor model, folate-conjugated liposomes exhibited targeted drug delivery with less systemic toxicity [37]. Folic acid was conjugated with Pullan backbone and poly (DL-lactide-*co*-glycolide) nanoparticles to target doxorubicin in cancer cells. This preparation was used to treat folate receptor overexpression in human carcinoma cells [38]. *Withania somnifera* is a common anticancer drug in India. It is an alcoholic extract of the leaves of *W. somnifera* that contains withaferin and withanone. Both can kill telomerase-negative cancer cells. Therefore, it can be used in aggressive and complex cancers. Conjugation of the alcoholic extract of *W. somnifera* with folate receptors forms a nanocomplex that has threefold tumor-suppression activities. It has proved to be a safe, efficient, and natural candidate for cancer treatment [39]. Folate receptors can be used in the imaging of cancer. Jin et al. performed dual targeting to improve tumor-killing efficiency and drug-originated toxic effects. A rhodamine probe was used with folic acid for imaging of SK-Hep-1 cell overexpression in folic acid receptors. This complex showed cytotoxic effects against overexpressed folic acid receptors. These can be adopted for imaging as well as serve as the basis for effective therapy [40].

2.3.2 *Anisamide-mediated targeting*

Anisamide, a low molecular weight benzamide, has high affinity and specificity for sigma-1 receptors. Sigma-1 receptors are expressed in normal tissue, whereas they are overexpressed in cancerous tissue, mainly in prostate, lung, and breast cancers. Sigma-binding ligands are prospects for targeted drug delivery both in diagnostics and therapeutics. Phospholipid liposomes are conjugated with anisamide via a PEG spacer. These ligands showed efficient interaction with human prostate carcinoma cells [41]. Anisamide can be used as a targeting ligan in gene delivery. RNA interference has been found to be effective in prostate cancer. The inclusion of octarginine in anisamide-targeted cyclodextrin nanoparticles increased the uptake and endocytosis of SiRNA in prostate cancer cell lines [42]. PEGylated anisamide-targeted gold nanoparticles loaded with epirubicin bound specifically to sigma receptors. This complex delivered the drug in endosome and induced antiproliferative effects in human prostate cancer cells [43].

2.3.3 *Sugar-mediated targeting*

Cancer cells have increased glucose uptake and metabolism. Glucose transport protein (GLUT) facilitates glucose uptake across plasma membranes and is abnormally expressed in cancer cells. GLUT 1, GLUT3, and GLUT 5 can be inhibited to reduce cancer growth [44]. According to the Warburg hypothesis, cancer cells alter their glucose metabolism, thus producing lactate in the presence of oxygen. Lactate production is approximately 10–100 times greater in cancer cells than it is in normal cells. Increased concentration of lactate decreases the pH in the tumor microenvironment, making the tumor more acidic (Fig. 2) [45]. Therefore, glucose targeting is a better opportunity for targeting drugs. Glucose-decorated poly (lactic-*co*-glycolic acid) (PLGA) and chitosan nanoparticles have been formulated and examined for anticancer activity. Results indicated that their use resulted in a 46% decrease of cancer cells in the HT-79 cell lines. The interaction between protein and glucose inhibits nutrition in the cells and thus cells begin to die after 24h of treatment. After 72h, the rate of apoptosis increases due to a lack of energy and inaccessibility of nutrients [46]. Glucose-conjugated nanoparticles loaded with doxorubicin were synthesized using succinic acid as linkers between glucosamine and chitosan. These glucose-conjugated

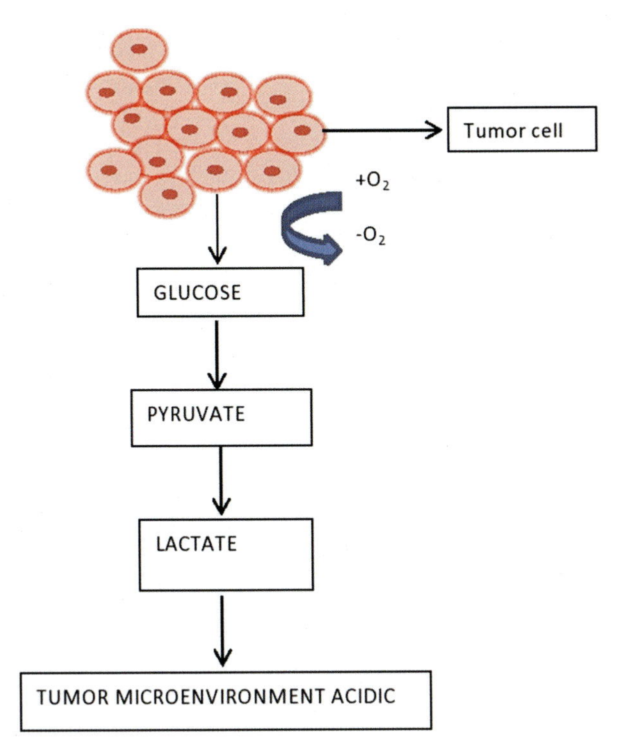

FIG. 2 The role of Warburg effect in tumorigenesis.

nanoparticles showed better cellular uptake and endocytosis activity and were found to be 4–5 times more effective than chitosan nanoparticles [47]. Poly (amidoamine) dendrimers loaded with doxorubicin were modified with glucose molecules for targeting of breast cancer in glucose deprived MCF-7 cell lines [48]. Glucose-coated iron oxide nanoparticles were formulated by metal vapor synthesis. The nanoparticles showed internalization of overexpressed GLUT-1 on the BxPC3 tumor line. Therefore, inhibition of GLUT1 overexpression can increase patient survival rate and open a new path for targeted drug delivery [49].

3 Protein-mediated targeting

3.1 Antibody-mediated targeting

Therapeutic monoclonal antibody (TMA)-conjugated drugs are effective in cancer therapy. The US FDA approved the first therapeutic monoclonal antibodies in September 1992. Currently, there are several TMAs marketed and used in cancer therapy [50]. Many antibodies are not used for targeting purposes due to their large size and higher density of antibodies on the surface of nanoparticles. Due to this limitation, some studies were performed using antibodies as ligands. Roth et al. developed immunoliposomes against CD166 expressed on culture prostate cells using an scFv antibody. Liposomes were loaded with the cytotoxic drug topotecan. The complex enhanced the overall efficacy of the delivered drug [51]. Perlecan (HSPG2), a cell surface protein, is overexpressed in human triple-negative breast cancer. HSPG2 is a large membrane protein that is highly glycosylated. Growth factors such as VEGF and FGF-2 bind with the perlecan protein. Paclitaxel-loaded poly (D, L-lactide-*co*-glycolide) nanoparticles were functionalized with antibodies, improving cell uptake, retention, and cytotoxicity [52]. Antibody (CD133)-conjugated gold nanoparticles showed peri-nuclear localization via endocytosis. Antibody-conjugated gold nanoparticles showed photodynamic as well as anticancer effects [53].

3.2 Transferrin-mediated targeting

Transferrin receptors are membrane-bound proteins expressed in almost all healthy cells for their requirement of iron. Transferrin is a single chain glycoprotein (70 kDa) which can bind to freely circulating two ferric (Fe^{3+}) atoms with high affinity. Interaction of transferrin with transferrin receptors is essential for iron absorption. Increased expression of transferrin receptors opens a path for drug delivery to malignant cells and blocks the natural process of iron uptake [54–56]. In one study, transferrin-conjugated poloxamer silica nanoparticles loaded with doxorubicin were formulated and evaluated on breast cancer cell lines, MDA-MB-231 (R). In vitro data suggested that the drug resistance can be minimized by drug accumulation in the nuclear region of cancer cells with minimum effect on healthy tissue [57]. Transferrin- and glucose-coupled chitosan vesicles were used for active and passive targeting to tumor cells. The vesicles were reached very rapidly and showed increased cellular uptake and cytotoxic effects in drug-resistant cells. These vesicles were found to be less active than the free drug [58]. Transferrin receptors targeting liposomes delivered doxorubicin into MDR cells via transferrin-mediated endocytosis. Liposomes increased cytotoxicity up to 3.5-fold more than free drugs, overcoming drug resistance [59]. Transferrin-mediated gene delivery provides cell-specific delivery of the DNA complex, which enhances the transfection efficiencies in vitro and in vivo in the human model. Transferrin-coupled liposomes–DNA complex has potential for use as systemic gene therapy in various human malignancies [60].

3.3 Lectin-mediated drug delivery

Lectins are a heterogeneous group of proteins or groups of glycoproteins that bind carbohydrates with high affinity and specificity to different carbohydrates such as mannose, sialic acid, *N*-acetylglucosamine, and galactosamine. Lectin interacts with oligosaccharide sites through van der Waals forces and hydrophobic interactions. Lectins can be screened from bacteria, viruses, animals, and plants [61]. Lectin-mediated drug delivery take the advantages of carbohydrate sites and acts as a site-specific drug delivery system. Anomalous protein glycosylation is a marker of the outgrowth of cancer cells. Alteration in glycosylation and production of glycan opens a new road for targeted drug delivery in cancer [62]. Wheat germ agglutinin is a common lectin used in targeted drug delivery. *Cis*-aconityl-linked doxorubicin wheat germ agglutinin showed 39% anti-proliferative effects on caco-2 cells as compared to the free drug [63]. In addition to wheat germ agglutinin, serum tuberosum lectin can be used as an excipient in caco-2 cells to target drugs to colonocytes. Sufficient adhesion and cell internalization occurred by both lectins [64]. Wheat germ agglutinin and *Ulex europaeus* agglutinin have the strongest interaction with *N*-acetyl-glucosamine, sialic acid, and alpha-L-fructose residue on the membrane surface [65]. Lectin of peanut agglutinin was conjugated with 5-flouorouracil, which revealed cytotoxicity, specificity, and selectivity towards human colorectal cancer cell lines [66]. Fungal Macrolepiota procera lectin (MpL) binds aminopeptidase N and integrin overexpressed on the membrane of tumor cells. MpL endocytosed upon binding and accumulated initially in the Golgi apparatus before finally reaching the lysosomes [67].

4 Carbohydrate-mediated drug delivery

Monosaccharide and polysaccharide carbohydrates are fundamental units of nucleotides. There is an abundant amount of carbohydrate moieties present for interaction with glycolipids and glycoproteins. Cancer cells have aberrant carbohydrate metabolism, which increases the need for glucose via glycolysis. Therefore, carbohydrate receptors can be used for targeting the molecule in malignant cells [68].

4.1 Hyaluronic-mediated targeted drug delivery

Hyaluronic acid is a linear, high molecular weight glycosaminoglycan composed of two saccharide units of glucuronic acid and *N*-acetyl glucosamine. Hyaluronic acid binds to CD44 receptors, which are overexpressed in malignant cells. Hyaluronic acid is biocompatible, nonimmunogenic, biodegradable, and easy to modify, making it a suitable candidate for drug targeting. Hyaluronic acid has been conjugated with various drug carriers such as nanoparticles, microspheres, and nanoemulsions to deliver drugs or genes to specific sites [69, 70]. Paclitaxel was covalently attached to hyaluronic acid with a disulfide bond. This complex increased the CD44-mediated endocytosis and tumor targeting of micelles in MCF cells [71]. The small molecule TH287 inhibits MTH1 protein in cells and can be used as an anticancer drug. HA-coupled mesoporous silica nanoparticles delivered drugs with controlled release and internalization of CAL27 cancer cells. HA-assembled silica-coated nanoparticles decreased the burden of tumor cells [72]. Hyaluronidase fragments hyaluronic acid into low molecular weight. Sulfate hyaluronic acid (SHA) synthesized by o-sulfation inhibited hyaluronidase activity up to 15-fold and affected the proliferation of osteoblasts, gene expression, and keratocytes. SHA blocked the proliferation of prostate cancer cells without weight loss and serum organ toxicity [73]. An HA-paclitaxel bioconjugate (ONCOFID) was developed for the treatment of ovarian and bladder cancer [74]. HA-modified chitosan nanoparticles loaded with SiRNA targeted the gene Bcl2. The B-lymphoma gene is involved in apoptosis and cell death. A549 cell lines expressing CD44 were used to study the cellular uptake and distribution of nanoparticles. This complex can silence Bcl genes that inhibit tumor growth [75]. Conjugated HA–bovine serum albumin (BSA) entrapped the hydrophobic cytotoxic drug paclitaxel, and imidazoacridinones having diameter less than 15 nm provided stability to the drug and facilitated receptor-mediated endocytosis. Drug-loaded nanoparticles showed internalization and cytotoxicity toward CD44-expressing cells rather than cells not expressing CD44 receptors [76]. HA-functionalized camptothecin-/curcumin-loaded polymeric nanoparticles showed a synergistic effect and had the greatest antitumor activity against colon cancer [77].

5 Physical targeting via EPR effects

In physical/passive targeting, abnormalities in the tumor microenvironment including hypervascularization, aberrant vascular architecture, extensive production of vascular permeability factors stimulating extravasation within tumor tissues, and lack of lymphatic drainage are exploited. Macromolecules accumulate in plasma and increase their half-life. Additionally, they extravasate in malignant cells due to changes in the vascularization of tumor tissue. The concentration of drugs reaches up to several-fold in tumor cells as compared to normal tissue due to enhanced permeability retention (EPR) effects [78]. EPR effects have three fundamental characteristics: (1) leaky vasculature, (2) abnormal lymphatic drainage, and (3) increased intratumor pressure. In physical targeting, nanoparticles should have optimal size (>40–50 kDa) to remain in the fast outflow of blood capillaries; however, they are smart enough to escape from the reticuloendothelial system such as the liver and spleen. The size of the gap between leaky endothelial vasculature ranges between 100 and 600 nm. Therefore, the size of nanoparticles should be 100 nm. Surface characteristics also play a key role in deciding the fate of nanoparticles. Ideally, nanoparticles should be coated with a hydrophilic polymer that protects against opsonization by antibodies, complements, and other proteins [79]. The macromolecule and drug carrier accumulates on the cancerous site, exploiting changes in the tumor microenvironment such as leaky vasculature, abnormal growth factors, problematic lymphatic drainage, increased interstitial pressure, elevated amount of bradykinin and prostaglandins, and so on [80]. Several pharmacological and physical means can be employed to enhance the tumor accumulation and efficacy of EPR-based nanomedicines. Among pharmacological strategies, the most prominent are treatments with drugs that modulate VEGF signaling, with angiotensin agonists and antagonists, with tumor necrosis factor-alpha (TNF-α), with vessel-promoting treatments, and with nitric oxide-producing agents [81]. PLGA nanoparticles extravasated through tumor vasculature and delivered drugs through EPR effects [82].

6 Triggered targeting

According to the Warburg theory, the tumor microenvironment differs from normal tissue in that there is a decrease in pH due to lactate formation. Taking advantage of the acidic pH, hyperthermia, and triggered drug delivery are used for targeting neoplastic cells. The drug from triggered drug delivery is released when exposed to specific environmental conditions such as a certain pH and temperature [45]. Ultrasound has been used in the diagnosis of many diseases. Ultrasonic-triggered drug delivery permits spatially confined delivery of therapeutic molecules in various ways, including triggering the drug release, promoting cellular uptake, and enhancing penetration of nanoparticles into malignant cells [83]. Ultrasound drug delivery increases efficacy and reduces toxic effects due to effects on the permeability of blood vessels. Ultrasound increased doxorubicin penetration into cancer cells, releasing the drug from carriers [84]. High-frequency ultrasound rays can increase cell membrane pore size, cavitation (microbubble formation), and hyperthermia on the cell. Drug-loaded microbubbles could be triggered by ultrasound, releasing drugs at the target site and reducing cytotoxic effects. At the same time, cavitation induced by acoustic waves destroys the malignant cells or changes their permeability, thus enhancing the therapeutic efficiency of drugs. A combination of microbubble and ultrasound can effectively deliver hydrophobic drugs such as paclitaxel, doxorubicin, and docetaxel. This therapy is particularly useful in the treatment of liver, brain, and lung cancers [85]. Nanobubble liposomes loaded with paclitaxel can be used for disease diagnosis and treatment. Sonoporation increased the drug concentration 3.5-fold. Therefore, ultrasound-triggered drug delivery can be used as a theranostic approach in the future [86].

The photo-triggered approach uses external light as stimuli to deliver the drug at the therapeutic site. In this case, the photosensitizer azobenzene is conjugated with mesoporous silica and a change in size from *trans* to *cis* regulated the transport of the molecule through the electrode under ultraviolet radiation or near-infrared radiation [87]. Polymer nanocarriers decorated with cell-targeting biomolecules were developed by introducing azobenzene into the backbone of the polymer [88]. In addition to controlled-release drug delivery, photo-triggered drug delivery penetrates the drug into deep tissues, which enhances biosafety and therapeutic efficacy.

Magnetically controlled drug delivery is used to overcome reticuloendothelial system clearance and drug concentration at non-target sites. It produces magnetic hyperthermia to induce apoptosis of tumor cells, making the malignant cells more sensitive to chemotherapy [89]. Polymerosome-encapsulated doxorubicin with superparamagnetic iron oxide nanoparticles released the drug at the target site due to an increase in hyperthermia upon application of a high magnetic field [90].

7 Concluding remarks

Cancer is a complex disease that involves many processes. Lack of specificity and selectivity of anticancer drugs to malignant cells is a major problem. Accumulation of drugs on non-target tissues increases drug-induced side effects. In addition, low drug concentration at target sites results in drug resistance. Targeted drug delivery is designed to overcome these limitations by improving the efficacy of drugs. There are several approaches to designing targeted drug delivery systems. These include active, passive, and triggered targeting. Active targeting contributes to surface modification or can be mediated via ligands like peptides, carbohydrates, lectins, and antibodies, whereas passive targeting leads to beneficial accumulation of anticancer drugs in the neoplastic cell because of EPR effects. Various nanoplatforms are designed to provide site-specific drug delivery, which provides a better therapeutic index. Triggered drug delivery is smart drug delivery utilizing high-frequency ultrasonic waves, magnetic fields, and UV or visible radiation. This chapter presents a major achievement in the area of targeting drug delivery. We hope it inspires more inventions in the future to improve and prolong the lives of cancer patients, which, after all, is the main goal of drug therapy.

References

[1] Siegel RL, Miller KD, Jemal A. Cancer statistics, 2020. CA Cancer J Clin 2020;70(1):7–30.

[2] Zajaczkowska R, et al. Mechanisms of chemotherapy-induced peripheral neuropathy. Int J Mol Sci 2019;20(6).

[3] Kang D, et al. Permanent chemotherapy-induced alopecia in patients with breast cancer: a 3-year prospective cohort study. Oncologist 2019;24(3):414–20.

[4] Bae YH, Park K. Targeted drug delivery to tumors: myths, reality and possibility. J Control Release 2011;153(3):198–205.

[5] Senapati S, et al. Controlled drug delivery vehicles for cancer treatment and their performance. Signal Transduct Target Ther 2018;3:7.

[6] Perez-Herrero E, Fernandez-Medarde A. Advanced targeted therapies in cancer: drug nanocarriers, the future of chemotherapy. Eur J Pharm Biopharm 2015;93:52–79.

[7] Muhamad N, Plengsuriyakarn T, Na-Bangchang K. Application of active targeting nanoparticle delivery system for chemotherapeutic drugs and traditional/herbal medicines in cancer therapy: a systematic review. Int J Nanomedicine 2018;13:3921–35.
[8] Ganta S, et al. Formulation development of a novel targeted theranostic nanoemulsion of docetaxel to overcome multidrug resistance in ovarian cancer. Drug Deliv 2016;23(3):968–80.
[9] Malarvizhi GL, et al. Transferrin targeted core-shell nanomedicine for combinatorial delivery of doxorubicin and sorafenib against hepatocellular carcinoma. Nanomedicine 2014;10(8):1649–59.
[10] Fang Y, et al. Aptamer-conjugated multifunctional polymeric nanoparticles as cancer-targeted, MRI-ultrasensitive drug delivery systems for treatment of castration-resistant prostate cancer. Biomed Res Int 2020;2020:9186583.
[11] Sharma R, et al. C-Type lectin receptor(s)-targeted nanoliposomes: an intelligent approach for effective cancer immunotherapy. Nanomedicine (Lond) 2017;12(16):1945–59.
[12] Cai LL, et al. RGD peptide-mediated chitosan-based polymeric micelles targeting delivery for integrin-overexpressing tumor cells. Int J Nanomed 2011;6:3499–508.
[13] Ndong C, et al. Antibody-mediated targeting of iron oxide nanoparticles to the folate receptor alpha increases tumor cell association in vitro and in vivo. Int J Nanomed 2015;10:2595–617.
[14] Qu D, et al. Anisamide-functionalized pH-responsive amphiphilic chitosan-based paclitaxel micelles for sigma-1 receptor targeted prostate cancer treatment. Carbohydr Polym 2020;229:115498.
[15] Kim K, et al. Hyaluronic acid-coated nanomedicine for targeted cancer therapy. Pharmaceutics 2019;11(7).
[16] Nicolas J, et al. Design, functionalization strategies and biomedical applications of targeted biodegradable/biocompatible polymer-based nanocarriers for drug delivery. Chem Soc Rev 2013;42(3):1147–235.
[17] Gilad Y, Firer M, Gellerman G. Recent innovations in peptide based targeted drug delivery to cancer cells. Biomedicine 2016;4(2).
[18] Ryvkin A, et al. Phage display peptide libraries: deviations from randomness and correctives. Nucleic Acids Res 2018;46(9):e52.
[19] Newton J, Deutscher SL. Phage peptide display. Handb Exp Pharmacol 2008;145–63 [185 Pt 2].
[20] Wu PH, et al. Targeting integrins in cancer nanomedicine: applications in cancer diagnosis and therapy. Cancers (Basel) 2019;11(11).
[21] Marelli UK, et al. Tumor targeting via integrin ligands. Front Oncol 2013;3:222.
[22] Sun Y, et al. RGD peptide-based target drug delivery of doxorubicin nanomedicine. Drug Dev Res 2017;78(6):283–91.
[23] Wang G, et al. RGD peptide-modified, paclitaxel prodrug-based, dual-drugs loaded, and redox-sensitive lipid-polymer nanoparticles for the enhanced lung cancer therapy. Biomed Pharmacother 2018;106:275–84.
[24] Yu X, et al. Enhanced tumor targeting of cRGD peptide-conjugated albumin nanoparticles in the BxPC-3 cell line. Sci Rep 2016;6:31539.
[25] Smaldone G, et al. gH625 is a viral derived peptide for effective delivery of intrinsically disordered proteins. Int J Nanomedicine 2013;8:2555–65.
[26] Perillo E, et al. Synthesis and in vitro evaluation of fluorescent and magnetic nanoparticles functionalized with a cell penetrating peptide for cancer theranosis. J Colloid Interface Sci 2017;499:209–17.
[27] Sun H, et al. Oligonucleotide aptamers: new tools for targeted cancer therapy. Mol Ther Nucleic Acids 2014;3:e182.
[28] Yang L, et al. Aptamer-conjugated nanomaterials and their applications. Adv Drug Deliv Rev 2011;63(14–15):1361–70.
[29] Wilson DS, Szostak JW. In vitro selection of functional nucleic acids. Annu Rev Biochem 1999;68(1):611–47.
[30] Ferlay J, et al. Estimates of worldwide burden of cancer in 2008: GLOBOCAN 2008. Int J Cancer 2010;127(12):2893–917.
[31] Ferreira CS, Matthews CS, Missailidis S. DNA aptamers that bind to MUC1 tumour marker: design and characterization of MUC1-binding single-stranded DNA aptamers. Tumour Biol 2006;27(6):289–301.
[32] Sayari E, et al. MUC1 aptamer conjugated to chitosan nanoparticles, an efficient targeted carrier designed for anticancer SN38 delivery. Int J Pharm 2014;473(1–2):304–15.
[33] Yu C, et al. Novel aptamer-nanoparticle bioconjugates enhances delivery of anticancer drug to MUC1-positive cancer cells in vitro. PLoS One 2011;6(9):e24077.
[34] Baek SE, et al. RNA aptamer-conjugated liposome as an efficient anticancer drug delivery vehicle targeting cancer cells in vivo. J Control Release 2014;196:234–42.
[35] Powell Gray B, et al. Tunable cytotoxic aptamer–drug conjugates for the treatment of prostate cancer. Proc Natl Acad Sci 2018;115(18):4761–6.
[36] Cheung A, et al. Targeting folate receptor alpha for cancer treatment. Oncotarget 2016;7(32):52553–74.
[37] Soe ZC, et al. Folate receptor-mediated celastrol and irinotecan combination delivery using liposomes for effective chemotherapy. Colloids Surf B Biointerfaces 2018;170:718–28.
[38] Lee SJ, et al. Folic-acid-conjugated pullulan/poly(DL-lactide-co-glycolide) graft copolymer nanoparticles for folate-receptor-mediated drug delivery. Nanoscale Res Lett 2015;10:43.
[39] Yu Y, et al. Folic acid receptor-mediated targeting enhances the cytotoxicity, efficacy, and selectivity of Withania somnifera leaf extract: in vitro and in vivo evidence. Front Oncol 2019;9:602.
[40] Jin X, et al. Folate receptor targeting and Cathepsin B-sensitive drug delivery system for selective Cancer cell death and imaging. ACS Med Chem Lett 2020;11(8):1514–20.
[41] Banerjee R, et al. Anisamide-targeted stealth liposomes: a potent carrier for targeting doxorubicin to human prostate cancer cells. Int J Cancer 2004;112(4):693–700.
[42] Evans JC, et al. Formulation and evaluation of anisamide-targeted amphiphilic cyclodextrin nanoparticles to promote therapeutic gene silencing in a 3D prostate cancer bone metastases model. Mol Pharm 2017;14(1):42–52.
[43] Wang L, et al. Development of anisamide-targeted PEGylated gold nanorods to deliver epirubicin for chemo-photothermal therapy in tumor-bearing mice. Int J Nanomedicine 2019;14:1817–33.
[44] Barron CC, et al. Facilitative glucose transporters: implications for cancer detection, prognosis and treatment. Metabolism 2016;65(2):124–39.
[45] Liberti MV, Locasale JW. The Warburg effect: how does it benefit cancer cells? Trends Biochem Sci 2016;41(3):211–8.
[46] Abolhasani A, et al. Investigation of the role of glucose decorated chitosan and PLGA nanoparticles as blocking agents to glucose transporters of tumor cells. Int J Nanomedicine 2019;14:9535–46.
[47] Li J, et al. Glucose-conjugated chitosan nanoparticles for targeted drug delivery and their specific interaction with tumor cells. Front Mater Sci 2014;8(4):363–72.
[48] Sztandera K, et al. Sugar modification enhances cytotoxic activity of PAMAM-doxorubicin conjugate in glucose-deprived MCF-7 cells - possible role of GLUT1 transporter. Pharm Res 2019;36(10):140.
[49] Barbaro D, et al. Glucose-coated superparamagnetic iron oxide nanoparticles prepared by metal vapour synthesis are electively

internalized in a pancreatic adenocarcinoma cell line expressing GLUT1 transporter. PLoS One 2015;10(4):e0123159.
[50] Firer MA, Gellerman G. Targeted drug delivery for cancer therapy: the other side of antibodies. J Hematol Oncol 2012;5:70.
[51] Roth A, et al. Anti-CD166 single chain antibody-mediated intracellular delivery of liposomal drugs to prostate cancer cells. Mol Cancer Ther 2007;6(10):2737–46.
[52] Khanna V, et al. Perlecan-targeted nanoparticles for drug delivery to triple-negative breast cancer. Future Drug Discov 2019;1(1):FDD8.
[53] Crous A, Abrahamse H. Effective gold nanoparticle-antibody-mediated drug delivery for photodynamic therapy of lung cancer stem cells. Int J Mol Sci 2020;21(11).
[54] Daniels TR, et al. The transferrin receptor and the targeted delivery of therapeutic agents against cancer. Biochim Biophys Acta 2012;1820(3):291–317.
[55] Daniels TR, et al. The transferrin receptor part I: biology and targeting with cytotoxic antibodies for the treatment of cancer. Clin Immunol 2006;121(2):144–58.
[56] Qian ZM, et al. Targeted drug delivery via the transferrin receptor-mediated endocytosis pathway. Pharmacol Rev 2002;54(4):561–87.
[57] Soe ZC, et al. Transferrin-conjugated polymeric nanoparticle for receptor-mediated delivery of doxorubicin in doxorubicin-resistant breast cancer cells. Pharmaceutics 2019;11(2).
[58] Dufes C, et al. Anticancer drug delivery with transferrin targeted polymeric chitosan vesicles. Pharm Res 2004;21(1):101–7.
[59] Kobayashi T, et al. Effect of transferrin receptor-targeted liposomal doxorubicin in P-glycoprotein-mediated drug resistant tumor cells. Int J Pharm 2007;329(1–2):94–102.
[60] Xu L, et al. Systemic tumor-targeted gene delivery by anti-transferrin receptor scFv-Immunoliposomes 1 this work was supported in part by National Cancer Institute Grant R01 CA45158 (to E. C.), National Cancer Institute small business technology transfer phase I Grant R41 CA80449 (to E. C.), and a grant from SynerGene therapeutics, Inc.<a id="xref-fn-1-1" class="xref-" href="#fn-1">1</a>. Mol Cancer Ther 2002;1(5):337–46.
[61] Coelho LC, et al. Lectins, interconnecting proteins with biotechnological/pharmacological and therapeutic applications. Evid Based Complement Alternat Med 2017;2017:1594074.
[62] Costa AF, et al. Targeting glycosylation: a new road for cancer drug discovery. Trends Cancer 2020;6(9):757–66.
[63] Wirth M, et al. Lectin-mediated drug targeting: preparation, binding characteristics, and antiproliferative activity of wheat germ agglutinin conjugated doxorubicin on Caco-2 cells. Pharm Res 1998;15(7):1031–7.
[64] Wirth M, Hamilton G, Gabor F. Lectin-mediated drug targeting: quantification of binding and internalization of wheat germ agglutinin and solanum tuberosum lectin using Caco-2 and HT-29 cells. J Drug Target 1998;6(2):95–104.
[65] Plattner VE, et al. Targeted drug delivery: binding and uptake of plant lectins using human 5637 bladder cancer cells. Eur J Pharm Biopharm 2008;70(2):572–6.
[66] Cai Q, Zhang ZR. Lectin-mediated cytotoxicity and specificity of 5-fluorouracil conjugated with peanut agglutinin (5-Fu-PNA) in vitro. J Drug Target 2005;13(4):251–7.
[67] Simon AU, et al. Fungal lectin MpL enables entry of protein drugs into cancer cells and their subcellular targeting. Oncotarget 2017;8(16):26896–910.
[68] Wang X, Guo Z. Targeting and delivery of platinum-based anticancer drugs. Chem Soc Rev 2013;42(1):202–24.
[69] Huang G, Huang H. Application of hyaluronic acid as carriers in drug delivery. Drug Deliv 2018;25(1):766–72.
[70] Alaniz L, et al. Interaction of CD44 with different forms of hyaluronic acid. Its role in adhesion and migration of tumor cells. Cell Commun Adhes 2002;9(3):117–30.
[71] Yin S, et al. Intracellular delivery and antitumor effects of a redox-responsive polymeric paclitaxel conjugate based on hyaluronic acid. Acta Biomater 2015;26:274–85.
[72] Shi XL, et al. Delivery of MTH1 inhibitor (TH287) and MDR1 siRNA via hyaluronic acid-based mesoporous silica nanoparticles for oral cancers treatment. Colloids Surf B Biointerfaces 2019;173:599–606.
[73] Benitez A, et al. Targeting hyaluronidase for cancer therapy: antitumor activity of sulfated hyaluronic acid in prostate cancer cells. Cancer Res 2011;71(12):4085–95.
[74] De Stefano I, et al. Hyaluronic acid-paclitaxel: effects of intraperitoneal administration against CD44(+) human ovarian cancer xenografts. Cancer Chemother Pharmacol 2011;68(1):107–16.
[75] Zhang W, et al. Antitumor effect of hyaluronic-acid-modified chitosan nanoparticles loaded with siRNA for targeted therapy for non-small cell lung cancer. Int J Nanomedicine 2019;14:5287–301.
[76] Edelman R, et al. Hyaluronic acid-serum albumin conjugate-based nanoparticles for targeted cancer therapy. Oncotarget 2017;8(15):24337–53.
[77] Xiao B, et al. Hyaluronic acid-functionalized polymeric nanoparticles for colon cancer-targeted combination chemotherapy. Nanoscale 2015;7(42):17745–55.
[78] Greish K. Enhanced permeability and retention (EPR) effect for anticancer nanomedicine drug targeting. Methods Mol Biol 2010;624:25–37.
[79] Cho K, et al. Therapeutic nanoparticles for drug delivery in cancer. Clin Cancer Res 2008;14(5):1310–6.
[80] Zhou Y, Kopecek J. Biological rationale for the design of polymeric anti-cancer nanomedicines. J Drug Target 2013;21(1):1–26.
[81] Nel A, Ruoslahti E, Meng H. New insights into "permeability" as in the enhanced permeability and retention effect of cancer nanotherapeutics. ACS Nano 2017;11(10):9567–9.
[82] Acharya S, Sahoo SK. PLGA nanoparticles containing various anticancer agents and tumour delivery by EPR effect. Adv Drug Deliv Rev 2011;63(3):170–83.
[83] Tharkar P, et al. Nano-enhanced drug delivery and therapeutic ultrasound for cancer treatment and beyond. Front Bioeng Biotechnol 2019;7:324.
[84] Thakkar D, et al. Overcoming biological barriers with ultrasound. AIP Conf Proc 2012;1481:381–7.
[85] Gong Q, et al. Drug-loaded microbubbles combined with ultrasound for thrombolysis and malignant tumor therapy. Biomed Res Int 2019;2019:6792465.
[86] Prabhakar A, Banerjee R. Nanobubble liposome complexes for diagnostic imaging and ultrasound-triggered drug delivery in cancers: a theranostic approach. ACS Omega 2019;4(13):15567–80.
[87] Lu J, et al. Light-activated nanoimpeller-controlled drug release in cancer cells. Small 2008;4(4):421–6.
[88] Mena-Giraldo P, et al. Photosensitive nanocarriers for specific delivery of cargo into cells. Sci Rep 2020;10(1):2110.
[89] Shen X, et al. PLGA-based drug delivery systems for remotely triggered cancer therapeutic and diagnostic applications. Front Bioeng Biotechnol 2020;8:381.
[90] Oliveira H, et al. Magnetic field triggered drug release from polymersomes for cancer therapeutics. J Control Release 2013;169(3):165–70.

Chapter 6

Material and strategies used in oncology drug delivery

Nitin Verma[a], Komal Thapa[a], and Kamal Dua[b]
[a]*Chitkara School of Pharmacy, Chitkara University, Solan, Himachal Pradesh, India,* [b]*Discipline of Pharmacy, Graduate School of Health, University of Technology Sydney, Ultimo, NSW, Australia*

1 Introduction

Cancer is a wide array of diseases that occurs due to unrestricted expansion of malignant cells that possesses the ability to invade or extend to rest of the parts of the body. Every year, 10 million cases of cancer occur with increased death rates and has been depicted to increase more in future in the world with 13.1 million by the year 2030 due to cancer. However, the death rate has been reduced during the past 5 years because of improved understanding of tumor pathology and improved treatment and methods for diagnosis. Currently, the treatment option involves chemotherapy, radiation therapy surgical intervention, or combination of these. The mechanism of conventional therapy is interference with the production of DNA and mitosis process and, thus, promoting death of fast multiplying cells. These agents may harm healthy tissue causing unwanted side effects such as appetite loss and nausea along with some adverse effects that is the foremost reason of higher death rates in cancer patients. In addition to this, higher doses of these drugs are needed due to their poor bioaccessibility of that increases the chance of multiple drug resistance. Consequently, chemotherapeutic agents with lower side effects that target cancerous cells either in active or passive means are needed to be developed [1]. However, from past few years, clear understanding of tumor biology has led to the advancement of delivering techniques for chemotherapeutics to tumor sites and also increased accessibility of adaptable resources, involving polymers [2–5]; lipids [6, 7]; inorganic carriers [8]; polymeric hydrogels [9, 10]; and biomacromolecular scaffolds [11]. In the last two decades, tremendous impact has been imposed by nanotechnology on clinical environment. Nanoscale drugs are comparatively better than conventional chemotherapeutic agents, as they do not provide toxic effects to normal cells because of their improved retention and permeability characteristic [12, 13]. Adhering nanocarriers carrying chemotherapeutics agent to molecules that bind over overexpressed antigen is an active approach. Among all nano-based carrier of drug, liposomes and micelles showed great clinical effect. Presently numerous nanocarriers are under clinical evaluation with some already approved. However, nanocarriers have many disadvantages such as poor bioavailability, lack of biodegradation, not enough tissue distribution, and possess toxicity for long-term administration. Resistance to chemotherapeutic agent in cancer is responsible for failure of cancer therapy in which cancerous cells become resistant to particular drug. Therefore, there is a need for better delivery systems with better targeting abilities for treating cancer and also toxicity and pain associated with cancer chemotherapy. This chapter has discussed the role of drug delivery vehicles in raising the therapeutic index of chemotherapeutic drugs used in cancer treatment. This book chapter holds recent challenges allied with chemotherapy, along with discussion on future trends in chemotherapy.

2 Materials and stratagem applied in cancer therapy

Various novel drug-releasing techniques have been developed for the treatment of cancer such as nanoscale composites based on proteins, lipids, organic and inorganic particles, and synthetic polymers. Encapsulation of drug in a carrier offers advantages like prevention of drug from blood stream degradation, improves drug solubility and stability, and delivers drug to the targeted site with less toxic side effects. Until now, different sizes of drug delivery vehicles have been developed that are either in clinical or preclinical stages of development.

2.1 Nanocarriers for drug delivery

Nanomedicine is fast emergent field in developing new cancer treatment and diagnostic method. Their exclusive biological characteristics such as minute size and huge surface area permit them to adhere and carry anticancer agents like drugs, DNA/RNA, or proteins. Nanocarriers employed in

Advanced Drug Delivery Systems in the Management of Cancer. https://doi.org/10.1016/B978-0-323-85503-7.00015-8

chemotherapy have been categorized into targeted or non-targeted drug delivery: The composition of organic nanocarriers is dendrimers, liposomes, lipids, carbon nanotubes, synthetic polymers, and emulsions (Scheme 1).

2.1.1 Inorganic nanocarriers

The inorganic nanocarriers serve as a great platform in development of novel treatment due to their improved drug loading capacity, minimum side effects and enhanced bioavailability. For the treatment of cancer, magnetic nanoparticles, carbon nanotubes, quantum dots, and mesoporous silica have been normally employed in treating cancer in several ways. A quantum dot is a powerful imaging probe that is employed for long-lasting, multifunction, diagnostic, and quantitative imaging [14–17]. On the other hand, QDs in biologics have major drawbacks such as high propensity of aggregation, inadequate adsorption, and its hydrophobic nature [18, 19]. The surface of QD is generally coated with polar species to make them more soluble in water, thereby enhancing their bioactivity and developing them into multifunction QDs [20, 21]. Gao et al. discovered encapsulated polymer and bioconjugated QD probes for cancerous cells and developed liposomes coated with D-α-tocopheryl polyethylene glycol 1000 succinate monoester (TPGS) for cancer imaging and treatment [22, 23]. Recently, multifunctional QDs have been regarded as a potential vehicle for anticancer drug release [24, 25]. A carbon-derived, synthetic one-dimensional (1D) nanomaterial known as carbon nanotubes (CNTs) is familiar for ideal near temperature within tumors [26, 27]. Another is water-soluble CNTs that were evaluated in drug delivery process as they can easily cross biological barriers without generating toxic effect [28, 29]. Conjugation of functional group with chemotherapeutic drug molecules on CNT surface improves humeral immune response cancerous cells [30, 31]. Layers of LDHs hold divalent or trivalent metal ion, such as Mg^{2+}, Ca^{2+}, Ni^{2+}, Zn^{2+}, and Fe^{3+}, and these charges are stabled by interlayer hydrated anions forming multiple layer with anions such as Cl^- and NO_3^- [32–35]. LDHs can inculcate various anionic biofunctional molecules, such as anticancer drugs and DNA, siRNA displaying persistent release, high bioactivity, and beneficial efficiency [36]. Fig. 1A demonstrates the fast release rate using phosphate-bound LDH drug (LP-R) while persistent release is attained with nitrate-based LDH (LN-R). Spectroscopic (XPS and UV-vis) and thermal studies (DSC) confirmed the contact between drug molecules and LDH host layers. Improved efficacies of anticancer drugs inserted in LDH have been established in in vitro anticancer studies (Fig. 1B and C). Various histogram of organs and parameters for biochemical analysis.

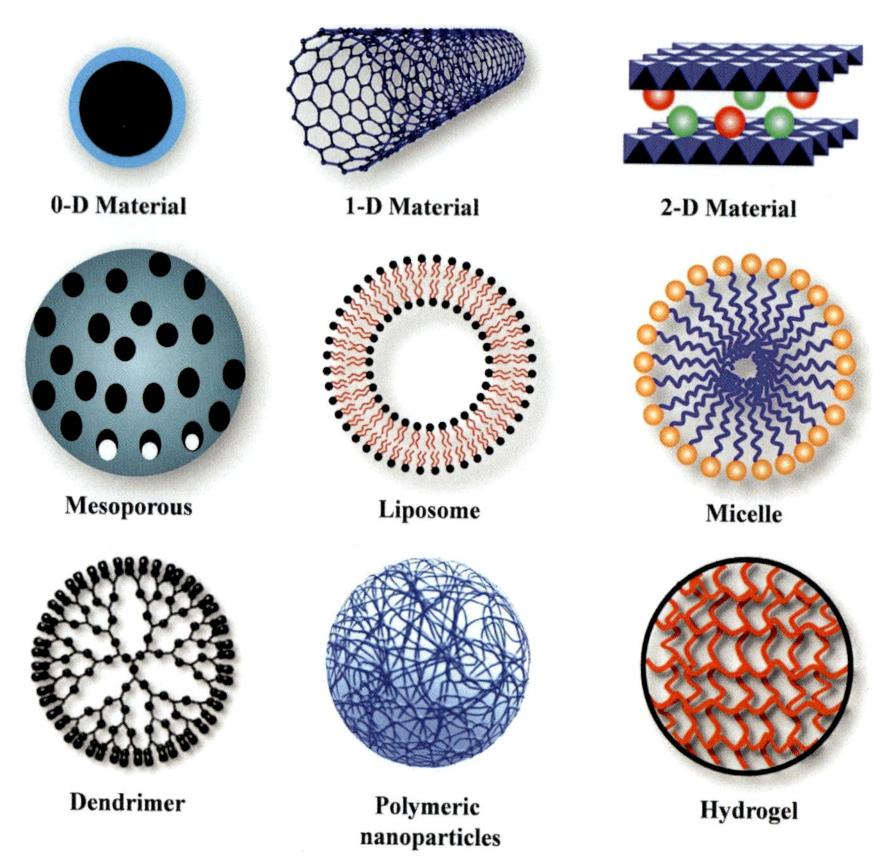

SCHEME 1 Different types of nanocarriers are used as controlled delivery vehicles for cancer treatment.

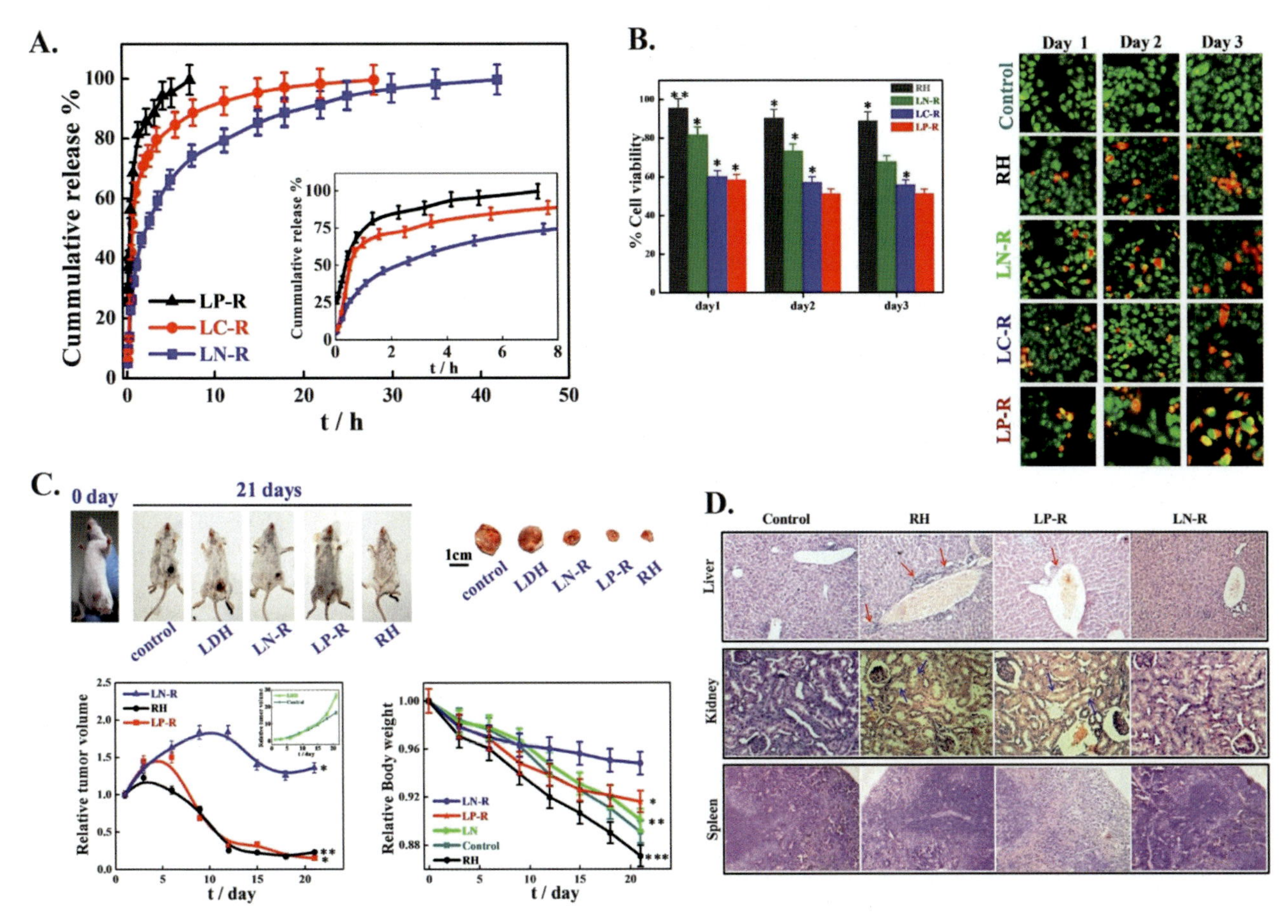

FIG. 1 In vitro and in vivo controlled release of drug using layered double hydroxides and its effects. (A) In vitro drug release profiles for drug intercalated nitrate, carbonate, and phosphate LDHs (LN-R, LC-R, and LP-R, respectively); inset figure describes the release pattern of the above-mentioned systems in a timeframe of 0–8 h; (B) in vitro cytotoxicity of free drug and drug intercalated LDHs against HeLa cells at different time intervals; (C) in vivo antitumor effect and systematic toxicity of pure RH and drug intercalated LDHs in comparison to control; and (D) histological analysis of liver, kidney, and spleen.

Liver cells damage with fast release vehicle (pure drug and LP-R), whereas mice liver cells treated with LN-R or slow-release vehicle did not show any damage (Fig. 1D). Additionally, LDH nanoparticles carrying positive charge could be simply introduced into cell membranes with negative charge via endocytosis pathway regulated by clathrin. Li et al. demonstrated the collective tactic that employed LDH to deliver anticancer drug (5-fluorouracil; 5-FU) and CD-siRNA providing higher cytotoxicity to cancerous cells [37]. Fullerenes are carbon allotropes with a huge spheroidal molecule bearing vacant cage of more than 60 atoms. They can easily react with electron-rich species and behave like electron-deficient alkenes [38]. Mesoporous silica nanoparticles (MSNs) are widely employed as drug delivery vehicles as they have distinctive properties such as larger fixed surface area, pore volume, and good biocompatibility, and provide physical protection of drugs and from degeneration or denaturation. MSNs have small pore size that makes it a better option for regulating drug-loading percentages and can distribute anticancer drugs in a targeted fashion [39]. MSNs have also come out as promising tool for both active and passive delivery of drugs on tumor tissues by means of increased permeation and retention (EPR) effect [40]. Calcium phosphate nanoparticles (CPN) are reported with outstanding biocompatibility, stability of colloids, and biodegradability and they can enclose negative charge-bearing therapeutic agents through Ca^{2+} chelation [41, 42]. CPN is the foremost element of bone and tooth enamel, and both Ca^{2+} and PO_4^{3-} are present in the blood stream at a comparatively high concentration (1–5 mM) [43, 44]. Lipid calcium phosphate (LCP) nanoparticles have been reported to deliver drug into lymphatic system and lymph node metastasis imaging [45]. PE glycated calcium phosphate hybrid micelles promote gene-silencing activity by increasing accumulation of siRNA in vivo in tumor tissues [46]. The organic or inorganic nanocarriers based on calcium phosphate have been employed in photodynamic therapy against acidic environments [47]. Nanoparticles of calcium phosphate embedded with Mn^{2+} are shown to act as a proficient magnetic resonance imaging (MRI) contrast agent in response to pH change and quickly increase the speed of magnetic resonance signals [48]. Superparamagnetic iron oxide nanoparticles (SPIONs) are gaining much interest in delivering drug to specific site using an external magnetic field. Polymer coating is required in SPIONs as it prevents in vivo particle agglomerations in blood plasma [49, 50]. Magnetic nanocarriers depend mainly on outer magnetic field that directs drug to reach their target site. In animal tumor model, better response of Adriamycin using "magnetic albumin microspheres (MM-ADR)" was reported than Adriamycin alone with respect to decrease in tumor size and survival of animal [51]. The high efficiency of "magnetic albumin microspheres" in delivering anticancer agent in rat model was mainly because of its magnetic effects [52]. SPION-based MRI is a very influential nonintrusive means in clinical diagnosis, biomedical imaging, and treatment. SPIONs efficiently provide advanced contrast development in MRI and is much eco-friendly than traditional contrast agents of Gd-based paramagnetic [53, 54]. Various method of SPION preparation for cancer treatment has been reported [55]. The most frequently used nanoparticles both in vivo and in vitro is MR157, which is monocrystalline iron oxide nanoparticles (MION) and cross-linked iron oxide nanoparticles (CLIO) [56]. Thermally cross-linked "super paramagnetic iron oxide nanoparticles (TCL-SPIONs)" coated by means of antibiofouling behaves as a new MR-based agent or in vivo cancer imaging [57], whereas "Cy 5.5-conjugated TCL-SPIONs behaves as double (MR/optical) cancer-imaging probe" [58]. SPION is the promising strategy to treat cancer via producing local heat under the influence of alternating magnetic field and cancerous cells are vulnerable to hyperthermia as hyperthermia induces apoptosis [59, 60]. Porphyrin-tethered, dopamine-oligoethylene, glycol ligand-coated, diamagnetic Fe/Fe_3O_4 nanoparticles behave as an important anticancer agent on B16-F10 mice on exposures of three small 10 min varying magnetic field (AMF) (Fig. 2) [61, 62]. Hyperthermia is inadequate for cancer treatment, therefore, new research has been focusing on combination of hyperthermia and chemotherapeutic agents by means of SPIONs. SPIONs coated with phospholipid-PEG can concurrently deliver doxorubicin with generation of heat for the treatment of advanced multimodal cancer [63].

2.1.2 *Organic nanocarriers*

Polymeric nanoparticles have achieved much consideration in nanomedicine owing to their easy synthesis and the ability of structural modifications that can improve the efficacy of loading drugs biodistribution, pharmacokinetic regulation, and therapeutic efficacy [63, 64]. Polymeric nanoparticles are prepared from synthetic polymers, e.g., poly (lactic acid) (PLA), poly (ε-caprolactone) (PCL), poly (lactic-*co*-glycolic acid), *N*-(2-hydroxypropyl)-meth acrylamide copolymer (HPMA), and poly (styrene-maleic anhydride) copolymer, or from natural polymers, for example, gelatin, dextran, guar gum, chitosan, and collagen. Encapsulation of drug is feasible by either dispersing it in a matrix of polymer or adhering it to the molecules of polymer. Polymeric nanoparticles have been evaluated for the last few decades as drug delivery systems with clinically accepted biodegradable polymeric nanoparticles, for example, PLA and PLGA. Conjugations of doxorubicin and dextran incorporated into a hydrogel showed effective cytotoxicity against solid tumors [65]. PLGA nanoparticles embedded with tamoxifen showed effective anticancer than the pure drug [66]. Multifaceted Taxol-loaded PLGA nanoparticles showed in vivo and in vitro chemotherapeutic effect and destruction of tumor cells via near-infrared photothermal [67]. Liposomes are closed structure with aqueous phase at the center and at least one concentric lipid bilayer. Since 1965, they are being used as drug delivery mediums because of their biodegradable and biocompatible properties, which make them multipurpose therapeutic carriers. Further, loading of amphiphilic drugs into the liposome inner aqueous core was done using, for example, ammonium sulfate technique for doxorubicin [68] or pH gradient technique for vincristine [69]. The major disadvantage of these conventional liposomes is quick removal from blood stream. The development of stealth liposomes is under developmental stage that utilizes hydrophilic polymer for surface coating; usually a lipid derivative of polyethylene glycol (PEG) that expands the liposomes half-life from less than a few minutes (conventional liposomes) to several hours (stealth liposomes) [70]. Liposomes target specific cells via both active and passive targeting strategies such as PEGylated specifically targeted cancer cells both inactive and passive strategies and exhibited high extent of nuclear transfection than conventional liposomes. Liposomal antisense oligonucleotides (ASO) are effectively shown to inhibit pump and nonpump multidrug-resistant tumors [71]. Liposomes targeted by ligand lead to deep drug interiorization into desired cells both in vitro and in vivo [72, 73]. Liposomes based on targeted zwitterionic oligopeptide exhibited improved tumor cell uptake, improved cytoplasmic distribution, and improved mitochondrial targeting [74]. Some clinically approved liposomal products are available in market and some are under clinical investigation. Ceramide liposomes showed inhibitory result on metastasis of peritoneal in murine xenograft model of human ovarian cancer [75]. Another interesting strategy is RNA interference (RNAi) in which disease-causing genes are silenced and liposome formulation such as chitosan-coated liposome is effective in developing siRNA delivery system that effectively silences disease-causing gene [76]. Paclitaxel enclosed in A7RC peptide-modified liposomes showed better antimitotic, antitumor, and antiangiogenic effect [77].

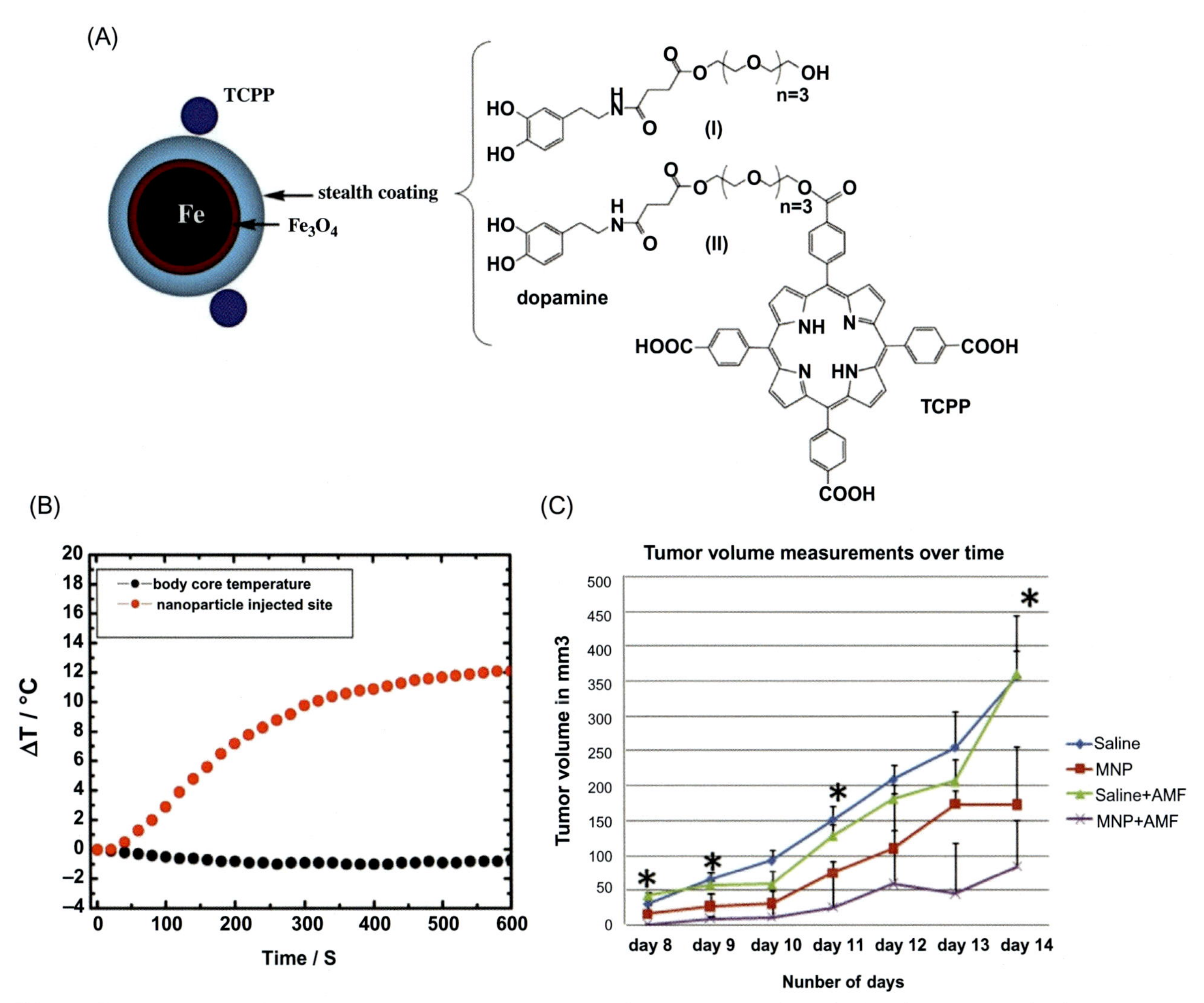

FIG. 2 Effect of surface modification on magnetic nanoparticle on hypothermia to reduce tumor size. (A) Schematic presentation showing the composition of the 4-tetracarboxyphenyl porphyrin (TCPP)-labeled, dopamine-anchored tetraethylene glycol ligands coated bimagnetic Fe/Fe_3O_4 nanoparticles; (B) graph illustrating the temperature profiles at the MNP injection site in the body core during alternating magnetic field (AMF) exposure, which is measured with a fiber optic temperature probe; (C) in vivo antitumor response after intratumoral injection of MNPs followed by AMF treatments. Graph demonstrates the relative changes in average tumor volumes overtime of B16–F10 tumor-bearing mice that were later injected with either saline or MNP intratumorally with or without AMF treatments [61].

2.2 Protein-based nanocarriers

The main sources of albumin protein include egg white (ovalbumin), bovine serum (bovine serum albumin, BSA), human serum (human serum albumin, HSA), soybeans, milk, and grains. The main advantages of albumin carrier are their easy preparation, elevated binding ability for several drugs, nonhazardous, nonimmunogenic, and is biodegradable as well as long half-life in circulating plasma. They are feasible in binding with targeting ligands and other modifications in surface due to functional groups existence like carboxylic and amino groups on nanoparticles surfaces of albumin [78]. Doxorubicin-fabricated human serum albumin (HSA) nanoparticles have been developed and showed enhanced in vitro anticancer effectiveness than pure drug in neuroblastoma cell lines (UKF-NB3 and IMR32) [79]. Nanoparticles based on paclitaxel-loaded bovine serum albumin (BSA) nanoparticles are made up by using desolvation technique accompanied by folic acid decoration to effect cancer cell line of human prostate efficiently [80]. Albumin through favorable, noncovalent, reversible binding is a hydrophobic carrier molecule (hormones, vitamins, and other plasma constituents) that helps in transportation of fluids in body and their release to cell surface. Furthermore, albumin binding to the glycoprotein (gp60) receptor promotes transcytosis of albumin-bound molecules [81, 82]. Abraxane (nab-paclitaxel; paclitaxel-albumin nanoparticles) with estimated diameter of 130 nm is the first FDA-approved commercial product that showed positive effects in metastatic breast carcinoma therapy.

2.3 Micelles as a drug carrier

Micelles are nanoscale system that is of globular or spherical created by self-relation between amphiphilic block copolymers in an aqueous solution that leads to formation of hydrophilic shell and hydrophobic core. Hydrophobic core acts as a reservoir hydrophobic drug as well as water-soluble drug, thus, forming suitable contender for i.v. administration. [63]. Paclitaxel is the first polymeric micelle formulation of cremophor-free polymeric micelle (Genexol-PM (PEG-poly (D, L-lactide)-paclitaxel) with less toxic effects and highly developed unmanageable malignancies [83]. Polymeric micelles of star shaped are based on four-arm disulfide-linked poly (ε-caprolactone) poly (ethylene glycol) amphiphilic copolymers that are in conjunction with folate ligands and exhibited high stability and persistent release, whereas an acidic environment is accompanied by prompt release [84]. Doxorubicin loaded into cationic 1,2-dioleoyl-3-trimethyl ammonium propane/methoxy poly (ethylene glycol) (DPP) nanoparticles results in micelles formation for intravesical drug delivery; and had shown effective antitumor effect against bladder cancer [85]. High encapsulation effectiveness against tumor was shown by cholesterol-modified mPEG-PLA micelles (mPEG-PLA-Ch) that considerably reduced size of tumor as compared to pure drug curcumin [86]. Phenyl boronic acid (PBA) can have high affinity for targeting sialylated epitope by selectively recognizing sialic acid (SA) and is highly expressed on cancer cells; and oxaliplatin-encapsulated micelles displayed enhanced targeting capacity on tumor by means of specific interactions with SA giving a striking strategy for cancer chemotherapy [87]. Gilbreth et al. developed lipid- and polyion complex-based micelles for improved therapeutic efficacy in tumor necrosis with the rapid production of multivalent agonists by binding with receptors [88].

2.4 Self-assembly as a drug carrier

Construction of nanoscale bioactive material via self-assembly is a remarkable approach due to its easy biomedical applications; which involves regenerative medicine, tissue engineering, and drug delivery. The main privilege of self-assembly is its unique structural feature that enables drug transformation by modification in the environmental conditions (pH, ionic strength, solvents, and temperature) and molecular chemistry [89]. In photosensitizer and chemotherapeutic agent, self-assembly decreases tumor relapse due to π-π stacking, hydrophobic, and electrostatic interactions (Fig. 3A). Intravenous administration of free Ce6 and NPs distributes drug throughout the body, whereas administration via self-assembly completely deposits drug at the tumor site (Fig. 3B). Ex vivo imaging further revealed elevated drug concentration in excised tumors with NPs as compared to free Ce6 solution (Fig. 3C) [90]. The production of switchable aptamer-diacyl lipid conjugates by the self-assembly of an aptamer

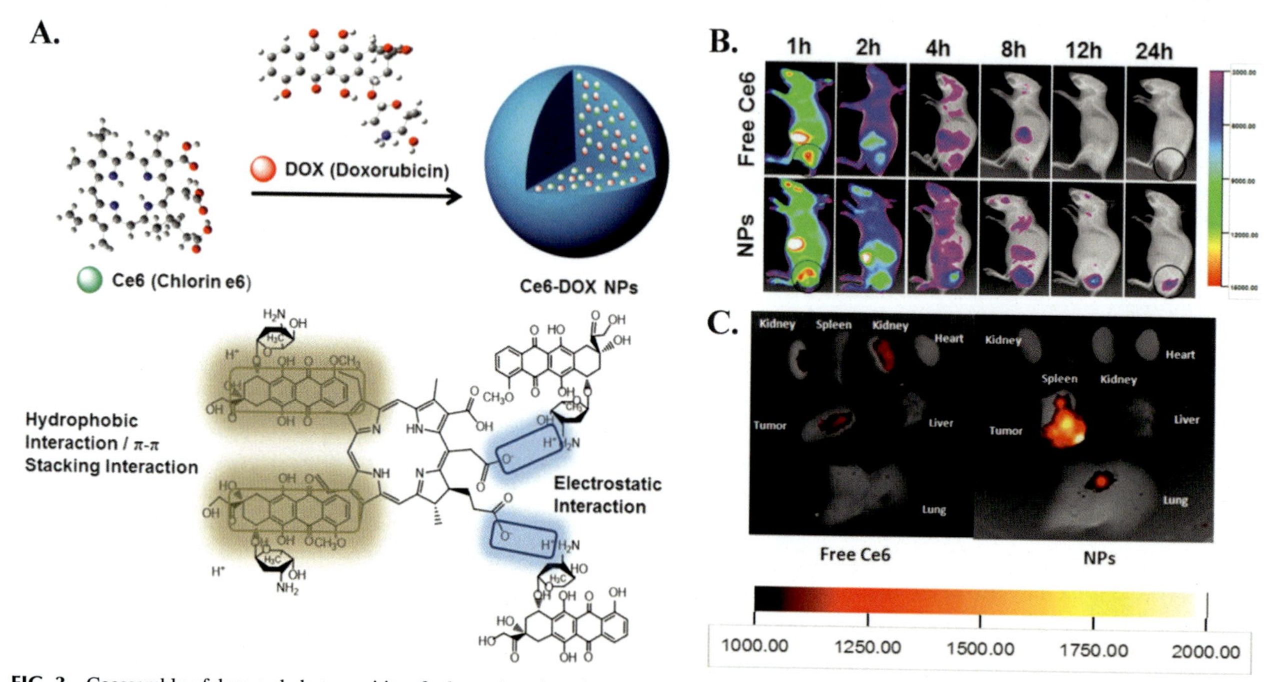

FIG. 3 Coassembly of drug and photosensitizer for better imaging of tumor size during treatment. (A) Schematic representation of carrier-free nanoparticles (NPs) via coassembly between DOX and Ce6; (B) in vivo fluorescence images of free Ce6 solution and DOX/Ce6 nanoparticles (NPs). The areas in the *black circles* represent tumor tissue; (C) representative ex vivo fluorescence imaging of tumor and organs excised from Balb/c nude mice-xenografted MCF-7 tumor at 24 h postinjection [90].

switch probe-diacyl lipid chimera discovered noteworthy results in molecular imaging for bioanalysis, detection of disease, and delivery of drugs [91]. Generation of intracellular nanofibers via enzyme-instructed self-assembly can change filaments of actin and increases cisplatin efficacy against drug-resistant ovarian carcinoma by controlling the effect of live cells [92]. The targeted nanoparticles is an advanced approach for inhibition of 40% tumor growth, and when combined with cisplatin, full inhibition lasted for 1 week [93].

2.5 Supramolecules as a delivery vehicle

Supramolecules can be formed by combining more than two molecular units and are stable due to noncovalent interactions like electrostatic, Van der Waals forces, hydrogen bonding, metal coordination, and π-π interactions. Therefore, they can be employed as drug carriers for delivery of drugs [94]. Amphiphilic dendrimers suppressed doxorubicin toxicity by creating supramolecular micelles for cancer therapy [95]. Supramolecular polymersomes stacked with DOX exhibited better antitumor effect [96]. The hydrogel of polymer-based supramolecular enclosed with cyclodextrin showed effective cytotoxic effects than non-CD containing polymer [97]. Another is the poly (*N*-isopropylacrylamide) supramolecular micelles that showed fast liberation of drug because of its cross-linking properties that aids tremendous biostability in most cell lines [98]. Dankers et al. demonstrated a novel idea about transient supramolecular networks that provided unique platform for modifying macroscopic rheological properties of drug via controlled microscopic supramolecular interactions that lead to discovery of promising carriers in medical field [99]. Drug delivery system based on "supramolecular interactions between polyamidation-polyethylene glycol-polyamidoamine (PAMAM-PEG-PAMAM) linear-dendritic copolymers" and "iron oxide-carbon nanotubes" can be used as possible systems in future for cancer treatment as small doses of drug are required [100]. Nanovesicles of porphysome are made from porphyrin-lipid bilayer that has photothermal and photoacoustic properties. Porphysome has high capability of absorbing light of infrared region and, therefore, has high capacity to destroy tumors by releasing heat. Incubation of porphysomes in detergent and lipase makes it vulnerable for enzymatic degradation. Tumors were destroyed in mice treated with porphysome and laser [101]. Muhanna et al. demonstrated the beneficial effect of porphysome and nanoparticles photothermal therapy (PTT) in hamster and rabbit model with oral cavity carcinomas [102]. Micelles loaded with drug showed better curative efficacy against triple-negative breast cancer via combined results of photothermal therapy, DOX, and TAX, with biostable porphyrin and DOX [103].

2.6 Hydrogel as a delivery vehicle

Hydrogels are regarded as three-dimensional hydrophilic polymer which can dissipate substantial volume of body fluids or water. These systems are thermodynamically water compatible which enable their swelling in presence of aqueous media [104]. These are extensively utilized for several purposes in the medical and pharmaceutical disciplines like artificial skin, biosensors, and materials for contact lenses as well as artificial hearts. Furthermore, these might also be applied as drug delivery carriers and three-dimensional cell culture [105, 106]. Hydrogels are successful mucoadhesive agents for controlled and targeted release of therapeutics since they can enclose hydrophilic as well as lipophilic drugs and biomacromolecules like deoxyribonucleic acid and proteins [107].

Hydrogel-based drug delivery technologies are being used for rectal, oral, skin, subcutaneous, and epidermal usage in diverse ways [104]. In situ gel formation has been the primary strength of hydrogel production. Hydrogels could be manufactured by self-assembly, reversible or nonreversible chemical processes, or ultraviolet/photopolymerization or irreversible covalent bonds. The phenomenon of gelation varies with time as well as concentration and, therefore, can be stimulated via an environmental stimulation like temperature, pH, and light [108]. Hydrogels are also identified for being biocompatible with insignificant cytotoxicity and are often used as the delivery system whenever the regular COS7 cell lines as well as the HepG2 or A549 cancer cell lines are monitored. A range of advanced semiinterpenetrating polymer matrix hydrogels comprised of salecan or poly (methacrylic acid) have been developed for controlled drug release by free-radical polymerization (Fig. 4A) [110]. In acidic environments (pH <5.5), released drug gets protonated and, therefore, interrupts the electrostatic contact among DOX and its hydrogel which facilitates drug delivery in contrary to high pH (~7.0) (Fig. 4B). In this aspect, pH ~5 is viewed as evidence of virtual cancer conditions. Successive responsiveness to pH 7.4 or 5.0 to a particular delivery medium induces pH-dependent "off-on" drug release switching. In A549 and HepG2 cells, cellular accumulation of DOX discharged from hydrogels was shown adequately, demonstrating tremendous potential for hydrogels being used as a device for cancer therapy (Fig. 4C). Fluorescent red copper-based nanoclusters stabilized by polyvinylpyrrolidone and cross-linked with polyvinyl alcohol could be translated into hydrogel nanomedicines, which could deliver cisplatin (CP) to HeLa (cervical cancer cells) via triggering cell apoptosis [111]. The greater encapsulating capacity is related to binding of molecule on surface and within the hydrogel polymer, accompanied with intense associations via different features like carboxylic acid. The gradual liberation of CP at physiological pH is

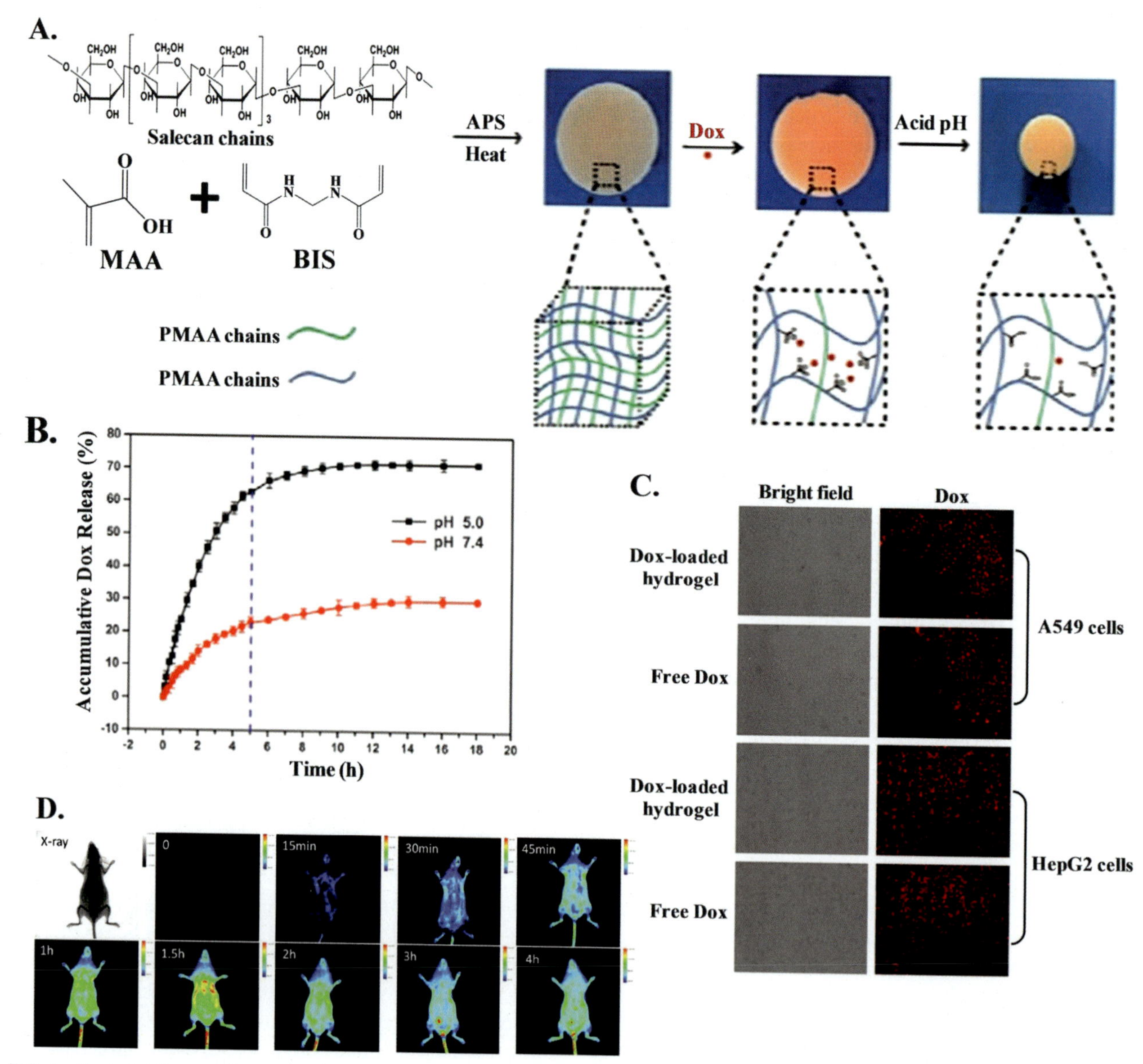

FIG. 4 Control delivery of drug using hydrogel as vehicle. (A) Illustration of the preparation and drug release of Salecan/PMAA semi-IPN hydrogels; (B) in vitro DOX release behaviors from the semi-IPN sample at two different pH values of 5.0 and 7.4; (C) Fluorescent microscopy images of A549 and HepG2 cells after 4 h of incubation with 6 μg/mL free DOX solutions and the extract liquid of Dox-loaded hydrogel; (D) intravital real-time fluorescence images of ICR mice injected with FITC-labeled PMAA nanohydrogels [109].

owing to greater coupling between drug and hydrogel that could be interrupted at acidic pH, promoting accelerated discharge. Doxorubicin-loaded PMAA hydrogel is proficient of intracellular decomposition as well as pH sensitivity through inserting cystamine bond inside polymeric network [112]. Biodegradable microgels synthesized with DOX-loaded polyvinyl caprolactam (PVCL) are being engineered to facilitate the release of drugs under acidic or alkaline conditions [113]. DOX-loaded microgels demonstrate powerful cytotoxic effects in contrast to blank microgels in HeLa cells. Polyethylene transformed gold nanocrystals and α-cyclodextrin-loaded supramolecular hybrid hydrogels manifested pH-dependent sustained DOX release by host-guest interrelationship [114]. Temperature reversible and pH-sensitive hydrogels synthesized using tetrapeptide were designed for sustained release of chemotherapeutic agents under physiological conditions [115]. Hexamethylene diisocyanate interacts with Pluronic F 127 as a copolymer-activating agent, and ultimately hyaluronic acid integration is being utilized to produce framework of matrix hydrogel with the property of sol gel transformation at 37°C contributing to development of injectable hydrogel from nanocomposites for controlled release delivery systems [116].

2.7 Hybrid materials for controlled drug delivery

Hydrogel structures of nanometer level are also referred to as "nanogels" and are manufactured by physical or chemical cross-linking with water-soluble polymers [10, 117]. Nanogels are currently used in numerous domains like chemical and biochemical sensing, diagnostics, cancer imaging, and tissue engineering, and particularly as drug delivery carriers [118–121]. Relative to current nanomedicines, nanogels provide many benefits in medicinal application like superior drug loading, improved stabilization during processing in contrast to liposomes and micelles, controlled or sustained drug delivery, simplicity of production, and enhanced sensitivity toward outside stimulus with relatively minimum toxicity [122, 123]. If subjected to aqueous media with excessive water content, nanogels behave like a soft substance [109]. Intravital real-time fluorescence image findings showed the rapid deposition and restoration of FITC-labeled PMAA nanohydrogels throughout the liver, kidney, and other organs like, lung, heart, and spleen following 30 min of therapy; extended in vivo blood supply lifespan was shown utilizing PEGylated FITC-labeled PMAA nanohydrogels (Fig. 4D). Yang et al. [118] produced P (NIPAM-ss-AA) (poly (*N*-isopropylacrylamide)-ss-acrylic acid) nanogels via cross-linking NIPAM and AA with the help of *N,N′*-bis-(acryloyl) cystamine (BAC) via precipitation polymerization that illustrated double reaction, i.e., pH or redox-triggered DOX liberation, during in vitro as well as tumor investigations. The improved passage of DOX-loaded nanogels with fewer adverse reactions was shown through animal research, suggesting a potential future forum for intracellular monitored release profile in cancer chemotherapy. Bovine serum albumin as well as chitosan nanogels formulated through an environmentally friendly strategy of self-assembly display sustained release with decreased cytotoxicity [119, 120]. As a prototype tumor-targeting delivery device, biocompatible as well as pH-responsive, self-assembled, chitosan-*graft*-poly (*N*-isopropylacrylamide) nanogels are being used and showed higher expression in acidic environment [121]. Alginate-PAMAM dendrimers-based hybrid nanogels efficiently targeted tumor cells and demonstrated pH-dependent sustained release [124]. In the field of photocatalysis, biomaterials, biosensors, electronic components, and targeted therapy, nanotechnology is a fast-emerging innovation domain [125–127]. The massive ratio and low surface-volume of nanomaterials, small size, capable of holding several molecules, retention properties, as well as their adsorption capacity makes them ideal alternative for application in biomedical field. Nanoparticles could moreover enhance bioavailability, restrain drugs from decomposition, as well as provide sustained drug release. Such specific features of nanomaterials, therefore, provide promising framework for their use as a reliable delivery device for medications [128]. Biodegradable nanohybrid hydrogel beads have been synthesized using carboxymethyl cellulose and graphene oxide followed by cross-linking with hydrated ferric chloride for controlled DOX release [129]. The drug release from hydrogels was largely pH dependent and revealed the quicker release at pH 6.8 than moderately basic medium (pH ~7.4). In addition, the higher levels of filler and graphene oxide decreased the release rate. Halloysite nanotubes containing polyhybrid hydrogels (hydroxyethyl methacrylate) with sodium hyaluronate have been very successful in delivering drugs for colon cancer [130]. Utilizing an equilibrium swelling technique, followed by dragging and splitting the vacuum, 5-fluorouracil (5-FU) was entrapped not just in hydrogels but also in Halloysite nanotubes. 5-FU in vitro release from nanohybrid hydrogels displayed pH-dependent controlled release. To develop a controlled release system, 5-FU was cross-linked inside montmorillonite clay and was reinforced with alginate accompanied by chitosan layering [131]. In gastrointestinal conditions, AlgCS/5-FU/Mt nanohybrids system was employed for release of 5-FU. Nanoscale and chitosan-polylactide hybrids discharged paclitaxel in a prominent fashion in basic conditions as compared to acidic conditions [132]. Biodegradable nanohybrids poly(ε-caprolactone) scaffolds containing organically transformed nanoclay developed by an electrospinning technique showed sustained delivery of an anticancer drug (dexamethasone) to basic polymer by establishing a maze or "tortuous path" that delayed the drug's diffusion from the matrix in the vicinity of a two-dimensional filler [133]. Sustained delivery system for anticancer drug was developed using aliphatic diisocyanate aliphatic chain extender with altering chain lengths, functionalized polyurethane nanohybrids, and 2 nanoclay [134]. Graphene-based polyurethane nanohybrids were discovered for anticancer drug (dexamethasone) delivery through transplanting long-chain polyurethane onto substrate of functionalized graphene oxide [135]. For the sustained release of dexamethasone, chemically marked amine and sulfonate-functionalized graphene have been formed throughout long-chain polyurethane molecules [136, 137]. Delayed drug release was due to hard segment of pure polyurethane, whereas diffusion of entrapped drug in nanohybrids was obstructed due to the self-assembled structure and graphene moieties. Numerous nanoparticles polymer has been discovered that offer sustained release platform for anticancer drugs, for example, DOX, 5-FU, and MTX [138–142].

3 Targeted delivery: Mechanistic pathway

The delivery of target-specific therapeutic is dependent on stimulus-reactive factors both intrinsic (pH, redox, and enzyme) or extracellular (temperature, acoustic, and light)

[143]. For oral insulin delivery, supramolecular gel obtained from pH-sensitive chitosan is used. The nanogels' pH sensitivity prevents insulin when it is in the intestine, and the bioadhesiveness of chitosan allows for enhanced insulin absorption through sustained interaction with the intestinal mucosa [144]. To accomplish multimodal-regulated release of drugs in combination of two, such as gemcitabine (GEM) and doxorubicin (DOX), the drug delivery platform was employed based on mechanized silica nanoparticles (MSNPs) that consisted of MSN vehicles, acid-cleavage intermediate linkages, and reversible supramolecular nanovalves, was designed by organizing the sequence of stimulation in series. The release of GEM at specific time and dose is regulated by applying external voltage, whereas the release of DOX due to subsequent treatment with acid causes breakdown of the transitional bonds comprising the ketal groups [145]. Active cross-linked supramolecular channels of poly (glycidyl methacrylate) derivative chains on mesoporous silica nanoparticles react well with dual pH and glutathione (GSH)-binding stimulation that regulates the release of anticancer drug doxorubicin hydrochloride (DOX) at intracellular tumor environment and improved drug release kinetics via disulfide bonds cleavage [146]. The oxidized form of glutathione (GSH) is glutathione disulfide (GSSG) is an endogenous antioxidant possesses specific antimetastatic function as glutathione disulfide (GSSG) dependent liposomes and completely prevents cell proliferation; cell separation and substantially decreases infiltration of cancer cell. Glutathione defends biological processes from reactive oxygen species [147]. For tumor-specific chemotherapy, thermally sensitive liposomes and amphiphilic polymer poly (EOEOVE)-OD4 have been employed. DOX-embedded liposomes administered intravenously in tumor-bearing mice substantially inhibited tumor growth [148]. Biocompatible functionalized cyclodextrin (CD) poly (*N*-(2-hydroxypropyl-methacrylamide) (PHPMA) is the key component that contains two guests, e.g., poly (*N*,*N*-dimethylacrylamide) (PDEAAm) and poly (*N*,*N*-dimethylacrylamide)(PDEAAm), formulated by polycondensation of RAFT (reversible addition-fragmentation chain transfer) and, therefore, can develop definite copolymer from supramolecular ABA that responds to UV light. Thermoresponsivity due to the negative enthalpy of complexation is demonstrated by CD-based host/guest complexes. In the case of PDEAAm, since heating just above cloud point of PDEAAm block [149], temperature-induced amplification is detected. Tripeptide Lys-Phe-Gly (KFG), a biologically significant tripeptide, gets self-assembled automatically in aqueous solutions into well-defined nanostructures, demonstrating an interesting mechanism which is reversible and concentration dependent, as depicted by dynamic light scattering, transmission electron microscopy, and atomic force microscopy experiments. The absence of the KFG series in the tyrosine kinase nerve growth factor (Trk NGF) receptor polypeptide chain greatly influences signaling cascade triggering. Examination of cell sorting via fluorescence, confocal microscopy, and fluorescence microscopy supports effective intracellular release of the drug [150]. For fluorescence and MRI detection, a combination of target-tracking aptamer and PLA2 (phosphatidylcholine 2-acetylhydrolase) enzyme for lipid bilayer breakup of uranin-containing liposomes and gadopentetic acid (DGTP) as signaling factors was examined. Aptamer-PLA2, thus, activates the release mechanism for signal transduction and preferential detection of biomolecules via the target-responsive liposome process [151].

3.1 Magnetic field for cancer treatment

Magnetic materials are the potential carriers discovered more than two decades ago for targeting drug to specific site. Recently, superparamagnetic Fe_3O_4 magnetic nanoparticles were produced via grafting with four-armed pentaerythritol poly (ε-caprolactone) for targeting tumor cells via magnetically controlled drug (DOX) delivery [152]. The release of nanoparticles carried with DOX was controlled by alternating magnetic fields of high frequency. The controlled DOX release is better than other conventional methods of drug delivery for patient compliance. The effectiveness of drug release increases depending on the intercellular uptake of drug under magnetic field. Another delivery system known as paclitaxel delivery system encapsulated with iron oxide has been discovered [153]. The hypothermic effects of magnetic nanoparticles enhance cell (MCF-7) mortality due to the presence of an external magnetic field, thus building biodegradable and biocompatible carrier for powerful delivery of anticancer agents. A striking transformation in quantity of drug discharge is initiated after silica magnetic nanocapsules were used as remote for switching "on" and "off" process. This magnetic capsule contained hydrophobic drug, camptothecin, and hydrophilic drug, doxorubicin, found in tumors of mouse breast and was effectual at reducing growth of tumor cell [154]. Superparamagnetic nickel ferrite is used as magnetic carriers in combination with poly (vinyl alcohol), poly (methacrylic acid) (PMAA), and poly (ethylene oxide) that encapsulated anticancer drug doxorubicin and this system of drug delivery significantly increased the release speed of drugs under magnetic fields [155].

3.2 Electric field for cancer therapy

Smart biomaterials have gained attention in the field of biomedicine and biotechnology [156–160]. Materials that respond to stimuli such as heat [161, 162], pH [163, 164], light [165, 166], enzymes [167, 168], and magnetic fields [169, 170] are extensively employed in biomedical field.

The generation of electric signal is easier than other stimuli and, therefore, they are employed to stimulate release of molecules through implantable electronic delivery devices or conductive polymeric bulk materials. Ge et al. created a drug delivery system responsive to electric field in which he employed nanoparticles made of conductive polymer polypyrrole and these nanoparticles served as drug reservoir and triggered the release of drug under electric field influence [171]. This type of gel is injectable and, when applied to an external DC, releases drug from these nanogels and permit its diffusion into surroundings. With each electric stimulus, 25 ng of drug is released in the solution. In nonavailability of an electric field, minimum drug is released; therefore, this sort of delivery system possesses more benefits than other conventional drug delivery systems. Weaver et al. demonstrated the electrically controlled drug delivery by using composite of graphene oxide with a polypyrrole scaffold, which facilitated linear release [172]. Another electrically controlled drug delivery system is a carbon nanotube (CNTs) which acts as nanoreservoirs for drugs and releases drug under electrical stimulations [173]. Similarly, CNT drug nanoreservoirs coated with polypyrrole effectively carry the drug and permit electrical stimulated drug with the voltage application [174]. With this, a dual stimulus responsive system of chitosan-gold nanocomposites (CGNC) has been developed for controlled release of anticancer drug "5-FU" to specific site of cancer cell at reduced pH [175].

3.3 Thermal treatment for cancer therapy

Photodynamic therapy is a highly developed strategy for delivering drug to a specific area. For example in Chitosan-functionalized MoS2 (MoS2-CS) nanosheets is such type of advancement, that is regarded as photo thermal-triggered drug delivery systems for chemotherapeutic [176]. This nanocarrier offers enhanced drug delivery system that enhances antitumor efficacy. Use of dual anticancer drug-loaded graphene oxide (GO) added with stabilizing agent, poloxamer 188, has been prepared to deliver anticancer drugs to cancerous cells under near-infrared (NIR) laser irradiation [177]. Nanoparticles attached with a photoactivatable o-nitro benzyl (ONB) derivative of 5-fluorouracil (5-FU) worked as a photocaging nanocarrier which absorbed NIR radiation and triggered the breakage of bonds between ONB-FU and release of 5-FU at specific site [178]. An amphiphilic block copolymer in combination with poly {γ-2-[2-(2-methoxyethoxy)-ethoxy] ethoxy-ε-caprolactone}-*b*-poly (γ-octyloxy-ε caprolactone) displayed controlled release of drug at low critical solution temperature (LCST) of 38°C [179]. Doxorubicin exhibited higher cytotoxic effects against MCF-7, when it was loaded into micelle. Another carrier is B-cyclodextrin-poly (*N*-isopropylacrylamide) star polymer that formed complex with PTX at room temperature and significantly improved the PTX solubilization [180]. The formation of nanoparticles is induced by phase transitions of poly (*N*-isopropylacrylamide) segments at body temperature (above LCST) that greatly enhance the cellular uptake of complex of polymer-drug leading to proficient thermoresponsive delivery of PTX. Formation of dual pH/light-responsive cross-linked polymeric micelles (CPM) is formed by assembling amphiphilic glycol chitosan-*o*-nitro benzyl succinate conjugates (GC-NBSCs) which then cross-links using glutaraldehyde (GA) and are employed as drug carrier that releases drugs quickly at low pH under light irradiation [181]. Thus, GC-NBSC CPMs have laid an encouraging platform for the construction of dual pH/light-responsive smart drug delivery systems (DDS) for treatment of cancer. Another nanoparticle discovered is a biodegradable plasmon resonant liposome gold that is synthesized through 1, 2-distearoyl-sn-glycero-3-phosphocholine (DSPC)-cholesterol coated with gold nanoparticles and are competent for attacking cancer cells via photothermal treatment.

4 Future challenges in cancer therapy

Development of novel delivery system for drug delivery is a great strategy for cancer treatment in coming future. Targeted nanocarriers have been evaluated for their better efficiency in treating cancer by employing several animal models. Current drug delivery systems developed are highly advanced, but their clinical success is limited. Therefore, data that are more clinical are needed for better understanding of the pros and cons of these vehicles. Targeting molecule in cancer treatment may improve chemotherapeutic index by identifying target in body while producing minimum adverse side effects. Therefore, for targeting cancer, ion beam therapy is the promising treatment in near future and is better than high-risk surgery; extensive damage due to radiation therapy, for example, X-rays injure healthy cells that comes their way. The other beams of proteins such as carbon and neon can be targeted to tumor cells accurately without harming healthy tissue adjacent to targeted site. Ion beam radiation has a major advantage as it specifically target any type of tumor and put down more energy in tumor tissue causing its destruction. Few doses of treatment are required in ion beam radiation; e.g., 30 days treatment of liver cancer using proton therapy, whereas 4 days in carbon therapy. Although proton therapy is normally used presently but it lead to heavier carbanions that leave additional energy in cancerous tissues. Carbon therapy offers maximum linear energy transfer (LET) among available form of clinical radiation therapy that results in double-stranded DNA destruction and is complicated for other conventional radiation treatment to achieve, as they are able to break single-stranded DNA.

5 Conclusions

The current chapter has discussed various approaches and potential strategies that can help to develop novel vehicles for delivering drugs in cancer treatment. The sole feature and characteristics of these novel systems have given way to clinicians to improve the therapeutic potency and efficacy of existing treatment (combined therapy or new treatment [monotherapy]). Some of these products have not shown appealing clinical translation, whereas some potential products are presently under clinical evaluation that could be of great promise and could develop into new therapeutic option in coming future.

References

[1] Boyle P, Bernard L. World cancer report 2008. Lyon: IARC Press; 2008.

[2] Soppimath KS, Aminabhavi TM, Kulkarni AR, Rudzinski WE. Biodegradable polymeric nanoparticles as drug delivery devices. J Control Release 2001;70:1–20.

[3] Su J, Chen F, Cryns VL, Messersmith PB. Catechol polymers for pH-responsive, targeted drug delivery to cancer cells. J Am Chem Soc 2011;133:11850–3.

[4] Kumar S, et al. Controlled drug release through regulated biodegradation of poly (lactic acid) using inorganic salts. Int J Biol Macromol 2017;104:487–97.

[5] Shim MS, Kwon YJ. Stimuli-responsive polymers and nanomaterials for gene delivery and imaging applications. Adv Drug Deliv Rev 2012;64:1046–59.

[6] Mo R, Jiang T, Gu Z. Recent progress in multidrug delivery to cancer cells by liposomes. Nanomedicine 2014;9:1117–20.

[7] Dong Y, et al. Lipid-like nanomaterials for simultaneous gene expression and silencing *in vivo*. Adv Healthc Mater 2014;3:1392–7.

[8] Gu FX, et al. Targeted nanoparticles for cancer therapy. Nano Today 2007;2:14–21.

[9] Shih H, Lin C-C. Photo click hydrogels prepared from functionalized cyclodextrin and poly (ethylene glycol) for drug delivery and *in situ* cell encapsulation. Biomacromolecules 2015;16:1915–23.

[10] Li Y, Maciel D, Rodrigues J, Shi X, Tomás H. Biodegradable polymer nanogels for drug/nucleic acid delivery. Chem Rev 2015;115:8564–608.

[11] Sun W, Gu Z. Engineering DNA scaffolds for delivery of anticancer therapeutics. Biomater Sci 2015;3:1018–24.

[12] Maeda H, Wu J, Sawa T, Matsumura Y, Hori K. Tumor vascular permeability and the EPR effect in macromolecular therapeutics: a review. J Control Release 2000;65:271–84.

[13] Koo H, et al. *In vivo* targeted delivery of nanoparticles for theranosis. Acc Chem Res 2011;44:1018–28.

[14] Bharali DJ, Lucey DW, Jayakumar H, Pudavar HE, Prasad PN. Folate-receptor-mediated delivery of InP quantum dots for bioimaging using confocal and two-photon microscopy. J Am Chem Soc 2005;127:11364–71.

[15] Zrazhevskiy P, Sena M, Gao X. Designing multifunctional quantum dots for bioimaging, detection, and drug delivery. Chem Soc Rev 2010;39:4326–54.

[16] Kairdolf BA, et al. Semiconductor quantum dots for bioimaging and bio-diagnostic applications. Annu Rev Anal Chem 2013;6:143–62.

[17] Bianco A, Kostarelos K, Prato M. Opportunities and challenges of carbon-based nanomaterials for cancer therapy. Expert Opin Drug Deliv 2008;5:331–42.

[18] Murray CB, Norris DJ, Bawendi MG. Synthesis and characterization of nearly monodisperse CdE (E = sulfur, selenium, tellurium) semiconductor nano crystallites. J Am Chem Soc 1993;115:8706–15.

[19] Bilan R, Nabiev I, Sukhanova A. Quantum dot-based nanotools for bioimaging, diagnostics, and drug delivery. Chembiochem 2016;17:2103–14.

[20] Halder GJ, Kepert CJ, Moubaraki B, Murray KS, Cashion JD. Guest-dependent spin crossover in a nano porous molecular framework material. Science 2002;298:1762–5.

[21] Zhao X, Li H, Lee RJ. Targeted drug delivery via folate receptors. Expert Opin Drug Deliv 2008;5:309–19.

[22] Gao X, Cui Y, Levenson RM, Chung LWK, Nie S. *In vivo* cancer targeting and imaging with semiconductor quantum dots. Nat Biotechnol 2004;22:969–76.

[23] Muthu MS, Kulkarni SA, Raju A, Feng S-S. Theranostic liposomes of TPGS coating for targeted co-delivery of docetaxel and quantum dots. Biomaterials 2012;33:3494–501.

[24] Huang C-L, et al. Application of paramagnetic graphene quantum dots as a platform for simultaneous dual-modality bioimaging and tumor-targeted drug delivery. J Mater Chem B 2015;3:651–64.

[25] Li Z, et al. Quantum dots loaded nanogels for low cytotoxicity, pH-sensitive fluorescence, cell imaging and drug delivery. Carbohydr Polym 2015;121:477–85.

[26] Huang N, et al. Single-wall carbon nanotubes assisted photothermal cancer therapy: animal study with a murine model of squamous cell carcinoma. Lasers Surg Med 2010;42:798–808.

[27] Ahmed N, Fessi H, Elaissari A. Theranostic applications of nanoparticles in cancer. Drug Discov Today 2012;17:928–34.

[28] Lin Y, et al. Advances toward bioapplications of carbon nanotubes. J Mater Chem 2004;14:527–41.

[29] Bianco A, Kostarelos K, Prato M. Applications of carbon nanotubes in drug delivery. Curr Opin Chem Biol 2005;9:674–9.

[30] Fadel TR, Fahmy TM. Immunotherapy applications of carbon nanotubes: from design to safe applications. Trends Biotechnol 2014;32:198–209.

[31] Villa CH, et al. Single-walled carbon nanotubes deliver peptide antigen into dendritic cells and enhance IgG responses to tumor-associated antigens. ACS Nano 2011;5:5300–11.

[32] Senapati S, et al. Layered double hydroxides as effective carrier for anticancer drugs and tailoring of release rate through interlayer anions. J Control Release 2016;224:186–98.

[33] Whilton NT, Vickers PJ, Mann S. Bioinorganic clays: synthesis and characterization of amino- and polyamino acid intercalated layered double hydroxides. J Mater Chem 1997;7:1623–9.

[34] Del Arco M, Gutiérrez S, Martín C, Rives V, Rocha J. Synthesis and characterization of layered double hydroxides (LDH) intercalated with non-steroidal anti-inflammatory drugs (NSAID). J Solid State Chem 2004;177:3954–62.

[35] Rives V, del Arco M, Martín C. Intercalation of drugs in layered double hydroxides and their controlled release: a review. Appl Clay Sci 2014;88-89:239–69.

[36] Tyner KM, Schiffman SR, Giannelis EP. Nanobiohybrids as delivery vehicles for camptothecin. J Control Release 2004;95:501–14.

[37] Li L, Gu W, Chen J, Chen W, Xu ZP. Co-delivery of siRNAs and anti-cancer drugs using layered double hydroxide nanoparticles. Biomaterials 2014;35:3331–9.

[38] Yadav BC, Kumar R. Structure, properties and applications of fullerenes. Int J Nanotechnol Appl 2008;2:15–24.
[39] Lai C-Y, et al. A mesoporous silica nanosphere-based carrier system with chemically removable CdS nanoparticle caps for stimuli-responsive controlled release of neurotransmitters and drug molecules. J Am Chem Soc 2003;125:4451–9.
[40] Mamaeva V, et al. Mesoporous silica nanoparticles as drug delivery systems for targeted inhibition of Notch signaling in cancer. Mol Ther 2011;19:1538–46.
[41] Okazaki M, Yoshida Y, Yamaguchi S, Kaneno M, Elliott JC. Affinity binding phenomena of DNA onto apatite crystals. Biomaterials 2001;22:2459–64.
[42] Lee MS, et al. Target-specific delivery of siRNA by stabilized calcium phosphate nanoparticles using dopa–hyaluronic acid conjugate. J Control Release 2014;192:122–30.
[43] Wang S, McDonnell EH, Sedor FA, Toffaletti JG. pH effects on measurements of ionized calcium and ionized magnesium in blood. Arch Pathol Lab Med 2002;126:947–50.
[44] Morgan TT, et al. Encapsulation of organic molecules in calcium phosphate nanocomposite particles for intracellular imaging and drug delivery. Nano Lett 2008;8:4108–15.
[45] Tseng Y-C, Xu Z, Guley K, Yuan H, Huang L. Lipid–calcium phosphate nanoparticles for delivery to the lymphatic system and SPECT/CT imaging of lymph node metastases. Biomaterials 2014;35:4688–98.
[46] Pittella F, et al. Systemic siRNA delivery to a spontaneous pancreatic tumor model in transgenic mice by PEGylated calcium phosphate hybrid micelles. J Control Release 2014;178:18–24.
[47] Nomoto T, et al. Calcium phosphate-based organic–inorganic hybrid nano-carriers with pH-responsive on/off switch for photodynamic therapy. Biomater Sci 2016;4:826–38.
[48] Mi P, et al. A pH-activatable nanoparticle with signal-amplification capabilities for non-invasive imaging of tumour malignancy. Nat Nanotechnol 2016;11:724–30.
[49] Raynal I, et al. Macrophage endocytosis of superparamagnetic iron oxide nanoparticles: mechanisms and comparison of ferumoxides and ferumoxtran-10. Investig Radiol 2004;39:56–63.
[50] Rogers WJ, Basu P. Factors regulating macrophage endocytosis of nanoparticles: implications for targeted magnetic resonance plaque imaging. Atherosclerosis 2005;178:67–73.
[51] Widder KJ, Senyei AE, Ranney DF. *In vitro* release of biologically active adriamycin by magnetically responsive albumin microspheres. Cancer Res 1980;40:3512–7.
[52] Gupta PK, Hung C-T. Targeted delivery of low dose doxorubicin hydrochloride administered via magnetic albumin microspheres in rats. J Microencapsul 1990;7:85–94.
[53] Aime S, et al. Insights into the use of paramagnetic Gd(III) complexes in MR-molecular imaging investigations. J Magn Reson Imaging 2002;16:394–406.
[54] Arbab AS, et al. Characterization of biophysical and metabolic properties of cells labeled with superparamagnetic iron oxide nanoparticles and transfection agent for cellular MR imaging. Radiology 2003;229:838–46.
[55] Laurent S, Saei AA, Behzadi S, Panahifar A, Mahmoudi M. Superparamagnetic iron oxide nanoparticles for delivery of therapeutic agents: opportunities and challenges. Expert Opin Drug Deliv 2014;11:1449–70.
[56] Josephson L, Tung C-H, Moore A, Weissleder R. High-efficiency intracellular magnetic labeling with novel superparamagnetic-Tat peptide conjugates. Bioconjug Chem 1999;10:186–91.
[57] Lee H, et al. Antibiofouling polymer-coated superparamagnetic iron oxide nanoparticles as potential magnetic resonance contrast agents for *in vivo* cancer imaging. J Am Chem Soc 2006;128:7383–9.
[58] Lee H, et al. Thermally cross-linked superparamagnetic iron oxide nanoparticles: synthesis and application as a dual imaging probe for cancer *in vivo*. J Am Chem Soc 2007;129:12739–45.
[59] Quinto CA, Mohindra P, Tong S, Bao G. Multifunctional superparamagnetic iron oxide nanoparticles for combined chemotherapy and hyperthermia cancer treatment. Nanoscale 2015;7:12728–36.
[60] Fortin J-P, et al. Size-sorted anionic iron oxide nanomagnets as colloidal mediators for magnetic hyperthermia. J Am Chem Soc 2007;129:2628–35.
[61] Balivada S, et al. A/C magnetic hyperthermia of melanoma mediated by iron(0)/iron oxide core/shell magnetic nanoparticles: a mouse study. BMC Cancer 2010;10:119.
[62] Hildebrandt B, et al. The cellular and molecular basis of hyperthermia. Crit Rev Oncol Hematol 2002;43:33–56.
[63] Park JH, et al. Polymeric nanomedicine for cancer therapy. Prog Polym Sci 2008;33:113–37.
[64] Parveen S, Sahoo SK. Polymeric nanoparticles for cancer therapy. J Drug Target 2008;16:108–23.
[65] Mitra S, Gaur U, Ghosh PC, Maitra AN. Tumour targeted delivery of encapsulated dextran–doxorubicin conjugate using chitosan nanoparticles as carrier. J Control Release 2001;74:317–23.
[66] Pandey SK, et al. Controlled release of drug and better bioavailability using poly(lactic acid-co-glycolic acid) nanoparticles. Int J Biol Macromol 2016;89:99–110.
[67] Cheng F-Y, Su C-H, Wu P-C, Yeh C-S. Multifunctional polymeric nanoparticles for combined chemotherapeutic and near-infrared photothermal cancer therapy *in vitro* and *in vivo*. Chem Commun 2010;46:3167–9.
[68] Bolotin EM, et al. Ammonium sulfate gradients for efficient and stable remote loading of amphipathic weak bases into liposomes and ligandoliposomes. J Liposome Res 1994;4:455–79.
[69] Boman NL, Masin D, Mayer LD, Cullis PR, Bally MB. Liposomal vincristine which exhibits increased drug retention and increased circulation longevity cures mice bearing P388 tumors. Cancer Res 1994;54:2830–3.
[70] Sapra P, Allen TM. Ligand-targeted liposomal anticancer drugs. Prog Lipid Res 2003;42:439–62.
[71] Pakunlu RI, et al. *In vitro* and *in vivo* intracellular liposomal delivery of antisense oligonucleotides and anticancer drug. J Control Release 2006;114:153–62.
[72] Jiang T, et al. Dual-functional liposomes based on pH-responsive cell-penetrating peptide and hyaluronic acid for tumor-targeted anticancer drug delivery. Biomaterials 2012;33:9246–58.
[73] Guo X, Szoka FC. Steric stabilization of fusogenic liposomes by a low-pH sensitive PEG-diortho ester-lipid conjugate. Bioconjug Chem 2001;12:291–300.
[74] Mo R, et al. Multistage pH-responsive liposomes for mitochondrial-targeted anticancer drug delivery. Adv Mater 2012;24:3659–65.
[75] Kitatani K, et al. Ceramide limits phosphatidylinositol-3-kinase C2β-controlled cell motility in ovarian cancer: potential of ceramide as a metastasis-suppressor lipid. Oncogene 2016;35:2801–12.
[76] Şalva E, Turan SÖ, Eren F, Akbuğa J. The enhancement of gene silencing efficiency with chitosan-coated liposome formulations of siRNAs targeting HIF-1α and VEGF. Int J Pharm 2015;478:147–54.
[77] Cao J, et al. A7RC peptide modified paclitaxel liposomes dually target breast cancer. Biomater Sci 2015;3:1545–54.

[78] Elzoghby AO, Samy WM, Elgindy NA. Albumin-based nanoparticles as potential controlled release drug delivery systems. J Control Release 2012;157:168–82.
[79] Dreis S, et al. Preparation, characterisation and maintenance of drug efficacy of doxorubicin-loaded human serum albumin (HSA) nanoparticles. Int J Pharm 2007;341:207–14.
[80] Zhao D, et al. Preparation, characterization, and *in vitro* targeted delivery of folate-decorated paclitaxel-loaded bovine serum albumin nanoparticles. Int J Nanomedicine 2010;5:669–77.
[81] Hawkins MJ, Soon-Shiong P, Desai N. Protein nanoparticles as drug carriers in clinical medicine. Adv Drug Deliv Rev 2008;60:876–85.
[82] Yardley DA. Nab-paclitaxel mechanisms of action and delivery. J Control Release 2013;170:365–72.
[83] Kim T-Y, et al. Phase I and pharmacokinetic study of Genexol-PM, a cremophor-free, polymeric micelle-formulated paclitaxel, in patients with advanced malignancies. Clin Cancer Res 2004;10:3708–16.
[84] Shi C, et al. Actively targeted delivery of anticancer drug to tumor cells by redox-responsive star-shaped micelles. Biomaterials 2014;35:8711–22.
[85] Jin X, et al. Efficient intravesical therapy of bladder cancer with cationic doxorubicin nanoassemblies. Int J Nanomedicine 2016;11:4535–44.
[86] Kumari P, et al. Cholesterol-conjugated poly(D, L-lactide)-based micelles as a nanocarrier system for effective delivery of curcumin in cancer therapy. Drug Deliv 2017;24:209–23.
[87] Deshayes S, et al. Phenylboronic acid-installed polymeric micelles for targeting sialylated epitopes in solid tumors. J Am Chem Soc 2013;135:15501–7.
[88] Gilbreth RN, et al. Lipid- and polyion complex-based micelles as agonist platforms for TNFR superfamily receptors. J Control Release 2016;234:104–14.
[89] Cui H, Webber MJ, Stupp SI. Self-assembly of peptide amphiphiles: from molecules to nanostructures to biomaterials. Biopolymers 2010;94:1–18.
[90] Zhang R, et al. Carrier-free, chemophotodynamic dual nanodrugs via self-assembly for synergistic antitumor therapy. ACS Appl Mater Interfaces 2016;8:13262–9.
[91] Wu C, et al. Engineering of switchable aptamer micelle flares for molecular imaging in living cells. ACS Nano 2013;7:5724–31.
[92] Zhou J, Du X, Li J, Yamagata N, Xu B. Taurine boosts cellular uptake of small D-peptides for enzyme-instructed intracellular molecular self-assembly. J Am Chem Soc 2015;137:10040–3.
[93] Li S-D, Chen Y-C, Hackett MJ, Huang L. Tumor-targeted delivery of siRNA by self-assembled nanoparticles. Mol Ther 2008;16:163–9.
[94] Yoon H-J, Jang W-D. Polymeric supramolecular systems for drug delivery. J Mater Chem 2010;20:211–22.
[95] Wei T, et al. Anticancer drug nanomicelles formed by self-assembling amphiphilic dendrimer to combat cancer drug resistance. Proc Natl Acad Sci U S A 2015;112:2978–83.
[96] Yu G, et al. Fabrication of a targeted drug delivery system from a Pillar[5]arene-based supramolecular diblock copolymeric amphiphile for effective cancer therapy. Adv Funct Mater 2016;26:8999–9008.
[97] Li J, Loh X. Cyclodextrin-based supramolecular architectures: syntheses, structures, and applications for drug and gene delivery. Adv Drug Deliv Rev 2008;60:1000–17.
[98] Cheng C-C, et al. Highly efficient drug delivery systems based on functional supramolecular polymers: *in vitro* evaluation. Acta Biomater 2016;33:194–202.
[99] Kaida S, et al. Visible drug delivery by supramolecular nanocarriers directing to single-platformed diagnosis and therapy of pancreatic tumor model. Cancer Res 2010;70:7031–41.
[100] Adeli M, Ashiri M, Chegeni BK, Sasanpour P. Tumor-targeted drug delivery systems based on supramolecular interactions between iron oxide–carbon nanotubes PAMAM–PEG–PAMAM linear-dendritic copolymers. J Iran Chem Soc 2013;10:701–8.
[101] Lovell JF, et al. Porphysome nanovesicles generated by porphyrin bilayers for use as multimodal biophotonic contrast agents. Nat Mater 2011;10:324–32.
[102] Muhanna N, et al. Phototheranostic porphyrin nanoparticles enable visualization and targeted treatment of head and neck cancer in clinically relevant models. Theranostics 2015;5:1428–43.
[103] Su S, Ding Y, Li Y, Wu Y, Nie G. Integration of photothermal therapy and synergistic chemotherapy by a porphyrin self-assembled micelle confers chemosensitivity in triple-negative breast cancer. Biomaterials 2016;80:169–78.
[104] Peppas N, Bures P, Leobandung W, Ichikawa H. Hydrogels in pharmaceutical formulations. Eur J Pharm Biopharm 2000;50:27–46.
[105] Peppas N, Langer R. New challenges in biomaterials. Science 1994;263:1715–20.
[106] Hoffman AS, Ratner BD. Synthetic hydrogels for biomedical applications. In: Hydrogels for medical and related applications. ACS symposium series. Washington, DC: American Chemical Society; 1976. p. 1–36.
[107] Lin C-C, Metters AT. Hydrogels in controlled release formulations: network design and mathematical modeling. Adv Drug Deliv Rev 2006;58:1379–408.
[108] Tomme SRV, Storm G, Hennink WE. *In situ* gelling hydrogels for pharmaceutical and biomedical applications. Int J Pharm 2008;355:1–18.
[109] Jin S, et al. Biodegradation and toxicity of protease/redox/pH stimuli-responsive PEGlated PMAA nanohydrogels for targeting drug delivery. ACS Appl Mater Interfaces 2015;7:19843–52.
[110] Qi X, et al. Fabrication and characterization of a novel anticancer drug delivery system: salecan/poly(methacrylic acid) semi-interpenetrating polymer network hydrogel. ACS Biomater Sci Eng 2015;1:1287–99.
[111] Ghosh R, Goswami U, Ghosh SS, Paul A, Chattopadhyay A. Synergistic anticancer activity of fluorescent copper nanoclusters and cisplatin delivered through a hydrogel nanocarrier. ACS Appl Mater Interfaces 2015;7:209–22.
[112] Xue B, et al. Intracellular degradable hydrogel cubes and spheres for anti-cancer drug delivery. ACS Appl Mater Interfaces 2015;7:13633–44.
[113] Wang Y, Nie J, Chang B, Sun Y, Yang W. Poly(vinylcaprolactam)-based biodegradable multiresponsive microgels for drug delivery. Biomacromolecules 2013;14:3034–46.
[114] Yu J, Ha W, Sun J-N, Shi Y-P. Supramolecular hybrid hydrogel based on host–guest interaction and its application in drug delivery. ACS Appl Mater Interfaces 2014;6:19544–51.
[115] Naskar J, Palui G, Banerjee A. Tetrapeptide-based hydrogels: for encapsulation and slow release of an anticancer drug at physiological pH. J Phys Chem B 2009;113:11787–92.
[116] Chen Y-Y, Wu H-C, Sun J-S, Dong G-C, Wang T-W. Injectable and thermoresponsive self-assembled nanocomposite hydrogel for long-term anticancer drug delivery. Langmuir 2013;29:3721–9.
[117] Kabanov AV, Vinogradov SV. Nanogels as pharmaceutical carriers: finite networks of infinite capabilities. Angew Chem Int Ed Eng 2009;48:5418–29.

[118] Yang H, et al. Smart pH/redox dual-responsive nanogels for on-demand intracellular anticancer drug release. ACS Appl Mater Interfaces 2016;8:7729–38.
[119] Wang Y, et al. Nanogels fabricated from bovine serum albumin and chitosan via self-assembly for delivery of anticancer drug. Colloids Surf B: Biointerfaces 2016;146:107–13.
[120] Mahanta AK, Senapati S, Maiti P. A polyurethane–chitosan brush as an injectable hydrogel for controlled drug delivery and tissue engineering. Polym Chem 2017;8:6233–49.
[121] Duan C, et al. Chitosan-g-poly(*N*-isopropylacrylamide) based nanogels for tumor extracellular targeting. Int J Pharm 2011;409:252–9.
[122] Eckmann DM, Composto RJ, Tsourkas A, Muzykantov VR. Nanogel carrier design for targeted drug delivery. J Mater Chem B 2014;2:8085–97.
[123] Chiang W-H, et al. Dual stimuli-responsive polymeric hollow nanogels designed as carriers for intracellular triggered drug release. Langmuir 2012;28:15056–64.
[124] Matai I, Gopinath P. Chemically cross-linked hybrid nanogels of alginate and PAMAM dendrimers as efficient anticancer drug delivery vehicles. ACS Biomater Sci Eng 2016;2:213–23.
[125] Lokina S, Stephen A, Kaviyarasan V, Arulvasu C, Narayanan V. Cytotoxicity and antimicrobial activities of green synthesized silver nanoparticles. Eur J Med Chem 2014;76:256–63.
[126] Wolfbeis OS. An overview of nanoparticles commonly used in fluorescent bioimaging. Chem Soc Rev 2015;44:4743–68.
[127] Rajendran S, et al. Ce^{3+}-ion-induced visible-light photocatalytic degradation and electrochemical activity of ZnO/CeO_2 nanocomposite. Sci Rep 2016;6:31641.
[128] Singh R, Lillard JW. Nanoparticle-based targeted drug delivery. Exp Mol Pathol 2009;86:215–23.
[129] Rasoulzadeh M, Namazi H. Carboxymethyl cellulose/graphene oxide bionanocomposite hydrogel beads as anticancer drug carrier agent. Carbohydr Polym 2017;168:320–6.
[130] Rao KM, Nagappan S, Seo DJ, Ha C-S. pH sensitive halloysite-sodium hyaluronate/poly(hydroxyethyl methacrylate) nanocomposites for colon cancer drug delivery. Appl Clay Sci 2014;97–98:33–42.
[131] Azhar FF, Olad A. A study on sustained release formulations for oral delivery of 5-fluorouracil based on alginate–chitosan/montmorillonite nanocomposite systems. Appl Clay Sci 2014;101:288–96.
[132] Nanda R, Sasmal A, Nayak PL. Preparation and characterization of chitosan–polylactide composites blended with Cloisite 30B for control release of the anticancer drug paclitaxel. Carbohydr Polym 2011;83:988–94.
[133] Singh NK, et al. Nanostructure controlled anti-cancer drug delivery using poly (ε-caprolactone) based nanohybrids. J Mater Chem 2012;22:17853–63.
[134] Mishra A, et al. Self-assembled aliphatic chain extended polyurethane nano-biohybrids: emerging hemocompatible biomaterials for sustained drug delivery. Acta Biomater 2014;10:2133–46.
[135] Patel DK, et al. Superior biomaterials using diamine modified graphene grafted polyurethane. Polymer (Guildf) 2016;106:109–19.
[136] Patel DK, et al. Graphene as a chain extender of polyurethanes for biomedical applications. RSC Adv 2016;6:58628–40.
[137] Patel DK, et al. Functionalized graphene tagged polyurethanes for corrosion inhibitor and sustained drug delivery. ACS Biomater Sci Eng 2017;3:3351–63.
[138] Dhanavel S, Nivethaa EAK, Narayanan V, Stephen A. *In vitro* cytotoxicity study of dual drug loaded chitosan/palladium nanocomposite towards HT-29 cancer cells. Mater Sci Eng C 2017;75:1399–410.
[139] Lei H, et al. Chitosan/sodium alginate modificated graphene oxide-based nanocomposite as a carrier for drug delivery. Ceram Int 2016;42:17798–805.
[140] Seema DM, Datta M. MMT-PLGA nanocomposites as an oral and controlled release carrier for 5-fluorouracil: a novel approach. Int J Pharm Pharm Sci 2013;5:332–41.
[141] Rasouli S, Davaran S, Rasouli F, Mahkam M, Salehi R. Synthesis, characterization and pH-controllable methotrexate release from biocompatible polymer/silica nanocomposite for anticancer drug delivery. Drug Deliv 2014;21:155–63.
[142] Zeynabad FB, et al. pH-controlled multiple-drug delivery by a novel anti-bacterial nanocomposite for combination therapy. RSC Adv 2015;5:105678–91.
[143] Liu D, Yang F, Xiong F, Gu N. The smart drug delivery system and its clinical potential. Theranostics 2016;6:1306–23.
[144] Saboktakin M, Maharramov A, Ramazanov M. pH sensitive chitosan-based supramolecular gel for oral drug delivery of insulin. J Mol Genet Med 2015;9:170.
[145] Wang T, Sun G, Wang M, Zhou B, Fu J. Voltage/pH-driven mechanized silica nanoparticles for the multimodal controlled release of drugs. ACS Appl Mater Interfaces 2015;7:21295–304.
[146] Li Q-L, et al. pH and glutathione dual-responsive dynamic cross-linked supra-molecular network on mesoporous silica nanoparticles for controlled anticancer drug release. ACS Appl Mater Interfaces 2015;7:28656–64.
[147] Sadhu SS, et al. *In vitro* and *in vivo* antimetastatic effect of glutathione disulfide liposomes. Cancer Growth Metastasis 2017;10, 117906441769525.
[148] Kono K, et al. Highly temperature-sensitive liposomes based on a thermosensitive block copolymer for tumor-specific chemotherapy. Biomaterials 2010;31:7096–105.
[149] Schmidt BVKJ, Hetzer M, Ritter H, Barner-Kowollik C. UV light and temperature responsive supramolecular ABA triblock copolymers via reversible cyclodextrin complexation. Macromolecules 2013;46:1054–65.
[150] Moitra P, Kumar K, Kondaiah P, Bhattacharya S. Efficacious anticancer drug delivery mediated by a pH-sensitive self-assembly of a conserved tripeptide derived from tyrosine kinase NGF receptor. Angew Chem Int Ed Eng 2014;53:1113–7.
[151] Xing H, et al. Multimodal detection of a small molecule target using stimuli-responsive liposome triggered by aptamer–enzyme conjugate. Anal Chem 2016;88:1506–10.
[152] Panja S, Maji S, Maiti TK, Chattopadhyay S. A smart magnetically active nanovehicle for on-demand targeted drug delivery: where van der waals force balances the magnetic interaction. ACS Appl Mater Interfaces 2015;7:24229–41.
[153] Mansouri M, Nazarpak MH, Solouk A, Akbari S, Hasani-Sadrabadi MM. Magnetic responsive of paclitaxel delivery system based on SPION and palmitoyl chitosan. J Magn Magn Mater 2017;421:316–25.
[154] Kong SD, et al. Magnetically vectored nanocapsules for tumor penetration and remotely switchable on-demand drug release. Nano Lett 2010;10:5088–92.
[155] Rana S, Gallo A, Srivastava RS, Misra RDK. On the suitability of nanocrystalline ferrites as a magnetic carrier for drug delivery: functionalization, conjugation and drug release kinetics. Acta Biomater 2007;3:233–42.
[156] Anderson DG. Materials science: smart biomaterials. Science 2004;305:1923–4.

[157] Stuart MAC, et al. Emerging applications of stimuli-responsive polymer materials. Nat Mater 2010;9:101–13.
[158] Guo X, Szoka FC. Chemical approaches to triggerable lipid vesicles for drug and gene delivery. Acc Chem Res 2003;36:335–41.
[159] LaVan DA, McGuire T, Langer R. Small-scale systems for *in vivo* drug delivery. Nat Biotechnol 2003;21:1184–91.
[160] Grayson ACR, et al. Multi-pulse drug delivery from a resorbable polymeric microchip device. Nat Mater 2003;2:767–72.
[161] Yavuz MS, et al. Gold nanocages covered by smart polymers for controlled release with near-infrared light. Nat Mater 2009;8:935–9.
[162] Choi S-W, Zhang Y, Xia Y. A temperature-sensitive drug release system based on phase-change materials. Angew Chem Int Ed Eng 2010;49:7904–8.
[163] Gillies ER, Jonsson TB, Fréchet JMJ. Stimuli-responsive supramolecular assemblies of linear-dendritic copolymers. J Am Chem Soc 2004;126:11936–43.
[164] Kim KT, Cornelissen JJLM, Nolte RJM, van Hest JCM. A polymersome nanoreactor with controllable permeability induced by stimuli-responsive block copolymers. Adv Mater 2009;21:2787–91.
[165] Kostiainen MA, Kasyutich O, Cornelissen JJLM, Nolte RJM. Self-assembly and optically triggered disassembly of hierarchical dendron–virus complexes. Nat Chem 2010;2:394–9.
[166] Dvir T, Banghart MR, Timko BP, Langer R, Kohane DS. Photo-targeted nanoparticles. Nano Lett 2010;10:250–4.
[167] Azagarsamy MA, Sokkalingam P, Thayumanavan S. Enzyme-triggered disassembly of dendrimer-based amphiphilic nanocontainers. J Am Chem Soc 2009;131:14184–5.
[168] Thornton PD, Heise A. Highly specific dual enzyme-mediated payload release from peptide-coated silica particles. J Am Chem Soc 2010;132:2024–8.
[169] Namiki Y, et al. A novel magnetic crystal–lipid nanostructure for magnetically guided *in vivo* gene delivery. Nat Nanotechnol 2009;4:598–606.
[170] Dames P, et al. Targeted delivery of magnetic aerosol droplets to the lung. Nat Nanotechnol 2007;2:495–9.
[171] Ge J, Neofytou E, Cahill TJ, Beygui RE, Zare RN. Drug release from electric-field-responsive nanoparticles. ACS Nano 2012;6:227–33.
[172] Weaver CL, LaRosa JM, Luo X, Cui XT. Electrically controlled drug delivery from graphene oxide nanocomposite films. ACS Nano 2014;8:1834–43.
[173] Luo X, Matranga C, Tan S, Alba N, Cui XT. Carbon nanotube nanoreservior for controlled release of anti-inflammatory dexamethasone. Biomaterials 2011;32:6316–23.
[174] Wadhwa R, Lagenaur CF, Cui XT. Electrochemically controlled release of dexamethasone from conducting polymer polypyrrole coated electrode. J Control Release 2006;110:531–41.
[175] Chandran PR, Sandhyarani N. An electric field responsive drug delivery system based on chitosan–gold nanocomposites for site specific and controlled delivery of 5-fluorouracil. RSC Adv 2014;4:44922–9.
[176] Yin W, et al. High-throughput synthesis of single-layer MoS2 nanosheets as a near-infrared photothermal-triggered drug delivery for effective cancer therapy. ACS Nano 2014;8:6922–33.
[177] Tran TH, et al. Development of a graphene oxide nanocarrier for dual-drug chemo-phototherapy to overcome drug resistance in cancer. ACS Appl Mater Interfaces 2015;7:28647–55.
[178] Fedoryshin LL, Tavares AJ, Petryayeva E, Doughan S, Krull UJ. Near-infrared-triggered anticancer drug release from upconverting nanoparticles. ACS Appl Mater Interfaces 2014;6:13600–6.
[179] Cheng Y, et al. Thermally controlled release of anticancer drug from self-assembled γ-substituted amphiphilic poly(ε-caprolactone) micellar nanoparticles. Biomacromolecules 2012;13:2163–73.
[180] Song X, et al. Thermo responsive delivery of paclitaxel by β-cyclodextrin-based poly(N-isopropylacrylamide) star polymer via inclusion complexation. Biomacromolecules 2016;17:3957–63.
[181] Meng L, et al. Chitosan-based nanocarriers with pH and light dual response for anticancer drug delivery. Biomacromolecules 2013;14:2601–10.

Chapter 7

Hydrogel-based drug delivery systems for cancer therapy

Brahmeshwar Mishra[a] **and Juhi Singh**[b]

[a]*Department of Pharmaceutical Engineering & Technology, Indian Institute of Technology (BHU), Varanasi, Uttar Pradesh, India,* [b]*NTU Institute for Health Technologies, Nanyang Technological University, Singapore, Singapore*

1 Introduction

According to the World Health Organization report, cancer is one of the leading causes of death globally, which led to about 9.6 million deaths in 2018, which is about 1 in 6 deaths worldwide due to cancer. Cancer not only reduces the quality of life but also decreases the life expectancy [1].

Two major characteristics of cancer are rapid cell proliferation and metastasis, leading to the recurrence of cancer and mortality. There are various treatment strategies available for cancer therapy, depending upon the type of cancer and the stage of diagnosis. The most commonly used strategies include surgical removal, chemotherapy, radiotherapy, gene therapy, immunotherapy, targeted therapy, hormonal therapy, etc. Chemotherapy is the most prevalent strategy used to control the tumor and prevent its recurrence. Chemotherapy works by killing the tumor cells by administering cytotoxic drug molecules. However, the application of chemotherapy is limited due to its side effects owing to the inability to limit the toxicity to tumor cells. Thus, lack of targeting ability of chemotherapeutic drugs leads to the toxicity of normal cells, thereby putting forward a clinical need to develop localized cancer therapies [2]. Advancement in nanoparticle development has led to the formulation of various drug delivery systems based on nanoparticles, micelles, etc. for cancer therapy. However, systemic circulation of these drug-loaded nanoparticles remains one of the concerns which has the potential to cause off-target toxicity, low dose at the target site, higher dose administration, reduced efficacy, and accompanied adverse effects. Development of localized therapy reduces toxicity to normal cells, prevents systemic circulation of drugs, prevents organ toxicity, reduces the dose required to be administered, provides sustained release at the target site, and enhances the efficiency of the therapy [3].

Hydrogels are crosslinked three-dimensional networks comprised of hydrophilic polymer chains. Hydrogels have been proposed to be used as drug depots for localized delivery to cancer sites providing prolonged release and reducing/avoiding off-target toxicity. Hydrogels can either be delivered orally in the form of nano/microparticles for targeting cancer in the gastrointestinal tract or can be injected locally to the tumor site. The hydrogels are generally delivered in the form of sol to the tumor site, which then converts into gel reacting to external or internal stimuli. The hydrogels can be designed according to the environment/nature of the targeted tumor site to avail of the internal responses. Apart from solving the issue of off-target toxicity, hydrogels offer advantages like reduced drug doses, increased amounts of drug delivered at the tumor site, thus overcoming the problem of low solubility in the case of hydrophobic chemotherapeutic drugs. Moreover, stimuli-sensitive hydrogels extend the opportunity to control the drug release kinetics based on the triggers and tumor position, further offering additional control over drug delivery. In this review, we give an overview of the hydrogels, their types, and different categories. Thereafter, the current research about the application of hydrogel for localized cancer therapy is discussed as summarized in Fig. 1. The primary focus is on the injectable hydrogel systems dependent on various internal (pH, temperature) and external stimuli (light, magnet, and ultrasound) as summarized in Fig. 2. In the end, a summary of the currently researched and patented systems is provided. The review gives an idea about the basic concepts behind the design of stimuli-sensitive hydrogels with examples from literature.

Advanced Drug Delivery Systems in the Management of Cancer. https://doi.org/10.1016/B978-0-323-85503-7.00011-0

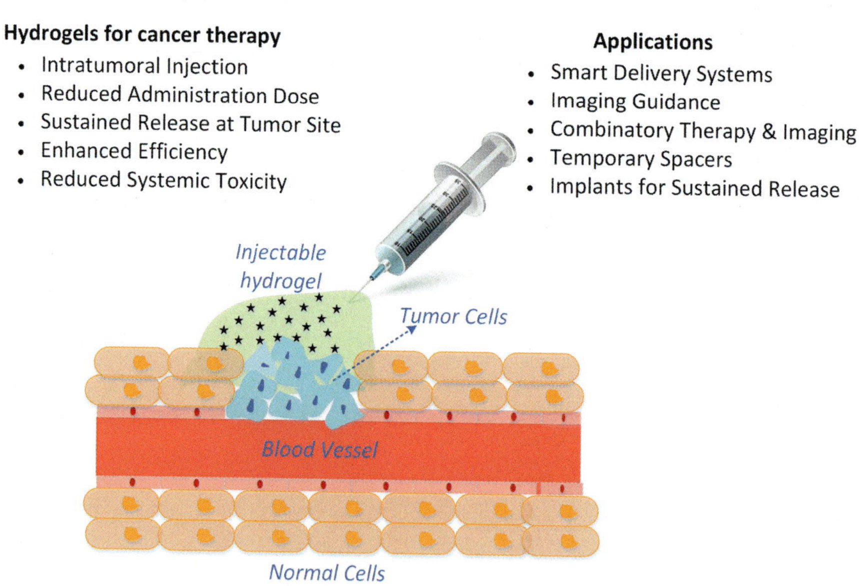

FIG. 1 Advantages and applications of injectable hydrogels in localized cancer therapy.

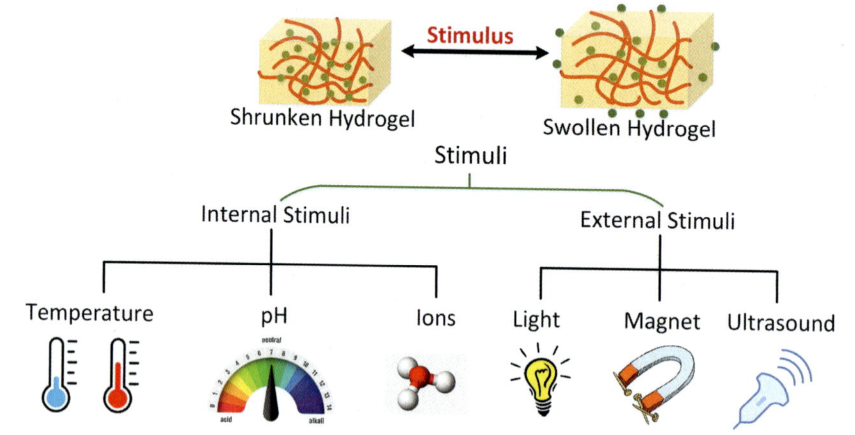

FIG. 2 Classification of hydrogels based on stimuli responsive for crosslinking as discussed in this chapter.

2 Hydrogels: An overview

Hydrogels can be described as 3D crosslinked networks composed of natural/synthetic polymers with the ability to absorb and retain water. The polymers are usually hydrophilic macromolecules consisting of hydrophilic functional groups like hydroxyl (-OH), carboxylic(-COOH), amide (-CONH-), primary amide (-$CONH_2$), and sulfonic groups (-SO_3H). These functional groups help in the formation of noncovalent and covalent bonds in inter and intrachain fashion, thereby helping in the physical and chemical crosslinking of hydrogels. Hydrogels are viscoelastic formulations comprising of both liquid-like (viscous) and solid (elastic) components, thereby often mimicking the mechanical properties of natural tissues. The polymer macromolecules can be modified by mixing with other polymers or chemical treatments to obtain a crosslinking type suitable for the desired application. The type of crosslinking, crosslinking density, synthesis methods are some of the factors that define the swelling behavior, pore density, pore size, and viscoelastic behavior of the hydrogels. Many times monomer units are used to synthesize large polymeric chains, wherein the reaction conditions, monomer concentration, initiator concentration, etc. affect the crosslinking density and the physical properties of hydrogels.

Hydrogels can be classified into several types based on factors like the source of origin (raw materials), nature of crosslinking, ionic charge, physical state, composition, physical structure, and stimuli required for activation. A brief introduction about the types of hydrogels is presented below:

(a) *Categorization based on the source of origin*
This category classifies hydrogels according to the source of origin of the raw materials/polymers [4].
- *Natural origin:* This comprises proteins and polysaccharide-based polymers derived from natural

sources like bacterial culture, plants, animals, algae, etc. Most exploited polymers are chitosan, alginate, gelatin, pectin, dextran, collagen, hyaluronic acid, fibrin, and adhesion proteins such as laminin, fibronectin. The hydrogels composed of natural polymers present properties, such as biodegradability, biocompatibility, and mimicking properties of biological extracellular matrix (ECM), thereby increasing their popularity among use in a wide variety of applications like drug delivery, wound healing, and tissue engineering.

- *Synthetic origin:* This comprises of human-made polymers which are chemically synthesized from monomers. Examples of such polymers are poly(ethyleneglycol) (PEG), polyacrylamide, polyacrylates, poloxamer, thiol-based polymers. Hydrogels formed from synthetic polymers are chemically more inert, stable, can be modified chemically easily, and biocompatible.
- *Hybrid polymers:* The polymers of natural origin are sometimes conjugated to synthetic polymers to modify the properties, thereby giving rise to hybrid hydrogels.

(b) *Categorization based on composition*

This category comprises of further classification of synthetic polymers. The categorization depends on the nature of the monomer used in the synthesis.

- *Homopolymer hydrogels:* As the name suggests, these hydrogels are formed from polymers synthesized from a single monomer type. The properties of hydrogels depend on the monomer property, polymerization initiator, and technique [5].
- *Copolymer hydrogels:* These hydrogels are produced from two types of monomers. One of the monomers essentially is responsible for absorbing water, thus responsible for forming hydrogel. The monomers are chemically crosslinked through polymerization or using initiator and physically crosslinked through chain aggregations, ionic interactions, complex formations, etc. [6].
- *Interpenetrating hydrogels*: This type of hydrogels are formed by a combination of two or more intertwined polymer networks. One of the polymers is linear in nature, and the other has a crosslinked network and is crosslinked in the immediate presence of other polymer networks [7].

(c) *Categorization based on nature of crosslinking*

- *Physical crosslinking*: The crosslinking of this type of polymer hydrogels occur by physical interactions like hydrogen bonds, crystallization, hydrophobic interactions, freeze-thawing, ionic interactions, heat-induced aggregation, etc. or environmental triggers like pH, temperature, etc. Physically crosslinked hydrogels do not require any chemical modifications or crosslinking initiators to form gels and are relatively easy to produce. However, it is difficult to tune the properties of these kinds of hydrogels like porosity, gelation time, degradation time, etc. [8].
- *Chemical crosslinking:* Chemical crosslinking usually occurs through covalent bonds among the polymer chains. The most used chemistries include the formation of Schiff base, secondary amines, sulfide bridges, hydrazine, etc. The hydrogels can be formed by adding small molecules to the prepolymer as a crosslinker or reaction of pre-functionalized polymers with reactive functional groups. Other methods of preparing hydrogels include chemical grafting, condensation, radical polymerization, enzymatic reaction, and high energy radiation [9].

(d) *Categorization based on ionic charge*

Hydrogels can be classified into three categories based on the final charge that exists on the crosslinked polymer chains:

- *Ionic hydrogels:* These hydrogels can be either cationic or anionic in nature. The anionic hydrogels contain negatively charged groups, such as carboxylic acid, sulfonic acid, whereas cationic hydrogels have positively charged groups such as amines. The presence of charged groups in hydrogels affects the swelling behavior and pH of the hydrogels [10].
- *Nonionic hydrogels:* These hydrogels do not carry any charged groups on their polymer chains or side chains [11].
- *Ampholytic hydrogels:* These hydrogels carry both positive and negative charge on the polymer chains [4].

(e) *Categorization based on physical properties*

The hydrogels can be divided into three types, such as solid, semisolid, and liquid based on their physical state, which further defines their biomedical applications.

- *Solid hydrogels:* As the name suggests, these hydrogels are in the solid state at room temperature due to their strong crosslinked network structure. The crosslinking can be either due to covalent or ionic interactions. These hydrogels can swell in the presence of buffers, water, or biological fluids. Solid hydrogels are engineered to mimic the physical, chemical, and biological properties of the tissues and thus have been widely evaluated for tissue engineering applications. Several attempts have been made to incorporate nanoparticles, inorganic metals, microparticles, metallic and ceramic nanomaterials, etc. into the hydrogel matrix to tailor the functionality and to meet the application requirements [12].
- *Liquid hydrogels:* These hydrogels are in liquid state at room temperature. Liquid hydrogels can also be classified as injectable or stimuli-sensitive hydrogels

wherein stimuli, such as pH, temperature, etc. lead to crosslinking and formation of the elastic hydrogel at the site of action/site of injection. Liquid hydrogels offer several advantages over other hydrogels like the ease of incorporation of drugs, active molecules, nanoparticles, etc., and the ease of injection without the requirement of surgery [13].

- *Semisolid hydrogels:* These hydrogels are semisolid in nature, adhesive in nature, and generally used as muco/bucoadhesive formulations. They are prepared using natural (alginate, cellulose derivatives, pectin, hyaluronic acid, other plant gums) or synthetic polymers, such as eudragits, carbopols, poloxamer, etc. The polymers used for preparation are hydrophilic macromolecules which can form noncovalent bonds (hydrogen bonds, ionic bonds, van der Waals bonds) with the exposed tissue proteins, thereby forming adhesion. The adhesion property is initiated upon wetting that occurs in contact with bodily fluids. This kind of hydrogels has been used as mucoadhesives for prolonged drug delivery to diseased sites [14].

(f) *Categorization based on stimuli required for activation* [15]

- *Physical stimulus-responsive hydrogels:* These are the hydrogels that respond to physical stimuli like temperature, light, pressure, etc. when applied to biological systems.
- *Chemical stimulus-responsive hydrogels:* These hydrogels respond to chemical stimuli like oxidants, pH, glucose, or other chemicals present at the site of action/application.
- *Biological stimulus-responsive hydrogels:* These hydrogels respond to a biochemical stimulus like enzymes, ligands, antigens, etc. and thus start crosslinking or releasing drugs, etc.

3 Hydrogel-based drug delivery systems for cancer therapy

3.1 Hydrogels for oral delivery

3.1.1 Active targeting hydrogels

The most used treatment modules for cancer therapy include chemotherapy, radiotherapy, surgery, or a combination of these. Chemotherapeutic agents work by obstructing the DNA synthesis in rapidly dividing cancer cells to cause cell death or hinder their growth. However, systemic delivery of chemotherapeutic agents has been shown to cause side effects and cellular toxicity to noncancerous cells leading to high mortality of cancer patients. The chemotherapeutic drugs are generally hydrophobic in nature and exhibit poor bioavailability leading to requirements of high dose administration. These high doses further lead to increased toxicity of healthy cells and the risk of development of drug resistance. These drawbacks associated with the systemic delivery of chemotherapeutics through the intravenous route have led researchers to work on the development of oral delivery platforms, which could target the cancer sites, especially the ones in the gastrointestinal tract. However, the delivery of anticancer drugs through the oral route poses challenges like low solubility of drugs, degradation of drugs when exposed to acidic, enzymatic environment of the stomach, low bioavailability, first-pass metabolism, and high binding to proteins leading to nonspecific biological proteins. To address these issues, several specialized biopolymers like gelatin, collagen, chitosan, PLGA, hyaluronic acid, etc. have been used to develop oral chemotherapeutics delivery platform which would prevent drug degradation and provide enhanced bioavailability to the target cancer site. Puraniket al. (2016) [16] presented one such system wherein hydrophobic therapeutics were loaded into a nanoscale polyanionic hydrogel for oral delivery. Nanoscale hydrogels were designed to overcome physicochemical challenges like low solubility and low permeability in the oral delivery of anticancer drug DOX. The polyanionic hydrogel system consists of a polymer backbone comprised of methacrylic acid (MAA) and a hydrophobic monomer chosen from *tert*-butyl methacrylate (t-BMA), *n*-butyl methacrylate (nBMA), *n*-butyl acrylate (nBA), and methyl methacrylate (MMA). The acidic monomer MAA is responsible for providing pH-responsive behavior to the hydrogels, whereas the hydrophobic part enables the loading of hydrophobic therapeutic drugs. PEG (Poly(ethylene glycol)) is then conjugated to the polymer backbone to provide colloidal stability to the nanoparticles, improve biocompatibility, provide mucoadhesion to the site and form hydrogel bonds in the stomach. The hydrogel nanoparticles exhibited good cytocompatibility, particle size ranging from 100 to 500 nm, and begin swelling at pH 4.9. Such pH-responsive systems can further be explored for oral chemotherapeutics delivery. Another system for oral delivery has been studied by Kunjiappan et al. [17] wherein a pH-sensitive hydrogel is formulated using zein-co-acrylic acid using *N,N*-methylene bisacrylamide for crosslinking, which is further loaded with chemotherapeutic drugs like rutin and 5-Fluorouracil. In this study, the -OH and -NH_2 groups present in the Zein is supposed to bind with the -COOH groups present in acrylic acid monomer and form acid-sensitive amide groups, thereby rendering pH-sensitive nature to the hydrogels. The hydrogels exhibited the highest swelling ratio in acidic environment leading to higher drug release due to larger surface area exposed due to swelling. Factors like the ratio of -NH_2 and -COOH groups into the copolymer, amount of crosslinking agent are the factors that could be modulated to change the swelling properties of the hydrogels and thus the drug release. The pH-sensitive hydrogels effectively carried the drugs and exhibited controlled release of both the drugs with enhanced cytotoxic effects against breast cancer cells and the least impact on the healthy cells.

3.2 Injectable hydrogels for local cancer therapy

Injectable hydrogels can be defined as 3D polymeric networks in liquid state, which can be injected into the body through a catheter or syringe. Injectable hydrogels have been proposed for biomedical applications such as therapeutics delivery and tissue engineering. The hydrogels, upon injection into the body, undergo physical or chemical crosslinking to form a 3D matrix at the site capable of controlled drug delivery. These hydrogels present several advantages specifically for tumor targeting and treatment, such as sustained/controlled delivery at the tumor site, thereby minimizing side effects reducing the systemic exposure to the body. Moreover, the chemotherapeutic drugs are hydrophobic in nature, thereby posing solubility issues which can be resolved with injectable hydrogels. The amount of drug required for targeted delivery to the tumor site is reduced while increasing the availability of the drug to the diseased site. The hydrogels also provide sensitive therapeutic agents protection from environmental forces, aid in tissue regeneration due to its physical and chemical properties, and can be delivered to any site because of its moldable nature. Apart from the therapeutic advantages, the injectable hydrogels also reduce the financial burden on the patient as the costs of implantation/delivery are relatively less compared with that of surgery. Moreover, as the surgical process is avoided, patient discomfort is also reduced.

Injectable hydrogels loaded with nanoparticles have been studied to provide localized delivery of one or more drugs, thereby overcoming biological barriers, reducing the dose, and avoiding systemic side effects. A similar system was presented by Bastiancich et al. (2017) [18], wherein injectable hydrogel made of Lauroyl-gemcitabine (GemC12)-loaded lipid nanocapsules is proposed for local glioblastoma therapy. The commonly used oral chemotherapy and radiotherapy after surgical removal of tissue are incapable of delaying the recurrences around the resection cavity. The intratumoral injection of the proposed hydrogel formulation was found to significantly increase the survival of the treatment group in an orthotopic human xenograft glioblastoma model. The presurgical administration of lipid nanocapsules hydrogel into the resection cavity also was shown to slow down the recurrences. Another similar approach is presented by Yang et al. [19], wherein two drugs, combretastatin-A4 phosphate (CA4P) and doxorubicin (DOX), are proposed to be delivered by means of injectable hydrogel to obtain synergistic antitumor therapy. The formulation comprises pH and redox-responsive nanogels of DOX and hydrogel of CA4P covalently bonded together to form injectable hydrogel. The in vitro release profiles suggested that DOX was released for 14 days and CA4P was released for 80 h, indicating the sequential controlled release of these drugs aimed at synergistic antiangiogenesis and cytotoxic activity. In vivo study in xenograft, tumor-bearing mice showed improved antitumor activity owing to the cell apoptosis and vascular collapse caused due to dual drug delivery systems.

As discussed above, injectable hydrogels have been studied for localized codelivery of chemotherapeutics. Researchers also have been focusing on the potential of hydrogels for delivery of multiple immunotherapeutics and combinations of immunotherapeutics and chemotherapeutics. One such approach is presented by Chen et al. [20] wherein localized injection of antiprogrammed cell death protein ligand 1 (αPDL1)-loaded hydrogel is proposed to reprogram the tumor microenvironment from pro-tumoral to antitumoral along with sustained delivery of αPDL1 to boost the immune system. The nanofiber hydrogel is based on betamethasone phosphate, an antiinflammatory steroid drug with noncovalent interactions like hydrogen bonds, hydrophobic interactions, coordination bonds, etc. as the crosslinking mechanism of the hydrogel. The inflammation present in the tumor microenvironment has been reported as one of the reasons promoting tumor development, the reduction of which has been proposed to improve the antitumor immune response and immune checkpoint blockade therapy. Taking this into consideration, an antiinflammatory steroid is used to form the hydrogel, and it has been found to reprogram the inflammation-related immunosuppressive tumor microenvironment due to antiinflammatory activity. The hydrogel also acts as a reservoir for localized sustained delivery of αPDL1, thereby reducing the systemic side effects. The combined effect of both these factors has led to inhibition of the growth of primary and distant tumors in studied CT26 tumor model. Similar systems targeting immunotherapy have been reported for cancer immunotherapy [21], obstructive nephropathy [22], immunochemotherapy [23], chronic kidney disease [24], etc. Several other types of injectable hydrogel systems which depend on stimuli like temperature, pH, laser, magnet, etc. will be discussed in detail in the next sections.

3.2.1 Local stimuli-sensitive hydrogels

- *pH-sensitive hydrogels:* As the name suggests, these hydrogels crosslink upon the change in the pH. Some of the examples of charged polymers forming pH-sensitive hydrogels are acrylates, Guar gum succinate, chitosan, dextran, xanthan, and alginates, etc. The change in pH leads to change in the charge present on the polymer chain, which is then responsible for phenomena like crosslinking, swelling, collapsing, and drug release. pH-sensitive behavior of hydrogels can be controlled by modulating factors like charge on the polymers, polymer concentration, dissociation constant, degree of ionization, and nature of pendant groups. The pendant groups get charged at a specific pH and develop repulsion among themselves leading to swelling of the hydrogel. Thus, pH-sensitive hydrogels are formulated considering the pH of the delivery site [25]. pH-sensitive hydrogels have also been evaluated in combination with nanoparticles for localized antitumor therapy. Injectable hydrogels loaded with nanoparticles have been reported to exhibit improved sustained drug release and therapeutic efficacy along with localized therapy. One such system has been reported by Liu et al. [26], wherein antineoplastic DOX is loaded into the micelles made from conjugated glycol chitosan-Pluronic F127 finally converted into injectable supramolecular hydrogel upon interaction with α-cyclodextrin. It was found that pH-controlled release of DOX led to enhanced tumor tissue targeting, cell uptake, therapeutic efficacy while also reducing systemic toxicity at the same administered dose of intravenous injection. Positively charged glycol chitosan was used here as it remains neutral in the pH range of 6–7, which is the normal pH range of the tumor microenvironment. The macrophages are known to easily phagocytize the nanoparticles with higher surface charge and large particle size. The neutral nature of the micelles in tumor microenvironment due to glycol chitosan prevents the rapid clearance of these micelles by macrophages. Thus, pH-sensitive nature here serves in the dual purpose of targeted drug delivery, increased efficacy, and reduced clearance by macrophages.

Polymers derived from biological sources like alginate, silk, fibrin, gelatin, etc. have gained interest in formulating hydrogels owing to their properties like stability, biocompatibility, and biodegradability. However, several limitations like minimal control over physical properties, reproducibility, degradation, etc. limit the translation of such systems into clinical applications. Silk is one such natural polymer which provides control to tune mechanical properties and ability to form drug delivery systems like hydrogels, films, composites, etc. Wu et al. [27] presented a silk-based injectable hydrogel with thixotropic properties to aid the injectability, loaded with DOX for local pH-sensitive delivery to the breast cancer site. The hydrogels were formed by self-assembly of silk nanofibers into neutral solution and exhibited thixotropic properties dependent on hydrophobic interactions and charge repulsions, i.e., behave as liquid upon application of stress and turn into solid after the removal of stress. It was reported that drug release upon local injection of these systems was dependent on pH, and maximum release was found at lower pH 4.5, thus presenting the required lysosomotropic delivery system. The sustained release of DOX from silk hydrogel was found to control the cytotoxicity of cancer cells during in vitro as well as in vivo studies. These kinds of systems can be further explored for localized therapy of cancer.

- *Temperature-sensitive hydrogels:* These hydrogels exhibit transitions from sol to gel or gel to sol based on the changes in temperature. Thermosensitive polymers exhibit both hydrophilic and hydrophobic portions interacting with each other, and the balanced interactions within these groups determine the response to temperature change. The interactions of these groups within the polymer structure with water molecules change upon variation in the temperature, leading to changes in the solubility of polymer. This change in solubility of the crosslinked polymer network leads to the formation of sol phase (flowing fluid) and gel phase (nonflowing fluid). The hydrogels exhibit sol-gel transition at lower critical solution temperature (LCST) or upper critical solution temperature (UCST). For gels exhibiting transition at LCST, the hydrogels remain as sol below LCST and convert to gel as the temperature is raised above LCST. Other types of hydrogels are cooling gels, wherein the sol phase exists above UCST, and as the cooling occurs below UCST, the gel phase is formed. Change in the ratio of hydrophilic and hydrophobic moieties in the polymer structure leads to a change in inherent UCST/LCST. Thermosensitive hydrogels exhibit several advantages like rapid sol-gel transition, avoidance of burst release due to gel, no requirement of crosslinking agent, controlled drug release, and tunable shape due to flowable nature. Poly(ε-caprolactone-*co*-lactide)-b-poly(ethylene gly*col*)-b-poly(ε-caprolactone-co-lactide) (PCLA-PEG-PCLA) is one such promising candidate for injectable hydrogels which show thermoresponsive sol-gel transitions.

The balance between the hydrophilic PEG moieties and hydrophobic PCLA moieties in the polymer structure determines the temperature-dependent behavior of the hydrogel. One such study is presented by Andrgie et al. [28], wherein PCLA-PEG-PCLA and heparin combination are used to formulate noncoagulant prodrug of heparin and evaluation of its antimetastatic activity. The thermosensitive hydrogel made up of triblock polymer was loaded with desulfated heparin and conjugated to the polymer via esterification reaction. The ratio of polymer conjugated heparin in the triblock polymer hydrogel has to be balanced to balance the hydrophilic nature of heparin with that of PEG and PCLA. The hydrogels were found to be cytocompatible to HeLa and HaCat cells, along with the reduction of tumor metastasis in vivo. Another study performed by Tsai et al. [29] wherein drug-loaded liposomes were incorporated into amphipathic thermosensitive hydrogels made up of Pluronic F127. In case of low temperature, polymer molecules form hydrogen bond interactions with water molecules, thus making the polymer soluble. Contradictory to this, in case of higher temperature, hydrophobic molecules tend to form micelles leading to the formation of gel. Immunotherapeutic anticancer drug imiquimod loaded into liposomes was mixed into temperature-sensitive Pluronic hydrogel and was evaluated for tumor suppressing activity upon in situ injection into breast cancer tissues. It was found that the drug-loaded hydrogel exhibited greater tumor suppression rate as observed by tumor size and animal survival rate compared with the intravenous administration of drug. The hydrogel also did not show significant toxicity, thus suggesting that topical hydrogel application can be used for localized breast cancer therapy.

3.2.2 External stimuli-responsive hydrogels

- *Light/Photoresponsive hydrogel:* As the name suggests, photosensitive hydrogels exhibit sol-to-gel transition upon exposure to light. Photodynamic therapy is widely studied for applications to cancer therapy because of the easy spatial and temporal control of the crosslinking and gelation, which further aids in targeted therapy. The crosslinking, gelation kinetics, drug release in these hydrogels can be controlled by tuning properties like light wavelength, light dose, exposure time, light intensity/power density, etc. The photosensitive hydrogels comprise a polymer network and a photosensitizer/photoreactive moiety as the light-responsive part. Upon irradiation with light, the photosensitive molecules in the hydrogel capture the light signal and convert the photo-irradiation to a chemical signal through a reaction like polymerization, dimerization, etc. leading to crosslinking of the hydrogel. Some other photoreactive agents absorb light and convert light energy to heat energy at the site of the tumor leading to the ablation of cancer cells [30]. Phototherapy has been evaluated in combination with other approaches like chemotherapy, radiotherapy, gene therapy, etc. Despite several efforts by researchers, photodynamic therapy has been limited due to the limitations in the light penetration depth. To solve this issue, the excitation wavelength has been shifted to the near-infrared range (NIR), which exhibits a penetration depth of several centimeters [31]. Phototherapy, in combination with other therapeutic approaches, can be used for the treatment of oral cancers, skin cancers, colon cancers, bladder cancers, and esophageal cancers. GhavamiNejad et al. (2016) [32] reported the formulation and evaluation of dual stimuli-responsive smart hydrogels loaded with DOX and bortezomib (BTZ) to utilize the different anticancer mechanisms of both drugs. The hydrogels were pH and NIR-responsive, thereby exhibiting pH-dependent drug release and NIR-dependent heat generation at the cancer site to achieve combined photo/chemotherapy. Mussel-inspired dopamine nanoparticles with catechol functional groups have been incorporated into hydrogel with dual purpose. Firstly, dopamine nanoparticles exhibit NIR-responsive photothermal characteristics, thus rendering the hydrogel photosensitive. Secondly, the catechol groups in dopamine can bind to the boron-containing antitumor drug BTZ and release the drug in a pH-controlled manner. It was found that when the hydrogel was exposed to NIR light, the dopamine nanoparticles absorbed light and generated heat, which was locally transmitted to the cancer cells, thereby causing hyperthermia. Simultaneously, the rise in the local temperature above LCST of the polymer network leads to deswelling, leading to the release of DOX. Thus, the system presented a combination of photothermal and chemotherapy, targeting localized cancer cell therapy through hyperthermia and on-demand multidrug release.

  Qiu et al. [33] presented a similar concept of controlling the drug release through external NIR light stimuli for treating subcutaneous breast cancer and melanoma without causing systemic side effects. Upon injection, the hydrogel undergoes phase transition as body temperature is lower than LCST and forms gel. Black phosphorus (BP) is used herein as the photosensitive agent to convert the exposed light to heat. The hydrogel made of agarose is loaded with BP nanostructures and DOX. Upon irradiation with NIR light, the BP nanostructures absorb light and increase the temperature of the hydrogel matrix. This increase in temperature leads to reversible hydrolysis and softening of the agarose gel triggering the release of DOX to the surrounding tissue environment. The drug release is tunable based on the concentration of BP nanostructures, agarose, light intensity, and light dose. Thus a light and temperature-sensitive hydrogel system with low toxicity and biodegradable nature with tunable release properties for localized treatment of breast cancers and melanoma is presented. Similar light-sensitive systems in combination with chemotherapy, thermosensitive, pH-sensitive, and glucose-responsive systems have been evaluated for localized therapy [34–37].
- *Magnetosensitive hydrogels:* These hydrogels consist of magnetic particles incorporated in the hydrogel network and responsible for causing hyperthermia upon exposure to the magnetic field. The magnetic particles, upon exposure to the field convert magnetic energy to heat due to several mechanisms like hysteresis loss, eddy current loss, and relaxation loss comprising Brownian and Neel relaxation. The conversion to heat energy can be modulated by controlling the magnetic field strength, frequency, properties of magnetic particles, concentration of these particles, particle size and distribution, and surrounding environment [38]. The magnetic particles undergo hysteresis loss leading to increased temperature crossing over the LCST of the hydrogel network leading to crosslinking and thus controlled release of drugs from the crosslinked matrix. There are various methods for incorporation of magnetic particles into hydrogels like (a) blending wherein particles are mixed into the hydrogel-forming polymer, blended and

then crosslinked to form hydrogel, (b) in situ synthesis, wherein magnetic particles are precipitated in the hydrogel upon crosslinking, and (c) grafting wherein magnetic particles are synthesized first and then grafted with functional groups to be further used for crosslinking purpose. Magnetic hyperthermia has been evaluated for use in real-time image guidance while delivering the hydrogel to deep cancerous tissues without the use of an invasive catheter. Tay et al. [39] presented the evaluation of theranostic platform based on magnetic hyperthermia combined with magnetic particle imaging (MPI) using superparamagnetic iron oxide nanoparticle (SPIONs). The SPIONs used for generating magnetic hyperthermia has been used here for magnetic particle imaging-guided spatial localization of hyperthermia to localized cancer region. This approach has advantages like reduced systemic side effects due to localization, accumulation in nonaffected organs, and damage due to heat to other areas. The experimental results indicated selective on-demand heating of the target sites. It was observed that even targets separated by short distance as 7mm exhibited no heating as desired. In vivo rodent model study exhibited localized heat therapy and sequential targeting ability using MPI guidance. The localized heat damage to tumor tissues was further confirmed by luciferase activity and histological assessment, suggesting no damage to healthy tissues. Further studies can be conducted to evaluate the drug release through MPI-guided magnetic hyperthermia and development of temperature feedback mechanism. Another SPIONs-based hydrogel system generating hyperthermia upon exposure to alternating magnetic field is evaluated by Lee et al. [40]. Poly (ethylene glycol) diacrylate (PEGDA)-based hydrogel system was used in combination with SPIONs magnetic particles. Upon application of external alternating magnetic field, the hydrogel undergoes crosslinking and gelation due to heat generation. The system is proposed to solve the issue with limited penetration of NIR light into deep tissues by using magnetic field with deeper penetration characteristics. In vitro gelation studies exhibited that the pregels undergo in situ gelation upon magnetic field application under thick tissues (2mm–20cm), which is higher than that of NIR wherein gelation occurs only up to 1–3mm of thickness. In vivo experiments in rats indicated the in situ formation of hydrogel upon magnetic field exposure in a controllable manner. The systems provide advantages of tunable gelation properties even in deep tissues, which can be utilized for localized therapy. Similar systems have also been presented in combination with phototherapy [41], chemotherapy [42], and nanoparticles for drug delivery to cancer sites [43].

- *Ultrasound-sensitive hydrogels*: High intensity focused ultrasound (HIFU) is one of the potential external stimuli actively being evaluated for localized therapy of tumors. It offers several advantages like cost-effectiveness, ease of access, noninvasive technique, and ease of combining therapeutic and imaging ability. Owing to these advantages, ultrasound-sensitive hydrogels have gained interest for their applicability to cancer therapy allowing spatial and temporal control over tissue penetration. Ultrasound uses high frequency sound waves leading to generation of mechanical/thermal effects which further lead to phenomena like cancerous cell death, stimulate drug delivery, enhance the permeability of endothelium to allow for drug entry, nucleate cavitation leading to mechanical effects like shear, shockwaves, which would lead to enhanced drug uptake at the target site. Several agents like microbubbles, nanoparticles, and nanoemulsions have been proposed to be used as ultrasound synergistic agents to improve the contrast of imaging, localize cavitation in the tumor site, and enhance the energy deposition at the tumor site, thereby leading to improved safety and efficacy of the ultrasound-based therapy. Ordeig et al. [44] presented one such implantable hydrogel system with the ability to release drug on demand upon ultrasound activation. Herein focused ultrasound is employed to initiate release from implantable thermally actuated gel with real-time dose tuning. The system also comprises of an imaging and actuation instrument which can be used in combination with fast ultrasound to image for imaging and temperature feedback control. The formulation comprises of thermally responsive hydrogel made up of N-isopropylacrylamide (NiPAAm)-co-acrylamide (AAm), which is imaged using real-time and placed onto the target tumor site noninvasively. Upon application of fast ultrasound (FUS), energy is imparted to the thermally responsive hydrogel leading to an increase in the temperature above the UCST of gels, ultimately leading to contraction of gel. It was found that FUS application leads to increase of the local temperature to 45°C in about 33s. The contraction of gel was found to initiate drug release from the capsule without causing structural damage to the capsule and no development of inflammation, necrosis or fibrous capsule around the implanted site in mice. Baghbani et al. [45] presented a similar system wherein ultrasound-sensitive doxorubicin-loaded alginate-stabilized perfluorohexane (PFH) nanodroplets were evaluated for their imaging and therapeutic properties in vitro and in vivo. It was reported that the nanodroplets exhibited pH-sensitive release of DOX, and higher release was observed in acidic environments similar to the tumor microenvironment. Upon application of ultrasound, the PFH nanodroplet undergoes acoustic droplet evaporation leading to the formation of nano/microbubbles triggering DOX release from nanodroplet. The reported acoustic droplet vaporization and simultaneous formation of microbubble are advantageous for modulating the drug delivery and improved contrast for imaging. Cellular uptake studies also revealed higher uptake upon FUS application, implying increased membrane permeability facilitating the drug uptake into the cells. Biodistribution study revealed that DOX remains tightly bound to the nanodroplets during circulation in blood at 7.4 pH, and significantly less release takes place. The study also showed significantly higher accumulation of DOX in tumor tissues and low presence in other organs, thereby reducing the toxicity. Thus encapsulation of DOX into alginate PFH nanodroplets enables ultrasound tunable drug release specifically at the tumor site with reduced systemic toxicity.

4 Marketed/patented hydrogel-based drug delivery systems

Despite a large amount of research in the field of stimuli-sensitive hydrogel for localized cancer therapy, very few of them have translated to the market owing to complex nature of delivery system, the difficulty of large-scale production, inability to pass clinical trials, and associated side effects. Regel, one such stimuli-sensitive injectable hydrogel, was developed by Macromed, Inc. The ReGel system comprises triblock copolymers PLGA-PEG-PLGA, which form gel when present in concentrations of 5%–30% by weight. The formulation stays in liquid form below room temperature, thus it can be injected either subcutaneously or intramuscularly. Upon injection to the body, the liquid converts to gel form due to the body temperature and act as a drug depot releasing drug slowly for weeks/months until the matrix degrades. Hydrophobic drugs are more suitable to be delivered using thermosensitive ReGel as they tend to remain attached to the polymer system as opposed to leaching out in case of hydrophilic drugs [46]. Anticancer drug paclitaxel was loaded onto the ReGel hydrogel system to develop an injectable intratumoral hydrogel OncoGel. It was found that OncoGel exhibited sustained release of drug paclitaxel up to 4–6 weeks upon intratumoral injection and has been clinically evaluated for primary brain and esophageal cancers [47]. SpaceOAR hydrogel is another polyethylene glycol-based injectable hydrogel system that is proposed to be injected between prostate and rectum for patients undergoing radiation therapy for prostate cancer. The hydrogel has been shown to reduce the radiation dose reaching to rectum and thus reduce rectal toxicity. However, the system is still under clinical trial owing to some adverse effects reported near the hydrogel injection site [48]. Atrigel delivery system is one such injectable system comprising of biodegradable polymers blended with pharmaceuticals in liquid form. This liquid is injected at the site of action wherein due to water present, it converts into a solid drug depot releasing the drug in a sustained manner. Eligard is one such extended-release formulation comprising of Atrigel and leuprolide acetate (analogue of luteinizing hormone-releasing hormone (LHRH)) prescribed for the treatment of prostate cancer [49]. Lupron Depot is another injectable PLGA microsphere system for delivery of LHRH for treatment of prostate cancer [50]. Apart from these, several patents regarding injectable hydrogels for localized cancer therapy have been filed. Some of them are listed below to give an idea about the approaches used for the development of such hydrogels:

- US9364545B2, Thermosensitive injectable hydrogel for drug delivery: The invention describes an in situ forming injectable hydrogel for the delivery of anticancer drugs. The thermosensitive injectable gel, which forms gel in the temperature range of 30°C–35°C is based on hyaluronic acid and copolymer of polyethylene oxide and polypropylene oxide [51].
- US20170209606A1, Hydrogels for localized radiotherapy: The invention describes the production and use of radioactive hydrogels to aid localized radiotherapy to the tumor site, thereby minimizing the toxicity to normal tissues. Microparticles loaded with radioisotope conjugated molecules are dispersed in the hydrogel. The hydrogel is expected to prevent the leaking radioisotopes, and thereby minimize the damage to healthy cells [52].
- US8003125B2, Injectable drug delivery systems with cyclodextrin-polymer-based hydrogels: The invention describes the formulation of cyclodextrin-based injectable hydrogel comprising of a secondary polymer which further conjugates to the loaded drug. The hydrogel is thermosensitive and thixotropic in nature, which can be further modulated by PEG or copolymers of PEG [53].
- US20060073281A1, Injectable hydrogel microspheres from aqueous two-phase system: The invention describes the development of injectable hydrogel microspheres, which can be loaded with protein drugs and used to deliver them in stable form. Emulsification technique is used using hydrogel precursors poly(ethylene glycol) diacrylate and N-isopropylacrylamide, with dextran solution as the aqueous phase [54].
- US20130266508A1, Thermosensitive hydrogel for coating radioisotope and chemotherapeutic agent to treat cancer and method for preparing the same: A thermosensitive hydrogel made of PCL-PEG-PCL which will be utilized to coat radioisotopes and anticancer drugs, thereby rendering them with temperature-sensitive behavior exhibited at body temperature. The formulation is aimed to concentrate radiation therapy and chemotherapeutics at the tumor site, thereby reducing adverse effects on the healthy cells [55].
- GR1009114, Pentablock polypeptidic and hybrid polymers to give in situ-forming injectable, self-healing, pH-and temperature-responsive hydrogels for localized, targeted delivery: pH and thermosensitive hydrogels are formed using pentagonal tripolypeptides, further loaded

with gemcitabine. The formulation is for the treatment of pancreatic cancers in a localized manner as the hydrogel only releases the drug at the tumor site owing to pH and temperature sensitivity [56].

- WO2019232114, Injectable thermoresponsive hydrogels as a combinatory modality for controlled drug delivery, biomaterial implant, and 3d printing bioink: Thermoresponsive hydrogels are formed using thermogelling polymer chitosan in combination with cellulose nanofibers/nanocrystals, further loaded with the drug for controlled delivery [57].
- EP2958933 B1, Crosslinked peptide hydrogels: Hydrogels comprising of amphiphilic peptides which have the ability to self-assemble into the form of nanofiber network capable of entrapping water molecules, which can be used an injectable in situ gel-forming agents for localized therapy [58].
- US20120100103A1, In situ forming hydrogel and biomedical use thereof: The invention describes the formulation of an in situ forming injectable hydrogel comprising of homogenous or heterogeneous polymers which are crosslinked with each other through the reaction of phenol and aniline molecules conjugated to their chains [59].

5 Conclusions and future perspectives

Hydrogels act as an excellent platform for localized therapy owing to its stimulus-dependent sol-to-gel conversion property. Stimulus-dependent crosslinking of hydrogels offers the opportunity to modulate the polymer network to achieve sensitivity based on the tumor microenvironment and the required drug release behavior. Here we discussed the basic working mechanisms of several types of stimuli-sensitive hydrogels like pH, light, magnet, ultrasound, temperature, ions, and combination of two stimuli in many cases. The injectable hydrogels have been used for various applications related to localized tumor therapy like imaging of tumor, drug delivery to the site, guided delivery using image-guiding property of hydrogel, and as temporary fillers/spacers to concentrate the effects of the therapy. These applications of hydrogels have exhibited advantages over chemotherapy, including targeted drug release controlled in spatially and temporally, reduced systemic circulation, reduced organ toxicity, reduced dose of administration, enhanced efficacy of treatment, increased cell apoptosis, inhibition of tumor growth, and enhanced dose at tumor site among few. Several clinical and experimental studies have highlighted the potential of injectable hydrogels for localized tumor therapy. As discussed, few of such hydrogels are available in the market like Eligard, Lupron Depot, OncoGel, ReGel, SpaceOAR, TraceIT, etc. prescribed tumor therapy. Hydrogels can also be evaluated for cancer therapy in combination with radiotherapy, immunotherapy, chemotherapy, gene therapy, etc. to further improve the efficiency of the treatment. Apart from imaging and drug delivery, hydrogels also extend the opportunity to understand the tumor microenvironment further, understand the drug uptake mechanism, cancer progression behavior, and changes in extracellular matrix, which will further help in improving the design of therapies against cancer. Despite the great potential of hydrogels, limitations like scaled-up production, sterilization of products, cost-effectiveness, patient compatibility, and adverse effects associated with hydrogels need to be addressed. Thus, the potential of hydrogels in localized treatment can further be explored in combination with other therapies and improved understanding of the cancer microenvironment and behavior.

References

[1] Bray F, Ferlay J, Soerjomataram I, Siegel RL, Torre LA, Jemal A. Global cancer statistics 2018: GLOBOCAN estimates of incidence and mortality worldwide for 36 cancers in 185 countries. CA Cancer J Clin 2018;68(6):394–424.

[2] Zhou J, Yu G, Huang F. Supramolecular chemotherapy based on host–guest molecular recognition: a novel strategy in the battle against cancer with a bright future. Chem Soc Rev 2017;46(22):7021–53.

[3] Andreyev J, Ross P, Donnellan C, Lennan E, Leonard P, Waters C, Wedlake L, Bridgewater J, Glynne-Jones R, Allum W, Chau I, Wilson R, Ferry D. Guidance on the management of diarrhoea during cancer chemotherapy. Lancet Oncol 2014;15(10):e447–60.

[4] Khansari MM, Sorokina LV, Mukherjee P, Mukhtar F, Shirdar MR, Shahidi M, Shokuhfar T. Classification of hydrogels based on their source: a review and application in stem cell regulation. JOM 2017;69(8):1340–7.

[5] Singhal R, Gupta K. A review: tailor-made hydrogel structures (classifications and synthesis parameters). Polym-Plast Technol Eng 2016;55(1):54–70.

[6] Singh NK, Lee DS. In situ gelling pH- and temperature-sensitive biodegradable block copolymer hydrogels for drug delivery. J Control Release 2014;193:214–27.

[7] Lipatov YS. Polymer blends and interpenetrating polymer networks at the interface with solids. Prog Polym Sci 2002;27(9):1721–801.

[8] Ullah F, Othman MBH, Javed F, Ahmad Z, Akil HM. Classification, processing and application of hydrogels: a review. Mater Sci Eng C 2015;57:414–33.

[9] Hennink WE, van Nostrum CF. Novel crosslinking methods to design hydrogels. Adv Drug Deliv Rev 2002;54(1):13–36.

[10] Ozmen MM, Okay O. Superfast responsive ionic hydrogels with controllable pore size. Polymer 2005;46(19):8119–27.

[11] Chujo Y, Sada K, Matsumoto K, Saegusa T. Synthesis of nonionic hydrogel, lipogel, and amphigel by copolymerization of 2-oxazolines and a bisoxazoline. Macromolecules 1990;23(5):1234–7.

[12] Varaprasad K, Raghavendra GM, Jayaramudu T, Yallapu MM, Sadiku R. A mini review on hydrogels classification and recent developments in miscellaneous applications. Mater Sci Eng C 2017;79:958–71.

[13] Gao Z, Kong L, Jin R, Liu X, Hu W, Gao G. Mechanical, adhesive and self-healing ionic liquid hydrogels for electrolytes and flexible strain sensors. J Mater Chem C 2020;8(32):11119–27.

[14] Abrego G, Alvarado H, Souto EB, Guevara B, Bellowa LH, Garduño ML, Garcia ML, Calpena AC. Biopharmaceutical profile of hydrogels containing pranoprofen-loaded PLGA nanoparticles for skin administration: in vitro, ex vivo and in vivo characterization. Int J Pharm 2016;501(1):350–61.

[15] Sood N, Bhardwaj A, Mehta S, Mehta A. Stimuli-responsive hydrogels in drug delivery and tissue engineering. Drug Deliv 2016;23(3):748–70.

[16] Puranik AS, Pao LP, White VM, Peppas NA. Synthesis and characterization of pH-responsive nanoscale hydrogels for oral delivery of hydrophobic therapeutics. Eur J Pharm Biopharm 2016;108:196–213.

[17] Kunjiappan S, Theivendran P, Baskararaj S, Sankaranarayanan B, Palanisamy P, Saravanan G, Arunachalam S, Sankaranarayanan M, Natarajan J, Somasundaram B, Wadhwani A. Modeling a pH-sensitive Zein-co-acrylic acid hybrid hydrogels loaded 5-fluorouracil and rutin for enhanced anticancer efficacy by oral delivery. 3 Biotech 2019;9(5):185.

[18] Bastiancich C, Bianco J, Vanvarenberg K, Ucakar B, Joudiou N, Gallez B, Bastiat G, Lagarce F, Préat V, Danhier F. Injectable nanomedicine hydrogel for local chemotherapy of glioblastoma after surgical resection. J Control Release 2017;264:45–54.

[19] Yang WJ, Zhou P, Liang L, Cao Y, Qiao J, Li X, Teng Z, Wang L. Nanogel-incorporated injectable hydrogel for synergistic therapy based on sequential local delivery of combretastatin-A4 phosphate (CA4P) and doxorubicin (DOX). ACS Appl Mater Interfaces 2018;10(22):18560–73.

[20] Chen M, Tan Y, Dong Z, Lu J, Han X, Jin Q, Zhu W, Shen J, Cheng L, Liu Z, Chen Q. Injectable anti-inflammatory Nanofiber hydrogel to achieve systemic immunotherapy post local administration. Nano Lett 2020.

[21] Chao Y, Chen Q, Liu Z. Smart injectable hydrogels for cancer immunotherapy. Adv Funct Mater 2020;30(2):1902785.

[22] Soranno DE, Lu HD, Weber HM, Rai R, Burdick JA. Immunotherapy with injectable hydrogels to treat obstructive nephropathy. J Biomed Mater Res A 2014;102(7):2173–80.

[23] Jiang L, Ding Y, Xue X, Zhou S, Li C, Zhang X, Jiang X. Entrapping multifunctional dendritic nanoparticles into a hydrogel for local therapeutic delivery and synergetic immunochemotherapy. Nano Res 2018;11(11):6062–73.

[24] Rodell CB, Rai R, Faubel S, Burdick JA, Soranno DE. Local immunotherapy via delivery of interleukin-10 and transforming growth factor β antagonist for treatment of chronic kidney disease. J Control Release 2015;206:131–9.

[25] Rizwan M, Yahya R, Hassan A, Yar M, Azzahari AD, Selvanathan V, Sonsudin F, Abouloula CN. pH sensitive hydrogels in drug delivery: brief history, properties, swelling, and release mechanism, material selection and applications. Polymers (Basel) 2017;9(4):137.

[26] Liu Z, Xu G, Wang C, Li C, Yao P. Shear-responsive injectable supramolecular hydrogel releasing doxorubicin loaded micelles with pH-sensitivity for local tumor chemotherapy. Int J Pharm 2017;530(1):53–62.

[27] Wu H, Liu S, Xiao L, Dong X, Lu Q, Kaplan DL. Injectable and pH-responsive silk nanofiber hydrogels for sustained anticancer drug delivery. ACS Appl Mater Interfaces 2016;8(27):17118–26.

[28] Andrgie AT, Mekuria SL, Addisu KD, Hailemeskel BZ, Hsu W-H, Tsai H-C, Lai J-Y. Non-anticoagulant heparin prodrug loaded biodegradable and injectable thermoresponsive hydrogels for enhanced anti-metastasis therapy. Macromol Biosci 2019;19(5):1800409.

[29] Tsai H-C, Chou H-Y, Chuang S-H, Lai J-Y, Chen Y-S, Wen Y-H, Yu L-Y, Lo C-L. Preparation of immunotherapy liposomal-loaded thermal-responsive hydrogel carrier in the local treatment of breast cancer. Polymers (Basel) 2019;11(10):1592.

[30] Tomatsu I, Peng K, Kros A. Photoresponsive hydrogels for biomedical applications. Adv Drug Deliv Rev 2011;63(14):1257–66.

[31] Ai F, Ju Q, Zhang X, Chen X, Wang F, Zhu G. A core-shell-shell nanoplatform upconverting near-infrared light at 808 nm for luminescence imaging and photodynamic therapy of cancer. Sci Rep 2015;5:10785.

[32] GhavamiNejad A, SamariKhalaj M, Aguilar LE, Park CH, Kim CS. pH/NIR light-controlled multidrug release via a mussel-inspired nanocomposite hydrogel for chemo-photothermal cancer therapy. Sci Rep 2016;6(1):33594.

[33] Qiu M, Wang D, Liang W, Liu L, Zhang Y, Chen X, Sang DK, Xing C, Li Z, Dong B, Xing F, Fan D, Bao S, Zhang H, Cao Y. Novel concept of the smart NIR-light–controlled drug release of black phosphorus nanostructure for cancer therapy. Proc Natl Acad Sci 2018;115(3):501.

[34] Lima-Sousa R, de Melo-Diogo D, Alves CG, Cabral CSD, Miguel SP, Mendonça AG, Correia IJ. Injectable in situ forming thermo-responsive graphene based hydrogels for cancer chemo-photothermal therapy and NIR light-enhanced antibacterial applications. Mater Sci Eng C 2020;117:111294.

[35] Xu X, Huang Z, Huang Z, Zhang X, He S, Sun X, Shen Y, Yan M, Zhao C. Injectable, NIR/pH-responsive nanocomposite hydrogel as long-acting implant for chemophotothermal synergistic cancer therapy. ACS Appl Mater Interfaces 2017;9(24):20361–75.

[36] Hao Y, Dong Z, Chen M, Chao Y, Liu Z, Feng L, Hao Y, Dong ZL, Chen MC, Chao Y, Liu Z, Feng LZ. Near-infrared light and glucose dual-responsive cascading hydroxyl radical generation for in situ gelation and effective breast cancer treatment. Biomaterials 2020;228:119568.

[37] Huang Z, Xiao H, Lu X, Yan W, Ji Z. Enhanced photo/chemo combination efficiency against bladder tumor by encapsulation of DOX and ZnPC into in situ-formed thermosensitive polymer hydrogel. Int J Nanomedicine 2018;13:7623–31.

[38] Ma M, Wu Y, Zhou J, Sun Y, Zhang Y, Gu N. Size dependence of specific power absorption of Fe3O4 particles in AC magnetic field. J Magn Magn Mater 2004;268(1–2):33–9.

[39] Tay ZW, Chandrasekharan P, Chiu-Lam A, Hensley DW, Dhavalikar R, Zhou XY, Yu EY, Goodwill PW, Zheng B, Rinaldi C, Conolly SM. Magnetic particle imaging-guided heating in vivo using gradient fields for arbitrary localization of magnetic hyperthermia therapy. ACS Nano 2018;12(4):3699–713.

[40] Lee H, Thirunavukkarasu GK, Kim S, Lee JY. Remote induction of in situ hydrogelation in a deep tissue, using an alternating magnetic field and superparamagnetic nanoparticles. Nano Res 2018;11(11):5997–6009.

[41] Wang Y, Wang L, Yan M, Dong S, Hao J. Near-infrared-light-responsive magnetic DNA microgels for photon- and magneto-manipulated cancer therapy. ACS Appl Mater Interfaces 2017;9(34):28185–94.

[42] Chen B, Xing J, Li M, Liu Y, Ji M. DOX@Ferumoxytol-medical chitosan as magnetic hydrogel therapeutic system for effective magnetic hyperthermia and chemotherapy in vitro. Colloids Surf B: Biointerfaces 2020;190:110896.

[43] Liu Y, Yang F, Yuan C, Li M, Wang T, Chen B, Jin J, Zhao P, Tong J, Luo S, Gu N. Magnetic nanoliposomes as in situ microbubble bombers for multimodality image-guided Cancer theranostics. ACS Nano 2017;11(2):1509–19.

[44] Ordeig O, Chin SY, Kim S, Chitnis PV, Sia SK. An implantable compound-releasing capsule triggered on demand by ultrasound. Sci Rep 2016;6(1):22803.

[45] Baghbani F, Chegeni M, Moztarzadeh F, Mohandesi JA, Mokhtari-Dizaji M. Ultrasonic nanotherapy of breast cancer using novel ultrasound-responsive alginate-shelled perfluorohexane nanodroplets: in vitro and in vivo evaluation. Mater Sci Eng C 2017;77:698–707.

[46] Shim MS, Lee HT, Shim WS, Park I, Lee H, Chang T, Kim SW, Lee DS. Poly(D,L-lactic acid-co-glycolic acid)-b-poly(ethylene glycol)-b-poly (D,L-lactic acid-co-glycolic acid) triblock copolymer and thermoreversible phase transition in water. J Biomed Mater Res 2002;61(2):188–96.

[47] Elstad NL, Fowers KD. OncoGel (ReGel/paclitaxel) — clinical applications for a novel paclitaxel delivery system. Adv Drug Deliv Rev 2009;61(10):785–94.

[48] Aminsharifi A, Kotamarti S, Silver D, Schulman A. Major complications and adverse events related to the injection of the SpaceOAR hydrogel system before radiotherapy for prostate cancer: review of the manufacturer and user facility device experience database. J Endourol 2019;33(10):868–71.

[49] Perez-Marrero R, Tyler RC. A subcutaneous delivery system for the extended release of leuprolide acetate for the treatment of prostate cancer. Expert Opin Pharmacother 2004;5(2):447–57.

[50] Jain A, Kunduru KR, Basu A, Mizrahi B, Domb AJ, Khan W. Injectable formulations of poly(lactic acid) and its copolymers in clinical use. Adv Drug Deliv Rev 2016;107:213–27.

[51] Jhan H-J, Ho H-O, Sheu M-T, Shen SC, Ho YS, Jun-Jen L. Thermosensitive injectable hydrogel for drug delivery [Google Patents]; 2016.

[52] Azab AK, de la Puente P, Azab F. Hydrogels for localized radiotherapy [Google Patents]; 2020.

[53] Li J, Yu H, Leong K. Injectable drug delivery systems with cyclodextrin-polymer based hydrogels [Google Patents]; 2011.

[54] Chu C-C, Zhang X-Z, Wu D-Q. Injectable hydrogel microspheres from aqueous two-phase system [Google Patents]; 2010.

[55] Luo T-Y, Peng C-L, Shieh M-J, Shih Y-H, Yeh C-H. Thermosensitive hydrogel for coating radioisotope and chemotherapeutic agent to treat cancer and method for preparing the same [Google Patents]; 2013.

[56] Dimitrios V, Panagioti IE, Sofokli DK, Panagiotis B, Vasileiou-Ioanni TC. Pentablock polypeptidic and hybrid polymers to give in situ-forming injectable, self-healing, pH-and temperature-responsive hydrogels for localised targeted delivery of gemcitabine. Patentscope, WIPO; 2017.

[57] Benhabbour SR, Maturavongsadit P. Injectable thermoresponsive hydrogels as a combinatory modality for controlled drug delivery, biomaterial implant and 3d printing bioink [Google Patents]; 2019.

[58] Charlotte H, Yang SW. In: Office EP, editor. Crosslinked peptide hydrogels. Google Patents; 2015.

[59] Park K-D, Joung Y-K, Park K-M. In situ forming hydrogel and biomedical use thereof [Google Patents]; 2014.

Chapter 8

Recent advances in drug formulation development for targeting lung cancer

Charles Gnanaraj, Ching-Yee Loo, Faizan Naeem Razali, and Wing-Hin Lee
Faculty of Pharmacy and Health Sciences, Royal College of Medicine Perak, Universiti Kuala Lumpur (RCMP UniKL), Ipoh, Perak, Malaysia

1 Introduction

Cancer contributes to the second highest mortality rate after cardiovascular disease globally. Among different type of cancers, lung cancer is the most lethal and statistically contributes to 1.3 million cases annually [1]. An increment of 14% new cases are generally reported each year [2]. The tumorigenesis of lung cancer has been closely associated with the inhalation or exposure of toxic gases such as tobacco smoke, radon gas, asbestos, and carcinogen compounds. As lung cancers develop in different locations of bronchus with high heterogeneity, specialized and personalized medication is often needed for a better clinical outcome of lung cancer therapy [3]. Based on the histological characterizations, lung cancer can be classified into small-cell lung cancer (SCLC) and non-small-cell lung cancer (NSCLC). NSCLC contributes to the majority (85%) of the total lung cancer cases [4]. NSCLC is prominently encountered in metastatic lung cancer patients, and adenocarcinoma is the most common form of NSCLC originated in peripheral lung tissue, and it make up 40% of the metastatic cases. Most of the adenocarcinoma cells are known to be derived from bronchial-alveolar stem cells, club cells, or alveolar type II cells [5]. Other forms of NSCLC are squamous-cell carcinoma, which make up 30% of NSCLC commonly detected in proximal airways in patients, and is known to be derived from basal cells that are found beneath bronchus epithelia or trachea, whereas large-cell carcinoma are rapidly dividing and quick spreading large cancer cells, which make up 9% of the NSCLC. Small-cell lung cancer (SCLC) are the less common type of cancer that is derived from pulmonary neuro-endocrine cells and could be found to be spread in circumference and submucosal of bronchia [5, 6]. Although SCLC contributed to relatively low (15%) incidence in lung cancer; however, this cell is highly aggressive and metastasize to other organs such as brain, bone, and liver. To worsen the scenario, patient diagnosed with SCLC is often have short median survival (> 1 year with appropriate treatment) and (< 4 months without any treatment). Therefore, combination of different type of treatments are necessarily to treat the lung tumor in different prospective.

The current lung cancer therapy that contributes to the highest clinical efficacy is still reliant on the use platinum-based chemotherapeutics administered either alone or in combination with other chemotherapeutics or biomolecules. It should be noted that intravenous administration remains the only approved route for application of chemotherapeutic to date. Despite the numerous side effects encountered with chemotherapy, this treatment option remains the most popular as most lung cancer is diagnosed in advanced state where surgery is no longer a feasible option. First-line chemotherapy treatment includes cisplatin, docetaxel, and paclitaxel, while second-line drugs (i.e., gemcitabine and pemetrexed) and combination therapy are used when patients failed to respond to the first-line drugs. Other treatment options include radiotherapy, surgery, and immunotherapies, which are often used in conjunction with chemotherapy for better clinical outcome. The treatment options vary according to the type and stages of lung cancer. Surgical option is recommended for early stages of tumor (Stage I and II for NSCLC) and is subjected to specific requirement such as (i) tumor localization, (ii) reduction of tumor size prior to surgery, (iii) patients' tolerance toward surgery the ability to reduce the tumor size prior to the surgery, and (iv) the tolerance of patient toward surgery. Meanwhile, a combination of chemotherapy and radiotherapy is recommended for Stage III NSCLC. It is proven that the possibility of recurrence of SCLC is significantly decreased when treated with combination of chemotherapy and radiotherapy. For advanced stage of lung cancer, chemotherapy and immunotherapy are the last treatment option.

Nevertheless, it is disheartening to acknowledge that the clinical success especially for advanced stage of lung cancer is relatively low despite the tremendous and continuous effort to develop novel effective drug formulations or discover new drugs. Chemotherapeutic drugs or biomolecules

Advanced Drug Delivery Systems in the Management of Cancer. https://doi.org/10.1016/B978-0-323-85503-7.00007-9

are conventionally administered via intravenous route (IV) as bolus or infusion. Systemic side effects associated with the drug alone or excipients that are used to increase the solubilities of these drugs in aqueous solutions are well documented. The revolution in drug delivery system has offered a new potential approach to improve the lung treatment outcome. The breakthrough includes various cutting-edge formulations that essentially deliver anticancer drugs to (i) increase biodistribution and bioavailability of drugs at targeted site, (ii) amplify cytotoxicity effect against tumor without compromising the viability of healthy cells, (iii) achieve targeted therapy with specific biomolecules (immunotherapies) and/or surface decoration of targeting moiety on carriers, and (iv) achieve simultaneous targeting of multiple oncogenic pathways and reduce drug resistance.

In this chapter, we summarize the current immunotherapies that are currently in development and approved for clinical use for lung cancer therapies. The advancement of chemotherapeutic formulations has been extensively reviewed elsewhere and therefore not a focus in this chapter [7–10]. Extensive innovative works have also been published, as well as recently reviewed in the field of aerosolized chemotherapy. Here, we presented the most recent technologies and clinical breakthrough in the treatment of lung cancer via pulmonary administration. Undoubtedly, chemotherapy remains the main option for the management of lung tumor via inhalation, and the trend is reflected in the higher number of clinical trials on aerosolized chemotherapy compared to other molecules (see Section 3.1).

1.1 Mechanisms of metastasis and drug resistance in lung cancer

Lung cancer treatment is addressed based on the type and stage of malignancy. Unfortunately, lung cancers are not traceable at early stages; hence, patients detected with lung cancer at late stages are majorly diagnosed with advanced metastatic cancer [11]. Metastasis of lung cancer spreads rapidly to other vital organs, pulmonary system, nervous system, adrenal glands, and bone, which is the main reason for reduced survival rate of lung cancer patients. Surgery or removal of tumor is performed for nonmetastatic cancers, which is not a feasible practice for most of the non-small-cell lung cancer (NSCLC) patients [12]. Metastasis of lung cancer in majority of patients makes radiotherapy and surgery nonviable methods to treat the condition, instead chemotherapy becomes the primary measure to control the progression of cancer, prolong the survival of patients, and enhance their quality of life.

There are several routes for drug administration in lung cancer treatment, which includes the inhalation route and the conventional systemic administration through intravenous and oral routes. Intravenous route is considered one of the most effective route of drug administration for metastatic cancer treatment as the localization of metastasis is unpredictable, hence the drug will be distributed to multiple sites throughout the body [13]. Drugs specified for anticancer treatment should have the tendency to penetrate the cancer cells to trigger elimination of the cells. Therefore, an effective concentration of the drug is needed to reach the cancer cells to exert the anticancer effects. But in the intravenous administration of chemotherapeutics, the drug concentration will be reduced when it reaches the metastatic tumors due to overall dispersion of the drugs [14]. Such situation facilitates metastatic cancer cells to progress despite encountering chemotherapeutic drugs, by developing resistance to drugs, causing relapse and failure in lung cancer chemotherapy.

Increasing the dose of chemotherapeutic drugs becomes the potential option to escalate their therapeutic effect on metastatic cancer cells. But upon administration of high dose, the human body becomes vulnerable to serious adverse effects, specifically at rapid dividing sites such as skin, hair, and active organs like liver and spleen [15]. These factors suppress the efficacy of metastatic lung cancer treatment using the conventional systemic drug administration. Moreover, there might be presence of genetically diversified subtypes of the metastatic NSCLC that complicates the treatment by exerting drug resistance [16]. Although there are several other minor reasons such as reduced bioavailability of drug, defect in apoptosis induction, reduced specificity of drug, DNA repair activation, and so on that indicate the relapse of lung cancer therapy, the prominent factors for failure of anticancer therapy are due to development of drug resistance and genetic diversity in NSCLC. Therefore, understanding the underlying mechanisms in metastasis and drug resistance development of NSCLC after drug administration are crucial at current to strategize effective measures to handle the issues of inefficient treatment for lung cancer.

2 Immunotherapy: Biomolecules targeting lung cancer

In the last five decades, the pioneer of cancer immuno-oncology, Lloyd J. Old, emphasized the pattern of cancer cell appearance compared to normal cells, which can be distinguished by the host's immune cells [17]. Ever since lots of intensive studies and clinical trials were carried out in supporting the development of immunotherapy against cancers, together with several other combination treatments such as invasive surgery, chemotherapy, and radiotherapy. Up to that extend, the manipulation and utilization of the host's immune components, along with their responses proving better prognosis among cancer patients. Immunotherapy or often recognized as biologic-material therapy is seen to expose slightly less hazardous compared to chemotherapy and radiation. This targeted approach involves immune-based molecules with the intent to regulate tumor growth [18].

2.1 Antibody-based biotherapies for lung cancer

In the early 1920s, researchers start to search for alternatives in replacing chemotherapy as it exerts various adverse effects on patient's recovery [19]. Biopsy and sampling were carried out upon lung tumor and began to wonder about the relationship between host antibody and cancer. Specific small molecules are upregulated in lung tumor microenvironment (TME) and are said to be antibodies. A strategy of antibody activity against cancer, for example, by specifically marking the cancer cells to be destroyed by specialized immune cells [20]. These biological molecules are directly targeting respective receptors that commonly be presented particularly on non-small-cell lung cancer (NSCLC) and some other type of cancer cells due to gene mutation and gene rearrangement. These receptors, for example, epidermal growth factor receptor (EGFR) and vascular endothelial growth factor (VEGF) receptor, are responsible for receiving growth signals for cells proliferation, metastasis, and angiogenesis enhancement. Technically, the growth of cancer cells can promisingly be terminated by blocking up these receptors. In addition, some type of biomolecules able to activate several pathways that lead to programmed cell death or apoptosis by binding to programmed cell death ligand 1 (PD-1) and inhibit cell growth by blocking anaplastic lymphoma kinase (ALK). Earlier on, some well-known antibody-based biomolecules have been approved to treat NSCLC, referring to adenocarcinoma, which includes monoclonal antibodies such as cetuximab, bevacizumab, nivolumab, pembrolizumab, and well-known protein kinase inhibitors example like erlotinib, gefitinib, crizotinib, and afatinib [18].

2.1.1 Approved biomolecules targeting epidermal growth factor receptor (EGFR)

The anti-EFGR monoclonal antibodies associate with its respective receptors that are commonly overexpressed on the surface of tumor cells. Ligand binding to receptor allows receptor dimerization and together with autophosphorylation of tyrosine residue at the tail side. It ensures the specific docking region for cytoplasmic proteins containing Src homology 2 and phosphotyrosine-binding domains. The binding of anti-EGFR to the phosphotyrosine residue activates related intracellular signaling pathways [21]. A monoclonal antibody (mAb), cetuximab, targets the EFGR that is expressed in most (80%–85%) of NSCLC patients. Cetuximab can be combined with chemotherapy agents in patients with advanced lung cancer conditions and the clinical testing revealed a better prognosis for patients receiving both combinations [22]. Toward recent clinical trials, the combination of cetuximab with popular anticancer drugs, namely, cisplatin and/or docetaxel is seen to be one of the promising strategies as neoadjuvant in combating NSCLC at a much earlier stage [23]. Throughout the assessment, patients reporting to experience skin complications, diarrhea, loss of appetite, and fatigue. Rarely, some of patients implicated respiratory reaction and blood pressure elevation. In much recent Phase I clinical study, determination for cetuximab regimen either alone or couple regiment was carried out and it was suggested that the ideal dosage regimen to administered cetuximab is 400 mg, once a week over 120 min of IV infusion [24].

Biological inhibitors erlotinib and gefitinib inhibit NSCLC tumor growth by blocking EGFR and inhibiting tyrosine kinase signaling. Some lung cancer cells expressed an altered version of extracellular protein ligands that ignite signals to support cell differentiation proliferation, motility, and survival. These tumorigenesis behaviors activated through MAPK and PI3K-AKT pathways, sustaining angiogenesis for tissue invasion and metastasis. Erlotinib is primarily been administered through oral, instead of parenteral for advanced lung cancer, and to overcome the mutation of the EGFR gene in 10%–15% of lung cancer patients [25–27]. In 2010, erlotinib is recognized as monotherapy for NSCLC after chemotherapy, approved by the FDA. A phase-III clinical trial on EGFR mutation-positive NSCLC patients was carried with a dosing regimen of 150 mg per day via oral administration and suggested that erlotinib is a promising candidate for first-line treatment of EGFR-mutation-derived NSCLC compared to other comparable chemotherapy agents [28]. Five years later, gefitinib was approved by the FDA to be the choice of NSCLC treatment for patients resulted from the EGFR gene mutation positive. Referring to a documented clinical report, gefitinib can be prescribed at a much lower dose than the standard dosing regimen (250 mg/day, oral) to NSCLC patients that frail to toxicity effects [29].

2.1.2 Approved biological molecules targeting vascular endothelial growth factor (VEGF)

VEGF antibodies promote the growth and proliferation of tumor through the enhancement of endothelial cells survival, facilitating cells migration and invasion, as well as increase permeability of vascular for metastasis activity. VEGF signaling was activated through three types of tyrosine kinase receptors (VEGFR-1, VEGFR-2, and VEGFR-3), which reported predominantly expressed on the surface of the epithelial cells [30]. The growth of tumor endothelial-derived cells is promoted by the response signals of the antibody-VEGF receptor association.

Bevacizumab is a selective anti-VEGF mAb allows to control the abnormal growth of blood vessel around TME, thus limiting the chances of cell migration and metastasis. In 2004, it was approved by the FDA to control metastasis for colorectal cancer, a combination treatment to chemotherapy agents. An extension to that, some clinical studies revealing positive outcomes for bevacizumab to be used as the first-line treatment for metastasized lung adenocarcinoma. Patients were administered with 15 mg/kg via IV, as a combination treatment between bevacizumab and other anticancer agents [31, 32]. However, it was reported to cause bleeding for prolong usage, elevating blood pressure, inducing fatigue, loss of appetite and headache, and as well as lowering the patient's white blood cells count [33–35].

2.1.3 Approved biological molecules targeting programmed cell death ligand (PD-1)

The expression of PD-1 on the surface of many tumor cells, including on squamous or nonsquamous NSCLC, is a coinhibitory receptor for T cells, B cells, monocytes, and natural killer cells. The system activation through the binding with complement molecules (PD-L2 and PD-L2) is proven to support cells survival through invading immunosurveillance strategy by preventing the activation of T-cell receptor signaling and demoting the upregulation of antiapoptotic molecules. The action of T cells is also further suppressed.

The two most popular biological molecules against lung cancer, namely, pembrolizumab and nivolumab, in a subclass of anti-PD-1 molecules in the action of promoting the apoptosis mechanisms in lung cancer cells [36]. Pembrolizumab used as the first-line treatment against untreated advanced NSCLC. The dose recommendation for lung cancer patients is 2 mg/kg or 200 mg, once for every 3 weeks via 30-min IV infusion [37]. Through series of clinical trials, pembrolizumab has been approved by the FDA. In 2015, pembrolizumab is used as the second-line treatment for patients with metastatic lung cancer. Along the way, patients responded well to the treatment with slight uncommon side effects such as body pain, chill sensation, bleeding, headache, fever, loss of voice, constipation, and rapid weight gain [38–40]. Going through a similar antineoplastic strategy, nivolumab was tested as the second-line treatment for both squamous and nonsquamous NSCLC with a recommended dose of 3 mg/kg or 240 mg, once for every 2 weeks administered via 30-min IV infusion [37]. In 2015, FDA has approved the use of nivolumab in the treatment for patients with metastatic squamous NSCLC that no longer responding to chemotherapy agents with adverse reactions similarly exerted by pembrolizumab, with addition of diarrhea, blister, and back pain [41–43].

2.1.4 Biomolecules under development as promising candidates for treating lung cancer

Till recent day, the effort on searching suitable candidates that have promising positive outcomes in cancer treatment, particularly against lung cancer are progressing, revealing some biological components possess the antitumor ability as good as chemotherapeutic agents but with slightly less exerting unwanted effects. Biological candidates such as figitumumab, nefotumomab, nimotuzumab ficlatuzumab, denosumab, ipilimumab, tremelimumab, veliparib, olaparib, and buparlisib are currently under trial to be suggested as part of lung cancer alternative treatments. These entities specifically interfere with biological systems in TME, executing specific binding interaction, channeling relevant signals, resulting in tumor inhibition behavior. It was well elucidated that receptors like insulin-like growth factor 1 (IGF-1), Fab fragment of murine mAb NR-LU-10, EGFR, hepatocyte growth receptor (HGFR), receptor activator of nuclear factor κB ligand (RANKL), cytotoxic T-lymphocyte-associated antigen (CTLA-4), receptor for poly-ADP ribose polymerase (PARP) enzyme production, and phosphatidylinositol 3-kinase (PI3k) are properly expressed onto lung cancer tumor cells (Table 1). Anticancer molecules bind to their relevant receptors and start to disrupt the endogenous signaling in TME exerting antitumor mechanisms.

3 Lung cancer therapy via inhalation

3.1 Clinical trials of inhaled chemotherapeutics: Success or failure?

Inhaled chemotherapy refers to the administration of drugs directly to the lung with the aid of aerosol generation devices such as nebulizer, dry powder inhaler, and soft mist inhaler (extensively reviewed in Refs. [70–72]. Emerging technologies on the design and improvement of these devices have been described in detail elsewhere [73–76]. Compared to other routes, local delivery of chemotherapeutics via inhalation is noninvasive in nature and deposits high concentrations of aerosolized drugs to solid tumors. In addition, inhalation therapy offers numerous advantages such as high bioavailability and rapid onset of action owing to its large surface area for drug absorption. The presence of limited metabolizing enzymes ensures limited degradation for drug delivered in such manner [77]. Apart from lung, inhaled drug has also been detected in lymph nodes, lymphatic tissues, and other peripheral airways, thus allowing the possibility of treating lung cancer that has metastasized to the other organs [78, 79]. This is beneficial as most lung cancers are detected in advanced metastatic state.

Effective deposition of aerosolized particles in the respiratory tract is determined primarily by (a) physicochemical

TABLE 1 Promising biomolecules for lung cancer treatment: under development.

Molecule	Treatment regimen	Receptor interaction	Mechanism of action	Status of development	References
Figitumumab	IV, 20 mg/kg in NS	*IGF-1R	Blocking the binding of IGF-1 to its receptor and inhibit cells growth by terminating anabolic effect	Combination anti-IGF-1R with carboplatin and paclitaxel passed Phase I and II clinical trials with promising outcome. The study was stop at phase-III for metastatic NSCLC	[44–46]
Nofetumomab	Coupled with radio-imaging probe	*NR-LU-10	Murine mAb is binding to tumor antigen, to be used as biomarker for bone metastatic detection from lung cancer	Preclinical assessment. No clinical trial recorded	[47, 48]
Nimotuzumab	IV,100–600 mg in 250 mL NS	*EGFR	Inhibit cells proliferation, enhance apoptosis, reduce angiogenesis and metastasis	Stop at Phase I clinical trial. Combine with radiotherapy against lung adenocarcinoma	[49–51]
Ficlatuzumab	IV, 10–20 mg/kg coupled with oral anti-EGFR	*HGFR	A humanized monoclonal antibody specifically binds to HGFR, promoting EGFR-inhibitors binding. Complement for anti-EGFR molecules	Phase-Ib clinical trial, combination with gefitinib	[52–54]
Denosumab	SC, 60–120 mg	*RANKL	Inhibit bone cancer by binding to osteoblast produced RANKL	Under evaluation for lung cancer as it increased overall survival of metastatic lung cancer patients at phase-III clinical trial	[55–57]
Ipilimumab	IV-infusion, 0.3–10 mg/kg	*CTLA-4	Human mAb-IgG1 binds to CTLA-4, inhibits critical negative T-Cell regulators	Phase-IIIb/IV clinical trial against NSCLC as monotherapy, combine therapy with nivolumab against SCLC	[58–60]
Tremelimumab	IV, 10–15 mg	*CTLA-4	Human Ab-IgG2 binds to CTLA-4, inhibits critical negative T-cell regulators, exerting anticancer effect on metastatic melanoma	Phase I clinical trials in combination therapy with durvalumab. Already proceed with phase-III clinical trial	[61–63]
Veliparib	Oral, 200 mg tab	$^{+}$PARP	Blocking DNA-damage repair mechanism for many types of tumor cells	Passed Phase I clinical trial, a combined maintenance treatment after chemotherapy, preventing lung cancer recurrent	[62, 64]
Olaparib	Oral, 100–200 mg tab	$^{+}$PARP	Blocking DNA-damage repair mechanism for many types of tumor cells	Passed Phase I and II clinical trials, coupled with oral temozolomide for SCLC treatment	[65–67]
Buparlisib	Oral, 100 mg tab	$^{+}$PI3K	Inhibition of PI3K-enzyme that promotes tumorigenesis of lung cancer	Stop at Phase I clinical trial due to disobey the primary criterion, as the activation of PI3K pathway is detected. Trial as phase-II continued as a combination treatment option with well-known chemotherapy agents	[68, 69]

(*): Antibody-type molecule. ($^{+}$): Biological inhibitor molecules. IV: Intravenous. SC: Subcutaneous. NS: Normal saline.

characteristics of particles (size, density, shape, and surface), (b) ventilatory parameters such as breath pattern and airflow velocity, which influenced the residence time in the respiratory tract [80–82]. From the point of view of developing an effective inhaled formulation, particle size and geometry are perhaps the most important criteria [82]. The optimal aerodynamic diameter for effective deposition ranges from 1 to 3 μm to avoid both deposition in the upper airways and exhalation [83]. This is achievable through specially designed formulation coupled with the use of suitable aerosol generation device.

The concept of nebulization is based on the generation of continuous aerosol droplets from solution or suspension using either compressed air (jet nebulizers) or ultrasonic nebulizers. The droplet size generally ranged between 2 and 5 μm. No specialized breath-actuation coordination is required as patients inhale the aerosol during normal breathing cycle [74]. For chemotherapeutics intended for delivery via nebulization, simple formulations are only needed by virtue of optimizing the drug solubility and stability in suspension/solution.

Nebulizers are the only device that used in clinical trials to produce aerosol droplets of chemotherapeutics and therapeutic cytokines for inhaled lung cancer therapy to date (extensively reviewed in Ref. [70]). Although dry powder inhalers and pressurized metered dose inhaler are available for other respiratory diseases, they are yet to be tested for aerosolized chemotherapy. The first clinical trial on aerosolized chemotherapy was reported in 1968 using 5-fluorouracil (5-FU) [84]. The authors demonstrated the retention of high 5-FU in tumor tissues for up to 4 h after administration. Since then, several on inhaled chemotherapy, involving cisplatin (CPT) [85, 86], doxorubicin (DOX) [87, 88], gemcitabine [89], and 9-nitrocamptothecin (9-NC) [90], have been initiated with moderate success in terms of safety and dose tolerability in phase I trials. Severe side effects associated with systemic toxicity were not found in the findings from the trials. However, it should be noted that though the safety and feasibility of inhaled chemotherapy had been demonstrated, majority of the aerosolized chemotherapy was unable to progress beyond phase II studies. From the cancer treatment perspective, high local chemotherapeutic dose should be administered to lung cancer patients to achieve clinical therapeutic drug levels. Pulmonary toxicity is correlated to the drug and its dose and remains one of the biggest challenges in aerosolized chemotherapy. As 10%–30% of chemotherapeutics in the lung is sufficient to induce toxicities, it is envisaged that high local delivery and retention of drugs following pulmonary administration would exacerbate the toxic effects [91]. The findings from phase I trials, however, disputed the idea as most nebulized chemotherapeutics are well tolerated. Based on the observation of Wittgen et al., the low cisplatin concentration in the formulation (1 mg/mL) restricted high aerosol deposition in lung (10%–15% nominal dose deposited) [85]. The poor aqueous solubility of cisplatin accounts for the low concentration in formulation for inhalation. As a consequence, it necessitated the increase in nebulization timing and frequency to achieve maximum delivered dose of 60 mg/m^2. The duration for a single nebulization is 20 min with three consecutive nebulization per session and a maximum of 3 sessions per day [85]. On the other hand, when the chemotherapeutic drug solubility is enhanced with the introduction of certain amount of ethanol, its nebulized formulation, however, induced local adverse effects (bronchospasm, dyspnea, etc.) [87]. From the safety point of view, the administration of nebulized chemotherapeutics in human trials should be conducted inside negative pressure units with high-efficiency particulate air (HEPA) filters and air extractor to capture aerosol particles that are lost during inspiration [71, 92]. The nebulizers used in clinical trials are equipped with additional features to reduce the aerosol losses in the air. For instance, Aero-Tech II nebulizer contains exhaled-collection filters and Pari LC Plus are meant for mouth-only inhalation [71, 92].

3.1.1 Conventional chemotherapeutic drugs for therapeutic purposes

In a study conducted to assess the dose-limiting toxicity (DLT) of aerosolized liposomal cisplatin on primary or metastatic lung carcinoma patients, the hematologic toxicity, nephrotoxicity, neurotoxicity, or ototoxicity was negligible. The adverse effects observed were associated with respiratory tract, which include pulmonary function loss, dyspnea, and hoarseness. However, it should be noted that DLT was not evident even at its maximum dose (60 mg/m^2) [85]. In another study, inhaled lipid cisplatin was generally well tolerated in recurrent osteosarcoma patients without showing specific side effects associated with intravenous cisplatin administration [86]. The same trend was seen in a phase I clinical trial for liposomal 9-NC for inhalation. Side effects higher than grade 2 were not reported in the enrolled patients with primary or metastatic lung cancer [90]. Similarly, the safety and pharmacokinetic profile of aerosolized gemcitabine were established with 11 patients with carcinoma localized in the lungs receiving dose-escalating inhaled chemotherapeutics once a week for 9 consecutive weeks [89]. Approximately 42% of the initial dose (between 1 and 4 mg/kg gemcitabine) was deposited successfully into lung with low systemic toxicity. The side effects of nebulized gemcitabine were mostly related to the respiratory tract, which is in accord with other published trials. DLT was observed at 4 mg/kg, whereby one patient with severe COPD complications developed acute respiratory failure [89]. To reduce the loss of aerosol to air, DOX was nebulized to patients using a mouth-only inhalation device (OncoMyst Model CDD-2a) in phase I and I/II clinical trials [87, 88]. Owing to the low aqueous solubility, DOX was dissolved in a co-solvent system containing four parts of water and 1 part of ethanol buffered at pH 3 prior to nebulization for these studies. As such, the presence of 20% ethanol in the formulation could be the attributing to the local adverse effects such as coughing observed in majority of the patients. The DLT for nebulized doxorubicin was 7.5 mg/m^2 [88]. Otterson et al. further demonstrated that nebulized doxorubicin is safe to be administered concurrently with systemic doses of i.v. docetaxel and cisplatin. However, the improvement in response rate failed to show significant differences [87].

3.1.2 *Nonchemotherapeutic drugs as cancer chemoprevention agent*

Several studies have demonstrated the potential use of nonchemotherapeutic drugs as chemoprevention agent that could delay lung tumorigenesis and disease progression (extensively reviewed elsewhere [4]). For example, in a long-term epidemiological study (1996–2001, $n = 10{,}474$) with COPD patients, those receiving inhaled corticosteroids is associated with reduced risk of lung cancer incidence [93]. One of the major limitation of the study is the small number of patients with lung cancer that matched the 80% compliance benchmark to be included in the analysis. As only 517 patients out of 10,474 were analyzed, the results were somewhat inconclusive to support the chemopreventive benefits of inhaled corticosteroids. However, several other independent studies also supported the findings of Parimon and colleagues. In a nested case–control study based on Korean national claim database, 9177 individuals with lung cancer out of a total of 792,687 eligible cohorts were assessed to correlate the use of inhaled corticosteroid and lung cancer incidence. The diagnosis of lung cancer is reduced in tandem with the regular use of inhaled corticosteroids (adjusted odds ratio (aOR), 0.79; 95% CI, 0.69–0.90) [94]. In the retrospective cohort study conducted by Kiri et al., the authors showed that only 6% COPD patients who were on regular use of inhaled corticosteroids/long-acting beta agonist developed lung cancer [95]. This study analyzed over 7000 ex-smoker COPD patients from 1989 to 2003 under regular medications of inhaled corticosteroids. The findings from the observational study indicated that inhaled corticosteroid acts to delay lung cancer progression in COPD patients [95]. A nationwide population-based cohort study in Taiwan was carried to determine the effectiveness of inhaled corticosteroid in female COPD patients ($n = 13{,}686$; 1997–2009) [96]. Interestingly, only 9.4% patients ($n = 1290$) were inhaled corticosteroid users. The main finding from this study is the reduced risk of female COPD patients prescribed with inhaled corticosteroid to develop lung cancer compared to noninhaled corticosteroid users (10.75 vs 9.68 years, $P < 0.001$) [96]. The lung cancer incidence rate per 100,000 person for noninhaled corticosteroid and inhaled corticosteroid users is 235.92 and 158.67, respectively (HR: 0.70, CI: 0.46–1.09) [96]. Sin et al. published a meta-analysis of seven randomized trials to associate the use of inhaled corticosteroid with decrease in all-cause mortality. Patients assigned to inhaled corticosteroid demonstrated lower risk of mortality compared to placebo (HR 0.75; 95% CI 0.57–0.99) [97].

However, findings from two separate randomized interventional trials contradicted the observational trial results, whereby inhaled corticosteroid failed to reduce the appearance of lesions or decrease the incidence of lung cancer [98, 99]. Lam et al. investigated the effectiveness of inhaled budesonide (Pulmicort Turbuhaler, 1600 μg) in a Phase IIb trial in heavy smokers ($n = 112$) with bronchial dysplasia. The 6-month intervention did not result in significant reduction of bronchial dysplasia or inhibit the appearance of new lesions [98]. In a randomized, double-blind, placebo-controlled trial, the treatment group between placebo (twice daily) and treated (inhaled budesonide 800 μg twice daily) arms showed negligible differences in nodule response [99]. Long-term monitoring (up to 5 years of follow-up) of solid nodules showed that inhaled budesonide was ineffective to reduce the incidence of lung cancer, as well as new nonsolid nodule appearances [100].

3.2 Current technological advances in inhalable drug delivery system (DDS)

3.2.1 *Representative inhalable DDS based on the drug encapsulation into microparticles*

As mentioned earlier, the optimal aerodynamic diameter should range from 1 to 3 μm for efficient delivery to the lung. Therefore, conventional inhalation formulations are focused on the development of micron-sized particles in the form of solution, suspension, or dried powder (Table 2). Large porous microparticles are characterized with high geometric diameter, low density, and aerodynamic diameter below 5 μm that are ideal for pulmonary administration. With appropriate geometric sizes (5–30 μm) and density, deposition of these microparticles into deep lung could be achieved. In addition, they are too large to be taken up by alveolar macrophages and thus increased their residence in the lung. Several porogens that are employed to create porous microparticles include ammonium bicarbonate, cyclodextrin, sodium chloride, and sugar [72, 111–114]. Li et al. studied the preparation of porous poly(cyclohexane-1,4-diyl acetone dimethylene ketal)/(PCADK)/polylactide-*co*-glycolide (PLGA) mixed matrix microspheres loaded with DOX using w/o/w emulsion method [111]. DOX was released in a sustained manner over 10 days and exhibited an increase in antiproliferative effect through enhanced tumor-related

TABLE 2 Inhalable microparticles used in lung cancer treatment.

Drugs	Dosage form	Carrier	Formulation into inhalable product	Target cells	Experimental setting	References
Cisplatin	DPI	Silk fibroin particle and mannitol	Spray drying or spray freeze drying with mannitol, silk fibroin and cisplatin	A549	In vitro	[101]
VEGF-siRNA	DPI	Chitosan	Spray freeze drying with siRNA 2%, chitosan 10%, mannitol 22%, L-leucine 66%	B16F10 cells	In vivo	[102]
PTX	Solution for intratracheal instillation	Polyethylene glycol (PEG)	Conjugation with 6kDa or 20kDa PEG	Lewis lung carcinoma mice model	In vivo	[103]
Topotecan	DPI	Trehalose and leucine	Spray drying at acidic condition (pH 3.5)	Sprague Dawley rats	In vivo	[104]
Oridonin	DPI	PLGA	Electrospraying to form porous PLGA microsphere loaded with orodinin	Lung cancer rat model induced with 3-methylcholanthrene (MCA) and diethylnitrosamine (DEN)	In vivo	[105]
IL-10	suspension	Polylactic acid microsphere	Phase inversion nanoencapsulation	LSLKras^{G12D} murine lung cancer model	In vivo	[106]
Cisplatin	DPI	PEGylated solid lipid microparticles	Spray drying	–	In vitro	[107]
Fisetin	DPI	Sulfobutylether-β-cyclodextrin complex	Spray drying	–	In vitro	[108]
Sodium hyaluronate	DPI	Stearylamine, stearyl alcohol, cetostearyl alcohol	Spray drying	A549 cells	In vitro	[109]
PTX	DPI	Superparamagnetic iron oxide nanoparticles (SPION) with solid lipid microparticles	Spray drying with different fatty acids	A549	In vitro	[110]

protein regulation, apoptosis, and cell cycle arrest [111]. Treatment of BALB/c mouse lung cancer model with DOX-loaded PCADK/PLGA microspheres (0.2mg/kg DOX) every 5days for a period of 20days reduced tumor mass and number. The predominant localization of microspheres in deep lung after delivery using tracheal insufflation demonstrated the good aerosolization efficiency of PCADK/PLGA microsphere [111]. Kim et al. reported similar findings, whereby inhalable DOX-loaded PLGA porous microspheres caused significant reduction in tumor mass for B16F10 melanoma bearing mice [112]. The micronization of chemotherapeutics into size range suitable for inhalation is a popular technique used extensively to produce DPI formulation. Recently, Verco et al. utilized the antisolvent precipitation method to prepare submicronized PTX crystals (NanoPac®) ranging from 600 to 800nm. This formulation is designed to be delivered in a reconstituted suspension in physiological saline and therefore could be delivered as IV and inhaled formulation [115, 116]. For inhalation purposes, aerosols of NanoPac® generated from jet nebulizers had MMAD of 1.8μm and were retained in the right lung lobes in vivo for more than 14days [116]. Following the inhalation of NanoPac® (0.5 or 1.0mg/kg) once or twice weekly for a duration of weeks into Calu-3 harboring nude male rats, signs of tumor regression and immune cell infiltration were observed [115].

3.2.2 Inhalable DDS based on nanotechnology approach

Inhalable nanoparticles have gained popularity owing to the versatile nature of various novel nanotechnology systems to deliver high drug payload to target cells [83] (Table 3). Unlike conventional inhalation therapy, these novel inhalable nanoformulation could be tailored to escape the rapid elimination from lungs and thus reducing the need for multiple inhalation dosing. In essence, the advantages of IV nanomedicine are translatable to inhaled nanomedicine, which include (a) enhanced retention and permeation effect, (b) increased local bioavailability, (c) reduced mucociliary and phagocytic clearance, (d) prolonged residence time locally, and (e) higher drug stability and solubility [77]. Despite these advantages, local delivery of individualized nanoparticles to the lung is not effective because their aerodynamic diameter is too small to be deposited efficiently at respiratory tract. As the aerodynamic diameter of nanoparticles are far below the optimal size range (1–5μm) for particle deposition via inhalation, these nanoparticles would be exhaled after inspiration. The high free energy in each nanoparticle would also cause irreversible particle aggregation and poor flowability [129]. To overcome this, the use of a carrier with suitable aerosolization profile for pulmonary delivery is needed.

For nebulization, the nanoparticles are dispersed in liquid solutions, which are then converted into aerosol droplets with suitable inhalable size range using commercial nebulizers (jet, ultrasonic, or mesh nebulizers). Assuming that the particles are spherical, a droplet of 5μm aerosol produced from nanodispersion contains approximately 50 100nm particles. Therefore, the nebulization of drug in the form of nanoparticle are postulated to show increased drug absorption rate compared to the nebulization of drug solution alone. However, it should be noted that the force of nebulization might lead to the destabilization of the nanoparticles [130]. In a study by Hureaux et al., the aerosol droplets of lipid nanocapsules produced from jet nebulizer underwent significant changes in size distribution [130]. Meanwhile, ultrasonic nebulizer was unable to produce adequate amount of aerosols. Only mesh nebulizers were able to comparatively produce stable droplets with good aerosol performance [130]. On the other hand, Lee et al. reported a stable aerosol formation of curcumin nanoparticles dispersed in phosphate-buffered saline using PARI LC Sprint® jet nebulizer [131]. The microscopic and quantitative analyses of the nanoparticles after nebulization revealed nonsignificant changes in both morphology and size distribution, respectively [131]. Therefore, each nanoformulation has different endurance capacity to the force of nebulization and hence the aerosol performance could be nebulizer-specific.

The design of formulation for the pulmonary delivery of nanotherapeutics in the form of DPI is technically challenging. Any carriers (excipients) used in the formulation must improve the aerosolization efficacy of DPI, as well as reduce the interparticle attraction forces that exist between nanoparticle-carrier and nanoparticle-nanoparticle. Similar to nebulization, destabilization of inhalable nanoparticle often occurs during the spray-drying process for lipid-based materials. Furthermore, the inability of dried powder to redisperse into individual nanoparticles during inspiration is also another problem of DPI-based nanoformulation. To date, two main approaches have been used to deliver inhalable nanoparticle in solid form. They are nano-in-microparticles and reversible nanoaggregates, whereby the former is held together by carriers and the latter is a result of van der Waals interactions between nanoparticles themselves [132]. The carriers used in nano-in-microparticles are often hydrophilic-based excipients (i.e., mannitol, leucine, inulin, etc.) that could be converted into inhalable particles with size suitable for inhalation (1–5μm) [122, 123, 127, 133]. Once deposited into lung, these hydrophilic excipients dissolve easily in alveolar fluid to redisperse into primary nanoparticles. Unlike nano-in-microparticles, nanoaggregates are drug-loaded nanoparticles that are agglomerated together and can be disintegrated into individualized nanoparticles after fluidization. One such example is hollow nanoaggregates with large geometric diameter and low density that exhibit smaller aerodynamic diameter and thus making them suitable for deep lung deposition [125].

TABLE 3 Representative preclinical studies of inhalable nanoparticles against lung cancer.

Drugs	Dosage form	Nanocarrier	Target	Outcome	References
Quercetin	Nebulized suspension	Nanoemulsion containing palm oil: ricinoleic acid, lecithin, Tween 80, glycerol	–	Nanoemulsion size: 106.1 nm Nebulized droplet size and FPF: 4.25 μm and 70.56%	[117]
Gemcitabine	Nebulized suspension	Gelatin nanocarriers cross-linked with genipin	A549 and H460 cell lines (in vitro)	Nanoparticle size: 178 nm. MMDA and FPF values are 2.0 μm and 75.2%, respectively. The nanoformulation showed comparable cytotoxic efficacy toward lung cancer cells with free gemcitabine solution	[118]
DOX	pMDI	pH-responsive, PEG-DOX NP	A549 (in vitro)	FPF ranged between 40% and 60% and drug release is higher at acidic environment	[119]
DOX	pMDI	PEGylated poly(amidoamine) dendrimer (PAMAM)	Calu-3 (in vitro)	High FPF (more than 82%)	[120]
siRNA	Nebulized suspension	Amide functionalized polyester nanoparticle	A549 cell line (in vitro), and orthotopic mice model with A549 cells	High local concentration of siRNA in the lung and hence significant gene silencing effect in tumor	[121]
Rapamycin and berberine	Nano-in-microparticles DPI	Primary nanoparticle: lipid nanocore and phospholipid shell Inhalable formulation: Spray drying with mannitol, leucine, PVP and/or maltodextrin	A549 cell lines (in vitro). Urethane induced lung cancer in albino mice (in vivo)	MMAD and FPF values are 3.28 μm and 55.5%, respectively. Reduction in VEGF expression coupled with increased caspase level. Low systemic toxicity in vivo. Higher inhibition in tumor regression and growth in vivo	[122]
Active: Pemetrxed and resveratrol Targeting moiety: Lactoferrin (LF) and chondroitin sulfate (CS)	Nano-in-microparticles DPI	Primary nanoparticulate liquid crystals Inhalable formulation: Spraying drying with mixtures of mannitol, inulin and leucine at a ratio of 1:1:1	Urethane induced lung cancer in BABL/c mice (in vivo)	MMAD and FPF values are 2.72 μm and 61.6%, respectively. Inhalable nanocomposite reduced the number and size of tumor. This is coupled increased expression of apoptotic protein (caspase-3) and reduced expression of VEGF-1	[123]

Curcumin	Nano-in-microparticles DPI	Primary nanoparticle: Curcumin-loaded PLGA NP Inhalable formulation: Spray drying of the NP in the presence of mannitol	A549 cell line (in vitro)	Curcumin acts as photosensitizer when activated by LED irradiating device. MMAD and FPF values are 3.02 μm and 64.94%, respectively	[124]
DOX and methotreaxate	Hollow nanoaggregates DPI	PEGylated paramagnetic hollow nanosphere	A549 cell line (in vitro)	The formulation showed appropriate aerosol performance (FPF of 22%)	[125]
STING agonist cyclic guanosine monophosphate-adenosine monophosphate (NP-c-GAMP)	Nebulized suspension	Liposomal of NP-c-GAMP	4-TI-luc lung metastasis mice model	Inhaled N-cGAMP acts synergistically with radiation therapy to induce systemic immunity as well as prolong survival in mice	[126]
Erlotinib	Nano-in-microparticles DPI	Erlotinib-loaded solid lipid nanoparticle	A549 cell line (in vitro)	Spray-dried erlotinib-SLN with mannitol resulted in microparticles with suitable aerosol performance	[127]
PTX	Suspension	Surface-modified SLN containing PEG and chitosan	HeLa adenocarcinoma ovarian cell line (in vitro) and murine M109-HiFR lung carcinoma cell subline (in vitro, in vivo)	Prolonged residence time of PTX in vivo. Preferential targeting delivery of PTX-SLN toward folate-receptors in both cell lines	[128]

3.2.2.1 Liposome

Liposome for inhalation is the first nanoformulation developed for the delivery of chemotherapeutics locally to the lung. In addition, inhaled liposome for lung cancer treatment remains the only nanocarrier that is evaluated in the clinical trials to date. In these clinical trials, liposome for inhalation is delivered to the lung via nebulization. In most cases, as dipalmitoylphosphatidylcholine (DPPC) accounts for 40% of lung surfactant; it is the most popular component used for the fabrication of liposome. As demonstrated in several preclinical studies, aerosolized liposomes harboring chemotherapeutic drugs such as CPT, 9-NC, paclitaxel (PTX), and DOX showed prolonged residency time in the lung, lower systemic translocation, and enhanced tumor inhibition characteristics [134–138]. Nebulized liposome containing 9-NC showed the highest deposition in the lung (310 ng/g of tissue) and reduced tumor nodules in osteosarcoma lung metastases mice [134, 135]. In an in vivo efficacy experiment of nebulized liposomal 9-NC formulation against lung cancer harboring mice xenograft model, it was observed that aerosolized liposomal 9-NC showed tumor inhibition activity. The growth of tumor size for treated groups after receiving aerosolized liposomal 9-NC for 31 days (8.1 μg/kg per day; 5 days per week) was sevenfold lower compared to control group [139]. Meanwhile, Zhang et al. developed a sustained release of liposomal formulation, which increased the mean residence time of 9-NC in the lung by at least threefold, as well as reduced the pulmonary damage [140]. Nebulization of liposomal formulation in the presence of 5% CO_2-enriched air is an effective strategy to increase drug deposition in lung [136]. The deposition of PTX liposome with or without the presence of CO_2-enriched air is 23.1 ± 4.3 and 5.5 ± 0.2 μg/g, respectively [136]. Furthermore, inhalation of PTX liposome (three times per week for 2 weeks; 5 mg/kg dose of PTX) to mice bearing murine renal carcinoma (Renca) pulmonary metastases decreased the tumor size from 0.88 to 0.59 mm [141].

A recent innovative approach has been published, which employed bacterial therapy as possible treatment of primary lung cancer [142]. The concept is based on the use of entrapment of drug harboring liposome into viable bacteria to mimic the nano-in-micro approach and subsequently delivered to lung via inhalation [142]. PTX-loaded liposome with mean particle size of 64 nm was introduced into *Escherichia coli* via electroporation with entrapment efficiency of 95%. Interestingly; both electroporation and PTX were not harmful to the bacterial cells as up to 90% cells remained viable. The pulmonary delivery of paclitaxel-in liposome-in bacteria exerted anticancer effects and immunostimulation simultaneously [142]. Higher apoptosis in cancer cells was noted, coupled with downregulation of VEGF and HIF-α expressions [142]. Repurposing established drug for new indication is another emerging approach for lung cancer treatment. Parvathaneni et al. reported the effectiveness of pirfenidone (PFD), an antifibrotic drug for NSLC by encapsulation into liposome [143]. The compositions of the cationic liposomal carrier are 5% 1,2-dioleoyl-3-trimethylammonium-propane (DOTAP), DPPC, and cholesterol (molar ratio of 5:5). Owing to the presence of cationic lipid, the PFD-loaded liposomes are positively charged (42.4 mV) with average size of 211.8 nm. The aerosol performance studies demonstrated that this liposomal formulation possess suitable characteristic for inhalation. Specifically, the mass median aerodynamic diameter (MMAD) and fine particle fraction (FPF) was 2.59 ± 0.04 μm and 76.88 ± 2.26%, respectively [143]. Hamzawy et al. demonstrated that the inhalation of gold nanoparticle embedded into liposome harboring temozolomide (TMZ)-induced apoptosis and necrosis in lung cancer cells possibly owing to the accumulation of intracellular reactive oxygen species (ROS) [144]. The use of gold nanoparticle-embedded liposome enhanced the stability and dispersion of TMZ [144]. Recent study described the use of antibody decorated liposome to target the receptors present on the surface of lung cancer cells such as carbonic anhydrase IX (CA IX) [145]. The conjugation of anticarbonic anhydrase IX to triptolide (TPL)-loaded liposomes promoted the cytotoxicity and drug uptake into tumor spheroid when delivered via pulmonary route [145].

The physical instability issue is often encountered for liposomal suspension for nebulization. The high shear force exerted on the liposomal suspension during nebulization destabilizes the structure of liposome and subsequently causes premature leakage of drugs. Additionally, the low long-term storage stability for liquid suspension is a main disadvantage and might cause drug degradation or irreversible fusions of liposomes. Therefore, dry liposomal formulation are popular options to overcome these limitations. Zhang et al. demonstrated the development of a stable and redispersible dried powder curcumin liposomal formulation via freeze-drying using mannitol as lyoprotectant [146]. The DPI formulation is suitable for inhalation with MMAD and FPF of 5.81 μm and 46.71%, respectively. Curcumin seems to be target-specific in its activity in which it is preferentially cytotoxic to cancer cell lines compared to healthy cell lines [131, 146]. In vivo administration of DPI curcumin liposome resulted in attenuation of pro-inflammatory cytokine and angiogenesis, as well as enhanced apoptotic action [146]. Zhu et al. recently reported that the conversion of liposomal suspension into DPI nanocomposite altered the pharmacodynamics and pharmacokinetics of docetaxel [147]. Co-spray drying of folic acid decorated docetaxel liposome with mannitol and leucine yielded microparticles with d_{50} of 3.94 μm. The MMAD and FPF of the dried powder formulation were 3.10 μm and 10.0%, respectively, which was not an optimized formulation for inhalation yet.

Although redispersible during fluidization, the liposomes underwent shape change from spherical into an irregular, distorted shape. Interestingly, the increase in particle size of liposome after redispersion enhanced the cellular uptake of docetaxel via micropinocytosis and consequently higher anticancer effect in vitro. Prolonged residence time and accumulation of docetaxel were observed for intratracheal administration of DPI folic acid conjugated docetaxel liposome compared to intravenous administration. Up to 50-fold higher concentration of docetaxel in lung was noted for administration of the DPI formulation via pulmonary route [147]. Recently. Fukushige et al. reported the pulmonary delivery of hyaluronic acid-coated-liposome-protamine-DNA complex prepared through spray-freeze drying method. The presence of hyaluronic acid reduced the toxic effect of siRNA, as well as protected the silencing effect of siRNA simultaneously [148].

3.2.2.2 Polymeric nanoparticle

Several studies have reported the potential of gelatin as nanocarrier of chemotherapeutics for delivery via inhalation [149–151]. Gelatin is a denatured protein with excellent biocompatibility with cells. It could be tailored to load both hydrophilic and hydrophobic drugs owing to its polyampholytic properties. Abdelraby et al. developed the DPI formulation of methotrexate GP using the concept nano-in-microparticles approach [149]. As the hydrophobic alkyl chain of leucine could act to reduce the cohesive interparticulate interactions between the microparticles, co-spraying leucine with methotrexate GP would result in particles with good aerosolization properties. The reported FPF and MMAD values for this DPI formulation were 49.53% and 2.59 μm [149]. In earlier reports, gelatin nanoparticles (GP) for inhalation were developed as nebulized formulation [150, 151]. A biotinylated epidermal growth factor (EGF)-modified GP induced negligible acute lung inflammation in nude mice following nebulization. The concentration of biotinylated EGF-GP was much higher in tumor cells compared healthy cells, which suggested the successful targeting toward EGFR-overexpressing cancer cells [150]. In another study, nebulization of cisplatin-loaded EGF-GP to mice reduced the tumor burden by 70% following 17 days posttreatment [151]. Long et al. demonstrated that DOX-loaded EGF-GP was selectively internalized by EGFR overexpressing cancer cells in vitro. In vivo administration of DOX-loaded EGF-GP resulted in 90% tumor inhibition, which was attributable to the high residence time of the DOX in lung (more than 24 h) [152]. In addition to EGR, other targeting ligands have been employed for treatment of lung cancer via pulmonary route [152–154]. These ligands mediated nanoformulations maintained its targeting specific behavior after inhalation and were generally more superior in terms of therapeutic efficacy. Owing to the specific and preferential uptake of chemotherapeutic drugs, ligand-mediated nanoformulation were also associated with reduced systemic side effects [152–154].

Two established antimalarial drugs, quinacrine and amodiaquine, were recently repositioned as potential anticancer agent, specifically developed as NP for inhalation to treat lung cancer [155, 156]. The high respirable fraction for amodiaquine-loaded polylactide-*co*-glycolide (PLGA) nanoparticle at 81% suggested that majority of emitted dose are deposited at the respirable region of the lung [155]. The same group also employed a similar PLGA NP carrier approach for the encapsulation of quinacrine. Slight modifications on the PLGA NP include the addition of a cationic polyethyleneimine (PEI) and bovine serum albumin (BSA) outer coating layer. PEI was introduced to confer positive charges to the NP, while BSA acted to attenuate the burst release and NP toxicities [156]. As expected, quinacrine NP was stable after nebulization and exhibited good aerosol performance. This nanoformulation displayed enhanced therapeutic efficacy against NSCLC in both in vitro and 3D spheroid models via induction of cell cycle arrest at G2/M phase and autophagy inhibition [156]. The same group also published the similar findings on the effectiveness of inhalable resveratrol-cyclodextrin complex-loaded PLGA NP against NSCLC [157]. Combination therapy of inhaled chemotherapeutic and photosentisizer was explored for the treatment of primary lung cancer recently [158]. Both in vitro and in vivo anticancer effects were investigated using gefitinib PLGA NP and 5-aminolevulinic acid (5-ALA) as photosensitizer. Remarkable synergism on the combinational therapy was noted as reflected in the significant decrease of tumor nodules and inhibition of inflammatory cell infiltration in Sprague–Dawley rates after treatment [158].

The development and evaluation of an aerosolized dry powder nanocomposite of PTX-loaded NP against air-interface 3D lung tumor spheroids was reported recently [159]. A single emulsion/solvent evaporation technique was employed for the encapsulation of PTX into acetylated dextran to form unimodal distribution nanoparticles with size of 200 nm. Co-spray drying the PTX-loaded NP with mannitol produced amorphous wrinkled particles exhibiting good aerosolization properties (MMAD = 2.44 μm, FPF = 66%, respirable fraction = 97%, and emitted dose = 88%) [159]. A 3D multicellular spheroid model grown in air-interface culture to mimic the solid lung tumor was used for a more realistic representation of solid tumor in relation to physiological responses, cell differentiation, and cell-drug interactions [159]. The exposure of 3D spheroid tumor cells with aerosolized PTX-loaded NP led to dramatic decrease in tumor size and disaggregation of cells. No appreciable growth in solid tumor was observed up to 15 days posttreatment [159]. In another study, etoposide/berberine loaded albumin nanoparticles was successfully converted into nanocomposite via spray drying with mannitol or leucine as carriers [160]. Mannitol was chosen over lactose to avoid the Maillard reaction between albumin and lactose [161]. Optimization of the mannitol to NP ratio to 3:1 during spray drying resulted in spherical rough particles with higher dried powder yield (78.4%) and excellent re-dispersibility [160]. On the other hand, the ratio of leucine:mannitol:NP at 1:2:1 produced hollow doughnut-shaped particles that are less cohesive and enhanced dispersibility. This is evident with the higher FPF for leucine/mannitol nanocomposite compared to mannitol nanocomposite (92.48% vs 77.86%). Mice treated with inhalable nanocomposite showed 1.89 reduction in lung weight, which corresponded to decrease of solid tumor nodules [160].

3.2.2.3 Other lipid-based nanocarrier

Nanostructured lipid carriers (NLC) are made up of a lipid phase (solid and liquid) and aqueous phase containing surfactants. Kaur et al. investigated the effect of surfactants (Cremophor, Tween 80, Tween 20) and spray-drying process parameter on the physicochemical and therapeutic properties of inhalable PTX/DOX-loaded NLC particles [162]. Inhalable dried powder nanoaggregates of these surfactant-differed-NLC prepared via spray drying exhibited excellent aerodynamic diameter ranging from 1.6 to 2.2 μm. The high dispersibility and thermal stability of cremophor resulted in the most stable nanoaggregates, highest DOX/PTX internalization into A549 cells, and in vivo distribution (retention and accumulation) in lung of Wistar rat [162]. In addition, Kamel et al. demonstrated that inhalable hybrid lipid nanocore-protein shell nanoparticles (HLPNP), encapsulating retinoic acid and genistein were superior compared to nanoparticle suspension against lung carcinoma [163].

4 Treatment of metastatic lung cancer

Metastasis in NSCLC is associated with modifications at molecular levels of the cancer cells, involving numerous gene expressions and signaling proteins. The subtypes of NSCLC can be identified through molecular manners since they possess unique gene expression and specific biomarkers, except for large-cell carcinoma, which is yet to exert specific biomarkers. Examples include adenocarcinoma expressing a p53 homologous nuclear protein, thyroid transcription factor-1, which is involved in basal cell commitment. On the other hand, SCLC exerts achaete-scute homolog 1, synaptophysin, and neural cell adhesion molecule, all accounting for neuroendocrine markers [164].

Distant metastasis of adenocarcinoma to the brain, adrenal glands, bone, and also to the ovary, which is a rare phenomenon, has been described by Chahin et al. and Wang et al. [165, 166]. It is known that adenocarcinomas and small-cell neuroendocrine carcinomas prefer metastasis to the brain, whereas squamous cell carcinoma metastasize to bones. The carcinomas leave the pulmonary circulation upon receiving specific signals for specific organs, for example, *E*-selectin, which is common in lung and breast cancer. There is a high probability for E-selectin to be overexpressed during systemic inflammation. The distant metastasis of lung carcinoma is further enabled through hyper-permeability mediated by endothelial cell FAK, which upregulates E-selectin for preferential homing of carcinoma at other organs. During hyperpermeability, blood flow is reduced, allowing the tumors to roll over endothelia allowing them to find a suitable space to grow. Metastasizing NSCLCs need to communicate with the new environment for the homing process to adapt and survive [167]. Most of singular NSCLC leaving the pulmonary circulation will not survive at distant metastasis and only a small percentage of tumors progress to form micronodules. The attachment of tumors to the endothelial cells are supported by several adhesion molecules such as vascular cell adhesion molecule-1 (VCAM-1) and vascular adhesion protein-1 (VAP-1), that is reliant on tumor cell-clot formation triggered by tissue coagulation factors. Modifications at the cell–cell junction of endothelia are necessary for the tumor to diffuse through inter-endothelial spaces. For the modification at cell–cell junction to take place, secretion of multiple permeability factors and upregulation of certain chemokine factors are required to enhance hyperpermeability [168, 169]. The permeability factors vary according to different models of tumor and the site of metastasis. Several factors that influence the homing and progression of metastatic NSCLC have been identified in various preclinical and clinical samples, which includes unique peptides like Semaphorin 5A, chemokine receptors like CXCR4, signaling proteins like matrix metalloproteinase (MMP), and at molecular level expression of microRNA (miRNA), which suppresses phosphatase and tensin homolog (PTEN) while activating the signals for Akt phosphorylation and NF-κβ activation [170]. These are the common factors influencing the distant metastasis and homing of NSCLC.

Recently, Lee et al. have reported the involvement of cluster of differentiation 109 (CD109), a glycosylphosphatidylinositol (GPI)-anchored protein belonging to the α2-macroglobulin/C3, C4, and C5 family, in metastasis of lung cancer by promoting invasiveness and growth of adenocarcinoma cells [171]. This protein was reported to be highly expressed by squamous cell carcinoma of NSCLC and was also found to be upregulated in several other malignancies. The mechanism of CD109 in triggering metastasis is largely undefined, but it has shown active involvement in regulating Janus kinase (JAK)/signal transducer and activator of transcription (STAT) signaling pathway in lung cancer. The JAK/STAT signaling pathway is an active regulator of the immune system, which includes activation of growth factors and various cytokines [172]. Expression of CD109 was also associated with the regulation of epidermal growth factor receptor (EGFR) pathway and its downstream signaling, including protein kinase B (Akt)/mammalian target of rapamycin (mTOR) pathway, which are important cell cycle regulators. Furthermore, CD109 is reportedly involved in the sensitivity to EGFR tyrosine kinase inhibitors (TKIs). Overexpression of the CD109 protein was associated with reduced survival of patients with lung adenocarcinoma. The driver mutations in metastasis were highlighted in the studies, and the metastatic NSCLC was linked to EGFR and echinoderm microtubule-associated protein-like 4 (EML-4)-anaplastic lymphoma kinase (ALK) rearrangement [173, 174]. Similar results were described by Mohapatra et al. stating concomitant mutations of EGFR/ALK in NSCLC of lung cancer patients from eastern India [175]. The aberrant EML4-ALK protein triggers the downstream signals of extracellular signal regulated kinase 1 and 2 (ERK 1/2), STAT3, and Akt, exhibiting the similar pattern of metastasis due to CD109. The ALK/ROS1 rearrangement mutations are prominent molecular drivers in metastasis, as exhibited in most of the primary carcinoma and also in metastasized NSCLC. It is noticeable that primary carcinoma and metastatic NSCLC have similar genetic or driver mutation in general, but some additional genetic modifications distinguish the primary carcinoma and metastatic NSCLC.

Metastasis of lung carcinoma is known to progress through blood stream and lymphatic systems. The NSCLCs are encountered at multiple sites and organs upon metastasizing, hence the targeted treatment should also be distributed to multiple sites in the patient's body. Considering the common driver mutations and proteins involved in the metastasis, tyrosine kinase inhibitor (TKI) drugs were introduced as an intervention to the metastatic progression [176]. Platinum-doublet chemotherapy was the basis of treatment for metastatic NSCLC in patients newly diagnosed with lung carcinoma. The conventional chemotherapy was replaced with targeted treatment using TKIs especially by targeting EGFR mutation and ALK rearrangement, which showed how positive indications on the patients' progression-free survival rate through reduction in metastasis and tumor growth inhibition [177]. The targeted protein or gene needs to be ascertained prior to treatment with TKIs to obtain a fruitful result. The life quality of patients improved in the initial stages upon administration of targeted therapy using TKIs such as erlotinib, crizotinib, and cabozatinib [178] (see Section 2). Other pathways involved in targeted therapy in metastatic lung cancer include RAS/MAPK, NTRK/ROS1, and PI3K/Akt/mTOR with specific drugs developed for each pathway and were clinically effective to replace chemotherapy as the first-line treatment, for instance entrectinib as NTRK/ROS1 pathway inhibitor and everolimus as PI3K/Akt/mTOR inhibitor. Concomitant mutations in metastasized NSCLC are treated with multi-TKI drugs, since the specific TKI for EGFR was not able to provide better progression-free survival rate for patients with simultaneous ALK mutation [177]. Though successful in the early periods of treatment, administration of targeted TKIs exerted signs of drug resistance in patients in the later stages.

Various inhalable formulations have been explored to increase the therapeutic efficacy against metastatic lung cancer (Table 4). Liu et al. recently reported the synthesis and efficacy of a stimuli-triggered formulation containing DOX-loaded sericin microparticles coated with metal–organic networks (DOX@SMPs-MON) [189]. A two-step synthesis method was described, whereby sericin microparticles with an average diameter of 4.6 μm were first prepared via single w/o emulsion method. This is followed with DOX loading and subsequent coating with a pH-sensitive metal–organic layer composed of tannic acid and ferric ions [189]. DOX is released in a pH-dependent manner and is favored in acidic conditions. Up to 86.3% of DOX release is achieved within 72 h in pH 5. In contrast, the cumulative release at pH 7.4 is significantly lower (37.2%). Intratracheal administration of DOX@SMPs-MON into metastatic lung model demonstrated significant decrease in mass size and number of metastatic foci [189]. The effect of initial particle size PTX-loaded albumin microparticles on the biodistribution, accumulation, and therapeutic efficacy against metastatic lung cancer was studied recently [190]. For this, three different sizes (0.5, 1.0, and 3.0 μm) of PTX-loaded albumin microparticles were converted into DPI with uniform size (5 μm) via spray drying. Similar with other findings, delivery of DPI samples via inhalation resulted in enhanced drug retention and efficacy compared to IV PTX solution. Although smaller microparticles (0.5 μm) facilitated faster drug release action, the 3.0 μm microparticle group exerted higher antitumor activity in vivo [190]. Curcumin as dried powder for inhalation based on mesoporous silica for metastatic cancer treatment reduced the number of B16F10 cells in C57BL/mice with the lung metastasis cancer [191]. In another work, PTX-loaded solid lipid nanoparticles (SLN) demonstrated more superior efficacy against MXT-B2 cells in both in vitro and in vivo settings compared to PTX solution [192]. Up to 20-fold higher in vitro antiproliferative effect of MXT-B2 cells was noted. Complete inhibition of lung metastasis and tumor growth was noted for MXT-B2 harboring B6D2F1 female mice receiving prolonged doses of PTX-SLN (1 mg/kg/dose) via pulmonary route. In contrast, for the group receiving IV PTX solution (2.4 mg/kg/dose), aggressive metastatic growth and lung regression were observed [192].

TABLE 4 Representative preclinical studies of inhalation therapy against metastatic lung cancers.

Drugs	Formulation (Carrier)	Particle size	Outcome	References
DOX	PLGA microparticles with surface-modified TRAIL	11.5±0.4 μm	The numbers and tumor size in BALB/c nu/nu mice model bearing H226 cells (metastatic lung cancer) were significantly reduced and smaller when treated with DOX-loaded PLGA microparticles decorated with TRAIL compared to TRAIL alone and nonmodified DOX-PLGA microparticles	[179]
DOX	Porous PLGA microparticles	14.1±2.1 μm	The microparticles could be retained in the lung for up to 14 days. The B16F10 tumor bearing mice that responded positively toward DOX-loaded PLGA microparticles as the mass and number of tumor were significantly reduced compared to control groups	[112]
DOX	PEGylated polylysine dendrimer	Not mentioned	Pulmonary instillation of this formulation was more effective in reducing the lung tumor burden (> 95%) compared to IV administration (30%–50% reduction). DOX encapsulated with dendrimer significantly reduced the drug toxicity-induced effects. Pulmonary instillation of naked DOX solution resulted in severe toxicity and death in MAT13762IIIB rat model	[180]
DOX	Poly(amidoamine) dendrimer (PAMAM)	4.7–9.7 nm	Inhaled formulation exerted stronger effect compared to intravenous administration. Higher retention of DOX in the lung and increased survival rate in rats	[181]
siRNA and Doxorubicin	pH sensitive nanoparticles	78.2±4.1 nm	In in vitro experiment, siRNA/DOX conjugated with PEI nanoparticle displayed more potent antiproliferative effect against B16F10 cells compared to siRNA or DOX alone. Higher drug residence time was observed following pulmonary administration compared to systemic IV route	[182]
Bcl2 siRNA and DOX	pH sensitive nanoparticles	76.0 nm	Local co-delivery into lung of B16F10 melanoma-bearing mice resulted in higher accumulation/retention of drug and gene. Higher antitumor activity was observed, and drugs were released from cargo in a dose-dependent manner	[183]
siRNA	Cationic nanoliposome	˂25 nm	Intrapulmonary delivery of the siRNA-loaded nanoliposome silenced the expressions of Mcl1 mRNA in two metastatic lung mouse models (B16F10 and Lewis carcinoma)	[184]
Interferon-β	Chitosan	Not mentioned	Intratracheal administration of dried powdered formulation pCMVMuβ was found more superior in maintaining the gene expression activity even at low dose (10 μg). Intravenous injection of interferon-β demonstrated a lack of activity In addition, the CT26 inoculated mice responded better toward inhaled therapy compared to IV therapy	[185]
Small hairpin osteopontin	–	Not mentioned	Delivery of lentivirus-based small hairpin osteopontin twice a week for 1–2 months reduced the metastatic level of lung tumors. Expressions of metastasis-related protein such as CD44v6 and VEGF were gradually reduced. Migration of breast cancer cells (MDA-MB231) toward lung tissues was inhibited	[186]
Curcumin	Theracurmin®	Not mentioned	Dose-dependent inhibition of B16F10 melanoma cells in C57BL/6J mice following pulmonary administration. Up to 60-fold higher systemic bioavailability compared to oral administration	[187]
DOX and Survivin siRNA	pH-sensitive hydrazine bond (3-maleimidopropionic acid hydrazide, BMPH) nanoparticle	84.0±3.7 nm	Release of drug in response to pH changes, i.e., higher release rate in acidic condition. Preferential accumulation of DOX/siRNA in tumor tissue of lung in vivo. Negligible uptake by healthy lung cells. Co-delivery of DOX/siRNA nanoparticle via pulmonary administration is more superior in terms of antitumor activity compared to monoformulation	[188]

5 The mechanism of drug resistance in lung cancer and its associated treatments

The development of drug resistance cases in patients undergoing chemotherapy are common for metastatic NSCLC. Mechanism of acquired drug resistance can be studied at three domains, namely, molecular, cellular, and pathological, given the type of cancer cell involved [178]. Alterations at molecular levels are the major causes for drug resistance, post-administration of chemotherapeutic or targeted therapy drugs, and such phenomena is known as acquired drug resistance. An example for developing this type of resistance in chemotherapy or platinum-based chemotherapy where paclitaxel is administered, inadequate amount of drug reaches the DNA, SRC-activated ERK pathway, and lack of DNA repair mechanism [193]. Drug resistance can be divided to on-target resistance, that is, when the direct molecular target of drug gets mutated causing poor or no response to the drug, and off-target resistance, that takes place by activation of signaling pathways parallel to or by-pass the signals downstream of targeted site. The development of drug resistance in tumors has been linked to rapid mutation of gene expression profile and multilevel epigenetic variations. Acquired drug resistance becomes the most common type of resistance besides intrinsic resistance because of evolving gene expression patterns under selection during therapy [194]. Intrinsic resistance occurs through epigenetic mutations in carcinomas to hinder host's immune reaction, which turns to develop resistance to therapeutic drugs as well. Development of drug resistance in tumor and metastasized NSCLC also depends on the interaction of the cancer cells with their surrounding environment, especially during homing. Mutation at gene due to acquired resistance to chemotherapy in NSCLC has brought to the generation of Multi-Drug Resistant gene-1 (MDR1), Multi-Drug Resistance associated Protein-1 (MRP1), ATP-binding cassette subfamily G member-2 (ABCG2), lung resistance protein (LRP), and many others [195].

Modified membrane transport is an initial step in multidrug resistance. Therapeutic drugs for cancer needs to penetrate the cellular membrane of the carcinoma to reach the DNA target site and other intracellular components. A membrane efflux pump protein that functions to expel the drug out of the cells is the phosphorylated and glycosylated p-glycoprotein (P-gp), or better known as multiple drug resistance protein, encoded by the MDR1 and ABCG2 genes [196]. P-gp prevents the anticancer drug from reaching its target site in the carcinoma, thus reducing the overall drug efficacy. It was found that P-gp were overexpressed in NSCLC cell lines from smokers. Moreover, the MDR1 gene can be triggered and upregulated in response to chemotherapeutic drugs, which explains the drug resistance behavior in most of NSCLC metastasis. Hence, to overcome the drug resistance due to P-gp, aiming MDR1 gene or P-gp is a strategic approach to develop drugs that targets these specific genes to be administered in combination with the chemotherapeutic drugs. A recent clinical study by Jethva et al. on molecular analysis of NSCLC from patients prior to drug administration also suggests that monitoring of the specific gene markers for multidrug resistance could be useful in developing a significant approach for lung cancer treatment [197].

DNA repair pathway is another mechanism of drug resistance. Chemotherapeutic drugs such as cisplatin tend to target the DNA of carcinoma, causing DNA damage or adduct formation, specifically causing mutation and damage to the tumors [198]. DNA damage induced by the chemotherapeutic drugs are encountered by DNA repair mechanisms, which includes nucleotide excision repair (NER) that is primarily involved in repairing the impaired DNA of tumor cells. Therefore, the gene encoding the protein for repair mechanism such as excision-repair cross complementation group 1 (ERCC1) is the molecular signaling reason for resistance toward chemotherapeutic drugs. Clinical evidence of platinum-based chemotherapy drug resistance was presented with upregulated ERCC1 mRNA levels in NSCLC [199]. Other DNA repair mechanisms that are normally upregulated in NSCLC to resist chemotherapy drugs include nonhomologous end joining (NHEJ), base excision repair pathway (BER), mismatch-repair pathway (MMR), and prevention of DNA alkylation by drugs to evade DNA targeting. Alkylation of DNA is performed by certain chemotherapeutic drugs at specific sites of DNA for easier targeting, but enzymes like O6-methylguanine DNA methyltransferase (MGMT) tend to excise the alkyl groups attached to DNA. Another enzyme responsible for DNA repair and synthesis, thymidylate synthase (TS) was also found to be highly expressed in NSCLCs that are under treatment with chemotherapeutic drug pemetrexed [200]. The target of chemotherapy drugs in NSCLC is to trigger apoptosis as one of the method to kill the cancer. Apoptosis pathway can be triggered by inducing DNA damage or triggering apoptosis pathway and related signals in the tumor. But the DNA repair mechanism and other resistance in tumors, causing an impaired apoptosis mechanism and with the deformed pathways, the tumor still survives. This causes stronger resistance in tumors toward chemotherapeutic drugs due to loss of normal apoptosis function. The apoptosis regulator proteins; caspases, Bax, SAPK/JNK, and antiapoptosis proteins; Bcl-2, inhibitor of apoptosis protein (IAP), and survivin are involved in the drug resistance mechanism in NSCLCs with impaired apoptosis pathways. Deranged caspases and apoptosis regulator proteins were also related to drug resistance in NSCLC [201]. Drug resistance has evidently escalated in targeted therapeutic drugs in the similar fashion to chemotherapy, at molecular levels. Targeted therapeutic drugs have specific molecular targets, but it was found that these targets could mutate or modify their configuration to suppress their expression to an untraceable level, thus causing targeted drugs to be ineffective. The antioxidant support within tumors such as reduced glutathione (GSH), metallothioneins (MTs), phase II metabolizing antioxidant enzymes, glutathione S-transferase (GST), superoxide dismutase (SOD) interacts with the therapeutic drugs to form conjugates and neutralizes the drugs [202]. These antioxidant defense system causes drug resistance in NSCLC as a preventive measure against oxidative stress and cell/DNA damaging effect of targeted or chemotherapy drugs. Targeted therapies have encountered major hurdle due to drug resistance in genetically mutated NSCLC. Majority of the patients undergoing targeted therapy such as TKI as the first-line treatment have developed drug resistance after 1 year of drug administration. Therefore, initial detection of molecular mutations and alterations are required in treating lung cancer based on targeted therapeutics [203]. Recent advances in molecular techniques using RT-PCR and immunohistochemical analysis have implicated the rapid detection of changes in gene expressions of NSCLC for an effective targeted drug administration [11, 16].

6 Conclusions

It is undeniable that the exploration in pathophysiological properties of lung cancer has significantly contributed to our understanding on developing novel and effective therapeutics agents. These include the discovery of novel immune-inhibitor molecules, repurposed drugs for new indications, and new drug designs. The exploitation of nanotechnology has driven the developments of precise, targeted, and personalized medication for cancer therapies (Table 5). Examples of FDA-approved nanoformulation include Doxil® (liposomal doxorubicin), Abraxane® (PTX nanoparticle), Sipuleucel-T (Provenge®), and Talimogene Laherparepvec (T-VEC). In addition, delivery of drugs directly into the lung via inhalation offers another option to increase drug payload, absorption, and bioavailability at target site while reducing the system-associated side effects simultaneously (Table 6). Although several novel formulations specifically for inhalation purposes with superior efficacy over IV or oral route have been designed, the translation into clinical practice is still in its infancy stage. In fact, the clinical trials for most inhaled chemotherapies are at Phase I stage only. Nebulization of aerosol into lung is time-consuming (often 20–30 min) and require multiple sessions per day. In addition, the air contamination as a result of aerosol escape during nebulization is a hazard to both the staffs and patients and thus requires special containment room for such exercise. Another major limitation is the nonachievable DLT due to low concentrations of drugs in nebulized solution/suspension. The trend of inhalation studies is shifted toward the engineering of more complex formulations, which could overcome the abovementioned disadvantages. The engineering of complex nano-in-microparticles or nanoaggregates in the form of DPI ensures a high drug payload in the formulation and hence increases the drug concentration in lung per inhalation. In addition, as DPI device is a breath-actuated device, the time required for drug delivery is short. In future, long-term safety evaluations of these complex sophisticated formulations for both pulmonary and systemic administration should be investigated. It is also obvious that personalized medication in future should be adapted to specific subpopulation of patients. Inhalation therapy perhaps could be proposed as adjuvant therapy in conjunction with other systemic modality (i.e., chemotherapy, radiotherapy, and immunotherapy) based on current preclinical and clinical findings.

TABLE 5 Limitations and advantages of nanotechnology in drug delivery for lung cancer treatment.

Advantages	Limitations
Significantly improves drug solubility	The loading efficiency is relatively low
Improves the drug residency in lung tumor tissue by using bioadhesive polymers	The excipient used in the production of nanoformulation could induce toxicity toward lung tissue
Increases the overall bioavailability of drugs in tumor tissue as the size of drugs has been reduced	The biodegradation of nanocarriers such as polymer materials is limited due to low enzymatic activity in the lung
The internalization of drugs into the cells can be easily regulated (i.e., decoration of active moiety onto nanoparticle to target specific surface receptors expressed on lung cancers)	Large-scale production is not easily done
The drugs could be easily deposited deep into the lung region. Physicochemical characteristics of nanoparticle can be alterable to provide sustained release behavior	Long-term storage stability of complex nanoformulation is an issue

TABLE 6 Limitations and advantages of pulmonary and systemic delivery in lung cancer treatment.

Pulmonary delivery	Systemic delivery
Advantages	*Advantages*
Higher absorption rate	Combination therapy is easily done
Minimizing drug toxicity effect	Higher dosage could be delivered
Rapid onset of action	Rapid onset of action
Limited intracellular and extracellular drugs metabolism enzymes	Enhanced retention and permeation effect
Limitations	*Limitations*
Not suitable for patients with impaired respiratory flow	Systemic toxicity
Particle size must range between 1 and 5 μm for efficient deposition in the lung	Large portion of drugs could be degraded by the enzymes in liver and gastrointestinal tract for oral chemotherapeutics
High clearance by ciliated cells and macrophages	Poor drug absorption
Local side effects could be enhanced if high dose of chemotherapeutics is delivered at once	
For patients with comorbidity such as COPD, the presence of thick mucus and restricted airway reduced the efficacy of inhaled nanoformulation	

Acknowledgments

The manuscript is funded by Fundamental Grant Research Scheme (FRGS/1/2018/STG03/UNIKL/02/2). C. Y. Loo is funded by Universiti Kuala Lumpur short-term research grant (UniKL/CoRI/str19078).

References

[1] Key statistics for lung cancer, http://www.cancer.org/cancer/lung-cancer-non-smallcell/detailedguide/non-small-cell-lung-cancer-key-statistics#; 2016. [updated 05/16/2016].

[2] Youlden DR, Cramb SM, Baade PD. The international epidemiology of lung Cancer: geographical distribution and secular trends. J Thorac Oncol 2008;3:819–31.

[3] Lemjabbar-Alaoui H, Hassan OU, Yang YW, Buchanan P. Lung cancer: biology and treatment options. Biochim Biophys Acta 2015;1856:189–210.

[4] Lee W-H, Loo C-Y, Ghadiri M, Leong C-R, Young PM, Traini D. The potential to treat lung cancer via inhalation of repurposed drugs. Adv Drug Deliv Rev 2018;133:107–30.

[5] Popper HH. Progression and metastasis of lung cancer. Cancer Metastasis Rev 2016;35:75–91.

[6] Kim GT, Hahn KW, Yoon SY, Sohn KY, Kim JW. PLAG exerts anti-metastatic effects by interfering with neutrophil elastase/PAR2/EGFR signaling in A549 lung cancer orthotopic model. Cancers (Basel) 2020;12.

[7] In GK, Nieva J. Emerging chemotherapy agents in lung cancer: nanoparticles therapeutics for non-small cell lung cancer. Transl Cancer Res 2015;4:340–55.

[8] Soares PIP, Romão J, Matos R, Silva JC, Borges JP. Design and engineering of magneto-responsive devices for cancer theranostics: Nano to macro perspective. Prog Mater Sci 2021;116:100742.

[9] Vanza JD, Patel RB, Patel MR. Nanocarrier centered therapeutic approaches: recent developments with insight towards the future in the management of lung cancer. J Drug Delivery Sci Technol 2020;60:102070.

[10] Sadhasivam J, Sugumaran A. Magnetic nanocarriers: emerging tool for the effective targeted treatment of lung cancer. J Drug Delivery Sci Technol 2020;55:101493.

[11] Sharma P, Mehta M, Dhanjal DS, Kaur S, Gupta G, Singh H, Thangavelu L, Rajeshkumar S, Tambuwala M, Bakshi HA, Chellappan DK, Dua K, Satija S. Emerging trends in the novel drug delivery approaches for the treatment of lung cancer. Chem Biol Interact 2019;309:108720.

[12] Dholaria B, Hammond W, Shreders A, Lou Y. Emerging therapeutic agents for lung cancer. J Hematol Oncol 2016;9:138.

[13] Zappa C, Mousa SA. Non-small cell lung cancer: current treatment and future advances. Transl Lung Cancer Res 2016;5:288–300.

[14] Jonna S, Reuss JE, Kim C, Liu SV. Oral chemotherapy for treatment of lung Cancer. Front Oncol 2020;10:793.

[15] Sosa Iglesias V, Giuranno L, Dubois LJ, Theys J, Vooijs M. Drug resistance in non-small cell lung cancer: a potential for NOTCH targeting? Front Oncol 2018;8:267.

[16] Liu WJ, Du Y, Wen R, Yang M, Xu J. Drug resistance to targeted therapeutic strategies in non-small cell lung cancer. Pharmacol Ther 2020;206:107438.

[17] Old LJ. Cancer immunology. Sci Am 1977;236:62–79.

[18] Silva AP, Coelho PV, Anazetti M, Simioni PU. Targeted therapies for the treatment of non-small-cell lung cancer: monoclonal antibodies and biological inhibitors. Hum Vaccin Immunother 2017;13:843–53.

[19] Herst PM, Berridge MV. Cell hierarchy, metabolic flexibility and systems approaches to cancer treatment. Curr Pharm Biotechnol 2013;14:289–99.

[20] History of vaccine. Cancer vaccine and immunotherapy; 2018.

[21] Scaltriti M, Baselga J. The epidermal growth factor receptor pathway: a model for targeted therapy. Clin Cancer Res 2006;12:5268–72.
[22] Janjigian YY, Azzoli CG, Krug LM, Pereira LK, Rizvi NA, Pietanza MC, Kris MG, Ginsberg MS, Pao W, Miller VA, Riely GJ. Phase I/II trial of cetuximab and erlotinib in patients with lung adenocarcinoma and acquired resistance to erlotinib. Clin Cancer Res 2011;17:2521–7.
[23] Hilbe W, Pall G, Kocher F, Pircher A, Zabernigg A, Schmid T, Schumacher M, Jamnig H, Fiegl M, Gächter A, Freund M, Kendler D, Manzl C, Zelger B, Popper H, Wöll E. Multicenter phase II study evaluating two cycles of docetaxel, cisplatin and Cetuximab as induction regimen prior to surgery in chemotherapy-naive patients with NSCLC stage IB-IIIA (INN06-study). PloS One 2015;10:e0125364.
[24] Azad N, Dasari A, Arcaroli J, Taylor GE, Laheru DA, Carducci MA, McManus M, Quackenbush K, Wright JJ, Hidalgo M, Diaz Jr LA, Donehower RC, Zhao M, Rudek MA, Messersmith WA. Phase I pharmacokinetic and pharmacodynamic study of cetuximab, irinotecan and sorafenib in advanced colorectal cancer. Invest New Drugs 2013;31:345–54.
[25] Bittner N, Ostoros G, Geczi L. New treatment options for lung adenocarcinoma- -in view of molecular background. Pathol Oncol Res 2014;20:11–25.
[26] Yeo WL, Riely GJ, Yeap BY, Lau MW, Warner JL, Bodio K, Huberman MS, Kris MG, Tenen DG, Pao W, Kobayashi S, Costa DB. Erlotinib at a dose of 25 mg daily for non-small cell lung cancers with EGFR mutations. J Thorac Oncol 2010;5:1048–53.
[27] Cohen MH, Johnson JR, Chattopadhyay S, Tang S, Justice R, Sridhara R, Pazdur R. Approval summary: erlotinib maintenance therapy of advanced/metastatic non-small cell lung cancer (NSCLC). Oncologist 2010;15:1344–51.
[28] Zhou C, Wu YL, Chen G, Feng J, Liu XQ, Wang C, Zhang S, Wang J, Zhou S, Ren S, Lu S, Zhang L, Hu C, Hu C, Luo Y, Chen L, Ye M, Huang J, Zhi X, Zhang Y, Xiu Q, Ma J, Zhang L, You C. Erlotinib versus chemotherapy as first-line treatment for patients with advanced EGFR mutation-positive non-small-cell lung cancer (OPTIMAL, CTONG-0802): a multicentre, open-label, randomised, phase 3 study. Lancet Oncol 2011;12:735–42.
[29] Satoh H, Inoue A, Kobayashi K, Maemondo M, Oizumi S, Isobe H, Gemma A, Saijo Y, Yoshizawa H, Hagiwara K, Nukiwa T. Low-dose gefitinib treatment for patients with advanced non-small cell lung cancer harboring sensitive epidermal growth factor receptor mutations. J Thorac Oncol 2011;6:1413–7.
[30] Rafii S, Lyden D, Benezra R, Hattori K, Heissig B. Vascular and haematopoietic stem cells: novel targets for anti-angiogenesis therapy? Nat Rev Cancer 2002;2:826–35.
[31] Claret L, Gupta M, Han K, Joshi A, Sarapa N, He J, Powell B, Bruno R. Prediction of overall survival or progression free survival by disease control rate at week 8 is independent of ethnicity: Western versus Chinese patients with first-line non-small cell lung cancer treated with chemotherapy with or without bevacizumab. J Clin Pharmacol 2014;54:253–7.
[32] Kubota T, Okano Y, Sakai M, Takaoka M, Tsukuda T, Anabuki K, Kawase S, Miyamoto S, Ohnishi H, Hatakeyama N, Machida H, Urata T, Yamamoto A, Ogushi F, Yokoyama A. Carboplatin plus weekly paclitaxel with bevacizumab for first-line treatment of non-small cell lung Cancer. Anticancer Res 2016;36:307–12.
[33] Herbst RS, Johnson DH, Mininberg E, Carbone DP, Henderson T, Kim ES, Blumenschein Jr G, Lee JJ, Liu DD, Truong MT, Hong WK, Tran H, Tsao A, Xie D, Ramies DA, Mass R, Seshagiri S, Eberhard DA, Kelley SK, Sandler A. Phase I/II trial evaluating the anti-vascular endothelial growth factor monoclonal antibody bevacizumab in combination with the HER-1/epidermal growth factor receptor tyrosine kinase inhibitor erlotinib for patients with recurrent non-small-cell lung cancer. J Clin Oncol 2005;23:2544–55.
[34] Tiseo M, Bartolotti M, Gelsomino F, Ardizzoni A. First-line treatment in advanced non-small-cell lung cancer: the emerging role of the histologic subtype. Expert Rev Anticancer Ther 2009;9:425–35.
[35] Su YL, Rau KM. Adding bevacizumab to chemotherapy effectively control radioresistant brain metastases in ALK-positive lung adenocarcinoma. J Thorac Oncol 2015;10:e21–2.
[36] Keir ME, Butte MJ, Freeman GJ, Sharpe AH. PD-1 and its ligands in tolerance and immunity. Annu Rev Immunol 2008;26:677–704.
[37] Schulze AB, Schmidt LH. PD-1 targeted immunotherapy as first-line therapy for advanced non-small-cell lung cancer patients. J Thorac Dis 2017;9:E384–6.
[38] Garon EB. Current perspectives in immunotherapy for non-small cell lung cancer. Semin Oncol 2015;42(Suppl 2):S11–8.
[39] Sul J, Blumenthal GM, Jiang X, He K, Keegan P, Pazdur R. FDA approval summary: pembrolizumab for the treatment of patients with metastatic non-small cell lung Cancer whose tumors express programmed death-ligand 1. Oncologist 2016;21:643–50.
[40] Garon EB, Rizvi NA, Hui R, Leighl N, Balmanoukian AS, Eder JP, Patnaik A, Aggarwal C, Gubens M, Horn L, Carcereny E, Ahn MJ, Felip E, Lee JS, Hellmann MD, Hamid O, Goldman JW, Soria JC, Dolled-Filhart M, Rutledge RZ, Zhang J, Lunceford JK, Rangwala R, Lubiniecki GM, Roach C, Emancipator K, Gandhi L. Pembrolizumab for the treatment of non-small-cell lung cancer. N Engl J Med 2015;372:2018–28.
[41] Reck M, Heigener D, Reinmuth N. Immunotherapy for small-cell lung cancer: emerging evidence. Future Oncol 2016;12:931–43.
[42] Gridelli C, Besse B, Brahmer JR, Crinò L, Felip E, de Marinis F. The evolving role of Nivolumab in non-small-cell lung Cancer for second-line treatment: a new cornerstone for our treatment algorithms. Results from an international experts panel meeting of the Italian Association of Thoracic Oncology. Clin Lung Cancer 2016;17:161–8.
[43] Kazandjian D, Suzman DL, Blumenthal G, Mushti S, He K, Libeg M, Keegan P, Pazdur R. FDA approval summary: Nivolumab for the treatment of metastatic non-small cell lung Cancer with progression on or after platinum-based chemotherapy. Oncologist 2016;21:634–42.
[44] Olmos D, Postel-Vinay S, Molife LR, Okuno SH, Schuetze SM, Paccagnella ML, Batzel GN, Yin D, Pritchard-Jones K, Judson I, Worden FP, Gualberto A, Scurr M, de Bono JS, Haluska P. Safety, pharmacokinetics, and preliminary activity of the anti-IGF-1R antibody figitumumab (CP-751,871) in patients with sarcoma and Ewing's sarcoma: a phase 1 expansion cohort study. Lancet Oncol 2010;11:129–35.
[45] Rossi A, Maione P, Bareschino MA, Schettino C, Sacco PC, Ferrara ML, Castaldo V, Gridelli C. The emerging role of histology in the choice of first-line treatment of advanced non-small cell lung cancer: implication in the clinical decision-making. Curr Med Chem 2010;17:1030–8.
[46] Maki RG. Small is beautiful: insulin-like growth factors and their role in growth, development, and cancer. J Clin Oncol 2010;28:4985–95.

[47] Machac J, Krynyckyi B, Kim C. Peptide and antibody imaging in lung cancer. Semin Nucl Med 2002;32:276–92.

[48] Straka MR, Joyce JM, Myers DT. Tc-99m nofetumomab merpentan complements an equivocal bone scan for detecting skeletal metastatic disease from lung cancer. Clin Nucl Med 2000;25:54–5.

[49] Xu S, Ramos-Suzarte M, Bai X, Xu B. Treatment outcome of nimotuzumab plus chemotherapy in advanced cancer patients: a single institute experience. Oncotarget 2016;7:33391–407.

[50] Kim SH, Shim HS, Cho J, Jeong JH, Kim SM, Hong YK, Sung JH, Ha SJ, Kim HR, Chang H, Kim JH, Tania C, Cho BC. A phase I trial of gefitinib and nimotuzumab in patients with advanced non-small cell lung cancer (NSCLC). Lung Cancer 2013;79:270–5.

[51] Lin S, Yan Y, Liu Y, Gao CZ, Shan D, Li Y, Han B. Sensitisation of human lung adenocarcinoma A549 cells to radiotherapy by Nimotuzumab is associated with enhanced apoptosis and cell cycle arrest in the G2/M phase. Cell Biol Int 2015;39:146–51.

[52] Tan EH, Lim WT, Ahn MJ, Ng QS, Ahn JS, Shao-Weng Tan D, Sun JM, Han M, Payumo FC, McKee K, Yin W, Credi M, Agarwal S, Jac J, Park K. Phase 1b trial of Ficlatuzumab, a humanized hepatocyte growth factor inhibitory monoclonal antibody, in combination with Gefitinib in Asian patients with NSCLC. Clin Pharmacol Drug Dev 2018;7:532–42.

[53] Scagliotti GV, Novello S, von Pawel J. The emerging role of MET/HGF inhibitors in oncology. Cancer Treat Rev 2013;39:793–801.

[54] D'Arcangelo M, Cappuzzo F. Focus on the potential role of ficlatuzumab in the treatment of non-small cell lung cancer. Biol Theory 2013;7:61–8.

[55] Chen Q, Hu C, Liu Y, Song R, Zhu W, Zhao H, Nino A, Zhang F, Liu Y. Pharmacokinetics, pharmacodynamics, safety, and tolerability of single-dose denosumab in healthy Chinese volunteers: a randomized, single-blind, placebo-controlled study. PLoS One 2018;13:e0197984.

[56] Carbone F, Crowe LA, Roth A, Burger F, Lenglet S, Braunersreuther V, Brandt KJ, Quercioli A, Mach F, Vallee JP, Montecucco F. Treatment with anti-RANKL antibody reduces infarct size and attenuates dysfunction impacting on neutrophil-mediated injury. J Mol Cell Cardiol 2016;94:82–94.

[57] Scagliotti GV, Hirsh V, Siena S, Henry DH, Woll PJ, Manegold C, Solal-Celigny P, Rodriguez G, Krzakowski M, Mehta ND, Lipton L, García-Sáenz JA, Pereira JR, Prabhash K, Ciuleanu TE, Kanarev V, Wang H, Balakumaran A, Jacobs I. Overall survival improvement in patients with lung cancer and bone metastases treated with denosumab versus zoledronic acid: subgroup analysis from a randomized phase 3 study. J Thorac Oncol 2012;7:1823–9.

[58] Sanghavi K, Zhang J, Zhao X, Feng Y, Statkevich P, Sheng J, Roy A, Vezina HE. Population pharmacokinetics of Ipilimumab in combination with Nivolumab in patients with advanced solid tumors. CPT Pharmacometrics Syst Pharmacol 2020;9:29–39.

[59] Golden EB, Demaria S, Schiff PB, Chachoua A, Formenti SC. An abscopal response to radiation and ipilimumab in a patient with metastatic non-small cell lung cancer. Cancer Immunol Res 2013;1:365–72.

[60] Rijavec E, Genova C, Barletta G, Burrafato G, Biello F, Dal Bello MG, Coco S, Truini A, Alama A, Boccardo F, Grossi F. Ipilimumab in non-small cell lung cancer and small-cell lung cancer: new knowledge on a new therapeutic strategy. Expert Opin Biol Ther 2014;14:1007–17.

[61] Centanni M, Moes D, Troconiz IF, Ciccolini J, van Hasselt JGC. Clinical pharmacokinetics and pharmacodynamics of immune checkpoint inhibitors. Clin Pharmacokinet 2019;58:835–57.

[62] Tarhini AA, Kirkwood JM. Tremelimumab, a fully human monoclonal IgG2 antibody against CTLA4 for the potential treatment of cancer. Curr Opin Mol Ther 2007;9:505–14.

[63] Antonia S, Goldberg SB, Balmanoukian A, Chaft JE, Sanborn RE, Gupta A, Narwal R, Steele K, Gu Y, Karakunnel JJ, Rizvi NA. Safety and antitumour activity of durvalumab plus tremelimumab in non-small cell lung cancer: a multicentre, phase 1b study. Lancet Oncol 2016;17:299–308.

[64] Werner TL, Sachdev J, Swisher EM, Gutierrez M, Kittaneh M, Stein MN, Xiong H, Dunbar M, Sullivan D, Komarnitsky P, McKee M, Tan AR. Safety and pharmacokinetics of veliparib extended-release in patients with advanced solid tumors: a phase I study. Cancer Med 2018;7:2360–9.

[65] Farago AF, Yeap BY, Stanzione M, Hung YP, Heist RS, Marcoux JP, Zhong J, Rangachari D, Barbie DA, Phat S, Myers DT, Morris R, Kem M, Dubash TD, Kennedy EA, Digumarthy SR, Sequist LV, Hata AN, Maheswaran S, Haber DA, Lawrence MS, Shaw AT, Mino-Kenudson M, Dyson NJ, Drapkin BJ. Combination Olaparib and Temozolomide in relapsed small-cell lung Cancer. Cancer Discov 2019;9:1372–87.

[66] Owonikoko TK, Zhang G, Deng X, Rossi MR, Switchenko JM, Doho GH, Chen Z, Kim S, Strychor S, Christner SM, Beumer J, Li C, Yue P, Chen A, Sica GL, Ramalingam SS, Kowalski J, Khuri FR, Sun SY. Poly (ADP) ribose polymerase enzyme inhibitor, veliparib, potentiates chemotherapy and radiation in vitro and in vivo in small cell lung cancer. Cancer Med 2014;3:1579–94.

[67] Owonikoko TK, Dahlberg SE, Khan SA, Gerber DE, Dowell J, Moss RA, Belani CP, Hann CL, Aggarwal C, Ramalingam SS. A phase 1 safety study of veliparib combined with cisplatin and etoposide in extensive stage small cell lung cancer: a trial of the ECOG-ACRIN Cancer research group (E2511). Lung Cancer 2015;89:66–70.

[68] Vansteenkiste JF, Canon JL, De Braud F, Grossi F, De Pas T, Gray JE, Su WC, Felip E, Yoshioka H, Gridelli C, Dy GK, Thongprasert S, Reck M, Aimone P, Vidam GA, Roussou P, Wang YA, Di Tomaso E, Soria JC. Safety and efficacy of Buparlisib (BKM120) in patients with PI3K pathway-activated non-small cell lung Cancer: results from the phase II BASALT-1 study. J Thorac Oncol 2015;10:1319–27.

[69] Massacesi C, Di Tomaso E, Urban P, Germa C, Quadt C, Trandafir L, Aimone P, Fretault N, Dharan B, Tavorath R, Hirawat S. PI3K inhibitors as new cancer therapeutics: implications for clinical trial design. Onco Targets Ther 2016;9:203–10.

[70] Okuda T, Okamoto H. Present situation and future Progress of inhaled lung Cancer therapy: necessity of inhaled formulations with drug delivery functions. Chem Pharm Bull 2020;68:589–602.

[71] Rosière R, Berghmans T, De Vuyst P, Amighi K, Wauthoz N. The position of inhaled chemotherapy in the care of patients with lung tumors: clinical feasibility and indications according to recent pharmaceutical progresses. Cancer 2019;11:329.

[72] Abdelaziz HM, Gaber M, Abd-Elwakil MM, Mabrouk MT, Elgohary MM, Kamel NM, Kabary DM, Freag MS, Samaha MW, Mortada SM, Elkhodairy KA, Fang JY, Elzoghby AO. Inhalable particulate drug delivery systems for lung cancer therapy: nanoparticles, microparticles, nanocomposites and nanoaggregates. J Control Release 2018;269:374–92.

[73] Kaur R, Garg T, Rath G, Goyal AK. Advanced aerosol delivery devices for potential cure of acute and chronic diseases. Crit Rev Ther Drug Carrier Syst 2014;31:495–530.

[74] Chan JG, Wong J, Zhou QT, Leung SS, Chan HK. Advances in device and formulation technologies for pulmonary drug delivery. AAPS PharmSciTech 2014;15:882–97.
[75] Ibrahim M, Verma R, Garcia-Contreras L. Inhalation drug delivery devices: technology update. Med Devices (Auckl) 2015;8:131–9.
[76] Pleasants RA, Hess DR. Aerosol delivery devices for obstructive lung diseases. Respir Care 2018;63:708–33.
[77] Lee W-H, Loo C-Y, Traini D, Young PM. Inhalation of nanoparticle-based drug for lung cancer treatment: advantages and challenges. Asian J Pharm Sci 2015;10:481–9.
[78] Zarogoulidis P, Darwiche K, Krauss L, Huang H, Zachariadis GA, Katsavou A, Hohenforst-Schmidt W, Papaiwannou A, Vogl TJ, Freitag L, Stamatis G, Zarogoulidis K. Inhaled cisplatin deposition and distribution in lymph nodes in stage II lung cancer patients. Future Oncol 2013;9:1307–13.
[79] Tatsumura T, Koyama S, Tsujimoto M, Kitagawa M, Kagamimori S. Further study of nebulisation chemotherapy, a new chemotherapeutic method in the treatment of lung carcinomas: fundamental and clinical. Br J Cancer 1993;68:1146–9.
[80] Vincent JH, Johnston AM, Jones AD, Bolton RE, Addison J. Kinetics of deposition and clearance of inhaled mineral dusts during chronic exposure. Br J Ind Med 1985;42:707–15.
[81] Byron PR, Patton JS. Drug delivery via the respiratory tract. J Aerosol Med 1994;7:49–75.
[82] Martonen TB, Katz IM. Deposition patterns of aerosolized drugs within human lungs: effects of Ventilatory parameters. Pharm Res 1993;10:871–8.
[83] Praphawatvet T, Peters JI, Williams RO. Inhaled nanoparticles–an updated review. Int J Pharm 2020;587:119671.
[84] Shevchenko IT, Resnik GE. Inhalation of chemical substances and oxygen in radiotherapy of bronchial cancer. Neoplasma 1968;15:419–26.
[85] Wittgen BP, Kunst PW, van der Born K, van Wijk AW, Perkins W, Pilkiewicz FG, Perez-Soler R, Nicholson S, Peters GJ, Postmus PE. Phase I study of aerosolized SLIT cisplatin in the treatment of patients with carcinoma of the lung. Clin Cancer Res 2007;13:2414–21.
[86] Chou AJ, Gupta R, Bell MD, Riewe KO, Meyers PA, Gorlick R. Inhaled lipid cisplatin (ILC) in the treatment of patients with relapsed/progressive osteosarcoma metastatic to the lung. Pediatr Blood Cancer 2013;60:580–6.
[87] Otterson GA, Villalona-Calero MA, Hicks W, Pan X, Ellerton JA, Gettinger SN, Murren JR. Phase I/II study of inhaled doxorubicin combined with platinum-based therapy for advanced non-small cell lung cancer. Clin Cancer Res 2010;16:2466–73.
[88] Otterson GA, Villalona-Calero MA, Sharma S, Kris MG, Imondi A, Gerber M, White DA, Ratain MJ, Schiller JH, Sandler A, Kraut M, Mani S, Murren JR. Phase I study of inhaled doxorubicin for patients with metastatic tumors to the lungs. Clin Cancer Res 2007;13:1246–52.
[89] Lemarie E, Vecellio L, Hureaux J, Prunier C, Valat C, Grimbert D, Boidron-Celle M, Giraudeau B, le Pape A, Pichon E, Diot P, el Houfia A, Gagnadoux F. Aerosolized gemcitabine in patients with carcinoma of the lung: feasibility and safety study. J Aerosol Med Pulm Drug Deliv 2011;24:261–70.
[90] Verschraegen CF, Gilbert BE, Huaringa AJ, Newman R, Harris N, Leyva FJ, Keus L, Campbell K, Nelson-Taylor T, Knight V. Feasibility, phase I, and pharmacological study of aerosolized liposomal 9-nitro-20(S)-camptothecin in patients with advanced malignancies in the lungs. Ann N Y Acad Sci 2000;922:352–4.
[91] Charpidou AG, Gkiozos I, Tsimpoukis S, Apostolaki D, Dilana KD, Karapanagiotou EM, Syrigos KN. Therapy-induced toxicity of the lungs: an overview. Anticancer Res 2009;29:631–9.
[92] Sardeli C, Zarogoulidis P, Kosmidis C, Amaniti A, Katsaounis A, Giannakidis D, Koulouris C, Hohenforst-Schmidt W, Huang H, Bai C, Michalopoulos N, Tsakiridis K, Romanidis K, Oikonomou P, Mponiou K, Vagionas A, Goganau AM, Kesisoglou I, Sapalidis K. Inhaled chemotherapy adverse effects: mechanisms and protection methods. Lung Cancer Manag 2020;8:LMT19.
[93] Parimon T, Chien JW, Bryson CL, McDonell MB, Udris EM, Au DH. Inhaled corticosteroids and risk of lung cancer among patients with chronic obstructive pulmonary disease. Am J Respir Crit Care Med 2007;175:712–9.
[94] Lee CH, Hyun MK, Jang EJ, Lee NR, Kim K, Yim JJ. Inhaled corticosteroid use and risks of lung cancer and laryngeal cancer. Respir Med 2013;107:1222–33.
[95] Kiri VA, Fabbri LM, Davis KJ, Soriano JB. Inhaled corticosteroids and risk of lung cancer among COPD patients who quit smoking. Respir Med 2009;103:85–90.
[96] Liu SF, Kuo HC, Lin MC, Ho SC, Tu ML, Chen YM, Chen YC, Fang WF, Wang CC, Liu GH. Inhaled corticosteroids have a protective effect against lung cancer in female patients with chronic obstructive pulmonary disease: a nationwide population-based cohort study. Oncotarget 2017;8:29711–21.
[97] Sin DD, Wu L, Anderson JA, Anthonisen NR, Buist AS, Burge PS, Calverley PM, Connett JE, Lindmark B, Pauwels RA, Postma DS, Soriano JB, Szafranski W, Vestbo J. Inhaled corticosteroids and mortality in chronic obstructive pulmonary disease. Thorax 2005;60:992–7.
[98] Lam S, leRiche JC, McWilliams A, Macaulay C, Dyachkova Y, Szabo E, Mayo J, Schellenberg R, Coldman A, Hawk E, Gazdar A. A randomized phase IIb trial of pulmicort turbuhaler (budesonide) in people with dysplasia of the bronchial epithelium. Clin Cancer Res 2004;10:6502–11.
[99] Veronesi G, Szabo E, Decensi A, Guerrieri-Gonzaga A, Bellomi M, Radice D, Ferretti S, Pelosi G, Lazzeroni M, Serrano D, Lippman SM, Spaggiari L, Nardi-Pantoli A, Harari S, Varricchio C, Bonanni B. Randomized phase II trial of inhaled budesonide versus placebo in high-risk individuals with CT screen-detected lung nodules. Cancer Prev Res (Phila) 2011;4:34–42.
[100] Veronesi G, Lazzeroni M, Szabo E, Brown PH, DeCensi A, Guerrieri-Gonzaga A, Bellomi M, Radice D, Grimaldi MC, Spaggiari L, Bonanni B. Long-term effects of inhaled budesonide on screening-detected lung nodules. Ann Oncol 2015;26:1025–30.
[101] Kim SY, Naskar D, Kundu SC, Bishop DP, Doble PA, Boddy AV, Chan H-K, Wall IB, Chrzanowski W. Formulation of biologically-inspired silk-based drug carriers for pulmonary delivery targeted for lung Cancer. Sci Rep 2015;5:11878.
[102] Miwata K, Okamoto H, Nakashima T, Ihara D, Horimasu Y, Masuda T, Miyamoto S, Iwamoto H, Fujitaka K, Hamada H, Shibata A, Ito T, Okuda T, Hattori N. Intratracheal administration of siRNA dry powder targeting vascular endothelial growth factor inhibits lung tumor growth in mice. Mol Ther Nucleic Acids 2018;12:698–706.
[103] Luo T, Loira-Pastoriza C, Patil HP, Ucakar B, Muccioli GG, Bosquillon C, Vanbever R. PEGylation of paclitaxel largely improves its safety and anti-tumor efficacy following pulmonary

delivery in a mouse model of lung carcinoma. J Control Release 2016;239:62–71.

[104] Kuehl PJ, Grimes MJ, Dubose D, Burke M, Revelli DA, Gigliotti AP, Belinsky SA, Tessema M. Inhalation delivery of topotecan is superior to intravenous exposure for suppressing lung cancer in a preclinical model. Drug Deliv 2018;25:1127–36.

[105] Zhu L, Li M, Liu X, Jin Y. Drug-loaded PLGA Electrospraying porous microspheres for the local therapy of primary lung Cancer via pulmonary delivery. ACS Omega 2017;2:2273–9.

[106] Li Q, Anderson CD, Egilmez NK. Inhaled IL-10 suppresses lung tumorigenesis via abrogation of inflammatory macrophage-Th17 cell Axis. J Immunol 2018;201:2842–50.

[107] Levet V, Rosière R, Merlos R, Fusaro L, Berger G, Amighi K, Wauthoz N. Development of controlled-release cisplatin dry powders for inhalation against lung cancers. Int J Pharm 2016;515:209–20.

[108] Mohtar N, Taylor KMG, Sheikh K, Somavarapu S. Design and development of dry powder sulfobutylether-β-cyclodextrin complex for pulmonary delivery of fisetin. Eur J Pharm Biopharm 2017;113:1–10.

[109] Martinelli F, Balducci AG, Kumar A, Sonvico F, Forbes B, Bettini R, Buttini F. Engineered sodium hyaluronate respirable dry powders for pulmonary drug delivery. Int J Pharm 2017;517:286–95.

[110] Reczyńska K, Marchwica P, Khanal D, Borowik T, Langner M, Pamuła E, Chrzanowski W. Stimuli-sensitive fatty acid-based microparticles for the treatment of lung cancer. Mater Sci Eng C 2020;111:110801.

[111] Li W, Chen S, Zhang L, Zhang Y, Yang X, Xie B, Guo J, He Y, Wang C. Inhalable functional mixed-polymer microspheres to enhance doxorubicin release behavior for lung cancer treatment. Colloids Surf B Biointerfaces 2020;196:111350.

[112] Kim I, Byeon HJ, Kim TH, Lee ES, Oh KT, Shin BS, Lee KC, Youn YS. Doxorubicin-loaded highly porous large PLGA microparticles as a sustained- release inhalation system for the treatment of metastatic lung cancer. Biomaterials 2012;33:5574–83.

[113] Arnold MM, Gorman EM, Schieber LJ, Munson EJ, Berkland C. NanoCipro encapsulation in monodisperse large porous PLGA microparticles. J Control Release 2007;121:100–9.

[114] Cai Y, Chen Y, Hong X, Liu Z, Yuan W. Porous microsphere and its applications. Int J Nanomedicine 2013;8:1111–20.

[115] Verco J, Johnston W, Frost M, Baltezor M, Kuehl PJ, Lopez A, Gigliotti A, Belinsky SA, Wolff R, diZerega G. Inhaled submicron particle paclitaxel (NanoPac) induces tumor regression and immune cell infiltration in an orthotopic athymic nude rat model of non-small cell lung cancer. J Aerosol Med Pulm Drug Deliv 2019;32:266–77.

[116] Verco J, Johnston W, Baltezor M, Kuehl PJ, Gigliotti A, Belinsky SA, Lopez A, Wolff R, Hylle L, diZerega G. Pharmacokinetic profile of inhaled submicron particle paclitaxel (NanoPac(®)) in a rodent model. J Aerosol Med Pulm Drug Deliv 2019;32:99–109.

[117] Arbain NH, Basri M, Salim N, Wui WT, Abdul Rahman MB. Development and characterization of aerosol Nanoemulsion system encapsulating low water soluble quercetin for lung Cancer treatment. Mater Today: Proc 2018;5:S137–42.

[118] Youngren-Ortiz SR, Hill DB, Hoffmann PR, Morris KR, Barrett EG, Forest MG, Chougule MB. Development of optimized, inhalable, gemcitabine-loaded gelatin nanocarriers for lung cancer. J Aerosol Med Pulm Drug Deliv 2017;30:299–321.

[119] Rao KSVK, Zhong Q, Bielski ER, da Rocha SRP. Nanoparticles of pH-responsive, PEG–doxorubicin conjugates: interaction with an in vitro model of lung adenocarcinoma and their direct formulation in propellant-based portable inhalers. Mol Pharm 2017;14:3866–78.

[120] Zhong Q, Humia BV, Punjabi AR, Padilha FF, da Rocha SRP. The interaction of dendrimer-doxorubicin conjugates with a model pulmonary epithelium and their cosolvent-free, pseudo-solution formulations in pressurized metered-dose inhalers. Eur J Pharm Sci 2017;109:86–95.

[121] Yan Y, Zhou K, Xiong H, Miller JB, Motea EA, Boothman DA, Liu L, Siegwart DJ. Aerosol delivery of stabilized polyester-siRNA nanoparticles to silence gene expression in orthotopic lung tumors. Biomaterials 2017;118:84–93.

[122] Kabary DM, Helmy MW, Abdelfattah E-ZA, Fang J-Y, Elkhodairy KA, Elzoghby AO. Inhalable multi-compartmental phospholipid enveloped lipid core nanocomposites for localized mTOR inhibitor/herbal combined therapy of lung carcinoma. Eur J Pharm Biopharm 2018;130:152–64.

[123] Abdelaziz HM, Elzoghby AO, Helmy MW, Abdelfattah E-ZA, Fang J-Y, Samaha MW, Freag MS. Inhalable lactoferrin/chondroitin-functionalized monoolein nanocomposites for localized lung cancer targeting. ACS Biomater Sci Eng 2020;6:1030–42.

[124] Baghdan E, Duse L, Schüer JJ, Pinnapireddy SR, Pourasghar M, Schäfer J, Schneider M, Bakowsky U. Development of inhalable curcumin loaded Nano-in-microparticles for bronchoscopic photodynamic therapy. Eur J Pharm Sci 2019;132:63–71.

[125] Nozohouri S, Salehi R, Ghanbarzadeh S, Adibkia K, Hamishehkar H. A multilayer hollow nanocarrier for pulmonary co-drug delivery of methotrexate and doxorubicin in the form of dry powder inhalation formulation. Mater Sci Eng C 2019;99:752–61.

[126] Liu Y, Crowe WN, Wang L, Lu Y, Petty WJ, Habib AA, Zhao D. An inhalable nanoparticulate STING agonist synergizes with radiotherapy to confer long-term control of lung metastases. Nat Commun 2019;10:5108.

[127] Bakhtiary Z, Barar J, Aghanejad A, Saei AA, Nemati E, Dolatabadi JEN, Omidi Y. Microparticles containing erlotinib-loaded solid lipid nanoparticles for treatment of non-small cell lung cancer. Drug Dev Ind Pharm 2017;43:1244–53.

[128] Rosière R, Van Woensel M, Gelbcke M, Mathieu V, Hecq J, Mathivet T, Vermeersch M, Van Antwerpen P, Amighi K, Wauthoz N. New folate-grafted chitosan derivative to improve delivery of paclitaxel-loaded solid lipid nanoparticles for lung tumor therapy by inhalation. Mol Pharm 2018;15:899–910.

[129] Zhang J, Wu L, Chan H-K, Watanabe W. Formation, characterization, and fate of inhaled drug nanoparticles. Adv Drug Deliv Rev 2011;63:441–55.

[130] Hureaux J, Lagarce F, Gagnadoux F, Vecellio L, Clavreul A, Roger E, Kempf M, Racineux JL, Diot P, Benoit JP, Urban T. Lipid nanocapsules: ready-to-use nanovectors for the aerosol delivery of paclitaxel. Eur J Pharm Biopharm 2009;73:239–46.

[131] Lee WH, Loo CY, Ong HX, Traini D, Young PM, Rohanizadeh R. Synthesis and characterization of inhalable flavonoid nanoparticle for lung Cancer cell targeting. J Biomed Nanotechnol 2016;12:371–86.

[132] Wan KY, Weng J, Wong SN, Kwok PCL, Chow SF, Chow AHL. Converting nanosuspension into inhalable and redispersible nanoparticles by combined in-situ thermal gelation and spray drying. Eur J Pharm Biopharm 2020;149:238–47.

[133] Gandhi M, Pandya T, Gandhi R, Patel S, Mashru R, Misra A, Tandel H. Inhalable liposomal dry powder of gemcitabine-HCl: formulation, in vitro characterization and in vivo studies. Int J Pharm 2015;496:886–95.
[134] Koshkina NV, Kleinerman ES, Waidrep C, Jia SF, Worth LL, Gilbert BE, Knight V. 9-Nitrocamptothecin liposome aerosol treatment of melanoma and osteosarcoma lung metastases in mice. Clin Cancer Res 2000;6:2876–80.
[135] Koshkina NV, Gilbert BE, Waldrep JC, Seryshev A, Knight V. Distribution of camptothecin after delivery as a liposome aerosol or following intramuscular injection in mice. Cancer Chemother Pharmacol 1999;44:187–92.
[136] Koshkina NV, Knight V, Gilbert BE, Golunski E, Roberts L, Waldrep JC. Improved respiratory delivery of the anticancer drugs, camptothecin and paclitaxel, with 5% CO2-enriched air: pharmacokinetic studies. Cancer Chemother Pharmacol 2001;47:451–6.
[137] Garbuzenko OB, Mainelis G, Taratula O, Minko T. Inhalation treatment of lung cancer: the influence of composition, size and shape of nanocarriers on their lung accumulation and retention. Cancer Biol Med 2014;11:44–55.
[138] Tagami T, Ando Y, Ozeki T. Fabrication of liposomal doxorubicin exhibiting ultrasensitivity against phospholipase A(2) for efficient pulmonary drug delivery to lung cancers. Int J Pharm 2017;517:35–41.
[139] Knight V, Koshkina NV, Waldrep JC, Giovanella BC, Gilbert BE. Anticancer effect of 9-nitrocamptothecin liposome aerosol on human cancer xenografts in nude mice. Cancer Chemother Pharmacol 1999;44:177–86.
[140] Zhang LJ, Xing B, Wu J, Xu B, Fang XL. Biodistribution in mice and severity of damage in rat lungs following pulmonary delivery of 9-nitrocamptothecin liposomes. Pulm Pharmacol Ther 2008;21:239–46.
[141] Koshkina NV, Waldrep JC, Roberts LE, Golunski E, Melton S, Knight V. Paclitaxel liposome aerosol treatment induces inhibition of pulmonary metastases in murine renal carcinoma model. Clin Cancer Res 2001;7:3258–62.
[142] Zhang M, Li M, Du L, Zeng J, Yao T, Jin Y. Paclitaxel-in-liposome-in-bacteria for inhalation treatment of primary lung cancer. Int J Pharm 2020;578:119177.
[143] Parvathaneni V, Kulkarni NS, Shukla SK, Farrales PT, Kunda NK, Muth A, Gupta V. Systematic development and optimization of inhalable Pirfenidone liposomes for non-small cell lung cancer treatment. Pharmaceutics 2020;12.
[144] Hamzawy MA, Abo-Youssef AM, Salem HF, Mohammed SA. Antitumor activity of intratracheal inhalation of temozolomide (TMZ) loaded into gold nanoparticles and/or liposomes against urethane-induced lung cancer in BALB/c mice. Drug Deliv 2017;24:599–607.
[145] Lin C, Wong BCK, Chen H, Bian Z, Zhang G, Zhang X, Kashif Riaz M, Tyagi D, Lin G, Zhang Y, Wang J, Lu A, Yang Z. Pulmonary delivery of triptolide-loaded liposomes decorated with anti-carbonic anhydrase IX antibody for lung cancer therapy. Sci Rep 2017;7:1097.
[146] Zhang T, Chen Y, Ge Y, Hu Y, Li M, Jin Y. Inhalation treatment of primary lung cancer using liposomal curcumin dry powder inhalers. Acta Pharm Sin B 2018;8:440–8.
[147] Zhu X, Kong Y, Liu Q, Lu Y, Xing H, Lu X, Yang Y, Xu J, Li N, Zhao D, Chen X, Lu Y. Inhalable dry powder prepared from folic acid-conjugated docetaxel liposomes alters pharmacodynamic and pharmacokinetic properties relevant to lung cancer chemotherapy. Pulm Pharmacol Ther 2019;55:50–61.
[148] Fukushige K, Tagami T, Naito M, Goto E, Hirai S, Hatayama N, Yokota H, Yasui T, Baba Y, Ozeki T. Developing spray-freeze-dried particles containing a hyaluronic acid-coated liposome-protamine-DNA complex for pulmonary inhalation. Int J Pharm 2020;583:119338.
[149] Abdelrady H, Hathout RM, Osman R, Saleem I, Mortada ND. Exploiting gelatin nanocarriers in the pulmonary delivery of methotrexate for lung cancer therapy. Eur J Pharm Sci 2019;133:115–26.
[150] Tseng CL, Wu SY, Wang WH, Peng CL, Lin FH, Lin CC, Young TH, Shieh MJ. Targeting efficiency and biodistribution of biotinylated-EGF-conjugated gelatin nanoparticles administered via aerosol delivery in nude mice with lung cancer. Biomaterials 2008;29:3014–22.
[151] Tseng CL, Su WY, Yen KC, Yang KC, Lin FH. The use of biotinylated-EGF-modified gelatin nanoparticle carrier to enhance cisplatin accumulation in cancerous lungs via inhalation. Biomaterials 2009;30:3476–85.
[152] Long JT, Cheang TY, Zhuo SY, Zeng RF, Dai QS, Li HP, Fang S. Anticancer drug-loaded multifunctional nanoparticles to enhance the chemotherapeutic efficacy in lung cancer metastasis. J Nanobiotechnol 2014;12:37.
[153] Varshosaz J, Hassanzadeh F, Mardani A, Rostami M. Feasibility of haloperidol-anchored albumin nanoparticles loaded with doxorubicin as dry powder inhaler for pulmonary delivery. Pharm Dev Technol 2015;20:183–96.
[154] Choi SH, Byeon HJ, Choi JS, Thao L, Kim I, Lee ES, Shin BS, Lee KC, Youn YS. Inhalable self-assembled albumin nanoparticles for treating drug-resistant lung cancer. J Control Release 2015;197:199–207.
[155] Parvathaneni V, Kulkarni NS, Chauhan G, Shukla SK, Elbatanony R, Patel B, Kunda NK, Muth A, Gupta V. Development of pharmaceutically scalable inhaled anti-cancer nanotherapy – repurposing amodiaquine for non-small cell lung cancer (NSCLC). Mater Sci Eng C 2020;115:111139.
[156] Vaidya B, Kulkarni NS, Shukla SK, Parvathaneni V, Chauhan G, Damon JK, Sarode A, Garcia JV, Kunda N, Mitragotri S, Gupta V. Development of inhalable quinacrine loaded bovine serum albumin modified cationic nanoparticles: repurposing quinacrine for lung cancer therapeutics. Int J Pharm 2020;577:118995.
[157] Wang X, Parvathaneni V, Shukla SK, Kulkarni NS, Muth A, Kunda NK, Gupta V. Inhalable resveratrol-cyclodextrin complex loaded biodegradable nanoparticles for enhanced efficacy against non-small cell lung cancer. Int J Biol Macromol 2020;164:638–50.
[158] Zhang T, Bao J, Zhang M, Ge Y, Wei J, Li Y, Wang W, Li M, Jin Y. Chemo-photodynamic therapy by pulmonary delivery of gefitinib nanoparticles and 5-aminolevulinic acid for treatment of primary lung cancer of rats. Photodiagnosis Photodyn Ther 2020;31:101807.
[159] Torrico Guzmán EA, Sun Q, Meenach SA. Development and evaluation of paclitaxel-loaded aerosol nanocomposite microparticles and their efficacy against air-grown lung cancer tumor spheroids. ACS Biomater Sci Eng 2019;5:6570–80.
[160] Elgohary MM, Helmy MW, Abdelfattah E-ZA, Ragab DM, Mortada SM, Fang J-Y, Elzoghby AO. Targeting sialic acid residues on lung cancer cells by inhalable boronic acid-decorated albumin nanocomposites for combined chemo/herbal therapy. J Control Release 2018;285:230–43.

[161] Aalaei K, Rayner M, Sjöholm I. Chemical methods and techniques to monitor early Maillard reaction in milk products; a review. Crit Rev Food Sci Nutr 2019;59:1829–39.

[162] Kaur P, Mishra V, Shunmugaperumal T, Goyal AK, Ghosh G, Rath G. Inhalable spray dried lipidnanoparticles for the co-delivery of paclitaxel and doxorubicin in lung cancer. J Drug Deliv Sci Technol 2020;56:101502.

[163] Kamel NM, Helmy MW, Abdelfattah E-Z, Khattab SN, Ragab D, Samaha MW, Fang J-Y, Elzoghby AO. Inhalable dual-targeted hybrid lipid nanocore–protein shell composites for combined delivery of genistein and all-trans retinoic acid to lung cancer cells. ACS Biomater Sci Eng 2020;6:71–87.

[164] Ranjan AP, Mukerjee A, Gdowski A, Helson L, Bouchard A, Majeed M, Vishwanatha JK. Curcumin-ER prolonged subcutaneous delivery for the treatment of non-small cell lung cancer. J Biomed Nanotechnol 2016;12:679–88.

[165] Chahin M, Krishnan N, Matthews-Hew T, Hew J, Pham D. Metastatic anaplastic lymphoma kinase rearrangement-positive adenocarcinoma of occult primary mimicking ovarian Cancer. Cureus 2020;12:e9437.

[166] Wang X, Wang Z, Pan J, Lu ZY, Xu D, Zhang HJ, Wang SH, Huang DY, Chen XF. Patterns of Extrathoracic metastases in different histological types of lung Cancer. Front Oncol 2020;10:715.

[167] Thakur C. Chapter 2—An overview, current challenges of drug resistance, and targeting metastasis associated with lung cancer. In: Kesharwani P, editor. Nanotechnology-based targeted drug delivery systems for lung cancer. Academic Press; 2019. p. 21–38.

[168] Yuan M, Huang L-L, Chen J-H, Wu J, Xu Q. The emerging treatment landscape of targeted therapy in non-small-cell lung cancer. Signal Transduct Target Ther 2019;4:61.

[169] Woodman C, Vundu G, George A, Wilson CM. Applications and strategies in nanodiagnosis and nanotherapy in lung cancer. Semin Cancer Biol 2020.

[170] Ning J, Li P, Zhang B, Han B, Su X, Wang Q, Wang X, Li B, Kang H, Zhou L, Chu C, Zhang N, Pang Y, Niu Y, Zhang R. miRNAs deregulation in serum of mice is associated with lung cancer related pathway deregulation induced by PM2.5. Environ Pollut 2019;254:112875.

[171] Lee KY, Shueng PW, Chou CM, Lin BX, Lin MH, Kuo DY, Tsai IL, Wu SM, Lin CW. Elevation of CD109 promotes metastasis and drug resistance in lung cancer via activation of EGFR-AKT-mTOR signaling. Cancer Sci 2020;111:1652–62.

[172] Kim MS, Haney MJ, Zhao Y, Mahajan V, Deygen I, Klyachko NL, Inskoe E, Piroyan A, Sokolsky M, Okolie O, Hingtgen SD, Kabanov AV, Batrakova EV. Development of exosome-encapsulated paclitaxel to overcome MDR in cancer cells. Nanomed Nanotechnol Biol Med 2016;12:655–64.

[173] Lim ZF, Ma PC. Emerging insights of tumor heterogeneity and drug resistance mechanisms in lung cancer targeted therapy. J Hematol Oncol 2019;12:134.

[174] Mangal S, Gao W, Li T, Zhou QT. Pulmonary delivery of nanoparticle chemotherapy for the treatment of lung cancers: challenges and opportunities. Acta Pharmacol Sin 2017;38:782–97.

[175] Mohapatra PR, Sahoo S, Bhuniya S, Panigrahi MK, Majumdar SKD, Mishra P, Patra S. Concomitant echinoderm microtubule-associated protein-like 4-anaplastic lymphoma kinase rearrangement and epidermal growth factor receptor mutation in non-small cell lung cancer patients from eastern India. J Cancer Res Ther 2020;16:850–4.

[176] Panchal R. Systemic anticancer therapy (SACT) for lung cancer and its potential for interactions with other medicines. Ecancermedicalscience 2017;11:764.

[177] Riccardo F, Barutello G, Petito A, Tarone L, Conti L, Arigoni M, Musiu C, Izzo S, Volante M, Longo DL, Merighi IF, Papotti M, Cavallo F, Quaglino E. Immunization against ROS1 by DNA electroporation impairs K-Ras-driven lung adenocarcinomas. Vaccines (Basel) 2020;8.

[178] Chen Y, Tang WY, Tong X, Ji H. Pathological transition as the arising mechanism for drug resistance in lung cancer. Cancer Commun 2019;39:53.

[179] Kim I, Byeon HJ, Kim TH, Lee ES, Oh KT, Shin BS, Lee KC, Youn YS. Doxorubicin-loaded porous PLGA microparticles with surface attached TRAIL for the inhalation treatment of metastatic lung cancer. Biomaterials 2013;34:6444–53.

[180] Kaminskas LM, McLeod VM, Ryan GM, Kelly BD, Haynes JM, Williamson M, Thienthong N, Owen DJ, Porter CJ. Pulmonary administration of a doxorubicin-conjugated dendrimer enhances drug exposure to lung metastases and improves cancer therapy. J Control Release 2014;183:18–26.

[181] Zhong Q, Bielski ER, Rodrigues LS, Brown MR, Reineke JJ, da Rocha SRP. Conjugation to poly(amidoamine) dendrimers and pulmonary delivery reduce cardiac accumulation and enhance antitumor activity of doxorubicin in lung metastasis. Mol Pharm 2016;13:2363–75.

[182] Xu C, Tian H, Sun H, Jiao Z, Zhang Y, Chen X. A pH sensitive co-delivery system of siRNA and doxorubicin for pulmonary administration to B16F10 metastatic lung cancer. RSC Adv 2015;5:103380–5.

[183] Xu C, Wang P, Zhang J, Tian H, Park K, Chen X. Pulmonary codelivery of doxorubicin and siRNA by pH-sensitive nanoparticles for therapy of metastatic lung cancer. Small 2015;11:4321–33.

[184] Shim G, Choi H-w, Lee S, Choi J, Yu YH, Park D-E, Choi Y, Kim C-W, Oh Y-K. Enhanced intrapulmonary delivery of anticancer siRNA for lung Cancer therapy using cationic Ethylphosphocholine-based Nanolipoplexes. Mol Ther 2013;21:816–24.

[185] Okamoto H, Shiraki K, Yasuda R, Danjo K, Watanabe Y. Chitosan–interferon-β gene complex powder for inhalation treatment of lung metastasis in mice. J Control Release 2011;150:187–95.

[186] Yu K-N, Minai-Tehrani A, Chang S-H, Hwang S-K, Hong S-H, Kim J-E, Shin J-Y, Park S-J, Kim J-H, Kwon J-T, Jiang H-L, Kang B, Kim D, Chae C-H, Lee K-H, Yoon T-J, Beck GR, Cho M-H. Aerosol delivery of small hairpin osteopontin blocks pulmonary metastasis of breast cancer in mice. PloS One 2010;5:e15623.

[187] Shimada K, Ushijima K, Suzuki C, Horiguchi M, Ando H, Akita T, Shimamura M, Fujii J, Yamashita C, Fujimura A. Pulmonary administration of curcumin inhibits B16F10 melanoma lung metastasis and invasion in mice. Cancer Chemother Pharmacol 2018;82:265–73.

[188] Xu C, Tian H, Wang P, Wang Y, Chen X. The suppression of metastatic lung cancer by pulmonary administration of polymer nanoparticles for co-delivery of doxorubicin and Survivin siRNA. Biomater Sci 2016;4:1646–54.

[189] Liu J, Deng Y, Fu D, Yuan Y, Li Q, Shi L, Wang G, Wang Z, Wang L. Sericin microparticles enveloped with metal-organic networks as a pulmonary targeting delivery system for intra-tracheally treating metastatic lung cancer. Bioact Mater 2021;6:273–84.

[190] Chaurasiya B, Huang L, Du Y, Tang B, Qiu Z, Zhou L, Tu J, Sun C. Size-based anti-tumoral effect of paclitaxel loaded albumin microparticle dry powders for inhalation to treat metastatic lung cancer in a mouse model. Int J Pharm 2018;542:90–9.

[191] Su W, Wei T, Lu M, Meng Z, Chen X, Jing J, Li J, Yao W, Zhu H, Fu T. Treatment of metastatic lung cancer via inhalation administration of curcumin composite particles based on mesoporous silica. Eur J Pharm Sci 2019;134:246–55.
[192] Videira M, Almeida AJ, Fabra A. Preclinical evaluation of a pulmonary delivered paclitaxel-loaded lipid nanocarrier antitumor effect. Nanomedicine 2012;8:1208–15.
[193] Huang SH, Huang AC, Wang CC, Chang WC, Liu CY, Pavlidis S, Ko HW, Chung FT, Hsu PC, Guo YK, Kuo CS, Yang CT. Front-line treatment of ceritinib improves efficacy over crizotinib for Asian patients with anaplastic lymphoma kinase fusion NSCLC: the role of systemic progression control. Thorac Cancer 2019;10:2274–81.
[194] Sgambato A, Casaluce F, Maione P, Gridelli C. Targeted therapies in non-small cell lung cancer: a focus on ALK/ROS1 tyrosine kinase inhibitors. Expert Rev Anticancer Ther 2018;18:71–80.
[195] Otoukesh S, Sanchez T, Mirshahidi S, Wallace D, Mirshahidi H. ASCEND-8 pharmacokinetic, safety, and efficacy data for ceritinib 450mg with food in patients with anaplastic lymphoma kinase-positive non-small cell lung cancer: a clinical perspective. Cancer Treat Res Commun 2019;20:100149.
[196] Wangari-Talbot J, Hopper-Borge E. Drug resistance mechanisms in non-small cell lung carcinoma. J Cancer Res Updates 2013;2:265–82.
[197] Jethva D, Desai U, Joshi J, Raval A, Shah F. Expression of multidrug resistance (MDR) genes in lung Cancer. GCSMC J Med Sci 2018;7:28–36.
[198] Young LC, Campling BG, Voskoglou-Nomikos T, Cole SP, Deeley RG, Gerlach JH. Expression of multidrug resistance protein-related genes in lung cancer: correlation with drug response. Clin Cancer Res 1999;5:673–80.
[199] Gomez-Cuadrado L, Tracey N, Ma R, Qian B, Brunton VG. Mouse models of metastasis: progress and prospects. Dis Model Mech 2017;10:1061–74.
[200] Iyer R, Nguyen T, Padanilam D, Xu C, Saha D, Nguyen KT, Hong Y. Glutathione-responsive biodegradable polyurethane nanoparticles for lung cancer treatment. J Control Release 2020;321:363–71.
[201] Singla AK, Downey CM, Bebb GD, Jirik FR. Characterization of a murine model of metastatic human non-small cell lung cancer and effect of CXCR4 inhibition on the growth of metastases. Onco Targets Ther 2015;2:263–71.
[202] Tew KD. Glutathione-associated enzymes in anticancer drug resistance. Cancer Res 2016;76:7–9.
[203] Ye Q, Liu K, Shen Q, Li Q, Hao J, Han F, Jiang RW. Reversal of multidrug resistance in Cancer by multi-functional flavonoids. Front Oncol 2019;9:487.

Chapter 9

Advanced drug delivery systems in lung cancer

Anil Philip and Betty Annie Samuel
School of Pharmacy, College of Pharmacy and Nursing, University of Nizwa, Nizwa, Oman

1 Introduction

Lung cancer is one of the most diagnosed cancers and the world's leading cause of death [1]. Lung cancer is a heterogeneous disease caused by genetic and pulmonary epithelial changes [2]. For right treatment, correct clinical diagnosis of lung cancer is critical. Lung cancer is a complex neoplasm, both histopathologically and biologically [3, 4]. Of the numerous histological forms, the most prevalent are the non-small-cell lung carcinoma (NSCLC) and small-cell lung carcinoma (SCLC) [5]. The choice of the lung cancer treatment regimen will depend upon the cancer's progression stage and on the physical state of the patient. In general, all SCLC phases undergo chemotherapy treatment as surgery is unable to manage the disease due to malignant cells easily double up and having metastatic properties [6]. Surgical removal is the preferred stage 1 NSCLC procedure since the tumor is principally situated in one lung area and not propagated to the draining lymph nodes. Chemotherapy either post-operative or pre-operative, and surgical procedures are necessary for the stage 2 and 3 NSCLC depending on the nature of the tumor infiltration in the chest cavity. Preoperative chemotherapy decreases tumors and regulates the incidence of the distal tissue. The only therapeutic approach available for patients with stage 4 lung cancer is chemoradiotherapy that ensures systemic transfer of metastatic lesions [7, 8].

For cellular survival, the respiratory and cardiovascular system work together to provide oxygen-rich blood with hormones, nutrients, and other factors. The cell should instantly remove the cellular respiration by-products, including carbon dioxide, via circulation and eventually outside the body through the respiratory system [9]. Several external and internal factors can contribute to the development of lung cancer. An elevated risk of lung disease is correlated with environmental contaminants such as tobacco smoke, ionizing radiation, air pollutants in the outside, such as small particulate matters and sulfate aerosols. Cigarette smoking is the leading cause of lung cancer growth, standing for 85% of all lung cancer cases [10]. These dangerous agents and other carcinogens manage mutations in essential genes that start and develop lung cancer.

2 Nanoparticle drug delivery

Nanoscale Technology has increased the therapeutic window, as well as the solubility of low water-solubility cancer drugs. Not only can nanoparticles improve the potency of drugs involved in chemotherapy but can also minimize non-specific toxic effects. Due to their biological compatibility, nanocarriers have made nanoparticles the perfect medium for in vivo drug delivery [11, 12].

2.1 Polymer-based nanoparticles

Polymer-based (natural or synthetic) nanoparticles that are biodegradable have lately become prevalent owing to their regulated, prolonged-release properties and biological compatibility. In supplement to the discomfort associated with administration, traditional anticancer agents have adverse reactions. For patients with lung cancer, effective oral administration of medicines was impossible due to inadequate therapeutic concentrations at the tumor site. Polymer-based nanoparticles provides a new method for the management of anticancer drugs because of their size, surface load, and form [13].

Using PEG–PLA block copolymers, Kim et al. created nanoparticles for lung cancer therapy [14], which entered the clinical trials for advanced NSCLC in phase II treatment. Using nanocarrier for treating metastatic lung cancer by gemcitabine is currently in a stage II clinical trial [15]. Jiang et al. have recently tried to build a nanoformulation specifically administrable via oral route to lung cancer patients. Researchers have developed a chitosan-altered polycaprolactone polymer nanoparticles, where the chitosan's mucoadhesive attributes are helpful for selective interaction

Advanced Drug Delivery Systems in the Management of Cancer. https://doi.org/10.1016/B978-0-323-85503-7.00024-9

with mucine in cancer cells [16]. Mehrotra et al. created chitosan-centered lomustine nanoparticles, which improved in vitro activity against lung cell L132 [17]. Recently, polymer nanoparticulates were used for the development of nanoformulations for early-stage lung cancer therapy wherever paclitaxel-laden nanoparticles postponed subcutaneous lesions locally [18].

2.2 Inhalable nanoparticles

The oral or nasal systemic drug delivery via inhalation is an appealing alternative to oral and parenteral delivery of medicinal items. Inhalation of nanosized carriers or delivery of respiratory chemotherapeutic drugs has recently gained increased acceptance as the drugs reach the lung or target site without negatively influencing the entire body. Consequently, overcoming the toxic effects associated with the systemic delivery may be a reality. Pulmonary nanoparticle deliveries often allows for a prolonged release of therapeutic medications in the lung. The nanoparticle delivery of drug to lungs have been primarily used for nebulization and aerosols [19]. In addition, pulmonary route for nanoparticles can avoid mucosal clearance and lung phagocytic ways and consequently remain in the respiratory tract for a longer period. The particle inhaled should either be tiny to go past the area of the upper lungs (to avoid mucosal trapping and clearance) or possess the necessary surface chemistry (for preventing adherence and/or mucopenetration to a mucosal layer). Inhaled self-constructed nanoparticles of human serum albumin together with doxorubicin and octyl aldehyde and adsorbed through apoptotic tumor necrosis factor-linked apoptosis-inducing ligand protein were developed by the Kim et al. to enhance the apoptosis-stimulating effects alongside drug-resilient lung tumors [20]. Table 1 represents examples of few drugs made into nanoformulations [21].

TABLE 1 Examples of drugs made into nanoformulations as dry powder inhalers (DPIs).

Drug/agent	Class	Condition	Route of administration
Vancomycin	Antibiotic	Infection	DPI
Clarithromycin	Antibiotic	Infection	DPI
Salmon calcitonin	Hormone	Hypocalcemia	DPI
Tacrolimus and Cyclosporine A	Immunosuppressant	Allograft rejection prevention in lung transplantation	DPI
Budesonide	Glucocorticoid	Asthma and COPD	DPI
Tranilast	Antiallergic agent	Bronchial asthma	DPI
Diatrizoic acid	Radio contrast agent	Imaging of airway	DPI
Ciprofloxacin	Antibiotic	Cystic fibrosis	DPI
Cyclosporine A	Immunosuppressant	Lung transplant rejection prevention	DPI
Paclitaxel	Microtubule inhibitor	Lung cancer	DPI
Tobramycin	Antibiotic	Infection	DPI
Azithromycin	Antibiotic	Infection	DPI
Rifampicin	Antibiotic	Tuberculosis	DPI
Ofloxacin	Antimicrobial	Infection	DPI
Moxifloxacin	Antibacterial	Infection	DPI
Doxorubicin	Anticancer agent	Lung cancer	DPI
Influenza virus	Antigen	Influenza	Nasal
Anthrax rPA	Antigen	Anthrax	Nasal
Fluticasone propionate/albuterol sulfate	Anti-inflammatory/β_2-agonist	Asthma and COPD	DPI
Salbutamol sulfate	β_2-agonist	Asthma	DPI

Abbreviation: *COPD*, chronic obstructive pulmonary disease; *rPA*, recombinant protective antigen.
Reproduced with permission from Elsevier Inc.

2.3 Lipid polymer hybrid nanoparticles

A new type of nanoparticle known as lipid polymer hybrid nanoparticles (LPHNPs) have recently become common in cancer chemotherapy. These nanoparticles normally include a biodegradable polymer inner core and a phospholipid shell (layer). The core delivers structure integrity and physical in-use stability, while the phospholipid layer helps in absorption and affinity at cellular level. The LPHNPs (natural, synthetic, and semisynthetic) can entrap both hydrophilic and hydrophobic drugs. The LPHNP's core holds the drug with hydrophilic nature, while the drugs of hydrophobic nature can be incorporated into the phospholipid layer [20]. The nanoparticle diameter and the phospholipid layer thickness determines the drug release from LPHNPs. The LPHNPs shell is controlling the H_2O flow into the nanoparticles, and therefore shell thickness determines the drug dissolution. An increase from 70% to 85% in drug release after 24h has been reported, with shell thickness decreasing from 300 to 100nm [22].

LPHNPs (Fig. 1) can offer a variety of advantages such as controlled drug release and high drug loading [23].

2.4 Hollow nanoparticles and nanoaggregates

Nanoparticles developed as large hollow or porous particles, enhance the geometric diameter, and reduce the aerodynamic diameter, creating particles more suitable for lung deposits (Figs. 2 and 3) [24]. A particle's geometric diameter adds minimal to deposition of the particle in the lung, while the aerodynamic diameter characterizes nanoparticles deep-seated lung deposition. A study by Edward et al. demonstrated better aerosolization effectiveness, sustained release, and bioavailability for porous drug particles [25].

FIG. 2 SEM image of the nanoaggregates indicating a significant presence of the fine particles. *Reproduced with permission from Elsevier Inc.*

In another study, the same authors found that it is not only the particle size but also the porosity that influences insulin bioavailability [26].

Nanoparticle aggregates constitute the drugs with accumulated nanoparticles that have the potential to dissociate in individual nanoparticles and release the drug into the lungs or the respiratory system. Larger hollow nanoparticulate aggregates with a geometric diameter of approximately 10μm have a restricted aerodynamic diameter (1–5μm) due to the low particle density. The larger particles reduces the particle propensity toward accumulation in the inhaler unit, in such a way as to ensure the powder is properly distributed while the smaller particles help prevent deposition anywhere within the respiratory system, except in the lungs [27].

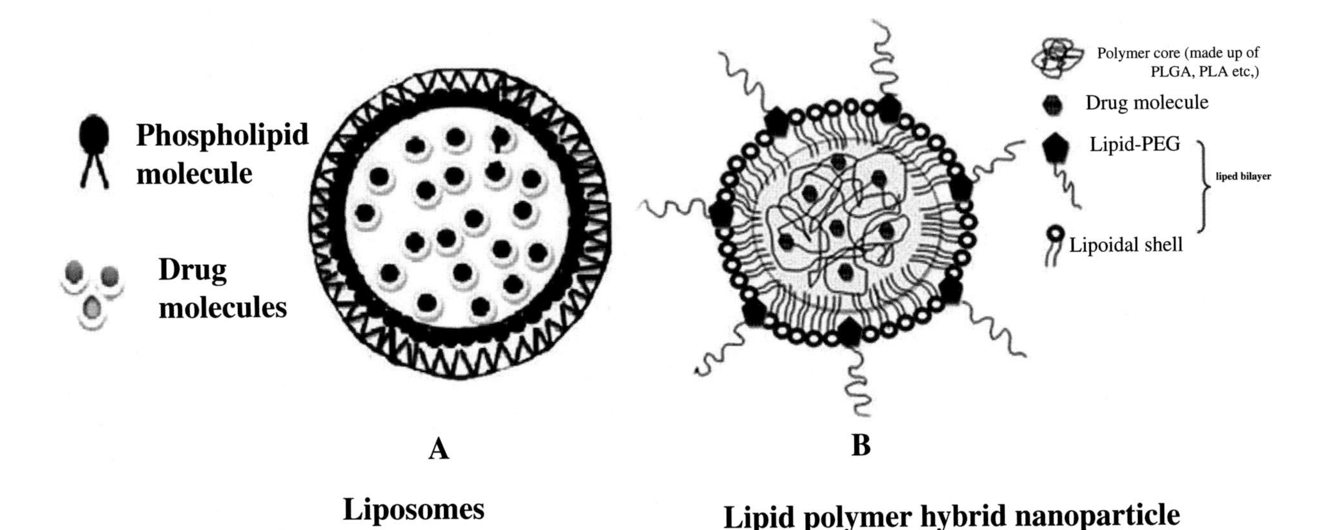

FIG. 1 (A) Liposomes bilayer structure, and (B) lipid polymer hybrid nanoparticles with inner polymer core containing drug molecule surrounded by lipid and lipid PEG layer. *Reproduced with permission from Elsevier Inc.*

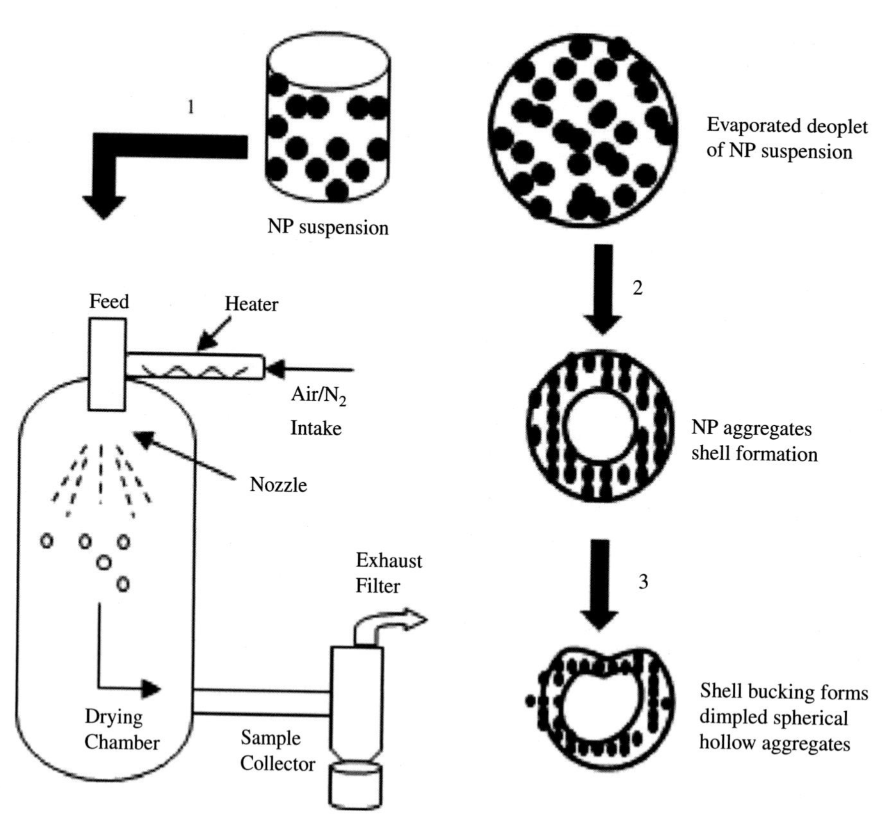

FIG. 3 Spray-drying formulation of the large hollow spherical nanoaggregates. *Reproduced with permission from Elsevier Inc.*

3 Peptide-based nanoparticles

In many physiological functions of the human body, small naturally occurring peptide-like growth factors and the neurotransmitters play a key role. As therapeutic agents, peptides can easily penetrate cells and release drugs in response to the tissue pH. Peptides liposomes conjugates were successfully incorporated into mitochondria (in a extracellular tumor atmosphere, pH ~ 6.8) by reversing their surface charge from negative to positive [28]. Therefore, peptide-based nanoparticles have become more important for targeting tissues as gene and drug transporters. The physical and chemical structures of peptides can be altered, which reduces the enzyme degradation and enhances the half-life of peptides in blood circulation. As the surface of the pulmonary mucosa is rich in antiprotease enzymes, small molecular peptides and proteins permeate naturally after inhalation. The delivery of pulmonary nanoparticles with peptide can therefore lead to faster absorption and increased local and systemic bioavailability compared with subcutaneous or intravenous injections [29].

3.1 Quantum dots

Quantum dots, due to their distinctive optical or electronic characteristics, are studied as a modern probe for in vivo and in vitro biomedical imaging. In supplement to the encouraging application of molecular in vitro imagery, quantum dots–based multifunctional probes have resulted in the development of anticancer drugs [30]. Quantum dots can be used to track complex changes to a tumor microenvironment that would help study the mechanism for cancer invasion and assist direct personalized clinical care. Quantum dots are nanocrystals comprising of a semiconductor material that is centered in the shell of another semiconductor with a wider spectral band gap. First manufactured in 1982, quantum dots are one of the latest novel methods for treating lung cancer. Quantum dots have distinctive features including photostability, high level of photobleaching, broad absorption spectrum, and a variety of fluorescence emissions that can be accomplished by altering their composition and dimensions. Quantum dots are being used in signal transmission, cell and molecular tracking, drug monitoring, and in vivo screening for cancer [31].

4 Micelles

Micelles are nanosized, self-assembled amphiphilic core structures that develop spontaneously through hydrophobic interactions in aqueous environments. They are obtained by dissolving specific polymer chains in an aqueous solution directly above the threshold and the solution temperature [32]. The majority of the micelles consist of amphiphilic polymers such as polyethylene glycol (PEG) and polyethylene oxide (PEO) [33]. In manufacture of micelles, polymer choice is critical because micelle size and form, stability, and drug retention are determined by the polymers. Integrating hydrogenated soy phosphatidylcholine modifies PEG-PE micelles and decreases drug retention inside the nanoparticle by reducing the electrostatic interactions between cargo (doxorubicin) and PEG-phatidylethanolamine [34]. Unimeric micelles are better suited for the drug delivery than diblock (hydrophilic-hydrophobic) or triblock (hydrophilic-hydrophilic-hydrophilic) particles, due to lower molecular mass and fast renal clearance of polymeric residues following micelle degradation. Micelles made from a hydrophobic center that enables the loading of paclitaxel, tamoxifen, and hydrophile shell, which serves as a barrier to the binding and opsonizing of particle aggregating proteins, degrade systemically before the target tissue is reached [35]. In micelles, two separate anticancer drugs can be co-encapsulated to boost the efficiency of the drug loading and synergize the anticancer effect [36].

5 Dendrimers

Dendrimers are nanoparticles that mimic a set of branches of the tree around a central nucleus. Dendrimers are known as dendritic polymers. The ability to alter the surface by acylation or PEGylation increases the bioavailability of dendrimers and increases the targeted delivery of drugs through the conjugation of target peptides and receptor protein [37]. The absorption efficiency and lung retention of dendrimers are determined by molecular weight and composition. Higher molecular weight (78 kDa) inhalations, for example, lead to lower absorption rate and improved drug retention in the lung tissue in contrast to low-molecular-weight dendrimers (< 22 kDa) [38].

6 Conclusion

Inhalation of anticancer drugs due to its noninvasive characteristics and ability to deliver patient-led drug delivery become an attractive treatment strategy. In general, lung cancer is diagnosed in a later phase of the disease. In advanced stage lung cancer patients, breathing difficulties and short breathing are common. A spacer and an energy source for producing and delivering nanosized molecules could be used to help people inhale medications safely. Breathing chemotherapeutic nanoparticle means the intact distribution of drugs to the lung is more efficiently done in contrast to the traditional intravenous delivery of drugs. Drug delivery through nanoparticles to only selected tissues presents a mechanism to create an efficient local concentration of drugs with decreased levels of cytotoxicity relative to topical application of free drug. Research has confirmed that cytotoxicity in nanocarriers is lower than in free medicines, if chemotherapeutics are encapsulated. Although lung cancer occurs from pulmonary epithelium, which is accessible to inhaled aerosol drugs, it can be found anywhere in the lung. Small-cell lung cancer of the lungs occurs in bronchus and the non-small-cell lung cancer, like adenocarcinoma and large cell carcinoma in peripheral bronchiole and alveolar epithelium. The drug carriers must therefore localize themselves to central airways for the treatment of lung cancer of small cells, while the location of adenocarcinomas should be peripheral. Nanoparticles in the aerosol form may be useful to treat small-cell lung cancer by impacting on the central epithelium of the airway. Due to the aerodynamic character of the nanoparticles, local and systemic drug delivery through the pulmonary route simultaneously poses a challenge. A good way to localize the delivery of drugs for anticancer is the use of multilamellar liposomes. Dendrimers are a better way of delivering systemic anticancer drugs due to its variety in particle size.

References

[1] Lopes Pegna A, Picozzi G. Lung cancer screening update. Curr Opin Pulm Med 2009;15:327–33.

[2] Larsen JE, Minna JD. Molecular biology of lung cancer: clinical implications. Clin Chest Med 2011;32(4):703–40.

[3] Minna JD, Gazdar A. Focus on lung cancer. Cancer Cell 2002;1:49–52.

[4] Herbst RS, Heymach JV, Lippman SM. Lung cancer. N Engl J Med 2008;359:1367–80.

[5] Travis WD, Brambilla E, Muller-Hermelink HK, Harris CC. Tumours of the lung. In: Travis WD, Brambilla E, Muller-Hermelink HK, Harris CC, editors. Pathology and genetics: tumours of the lung, pleura, thymus and heart. Lyon: International Agency for Research on Cancer (IARC); 2004. p. 9–124.

[6] Neal JW, Gubens MA, Wakelee HA. Current management of small cell lung cancer. Clin Chest Med 2011;32:853–63.

[7] Lemjabbar-Alaoui H, Hassan OU, Yang Y-W, Buchanan P. Lung cancer: biology and treatment options. Biochim Biophys Acta 2015;1856(2):189–210.

[8] Dietrich MF, Gerber DE. Chemotherapy for advanced non-small cell lung cancer. In: Reckamp K, editor. Lung cancer. Cancer treatment and research, Vol. 170. Cham: Springer; 2016. p. 119–49.

[9] Aung HH, Sivakumar A, Gholami SK, Venkateswaran SP, Gorain B, Shadab. An overview of the anatomy and physiology of the lung. In: Nanotechnology-based targeted drug delivery systems for lung cancer. Elsevier Inc.; 2019. p. 1–20.

[10] Furrukh M. Tobacco smoking and lung cancer: perception-changing facts. Sultan Qaboos Univ Med J 2013;13(3):345–58.

[11] Ahmad K. Gene delivery by nanoparticles offers cancer hope. Lancet Oncol 2002;3(8):451.

[12] Ramesh R. Nanoparticle-mediated gene delivery to the lung. Methods Mol Biol 2008;434:301–31.

[13] Jung J, Park SJ, Chung HK, Kang H-W, Lee S-W, Seo MH, Park HJ, Song SY, Jeong SY, Choi EK. Polymeric nanoparticles containing taxanes enhance chemoradiotherapeutic efficacy in non-small cell lung cancer. Int J Radiat Oncol 2012;84:e77–83.

[14] Kim DW, Kim SY, Kim HK, Kim SW, Shin SW, Kim JS, Park K, Lee MY, Heo DS. Multicenter phase II trial of genexol-pm, a novel cremophor-free, polymeric micelle formulation of paclitaxel, with

cisplatin in patients with advanced non-small-cell lung cancer. Ann Oncol 2007;18:2009–14.

[15] Cho EK. A phase II trial of genexol-pm and gemcitabine in patients with advanced non-small-cell lung cancer. Available from: http://clinicaltrials.gov/show/NCT01770795. [Accessed 22 September 2020].

[16] Zhao T, Chen H, Yang L, Jin H, Li Z, Han L, Lu F, Xu Z. DDAB-modified tpgs- b - (pcl- ran -pga) nanoparticles as oral anticancer drug carrier for lung cancer chemotherapy. Nano 2013;8:1350014 [10 pages].

[17] Mehrotra A, Nagarwal RC, Pandit JK. Lomustine loaded chitosan nanoparticles: Characterization and in-vitro cytotoxicity on human lung cancer cell line 1132. Chem Pharm Bull (Tokyo) 2011;59:315–20.

[18] Liu R, Khullar OV, Griset AP, Wade JE, Zubris KAV, Grinstaff MW, Colson YL. Paclitaxel-loaded expansile nanoparticles delay local recurrence in a heterotopic murine non-small cell lung cancer model. Ann Thorac Surg 2011;91:1077–84.

[19] Elgindy N, Elkhodairy K, Molokhia A, ElZoghby A. Biopolymeric nanoparticles for oral protein delivery: design and in vitro evaluation. J Nanomed Nanotechnol 2011;02:1–8.

[20] Kim I, Byeon HJ, Kim TH, Lee ES, Oh KT, Shin BS, Lee KC, Youn YS. Doxorubicin-loaded porous PLGA microparticles with surface attached TRAIL for the inhalation treatment of metastatic lung cancer. Biomaterials 2013;34:6444–53.

[21] Muralidharan P, Malapit M, Mallory E, Hayes D, Mansour HM. Inhalable nanoparticulate powders for respiratory delivery. Nanomed Nanotech Biol Med 2015;11:1189–99.

[22] Hadinoto K, Sundaresan A, Cheow WS. Lipid–polymer hybrid nanoparticles as a new gneration therapeutic delivery platform: a review. Eur J Pharm Biopharm 2013;85:427–43.

[23] Dave V, Tak K, Sohgaura A, Gupta A, Sadhu V, Reddy KR. Lipid-polymer hybrid nanoparticles: synthesis strategies and biomedical applications. J Microbiol Methods 2019;160:130–42.

[24] Cheow WS, Li S, Hadinoto K. Spray drying formulation of hollow spherical aggregates of silica nanoparticles by experimental design. Chem Eng Res Des 2010;88:673–85.

[25] Edwards DA. Large porous particles for pulmonary drug delivery. Science 1997;276:1868–72.

[26] Primavera R, Kevadiya BD, Swaminathan G, Wilson RJ, De Pascale A, Decuzzi P, Thakor AS. Emerging nano- and micro-technologies used in the treatment of type-1 diabetes. Nanomaterials 2020;10(789):1–27.

[27] Hadinoto K, Phanapavudhikul P, Kewu Z, Tan RB. Dry powder aerosol delivery of large hollow nanoparticulate aggregates as prospective carriers of nanoparticulate drugs: effects of phospholipids. Int J Pharm 2007;333:187–98.

[28] Hirn S, Wenk A, Schleh C, Schäffler M, Haberl N, Gibson N, Schittny JC. Age-dependent rat lung deposition patterns of inhaled 20 nanometer gold nanoparticles and their quantitative biokinetics in adult rats. ACS Nano 2018;12:7771–90.

[29] Ding Y, Kutschke D, Möller G, Schittny JC, Burgstaller G, Hofmann W, Stoeger T, Razansky D, Walch A, Schmid O. Three-dimensional quantitative co-mapping of pulmonary morphology and nanoparticle distribution with cellular resolution in nondissected murine lungs. ACS Nano 2018;13:1029–41.

[30] Medarova Z, Pham W, Farrar C, Petkova V, Moore A. In vivo imaging of siRNA delivery and silencing in tumors. Nat Med 2007;13:372–7.

[31] Mishra V, Gurnany E, Mansoori MH. Quantum dots in targeted delivery of bioactives and imaging. In: Nanotechnology-based approaches for targeting and delivery of drugs and genes. Academic Press; 2017. p. 427–50.

[32] Jakobsson JKF, Aaltonen HL, Nicklasson H, Gudmundsson A, Rissler J, Wollmer P, Löndahl J. Altered deposition of inhaled nanoparticles in subjects with chronic obstructive pulmonary disease. BMC Pulm Med 2018;18:1–11.

[33] Torchilin VP. Structure and design of polymeric surfactant-based drug delivery systems. J Control Release 2001;73:137–72.

[34] Wei X, Wang Y, Zeng W, Huang F, Qin L, Zhang C, Liang W. Stability influences the biodistribution, toxicity, and anti-tumor activity of doxorubicin encapsulated in PEG- PE micelles in mice. Pharm Res 2012;29:1977–89.

[35] Batrakova EV, Bronich TK, Vetro JA, Kabanov AV. Polymer micelles as drug carriers. In: Nanoparticulates as drug carriers. Imperial College Press; 2006. p. 57–93.

[36] Han W, Shi L, Ren L, Zhou L, Li T, Qiao Y, Wang H. A nanomedicine approach enables co- delivery of cyclosporin A and gefitinib to potentiate the therapeutic efficacy in drug-resistant lung cancer. Signal Transduct Target Ther 2018;3:16.

[37] Myung JH, Roengvoraphoj M, Tam KA, Ma T, Memoli VA, Dmitrovsky E, Freemantlec SJ, Hong S. Effective capture of circulating tumor cells from a transgenic mouse lung cancer model using dendrimer surfaces immobilized with anti-EGFR. Anal Chem 2015;87:10096–102.

[38] Ryan GM, Kaminskas LM, Kelly BD, et al. Pulmonary administration of PEGylated polylysine dendrimers: absorption from the lung versus retention within the lung is highly size-dependent. Mol Pharm 2013;10:2986–95.

Chapter 10

Advanced drug delivery systems in breast cancer

Samipta Singh, Priya Singh, Nidhi Mishra, Priyanka Maurya, Neelu Singh, Raquibun Nisha, and Shubhini A. Saraf
Department of Pharmaceutical Sciences, Babasaheb Bhimrao Ambedkar University, Lucknow, Uttar Pradesh, India

1 Introduction to breast cancer

Breast anatomy outlines three essential parts: (1) lobules (milk-producing glands), (2) ducts (tubes that carry milk to the nipple), and (3) connective tissue. Most breast cancers instigate in either ducts or lobules. They can even metastasize outside the breast region via blood vessels or lymph vessels to other body tissues, Fig. 1 [2]. Breast cancer is a kind of tissue cancer involving mainly the inner layer of lobules and ducts [3].

Breast cancer has been recorded as a prevalent type of malignant neoplasms among women with more than one million new cases annually. It is known to be the second leading cancer-associated source of demise and a prime cause of mortality among women aged 45–55 years. The incidence in women being 1/8, often requires complete removal of tissue, chemo/hormone therapy, or radiotherapy most of the time [4]. Prevention is possible by working on the risk factors that may lead to breast cancer. These risk factors may be classified into changeable and unchangeable risk factors. Nonchangeable risk factors include increasing age, mutations, family history, dense breasts, personal history of either breast carcinoma or specific noncancerous breast-associated disease, any previous radiation therapy-based treatment, and women on diethylstilbestrol. People with strong family history of breast cancer or inherited changes in BRCA1 gene and BRCA 2 gene are likely to suffer. The changeable risk factors include being overweight or obese, physically not being active, hormonal intake, reproductive factors, and drinking alcohol [5]. Upholding a healthy weight, exercising regularly, avoiding alcohol, being careful while on hormone replacement therapy or oral contraceptives, or while breastfeeding children, if possible, are some preventive approaches [6]. However, once breast cancer occurs, it must be treated to be defeated.

2 Types of breast cancer

Breast cancer is diverse in characteristics and content and, consequently, may be classified based on several considerations, such as hormonal (in a broad sense), histology, and molecular basis.

2.1 Breast cancer types based on histology

Widely used histological classification is WHO's classification [7], which includes

I. An invasive type
II. A noninvasive type

A noninvasive type may be lobular–ductal carcinoma in situ (LCIS) or ductal carcinoma in situ (DCIS)

Invasive carcinoma can be further sub-classified into:

(a) *Invasive ductal carcinoma (IDC):* Commonest type of breast carcinoma, has ductal malignant proliferation, with stromal invasion in either DCIS presence or absence, shows variable morphologic ductal differentiation [8].
(b) *Invasive ductal carcinoma with prominent intraductal component (IDCPIC):* Presence of an IDC allied with an intraductal component, as a minimum of four times bigger than the invasive one; represents 5% of breast cancers [9].
(c) *Invasive lobular carcinoma (ILC):* Lobular carcinoma that invades the adjacent tissues and is characterized by monotonous round, small, discohesive cells, assembled into clusters [10].
(d) *Papillary carcinoma:* Papillary carcinoma represents around 0.5% of newer breast cancer cases, presents with bloody discharge from the nipple, abnormal mass, abnormalities in radiography, the arrangement of cell proliferation around fibrovascular cores, establishing a circumscribed mass [11].

Advanced Drug Delivery Systems in the Management of Cancer. https://doi.org/10.1016/B978-0-323-85503-7.00028-6

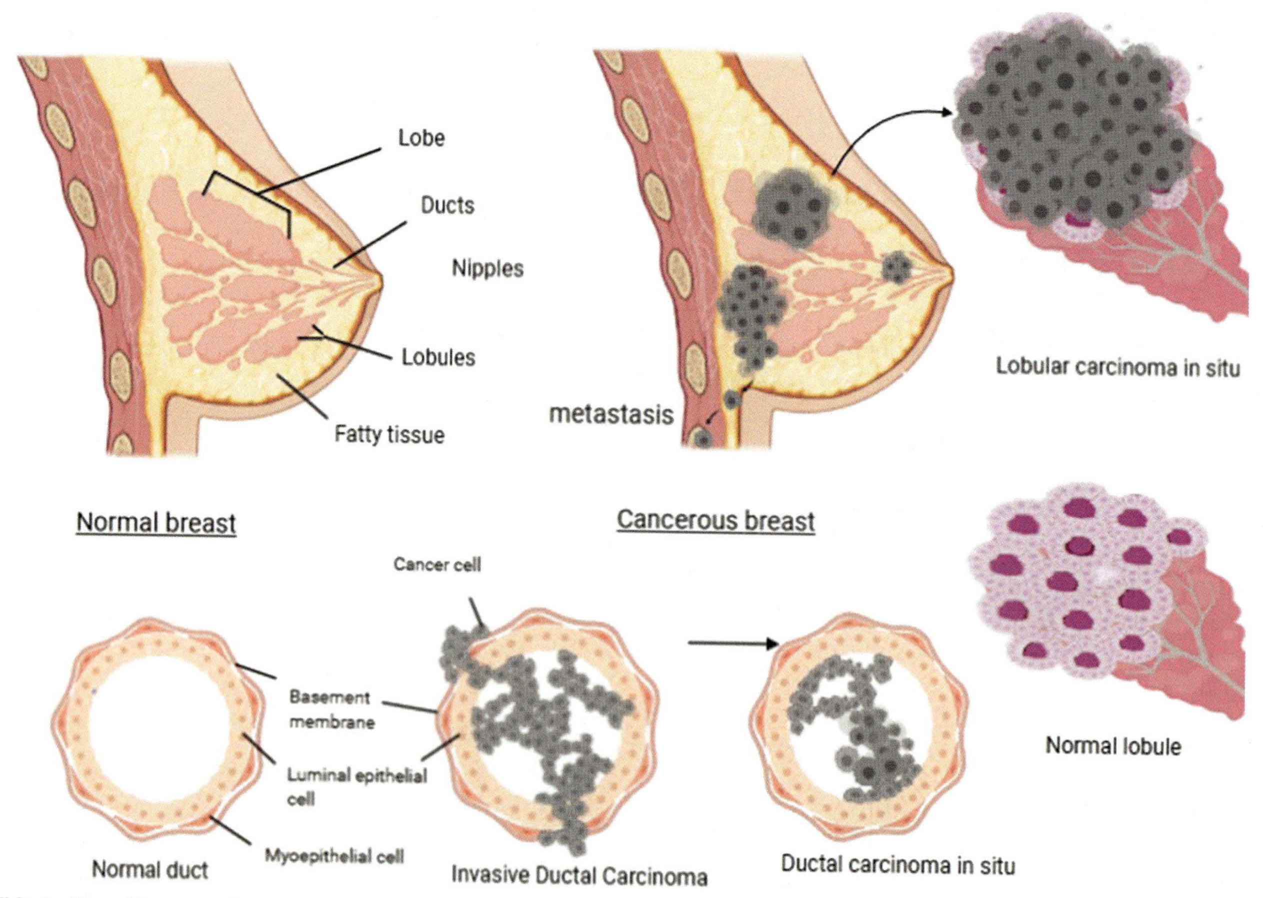

FIG. 1 Normal breast and few breast cancer types (lobular [1] and ductal). Created using biorender.com

(e) *Medullary carcinoma:* Rare (< 5%), "carcinomas with medullary features." [12]

(f) *Mucinous carcinoma:* Rare subtype (2%), may either be pure or mixed type based on cellularity. In pure type, 90% of tumor produces extracellular mucin exclusively, whereas the mixed type also comprises an infiltrating ductal epithelial component [13].

(g) *Tubular carcinoma:* Rare (1%–2% of invasive ones), these are comprised of tubular structures, with lumina open, accompanied by stroma [14].

(h) *Apocrine carcinoma:* Another rare subtype, a characteristic feature of which is the presence of apocrine cells [15].

(i) *Secretory (juvenile) carcinoma:* The incidence of this subtype is very low (0.015%). Histological analysis shows copious extra- and intra-cellular secretory material [16].

(j) *Adenoid cystic carcinoma:* Also a rare breast cancer subtype that is marked by the existence of a luminal and basaloid cellular population organized in a definite growth pattern [17].

(k) Carcinoma with metaplasia: It encompasses 0.25%–1% of annually diagnosed patients and can be subdivided into squamous type, spindle-like, matrix-producing, carcinosarcoma, metaplastic with giant osteoclastic cells [18].

(l) *Paget's disease of the nipple:* This subtype is also rare in which the nipple skin and the darker area that surrounds the nipple, both become eczema-like [19].

2.2 Molecular classification of breast cancer

Molecular level enables us to understand the remarkable counterparts between normal tissue development and cancer progression.

Where normal cell development is in tight control by complex signaling pathways, many of these become dysregulated or usurped by cancer cells or cancer stem cells. A genetic or epigenetic change allows the cells to bypass these mechanisms which drive cancer to take place. Therefore, activating proto-oncogenes or inactivating tumor suppressors may result in hyperactivation or elimination of critical negative regulators, respectively. Estrogen receptor (ER) signaling, human epithelial growth factor receptor (HER2) signaling, and canonical Wnt signaling are major signaling pathways that regulate normal glandular development and stem cell functions of the breast. Apart from these pathways,

TABLE 1 Breast cancer primary molecular classification [21].

	ER	PR	HER2	Other important features	Responsiveness
Luminal A	+	+/−	−	Ki67 low	Hormonal therapy or chemotherapy
Luminal B	+	+/−	+	Ki67 high	Usually endocrine or trastuzumab-responsive, variable to chemotherapy
HER2	−	−	+	Ki67 high	Trastuzumab and chemotherapy responsive
Basal	−	−	−	Ki67 high, cytokeratin 5/6 +	Nonresponsive to endocrine therapy, responsive to chemotherapy
Claudin-low	−	−	−	Ki67 high, E-cathedrin, claudin-3, claudininin-4, and 7 low	Intermediate response to chemotherapy

many other pathways also play significant roles but are out of the scope of this chapter [20]. Primary molecular breast cancer classification is mentioned in Table 1.

However, in a comprehensive sense, breast cancer can be ER +, HER2 +, or triple-negative (absence of ER, PR, and HER2) type [22]. Treatment is also dependent upon the kind of breast cancer.

3 Drugs approved/nonapproved for treating breast cancer

3.1 FDA-approved chemotherapeutic agent/hormonal/targeted drugs

Drugs for preventing breast cancer include tamoxifen citrate and raloxifene hydrochloride. Drugs approved by the United States Food and Drugs Administration for treating breast cancer are capecitabine, docetaxel, doxorubicin hydrochloride, epirubicin hydrochloride, cyclophosphamide, eribulin mesylate, paclitaxel, 5-fluorouracil, everolimus, fulvestrant, toremifene, gemcitabine hydrochloride, goserelin acetate, ixabepilone, methotrexate, vinblastine sulfate, thiotepa, and letrozole. Some specific drugs includes hormonal drugs (tamoxifen citrate (for HR +) [23], anastrozole (for HR +), exemestane (for ER +)), lapatinib (for HER2 +) [24], olaparib (for BRCA mutation carriers) [25], megestrol acetate (for HR +) [26], neratinib maleate (for HER2 +) [27], pamidronate disodium (breast cancer with osteolytic bone metastases) [28], talazoparib tosylate (for gBRCAm; HER2-) [29], toremifene (for HR +) [30], ribociclib (for HR +) [31], tucatinib (for HER2 +) [32], palbociclib (for HR +) [33], abemaciclib (for metastatic HR +/HER2- breast cancers) [34], alpelisib (for HR +) [35], etc. [36]

3.2 Monoclonal antibodies

Monoclonal antibodies are lab-generated proteins which, like our own antibodies, can also recognize specific targets (like breast tumor antigens) and provide improved therapeutic benefit [37, 38]. Some known FDA-approved monoclonal antibodies aimed to cure breast cancer are: Trastuzumab (for HER2 + breast cancer), Pertuzumab (for HER2 + breast cancer) [39], Atezolizumab (where tumors express PD-L1) [40].

3.3 Other chemotherapeutic agent/phytocomponent/extracts

Recently, aspirin (analgesic) has been under clinical trial for preventing recurrence in stage II-III, HER2-, node + breast cancer [41]. Curcumin [42], quercetin [43], and resveratrol [44] have also shown some potential against breast cancer. Some common herbs or herbal extracts that show some potentiality against breast cancer are (dried roots of *P. ginseng*, Black cohosh) [45], *Nelumbo nucifera* leaves extract [46], etc. Some essential oils, such as frankincense, geranium, and pine needle essential oil also show effectiveness against breast cancer [47].

3.4 Nanoparticle itself showing anticancer activity against breast cancer

There are some nanoparticles that themselves have exhibited some activity against breast cancer. Thus, they hold the potential to be utilized as a drug themselves. Examples include, fullerenes [48], graphene oxide [49], gold nanoparticles [50], silver nanoparticles [51], selenium nanoparticles [52], platinum nanoparticles [53, 54], copper oxide nanoparticles [55], Nanobacteria from calcified placental tissue [56], etc. However, these hold some drawbacks due to which clinical application remains a challenge.

3.5 Genes therapy product

A gene therapy product is fundamentally a part of nucleic acid, which might be either prophylactic, diagnostic, or therapeutic, and can be administered either via in vivo method or ex vivo method [57]. Gene therapy based on RNA interference (especially short interfering RNAs (siRNAs)) has gained the highest attention. However, unmodified siRNA has its limitations, such as intravascular instability, susceptibility to RNase A-type nuclease, quick renal clearance, short half-life, potential toxicities [58]. Other examples of gene therapy in breast cancer are: shRNA, pTRAIL DNA, virus-based gene therapy (rodent parvovirus-H1 (H1-PV), HF10 clone), etc. [57]

4 Conventional delivery systems

Conventional approaches for drug delivery (Fig. 2) in the treatment of breast cancer are still dominating the market. Most breast cancer therapeutics are given by IV injection (either bolus or infusion) [59]. Some drugs (for example, hormonal: exemestane [60], tamoxifen [61], letrozole [62, 63], anastrazole [62, 64], megestrol acetate [65] are given orally in tablet form. Palbociclib [66] and talazoparib [67] are available as a capsule. Nowadays, conventional delivery systems, including tablets, capsules are influenced by a controlled drug delivery system and it gives therapeutic advantages to the patients by altering the release of pharmaceutical agents and thus, rescheduling their pharmacological action [68].

5 Advanced drug delivery system in breast cancer

Despite being successful, delivering anticancer drugs via conventional drug therapy system (CDDS) holds some main drawbacks such as adverse or unwanted side effects, low therapeutic indices, poor bioavailability, requirements of high dose, multiple drug resistance (MDR) development, and nonspecific targeting. Thus, the major objective in developing a formulation against breast cancer is to properly address these issues and to carry the drug to the desired sites, also reducing adverse or side effects [69]. Advanced drug delivery systems (ADDS) have been introduced to overcome the limitation of the CDDS [70].

6 Introduction of nanotechnology to breast cancer therapeutics

The arrival of nanotechnology provided conceivable solutions to the glitches in breast cancer conventional "drug" based treatment that was generally difficult to deal with, including its biological diversity in the patient population, the multiple disease progressive pathways, the inception of "resistance" to established drugs, and the severity of side or adverse effects of medicament, which generally result from very poor distribution of therapeutics into the body [71]. The overall nanotechnology utilization in breast cancer can be broadly divided into 1. therapy, 2. diagnostics, and 3. diagnostics and therapeutics both [72].

Nano-based strategies often involve a build-up because a greater permeation capability is noticed in all cancer cells. Cancer cells are known to grow rapidly. This rapid growth leads to a poorly formed vasculature. This, in turn, results in poor flow of blood which would have normally carried out the nanoparticles from the tumor but is unable to do so in such cells. Poorly formed lymphatics also contribute to this phenomenon. The positive side of this build-up of nanoparticles is that the drug is passively targeted to the required site. Alternatively, the nanoparticles can be surface decorated by a ligand or an alternative molecule that is able to actively reach the target site because the tumor has an enhanced requirement of this molecule. This strategy is called active targeting [73].

7 Nano-based advanced DDS in breast cancer

Several nano-carriers have shown promising results at a preclinical level. Even at a clinical level, some drug-loaded nanoparticles have shown hopefulness. A variety of

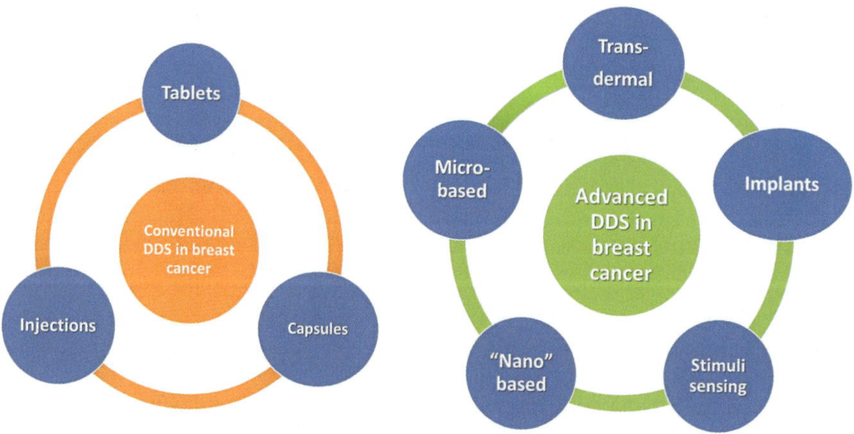

FIG. 2 Conventional and advanced DDS in breast cancer: an overview.

substances, for example, polymers, proteins, polysaccharides, lipids, or inorganic materials can be used to fabricate these nano-carriers. Their selection depends on multiple factors such as desired size, drug properties (e.g., solubility in aqueous solvent and stability), drug release characteristics, desired surface properties such as charge or permeability; biodegradability, biocompatibility and toxicity; and regulatory challenges [74]. Some important advanced "nano" based system for delivering drugs to the cancer of the breast are summarized below:

7.1 Vesicular drug delivery system

In vesicular drug delivery systems (VDDS), a vesicular structure encapsulates active moieties [75]. There are a plethora of VDDS types, while the types investigated in breast cancer are limited and are given below:

7.1.1 Liposomes

Liposomes are artificial spherical vesicles that can easily be fabricated using natural nontoxic phospholipids and cholesterol [76]. They may vary in size (25 nm to 2.5 μm) and can be categorized into unilamellar, multilamellar, or oligolamellar according to their lamellae [77]. Phospholipid bilayer presence in liposomes prevents breaking down of the encapsulated active drug form before reaching the tumor tissue and is expected to minimize the drug exposure to healthy sensitive tissue [78]. Though conventional liposomes reduced the drug toxicity in vivo, they were prone to rapid elimination from the bloodstream, thus limiting its therapeutic efficacy. Sterically stabilized liposomes were introduced to ameliorate the stability of liposomes and extend their time of circulation in blood. Polyethylene glycol (PEG) is usually the optimal choice [79].

In breast cancer pharmaceuticals development, doxorubicin liposomes such as Doxil, Myocet, ThermoDox have made their presence felt at a clinical level [80]. Doxil is PEGylated and contains HSPC, cholesterol, and PEG 2000-DSPE. Myocet (150–250 nm) is a non-PEGylated liposomal preparation, manufactured by Elan Pharmaceuticals, and contains acidic egg phosphatidylcholine and cholesterol in 55:45 molar ratio, the drug-to-lipid ratio is around 0.27. Mononuclear phagocyte system recognizes them easily due to their larger size, thereby lessening their exposure to normal tissues [80, 81]. In a study, the occurrence of cardiac adverse events was also lowered significantly with Myocet in metastatic breast cancer patients [80, 82].

7.1.2 Niosomes

Niosomes are also a bilayered structure primarily encompassing the self-association of nonionic surfactants and cholesterol dispersed in an aqueous phase [83]. Easy derivatization and versatile structure of surfactants lead to enhanced chemical and better storage capabilities than liposomes [84]. Niosomes containing tamoxifen citrate, formulated using Span 60: cholesterol (1: 1) gave 2.8 fold enhancements in cellular uptake in MCF-7 cells. Tumor volume was also found to be reduced as compared to free tamoxifen citrate in vivo [85]. Carum carvil extract-loaded Niosomes (containing Tween 60, Span 60, ergosterol, and Carum in a 30:30:30:10 molar ratio) showed better anticancer activity against MCF-7 cells [86]. PEGylated gingerol nano-niosomes (enclosing more than 76% of the Gingerol) exhibited less IC50 (2.97 ± 4.7 μM) than the standard drug (26.67 ± 1.5 μM) in T47D cell lines [87].

7.1.3 Phytosomes

Phytosomes are unique for having bioactive compounds as their core construction besides phospholipid. They are constructed by the conjugation of phytoconstituents such as terpenoids, flavonoids, and tannins or herbal extract or to phospholipids in a nonpolar solvent [88]. PEGylated surfaces can protect nanocarriers from opsonization and reticuloendothelial system clearance by repelling plasma proteins. Greenselect phytosome (green tea catechin-based formulation) exerted high bioavailability of catechin and antiproliferative effects on breast cancer tissue [89]. Phytosomal-curcumin perturbated AMP-activated protein kinase, reduced MCF-7 invasiveness through E-cadherin perturbation, inhibited the tumor growth in vivo [90]. Codelivery of monascin and ankaflavin incorporated in phytosomal casein micelles, along with resveratrol, showed in vivo tumor volume reduction [91]. Luteolin nano-phytosomes enhanced the bioavailability and improved passive targeting of luteolin in MDA-MB 231 breast cancer cells. Its cotreatment with doxorubicin resulted in the highest cell death [92].

7.1.4 Exosomes

Despite the role of exosomes (an important class of extracellular vesicles) in the progress of the breast cancer, exosomes also displayed their role in the delivery of drug as a carrier to breast cancer [93]. Exosomes have begun to be explored for their usage as drug delivery nano vehicles for the delivery of small molecule drugs or genes. They have exceptional interaction ability with cellular membranes and deliver their cargo to target cells [94]. They fuse with the targeted-cellular membrane, open up and get their payload delivered to the recipient cell and mediate many processes [95]. However, exosome uptake needs further investigation. Exosomes as a carrier have been utilized to deliver miRNA (Let-7a miRNA, miR-379), other Ra (Cas9 mRNA), proteins (MHC-I/peptide complexes), and other chemical drugs, like doxorubicin for treating breast cancer either in vitro or in vivo or both [96]. Pullan et al. loaded delivered doxorubicin using exosomes isolated from immature dendritic cells (imDCs) of the mouse as a targeted DDS against human breast cancer cells. Doxorubicin carrying exosomes inhibited the growth of tumor cells in breast, and had no toxic outcome on mice. Further modification with specific iRGD peptides endowed their targeting ability [97].

7.1.5 Polymersomes

Polymersomes are spherical nanovesicular systems, composed mainly of amphiphilic block or triblock copolymers held together in water because of robust physical interactions creating a hydrophobic bilayer and an aqueous core. They can be engineered to make the delivery responsive to pH, temperature, light, enzymes, or any other stimuli [98–100]. They are more stable than liposomes and can be effectively used for the treatment of breast cancer targeting [101]. In a study, doxorubicin-loaded polymersomes were utilized against breast cancer [102]. In another study, doxorubicin-loaded redox-sensitive polymersomes gave about 85% tumor regression in contrast to 42% displayed by plain doxorubicin [103].

7.2 Nanoparticulate systems

7.2.1 Polymeric

The most common investigated polymeric nanoparticles in breast cancer are poly(lactic-co-glycolic acid) (PLGA) nanoparticles. PLGA nanoparticles possess moderate MDR reversal action on their own and are biodegradable. Katiyar et al. codelivered rapamycin and piperine (a chemosensitizer) via PLGA nanoparticles and found it to be a hopeful approach [104]. Other common nanoparticles are chitosan [105–107] and polycaprolactone (PCL) commonly used with PEG [108, 109] holding a plethora of literatures.

Polymeric nanoparticles are broadly classified into nanospheres (solid) and nanocapsules (hollow) [110]. Nanospheres and nanocapsules have different architecture and morphology, where nanospheres are closed packed matrix systems, whereas nanocapsules are core-shell structures where the drug is amalgamated into the inner core surrounded by a nanoshell [111]. Polymersome is one of the specific types of nanocapsule [112]. As per the current trend, nanoparticles are often functionalized to get desired active targeting. Some examples that have shown positive results against breast cancer are calciferol-loaded nanocapsules, [113] mesoporous polymeric nanospheres for siRNA delivery [114], etc.

7.2.2 Protein nanoparticles

The emergence of Abraxane (a protein-based nanoparticle of paclitaxel) is a boon to breast cancer research at a clinical level [115]. Human serum albumin (HSA (66.5 kDa)) is a very abundant serum protein and a versatile carrier facilitating the transport of various substances throughout the systemic circulation. Drug binding to albumin may affect its biodistribution, bioactivity, and metabolism [116]. Research on albumin nanoparticles is still ongoing. Lee et al. incorporated paclitaxel in HSA-PEG nanoparticles using film casting and re-hydration method, without chemical crosslinking or high pressure/shear application. As per cytotoxicity assay against various cancer cells of breast, the obtained HSA-PEG nanoparticles (280 nm) were equivalent or even better than Abraxane. Its systemic administration in SK-BR-3 xenograft mouse model also exhibited a greater intratumoral accumulation and systemic circulation of >96 h [116]. Other examples of protein nanoparticles that have shown some positive outcome in breast cancer research are: 7% gliadin nanoparticles carrying cyclophosphamide [117], Tobacco mosaic virus protein nanoparticles carrying doxorubicin [118], whey protein isolate protein nanoparticles carrying lycopene [119], shell-crosslinked zein nanocapsules for resveratrol and exemestane oral delivery, [120] etc.

7.2.3 Solid lipid nanoparticles and nanostructured lipid carriers

Solid lipid nanoparticles (SLN; first-generation lipidic carriers) are a promising approach against breast cancer. SLN loaded with mitoxantrone improved lymph node metastases [121]. Solid lipid is also able to overcome drug resistance against adriamycin-resistant breast cancer [122], deliver microRNA-200c (enhancing paclitaxel ability) [123], and reverse tamoxifen resistance [124]. However, SLN holds some disadvantages (such as drug expulsion), due to which second-generation lipidic nanocarrier also exists, i.e., nanostructured lipid carriers (NLCs). NLCs displayed high encapsulation efficiency for quercetin, enhanced its stability in an aqueous solution, and showed high uptake in breast cancer cells [125]. NLCs can be made dual-functional against breast cancer (antimetastatic and photothermal) [126]. They have also proven to be effective against adriamycin-resistant breast cancer [127]. NLCs are also a good carrier for the oils that exhibit activity against breast cancer, for example, Perilla oil [128, 129].

7.3 Micellar drug delivery system

Polymeric micelles are basically composed of amphiphilic block copolymers entailing hydrophobic and hydrophilic blocks in an aqueous medium. Poly (ethylene glycol) is the most extensively used hydrophilic block. Selection of hydrophobic blocks rings versatility which provides predictive properties of polymeric micelles [130]. Genexol-PM, paclitaxel-loaded cremophor EL-free PEG-b- poly(D,L-lactic acid) polymeric micelle showed more favorable efficacy (45.3%–72.3% response rate) than conventional cremophor EL-based paclitaxel (21%–54% response rate) in metastatic cancer patients [130, 131]. Amphiphilic polymer Pluronic F127-based polymeric micelles with an inner core containing Zileuton accumulated in the tumor region; however, liver and spleen also retained it [132]. Alendronate-modified docetaxel-loaded micelles were valuable against breast cancer with bone metastasis [133].

7.4 Dendrimers

These highly branched polymeric macromolecules [134] have also significantly improved the antiproliferative activity of doxorubicin against triple-negative (MDA-MB-231) cells. Dendrimer modification with EBP-1 and TAT peptides proved advantageous against EGFR and gave dual functional benefits [135]. Owing to its structure, chemical modification of dendrimers to give dendrimer-drug conjugate is possible [136], for example, curcumin-dendrimer conjugate (which proved efficacious against SKBr3 and BT549 breast cancer cells) [137], polyamidoamide (PAMAM) dendrimer conjugated with cisplatin and doxorubicin for their codelivery (which proved efficacious against MCF-7 and MDA-MB-231 cells at relatively low concentration) [138], etc. PAMAM G4.5 carboxylate dendrimers showed good uptake efficiency in 4T1 tumor cells [139].

7.5 Nanoemulsions

Nanoemulsions are emulsions with nano-size range, o/w being more popular. Oils exhibiting anticancer activity against breast cancer can easily be delivered via nanoemulsions, for example, spearmint oil [140]. Microemulsions are also nano-in-size but exhibit thermodynamic stability [141]. Lipophilic drugs can also be delivered as nanoemulsion, with or without any synergistic or additive effect. Tripathi et al. developed α-linolenic acid nanoemulsion and functionalized it with folic acid for better activity of doxorubicin [142]. Phospholipid-based lipid nanoemulsion (LN) has also been formulated to coencapsulate and concurrently deliver doxorubicin and W198 to MDR breast cancer cells [143].

7.6 Nanofibers

Fine pores, high surface area of nano/micro-fibers, and porous structure allow the drugs to be integrated into the nanofibers. Changing the electrospinning parameters may be helpful in achieving a controlled release pattern of drugs from nanofibers, followed by size adjustment and nanofibers morphology [144]. Curcumin- and chrysin-loaded PLGA/PEG electrospun nanofibers exhibited excellent synergistic antiproliferative activities and excellent inhibitory capacity against T47D cancer cells of breast and hold the application in decreasing locoregional recurrence of breast cancer [145].

7.7 Layer-by-layer nanoparticles

Layer-by-Layer (LbL) deposition is a beneficial technique, which creates multi-layered core-shell nanocapsule. A variety of components for shell may be used, such as polyelectrolytes, enzymes, lipids, colloids, viruses, or any other active compounds. The multiple layered films allow the combination of drugs in a solitary unit at "nano" level, bringing on the multifunctional architectures. It also provides protection to drugs and the possibility to load a higher amount of drugs [146]. LbL nanoparticles forming by depositing poly-l-arginine and siRNA alternately, exhibited protracted 28 h of serum half-life, reduced up to 80% tumor target gene expression [147].

7.8 Magnetic systems for drug delivery

Due to numerous challenges in oncology, magnetic nanoparticles (MNPs) were introduced for more efficient release of drugs in the desired manner and drug penetration into solid tumors [148]. The unique magnetic characteristics of MNPs enable them to be utilized for detecting cancer lesions and deliver drugs into the desired target. Iron oxide is extensively used for this purpose because of its superparamagnetic properties. Magnetic field exposure leads to the generation of heat by these nanoparticles, which aids in the destruction of tumors [149]. Violamycin B1-coated iron oxide nanoparticles were tested on MCF-7 cells, which showed better delivery outcome and was efficient [150].

7.9 Hybrid nanoparticles

Hybrid nanoparticles often combine two or more nanoparticulate systems to form a single nano-based system. Lipid polymer hybrid is most common, such as for loading methotrexate and beta carotene (gave apoptosis index of 0.89 against MCF-7 cells) [151], psoralen (similar antitumor ability as that of doxorubicin but lower toxicity than doxorubicin), and several other similar investigations. The outer lipid layer improves biocompatibility [152]. Other are calcium phosphate-polymer hybrid nanoparticles [153], hybrid magnetic [154], etc.

7.10 Miscellaneous

There are also some other systems, which may carry on the drug after conjugating it to the nanosystems, such as, metallic nanoparticles (silver [155], gold [156]), carbon nanotropes (graphene oxide [157], carbon nanotube [158], fullerenes [159]), and ceramic (mesoporous silica [160]). However, due to some drawbacks, their application as a DDS is limited.

8 Other advanced DDS

8.1 Transdermal patches in breast cancer

Transdermal patches are attractive as they are noninvasive, easy to use, and can also deliver multiple doses for an extended time period, while also showing patient compliance and cost-effectiveness. The approach to transdermal patches may either be matrix or reservoir approach polymeric systems that house the drug such that it diffuses through the polymer by the way of skin beneath for reaching the systemic circulation in a controlled manner [161–164].

A transdermal patch is an interesting approach for breast cancer since it can act as a reservoir for the drug at the required site. On an animal observation basis, the transdermal patch of anastrazole gave high anastrazole accumulation in the area beneath the patch application site compared to oral administration [165]. Apart from anastrazole, there are also positive results promoting site-specific delivery and high drug concentration obtained from a transdermal patch of Letrozole [166]. Mean residence time (MRT) was significantly enhanced following transdermal administration (97.5 ± 2.49 h) compared with oral administration (39.8 ± 1.93 h) [164].

The application of transdermal patches is not just limited to direct treatment or prevention but may also show some significance indirectly. The transdermal patch of estradiol is mainly used for reducing menopausal symptoms (such as hot flashes and vaginal dryness) [167]. In a study named ULTRA, transdermal patches of estradiol resulted in no increment in breast density which is a positive sign as heavy breasts are often associated with an increased risk of cancer [168]. However, estradiol may show interaction with aromatase inhibitors, fulvestrant, tamoxifen, and other hormonal medications [167].

8.2 Microneedles

As an alternative to syringe-based therapeutic administration, microneedle-devices (Fig. 3) are promising [169]. These are the device composed of needles which are very thin and are arranged in single lines depending upon the surface area to be covered for the therapy. The dimensions of these microprojections are generally around 20–250 μm wide and as long as 1500 μm. These are compatible with most small and macromolecular therapeutics, or even nanomedicines. These are versatile, not as painful as conventional needles, and are safer to use. They can be made using various materials such as stainless steel or polymers. Microneedles designs may either be solid or hollow [170]. Dissolvable microneedles are of high interest owing

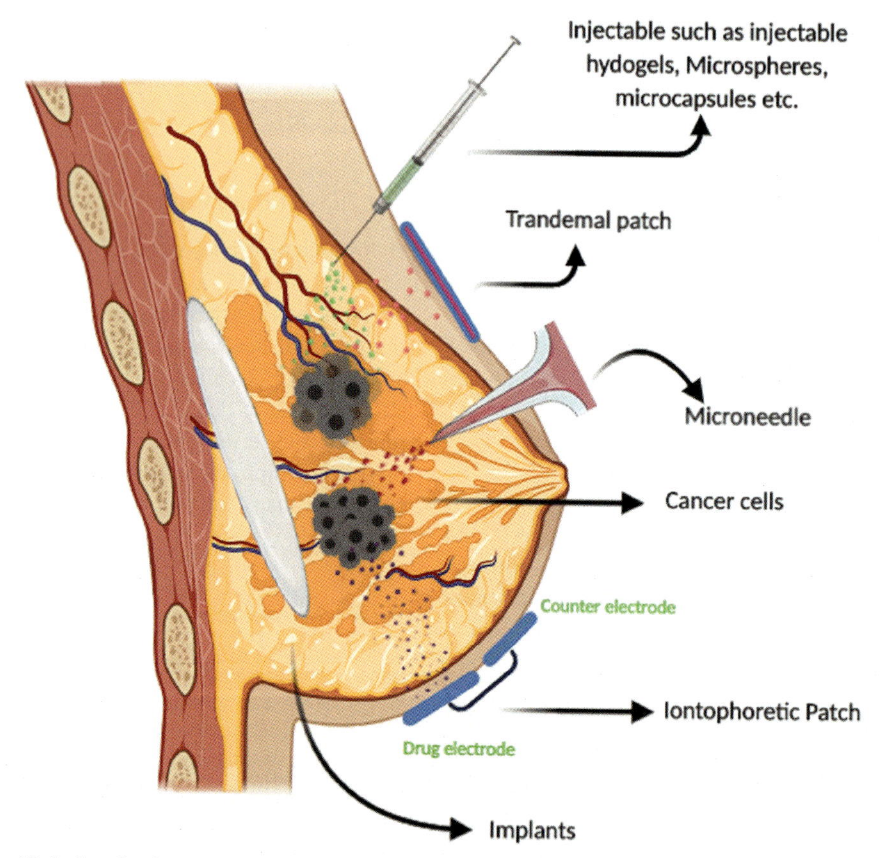

FIG. 3 Implants, injectable hydrogel and transdermal approaches for localized drug delivery to breast cancer. Created by biorender.com

to their safety, high payload, and low device complexity [169, 171]. Bhatnagar et al. constructed dissolvable polymeric microneedles for the codelivery of docetaxel and doxorubicin HCl in 4T1 xenograft breast cancer model. Microneedles dissolved into the excised skin within an hour of insertion and brought about greater survival in comparison to those administered via intratumoral injections [169]. Desimone and MOGA prepared a rapidly dissolving microneedle ("PRINT microneedles"). They loaded it with docetaxel to determine the application in breast cancer, and administered the patches to SUM149 xenograft mice (inflammatory breast cancer), despite penetrating the skin, the microneedles with 400 μm tall needles could not deliver docetaxel into the tumor when the cancer presented as a lump, indicating the requirement of longer microneedles [171]. Zein microneedles (ZMN) are plastic; undissolvable in an aqueous solvent or skin tissue, however, can imbibe water and swell upon prolonged incubation in aqueous medium [172].

8.3 Other transdermal approaches

There are several other transdermal approaches investigated in breast cancer treatment. Some of them are given below:

Iontophoresis utilizes an electric field to deliver drugs to the tissue [173]. It helps in the transport of charged compounds and compounds with high molecular weights, which are normally unable to be delivered via passive transport. Komuro et al. delivered miproxifen phosphate (TAT-59; formulated in the form of a gel) [174].

Electroporation is another technique for enhancing the permeability of a drug by exposing the cells to electric pulses [175]. Reversible electroporation is most commonly exploited for increasing the concentration of drugs at the required site [176]. Rodríguez-Cuevas et al. in 2001 applied electric pulses post 10 min after bleomycin injection (100 microsec long; field strength: 1,300 V/cm; frequency: 1 Hz). They anticipated the technique to be useful for those breast cancer patients, whose treatment failed from chemotherapy or radiotherapy [177].

Apart from the above-mentioned complex techniques, simple transdermal gels have also been tried. The basic difference between conventional topical gel and transdermal gel is that, in transdermal gels, one or more permeation enhancers are added to enhance the permeability [178]. Lee et al. conducted a study in which 4-hydroxytamoxifen (4-HT) transdermal gel (by BHR Pharma; prepared by Enhanced Hydroalcoholic Gel technology (patchless transdermal delivery system)) were taken and compared with oral 4-HT in breast cancer patients. They found similarity in the antiproliferative effect of 4-HT gel as that of oral 4-HT; however, there were reduced effects on coagulation and endocrine parameters [179].

8.4 Implants

Breast Implants (Fig. 3) have already been well utilized by women for either aesthetic purpose or after mastectomy or lumpectomy as a process of breast augmentation [180]. This concept could be used as a precise approach against cancer tissues. However, its application remains underutilized because of the surgical procedure requisite. Nevertheless, drug-loaded implants hold some advantages, such as drug administration frequency is reduced, delivery efficacy is increased, and systemic toxicity is minimal. Yang et al. developed implantable scaffolds (of poly-lactic-co-glycolic acid) via 3D printing for immobilizing NVP-BEZ235 and 5-fluorouracil, for orthotopic breast cancer therapy. These scaffolds lessened the requisite drug dosage significantly, ensured therapeutic drug level at desired site for a longer period, long-term drug release, reduced the drug administration frequency, keeping exposure of drug in the normal tissue limited [181]. Talazoparib-loaded PLGA implants provided a sustained release for over 28 days and significantly inhibited tumor growth in Brca1Co/Co;MMTV-Cre;p53 +/− mice treated with these implants in comparison to free talazoparib [182].

8.5 Injectable hydrogels

Injectable hydrogels are nothing but highly crosslinked 3D structures because of the intercrosslinked mesh-like structure that is formed by the polymers. These structures are capable of absorbing a large quantity of water. This water gets entrapped in the inter-penetrating network (IPN). This IPN is capable of entrapping within itself. Proteins, peptides, genetic material like RNA as well as anticancer drugs. Peptide-based hydrogels are well received by the biological system due to the physiological similarity between the system and also because these can break down into smaller units. Qi et al. formulated hexapeptide-based hydrogel with doxorubicin for breast cancer chemotherapy [183]. In another drug that utilized the same drug but silk hydrogel as a dosage form was able to display sustained release almost up to one month. Excellent response was observed against MDA-MB-231- induced xenograft mice and outperformed doxorubicin (equivalent amount) delivered intravenously [184].

8.6 Microcapsules and microspheres

Microspheres and microcapsules aid in prolonging drug release and targeting the therapeutic or a particular site. Microspheres are considered as the matrix system in which the drug is dispersed homogeneously, either dissolved or suspended. Characteristically microspheres are free-flowing powders (1–1000 μm size range). Microcapsules are heterogeneous particles in which a membrane shell surrounds the core creating a reservoir [185]. Microencapsulation is a process of coating solid, liquid, or gaseous particles with a thin layer of polymeric solution to yield micron-sized capsules [186–188].

Pal et al. prepared gum acacia (GA) microspheres by co-precipitation method, encapsulated curcumin, and evaluated their cytotoxicity on TNBC cell lines. They perturbed MMP and induced apoptosis. Folic acid conjugation resulted in even higher tumor regression BALB/C mice model [189]. Microspheres offer the benefits of localized and prolonged drug release. Drug-loaded albumin microspheres improved survival in a murine breast cancer model [190]. Since the internal artery of the breast perfuses the adjoining areas, the delivered microspheres would reach the abdominal wall which is not desired [191]. Thiolated polymethacrylic acid (PMA)-based redox active microcapsules containing doxorubicin internalized in SKBR3 cells and 3D spheroids [192].

9 Antibody-drug conjugates

Antibody-drug conjugates (ADCs) exploit the monoclonal antibody (mAb) usage by chemically conjugating mAb to the drug. These deliver the drug to the cells that express antigen specifically by the aid of mAb. This lessens the drug exposure to the cells that do not express the target antigen [193]. The regulatory sanction of a few ADCs is a clear indication of their clinical potentiality. Few FDA-approved ADCs are Trastuzumab emtansine (for HER2 + metastatic breast cancer) [194], Sacituzumab govitecan (Trop-2-directed antibody conjugated to a topoisomerase I inhibitor (SN-38) for TNBC) [195], and glembatumumab vedotin (for glycoprotein NMB (gpNMB) expressing breast cancer). However, glembatumumab vedotin failed to encounter the primary endpoint in the METRIC study [196].

10 Physiological stimuli responsive-based ADD approach

Existing delivery systems can be modified using smart polymer/other material which hold some sensitivity to some stimuli, such as, pH, redox, ROS, ATP, and enzymes. An external or physical approach may also be carried out, where an external stimulus is applied to enhance drug delivery outcome (discussed in Section 11.4). pH-responsive systems are of utmost importance in breast cancer due to the accumulation of acidic metabolites in tumor cells, because of anaerobic glycolysis. Thus, it is the feature of the tumor microenvironment that they are weakly acidic (pH 6.5–7.0). Thus, several pH-sensitive carriers have been designed, so that it degrades in the low acidic condition, whereas their stability stays retained in normal physiological pH (i.e., 7.4) [197]. Redox-responsiveness is another strong stratagem because of the GSH concentration difference between the intracellular reducing space and extracellular mildly oxidizing space [197, 198]. There is evidence of higher GSH level in tumor tissues than that in normal tissues. ROS responsive systems are developed due to the engrossment of cancer cells with intrinsic oxidative stress with a fairly higher H_2O_2 level than healthy tissues and increased ROS in the tumor. An enzyme-responsive system takes advantage of tumor-associated enzyme dysregulations. Hypoxia has been known to be a prime cause of restricting the growth of cancer cells via the pro-drug approach. Through this approach, the pro-drug molecules are able to act like electron acceptors in hypoxic conditions, thus, leading to the release of the drug [197] (Table 2).

11 Delivering micro/nanoparticles (drug-loaded or blank)

11.1 Delivering in the form of conventional DDS

They can be served as a dried lyophilized powder (example, Abraxane) and later be reconstituted into the desirable solvent or directly in a suspension form. Orally, they can be packed into capsules or served in a liquid form. Topically, the nanoparticles can be incorporated into the gel, to form emulgel [207], SLN or NLC-based gel [208], or nanoparticulate gel [209].

11.2 Delivering as advanced transdermal approach

Nanoparticles/emulsions or microemulsions themselves enhance the transdermal ability [210, 211]. However, they can be combined with other approaches such as transdermal patch [212] or microneedles [213] or iontophoresis [214] for more efficient delivery. Mehnath et al. combined paclitaxel carrying micelles with core-shell nanofibers-based patch, which aided in releasing the drug at target sites [215].

11.3 Surface modification of nanoparticles

Nanoparticles, whether with drug or without including any drug may be surface modified/ surface functionalized/ surface engineered by a ligand or a polymer or antibody or a suitable substance, depending on the reason to modify. This may be done either by adsorption, chemically by the help of linkers, or doping. The major reasons for changes in the nanoparticle surface are to reduce the toxicity of metallic nanoparticles and make them biocompatible, make the nanoparticle dispersible in water, or facilitate active or passive targeting.

11.3.1 To solve issues of nanoparticles

Major issues faced by nanoparticles are clearance, toxicity, and stability of some nanoparticles. PEGylation helps in prolonged circulation [216]. Engineered capsid has also helped

TABLE 2 Stimuli-responsive nano/micro systems for drug delivery to breast cancer.

Responsive systems	Smart key composition	Drug	Outcome
pH-responsive micelles	Diblock copolymer, poly(ethylene glycol)-block-poly(2-(diisopropylamino)ethyl methacrylate)	Doxorubicin	- pH-triggered micelle dissociation - Targeted mitochondrial organelles, reduced MMP - Reversed doxorubicin resistance [199]
Polymeric nanoparticles (pH-responsive)	Chitosan	Tamoxifen	- Increased intracellular Tamoxifen concentration - Enhance anticancer efficiency of tamoxifen - Induced apoptosis in a caspase-dependent manner [105]
pH-responsive zinc oxide nanoparticle	Phenylboronic acid (PBA)	Quercetin	- Apoptotic cell (MCF-7) death via enhanced oxidative stress - Reduced tumor growth in EAC tumor-bearing mice [200]
pH-Sensitive cleavable liposomes	PEG5K-Hydrazone-PE and DSPE-PEG2K-R8	Paclitaxel	High tumor penetration depletion of collagen I [201]
pH-responsive microspheres	Gum acacia	Curcumin	- pH-responsive curcumin release [189]
Layer-by-layer microneedle (pH sensitivity)	Chitosan	Doxorubicin	- Control the doxorubicin release [202]
Redox-responsive micelles	Indomethacin and dextran	Doxorubicin	- Good micellar stability in normal physiological condition - Depolymerized, released doxorubicin in a reducing condition. - Intracellular doxorubicin accumulation and retention - Synergism [198]
Redox-responsive Micelles	Built-in disulfide bonds, poly(Tyrosine(alkynyl)-OCA) conjugate and monomethoxy poly(ethylene glycol)-b-poly(Tyrosine(alkynyl)-OCA)	Camptothecin	- Excellent stability under normal conditions - Rapid dissociation in reduced condition - Cytotoxicity enhancement [203]
ROS responsive prodrug nanoparticles	Thioether linker to conjugate 6-maleimidocaproic acid with drug; Nanoparticles coated with polydopamine	Paclitaxel	- High sensitivity [204]
Enzyme-Responsive	mPEGylated peptide dendrimer via a tetra-peptide linker GFLG	Doxorubicin	- Improved antitumor efficacy [205]
Hypoxia-responsive	Micellar nanoparticles of γ-Propargyl-L-glutamate N-carboxyanhydride	Doxorubicin	- Presented hypoxia sensitive drug release behavior - Significant anticancer activities under low-oxygen conditions - Longer circulation time [206]

in prolonging circulation [217]. Modification is also carried out to reduce the toxicity of nanoparticles that have shown some toxicity but are effective in treating breast cancer or themselves show antibreast cancer activity. For example, polyethylene glycol reduced the toxicity of gold nanorods [218] and mesoporous platinum nanoparticles [219]. PVP reduced toxicity of platinum nanoparticles [220] and so on. Surface modification also helps in improving water dispersibility. For example, PEG-phospholipids improved water dispersibility of iron oxide nanoparticles [221]

11.3.2 Surface modification to facilitate active targeting

Passive targeting has already been discussed in Section 6 of this chapter. Different surface substances can be applied to facilitate active targeting. Basic approaches include antibodies–antigen interaction, aptamers–protein interaction, carbohydrate–lectin interaction, ligand–receptor interaction [101]. List of active targeting ligands (not just limited to) are summarized in Table 3.

11.4 Via external stimuli (ultrasound, magnetic field)

Apart from delivering the nanoparticles via a normal way, nanoparticles with or without surface modification or functionalization may also be delivered to the target cells or organs via the addition of specific external stimuli. These stimuli are either given as an add on (such as thermal therapy: nanoparticle will kill breast cancer cells via generation of heat inside the cells, after the nanoparticles come under the exposure of a certain factor, at a certain wavelength, such as NIR (Fig. 4)) or help in targeting. These are summarized in Table 4.

12 Conclusion

It is evident that there are numerous strategies to develop advanced systems to deliver drugs to breast cancer cells. While most systems focus on localized or targeted delivery, it is absolutely essential that the delivery system fulfills therapeutic purposes. These delivery systems should be explored to translate into clinical application. Negative aspects of each delivery system should be encouraged to be explored well and extensively reported so that concerted efforts can be

TABLE 3 Various targeting agents w.r.t. hormonal classification of breast cancer.

Type of breast cancer	Nanoparticle	Targeting agent	Targeted
HR +	Liposomes	Biotin	SMVT receptors [222]
	Polyion complex micelles	Estrone	Estrogen receptors (ERα) [223]
	Gold NPs	Estradiol	Estrogen receptor [224]
	Gemcitabine-loaded albumin NPs	Folic acid	Folate receptor [225]
	Doxorubicin-loaded albumin NPs	AS1411 aptamer	Nucleolin [226]
HER2 +	Dasatinib-loaded polyethyleneimine (PEI)-coated polylactide NPs	Trastuzumab	HER2 receptor [227]
	Gold-nanoshelled PLGA magnetic hybrid NPs	HER2 antibody	HER2 receptor [228]
	Curcumin-loaded HSA NPs	HER2 aptamer	HER2 receptor [229]
	Daunorubicin-loaded DNA nanospindles	NRG-1	HER2/neu receptors [230]
TNBC	MWCNT	Hyaluronic acid	CD44 [231]
	Nanoshell	FZD7 antibody	Frizzled7 transmembrane receptors inhibited Wnt signaling [232]
	Chitosan NPs	α-Santalol	Polo-like kinase 1 [233]
	ABT-737-loaded PLGA NPs	Notch-1 antibodies	Notch-1 receptors [234]
	PEG-coated magnetite nanoparticles	Luteinizing hormone-releasing hormone	Luteinizing hormone-releasing hormone receptors [235]
	PEG5000-b-PAE10000 micelles	Glutamine	SLC1A5 receptor [236]
	Micelles	Folic acid	Folate receptor [237]

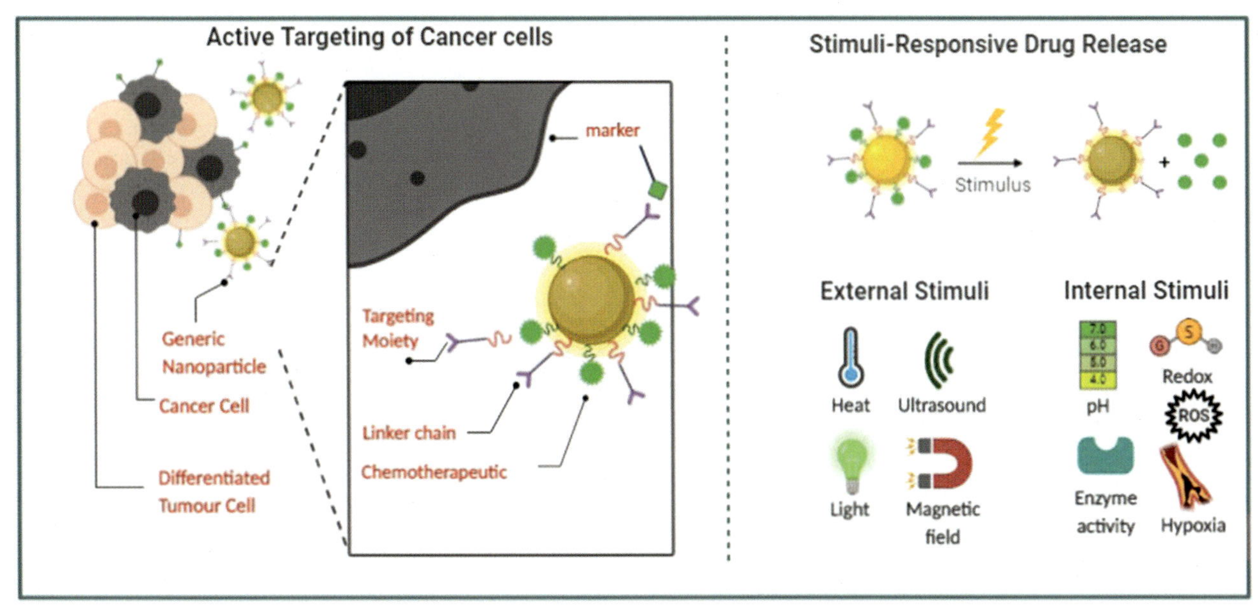

FIG. 4 Ligand aided (*left*) and external stimuli aided (*right*) drug delivery via nano-carrier to breast cancer. Created by biorender.com

TABLE 4 Use of external stimuli to deliver micro/nanosystems.

Nanoparticle	External stimuli	Purpose	Outcome
Mesoporous platinum nanoparticle-(mesoPt)	Laser	Combined chemo-photothermal breast cancer therapy	Improved anticancer effect after irradiation by 808 nm laser and 84% cancer cells were killed (Dox concentration is 8 μg/mL) [219]
Hyaluronic Acid- platinum nanoparticles	NIR	Targeted photothermal therapy (PTT)	More efficient inhibition of tumor growth via PTT [238]
Avidin-conjugated titanium dioxide nanoparticles	Ultrasound	Sonodynamic therapy	Till 24 h, there was no cell injury, but the viable concentration of MCF-7 cells dropped to 68% after 96 h [239]
Estrone-modified liposomes	Ultrasound (US) waves	Ultrasonically controlled therapy against breast cancer	Ultrasound improved uptake [240]
Magnetoliposomes	Magnetic field	Magnetic drug delivery	Reduced breast tumor cell viability [241]

made in the right direction. Because of being a heterogeneous disease, more understanding of breast cancer and its regulation pathway of each delivery system is required not just for marker determination but also for better development of the targeted delivery system. Dovetailing of the multifaceted researches shall lead to better outcomes in the future.

References

[1] Lobular Carcinoma In Situ, https://siteman.wustl.edu/glossary/cdr0000046315/; 2012.

[2] Breast Cancer 2018.What Is Breast Cancer?; https://www.cdc.gov/cancer/breast/basic_info/what-is-breast-cancer.htm. Accessed September 11, 2018.

[3] Sariego J, Zrada S, Byrd M, Matsumoto T. Breast cancer in young patients. Am J Surg 1995;170(3):243–5.

[4] Ataollahi M, Sharifi J, Paknahad M, Paknahad A. Breast cancer and associated factors: a review. J Med Life 2015;8(Spec Iss 4):6.

[5] Breast Cancer 2018.What are the risk factors for breast cancer?; https://www.cdc.gov/cancer/breast/basic_info/risk_factors.htm. Accessed Sept 14, 2020.

[6] Breast Cancer 2018; What can i do to reduce my risk of breast cancer? https://www.cdc.gov/cancer/breast/basic_info/prevention.htm. Accessed Sept 14, 2020.

[7] Deshpande TM, Pandey A, Shyama S. Breast cancer and etiology. Trend Med 2017;17:3–5.

[8] Makki J. Diversity of breast carcinoma: histological subtypes and clinical relevance. Clin Med Insights Pathol 2015;8:23–31.

[9] Sabatier R, Sabiani L, Zemmour C, et al. Invasive ductal breast carcinoma with predominant intraductal component: clinicopathological features and prognosis. Breast (Edinburgh, Scotland) 2016;27:8–14.

[10] Chen Z, Yang J, Li S, et al. Invasive lobular carcinoma of the breast: A special histological type compared with invasive ductal carcinoma. PLoS One 2017;12(9), e0182397.

[11] Pal SK, Lau SK, Kruper L, et al. Papillary carcinoma of the breast: an overview. Breast Cancer Res Treat 2010;122(3):637–45.

[12] Limaiem F, Mlika M. Medullary Breast Carcinoma. StatPearls [Internet]. StatPearls Publishing; 2019.

[13] Limaiem F, Ahmad F. Cancer, Mucinous Breast Carcinoma. *StatPearls [Internet]*: StatPearls Publishing; 2019.

[14] Limaiem F, Mlika M. *Tubular Breast Carcinoma*. In: StatPearls [Internet]. Treasure Island (FL): StatPearls Publishing; 2020.

[15] Vranic S, Feldman R, Gatalica Z. Apocrine carcinoma of the breast: a brief update on the molecular features and targetable biomarkers. Bosn J Basic Med Sci 2017;17(1):9.

[16] Joseph B, Jack M, Alan H, Edna K. Secretory carcinoma of the breast: an elusive presentation of this rare pathology. Int J Oncol Res 2017;1.

[17] Miyai K, Schwartz MR, Divatia MK, et al. Adenoid cystic carcinoma of breast: recent advances. World J Clin Cases 2014;2(12):732.

[18] Shah DR, Tseng WH, Martinez SR. Treatment options for metaplastic breast cancer. ISRN Oncol. 2012/06/21 2012;2012:706162.

[19] Karakas C. Paget's disease of the breast. J Carcinog 2011;10:31.

[20] Feng Y, Spezia M, Huang S, et al. Breast cancer development and progression: risk factors, cancer stem cells, signaling pathways, genomics, and molecular pathogenesis. Genes Dis 2018;5(2):77–106.

[21] Yersal O, Barutca S. Biological subtypes of breast cancer: prognostic and therapeutic implications. World J Clin Oncol 2014;5(3):412–24.

[22] Ferrari N, Mohammed ZM, Nixon C, et al. Expression of RUNX1 correlates with poor patient prognosis in triple negative breast cancer. PLoS One 2014;9(6), e100759.

[23] Maji R, Dey NS, Satapathy BS, Mukherjee B, Mondal S. Preparation and characterization of Tamoxifen citrate loaded nanoparticles for breast cancer therapy. Int J Nanomedicine 2014;9:3107–18.

[24] Moy B, Kirkpatrick P, Kar S, Goss P. Lapatinib. Nat Rev Drug Discov. 2007/06/01 2007;6(6):431-432.

[25] Caulfield SE, Davis CC, Byers KF. Olaparib: a novel therapy for metastatic breast cancer in patients with a BRCA1/2 mutation. J Adv Pract Oncol 2019;10(2):167–74.

[26] Bines J, Dienstmann R, Obadia R, et al. Activity of megestrol acetate in postmenopausal women with advanced breast cancer after nonsteroidal aromatase inhibitor failure: a phase II trial. Ann Oncol 2014;25(4):831–6.

[27] Kotecki N, Gombos A, Awada A. Adjuvant therapeutic approaches of HER2-positive breast cancer with a focus on neratinib maleate. Expert Rev Anticancer Ther 2019;19(6):447–54.

[28] Glover D, Lipton A, Keller A, et al. Intravenous pamidronate disodium treatment of bone metastases in patients with breast cancer. A dose-seeking study. Cancer: Interdisciplinary Int J Am Cancer Soc 1994;74(11):2949–55.

[29] FDA Database: Drug Approvals and Databases. FDA approves talazoparib for gBRCAm HER2-negative locally advanced or metastatic breast cancer. U. S. Food and Drugs Administration; 2018.

[30] Fareston Hormonal Therapy, https://www.breastcancer.org/treatment/hormonal/serms/fareston; 2020. [Accessed 28 March 2020].

[31] Shah A, Bloomquist E, Tang S, et al. FDA approval: ribociclib for the treatment of postmenopausal women with hormone receptor–positive, HER2-negative advanced or metastatic breast cancer. Clin Cancer Res 2018;24(13):2999–3004.

[32] Lee A. Tucatinib: first approval. Drugs 2020;80(10):1033–8.

[33] Serra F, Lapidari P, Quaquarini E, Tagliaferri B, Sottotetti F, Palumbo R. Palbociclib in metastatic breast cancer: current evidence and real-life data. Drugs Context 2019;8:212579.

[34] Verzenio®abemaciclib 50 I 100 I 150 I 200 mg tablets twice a day. https://www.verzenio.com/.

[35] PIQRAY® (alpelisib) tablets. 2020; https://www.hcp.novartis.com/products/piqray/metastatic-breast-cancer/, 2020.

[36] Drugs Approved for Breast Cancer, https://www.cancer.gov/about-cancer/treatment/drugs/breast; 2020. [Accessed 16 July 2020].

[37] Monoclonal Antibodies Immunotherapy 2019; https://www.cancer.gov/about-cancer/treatment/types/immunotherapy/monoclonal-antibodies. Accessed Sept 21, 2019.

[38] Weiner LM, Surana R, Wang S. Monoclonal antibodies: versatile platforms for cancer immunotherapy. Nat Rev Immunol. 2010/05/01 2010;10(5):317-327.

[39] FDA Database: Drug Approvals and Databases. FDA approves combination of pertuzumab, trastuzumab, and hyaluronidase-zzxf for HER2-positive breast cancer. US Food and Drug Administration; 2020.

[40] FDA Database: Drug Approvals and Databases. FDA approves atezolizumab for PD-L1 positive unresectable locally advanced or metastatic triple-negative breast cancer. U. S. Food and Drug Administration; 2019.

[41] Anon. Aspirin in Preventing Recurrence of Cancer in Patients With HER2 Negative Stage II-III Breast Cancer After Chemotherapy, Surgery, and/or Radiation Therapy, https://clinicaltrials.gov/ct2/show/NCT02927249#eligibility; 2020. [Accessed 23 July 2020].

[42] Atlan M, Neman J. Targeted transdermal delivery of curcumin for breast cancer prevention. Int J Environ Res Public Health 2019;16(24):4949.

[43] Kikuchi H, Yuan B, Hu X, Okazaki M. Chemopreventive and anticancer activity of flavonoids and its possibility for clinical use by combining with conventional chemotherapeutic agents. Am J Cancer Res 2019;9(8):1517–35.

[44] Vinod BS, Nair HH, Vijayakurup V, et al. Resveratrol chemosensitizes HER-2-overexpressing breast cancer cells to docetaxel chemoresistance by inhibiting docetaxel-mediated activation of HER-2-Akt axis. Cell Death Dis 2015;1:15061.

[45] Shareef M, Ashraf MA, Sarfraz M. Natural cures for breast cancer treatment. Saudi Pharm J 2016;24(3):233–40.

[46] Wang C-J, Yang M-Y, Inventors; Chung Shan Medical University, assignee. Extracts of sacred water lotus for the treatment of cancer. 2010.

[47] Ren P, Ren X, Cheng L, Xu L. Frankincense, pine needle and geranium essential oils suppress tumor progression through the regulation of the AMPK/mTOR pathway in breast cancer. Oncol Rep 2018;39(1):129–37.

[48] Bianco A, Da Ros T. Biological applications of fullerenes. In: Lang F, Nierengarten JF, editors. Fullerenes: Principles and Applications; 2007. p. 301–28.

[49] Mahanta S, Paul S. Bovine α-lactalbumin functionalized graphene oxide nano-sheet exhibits enhanced biocompatibility: A rational strategy for graphene-based targeted cancer therapy. Colloids Surf B Biointerfaces 2015;134:178–87.

[50] Jain S, Hirst DG, O'Sullivan JM. Gold nanoparticles as novel agents for cancer therapy. Br J Radiol 2012;85(1010):101–13.

[51] Jeyaraj M, Sathishkumar G, Sivanandhan G, et al. Biogenic silver nanoparticles for cancer treatment: An experimental report. Colloids Surf B Biointerfaces 2013;106:86–92.

[52] Yazdi MH, Mahdavi M, Varastehmoradi B, Faramarzi MA, Shahverdi AR. The immunostimulatory effect of biogenic selenium nanoparticles on the 4T1 breast cancer model: an in vivo study. Biol Trace Elem Res 2012;149(1):22–8.

[53] Baskaran B, Muthukumarasamy A, Chidambaram S, Sugumaran A, Ramachandran K, Manimuthu TR. Cytotoxic potentials of biologically fabricated platinum nanoparticles from Streptomyces sp. on MCF-7 breast cancer cells. IET Nanobiotechnol 2016;11(3):241–6.

[54] Rokade SS, Joshi KA, Mahajan K, et al. Gloriosa superba mediated synthesis of platinum and palladium nanoparticles for induction of apoptosis in breast cancer. Bioinorg Chem Appl 2018;2018:1–9.

[55] Ramaswamy SVP, Narendhran S, Sivaraj R. Potentiating effect of ecofriendly synthesis of copper oxide nanoparticles using brown alga: antimicrobial and anticancer activities. Bull Mater Sci 2016;39(2):361–4.

[56] M-j Z, S-n L, Xu G, Guo Y-N, Fu J-N, Zhang D-C. Cytotoxicity and apoptosis induced by nanobacteria in human breast cancer cells. Int J Nanomedicine 2014;9:265–71.

[57] McCrudden CM, McCarthy HO. Current status of gene therapy for breast cancer: progress and challenges. Appl Clin Genet 2014;7:209–20.

[58] Huynh A, Madu C, Lu Y. siRNA: a promising new tool for future breast cancer therapy. Oncomedicine 2018;3:74–81.

[59] Breast Cancer 2020 Chemotherapy for Breast Cancer; https://www.cancer.org/cancer/breast-cancer/treatment/chemotherapy-for-breast-cancer.html. Accessed 2020.

[60] Annapurna MM, Swati B, Ramya DN, Pramadvara K, Ram A, Aslesha N. A validated stability-indicating liquid chromatographic method for the determination of exemestane. Chem Sci 2014;3(3):961–8.

[61] Adam HK, Patterson JS, Kemp JV. Studies on the metabolism and pharmacokinetics of tamoxifen in normal volunteers. Cancer Treat Rep 1980;64(6–7):761–4.

[62] Badawy A, Mosbah A, Shady M. Anastrozole or letrozole for ovulation induction in clomiphene-resistant women with polycystic ovarian syndrome: a prospective randomized trial. Fertil Steril 2008;89(5):1209–12.

[63] Drugs & Medications Letrozole. https://www.webmd.com/drugs/2/drug-4297/letrozole-oral/details. Accessed 2020.

[64] Drugs & Medications 2020; Anastrozole. https://www.webmd.com/drugs/2/drug-1555/anastrozole-oral/details. Accessed 2020.

[65] USP. Megestrol Acetate Tablets, https://www.rxlist.com/megestrol-acetate-tablets-drug.htm#indications; 2020.

[66] Laderian B, Fojo T. CDK4/6 Inhibition as a therapeutic strategy in breast cancer: palbociclib, ribociclib, and abemaciclib. Semin Oncol 2017;44(6):395–403.

[67] Hoy SM. Talazoparib: first global approval. Drugs 2018;78(18):1939–46.

[68] Murali S, Nair SC, Priya A. Drug delivery research: current status ad future prospects. Int J Pharm Sci Rev Res 2016;40(1):94–9.

[69] Senapati S, Mahanta AK, Kumar S, Maiti P. Controlled drug delivery vehicles for cancer treatment and their performance. Signal Transduct Target Ther 2018;3:7.

[70] Bhagwat R, Vaidhya I. Novel drug delivery systems: an overview. Int J Pharm Sci Res 2013;4(3):970.

[71] Tanaka T, Decuzzi P, Cristofanilli M, et al. Nanotechnology for breast cancer therapy. Biomed Microdevices 2009;11(1):49–63.

[72] Avitabile E, Bedognetti D, Ciofani G, Bianco A, Delogu LG. How can nanotechnology help the fight against breast cancer? Nanoscale 2018;10(25):11719–31.

[73] Marta T, Luca S, Serena M, Luisa F, Fabio C. What is the role of nanotechnology in diagnosis and treatment of metastatic breast cancer? Promising Scenarios for the Near Future. J Nanomat 2016;2016:5436458.

[74] Vega-Vásquez P, Mosier NS, Irudayaraj J. Nanoscale drug delivery systems: from medicine to agriculture. Front Bioeng Biotechnol 2020;8.

[75] Buchiraju R, Sreekanth DN, Bhargavi S, et al. Vesicular drug delivery system–an over view. Res J Pharm Biol Chem Sci 2013;4(3):462–74.

[76] Akbarzadeh A, Rezaei-Sadabady R, Davaran S, et al. Liposome: classification, preparation, and applications. Nanoscale Res Lett 2013;8(1):102.

[77] Olusanya TO, Haj Ahmad RR, Ibegbu DM, Smith JR, Elkordy AA. Liposomal drug delivery systems and anticancer drugs. Molecules 2018;23(4):907.

[78] Brown S, Khan DR. The treatment of breast cancer using liposome technology. J Drug Deliv 2012;2012:212965.

[79] Sercombe L, Veerati T, Moheimani F, Wu SY, Sood AK, Hua S. Advances and challenges of liposome assisted drug delivery. Front Pharmacol 2015;6:286.

[80] Bulbake U, Doppalapudi S, Kommineni N, Khan W. Liposomal formulations in clinical use: an updated review. Pharmaceutics 2017;9(2):12.

[81] Kanter P, Bullard G, Pilkiewicz F, Mayer L, Cullis P, Pavelic Z. Preclinical toxicology study of liposome encapsulated doxorubicin (TLC D-99): comparison with doxorubicin and empty liposomes in mice and dogs. In Vivo (Athens, Greece) 1993;7(1):85–95.

[82] Harris L, Batist G, Belt R, et al. Liposome-encapsulated doxorubicin compared with conventional doxorubicin in a randomized multicenter trial as first-line therapy of metastatic breast carcinoma. Cancer 2002;94(1):25–36.

[83] Ag Seleci D, Seleci M, Walter J-G, Stahl F, Scheper T. Niosomes as nanoparticular drug carriers: fundamentals and recent applications. J Nanomater 2016;2016(3):1–13.

[84] Xu Y-Q, Chen W-R, Tsosie JK, et al. Niosome encapsulation of curcumin: characterization and cytotoxic effect on ovarian cancer cells. J Nanomater 2016;2016:1–9.

[85] Shaker DS, Shaker MA, Hanafy MS. Cellular uptake, cytotoxicity and in-vivo evaluation of Tamoxifen citrate loaded niosomes. Int J Pharm 2015;493(1):285–94.
[86] Barani M, Mirzaei M, Torkzadeh-Mahani M, Adeli-sardou M. Evaluation of carum-loaded niosomes on breast cancer cells:physicochemical properties, in vitro cytotoxicity, flow cytometric, dna fragmentation and cell migration assay. Sci Rep 2019;9(1):7139.
[87] Behroozeh A, Mazloumi Tabrizi M, Kazemi SM, et al. Evaluation the anti-cancer effect of pegylated nano-niosomal gingerol, on breast cancer cell lines (T47D) in-vitro. Asian Pac J Cancer Prev 2018;19(3):645–8.
[88] Babazadeh A, Zeinali M, Hamishehkar H. Nano-phytosome: a developing platform for herbal anti-cancer agents in cancer therapy. Curr Drug Targets 2018;19(2):170–80.
[89] El-Far SW, Helmy MW, Khattab SN, Bekhit AA, Hussein AA, Elzoghby AO. Folate conjugated vs PEGylated phytosomal casein nanocarriers for codelivery of fungal-and herbal-derived anticancer drugs. Nanomedicine 2018;13(12):1463–80.
[90] Hashemzehi M, Behnam-Rassouli R, Hassanian SM, et al. Phytosomal-curcumin antagonizes cell growth and migration, induced by thrombin through AMP-Kinase in breast cancer. J Cell Biochem 2018;119(7):5996–6007.
[91] El-Far SW, Helmy MW, Khattab SN, Bekhit AA, Hussein AA, Elzoghby AO. Phytosomal bilayer-enveloped casein micelles for codelivery of monascus yellow pigments and resveratrol to breast cancer. Nanomedicine 2018;13(5):481–99.
[92] Sabzichi M, Hamishehkar H, Ramezani F, et al. Luteolin-loaded phytosomes sensitize human breast carcinoma MDA-MB 231 cells to doxorubicin by suppressing Nrf2 mediated signalling. Asian Pac J Cancer Prev 2014;15(13):5311–6.
[93] Mughees M, Kumar K, Wajid S. Exosome vesicle as a nanotherapeutic carrier for breast cancer. J Drug Target 2020;1–10.
[94] Kim MS, Haney MJ, Zhao Y, et al. Development of exosome-encapsulated paclitaxel to overcome MDR in cancer cells. Nanomed Nanotechnol Biol Med 2016;12(3):655–64.
[95] Bunggulawa EJ, Wang W, Yin T, et al. Recent advancements in the use of exosomes as drug delivery systems. J Nanobiotechnol 2018;16(1):1–13.
[96] Zhao X, Wu D, Ma X, Wang J, Hou W, Zhang W. Exosomes as drug carriers for cancer therapy and challenges regarding exosome uptake. Biomed Pharmacother 2020;128:110237.
[97] Pullan JE, Confeld MI, Osborn JK, Kim J, Sarkar K, Mallik S. Exosomes as drug carriers for cancer therapy. Mol Pharm 2019;16(5):1789–98.
[98] Onaca O, Enea R, Hughes DW, Meier W. Stimuli-responsive polymersomes as nanocarriers for drug and gene delivery. Macromol Biosci 2009;9(2):129–39.
[99] Sharma AK, Prasher P, Aljabali AA, et al. Emerging era of "somes": polymersomes as versatile drug delivery carrier for cancer diagnostics and therapy. Drug Deliv Transl Res 2020;10:1171–90.
[100] Lee JS, Feijen J. Polymersomes for drug delivery: design, formation and characterization. J Control Release 2012;161(2):473–83.
[101] Singh S, Maurya P, Saraf SA. Cutting edge targeting strategies utilizing nanotechnology in breast cancer therapy. Front Anticancer Drug Discov 2019;*10*:180.
[102] Upadhyay KK, Bhatt AN, Mishra AK, et al. The intracellular drug delivery and anti tumor activity of doxorubicin loaded poly (γ-benzyl l-glutamate)-b-hyaluronan polymersomes. Biomaterials 2010;31(10):2882–92.
[103] Kumar A, Lale SV, Mahajan S, Choudhary V, Koul V. ROP and ATRP fabricated dual targeted redox sensitive polymersomes based on pPEGMA-PCL-ss-PCL-pPEGMA triblock copolymers for breast cancer therapeutics. ACS Appl Mater Interfaces 2015;7(17):9211–27.
[104] Katiyar SS, Muntimadugu E, Rafeeqi TA, Domb AJ, Khan W. Co-delivery of rapamycin- and piperine-loaded polymeric nanoparticles for breast cancer treatment. Drug Deliv 2016;23(7):2608–16.
[105] Vivek R, Babu VN, Thangam R, Subramanian K, Kannan S. pH-responsive drug delivery of chitosan nanoparticles as Tamoxifen carriers for effective anti-tumor activity in breast cancer cells. Colloids Surf B Biointerfaces 2013;111:117–23.
[106] Liu Z, Lv D, Liu S, et al. Alginic acid-coated chitosan nanoparticles loaded with legumain DNA vaccine: effect against breast cancer in mice. PLoS One 2013;8(4), e60190.
[107] Esfandiarpour-Boroujeni S, Bagheri-Khoulenjani S, Mirzadeh H, Amanpour S. Fabrication and study of curcumin loaded nanoparticles based on folate-chitosan for breast cancer therapy application. Carbohydr Polym 2017;168:14–21.
[108] Shenoy DB, Amiji MM. Poly (ethylene oxide)-modified poly (ε-caprolactone) nanoparticles for targeted delivery of tamoxifen in breast cancer. Int J Pharm 2005;293(1-2):261–70.
[109] Danafar H, Sharafi A, Kheiri Manjili H, Andalib S. Sulforaphane delivery using mPEG–PCL co-polymer nanoparticles to breast cancer cells. Pharm Dev Technol 2017;22(5):642–51.
[110] Singh AK. Introduction to Nanoparticles andNanotoxicology. In: Singh AK, editor. Engineered Nanoparticles. Boston: Academic Press; 2016. p. 1–18.
[111] Moshiri A. The role of nanomedicine, nanotechnology, and nanostructures on oral bone healing, modeling, and remodeling. In: Baghaban-Eslaminejad M, Oryan A, Kamali A, Andronescu E, Grumezescu AM, editors. Nanostructures for Oral Medicine. Elsevier; 2017. p. 777–832.
[112] Zhu X, Anquillare ELB, Farokhzad OC, Shi J. Chapter 22–Polymer- and Protein-Based Nanotechnologies for Cancer Theranostics. In: Chen X, Wong S, editors. Cancer Theranostics. Oxford: Academic Press; 2014. p. 419–36.
[113] Nicolas S, Bolzinger M-A, Jordheim LP, Chevalier Y, Fessi H, Almouazen E. Polymeric nanocapsules as drug carriers for sustained anticancer activity of calcitriol in breast cancer cells. Int J Pharm 2018;550(1):170–9.
[114] Wu X, Zheng Y, Yang D, et al. A strategy using mesoporous polymer nanospheres as nanocarriers of Bcl-2 siRNA towards breast cancer therapy. J Mater Chem B 2019;7(3):477–87.
[115] Miele E, Spinelli GP, Miele E, Tomao F, Tomao S. Albumin-bound formulation of paclitaxel (Abraxane® ABI-007) in the treatment of breast cancer. Int J Nanomedicine 2009;4:99.
[116] Lee JE, Kim MG, Jang YL, et al. Self-assembled PEGylated albumin nanoparticles (SPAN) as a platform for cancer chemotherapy and imaging. Drug Deliv 2018;25(1):1570–8.
[117] Gulfam M, Kim J-E, Lee JM, Ku B, Chung BH, Chung BG. Anticancer drug-loaded gliadin nanoparticles induce apoptosis in breast cancer cells. Langmuir 2012;28(21):8216–23.
[118] Bruckman MA, Czapar AE, VanMeter A, Randolph LN, Steinmetz NF. Tobacco mosaic virus-based protein nanoparticles and nanorods for chemotherapy delivery targeting breast cancer. J Control Release 2016;231:103–13.
[119] Jain A, Sharma G, Ghoshal G, et al. Lycopene loaded whey protein isolate nanoparticles: An innovative endeavor for enhanced bioavailability of lycopene and anti-cancer activity. Int J Pharm 2018;546(1):97–105.

[120] Elzoghby AO, El-Lakany SA, Helmy MW, Abu-Serie MM, Elgindy NA. Shell-crosslinked zein nanocapsules for oral codelivery of exemestane and resveratrol in breast cancer therapy. Nanomedicine 2017;12(24):2785–805.

[121] Lu B, Xiong S-B, Yang H, Yin X-D, Chao R-B. Solid lipid nanoparticles of mitoxantrone for local injection against breast cancer and its lymph node metastases. Eur J Pharm Sci 2006;28(1-2):86–95.

[122] Kang KW, Chun M-K, Kim O, et al. Doxorubicin-loaded solid lipid nanoparticles to overcome multidrug resistance in cancer therapy. Nanomed Nanotechnol Biol Med 2010;6(2):210–3.

[123] Liu J, Meng T, Yuan M, et al. MicroRNA-200c delivered by solid lipid nanoparticles enhances the effect of paclitaxel on breast cancer stem cell. Int J Nanomedicine 2016;11:6713.

[124] Eskiler GG, Cecener G, Dikmen G, Egeli U, Tunca B. Solid lipid nanoparticles: Reversal of tamoxifen resistance in breast cancer. Eur J Pharm Sci 2018;120:73–88.

[125] Sun M, Nie S, Pan X, Zhang R, Fan Z, Wang S. Quercetin-nanostructured lipid carriers: characteristics and anti-breast cancer activities in vitro. Colloids Surf B Biointerfaces 2014;113:15–24.

[126] Li H, Wang K, Yang X, et al. Dual-function nanostructured lipid carriers to deliver IR780 for breast cancer treatment: anti-metastatic and photothermal anti-tumor therapy. Acta Biomater 2017;53:399–413.

[127] Zhang X-G, Miao J, Dai Y-Q, Du Y-Z, Yuan H, Hu F-Q. Reversal activity of nanostructured lipid carriers loading cytotoxic drug in multi-drug resistant cancer cells. Int J Pharm 2008;361(1-2):239–44.

[128] Kumar K, Teotia D, Al-kaf AGA. Clinical application of perilla oil in breast cancer. Pharml Bioprocess 2018;6(2):59–63.

[129] Tripathi CB, Parashar P, Arya M, et al. Biotin anchored nanostructured lipid carriers for targeted delivery of doxorubicin in management of mammary gland carcinoma through regulation of apoptotic modulator. J Liposome Res 2020;30(1):21–36.

[130] Cho H, Lai TC, Tomoda K, Kwon GS. Polymeric micelles for multi-drug delivery in cancer. AAPS PharmSciTech 2015;16(1):10–20.

[131] Lee KS, Chung HC, Im SA, et al. Multicenter phase II trial of Genexol-PM, a Cremophor-free, polymeric micelle formulation of paclitaxel, in patients with metastatic breast cancer. Breast Cancer Res Treat 2008;108(2):241–50.

[132] Gener P, Montero S, Xandri-Monje H, et al. Zileuton™ loaded in polymer micelles effectively reduce breast cancer circulating tumor cells and intratumoral cancer stem cells. Nanomed Nanotechnol Biol Med 2020;24, 102106.

[133] Liu T, Romanova S, Wang S, et al. Alendronate-modified polymeric micelles for the treatment of breast cancer bone metastasis. Mol Pharm 2019;16(7):2872–83.

[134] Palmerston Mendes L, Pan J, Torchilin VP. Dendrimers as nanocarriers for nucleic acid and drug delivery in cancer therapy. Molecules 2017;22(9):1401.

[135] Liu C, Gao H, Zhao Z, et al. Improved tumor targeting and penetration by a dual-functional poly (amidoamine) dendrimer for the therapy of triple-negative breast cancer. J Mater Chem B 2019;7(23):3724–36.

[136] Castro RI, Forero-Doria O, Guzman L. Perspectives of dendrimer-based nanoparticles in cancer therapy. An Acad Bras Cienc 2018;90(2):2331–46.

[137] Debnath S, Saloum D, Dolai S, et al. Dendrimer-curcumin conjugate: a water soluble and effective cytotoxic agent against breast cancer cell lines. Anti-Cancer Agents Med Chem 2013;13(10):1531–9.

[138] Guo X-L, Kang X-X, Wang Y-Q, et al. Co-delivery of cisplatin and doxorubicin by covalently conjugating with polyamidoamine dendrimer for enhanced synergistic cancer therapy. Acta Biomater 2019;84:367–77.

[139] Oddone N, Lecot N, Fernández M, et al. In vitro and in vivo uptake studies of PAMAM G4. 5 dendrimers in breast cancer. J Nanobiotechnol 2016;14(1):1–12.

[140] Periasamy VS, Athinarayanan J, Alshatwi AA. Anticancer activity of an ultrasonic nanoemulsion formulation of Nigella sativa L. essential oil on human breast cancer cells. Ultrason Sonochem 2016;31:449–55.

[141] Anton N, Vandamme TF. Nano-emulsions and micro-emulsions: clarifications of the critical differences. Pharm Res 2011;28(5):978–85.

[142] Tripathi CB, Parashar P, Arya M, et al. QbD-based development of α-linolenic acid potentiated nanoemulsion for targeted delivery of doxorubicin in DMBA-induced mammary gland carcinoma: in vitro and in vivo evaluation. Drug Deliv Transl Res 2018;8(5):1313–34.

[143] Cao X, Luo J, Gong T, Zhang Z-R, Sun X, Fu Y. Coencapsulated doxorubicin and bromotetrandrine lipid nanoemulsions in reversing multidrug resistance in breast cancer in vitro and in vivo. Mol Pharm 2015;12(1):274–86.

[144] Banihashem S, Nikpour Nezhati M, Panahi HA, Shakeri-Zadeh A. Synthesis of novel chitosan-g-PNVCL nanofibers coated with gold-gold sulfide nanoparticles for controlled release of cisplatin and treatment of MCF-7 breast cancer. Int J Polym Mater Polym Biomater 2019;1–12.

[145] Rasouli S, Montazeri M, Mashayekhi S, et al. Synergistic anticancer effects of electrospun nanofiber-mediated codelivery of Curcumin and Chrysin: Possible application in prevention of breast cancer local recurrence. J Drug Delivery Sci Technol 2020;55:101402.

[146] Santos AC, Caldas M, Pattekari P, et al. Layer-by-Layer coated drug-core nanoparticles as versatile delivery platforms. In: Grumezescu AM, editor. Design and Development of New Nanocarriers. William Andrew Publishing; 2018. p. 595–635.

[147] Deng ZJ, Morton SW, Ben-Akiva E, Dreaden EC, Shopsowitz KE, Hammond PT. Layer-by-layer nanoparticles for systemic codelivery of an anticancer drug and siRNA for potential triple-negative breast cancer treatment. ACS Nano 2013;7(11):9571–84.

[148] Tran LA, Wilson LJ. Nanomedicine: making controllable magnetic drug delivery possible for the treatment of breast cancer. Breast Cancer Res 2011;13(2):303.

[149] Attari E, Nosrati H, Danafar H, Manjili HK. Methotrexate anticancer drug delivery to breast cancer cell lines by iron oxide magnetic based nanocarrier. J Biomed Mater Res A 2019;107(11):2492–500.

[150] Marcu A, Pop S, Dumitrache F, et al. Magnetic iron oxide nanoparticles as drug delivery system in breast cancer. Appl Surf Sci 2013;281:60–5.

[151] Jain A, Sharma G, Kushwah V, et al. Methotrexate and beta-carotene loaded-lipid polymer hybrid nanoparticles: a preclinical study for breast cancer. Nanomedicine 2017;12(15):1851–72.

[152] Jadon RS, Sharma M. Docetaxel-loaded lipid-polymer hybrid nanoparticles for breast cancer therapeutics. J Drug Delivery Sci Technol 2019;51:475–84.

[153] Zhou Z, Kennell C, Lee J-Y, Leung Y-K, Tarapore P. Calcium phosphate-polymer hybrid nanoparticles for enhanced triple negative breast cancer treatment via co-delivery of paclitaxel and miR-221/222 inhibitors. Nanomed Nanotechnol Biol Med 2017;13(2):403–10.

[154] Larson TA, Bankson J, Aaron J, Sokolov K. Hybrid plasmonic magnetic nanoparticles as molecular specific agents for MRI/optical imaging and photothermal therapy of cancer cells. Nanotechnology 2007;18(32):325101.
[155] Rasheed M, ALIa A, Kanwal S, Ismail M, Sabir N, Amin F. Synergy of green tea reduced tamoxifen-loaded silver nanoparticles exhibit ogt downregulation in breast cancer cell line. Dig J Nanomater Biostruct 2019;14(3):695–704.
[156] Teixeira RA, Lataliza AA, Raposo NR, Costa LAS, Sant'Ana AC. Insights on the transport of tamoxifen by gold nanoparticles for MCF-7 breast cancer cells based on SERS spectroscopy. Colloids Surf B Biointerfaces 2018;170:712–7.
[157] Huang YP, Hung CM, Hsu YC, et al. Suppression of breast cancer cell migration by small interfering RNA delivered by polyethylenimine-functionalized graphene oxide. Nanoscale Res Lett 2016;11(1):247.
[158] Al Faraj A, Shaik AP, Shaik AS. Magnetic single-walled carbon nanotubes as efficient drug delivery nanocarriers in breast cancer murine model: noninvasive monitoring using diffusion-weighted magnetic resonance imaging as sensitive imaging biomarker. Int J Nanomedicine 2015;10:157.
[159] Raza K, Thotakura N, Kumar P, et al. C60-fullerenes for delivery of docetaxel to breast cancer cells: a promising approach for enhanced efficacy and better pharmacokinetic profile. Int J Pharm 2015;495(1):551–9.
[160] Meng H, Mai WX, Zhang H, et al. Codelivery of an optimal drug/siRNA combination using mesoporous silica nanoparticles to overcome drug resistance in breast cancer in vitro and in vivo. ACS Nano 2013;7(2):994–1005.
[161] Prausnitz MR, Langer R. Transdermal drug delivery. Nat Biotechnol 2008;26(11):1261–8.
[162] Moses MA, Brem H, Langer R. Advancing the field of drug delivery: taking aim at cancer. Cancer Cell 2003;4(5):337–41.
[163] Ahn JS, Lin J, Ogawa S, et al. Transdermal buprenorphine and fentanyl patches in cancer pain: a network systematic review. J Pain Res 2017;10:1963.
[164] Li L, Fang L, Xu X, Liu Y, Sun Y, He Z. Formulation and biopharmaceutical evaluation of a transdermal patch containing letrozole. Biopharm Drug Dispos 2010;31(2-3):138–49.
[165] Xi H, Yang Y, Zhao D, et al. Transdermal patches for site-specific delivery of anastrozole: In vitro and local tissue disposition evaluation. Int J Pharm. 2010/05/31 2010;391(1):73-78.
[166] Li L, Xu X, Fang L, et al. The transdermal patches for site-specific delivery of letrozole: a new option for breast cancer therapy. AAPS PharmSciTech 2010;11(3):1054–7.
[167] Drugs & Medications. Estradiol Transdermal Patch Patch, Weekly, https://www.webmd.com/drugs/2/drug-32968/estradiol-transdermal-patch/details. [Accessed 21 August 2020].
[168] Grady D, Vittinghoff E, Lin F, et al. Effect of ultra-low-dose transdermal estradiol on breast density in postmenopausal women. Menopause 2007;14(3):391–6.
[169] Bhatnagar S, Bankar NG, Kulkarni MV, Venuganti VVK. Dissolvable microneedle patch containing doxorubicin and docetaxel is effective in 4T1 xenografted breast cancer mouse model. Int J Pharm 2019;556:263–75.
[170] Moreira AF, Rodrigues CF, Jacinto TA, Miguel SP, Costa EC, Correia IJ. Microneedle-based delivery devices for cancer therapy: A review. Pharmacol Res 2019;148:104438.
[171] Desimone J, MOGA KA, Inventors; The University Of North Carolina At Chapel Hill, assignee. Rapidly dissolvable microneedles for the transdermal delivery of therapeutics. 2017.
[172] Bhatnagar S, Kumari P, Pattarabhiran SP, Venuganti VVK. Zein microneedles for localized delivery of chemotherapeutic agents to treat breast cancer: drug loading, release behavior, and skin permeation studies. AAPS PharmSciTech 2018;19(4):1818–26.
[173] Byrne JD, Jajja MRN, O'Neill AT, et al. Local iontophoretic administration of cytotoxic therapies to solid tumors. Sci Transl Med 2015;7(273). 273ra214–273ra214.
[174] Komuro M, Suzuki K, Kanebako M, et al. Novel iontophoretic administration method for local therapy of breast cancer. J Control Release 2013;168(3):298–306.
[175] Neal RE, Singh R, Hatcher HC, Kock ND, Torti SV, Davalos RV. Treatment of breast cancer through the application of irreversible electroporation using a novel minimally invasive single needle electrode. Breast Cancer Res Treat 2010;123(1):295–301.
[176] Pehlivanova VN, Tsoneva IH, Tzoneva RD. Multiple effects of electroporation on the adhesive behaviour of breast cancer cells and fibroblasts. Cancer Cell Int 2012;12(1):9.
[177] Rodriguez-Cuevas S, Barroso-Bravo S, Almanza-Estrada J, Cristóbal-Martinez L, González-Rodriguez E. Electrochemotherapy in primary and metastatic skin tumors: phase II trial using intralesional bleomycin. Arch Med Res 2001;32(4):273–6.
[178] Wilbur RL. The Difference Between Topical and Transdermal Medications. Miami: Gensco Pharma; 2017. p. 1–2.
[179] Lee O, Page K, Ivancic D, et al. A randomized phase ii presurgical trial of transdermal 4-hydroxytamoxifen gel versus oral tamoxifen in women with ductal carcinoma in situ of the breast. Clin Cancer Res 2014;20(14):3672–82.
[180] Mehrabani D, Manafi A. Breast implants and breast cancer. World J Plast Surg 2012;1(2):62–3.
[181] Yang Y, Qiao X, Huang R, et al. E-jet 3D printed drug delivery implants to inhibit growth and metastasis of orthotopic breast cancer. Biomaterials 2020;230:119618.
[182] Belz JE, Kumar R, Baldwin P, et al. Sustained release talazoparib implants for localized treatment of BRCA1-deficient breast cancer. Theranostics 2017;7(17):4340–9.
[183] Qi Y, Min H, Mujeeb A, et al. Injectable hexapeptide hydrogel for localized chemotherapy prevents breast cancer recurrence. ACS Appl Mater Interfaces 2018;10(8):6972–81.
[184] Seib FP, Pritchard EM, Kaplan DL. Self-assembling doxorubicin silk hydrogels for the focal treatment of primary breast cancer. Adv Funct Mater 2013;23(1):58–65.
[185] Lengyel M, Kállai-Szabó N, Antal V, Laki AJ, Antal I. Microparticles, microspheres, and microcapsules for advanced drug delivery. Sci Pharm 2019;87(3):20.
[186] Prasanth V, Moy AC, Mathew ST, Mathapan R. Microspheres-an overview. Int J Res Pharmaceut Biomed Sci 2011;2(2):332–8.
[187] Tiwari S, Verma P. Microencapsulation technique by solvent evaporation method (Study of effect of process variables). Int J Phar Life Sci 2011;2(8):998–1005.
[188] Ramteke K, Jadhav V, Dhole S. Microspheres: as carrieres used for novel drug delivery system. Iosrphr 2012;2(4):44–8.
[189] Pal K, Roy S, Parida PK, et al. Folic acid conjugated curcumin loaded biopolymeric gum acacia microsphere for triple negative breast cancer therapy in invitro and invivo model. Mater Sci Eng C 2019;95:204–16.

[190] Almond BA, Hadba AR, Freeman ST, et al. Efficacy of mitoxantrone-loaded albumin microspheres for intratumoral chemotherapy of breast cancer. J Control Release 2003;91(1-2):147–55.
[191] Doughty J, Anderson J, Willmott N, McArdle C. Intra-arterial administration of adriamycin-loaded albumin microspheres for locally advanced breast cancer. Postgrad Med J 1995;71(831):47–9.
[192] Colone M, Kaliappan S, Calcabrini A, Tortora M, Cavalieri F, Stringaro A. Redox-active microcapsules as drug delivery system in breast cancer cells and spheroids. J Mol Genet Med 2016;10(1):1–6.
[193] Trail PA, Dubowchik GM, Lowinger TB. Antibody drug conjugates for treatment of breast cancer: Novel targets and diverse approaches in ADC design. Pharmacol Ther 2018;181:126–42.
[194] Zhao P, Zhang Y, Li W, Jeanty C, Xiang G, Dong Y. Recent advances of antibody drug conjugates for clinical applications. Acta Pharm Sin B 2020;10(9):1589–600.
[195] Syed YY. Sacituzumab govitecan: first approval. Drugs 2020;80(10):1019–25.
[196] Yardley DA, Weaver R, Melisko ME, et al. EMERGE: a randomized phase II study of the antibody-drug conjugate glembatumumab vedotin in advanced glycoprotein NMB–expressing breast cancer. J Clin Oncol 2015;33(14):1609–19.
[197] Xie A, Hanif S, Ouyang J, et al. Stimuli-responsive prodrug-based cancer nanomedicine. EBioMedicine 2020;56:102821.
[198] Zhou Y, Wang S, Ying X, et al. Doxorubicin-loaded redox-responsive micelles based on dextran and indomethacin for resistant breast cancer. Int J Nanomedicine 2017;12:6153–68.
[199] Yu P, Yu H, Guo C, et al. Reversal of doxorubicin resistance in breast cancer by mitochondria-targeted pH-responsive micelles. Acta Biomater 2015;14:115–24.
[200] Sadhukhan P, Kundu M, Chatterjee S, et al. Targeted delivery of quercetin via pH-responsive zinc oxide nanoparticles for breast cancer therapy. Mater Sci Eng C 2019;100:129–40.
[201] Zhang L, Wang Y, Yang Y, et al. High tumor penetration of paclitaxel loaded ph sensitive cleavable liposomes by depletion of tumor collagen i in breast cancer. ACS Appl Mater Interfaces 2015;7(18):9691–701.
[202] Moreira AF, Rodrigues CF, Jacinto TA, Miguel SP, Costa EC, Correia IJ. Poly (vinyl alcohol)/chitosan layer-by-layer microneedles for cancer chemo-photothermal therapy. Int J Pharm 2020;576:118907.
[203] Wang H, Tang L, Tu C, et al. Redox-responsive, core-cross-linked micelles capable of on-demand, concurrent drug release and structure disassembly. Biomacromolecules 2013;14(10):3706–12.
[204] Yang B, Wang K, Zhang D, et al. Polydopamine-modified ROS-responsive prodrug nanoplatform with enhanced stability for precise treatment of breast cancer. RSC Adv 2019;9(16):9260–9.
[205] Zhang C, Pan D, Luo K, et al. Peptide dendrimer–doxorubicin conjugate-based nanoparticles as an enzyme-responsive drug delivery system for cancer therapy. Adv Healthc Mater 2014;3(8):1299–308.
[206] Zhang P, Yang H, Shen W, Liu W, Chen L, Xiao C. Hypoxia-responsive polypeptide nanoparticles loaded with doxorubicin for breast cancer therapy. ACS Biomater Sci Eng 2020;6(4):2167–74.
[207] Mohamed MI. Optimization of chlorphenesin emulgel formulation. AAPS J 2004;6(3):81–7.
[208] Vohra T, Kaur I, Heer H, Murthy RR. Nanolipid carrier-based thermoreversible gel for localized delivery of docetaxel to breast cancer. Cancer Nanotechnol 2013;4(1-3):1–12.
[209] Luo Y, Li J, Hu Y, et al. Injectable thermo-responsive nano-hydrogel loading triptolide for the anti-breast cancer enhancement via localized treatment based on "two strikes" effects. Acta Pharm Sin B 2020;10(11):2227–45.
[210] Yehia R, Hathout RM, Attia DA, Elmazar MM, Mortada ND. Antitumor efficacy of an integrated methyl dihydrojasmonate transdermal microemulsion system targeting breast cancer cells: In vitro and in vivo studies. Colloids Surf B Biointerfaces 2017;155:512–21.
[211] Abdel-Hafez SM, Hathout RM, Sammour OA. Curcumin-loaded ultradeformable nanovesicles as a potential delivery system for breast cancer therapy. Colloids Surf B Biointerfaces 2018;167:63–72.
[212] Adhyapak AA, Desai BG. Formulation and evaluation of liposomal transdermal patch for targeted drug delivery of tamoxifen citrate for breast cancer. Indian J Health Sci Biomedl Res (KLEU) 2016;9(1):40.
[213] Lan X, She J, Lin D-A, et al. Microneedle-mediated delivery of lipid-coated cisplatin nanoparticles for efficient and safe cancer therapy. ACS Appl Mater Interfaces 2018;10(39):33060–9.
[214] Chang C-C, Yang W-T, Ko S-Y, Hsu Y-C. Liposomal curcuminoids for transdermal delivery: iontophoresis potential for breast cancer chemotherapeutics. Dig J Nanomater Biostruct 2012;7(1):59–71.
[215] Mehnath S, Chitra K, Karthikeyan K, Jeyaraj M. Localized delivery of active targeting micelles from nanofibers patch for effective breast cancer therapy. Int J Pharm 2020;584:119412.
[216] Kaul G, Amiji M. Long-circulating poly(ethylene glycol)-modified gelatin nanoparticles for intracellular delivery. Pharm Res 2002;19(7):1061–7.
[217] Aanei IL, ElSohly AM, Farkas ME, et al. Biodistribution of antibody-MS2 viral capsid conjugates in breast cancer models. Mol Pharm 2016;13(11):3764–72.
[218] Alkilany AM, Shatanawi A, Kurtz T, Caldwell RB, Caldwell RW. Toxicity and cellular uptake of gold nanorods in vascular endothelium and smooth muscles of isolated rat blood vessel: importance of surface modification. Small 2012;8(8):1270–8.
[219] Fu B, Dang M, Tao J, Li Y, Tang Y. Mesoporous platinum nanoparticle-based nanoplatforms for combined chemo-photothermal breast cancer therapy. J Colloid Interface Sci 2020;570:197–204.
[220] Teow Y, Valiyaveettil S. Active targeting of cancer cells using folic acid-conjugated platinum nanoparticles. Nanoscale 2010;2(12):2607–13.
[221] Choi JY, Lee SH, Na HB, An K, Hyeon T, Seo TS. In vitro cytotoxicity screening of water-dispersible metal oxide nanoparticles in human cell lines. Bioprocess Biosyst Eng 2009;33(1):21.
[222] Tang B, Peng Y, Yue Q, et al. Design, preparation and evaluation of different branched biotin modified liposomes for targeting breast cancer. Eur J Med Chem 2020;193:112204.
[223] Raveendran R, Chen F, Kent B, Stenzel MH. Estrone-decorated polyion complex micelles for targeted melittin delivery to hormone-responsive breast cancer cells. Biomacromolecules 2020;21(3):1222–33.
[224] Lara-Cruz C, Jiménez-Salazar JE, Arteaga M, et al. Gold nanoparticle uptake is enhanced by estradiol in MCF-7 breast cancer cells. Int J Nanomedicine 2019;14:2705–18.
[225] Dubey RD, Alam N, Saneja A, et al. Development and evaluation of folate functionalized albumin nanoparticles for targeted delivery of gemcitabine. Int J Pharm 2015;492(1):80–91.
[226] Baneshi M, Dadfarnia S, Shabani AMH, Sabbagh SK, Haghgoo S, Bardania H. A novel theranostic system of AS1411 aptamer-functionalized albumin nanoparticles loaded on iron oxide and gold nanoparticles for doxorubicin delivery. Int J Pharm 2019;564:145–52.

[227] Niza E, Noblejas-López MM, Bravo I, et al. Trastuzumab-targeted biodegradable nanoparticles for enhanced delivery of Dasatinib in HER2+ metastasic breast cancer. Nanomaterials 2019;9(12):1793.

[228] Dong Q, Yang H, Wan C, et al. HER2-functionalized gold-nanoshelled magnetic hybrid nanoparticles: a theranostic agent for dual-modal imaging and photothermal therapy of breast cancer. Nanoscale Res Lett 2019;14(1):235.

[229] Saleh T, Soudi T, Shojaosadati SA. Aptamer functionalized curcumin-loaded human serum albumin (HSA) nanoparticles for targeted delivery to HER-2 positive breast cancer cells. Int J Biol Macromol 2019;130:109–16.

[230] Baig MMFA, Lai W-F, Mikrani R, et al. Synthetic NRG-1 functionalized DNA nanospindels towards HER2/neu targets for in vitro anti-cancer activity assessment against breast cancer MCF-7 cells. J Pharm Biomed Anal 2020;182:113133.

[231] Singhai NJ, Maheshwari R, Ramteke S. CD44 receptor targeted 'smart' multi-walled carbon nanotubes for synergistic therapy of triple-negative breast cancer. Colloids Interface Sci Commun 2020;35:100235.

[232] Riley RS, Day ES. Frizzled7 antibody-functionalized nanoshells enable multivalent binding for Wnt signaling inhibition in triple negative breast cancer cells. Small 2017;13(26):1700544.

[233] Zhang J, Wang Y, Li J, Zhao W, Yang Z, Feng Y. α-Santalol functionalized chitosan nanoparticles as efficient inhibitors of polo-like kinase in triple negative breast cancer. RSC Adv 2020;10(9):5487–501.

[234] Valcourt DM, Dang MN, Scully MA, Day ES. Nanoparticle-mediated co-delivery of Notch-1 antibodies and ABT-737 as a potent treatment strategy for triple-negative breast cancer. ACS Nano 2020;14(3):3378–88.

[235] Hu J, Obayemi JD, Malatesta K, Košmrlj A, Soboyejo WO. Enhanced cellular uptake of LHRH-conjugated PEG-coated magnetite nanoparticles for specific targeting of triple negative breast cancer cells. Mater Sci Eng C 2018;88:32–45.

[236] Zhu Y-Z, Xu D, Liu Z, et al. The synthesis of glutamine-functionalized block polymer and its application in triple-negative breast cancer treatment. J Nanomat 2020;2020:4943270.

[237] Paulmurugan R, Bhethanabotla R, Mishra K, et al. Folate receptor–targeted polymeric micellar nanocarriers for delivery of orlistat as a repurposed drug against triple-negative breast cancer. Mol Cancer Ther 2016;15(2):221–31.

[238] Zhu Y, Li W, Zhao X, et al. Hyaluronic acid-encapsulated platinum nanoparticles for targeted photothermal therapy of breast cancer. J Biomed Nanotechnol 2017;13(11):1457–67.

[239] Ninomiya K, Fukuda A, Ogino C, Shimizu N. Targeted sonocatalytic cancer cell injury using avidin-conjugated titanium dioxide nanoparticles. Ultrason Sonochem 2014;21(5):1624–8.

[240] Salkho NM, Paul V, Kawak P, et al. Ultrasonically controlled estrone-modified liposomes for estrogen-positive breast cancer therapy. Artif Cells Nanomed Biotechnol 2018;46(sup. 2):462–72.

[241] Szuplewska A, Rękorajska A, Pocztańska E, Krysiński P, Dybko A, Chudy M. Magnetic field-assisted selective delivery of doxorubicin to cancer cells using magnetoliposomes as drug nanocarriers. Nanotechnology 2019;30(31):315101.

Chapter 11

Advanced drug delivery systems in the treatment of ovarian cancer

Santwana Padhi[a] and Anindita Behera[b]

[a]KIIT Technology Business Incubator, KIIT Deemed to be University, Bhubaneswar, Odisha, India, [b]School of Pharmaceutical Sciences, Siksha 'O' Anusandhan Deemed to be University, Bhubaneswar, Odisha, India

1 Introduction

Ovarian cancer (OC) is known to be a deadly ailment that counts for many deaths in female population every year, rendering it a major health concern worldwide [1]. OC is the most deadly reproductive cancer in females in the world across the countries as per the World Health Organization (WHO) and is generally an asymptomatic condition intensified by the lack of early diagnosis and proximity to effective and expensive chemotherapeutics. The treatment modality of OC care usually involves invasive surgery, radiotherapy or conventional chemotherapy following the surgical method, based on the stage of diagnosis of the OC disease [2]. Owing to the emergence of multidrug resistance (MDR), traditional therapy regimen has its specific set of associated shortcomings, comprising toxicity and the resultant reversion of the disease. Furthermore, the chemotherapeutic drugs are not restricted to OC depletion and thus demonstrate enhanced toxicity to surrounding healthy cells. In addition, the long-term condition with the presentation and growth of chemoresistant tumors is typically aversive. Patients experience a number of side effects, comprising persistent fatigue, loss of hair, and deteriorating number of plasma cells concurrent with the administration of chemotherapeutics for OC treatment [3]. Several targeted drug delivery systems have been designed for site-specific distribution of the antineoplastics to particular tumor sites with an aim to overcome adverse effects related to the usage of conventional chemotherapeutic drugs [4].

The nanodrug delivery systems are fabricated to encapsulate an appreciable amount of drug load in the core, protect the drugs from degradation in the biological milieu, and thereby release the entrapped drug at the targeted site to enhance the therapeutic efficacy and decrease the associated off-target side effects [5]. The design framework of these delivery systems aims to boost in vivo blood circulation with adequate compatibility and sufficient retention time. In addition, the polymers employed in the formulation process are usually nontoxic, biocompatible, and noninteractive with inflammatory invoking molecules.

This chapter details about the formulation details of various passive and active targeted nanodrug delivery systems encasing chemotherapeutic drugs for the amelioration of ovarian cancer. A particular focus is laid on the optimization protocols followed by the formulation methods, optimization, in vitro attributes along with the in vivo efficacy of the nanosystems entrapping suitable amount of the drugs.

2 Pathophysiology of ovarian cancer

Ovarian cancer can be categorized as per the origin and the histology involved in the cancerous tissues. As per the origin of the cancer, the ovarian cancer can be classified as epithelial and nonepithelial ovarian cancer as illustrated in Fig. 1. Majority incidence of epithelial ovarian cancer (EOC) surpasses the nonepithelial ovarian cancer which accounts for only 10%. EOC is again of different subtypes depending on their origin, pathophysiology, cellular profile of the cancerous tissues, and the risk factors responsible for development [6]. Again EOC is classified into a major class of nonmucinous (97%) and a minor class of mucinous type (3%). Histopathologically mucinous origin of ovarian cancer can be differentiated into serous carcinoma (~70%), endometroid carcinoma (~10%), clear cell carcinoma (~10%), and unspecified (~10%) as illustrated in Fig. 1. Serous-type ovarian cancer is further classified as low-grade serous carcinoma (LGSC) and high-grade serous carcinoma (HGSC) [7, 8] (Fig. 1).

EOC is also classified as type I and type II depending on the origin, molecular level, and stage at which it is generally diagnosed. Type I EOC develops as a sequential growth of marginal serous tumors or sometimes from endometriosis. The endometrioid, LGSCs, and clear cell carcinoma are diagnosed at an early stage and are mostly low-grade carcinoma [9–11]. However, type II EOC are high grade,

Advanced Drug Delivery Systems in the Management of Cancer. https://doi.org/10.1016/B978-0-323-85503-7.00020-1

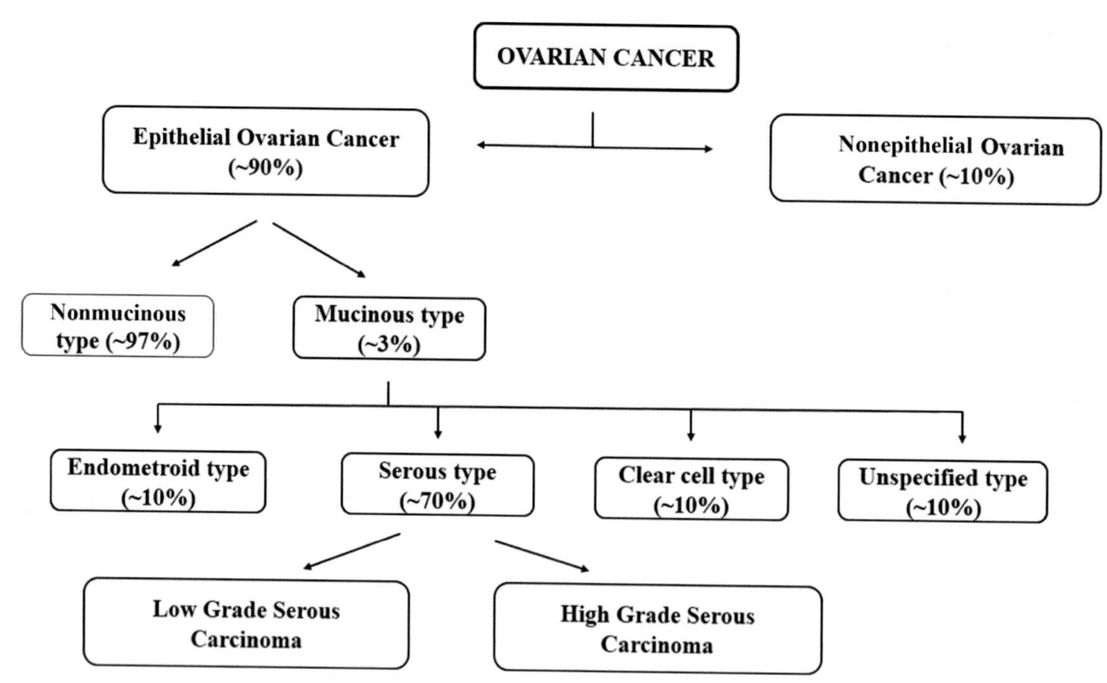

FIG. 1 Classification of ovarian cancer.

occur frequently, and developed from fimbrial epithelium diagnosed at stages 3 and 4 [12].

Several theories explain the occurrence of ovarian cancer and illustrated in Fig. 2; mainly genetic mutations play an important role in uncontrolled cellular growth leading to formation of tumor and subsequently converted to carcinoma. Incessant ovulation theory explains the origin of the ovarian cancer due to the physical trauma experienced by the epithelial cells on the surface of ovary. In each ovulation, the epithelium has number of repetitive physical trauma which results in DNA damage at cellular level. The epithelial cells invaginate into the cortical stromatal layer and get trapped to form spherical cortical inclusion cysts. While inside the ovary, the epithelial cells are continuously exposed to ovarian hormones which facilitates the cell proliferation and transforms the cells into cancerous one [13, 14]. But this theory was unable to explain the origin of different types of EOC according to their origin or prognosis [15]. It also failed to explain the high risk of EOC in patients with polycystic ovarian syndrome, where the number of ovulation is less [13, 14].

It was assumed that the origin of ovarian cancer was only in ovary, but later few researchers explored the origin other than ovary, i.e., in the fallopian tubes. Fallopian tubes theory explains the genes responsible for the EOC. *BRCA1* and *BRCA2* are two oncosuppressor genes found on the chromosome 17q21 and 13q, respectively [16]. Mutation in these genes causes premature and altered protein generation, which causes the uncontrolled growth of cells leading to cancer. Those women having prophylactic salpingo-oophorectomy mostly face these *BRCA1* or *BRCA2* gene mutation, which results in epithelial dysplasia in the fallopian tubes and later it is converted into tubal intraepithelial carcinoma (TIC). So fallopian tubes can be the origin of primary lesion and subsequently it spreads into the ovaries. Another gene *TP53* also undergoes mutation and found in TIC. The *TP53* gene expression in normal fallopian tubes secretory cells is identical to the *TP53* in mutated form in serous-type EOC. Expression of *TP53* is a sequential result of DNA damage in the epithelial cells of the fallopian tubes when it comes across with cytokines or oxidants. About 50% of *TP53* gene mutation results in growth of cancerous cells [17, 18]. And 70%–90% of TIC are also found in fimbriae, as it is the distal part of the fallopian tubes and it is located close to the ovary. Fimbriae come across the same kind of stress conditions similar to ovary and rich with blood vasculature, so they spread the metastasis to ovary very fast through blood stream [13].

Kurman and Shih [13] proposed the two-pathway theory in 2004 to explain the phenomenon of EOC. They classified the EOC into type I and type II ovarian cancer and differentiated the difference in their location of origin, genetic stability, and the type of gene mutations involve in the development. Type I EOC is generally originated inside the ovary, grow slowly, become benign, and genetically stable [15]. Type I ovarian cancer undergoes some morphological changes and later crosses the intermediate phase and becomes cancerous. The genetic mutations involved in type I ovarian cancer are *KRAS*, *BRAF*, *ERBB2*, *CTNNB1*, *PTEN*, *PIK3CA*, *ARID1A*, *PPP2R1A*, and *BCL2* [11]. These genes activate the MAPK pathway which converts the benign tumors into cancerous mass [13, 17]. However, type II ovarian

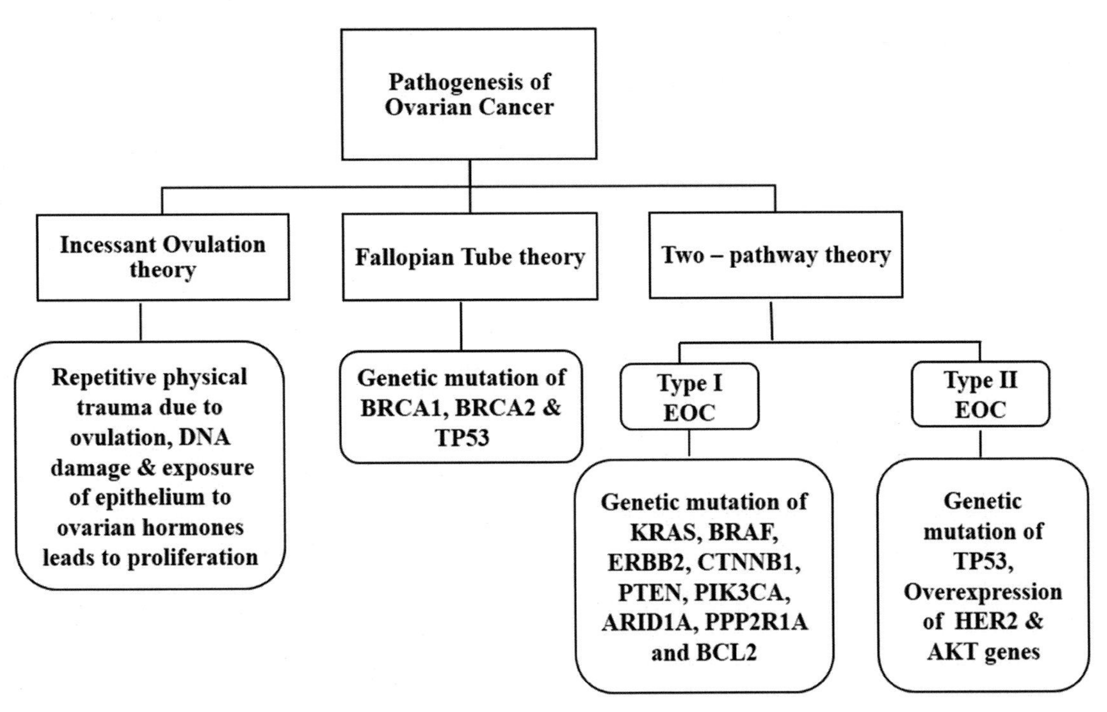

FIG. 2 Theories explaining pathogenesis of ovarian cancer.

cancer originates outside the ovaries, i.e., in the fallopian tubes. Type II ovarian cancer grow very fast, genetically unstable, and are diagnosed at advanced stage. Mostly mutation of *TP53* and overexpression of human epidermal growth factor receptor 2 (*HER2*/neu) and *AKT* genes. *BRCA1* and *BRCA2* gene mutation accounts for development of ~50% of ovarian cancer. As the type II ovarian cancer originates in fallopian tubes, the environmental stress due to inflammatory cytokines and ROS (reactive oxygen species) combines with *TP53* gene mutation and brings about neoplastic changes in the epithelial cells of fallopian tubes and subsequently the uncontrolled gene stability leads to development and growth of cancerous tissue [13, 15]. The theory was able to explain the pathogenesis of ovarian cancer as compared to other theories, but lacked to explain the pathogenesis of nonovarian origin of cancer [15] (Fig. 2).

3 Nanotools for the treatment of ovarian cancer

Polymeric nanoparticles, polymeric nanomicelles, nanoconjugates, dendrimers, liposomes, and nanostructured lipid formulations are the major nanodrug delivery systems that are suitably explored for the amelioration of ovarian cancer [19]. The above-mentioned drug delivery systems offer several benefits, including the improvement in the delivery of therapeutic drugs along with the fulfillment of an array of biopharmaceutical criteria, such as a considerable improvement in therapeutic efficacy as compared to native drugs, good biodegradability and biocompatibility, nontoxic and noninflammatory properties, as well as potential prospects for scaling up production [20]. Chemotherapeutics entrapped nanoformulations must possess certain unique characteristics such as higher drug entrapment efficiency, potential to incorporate drugs inside the internal core of the fabricated structure, and preferentially deliver the encased drug in tumor tissue through enhanced permeability and retention effect [21]. Furthermore, the nanoformulation engineered with particular ligands promotes better accumulation in OC tumors thereby consequently boosts the therapeutic efficacy as compared to nonfunctionalized nanoformulations [22].

There are a myriad of possible advantages conferred by the targeted nanosystems [23]. Targeting is typically accomplished using two distinct delivery mechanisms: (i) passive targeting with enhanced vascular permeability and absorption and (ii) active targeting by attaching specific ligands for the overexpressed receptors. The following section entails a brief description about the passive and active targeting approaches.

3.1 Passive targeting

The ability of the nanoformulations confers a large continuous blood circulation duration and preferentially gets deposited in tumor tissue by enhanced permeation and retention effect (EPR) and forms the basis of passive targeting [24]. The therapeutic payloads are dispersed in tumor extracellular matrix and further transported into the tumor cells. EPR targeting is due to tumor pathophysiological characteristics which are absent in healthy tissues [25].

3.2 Active targeting

The active targeting strategy involves binding of functional ligands to the surface of the nanoparticulate systems. These chosen ligands recognize overexpressed receptors present on the plasma membranes of tumor cells, leading to an increased absorption followed by cellular internalization through the receptor-mediated endocytosis pathway [26]. The widely used ligands for tumor targeting includes nucleic DNA/RNA aptamers, sugar groups, folates, monoclonal antibodies (mAb), oligopeptides [27].

Ehrlich proposed that antibody-conjugated nanocarriers have advanced into a prototype employing three constituents: a therapeutic agent, a copolymer, and active targeting moieties linked to a single formulation. This therapeutic targeting approach offers benefits, including high target specificity to the pathological site with minimal toxicity to surrounding healthy cells. In addition, this clinical method also increases tumor therapy, the chemotherapy of initial stage carcinoma (metastatic cancer) when major papillary fallopian tubes are still premature.

3.2.1 Liposomes

Liposomes are microscopic spherical structures comprising a lipid bilayer enclosing an aqueous compartment. Nonpolar lipid soluble drugs are integrated in the lipid bilayer, while aqueous-soluble drugs remain inside vesicle. The suitable encapsulation of chemotherapeutic drugs in liposomes results in alteration in pharmacological attributes, resulting in a decrease in rate of deterioration of drugs with an enhanced cell death at the administered doses [28]. These constructed lipophilic nanocarriers can be used for precise tumor tissue targeting along with imaging, but ligands decorated on to the surface (phospholipid coating) layer enhances cellular internalization, allowing a marked therapeutic impact at the targeted specific tumor sites.

On the basis of abovementioned concept, for the effective and successful treatment of ovarian cancer, paclitaxel-topotecan at a concentration of 20:1 w/w was loaded into folate-conjugated PEGylated liposomes (FPL-Pac-Top). In order to evaluate the potential of cytotoxicity of the fabricated liposomes in terms of apoptosis and necrosis, flow cytometry analysis was performed. The findings indicated the fact that 49.5% of cells in the control group were noted to be in the live stage while those cells interacted with Pac-Top, 48% cells showed apoptosis, and 45% of cell death occurred due to necrosis. The results exhibited the fact that Pac-Top is also proposed to be significantly toxic to cells, while Pac-Top liposomes, PEGylated Pac-Top liposomes, and FPL-Pac-Top demonstrated $12.6 \pm 0.7\%$, $3.5 \pm 0.2\%$, and $10.7 \pm 0.8\%$ cell death by necrosis, respectively. Reduced rate of necrosis in the PEGylated Pac-Top liposomes may be attributed to hydrophilic nature of PEG. In vivo studies demonstrated that a significant percentage of liposomal Pac-Top was noted in the blood after first hours of intravenous administration indicating a preferential uptake by liver and spleen when compared to PEGylated liposomal formulations. The observed result discovered that the functionalized nanosystems were more effective in retaining the drug concentration in blood as compared to the nonfunctionalized ones. In addition, the FPL-Pac-Top provided dual advantage of enhanced retention and target to the overexpressed folate receptors on ovarian cancer cells. The noted effects may be due to thermosensitive, long circulatory characteristics and targetability properties of the FPL-Pac-Top [29].

A critical factor responsible for paclitaxel-resistant ovarian cancer is the overexpression of permeability glycoprotein (P-gp), an ABC transporter associated with the removal of chemotherapeutic drugs from cells. In such a scenario, codelivery of P-gp inhibitor along with the anticancer drugs has not significantly contributed to an effective reversal of drug resistance in the clinical trials. Tariquidar (XR-9576, XR) is a third-generation P-gp inhibitor, which alone or along with paclitaxel (PCT) was encapsulated in liposomes (LP). The liposomal formulation illustrated advantageous physicochemical features and the potential to overcome chemoresistance in chemosensitive/chemoresistant ovarian cancer cell lines. In vitro studies revealed that retention of rhodamine 123, a P-gp substrate in the cells, was significantly increased by LP (XR). Further, LP (XR, PCT) synergistically produced appreciable cytotoxicity, disrupted proliferation rate, and induced G2-M arrest that overexpressed P-gp when evaluated in paclitaxel-resistant SKOV3-TR and HeyA8-MDR cell lines. In vivo studies were conducted on mice with orthotopic HeyA8-MDR ovarian tumors, and LP (XR, PCT) illustrated substantially decreased tumor weight (43.2% vs 16.9%) and number of metastases (44.4% vs 2.8%) as compared to LP (PCT). Apoptosis and compromised proliferation were effectively triggered in the xenografts by LP (XR, PCT). The present study results indicate that paclitaxel-resistant ovarian tumor cells can be sensitized to paclitaxel by the codelivery of a P-gp inhibitor along with paclitaxel utilizing a liposomal platform. Hence, it becomes imperative to state that this potential formulation may be accounted for clinical research in patients with P-gp overexpressing tumors [30].

Rapid advancement in the field of genomics and proteomics has driven patient stratification with specified treatment regimen tailor-made to the genetic contour of cancer in patients. In such a context, the usage of poly(ADP-ribose) polymerase (PARP) inhibitors showed an outstanding output for *BRCA1/2* mutations in breast and ovarian cancers. Using proteomic analysis, it was inferred that a cytolinker protein known as plectin is found to be mislocalized on the cell surface during progression of malignancy. In view of the stated theoretical concept, plectin (PTP)-conjugated liposomes were formulated entrapping PARP inhibitor

(AZ7379). In vivo tumor growth studies were conducted in mice with OVCAR8 cancer cells (high-grade epithelial ovarian tumors). The fabricated formulation improved PARP inhibition with a decelerated tumor growth. The conjugated and unconjugated AZ7379-entrapped liposomes resulted in a 3- and 1.7-fold reduction in tumor volume, respectively, as compared to neat form of the drug [31].

In another piece of work, sphingomyelin-cholesterol liposomal formulation (CPD100Li) entrapping a prodrug of vinblastine, mon-N-oxide (CPD100), which is activated by hypoxia, was prepared for evaluating its efficacy in hypoxic tumor microenvironment. CPD100Li with ionophore were noted to be stable for a time period of 48 h at a temperature of 4°C and the lyophilized formulation maintained the stability for 3 months in the presence of argon at 4°C. Upon evaluation in ES2 cells under conditions of hypoxia, CPD100Li demonstrated 9.2-fold lesser IC_{50} value when compared to CPD100Li in normoxic setting and the observed effect may be attributed to the hypoxia-dependent activation of CPD100. The dose-limiting toxicity was found to be 45 mg/kg for both CPD100 and CPD100Li, while the maximum tolerated dose corresponded to 40 mg/kg. The preliminary results showed in vivo stability of CPD100 is affected by ionophore. CPD100Li represented a remarkable platform to build a first such type of chemotherapy treatment focused on the preferential transformation of drugs via the solid tumors' hypoxic physiological microenvironment [32].

One of the most common therapies for ovarian cancer is the use of combinatorial therapy involving carboplatin (CPT) and paclitaxel (PTX). However it has limited application owing to the differential pharmacokinetics profile of the individual drugs. The drugs CPT/PTX at a molar ratio of 1:1 produced superior antitumor synergism; hence this ratio of the drugs was suitable to incorporate in multilamellar liposome vesicle (cMLV). Combination therapy with cMLVs (CPT + PTX) was evaluated in OVCAR8 xenograft nude mice and cMLVs (CPT) (1.8 mg/kg), cMLVs (PTX) (4.0 mg/kg), and cMLVs (CPT + PTX) (1.8 mg + 4.0 mg/kg) were administered to the group of mice when the tumors reached approximately 100 mm^3 in size. Mice injected with only PBS served as control. The formulations were given to the group of mice intravenously on days 0, 3, 6, and 9. In the tumor inhibition assay, cMLVs entrapping CPT + PTX at an administered dose of 1.8 mg CPT + 4.0 mg PTX per kg enabled higher tumor regression effects as compared to cMLVs (CPT) and cMLVs (PTX) at a dose of 1.8 mg CPT per kg and 4.0 mg PTX per kg, respectively. The superior antitumor efficacy may be attributed to the synergistic effects of both CPT and PTX. The combinatorial drug-loaded cMLV resulted in a diminished systemic toxicity with a superior antitumor effect when compared to combinations involving both the neat drugs and individual drug-loaded cMLVs, upon evaluation in an OVCAR8 ovarian cancer xenograft mouse model. The observed results indicated the suitability in using the combination of drugs for the treatment of solid tumors [33].

3.2.2 Nanoemulsion

Nanoemulsions are novel drug delivery systems, which are single-phase systems comprising emulsified oil, water, and amphiphilic molecules. The use of nanoemulsions is an appealing platform, in particular, because of their advantageous properties for solubilizing hydrophobic chemotherapeutic drugs and offering thermodynamic stability, biocompatibility, as well as targetability. Owing to their submicron dispersed colloids, nanoemulsions further helps to overcome biological barriers associated with drug delivery of conventional chemotherapeutic drugs to reach the targeted tumor site [34]. Surface-engineered nanoemulsion with anchored ligands has the ability to revolutionize the treatment of ovarian cancer by tailoring it according to the need for advanced therapy to address targeted delivery in the microenvironment of cancer cells [35].

In order to enhance the therapeutic efficacy and to reduce the associated toxicity, active targeting strategy was employed to encapsulate paclitaxel in hyaluronan-coated solid nanoemulsions forming paclitaxel-loaded hyaluronan solid nanoemulsions (PTX-HSNs). The particle size of the actively targeted nanoemulsion was observed to be 100 nm with narrow particle size distribution and encapsulation efficiency close to 100%. A 100-fold enhanced targeting capability was shown by the PTX-HSNs as compared to the PTX-loaded solid nanoemulsions (PTX-SNs) when evaluated in SK-OV-3 and OVCAR-3 cells. Both the formulations sustained the release of entrapped drug, PTX, for more than 6 days with no burst effect, indicating a superior effect on the tumor cells as compared to the marketed version of the drug, Taxol. In vivo studies showed that post PTX-HSN administration, the circulation longevity was enhanced with a better retention of the drug at the tumor site. The maximum tolerated dose (MTD) for PTX-HSNs was 2.5-fold more than that of Taxol. The noted results support the application of PTX-HSN as an efficacious nanosystem for delivery of PTX to ovarian cancer [36].

Pyridoclax, a BCS II drug, is appended with unfavorable physicochemical attributes which makes it unsuitable for preclinical evaluation. In such a prevailing context, researchers report the formulation of a pyridoclax-entrapped nanoemulsion for evaluating its in vitro and ex vivo efficacy. A 1000-fold increase in apparent solubility of the drug was noted upon its nanoformulation. The particle size was approximately 100 nm and a high encapsulation efficiency ($>95\%$). Pyridoclax-encapsulated nanoemulsion (PNEs) demonstrated a 2.5-fold higher antitumor activity than free pyridoclax when evaluated on chemoresistant ovarian cancer cells. This effect was supported by a sharp increase in activation of caspase 3/7. Based on the observed preliminary results, it may be stated that NEs can be explored as a potential nanocarrier system for preserving the antitumor activity of encased pyridoclax [37].

Second- and third-line treatments of chemotherapy cause development of MDR leading to recurrence of ovarian cancer. To resolve the stated problems, nanoemulsions (NEs) are progressing as an appealing drug delivery system in view of the fact that exposure of therapeutic cargo to normal tissues may also be reduced by NEs leading to potentially reduced side effects. Folate receptor-α (FR-α) is generally found to be overexpressed in ovarian cancer so a nanoemulsion of docetaxel with gadolinium (Gd) was fabricated. The nanoemulsion had a particle size of 120–150 nm and zeta potential around −45 mV. In vivo study was conducted in spontaneous transgenic ovarian cancer model, which showed that NMI-500 was able to inhibit tumor growth and this growth inhibition corresponded with the images obtained from MRI. These findings show that NMI-500 justifies progress and development in clinical trials [38].

Linum usitatissimum seed essential oil (LSEO) has proven to be potential in reducing the risk associated with prostate and colon cancers. The cytotoxicity, apoptotic, and antiangiogenic activities were evaluated on the human ovarian cancer cell line A2780 based on its and biocompatibility. LSEO nanoemulsions (LSEO-NEs) were synthesized using the ultrasound-based method, which were further evaluated for its cytotoxicity efficacy in A2780 cells. A significant dose-dependent cytotoxicity was shown by LSEO-NE in the A2780 cell line and the corresponding gene expression for apoptosis confirmed the cell death. Chick chorioallantoic membrane (CAM) assay also displayed a shortening of dimension and number of blood vasculature which assured its antiangiogenic potential. The positive results by ex vivo assays imply LSEO as a promising anticancer compound. An in-depth cell line and in vivo studies need to be carried out to support the current findings [39].

3.2.3 Micelle

Owing to their prolonged circulatory spans, increased drug stability, and precise targeting capability in tumor tissue, nanomicelles are perceived as promising carrier systems for imaging agents and therapeutics. Micelles are primarily delivered intravenously (IV), and prior to reaching the peritoneal cavity, they are typically exposed to the biological milieu leading to cytotoxicity [40]. The predominant site of OC is the intraperitoneal (IP) cavity, therefore, the preferred route of administration for OC treatment is IP route, which may lead to better patient compliance relative to intravenous (IV) treatment [41].

The biggest hindrance in the management and therapy of multidrug-resistant tumors is to provide large doses of chemotherapeutic drugs directly to malignant cells across an extended span of time. Established P-glycoprotein inhibitors are poloxamer and D-α-tocopheryl polyethylene glycol 1000 succinate (TPGS). Hence, it is pertinent to state that mixed micelles incorporating poloxamer 407 and TPGS can maximize the efficacy by delivery of the entrapped drug at a high concentration of the drug within the cancer cells and by inhibiting P-glycoprotein. In view of the concept, curcumin (CUR) was formulated as a mixed micelle with poloxamer 407 and TPGS and evaluated for its efficacy against multidrug-resistant ovarian cancer. The formulated nanosystems had a particle size of 21.4 ± 0.3 nm and a zeta potential of − 11.56 ± 0 7 mV with an entrapment efficiency in the range of 86%–95%. The formulation released curcumin for a period of 9 days. The cytotoxicity and cellular internalization studies were conducted in multidrug-resistant ovarian cancer (NCI/ADR-RES) cells. CUR-loaded poloxamer 407/TPGS mixed micelles were taken up by the cells in a more rapid manner resulting in threefold more cytotoxicity as compared to neat CUR after a time period of 48 h. The noted results imply the suitability of utilizing micelles as a potential nanocarrier for the treatment of ovarian cancer [42].

Huge focus has been drawn for the treatment of multidrug-resistant cancers with combination of chemotherapeutic drugs. However, the low water solubility of certain anticancer drugs limits their clinical implementation. A group of researchers reported the preparation and evaluation of poly(ethylene glycol)-poly(ε-caprolactone) (MPEG-PCL) micelles as a codelivery nanosystem entrapping the chemotherapeutic drug paclitaxel (PTX) and tacrolimus (FK506), a multidrug-resistant reversing agent. The PTX- and FK506-coloaded MPEG-PCL micelles (P-F/M) had a particle size of around 28.7 ± 3.2 nm with 99.3 ± 0.5% of encapsulation efficiency. A superior cytotoxicity with an efficient internalization was conferred by the P-F/M as compared to PTX-loaded micelles when evaluated in a PTX-sensitive human ovarian cancer cell line, A2780. In addition, the codelivery system imparted a substantial G2/M arrest and induced apoptosis. The observed results implicated the suitability of the use of codelivery micelles entrapping PTX and FK506 against MDR human ovarian cancer [43].

Although doxorubicin (DOX) is choice of drug for first-line treatment in ovarian cancer, severe adverse effects like toxicity to cardiac cells limit its potential clinical usage. A publication reported the utilization of micelle-former biomaterials such as Soluplus®, Pluronic F127, Tetronic T1107, and D-α-tocopheryl polyethylene glycol 1000 succinate (TPGS) in the formulation of micelle for the delivery of DOX. In vitro release of the formulation was noted to be higher at a pH of 5.5, which corresponded to the acidic tumor microenvironment pH value when compared to the release profile at physiological counterpart (7.4). The size of the formulation corresponded to 10.7 nm with a PDI of 0.239. Superior cytotoxicity was conferred by the formulation against ovarian (SKOV-3) as compared to Doxil.

The noted effect may be due to significant uptake of the formulation by SKOV-3 cells post 2, 4, and 6h of incubation. These potential results conclude the fact that T1107:TPGS (1:3) mixed micelles could be utilized for targeted delivery of DOX [44].

There has been a bunch of reports supporting the efficacy of phenolic compounds from different dietary sources in altering the pathophysiological attributes of tumor cells. In such a pursuit, ellagic acid (EA), a natural polyphenolic compound, has poor water solubility leading to insufficient oral bioavailability which restricts its clinical application. In order to solve the pertinent issue, film-hydration method was used to formulate the micellar delivery system entrapping EA with D-α-tocopheryl polyethylene glycol succinate (TPGS). The delivery system entrapping EA were of spherical shape with an appreciable size of 113.2 ± 23 nm with 0.260 ± 0.038 PDI and drug entrapment efficiency of $88.67 \pm 3.21\%$. The formulation sustained the release of entrapped EA with 67.8% cumulative release within a time period of 12h. An IC_{50} value of 12.36μM was conferred by the micellar formulation in OVCAR-3 cells. EA-TPGS induced cell death by cell cycle arrest of G1 phase leading to apoptosis through p15 and p21 inhibition. The positive results further strengthened the underlying reason for the usage of EA-TPGS micelles in overcoming the hurdles in targeted delivery of numerous hydrophobic chemotherapeutic agents by oral administration [45].

3.2.4 Metal nanoparticles

Metal nanoparticles are the most useful nanodrug delivery system in application of therapeutic and diagnostic fields. Monometallic and bimetallic nanoparticles show enhanced physicochemical properties according to their size and shape as compared to bulk metal [46]. The metal nanoparticles due to their specific optical, thermal, magnetic, electrical, and catalytic behavior can interact with the cellular proteins, enzymes, or receptors to exhibit their therapeutic activity in cancerous cells [47].

Toubhans et al. reported the anticancer activity of selenium nanoparticles (Se NPs) on two cancer cell lines, i.e., SKOV-3 and OVCAR-3. The Se NPs showed contrast results on the two cell lines. OVCAR-3 cells showed enhancement of irregularity on the surface of the cell membrane and toughness and formation of intracellular vesicles which behaves as autophagosomes. The autophagosomes induce autophagy causing death of cancerous cells. While SKOV-3 showed declination in the irregularity and toughness of the surface of cell membrane leading to decrease in metastasis and proliferation of cancerous cells. So the study concluded that the efficacy of the anticancer activity of Se NPs is dependent on the type of cell. Since the in vivo study of Se NPs proved to be less toxic, the Se NPs are also useful for high-grade serous ovarian cancer with potential anticancer activity [48].

Ramalingam et al. prepared the hematite (α-Fe_2O_3) nanoparticles by wet chemical method and studied the anticancer activity on metastatic ovarian cancer. The synthesized α-Fe_2O_3 nanoparticles increased the intracellular concentration of reactive oxygen species (ROS), damaging the mitochondrial membrane potential. The loss of mitochondrial membrane potential retards the cancer cell proliferation and cell death by induction of apoptosis. On the other hand, α-Fe_2O_3 nanoparticles also overexpress the p53 and caspase 3 proteins which regulate the cell cycle and initiate the apoptosis. The overexpression of these proteins induce the apoptosis through the mitochondria-mediated signaling pathway. The apoptosis regulates the metastatic ovarian cancer [49].

Han et al. prepared the microbubbles coated copper and selenium nanoparticles and studied the anticancer activity on ovarian cancer cells like A2780 and CisRA2780. The efficacy of Cu–Se NPs, MB@ Cu–Se NPs, and MB@ Cu–Se NPs with ultrasound was evaluated for the cancer cell death by induction of apoptosis. The MB@ Cu–Se NPs with ultrasound showed highest capability of induction of apoptosis in the ovarian cancer cells and was confirmed by different staining assay methods like AO-EB and Hoechst 33258 nuclear staining method. Apoptosis of the cells in contact with MB@ Cu–Se NPs and ultrasound occurred due to the damage in the DNA and chromatin condensation. Therefore, the nanoformulation with ultrasound showed the maximum efficacy as compared to Cu–Se NPs or MB@ Cu–Se NPs [50].

Ramezani et al. synthesized biogenic curcumin-coated silver nanoparticles (cAg NPs) and studied its effect in returning the sensitivity of cisplatin to cisplatin-resistant A2780 cell line. The efficacy of cAg NPs was assessed by MTT assay, acridine orange/propidium iodide (AO/PI), DAPI staining, Annexin V/PI assay, and caspase 3/9 activation assay. Similarly, the gene expression of *p53* and *MPP-9* genes were assessed by reverse transcription polymerase chain reaction (RT-PCR). The caspase 3/9 activity was increased by synergistic action of cisplatin with cAg NPs and the *p53* gene was upregulated, whereas the *MPP-9* gene was downregulated. The combination of cisplatin and cAg NPs showed increased efficacy in inducing the apoptosis in A2780 cancer cell line as compared to cisplatin or cAg NPS individually [51].

Maity et al. used the polyphenols of tea leaves and green synthesized gold nanoparticles conjugated with theaflavin. The nanoconjugates Au NP@TfQ showed increased anticancer effect more than only theaflavin as the theaflavin gets oxidized to its quinone form when gets conjugated with Au NPs. The quinones present in the nanoconjugates increased the ROS generation by depolarizing the mitochondrial complex and activating the caspase-regulated apoptosis and cell death of cancerous cells. PARP enzyme is activated for the repair of DNA during the damage of DNA by ROS. In this study, the PARP enzyme was inactivated by the nanoconjugates and the nanoconjugates also had low hemolytic property so it can be targeted safely for in vivo application [52].

Gurunathan et al. synthesized biogenic palladium nanoparticles (Pd NPs) from the extracts of the leaves of *Evolvulus alsinoides* and studied the anticancer activity of the biosynthesized NPs on human ovarian cancer cell lines A2780 and SKOV-3. In both the cell lines, the Pd NPs showed dose- or concentration-dependent cytotoxicity, oxidative stress leading to DNA damage, and apoptosis. Pd NPs showed increased toxicity to the cells by leakage of lactase dehydrogenases and increased malondialdehyde level and ROS and decreased other antioxidant levels like glutathione and superoxide dismutase. On further study, the Pd NPs caused dysfunction of mitochondrial complex by changing the mitochondria membrane potential, reduction of ATP, induction of DNA damage, and activation of caspase 3 pathway. All phenomenon lead to apoptosis in both the cell lines, A2780 and SKOV-3 [53,54].

Padmanabhan et al. synthesized variable-sized zinc oxide nanoparticles (ZnO NPs) from 15 to 55 nm. The ZnO NPs showed cytotoxicity by inducing oxidative and proteotoxic stress conditions to the cancer cells and induce the cell death by apoptosis. The cytotoxic efficacy of ZnO NPs was found to be inversely proportional to the size of the NPs and independent of mutation status of *p53* genes in the cancer cells. So the ZnO NPs can be used for drug-resistant cancer cells [55]. Bai et al. also synthesized the ZnO NPs of size 20 nm and studied the anticancer activity of the NPs. The ZnO NPs showed dose-dependent cytotoxicity and apoptosis by increased ROS generation, change in mitochondria membrane potential, and damage of double strands of DNA. The DNA damage was associated with expression of γ-H_2AX and Rad51 cells. ZnO NPs induced the apoptosis and autophagy by upregulating the *p53* and *LC3* genes. Western blot analysis showed the upregulated *Bax*, *caspase 9*, *Rad51*, γ-H_2AX, *p53*, and *LC3* genes and downregulated *BCL-2* genes when ZnO NPs were exposed to human ovarian cancer cell line SKOV-3 [56].

Vassie et al. synthesized the nanosized cerium oxide by flame spray pyrolysis method, having diameter of 7 nm and 94 nm did not affect the proliferation of ovarian cancer cells and colon cancer cell lines. The nanoceria was capable of entering into the cells by nonspecific pathways and energy dependent and got internalized into the cytoplasm and lysosomes. The nanoceria was an effective ROS scavenger and the scavenging activity was found to be directly proportional to the size. The bigger nanoceria was a better scavenger than the smaller one. So it could be an option for therapeutic activity in lowering the oxidative stress in the cancerous cells [57].

Pašukonienė et al. synthesized superparamagnetic cobalt ferrite nanoparticles (Co-SPIONs) and studied its cytotoxity and anticancer activity in expression of marker compounds. The Co-SPIONs were investigated on A2780 cell line for its clonogenicity and biomarker ESA expression and showed a dose-dependent response. Higher dose of the Co-SPIONs reduced the clonogenicity and ESA expression. ESA expression facilitates the metastasis, so the Co-SPIONs hinder the metastatic growth of cancer cells [58].

3.2.5 Polymeric nanoparticles

Polymeric nanoparticles have attained more interest in the last few years due to its smaller size and modification of the drug molecules as per the requirement by different types of polymers. The active molecules are encapsulated into the biocompatible and biodegradable polymers to elicit their therapeutic action in a desired manner like controlled release or sustained release and targeted drug delivery [59]. So many bioavailability issues could be resolved by formulating the desired therapeutic or diagnostic agents into polymeric nanoparticles. The properties of the polymeric nanoparticles can be easily modified by controlling their size and shape and can enhance their therapeutic index [60].

Sanchez-Ramirez et al. designed polymeric nanoparticles with combination therapy involving chemotherapy with phototherapy. The combination therapy showed combination of mechanisms including damage to DNA and damage to cell membrane and cytoskeleton by thermal means. Poly(lactic-*co*-glycolic) nanoparticles (PLGA NPs) contained carboplatin and a near infrared photosensitizer indocyanine green (ICG), and the efficacy was tested on SKOV-3 cell line. When the cells were exposed to a laser irradiation of 800 nm, the cytotoxicity was enhanced due to the synergistic action of the anticancer drug carboplatin and the photo-induced damage by ICG. The phototherapy caused the increase in the temperature of the cancerous cells and along with the chemotherapeutic agent, the damage of DNA and cytoskeleton reduces the chances of the development of the resistance. The biodegradability, photoresponsiveness, and synergistic action by the polymeric nanoparticles can be an option for the clinical trial for the therapeutic activity in ovarian cancer [61].

Domínguez-Ríos et al. targeted the overexpressed HER2 receptors for the treatment of ovarian cancer by PLGA NPs. Cisplatin-loaded chitosan-coated PLGA NPs were conjugated with anti-HER2 monoclonal antibody trastuzumab (TZ) and its cellular uptake and cytotoxicity were evaluated on epithelial ovarian cancer cell line SKOV-3. The cisplatin loaded in the NPs system was found to be released in a better way at acidic conditions in a continuous manner and the conjugation to the monoclonal antibody gave an idea of the biofunctionalization of the PLGA NPs modified with chitosan. In the SKOV-3 cell line, the cytotoxity of the NPs was found to be the highest as compared to free cisplatin. The conjugated NPs showed decrease in the cell viability to about 38% and the internalization of the NPs occurred by endocytosis. Trastuzumab alone causes development of resistance, whereas when conjugated with the PLGA NPs, the MDR was avoided [62].

Abriata et al. prepared paclitaxel (PCX) loaded poly-ε-caprolactone (PCL) nanoparticles and studied its anticancer activity with SKOV-3 cell line. PCL was selected as a polymer to deliver the PCX for intravenous administration for a controlled release formulation. The PCX-PCL polymeric nanoparticle was evaluated for cytotoxicity by XTT assay and the cellular uptake was assessed by flow cytometry and confocal laser scanning microscopy. The cytotoxicity study showed the controlled release of PCX leading to decrease in cell viability and the cellular uptake was time dependent and the internalization of the polymeric NPs got confirmed [63].

Jayawardhanaa et al. designed an organoplatinum polymeric nanoparticle to overcome the drug resistance in ovarian cancer cells. They prepared the polymeric nanoparticle by assembling the organoplatinum with an anionic block copolymer, i.e., methoxy polyethylene glycol-*block*-polyglutamic acid. The polymeric nanoparticle increases the properties of hydrophobic organoplatinum by enhancing the biocompatibility and solubility. The endocytic internalization of these NPs exhibited a dual action targeting the damage of the nuclear DNA with damage of mitochondria. So it can be an option of therapy for cisplatin-resistant ovarian cancer cells [64].

Luiz et al. conducted a comparative study on the anticancer activity of PCX by two nanoformulations on SKOV-3 cell line of ovarian cancer. The polymeric nanoparticle of PCX was prepared with PLGA, and the PLGA NPs was modified with folate for comparative study in cancer cells. The folate-modified polymeric NPs showed higher cytotoxity as compared to PCX-PLGA NPs and 3.6-fold higher uptake by the cells. The folate modification of the polymeric nanoparticle had better internalization and better therapeutic activity for ovarian cancer cells [65].

Zheng et al. coformulated verapamil and doxorubicin loaded with methoxy poly(ethylene glycol)-poly(L-lactic acid) (MPEG-PLA) nanoparticles. Verapamil was coformulated as it is a multidrug-resistant protein inhibitor and the doxorubicin showed an improved pharmacokinetics with lesser systemic toxicity. Doxorubicin as MPEG-PLA nanoparticle was not efficient in reversing the drug resistance and tumor suppression efficiency. The anticancer activity was tested in A2780 and SKOV-3 cell lines. Doxorubicin in MPEG-PLA decreased the tumor growth and prolonged the survival time, whereas the doxorubicin coencapsulated with verapamil showed suppression of tumor [66].

Ghassami et al. synthesized poly(butylene adipate-*co*-butylene terephthalate)-loaded docetaxel by electrospraying technique. Then aptamer molecules were attached to the polymeric nanoparticles by covalent bond. A comparative study was conducted with intravenous administration of Taxotere®, docetaxel nanoparticles, and aptamers-docetaxel nanoparticle in female Balb/c and *HER2* overexpressed tumors. The aptamers-docetaxel nanoparticle had showed significant increase in in vitro cytotoxicity and uptake by *HER2* overexpressed tumor cells as compared to docetaxel nanoparticles and free docetaxel. The pharmacokinetic profile, cytotoxicity, and antitumor efficacy were improved due to targeted delivery of docetaxel at the site of cancer cells [67].

Zhang et al. formulated the drug of second-line treatment for platinum-resistant ovarian cancer, i.e., cisplatin along with wortmannin as a polymeric nanoparticle. In case of platinum-resistant ovarian cancer, cisplatin efficacy is reduced as the metal transporting proteins are downregulated and the damaged DNA starts repairing. Wortmannin, a phosphoinositide 3-kinase inhibitor, blocks or downregulates the DNA repair in platinum-resistant ovarian cancer. So the combination of cisplatin and wortmannin was encapsulated in poly(lactic-*co*-glycolic acid)-poly(ethylene glycol) (PLGA-PEG) NPs. The nanoformulation enhanced the uptake of cisplatin, reduced the systemic toxicity, improved stability of wortmannin, and enhanced therapeutic efficacy as anticancer agent. In vivo efficacy of the polymeric NPs was examined in A2780 and A2780cis xenograft murine models and found to be efficacious as chemotherapeutic agent with complex mechanisms other than direct killing of cancerous cells [68].

Zou et al. prepared paclitaxel and borneol combination as PEG-PAMAM nanoparticles to reverse the drug-resistant ovarian cancer. Borneol is a natural p-glycoprotein inhibitor combined with paclitaxel and the anticancer activity was tested in A2780 cancer cell line. The paclitaxel-borneol polymeric nanoparticle showed enhanced cellular uptake and cytotoxicity, decreased the mitochondria membrane potential, and increased the rate of apoptosis. The paclitaxel-borneol NPs significantly decreased the tumor growth in A2780 in comparison to paclitaxel alone [69].

Poon et al. synthesized core-shell-type carboplatin and gemcitabine polymeric nanoparticles and evaluated against platinum-resistant ovarian cancer cell lines SKOV-3 and A2780/CDDP tumors. The combined core-shell nanoparticles had prolonged half-life with improved cellular uptake by tumor cells. This formulation resulted in 71% cell deterioration and inhibition of tumor growth of about 80%. So the combination polymeric nanoparticle was found to be a better option for therapy of drug-resistant cancer [70].

3.2.6 Combination nanodrug therapy

Combination or amalgamation of two or more anticancer drugs has extensive research and advantages for treatment of cancers. The combination of drugs shows synergistic or additive effect to treat the solid tumors more effectively. The combination drug therapy enhances the drug efficacy as compared to monotherapy and potentially reduces the drug resistance or can be effective in reversal of drug resistance of tumor cells. The combination of drugs in nano form aims for more targeting capability either by passive or active way of treatment. These are more beneficial in reduction of tumor growth, metastasis, stem cell population, arrest mitotic active cell count, and induce apoptosis and cell death [71,72].

Huang et al. used the concept of amphiphilic drug-drug conjugate (ADDC) for designing the delivery of two chemotherapeutic drugs in a single formulation. They linked a water-soluble chemotherapeutic drug floxuridine and water-insoluble drug chlorambucil by an esterification reaction. The ADDC technology has so many advantages of easy preparation, precise dimension, high drug entrapment, and self-delivery of different chemotherapeutic drugs without any carrier. The floxuridine-chlorambucil conjugate formed a nanodrug of average size 103 nm. Once the drug conjugate entered the tumor cells, it degrades by breakage of ester bond and the free floxuridine and chlorambucil are released and showed synergistic action on OVCAR-3 cells. The drug conjugate showed 4.7-fold increase in expression of caspase 3 in OVCAR-3 cells leading to increase in apoptosis of about 18.76% as compared to other drug combinations in the similar conditions [73].

Fathi et al. synthesized methotrexate-conjugated chitosan-grafted magnetic nanoparticles for targeted delivery of erlotinib to OVCAR-3 ovarian cancer cell line via overexpressed folate receptor in solid tumors. The chitosan was engineered and magnetic nanoparticles were prepared and conjugated with methotrexate to resemble the folic acid receptor-mediated drug delivery. The erlotinib was encapsulated into the methotrexate-conjugated chitosan-grafted magnetic nanoparticles with entrapment efficiency of 86% and the formulation was pH and thermo responsive, effective at pH 5.5 and 40°C [74].

Gawde et al. designed folic acid-decorated bovine serum albumin nanoparticles of paclitaxel and difluorinated curcumin (FA-BSA-PTX and FA-BSA-CDF) and studied anticancer activity on intravenous administration on ovarian (SKOV-3 cells) and cervical (HeLa cells) tumors. The combined application of FA-BSA-PTX and FA-BSA-CDF showed synergistic anticancer effect as enhancement in cellular uptake and induction of apoptosis causing cell death [75].

Catanzaro et al. investigated on liposomal formulation of cisplatin with 6-amino nicotinamide on cisplatin-resistant ovarian cancer. The demand for glucose by the resistant cells triggers the pentose phosphate pathway and 6-amino nicotinamide is an inhibitor of glucose-6-phosphate dehydrogenase. So combined administration of cisplatin with 6-amino nicotinamide indicated greater cytotoxicity in drug-resistant cells with improved pharmacokinetic profile [76].

Shen et al. designed a thermosensitive hydrogel containing cisplatin and paclitaxel with a thermosensitive copolymer Bi (mPEG-PLGA)-Pt (IV). The thermosensitive copolymer showed a temperature-dependent sol-gel transition in concentrated aqueous solution. In vitro study of the dual drug showed a synergistic anticancer effect on SKOV-3 cell line in a sustained release form for 2.5 months [77]. Similarly, Cai et al. designed a telodendrimer nanocarrier for administration of these two drugs. They developed three-layered linear dendritic telodendrimer micelles of cisplatin and encapsulated the paclitaxel. The low dose of paclitaxel with cisplatin acted synergistically at a ratio of 1:2 in SKOV-3 cell line. The coencapsulation of paclitaxel and cisplatin was found to be an efficacious targeted delivery for chemotherapy reducing the cytotoxicity and synergistic antitumor effect as compared to individual free or single drug-loaded telodendrimer formulation [78].

Other than drug-drug conjugates, drug-silver NPs have also been found to be in application for combination therapy for ovarian cancer or drug-resistant ovarian cancer treatment. Zhang et al. studied the effect of lycopene-reduced graphene oxide Ag NPs with trichostatin A on human ovarian cancer cells SKOV-3. The combination of lycopene-reduced graphene oxide Ag NPs and trichostatin A showed a dose-dependent cytotoxicity and the cytotoxicity was enhanced by enhancing the malondialdehyde level, reducing glutathione, and causing dysfunction of mitochondria. The combination had an effect on DNA damage inducing apoptosis [79]. Yuan et al. [80] reported the enhanced cytotoxic and apoptotic effects of gemcitabine with resveratrol reduced Ag NPs on A2780 cell line. Zhang et al. reported the effect of combination of salinomycin and Ag NPs synthesized from *Bacillus clausii* on human ovarian cancer cell line A2780. The combination showed a significant cytotoxic effect with expression of genes responsible for apoptosis. They also caused deposition of autophagolysosomes leading to dysfunction of mitochondria and loss of cell viability. The combination of salinomycin and Ag NPs exhibited a synergistic anticancer activity on A2780 cell line [81].

4 Conclusion

The promising results of the preclinical studies propels the pharmaceutical researchers to gain deeper insights into the said disease and design drug-loaded nanotools to maximize therapeutic efficacy. Before approval for clinical usage, substantial distinction is demanded in physicochemical, pharmacological, and immunological contexts. The efficacy of most of the anticancer drugs in nanoformulations has not progressed to a sufficient grade toward potential clinical application of the fabricated nanotherapeutics. Nanotools entrapping a combination of specific drugs may provide potential efficacy as compared to those provided by a single drug. Future developments in the development and delivery of nanosystems need to include circulatory computational assessments, simulation of pathogenic and cell lines environments derived from cancer patients, induced pluripotent stem cell (iPSC) technology, 3D cell culture, organotypic systems, advanced cell scanning systems, microfluidics, nanotechnological, and editing of genes technologies. Nanotheranostics are pharmacotherapeutic interventions that are triggered by diagnosis of ovarian cancer in early stage and hence can be employed in the treatment regimen in the future.

References

[1] Pantshwa J, Kondiah P, Choonara Y, Marimuthu T, Pillay V. Nanodrug delivery systems for the treatment of ovarian cancer. Cancer 2020;12(1):213.

[2] Padhi S, Kapoor R, Verma D, Panda A, Iqbal Z. Formulation and optimization of topotecan nanoparticles: *in vitro* characterization, cytotoxicity, cellular uptake and pharmacokinetic outcomes. J Photochem Photobiol B Biol 2018;183:222–32.

[3] Napoletano C, Ruscito I, Bellati F, Zizzari I, Rahimi H, Gasparri M, Antonilli M, Panici P, Rughetti A, Nuti M. Bevacizumab-based chemotherapy triggers immunological effects in responding multi-treated recurrent ovarian cancer patients by favoring the recruitment of effector T cell subsets. J Clin Med 2019;8(3):380.

[4] Chishti N, Jagwani S, Dhamecha D, Jalalpure S, Dehghan MH. Preparation, optimization, and in vivo evaluation of nanoparticle-based formulation for pulmonary delivery of anticancer drug. Medicina 2019;55(6):294.

[5] Verma D, Thakur P, Padhi S, Khuroo T, Talegaonkar S, Iqbal Z. Design expert assisted nanoformulation design for co-delivery of topotecan and thymoquinone: optimization, in vitro characterization and stability assessment. J Mol Liq 2017;242:382–94.

[6] Holschneider C, Berek J. Ovarian cancer: epidemiology, biology, and prognostic factors. Semin Surg Oncol 2000;19:3–10.

[7] Gaona-Luviano P, Medina-Gaona L, Magaña-Pérez K. Epidemiology of ovarian cancer. Chin Clin Oncol 2020;9:47.

[8] Momenimovahed Z, Tiznobaik A, Taheri S, Salehiniya H. Ovarian cancer in the world: epidemiology and risk factors. Int J Womens Health 2019;11:287–99.

[9] Karst A, Levanon K, Drapkin R. Modeling high-grade serous ovarian carcinogenesis from the fallopian tube. Proc Natl Acad Sci USA 2011;108:7547–52.

[10] Kurman R. Origin and molecular pathogenesis of ovarian high-grade serous carcinoma. Ann Oncol 2013;24:x16–21.

[11] Nezhat F, Pejovic T, Reis F, Guo S. The link between endometriosis and ovarian cancer: clinical implications. Int J Gynecol Cancer 2014;24:623–8.

[12] Przybycin C, Kurman R, Ronnett B, Shih I, Vang R. Are all pelvic (nonuterine) serous carcinomas of tubal origin? Am J Surg Pathol 2010;34:1407–16.

[13] Kurman R, Shih I. The origin and pathogenesis of epithelial ovarian cancer: a proposed unifying theory. Am J Surg Pathol 2010;34:433–43.

[14] Kyo S, Ishikawa N, Nakamura K, Nakayama K. The fallopian tube as origin of ovarian cancer: change of diagnostic and preventive strategies. Cancer Med 2019;9:421–31.

[15] Koshiyama M, Matsumura N, Konishi I. Recent concepts of ovarian carcinogenesis: type I and type II. Biomed Res Int 2014;2014:1–11.

[16] Koh S, Huak C, Lutan D, Marpuang J, Ketut S, Budiana N, Saleh A, Aziz M, Winarto H, Pradjatmo H, Hoan N, Thanh P, Choolani M. Combined panel of serum human tissue kallikreins and CA-125 for the detection of epithelial ovarian cancer. J Gynecol Oncol 2012;23:175.

[17] Erickson B, Conner M, Landen C. The role of the fallopian tube in the origin of ovarian cancer. Am J Obstet Gynecol 2013;209:409–14.

[18] Gross A, Kurman R, Vang R, Shih I, Visvanathan K. Precursor lesions of high-grade serous ovarian carcinoma: morphological and molecular characteristics. J Oncol 2010;2010:1–9.

[19] Basso J, Miranda A, Nunes S, Cova T, Sousa J, Vitorino C, Pais A. Hydrogel-based drug delivery nanosystems for the treatment of brain tumors. Gels 2018;4(3):62.

[20] Larrañeta E, Stewart S, Ervine M, Al-Kasasbeh R, Donnelly R. Hydrogels for hydrophobic drug delivery. Classification, synthesis and applications. J Funct Biomater 2018;9(1):13.

[21] Khuroo T, Verma D, Talegaonkar S, Padhi S, Panda A, Iqbal Z. Topotecan–tamoxifen duple PLGA polymeric nanoparticles: investigation of in vitro, in vivo and cellular uptake potential. Int J Pharm 2014;473(1–2):384–94.

[22] Bhise K, Sau S, Alsaab H, Kashaw S, Tekade R, Iyer A. Nanomedicine for cancer diagnosis and therapy: advancement, success and structure–activity relationship. Ther Deliv 2017;8(11):1003–18.

[23] Padhi S, Mirza M, Verma D, Khuroo T, Panda A, Talegaonkar S, Khar R, Iqbal Z. Revisiting the nanoformulation design approach for effective delivery of topotecan in its stable form: an appraisal of its in vitro behavior and tumor amelioration potential. Drug Deliv 2015;23(8):2827–37.

[24] Chavoshy F, Makhmalzade B. Polymeric micelles as cutaneous drug delivery system in normal skin and dermatological disorders. J Adv Pharm Technol Res 2018;9(1):2.

[25] Behera A, Padhi S. Passive and active targeting strategies for the delivery of the camptothecin anticancer drug: a review. Environ Chem Lett 2020;18(5):1557–67.

[26] Padhi S, Behera A. Nanotechnology based targeting strategies for the delivery of camptothecin. In: Saneja A, Panda A, Lichtfouse E, editors. Sustainable agriculture reviews, vol. 44. Cham: Springer; 2020. p. 243–73.

[27] Cha H, Song K. Effect of MUC8 on airway inflammation: a friend or a foe? J Clin Med 2018;7(2):26.

[28] Trucillo P, Campardelli R, Reverchon E. Supercritical CO_2 assisted liposomes formation: optimization of the lipidic layer for an efficient hydrophilic drug loading. J CO_2 Util 2017;18:181–8.

[29] Jain A, Jain S. Multipronged, strategic delivery of paclitaxel-topotecan using engineered liposomes to ovarian cancer. Drug Dev Ind Pharm 2015;42(1):136–49.

[30] Zhang Y, Sriraman S, Kenny H, Luther E, Torchilin V, Lengyel E. Reversal of chemoresistance in ovarian cancer by co-delivery of a P-glycoprotein inhibitor and paclitaxel in a liposomal platform. Mol Cancer Ther 2016;15(10):2282–93.

[31] Dasa S, Diakova G, Suzuki R, Mills A, Gutknecht M, Klibanov A, Slack-Davis J, Kelly K. Plectin-targeted liposomes enhance the therapeutic efficacy of a PARP inhibitor in the treatment of ovarian cancer. Theranostics 2018;8(10):2782–98.

[32] Shah V, Nguyen D, Al Fatease A, Patel P, Cote B, Woo Y, Gheewala R, Pham Y, Huynh M, Gannett C, Rao D, Alani A. Liposomal formulation of hypoxia activated prodrug for the treatment of ovarian cancer. J Control Release 2018;291:169–83.

[33] Zhang X, Liu Y, Kim Y, Mac J, Zhuang R, Wang P. Co-delivery of carboplatin and paclitaxel via cross-linked multilamellar liposomes for ovarian cancer treatment. RSC Adv 2017;7(32):19685–93.

[34] Kassem M, Ghalwash M, Abdou E. Development of nanoemulsion gel drug delivery systems of cetirizine; factorial optimization of composition, in vitro evaluation and clinical study. J Microencapsul 2020;37(6):413–30.

[35] Sahu P, Das D, Mishra V, Kashaw V, Kashaw S. Nanoemulsion: a novel Eon in cancer chemotherapy. Mini Rev Med Chem 2017;17(18):1778–92.
[36] Kim J, Park Y. Paclitaxel-loaded hyaluronan solid nanoemulsions for enhanced treatment efficacy in ovarian cancer. Int J Nanomedicine 2017;12:645–58.
[37] Groo A, Hedir S, Since M, Brotin E, Weiswald L, Paysant H, Nee G, Coolzaet M, Goux D, Delépée R, Freret T, Poulain L, Voisin-Chiret A, Malzert-Fréon A. Pyridoclax-loaded nanoemulsion for enhanced anticancer effect on ovarian cancer. Int J Pharm 2020;587:119655.
[38] Patel N, Piroyan A, Ganta S, Morse A, Candiloro K, Solon A, Nack A, Galati C, Bora C, Maglaty M, O'Brien S, Litwin S, Davis B, Connolly D, Coleman T. *In vitro* and *in vivo* evaluation of a novel folate-targeted theranostic nanoemulsion of docetaxel for imaging and improved anticancer activity against ovarian cancers. Cancer Biol Ther 2018;19(7):554–64.
[39] Keykhasalar R, Tabrizi M, Ardalan P, Khatamian N. The apoptotic, cytotoxic, and antiangiogenic impact of *Linum usitatissimum* seed essential oil nanoemulsions on the human ovarian cancer cell line A2780. Nutr Cancer 2020;72:1–9.
[40] Lengyel E. Ovarian cancer development and metastasis. Am J Pathol 2010;177(3):1053–64.
[41] Jahangirian H, Ghasemian Lemraski E, Webster T, Rafiee-Moghaddam R, Abdollahi Y. A review of drug delivery systems based on nanotechnology and green chemistry: green nanomedicine. Int J Nanomedicine 2017;12:2957–78.
[42] Saxena V, Hussain M. Polymeric mixed micelles for delivery of curcumin to multidrug resistant ovarian cancer. J Biomed Nanotechnol 2013;9(7):1146–54.
[43] Wang N, He T, Shen Y, Song L, Li L, Yang X, Li X, Pang M, Su W, Liu X, Wu Q, Gong C. Paclitaxel and tacrolimus co-encapsulated polymeric micelles that enhance the therapeutic effect of drug-resistant ovarian cancer. ACS Appl Mater Interfaces 2016;8(7):4368–77.
[44] Cagel M, Bernabeu E, Gonzalez L, Lagomarsino E, Zubillaga M, Moretton M, Chiappetta D. Mixed micelles for encapsulation of doxorubicin with enhanced in vitro cytotoxicity on breast and ovarian cancer cell lines versus Doxil®. Biomed Pharmacother 2017;95:894–903.
[45] Md S, Alfaifi M, Elbehairi S, Shati A, Fahmy U, Alhakamy N. Ellagic acid loaded TPGS micelles for enhanced anticancer activities in ovarian cancer. Int J Pharmacol 2020;16(1):63–71.
[46] Behera A, Mittu B, Padhi S, Patra N, Singh J. In: Abd-Elsalam K, editor. Multifunctional hybrid nanomaterials for sustainable agrifood and ecosystems. Amsterdam: Elsevier; 2020. p. 639–82.
[47] Sharma A, Goyal A, Rath G. Recent advances in metal nanoparticles in cancer therapy. J Drug Target 2017;26:617–32.
[48] Toubhans B, Gazze S, Bissardon C, Bohic S, Gourlan A, Gonzalez D, Charlet L, Conlan R, Francis L. Selenium nanoparticles trigger alterations in ovarian cancer cell biomechanics. Nanomed Nanotechnol Biol Med 2020;29:102258.
[49] Ramalingam V, Harshavardhan M, Dinesh Kumar S, Malathi Devi S. Wet chemical mediated hematite α-Fe_2O_3 nanoparticles synthesis: preparation, characterization and anticancer activity against human metastatic ovarian cancer. J Alloys Compd 2020;834:155118.
[50] Han W, Liu X, Wang L, Zhou X. Engineering of lipid microbubbles-coated copper and selenium nanoparticles: ultrasound-stimulated radiation of anticancer activity in human ovarian cancer cells. Process Biochem 2020;98:113–21.
[51] Ramezani T, Nabiuni M, Baharara J, Parivar K, Namvar F. Sensitization of resistance ovarian cancer cells to cisplatin by biogenic synthesized silver nanoparticles through p53 activation. Iran J Pharm Res 2019;18(1):222–31.
[52] Maity R, Chatterjee M, Banerjee A, Das A, Mishra R, Mazumder S, Chanda N. Gold nanoparticle-assisted enhancement in the anticancer properties of theaflavin against human ovarian cancer cells. Mater Sci Eng C 2019;104:109909.
[53] Gurunathan S, Kim E, Han J, Park J, Kim J. Green chemistry approach for synthesis of effective anticancer palladium nanoparticles. Molecules 2015;20:22476–98.
[54] Gurunathan S, Qasim M, Park C, Arsalan Iqbal M, Yoo H, Hwang J, Uhm S, Song H, Park C, Choi Y, Kim J, Hong K. Cytotoxicity and transcriptomic analyses of biogenic palladium nanoparticles in human ovarian cancer cells (SKOV3). Nanomaterials 2019;9:787.
[55] Padmanabhan A, Kaushik M, Niranjan R, Richards J, Ebright B, Venkatasubbu G. Zinc oxide nanoparticles induce oxidative and proteotoxic stress in ovarian cancer cells and trigger apoptosis independent of p53-mutation status. Appl Surf Sci 2019;487:807–18.
[56] Bai D, Zhang X, Zhang G, Huang Y, Gurunathan S. Zinc oxide nanoparticles induce apoptosis and autophagy in human ovarian cancer cells. Int J Nanomed 2017;12:6521–35.
[57] Vassie J, Whitelock J, Lord M. Endocytosis of cerium oxide nanoparticles and modulation of reactive oxygen species in human ovarian and colon cancer cells. Acta Biomater 2017;50:127–41.
[58] Pašukonienė V, Mlynska A, Steponkienė S, Poderys V, Matulionytė M, Karabanovas V, Statkutė U, Purvinienė R, Kraśko J, Jagminas A, Kurtinaitienė M, Strioga M, Rotomskis R. Accumulation and biological effects of cobalt ferrite nanoparticles in human pancreatic and ovarian cancer cells. Medicina 2014;50:237–44.
[59] Karlsson J, Vaughan H, Green J. Biodegradable polymeric nanoparticles for therapeutic cancer treatments. Annu Rev Chem Biomol Eng 2018;9:105–27.
[60] Crucho C, Barros MT. Polymeric nanoparticles: a study on the preparation variables and characterization methods. Mater Sci Eng C 2017;80:771–84.
[61] Sánchez-Ramírez D, Domínguez-Ríos R, Juárez J, Valdés M, Hassan N, Quintero-Ramos A, del Toro-Arreola A, Barbosa S, Taboada P, Topete A, Daneri-Navarro A. Biodegradable photoresponsive nanoparticles for chemo-, photothermal- and photodynamic therapy of ovarian cancer. Mater Sci Eng C 2020;116:111196.
[62] Domínguez-Ríos R, Sánchez-Ramírez D, Ruiz-Saray K, Oceguera-Basurto P, Almada M, Juárez J, Zepeda-Moreno A, del Toro-Arreola A, Topete A, Daneri-Navarro A. Cisplatin-loaded PLGA nanoparticles for HER2 targeted ovarian cancer therapy. Colloids Surf B Biointerfaces 2019;178:199–207.
[63] Abriata J, Turatti R, Luiz M, Raspantini G, Tofani L, do Amaral R, Swiech K, Marcato P, Marchetti J. Development, characterization and biological in vitro assays of paclitaxel-loaded PCL polymeric nanoparticles. Mater Sci Eng C 2019;96:347–55.
[64] Jayawardhana A, Qiu Z, Kempf S, Wang H, Miterko M, Bowers D, Zheng Y. Dual-action organoplatinum polymeric nanoparticles overcoming drug resistance in ovarian cancer. Dalton Trans 2019;48:12451–8.
[65] Luiz M, Abriata J, Raspantini G, Tofani L, Fumagalli F, de Melo S, Emery F, Swiech K, Marcato P, Lee R, Marchetti J. In vitro evaluation of folate-modified PLGA nanoparticles containing paclitaxel for ovarian cancer therapy. Mater Sci Eng C 2019;105:110038.

[66] Zheng W, Li M, Lin Y, Zhan X. Encapsulation of verapamil and doxorubicin by MPEG-PLA to reverse drug resistance in ovarian cancer. Biomed Pharmacother 2018;108:565–73.

[67] Ghassami E, Varshosaz J, Jahanian-Najafabadi A, Minaiyan M, Rajabi P, Hayati E. Pharmacokinetics and in vitro/in vivo antitumor efficacy of aptamer-targeted Ecoflex® nanoparticles for docetaxel delivery in ovarian cancer. Int J Nanomedicine 2018;13:493–504.

[68] Zhang M, Hagan IV CT, Min Y, Foley H, Tian X, Yang F, Mi Y, Au KM, Medik Y, Roche K, Wagner K, Rodgers Z, Wang AZ. Nanoparticle co-delivery of wortmannin and cisplatin synergistically enhances chemoradiotherapy and reverses platinum resistance in ovarian cancer models. Biomaterials 2018;169:1–10.

[69] Zou L, Wang D, Hu Y, Fu C, Li W, Dai L, Yang L, Zhang J. Drug resistance reversal in ovarian cancer cells of paclitaxel and borneol combination therapy mediated by PEG-PAMAM nanoparticles. Oncotarget 2017;8:60453–68.

[70] Poon C, Duan X, Chan C, Han W, Lin W. Nanoscale coordination polymers co-deliver carboplatin and gemcitabine for highly effective treatment of platinum-resistant ovarian cancer. Mol Pharm 2016;13(11):3665–75.

[71] Mokhtari R, Homayouni T, Baluch N, Morgatskaya E, Kumar S, Das B, Yeger H. Combination therapy in combating cancer. Oncotarget 2017;8:38022–43.

[72] Yap TA, Omlin A, de Bono JS. Development of therapeutic combinations targeting major cancer signaling pathways. J Clin Oncol 2013;31:1592–605.

[73] Huang P, Wang G, Wang Z, Zhang C, Wang F, Cui X, Guo S, Huang W, Zhang R, Yan D. Floxuridine-chlorambucil conjugate nanodrugs for ovarian cancer combination chemotherapy. Colloids Surf B Biointerfaces 2020;194:111164.

[74] Fathi M, Barar J, Erfan-Niya H, Omidi Y. Methotrexate-conjugated chitosan-grafted pH- and thermo-responsive magnetic nanoparticles for targeted therapy of ovarian cancer. Int J Biol Macromol 2020;154:1175–84.

[75] Gawde K, Sau S, Tatiparti K, Kashaw S, Mehrmohammadi M, Azmi A, Iyer A. Paclitaxel and di-fluorinated curcumin loaded in albumin nanoparticles for targeted synergistic combination therapy of ovarian and cervical cancers. Colloids Surf B Biointerfaces 2018;167:8–19.

[76] Catanzaro D, Nicolosi S, Cocetta V, Salvalaio M, Pagetta A, Ragazzi E, Montopoli M, Pasut G. Cisplatin liposome and 6-amino nicotinamide combination to overcome drug resistance in ovarian cancer cells. Oncotarget 2018;9:16847–60.

[77] Shen W, Chen X, Luan J, Wang D, Yu L, Ding J. Sustained codelivery of cisplatin and paclitaxel via an injectable prodrug hydrogel for ovarian cancer treatment. ACS Appl Mater Interfaces 2017;9(46):40031–46.

[78] Cai L, Xu G, Shi C, Guo D, Wang X, Luo J. Telodendrimer nanocarrier for co-delivery of paclitaxel and cisplatin: a synergistic combination nanotherapy for ovarian cancer treatment. Biomaterials 2014;37:456–68.

[79] Zhang X, Huang F, Zhang G, Bai D, Massimo D, Huang Y, Gurunathan S. Novel biomolecule lycopene-reduced graphene oxide-silver nanoparticle enhances apoptotic potential of trichostatin Ain human ovarian cancer cells (SKOV3). Int J Nanomedicine 2017;12:7551–75.

[80] Yuan Y, Peng Q, Gurunathan S. Silver nanoparticles enhance the apoptotic potential of gemcitabine in human ovarian cancer cells: combination therapy for effective cancer treatment. Int J Nanomedicine 2017;12:6487–502.

[81] Zhang X, Gurunathan S. Combination of salinomycin and silver nanoparticles enhances apoptosis and autophagy in human ovarian cancer cells: an effective anticancer therapy. Int J Nanomedicine 2016;11:3655–75.

Chapter 12

Advanced drug delivery systems in blood cancer

Ashish Garg[a], Sweta Garg[b], Neeraj Mishra[c], Sreenivas Enaganti[d], and Ajay Shukla[e]

[a]*Department of P.G. Studies and Research in Chemistry and Pharmacy, Rani Durgavati University, Jabalpur, MP, India,* [b]*Shri Ram Institute of Pharmacy, Jabalpur, MP, India,* [c]*Amity Institute of Pharmacy, Amity University Madhya Pradesh, Gwalior, MP, India,* [d]*Bioinformatics Division, Averin Biotech Pvt. Ltd., Nallakunta, Hyderabad, TG, India,* [e]*Department of Pharmacy, Institute of Technology and Management (ITM), Gida, Gorakhpur, UP, India*

1 Introduction

Blood is a fluid connective tissue involved in the transportation of hormones, respiratory gases, nutrients, etc. It plays a vital role in regulating pH, the temperature of the body, and other processes involved in thermoregulation. Human blood is composed of RBC, WBC, blood platelets, and plasma. Hematologic cancer is developed when the cells of bone marrow and the blood cells are affected, resulting in malfunction of the cells which results in the outgrowth of tumor tissue. According to the WHO, lung and bronchial cancers were found to be the most frequently manifesting cancers, whereas leukemia stands at ninth position in India and sixth in the United States with a total of 347,583 leukemia deaths in 2017 https://ourworldindata.org/cancer. In addition, majority of population of the world suffer from cancer, of them 0.31% in India, 5.42% in the United States, and 3.04% in Australia [1]. As per 2018 estimations, 62% increased death rate is expected by 2040.

Hematologic malignancies are the type of cancer that are commonly observed in young adults and children, comprising myeloma, lymphoma, and leukemia, affecting the lymphatic system, bone marrow, and blood cells (Fig. 1). The sequential stages of hematopoietic differentiation provide multiple opportunities for mutations to occur together with other disruptive events that lead to distinct tumor subtypes and clinical presentations [2]. The tumor heterogeneity in the tissues of lymphoid and hematopoietic organs is distinctive challenges for treatment and diagnosis of cancer [3–6].

2 Types of blood cancer

2.1 Leukemia

In leukemia condition, patient produces a large amount of poorly differentiated WBCs, which have an effect on cells of bone marrow and leukocytes. Leukemia is divided into four types based on the cells affected (lymphoid or myeloid) and the level of cell multiplication (chronic or acute). They are (a) chronic lymphocytic leukemia (CLL), (b) chronic myeloid leukemia (CML), (c) acute lymphoblastic leukemia (ALL), and (d) acute myeloid leukemia (AML) [2]. Acute myeloid leukemia (AML) has an incident rate of 4.2 cases per 100,000 with a 2.8 mortality rate per 100,000 adult individuals. Similarly, the incidence rates in ALL and CML are, respectively, 1.7 and 1.8 incident cases with 0.4 mortality rate in 100,000 individuals, and they are prevalent in children and adults [7, 8].

2.2 Lymphoma

In lymphoma, lymphatic system, bone marrow, spleen, and blood cells are affected and are classified into (a) non-Hodgkin lymphoma and (b) Hodgkin lymphoma. Lymphoma originates in lymph nodes where the lymphoid lineage of hematopoiesis differentiates into B-cells, T-cells, or natural killer cells. Abnormal events include extensive cell proliferation, somatic mutations, and antibody class switching, that ultimately impair the immune system as a

Advanced Drug Delivery Systems in the Management of Cancer. https://doi.org/10.1016/B978-0-323-85503-7.00008-0

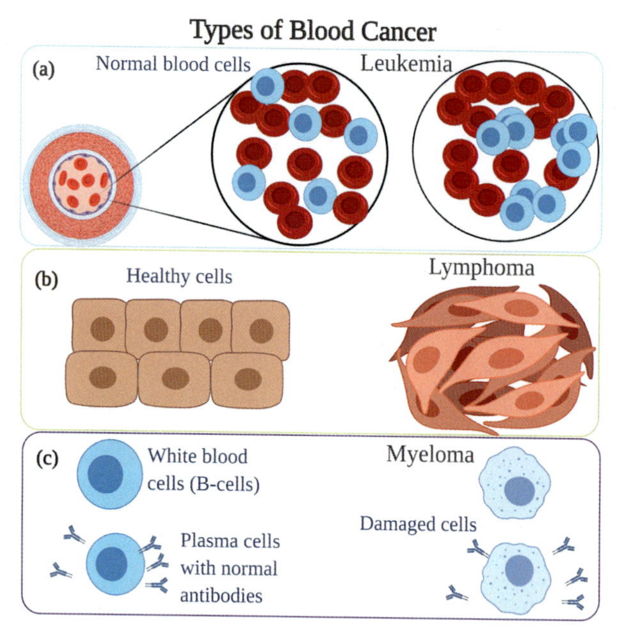

FIG. 1 Different types of blood cancer: (A) leukemia, (B) lymphoma, and (C) myeloma. *Created by Biorender.com.*

whole and the adaptive immune response in particular [9]. The annual hematological cancer incidence rate of 19.5 cases for 100,000 persons was observed in Western countries along with 5.9 death rate in 100,000 individuals per year, and most of them accompanied with aberrant B cells.

2.3 Myeloma

Myeloma is a tumor of plasma cells that hinders the antibody production and thus weakens the immune system.

3 Symptoms of blood cancer

Different blood cancer symptoms were observed in women and men. The general symptoms of blood cancer include difficulty in breathing, sudden weight loss, fever, stomachache, severe headache, itchy skin, wheezing, fatigue, continuous cough, difficulty in swallowing, back pain, frequent infections, weakness, swollen breasts (in women), pelvic pain, etc. Fig. 2A represents all the symptoms related to blood cancer including the various stages of blood cancer (Fig. 2B).

4 Causes of blood Cancer

No specific reasons are accompanied with the occurrence of blood cancer. The factors that are related for the occurrence of cancer are age, excessive alcohol consumption, usage of hair dyes, hereditary, immune system, AIDS, infections, smoking, organ transplantation, exposure to chemotherapy, etc.

5 Pathophysiology in blood cancer

Conditions that appear in the patients during pathophysiology include the following:

In *anemia*, patients suffer from chest pain, dizziness, tiredness, hypercalcemia, pale skin, and shortness of breath. Due to *poor clotting of blood*, patients show bleeding gums, red and black bowel movements, red dots, and bruising. During pathophysiology, patients suffer with bone pains, lump formation in arm pit and neck, loss of weight, damaged nerves, pain in the chest, numbness in arms and legs, etc.

Factors that are responsible for developing blood cancer are as follows:

Age-related factor: Children are mostly affected with ALL and adults with AML.

Genetic factor: Individuals suffering from Bloom and Down's syndrome are at high risk of developing blood cancer. Moreover, genetic inheritance may also lead to the development of cancer. In addition, people with HIV and other immunodeficiency diseases easily develop blood cancer.

Chemotherapy: Patients who are being treated with topoisomerase 2 inhibitors, platinum, and alkylating agents are prone to blood cancer.

Radiation therapy: Blood cancer is easily developed in patients who are under radiation therapy.

6 Blood cancer treatment

Treatment of blood cancer with conventional therapies not only affects the cancerous tissue but also damages the surrounding tissues. The side effects observed in the conventional therapies include muscle pain, disorders in blood, nervous system malfunction, vomiting, mouth sores, headache, etc. Thinking and memory of the patient were severely affected during chemotherapy (chemobrain). Few side effects persist for a longer time that include kidney failure, lung defects, damage to the reproductive system, and heart. It takes a very long duration for complete curing of blood cancer using conventional therapies. Changes occur rapidly during the cancer treatment, and the effects are similar in adults and children. Personalized medicine helps the doctor to treat the patient in a more effective manner. Fig. 3 represents the treatment strategy for blood cancer managementwww.cancer.net.

6.1 Chemotherapy

In chemotherapy, the administration of drugs such as antimetabolites was employed for the treatment of cancer. In advanced chemotherapy, two drugs that act in a synergetic way and stable were administered to patients. For more effective treatment, therapeutic agents that cannot be taken orally are directly interposed into the cerebrospinal fluid (intrathecal chemotherapy). In addition, antimitotic,

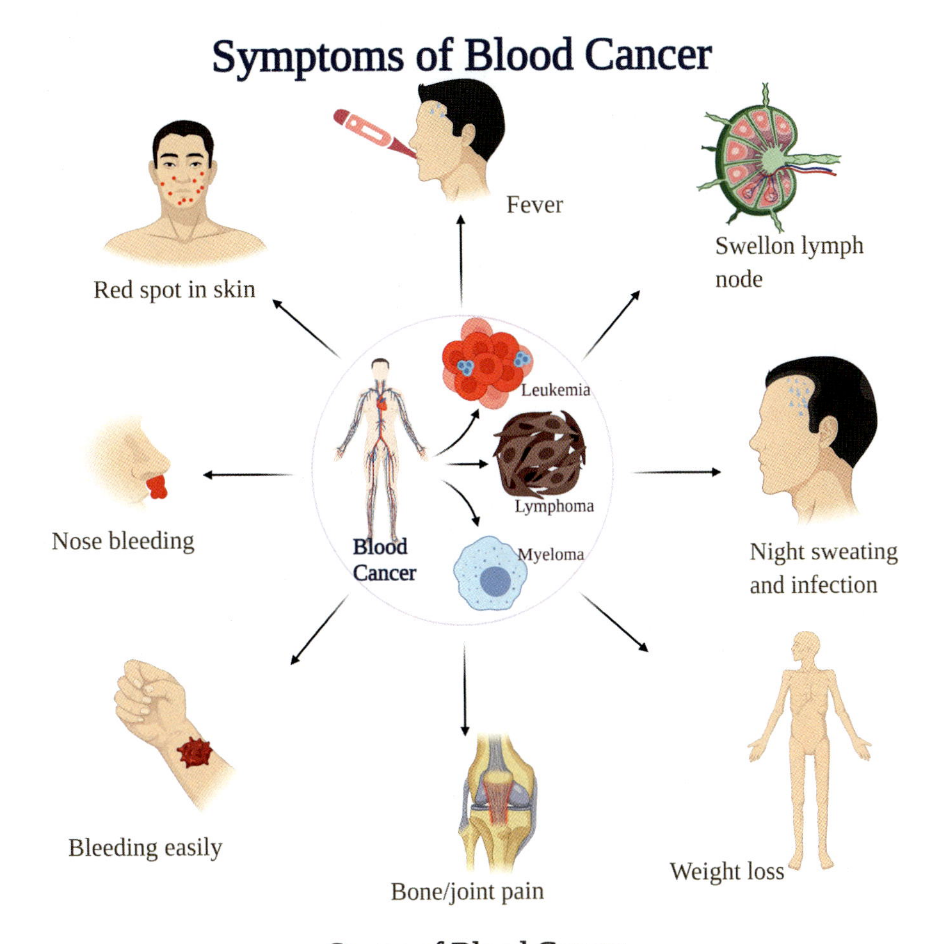

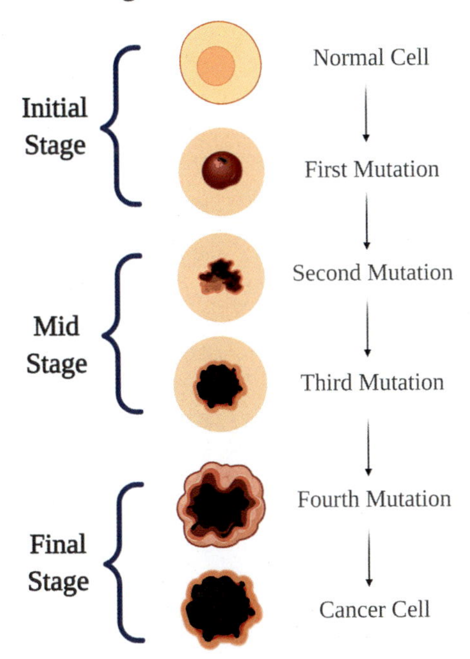

FIG. 2 (A) Symptoms of blood cancer. (B) Stages of blood cancer. *Created by Biorender.com.*

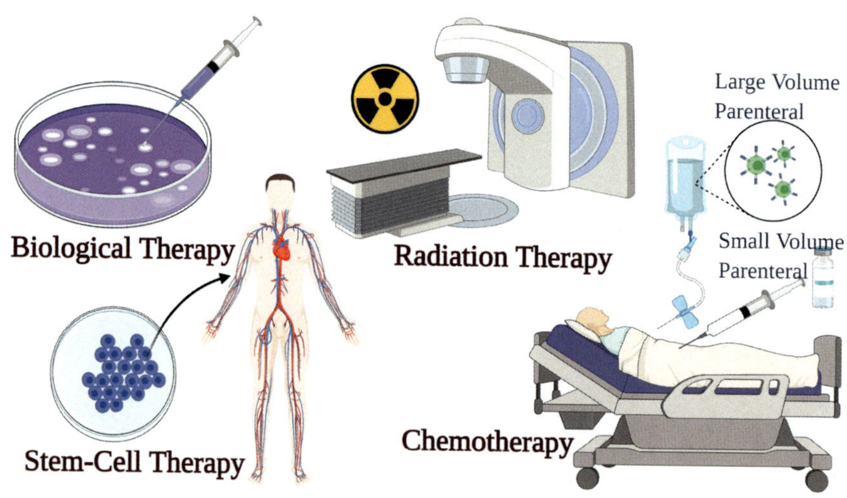

FIG. 3 Current available treatment strategy for blood cancer. *Created by Biorender.com.*

alkylating agents, and antitumor drugs were administered to prevent the cancerous cells from cell division. Asparagine-specific enzymes help in increasing thickness of the bones and reduce the pains in the bone and risk of bone fracture. In chemotherapy, the drugs employed include tyrosine kinase, histone deacetylase and DNA-repair enzyme inhibitors, immunomodulators, hypomethylating, bisphosphonates, and monoclonal antibodies.

6.2 Radiation therapy

In radiation therapy, cancerous cells are destroyed by treating them with high-energy X-rays. In intensity-modulated radiotherapy, early-stage tumors were exposed to a threshold possible dose of radiation, taking care that the surrounding tissues were not damaged. In image-maneuvered radiation therapy, using MRI, a laser was employed to focus exactly onto the affected areas, whereas in stereotactic radiosurgery, a high dose of radiations is delivered to a small area in the brain and spine.

6.3 Immunotherapy

In immunotherapy, monoclonal antibodies bind to antigens and assist in the detection and destruction of tumor cells. In addition, checkpoint inhibitors were employed to find the hidden cancerous tissues. Excess generation of programmed death-ligand 1 (PD-L1) and programmed death-ligand 2 (PD-L2) may interfere in primary treatment, under such circumstances employing immune checkpoint inhibitors results in blocking of PD-L1, assuring positive results in the treatment. In chimeric antigen receptor (CAR) T-cell therapy, T cells are employed to provide the immunity to patients against viruses and T cells are administered in the form of vaccines to prevent cancer and enhance the immunity against cancer.

6.4 Gene therapy

Genetic makeup of the patient determines the formulation of personalized medicine. In advanced treatments, the targets like blood, proteins, and genes are selected, and the treatment of cancer using such therapies leaves behind many side effects. For example, trastuzumab is used for treating blood cancer, colorectal and lung cancers are treated with afatinib and cetuximab, and melanoma is treated with dabrafenib drug.

7 Challenges in the blood cancer management

Nanomedicines are efficient with high efficacy with decreased morbidity of anticancer therapeutic drugs and help in checking the treatment and diagnosis of liquid tumors. The successful application of nanomedicine depends on the accessibility of models that simulate the similar circumstances in patients. Various leukemia and lymphoma models that are used currently have several limitations. In most of the cases, the murine parthenogenesis does not reflect the human complex environment in cancer, thus making the diagnoses difficult.

7.1 Biological barriers

The blood–brain barrier (BBB) comprises endothelial cells and basal lamina that form a tight impermeable junction, thus making the treatment of blood cancer more difficult. Advanced techniques like infusion, implantation, and intraventricular or intracerebral injection were employed for treating blood cancer. Moreover, these techniques have high risk of toxicity due to nonuniform dispersions of the drug [10]. Attaching poloxamer 188, polysorbate 80 like surfactants with the drug molecule increases the drug penetration through BBB [11]. Copper, aluminum, and silver-based nanoparticles are not employed in the treatment as they cause degradation of BBB and increase the risk of neurotoxicity [12].

7.2 Reticuloendothelial system

RES consists of both cellular and noncellular components. Binding of nanoparticles to the phagocytes releases the cytokines, resulting in enhanced particle clearance in the blood [10]. Modification of hydrophilic moieties on the surface of nanoparticles (like cysteine and glutathione, PEG) prevents them from exposure to tissues and also avoids the uptake by RES [13]. In addition, coating of nanoparticles with leucocytes or erythrocyte derivatives helps the nanoparticles to circulate for more time in the blood.

7.3 Renal system

The renal system consists of glomerular basement membrane through which the blood flows for the filtration. The cationic particles are easily cleared as the basement is negatively charged [14]. Nanoparticle size, shape, and charge are important for the clearance. Size of nanoparticles less than 6 nm is cleared easily from the renal system [15]. A patient's renal deficiency allows circulation of optimum-sized particles for a longer time in the blood, leading to increased side effects [10].

7.4 Management

Conventional chemotherapy has more drawbacks in treating hematological cancers due to low efficiency and high adverse effects. The innovative approaches using nanoparticles showed good results in treating hematological cancers. In a nanotechnology, biomarker assistance and targeted treatment proved to treat leukemia and lymphoma. Tumor profile of a patient can be predicted by liquid biopsy, which suggests the therapeutic regimen. Tumors that are liquid in nature circulate continuously in the blood stream and require advanced strategies for identification and treatment, whereas nanoparticles can be used for solid tumors. The tumors can be targeted by stimulated, active, and passive targeting. In passive targeting, accumulation of particles occurs at the leakage region of the blood vasculature of tumors by EPR, whereas ligands like engineered antibodies, folic acid, enzymes, and transferrin were employed for active targeting [15]. Success in releasing the drug by stimulative targeting depends on internal and external factors.

Ionic strength, stress, redox, and pH of the cells are included under internal factors, which regulate the drug release at target site. Iron peroxide particles coated with sodium alginate release the drug due to the changes in the pH. Nanoparticles containing disulfide bonds catalyze the oxidation of glutathione in the tumor cells, leading to enhanced cellular apoptosis [15, 16]. External factors including magnetic force, temperature, electric force, light, and ultrasound also showed effect on drug release. Temperature between 37 °C and 42 °C increases the blood vessel permeability [17], and near infrared increases the penetration of nanoparticles [18], whereas magnetic and electric fields enhance the aggregation of nanoparticles at the target sites [19].

8 Blood cancer diagnosis

Treatment of blood cancer depends on the kind, stage, treatment history, patient age, and genetic background of the patient. Radiotherapy, chemotherapy, cytology and molecular methodologies, targeted medical aid, and stem cell transplantation were employed for treating blood cancer. The strategies like flow cytometry with fluorescent markers, morphological analysis, antibody microarray assay, in situ hybridization, immune histochemistry, and DNA sequencing were used for detecting leukemia and lymphoma. The effective treatment depends on accuracy and sensitivity of the target tissue.

Nanoparticles coupled with signal amplification were used for early detection of cancer. Molecular management and diagnosis allow the genetic and histological tumor characterization in biopsies, which helps to compare the therapy procedure for the detection of patient responses to therapeutics. Biopsies of solid tumors at a specific time does not give any information about the whole cancer or the pattern of variations in genetic mutations, and executing multiple biopsies on a single patient is expensive, difficult, and risky for the patient.

9 Current theranostic approach

Theranostics includes both therapy and diagnostics. It is used for observing the drug delivery, response, and diagnosis. Due to their versatile structures, functional properties, sensitivity, specificity, and rapid diagnosis, nanoparticles are widely used in treatment of cancer. The size of nanocarriers makes them cross biological barriers efficiently, and moreover, the surface ligand helps in specific tumor targeting. Theranostics is efficiently applied in the detection of liquid tumors. Preclinical studies on lymphoma were done using nanotheranostic methods with diatomite nanoparticles, nanoantibodies, and metal nanoparticles. Nanotheranostic performance was found to be effective and more efficient by using albumin-bound paclitaxel nanoparticles linked to rituximab [20] in combination with the imaging of tumors with Alexa Fluor 750 linked with abraxane than the use of two antibodies individually.

10 Benefits and features of advanced drug delivery system

The physicochemical properties like surface, shape, and size of the nanocarriers can be altered based on the requirement. Particle size determines the distribution and site-specific delivery, whereas particle shape determines the fluid impact dynamics. The nanoparticle surface charge can be changed by adding ligands, thus improving the clearance, degradation, solubility, and cellular uptake of drugs [21]. The encapsulated hydrophilic nanoparticles have increased solubility, thereby enhancing the drug accessibility and effectiveness [22]. Reticuloendothelial system (RES) detects hydrophobic materials and is discharged, so PEG-coated carrier molecules enhance the hydrophilic nature of carriers; thereby, the circulation time is increased in the patient blood [23].

CAR T-cell therapy is an advanced therapy for treating blood cancer, which uses CD19 cells to target the receptor of chimeric antigen. For CAR T-cell therapy, ZUMA-1 (large B-cell lymphoma) was used, and 82% of the patients completely recovered and 54% of them were cured from the disease completely. Leukapheresis separates T cells from WBC, and the T cells are engineered with CAR T cells to assail the cancerous T cells. In the stem cell therapy, monoclonal antibodies and other drugs were engineered to attack CD20 antigen in lymphoma condition to stop the cancerous cell growth and multiplication. For leukemia, drugs are developed to inhibit FLT3 gene by developing mutations in the gene. Daratumumab, elotuzumab, and ixazomib were used for treating myeloma, which target the myeloma cells and inhibit the growth of the cells.

11 Advance drug delivery tool in the blood cancer management

The success of the treatment depends on the ability of drug to target the cancerous cells. Nanoparticles inhibit the growth of metastatic tumors and also cause apoptosis of cancerous cells [24]. Various nanodevices were used in the management of blood cancer and are depicted in Fig. 4.

Multiple therapeutic agents were carried out by nanoparticles, thus increasing the effectiveness of treatment [25]. Theranostics helps in visualizing the effects of drugs on cancer cells [26]. The ability of the drug to target and the required concentration at the action site enhance the effectiveness of the drug and also minimize the toxicity. The advanced nanotechnologies and nanomedicines decrease the patient's risk and enhance the survivability of the patient. Advance therapeutics developed using nanotechnology are more efficient in targeting the cancer cells with minimized chemotherapy side effects. Large surface area provided by the smaller particles of nanomedicine can be utilized to attach DNA, RNA, ligands, aptamers, small molecules, and peptides. Such a combination contributes an advantage of theranostic action. The treatment strategies with nanotechnology showed improved pharmacokinetics, selective targeting, and reduced toxicity of the drugs. The advantages of chemotherapy in combination with nanotechnology are encapsulation and conjugation of drugs, targeting with

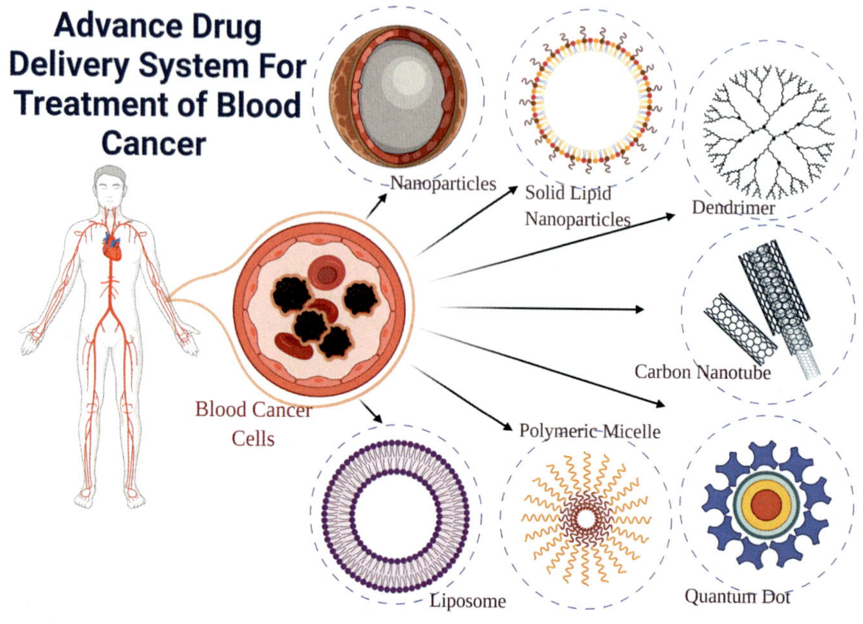

FIG. 4 Nano-based system for the management of blood cancer. *Created by Biorender.com.*

the help of heat, ultrasound and pH, and passive targeting, leading to decreased threat and increased efficacy of the therapeutics.

Mesoporous silica-containing gemcitabine and paclitaxel were used as nanomedicines. In immunotherapy, the cancer type is identified based on functional and molecular analyses of single tumor. Nanodevices help in image characterization, identification of T cells, and synergy with nanomedicines. Nanomedicines engage close to the tumor cells and destroy the cancerous cells. In radiation therapy, patients are treated with high radiation that damages the DNA in the cancerous cells. The photoelectric effects of radiation therapy can be enhanced by using nanotechnology-based approaches. This therapy includes electromagnetic radiation for superficial tumors, which works by photosensitizing the cells and, thus, the activation of reactive oxygen species takes place. Lanthanides or hafnium are injected, which are high-molecular weight atoms that are irradiated externally, and, thus, these atoms emit visible light photons and singlet oxygen forms, leading to the destruction of cancerous cells. Different methods employed for the preparation of nanoparticles include bio-aggregation [27], electrostatic deposition [28], nanomanipulation [29], imprinting, vapor deposition, and photochemical patterning [30]. Various advanced delivery systems that were utilized in treating blood cancer are shown in Table 1.

11.1 Nanoparticle-based system for blood cancer management

Various nanocarriers used in cancer treatment are lipid nanocarriers (liposome, solid lipid, and liposome nanoparticles), polymer-based nanoparticles (albumin bound nanoparticles and micelles), and polymeric nanoparticles. Conjugated nanoparticles include antibody and polymer drug, polymer protein, and inorganic nanoparticles, viz., silica, metal, and hafnium oxide [22].

Advancements in technology led to the rapid development, characterization, and applications of nanoparticles in cancer treatment. The meliorated permeability and retention (EPR) effect developed in therapeutic agents play a role in facilitating the accumulation of nanoparticles at the tumor tissues than normal tissues [51, 52]. In general, nanoparticles encounter five different barriers in vivo, viz., adsorption of serum opsonin onto the surface of NPs, interaction of nanoparticles with the immune system [53], action of nanoparticles at the site of tumor, penetration of NPs into the tumor, and internalization of NPs. The complex biological barriers make the drug delivery more difficult with a 0.7% (median) of administered nanoparticles to reach target sites [54, 55].

Polymeric nanoparticles are made up of dense matrices and synthesized by emulsification, electrospray technique, and nanoprecipitation. These polymers were employed in

TABLE 1 Various advanced drug delivery systems for the management of blood cancer.

S. No.	Types of advanced drug delivery system	Approach	Activity, targeting, and applications	References
1	Silica NPs	Rituximab-conjugated mesoporous silica nanoparticles	Targeted delivery of lymphoma B cell	[31]
2	NPs	B-cell activating factor receptor (BAFF-R) aptamer-siRNA delivery system	Gene silencing of STAT3 mRNAs in a variety of B-cell lines	[32]
3	NPs	Barasertib (AZD2811) nanoparticles targeting Aurora b kinase	Treatment of diffuse large B-cell lymphoma	[33]
4	Iron oxide NPs	Rituximab-conjugated iron oxide nanoparticles	Treatment of CD20-positive lymphoma	[34]
5	Selenium nanoparticles (SeNPs)	Carboxylic group-induced synthesis of selenium nanoparticles (SeNPs)	Treatment of Dalton's lymphoma cells	[35]
6	NPs	Nisin and nisin-loaded PLA–PEG–PLA nanoparticles	Cytotoxicity of nisin-loaded nanoparticles was more than that of the free nisin against K562 blood cancer cells	[36]
7	NPs	Aptamer-conjugated Au-coated magnetic NPs	Aptamer (sgc8) recognizes PTK7 on screen-printed graphene-nitrogen nanosheet electrode (N-GN) in vitro cells and human blood plasma	[37]

Continued

TABLE 1 Various advanced drug delivery systems for the management of blood cancer.—cont'd

S. No.	Types of advanced drug delivery system	Approach	Activity, targeting, and applications	References
8	Nanoconjugates	QD-CdTe@Wogonin	In vitro/in vivo multidrug-resistant leukemia (K562-A02 cell line; leukemia-bearing mice)	[38]
9	Nanoconjugates	MNP-Fe_3O_4@Wogonin	MNPs as contrast agents for MRI. In vitro multidrug-resistant leukemia (K562-A02 cell line)	[39]
10	Nanoconjugates	Rituximab-ABX-Alexa750 (rituximab binds to albumin of ABX, NP-albumin-bound PTX)	In vitro/in vivo B-cell lymphoma (Daudi cells; human B-cell lymphoma murine model)	[40]
11	Nanoconjugates	AgNP@p-MBA@Rituximab	p-Mercaptobenzoic acid (p-MBA) (Raman reporter and linker molecule). In vitro Burkitt's lymphoma (Daudi and Raji cell lines)	[41]
12	Liposome	Nonpegylated liposomal doxorubicin	Management of anthracycline-induced cardiotoxicity in lymphoma patients	[42]
13	Liposome	Dual-mode dark-field and surface-enhanced Raman scattering liposomes	Lymphoma and leukemia cell imaging	[43]
14	Liposome	Liposome-encapsulated doxorubicin in combination with cyclophosphamide, vincristine, prednisone, and rituximab	Treatment of patients with lymphoma and concurrent cardiac diseases or pretreated with anthracyclines	[44]
15	Microspheres	Puerarin-loaded carboxymethyl chitosan microspheres (Pue-CCMs)	Pue-CCMs downregulated the expression of the Bcl-2 associated X protein (Bax) and upregulated B-cell lymphoma-2 (Bcl-2) expression	[45]
16	Microspheres	Vincristine sulfate loaded dextran microspheres	Enhanced cytotoxicity in THP-1, human leukemia cells	[46]
17	Microspheres	Autologous tumor antigen-coated microbeads	Treatment and immunotherapy of B-cell lymphoma	[47]
18	QDs	Immunofluorescent labeling of CD20 tumor marker with quantum dots	Bioimaging tool; rapid and quantitative detection of diffuse large B-cell non-Hodgkin's lymphoma	[48]
19	QDs	T-lymphoma labeled with fluorescent quantum dots	Bioimaging tools for tracing target cells	[49]
20	QDs	Folate-mediated tumor cell uptake of quantum dots entrapped in lipid nanoparticles	Folate-targeted lipodots represent an attractive approach for tumor cell labeling both in vitro and in vivo	[50]

NPs, nanoparticles; QDs, quantum dots.

delivering the imaging, diagnostic, and chemotherapeutic agents in the treatment of cancer [56]. Genes, vaccines, proteins, and drugs were delivered using polylactic acid, copolymer, and polyglycolic acid polymers [57, 58]. Metal-based polymers are hollow metal nanoshells prepared using palladium, silver, gold, and platinum [59]. Previous studies elucidate that to enhance the efficacy of newly developed therapies, the strength of metal nanoparticles was also involved. These metal nanoparticles are biodegradability, low toxicity, and biocompatibility in nature. Silver nanoparticles, in the acidic pH, release reactive oxygen species leading to the destruction of cellular components and finally apoptosis of the cell. Cancerous cells are detected using the optical absorbance obtained from the gold nanoparticles, and cancer therapy is done using the photo thermal effect of gold particles. Doxorubicin conjugated with H-sensitive mesoporous silica nanoparticles (MSNs) controls the release of drug even at low physiological pH [60].

DMSA-Fe_3O_4 magnetic nanoparticles (MNPs) covalently linked with ADM and As_2O_3 as a drug delivery system showed higher anticancer efficacy in non-Hodgkin's lymphoma (NHL) cell line—Raji [61]. Fe_3O_4@MTX magnetic nanoparticles demonstrated in vivo and in vitro that the combined thermo-chemotherapy of Fe_3O_4@MTX MNPs promotes apoptosis of diffuse large B-cell lymphoma (DLBCL) cells through the mitochondrial apoptotic pathway and provides an advanced treatment for primary central nervous system lymphomas (PCNSL) [62]. Reneeta [63] formulated protein NPs by conjugating 5-fluorouracil with silkworm pupa (5FU-PpNPs) in treating lymphoma, and the results suggest that 5-FU-PpNPs played a critical role in decreasing the tumor volume and the viability of tumor cell. Methotrexate (MTX)-loaded hydrogel nanoparticles prepared by treating primary CNS lymphoma via intranasal (IN) administration showed significantly higher concentration of MTX in the brain but not in the plasma. Further, intranasal administration of nanogel significantly increased the MTX concentration in the brain [64]. Guorgui et al. [65] prepared curcumin in solid lipid nanoparticles (SLN-curc) for the treatment of Hodgkin's lymphoma (HL).

11.2 Nanoconjugated system

In protein drug conjugates, a biodegradable bond binds to both protein and drug. Degradation of the linkage by proteases or redox altering agents at nontarget site leads to decreased drug concentration at the target sites. Newer linkers that are stable may increase the accuracy of drug delivery and the circulation time with reduced cytotoxicity. The drug targeting ability can be enhanced by conjugating the drugs with antibodies [66].

11.3 Liposome-based system

Liposomes are spherical lipid bilayer systems. The liposomes of 50- and 500-nm size can be synthesized using extrusion, sonication, reverse-phase, solvent injection, and evaporation [56]. Liposomes are very sensitive and are easily destroyed by ultrasound microwaves, heat, and radio frequencies [67] and can also be easily formulated according to the requirements [56]. Some antibodies are also used in targeting due to their binding ability to antigens of the tumor cells. Monoclonal antibodies were employed as imaging vehicles, drug targeting, and carriers [68].

Gemcitabine (G), vinorelbine (V), and liposomal doxorubicin (GVDoxil) are having high response rates, high toxicity, and price. Ganesan and his researcher group [69] suggested that GVDex is an effective regimen than GVDoxil in relapsed/refractory HL. UV-responsive liposome was developed to overcome chemo-resistance in non-Hodgkin lymphoma. The antitumor activities of the nanoparticles lead to the accumulation of therapeutic agents and retention of drugs by the malignant tissues and cells. Further, the release was in co-operation of active targeting via antibody–antigen reaction and passive targeting with enhanced permeability and retention effect [70]. Another group of researchers suggested the efficacy and safety assessment of vincristine sulfate injection for acute lymphocytic leukemia [71].

Some of the therapeutic agents synthesized and studied against adult Burkitt lymphoma and aggressive B-cell lymphoma are vincristine, doxorubicin, methotrexate (CODOX-M) ifosfamide, rituximab (R), liposome-encapsulated cytarabine (D), cyclophosphamide, and etoposide with intrathecal liposomal cytarabine and rituximab [72]. Treatment of lymphoma using quercetin-decorated curcumin liposome exhibited increased body mass and life span in mice inflicted with lymphoma and showed the cytotoxic activity toward HT-29 cells and HCT-15 cells due to the presence of quercetin [73]. Phase II clinical studies on advanced-stage DLBCL and MCL with rituximab and vincristine sulfate liposome injection (VSLI) showed positive response in patients [74]. Scientific investigations on interleukin-2 (IL-2) and interleukin-6 (IL-6) gene therapy evidenced the activation of antitumor response and macrophage suppression in the mice with lymphoma [75].

11.4 Dendrimeric system for blood cancer treatment

Dendrimers are branched macromolecules with flexibility, density, size, shape, and water soluble used for aiming and treatment of tumors [76]. Gothwal [77] developed bendamustine-PAMAM conjugates that are very effective on cancer at a very low dose with more stability and with easy uptake by THP-1 cells. In addition, PAMAM-bendamustine conjugate was effective on tumor-bearing BALB/C mice with higher accumulation in the tumor. Lidický et al. [78] elucidated the antilymphoma activity of anti-Cd20 targeted against monoclonal antibodies and nontargeted polymer-prodrug conjugates. PEG-oligocholic acid telodendrimer micelles were formulated by Xiao et al. [79] and showed that DOX-PEG micelles exhibited enhanced efficacy and increased survival in mice with Raji lymphoma than free DOX and Doxil® with the same dose.

11.5 Carbon-based drug delivery system

11.5.1 Quantum dots

Quantum dots were utilized as nanocarriers to enhance the efficacy of antitumoral therapeutics. A natural antiproliferative active flavonoid (4-nm (diameter) cadmium telluride quantum dots) acts against several cancers and in conjugation with wogonin causes apoptosis of cancerous cells. Zhou et al. [80] developed daunorubicin and gambogic acid co-loaded cysteamine-CdTe quantum dots for reducing the multidrug resistance to lymphoma. In the same way, another group of scientists utilized multicolor QDs for recognition of Hodgkin's lymphoma [81]. Belletti et al. [82] prepared anticancer drug (curcumin) containing quantum dots engineered polymeric (PLGA) nanoparticles for imaging and treatment of primary effusion lymphoma (PEL) in BCBL-1 and HBL-6 cell lines.

11.5.2 Carbon nanotubes

The carbon nanotubes (CNTs) were developed as nanosuspensions with single-walled forms that bind and taken up by endocytosis. The disulfide bonds in the carbon nanotube are cleaved to release the drugs in the presence of enzymes. The photochemical destruction of tumor cells occurs by carbon nanotubes [83]. Employing Raman signature technique, CNTs can also be used in imaging of tumors [84]. A medical-grade, multiwalled carbon nanotubes were developed for hematological malignance treatment [85]. Monoclonal antibodies conjugated with CNTs can specifically bind and target the cancerous cells by NIR using anti-CD22, anti-CD25, CD22-CD25+ normal human mononuclear cells or CD22+ CD25-daudi cells [86].

12 Clinical trials and marketed treatments available for blood cancer

Among many nanomedicines, liposomal nanoformulations exhibited an effective impact on malignancies; still three liposomal formulations reached clinical trials in treating leukemia. FDA approved a new nanoformulation, liposomal vincristine sulfate (Marqibo®) for the treatment of Ph+ALL in adults. Vincristine inhibits the formation of microtubule during mitotic spindle formation, thus arresting the division of metaphase cancerous cells and used as an essential constituent for treating other lymphoid malignancies and ALL using chemotherapy regimens. Encapsulation using sphingomyelin/cholesterol-based liposome successfully functioned in slow release of the drug from the vector and efficient delivering of high doses of therapeutics to tissues [71]. Various nanomedicines are in the advanced development stage, viz., (i) cytarabine and daunorubicin (CPX-351) liposomal formulations, which showed positive output in clinical trials (phase III) in treating high-risk AML (ClinicalTrials.gov Identifier: NCT02286726, [87]); (ii) encapsulated liposome with an anthracycline (annamycin) was employed in treating relapsed or refractory leukemia in phase I/II clinical trials.

For the treatment of cancer, nucleic acid nanoconstructs were developed by Northwestern University at the University Cancer Centre. DNMT3A gene mutation was observed in AML patients, and it was identified that high doses of anthracyclines showed better results, whereas stem cell transplants exhibited positive results in patients with RVNX1 gene mutation. New drugs and treatment regimens with modified therapies are stated in the following. FLT3 inhibitors (sorafenib, quizartinib, and crenolanib) develop mutations in FLT3 gene, resulting in arrested cell division of AML cells.

BCL-2 gene inhibitors cause programmed cell death of cells. Cytarabine and venetoclax used at low dose exhibited apoptosis of leukemia cells. Enasidenib, inhibitor or isocitrate dehydrogenase, develops mutations in the tumor cell and arrests the cell from maturation. Similarly, volasertib, inhibitor of serine/threonine protein kinase, arrests the cell division and finally leads to the cell death of tumor by inhibiting the PLK1 enzyme. CAR T-cell therapy, nanoparticle vaccines, and monoclonal antibodies are few types of immunotherapies that are employed for treating different types of cancers. The various advanced drug delivery systems which are currently available in market and are also approved in preclinical and clinical trials, are depicted in Table 2.

13 Conclusion and future perspective

Blood cancer or HM treatment was executed in the clinic using radiotherapy, stem cell transplantation, immunotherapy, and chemotherapy. Employing two or more drugs in combination to target the cancerous cells by altering the various oncogenic pathways resulted in enhanced molecular and cytogenetic responses and an increased rate of survival. Chemotherapeutic drugs have limitations like poor pharmacokinetic profile, reduced bioavailability, lack of specificity, development of multidrug resistance, and dose-limiting toxicities. The small drug molecules are transported by passive diffusion through the cell membrane into malignant cells, and the drug molecules combat different drug efflux pumps leading to reduced potential concentration at the target site resulting in MDR [110–112].

Nanomedicines such as polymeric micelles and liposomes showed better pharmacokinetic properties and with increased retention in the blood circulation with reduced side effects. The value of therapeutic nanoparticles in the treatment of HM was evidenced by the success of VYXEOS® and Doxil®, while liposomal formulation increased the chemotherapeutic efficiency and safety. Classical nanocarrier systems showed high improvement in overcoming MDR in various in vitro experiments. The available information has to be carefully utilized as in vitro models lack to mimic many events like EPR effect, angiogenesis, tumor microenvironment, etc. even after a thorough understanding the disease pathology physiology; only few formulations were employed successfully in treating HM.

TABLE 2 List of advanced drug delivery system available in market/clinical trials.

S. No.	Types of advanced drug delivery system	Product name	Company name	References
1	Nanoparticles	Abraxane®	Mayo Clinic (Phase 2)	[88]
		CD19-DOX-NPs	Preclinical	[89]
2	Micelles	SP1049C	Supratek Pharma (Phase 3)	[90, 91]
		NK012	Nippon Kayaku Co., Ltd. (Market)	[92]
3	Liposome	Doxil®/Caelyx®	Johnson & Johnson (Market)	[93–95]
		Marqibo® (Talon)	Spectrum Pharmaceuticals (Market)	[96–98]
		DaunoXome®	Gilead Sciences (Market)	[94]
		CPX-351	Celator (Market)	[96, 99]
		Liposomal tretinoin (ATRA-IV)	MD Anderson/NCI (Phase II)	[89]
		Liposomal Grb-2	MD Anderson/Bio-Path (Preclinical)	[100]
4	Nanoconjugates	Ibritumomab tiuxetan, Zevalin®	IDEC/spectrum (Market)	[101]
		Brentuximab vedotin (Adcetris®)	Seattle Genetics (Market)	[102, 103]
		PegAsys	Hoffmann-La Roche (Phase III)	[104]
		Elacytarabine	Clavis Pharma (Phase III)	[103, 105–107]
		OncasparTM (PEG)	Baxalta (Market)	[108, 109]

The reality lies in the explanation that encapsulation of therapeutic molecules into nanoparticles, nanoparticle stability, and discharge of drug at the tumor site are few challenges that are faced in the development of therapeutic nanomedicine. Nevertheless, nanomedicines are effective and potential drug delivery systems, and various formulations are clinically approved for the treatment of malignancies and chronic inflammatory diseases. Advanced and extensive research may develop novel nanoparticle formulations for the treatment of blood cancer in patients.

References

[1] Siegel RL, Miller KD, Fedewa SA, Ahnen DJ, Meester RG, Barzi A, et al. Colorectal cancer statistics, 2017. CA Cancer J Clin 2017;67(3):177–93.

[2] Hu D, Shilatifard A. Epigenetics of hematopoiesis and hematological malignancies. Genes Dev 2016;30:2021–41.

[3] Cazzola M. Introduction to a review series: the 2016 revision of the WHO classification of tumors of hematopoietic and lymphoid tissues. Blood 2016;127:2361–4.

[4] Den Boer ML, van Slegtenhorst M, De Menezes RX, Cheok MH, Buijs-Gladdines JG, Peters ST, et al. A subtype of childhood acute lymphoblastic leukaemia with poor treatment outcome: a genome-wide classi-fication study. Lancet Oncol 2009;10:125–34.

[5] Freeman CL, Gribben JG. Immunotherapy in chronic lympho-cytic leukaemia (CLL). Curr Hematol Malig Rep 2016;11:29–36.

[6] Tiacci E, Park JH, De Carolis L, Chung SS, Broccoli A, Scott S, et al. Targeting mutant BRAF in relapsed or refractory hairy-cell leukemia. N Engl J Med 2015;373:1733–47.

[7] Jabbour E, Kantarjian H. Chronic myeloid leukemia: 2012 update on diagnosis, monitoring, and management. Am J Hematol 2012;87:1037–45.

[8] Siegel RL, Miller KD, Jemal A. Cancer statistics. CA Cancer J Clin 2016;66:7–30.

[9] Young RM, Staudt LM. Targeting pathological B cell receptor signalling in lymphoid malignancies. Nat Rev Drug Discov 2013;12:229–43.

[10] von Roemeling C, Jiang W, Chan CK, Weissman IL, Kim BY. Breaking down the barriers to precision cancer nanomedicine. Trends Biotechnol 2017;35(2):159–71.

[11] Kreuter J. Mechanism of polymeric nanoparticle-based drug transport across the blood_brain barrier (BBB). J Microencapsul 2013;30(1):49–54.

[12] Shanker Sharma H, Sharma A. Neurotoxicity of engineered nanoparticles from metals. CNS Neurol Disord Drug Targets 2012;11(1):65–80.

[13] Garcia KP, Zarschler K, Barbaro L, Barreto JA, O'Malley W, Spiccia L, et al. Zwitterionic-coated "stealth" nanoparticles for biomedical applications: recent advances in countering biomolecular corona formation and uptake by the mononuclear phagocyte system. Small 2014;10(13):2516–29.

[14] Liu J, Yu M, Zhou C, Zheng J. Renal clearable inorganic nanoparticles: a new frontier of bionanotechnology. Mater Today 2013;16(12):477–86.

[15] Cho K, Wang X, Nie S, Chen ZG, Shin DM. Therapeutic nanoparticles for drug delivery in cancer. Clin Cancer Res 2008;14(5):1310–6.
[16] Yang J, Duan Y, Zhang X, Wang Y, Yu A. Modulating the cellular microenvironment with disulfide-containing nanoparticles as an auxiliary cancer treatment strategy. J Mater Chem B 2016;4(22):3868–73.
[17] Chen KJ, Liang HF, Chen HL, Wang Y, Cheng PY, Liu HL, et al. A thermoresponsive bubble generating liposomal system for triggering localized extracellular drug delivery. ACS Nano 2012;7(1):438–46.
[18] Rapoport N, Gao Z, Kennedy A. Multifunctional nanoparticles for combining ultrasonic tumor imaging and targeted chemotherapy. J Natl Cancer Inst 2007;99(14):1095–106.
[19] Guduru R, Liang P, Runowicz C, Nair M, Atluri V, Khizroev S. Magneto-electric nanoparticles to enable field-controlled high-specificity drug delivery to eradicate ovarian cancer cells. Sci Rep 2013;3:2953.
[20] Vinhas R, Mendes R, Fernandes AR, Baptista PV. Nanoparticles-emerging potential for managing leukemia and lymphoma. Front Bioeng Biotechnol 2017;5:1–10.
[21] Stylianopoulos T, Poh M-Z, Insin N, Bawendi MG, Fukumura D, Munn LL, et al. Diffusion of particles in the extracellular matrix: the effect of repulsive electrostatic interactions. Biophys J 2010;99(5):1342–9.
[22] Wicki A, Witzigmann D, Balasubramanian V, Huwyler J. Nanomedicine in cancer therapy: challenges, opportunities, and clinical applications. J Control Release 2015;200:138–57.
[23] Bregoli L, Movia D, Gavigan-Imedio JD, Lysaght J, Reynolds J, Prina-Mello A. Nanomedicine applied to translational oncology: a future perspective on cancer treatment. Nanomed Nanotechnol Biol Med 2016;12(1):81–103.
[24] Hood JD, et al. Tumor regression by targeted gene delivery to the neovasculature. Science 2002;296:2404–7.
[25] Wang L, Shi C, Wright FA, Guo D, Wang X, Wang D, et al. Multifunctional telodendrimer nanocarriers restore synergy of bortezomib and doxorubicin in ovarian cancer treatment. Cancer Res 2017;77(12):3293–305.
[26] Ahmed N, Fessi H, Elaissari A. Theranostic applications of nanoparticles in cancer. Drug Discov Today 2012;17(17):928–34.
[27] Mirkin CA, Letsinger RL, Mucic RC, Storhoff JJ. A DNA method for rationally assembling nanoparticles into macroscopic materials. Nature 1996;382:607–9.
[28] Ai H. Electrostatic layer-by-layer nanoassembly on biological microtemplates: platelets. Biomacromolecules 2002;3:560–4.
[29] Hansma HG, Kasuya K, Oroudjev E. Atomic force microscopy imaging and pulling of nucleic acids. Curr Opin Struct Biol 2004;14:380–5.
[30] Cui D. Advance and prospect of bionanomaterials. Biotechnol Prog 2003;19:683–92.
[31] Zhou S, Wu D, Yin X, Jin X, Zhang X, Zheng S, et al. Intracellular pH-responsive and rituximab-conjugated mesoporous silica nanoparticles for targeted drug delivery to lymphoma B cells. J Exp Clin Cancer Res 2017;36(1):24.
[32] Zhou J, Rossi JJ, Shum KT. Methods for assembling B-cell lymphoma specific and internalizing aptamer-siRNA nanoparticles via the sticky bridge. Methods Mol Biol 2015;1297:169–85.
[33] Floc'h N, Ashton S, Ferguson D, Taylor P, Carnevalli LS, Hughes AM. Modeling dose and schedule effects of azd2811 nanoparticles targeting aurora B kinase for treatment of diffuse large B-cell lymphoma. Mol Cancer Ther 2019;18(5):909–19.
[34] Song L, Chen Y, Ding J, Wu H, Zhang W, Ma M. Rituximab conjugated iron oxide nanoparticles for targeted imaging and enhanced treatment against CD20-positive lymphoma. J Mater Chem B 2020;8(5):895–907.
[35] Kumar S, Tomar MS, Acharya A. Carboxylic group-induced synthesis and characterization of selenium nanoparticles and its anti-tumor potential on Dalton's lymphoma cells. Colloids Surf B Biointerfaces 2015;126:546–52.
[36] Goudarzi F, Asadi A, Afsharpour M, Jamadi RH. In vitro characterization and evaluation of the cytotoxicity effects of nisin and nisin-loaded pla-peg-pla nanoparticles on gastrointestinal (AGS and KYSE-30), hepatic (HepG2) and blood (K562) cancer cell lines. AAPS PharmSciTech 2018;19(4):1554–66.
[37] Khoshfetrat SM, Mehrgardi MA. Amplified detection of leukemia cancer cells using an aptamer-conjugated gold-coated magnetic nanoparticles on a nitrogen-doped graphene modified electrode. Bioelectrochemistry 2017;114:24–32.
[38] Huang B, Liu H, Huang D, Mao X, Hu X, Jiang C, et al. Apoptosis induction and imaging of cadmium-telluride quantum dots with wogonin in multidrug-resistant leukemia K562/A02 cell. J Nanosci Nanotechnol 2016;16:2499–503.
[39] Peng M-X, Wang X-Y, Wang F, Wang L, Xu P-P, Chen B. Apoptotic mechanism of human leukemia K562/A02 cells induced by magnetic ferroferric oxide nanoparticles loaded with wogonin. Chin Med J (Engl) 2016;129:2958.
[40] Nevala WK, Butterfield JT, Sutor SL, Knauer DJ, Markovic SN. Antibody-targeted paclitaxel loaded nanoparticles for the treatment of CD20+ B-cell lymphoma. Sci Rep 2017;7:45682.
[41] Yao Q, Cao F, Feng C, Zhao Y, Wang X. SERS detection and targeted ablation of lymphoma cells using functionalized Ag nanoparticles. In: Proc. SPIE 9724, Plasmonics in biology and medicine XIII; 2016. p. 972407.
[42] Olivieri J, Perna GP, Bocci C, Montevecchi C, Olivieri A, Leoni P, Gini G. Modern management of anthracycline-induced cardiotoxicity in lymphoma patients: low occurrence of cardiotoxicity with comprehensive assessment and tailored substitution by nonpegylated liposomal doxorubicin. Oncologist 2017;22(4):422–31.
[43] Ip S, MacLaughlin CM, Joseph M, Mullaithilaga N, Yang G, Wang C. Dual-mode dark field and surface-enhanced Raman scattering liposomes for lymphoma and leukemia cell imaging. Langmuir 2019;35(5):1534–43.
[44] Rigacci L, Mappa S, Nassi L, Alterini R, Carrai V, Bernardi F, Bosi A. Liposome-encapsulated doxorubicin in combination with cyclophosphamide, vincristine, prednisone and rituximab in patients with lymphoma and concurrent cardiac diseases or pre-treated with anthracyclines. Hematol Oncol 2007;25(4):198–203.
[45] Song X, Bai X, Liu S, Dong L, Deng H, Wang C. A novel microspheres formulation of puerarin: pharmacokinetics study and in vivo pharmacodynamics evaluations. Evid Based Complement Alternat Med 2016;2016:4016963.
[46] Thakur V, Kush P, Pandey RS, Jain UK, Chandra R, Madan J. Vincristine sulfate loaded dextran microspheres amalgamated with thermosensitive gel offered sustained release and enhanced cytotoxicity in THP-1, human leukemia cells: in vitro and in vivo study. Korean J Couns Psychother 2016;61:113–22.
[47] Henson MS, Curtsinger JM, Larson VS, Klausner JS, Modiano JF, Mescher MF, Miller MF. Immunotherapy with autologous tumour antigen-coated microbeads (large multivalent immunogen),

IL-2 and GM-CSF in dogs with spontaneous B-cell lymphoma. Vet Comp Oncol 2011;9(2):95–105.
[48] Shariatifar H, Hakhamaneshi MS, Abolhasani M, Ahmadi FH, Roshani D, Nikkhoo B. Immunofluorescent labeling of CD20 tumor marker with quantum dots for rapid and quantitative detection of diffuse large B-cell non-Hodgkin's lymphoma. J Cell Biochem 2019;120(3):4564–72.
[49] Hoshino A, Hanaki KI, Suzuki K, Yamamoto K. Applications of T-lymphoma labeled with fluorescent quantum dots to cell tracing markers in mouse body. Biochem Biophys Res Commun 2004;314(1):46–53.
[50] Schroeder JE, Shweky I, Shmeeda H, Banin U, Gabizon A. Folate-mediated tumor cell uptake of quantum dots entrapped in lipid nanoparticles. J Control Release 2007;124(1–2):28–34.
[51] Bae YH, Park K. Targeted drug delivery to tumors: myths, reality and possibility. J Control Release 2011;153:198–205.
[52] Torchilin V. Tumor delivery of macromolecular drugs based on the EPR effect. Adv Drug Deliv Rev 2011;63:131–5.
[53] La-Beck NM, Gabizon AA. Nanoparticle interactions with the immune system: clinical implications for liposome-based cancer chemotherapy. Front Immunol 2017;8:416.
[54] Behzadi S, Serpooshan V, Tao W, Hamaly MA, Alkawareek MY, Dreaden EC, et al. Cellular uptake of nanoparticles: journey inside the cell. Chem Soc Rev 2017;46:4218–44.
[55] Chen H, Zhang W, Zhu G, Xie J, Chen X. Rethinking cancer nanotheranostics. Nat Rev Mater 2017;2:17024.
[56] Sun T, Zhang YS, Pang B, Hyun DC, Yang M, Xia Y. Engineered nanoparticles for drug delivery in cancer therapy. Angew Chem Int Ed Engl 2014;53(46):12320–64.
[57] Panyam J, Dali MM, Sahoo SK, Ma W, Chakravarthi SS, Amidon GL, et al. Polymer degradation and in vitro release of a model protein from poly (d, l-lactide-coglycolide) nano- and microparticles. J Control Release 2003;92:173–87.
[58] Katare YK, Panda AK, Lalwani K, Haque IU, Ali MM. Potentiation of immune response from polymerentrapped antigen: toward development of single dose tetanus toxoid vaccine. Drug Deliv 2003;10:231–8.
[59] Sun Y, Mayers BT, Xia Y. Template engaged replacement reaction: a one-step approach to the large scale synthesis of metal nanostructures with hollow interiors. Nano Lett 2002;2:481–5.
[60] Chen JF, Ding HM, Wang JX, Shao L. Preparation and characterization of porous hollow silica nanoparticles for drug delivery application. Biomaterials 2004;25:723–7.
[61] Cai X, Yu X, Qin W, Wang T, Jia Z, Xiao R, Qi C. Preparation and anti-Raji lymphoma efficacy of a novel pH sensitive and magnetic targeting nanoparticles drug delivery system. Bioorg Chem 2020;20:103375.
[62] Dai X, Yao J, Zhong Y, Li Y, Lu Q, Zhang Y, et al. Preparation and characterization of Fe_3O_4@MTX magnetic nanoparticles for thermochemotherapy of primary central nervous system lymphoma in vitro and in vivo. Int J Nanomedicine 2019;14:9647–63.
[63] Reneeta NP, Thiyonila B, Aathmanathan VS, Ramya T, Chandrasekar P, Subramanian N. Encapsulation and systemic delivery of 5-fluorouracil conjugated with silkworm pupa derived protein nanoparticles for experimental lymphoma cancer. Bioconjug Chem 2018;29(9):2994–3009.
[64] Jahromi LP, Mohammadi-Samani S, Heidari R, Azadi A. In vitro- and in vivo evaluation of methotrexate-loaded hydrogel nanoparticles intended to treat primary cns lymphoma via intranasal administration. J Pharm Pharm Sci 2018;21(1):305–17.
[65] Guorgui J, Wang R, Mattheolabakis G, Mackenzie GG. Curcumin formulated in solid lipid nanoparticles has enhanced efficacy in Hodgkin's lymphoma in mice. Arch Biochem Biophys 2018;648:12–9.
[66] Alley SC, Okeley NM, Senter PD. Antibody-drug conjugates: targeted drug delivery for cancer. Curr Opin Chem Biol 2010;14(4):529–37.
[67] Frenkel V. Ultrasound mediated delivery of drugs and genes to solid tumors. Adv Drug Deliv Rev 2008;60(10):1193–208.
[68] Brongersma ML. Nanoscale photonics: nanoshells: gifts in a gold wrapper. Nat Mater 2003;2:296–7.
[69] Ganesan P, Mehra N, Joel A, Radhakrishnan V, Dhanushkodi M, Kalayarasi JP, et al. Gemcitabine, vinorelbine and dexamethasone: a safe and effective regimen for treatment of relapsed/refractory hodgkin's lymphoma. Leuk Res 2019;84:106188.
[70] Li H, Guo K, Wu C, Shu L, Guo S, Hou J, Zhao N. Controlled and targeted drug delivery by a UV-responsive liposome for overcoming chemo-resistance in non-hodgkin lymphoma. Chem Biol Drug Des 2015;86(4):783–94.
[71] Douer D. Efficacy and safety of vincristine sulfate liposome injection in the treatment of adult acute lymphocytic leukemia. Oncologist 2016;21(7):840–7.
[72] Corazzelli G, Frigeri F, Russo F, Frairia C, Arcamone M, Esposito G, et al. RD-CODOX-M/IVAC with rituximab and intrathecal liposomal cytarabine in adult Burkitt lymphoma and 'unclassifiable' highly aggressive B-cell lymphoma. Br J Haematol 2012;156(2):234–44.
[73] Ravichandiran V, Masilamani K, Senthilnathan B, Maheshwaran A, Wong TW, Roy P. Quercetin-decorated curcumin liposome design for cancer therapy: in-vitro and in-vivo studies. Curr Drug Deliv 2017;14(8):1053–9.
[74] Kaplan LD, Deitcher SR, Silverman JA, Morgan G. Phase II study of vincristine sulfate liposome injection (Marqibo) and rituximab for patients with relapsed and refractory diffuse large B-cell lymphoma or mantle cell lymphoma in need of palliative therapy. Clin Lymphoma Myeloma Leuk 2014;14(1):37–42.
[75] Wang Q, Cao X, Wang J, Zhang W, Tao Q, Ye T. Macrophage activation of lymphoma-bearing mice by liposome-mediated intraperitoneal IL-2 and IL-6 gene therapy. Chin Med J (Engl) 2000;113(3):281–5.
[76] Namazi H, Adeli M. Dendrimers of citric acid and poly (ethylene glycol) as the new drugdelivery agents. Biomaterials 2004;26:1175–83.
[77] Gothwal A, Khan I, Kumar P, Raza K, Kaul A, Mishra AK, Gupta U. Bendamustine-PAMAM conjugates for improved apoptosis, efficacy, and in vivo pharmacokinetics: a sustainable delivery tactic. Mol Pharm 2018;15(6):2084–97.
[78] Lidický O, Janoušková O, Strohalm J, Alam M, Klener P, Etrych T. Anti-lymphoma efficacy comparison of anti-cd20 monoclonal antibody-targeted and non-targeted star-shaped polymer-prodrug conjugates. Molecules 2015;20(11):19849–64.
[79] Xiao K, Luo J, Li Y, Lee JS, Fung G, Lam KS. PEG-oligocholic acid telodendrimer micelles for the targeted delivery of doxorubicin to B-cell lymphoma. J Control Release 2011;155(2):272–81.
[80] Zhou Y, Wang R, Chen B, Sun D, Hu Y, Xu P. Daunorubicin and gambogic acid coloaded cysteamine-CdTe quantum dots minimizing the multidrug resistance of lymphoma in vitro and in vivo. Int J Nanomedicine 2016;11:5429–42.
[81] Liu J, Lau SK, Varma VA, Kairdolf BA, Nie S. Multiplexed detection and characterization of rare tumor cells in Hodgkin's lymphoma with multicolor quantum dots. Anal Chem 2010;82(14):6237–43.

[82] Belletti D, Riva G, Luppi M, Tosi G, Forni F, Vandelli MA, et al. Anticancer drug-loaded quantum dots engineered polymeric nanoparticles: diagnosis/therapy combined approach. Eur J Pharm Sci 2017;107:230–9.
[83] Son KH, Hong JH, Lee JW. Carbon nanotubes as cancer therapeutic carriers and mediators. Int J Nanomedicine 2016;11:5163.
[84] Rao A, Richter E, Bandow S, Chase B, Eklund P, Williams K, et al. Diameter-selective Raman scattering from vibrational modes in carbon nanotubes. Science 1997;275(5297):187–91.
[85] Falank C, Tasset AW, Farrell M, Harris S, Everill P, Marinkovic M, Reagan MR. Development of medical-grade, discrete, multi-walled carbon nanotubes as drug delivery molecules to enhance the treatment of hematological malignancies. Nanomedicine 2019;20:102025.
[86] Marches R, Chakravarty P, Musselman IH, Bajaj P, Azad RN, Pantano P. Specific thermal ablation of tumor cells using single-walled carbon nanotubes targeted by covalently-coupled monoclonal antibodies. Int J Cancer 2009;125(12):2970–7.
[87] ClinicalTrials.gov ID: NCT02286726.
[88] Bernacki RJ, Veith J, Pera P, Kennedy BJ, Manzotti C, Morazzoni C, et al. A novel nanoparticle albumin bound thiocolchicine dimer (nab-5404) with dual mechanisms of action on tubulin and topoisomerase-1: evaluation of in vitro and in vivo activity. Cancer Res 2005;65. 560-560.
[89] Krishnan V, Xu X, Kelly D, Snook A, Waldman S, Mason RW, et al. CD19-targeted nanodelivery of doxorubicin enhances therapeutic efficacy in bcell acute lymphoblastic leukemia. Mol Pharm 2015;12:2101–11.
[90] Alakhova DY, Zhao Y, Li S, Kabanov AV. Effect of doxorubicin/pluronic SP1049C on tumorigenicity, aggressiveness, DNA methylation and stem cell markers in murine leukemia. PLoS One 2013;8, e72238.
[91] Sharma AK, Zhang L, Li S, Kelly DL, Alakhov VY, Batrakova EV, Kabanov AV. Prevention of MDR development in leukemia cells by micelle-forming polymeric surfactant. J Control Release 2008;131:220–7.
[92] Miyazaki O, Sekine K, Nakajima N, Ichimura E, Ebara K, Nagai D, et al. Antimyeloma activity of NK012, a micelle-forming macromolecular prodrug of SN-38, in an orthotopic model. Int J Cancer 2014;134:218–23.
[93] Visani G, Loscocco F, Isidori A. Nanomedicine strategies for hematological malignancies: what is next? Nanomedicine (Lond) 2014;9:2415–28.
[94] Lutzny G, Kocher T, Schmidt-Supprian M, Rudelius M, Klein-Hitpass L, Finch AJ, et al. Protein kinase c-beta-dependent activation of NF-kappa B in stromal cells is indispensable for the survival of chronic lymphocytic leukemia B cells in vivo. Cancer Cell 2013;23:77–92.
[95] Porter CJ, Moghimi SM, Illum L, Davis SS. The polyoxyethylene/polyoxypropylene block co-polymer poloxamer-407 selectively redirects intravenously injected microspheres to sinusoidal endothelial cells of rabbit bone marrow. FEBS Lett 1992;305:62–6.
[96] Tardi P, Wan CP, Mayer L. Passive and semi-active targeting of bone marrow and leukemia cells using anionic low cholesterol liposomes. J Drug Target 2016;24:797–804.
[97] Anada T, Takeda Y, Honda Y, Sakurai K, Suzuki O. Synthesis of calcium phosphate binding liposome for drug delivery. Bioorg Med Chem Lett 2009;19:4148–50.
[98] Bangham AD, Standish MM, Watkins JC. Diffusion of univalent ions across the lamellae of swollen phospholipids. J Mol Biol 1965;13:238–52.
[99] Orlowski RZ, Nagler A, Sonneveld P, Blade J, Hajek R, Spencer A, et al. Final overall survival results of a randomized trial comparing bortezomib plus pegylated liposomal doxorubicin with bortezomib alone in patients with relapsed or refractory multiple myeloma. Cancer 2016;122:2050–6.
[100] Tari AM, Gutierrez-Puente Y, Monaco G, Stephens C, Sun T, Rosenblum M, et al. Liposome-incorporated Grb2 antisense oligodeoxynucleotide increases the survival of mice bearing bcr-abl-positive leukemia xenografts. Int J Oncol 2007;31:1243–50.
[101] Morschhauser F, Radford J, Van Hoof A, Vitolo U, Soubeyran P, Tilly H, et al. Phase III trial of consolidation therapy with yttrium-90-ibritumomab tiuxetan compared with no additional therapy after first remission in advanced follicular lymphoma. J Clin Oncol 2008;26:5156–64.
[102] Younes A, Gopal AK, Smith SE, Ansell SM, Rosenblatt JD, Savage KJ, et al. Results of a pivotal phase II study of brentuximab vedotin for patients with relapsed or refractory Hodgkin's lymphoma. J Clin Oncol 2012;30:2183–9.
[103] Keane N, Freeman C, Swords R, Giles FJ. Elacytarabine: lipid vector technology under investigation in acute myeloid leukemia. Expert Rev Hematol 2013;6:9–24.
[104] ClinicalTrials.gov ID: NCT02736721.
[105] ClinicalTrials.gov ID: NCT01147939.
[106] Roboz GJ, Rosenblat T, Arellano M, Gobbi M, Altman JK, Montesinos P, et al. International randomized phase III study of elacytarabine versus investigator choice in patients with relapsed/refractory acute myeloid leukemia. J Clin Oncol 2014;32:1919–26.
[107] DiNardo CD, O'Brien S, Gandhi VV, Ravandi F. Elacytarabine (CP-4055) in the treatment of acute myeloid leukemia. Future Oncol 2013;9:1073–82.
[108] Hare JI, Lammers T, Ashford MB, Puri S, Storm G, Barry ST. Challenges and strategies in anti-cancer nanomedicine development: an industry perspective. Adv Drug Deliv Rev 2017;108:25–38.
[109] Dinndorf PA, Gootenberg J, Cohen MH, Keegan P, Pazdur R. FDA drug approval summary: pegaspargase (oncaspar) for the first-line treatment of children with acute lymphoblastic leukemia (ALL). Oncologist 2007;12:991–8.
[110] Kunjachan S, Blauz A, Mockel D, Theek B, Kiessling F, Etrych T, et al. Lammers, overcoming cellular multidrug resistance using classical nanomedicine formulations. Eur J Pharm Sci 2012;45:421–8.
[111] Kunjachan S, Rychlik B, Storm G, Kiessling F, Lammers T. Multidrug resistance: physiological principles and nanomedical solutions. Adv Drug Deliv Rev 2013;65:1852–65.
[112] Tezcan O, Ojha T, Storm G, Kiessling F, Lammers I. Targeting cellular and microenvironmental multidrug resistance. Expert Opin Drug Deliv 2016;13:1199–202.

Chapter 13

Advanced drug delivery systems in kidney cancer

Nimisha, Apoorva Singh, and Kalpana Pandey
Amity Institute of Pharmacy Lucknow, Amity University, Uttar Pradesh, Noida, India

1 Introduction

Vertebrate kidneys are two bean-shaped organs that are an essential part of the renal system. In humans, this set is placed on the left and right side of the retroperitoneal space, which means that both are located below the diaphragm. The right kidney is posterior to the liver and the left one is posterior to the spleen. The kidney consists of a unique collecting system and a functional substance called parenchyma. The parenchyma further includes an outer renal cortex, a renal medulla on the inner side (Fig. 1A), and around 1 to 1.5 million microscopic filtering units called nephrons [1]. The basic structure of nephron is shown in Fig. 1B. Nephrons are complex structures and can be fundamentally divided into two parts: the renal corpuscle and renal tubule, lined by endothelial cells [2]. Due to its salient features, kidney is a vital organ for the proper functioning of the human body. Thus, if it gets associated with any kind of disease, our body's balance gets seriously affected. There are a varied number of diseases that may affect the overall functioning of our renal system, and "kidney cancer" is among the most dreadful ones. People commonly feel apprehensive by the word cancer which starts with an abnormal cellular growth but has the potential to invade different parts of our body, except for benign tumors [3].

Kidney cancer may occur in one or both kidneys and eventuate when healthy cells grow out of control and form a tumor. Some of the symptoms which become evident are lower back pain, lump formation in the abdomen, loss of appetite, reiterative fever, anemia, loss of appetite, fatigue, and fluctuations in blood pressure. The factors which lead to such states can only define the likelihood of getting this disease because the actual causes are not known. Some of these risk factors can be obesity, cadmium exposure, smoking, genetic, long-term dialysis, high blood pressure issues, use of diuretics, or exposure to asbestos (naturally occurring silicate minerals) [4] (Fig. 2).

There are many forms of cancer, but kidney cancer is not a single uniform type. In fact, there are different forms in which kidney cancer may influence the kidney. These discrete forms can be distinguished by evaluating their clinical course, histologies, response to therapeutic agents, tumor suppressor genes, and oncogenes. There are around 12 genes which can be linked to the progression of kidney cancer. These are:

- **i.** Von Hippel-Lindau (VHL)
- **ii.** MET (Proto-Oncogene, Receptor Tyrosine Kinase)
- **iii.** Melanocyte Inducing Transcription Factor (MITF)
- **iv.** Transcription Factor EB (TFEB)
- **v.** Tuberous sclerosis Complex 1 (TSC1)
- **vi.** Tuberous Sclerosis Complex 2 (TSC2)
- **vii.** Transcription Factor Binding To IGHM Enhancer 3 (TFE3)
- **viii.** Phosphatase and tensin homolog (PTEN)
- **ix.** Succinate dehydrogenase B (SDHB)
- **x.** Folliculin (FLCN)
- **xi.** Fumarate hydratase (FH)
- **xii.** Succinate dehydrogenase D (SDHD)

These genes get involved with the single cells activity and metabolism whenever there is a lack of nutrients around them or in case the surrounding environment is depriving them. They regulate the cell's mechanism in such a way that it becomes more responsive toward alterations in levels of iron, oxygen, or growth-limiting factors like energy and nutrients [5, 6].

1.1 Types of kidney cancer

Grouping kidney cancers into levels and sublevels provides assistance in understanding them in more descriptive ways (like on the basis of their clinical behavior, pathologies, or molecular origins) [7, 8].

Advanced Drug Delivery Systems in the Management of Cancer. https://doi.org/10.1016/B978-0-323-85503-7.00018-3

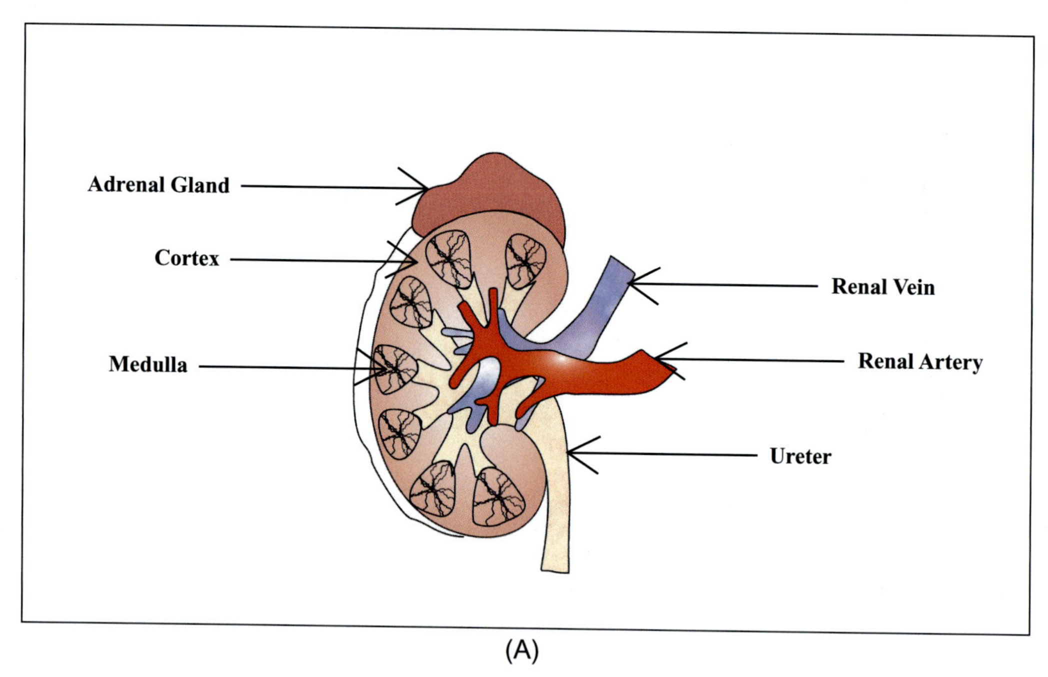

(A)

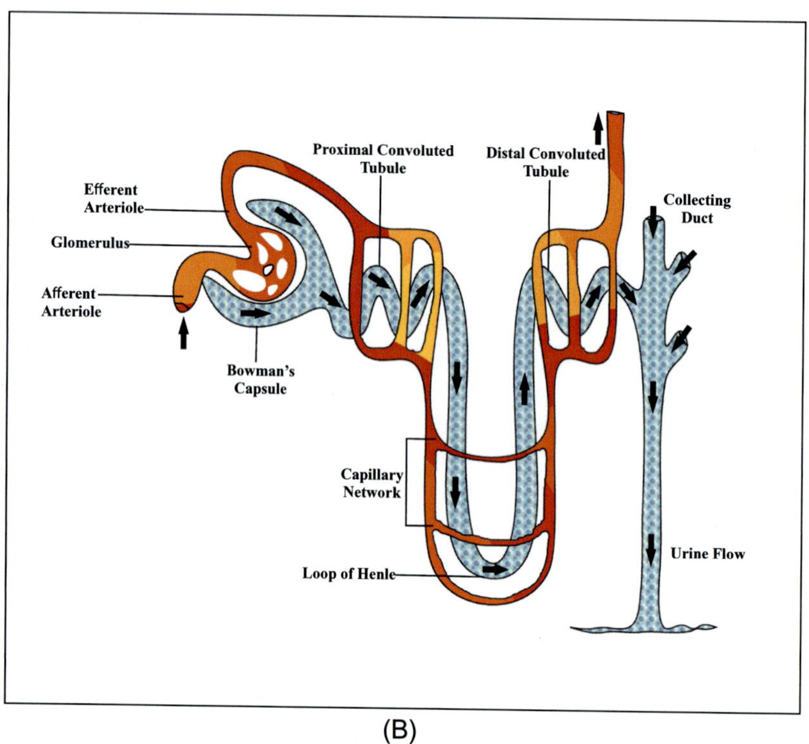

(B)

FIG. 1 General structure of (A) kidney and (B) nephron.

1. Renal cell carcinoma or adenocarcinoma
   RCC, the most common type of kidney cancer, belongs to a continuously evolving area of tumor research and oncology. About 9 out of the 10 kidney cancers diagnosed belong to the RCC category. The functional part of kidney—parenchyma—is the site from where RCC arises and proceeds to form a mass of abnormally grown cells—tumor. The cells grow in the renal epithelial lining of the tubules positioned inside the kidneys. The tumor grows over time and can form singular or multiple manifestations in one or both the kidneys at the same time.
   Since the past decades researchers are aiming more and more of their efforts toward developing targeted formulations. Thus, it becomes important to divide RCC into further subtypes to enhance the functionality of cancer therapy. The classification of RCC resolves the

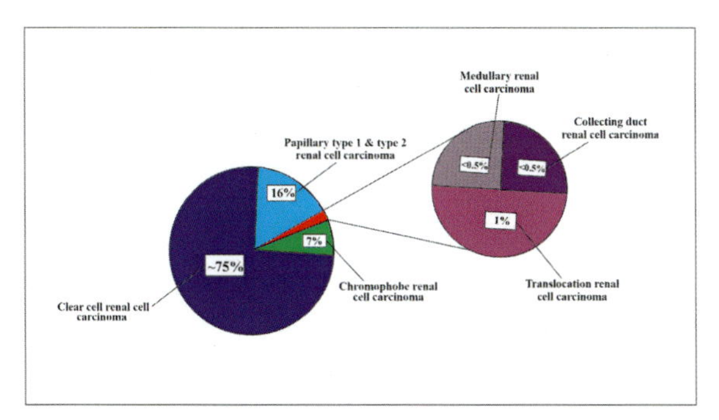

FIG. 2 Pie chart showing distribution subtypes of renal cell carcinoma.

complexity of understanding the clinical constructs and management of kidney cancer [9, 10] (Fig. 3).

Subtypes of renal cell carcinoma

i. Clear cell renal cell carcinoma (ccRCC)

Among the various subtypes of RCC, clear cell RCC is found the most abundantly. It is found in 70%–75% of RCC-affected patients and detected by necrosis or hemorrhage affected areas, seeming golden yellow in appearance and well confined. The microscopic evaluations of these cells reveal ample amounts of clear cytoplasm that is developed due to the accumulation of glycogen and lipid molecules. The larger ccRCC-originated tumors reveal presence of perinephric fat (swelling within the fat of the perirenal space), renal sinus fat invasion and/or renal vein [11]. The ccRCCs display hypervascular behavior and its tumor cells exhibit coagulative necrosis. There is improper functioning of a few genes like *VHL* gene and the chromatin remodeling genes like JARIDIA, Polybromo 1 (*PBMR1*), SET domain containing 2 (*SETD2*), and *BRCA1* associated protein-1 (ubiquitin carboxyterminal hydrolase) (*BAP1*) [12].

ii. Papillary renal cell carcinoma

The second most common kind of kidney cancer, papillary RCC, covers around 10%–16% of cases. Among these cases, almost 10% are coetaneous bilateral and multifocal papillary carcinoma cases. During operative surgeries, it is observed that these tumors show cystic, solid, or mixed textures, and the lesions formed are either reddish or brownish but usually well delineated. When viewed under the microscope, these tumors have a papillary or tubule-like frame with histologic features like an accumulation of calcium in tissue area, necrosis, or infiltration of macrophages [13].

Based on the morphological features, there are two types of papillary RCC. The International Society of Urological Pathology (ISUP) provides a well-recognized distinction between the papillary RCC subtypes [14].

Type I: The tumor cells in type I are present lining the papillae, contains light-colored cytoplasm, and upon C.I. 75,290 and eosin staining, it can be viewed as basophilic. This appearance is due to the presence of small nuclei of the Fuhrman nuclear grade I–II class (a method employed for grading RCC).

Type II: In comparison with type I, type II is far more heterogeneous, and the tumor cells are aligned on thicker papillae. These papillae have proliferative eosinophilic or granular cytoplasm. The nuclei are larger and of pleomorphic type (Fuhrman nuclear grade II–III) [15].

iii. Chromophobe renal cell carcinoma

Among all malignant RCCs, chRCC covers around 5% cases. The chRCC cells have well-defined borders and prolific cytoplasms, which collaborate to form tumors with an overall brownish appearance. The size of the solid tumors may range from 6 to 16 cm with koilocyte kind of cells containing uneven raisinoid nuclear membrane and recurring binucleation [16].

iv. Oncocytoma

Oncocytoma is the benign form of RCC, which is very similar to the eosinophilic renal malignancies. This makes it difficult for the doctors to distinguish it from subtypes like chRCC. Microscopic analysis depicts these tumors as combination of abnormal cells with polygonal cell membrane, voluminous cytoplasm, and well-defined round nuclei. The overall morphology of these tumors describes amber or tan colored, well-circumscribed cells that do not suffer necrosis by autolysis. With the

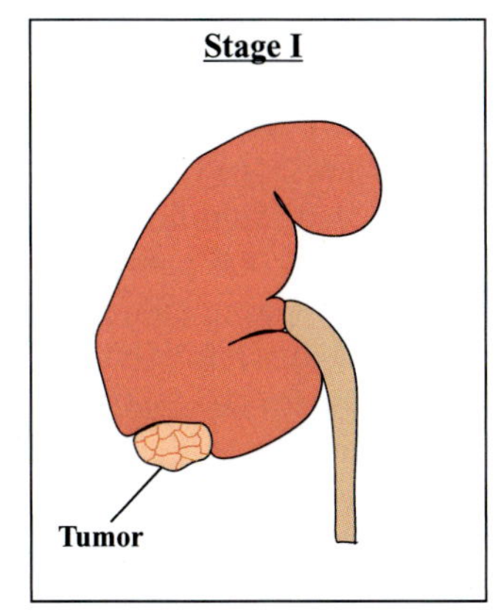

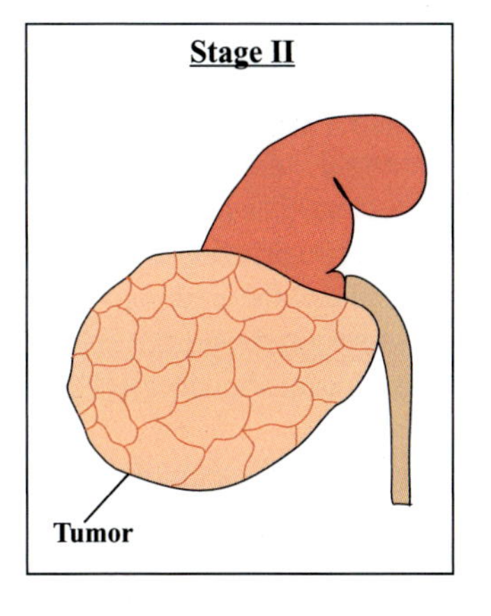

FIG. 3 (A) Stage I tumor and (B) Stage II tumor.

identification via imaging techniques like CT scan, doctors can interject the size of tumors that may range from small abnormal tissues without natural scarring and large tumors in case of last-stage cancers [17].

v. Collecting duct renal cell carcinoma
Collecting duct RCC originates from the innermost part of the kidney—renal medulla. The number of cases under this subtype is not more than 0.5%. On closer morphological examination, the tumor cells appear as white or gray, partially cystic, and often invade the renal vein or renal sinus. There is an advanced growth of dense connective tissues along with tubulopapillary architecture surrounded by mucin—the main component of mucus. According to immunohistochemical studies, the carcinoma cells of this class arise from the distal collecting duct and show hostile behavior [18].

vi. Medullary renal cell carcinoma
Medullary RCC is known to be more apparent in young black patients who suffer from blood disorders like hemoglobinopathy with fewer exceptions [19, 20].
For some unknown reason(s), this RCC subtype is mostly diagnosed in the right kidney. The survival rate is low here, as the tumor reaches its adversity within a few months only. The basic tumor progression starts from the epithelium of renal calyces. The appearance of these tumors ranges from white to tan color and grossly they do not possess well-defined structures. Unlike oncocytoma, medullary RCC exhibits extreme necrosis with an additional disadvantage of severe hemorrhage. Although a violent subtype, it is rare and thus little literature is available regarding its molecular biology [21].

vii. Microphthalmia transcription factor family translocation renal cell carcinoma (tRCC)
The translocation RCC consists of chromosomal translocation of transcription factors that come under the microphthalmia transcription factor family (MiT). MiT includes factors like, TFEC (transcription Factor EC), TFE3, MITF, and TFEB. The uncommon t(6;11) RCC retains the metastasis-associated lung adenocarcinoma transcript 1 and transcription Factor EB (MALAT1–TFEB) gene fusion through TFEB rearrangements. On the other hand, Xp11 tRCC retains TFE3 gene fusions. The basic understanding of how these translocations and gene fusions results in abnormal tumor growth is still undergoing research but it is clear that these genes promote nuclear localization by overexpressing themselves. Microscopic observations reveal that these tumors exhibit calcifications around a nested structure and the cells are eosinophilic/granular with cytoplasm abundance [22].

2. Urothelial carcinoma
Commonly known as transitional cell carcinoma, TCC, urothelial carcinoma covers around 5%–10% of overall diagnosed kidney cancer cases. It originates from the parts where urine resides before proceeding toward the bladder. This area is the renal pelvis. There are many variants of urothelial carcinoma. Some show glandular or squamous differentiation, while others may have nested superficial and deep portions, numerous microcysts, or prominent tubule-like components [23].

3. Renal sarcoma
Renal sarcoma accounts for a rare percentage of kidney cancer cases (< 1%) and is known to originate from the soft connective tissues or blood vessels, which line the kidney. Conventionally, sarcoma of the kidney is managed with the aid of surgery or chemotherapy. Both radical nephrectomy and adjuvant chemotherapy can be employed to treat sarcoma with a 29%–36% chance of curability [24].

4. Wilms tumor
Nephroblastoma or Wilms tumor is commonly diagnosed in the early stages of childhood and accounts for around 1% of kidney cancer cases. The type of treatment given here is basically different from that given to adults and is most positively treated with the help of radiation therapy and chemotherapy. Since some cases may suffer from radiosensitivity, chemotherapeutic agents are more favorably employed for attaining maximum survival rates. Agents like dactinomycin, vincristine, and doxorubicin are mostly used and have shown positive results since 1950. In fact, treatment of Wilms tumor is considered a major success in the field of oncology [25].

5. Lymphoma
Renal lymphomas are known to arise from the renal parenchyma and cause the enlargement of lymph nodes in parts like chest, abdomen, or neck, called lymphadenopathy. Diagnosis is mostly done by biopsy and doctors usually suggest chemotherapy rather than operative surgeries. Common nodular masses may be observed, and it may mimic renal cell carcinoma (as observed clinicoradiologically) [26].

2 Pathophysiology of renal carcinoma

There is a wide range of subtypes of carcinomas but the most common type of carcinoma found in kidney is renal cell carcinoma (RCC). A huge percentage, about 85%, accounts for cancers caused in kidney i.e., renal cell cancers. Pathologists diagnose and identify various subtypes of RCC by monitoring the cancer tissue cells under microscope followed by either biopsy or sometimes surgery also. Significant improvement has been seen in understanding the pathophysiology and biology of these cancer subtypes in kidney over the past few decades. The newer insights into the disease progression have helped a lot in the development of new approach of therapeutics for the purpose of diagnosis and treatment of these renal cell cancers.

There are various mechanisms related to genetic and immune responses that contribute to the development and pathogenesis of these types of cancers, and also it is very helpful in developing newer therapeutics that are specifically targeted onto the renal cancer cells and helps in the RCC treatment.

Herein, we focus on finding the alterations in the genetic contents of renal cancer cells and also the mechanism of action of various therapeutic drugs for targeting these pathways involved in the progression of disease. The immune system plays an important role in the growth and fast development of these renal cell carcinoma and the focus of this review is on manipulation of immune system to prevent its growth and development by reactivating cytotoxic immunity that is proven hostile to renal cell carcinoma. Polybromo-1 (*PBRM-1*) gene and Von Hippel Lindau (*VHL*) are two types of genes that play a vital role in the pathogenesis of renal cell carcinoma. Hypertension, obesity, smoking, diabetes, and excess analgesic use are few common risk factors that can easily contribute to the aggressiveness of the disease condition [27] (Table 1).

2.1 Changes in the genetic make-up

Losing of 3p chromosome, i.e., short arm of chromosome 3, is the most common genetic alteration found associated with developing cRCC and this genetic alteration is found in appropriately into 95% of the total found cases of renal cell carcinoma (RCC).

MTOR, *VHL*, *BAP-1*, *PBRM-1*, *KDM5C*, and *SETD2* are some class of genes that are involved in RCC pathogenesis. There are different other genetic alterations that consist of partially losing of 14q (42%) and 9p (29%); gain of 7q (20%) and 5q (69%) and deletion of 8p (32%) as well [28].

2.2 Von Hippel–Lindau gene

It is one of the class of genes that plays a vital role in the progression of renal cell carcinomas. VHL can be either transmitted or altered in a sporadic or autosomal dominant way like in the case of Von Hippel Lindau disease.

TABLE 1 Type of renal cell carcinoma and their percent occurrence of patients.

S. no.	Types of renal cell carcinoma (RCC)	Percent occurrence of RCC patients (%)
1.	Renal medullar carcinoma (RMC)	1
2.	Unclassified renal cell carcinomas	3–4
3.	Translocation renal cell carcinomas (tRCC)	3–4
4.	Chromophobe renal cell carcinoma(chRCC)	5
5.	Papillary renal cell carcinoma	10–15
6.	Clear cell renal cell carcinoma (ccRCC)	70

There are rare cases of inherited VHL disease that come into account. Understanding the approach for identifying VHL gene suppressor and also knowing its molecular basis of disease progression and in the pathogenesis of sporadic disease. The protein pVHL is a cancer suppressor protein and it is product of *VHL* gene. It forms complexes with many other proteins like cellulin 2, elongin B, and elongin C.

The resulting VBC complex is helpful in proteasomal deterioration of various intercellular proteins. Regulation of levels of various intercellular proteins like alpha (HIF1A and HIF2A) and hypoxia-inducible factor 1 and 2 alpha is the major key responsibility of VHL gene product. *VHL* gene known to be major cancer cell suppressor gene in sporadic renal cell carcinoma (RCC), as it helps in minimizing the defects in *VHL* gene that arise due to various factors like epigenetic slicing in approximately 60% of the cases, mutation, or deletion. VHL-protein regulation pathway gets affected by inactivation of this *VHL* gene.

Genes that are functional in the process of angiogenesis, proliferation, and formation of extracellular matrix get activated because there is negative regulation of hypoxia-inducible factors (HIFs) 1a and 2a that are transcriptional activators by VHL protein. Chromosome 3 is the site for addition of cancer suppressor genes involved in sporadic and inherited clear cell renal cell carcinomas [29, 30].

2.3 Protein polybromo-1 gene

In the process of pathogenesis of clear cell renal cell carcinoma, *PBRM-1* plays a vital role as cancer suppressor gene. A protein known as BAF180 that is a part of nucleosomes remodeling complex is encoded by *PBRM-1*. The nucleosomes remodeling complex is a combination of several other proteins that help to control certain gene expression by retrieving the condensed DNA part [31, 32].

2.4 BRCA1-associated protein-1

3p chromosome is the site where *BAP-1* (i.e., BRCA-1 associated protein-1, cancer suppressor gene) can be found. In almost 15% of the total cases of clear cell renal cell carcinomas, this *BAP-1* gene is mutated. The mutation of cancer cells via *BAP-1* contributes to the rapid worsening and aggressiveness of the cancer prognosis [25]. *BAP-1* plays a vital role in cell proliferation suppression as done by other cancer suppressor genes. This suppression takes place when there is interaction with host cell factor-1 (HCF-1) also known as transcription protein. Thereafter, this HCF-1 binds with various transcription factors that helps in inhibiting cell proliferation [33].

The investigation is under process hoping to achieve novel targets and strategies so that it can provide sustained and more stable responses and improve the chances of survival. In summary, we can conclude that there is improvement in the understanding and detection of pathogenesis of renal cell carcinoma. Many therapeutics that are proven effective for metastatic diseases have been incorporated into the list in the past decade.

This improved progress results in increased rate of patients suffering with severe renal cell carcinoma. Inhibition of MTOR and VEGFR pathways is the main area of focus for research and scientist by developing better therapeutics. Current research also emphasizes on developing new combined approaches, immunotherapy targets, and metabolic targeting for diagnosis and treatment of disease condition [34] (Table 2).

3 Diagnosis

3.1 Stages I and II kidney tumor open pop-up dialog box

The primary treatment for the beginning phase of kidney cancer is radical nephrectomy. This activity incorporates total resection of the kidney including Gerota's fascia, and of the related adrenal gland, renal vein, and ureter. Most relapses after revolutionary nephrectomy are systemic (lung, distant nodes, bone, and liver); within the operative site, relapses can occur around 5%–10% of cases and territorial hubs of around 5%–10% of cases. Abdominal imaging is necessary for local relapses occurring in advanced-stage and high-grade tumor in every duration of 6–12 months for the duration of about 5 years. There are no guidelines validated for the procedure and are solely based on clinical experience. Surgery is the successful mode for the treatment of local relapses. Radical nephrectomy is beyond the realm of imagination, in any case, if the patient would be left with no or exceptionally undermined renal function. Such patients incorporate those with bilateral synchronous tumors, carcinoma in an anatomically or practically solitary kidney, carcinoma in addition to another disorder that could unfavorably influence renal function, carcinoma and working however impeded contralateral kidney, and carcinoma with hidden VHL disease. Under these conditions, partial nephrectomy ought to be done to save however much typical kidney as could reasonably be expected. The specialized difficulties of partial nephrectomy, which incorporate satisfactory intraoperative identification of the lesion, identification and vascular supply control, and shirking of ischemic injury to the typical kidney tissue, require counseling with an accomplished urological specialist, particularly for patients with VHL ailment, since these patients have several renal cysts of variable malignancy potential [35]. By the by, results for patients who have gone through partial nephrectomy are close to as great as those for patients with a comparable phase of cancer treated with standard radical nephrectomy and dissection of lymph node. Since fractional or partial nephrectomy permits the chance of continuous and repeat, cautious radiological follow-up each 3–6 months for 5–10 years is required. A salvage radical nephrectomy is remedial in most of these patients (Fig. 3).

3.2 Stage III kidney tumor open pop-up dialog box

Patients with stage 3 kidney malignant growth incorporate those with tumor expansion to the adrenal organ or perinephric tissues however confined to Gerota's fascia, those with tolerably expanded (1–2 cm) abdominal lymph nodes on registered tomography, and those with net tumor augmentation into the renal vein or inferior rate vena cava. The first group of patients is generally recognized on pathological investigation and is treated with standard radical nephrectomy. Although prognosis is more regrettable than for patients with a less progressed malignant growth, no change in surgical method is required. Relapse, both distant and local, is normal, and if a clinical and radiologically secluded distant or local relapse is distinguished, surgical cure ought to be endeavored. The second group of patients ought to likewise be treated with a standard radical nephrectomy, since over half of extended nodes on registered tomography (1·0–2·0 cm) are auxiliary to responsive changes. A patient ought not be denied the opportunity for curative resection based on tolerably expanded retroperitoneal nodes on computed tomography. However, there is no clear therapeutic benefit of extended lymphadenectomy (formal dissection of the primary retroperitoneal landing site) for patients with clinically localized disease: we and most other urological oncologists do not recommend this procedure. Gross tumor involvement of the renal vein or the inferior vena cava is unusual and is difficult to treat. Most of these patients have frank metastatic disease and should be treated as such. If a thorough assessment for metastatic disease or bulky regional nodal disease is negative, the patient may be suitable for attempted curative resection. Before any surgery, the extent of tumor in the vena cava must be assessed by magnetic resonance imaging or venography of the vena cava [36]. If intrahepatic or supradiaphragmatic involvement is suggested, transoesophageal echocardiography will show the presence and extent of any intracardiac tumor. Most patients with a tumor thrombus in the vena cava do not have direct invasion into the vessel wall. Thus, minor extension into the vena cava can be easily removed through a renal vein or vena-cava incision. If there is invasion of the vessel wall, resection of the vena cava with or without graft replacement can be done. More extensive disease, including thrombus that extends supradiaphragmatically, requires thoracotomy and hypothermic circulatory arrest for successful removal of the tumor. In specialized centers, these procedures have an operative mortality of 2%–13% and a postoperative complication rate of 30%–60%. Long-term outcomes, however, are reasonably good, with 56% survival at 5 years and 45% survival at 10 years.

TABLE 2 Main histological subtypes of RCC—Epidemiology, histology, and imaging characteristics.

S. no.		Subtypes					
		Clear cell	Papillary	Cystic solid	Collecting ducts (Bellini)	Xp11 translocation	Nonclassified
1.	Incidence	75%	10%	1%–4%	1%	Rare	4%–6%
2.	Origin, histology	Proximal nephron, tubular epithelium	Distal nephron, tubular epithelium	Similar to renal clear cell cancer, without solid nodules	Collecting tubules	Distal/proximal nephron, may be similar to papillary or clear cell carcinoma	Variable
3.	Patient's age	> 50 years	> 50 years	Fourth and fifth decades of life	> 50 years	Children (early childhood)	Variable
4.	Signal/density pattern	Heterogeneous density/signal	Low T2 signal, hypodense	High T2 signal intensity, fluid density	Low T2 signal, heterogeneous	Hypodense, intermediate T2 signal intensity	Variable
5.	Biological behavior	Aggressive, according to the stage, Furhman grade and sarcomatoid transformation	Aggressive, according to the stage, Furhman grade and sarcomatoid transformation	Indolent, without metastases	Very aggressive, mortality: 70% in 2 years	Indolent	High mortality
6.	Postcontrast hemodynamic pattern	Hypervascular	Hypovascular	Septal and solid portions enhancement	Hypovascular	Hypovascular	Variable
7.	Associations and predispositions	Von Hippel-Lindau (25%–45%), tuberous sclerosis (2%)	Hereditary papillary RCC	Predominance in men	Subtle predominance in men	*TFE3* gene involved in the tumori genesis	–

3.3 Stage IV tumors

There is little sign for careful resection of neighborhood malignant growth if there is clinically clear tumor augmentation past Gerota belt (T4, board), broad nodal malady (N2), or straight-to-the-point metastatic infection. Urology specialists have embraced nephrectomy in these settings if the patient has critical uncontrolled nearby side effects, for example, torment or bonding safe hematuria. Even though nephrectomy may sporadically be valuable under such conditions, transarterial angioinfarction may likewise be utilized. Resection of the essential tumor is demonstrated if the patient is to have immunotherapy for metastatic ailment, in light of the fact that a huge nearby tumor may restrain the insusceptible reaction. This perception has not, notwithstanding, been affirmed in randomized preliminaries. A 15% reaction rate to immunotherapy, the related horribleness of nephrectomy, and the finding that up to 40% of patients with metastatic sickness who go through nephrectomy never get immunotherapy auxiliary to quick infection progression, lead us to suggest that patients with metastatic ailment should initially be treated with immunotherapy or other fundamental treatment. In the event that all the metastatic injuries react totally or totally, nephrectomy can impact a total fix [35, 36] (Fig. 4).

Tests and methodology used to analyze kidney disease include:

- *Test for blood and urine:* Blood and urine test may give primary care physician hints about what is causing signs and side-effects.
- *Imaging tests:* Imaging tests permits to visualize a kidney tumor or irregularity. Imaging tests may incorporate ultrasound, X-beam, CT or MRI.
- *Removing a sample of kidney tissue (biopsy):* In certain circumstances, physician may prescribe a technique to eliminate a sample of cells (biopsy) from a dubious region of kidney. The sample is examined in a lab to search for indications of malignant growth. This technique is not constantly required.

3.4 Kidney cancer staging

When your primary care physician distinguishes a kidney cancer that may be kidney cancer, the following stage is to decide the degree (phase) of the disease. Staging tests for kidney disease may incorporate extra CT scans or other imaging tests which the physicians feel are suitable.

The phases of kidney malignancy are demonstrated by Roman numerals that range from I to IV, with the most minimal stages showing disease that is bound to the kidney. By stage IV, the disease is viewed as cutting edge and may have spread to the lymph hubs or to different zones of the body (Fig. 5).

3.5 Treatment with nonsurgical medicines

Small kidney tumors are some of the time obliterated utilizing nonsurgical medicines, for example, heat and cold. These techniques might be an alternative in specific

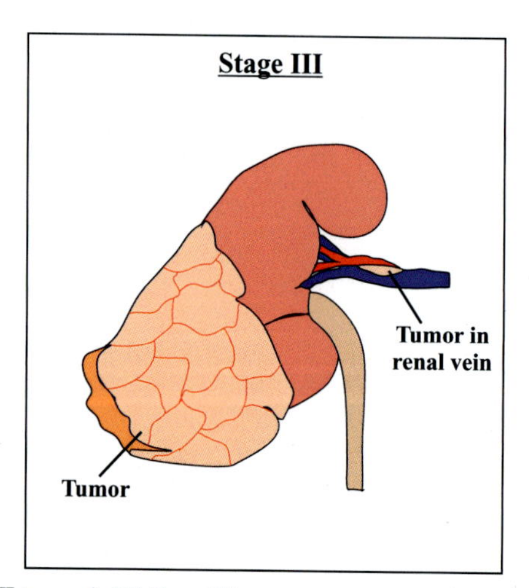

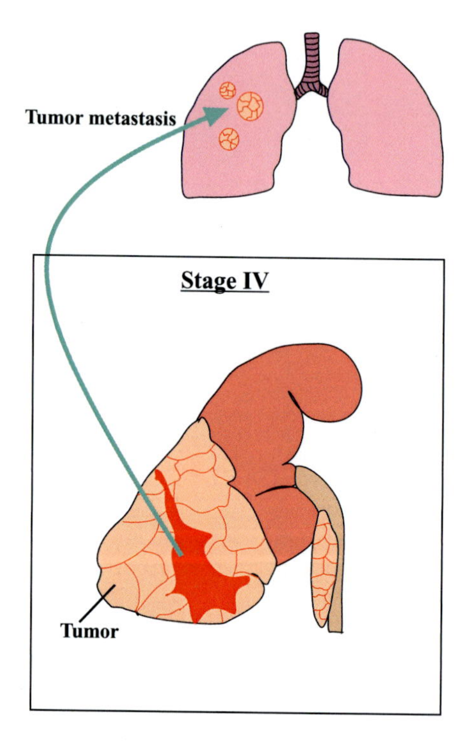

FIG. 4 (A) Stage III tumor & (B) Stage IV tumor.

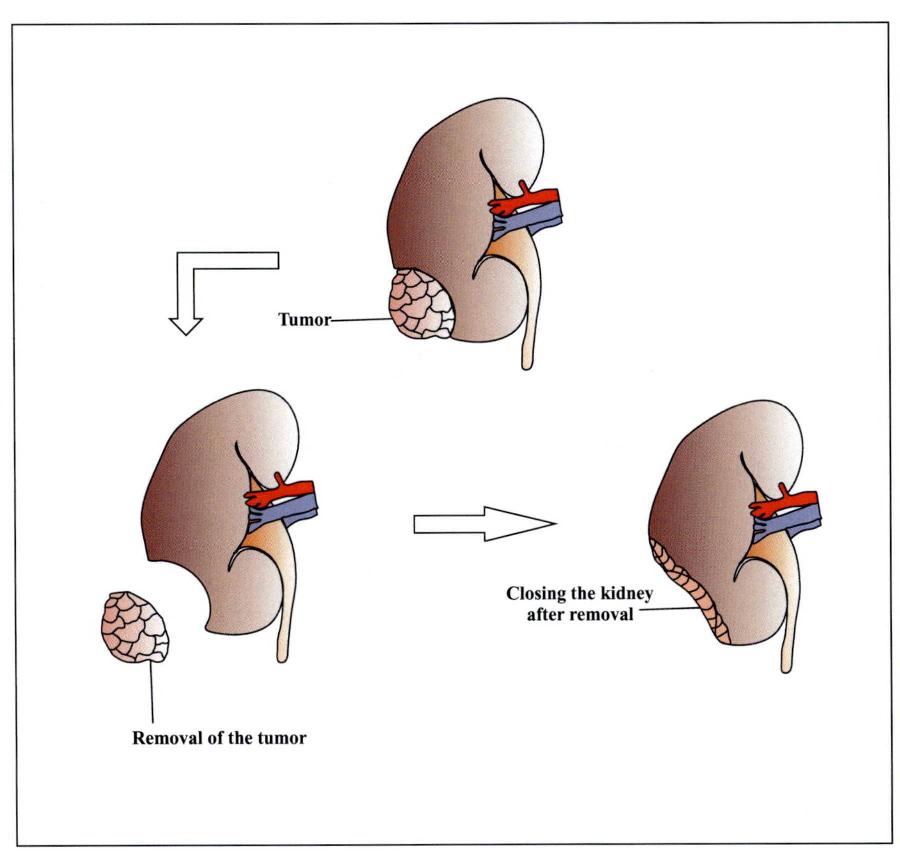

FIG. 5 Partial nephrectomy.

circumstances, for example, in individuals with other medical issues that make medical procedure hazardous.

Alternatives may include:

- *Treatment to freeze disease cells (cryoablation).* During cryoablation, a special empty needle is embedded through the skin and into the kidney tumor utilizing ultrasound or other image direction. Cold gas in the needle is utilized to freeze the malignancy cells.
- *Treatment to heat cancer cells (radiofrequency ablation).* During radiofrequency ablation, a special probe is inserted through the skin and into the kidney tumor utilizing ultrasound or other imaging to guide placement of the probe. An electrical current is run through the needle and into the cancer cells, causing the cells to heat up or burn [37, 38].

4 Epidemiology

Kidney cancer is the thirteenth most normal disease around the world, representing 2.4% among all cancers, with in excess of 330,000 new cases analyzed yearly. It positions higher in Europe, North America, Australia/New Zealand, and Japan, where it is on the seventh most basic cancer. With the end goal of likeness across nations and after some time, unmistakably the study of disease transmission information bases frequently bunch renal cell carcinomas along with other upper urinary tract cancers [39, 40].

4.1 Genetic and molecular epidemiology

4.1.1 Inherited susceptibility

Although most renal cell carcinomas are sporadic, several genetic diseases are associated with RCC, including von Hippel-Lindau syndrome, hereditary papillary renal carcinoma, tuberous sclerosis, Birth Hogg-Dubé syndrome and hereditary leiomyoma renal cell carcinoma. The genes underlying each of these conditions have been cloned and germline mutations in affected patients have been identified. Strong correlations have been found between some of these genes involved in the pathogenesis of renal tumors and the histopathological and clinical behavioral features *[41–43]*.

Few analytical epidemiologic studies have reported on a family history of RCC. One investigation found a 1.6-crease expanded danger for RCC [42] and another a 2.5-overlap hazard [43] if a first-degree relative was influenced with the disease, though no affiliation was found in a later report [44]. One ongoing library study found a 1.6-crease increment for RCC in posterity by parental disease and a 4.9-overlay increment for RCC by sibling's cancer [45].

4.1.2 Age and sex

Frequency paces of kidney cancer growth increases consistently with age, with a pinnacle of occurrence at roughly age 75 years. Around the world, roughly one portion of all cases are analyzed before age 65 years. The rate of kidney cancer is in men than ladies. The soundness of the sex proportion shows that biologic contrasts among people, as opposed to way of life contrasts, for example, tobacco smoking, are probably going to represent a great part of the frequency inconsistencies.

4.1.3 Geography and ethnicity

There is enormous variety in incidence rates far and wide, with the most noteworthy rates at the nation level found in the Czech Republic (age-normalized rate, 21.9/100,000 in guys) and Lithuania (age-normalized rate, 18.7/100,000 in guys). Occurrence rates are under 2/100,000 in generally safe nations, for example, Thailand, and African nations.

In the United States, rates are higher in black men (age-normalized rate, 15.6/100,000) than in white men (age-normalized rate, 14.0/100,000). Hispanics and non-Hispanics show comparable rates. Native Americans in the United States have transitional rates (age-normalized rate, 10.9/100,000 in guys), though Asians in the United States have low occurrence rates (age normalized rate, 6.4/100,000 in guys). In Europe, enormous intra nation provincial varieties have been portrayed in certain nations, strikingly in Germany (higher occurrence rates in the eastern locales of the nation) and Italy (higher rate rates in the north) [46].

4.2 Lifestyle risk factors

4.2.1 Tobacco smoking

Tobacco smoking has been named cancer causing for the kidney by the International Agency for Research on Cancer and the United States Department of Health and Human Services. The impact size on kidney cancer is unassuming, with an inexact 30% expanded danger in current smokers and a 15% expanded danger in previous smokers contrasted and never smokers. Epidemiologic proof for a causal function for tobacco smoking incorporates a dose-response connection among hazard and the amount of tobacco smoked every day, just as diminished danger with expanding long stretches of smoking suspension. In developed nations, it is assessed that 6% of kidney cancer deaths are a consequence of tobacco smoking [47].

4.2.2 Excess body weight

The relationship between abundance body weight and danger of kidney malignancy has been broadly detailed in enormous, forthcoming associates. Excess body weight has been overwhelmingly evaluated through a height of weight list (BMI; in kilograms per square meter). According to the studies on BMI class (18.5 # 25 kg/m^2), overweight (BMI, 25 # 30 kg/m^2) and fat (BMI, 30 kg/m^2) people have an expected 28% and 77% expanded danger, respectively. The affiliation was shown to be direct in a few investigations, with a 4% expansion in hazard for every 1-kg/m^2 increase in BMI or a 25% expansion in hazard for every 5-kg/m^2 increase. An imminent investigation of male young people with BMI estimated at age 16–19 years revealed that excessive body weight prior in life is related to the higher risk of renal carcinoma. High BMI is assessed to be liable for 26% of occurrence of renal carcinoma cases around the world [48].

Obesity has been evaluated utilizing waist circumference and waist-to-hip proportion. Results obtained from different studies were predictable, showing a critical increment in hazard with expanding waist circumference and waist-to-hip proportion. Weight cycling during adulthood has not been shown to altogether adjust kid hazard in the wake of representing standard BMI. There is additionally no information accessible on the advantage of weight reduction or potentially long-haul adjustment of lower BMI in relationship with the danger of kidney disease. Systems included have not yet been illustrated, and research is progressing on the part of incendiary status, sex and development hormone levels, metabolic state (insulin resistance and insulin-like development factor-1 levels), and adipokine levels. It will be imperative to comprehend the instruments in play, especially with regard to the stoutness oddity wonder, whereby overweight and hefty patients were shown in different arrangement to have a prognostic bit of leeway contrasted and patients with typical BMI [49].

4.2.3 Liquor consumption

A few metaexaminations and enormous planned partner consideration were directed on the relationship between liquor utilization and danger of kidney malignancy. All investigations revealed diminished danger in consumers contrasted with nondrinkers or light consumers. Consumers commonly have a 20% decrease in hazard contrasted with nondrinkers and light consumers. As there is no relationship between nonalcoholic drink admission, nor complete liquid admission and kidney malignant growth hazard, and on the grounds that the impact of liquor utilization does not vary by sort of mixed refreshment, ethanol presentation probably has an unthinking job. Hyperinsulinemia and insulin obstruction as a rule have been related to kidney malignancy hazard. Liquor utilization has been shown to expand insulin affectability and could be related to a decrease in kidney disease hazard through this backhanded course [50].

4.2.4 Physical activity

Physical action has been related to an unobtrusive decrease in danger in a huge metainvestigation of planned partners.

While considering a wide range of physical exercises together, the most elevated action classification had a 13% diminished danger contrasted and the least classification. Essentially, a huge, pooled investigation of accomplice contemplates detailed a decreased danger related to more significant levels of relaxation time physical movement. Inactive conduct as estimated by time spent sitting does not appear to expand the danger of kidney malignancy. How physical movement may impact the danger of kidney malignancy is indistinct, and physical action as a danger factor that is autonomous of abundance body weight and hypertension has not been illustrated. In its 2015 report, the World Cancer Research Fund and American Institute for Cancer Research inferred that there was restricted to no proof of a connection between physical action and kidney malignancy hazard [49].

4.2.5 Diet

Convincing epidemiologic proof on diet and kidney malignant growth hazard is inadequate in the literature. Results on products of the soil admission from imminent companions showed generally invalid or nonsignificant affiliations, or an unassuming decrease in danger in the most noteworthy leafy foods consumption classes. Information on supplement explicit affiliations is inadequate. Besides, examiners that explored the admission of fat and protein in relationship with the danger of kidney malignancy generally detailed invalid affiliations [51].

4.2.6 Regenerative factors and hormones

There is some proof that specific hormone-related variables are related to the danger of RCC. Not many logical epidemiologic investigations of RCC have zeroed in on regenerative variables or exogenous hormones and those which have, for the most part discovered little proof that components like equality and hysterectomy/oophorectomy are of significance. Nonetheless, positive affiliations, regardless of whether feeble, have been accounted for once in a while for oral contraceptives and for utilization of substitution estrogens. Interestingly, a multicenter case-control study found an essentially diminished danger of RCC following oral prophylactic use among ladies who did not smoke, with a proposal of expanded decrease with span of utilization. Moreover, a few investigations have discovered an expanded danger related to number of births. The discoveries from the generally scarcely any examinations managing conceptive and hormonal components stay confounding and conflicting [44].

4.2.7 Ionizing radiation

Different sorts of ionizing radiation have been related to an abundance danger of kidney malignancy. Ladies with cervical malignant growth, treated with radiotherapy, encountered a little however altogether expanded danger as did men treated for testicular disease. Among patients with ankylosing spondylitis who had gotten X-beam therapy the mortality from kidney malignancy was essentially expanded. Thorotrast, a α-radiating differentiation medium, has been connected to kidney disease in patients who had gone through retrograde pyelography with thorotrast; be that as it may, the malignancies are generally urothelial malignancy of the renal pelvis. Just one case-control study found a noteworthy positive relationship among ladies between any lifetime radiation treatment got and RCC [52].

4.2.8 Occupation

Aside from asbestos, no word-related presentation or occupation has been reliably connected with RCC, and even asbestos is probably not going to be answerable for any significant increment in RCC hazard. Renal cell carcinoma, as opposed to bladder disease, is commonly not considered an occupation-related malignant growth. Nonetheless, affiliations have been accounted for with introduction or business related to presentation with asbestos in a few investigations, with fuel and other oil-based goods in a few; however, not all examinations, and further with hydrocarbons, lead, cadmium, and work or introduction identified with cleaning and clothing. Arsenic has been accounted for to be related to expanded danger for RCC in zones with interminable introduction from drinking water.

Worldwide Agency for Research on Cancer has considered both trichloroethylene utilized chiefly in metal degreasing and tetrachloroethylene utilized in cleaning as cancer-causing in creatures, and most likely additionally to people. In any case, basic audits of both accomplice and case-control examiners presumed that there is no persuading proof that these solvents represent a danger in people; discoveries from companion contemplates contend against a causal relationship. In any case, another ongoing audit has an alternate view, and finds the proof supporting a relationship among trichloroethylene and RCC more grounded than for some other malignant growth site. There is additionally proof of a relationship between frequency of RCC among laborers presented to degreasing operators and solvents and to those in both iron and steel and cleaning and clothing work ventures, although direct causality cannot be surveyed. RCC patients with high, combined presentation of trichloroethylene have been shown to have more regular substantial VHL changes, however this discovering needs further affirmation [53].

4.3 Clinical history

4.3.1 Hypertension

Hypertension inclines kidney to malignant growth. In the United States, a background marked by hypertension has been assessed to twofold the danger of kidney malignancy in white patients and triple the danger in dark patients. Planned partner concentrates reliably report portion reaction relationship between pulse at pattern and kidney disease hazard, in any event, while confining the danger examination to over 5 years after circulatory strain estimation when turn around causation is more uncertain.

Antihypertensive drug has additionally been related to an expanded danger of kidney malignant growth, yet it is hard to unravel the impact of the condition from the impact of treatment. In an investigation with rehashed proportions of circulatory strain after some time, a diminished danger was seen with the decrease of pulse. This demonstrates the hypertensive condition, as opposed to the treatment, is bound to be the risk factor. Moreover, controlling the condition using hypertensive drug might be a compelling remedial intercession in the counteraction of kidney disease. Hypertension likewise is by all accounts organically free from corpulence in expanding the danger of kidney disease, with a total impact among people who present with the two conditions [54].

5 Management of kidney cancer

The paradigms for the treatment of kidney cancer exist in a constantly evolving state owing to the fact that each year different delivery methods and active agents get researched. Kidney cancer alone accounts for around 2.4% cases among the vast pool of distinctive cancers, and this exerts a considerable financial burden over the healthcare system worldwide. Thus, it is important that the advancements in cancer therapy should focus on both present and future priorities. This also included developing management options which have well-established clinical benefits, significant molecular biomarkers and better understanding of therapies toward which patients may develop resistance. Advancements in any field of cancer management marks a new level of improving the affected populations' quality of life [3].

5.1 Conventional approaches for the treatment of kidney cancer

The conventional management of kidney cancer relies on various factors like stages of cancer, basic side effects, health of patient and his/her preferences. The objective of any kind of treatment option should be well analyzed by both the doctor and the patient seeking it. The shared decision making helps in utilizing the best option among the variety of treatment options available. The main intent of therapy in kidney cancer is to increase or improve patient's lifespan and prohibit the spreading of tumor to the other parts of the body.

Common treatments used for kidney cancer are:

1. Surgical procedure
Regardless of the development of fundamental treatments like immunotherapy or targeted drug delivery systems, surgery still stays as the main pillar of treatment for patients with existing metastatic renal cell carcinoma. Surgery refers to the excision of tumor or the tissues encompassing it. In the event that the disease has not spread past the kidneys, medical procedure to remove part or the entire kidney and lymph nodes might be the main therapy required. One of the many curative options is blocking the renal vessels especially the artery, and further ejecting the tumor mass adjoining the peritoneum, lymphatic drainage fluid, perinephric fat (situated between the kidney and the Gerota fascia). The blockade is created to obstruct the tumor from spreading into the bloodstream. The different types of surgery options opted for kidney cancer mainly depend upon the stage of malignancy.

i. Radical nephrectomy
During radical nephrectomy the surgical procedures involves the basic removal of the tumor, borderline tissues or the entire kidney. The procedure may also be extended to remove the affected lymph node through radical dissection. A major subtype of radical nephrectomy is adrenalectomy in which the tumor invaded adrenal gland(s) and blood vessels are also excised. This type of nephrectomy is used frequently in case of large tumors or when only a small fraction of healthy or normal cells is remaining. In severe cases, when the renal tumor progresses toward the heart through vena cava, complex techniques are maneuvered to eradicate the manifestations [55].

ii. Partial nephrectomy
While performing partial nephrectomy only the tumor or infected tissue is taken out (focused removal), and the healthy tissues are spared. This procedure focuses on saving the healthy parts as much as possible; hence, it is also referred to as "kidney-sparing" medical procedure. This procedure is appreciated for its kidney preservation aspects and for the fact that it significantly reduces chronic morbidities. It shows effective removal of abnormal cells in case of T1 tumors, and the additional side effects can be depleted by using smaller surgical incisions. Some of the factors which may trifle the glomerular filtration rate after partial nephrectomy are old age, presence of solitary kidney, size of tumor, longer ischemic intervals, and lower preoperative glomerular filtration rates [56].

iii. Laparoscopic and robotic surgery (minimally invasive surgery)
Laparoscopy has risen as a favored choice for the careful removal of kidney cancers. Numerous reports have been distributed in regard to the operative result of renal cell carcinoma (RCC) and upper-tract transitional cell carcinoma (TCCA) treated laparoscopically. During this procedure, surgeons make few smaller cuts instead of the one bigger cut opted during a conventional surgery. Surgeons proceed by inserting telescoping equipment into these little keyhole cuts to totally eliminate the tumor masses or to operate through partial nephrectomy. This medical procedure

may take longer; however, might be less difficult. Robotic instruments are also equipped in some of the cases by surgeons, but every aspect of these procedures should be clearly discussed with the patient. It is essential to examine the likely advantages and risks/dangers of these kinds of medical procedure with the surgical team [57].

2. Therapies using medication
While using systemic therapy the cancer cells are interdicted by administering the drug directly into the bloodstream through oral consumption or through direct delivery into the blood with the help of intravenous (IV) tube placed into a vein using a needle. Systemic therapies can be prescribed singularly or in combination with radiation therapy and/or surgical operations. The drugs used to treat renal carcinomas are constantly being assessed. As scientists get familiar with the evolution in cells that cause malignancies, they focus toward creating options which can be targeted toward the cancerous cells only without creating any damage to the normal healthy cells. As compared with chemotherapy, targeted therapy has now become more efficient tool for the treatment of kidney cancer. Apart from this, immunotherapy is also utilized to enhance body's natural defenses to fight cancer. Chemotherapy, immunotherapy, and targeted therapy are further described below.

i. Chemotherapy
Chemotherapy involves the prevention of cancer cells from growing, dividing and producing more cells, by destroying the cancer cells, through the usage of drugs. The schedule of chemotherapy involves a specific number of cycles that is given over a set period of time. Under this, a patient may be given one drug at a time or at the same time a combination of various drugs.
Although proven beneficial for curing different kinds of cancer, it still lags behind in most cases of kidney cancer. The researchers are continuously studying new drugs or the combinations of different existing active agents to enhance their chemotherapeutic effects. A combination of gemcitabine (Gemzar) with capecitabine (Xeloda) or fluorouracil (5-FU) has helped few patients to temporarily reduce the size of tumor.
The chemotherapy has been proven to be more successful for the treatment of transitional cell carcinoma, also called urothelial carcinoma, and Wilms tumor. It does involve many side-effects like fatigue, risk of infection, nausea and vomiting, hair loss, loss of appetite, and diarrhea, but this depends on the patient and the dosage involved. However, these are reversed after the treatment is finished [58, 59].

ii. Immunotherapy
Also known as biologic therapy, immunotherapy is created to boost the natural defenses of the body to fight cancer. It makes use of materials made by the body or in a laboratory which targets, restores or improves the immune system function.

- Interleukin-2 (IL-2, Proleukin): This is a kind of immunotherapy which is utilized for the treatment of later stage of kidney cancer. It is a protein that is generated by white blood cells known as cytokine. IL-2 has its importance in immune system functioning and also includes the devastation of tumor cells.
  Various severe side-effects are caused due to high dosage of IL-2 like kidney damage, bleeding, heart attack, chills and fever, low blood pressure, excess fluid in the lungs, etc. A 10-day treatment is required for the patients in such cases. On the other hand, low doses of IL-2 have lower side-effects but are not very effective. However, high dosage IL-2 can help for treating small percentage of people with metastatic kidney cancer. The centers which have the expertise in dealing with high dosage of IL-2 treatment for kidney cancer should have the authority to recommend IL-2.
- Alpha-interferon: It is useful in treatment of kidney cancer that has already spread. It functions by changing the proteins on the surface of cancer cells and slows down the growth of cancer cells. Alpha-interferon has not proved to be as effective as IL-2, but on the other hand it has showed to increase the lives when it is compared with the other treatment using megestrol acetate (Megace).
- Immune checkpoint inhibitors: It is a kind of immunotherapy which is being tested in kidney cancer. A combination of two immune checkpoint inhibitors, i.e., nivolumab (Opdivo) and ipilimumab (Yervoy), has been approved by the FDA for the treatment of specific patients who suffer from advanced renal cell carcinoma that has not been treated before. Previously the research indicated that nivolumab when injected through the vein for every 2 weeks proved helpful for certain people who had received treatment previously and have longer life than the patients treated with the targeted therapy. Also, the combinations of checkpoint inhibitor that includes the pembrolizumab (Keytruda) or avelumab (Bavencio), plus a targeted therapy, axitinib has been approved by the FDA, for the people suffering from advanced renal cell carcinoma. These types of drugs are being investigated by many clinical trials for curing of kidney cancer.
  There are various side-effects involved with various kinds of immunotherapy. The most common ones are diarrhea, flu-like symptoms, weight changes and skin reactions [60].

iii. Targeted therapy
With the relative evolution in the cells that cause cancer, the researchers have developed drugs which target some of the most relevant changes. The targeted drugs that have been developed so far are distinct from the standard chemotherapy drugs. Even though they have different kinds of side effects, these drugs work efficiently in

most cases, especially when compared with the standard chemo drugs. In case of kidney cancer, where the chemotherapy could not leave its mark, these targeted drugs are proving to be very effective.

Under targeted therapy, the treatment targets the genes and proteins relating to cancer or the tissue environment that facilitates the growth and survival of cancer. Along with limiting the damage to the healthy cells, this treatment also blocks the growth and spread of cancer cells. The studies by the researchers aim at finding more about specific molecular targets and the new treatments which are directed toward them as all tumors do not have the same targets [61].

Targeted therapy for kidney cancer includes:

- *Antiangiogenesis therapy:* The major focus of this treatment is to prevent angiogenesis, i.e., the process of making new blood vessels. In most cases, the kidney cancer cells are mutated by the *VHL* gene which causes the cancer cells to promote the production of certain proteins called vascular endothelial growth factors (VEGF). This protein is found responsible for the formation of new blood vessels and it can be stopped with the help of specific drugs. The aim of antiangiogenesis therapies is to block the vessels to control the tumor by starving it because the nutrients are delivered to the tumor through the blood vessels. To block the VEGF, there are two methods that can be used. These are: utilizing the small molecule inhibitors of the VEGF receptors i.e. VEGFR or by equipping certain antibodies against these receptors. For people who have metastatic renal cell carcinoma, it has been proven that an antibody called bevacizumab (Avastin) can slow growth of tumor and when combined with interferon, it slows the growth and spread of tumor. To treat the metastatic kidney cancer, the U.S. Food and Drug Administration (FDA) have approved two drugs called bevacizumab-awwb (Mvasi) and bevacizumab-bvzr (Zirabev) which are alike hence they are called biosimilars. They are called so due to similarities in their function to the original bevacizumab antibody.
- There are other methods to block VEGF such as with tyrosine kinase inhibitors (TKIs). Some of these TKIs that can be useful to cure clear cell kidney cancers are Axitinib (Inlyta), cabozantinib (Cabometyx, Cometriq), pazopanib (Votrient), sorafenib (Nexavar), and sunitinib (Sutent). These TKIs also have some common side effects such as high blood pressure, diarrhea and tenderness and sensitivity in the hands and feet [62].
- *mTOR inhibitors:* Drugs like everolimus (Afinitor) and temsirolimus (Torisel) target mTOR i.e., protein which help in growth of kidney cancer cells. The studies indicate that these drugs slow down such kind of growth [63].
- *Combined therapies:* For primary treatment of advanced renal cell carcinoma, the FDA, in the year 2019, gave approval of two combination treatments. Under the first combination, there are axitinib and pembrolizumab (Keytruda), which act as immune checkpoint inhibitor. The second combination includes axitinib and avelumab (Bavencio) that is also an immune checkpoint inhibitor. The PD-1 pathway is targeted by pembrolizumab and avelumab which help to activate the immune system to attack cancer cells. Axitinib is an antiangiogenesis therapy. The treatment combination works whether or not the tumor expresses the PD-L1 protein, therefore patients who go through this treatment are not required to be tested for PD-L1 [61].

3. Radiation therapy

   Under this, to destroy the cancer cells, there is usage of high-energy X-rays or other particles. In case of kidney cancer, it is not considered as a primary treatment.

   Radiation therapy is opted only in cases where the patient cannot cope up with the operative procedures. This is because radiations have the drawback of affecting the healthy kidney while treating the damaged one. This rare therapy is only used for the areas where the cancer has already spread and not on the primary kidney tumor. Hence, it is prescribed when the cancer has already spread. Radiation therapy helps to reduce the symptoms like bone pain or swelling in the brain [64].

5.2 Novel treatment

Various drug delivery methods can target both cancer and healthy cells creates the need of targeted active and passive drug delivery system. The physicochemical properties of the nanoparticles are altered so that it can be used as both targeted and controlled drug release system. Many antibodies, macromolecules can be targeted through nanoparticles to the site of action thereby avoiding side effects and harming the other healthy cells. Encapsulation can, though, enhance the property of suspension by protecting the drug from degradation [65].

Pharmacokinetics is poor in many therapeutic molecules which are used for kidney though many small molecules are filtered from the kidney and also cleared by hepatic clearance before they have shown their therapeutic effect which gives less exposure to kidney nanoparticles; and thus it has to be biocompatible, should have a specific pH and should protect the premature delivery of drug, controlled drug interaction with cells, and extended circulation of drug in the body and should be easily functioned and disposable [66].

Treatments focused against vascular endothelial growth factor (VEGF) changed the course of treatment of metastatic renal cell carcinoma (mRCC) years ago. Recently, the immune checkpoint inhibitors have changed treatment designs for a subsequent time. Complete reduction is turning into a practical objective, rising up to 10% of patients with invulnerable checkpoint inhibitors. 60 months survival is overall observed as median. To accomplish these objectives, new treatment strategies and novel combination treatment will be required, expanding on the establishment of antiangiogenic or potentially immunotherapeutic agents. Various new agents have been grown, some of which have empowering stage II data in RCC as monotherapy or in mix with different specialists. Choosing patients for explicit treatments additionally requires examination. Customized treatment has been deficient in RCC. More strongly randomized stage III information shows that customized passing ligand 1 (PDL1) can be utilized to choose patients for treatment (atezolizumab and bevacizumab). The use of rising operators is changing the methodology and outlook of doctors and patients. Moreover, planning preliminaries in this setting has gotten more earnestly on the grounds that norm of care is quickly developing and set up end focuses are progressively unsure. Ongoing information recommends that general endurance (OS) will be the favored as essential endpoint with safe treatment studies, and movement free endurance (PFS), which has been the administrative end point for focused treatment, may not be ideal. Thus, the distinguishing proof of the following ages of treatments/blends will require serious examination before any choice to go to stage III and will be related to critical danger. Despite the fact that there are countless expected focuses, this chapter centers around a modest number that are generally applicable to RCC. It additionally centers around metastatic clear cell RCC histology, where most of improvements are happening. Advancement in nonclear cell RCC will be secured inside a particular segment of this extraordinary version [67].

Various drug delivery methods can target both cancer and healthy cells, creating the need for targeting active and passive drug delivery system. The physicochemical properties of the nanoparticles are altered so that it can be used as both targeted and controlled drug release system. Many antibodies, macromolecules can be targeted through nanoparticles to the site of action hence avoiding side effects and harming of the other cells. Encapsulation can, though, enhance the property of suspension by protecting the drug from degradation [68].

Pharmacokinetics is poor in many therapeutic molecules which are used for kidney though many small molecules are filtered from the kidney and also cleared by hepatic clearance before they have shown their therapeutic effect which gives less exposure to kidney nanoparticles and thus it has to be biocompatible, should have a specific pH and should protect against the premature delivery of drug, ensure controlled drug interaction with cells, and facilitate extended circulation of drug in the body, and also should be easily functional and disposable [69].

Pharmacokinetic profiling of nanoparticles has been found as unique and does include renal filtration having high surface area nanoparticles has high surface-to-volume ratio, which can be modified by various functional groups so as to stabilize and internalize the therapeutic agents. Drug distribution is more in the targeted tissues than in the healthy cells. In addition to the properties nanoparticles have unique optical, electromagnetic and thermal properties such as in gold, silver, SPIONs which are being used for the treatment of cancer through photothermal killing of the tumor cells [70].

Multifactorial diseases like renal cancer can need multitargeted inhibitors for development of resistance. Both single drug and multidrug inhibitors are used but at present, there is no cure for metastatic RCC and hence the use of nanoparticles is limited. Nanoparticles have potential for early diagnostic devices like imaging, sensing and remote actuation that can be used for dual advantage therapy [71].

Nanoparticles are novel systems having size in nanometer range and were initially used in scanning probe microscopy molecular structure discovery. Various methods are involved in the production of different sizes and shapes of microsponges, the size of nanoparticles at which it enables the living tissue functioning is 10^{-9}.

Nanoparticles have various applications in medicine field, it can accumulate in the tissue and also has retention effect under certain circumstances specially for tumors as they have endothelial lining of the cell. Nanoparticles can be used for the encapsulation of poorly soluble drugs which modifies drug circulation, its permeability, tissue distribution and facilitation.

Many nanoparticles like liposomes, polymer microspheres, carriers, polymer drug conjugates, micelles and ligand targeted systems are under various stages of drug development systems. Moreover, nanoparticles play an important part in cancer diagnosis and treatment of the tumors and are a potentially effective for management of cancer.

Nanotechnology has shown more effective response over traditional methods, different types of nanoparticles are used for the solubility of insoluble drug and they have specialty toward prostate, bladder, and renal cancers. By combining with other new technologies and enhancing their effectiveness, the nanoparticles also show various effect toward urological cancer also. So, in the above discussion, we can conclude that there are many applications of nanoparticles and there are various NP-based strategies which will be implemented in future [72].

5.3 Nanotechnology in renal cancer

Renal cell carcinoma (RCC) is certifiably not a solitary element, yet incorporates a gathering of tumors with an exceptionally heterogeneous epithelium starting from the renal tubules. There are three prevailing histopathological sorts of RCC: clear cell (65%), papillary (15%–20%), and chromophobe (5%). RCC is the third most common urinary tract disease, with an occurrence pace of around 5–10 cases for every 100,000 individuals, and records for 2%–3% of every harmful tumor. Despite the fact that chemotherapy is one of the principal methods of malignant growth treatment, its adequacy is restricted by drug resistance. For instance, sunitinib is at present the standard first-line treatment for cutting edge RCC, and patients definitely become impervious to sunitinib treatment. Likewise, the clinical results in patients with RCC stay poor. Hence, fast advancement of nanotechnologies may improve remedial techniques for the finding and treatment of RCC [73].

5.3.1 Imaging in renal cancer

Pinpoint imaging of RCC is one of the fundamental issues for the conclusion, arranging, and clinical treatment of patients. In current oncology, a progression of NPs for demonstrative examination have been created, and these NPs are generally utilized as differentiation specialists for MRI and CT imaging. An in vitro MRI investigation of ccRCC and control cells affirmed that the manufactured mAb G250-SPIO nanoprobe could be utilized as a particular marker for ccRCC cells. Additionally, the nanoprobe could be effortlessly formed with other antitumor medications and imaging materials to accomplish multimodal imaging and symptomatic treatment combination. For instance, the blend of Cy5-labeled glycosylated bleomycin and the nanoprobe may give another heading to the plan of novel tumor imaging and helpful specialists. Also, GNPs or quantum specks as imaging materials could likewise be created with the mAb G250 nanoprobe to advance exceptionally multiplexed equal recoloring, coming about best imaging. Numerous investigations have exhibited that the conclusion of RCC can be performed by diverse NP imaging. Be that as it may, urologists are frequently confronted with imaging difficulties, remembering absence of conclusive responses for clinical work on during the finding of RCC histology [74].

5.3.2 Need of multitarget inhibitors (MTIs)

In multifactorial ailments complexities that emerge are advancement of obstruction and harmfulness due to nonexplicitness of medication conveyance. Renal malignancy being multifactorial illness needs MTIs utilizing either single-drug inhibitor (SDIs) or multidrug inhibitors (MDI). SDIs, like, kinase inhibitor, at the same time influence various pathways, though MDIs hinder numerous pathways in light of the fact that numerous combinatorial operators are utilized that demonstration synergistically. Medications for RCC that are in preclinical or clinical stage are sorafenib, sunitinib, pazopanib, and cabozantinib. These medications restrain vascular endothelial growth factor receptor; platelet-derived growth factor receptor and mitogen-activated protein pathway. MTIs restrain various flagging pathways in this way and can possibly cause fundamental harmfulness. Utilization of nanoparticles for focused conveyance of MTI guarantees its gathering in the tumor vasculature and decreases the foundational side effects. MDI utilizes synergistic medication mixes, which upgrades the adequacy. Berenbaum had proposed the utilization of blend operators as an answer for malignant growth treatment. Utilizing nanotechnology, the proportion of various specialists can be custom-made to accomplish ideal for various targets [75] (Fig. 6).

5.3.3 Appropriateness of nanoparticles for targeted drug delivery

Utilization of nanotechnology for conveying medications to renal disease is still in its earliest stages. Kidneys channel blood through a structure known as the glomerular hairlike divider. Nanoparticles <10 nm breadth faces first-pass renal filtration, and consequently, size of nanoparticles is very important. Dreher et al. have indicated that particles >100 nm distance across can collect in the tumor tissue, while being latently focused on, which is totally reliant on diffusion-mediated transport into the tumor. Stable nanoparticles that are available for use for longer period bring about improved medication take-up by tumor and by staying away from reticuloendothelial framework, it limits the poisonousness [76].

5.3.4 Nanoparticles

Corsi et al. surveyed the biological impact of cerium oxide NPs. In spite of cerium oxide NPs (CNPs) being found to exert strong anticancer activities, which is a characteristic generally attributed to CNPs redox activity, other studies, however, reported nonredox mechanisms. The authors recently demonstrated that the radio-sensitizing effect of CNPs on human keratinocytes is independent from the redox switch. Mechanisms involving particle dissolution with release of toxic Ce4+ atoms, or differential inhibition of the catalase vs superoxide dismutase mimetic activity with accumulation of H2O2 have been proposed, explaining such intriguing findings only partially.

Jia et al. summarized the recent progress in targeting DCs using NPs as a drug delivery carrier in cancer immunotherapy, the recognition of NPs by DCs, and the ways the physicochemical properties of NPs affect functions of DCs. The molecular pathways in DCs that are affected by NPs were also discussed [76].

5.3.5 Nanospheres

Nanospheres can self-assemble, are biodegradable, their matrix is made of bovine serum albumin or other

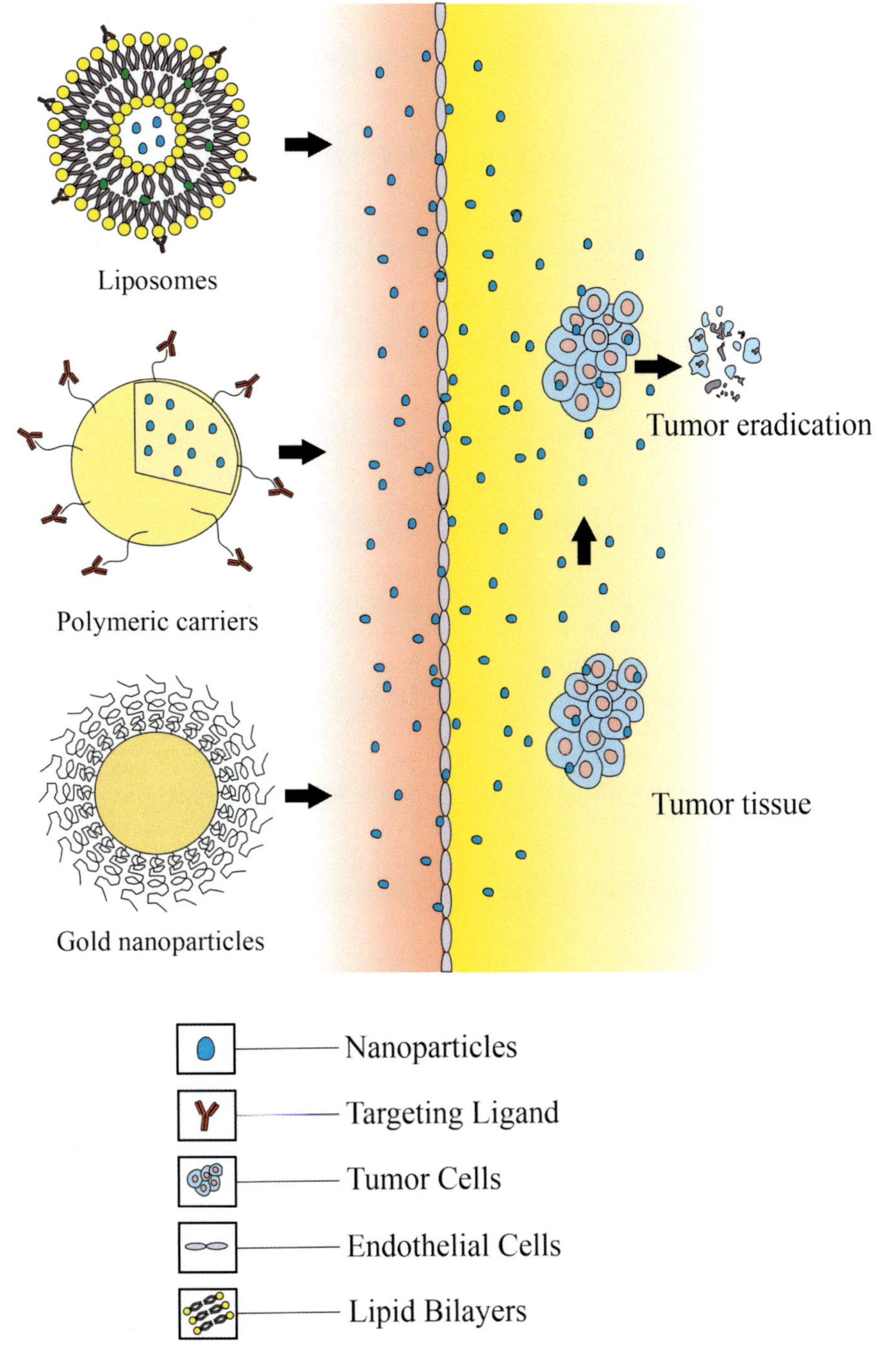

FIG. 6 Application of NPs in urological cancer diagnosis and therapeutics.

amphiphilic copolymers, in which drug molecules are uniformly distributed. Nanospheres, when used for delivering antiangiostatin and endostatins (anti-VEGF), were found to be long circulating. Composition of the polymer matrix and its ability to imbibe fluids determine how rapidly the drug will be released. These properties indicate that if tried, nanospheres may be suitable for delivering drugs to renal cancer tumor [77].

5.3.6 Nanocapsules

Nanocapsules are composed of polymeric membranes and core of oil, which encompasses drugs that can diffuse out under appropriate conditions, by responding to environmental, chemical, thermal, or biological triggers. Nanocapsules are colloidal and synthesized by interfacial deposition of preformed polymers (PLA, PLGA, PCL, and PEG). It is suitable for the delivery of hydrophobic drugs. Lipid nanocapsules have been used for multidrug resistant cases in rat tumor. Though not much work has been done on drug delivery to RCC using nanocapsules, Hureaux et al. have suggested a novel hybrid protein-lipid polymer nanocapsule of 180 nm as nontoxic drug for codelivery of transcription factor p53 and lipophilic drug paclitaxel to induce HeLa cell apoptosis; this gives incentive to try using nanocapsules for delivering drugs to RCC [78] (Fig. 7).

5.3.7 Micelles

They are amphiphilic block of copolymers that are biocompatible, biodegradable, and capable of self-assembly in aqueous solution. They are aggregate of molecules in a colloidal solution, that can be anionic (sodium dodecyl sulfate), cationic (cetylpyridinium bromide), zwitterionic (lecithin) or nonionic (polyoxyethylene lauryl ether). There is reverse and bilayer micelle (Fig. 1C). Drugs entrapped in micelles have shown increased vascular permeability and impaired lymphatic drainage, causing EPR effect, suitable for passive targeting and can be used for controlled drug release, thus showing a potential for its use in RCC therapy. Suitability of micelle for MTI drug delivery; for concurrent delivery of two or more MTI modalities; as stimulus-responsive targeted; easy tailoring to optimize delivery to solid tumors and delivery of poorly soluble drugs have been observed [79].

5.3.8 Nanoliposomes

They are spherical, concentric bilayered vesicles having phospholipid membrane. The amphiphilic nature of liposomes, their ease of surface modification, and a good biocompatibility make them an appealing solution for increasing the circulating half-life of proteins and peptides. The hydrophilic compounds remain encapsulated in the aqueous interior and hydrophobic compounds can escape encapsulation by diffusion through phospholipids membrane. Nanoliposomes can be designed to adhere to cellular membranes to transfer drugs following endocytosis. 100–200 nm nanoliposomes can rapidly enter tumor through vasculature. In nanoliposomes at a time many different molecules can be attached to the surface. PEG linked nanoliposomes are highly stable and have been used to deliver drugs to brain and colorectal cancer but not for renal cancer. Since renal cancer has multiple signaling pathway, many nanoliposomal formulations containing therapeutic agents, antisense oligodeoxynucleotides, siRNA, DNA or radioactive particles that can target multiple signaling pathway are being used for cancer treatment and they can be of potential use for renal cancer also. Use of nanoliposomes to deliver iron-oxide nanoparticles for photothermal therapy of tumors and conjugation of nanoparticles of gold with liposomes for photothermal therapy and multimodel imaging have also generated hopes for its use in renal cancer treatment [80] (Fig. 8).

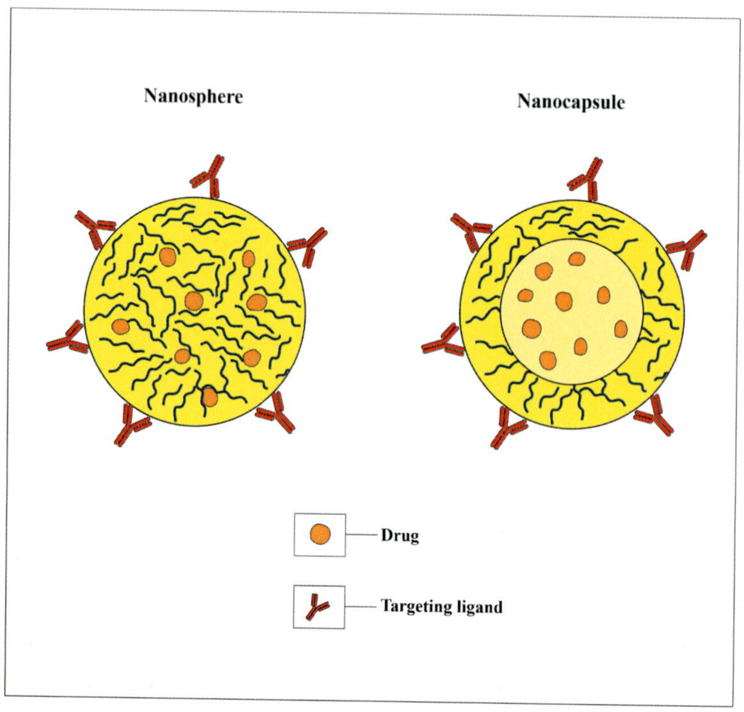

FIG. 7 Diagrammatic representation of nanosphere and nanocapsule.

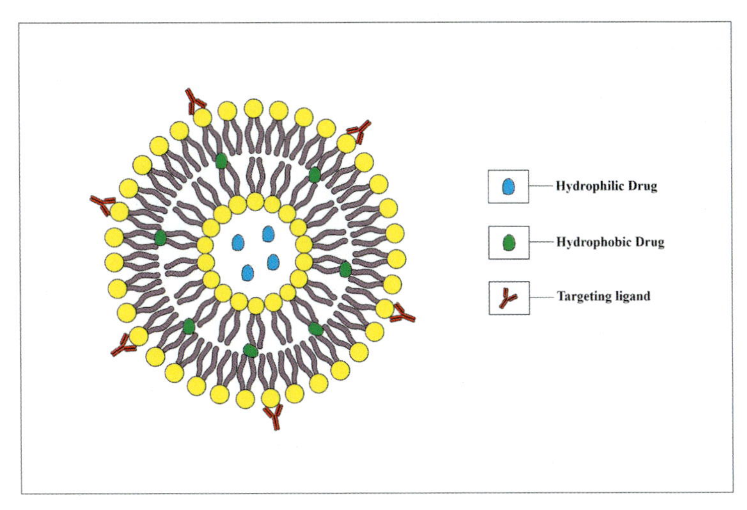

FIG. 8 Diagrammatic representation of liposome.

5.3.9 Application of liposome in RCC

Singh et al. [81] formulated monoclonal antibody-linked small unilamellar vesicles containing methotrexate [(MTX) SUVs) in cancer chemotherapy. (MTX)SUVs were prepared by probe sonication and were linked covalently to Dal K29 (an IgGl monoclonal antibody against human renal cancer), normal mouse IgG, or a nonspecific mouse myeloma IgGl. A colony inhibition assay showed that Dal K29-linked (MTX) SUVs were 5 and 40 times more potent than Dal K29-MTX and equimolar amounts of free untrapped MIX in inhibiting the growth of the target CaKi-1 cells but were not toxic to a human melanoma line (that did not react with Dal K29) [81].

Zhang et al. [82] formulated sertraline-loaded PLGA particles and HM-coated DPPC liposomes exhibited significantly improved cell kill compared with sorafenib alone at lower concentrations, namely 10–15 mM and 5–15 mM from 24 to 96 h, respectively. PLGA and HMC-coated liposomes are promising platforms for drug delivery of sorafenib. Likewise, due to different particle characteristics of PLGA and liposomes, each model can be further developed for unique clinical modalities and paired with ablative particles for synergistic treatments [82].

6 Herbal remedies

Kidney cancer or renal carcinoma is the disease characterized by the proliferation of the renal cells or the tubules which lead to malignancy in the kidney and account for 85% cases of malignant renal cell carcinoma. The number of registered cases and deaths due to renal carcinoma increases at the rate of 2%–4% globally [83]. The understanding and research in genomics and molecular biology have increased and have revealed several pathways of carcinoma progression. The US Food and Drug Administration (FDA) provides significant data in this respect from its approved agents. The drugs of choice have become first choice as they provide clinical benefit and help in remission of the disease. However, the signs and symptoms relapse after the treatment; therefore, further search for an effective therapy is always on the way to seek better treatment measures. The natural products or herbal remedies are also used as anticancer agents in the prevention and treatment of renal carcinoma alone or in combination with chemotherapy. The renal cell carcinoma (RCC) is the disease of the kidney tubules that results in various symptoms like fever, hypertension, weight loss, hypercalcemia, malaise, and night sweats. Although rare, but still RCC is among top ten cancers, commonly affecting the population of an average age of more than 45 years [84] with an average of more men over women with more than 60 years of age [85, 86].

6.1 Treatment of renal carcinoma

The current treatment of renal carcinoma includes various therapies like radiotherapy, metastasectomy, and modalities. Metastasectomy provides better survival in comparison with those who did not undergo metastasectomy. Also, total procedure of cure through metastasectomy is under conflict [87]. Last 30 years show little progress with no chemotherapeutic agents presently available that are totally effective against renal carcinoma [88]. The average life of survival of metastatic patients is 10 months, but only less than 2% have survival life greater than 5 years. Therefore, the search for treatment of renal malignancy to provide effective therapy is the need of the hour.

The last decade shows usage of natural or herbal remedies for the varied ailments at a growing pace [89]. Diabetes, infertility and other diseases are effectively treated through natural remedies [90, 91]. The research on alternative and complementary medicine gives greater emphasis on cancer management.

6.2 Natural products into renal carcinoma

Natural products are already being used for hundreds and thousands of years in the treatment of the disease due to their medicinal properties. A wide variety of Ayurveda and siddha traditional medicines are used in the treatment and cure of various diseases. The primary health of the 60%–70% rural population is still cared and cured by these traditional medicines. Therefore, anticancer drugs have been derived from natural products and utilized for the same. For example, taxanes and camptothecins show pioneer utilization in the development of medicaments with anticancer or anti solid tumor effect [92, 93]. As the main therapy for treatment of kidney carcinoma is surgery, it becomes necessary to research and examine a natural agent to treat renal carcinoma.

i. *Englerin:* The extract of Englerin, obtained from *Phyllanthus engleri*, (Fam. Euphorbiaceae), selectively inhibits growth of renal cancer cells in different cell lines [94].
ii. *Bullatacin and Bullatacinon:* Both the bioactive compounds are highly potent that are obtained from the bark of stem of *Annona bullata.* At low concentrations of 1 ppm and less, Bullatacin shows high cytotoxic action that indicates Bullatacin as promising drug candidate against renal cancer [95].
iii. *Lapachol:* Isolation of Lapachol is done using dichloromethane from the fruits of *Kigelia pinnata.* The compound has inhibitory effect on (Caki-2) renal cell carcinoma cell line. It shows dose-dependent activity [96].
iv. *Honey:* Honey is traditional medicine and is used as herbal remedy for many diseases since time immortal. Increase in cell apoptosis rate and decrease in viability of cell are shown at concentrations of 2.5%, 5%, 10% and 20% in the ACHN (Human Renal Cancer cell lines) [97].
v. *Coumarin:* The isolation of coumarin is done from *Dipteryx odorata,* (Fam. Fabiaceae). Its hierarchical order goes from Fabaceae > Benzopyrone > Flavonoids > Coumarin. Throns et al. in their work showed that coumarins on being given orally at 100 mg/day dosing for 15 days shows reduction in metastatic development of carcinoma among the patients along with zero side effects [98]. The in vitro and in vivo studies gave similar results on treating patients with 600–5000 mg of Coumarin [99].
vi. *Annonaceous Acetogenins:* These are isolated from *Annona bullata* bark using ethanolic extraction. These acetogenins have shown cytotoxic effect on human kidney tumor which is comparatively better than adriamycin [100].
vii. *Resveratrol:* This is a natural polyphenol found in grapes altogether with other 72 species and in abundance in red wine. Resveratrol has shown induction in apoptosis, inhibition of tumor growth, and suppression in metastasis and angiogenesis in a variety of malignant cancer and renal carcinoma. Induction of apoptosis and inhibition of cell growth are done by resveratrol that modulates gene expression [101].

6.3 Patented treatment for kidney stone and UTI

Patankar [102] developed patented synergistic formulation of herbal composition, that has therapeutic amount of Kadalikshar (15%–85%), Varunbhavit (15%–85%), Yavkshar (15%–85%), Apamargkshar (15%–85%), and Gomutrakshar (15%–85%) as an option with pharmaceutical additives. Accordingly, this formulation is useful for urinary and kidney-related disorders, such as UTI, renal calculi, and urinary stenting. The synergistic herbal-based formulation is made in the form of tablets, cachets, capsules, emulsion, suspension, powder, or paste that can be delivered through nasal, oral, topical, parenteral, or vaginal route.

The evidence from scientific research describes that the herbal compounds (banana, varun, yav, and aghada extract) used herein are helpful in reducing the size of renal calculi. This treatment is also helpful in reducing hematuria, causing pain reduction and shows antiinflammatory effect. The preferred formulation ingredients have been considerably emphasized upon. If added, changes in the dosage form should be apparent in regard to the patented formulation [102].

6.4 Traditional Chinese medicine for renal carcinoma

Chinese herbs have been used as traditional medicine for the thousands of years in the treatment of RCC that has advantages of lowering the toxicity and showing less adverse drug reactions. Combination therapy along with herbal components shows long-lasting therapeutic efficacy. Herbs have sensitizing effect on the body by reducing radio and chemotherapy, increasing normal tissue protection and prevent recurrence of tumor and metastasis. Hence, herbs have special advantages by showing improvement in life quality and increasing life expectancy of patients [103, 104].

Herbs used from the Chinese medicine are:

- *Oldenlandiadiffusa*
- *Curcuma longa*
- *Astragalus membranaceus*
- *Panax ginseng*

- *Ganoderma lucidum*
- *Angelica sinensis*
- *Panax notoginseng*
- *Scutellariabarbata*
- *Liquorice*
- *Radix Salvia miltiorrhiza*

Research on Chinese medicine shows that this medication can be used to change the gene expression of carcinoma cells. Further clinical trials are still required to prove the efficacy. The advantage of herbal treatment for cancer is that these herbal medications are easily available at cheaper prices than synthetic medicines. Effort is reduced to half by these Chinese herbs, thereby doubling the therapeutic effect. It is also seen that the traditional medicine causes hepatotoxicity due to overdose effects [105].

6.5 Recent *Berberine*-associated therapy

Lopes et al. [107] performed a study on Berberine (BBR), as an anticancer agent, which is an alkaloid categorized as an isoquinoline type. It is majorly found in roots, rhizomes, barks, and stem of various plant species, and has been used extensively in Chinese medicine [106]. Berberine has antiinflammatory, antioxidant, antidiabetic, and antimicrobial activity. The procured berberine was tested in vitro and in vivo for its performance [107].

These tests include:

- Cell line culture conditions
- Cell viability assay
- Cellular uptake of Berberine
- Cell viability and phototoxicity
- Measurement of ROS assay (Reactive Oxygen Species)
- Autophagy assay
- Apoptosis assay
- PCR array
- Metabolite foot printing analysis
- NMR Spectroscopy
- Statistical analysis

6.6 Esculetin in renal carcinoma

Esculetin is the main ingredient of herbal Chinese medicine, *Fraxini,* that possesses antitumor activity. Duan et al. [108] studied esculetin for its therapeutic effect on renal carcinoma. It showed that esculetin decreases the cell viability or cell life of tumor cells of SN12-PM6 and 786-O cells. To detect cell apoptosis in ccRCC cells, flow cytometry analysis is done. Esculetin inhibits tumor growth by arresting the cell cycle phases (Table 3).

For study to be performed in the cell lines, ccRCC and SN12-PM6 were procured and cultured in the laboratory. The other reagents were fetal bovine serum (10%), penicillin-streptomycin (1%), CO_2 (5%) at 37°C and the drug esculetin was dissolved in dimethyl sulfoxide (DMSO) [108].

The methods used for evaluation and analysis are:

- MTS Assay
- Colony formation assay
- Flow cytometry
- Western blot analysis
- Wound-healing assay
- Transwell assay
- Statistical analysis

7 Future perspectives

Besides having outstanding potential for the treatment, diagnosis as well as detection of urological cancer, nanoparticles still need more research. The biocompatibility of nanocarriers must be known for the safety priority. Safety of gold nanoparticles and gold nanoshells are still undergoing clinical trials. Nanoparticles containing iron oxide are used for MRI agent approved by US FDA. Different nanoparticles have different pharmacokinetic properties that should be stored in database.

Another additional research should be done for single nanoparticle having multiple functions targeted therapy or combined therapy. Multifunctional nanoparticles can also be used in urological cancer with respect to the targeted loaded agents and in PTTs.

The another widely used agent for nanoparticles is dosimetry, which is used to estimate its concentration in tumors; dosimetry is used to guide in radiation therapy but not in the case of PTT because of undistributed particles in tissues; lastly, more research has to be done for manufactured nanoparticles environmental influence since these nanoparticles have been shown to enter air, water, and soil, so there is an urgent need for safety guidelines of nanoparticles.

Since nanotherapy of renal cancer is still at its infancy stage, different types of renal cancer, renal clearance, and multifactorial nature of cancer are still a challenge. Nanoparticles' suitability for drug carriers like CNTs has already been used for hyperthermia treatment as well as in drug delivery, and a consideration which has to be noticed is that biodegradable polymer should be used and accumulation size range for particles should not cross BBB. In future, the treatment of metastatic cancer is a challenge for nanoparticles due to the size of tumor cells in metastatic cancer and its diverse spreading in the organ.

8 Conclusion

During the decades, many combined drug delivery systems have been made for the treatment of cancer. Nanoparticles are one of the most promising drug delivery systems for

TABLE 3 Patents depicting the novel and conventional inventions toward treatment of kidney cancer/renal cell carcinoma (in ascending order of year).

S. no.	Inventors(s) name	Patent no. and date	Title of the patent	Description	Ref. no.
1.	Alton Kremer	WO2017179739A1, 2017	Treatment of renal cell carcinoma with lenvatinib and everolimus	This invention provides various methods and compositions for treatment of kidney cancer along with combination of pharmaceutical salts or lenvatinib and therefore, everolimus being provided.	[109]
2.	You-wen HE, Yu Wang	WO2017087784A1, 2016	Tumor-infiltrating lymphocytes for treatment of cancer	Lymphocytes infiltered by tumor are WBCs that are located at cancer tissue site to build up an immune response for opposing growth and metastasis of cancer.	[110]
3.	Xiangjun Zhou, Jin Li, Yanyan HAN, Dongyun WU, Junyun LIU, Ran TAO, Longqing TANG	WO2016146035A1, 2016	Methods of cancer treatment using activated t cells	This particular invention specifically gives compositions, methods, and kits for treatment of tumor with application of activated T cells.	[111]
4.	Michael Bausch, Wetzler Rainer Christian Muller	US8465736B2, 2007	Glutadon	This invention reflects regarding combined product for tumor therapy. These consist of glutaminase and 6 –diazo- 5 –oxo-L-norleucine (DON)	[112]
5.	BörjeHaraldsson Ulf Nilsson Lisa Buvall Jenny Nyström	US20100152243A1, 2009	Treatment of renal cell carcinoma	This invention provides methods for kidney cancer treatment by incorporating some compounds like 3,3′,4,4′-tetrahydroxy-2,2′-bipyridine-N,N′-dioxide by means of standard protocols for administration	[113]
6.	Suresh Balkrishna Patankar	US20130337057A1, 2012	Novel herbal composition for the treatment of kidney stone and other urinary tract disorders	This invention discloses information regarding combination composition of Varunbhavit, Kadalikshar, Apamargkshar, Yavkshar with additives that are pharmaceutically active	[114]
7.	Olavi E. Kajander, K. M. Aho, Neva Ciftcioglu, Brady H. Millican	EP1796683A1, 2007	Methods and compositions for the administration of calcium chelators, bisphosphonates and/or citrate compounds and their pharmaceutical uses	Providing composition that consists of citrate compounds, calcium chelators, and bisphosphonates for the purpose of treatment and helpful in decreasing metal poisoning in different patients	[115]
8.	Christine Kritikou, Kevin J. Saliba	US20120195961A1, 2010	Use of spinosyns and spinosyn Compositions against disease caused by protozoans, viral infections and cancer	This invention depicts cancer in immune system cells	[116]
9.	Shuhua Gu, Xuecheng Wang	US20120039995A1, 2012	Arsenic compounds, their preparation methods and uses thereof	This invention discloses compounds with general formula of Arsenic and its uses in manufacturing of formulations for cancer treatment	[117]
10.	Joseph A. VETRO, Sam D. Sanderson	WO2013082535A2, 2012	Controlled-release peptide compositions and uses thereof	This depicts information regarding tumor that occur mainly in cartilage, bone, muscle, fat, blood vessel	[118]

11.	Stephen C. Suffin, W. Hamin Emory, Leonard Brandt	EP1682152A2, 2003	Compositions and methods for treatment of nervous system disorders	This invention depicts treatment methods for patients with CNS disorders providing with anticonvulsant and neuroactive modulator	[119]
12.	BörjeHaraldsson, Ulf Nilsson, Lisa Buvall, Jenny Nyström	US8349823B2, 2009	Treatment of renal cell carcinoma	This invention provides standard protocol and method for administration of various 3,3′,4,4′-tetrahydroxy-2,2′-bipyridine-N,N′-dioxide compounds, especially 3,3′,4,4′-tetrahydroxy-2,2′-bipyridine-N,N′-dioxide (Orellanine)	[120]
13.	Laurence Elias, Gary Witherell	US8012465B2, 2006	Methods for treating renal cell carcinoma	This invention discloses various methods for renal cancer treatment along with minimum doses of IL-2	[121]
14.	Roland Kellner, Rudolf Lichtenfels, Siegfried Matzku, Barbara Seliger	WO2002082076A3, 2003	Renal cell carcinoma tumor markers	This invention takes tumor markers into consideration for the purpose of prognosis, diagnosis and subtypes identification in kidney cancer	[122]
15.	Peter Hulick, Othon Iliopoulos	WO2008128043A3, 2008	Diagnostic and prognostic methods for renal cell carcinoma	This invention reflects different methods for prognosis and diagnosis of kidney cancer by using groups of genes for expression analysis by collection of biological sample from the patient	[123]
16.	Gary Dukart, James Joseph Gibbons, Jr. Anna Berkenblit Jay Marshall Feingold	US8791097B2, 2008	Antitumor activity of CCI-779 in papillary renal cell cancer	This invention reflects CCI-779 use for treatment of papillary types of kidney cancer	[124]
17.	Esther Helen Rose, Mary Ellen Rybak	EP1535623A1, 2000	Renal cell carcinoma treatment	This invention reflects enhanced therapy for kidney cancer by incorporating pegylated interferon alpha to gain minimal cancer responses	[125]
18.	Esther Helen Rose Mary Ellen Rybak	EP1043025B1, 2000	Renal cell carcinoma treatment	The present invention provides the use of pegylated interferon-alpha for the manufacture of a medicament for use in a method of treating a patient with renal cell carcinoma	[126]
19.	Yusuke	EP1907580A2, 2008	Method for diagnosing and treating renal cell carcinoma	This invention describes the procedure for detection and diagnosis of kidney cancer	[127]
20.	Richard William Wyatt Childs, Yoshiyuki Takahashi, Sachiko Kajigaya, Nanae Harashima	US8921050B2, 2007	Methods of diagnosing renal cell carcinoma	This invention depicts methods and composition for prevention and treatment tumor related effects in mammalian patients	[128]

the treatment of urological cancers. Many clinical trials have proved that the nanoparticles can overcome several limitations of current medical approaches. Various metallic nanoparticles are used for the diagnosis with MRI and CT imaging providing higher resolution to early diagnosis of cancer. Gold nanoparticles are one of the most relevant delivery systems to be used because of its potential to generate more heat efficiently. Researchers have clearly demonstrated that nanoparticles have the improved safety and efficacy because of its size range, which allows the drugs to release from the system, and polymeric nanoparticles have many functions for imagining and diagnosis of the tumor cells. Multifunctional nanoparticles have the most advantage because of its multifunctional action and target based on cancer cells; however, there are many issues still which must be clarified.

For the treatment of RCB and antiangiogenesis, there are several immunotherapy drugs available. One of the deadliest cancers is renal carcinoma which does not have any cure in the advanced stage. Some compounds show adverse effects during the treatment of tumors and therefore many other compounds have been discovered for the proper treatment.

Many compounds have been discovered, which do not show adverse drug reactions and in recent studies, it has been proved by many in vitro and in vivo studies. Some of these compounds are quercetin, EGCG, englerin A, curcumin, and resveratrol, which show beneficial effect in the preclinical studies of RCB.

References

[1] Rouiller C. General anatomy and histology of the kidney. In: The kidney. Academic Press; 1969. p. 61–156.

[2] Oni MO, Oguntibeju OO. Clinical and diagnostic importance of proteinuria: a review. Afr J Biotechnol 2008;7(18).

[3] Linehan JA, Nguyen MM. Kidney cancer: the new landscape. Curr Opin Urol 2009;19(2):133–7.

[4] Linehan WM, Zbar B. Focus on kidney cancer. Cancer Cell 2004;6(3):223–8.

[5] Lindblad P, Adami HO. Kidney cancer. Textbook of cancer epidemiology. New York: Oxford University Press; 2002.

[6] Sudarshan S, Linehan WM. Genetic basis of cancer of the kidney. In: Seminars in oncology, vol. 33. WB Saunders; 2006. p. 544–51 [No. 5].

[7] Weiss RH, Lin PY. Kidney cancer: identification of novel targets for therapy. Kidney Int 2006;69(2):224–32.

[8] Störkel S, Van Den Berg E. Morphological classification of renal cancer. World J Urol 1995;13(3):153–8.

[9] Karumanchi SA, Merchan J, Sukhatme VP. Renal cancer: molecular mechanisms and newer therapeutic options. Curr Opin Nephrol Hypertens 2002;11(1):37–42.

[10] Shuch B, Amin A, Armstrong AJ, Eble JN, Ficarra V, Lopez-Beltran A, Martignoni G, Rini BI, Kutikov A. Understanding pathologic variants of renal cell carcinoma: distilling therapeutic opportunities from biologic complexity. Eur Urol 2015;67(1):85–97.

[11] Linehan WM, Ricketts CJ. The metabolic basis of kidney cancer. In: Seminars in cancer biology, vol. 23. Academic Press; 2013. p. 46–55 [No. 1].

[12] Cancer Genome Atlas Research Network. Comprehensive molecular characterization of clear cell renal cell carcinoma. Nature 2013;499(7456):43–9.

[13] https://radiopaedia.org/articles/papillary-renal-cell-carcinoma?lang=us.

[14] Warren AY, Harrison D. WHO/ISUP classification, grading, and pathological staging of renal cell carcinoma: standards and controversies. World J Urol 2018;36(12):1913–26.

[15] Mikhaylenko DS, Klimov AV, Matveev VB, Samoylova SI, Strelnikov VV, Zaletaev DV, Lubchenko LN, Alekseev BY, Nemtsova MV. Case of hereditary papillary renal cell carcinoma type I in a patient with a germline MET mutation in Russia. Front Oncol 2019;9:1566.

[16] Vera-Badillo FE, Conde E, Duran I. Chromophobe renal cell carcinoma: a review of an uncommon entity. Int J Urol 2012;19(10):894–900.

[17] Wobker SE, Williamson SR. Modern pathologic diagnosis of renal oncocytoma. J Kidney Cancer VHL 2017;4(4):1.

[18] Chao D, Zisman A, Pantuck AJ, Gitlitz BJ, Freedland SJ, Said JW, Figlin RA, Belldegrun AS. Collecting duct renal cell carcinoma: clinical study of a rare tumor. J Urol 2002;167(1):71–4.

[19] Srigley JR, Delahunt B. Uncommon and recently described renal carcinomas. Mod Pathol 2009 Jun;22(2):S2–3.

[20] Davis Jr CJ, Mostofi FK, Sesterhenn IA. Renal medullary carcinoma the seventh sickle cell nephropathy. Am J Surg Pathol 1995;19(1):1.

[21] Carlo MI, Chen Y, Chaim J, Coskey DT, Woo K, Hsieh J, et al. Medullary renal cell carcinoma (RCC): Genomics and treatment outcomes. J Clin Oncol 2016;34(15). 4556-4556.

[22] Xie L, Zhang Y, Wu CL. Microphthalmia family of transcription factors associated renal cell carcinoma. Asian J Urol 2019;6(4):312–20.

[23] Amin MB. Histological variants of urothelial carcinoma: diagnostic, therapeutic and prognostic implications. Mod Pathol 2009;22(2):S96–118.

[24] Vogelzang NJ, Fremgen AM, Guinan PD, Chmiel JS, Sylvester JL, Sener SF. Primary renal sarcoma in adults. A natural history and management study by the American Cancer Society, Illinois division. Cancer 1993;71(3):804–10.

[25] Metzger ML, Dome JS. Current therapy for Wilms' tumor. Oncologist 2005;10(10):815–26.

[26] Cyriac S, Rejiv R, Shirley S, Sagar GT. Primary renal lymphoma mimicking renal cell carcinoma. Indian J Urol: J Urolo Soc India 2010 Jul;26(3):441.

[27] Motzer RJ, Jonasch E, Agarwal N, Bhayani S, Bro WP, Chang SS, Choueiri TK, Costello BA, Derweesh IH, Fishman M, Gallagher TH. Kidney cancer, version 2.2017, NCCN clinical practice guidelines in oncology. J Natl Compr Cancer Netw 2017;15(6):804–34.

[28] Beroukhim R, Brunet JP, Di Napoli A, Mertz KD, Seeley A, Pires MM, Linhart D, Worrell RA, Moch H, Rubin MA, Sellers WR. Patterns of gene expression and copy-number alterations in von-hippellindau disease-associated and sporadic clear cell carcinoma of the kidney. Cancer Res 2009;69(11):4674–81.

[29] Foster K, Prowse A, van den Berg A, Fleming S, Hulsbeek MM, Crossey PA, Richards FM, Cairns P, Affara NA, Ferguson-Smith MA, Buys CH. Somatic mutations of the von Hippel—Lindau disease tumor suppressor gene in non-familial clear cell renal carcinoma. Hum Mol Genet 1994;3(12):2169–73.

[30] Gnarra JR, Tory K, Weng Y, Schmidt L, Wei MH, Li H, Latif F, Liu S, Chen F, Duh FM, Lubensky I. Mutations of the VHL tumour suppressor gene in renal carcinoma. Nat Genet 1994;7(1):85–90.
[31] Brugarolas J. PBRM1 and BAP1 as novel targets for renal cell carcinoma. Cancer J (Sudbury, Mass) 2013;19(4):324.
[32] Thompson M. Polybromo-1: the chromatin targeting subunit of the PBAF complex. Biochimie 2009;91(3):309–19.
[33] Vega-Rubin-de-Celis P-LS, Liao S, Leng A, Pavia-Jimenez N, Wang A, Yamasaki S, Zhrebker T, Sivanand L, Spence S, et al. [Submitted]. Novel tumor suppressor, BAP1, sets foundation for molecular genetic classification of renal cell carcinoma; 2012.
[34] Nabi S, Kessler ER, Bernard B, Flaig TW, Lam ET. Renal cell carcinoma: a review of biology and pathophysiology. F1000Res 2018;7:307.
[35] Niceforo J, Coughlin BF. Diagnosis of renal cell carcinoma: value of fine-needle aspiration cytology in patients with metastases or contraindications to nephrectomy. Am J Roentgenol 1993;161(6):1303–5.
[36] Press GA, McClennan BL, Melson GL, Weyman PJ, Mauro MA, Lee JK. Papillary renal cell carcinoma: CT and sonographic evaluation. Am J Roentgenol 1984;143(5):1005–9.
[37] Nadel L, Baumgartner BR, Bernardino ME. Percutaneous renal biopsies: accuracy, safety, and indications. Urol Radiol 1986;8(1):67.
[38] Dean A. Treatment of solitary cyst of the kidney by aspiration. Trans Am Assoc Genitourin Surg 1939;32:91–5.
[39] Gago-Dominguez M, Yuan JM, Castelao JE, Ross RK, Mimi CY. Family history and risk of renal cell carcinoma. Cancer Epidemiol Prev Biomark 2001;10(9):1001–4.
[40] Choyke PL, Glenn GM, Walther MM, Zbar B, Linehan WM. Hereditary renal cancers. Radiology 2003;226(1):33–46.
[41] Linehan WM, Walther MM, Zbar B. The genetic basis of cancer of the kidney. J Urol 2003 Dec;170(6):2163–72.
[42] Hwang JJ, Uchio EM, Linehan WM, Walther MM. Hereditary kidney cancer. Urol Clin 2003;30(4):831–42.
[43] Schlehofer B, Pommer W, Mellemgaard A, Stewart JH, McCredie M, Niwa S, Lindblad P, Mandel JS, McLaughlin JK, Wahrendorf J. International renal-cell-cancer study. VI. The role of medical and family history. Int J Cancer 1996;66(6):723–6.
[44] Kreiger N, Marrett LD, Dodds L, Hilditch S, Darlington GA. Risk factors for renal cell carcinoma: results of a population-based case-control study. Cancer Causes Control 1993 Mar 1;4(2):101–10.
[45] Czene K, Hemminki K. Kidney cancer in the Swedish family cancer database: familial risks and second primary malignancies. Kidney Int 2002;61(5):1806–13.
[46] Li P, Znaor A, Holcatova I, Fabianova E, Mates D, Wozniak MB, Ferlay J, Scelo G. Regional geographic variations in kidney cancer incidence rates in European countries. Eur Urol 2015;67(6):1134–41.
[47] Scelo G, Li P, Chanudet E, Muller DC. Variability of sex disparities in cancer incidence over 30 years: the striking case of kidney cancer. Eur Urol Focus 2018;4(4):586–90.
[48] Bhaskaran K, Douglas I, Forbes H. dos-Santos-Silva I, Leon DA, Smeeth L. body-mass index and risk of 22 specific cancers: a population-based cohort study of 5· 24 million UK adults. Lancet 2014;384(9945):755–65.
[49] Stevens VL, Jacobs EJ, Patel AV, Sun J, McCullough ML, Campbell PT, Gapstur SM. Weight cycling and cancer incidence in a large prospective US cohort. Am J Epidemiol 2015;182(5):394–404.
[50] Karami S, Daugherty SE, Purdue MP. A prospective study of alcohol consumption and renal cell carcinoma risk. Int J Cancer 2015 Jul 1;137(1):238–42.
[51] Nicodemus KK, Sweeney C, Folsom AR. Evaluation of dietary, medical and lifestyle risk factors for incident kidney cancer in postmenopausal women. Int J Cancer 2004;108(1):115–21.
[52] Kleinerman RA, Boice Jr JD, Storm HH, Sparen P, Andersen A, Pukkala E, Lynch CF, Hankey BF, Flannery JT. Second primary cancer after treatment for cervical cancer. An international cancer registries study. Cancer 1995;76(3):442–52.
[53] Asal NR, Risser DR, Kadamani S, Geyer JR, Lee ET, Cherng N. Risk factors in renal cell carcinoma: I. methodology, demographics, tobacco, beverage use, and obesity. Cancer Detect Prev 1988;11(3–6):359–77.
[54] Colt JS, Schwartz K, Graubard BI, Davis F, Ruterbusch J, DiGaetano R, Purdue M, Rothman N, Wacholder S, Chow WH. Hypertension and risk of renal cell carcinoma among white and black Americans. Epidemiology (Cambridge, Mass) 2011;22(6):797.
[55] Robson CJ. Radical nephrectomy for renal cell carcinoma. J Urol 1963;89(1):37–42.
[56] Lane BR, Babineau DC, Poggio ED, Weight CJ, Larson BT, Gill IS, Novick AC. Factors predicting renal functional outcome after partial nephrectomy. J Urol 2008;180(6):2363–9.
[57] Al-Qudah HS, Rodriguez AR, Sexton WJ. Laparoscopic management of kidney cancer: updated review. Cancer Control 2007;14(3):218–30.
[58] George CM, Stadler WM. The role of systemic chemotherapy in the treatment of kidney cancer. In: Kidney cancer. Boston, MA: Springer; 2003. p. 173–82.
[59] Lilleby W, Fosså SD. Chemotherapy in metastatic renal cell cancer. World J Urol 2005;23(3):175–9.
[60] Bedke J, Stühler V, Stenzl A, Brehmer B. Immunotherapy for kidney cancer: status quo and the future. Curr Opin Urol 2018;28(1):8–14.
[61] Allory Y, Culine S, De La Taille A. Kidney cancer pathology in the new context of targeted therapy. Pathobiology 2011;78(2):90–8.
[62] Gurevich F, Perazella MA. Renal effects of anti-angiogenesis therapy: update for the internist. Am J Med 2009;122(4):322–8.
[63] Sawyers CL. Will mTOR inhibitors make it as cancer drugs? Cancer Cell 2003;4(5):343–8.
[64] Blanco AI, Teh BS, Amato RJ. Role of radiation therapy in the management of renal cell cancer. Cancer 2011;3(4):4010–23.
[65] Wang S, Luo J, Lantrip DA, Waters DJ, Mathias CJ, Green MA, Fuchs PL, Low PS. Design and synthesis of [111In] DTPA– folate for use as a tumor-targeted radiopharmaceutical. Bioconjug Chem 1997;8(5):673–9.
[66] Singh R, Lillard Jr JW. Nanoparticle-based targeted drug delivery. Exp Mol Pathol 2009;86(3):215–23.
[67] Wang H, Li X, Ma Z, Wang D, Wang L, Zhan J, She L, Yang F. Hydrophilic mesoporous carbon nanospheres with high drug-loading efficiency for doxorubicin delivery and cancer therapy. Int J Nanomedicine 2016;11:1793.
[68] Duncan R, Gaspar R. Nanomedicine (s) under the microscope. Mol Pharm 2011;8(6):2101–41.
[69] El-Sayed IH, Huang X, El-Sayed MA. Selective laser photo-thermal therapy of epithelial carcinoma using anti-EGFR antibody conjugated gold nanoparticles. Cancer Lett 2006;239(1):129–35.

[70] Ojha T, Pathak V, Shi Y, Hennink WE, Moonen CT, Storm G, Kiessling F, Lammers T. Pharmacological and physical vessel modulation strategies to improve EPR-mediated drug targeting to tumors. Adv Drug Deliv Rev 2017;119:44–60.
[71] Bertrand N, Wu J, Xu X, Kamaly N, Farokhzad OC. Cancer nanotechnology: the impact of passive and active targeting in the era of modern cancer biology. Adv Drug Deliv Rev 2014;66:2–5.
[72] Mieszawska AJ, Mulder WJ, Fayad ZA, Cormode DP. Multifunctional gold nanoparticles for diagnosis and therapy of disease. Mol Pharm 2013;10(3):831–47.
[73] Zrazhevskiy P, Gao X. Quantum dot imaging platform for single-cell molecular profiling. Nat Commun 2013;4(1):1–2.
[74] Escudier B, Eisen T, Stadler WM, Szczylik C, Oudard S, Siebels M, Negrier S, Chevreau C, Solska E, Desai AA, Rolland F. Sorafenib in advanced clear-cell renal-cell carcinoma. N Engl J Med 2007;356(2):125–34.
[75] Vaishampayan U. Cabozantinib as a novel therapy for renal cell carcinoma. Curr Oncol Rep 2013;15(2):76–82.
[76] Haley B, Frenkel E. Nanoparticles for drug delivery in cancer treatment. In: Urologic oncology: seminars and original investigations, vol. 26. Elsevier; 2008. p. 57–64 [No. 1].
[77] Santhi K, Dhanaraj SA, Joseph V, Ponnusankar S, Suresh B. A study on the preparation and anti-tumor efficacy of bovine serum albumin nanospheres containing 5-fluorouracil. Drug Dev Ind Pharm 2002;28(9):1171–9.
[78] Fessi HP, Puisieux F, Devissaguet JP, Ammoury N, Benita S. Nanocapsule formation by interfacial polymer deposition following solvent displacement. Int J Pharm 1989;55(1):R1–4.
[79] Yokoyama M, Kwon GS, Okano T, Sakurai Y, Kataoka K. Development of micelle-forming polymeric drug with superior anticancer activity. Polym Drugs Drug Admin 1994;10:126–34.
[80] Torchilin VP. Recent advances with liposomes as pharmaceutical carriers. Nat Rev Drug Discov 2005;4(2):145–60.
[81] Singh M, Ghose T, Faulkner G, Kralovec J, Mezei M. Targeting of methotrexate-containing liposomes with a monoclonal antibody against human renal cancer. Cancer Res 1989;49(14):3976–84.
[82] Lei Y, Zeng L, Xie S, Fan K, Yu Y, Chen J, Zhang S, Wang Z, Zhong L. Sertraline/ICG-loaded liposome for dual-modality imaging and effective chemo-photothermal combination therapy against metastatic clear cell renal cell carcinoma. Chem Biol Drug Des 2020;95(3):320–31.
[83] Shroff EH, Eberlin LS, Dang VM, Gouw AM, Gabay M, Adam SJ, Bellovin DI, Tran PT, Philbrick WM, Garcia-Ocana A, et al. MYC oncogene overexpression drives renal cell carcinoma in a mouse model through glutamine metabolism. Proc Natl Acad Sci U S A 2015;112:6539–44.
[84] Malouf GG, Camparo P, Oudard S, Schleiermacher G, Theodore C, Rustine A, Dutcher J, Billemont B, Rixe O, Bompas E, et al. Targeted agents in metastatic Xp11 translocation/TFE3 gene fusion renal cell carcinoma (RCC): a report from the juvenile RCC network. Ann Oncol 2010;21:1834–8.
[85] Monteiro MS, Barros AS, Pinto J, Carvalho M, Pires-Luis AS, Henrique R, Jeronimo C, Bastos ML, Gil AM, Guedes de Pinho P. Nuclear Magnetic Resonance metabolomics reveals an excretory metabolic signature of renal cell carcinoma. Sci Rep 2016;6:37275.
[86] Qu Y, Chen H, Gu W, Gu C, Zhang H, Xu J, Zhu Y, Ye D. Age-dependent association between sex and renal cell carcinoma mortality: a population-based analysis. Sci Rep 2015;5:9160.
[87] Dabestani S, Marconi L, Hofmann F, Stewart F, Lam TB. Local treatments for metastases of renal cell carcinoma: a systematic review. Lancet Oncol 2014;15:e549–61.
[88] Sosman JA. Targeting of the VHL-hypoxia-inducible factor-hypoxia-induced gene pathway for renal cell carcinoma therapy. J Am Soc Nephrol 2003;14:2695–702.
[89] Ji HF, Li XJ, Zhang HY. Natural products and drug discovery. Can thousands of years of ancient medical knowledge lead us to new and powerful drug combinations in the fight against cancer and dementia? EMBO Rep 2009;10:194–200.
[90] Rangika BS, Dayananda PD, Peiris DC. Hypoglycemic and hypolipidemic activities of aqueous extract of flowers from Nycantusarbortristis L in male mice. BMC Complement Altern Med 2015;15:289.
[91] Peiris LD, Dhanushka MA, Jayathilake TA. Evaluation of aqueous leaf extract of *Cardiospermum halicacabum* (L.) on fertility of male rats. Biomed Res Int 2015;2015:175726.
[92] Wani MC, Taylor HL, Wall ME, Coggon P, McPhail AT. Plant antitumor agents. VI. The isolation and structure of taxol, a novel antileukemic and antitumor agent from Taxus brevifolia. J Am Chem Soc 1971;93:2325–7.
[93] Wall ME, Wani MC, Cook CE, et al. Plant anti-tumor agents: I. the isolation and structure of camptothecin, a novel alkaloidal leukemia and tumor inhibitor from Camptotheca acuminata. J Am Chem Soc 1966;88:3888–90.
[94] Ratnayake R, Covell D, Ransom TT, Gustafson KR, Beutler JA. Englerin a, a selective inhibitor of renal cancer cell growth, from Phyllanthus engleri. Org Lett 2009;11:57–60.
[95] Hui YH, Rupprecht JK, Liu YM, Anderson JE, Smith DL. Bullatacin and bullatacinone: two highly potent bioactive acetogenins from Annona bullata. J Nat Prod 1989;52:463–77.
[96] Houghton PJ, Photiou A, Uddin S, Shah P, Browning M, et al. Activity of extracts of Kigelia pinnata against melanoma and renal carcinoma cell lines. Planta Med 1994;65:430–3.
[97] Samarghandian S, Afshari JT, Davoodi S. Honey induces apoptosis in renal cell carcinoma. Pharmacogn Mag 2011;7:46–52.
[98] Thornes RD, Lynch G, Sheehan MV. Cimetidine and coumarin therapy of melanoma. Lancet 1982;2:328.
[99] Marshall ME, Butler K, Fried A. Phase I evaluation of coumarin (,2-benzopyrone) and cimetidine in patients with advanced malignancies. Mol Biother 1991;3:170–8.
[100] Hui Y, Wood KV, McLaughlin JL. Bullatencin, 4-deoxyasimicin, and the uvariamicins: additional bioactive annonaceousacetogenins from Annona bullata rich. (Annonaceae). Nat Toxins 1992;1:4–14.
[101] Shi T, Liou LS, Sadhukhan P, Duan ZH, Novick AC, Hissong JG, Almasan A, DiDonato JA. Effectsof resveratrol on gene expression in renal cell carcinoma. Cancer Biol Ther 2004;3:882–8.
[102] Patankar SB. Novel herbal composition for the treatment of kidney stone and other urinary tract disorders, US20130337057A1; 2012.
[103] Zulfiker HM, Hashimi SM, Good DA, Grice ID, Wei MQ. Cane toad skin extract-induced upregulation and increased interaction of serotonin 2A and D2 receptors via Gq/11 figure 10: photo of dry root Salvia miltiorrhiza. Evidence-based complementary and alternative medicine 7 signaling pathway in CLU213 cells. J Cell Biochem 2017;118(5):979–93.
[104] Zulfiker HM, Sohrabi M, Qi J, Matthews B, Wei MQ, Grice ID. Multi-constituent identification in Australian cane toad skin extracts using high-performance liquid chromatography high-resolution tandem mass spectrometry. J Pharm Biomed Anal 2016;129:260–72.

[105] Liu W, Yang B, Yang L, Kaur J, Jessop C, Fadhil R, Wei MQ. Therapeutic effects of ten commonly used Chinese herbs and their bioactive compounds on cancers. Evid Based Complement Alternat Med 2019;2019:1–10.
[106] Sun Y, Xun K, Wang Y, Chen X. A systematic review of the anticancer propertiesof berberine, a natural product from Chinese herbs. Anti-Cancer Drugs 2009;20(9):757–69.
[107] Lopes TZ, de Moraes FR, Tedesco AC, Arni RK, Rahal P, Calmon MF. Biomed Pharmacother 2020;123:109794.
[108] Duan J, Shi J, Ma X, Xuan Y, Li P, Wang H, Fan Y, Gong H, Wang L, Pang Y, Pang S, Yan Y. Esculetin inhibits proliferation, migration, and invasion of clear cell renal cell carcinoma cells. Biomed Pharmacother 2020;125:110031.
[109] https://patents.google.com/patent/WO2017179739A1/en.
[110] https://patents.google.com/patent/WO2017087784A1/ar.
[111] https://patents.google.com/patent/WO2016146035A1/ru.
[112] https://patents.google.com/patent/US8465736.
[113] https://patents.google.com/patent/US20100152243A1/en.
[114] https://patents.google.com/patent/US20130337057.
[115] https://patents.google.com/patent/EP1796683A1.
[116] https://patents.google.com/patent/US20120195961A1.
[117] https://patents.google.com/patent/US20120039995A1.
[118] https://patents.google.com/patent/WO2013082535A2.
[119] https://patents.google.com/patent/EP1682152A2.
[120] https://patents.google.com/patent/US8349823.
[121] https://patents.google.com/patent/US8012465B2.
[122] https://patents.google.com/patent/WO2002082076A3.
[123] https://patents.google.com/patent/WO2008128043A3.
[124] https://patents.google.com/patent/US8791097B2.
[125] https://patents.google.com/patent/EP1535623A1.
[126] https://patents.google.com/patent/EP1043025B1.
[127] https://patents.google.com/patent/EP1907580A2.
[128] https://patents.google.com/patent/US8921050B2.

Chapter 14

Advanced drug delivery systems for glioblastoma

Ganesh B. Shevalkar[a], Nisha R. Yadav[b], Chandrakantsing V. Pardeshi[a], and Sanjay J. Surana[a]
[a]*Department of Pharmaceutics and Pharmaceutical Technology, R. C. Patel Institute of Pharmaceutical Education and Research, Shirpur, India,*
[b]*Edhaa Innovations Pvt. Ltd., Mumbai, India*

1 Introduction

A glioblastoma (GBM) is a rapidly growing tumor of the star-shaped glial cells, i.e., astrocytes. These astrocytes are abundantly present in the brain and have the primary function to support the health of the nerve cells. Besides this, they also regulate the transmission of electrical impulses within the brain. These cells in this GBM tumors become pleomorphic, i.e., they show variable size and shape; hence, GBM is also called as glioblastoma multiforme [1]. GBM is the most intrusive type of glial tumor that proliferates to make their blood vessels to increase their blood supply and spread into nearby brain tissue. The glioma related to glial cells invades astrocytes, oligodendrocytes, and ependymal cells in the brain [2]. On that basis, it is categorized as astrocytoma, oligodendroglioma, and ependymoma, respectively. World Health Organization (WHO) has given grading to these brain tumors from I to IV based on their pathological evaluation. Grade I indicates slow tumor growth, whereas Grade IV indicates rapid growth. As glioblastoma shows aggressive growth in the brain, it is categorized as a Grade IV brain tumor [3].

Epidemiology: Glioblastoma is considered the most aggressive malignant tumor with a rare incidence rate (3.19 per 100,000 people per annum); however, its poor diagnosis and minimal survival rate make it a significant health issue [4, 5]. After the diagnosis of GBM, most of the patients survive only for 2.5 years, whereas only 5% survive for 5 years [6]. The incidence of GBM is more significant in males than in females. In Caucasians, the incidence rate is double than of Africans and Afro-Americans, whereas it is rarely found in American Indians and Asians [4]. The supratentorial area (frontal, temporal, parietal, and occipital lobes) is more prone to GBM than the infratentorial area (cerebellum) [7]. In the coming time, understanding the role of genes using multicenter clinical studies will help to decide the epidemiology of this mortifying disease.

Causes: The exact reason for the development of glioblastoma is unidentified until today. However, environmental and genetic factors were examined in GBM [8]. The factors that could be considered as causes for GBM include prior ionizing radiation therapy for treating cancer, reduced sensitivity to allergy, immune factors and genes, and some single-nucleotide polymorphisms recognized by a genomic study [9]. Other factors like working in petroleum refinery, synthetic rubber industry, and longtime exposure to pesticides are also contemplated causes of GBM. However, it is worth noting that individuals diagnosed with glioblastoma may not have any risk, as mentioned above; similarly, those who are associated with these factors may never develop GBM throughout their life. The involvement of these factors in the occurrence of GBM has not been proven yet; hence, more research is required in this direction. The hereditary diseases like Turcot syndrome, Li-Fraumeni syndrome, and neurofibromatosis are also considered as a risk of GBM; however, their occurrence is hardly diagnosed in the case of GBM [10].

Symptoms and Diagnosis: GBM shows symptoms like headache, cognitive impairment, posture imbalances, loss of senses, disturbance in vision, delirium, and seizures [1]. As many of these symptoms are familiar to other neurological and psychological disorders like epilepsy, dementia, or stroke, the chances of misdiagnosis are more. Various techniques, like the neurological exam, sophisticated imaging tests, biopsy, and molecular testing, are used to diagnose glioblastoma [11]. In a neurological exam, the patient is observed for posture imbalance, strength, muscle coordination, and the senses, including hearing or vision problems [12]. Sophisticated imaging techniques like computed tomography (CT) scan and magnetic resonance imaging (MRI) have great accuracy in locating the brain tumors with their size [13]. Intraoperative MRI could also be used during surgery for finding and removal of the tumor as well as for tissue biopsies. Magnetic resonance spectroscopy (MRS) is used

Advanced Drug Delivery Systems in the Management of Cancer. https://doi.org/10.1016/B978-0-323-85503-7.00025-0

to study the chemical profile of tumors, whereas a positron emission tomography (PET) scan helps to identify the recurrence of tumors [14].

2 Glioblastoma and blood–brain barrier

The main hurdle in treating all brain disorders and tumors is the blood–brain barrier (BBB) as it prevents the drug from getting into the brain. Hence, it is a very challenging task to design the therapy that could target the brain for treating the disease like GBM [15]. Considerable effort has been taken so far for understanding the mechanism behind the transportation of drugs across this barrier. The BBB comprises endothelial cells that surround the brain and spinal cord [16]. These endothelial cells are linked by the continuous tight junctions that act as barriers for extracellular fluids. This barrier plays a vital role in maintaining the brain homeostasis and shielding the brain from exposure to extraneous materials [17]. This tight junction allows only small molecules, fat-soluble vitamins, and some gases to enter the brain tissue through paracellular transport.

Apart from this paracellular transport, drugs are also transported through BBB employing other routes. The transcellular route is one of that kind, where drugs having a low molecular weight (<500Da) and high lipophilicity easily transported across BBB [18]. However, extensive plasma protein binding of lipophilic drugs results in less free drug available for brain uptake [19]. The polar nature of the drug molecule is also the reason behind less brain uptake, as inverse correlation is described for the polar surface of drug and brain permeability [20, 21]. Besides these paracellular and transcellular routes, drug molecules are also transported via facilitated diffusion or receptor-mediated transport [22]. However, these routes favor permeation of water, ions, amino acids, peptides, and some small lipophilic molecules across BBB. Most drugs and large molecules cannot cross the barrier and could not exhibit their pharmacological action [23, 24]. Reports have shown that nearly 98% of small and 100% of large molecules could not pass through the BBB [25].

The integrity of this protective barrier, however, affected in some clinical conditions like brain tumor, autoimmune deficiency syndrome, dementia, multiple sclerosis, and stroke. Studies have shown that patient with GBM or other brain tumor possesses disrupted and highly permeable vascular network of BBB. However, it does not offer a considerable rise in the amount of drugs entering the brain. Moreover, in the outer region of disrupted BBB, the intrusive nature of glial tumors produces malignant cells [26]. In GBM, tumor vessels developed a compact network around tumor tissue, which acts as another barrier for the chemotherapeutic agent and prevents it from getting access to tumor tissue. This barrier is known as the blood–brain tumor barrier (BBTB). Besides this, the endothelial cells of the GBM tumor exhibit efflux transporter on its surface that expel the drug out from the cell and thereby develop resistance to chemotherapy. All these factors make the treatment of GBM very challenging. Very few cytotoxic drugs like temozolomide (TMZ) that can permeate through BBB in an acceptable range are employed to manage GBM [27, 28]. In vitro cytotoxicity study using drugs like paclitaxel, doxorubicin, and cisplatin has also shown sufficient activity against the GBM cell line; however, their insufficient brain uptake limits their use for the treatment of GBM [29–31]. Such type of situation demands higher doses of the drug to achieve therapeutic concentration at the target site. However, it develops other adverse effects as healthier cells are also exposed to that high concentration [32].

Hence, researchers have developed various innovative approaches for delivering the drug to the target site of the brain. The most common techniques that are used for this purpose are pharmacological and neurosurgical approaches [17]. In the pharmacological process, nanotherapeutics like polymeric or lipid-based drug delivery systems are used. Whereas in the neurosurgical procedure, invasive techniques like convection-enhanced diffusion (CED), intraventricular infusion, and intracerebral delivery are used [17]. The neurosurgical-based approach delivers the drug at the specific target site in the brain, thereby increasing the bioavailability and effectiveness of the chemotherapy. However, to achieve such drug delivery via this approach is technically challenging. This chapter deals with the understanding of various advanced therapeutic modalities for the treatment or management of GBM. It also elaborates on different promising technologies and the loopholes of current therapies. This chapter emphasizes the correlation of GBM biology with the development of emerging therapies.

3 Management of glioblastoma (marketed or FDA-approved products/technologies) and challenges

Therapeutic management (antitumor therapy) and supportive care are the current strategies that are being utilized for the management of GBM in patient [9].

3.1 Supportive care

General signs and symptoms of the GBM are controlled in supportive care [33]. Neurological symptoms are relieved by the administration of corticosteroids. Dexamethasone is generally preferred in patients with substantial side effects because of its low mineralocorticoid activity [1]. At the same time, levetiracetam is given to the patient having seizures due to its minimal toxicity and no interactions with antitumor drugs [1, 9].

3.2 Currently approved therapies

In the therapeutic management of GBM, surgical resection of the tumor is performed, if required, and is safe to perform. After that, radiotherapy and chemotherapy are given simultaneously to the patient [34]. Nowadays, after the removal of the tumor, multimodal therapy is given to the patient, which consists of local treatment with Gliadel® (Guilford Pharmaceuticals, Baltimore [EEUU]), which is followed by radiation therapy and systemic administration of TMZ [35]. USFDA has approved a new approach for treating recurrent and newly diagnosed GBM, i.e., tumor treating fields (TTFields; Optune). In this approach, an electric field is applied at the tumor location for disrupting cell division of tumor cells [34].

3.2.1 Surgery

Surgery is the main part of standard care, where the maximum part of the tumor is removed and shows subsequent enhancement in progression-free survival and overall survival rate [9, 34]. The time required for surgery depends on the location and area acquired by the tumor in the brain. Although the tumor is removed using surgery, its recurrence is found in approximately 80% of cases [9].

3.2.2 Radiation therapy

As chances of GBM recurrence after surgery are more, hence to prevent its recurrence and killing of remaining tumor cells, radiation therapy at a total dose of 60Gy is employed after surgery [9]. Radiation therapy/radiotherapy can also be used as a primary treatment strategy for a brain tumor [34]. Radiation therapies like intensity-modulated and boron neutron capture therapy are newly employed treatment modalities for GBM. In these newer techniques, exposure of healthy tissues to radiation is significantly less and hence reduces the subsequent toxicity also. However, necrosis of tissues, permanent neuronal damage, resistance of some tumor tissues to radiation, and the invasive nature of GBM are the limitations of radiotherapy [9, 33].

3.2.3 Tumor-treating fields

Tumor-treating fields (TTF) is the noninvasive technique approved by the USFDA in 2011 to treat newly diagnosed and recurrent GBM. In this technique, the noninvasive transducer arrays are attached to the scalp of patients for delivering an alternating electric field of low intensity and immediate frequency around the anatomic region of the tumor. This alternating electric field disturbs the mitosis of cells where it is applied, and this ultimately results in the killing of the cells. It is generally employed after surgery and radiotherapy [36]. If the continuous application is required, the patient can use it at home also, where the field generator (NovoTTF System, Novocure Ltd.) is needed to connect with a portable battery (total weight of 1.2kg) to activate the device. For better results, the device should be used daily for 18h with short personal breaks [36]. However, the high cost and requirement of the specialized instrument are the main limitations associated with this therapy [37].

3.2.4 Chemotherapy

Various chemotherapeutic agents were investigated until today for their effectiveness against GBM. Among them, very few like TMZ, carmustine (BCNU), and lomustine (CCNU) were found effective against GBM. However, BCNU and CCNU are associated with adverse effects like harsh cytotoxicity and early development of resistance, which limits their use in the treatment of GBM. Table 1 shows the list of chemotherapeutic agents used for the treatment of GBM along with their associated side effects. Before TMZ approval, procarbazine and vincristine, along with lomustine, were used as the prime drugs for the management of GBM. Nowadays, TMZ, along with radiotherapy, is used as a standard therapeutic regimen for GBM patients [9]. This standard treatment regimen has increased the survival rate of the patient; however, the patient who has developed resistance to TMZ shows the progression and recurrence of the tumor [34]. If the patient does not respond to these drugs, then drugs like carboplatin, oxaliplatin, etoposide, and irinotecan are used as second-line treatment [9].

TABLE 1 List of chemotherapeutic agents used to treat GBM and their associated side effects [9].

Agent	Side effects
Carmustine	Nausea, myelosuppression, pulmonary fibrosis
Lomustine	Nausea, myelosuppression, pulmonary fibrosis
Temozolomide	Nausea, fatigue, headache, constipation, myelosuppression
Vincristine	Peripheral neuropathy, constipation
Cisplatin	Nausea, renal insufficiency, peripheral neuropathy, myelosuppression
Bevacizumab	Convulsions, chest pain, bleeding gums, cough, difficult breathing, burning, tingling, numbness, chills, body pain, cracks in the skin, dilated neck veins
Etoposide	Swelling of the face, eyes, tongue, and lips; sweating, difficulty in swallowing, cough, nervousness, itching, dizziness
Procarbazine	Convulsions, hallucinations, shortness of breath, confusion, thick bronchial secretions, tiredness

4 Treatment modalities for glioblastoma

From the last few decades, nanomedicines have emerged as a promising alternative to conventional chemotherapy because of its potential to overcome the limitations associated with conventional chemotherapy [35]. Several nanomaterials, including polymeric nanoparticles (micelles, dendrimers, polymersomes, and nanogels), lipid nanoparticles (nanoemulsion, solid lipid nanoparticles, liposomes, and nanostructured lipid carriers), inorganic nanoparticles (silica and gold nanoparticles, nanorods, nanowires, and quantum dots), and organic nanoparticles (carbon nanotubes and nanographene), have been studied as carriers for the drugs, which are used in the treatment of GBM. Table 2 provides a short description of different nanocarriers used for delivering the drug to the brain. Nanomedicines are employed in two ways, i.e., local treatment and systemic treatment. In the case of systemic treatment, nanoparticles are delivered to the tumor site either by active (ligand/receptor) or passive (enhanced permeation and retention) targeting [38].

TABLE 2 Description of nanocarriers evaluated for their application in the management of GBM [36,39–45].

Drug delivery systems	Short description/characteristic	Advantages	Commonly used components
Polymeric nanoparticles	• Biocompatible or biodegradable • Natural (e.g., gelatin, alginate) or synthetic (e.g., PLA, PLGA) polymers are used as carriers • Used for hydrophilic as well as hydrophobic drugs • Size: 10–1000 nm	• Simple preparation techniques • Surface modification for enhanced permeability across BBB • Good stability • Controlled release of drugs	Poly-(butyl cyanoacrylate) (PBCA), poly-(lactic-*co*-glycolic acid) (PLGA), chitosa
Liposomes	• Vesicular system made up of phospholipid having an aqueous inner core enclosed by one or more bilayers • This is the first nanomedicine which is approved by the FDA for clinical use • E.g., Doxil®, doxorubicin hydrochloride liposomal injection • Size: 50–300 nm	• Suitable for hydrophilic as well as lipophilic therapeutics • Good biocompatibility • Simple preparation technique • Available commercially • Half-life can be improved by surface modification using PEG	Hydrogenated soy phosphatidylcholine (HSPC), cholesterol, 1,2-distearoyl-sn-glycero-3-phosphocholine, DSPE-PEG2000
Dendrimers	• Highly branched polymeric macromolecules • Has three main components: a central core, an interior dendritic structure (the branches), and an exterior surface with functional surface groups • Size: 2–15 nm	• Small size suitable for escaping from RES, but retention in the brain is weak • Suitable for delivery of drug and genes • Used for imaging of brain tumors • Easily functionalized • Most versatile nanocarrier	Polyamidoamine (PAMAM)
Micelles	• Self-assembled structure obtained from the dispersion of amphiphilic copolymer in water • Dispersion of such polymer in water forms an aggregate with a hydrophobic core and hydrophilic surface in contact with water • Size: 20–200 nm	• Can entrap a large amount of hydrophobic drug	Poly (D, L-lactide) (PLA), poly (glycolic acid) (PGA), poly (ε-caprolactone) (PCL), etc.

TABLE 2 Description of nanocarriers evaluated for their application in the management of GBM [36,39–45]—cont'd

Drug delivery systems	Short description/characteristic	Advantages	Commonly used components
Solid lipid nanoparticles (SLNs)	• First generation of lipid-based nanoparticles • Lipids which remain solid at body temperature are used for drug loading • Used for controlled drug release • Due to the tendency toward aggregation and gelation, solid lipid matrices stabilized by surfactants • Size: 10–1000 nm	• Overcome the problems of drug solubility, permeability, and toxicity • Suitable alternative to liposomes and polymeric systems • Biocompatible and degradable system • Large area available for surface functionalization	Stearic acid, stearic amine, glyceryl monostearate, glyceryl distearate, glyceryl dibehenate, cetyl alcohol, cetyl palmitate, tripalmitin, trimyristin, tristearin, etc.
Nanostructured lipid carrier (NLC)	• Second generation of lipid-based nanoparticles • Formulated via blending of solid lipids and oils • Formulated for enhancement of drug loading capacity • Size: 10–1000 nm	• Overcome the drug loading and stability problems associated with the SLN system • Low cytotoxicity	• Solid lipids, as mentioned in SLN • Oils: almond oil, corn oil, oleic acid, olive oil, peanut oil, sesame oil, soybean oil, medium-chain triglycerides, etc. • Surfactants: Lutrol F68, Span®85 Cremophor®EL, Cremophor®RH, Eumulgin, Tween® 20, Tween® 80, Pluronic® F68, etc.
Gold nanoparticles	• Are small *gold* particles • Modified chemically for selective targeting of the tumor • Size: 1–100 nm	• Biocompatible • Less toxic • Used for diagnostic applications • Used as an aid in radiation therapy because of its ability to absorb energy • Last for a longer time in tumor tissue when injected as a single injection	• Gold nanoparticles (AuNP) coated with PEG, PCL, gelatin, polyacrylic acid (PAA)
Polymersomes	• Are composed of synthetic amphiphilic or ionic block copolymers that form self-assembled polymeric vesicles • Structure is similar to liposomes	• Bilayer synthetic polymers are more stable to shear stresses than lipid membranes • Provide high mechanical stability, elastic behavior, higher membrane viscosity, biodegradability, and biocompatibility	• Hydrophobic polymers: polybutadiene (PBD), dimethylsiloxane (DMS), polystyrene (PS), and polylactide (PLA) • Hydrophilic polymers: polyethylene glycol, polyacrylic acid, and poly L-glutamic acid
Nanogels	• A new class of nanosystems characterized by their softness and a macromolecular structure resulting from the 3D cross-linking that occurs between water-soluble polymeric units in a supramolecular framework	• Structure allows numerous bioactive components (such as drugs, proteins, peptides, and DNA/RNA) to be encapsulated, as well as stimulating the release of drugs under specific conditions in both organs and cell lines	Polyacrylamide, polyvinyl alcohol [PVA], alginate
Silica nanoparticles	• Among all the nanomaterials studied, silica NPs have unique properties, including good biocompatibility, a large surface area, and variable pore size	• Used to improve biocompatibility in the nervous tissue of other nanoformulations, such as iron oxide NPs	• Mesoporous silica NPs (MSNs), the most widely used for the transport and release of drugs

Continued

TABLE 2 Description of nanocarriers evaluated for their application in the management of GBM [36,39–45]—cont'd

Drug delivery systems	Short description/characteristic	Advantages	Commonly used components
Carbon nanotubes	• These are nanosized cylindrical structures made of carbon atoms organized into aromatic rings	• Good physical and chemical characteristics • Ability to cross BBB	• Modified by polymers like polyethylene glycol (PEG) or poly amino benzene sulfonic acid (PABS) for increasing water solubility
Nanographene	• A 2D material formed by layers of carbon atoms disposed of in a regular hexagonal pattern	• Large surfaces available for transporting molecules and many functional groups that allow its functionalization	
Nanorods and nanowires	• A new type of NPs which is originated from semiconducting materials or metals • Among them, gold nanorods (GNRs), characterized by size: 1–100 nm • Nanowires are structures with diameters of only a few nanometers and extended lengths and are principally used for diagnosis	• Low toxicity • GNRs are used as contrast agents for imaging of tumors • Used in photothermal therapy • Can monitor brain electrical activity without the use of a probe	• Nanowires, developed from silicon, germanium, carbon, gold, and copper
Quantum dots	• These are optical semiconductor nanocrystals that have unique optical and electrical properties • Size: 2–10 nm	• Easily detectable since they only emit at a characteristic wavelength • Can absorb photons over a wide range of wavelengths	
Magnetic NPs	• Versatile surface functionalization, biocompatibility, and magnetic characteristics • Surface functionalization is required to incorporate targeting ligands to further surface conjugations, to prevent surface oxidation and agglomeration, and to improve biocompatibility • Magnetic force generated by small size may be insufficient to disrupt key cell components	• Theranostic applications include cell labeling, sorting and manipulation, magnetic guidance for drug delivery, hyperthermia, and magnetic resonance imaging (MRI) • Ability to cross the BBB under the control of a magnetic field • New strategies to associate Zn, Ni, and Co metal elements into the NPs can enhance the magnetization	• FDA-approved magnetic materials for MNP synthesis are maghemite (γ-Fe_2O_3) and magnetite (Fe_3O_4)

4.1 Local administration of nanotherapeutics

Local administration of nanotherapeutics directly delivers the drug at the tumor site. This approach includes various invasive techniques like convection-enhanced delivery (CED), locally implanted systems, intraventricular, intraparenchymal, or intrathecal strategies to deliver the drug. These techniques bypass the BBB and directly deliver the drug molecules at the tumor site resulting in the enhancement of bioavailability and reduced systemic toxicity of drug molecules [35].

4.1.1 Direct brain injection

The nanoparticles formulated by employing nanotechnology have an adequate ability to encapsulate the drug molecule and control its release. Hence, injecting the anticancer drug in the form of nanoformulations at the tumor site could exhibit sustained release of drugs with a significant reduction of tumor volume. Clinical trials are under evaluation for this strategy, where nanotechnology-based liposomal formulation of cytarabine was intrathecally administered to the patient (Table 3) [35].

4.1.2 Brain infusion

CED method was developed in the early 1990s to administer nanocarriers through brain infusion. In this technique, a small-caliber catheter is introduced through a burr hole into the brain parenchyma. After insertion, it generates a convective flow under pressure and actively pumps the diffusate

TABLE 3 List of drug delivery system under clinical evaluation for the management of GBM [35].

System for drug delivery	Active	Route of administration	Active combined with	Phase of clinical trial	Identifier
Local treatments					
Liposomes	Cytarabine	Intrathecal	Temozolomide	Phase I/II	NCT01044966
Carmustine sustained release implant (CASANT)	Carmustine	Locally	None	Phase III	NCT01637753 NCT01656980
Wafers	Carmustine	Locally	Radiation therapy, bevacizumab, temozolomide	Phase II	NCT01186406
Wafers	Carmustine	Locally	5-amino-levulinic acid	Phase II	NCT01310868
Systemic treatments					
Liposomes	Irinotecan	IV	None	Phase I	NCT00734682
Nanoparticles	Etirinotecan pegol	IV	None	Phase II	NCT01663012

into the brain irrespective of drug solution diffusivity. The diffusivity of the drug from nanocarrier is based on its size, surface charge, and surface coating. These parameters play a very crucial role in the diffusivity of the drug from nanocarriers; hence are needed to be optimized during their administration through CED. For example, an optimized glucosamine conjugate coating on the liposome can maintain its in vivo safety for healthy brain tissues. Hence, the application of such nanocarriers for CED requires extensive optimization for achieving the desired delivery, along with safety [40].

CED technique has been widely studied in animal models using numerous liposomal formulations containing antitumor drugs. It was found that drug-containing liposomes were successfully distributed in tumor tissue and surrounding invasive tumor cells [41]. Additionally, surface-modified liposome with ability to cause "selective" toxicity to glioma cell without affecting the healthy brain cells/tissue can be an exciting approach for delivery of therapeutics [46]. Besides this, CED technology can also deliver the drugs with different antineoplastic mechanisms, or it can be given in combination with radiotherapy to achieve a synergic effect [42,43].

4.1.3 Local implantation

Local implantation of nanocarriers is an alternative technique to brain infusion and direct injection systems for the administration of anticancer agents. Polymeric nanocarriers like wafers, gels, or microchips are administered by this technique.

Gliadel® (Guilford Pharmaceuticals, Baltimore [EEUU]) is the first USFDA-approved polymer wafer that contains carmustine as an anticancer drug. This wafer is used during tumor surgery and placed near the tumor bed. However, its magnitude of benefit for a patient was relatively modest, leading to the clinical evaluation of new systems for the treatment of newer as well as recurrent malignant glioma (Table 3).

A biocompatible microcapsule device loaded with the therapeutic molecule, gel polymer-based system incorporated with controlled release formulation of chemotherapeutic drugs, microchips, and micro-electro-mechanical system (MEMS) as a source of controlled release of drugs are innovative methods currently under evaluation. The important benefit of these systems is that they can combine different drugs or therapies. This strategy combines various treatment modalities to take advantage of diverse pharmacologic mechanisms of the drugs. The synergic effect of combination therapy also makes it the right choice against tumor cells exhibiting different resistance mechanisms.

The liquid crystal polymer is used for the fabrication of microcapsule device. As compared with polymer wafers, these devices can load double amounts of a drug in a similar volume. A gelled polymer system loaded with chemotherapeutic drugs is composed of thermosensitive, biodegradable polymers. The OncoGel™ (MacroMed, Inc., Sandy [EEUU]) system is one of that kind where paclitaxel is loaded into the polymer base, i.e., ReGel™ (MacroMed, Inc., Sandy [EEUU]). This injectable formulation has shown the controlled release of paclitaxel from the polymer base and found safe and effective when evaluated in preclinical and clinical studies (NCT00479765). PLGA and PEG are the main components of this thermosensitive, biodegradable polymeric system. Microchip-based system is implanted in the cavity after brain tumor resection and act as controlled release systems for loaded drugs. Another advantage of such systems is their capability to combine and control the release of different types of drugs through various small reservoirs inside a small size. Comparative efficacy testing of microchips containing BCNU and locally implanted wafers has shown similar efficacy. Similarly, to control the drug release after implantation, the MEMS applies a short electrical pulse. Regardless of the advantages, this local delivery of the nanocarrier system has some limitations, as mentioned below:

- The implant size limits drug dosage.
- Distance from the resection cavity leads to an exponential decrease in drug diffusion. Hence, the lower drug concentration is observed in cells situated away from the insert.
- Need hospitalization of the patient.
- Associated with the risk of neurotoxicity, cerebral edema or infection, cerebrospinal fluid leak, poor wound healing, and tumor cyst formation.
- The patient needs to be anesthetized for the implantation of devices or catheters (e.g., microchips and CED)
- An experienced skill set is required for implantation.

4.2 Systemic administration of nanotherapeutics

Chemotherapeutic agents used for the treatment of GBM can cause harm to the healthier cells because they cannot differentiate between the tumor cell and healthy cells. Besides this, it is difficult for the chemotherapeutic agents to achieve significant therapeutic concentration in the brain because of their inability to cross BBB, or they become ineffective if tumor cells become resistant to it. In contrast, the systemic administration of nanocarriers enhanced the bioavailability and therapeutic concentration of drugs in the brain as they are capable of crossing the BBB and targeting the tumor cells. Nanocarriers also protect the drugs from degradation and maintain their therapeutic efficacy. Nanocarrier will be selected based on its ability to encapsulate a significant amount of drug and controlling its release in a stable form [35].

4.2.1 *Passive targeting*

Chemotherapeutics encapsulated into the nanocarriers were supposed to target the tumor site by the enhanced permeability and retention (EPR) effect. The speedy and inadequate tumor angiogenesis results in increased permeability of tumor vasculature that helps to deliver drugs at tumor cells. After intravenous administration, injected nanoparticles extravasate through the leaky gaps into the tumor tissue, and dysfunctional lymphatic drainage causes the collection of these nanoparticles in the tumor bed. However, this extravasation and retention of nanoparticles are primarily governed by their physical properties like size, shape, surface charge, and hydrophobicity. Nanoparticles having a particle size of <400 nm are usually considered to extravasate and accumulate in the interstitial space. The permeability of blood vessels in the tumor is mostly nonhomogenous and may not have control over the diffusion and spread of a nontherapeutic agent into the tumor mass.

Moreover, the impedance of diffusion by the brain extracellular space (ECS) and elevated interstitial pressure results in diminished EPR effect in tumors [44]. These are some of the probable causes of the disappointing performance of EPR-centered nanosystems in clinical use. These nanoparticles are generally unable to improve response rates or survival periods [35]. The minimal increment in tumor accumulation by passive targeting approach resulted in the development of alternative strategies: active targeting, which may cause significant enhancements in drug delivery to GBM tumor cells.

4.2.2 *Active targeting*

Nanoformulations are targeted actively by functionalizing their surface using the targeting moieties like ligands, antibodies, or peptides. These targeting moieties recognize the key structures expressed on the tumor surface or its associated endothelium. Active targeting (surface-functionalized nanoparticles) provides a significant accumulation of drug at the tumor region, and less in nontargeted organs. This ultimately reduces the associated side effects of chemotherapeutic agents on normal tissues. Identification of a suitable target and selection of highly specific targeting moiety are crucial for designing effective targeted nanoparticle [35,44].

4.2.2.1 Functionalization of nanocarriers

The surface of the nanocarriers is functionalized with an objective to enhance the concentration of drug in GBM. Functionalization for nanocarrier involves recognition of receptors, which are preferentially expressed on the surface of GBM tumor cells compared with nonneoplastic brain cells and then modification of the surface of nanoparticle with target-specific ligands, which eventually increase the affinity of nanoparticle for the targeted receptors overexpressed on the target cells. After binding to the targeted receptor, nanoparticles get internalized via receptor-mediated endocytosis to promote their cellular uptake and to enhance the therapeutic activity of an encapsulated drug. Moreover, considering the fact that metastatic cells may also express the targeted receptors, this strategy can also be applied for targeting such metastatic cells that have moved away from the tumor site [44]. Another type of active targeting is dual targeting that involves the functionalization of nanoparticle surface with moieties that bind to target expressed on tumor cells and the neovasculature around them [35]. This strategy is a step forward in developing an advanced drug delivery system to target BBB and tumors. In fact, in vivo evaluation of such multifunctional nanocarrier in animal studies exhibited enhanced accumulation in the tumor and improved survival time with a reduction in toxicities [35].

4.2.2.2 Systemic nanocarriers currently under clinical trial

Nanocarriers loaded with antitumor drugs provide a novel strategy for enhancing the bioavailability of the drug. Such systems possess the capability to deliver the drug to the brain in a noninvasive manner and also overcome many limitations presented by conventional dosage forms. Table 3 provides a brief description of systemic nanocarriers currently under clinical trials for brain tumor patients.

4.3 Other advanced technologies for GBM management

Although numerous approaches have been developed for delivering the drug at the tumor site, their efficacy remains an issue. To overcome this, studies are ongoing to evaluate other advanced technologies that explore different routes to bypass BBB (nose to brain) or different types of stimuli (internal and external) for enhanced accumulation.

4.3.1 Stimuli-responsive therapy

The recent advancement in materials chemistry and drug delivery has opened a new window for delivering drugs in response to stimuli. This stimuli-responsive drug delivery system has the ability to deliver the drug in spatial-, temporal-, and dosage-controlled fashions. Biocompatible materials that are subject to a specific physical stimulation or that, in response to a specific stimulus, show protonation, hydrolytic cleavage, or molecular conformational change are required for the development of stimuli-responsive systems. The stimuli used for this purpose are either external (variations in temperature, magnetic field, ultrasound intensity, light or electric pulses) or internal (changes in pH, enzyme concentration, or redox gradients). This stimuli-responsive system facilitates the on-demand or controlled release of the drug, signaling in specific positions for the activation of drug, exposure of ligand, enhanced accumulation at tumor cells, conversion of charge, and sensing of unique pathological factors [45,47,48].

4.3.2 Nose to brain delivery

Intranasal delivery of a drug is one of the promising approaches for the treatment of GBM. This route provides a safe and noninvasive route for delivering a drug into the brain [49]. Due to the local delivery of therapeutic agents across the olfactory mucosa and connected tissues, BBB is bypassed, resulting in reduced side effects associated with systemic drug delivery. It also holds advantages such as a simple route of administration, ease of repetitive, and programmed administration of the drug, and there is no need for invasive surgical procedures [49].

Various molecules have been evaluated until today to deliver drugs into the brain via the intranasal route. However, very few of them have shown promising effects. Perillyl alcohol (POH) is one of such kinds. It is a naturally occurring hydroxylated monoterpene having both hydrophilic and lipophilic properties. When administered intranasally, it solubilizes readily in biological membranes and interacts with bilayers of glioma cells and thereby effectively delivers into the tumor cells. Its intranasal formulation has shown promising results against a brain tumor. This formulation has now reached to the clinical trials of phase 1/2 [50].

5 Emerging therapies for targeting GBM

GBM is an aggressive tumor exhibiting neoangiogenesis and intratumor heterogeneity. Considering this fact, different types of genetic and epigenetic modifications were studied in GBM (Table 4). The studies have shown that the three main signaling pathways are dysregulated in GBM. In approximately 88% of cases, the receptor tyrosine kinase (RTK)/Ras/phosphoinositide 3-kinase (PI3K) pathway gets activated. However, inhibition was found in p53 (approx. 87% cases) and retinoblastoma protein (Rb) (approx. 78% cases) signaling pathways. Hence, it is proposed that drug targeting these alterations could be the potential targeted therapies for GBM [34].

5.1 Inhibition of EphA3 receptor

The tumor that initiates the cell population of glioma shows overexpression of the EphA3 receptor. This receptor maintains the tumor cells in a less differentiated and stem cell-like state. Additionally, in vitro study pointed that EphA3 facilitates tumorigenic potential in GBM cells, suggesting its candidature as a potential target for GBM [34].

Ifabotuzumab is the EphA3 monoclonal antibody (IIIA4) for human use, and it acts by binding to the EphA3 globular ephrin-binding domain. It is being investigated in the Phase 0/1 clinical trial in recurrent GBM patients. This study was aimed to find out the optimal dose for tumor penetration. In another study, a bispecific antibody was evaluated against EphA2/A3, and it showed in vitro reduction in clonogenicity and in vivo decline in tumor burden. Thus, it indicates the application of the antibodies for effective targeting and inhibition of the EphA3 receptor. After that, different studies were carried out in this direction and that have proposed the potential of EphA3 inhibitors for the treatment of GBM, which exhibits an amplified EphA3 receptor. However, only detailed clinical testing can give conclusive data for its use in treatment [34].

TABLE 4 Commonly identified genetic alteration in GBM [34].

Name	Function	Expression status	Prevalence
EPHA3	Regulates cell motility and adhesion properties	Overexpressed	40%–60%
EGFR	Regulates the process of cell division, its growth, and survival	Overexpressed	40%–60%
MGMT	Prevention of mismatch errors	Methylated	40%–60%
CDKN2A	Regulation of cell cycle and retinoblastoma activation	Decreased	49%–52%
PTEN	Regulates the signaling pathway involved in the proliferation of cells	Deleted and/or mutated	34%
PIK3CA	Regulates the process of cell division, its growth, and survival	Overexpressed and/or mutated	15%
PDGFRA	Regulates the process of cell division, its growth, and survival	Overexpressed	13%
IDH1	Production of NADPH	Mutated	5%–10%
MDM2	Regulation of p53 activity	Overexpressed	8%–9%
MET	Regulation of proliferation, survival, and motility	Overexpressed and/or mutated	4%–6%
SF/HGF	Activates tumor growth and angiogenesis	Overexpressed	1.6%–4%
VEGF	Promotion of angiogenesis	Overexpressed and/or mutated	40%–60%
Fn14	Intracellular signaling and promote migration, invasion, and cell survival	Overexpressed	70%–85%

5.2 Inhibition of epidermal growth factor receptor (EGFR)

The overexpression or mutation of EGFR is responsible for downstream signaling, resulting in reduced apoptosis leading to enhance proliferation and angiogenesis. In-frame deletion of 801 base pairs from the extracellular domain results in developing a mutated form of 1EGFR (EGFRvIII or de2-7EGFR), and this mutant is most commonly found in GBM. The high prevalence of EGFR mutation led to a thorough preclinical and clinical examination of EGFR inhibitors [34]. Some of the inhibitors, small molecule, and monoclonal antibodies are briefly discussed in the following sections.

5.2.1 Small molecule inhibitors

Tyrosine kinase inhibitors, including erlotinib, gefitinib, and lapatinib, are the well-studied EGFR inhibitors for GBM. The study performed on the GBM patient-derived xenograft (GBM-PDX) model showed the sensitivity of phosphate and tensin homolog (PTEN) against erlotinib. Another study demonstrated that combining erlotinib with TMZ and radiotherapy results in improved survival period (19.3 vs 14.1 months) in comparison with monotherapy [34].

5.2.2 Monoclonal antibodies

Cetuximab is the monoclonal antibody that shows its action against wild-type EGFR and 1EGFR in GBM. Preclinical studies with cetuximab alone or in combination with radiotherapy have demonstrated an increase in survival as well as complete tumor elimination in EGFR-amplified PDX models. A phase II trial in a patient with recurring GBM showed limited activity of cetuximab but established its safety in the patient population [34].

5.3 Inhibition of vascular endothelial growth factor (VEGF)

VEGF is overexpressed in GBM and is responsible for angiogenesis and cell proliferation. GBM tumors are commonly hypoxic, and VEGF-induced angiogenesis counteracts hypoxia, which explain increased VEGF expression and irregular vasculature in GBM. The function of VEGF and its overexpression in glioma highlighted an analysis of VEGF as a potential therapeutic target [34]. Bevacizumab is a monoclonal antibody that acts against VEGF and is specifically used in humans. In studies with the GBM mouse model, bevacizumab alone, or in combination with radiotherapy, has found useful in blocking the angiogenesis and subsequent tumor growth. Its similar encouraging performance in the clinical investigation led to approval from FDA [34].

5.4 Inhibition of multiple RTKs

Coactivation of multiple RTKs in glioma leads to redundancy and limited efficacy of single RTK targeting therapies. Co-expression of EGFR and platelet-derived growth factor receptor A (PDGFRA) occur in 37% of GBM.

Sunitinib is an oral, small-molecule, multitargeted receptor tyrosine kinase (RTK) inhibitor that inhibits PDFR and VEGFR. Inhibition of these receptors results in decreased vascularization, followed by an increase in apoptosis and tumor reduction. In preclinical models, sunitinib exhibited improved survival in an intracerebral GBM mouse model. Similarly, when it is used alone or combined with low-dose radiotherapy against the PDGFF-driven mouse model, it showed retardation in tumor growth and enhanced survival time [34].

5.5 Inhibition of PI3K

In GBMs, activation of the PI3K/Akt signaling pathway plays a vital role in the regulation of signal transduction and thereby facilitates the cellular processes like proliferation, migration, survival, and angiogenesis. PI3K pathway inhibitors may prove beneficial in GBM because of their role in downstream activation of the PI3K/Akt pathway. More than 50 PI3K inhibitors have shown their effectiveness against GBM in preclinical studies, and many of them are now under clinical evaluation [34].

5.6 Inhibition of HGFR/MET

Through activation of MET, the HGF receptor shows an increase in tumor growth and angiogenesis. An active variant of MET, MET17–8, has been recognized to enhance downstream signaling to boost tumor progression and angiogenesis in GBM patients.

SGX-523 is a small molecule used for the inhibition of HGFR/MET. It acts by inhibiting the tumor cellular processes like cell growth, relocation, and invasion of glioma cells. It has shown decreased tumor growth in a murine xenograft model of GBM, developed through the U87MG GBM cell line. Similarly, another small-molecule inhibitor of multiple tyrosine kinases, amuvatinib (MP470), has shown radiosensitivity for GBM cell lines [34].

5.7 Inhibition of fibroblast growth factor-inducible 14 (Fn14)

Fn14 is a member of the superfamily of tumor necrosis factor receptor (TNFR). In GBM, an increase in the Fn14 gene transcription and protein levels is observed at the tumor margin. It is an emerging molecular target because: (i) it is minimally expressed in the normal brain tissues, whereas highly expressed (~70%–85%) in most of the GBM tumors, (ii) its high expressiveness is also found in stationary as well as malignant glioma cell, and (iii) it undergoes constitutive receptor internalization for enabling therapeutic agent delivery. These findings indicate the suitability of Fn14 for targeting therapeutics to GBM cells [44].

6 Future perspectives

Application of microdevices has shown tremendous potential in the management of GBM as they directly inject the therapeutic agent into the brain. These strategies have advantages like bypassing the BBB, direct delivery to the tumor, safety of drugs from systemic degradation, and control over the amount of drug to be delivered. CED also has shown its usefulness in facilitating the local delivery of various types of nanocarrier in the broad area of the brain. However, limited drug penetration depth calls for identifying new targets to facilitate the development of effective nanotherapeutics and other advanced delivery systems. Nanotherapeutics, like active targeting nanocarriers, need to be studied in detail for its safety, biodistribution, and tumor targeting. Furthermore, nanotechnology can provide the scope of developing a combination of therapies resulting in unexpected synergies that can arise. It also offers considerable knowledge about transport mechanisms, biological processes, and cellular mechanisms that will help in the development of advance and effective drug delivery systems for GBM. Along with effectiveness, long-term toxicological examinations should be the prerequisite to ensure their safety. Hopefully, continued research efforts will improve the chances of recovery of patients suffering from GBM and thereby help in combating the disease in the future.

7 Conclusion

Regardless of extensive global efforts, complete management of GBM is still considered the most impenetrable task. From several decades, different types of treatment strategies were examined for GBM but have shown minimal success. Among these strategies, nanotechnology has emerged as a powerful strategy that can resolve many limitations of the current therapies, such as crossing the BBB or overcoming tumor-cell drug resistance. Some nanoformulations, such as liposomes, are under evaluations in clinical trials to treat GBM. However, many of these nanotechnology-based formulations are associated with various issues, like drug encapsulation, specificity, selectivity, toxicity, size, and composition. These problems need to be solved to ensure that nanoformulations could reach to the clinical application. Apart from formulation-related problems, the performance-related problems may get a probable solution through the genetic profiling of GBM biopsies. The recently approved bevacizumab immunotherapy has shown that a basic understanding of GBM biology and the identification of additional targets on tumor will help in the development of a clinically useful strategy for the management of GBM.

References

[1] Omuro A. Glioblastoma and other malignant gliomas. JAMA 2013;310:1842. https://doi.org/10.1001/jama.2013.280319.

[2] Goodenberger ML, Jenkins RB. Genetics of adult glioma. Cancer Genet 2012;205:613–21. https://doi.org/10.1016/j.cancergen.2012.10.009.

[3] Louis DN, Ohgaki H, Wiestler OD, Cavenee WK, Burger PC, Jouvet A, Scheithauer BW, Kleihues P. The 2007 WHO classification of tumours of the central nervous system. Acta Neuropathol 2007;114:97–109. https://doi.org/10.1007/s00401-007-0243-4.

[4] Iacob G, Dinca EB. Current data and strategy in glioblastoma multiforme. J Med Life 2009;2:386–93. http://www.ncbi.nlm.nih.gov/pubmed/20108752.

[5] Gutkin A, Cohen ZR, Peer D. Harnessing nanomedicine for therapeutic intervention in glioblastoma. Expert Opin Drug Deliv 2016. https://doi.org/10.1080/17425247.2016.1200557.

[6] Fernandes C, Costa A, Osório L, Lago RC, Linhares P, Carvalho B, Caeiro C. Current standards of care in glioblastoma therapy. In: Glioblastoma. Codon Publications; 2017. p. 197–241. https://doi.org/10.15586/codon.glioblastoma.2017.ch11.

[7] Rani V, Venkatesan J, Prabhu A. Nanotherapeutics in glioma management: advances and future perspectives. J Drug Deliv Sci Technol 2020;57:101626. https://doi.org/10.1016/j.jddst.2020.101626.

[8] Reilly KM. Brain tumor susceptibility: the role of genetic factors and uses of mouse models to unravel risk. Brain Pathol 2009;19:121–31. https://doi.org/10.1111/j.1750-3639.2008.00236.x.

[9] Hanif F, Muzaffar K, Perveen K, Malhi SM, Simjee SU. Glioblastoma multiforme: a review of its epidemiology and pathogenesis through clinical presentation and treatment. Asian Pac J Cancer Prev 2017;18:3–9. https://doi.org/10.22034/APJCP.2017.18.1.3.

[10] Huson SM. Neurofibromatosis: phenotype, natural history and pathogenesis. J Med Genet 1993;30:351. https://doi.org/10.1136/jmg.30.4.351.

[11] Nam L, Coll C, Erthal L, de la Torre C, Serrano D, Martínez-Máñez R, Santos-Martínez M, Ruiz-Hernández E. Drug delivery nanosystems for the localized treatment of glioblastoma multiforme. Materials (Basel) 2018;11:779. https://doi.org/10.3390/ma11050779.

[12] Fakhoury M. Drug delivery approaches for the treatment of glioblastoma multiforme. Artif Cells Nanomed Biotechnol 2016;44:1365–73. https://doi.org/10.3109/21691401.2015.1052467.

[13] Park SS, Chunta JL, Robertson JM, Martinez AA, Oliver Wong C-Y, Amin M, Wilson GD, Marples B. MicroPET/CT imaging of an orthotopic model of human glioblastoma multiforme and evaluation of pulsed low-dose irradiation. Int J Radiat Oncol 2011;80:885–92. https://doi.org/10.1016/j.ijrobp.2011.01.045.

[14] Adamson C, Kanu OO, Mehta AI, Di C, Lin N, Mattox AK, Bigner DD. Glioblastoma multiforme: a review of where we have been and where we are going. Expert Opin Investig Drugs 2009;18:1061–83. https://doi.org/10.1517/13543780903052764.

[15] Fakhoury M, Takechi R, Al-Salami H. Drug permeation across the blood-brain barrier: applications of nanotechnology. Br J Med Med Res 2015;6:547–56. https://doi.org/10.9734/BJMMR/2015/15493.

[16] Khalin I, Alyautdin R, Ismail NM, Haron MH, Kuznetsov D. Nanoscale drug delivery systems and the blood–brain barrier. Int J Nanomedicine 2014;795. https://doi.org/10.2147/IJN.S52236.

[17] Serwer LP, James CD. Challenges in drug delivery to tumors of the central nervous system: an overview of pharmacological and surgical considerations. Adv Drug Deliv Rev 2012;64:590–7. https://doi.org/10.1016/j.addr.2012.01.004.

[18] Clark DE. In silico prediction of blood–brain barrier permeation. Drug Discov Today 2003;8:927–33. https://doi.org/10.1016/S1359-6446(03)02827-7.

[19] Lalatsa A, Schätzlein AG, Uchegbu IF. Drug delivery across the blood–brain barrier. In: Compr. Biotechnol. Elsevier; 2011. p. 628–37. https://doi.org/10.1016/B978-0-444-64046-8.00313-X.

[20] Gleeson MP. Generation of a set of simple, interpretable ADMET rules of thumb. J Med Chem 2008;51:817–34. https://doi.org/10.1021/jm701122q.

[21] Serrano Lopez DR, Lalatsa A. Peptide pills for brain diseases? Reality and future perspectives. Ther Deliv 2013;4:479–501. https://doi.org/10.4155/tde.13.5.

[22] Hervé F, Ghinea N, Scherrmann J-M. CNS delivery via adsorptive transcytosis. AAPS J 2008;10:455–72. https://doi.org/10.1208/s12248-008-9055-2.

[23] Liu W-Y, Wang Z-B, Zhang L-C, Wei X, Li L. Tight junction in blood-brain barrier: an overview of structure, regulation, and regulator substances. CNS Neurosci Ther 2012;18:609–15. https://doi.org/10.1111/j.1755-5949.2012.00340.x.

[24] Ronaldson PT, Davis TP. Targeting blood–brain barrier changes during inflammatory pain: an opportunity for optimizing CNS drug delivery. Ther Deliv 2011;2:1015–41. https://doi.org/10.4155/tde.11.67.

[25] Patel MM, Goyal BR, Bhadada SV, Bhatt JS, Amin AF. Getting into the brain: approaches to enhance brain drug delivery. CNS Drugs 2009;23:35–58. https://doi.org/10.2165/0023210-200923010-00003.

[26] van Tellingen O, Yetkin-Arik B, de Gooijer MC, Wesseling P, Wurdinger T, de Vries HE. Overcoming the blood–brain tumor barrier for effective glioblastoma treatment. Drug Resist Updat 2015;19:1–12. https://doi.org/10.1016/j.drup.2015.02.002.

[27] Ostermann S. Plasma and cerebrospinal fluid population pharmacokinetics of temozolomide in malignant glioma patients. Clin Cancer Res 2004;10:3728–36. https://doi.org/10.1158/1078-0432.CCR-03-0807.

[28] Laquintana V, Trapani A, Denora N, Wang F, Gallo JM, Trapani G. New strategies to deliver anticancer drugs to brain tumors. Expert Opin Drug Deliv 2009;6:1017–32. https://doi.org/10.1517/17425240903167942.

[29] Lesniak MS, Upadhyay U, Goodwin R, Tyler B, Brem H. Local delivery of doxorubicin for the treatment of malignant brain tumors in rats. Anticancer Res 2005;25:3825–31. http://www.ncbi.nlm.nih.gov/pubmed/16312042.

[30] Zhan C, Gu B, Xie C, Li J, Liu Y, Lu W. Cyclic RGD conjugated poly(ethylene glycol)-co-poly(lactic acid) micelle enhances paclitaxel anti-glioblastoma effect. J Control Release 2010;143:136–42. https://doi.org/10.1016/j.jconrel.2009.12.020.

[31] Kondo Y, Kondo S, Tanaka Y, Haqqi T, Barna BP, Cowell JK. Inhibition of telomerase increases the susceptibility of human malignant glioblastoma cells to cisplatin-induced apoptosis. Oncogene 1998;16:2243–8. https://doi.org/10.1038/sj.onc.1201754.

[32] Wang PP, Frazier J, Brem H. Local drug delivery to the brain. Adv Drug Deliv Rev 2002;54:987–1013. https://doi.org/10.1016/S0169-409X(02)00054-6.

[33] Norden AD, Wen PY. Glioma therapy in adults. Neurologist 2006;12:279–92. https://doi.org/10.1097/01.nrl.0000250928.26044.47.

[34] Taylor OG, Brzozowski JS, Skelding KA. Glioblastoma multiforme: an overview of emerging therapeutic targets. Front Oncol 2019;9:1–11. https://doi.org/10.3389/fonc.2019.00963.

[35] Saenz del Burgo L, Hernández RM, Orive G, Pedraz JL. Nanotherapeutic approaches for brain cancer management. Nanomed Nanotechnol Biol Med 2014;10:e905–19. https://doi.org/10.1016/j.nano.2013.10.001.

[36] Mun EJ, Babiker HM, Weinberg U, Kirson ED, Von Hoff DD. Tumor-treating fields: a fourth modality in cancer treatment. Clin Cancer Res 2018;24:266–75. https://doi.org/10.1158/1078-0432.CCR-17-1117.

[37] Cha GD, Kang T, Baik S, Kim D, Choi SH, Hyeon T, Kim D-H. Advances in drug delivery technology for the treatment of glioblastoma multiforme. J Control Release 2020;328:350–67. https://doi.org/10.1016/j.jconrel.2020.09.002.

[38] Rezaei V, Rabiee A, Khademi F. Glioblastoma multiforme: a glance at advanced therapies based on nanotechnology. J Chemother 2020;32:107–17. https://doi.org/10.1080/1120009X.2020.1713508.

[39] Ortiz R, Cabeza L, Perazzoli G, Jimenez-Lopez J, García-Pinel B, Melguizo C, Prados J. Nanoformulations for glioblastoma multiforme: a new hope for treatment. Future Med Chem 2019;11:2459–80. https://doi.org/10.4155/fmc-2018-0521.

[40] Yadav N, Rajendra J, Acharekar A, Dutt S, Vavia P. Effect of glucosamine conjugate-functionalized liposomes on glioma cell and healthy brain: an insight for future application in brain infusion. AAPS PharmSciTech 2020;21:24. https://doi.org/10.1208/s12249-019-1567-9.

[41] Saito R, Krauze MT, Noble CO, Drummond DC, Kirpotin DB, Berger MS, Park JW, Bankiewicz KS. Convection-enhanced delivery of Ls-TPT enables an effective, continuous, low-dose chemotherapy against malignant glioma xenograft model1. Neuro Oncol 2006;8:205–14. https://doi.org/10.1215/15228517-2006-001.

[42] Yamashita Y, Krauze MT, Kawaguchi T, Noble CO, Drummond DC, Park JW, Bankiewicz KS. Convection-enhanced delivery of a topoisomerase I inhibitor (nanoliposomal topotecan) and a topoisomerase II inhibitor (pegylated liposomal doxorubicin) in intracranial brain tumor xenografts1. Neuro Oncol 2007;9:20–8. https://doi.org/10.1215/15228517-2006-016.

[43] Allard E, Jarnet D, Vessières A, Vinchon-Petit S, Jaouen G, Benoit J-P, Passirani C. Local delivery of ferrociphenol lipid nanocapsules followed by external radiotherapy as a synergistic treatment against intracranial 9L glioma xenograft. Pharm Res 2010;27:56–64. https://doi.org/10.1007/s11095-009-0006-0.

[44] Wadajkar AS, Dancy JG, Hersh DS, Anastasiadis P, Tran NL, Woodworth GF, Winkles JA, Kim AJ. Tumor-targeted nanotherapeutics: overcoming treatment barriers for glioblastoma. Wiley Interdiscip Rev Nanomed Nanobiotechnol 2017;9. https://doi.org/10.1002/wnan.1439, e1439.

[45] Mi P. Stimuli-responsive nanocarriers for drug delivery, tumor imaging, therapy and theranostics. Theranostics 2020;10:4557–88. https://doi.org/10.7150/thno.38069.

[46] Yadav N, Rajendra J, Acharekar A, Dutt S, Vavia P. Effect of glucosamine conjugate-functionalized liposomes on glioma cell and healthy brain: an insight for future application in brain infusion. AAPS PharmSciTech 2020;21(1). https://doi.org/10.1208/s12249-019-1567-9.

[47] Yu J, Chu X, Hou Y. Stimuli-responsive cancer therapy based on nanoparticles. Chem Commun 2014;50:11614–30. https://doi.org/10.1039/C4CC03984J.

[48] Yao J, Feng J, Chen J. External-stimuli responsive systems for cancer theranostic. Asian J Pharm Sci 2016;11:585–95. https://doi.org/10.1016/j.ajps.2016.06.001.

[49] Zhang H, Wang R, Yu Y, Liu J, Luo T, Fan F. Glioblastoma treatment modalities besides surgery. J Cancer 2019;10:4793–806. https://doi.org/10.7150/jca.32475.

[50] Bruinsmann FA, Richter Vaz G, de Cristo Soares Alves A, Aguirre T, Pohlmann AR, Guterres SS, Sonvico F. Nasal drug delivery of anticancer drugs for the treatment of glioblastoma: preclinical and clinical trials. Molecules 2019;24:4312. https://doi.org/10.3390/molecules24234312.

Chapter 15

Advanced drug delivery systems in prostate cancer

C. Sarath Chandran, Alan Raj, and T.K. Shahin Muhammed
College of Pharmaceutical Sciences, Government Medical College Kannur, Kerala, India

1 Introduction

According to global death statics, cancer is in the second position. One in six death in the world is reported due to cancer. Day by day the death rate due to cancer is increasing, which may be due to several reasons including poor diagnosis, improper treatment strategies, etc. The main cause of cancer is due to the defect or mutation in genetic materials inside the cell, which is caused either by environment factors or by inherited factors. The proto-oncogenes are involved in protein synthesis, cell proliferation, and differentiation. Any mutation in this gene may lead to the production of inhibitory signals for cell division and induce apoptosis [1, 2]. There are the few exogenous and endogenous factors that contribute to cancer pathogenesis, which includes lifestyle, exposure to radiations, chemical agents, tobacco smoking, nutritional habits, etc. The viruses are reported to cause cancer, i.e., human papillomavirus (HPV) induces cervical cancer [3].

The prostate cancer is mainly reported in men, which is characterized by abnormal prostate gland with uncontrollable proliferation. Based on global cancer statics, prostate cancer is in second position among other cancer in men [4]. When compared with other cancers, the mortality risk is less [5, 6]. According to US National Cancer Institute, prostate cancer is developed in four stages, they are early stage or stages I and II, where tumor is concise in the prostate gland. The stage III or locally advanced stage, i.e., when spreading starts, and the last one is advanced stage or stage IV, in which the cancer spreads outside the prostate gland and affect other parts including lymph nodes, bones, liver, or lungs [7]. Men below 65 years with mutation in BRCA-2 gene are more prone to prostate cancer, also individuals with a previously reported urinary tract infection are more susceptible to this tumor [8, 9].

The diagnostic tools for the detection of prostate cancer include digital rectal examination, blood level of prostate-specific antigen (PSA), and transrectal ultrasound biopsies. The abnormality in digital rectal examination or elevated level of PSA helps in accurate diagnosis. When comparing digital rectal examination and transrectal ultrasound, the PSA level examination helps to predict prostate cancer more accurately [10]. The PSA level of less than 4 ng/mL is fixed as normal, over 10 ng/mL considered as high, and in between 4 and 10 ng/mL as intermediate. If the PSA level is high, there is a higher chance of spreading of cancer to other organs. In some individuals, the PSA level is less than 4 ng/mL, even if in cancerous condition, and the PSA level is high in noncancerous conditions like benign prostate hyperplasia [11, 12].

Early detection and treatment can bring back the patient to normal condition. The symptoms of prostate cancer included difficulty in urination, blood in urine or semen, and pain or discomfort when sitting. The treatment options for prostate cancer depended upon the stage of cancer progression. The current treatment strategies are surgery and radiation therapy which included: brachytherapy, conformal radiation therapy, chemotherapy, hormonal therapy, etc. But these treatment led the patient to infertility in future. The removal of prostate glands through surgery may affect the semen production, and radiation therapy may damage the sperm, normal urinary, and bowel function [13]. Hence, these problems may be overcome by introducing novel drug delivery systems like nanoparticles for the effective delivery of drug to the tumor cell thereby reducing the cytotoxicity of normal cells.

2 Targeted drug delivery system

A drug delivery system can be defined as a formulation or a device which helps to deliver a therapeutic substance in the body to improve its efficacy, safety, and control the release rate, time, and place of release of the drug in the body [14–16]. Targeted drug delivery system is considered as a type of novel drug delivery system, in which the drug/biological agent is specifically targeted to an organ/tissue and release

Advanced Drug Delivery Systems in the Management of Cancer. https://doi.org/10.1016/B978-0-323-85503-7.00034-1

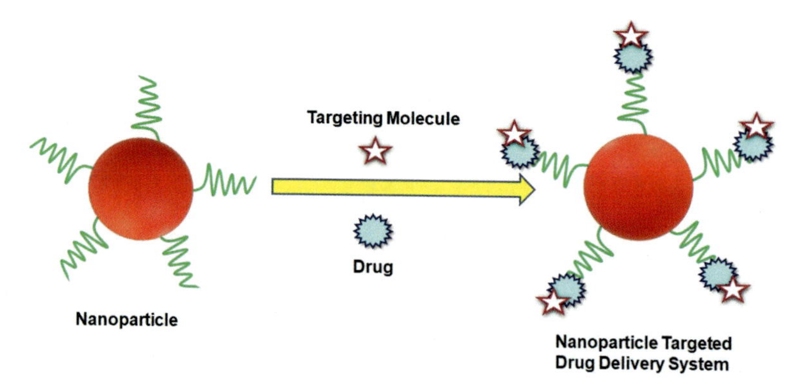

FIG. 1 Drug targeting via nanoparticle.

the active therapeutic agent. It has many advantages over the conventional route of drug delivery. It enhances the stability and efficient delivery of the drug, reduce unwanted side effects, reduce the dose of drug, and protect the nontargeted organs/tissues. The application nanotechnology is widely utilized to deliver the molecule to a specific target organ without any side effects [17, 18]. Fig. 1 shows a schematic representation of drug delivery with the aid of nanoparticle. The release of drug from the dosage form is important, it should be in a controlled manner in order to obtain a better therapeutic activity. There were various nanoparticles available for the effective delivery of drugs. The drug/s are delivered to the target organ with the help of nanoparticle via passive or active approach.

3 Passive and active targeting

Passive targeting of drug to cancer cells by the enhanced permeation and retention effect (EPR) is one of the best approaches in cancer therapy. It is a unique character of cancer cells in which the cancer cells have high permeability toward macromolecules. Due to the absence of lymphatic drainage in cancer cells, these particles will accumulate in these cells [19]. It has to be noted that, for utilizing the EPR effect, the nanoparticles should circulate in the systemic circulation for longer time. Surface modification with polyethylene glycol (PEG) made them more hydrophilic, which enhances the circulation time in blood [20]. The reticuloendothelial system showed a faster uptake of negatively charged nanoparticles, so modification in surface charge (positive/neutral) will increase the circulation in blood. Apart from this, not all tumors exhibit the EPR effect, especially the metastatic tumor cells because of extreme hypoxia. The EPR effect in solid tumors is enhanced externally by the following means: by slow infusion of angiotensin-converting enzyme II, applying nitric oxide, and releasing agent topically. It is not just the molecular size and weight that determine the EPR effect but also the surface charge, and biocompatibility should be considered while formulating the nanoparticles for cancer cell targeting [21].

The passive targeting is mainly focused in EPR effect of cancerous cells, but it is heterogeneous, which means its strength varies with different tumors and individuals. Hence, active targeting to the cancer cell is one of the best approaches to achieve better stability and therapeutic efficacy of a drug. In active targeting approach, the drug is delivered to the subcellular components [22]. The surface of nanoparticle was modified with polymers, and biological agents like antibodies, aptamers, which have more affinity toward the target site [23, 24]. The nanoparticle will deliver the drug/s to extracellular matrix of tumor cells and diffuse into the cell via receptor-mediated endocytosis [25]. Fig. 2 gives a detailed overview of passive and active approach.

4 Nanoparticle delivery systems in prostate cancer

Nanoparticles have great possibilities in cancer therapy. Their unique properties in terms of size, surface characteristics, and shape play a key role in cancer therapy. The size of nanoparticles are within the range 1–1000 nm. The nano size range and unique properties of nanoparticles are explored to deliver the poorly water-soluble drugs (anticancer drugs) for site-targeted delivery of chemical and biological substances. The drug loaded nanoparticles are widely applied in cancer therapy as a delivery tool, for diagnosis and for both (theranostic agent) [26, 27]. The nanoparticle carrier system is composed of either synthetic material or natural origin. The natural compounds like hyaluronic acid, heparin, dextran, chitosan, cellulose, and derivatives are frequently used because they are nontoxic and nonimmunogenic [28]. Recently, gold nanoparticles are extensively used in cancer therapy. Large molecules like DNA, RNA, or proteins can conjugate with gold nanoparticles for the effective delivery [29]. The targeting ability of nanoparticle can be enhanced by conjugating it with certain ligands, aptamers, etc. The most common nanoparticles used in prostate cancer are liposome, polymeric nanoparticles, micelles, dendrimers, and carbon-based nanoparticles (Fig. 3).

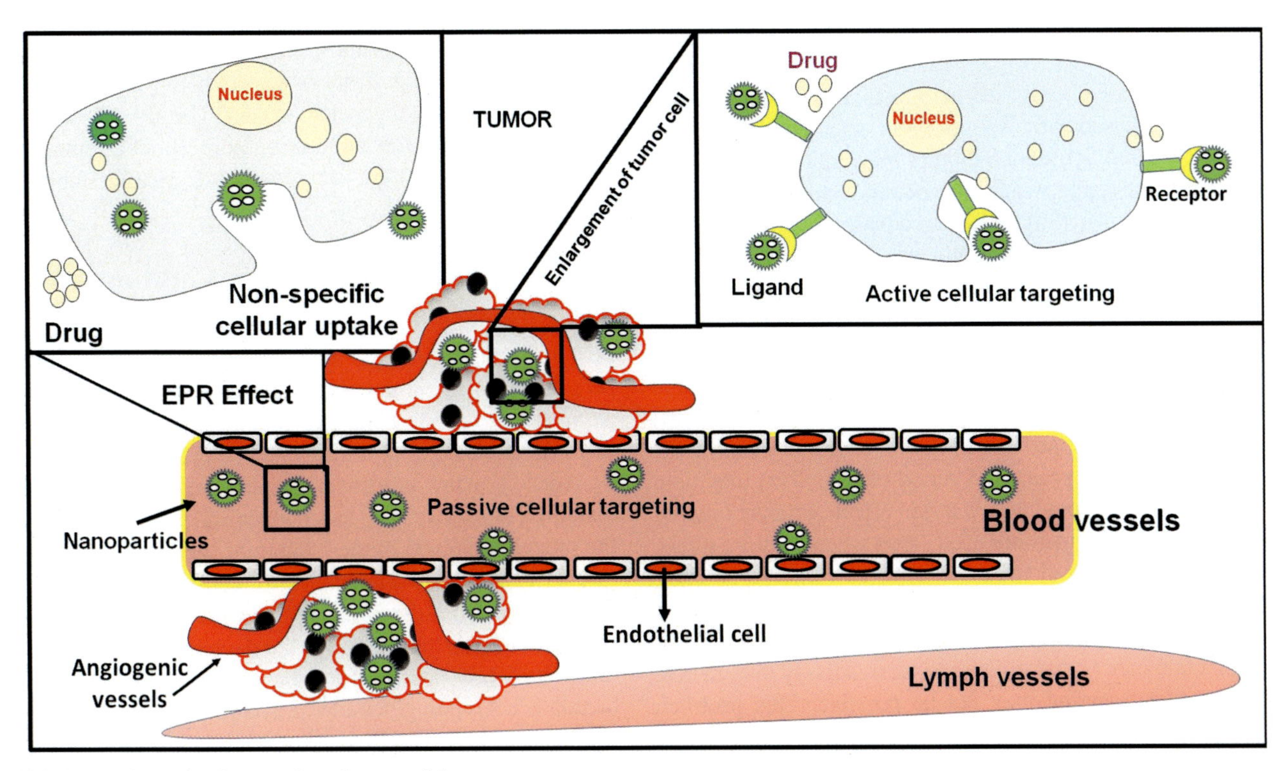

FIG. 2 Passive and active targeting of nanoparticles.

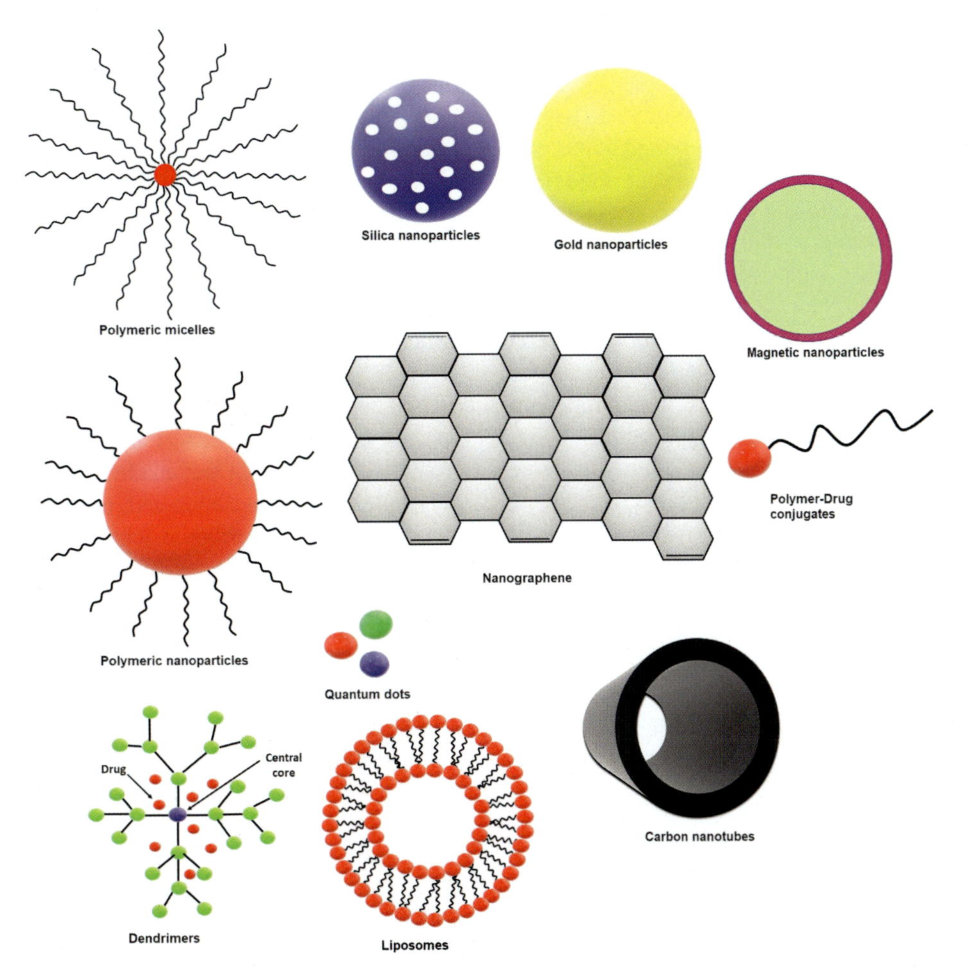

FIG. 3 Nanoparticle-based targeting.

5 Liposome

Liposomes are lipid-based nanovesicular systems. It is mainly composed of phospholipids, in which it had a hydrophilic core and hydrophobic outer bilayer. The hydrophilic drugs are entrapped within the core and hydrophobic drugs in the outer layer [30]. It has an advantage in biocompatibility and biodegradability, and also it can protect the drug from degradation due to enzymes that enhance the bioavailability [31]. The anticancer liposomes can be passively targeted to the prostate cancer cells by EPR effect because the tumor cells have high permeability to macromolecules due to the presence of vascular mediators. But the lack of lymphatic drainage helped the liposome to accumulate in the prostate gland for prolonged period of time. This may help to reduce the dose and possible side effects when compared with nontargeted methods [32]. The vascularity to tumor cells is differentiated from one type of tumor to another, thus the EPR effect is not preferable and the drug-loaded nanoparticles cannot accumulate in tumor cells [33]. In case of prostate cancer, the tumor vascularization is low; hence, the passive targeting may not be possible. This problem may be overcome by active targeting approach [34].

The targeting efficiency of liposome may be further enhanced by incorporating PEG, aptamers, antibodies, proteins, peptides, ligands, and carbohydrates. The PEG-liposomes have the ability to overcome the detection of mononuclear phagocytic system to reduce the cellular toxicity. The liposomes may enhance the stability and targeting property of entrapped drug, but it was reported with a major drawback of not delivering the drugs in a controlled rate.

The prostate cancer cells are overexpressed by gastrin-releasing peptide receptors (GRPR). Docetaxel is an anticancer drug which is extensively used to treat prostate cancer. The hydrophobic nature of docetaxel limits its therapeutic use, and Zhang et al. made an attempt to overcome the limitation and improved the treatability by incorporating the docetaxel in self-assembling hybrid ELP/liposome. The targeting property was modified by attaching gastric-releasing peptide (GRP) on the surface of hybrid liposomes. It was evaluated in vitro by MTT assay. The reports suggested that the hybrid liposome specifically bound to GRP receptors and cell viability at very low dose [35].

Bauhinia purpurea agglutinin (BPA) is a lecithin that extensively bind to prostate cancer cells rather than normal cells. Ikemoto et al. conjugated the targeting ability of BPA with PEG-modified liposomes. This BPA-modified liposome was evaluated in terms of targeting property. The study suggested that the modified liposome significantly suppressed the tumor growth in prostate cancer cells [36].

High concentrations of specific enzyme were detected in prostate cancer cells, which included secretory phospholipase A2 (sPLA2) and prostate-specific antigen (PSA) and cathepsin [37–39]. Recently, a new approach of incorporating drugs into enzyme-responsive liposomes was adopted for the delivery of chemotherapy agents for the effective treatment of prostate cancer. The sPLA2 enzyme-binding liposome loaded doxorubicin showed enhanced tumor growth reduction activity than sterically stabilized liposomes [40].

6 Polymeric nanoparticles and micelles

Polymeric nanoparticles are small (10–1000 nm), spherical, branched, or core-shelled structures with great potential in delivering the chemotherapeutic agents [41]. The polymeric nanoparticle has the ability in specific targeting as well as controlled releasing of drug, which improves the therapeutic index of many anticancer agents. The nanoparticles are classified as nanospheres and nanocapsules depending upon the process involved in the formation. The targeting ability of polymeric nanoparticle helps to reduce the cytotoxic effect of anticancer drugs and protect the normal cells. The examples of polymers used for the development of nanoparticles included poly(lactide-*co*-glycolide) (PLGA), polylactide (PLA), polyglycolide, polycaprolactone (PCL), and poly(D,L-lactide). The PLGA is extensively used in prostate cancer therapy [42]. A new multifunctional nanoparticle was designed to treat prostate cancer, in which PLGA, PEG, and an aptamer A10 were linked together to form an amphiphilic triblock copolymer. These multifunctional nanoparticles have the ability to bind with prostate-specific membrane antigen (PSMA) which is overexpressed on the surface of prostate cancer cells. The Pegaptanib, an aptamer, has a single-stranded nucleic acid with an ability to bind specifically to the 165 amino acid form of vascular endothelial growth factor VEGF165 protein in prostate cancer cells.

The newly developed, multifunctional, polymer-lipid hybrid, nanoparticles are able to deliver anticancer agents efficiently to the prostate cancer cells. It had both the advantages of liposome and polymeric nanoparticles [43]. The US FDA-approved anticancer drug, docetaxel, for the treatment of metastatic prostate cancer helps to improve the disease condition. An attempt was made by Yan et al. to formulate docetaxel-curcumin-conjugated, polymer-lipid hybrid type, nanoparticles. The proposed formulation had the ability to overcome the multidrug resistance by anticancer drugs. The prepared formulation was evaluated ex vivo and in vivo, in human prostate carcinoma cell lines by MTT assay and PC3 tumor xenografts in mice, respectively. It showed high antitumor activity against tumor cells (ex vivo and in vivo) without any side effects [44].

The continuous delivery of chemotherapeutic agent to tumor cells may lead to possible resistance. This tumor resistance is due to the changes that occurred in DNA structure, and one of the major reasons may be the binding of drug to the macromolecules such as fibrous proteins, collagen, etc., present on the extracellular matrix.

The polymer-lipid hybrid nanoparticles have the ability to overcome the chemotherapeutic resistance [45]. The sphingosine kinase 1 is a protoenzyme, mainly seen in prostate tumor cells, that causes chemoresistance. The fingolimod (FTY720) is a potential inhibitor of sphingosine kinase 1 which helped in reducing the chemoresistance to docetaxel. Wang et al. incorporated these two drugs in hybrid-lipid-polymer nanoparticle to reduce the unwanted systemic absorption and enhance the targeting ability of anticancer drug docetaxel. The evaluation results suggested that the multifunctional nanoparticles have better tumor suppressing activity even in low dose when compared with free drug [46].

The polymeric micelles have a vital role in delivering chemotherapeutic agents in prostate cancer. They are self-assembled colloidal particulates within a size range 5–100 nm. It had an ability to carry hydrophobic drugs in its inner core [47, 48]. Due to its small size, it can passively target tumor cells, especially in poorly vascularized cells. The poly(ethylene glycol) (PEG)-based micelle was extensively used in the field of prostate cancer management. It is highly water soluble, neutral, and nontoxic in nature. The presence of hydrophilic molecule in its surface helps to block the nonspecific binding of micelles with blood components, and results in prolonged circulation time of micelle in blood [49]. Other polymers used to prepare micelles include poly(*N*-vinyl pyrrolidone) (PVP) and poly(*N*-isopropylacrylamide) (PNIPAm) [50].

Cabazitaxel is a US FDA-approved anticancer drug in the management of prostate cancer. It has low aqueous solubility and poor targeting property. The enzyme-responsive polymeric micelle was composed of two amphiphilic block copolymers. The enzyme in the micelle had the ability to bind with matrix metalloproteinase-2 (MMP-2), which was highly expressed in the surface of cancer cells. The incorporation of Cabazitaxel in an enzyme-responsive polymeric micelle had better results in inhibiting tumor growth when compared with normal micelle and free cabazitaxel [51].

7 Dendrimers

Dendrimers are polymeric nanoparticles in which it looks like a tree. They are hyperbranched macromolecular substances and its size ranges from 1 nm to 10 nm [52, 53]. It has some unique properties in terms of size, monodispersity, long or short branches around a central core, and it was synthesized quickly. The interesting fact is that the relative molecular mass and chemical composition of dendrimer can easily regulate, thereby, we can predict its pharmacokinetics and biodistribution [54]. The drug is encapsulated within the center of the dendrimer which helps to deliver the drug in a controlled manner [55]. The dendrimers are synthesized by two methods, i.e., divergent method and convergent method. In divergent method, the repeating molecules were bound directly to the inner core of the dendrimer. The disadvantages of this method are low production yield and difficulty in purification, which may be due to the large quantity of reagents used for the production [56, 57]. In convergent method, initially the polymer segments are prepared and finally joined together. The purification process is easier in this step, i.e., each segment can be subsequently purified. Following polymers are extensively used in the field of dendrimer preparation: poly(amidoamine) and poly(propylene imine), other polymers include polypeptide scaffolds, polyester, etc., which have improved biodegradable property [58, 59].

Dendrimer consists of three parts: an inner core, monomer branches, and a terminal group. The monomers are connected to central core (G_o) called as first-generation dendrimers (G_1); and in second generation (G_2), the second type monomers were bound to the G_0. The successive addition of these monomers to the previous generation will get next generations. The molecular weight of dendrimer will be doubled in each addition [60]. Modification in the terminal part of dendrimer will get charged (positive or negative), hydrophilic or hydrophobic dendrimers for desired action [61, 62].

As we know that PMSA is highly expressed in the cell surface of prostate cancer, glutamate urea ligand has a high affinity toward PMSA. Lim et al. made an attempt to conjugate urea-based ligand 2-[3-(1,3-dicarboxy propyl)-ureido] pentanoic acid (DUPA) to triazine dendrimer. The triazine dendrimer was prepared by five steps starting with dichlorotriazine with DUPA naming G_1, G_2, and G_5. In the ex vivo, in vivo data suggested that G_1 has smaller size and possesses low molecular weight. This dendrimer conjugate showed better targeting property toward prostate cancer cells [63].

Monoclonal antibodies (mAb) are currently used to treat prostate cancer. PSMA is a type II membrane glycoprotein that is present on the surface of prostate cancer cells. Monoclonal antibody J591, which is under clinical trial, has high membrane targeting toward PSMA. It is a deimmunized IgG monoclonal antibody [64]. PAMAM dendrimers are extensively used in the area of prostate cancer. Conjugating anti-PSMA mAb J591 dendrimer effectively delivers the mAb at the target site [65, 66].

8 Gold nanoparticles

Gold nanoparticles (AuNp) are nanoparticles of gold with an average size of 1–100 nm. If it is in aqueous medium, we get colloidal gold. The nanoparticles are small and have high surface energy, which help to conjugate wide variety of molecules. The size of particle is important because small size of gold nanoparticles have targeting capacity [67, 68]. Because of these properties, it is extensively used to treat cancer, especially as a targeting moiety in cancer therapy. The application of gold nanoparticles is extensively used in photon-based therapy in cancer management [69].

While designing gold nanoparticles, some factors such as size, charge of the molecules, and surface chemistry shall be considered. The type of interaction between the gold nanoparticles and drug should be known (covalent/noncovalent binding) [70]. The possible challenge in the delivery of gold nanoparticles is the barrier properties of epithelial and endothelial cells. This problem was covered by applying penetration enhancers. Apart from these factors, drug released from the gold nanoparticles is depending on the strength of the drug-nanoparticle complex and the time of drug release from the nanoparticle [71]. Chemical methods like Turkevich, Frens, Brust biphasic, microemulsion, seeding, etc. are the commonly used methods for the preparation of gold nanoparticles.

Gold nanoparticles is a strong backbone for the success of prostate cancer therapy. Tsai et al. formulated 67 kDa laminin receptor (67LR)-MMP responsive nanoparticle of Doxorubicin for prostate cancer treatment. Initially, the Doxorubicin is combined with gelatine, then this molecule is conjugated with epigallocatechin gallate (EGCG)-AuNp complex, which is prepared by layer-by-layer assembly method. The results revealed that the gold nanoparticle complex inhibited PC-3 cell proliferation. The 67-LR-mediated delivery of Doxorubicin showed better cellular uptake and induced apoptosis of PC-3 cells [72].

Butterworth et al. designed nanoparticle by reduction in Diethylenetriaminepentaacetic acid (DTDTPA)-conjugated gold nanoparticles (AuNp-DTDTPA), which was utilized as theranostic agent for the management of prostate cancer. The final data revealed that the AuNp-DTDTPA was able to reduce tumor cell growth from 16 to 38 days compared with normal [73]. The targeting capacity of gold nanoparticles can be improved by modifying the surface charge for conjugating siRNA via electrostatic interaction. The cationic gold nanoparticles will protect the siRNA from enzymatic degradation. Fitzgerald revealed the potential use of gold nanoparticles for the effective delivery of anisamide-labeled gold nanoparticles of siRNA. Anisamide (AA) had the ability to bind with the sigma receptor present on prostate cancer cells. The results revealed that the modified gold nanoparticle successfully bind with the PC-3 cell lines to reduce the cancer progression in vitro [74].

9 Carbon-based nanoparticles

Various applications were reported for carbon-based nanoparticles (CBN) as theranostic agent, which means it is used for both imaging and treatment purpose. The carbon-based nanoparticles include fullerenes, carbon nanotubes (CNT), graphene derivatives, carbon quantum dots (QDs), and nanodiamonds (NDs) [75]. It has the ability to stimulate the reactive oxygen species (ROS) in cancer cells, which led to the damage of DNA and lipid components in the cancer cells [76, 77]. Carbon-based nanoparticle graphene causes cell death by similar mechanism with slight modification. It damages mitochondria, resulting in the production of ROS, which lead to DNA damage followed by apoptosis [78].

Carbon-based nanotubes have potential role in the treatment of prostate cancer. Carbon nanotubes are single-atom thickness sheets which can be rolled into small cylindrical tubes. Based on the number of carbon layers, they are classified as: single-walled carbon nanotubes and multiwalled carbon nanotubes. Their diameter varies from 0.4 to 2.5 nm for single-walled carbon nanotubes and up to 100 nm for multiwalled carbon nanotubes [79]. Due to its unique characteristics in size, biodistribution, and geometry, CNT can easily deliver wide variety of drug molecules. But its toxic nature and tendency to aggregate limits its therapeutic application. This can be minimized by conjugating it with polyethylene glycol (PEG) [80]. The conjugation of targeting ligand to phospholipid-poly(ethylene glycol) (PL-PEG), single-walled, carbon nanotube complex made them more site specific and can be used for cancer treatment [81]. The targeting capacity can be enhanced by conjugating it with aptamers or certain antibodies, etc., which specifically bind to the target cells. Beckler et al. [82] made an attempt to formulate multiwalled, carbon nanotube antibody complex by a diimide-activated amidation process. The surface of prostate cancer cell is overexpressed by cadherin 4 (CDH4) cells. The combination of anti-CDH4 with carbon nanotubes will efficiently bind with prostate cancer cells. When applying microwaves externally, there was a hyperthermic environment created at the tumor cells which cause cell death. Hyperthermic effect is an unique property of carbon nanotubes [82].

The carbon nanotubes are extensively used as contrasting agent for cancer diagnosis. Because of the rapid and early detection of prostate cancer, the mortality rate can be reduced significantly. For the effective binding of nanoparticle into the prostate cancer cells, some surface modifications has to be done. Conjugation of nanoparticle with ligands like monoclonal antibodies, programmed antibodies, and aptamers shows high affinity toward the surface cell protein in tumor cells [83]. Gu et al. [84] reported the application of carbon nanotubes as a contrasting agent. They formulated a novel ultrasound-based contrasting agent by slight modification in the surface of multiwalled carbon nanotubes by combination of PEG and PSMA aptamer A10-3.2 [84].

Quantum dots are also a carbon-based nanoparticle that is unique in their shape and size, with high stability [85]. Quantum dots are reported with low solubility with almost all common solvents. The surface modifications with oxygen containing functional groups made them water soluble. The surface functional groups determine their physical properties [86]. The cancer cell suppression behavior of quantum dots was reported in several literatures [87, 88]. The targeting ability of quantum dots can be enhanced by conjugating with polymers like polyethylene glycol (PEG) and some biological agents like antibodies. Quantum dots absorb via receptor-mediated endocytosis and accumulate in the cancer cells [89, 90]. The carbon quantum dots can be prepared by two methods, i.e., top-down method and bottom-up method. In top-down process, the macromolecules are reduced into small fragments by means of either physical or chemical methods, on the other hand the bottom-up method is done by polymerization and carbonization of series of small molecules through chemical reactions [91, 92].

Carbon dots are mainly used for imaging purpose. Gao et al. formulated antibody-labeled carbon dots, which specifically bind to PMSA present in target cell [93]. Bagalkot et al. formulated carbon dots as a theranostic agent. A10 aptamer was chosen and conjugated with nanoparticle for the delivery of Doxorubicin. The prepared formulation effectively binds with prostate cancer cell and is taken via receptor-mediated endocytosis [94].

10 Magnetic nanoparticles

They are special class of nanoparticles which was prepared from pure metals like Fe, Co, and Ni, or a mixture of metals which is compatible with human tissues [95]. Due to its unique properties like controlled release of drug, hyperthermia effect, bio sensing, etc., it is commonly used in prostate cancer therapy. One of the main advantage of magnetic nanoparticle is that it can control with the help of an external magnetic field [96]. Applying a biocompatible polymer on the surface of these nanoparticles will enhance its stability and improve the targeting property to various parts of the body [97]. Modifications in surface characteristics of nanoparticle led to multifunction property, i.e., hyperthermia effect along with delivery of drug [98].

Several investigations were made on magnetic nanoparticles for the effective delivery of antineoplastic agents. Paclitaxel is an antiprostate cancer drug, which was conjugated with poly[aniline-co-sodium-*N*-(1-one butyric acid) aniline] in the center of magnetic nanoparticle, which improves cell cytotoxicity against PC-3 cells in vitro [99, 100]. The surface of nanoparticle micelle was conjugated with polyethylene glycol for the improvement of anticancer activity against C4-2 cells in a xenograft mice models [101]. Another attempt was made by Sato et al. in magnetic nanoparticle for the effective targeting of docetaxel. Magnetic nanoparticle was conjugated with Fe_3O_4, which reduced the cell viability at low dose of docetaxel in vitro [102]. Mitoxantrone is a US FDA-approved drug for the prostate cancer treatment. The anticancer activity was enhanced by using magnetic nanoparticle prepared by chemical coprecipitation technique. The drug-nanoparticle complex inhibited the cell proliferation and showed cytotoxic activity against DU145 (DU-145) prostate cancer cell lines ex vivo [103].

11 Conclusion

Nanoparticles have a key role in the delivery of active therapeutic agent to the target tissues. It controls the release of active ingredients, protects the entrapped molecules from degradation due to various enzymes, and protects the drug from causing unwanted side effects. Moreover, nanoparticle delivery systems coupled with drug have more stability and can overcome the solubility issues of hydrophobic anticancer drugs. The targeted drug delivery ability of liposomes, polymeric nanoparticles and micelles, and dendrimers may be used for the effective management of prostate cancer. The multidrug resistance in cancer cells is the major drawback of conventional anticancer drug delivery, which can be countered by the advanced nanoparticle-based drug delivery systems. Nanoparticles were not only used as a delivery agent but it also has a potential application in the imaging sector. The nanoparticle-based systems have proved their application in the field of diagnosis in prostate cancer. The utilization of nanoparticle drug delivery systems can overcome various drawbacks of current prostate cancer therapy to effectively reduce the mortality rate.

References

[1] Kufe DW, Pollock RE, Weichselbaum RR, Robert C, Bast J, Gansler TS, Holland JF, et al. Holland-Frei cancer medicine. [cited 2020 Aug 29]; Available from: https://www.ncbi.nlm.nih.gov/books/NBK12354/; 2003.

[2] Krug U, Ganser A, Koeffler HP. Tumor suppressor genes in normal and malignant hematopoiesis. Oncogene 2002;21(21):3475–95.

[3] Walboomers JM, Jacobs MV, Manos MM, Bosch FX, Kummer JA, Shah KV, et al. Human papillomavirus is a necessary cause of invasive cervical cancer worldwide. J Pathol 1999;189(1):12–9.

[4] Hema S, Thambiraj S, Shankaran DR. Nanoformulations for targeted drug delivery to prostate cancer: an overview. J Nanosci Nanotechnol 2018;18(8):5171–91.

[5] Jemal A, Thomas A, Murray T, Thun M. Cancer statistics, 2002. CA Cancer J Clin 2002;52(1):23–47.

[6] Jemal A, Siegel R, Xu J, Ward E. Cancer statistics, 2010. CA Cancer J Clin 2010;60(5):277–300.

[7] Mulla JAS. Drug delivery and therapeutic approaches to prostate. Cancer 2018;12.

[8] Cuzick J, Thorat MA, Andriole G, Brawley OW, Brown PH, Culig Z, et al. Prevention and early detection of prostate cancer. Lancet Oncol 2014;15(11):e484–92.

[9] Castro E, Eeles R. The role of BRCA1 and BRCA2 in prostate cancer. Asian J Androl 2012;14(3):409–14.

[10] Cooperberg MR, Moul JW, Carroll PR. The changing face of prostate cancer. J Clin Oncol 2005;23(32):8146–51.

[11] Nogueira L, Corradi R, Eastham JA. Prostatic specific antigen for prostate cancer detection. Int Braz J Urol 2009;35(5):521–31.

[12] Thompson IM, Lucia MS, Lippman SM. Prevalence of prostate cancer among men with a prostate-specific antigen level $\leq$ 4.0 ng per milliliter. N Engl J Med 2004;350(22):2239–46.

[13] Malinowski B, Wiciński M, Musiała N, Osowska I, Szostak M. Previous, current, and future pharmacotherapy and diagnosis of prostate cancer—a comprehensive review. Diagnostics 2019;9(4):161.

[14] Jain KK. Drug delivery systems—an overview. In: Jain KK, editor. Drug delivery systems Totowa, NJ: Humana Press; 2008. p. 1–50. (Methods in Molecular Biology™; vol. 437). Walker J, editor. [Internet]. [cited 2020 Sep 16] Available from: http://link.springer.com/10.1007/978-1-59745-210-6_1.
[15] Tiwari G, Tiwari R, Bannerjee S, Bhati L, Pandey S, Pandey P, et al. Drug delivery systems: an updated review. Int J Pharm Investig 2012;2(1):2.
[16] Paolino D, Sinha P, Fresta M, Ferrari M. Drug delivery systems. In: Webster JG, editor. Encyclopedia of medical devices and instrumentation. Hoboken, NJ, USA: John Wiley & Sons, Inc.; 2006. p. emd274. [Internet]. [cited 2020 Sep 17]. p Available from: http://doi.wiley.com/10.1002/0471732877.emd274.
[17] Ravichandran R. Pharmacokinetic study of nanoparticulate curcumin: oral formulation for enhanced bioavailability. J Biomater Nanobiotechnol 2013;4(3):291–9.
[18] Wong PT, Choi SK. Mechanisms of drug release in nanotherapeutic delivery systems. Chem Rev 2015;115(9):3388–432.
[19] Torchilin V. Tumor delivery of macromolecular drugs based on the EPR effect. Adv Drug Deliv Rev 2011;63(3):131–5.
[20] Zhao Y, Chen G, Meng Z, Gong G, Zhao W, Wang K, et al. A novel nanoparticle drug delivery system based on PEGylated hemoglobin for cancer therapy. Drug Deliv 2019;26(1):717–23.
[21] Prabhakar U, Maeda H, Jain RK, Sevick-Muraca EM, Zamboni W, Farokhzad OC, et al. Challenges and key considerations of the enhanced permeability and retention effect for nanomedicine drug delivery in oncology. Cancer Res 2013;73(8):2412–7.
[22] Torchilin VP. Passive and active drug targeting: drug delivery to tumors as an example. In: Schäfer-Korting M, editor. Drug delivery. Handbook of experimental pharmacology, vol. 197. Berlin, Heidelberg: Springer Berlin Heidelberg; 2010. p. 3–53. [Internet]. [cited 2020 Sep 17].Available from: http://link.springer.com/10.1007/978-3-642-00477-3_1.
[23] Attia MF, Anton N, Wallyn J, Omran Z, Vandamme TF. An overview of active and passive targeting strategies to improve the nanocarriers efficiency to tumour sites. J Pharm Pharmacol 2019;71(8):1185–98.
[24] Dadwal A, Baldi A, Kumar Narang R. Nanoparticles as carriers for drug delivery in cancer. Artif Cells Nanomed Biotechnol 2018;46(sup2):295–305.
[25] Yoo J, Park C, Yi G, Lee D, Koo H. Active targeting strategies using biological ligands for nanoparticle drug delivery systems. Cancer 2019;11(5):640.
[26] Sanna V, Sechi M. Nanoparticle therapeutics for prostate cancer treatment. Maturitas 2012;73(1):27–32.
[27] Alexis F, Pridgen E, Molnar LK, Farokhzad OC. Factors affecting the clearance and biodistribution of polymeric nanoparticles. Mol Pharm 2008;5(4):505–15.
[28] Prajapati VD, Jani GK, Moradiya NG, Randeria NP, Nagar BJ, Naikwadi NN, et al. Galactomannan: a versatile biodegradable seed polysaccharide. Int J Biol Macromol 2013;60:83–92.
[29] Paciotti GF, Myer L, Weinreich D, Goia D, Pavel N, McLaughlin RE, et al. Colloidal gold: a novel nanoparticle vector for tumor directed drug delivery. Drug Deliv 2004;11(3):169–83.
[30] Kumari P, Ghosh B, Biswas S. Nanocarriers for cancer-targeted drug delivery. J Drug Target 2016;24(3):179–91.
[31] Bamrungsap S, Zhao Z, Chen T, Wang L, Li C, Fu T, et al. Nanotechnology in therapeutics: a focus on nanoparticles as a drug delivery system. Nanomedicine 2012;7(8):1253–71.
[32] Yamashita F, Hashida M. Pharmacokinetic considerations for targeted drug delivery. Adv Drug Deliv Rev 2013;65(1):139–47.
[33] Araki T, Ogawara K, Suzuki H, Kawai R, Watanabe T, Ono T, et al. Augmented EPR effect by photo-triggered tumor vascular treatment improved therapeutic efficacy of liposomal paclitaxel in mice bearing tumors with low permeable vasculature. J Control Release 2015;200:106–14.
[34] Dostalova S, Polanska H, Svobodova M, Balvan J, Krystofova O, Haddad Y, et al. Prostate-specific membrane antigen-targeted site-directed antibody-conjugated apoferritin nanovehicle favorably influences in vivo side effects of doxorubicin. Sci Rep 2018;8(1):8867.
[35] Zhang W, Song Y, Eldi P, Guo X, Hayball J, Garg S, et al. Targeting prostate cancer cells with hybrid elastin-like polypeptide/liposome nanoparticles. Int J Nanomed 2018;13:293–305.
[36] Ikemoto K, Shimizu K, Ohashi K, Takeuchi Y, Shimizu M, Oku N. *Bauhinia purprea* agglutinin-modified liposomes for human prostate cancer treatment. Cancer Sci 2016;107(1):53–9.
[37] Graff CL, Pollack GM. Nasal drug administration: potential for targeted central nervous system delivery. J Pharm Sci 2005;94(6):1187–95.
[38] Denmeade SR, Sokoll LJ, Chan DW, Khan SR, Isaacs JT. Concentration of enzymatically active prostate-specific antigen (PSA) in the extracellular fluid of primary human prostate cancers and human prostate cancer xenograft models. Prostate 2001;48(1):1–6.
[39] Gondi CS, Rao JS. Cathepsin B as a cancer target. Expert Opin Ther Targets 2013;17(3):281–91.
[40] Mock JN, Costyn LJ, Wilding SL, Arnold RD, Cummings BS. Evidence for distinct mechanisms of uptake and antitumor activity of secretory phospholipase A2 responsive liposome in prostate cancer. Integr Biol 2013;5(1):172–82.
[41] Farokhzad OC, Langer R. Impact of nanotechnology on drug delivery. ACS Nano 2009;3(1):16–20.
[42] Gref R, Lück M, Quellec P, Marchand M, Dellacherie E, Harnisch S, et al. 'Stealth' corona-core nanoparticles surface modified by polyethylene glycol (PEG): influences of the corona (PEG chain length and surface density) and of the core composition on phagocytic uptake and plasma protein adsorption. Colloids Surf B Biointerfaces 2000;18(3–4):301–13.
[43] Zhang L, Chan JM, Gu FX, Rhee J-W, Wang AZ, Radovic-Moreno AF, et al. Self-assembled lipid–polymer hybrid nanoparticles: a robust drug delivery platform. ACS Nano 2008;2(8):1696–702.
[44] Yan J, Wang Y, Zhang X, Liu S, Tian C, Wang H. Targeted nanomedicine for prostate cancer therapy: docetaxel and curcumin co-encapsulated lipid–polymer hybrid nanoparticles for the enhanced anti-tumor activity *in vitro* and *in vivo*. Drug Deliv 2016;23(5):1757–62.
[45] Salvador-Morales C, Gao W, Ghatalia P, Murshed F, Aizu W, Langer R, et al. Multifunctional nanoparticles for prostate cancer therapy. Expert Rev Anticancer Ther 2009;9(2):211–21.
[46] Wang Q, Alshaker H, Böhler T, Srivats S, Chao Y, Cooper C, et al. Core shell lipid-polymer hybrid nanoparticles with combined docetaxel and molecular targeted therapy for the treatment of metastatic prostate cancer. Sci Rep 2017;7(1):5901.
[47] Oerlemans C, Bult W, Bos M, Storm G, Nijsen JFW, Hennink WE. Polymeric micelles in anticancer therapy: targeting, imaging and triggered release. Pharm Res 2010;27(12):2569–89.
[48] Siddiqui IA, Adhami VM, Bharali DJ, Hafeez BB, Asim M, Khwaja SI, et al. Introducing nanochemoprevention as a novel approach for cancer control: proof of principle with green tea polyphenol epigallocatechin-3-gallate. Cancer Res 2009;69(5):1712–6.

[49] Suk JS, Xu Q, Kim N, Hanes J, Ensign LM. PEGylation as a strategy for improving nanoparticle-based drug and gene delivery. Adv Drug Deliv Rev 2016;99:28–51.

[50] Benahmed A, Ranger M, Leroux J-C. Novel polymeric micelles based on the amphiphilic diblock copolymer poly(N-vinyl-2-pyrrolidone)-block-poly(D,L-lactide). Pharm Res 2001;18(3):323–8.

[51] Barve A, Jain A, Liu H, Zhao Z, Cheng K. Enzyme-responsive polymeric micelles of cabazitaxel for prostate cancer targeted therapy. Acta Biomater 2020;113(1):501–11.

[52] Boas U, Heegaard PMH. Dendrimers in drug research. Chem Soc Rev 2004;33(1):43.

[53] Cheng Y. Pharmaceutical applications of dendrimers: promising nanocarriers for drug delivery. Front Biosci 2008;13(13):1447.

[54] Lee CC, MacKay JA, Fréchet JMJ, Szoka FC. Designing dendrimers for biological applications. Nat Biotechnol 2005;23(12):1517–26.

[55] Gupta U, Agashe HB, Asthana A, Jain NK. A review of in vitro–in vivo investigations on dendrimers: the novel nanoscopic drug carriers. Nanomed Nanotechnol Biol Med 2006;2(2):66–73.

[56] Hawker CJ, Frechet JMJ. Preparation of polymers with controlled molecular architecture. A new convergent approach to dendritic macromolecules. Am Chem Soc 1990;112(21):7638–47.

[57] Sowinska M, Urbanczyk-Lipkowska Z. Advances in the chemistry of dendrimers. New J Chem 2014;38(6):2168.

[58] Boyd BJ, Kaminskas LM, Karellas P, Krippner G, Lessene R, Porter CJH. Cationic poly-L-lysine dendrimers: pharmacokinetics, biodistribution, and evidence for metabolism and bioresorption after intravenous administration to rats. Mol Pharm 2006;3(5):614–27.

[59] Abbasi E, Aval S, Akbarzadeh A, Milani M, Nasrabadi H, Joo S, et al. Dendrimers: synthesis, applications, and properties. Nanoscale Res Lett 2014;9(1):247.

[60] Tomalia DA. Birth of a new macromolecular architecture: dendrimers as quantized building blocks for nanoscale synthetic polymer chemistry. Prog Polym Sci 2005;30(3–4):294–324.

[61] Lammers T, Hennink WE, Storm G. Tumour-targeted nanomedicines: principles and practice. Br J Cancer 2008;99(3):392–7.

[62] Mousa S, Bharali D, Khalil M, Gurbuz M, Simone T. Nanoparticles and cancer therapy: a concise review with emphasis on dendrimers. Int J Nanomed 2009;4(1):1–7.

[63] Lim J, Guan B, Nham K, Hao G, Sun X, Simanek EE. Tumor uptake of triazine dendrimers decorated with four, sixteen, and sixty-four PSMA-targeted ligands: passive versus active tumor targeting. Biomolecules 2019;9(9):421.

[64] Tagawa ST, Beltran H, Vallabhajosula S, Goldsmith SJ, Osborne J, Matulich D, et al. Anti-prostate-specific membrane antigen-based radioimmunotherapy for prostate cancer. Cancer 2010;116(S4):1075–83.

[65] Patri AK, Myc A, Beals J, Thomas TP, Bander NH, Baker JR. Synthesis and in vitro testing of J591 antibody–dendrimer conjugates for targeted prostate cancer therapy. Bioconjug Chem 2004;15(6):1174–81.

[66] Wolinsky J, Grinstaff M. Therapeutic and diagnostic applications of dendrimers for cancer treatment☆. Adv Drug Deliv Rev 2008;60(9):1037–55.

[67] Shah M. Gold nanoparticles: various methods of synthesis and antibacterial applications. Front Biosci 2014;19(8):1320.

[68] Brown SD, Nativo P, Smith J-A, Stirling D, Edwards PR, Venugopal B, et al. Gold nanoparticles for the improved anticancer drug delivery of the active component of oxaliplatin. J Am Chem Soc 2010;132(13):4678–84.

[69] Wang S, Lu G. Applications of gold nanoparticles in cancer imaging and treatment. In: Seehra MS, Bristow AD, editors. Noble and precious metals—properties, nanoscale effects and applications. InTech; 2018. [Internet]. [cited 2020 Sep 2]. Available from: http://www.intechopen.com/books/noble-and-precious-metals-properties-nanoscale-effects-and-applications/applications-of-gold-nanoparticles-in-cancer-imaging-and-treatment.

[70] Duncan B, Kim C, Rotello VM. Gold nanoparticle platforms as drug and biomacromolecule delivery systems. J Control Release 2010;148(1):122–7.

[71] Lim Z-ZJ, Li J-EJ, Ng C-T, Yung L-YL, Bay B-H. Gold nanoparticles in cancer therapy. Acta Pharmacol Sin 2011;32(8):983–90.

[72] Tsai L-C, Hsieh H-Y, Lu K-Y, Wang S-Y, Mi F-L. EGCG/gelatin-doxorubicin gold nanoparticles enhance therapeutic efficacy of doxorubicin for prostate cancer treatment. Nanomedicine 2016;11(1):9–30.

[73] Butterworth KT, Nicol JR, Ghita M, Rosa S, Chaudhary P, McGarry CK, et al. Preclinical evaluation of gold-DTDTPA nanoparticles as theranostic agents in prostate cancer radiotherapy. Nanomedicine 2016;11(16):2035–47.

[74] Fitzgerald KA, Rahme K, Guo J, Holmes JD, O'Driscoll CM. Anisamide-targeted gold nanoparticles for siRNA delivery in prostate cancer—synthesis, physicochemical characterisation and in vitro evaluation. J Mater Chem B 2016;4(13):2242–52.

[75] Lin J, Chen X, Huang P. Graphene-based nanomaterials for bioimaging. Adv Drug Deliv Rev 2016;105:242–54.

[76] Alarifi S, Ali D. Mechanisms of multi-walled carbon nanotubes-induced oxidative stress and genotoxicity in mouse fibroblast cells. Int J Toxicol 2015;34(3):258–65.

[77] Muthu MS, Abdulla A, Pandey BL. Major toxicities of carbon nanotubes induced by reactive oxygen species: should we worry about the effects on the lungs, liver and normal cells? Nanomedicine 2013;8(6):863–6.

[78] Schinwald A, Murphy FA, Jones A, MacNee W, Donaldson K. Graphene-based nanoplatelets: a new risk to the respiratory system as a consequence of their unusual aerodynamic properties. ACS Nano 2012;6(1):736–46.

[79] Shokrieh MM, Rafiee R. A review of the mechanical properties of isolated carbon nanotubes and carbon nanotube composites. Mech Compos Mater 2010;46(2):155–72.

[80] Liu Z, Tabakman S, Welsher K, Dai H. Carbon nanotubes in biology and medicine: in vitro and in vivo detection, imaging and drug delivery. Nano Res 2009;2(2):85–120.

[81] Liu Z, Tabakman SM, Chen Z, Dai H. Preparation of carbon nanotube bioconjugates for biomedical applications. Nat Protoc 2009;4(9):1372–82.

[82] Beckler B, Cowan A, Farrar N, Murawski A, Robinson A, Diamanduros A, et al. Microwave heating of antibody-functionalized carbon nanotubes as a feasible cancer treatment. Biomed Phys Eng Express 2018;4(4), 045025.

[83] Wang L, Li L, Guo Y, Tong H, Fan X, Ding J, et al. Construction and in vitro/in vivo targeting of PSMA-targeted nanoscale microbubbles in prostate cancer: PSMA-targeted nanoscale microbubbles. Prostate 2013;73(11):1147–58.

[84] Gu F, Hu C, Xia Q, Gong C, Gao S, Chen Z. Aptamer-conjugated multi-walled carbon nanotubes as a new targeted ultrasound contrast agent for the diagnosis of prostate cancer. J Nanopart Res 2018;20(11):1–10.

[85] Zheng XT, Ananthanarayanan A, Luo KQ, Chen P. Glowing graphene quantum dots and carbon dots: properties, syntheses, and biological applications. Small 2015;11(14):1620–36.
[86] Deka MJ, Dutta P, Sarma S, Medhi OK, Talukdar NC, Chowdhury D. Carbon dots derived from water hyacinth and their application as a sensor for pretilachlor. Heliyon 2019;5(6), e01985.
[87] Li C-L, Ou C-M, Huang C-C, Wu W-C, Chen Y-P, Lin T-E, et al. Carbon dots prepared from ginger exhibiting efficient inhibition of human hepatocellular carcinoma cells. J Mater Chem B 2014;2(28):4564.
[88] Hsu P-C, Chen P-C, Ou C-M, Chang H-Y, Chang H-T. Extremely high inhibition activity of photoluminescent carbon nanodots toward cancer cells. J Mater Chem B 2013;1(13):1774.
[89] Yu B, Tai HC, Xue W, Lee LJ, Lee RJ. Receptor-targeted nanocarriers for therapeutic delivery to cancer. Mol Membr Biol 2010;27(7):286–98.
[90] Zhang Q, Li F. Combating P-glycoprotein-mediated multidrug resistance using therapeutic nanoparticles. Curr Pharm Des 2013;19(37):6655–66.
[91] Mosconi D, Mazzier D, Silvestrini S, Privitera A, Marega C, Franco L, et al. Synthesis and photochemical applications of processable polymers enclosing photoluminescent carbon quantum dots. ACS Nano 2015;9(4):4156–64.
[92] Wang X, Feng Y, Dong P, Huang J. A mini review on carbon quantum dots: preparation, properties, and electrocatalytic application. Front Chem 2019;7:671.
[93] Gao X, Cui Y, Levenson RM, Chung LWK, Nie S. In vivo cancer targeting and imaging with semiconductor quantum dots. Nat Biotechnol 2004;22(8):969–76.
[94] Bagalkot V, Zhang L, Levy-Nissenbaum E, Jon S, Kantoff PW, Langer R, et al. Quantum dot–aptamer conjugates for synchronous cancer imaging, therapy, and sensing of drug delivery based on bi-fluorescence resonance energy transfer. Nano Lett 2007;7(10):3065–70.
[95] Alagiri M, Muthamizhchelvan C, Ponnusamy S. Structural and magnetic properties of iron, cobalt and nickel nanoparticles. Synth Met 2011;161(15–16):1776–80.
[96] Farzin A, Etesami SA, Quint J, Memic A, Tamayol A. Magnetic nanoparticles in cancer therapy and diagnosis. Adv Healthc Mater 2020;9(9):1901058.
[97] Gupta AK, Gupta M. Synthesis and surface engineering of iron oxide nanoparticles for biomedical applications. Biomaterials 2005;26(18):3995–4021.
[98] Kumar S, Balasubramanian RG, Nagaoka Y, Iwai S, et al. Curcumin and 5-fluorouracil-loaded, folate- and transferrin-decorated polymeric magnetic nanoformulation: a synergistic cancer therapeutic approach, accelerated by magnetic hyperthermia. Int J Nanomedicine 2014;9(1):437–59.
[99] Taylor RM, Sillerud LO. Paclitaxel-loaded iron platinum stealth immunomicelles are potent MRI imaging agents that prevent prostate cancer growth in a PSMA-dependent manner. Int J Nanomedicine 2012;7(1):4341–52.
[100] Hua M-Y, Yang H-W, Chuang C-K, Tsai R-Y, Chen W-J, Chuang K-L, et al. Magnetic-nanoparticle-modified paclitaxel for targeted therapy for prostate cancer. Biomaterials 2010;31(28):7355–63.
[101] Wu M, Huang S. Magnetic nanoparticles in cancer diagnosis, drug delivery and treatment. Mol Clin Oncol 2017;7(5):738–46.
[102] Watanabe M, Sato, Itcho, Ishiguro, Kawai, Kasai, et al. Magnetic nanoparticles of Fe_3O_4 enhance docetaxel-induced prostate cancer cell death. Int J Nanomedicine 2013;8(1):3151–60.
[103] Lee K-J, An JH, Chun J-R, Chung K-H, Park W-Y, Shin J-S, et al. In vitro analysis of the anti-cancer activity of mitoxantrone loaded on magnetic nanoparticles. J Biomed Nanotechnol 2013;9(6):1071–5.

Chapter 16

Nanomedicine-based doxorubicin delivery for skin cancer with theranostic potential

Ummarah Kanwal[a], Abida Raza[b], Muzaffar Abbas[c], Nasir Abbas[a], Khalid Hussain[a], and Nadeem Irfan Bukhari[a]

[a]*Punjab University College of Pharmacy, University of Punjab, Allama Iqbal Campus, Lahore, Pakistan,* [b]*NILOP Nanomedicine Research Laboratories, National Institute of Lasers and Optronics College, PIEAS, Islamabad, Pakistan,* [c]*Faculty of Pharmacy, Capital University of Science & Technology, Islamabad, Pakistan*

1 Introduction

Cancer is a leading cause of morbidity and mortality and one of the major challenging illnesses worldwide, despite all advancement in field of medicine [1]. Skin cancer is a fatal disease and its prevalence is an important global public health problem. The World Health Organization (WHO) has ranked the skin cancer as one of the most common ailments of white population [2]. Detection and diagnosis of the skin cancer at earlier stage are importantly required for better treatment outcomes [3]. Several options are available for cancer treatment, including surgery, chemotherapy, radiotherapy, immunotherapy, photothermal therapy, and photodynamic therapy [4]. There is a growing interest in developing anticancer moieties, which can concurrently treat and diagnose or assess the therapeutic prognosis. A therapeutic agent, which has the ability to simultaneously monitor therapeutic effect, as well as its progression (diagnosis), is termed as theranostic [5]. A limited number of radiopharmaceuticals are capable simultaneously of treating as well as diagnosing, i.e., acting as theranostic. For instance, the radioiodine is a radiotheranostic, which acts as diagnostic and therapeutic radiopharmaceutical for thyroid cancer. However, it is not always advisable to expose the patients to the radiations emitted from the radiopharmaceuticals [6]. None of the nonradiopharmaceutical options is available which can serve as a theranostic. Similarly, the available cancer chemotherapy modalities lack the capability to monitor therapeutic response, systemic toxicity, and ineffective drug concentrations within a tumor [7]. The majority of conventional anticancer agents are unable to achieve optimum pharmacokinetics and biodistribution in tumor tissues versus host tissues on account of their rapid systemic clearance due to lower molecular weights. Such nonaccumulation of anticancer agent in the region of interest (cancerous/tumor tissue) and nonspecific biodistribution result in toxicity for normal tissues and side effects with reduced therapeutic efficacy [8]. Nonaccumulation of anticancer agent, on the other hand, in tumor makes it incapable of monitoring the progression of anticancer treatment. The current chapter provides a brief overview of an underresearched nanotechnology-based formulation of DOX as a potential theranostic option.

2 Doxorubicin—Limitations

Doxorubicin (DOX), a member of anthracycline family, is a broad-spectrum anticancer drug used in the treatment of solid tumors, hepatocellular carcinoma, squamous cell carcinoma, and lymphomas [9]. However, its clinical use has been limited by its cardiotoxicity, hydrolytic degradation in plasma, and enzyme-led rapid elimination, following intravenous administration [10]. The currently FDA-approved liposomal DOX nanoformulations, Doxil® and Myocet®, which are in use for treatment of breast and ovarian cancer without cardiotoxicity [11], nevertheless cause other toxicities such as hand-foot syndrome [12]. However, DOX is still in use for cancer treatment due to its broad-spectrum anticancer activity [11]. Furthermore, DOX's autofluorescence has made it a model drug for skin tumors facilitating the distribution monitoring by optical imaging [3].

3 Nanomedicines—A solution

Attempts have been made to sustain the broad-spectrum anticancer efficiency and exclude toxicity associated with DOX, with a dominant emphasis on developing delivery systems based on nanotechnology. A nanomedicine is nanosized system comprising nanocarriers and the active moieties that can be used for treatment of diseases, as well as for their diagnosis. The nanomedicine does not contain excipients, and demonstrates enhanced bioavailability and drug delivery at

Advanced Drug Delivery Systems in the Management of Cancer. https://doi.org/10.1016/B978-0-323-85503-7.00041-9

site of action without excipient-associated side effects and with reduced toxicity [13]. Nanomedicines hold significant advantages over current chemotherapeutic approaches. A well-designed nanoparticulate system helps to achieve a control rate of drug release in tumor. Nanoparticles (NPs) of appropriate size (≤200 nm) and surface properties have potential to target tumor tissues passively based upon enhanced permeation and retention (EPR) effect [14]. NP formulations enhance drug efficacy by protecting their cargo drugs from the severe physiological environment and metabolic effects of enzymes. As a result of recent research, a number of NPs have been developed for the utilization as drug delivery moieties, imaging probes, and as multifunctional particulates [5]. The drug delivery with reduced side effects has been achieved by a variety of nanodelivery systems including polymeric, metallic, organic, inorganic nanoparticles, and lipidic nanoformulations. Multifunctional NPs are developed for simultaneous imaging and treatment of tumor tissues by combination of different imaging and targeted drug delivery modalities in single particle [15]. For DOX, polymeric nanoformulations, developed using nontoxic, biocompatible, and biodegradable materials, are a promising approach to achieve enhanced efficacy and lesser side effects and cytotoxicity by its delivery at the intended site [16]. Such systems distinctively exhibit a longer circulation half-life, enhanced permeability, and EPR effect and lesser hepatic uptake [17]. The above-mentioned characteristics have made polymeric nanomaterials the efficient nanosystems for drug transport [17].

4 Desired features of a DOX nanocarrier

One of the desired features in a nanocarrier of DOX is to effectively deliver this chemotherapeutic agent to tumor and ability to simultaneous monitor the progression of anticancer therapy and tumor regression [15]. Nanotheranostics is now an emerging field, encompassing the development of nanomedicine-based approaches for theranostic applications. Under this field, various nanocarriers have been developed for codelivery of diagnostic and therapeutic moieties in a controlled, sustained, and targeted manner to achieve improved biodistribution and enhanced theranostic effects with reduced side effects. Nanocarriers include polymer conjugations, dendrimers, micelles, liposomes, metal, and inorganic nanoparticles and carbon nanotubes [18–20]. The developed system, as a whole, drug or the carrier, must have the ability to relay the information to an external source, so that the biodistribution in an appropriate animal model could be imaged by appropriate device externally [17]. The biodistribution of nanomedicine in tissues, tumors, and organs could be optically imaged via fluorescence ligand-coupled polymeric nanoparticles [21, 22]. In vivo optical imaging of tumor tissues by the polymeric fluorescent nanoparticles at cell and organ level has already been reported [23]. The in vivo tumor imaging can be used for earlier cancer detection and monitoring of cancer therapy. DOX's self-fluorescence property enables its monitoring in tissues via optical imaging, in addition to its anticancer potential, and the above-mentioned characteristics have made it a popular theranostic moiety [24–26]. DOX-fluorescent nanomedicine-guided cancer therapy can be used for imaging and diagnostic purpose along with the therapeutic intervention in clinical conditions [25].

5 Optimum design of nanoparticulate drug delivery system

For the rational design of a nanoparticulate drug delivery system, various properties (attributes) are accomplished based on the manipulation of several parameters as shown in Fig. 1. New formulations are quickly optimized, usually based on the few critical quality attributes to keep them as simple as possible. Once optimized, such a formulation is subjected to detailed characterization wherein the majority of the characteristics of a newly developed optimized formulation are tested and matched with the desired quality attributes, features of a reference formulation, the desired features or that of a control, which may be free drug (without encapsulation in nanoformulation) or a blank nanocarrier (without drug) [27, 28].

6 Methods for development of polymeric nanoparticles

Literature cites several methods for the preparation of polymeric NPs such as physical, chemical, and biological

FIG. 1 Nanoparticulate parameters for drug delivery design.

methods. The commonly used methods for polymeric NPs production are multistep, which require organic solvents, and agents to reduce toxicity and stabilizers. The preparatory steps of NPs development also need to control several factors for optimized outputs [29]. The chemical substances and organic solvents have the potential biological and environmental risks [30]. Contrary to the conventional methods, sonication is a single-step, simple, convenient, relatively economical and a green process, categorized under physical method [29]. While sustaining the main merits of the sonochemical process, i.e., material exposure to the milder conditions during sonication, shorter reaction time, and the higher yields, probe-sonication provides several advantages in process development and manufacturing and product design [31]. Thus, the method could be implemented easily for any scientific and/or industrial applications, and the DOX-GCPQ nanoformulation could be prepared using simpler sonication approach without incorporating any chemical substances or excipients.

7 Development of DOX-GCPQ nanoformulation

DOX-GCPQ is prepared by loading DOX with GCPQ in normal saline as solution, using probe sonication (Sonics, Vibra cell, USA) [32]. The light-protected sample of the above-mentioned solution kept on ice is sonicated by immersing probe in the middle of sample at amplitude (60%) using pulsed mode of 50 on and 10 off positions, for 500 s. The prepared DOX-GCPQ formulation is characterized for particle size, zeta potential by dynamic light scattering at 25°C (Microtrac Nanotrac Wave II, USA), and surface morphology could be studied by atomic force microscopy (Alpha Contec, Germany) [33]. Stability of optimized nanoformulation DOX-GCPQ is analyzed based on the modification of particle size, zeta potential, and encapsulation efficiency after every month for 3 months (i.e., on 1st, 2nd, and 3rd, months) while storing at two different temperatures, i.e., 4°C and 25°C [34]. The two-way ANOVA could be used for revealing any difference, at *P* value ˂0.05 between the stability data of free DOX and DOX-GCPQ at various time intervals.

Balb/c mice of appropriate age and weight are the tumor-induced mice model to study biodistribution of nanoformulation in optical imaging. The skin tumor is induced in Balb/c mice using chemical method after an ethical approval and using the guidelines of the International Conference on Harmonization ICH [35]. The schematics given in Fig. 2 provide general guidelines for in vivo evaluation of DOX-GCPQ nanoformulation.

The animals were divided into treatment and control groups. The dorsal skin of animals of both groups is shaved, and the animals are confined in disposable biohazard cages whenever required. Two days after skin shaving, dimethylbenz(*a*)anthracene (DMBA) solution is applied to dorsal skin area for tumor initiation. In the control group, acetone is applied. Following DMBA application, the mice are isolated for 2 weeks, after which the tetradecanoyl phorbol (TPA) solution in acetone is applied twice in a week for 5 weeks. After 6 weeks of TPA treatment for tumor promotion, back of each mouse is palpated once in a week to detect formation of tumor and is recorded for presence of palpable mass ≥1 mm in size.

The body weight gain at regular intervals is compared weight in both test and control mice. The mice are under continuous monitoring for the papillomas development and harvested for histological verification. The tumors with skin flap are cryopreserved or fixed with formalin or ethanol according to experimental requirements. When the tumor size reaches to 50 mm^3, the mice are allocated into two groups ($n=6$). DOX and DOX-GCPQ nanoformulation at dose of 6 mg/kg [36–38] are administered to Group I and Group II intravenously, respectively, and are sacrificed after 24 h. The fluorescence intensity of the DOX-GCPQ nanoformulation in tumor is examined using I-Box Explorer^2 (iBox® Explorer^2 Imaging Microscope, UVP Ltd., UK). System excitation filter is to be set at 535/45 and emission filter 605/50 with automated BioLite™ MultiSpectral Light Source. Images are recorded at 0.17× magnification by 3.2 MP OptiChemi 610 camera and image is interpreted using Vision Works® LS Acquisition software.

8 Characteristics and features of DOX-GCPQ nanoformulation

With the earlier mentioned probe sonication method, DOX-GCPQ nanoformulation, as micelles are produced, which has the particle size of <100 nm with zeta potential of −26 mV, measured at 180 degree angle back scattering with diode laser (ʎ 780. 3 mW), using Nanotrac wave™ II, Microtrac [32]. DOX-GCPQ nanoformulation remains stable at 4°C for 3 months (Fig. 3). The above-mentioned characteristic is in line with the literature citation, which reports that the synthesized GCPQ self-assembles into the stable nano (colloidal)-sized particles [30]. For DOX-GCPQ, no chemical cross-linkers and/or ionic gelation agents are required, in accordance with the previous reports [39, 40].

The nanoformulation is considered stable when there no remarkable change occurs in particle, zeta charge, and entrapment efficiency over a time. This change is studied over a period of 3 months at two different temperatures, viz., 4°C and 25°C. The DOX-GCPQ is stable at 4°C, as there is no significant change except slight (nonsignificant, $P>0.05$) increase in particle size at 3 months as compared to that at 1 month (Fig. 4).

With the method adopted for induction of tumor, tumor of about 50 mm is observable after 6 weeks of chemical administration as shown in Fig. 5A. Histological examination of the section reveals the superficial epidermis showing

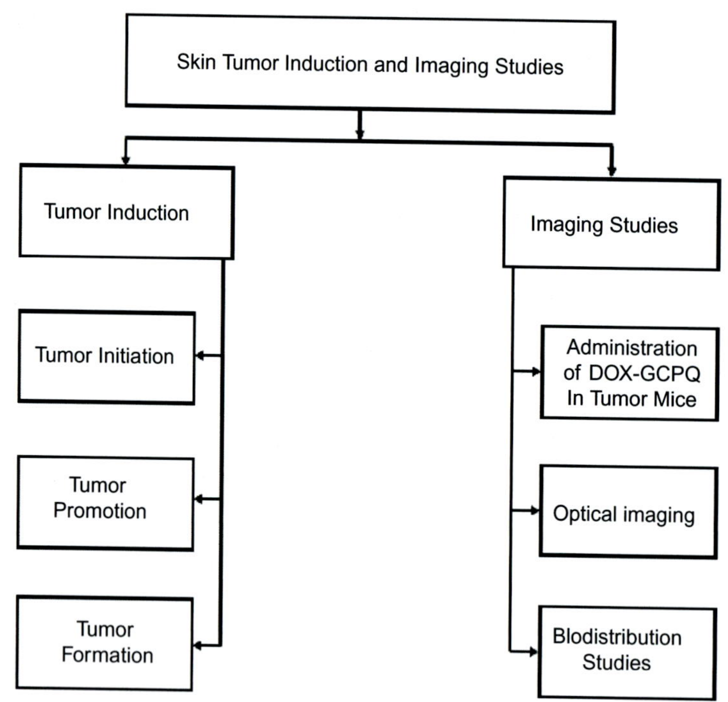

FIG. 2 General guidelines for in vivo evaluation of DOX-GCPQ nanoformulation.

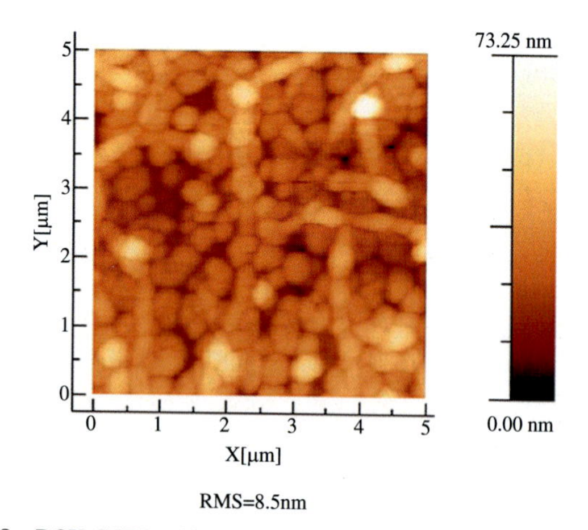

FIG. 3 DOX-GCPQ self-assembled micelle nanoformulation.

hyperkeratosis and acanthosis and the dermis shows dense acute and chronic inflammation and an evidence of the focal papillomas in the skin segment as given in Fig. 5B.

Optical imaging can noninvasively monitor biodistribution of a florescent nanomedicine properties [22]. The measurement of fluorescence is qualitative and semiquantitative, presenting relative uptake of nanoformulation by organs. The imaging of biodistribution of a nanomedicine is helpful in an earlier cancer detection and enables therapy monitoring simultaneously [41]. Being the self-fluorescent, DOX imparted florescence property in the DOX-GCPQ micelles, which enables its noninvasive optical imaging. For optical imaging of the DOX biodistribution, the DOX, loaded GCPQ nanoparticles, is injected intravenously in tail of mice models. Real-time in vivo imaging of DOX-loaded polymeric nanoparticles in skin tumor mouse model is accomplished using iBox explorer at 0.17× level of magnification as shown in Fig. 6. In tumor-bearing mice, strong florescent signals could be recorded from the tumor site, showing that the nanoparticles have distributed throughout the circulation system and accumulated in tumor within 60 min. An improved tumor accumulation in comparison with the free DOX [33] that has been reported to accumulate in heart, liver, spleen. Furthermore, the DOX-GCPQ has shown to be accumulated progressively in the tumor and lasts there up to 10 days. A significantly higher DOX-GCPQ accumulation in tumor has been linked with the enhanced fluorescence signal than in other organs as compared to the free drug. DOX florescence could also be imaged in the excised tumor tissue (Fig. 7). The improved accumulation of DOX from DOX-GCPQ nanoformulation is possibly due to EPR effect and could be attributed to feature of GCPQ's, as reported earlier which is not uptaken by spleen and liver [33, 42, 43]. The lasting of DOX from DOX-GCPQ could be attributed to the controlled release of GCPQ micelle. The controlled release systems smoothen the drug systemic levels for longer time and also enhances the efficacy and lowers side effects, leading to improved patient compliance [10].

The DOX encapsulation in biocompatible nanodrug delivery dosage form not only facilitates effective tumor targeting, actively and passively, but also reduces its adverse effects [44], particularly cardiotoxicity due to its lower accumulation in the heart. In short, the recently developed DOX-GCPQ nanoformulation has demonstrated simultaneous possibility of an in vivo localization in the tumor and optical imaging of the tumor, as reported earlier for other polymeric fluorescent

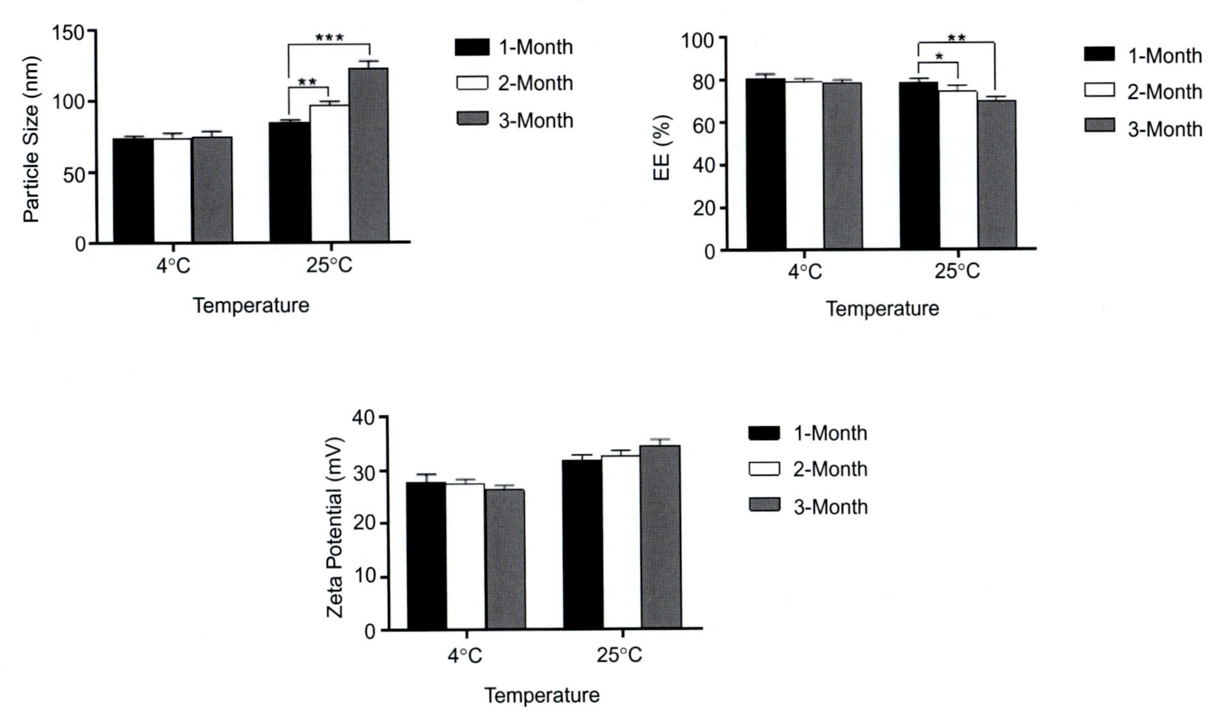

FIG. 4 Properties of DOX and DOX-GCPQ formulation during 3-month stability study at 4°C and 25°C.

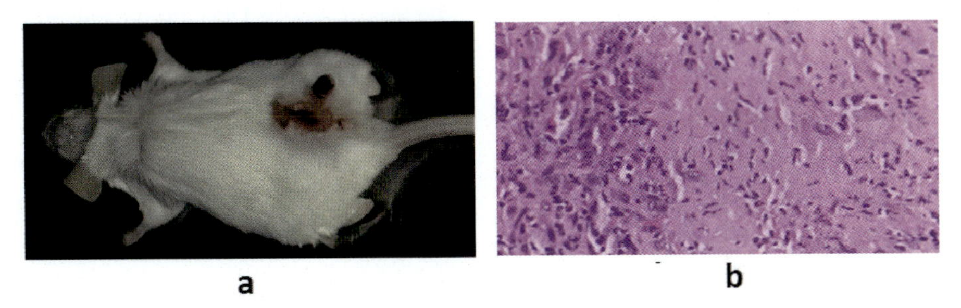

FIG. 5 (A) Skin tumor formation in mice. (B) Histopathology of tumor tissue.

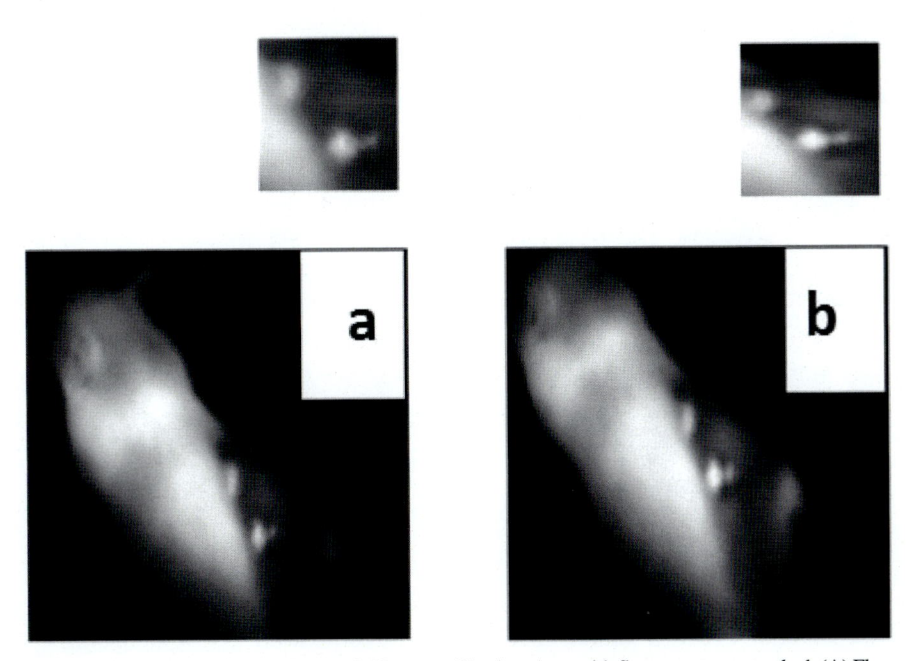

FIG. 6 Florescent signals were imaged with I-box explorer at 0.17× magnification. Area with florescence was marked. (A) Florescent imaging of tumor model mice with particle accumulation at tumor site after 20 min of dose. (B) Florescent imaging of tumor model mice with particle accumulation at tumor site after 60 min.

FIG. 7 Ex vivo imaging of excised tumor tissue (A) untreated tumor and (B) doxorubicin treated.

nanoparticles [23]. Thus, DOX-GCPQ can be explored for its theranostic potential after detailed kinetics and in vivo imaging studies. The previously mentioned methodology could be used to evaluate the performance of other nanoscale drug carriers in cancer therapy, without sacrificing a large number of animals. Targeted chemotherapy treatment with minimal off-target biodistribution to reduce side effects could be accomplished by studying biodistribution of drug from the delivery systems (Fig. 8).

9 Commercial viability and marketability of DOX-GCPQ

The raw materials of GCPQ, chitosan, is available abundantly, but the laboratory-scale production of such a material could be expensive. The direct cost for preparing, for instance, 300mg of GCPQ, as shown in Table 1, is 241 Euro. However, since development of a single DOX dose requires 10mg of GCPQ, the cost of DOX-GCPQ production is 8 €. The overall production of DOX-GCPQ is a single-step sonication followed by sterilization, which can be done easily by filtration [45]. Thus, a mass-level DOX-GCPQ production would be further economical and commercially viable. Nevertheless, the loss, if any, of DOX during encapsulation in GCPQ and sterilization (filtration) should be taken into consideration for examining its cost and commercial viability.

10 Potential applications of DOX-GCPQ nanoformulation

Recently, several types of in vivo animal imaging, including imaging of tumor tissues [26], angiogenic vasculature

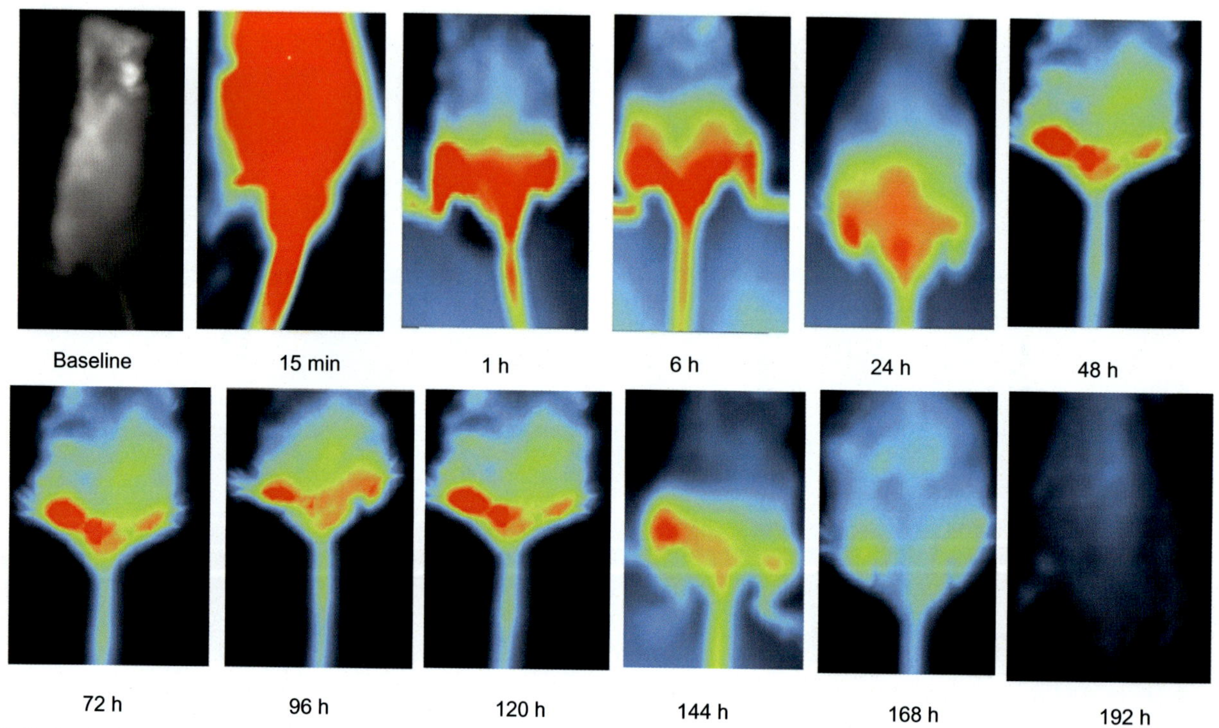

FIG. 8 In vivo imaging of mice.

TABLE 1 Direct cost incurred for the preparation of DOX-GCPQ nanoformulation.

Name of chemical required	Make	Cat no.	Pack size	Price in Euro	Quantity required in the formulation	Cost in Euro
Glycol chitosan (GC)	Sigma-Aldrich	G7753	500 mg	106	500 mg	106.00
HCL	Sigma-Aldrich	H1758	100 mL	47	37.5 mL	17.63
Sodium iodide	Sigma-Aldrich	383112	1000 g	82	45 mg	3.69
Sodium hydroxide	Sigma-Aldrich	S8045	5000 g	63	40 mg	0.50
Methyl iodide	Sigma-Aldrich	456756	100 mL	119	2 mL	2.38
Sodium bicarbonate	Sigma-Aldrich	792519	5000 g	45	376 mg	3.38
Palmitic acid *N*-hydroxysuccinimide	Sigma-Aldrich	P1162)	1000 g	85	475 mg	40.38
N-Methyl-2-pyrrolidone	Sigma-Aldrich	270458	1000 mL	120	25 mL	1.68
Diethyl ether	Sigma-Aldrich	296082	100 mL	48	100 mL	48.00
Acetone	Sigma-Aldrich	65050	1000 L	54	300 mL	16.20
DOX	Fischer scientific	25316-40-9-	100 mg	75	2 mg	1.50
Total cost						**241.333**

[46], and sentinel lymph nodes [47], have been reported with the fluorescent NPs. With the possible application of DOX-GCPQ in imaging-guided therapy that can noninvasively monitor the progression of disease and therapy, an early detection of cancer and monitoring of cancer therapy could be achieved by in vivo tumor imaging. With this promising progress in the development of fluorescent NP-guided cancer therapy, it is imperative to constantly improve the fundamental understanding of designing and applying NPs for diagnosis, treatment, or the combination of imaging and therapeutics in different clinical situations [48]. The following properties of DOX-GCPQ could expand its applications. The GCPQ self-assembles into stable nano colloidal-sized particles [30] without a cross linker [39] and/or ionic gelation agents [40], with a lower critical micellar concentration; thus, the nanoparticles do not prematurely disaggregate in the body fluid, even after extreme dilution conditions (i.e., are more stable). Furthermore, the GCPQ NPs possess enhanced permeation properties and thus reach the target site [49]. GCPQ has high drug incorporation efficacy, and shows enhanced bioavailability of hydrophobic drugs, such as DOX. It is available as dispersion with low viscosity (dynamic viscosity less than 5mPas) when its concentration remained <40 mg/mL; thus, it is favorable for injectable formulations [30]. GCPQ is biodegradable, biocompatible, and nontoxic polymeric nanocarrier, which avoids uptake by liver and spleen and target tumor passively [30, 50]. GCPQ is also capable of passively targeting the tumor site based on the EPR effect [30]. GCPQ is reported to encapsulate and increase the transport across cornea, blood-brain barrier, and gastrointestinal tract for both the hydrophilic and hydrophobic drugs. Drug encapsulation inside GCPQ polymer protects drug's degradation from gastric and intestinal environment, leading to improved gut absorption of active therapeutic agents. The GCPQ NPs also promote drug dissolution by providing large surface area [51]. Upon oral administration, GCPQ promotes adhesion of loaded drug to the mucous allowing prolonged contact between NPs and absorptive enterocytes in the gastrointestinal tract [52]. GCPQ NPs were also reported to promote brain delivery of neuropeptides known as leucine5-enkephalin [51]. Based on these properties, GCPQ should provide useful nanoparticles for delivery of drugs from the oral or intravenous route of administration into the tumors.

11 Conclusion

Probe sonication, a single-step, excipient-free, and a green physical method, is used for the development of DOX-GCPQ nanoformulation. The DOX-loaded GCPQ nanoformulation enables an effective DOX delivery and noninvasive monitoring of DOX biodistribution and accumulation at tumor site utilizing self-florescent property of doxorubicin. In the animal model, the system is proved to be without major side effects. The system is industrially and commercially viable. In future, the potential for tumor regression of this design could be studied further. The cancer theranostic with the recently developed DOX-GCPQ micelles is expected to present a new approach in cancer treatment coupled with an early-stage diagnosis, drug delivery, and real-time noninvasive imaging for therapeutic efficacy simultaneously.

References

[1] Blanco E, Shen H, Ferrari M. Principles of nanoparticle design for overcoming biological barriers to drug delivery. Nat Biotechnol 2015;33(9):941–51.

[2] Gupta AK, Bharadwaj M, Mehrotra R. Skin cancer concerns in people of color: risk factors and prevention. Asian Pac J Cancer Prev 2016;17(12):5257.

[3] Huber LA, Pereira TA, Ramos DN, Rezende LC, Emery FS, Sobral LM, et al. Topical skin cancer therapy using doxorubicin-loaded cationic lipid nanoparticles and iontophoresis. J Biomed Nanotechnol 2015;11(11):1975–88.

[4] Brannon-Peppas L, Blanchette JO. Nanoparticle and targeted systems for cancer therapy. Adv Drug Deliv Rev 2012;64:206–12.

[5] Wang Z, Niu G, Chen X. Polymeric materials for theranostic applications. Pharm Res 2014;31(6):1358–76.

[6] Buscombe J, Hirji H, Witney-Smith C. Nuclear medicine in the management of thyroid disease. Expert Rev Anticancer Ther 2008;8(9):1425–31.

[7] Chidambaram M, Manavalan R, Kathiresan K. Nanotherapeutics to overcome conventional cancer chemotherapy limitations. J Pharm Pharm Sci 2011;14(1):67–77.

[8] Leonard R, Williams S, Tulpule A, Levine A, Oliveros S. Improving the therapeutic index of anthracycline chemotherapy: focus on liposomal doxorubicin (Myocet™). Breast 2009;18(4):218–24.

[9] González-Fernández Y, Imbuluzqueta E, Zalacain M, Mollinedo F, Patiño-García A, Blanco-Prieto MJ. Doxorubicin and edelfosine lipid nanoparticles are effective acting synergistically against drug-resistant osteosarcoma cancer cells. Cancer Lett 2017;388:262–8.

[10] Jayakumar R, Nair A, Rejinold NS, Maya S, Nair S. Doxorubicin-loaded pH-responsive chitin nanogels for drug delivery to cancer cells. Carbohydr Polym 2012;87(3):2352–6.

[11] Kanwal U, Irfan Bukhari N, Ovais M, Abass N, Hussain K, Raza A. Advances in nano-delivery systems for doxorubicin: an updated insight. J Drug Target 2018;26(4):296–310.

[12] Kwakman JJ, Elshot YS, Punt CJ, Koopman M. Management of cytotoxic chemotherapy-induced hand-foot syndrome. Oncol Rev 2020;14(1).

[13] Liu Y, Miyoshi H, Nakamura M. Nanomedicine for drug delivery and imaging: a promising avenue for cancer therapy and diagnosis using targeted functional nanoparticles. Int J Cancer 2007;120(12):2527–37.

[14] Peer D, Karp JM, Hong S, Farokhzad OC, Margalit R, Langer R. Nanocarriers as an emerging platform for cancer therapy. Nat Nanotechnol 2007;2(12):751–60.

[15] Choi KY, Liu G, Lee S, Chen X. Theranostic nanoplatforms for simultaneous cancer imaging and therapy: current approaches and future perspectives. Nanoscale 2012;4(2):330–42.

[16] Manivasagan P, Bharathiraja S, Bui NQ, Jang B, Oh Y-O, Lim IG, et al. Doxorubicin-loaded fucoidan capped gold nanoparticles for drug delivery and photoacoustic imaging. Int J Biol Macromol 2016;91:578–88.

[17] Masood F. Polymeric nanoparticles for targeted drug delivery system for cancer therapy. Mater Sci Eng C 2016;60:569–78.

[18] Sumer B, Gao J. Theranostic nanomedicine for cancer. Nanomedicine 2008;3(2):137–40.

[19] Deveza L. Therapeutic angiogenesis for treating cardiovascular diseases. Department of Orthopaedic Surgery and Bioengineering, Stanford University; 2012.

[20] Janib SM, Moses AS, MacKay JA. Imaging and drug delivery using theranostic nanoparticles. Adv Drug Deliv Rev 2010;62(11):1052–63.

[21] Kim J, Lee JE, Lee SH, Yu JH, Lee JH, Park TG, et al. Designed fabrication of a multifunctional polymer nanomedical platform for simultaneous cancer-targeted imaging and magnetically guided drug delivery. Adv Mater 2008;20(3):478–83.

[22] Kunjachan S, Ehling J, Storm G, Kiessling F, Lammers T. Noninvasive imaging of nanomedicines and nanotheranostics: principles, progress, and prospects. Chem Rev 2015;115(19):10907–37.

[23] Vollrath A, Schubert S, Schubert US. Fluorescence imaging of cancer tissue based on metal-free polymeric nanoparticles—a review. J Mater Chem B 2013;1(15):1994–2007.

[24] Mohan P, Rapoport N. Doxorubicin as a molecular nanotheranostic agent: effect of doxorubicin encapsulation in micelles or nanoemulsions on the ultrasound-mediated intracellular delivery and nuclear trafficking. Mol Pharm 2010;7(6):1959–73.

[25] Jiang S, Gnanasammandhan MK, Zhang Y. Optical imaging-guided cancer therapy with fluorescent nanoparticles. J R Soc Interface 2010;7(42):3–18.

[26] Miki K, Kuramochi Y, Oride K, Inoue S, Harada H, Hiraoka M, et al. Ring-opening metathesis polymerization-based synthesis of ICG-containing amphiphilic triblock copolymers for in vivo tumor imaging. Bioconjug Chem 2009;20(3):511–7.

[27] Navya P, Daima HK. Rational engineering of physicochemical properties of nanomaterials for biomedical applications with nanotoxicological perspectives. Nano Converg 2016;3(1):1.

[28] ICH. Pharmaceutical development. Q8 current step; 2009. p. 4.

[29] Vauthier C, Bouchemal K. Methods for the preparation and manufacture of polymeric nanoparticles. Pharm Res 2009;26(5):1025–58.

[30] Uchegbu IF, Carlos M, McKay C, Hou X, Schätzlein AG. Chitosan amphiphiles provide new drug delivery opportunities. Polym Int 2014;63(7):1145–53.

[31] Uchegbu IF. Pharmaceutical nanotechnology: polymeric vesicles for drug and gene delivery. Expert Opin Drug Deliv 2006;3(5):629–40.

[32] Ummarah K, Nadeem IB, Nosheen F, Mahreen R, Khalid H, Nasi RA, et al. Doxorubicin loaded quaternary ammonium palmitoyl glycol chitosan polymeric nanoformulation: uptake by cells and organs. Int J Nanomed 2019;14:1–15.

[33] Kanwal U, Bukhari NI, Rana NF, Rehman M, Hussain K, Abbas N, et al. Doxorubicin-loaded quaternary ammonium palmitoyl glycol chitosan polymeric nanoformulation: uptake by cells and organs. Int J Nanomed 2019;14:1.

[34] Jain A, Thakur K, Kush P, Jain UK. Docetaxel loaded chitosan nanoparticles: formulation, characterization and cytotoxicity studies. Int J Biol Macromol 2014;69:546–53.

[35] ICH. http://www.ich.org/products/guidelines/safety/article/safety-guidelines.html; 2018 [Accessed on September 13, 2018].

[36] Z-h J, M-j J, C-g J, X-z Y, S-x J, X-q Q, et al. Evaluation of doxorubicin-loaded pH-sensitive polymeric micelle release from tumor blood vessels and anticancer efficacy using a dorsal skin-fold window chamber model. Acta Pharmacol Sin 2014;35(6):839–45.

[37] Anders CK, Adamo B, Karginova O, Deal AM, Rawal S, Darr D, et al. Pharmacokinetics and efficacy of PEGylated liposomal doxorubicin in an intracranial model of breast cancer. PLoS One 2013;8(5):61359–69.

[38] Gustafson DL, Merz AL, Long ME. Pharmacokinetics of combined doxorubicin and paclitaxel in mice. Cancer Lett 2005;220(2):161–9.

[39] Thanoo BC, Sunny M, Jayakrishnan A. Cross-linked chitosan microspheres: preparation and evaluation as a matrix for the controlled release of pharmaceuticals. J Pharm Pharmacol 1992;44(4):283–6.

[40] Shiraishi S, Imai T, Otagiri M. Controlled release of indomethacin by chitosan-polyelectrolyte complex: optimization and in vivo/in vitro evaluation. J Control Release 1993;25(3):217–25.

[41] Yang C, Qin Y, Tu K, Xu C, Li Z, Zhang Z, et al. Star-shaped polymer of β-cyclodextrin-g-vitamin E TPGS for doxorubicin delivery and multidrug resistance inhibition. Colloids Surf B: Biointerfaces 2018;169:10–9.

[42] Lalatsa A, Schätzlein AG, Mazza M, Le TBH, Uchegbu IF. Amphiphilic poly (L-amino acids)—new materials for drug delivery. J Control Release 2012;161(2):523–36.

[43] Wang Y, Mei X, Yuan J, Lu W, Li B, Xu D. Taurine zinc solid dispersions attenuate doxorubicin-induced hepatotoxicity and cardiotoxicity in rats. Toxicol Appl Pharmacol 2015;289(1):1–11.

[44] Piktel E, Niemirowicz K, Wątek M, Wollny T, Deptuła P, Bucki R. Recent insights in nanotechnology-based drugs and formulations designed for effective anti-cancer therapy. J Nanobiotechnol 2016;14(1):39.

[45] Vetten MA, Yah CS, Singh T, Gulumian M. Challenges facing sterilization and depyrogenation of nanoparticles: effects on structural stability and biomedical applications. Nanomedicine 2014;10(7):1391–9.

[46] Smith RE. Trends in recommendations for myelosuppressive chemotherapy for the treatment of solid tumors. J Natl Compr Cancer Netw 2006;4(7):649–58.

[47] Zhang J, Chen XG, Li YY, Liu CS. Self-assembled nanoparticles based on hydrophobically modified chitosan as carriers for doxorubicin. Nanomedicine 2007;3(4):258–65.

[48] Zhang J, Sun Y, Tian B, Li K, Wang L, Liang Y, et al. Multifunctional mesoporous silica nanoparticles modified with tumor-shedable hyaluronic acid as carriers for doxorubicin. Colloids Surf B: Biointerfaces 2016;144:293–302.

[49] Qu X, Khutoryanskiy VV, Stewart A, Rahman S, Papahadjopoulos-Sternberg B, Dufes C, et al. Carbohydrate-based micelle clusters which enhance hydrophobic drug bioavailability by up to 1 order of magnitude. Biomacromolecules 2006;7(12):3452–9.

[50] Lalatsa A, Garrett N, Ferrarelli T, Moger J, Schatzlein A, Uchegbu I. Delivery of peptides to the blood and brain after oral uptake of quaternary ammonium palmitoyl glycol chitosan nanoparticles. Mol Pharm 2012;9(6):1764–74.

[51] Lalatsa A, Lee V, Malkinson JP, Zloh M, Schatzlein AG, Uchegbu IF. A prodrug nanoparticle approach for the oral delivery of a hydrophilic peptide, leucine5-enkephalin, to the brain. Mol Pharm 2012;9(6):1665–80.

[52] Siew A, Le H, Thiovolet M, Gellert P, Schatzlein A, Uchegbu I. Enhanced oral absorption of hydrophobic and hydrophilic drugs using quaternary ammonium palmitoyl glycol chitosan nanoparticles. Mol Pharm 2011;9(1):14–28.

Chapter 17

Advanced drug delivery systems in liver cancer

Devaraj Ezhilarasan and Roy Anitha

Department of Pharmacology, Biomedical Research Unit and Laboratory Animal Centre, Saveetha Dental College and Hospitals, Saveetha Institute of Medical and Technical Sciences, Chennai, TN, India

1 Introduction

Hepatocellular carcinoma (HCC) is responsible for more than 90% of primary liver cancers and it is the fifth leading cause of cancer-related mortality and the seventh-most commonly occurring cancer worldwide [1, 2]. There is a 70% of tumor recurrence and ablation was reported after surgery in HCC patients [2]. Chronic hepatitis B virus (HBV) and hepatitis C virus (HCV) infections, high alcohol intake, metabolic syndromes, such as diabetes and obesity, non-alcoholic fatty liver diseases (NAFLD), aflatoxin B1-contaminated food, and tobacco are considered as the primary risk factors for HCC [1]. Viral-related HCC progression is under control in some countries due to the emergence of antihepatitis vaccination. However, numbers of HCC cases are rapidly increasing in the United States, European countries, and India [1, 3].

Targeted therapy using tyrosine kinase inhibitors (TKI) like sorafenib and lenvatinib is considered as the first-line treatment [4] and regorafenib, ramucirumab, and cabozantinib are also employed as the second-line treatment in patients with HCC [5, 6]. In phase II clinical trials, monotherapy with antiprogrammed cell death protein 1 (PD-1) antibodies such as nivolumab and pembrolizumab exhibited anticancer effects in patients with advanced HCC [7, 8]. However, these drugs are failed in phase III clinical trials [9]. Further, adverse events such as hand-foot syndrome, hypertension, rashes, gastrointestinal distress (diarrhea), eosinophilia, and fatigue are commonly observed in HCC patients treated with TKI [10, 11]. Though TKIs are well tolerated and with good efficacy, their pharmacokinetics shows variation among HCC patients and their bioavailability is approximately less than 60% [12]. TKIs also undergo enterohepatic cycling. Chemotherapeutic drugs like doxorubicin (DOX) and paclitaxel (PTX) are commonly prescribed drugs in HCC patients. However, these agents often cause nausea, vomiting, and several off-target effects (Fig. 1) [13]. Therefore, several strategies are used to enhance their bioavailability in liver cancer patients [14]. For instance, nanoparticle encapsulation is often used strategy to enhance the bioavailability of TKIs and chemotherapeutic drugs.

2 Nanoparticles-mediated drug delivery in HCC

Nanoparticles are used as monotherapy in HCC and are used to deliver poorly soluble anticancer drugs. The chemotherapeutic drugs or targeted therapies with TKIs for HCC have high systemic toxicity, low specificity, and often induce severe side effects such as nausea, vomiting, and hair loss. To reduce the chemotherapy-induced side effects on nontumor tissues, targeted drug delivery systems using nanotechnology need to be explored. Nanoparticles possess several advantages as drug carriers, have high encapsulation efficiency for drug coating, increase cellular uptake of drugs, improve solubility and bioavailability, etc. [15]. Nanoparticles-based anticancer drug delivery often enhances the bioavailability, cancer cell internalization due to targeted drug delivery, and reduces side effects [16]. Currently, sorafenib is the only therapy available for HCC patients to improve the overall survival rate. However, its poor oral bioavailability and side effects as aforementioned limit the efficacy of this drug. Several nano-based drug delivery strategies have been tried with sorafenib due to its uncontrolled delivery-mediated drug resistance. For instance, thermosensitive hydrogels are nowadays frequently used as a drug delivery carrier due to their controlled and sustained drug release, biodegradable property, low toxicity, high drug loading capacity, and site-specificity. The selenium and sorafenib thermosensitive hydrogel nanosystem showed

Advanced Drug Delivery Systems in the Management of Cancer. https://doi.org/10.1016/B978-0-323-85503-7.00005-5

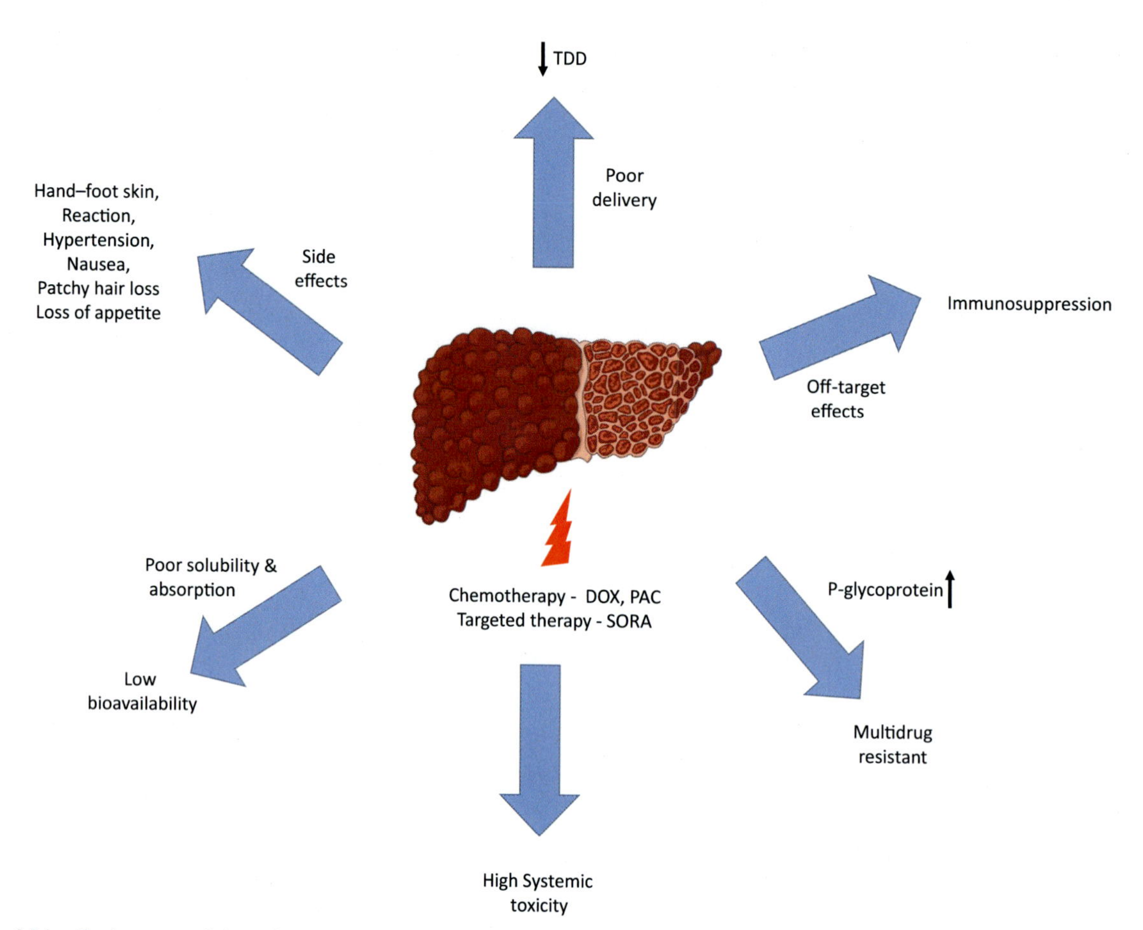

FIG. 1 Major disadvantages of chemo/targeted therapy in HCC patients. TDD, targeted drug delivery; DOX, doxorubicin; PAC, paclitaxel.

long-term and site-specific anticancer potential in a mouse bearing HepG2 cells even after 21 days of single subcutaneous injection [17]. Codelivery of sorafenib and metapristone encapsulated by C-X-C chemokine receptor type 4 (CXCR4) targeted PEGylated poly (lactic-*co*-glycolic acid) nanoparticles conjugated with a peptide inhibitor of CXCR4 (LFC131) accumulated at tumor sites and inhibited the tumor growth in HCC-bearing xenograft nude mice model [18].

When locally injected, DOX and paclitaxel PTX gel formed nanoparticles reservoir at tumor through sol–gel transformation. These drug-gel preparations showed long-lasting anticancer potential on tumor-bearing mice [19]. Cotreatment with DOX and verapamil and ginger extract loaded chitosan nanoparticles enhanced apoptosis via downregulation of B-cell lymphoma-2 (BCL-2) and inhibited angiogenesis marker (vascular endothelial growth factor (VEGF)) and cardiotoxicity in diethylnitrosamine-induced HCC in mice [20, 21]. Superparamagnetic iron oxide nanoparticles (SPIONs) have target specificity and, hence, sorafenib was used along with SPION nanoparticles for targeted drug delivery. MRI-visible SPION and pH-sensitive micelles conjugated with DOX have high blood stability and tumor extravasation that prolonged the survival rate of HCC-bearing mice [22]. Several studies have shown that chemotherapeutic drugs have a thermosensitive effect that could enhance tumor cells sensitivity to several anticancer drugs. For example, between 40.5 °C and 43.0 °C, DOX had a synergistic thermotherapy and chemotherapy effect on tumor cells [23]. Meanwhile, CD147 protein is highly expressed in liver cancer cells. Therefore, magnetothermally responsive manganese zinc (Mn–Zn), ferrite magnetic nanocarriers conjugated with DOX and CD147 monoclonal antibodies were prepared and targeted against liver tumor in mice. The elevated temperature of Mn–Zn ferrite magnetic nanoparticles causes structural changes in the thermosensitive copolymer drug carriers, thereby releasing DOX at tumor sites [24]. There are several nanoparticles that are used to enhance the anticancer potential of chemotherapeutic drugs. Especially, thermotherapy using nanoparticle and chemotherapeutic drug conjugation induces local heating effect at tumor surface which could enhance the cytotoxic potential of chemotherapeutic drugs. Thermotherapy also enhances the blood flow and thereby enhances the chemotherapeutic drug absorption and drug delivery to the tumor cells.

After chemotherapy, thermotherapy reportedly impedes DNA polymerase-associated DNA damage repair in cancer cells.

3 Nucleic acids-mediated drug delivery

Biodegradable polyphosphoester nanoparticles are used for the delivery of several chemotherapeutic drugs such as sorafenib and paclitaxel and nucleic acids [25]. Small interfering RNAs (siRNAs) are known to regulate gene expression by RNA interference (RNAi). In general, RNAi, a post-transcriptional process involving specific inhibition of gene expression by inducing cleavage on a specific place of mRNA [26]. siRNA-mediated drug delivery is applied in cancer therapeutics to specifically inhibit cancer cell survival by inducing the interference of specific carcinogenic gene expressions [27]. Often, siRNAs have been codelivered along with a standard chemotherapeutic agent to enhance the anticancer potential [28]. Anti-BCL-2 (antiapoptotic gene) siRNA and DOX along with a copolymer of poly (ethylene glycol)-*block*-poly (L-lysine)-*block*-poly aspartyl (N-(N′,N′-diisopropylaminoethyl)) caused synergistic anticancer potential by reducing tumor volume in HepG2/ADM cells-xenografted mice model. This siRNA/nanoparticle/DOX-mediated complex could bypass the drug efflux system, reduce chemoresistance, and suppress the BCL-2 gene expression [28]. Glypican 3 (GPC3), a heparan sulfate proteoglycan siRNA, was also studied against an experimental HCC model [29]. GPC3 is a cell surface oncofetal protein highly expressed on a variety of solid tumors including HCC [30, 31]. Intravenous injection of sorafenib, siRNA-targeting GPC3-loaded liposomes, effectively downregulated the GPC3 expression and increased the tumor sensitivity to sorafenib. This drug and siRNA-loaded liposomal system also inhibited tumor growth and increased the survival of nude mice bearing HepG2 xenografts [29]. UBC9 is an E2-conjugating enzyme upregulated in HCC progression [32]. Interestingly, UBC9 downregulation by small hairpin RNA (shRNA) enhanced the sensitivity of HCC to DOX by regulation of extracellular signal-regulated kinase 1/2 and P38 activation and also reduced the expression of BCL-2 and increased the expression of caspase 3 [33]. Multidrug resistance (MDR) is one of the main obstacles that hamper the treatment of HCC due to the enhanced expression of P-glycoprotein [34]. A variety of miRNAs have been reported to get dysregulated during the progression of HCC [35]. The overexpression of miR-375 has been reported to inhibit P-glycoprotein expression in HCC [36]. Therefore, in a study, lipid-coated hollow mesoporous silica nanoparticles have been developed to deliver DOX and miR-375. This nano/miR-375/DOX complex could overcome the drug efflux and specifically delivered miR-375/DOX into MDR HCC tissues, and this combination also exhibited significant antitumor effects in xenografts and primary tumors [37]. miRNA-17 family has been upregulated in HCC cells such as HepG2 and Hep3B [38]. The lipid nanoparticle encapsulated antagomir-17 or antagomir-17 administration alone in vivo resulted in tumor growth retardation and increased apoptosis [38, 39]. miRNA-539 has been downregulated during HCC progression. The intratumoral delivery of miR-539 significantly retarded the xenograft tumor growth [40]. Undoubtedly, siRNA/miRNA/shRNA is used for the treatment of HCC and nanoparticles and stem cells are mainly used to deliver these nucleic acids.

4 Mesenchymal stem cells (MSCs)-mediated drug delivery

Stem cell therapy is one of the emerging and promising strategies involved in targeted drug delivery [41]. Mesenchymal stem cells exhibit tropism toward tumor sites. The stem cells coated with chemotherapeutic drugs or nanoparticles deliver drugs precisely into the cancer tissue due to the cancer tropism properties of stem cells [42, 43]. MiR-199a-3p has been shown to enhance chemosensitivity and inhibit the growth of HCC [44, 45]. Therefore, exosomes from adipose tissue-derived mesenchymal stem cells (MSCs) were used as a vehicle to deliver miR-199a (AMSC-Exo-199a) and improve HCC chemosensitivity. In an orthotopic HCC mouse model, tail vein injection of AMSC-Exo-199a and DOX caused tumor growth retardation via mechanistic target of rapamycin and p21 activated kinase 4 modulation and increased the levels of miR-199a in the HCC sample. This study demonstrated that AMSC-Exo-199a administration could increase the chemosensitivity and growth inhibitory effect of DOX [46]. Genetically, engineered MSCs with Apoptin, a protein with an ability to selectively destroy cancer cells, reduced tumor volume in HepG2 cells transplanted mice. In this study, MSCs have been used to deliver apoptin to selectively kill cancer cells [47]. Site-specific delivery of chemotherapeutic drugs to tumor sites is often a problem. Hence, MSCs incorporated chemotherapeutic drugs or MSCs conditioned cytotoxic drugs are employed against HCC. For instance, in xenograft mice models of HCC, human placenta–derived MSCs and sorafenib intravenous administration induced significant tumor tissue necrosis and inhibited angiogenesis and tumor growth as compared to sorafenib monotherapy [48, 49]. MSCs labeled with iron oxide nanoparticles significantly reduced the tumor mass. These MSCs and iron oxide nanoparticle complexes were traced by MRI and it revealed the homing and localization of MSCs in the tumor sites [50]. The stem cells are used as a monotherapy and also used to deliver nucleic acids, nanoparticles, and anticancer agents due to their high tumor homing and tropism properties. In stem cell monotherapy, stem cells sense the tumor sites and modulate the tumor milieu by secreting a variety of anticancer soluble factors.

4.1 Oncolytic virus and viral-like particles-based targeted therapy in HCC

One of the hallmarks of targeted therapy is to kill the cancer cells without affecting the normal tissue [51]. Oncolytic viral therapy involves the selective killing of cancer cells by infection [52]. In preclinical studies, adenoviruses and vesicular stomatitis virus are used as vectors due to their ability to kill the HCC cells. The vesicular stomatitis virus has the ability to infect the HCC cells, while adenoviruses have tropism properties toward hepatocytes. The oncolytic virus-like JX-594 replicates in tumor cells that cause oncolysis. Mechanistically, once injected, the oncolytic virus infects HCC cells, and they replicate inside the HCC cells and cause cancer cell rupture. The cancer cell rupture causes the release of viral particles and several cytokines further infect surrounding cancer cells. The cancer cell-derived cytokines further activate the T cells that are responsible for HCC cell death [53]. During oncolytic virus therapy, interferons are stimulated, and this may cause the immune clearance of viruses, and this has been considered as one of the main disadvantages [54]. Virus-like particles are nanoparticles and they do not have genetic material and structurally they resemble viruses [55]. MS2 virus-like particles structurally resemble the MS2 bacteriophage. The MS2 virus-like particles have been used to specifically deliver anticancer agents in HCC cells. In a previous study, the MS2 virus-like particles were cross-linked with GE11, an epidermal growth factor receptor (EGFR) targeting peptide used to target EGFR. These cross-linked particles had a high affinity to EGFR and delivered tumor suppressor [56]. The MSP (SP94) viruses-like particles have a higher affinity to hepatocytes and have been shown to specifically bind to HCC cells. These particles are specifically used to deliver anticyclin siRNAs in Hep3B cells and thereby could induce growth arrest apoptosis [57]. The viruses-like particles are considered as one of the rapid, safe, and very efficient way of drug delivery systems in HCC therapy. The different drug delivery systems employed in HCC are presented in Fig. 2.

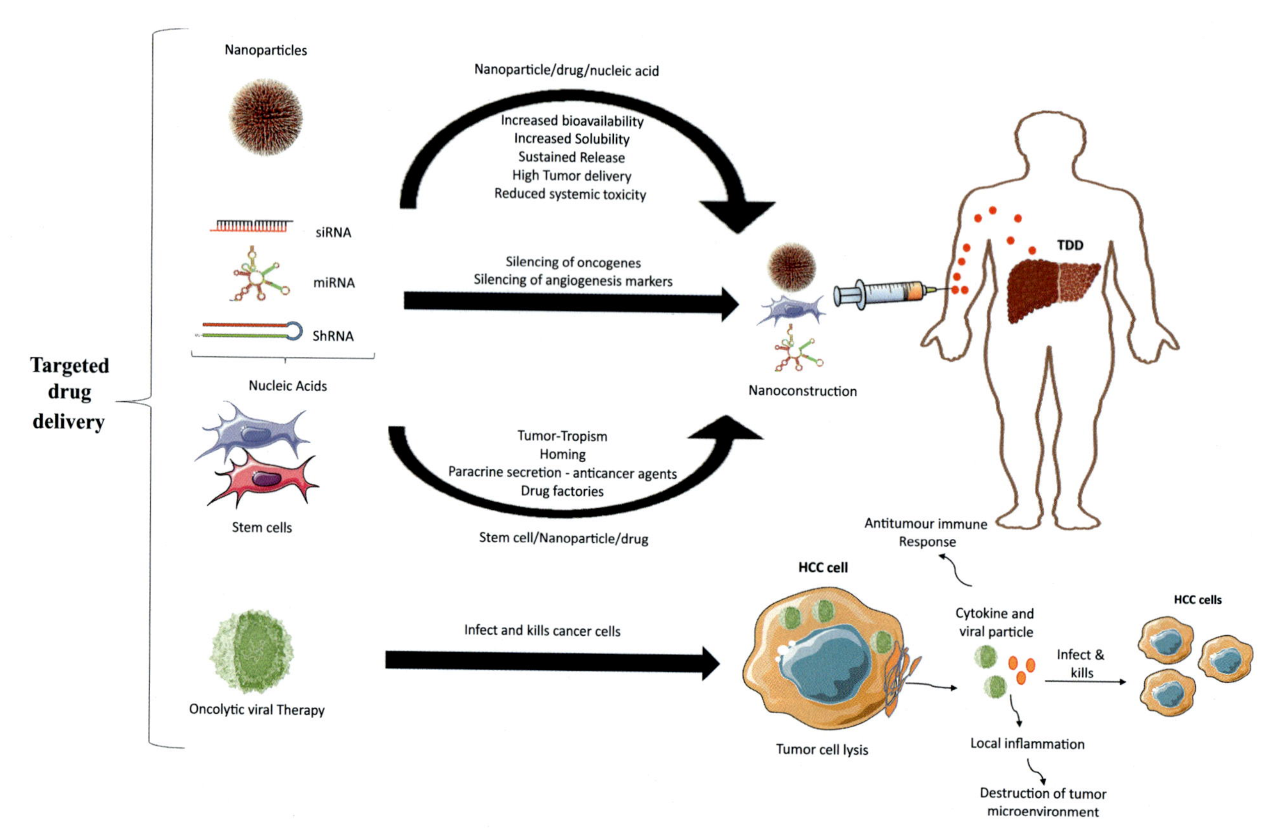

FIG. 2 Different drug delivery systems used in HCC. siRNA, Small interfering RNAs; miRNA, microRNA; ShRNA, small hairpin RNA; HCC, hepatocellular carcinoma; TDD, targeted drug delivery.

5 Conclusion

Chemotherapy often failed due to side effects, low solubility, poor bioavailability, and off-target effects of the drug used. Therefore, several drug delivery systems have been used to deliver drugs precisely into the tumor sites. The nanoparticles are used alone and also they are used as a vehicle to deliver the poorly soluble anticancer agents against solid cancers. The nanoparticles possess several advantages. They act as drug carriers, have high encapsulation efficiency, can increase cellular uptake of drugs, and improve solubility and bioavailability. Therefore, encapsulation of nanoparticles and anticancer agents can improve the bioavailability and drug delivery to tumor sites. Stem cells, oncolytic virus, and viral-like particles are called biotherapeutics involved in HCC therapy. The stem cells have unique homing and tumor tropism properties, and their transplantation can be used to modulate the HCC microenvironments. Once injected, these stem cells secrete several cytokines and soluble factors that control cancer cell proliferation and halt their cell cycle. Stem cells are also used as a carrier vehicle to deliver the chemotherapeutic drugs/siRNAs due to their homing properties. Oncolytic viruses are used to infect and kills cancer cells. The oncolytic viruses selectively enter, lyse the HCC cells, and modulate the cancer microenvironment via proinflammatory action. The virus-like particles are also used as an anticancer delivery vehicle and they offer a rapid, safe, and very efficient way of drug delivery system in HCC therapy.

References

[1] McGlynn KA, Petrick JL, El-Serag HB. Epidemiology of hepatocellular carcinoma. Hepatology 2020. https://doi.org/10.1002/hep.31288.

[2] Zhu XD, Li KS, Sun HC. Adjuvant therapies after curative treatments for hepatocellular carcinoma: current status and prospects. Genes Dis 2020;7(3):359–69.

[3] Singal AG, Lampertico P, Nahon P. Epidemiology and surveillance for hepatocellular carcinoma: new trends. J Hepatol 2020;72(2):250–61.

[4] Liu Z, Lin Y, Zhang J, Zhang Y, Li Y, Liu Z, Li Q, Luo M, Liang R, Ye J. Molecular targeted and immune checkpoint therapy for advanced hepatocellular carcinoma. J Exp Clin Cancer Res 2019;38(1):447.

[5] Abou-Alfa GK, Meyer T, Cheng AL, El-Khoueiry AB, Rimassa L, Ryoo BY, Cicin I, Merle P, Chen Y, Park JW, Blanc JF, Bolondi L, Klümpen HJ, Chan SL, Zagonel V, Pressiani T, Ryu MH, Venook AP, Hessel C, Borgman-Hagey AE, Kelley RK. Cabozantinib in patients with advanced and progressing hepatocellular carcinoma. N Engl J Med 2018;379(1):54–63.

[6] Bruix J, Qin S, Merle P, Granito A, Huang YH, Bodoky G, Pracht M, Yokosuka O, Rosmorduc O, Breder V, Gerolami R, Masi G, Ross PJ, Song T, Bronowicki JP, Ollivier-Hourmand I, Kudo M, Cheng AL, Llovet JM, Finn RS, RESORCE Investigators. Regorafenib for patients with hepatocellular carcinoma who progressed on sorafenib treatment (RESORCE): a randomised, double-blind, placebo-controlled, phase 3 trial. Lancet 2017;389(10064):56–66.

[7] El-Khoueiry AB, Sangro B, Yau T, Crocenzi TS, Kudo M, Hsu C, Kim TY, Choo SP, Trojan J, Welling THR, Meyer T, Kang YK, Yeo W, Chopra A, Anderson J, Dela Cruz C, Lang L, Neely J, Tang H, Dastani HB, Melero I. Nivolumab in patients with advanced hepatocellular carcinoma (CheckMate 040): an open-label, non-comparative, phase 1/2 dose escalation and expansion trial. Lancet 2017;389(10088):2492–502.

[8] Zhu AX, Finn RS, Edeline J, Cattan S, Ogasawara S, Palmer D, Verslype C, Zagonel V, Fartoux L, Vogel A, Sarker D, Verset G, Chan SL, Knox J, Daniele B, Webber AL, Ebbinghaus SW, Ma J, Siegel AB, Cheng AL, KEYNOTE-224 investigators. Pembrolizumab in patients with advanced hepatocellular carcinoma previously treated with sorafenib (KEYNOTE-224): a non-randomised, open-label phase 2 trial. Lancet Oncol 2018;19(7):940–52.

[9] Finn RS, Ryoo BY, Merle P, Kudo M, Bouattour M, Lim HY, Breder V, Edeline J, Chao Y, Ogasawara S, Yau T, Garrido M, Chan SL, Knox J, Daniele B, Ebbinghaus SW, Chen E, Siegel AB, Zhu AX, Cheng AL, KEYNOTE-240 investigators. Pembrolizumab as second-line therapy in patients with advanced hepatocellular carcinoma in KEYNOTE-240: a randomized, double-blind, phase III trial. J Clin Oncol 2020;38(3):193–202.

[10] Brose MS, Frenette CT, Keefe SM, Stein SM. Management of sorafenib-related adverse events: a clinician's perspective. Semin Oncol 2014;41(Suppl 2):S1–S16.

[11] Salame N, Chow ML, Ochoa MT, Compoginis G, Crew AB. Sorafenib toxicity mimicking drug reaction with eosinophilia and systemic symptoms (DRESS) syndrome. J Drugs Dermatol 2019;18(5):468–9.

[12] Singh RP, Patel B, Kallender H, Ottesen LH, Adams LM, Cox DS. Population pharmacokinetics modeling and analysis of foretinib in adult patients with advanced solid tumors. J Clin Pharmacol 2015;55(10):1184–92.

[13] Ezhilarasan D. Herbal therapy for Cancer: Clinical and experimental perspectives. In: Timiri Shanmugam P, editor. Understanding cancer therapies. Boca Raton: CRC Press; 2018. p. 129–66.

[14] Scheffler M, Di Gion P, Doroshyenko O, Wolf J, Fuhr U. Clinical pharmacokinetics of tyrosine kinase inhibitors: focus on 4-anilinoquinazolines. Clin Pharmacokinet 2011;50(6):371–403.

[15] Usmani A, Mishra A, Ahmad M. Nanomedicines: a theranostic approach for hepatocellular carcinoma. Artif Cells Nanomed Biotechnol 2018;46(4):680–90.

[16] Samie H, Saeed M, Faisal SM, Kausar MA, Kamal MA. Recent findings on nanotechnology-based therapeutic strategies against hepatocellular carcinoma. Curr Drug Metab 2019;20(4):283–91.

[17] Zheng L, Li C, Huang X, Lin X, Lin W, Yang F, Chen T. Thermosensitive hydrogels for sustained-release of sorafenib and selenium nanoparticles for localized synergistic chemoradiotherapy. Biomaterials 2019;216:119220.

[18] Zheng N, Liu W, Li B, Nie H, Liu J, Cheng Y, Wang J, Dong H, Jia L. Co-delivery of sorafenib and metapristone encapsulated by CXCR4-targeted PLGA-PEG nanoparticles overcomes hepatocellular carcinoma resistance to sorafenib. J Exp Clin Cancer Res 2019;38(1):232.

[19] Fu H, Huang L, Xu C, Zhang J, Li D, Ding L, Liu L, Dong Y, Wang W, Duan Y. Highly biocompatible thermosensitive nanocomposite gel for combined therapy of hepatocellular carcinoma via the enhancement of mitochondria related apoptosis. Nanomed Nanotechnol Biol Med 2019;21:102062.

[20] Abo Mansour HE, El-Batsh MM, Badawy NS, Mehanna ET, Mesbah NM, Abo-Elmatty DM. Effect of co-treatment with doxorubicin and verapamil loaded into chitosan nanoparticles on diethylnitrosamine-induced hepatocellular carcinoma in mice. Hum Exp Toxicol 2020;39(11):1528–44.
[21] Abo Mansour HE, El-Batsh MM, Badawy NS, Mehanna ET, Mesbah NM, Abo-Elmatty DM. Ginger extract loaded into chitosan nanoparticles enhances cytotoxicity and reduces cardiotoxicity of doxorubicin in hepatocellular carcinoma in mice. Nutr Cancer 2020;1–16. Advance online publication https://doi.org/10.1080/01635581.2020.1823436.
[22] Li B, Cai M, Lin L, Sun W, Zhou Z, Wang S, Wang Y, Zhu K, Shuai X. MRI-visible and pH-sensitive micelles loaded with doxorubicin for hepatoma treatment. Biomater Sci 2019;7(4):1529–42.
[23] Urano M, Kuroda M, Nishimura Y. For the clinical application of thermochemotherapy given at mild temperatures. Int J Hyperthermia 1999;15(2):79–107.
[24] Li M, Deng L, Li J, Yuan W, Gao X, Ni J, Jiang H, Zeng J, Ren J, ang, P. Actively targeted magnetothermally responsive nanocarriers/doxorubicin for thermochemotherapy of hepatoma. ACS Appl Mater Interfaces 2018;10(48):41107–17.
[25] Elzeny H, Zhang F, Ali EN, Fathi HA, Zhang S, Li R, El-Mokhtar MA, Hamad MA, Wooley KL, Elsabahy M. Polyphosphoester nanoparticles as biodegradable platform for delivery of multiple drugs and siRNA. Drug Des Devel Ther 2017;11:483–96.
[26] Mahmoodi Chalbatani G, Dana H, Gharagouzloo E, Grijalvo S, Eritja R, Logsdon CD, Memari F, Miri SR, Rad MR, Marmari V. Small interfering RNAs (siRNAs) in cancer therapy: a nano-based approach. Int J Nanomedicine 2019;14:3111–28.
[27] Singh A, Trivedi P, Jain NK. Advances in siRNA delivery in cancer therapy. Artif Cells Nanomed Biotechnol 2018;46(2):274–83.
[28] Sun W, Chen X, Xie C, Wang Y, Lin L, Zhu K, Shuai X. Co-delivery of doxorubicin and anti-BCL-2 siRNA by pH-responsive polymeric vector to overcome drug resistance in in vitro and in vivo HepG2 hepatoma model. Biomacromolecules 2018;19(6):2248–56.
[29] Sun W, Wang Y, Cai M, Lin L, Chen X, Cao Z, Zhu K, Shuai X. Codelivery of sorafenib and GPC3 siRNA with PEI-modified liposomes for hepatoma therapy. Biomater Sci 2017;5(12):2468–79.
[30] El-Saadany S, El-Demerdash T, Helmy A, Mayah WW, El-Sayed Hussein B, Hassanien M, Elmashad N, Fouad MA, Basha EA. Diagnostic value of glypican-3 for hepatocellular carcinomas. Asian Pac J Cancer Prev 2018;19(3):811–7.
[31] Montalbano M, Georgiadis J, Masterson AL, McGuire JT, Prajapati J, Shirafkan A, Rastellini C, Cicalese L. Biology and function of glypican-3 as a candidate for early cancerous transformation of hepatocytes in hepatocellular carcinoma (review). Oncol Rep 2017;37(3):1291–300.
[32] Yang H, Gao S, Chen J, Lou W. UBE2I promotes metastasis and correlates with poor prognosis in hepatocellular carcinoma. Cancer Cell Int 2020;20:234.
[33] Fang S, Qiu J, Wu Z, Bai T, Guo W. Down-regulation of UBC9 increases the sensitivity of hepatocellular carcinoma to doxorubicin. Oncotarget 2017;8(30):49783–95.
[34] Kong XB, Yang ZK, Liang LJ, Huang JF, Lin HL. Overexpression of P-glycoprotein in hepatocellular carcinoma and its clinical implication. World J Gastroenterol 2000;6(1):134–5.
[35] Mizuguchi Y, Takizawa T, Yoshida H, Uchida E. Dysregulated miRNA in progression of hepatocellular carcinoma: a systematic review. Hepatol Res 2016;46(5):391–406.
[36] Yang T, Zhao P, Rong Z, Li B, Xue H, You J, He C, Li W, He X, Lee RJ, Ma X, Xiang G. Anti-tumor efficiency of lipid-coated cisplatin nanoparticles co-loaded with MicroRNA-375. Theranostics 2016;6(1):142–54.
[37] Xue H, Yu Z, Liu Y, Yuan W, Yang T, You J, He X, Lee RJ, Li L, Xu C. Delivery of miR-375 and doxorubicin hydrochloride by lipid-coated hollow mesoporous silica nanoparticles to overcome multiple drug resistance in hepatocellular carcinoma. Int J Nanomedicine 2017;12:5271–87.
[38] Dhanasekaran R, Gabay-Ryan M, Baylot V, Lai I, Mosley A, Huang X, Zabludoff S, Li J, Kaimal V, Karmali P, Felsher DW. Anti-miR-17 therapy delays tumorigenesis in MYC-driven hepatocellular carcinoma (HCC). Oncotarget 2017;9(5):5517–28.
[39] Huang X, Magnus J, Kaimal V, Karmali P, Li J, Walls M, Prudente R, Sung E, Sorourian M, Lee R, Davis S, Yang X, Estrella H, Lee EC, Chau BN, Pavlicek A, Zabludoff S. Lipid nanoparticle-mediated delivery of anti-miR-17 family oligonucleotide suppresses hepatocellular carcinoma growth. Mol Cancer Ther 2017;16(5):905–13.
[40] Zhu C, Zhou R, Zhou Q, Chang Y, Jiang M. microRNA-539 suppresses tumor growth and tumorigenesis and overcomes arsenic trioxide resistance in hepatocellular carcinoma. Life Sci 2016;166:34–40.
[41] Elkhenany H, Shekshek A, Abdel-Daim M, El-Badri N. Stem cell therapy for hepatocellular carcinoma: future perspectives. Adv Exp Med Biol 2020;1237:97–119.
[42] Ai J, Ketabchi N, Verdi J, Gheibi N, Khadem Haghighian H, Kavianpour M. Mesenchymal stromal cells induce inhibitory effects on hepatocellular carcinoma through various signaling pathways. Cancer Cell Int 2019;19:329.
[43] Devaraj E, Rajeshkumar S. Nanomedicine for hepatic fibrosis. In: Shukla A, editor. Nanoparticles and their biomedical applications. Singapore: Springer; 2020.
[44] Callegari E, D'Abundo L, Guerriero P, Simioni C, Elamin BK, Russo M, Cani A, Bassi C, Zagatti B, Giacomelli L, Blandamura S, Moshiri F, Ultimo S, Frassoldati A, Altavilla G, Gramantieri L, Neri LM, Sabbioni S, Negrini M. miR-199a-3p modulates MTOR and PAK4 pathways and inhibits tumor growth in a hepatocellular carcinoma transgenic mouse model. Mol Ther–Nucleic Acids 2018;11:485–93.
[45] Lou G, Song X, Yang F, Wu S, Wang J, Chen Z, Liu Y. Exosomes derived from miR-122-modified adipose tissue-derived MSCs increase chemosensitivity of hepatocellular carcinoma. J Hematol Oncol 2015;8:122.
[46] Lou G, Chen L, Xia C, Wang W, Qi J, Li A, Zhao L, Chen Z, Zheng M, Liu Y. MiR-199a-modified exosomes from adipose tissue-derived mesenchymal stem cells improve hepatocellular carcinoma chemosensitivity through mTOR pathway. J Exp Clin Cancer Res 2020;39(1):4.
[47] Zhang J, Hou L, Wu X, Zhao D, Wang Z, Hu H, Fu Y, He J. Inhibitory effect of genetically engineered mesenchymal stem cells with Apoptin on hepatoma cells in vitro and in vivo. Mol Cell Biochem 2016;416(1–2):193–203.
[48] Hajighasemlou S, Nikbakht M, Pakzad S, Muhammadnejad S, Gharibzadeh S, Mirmoghtadaei M, Zafari F, Seyhoun I, Ai J, Verdi J. Sorafenib and mesenchymal stem cell therapy: a promising approach for treatment of HCC. Evid Based Complement Alternat Med 2020;2020:9602728.
[49] Seyhoun I, Hajighasemlou S, Muhammadnejad S, Ai J, Nikbakht M, Alizadeh AA, Hosseinzadeh F, Mirmoghtadaei M, Seyhoun SM, Verdi J. Combination therapy of sorafenib with mesenchymal stem cells as a novel cancer treatment regimen in xenograft models of hepatocellular carcinoma. J Cell Physiol 2019;234(6):9495–503.

[50] Faidah M, Noorwali A, Atta H, Ahmed N, Habib H, Damiati L, Filimban N, Al-Qriqri M, Mahfouz S, Khabaz MN. Mesenchymal stem cell therapy of hepatocellular carcinoma in rats: detection of cell homing and tumor mass by magnetic resonance imaging using iron oxide nanoparticles. Adv Clin Exp Med 2017;26(8):1171–8.

[51] Fukuhara H, Ino Y, Todo T. Oncolytic virus therapy: a new era of cancer treatment at dawn. Cancer Sci 2016;107(10):1373–9.

[52] Taguchi S, Fukuhara H, Todo T. Oncolytic virus therapy in Japan: progress in clinical trials and future perspectives. Jpn J Clin Oncol 2019;49(3):201–9.

[53] Daher S, Massarwa M, Benson AA, Khoury T. Current and future treatment of hepatocellular carcinoma: an updated comprehensive review. J Clin Transl Hepatol 2018;6(1):69–78.

[54] Geoffroy K, Bourgeois-Daigneault MC. The pros and cons of interferons for oncolytic virotherapy. Cytokine Growth Factor Rev 2020;56:49–58.

[55] Reghupaty SC, Sarkar D. Current status of gene therapy in hepatocellular carcinoma. Cancer 2019;11(9):1265.

[56] Chang L, Wang G, Jia T, Zhang L, Li Y, Han Y, Zhang K, Lin G, Zhang R, Li J, Wang L. Armored long non-coding RNA MEG3 targeting EGFR based on recombinant MS2 bacteriophage virus-like particles against hepatocellular carcinoma. Oncotarget 2016;7(17):23988–4004.

[57] Ashley CE, Carnes EC, Phillips GK, Durfee PN, Buley MD, Lino CA, Padilla DP, Phillips B, Carter MB, Willman CL, Brinker CJ, Caldeira J, Chackerian B, Wharton W, Peabody DS. Cell-specific delivery of diverse cargos by bacteriophage MS2 virus-like particles. ACS Nano 2011;5(7):5729–45.

Chapter 18

Advanced drug delivery systems in hepatocellular carcinoma

Dhrubojyoti Mukherjee[a] and Shvetank Bhatt[b]

[a]*Department of Pharmaceutics, Faculty of Pharmacy, Ramaiah University of Applied Sciences, Bengaluru, Karnataka, India,* [b]*Amity Institute of Pharmacy, Amity University Madhya Pradesh (AUMP), Gwalior, Madhya Pradesh, India*

1 Introduction

One of the predominant causes of cancer-associated death worldwide is hepatocellular carcinoma (HCC). It is the fifth-most commonly diagnosed cancer globally and is the second- and sixth-most significant reason for death related to cancer in males and females, respectively. Patients when first diagnosed are in a transitional or advanced stage because of the lengthy latent periods of HCC. The most frequently used mode of treatment, chemotherapy, is associated with severe side effects and a low response rate. Conventional chemotherapeutic drugs have restricted roles in the improvement of the disease in long term as drug resistance and systemic toxicity cause potential threat in the treatment approach. However, locoregional therapies are often recommended as a treatment option clinically, but only an insignificant portion of the patients are potential candidates for the therapy when diagnosed [1, 2].

HCC can be cured in its earlier stages with a different mode of therapies, but for the patient with intermediate stages, the ideal treatment strategy has not been well established. Over the course of the disease, a small number of patients receive any substantial treatment, even though treatment of HCC at any stage has been associated with survival benefits. The most curative intervention for HCC is liver transplant but the opportunities are restricted by the progressive stage at diagnosis, comorbid conditions, scarce social support, and limited resources [3]. The concept of drug delivery can be explained as the method and route followed to administer an active pharmaceutical ingredient, which can stimulate its pharmacological activity and also reduce the adverse effects associated with it. On the other hand, drug delivery systems are the device or the formulations that assist in the delivery of the API to a specific physiological site as well as promote the timely release of the API [4, 5].

Drug delivery technology (DDT) is developed to reconnoitre novel delivery routes. In the past decades, several DDT has been explored, besides oral and transdermal, the concept of drug delivery has expanded to pulmonary, ocular, vaginal, nasal, and rectal to render local and systemic effects. Pharmaceutical industries started focusing on the design and development of ADDSs to overcome the limitations of conventional drug delivery systems. Advanced/novel drug delivery systems are designed based on rationale engineering which can promote delivery and improve the performance of existing benchmark drugs. Novel or ADDSs can combine superior techniques and modulate the novel dosage forms to target and control the delivery of drugs sometimes being site specific. The two prerequisites are fulfilled by the advanced drug carriers: first, the delivery of the API to a specific target site at an extent and pace which fulfils the demand of the body, and second is monitoring the active unit during the treatment [6]. Both of these conditions are important during the treatment approach of cancer like HCC.

2 Morphology, anatomy, and physiology of normal liver

2.1 Morphology and anatomy of liver

The liver is the cone-shaped or triangular, second largest organ of our body that stands after the skin. The average weight of the liver is approximately 3 pounds and it is dark brown in color. It is located just under the diaphragm in the right upper abdomen and mid-abdomen and lengthens to the left upper abdomen. Two distinct sources of blood circulation to the liver are as follows: (i) hepatic artery: which supplies the oxygenated, pure blood and (ii) hepatic portal vein: which connects the duodenum to the liver and supplied with nutrient-rich blood. On the other hand, the blood is collected by hepatic vein that leads to the vena cava and returns to the heart.

Since the liver has a big size, at any given moment liver contains about 1 pint (13%) of total blood supply with

Advanced Drug Delivery Systems in the Management of Cancer. https://doi.org/10.1016/B978-0-323-85503-7.00023-7

respect to body. The liver is made up of four lobes, namely left, right, caudate, and quadrate lobes. The right and left lobes of the liver are considered as main lobes and they are bigger in size as compared with the other two lobes. These lobes are parted by the falciform ligament. The right lobe is around 5–6 times bigger than the tapered left lobe [7]. The small caudate lobe wraps around the inferior vena cava and extends from the posterior side. The quadrate lobe is also small and found below the caudate lobe and extends from the posterior side of the right lobe and covers around the gallbladder. The right and left lobes consist of eight segments comprising of small size lobules (small lobes). These lobules are linked to small tubes or ducts that further connect with larger tubes to form the common hepatic duct. The common hepatic duct involves in the transportation of the bile synthesized by the hepatic cells to the gallbladder and duodenum via the common bile duct [8]. Anatomical structure of normal liver has been shown in Fig. 1.

The interior assembly of the liver is made up of lobules. The interior assembly has 100,000 small hexagonal functional units of lobules [7]. Each lobule contains a central vein bounded by six hepatic portal veins and six hepatic arteries. These blood vessels are connected with many small capillary-like tubes named as sinusoids. Each sinusoid that passes through the liver contains two major cell types, i.e., Kupffer cells and hepatocytes. The macrophage cells of the liver are known as Kupffer cells involved in the breakdown of old and worn-out RBC. Hepatocytes are the cuboidal cells that lined the sinusoids and make up the majority of the portion of the liver. Hepatocytes are responsible for most of the liver's functions—biotransformation of drugs and nutrients, storage, digestion, and production of bile [9].

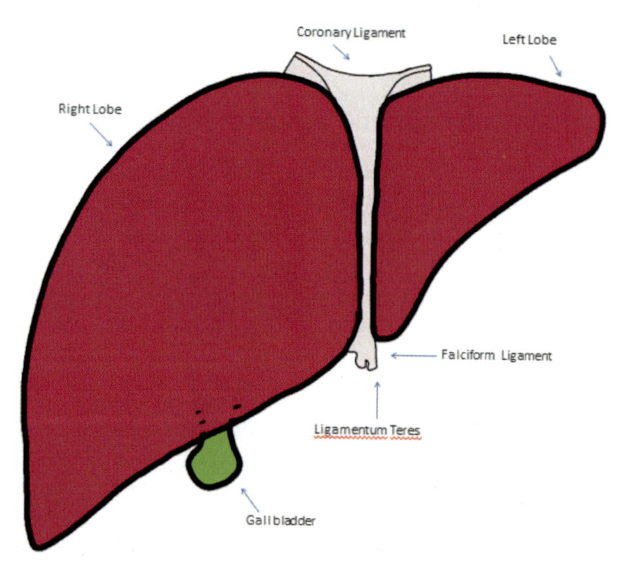

FIG. 1 Normal liver.

2.2 Physiology of liver

The liver is the main metabolizing organ involved in metabolic transformation or inactivation of the drug molecules [10]. The bile synthesized in the liver is involved in the emulsification of the fat. It is also important to make certain plasma proteins and cholesterol [11]. It is also involved in the processing of hemoglobin as well as the storage of iron and conversion of glucose into glycogen [12]. It is also involved in the conversion of toxic ammonia into urea. Formation of bile pigments and important clotting factors is also some important functions of the liver. It is also involved in the processing of RBC and to form the bile pigments line bilirubin and biliverdin [13].

3 Pathophysiology of HCC

HCC, associated with poor prognosis, is one of the primary causes of death in the globe. The infection with Hepatitis B or C virus (HBV or HCV) and chronic abuse of alcohol is the main reason for cirrhosis of the liver [14]. Patients with liver cirrhosis are more often affected by HCC. Cirrhosis is responsible for oncogenic variations and is observed in approximately 90% of the patients detected with HCC. In the remaining 10% of the patients, noncirrhotic mechanism of tumor progression is accountable for the malignancy of the disease [15]. In liver cirrhosis, metabolic stress and reactive oxygen species cause cyclic inflammation, tissue necrosis, and elevated turnover of hepatic which leads to the accumulation of genetic errors and mutations such as telomere erosion; telomerase reactivation; point mutations; chromosomal gains; and deletions in TP53, AXIN1, and CTNNB1. They are also associated with the activation of proto-oncogenes such as the RAS-RAF-MEK-MAP kinase pathway and β-catenin signaling. The risk of getting affected with HCC is more in the patients infected with viral hepatitis as compared to nonviral causes [16].

3.1 Tumor microenvironment (TME)

HCC belongs to the cold tumor category as per the presence of TME. It (TME) is a complex network of tumor cells, stromal, angiogenic cells, cells targeting the immune system, and cancer-related fibroblastic cells, in which transduction signaling pathways and production of molecules and other factors promote tumor progression [17]. Chronic hepatic injury leads to hypoxia and fibrosis due to the deposition of extracellular matrix (ECM) [18]. This hypoxia and ECM lead to activation and secretion of proangiogenic factors by stromal cells, and hypoxia-inducible factor-1α (HIF-1α) production is enhanced as well as tumor growth, cell proliferation, and cell survival are stimulated. The involvement of a fibrotic environment, i.e., tumor-associated fibroblasts (TAFs), secretes factors that support the growth of tumor and angiogenesis and is involved in cross talk with tumor cells [19].

Immune cells such as tumor-infiltrating lymphocytes are important for initiating the antitumor response and T-cell activation cascade [20]. On the other hand, circulatory regulatory T cells (Tregs) and myeloid-derived suppressive cells (MDSCs) are associated with the suppression of the immune response. Moreover, tumor-associated macrophages (TAMs) release various chemokines and growth factors that suppress antitumor immunity [21].

3.2 Angiogenesis

Angiogenesis is also one of the important factors playing a significant role in the pathological development of HCC. High levels of angiogenic factors produced by cancerous cells, vascular endothelial cells, immune cells, and surrounding TME are linked with HCC [22]. The vascular endothelial growth factor (VEGF) is one of the utmost important growth factors involved in the progression and angiogenesis of HCC. Hypoxic environments lead to overexpression of VEGF due to oncogenic gene mutations, hormones, and cytokine production [23]. VEGF through VEGR receptor acts on the neighboring stromal environment consisting of hepatic stellate cells and Kupffer cells.

4 Immunobiology of liver in HCC

Recent reports have revealed the promising data with various immunotherapeutic approaches for the treatment of HCC suggest the involvement of the immune environment in the progression of HCC. However, the roles of several immune components and factors, such as cytotoxic T cells, regulatory T cells, gamma delta T cells, and various cytokines and chemokines, have yet to be explored in detail [24]. An important role of CD8 + T cells in antitumor immunity by inhibiting the growth of the tumor has been identified. On the other hand, TAMs, MDSC, Tregs, Th17 cells, and related cytokines such as IL-6, TNF-α, IL-1β, IL-23, and TGF-β are assumed to play significant roles in promoting the progress and survival of cancer [25].

However, the role of immunotherapies for the treatment of HCC is still not confirmed. The signaling transduction pathway by which TAMs and TGF-β regulate the generation and function of Tregs in HCC is still poorly understood and debatable. Moreover, a better understanding of whether TGF-β production favorably induces Tregs or stimulates the production of Th17 cells within the TME is needed [26]. Development of novel and potential antitumor drugs targeting or utilizing immunological mechanisms warrants additional research into improved understanding of the balance between all immune machineries at various phases of tumor development [27]. In recent times the research is moving on around the success of immune checkpoint inhibitors (ICIs) for the cure of various solid tumors so on for HCC. The role of other costimulatory and coinhibitory molecules still need to explore in HCC.

5 Cell surface proteins/carriers involved in HCC

Cell signaling proteins are connecting links between the extracellular environment and the intracellular environment and assist in transmitting the signaling data to the intracellular signaling proteins from the extracellular environment. They are dysregulated during cancer development including HCC. These proteins may serve as a biomarker for the disease or as a target for the vaccines for the treatment of HCC. Some of the important proteins are discussed in the next section.

5.1 Epidermal growth factor receptor (EGFR)

This is an important transmembrane glycoprotein belonging to enzyme-linked receptor family [28]. Epithelial tumors have expressed EGFR very frequently. EGFR has a role in tumor proliferation, intrahepatic metastasis, and poor disease-free survival during the course of the disease. The expression of EGFR is found in 68% of HCC analyzed. Several EGFR inhibitors entered in clinical setup for the treatment of HCC [29, 30].

5.2 G-protein coupled receptor (GPCR)

GPCR belongs to seven transmembranes (7-TM) superfamily. Numerous shreds of evidence indicated that unusual presence of GPCR, representing to class A (rhodopsin-like) and class F (frizzled/smoothened), had a vital role in HCC tumor progression [31, 32]. Adrenergic receptors, frizzled receptors (FZD), Adhesion Receptors, and G-protein receptor kinases (GRKs) belong to the GPCR receptor category and have a role in the pathological progression of HCC [33, 34].

5.3 Chemokine receptors (CR)

The CR expressed in HCC cells are involved in cell proliferation, migration apoptosis, and invasion of cells. These chemokine receptors work through the activation of the phosphatidylinositol-4,5-bisphosphate 3-kinase (PI3K)/protein kinase B (Akt) transduction pathway. CC chemokine receptor 1 (CCR) is vastly presented in tissues involved in HCC [35]. On the other hand, tumoral and nontumoral cells within HCC tissues have expression of CCR2 and its ligand CCL2. Besides, the role of CCR5, CCR7, and CCR9 chemokine receptors is also established in HCC. Also, high levels of C-X-C chemokine receptor (CXCR) 2 are associated with poor prognosis and progression of

HCC in humans. The CXCL5/CXCR2 axis induced EMT and supported the invasion and migration of HCC cells by activating the PI3K/Akt/GSK-3β/Snail and PI3K/Akt/ERK1/2 transduction signaling pathways. The clinical trials of these chemokine receptors in combination with immunotherapy and targeted therapy are underway and showed promising results [36, 37].

5.4 E-prostanoid (EP) receptors

EP receptors EP1, EP2, EP3, and EP4 belong to GPCR that have specificity toward prostaglandin E2 (PGE2). These receptors are participated in the advancement and progression of tumorigenesis in HCC by modulating the invasion, adhesion, growth, metastasis, and angiogenesis of tumor cells [38, 39].

5.5 Cluster of Differentiation 44 (CD44)

A transmembrane glycoprotein receptor, CD44, for various ligands such as hyaluronic acid, collagen, fibronectin, osteopontin, and a co-receptor for EGFR and c-Met. Kupffer cells and lymphocytes which are myeloid cells in the liver expressed CD44 glycoprotein in normal physiologic conditions. Increased expression of CD44 has been observed in HCC progenitor cells (HcPCs) [40, 41].

5.6 Endoglin (CD105)

Endoglin is a TGF-β coreceptor, which is a glycoprotein in nature that is expressed on activated sinusoidal endothelial cells of liver and has angiogenic, migratory, and antiapoptotic effects [42].

5.7 Integrins

These are glycoproteins associated with the cell adhesion to the ECM. They are linked with the cellular pathways involved in cell differentiation and proliferation as well as motility [43]. Elevated levels of β1 integrin (ITGB1) were observed in tissues and multicellular spheroids of HCC cells. In addition to above-mentioned protein carrier molecules mucin, solute carrier transporters (SLCs), annexins (ANXs), glypican-3, cluster of differentiation 147 (CD147), glucose-regulated protein 78 (GRP78), and ezrin are the important proteins that have a significant role in the pathological progression of HCC [44, 45]

6 Available treatment for HCC

6.1 Approved drugs for the treatment of HCC

Presently, four drugs have been approved by FDA for the treatment of HCC as the first-line or second-line agents. An orally administered small molecule, sorafenib, was shown to inhibit a couple of serine/threonine and tyrosine kinase. The drug was approved in 2007 and considered as the first-line treatment of HCC. One more tyrosine kinase inhibitor (TKI), regorafenib, having a mechanism similar to sorafenib was approved by FDA in 2017 as a second-line treatment for patients who tolerated but progressed with sorafenib [46, 47]. In 2018, FDA has approved one more TKI inhibitor, levantinib, which selectively inhibits fibroblast growth factor receptor 1–4 (FGFR 1–4), vascular endothelial growth factor receptor 1–3 (VEGFR 1–3), platelet-derived growth factor receptor alpha (PDGFRα), and tyrosine-protein kinase (KIT) proto-oncogenes. The drug was approved as first-line treatment for patients with unresectable HCC [48]. In 2019, carbozantinib was approved as a second-line treatment of progressive HCC subsequent to Sorafenib administration. The approved drug can block several angiogenic and oncogenic pathways associated with tumor progression and metastasis like hepatocyte growth factor receptor (HGFR), PDGFR, AXL receptor tyrosine kinase (AXL), VEGFR2, and FMS-like tyrosine kinase 3 (FLT3) [49].

Immunotherapeutic tactics are one of the prominent ways for management and control of HCC. The treatment approach is based on programmed cell death protein-1 in HCC by tumor-infiltrating lymphocytes (TIL) expression. Two drugs, nivolumab and pembolizumab, were the ICIs and approved as second-line agents that exhibited to progressively improve the chance of survival of patients with HCC [50, 51].

Recently, several natural compounds have been observed to be active against HCC. Piperine, extracted from the pepper, is an alkaloid and has been found to have anticancer activities [52]. Piperine can show selectivity to cancer cells, evidenced by concentration of piperine, lethal to cancer cells but exhibited nontoxicity to hepatocytes. The alkaloid also has capabilities to stimulate mitochondria-mediated apoptosis by increasing caspase-3 and caspase-9 activity with inhibition of catalase. Piperine also has proven activity of inhibition to human HCC progression by TKI. Thus, piperine could be a potential treatment approach for HCC [53]. Curcumin, a constituent of turmeric has shown biological efficacy in HCC. Research studies demonstrated the synergistic effect of curcumin when administered with piperine [54].

Liver transplantation has a strong curative impact in HCC, but it comes with a disadvantage of organ shortage and morbidity, which impedes the medical practitioner to look out for alternative treatment options. Radiofrequency ablation (RFA), on the other hand, has come out fruitful for some specific cases with very low morbidity and mortality associated with it. Another alternative for RFA is liver resection (LR), it allows better results of treatments and lets medical practitioners to analyze tumor pathology and risk factors associated with it [55]. In liver cirrhotic patients, laparoscopy is gaining popularity recently, which is a minimally invasive approach accompanied by a substantial decrease in complications associated with perioperative procedures [56].

7 Significance of physicochemical properties of a drug delivery system in HCC

An API carried by a DDS needs to physically contact the physiological target, e.g., receptors in the liver cells in the case of HCC. Targeted or site-specific drug delivery in this context is significant as it ensures interactions with the target site. The following criteria must be fulfilled by the DDS in the treatment of HCC: (a) it should be efficient enough to cross the biological barrier; (b) in the case of HCC, the DDS should be selectively identified by the liver receptors; (c) exogenous and endogenous ligands should compete with each other for targeting; (d) delivery system designed in the lab scale should be biodegradable, biocompatible, nontoxic, and should be stable physiochemically in the liver cells; (e) should have uniform sinusoid capillary distribution; (f) the rate of drug release from the delivery system should be controlled and predictable, which could assists in therapeutic amount of drug release in the liver cells; (g) drug distribution should not get affected due to drug release; (h) ideally there should not be any drug leakage from the delivery system during the passage through various parts of the body; (i) the drug delivery system as a whole should be inert in nature and should not possess any therapeutic activity or toxicity and should get eliminated from the body after without modulating the disease state; and (j) the DDS should be easy to develop and cost-effective [57].

8 Strategies to design ADDS for HCC

8.1 General consideration

ADDSs are the device or the formulations which allow target-oriented delivery of therapeutic agent in a way that augment its safety and efficacy. ADDS, which are capable of targeting a drug to the action site, have influenced the treatment of cancer as a drug delivery system. Targeted drug delivery systems as a subpart of ADDS mostly comprise of carriers containing lipids and polymers as therapeutics. The major criteria of an efficient ADDS should be such that it should be capable of retaining the drugs, eluding the macrophage-oriented clearance to reach the target site, and capable of delivering the therapeutics to the site of action [58]. The process of transport of the DDS to the intended location in cases of HCC without getting distributed all through the body should be considered and is a challenging task. The DDS in the form of drug carriers should be specifically targeted to the cancerous cells and this could be a potential approach to reduce the side effects of anticancerous agents [59, 60]. Selectivity and specificity of ligands to the tumor cells is one of the major concerns in drug targeting to HCC as the site of a target is not limited to the tumor cells but also to the adjoining normal cells with a similar or same affinity. This results in the nonspecific toxicity of the normal cells by the ligand-oriented DDS set for cancer cells. Therefore, the approach of targeting the tumor cells should be targeting the highly overexpressed receptors on cancer cells. This approach is idealistic to achieve superior selectivity and therapeutic efficacy. The mechanism by which the loaded drug releases into the intracellular region when the DDS is exposed is an important parameter to be considered. The factors related to the design of DDS which influence the drug release to be considered are physicochemical, structural, and mechanical properties [61].

8.2 Design and development of DDS/ADDS based on biological properties of HCC with a molecular dynamics' insight

For transvascular transport of macromolecules in tumors, one of the significant factors is molecular size. The delivery of the drugs depends upon the vascular pore cut-off size, and DDS with a particle size range of below 200 nm diameter could effectively pass onto parenchymal cells of the liver. However, the particle size more than 200 nm will be destroyed by monocytes and reticuloendothelial system-oriented scavenging which signifies that the ideal size range for liver targeting would be 100–200 nm [62, 63]. Another important parameter playing a major part in drug delivery is tumor vasculature. Enhanced permeability and retention (EPR)-mediated drug delivery is considered as a useful way of transporting macromolecules (nanoparticles, liposomes, etc.) passively in tumors. In HCC, EPR-mediated drug delivery help in increased accumulation of macromolecules in tumor tissues. The macromolecular accumulation also may be augmented by angiotensin II-induced hypertension. However, drug targeting through EPR does not guarantee therapeutic success as the major determining step for anticancer activity is effective drug uptake intracellularly. This phenomenon has drawn greater attention in drug delivery precise to receptors achieved by altering the surface properties of DDS with targeting ligands [64, 65].

Different strategies related to targeting HCC using receptors precise to the cell membrane is a useful approach toward potential development of ADDS. Asialoglycoprotein receptor (ASGP-R) is an appealing receptor to target the hepatocarcinomatous cells with ADDS. To improve the efficiency of chemotherapeutic drugs associated with ADDS, approaches associated with receptor-mediated targeting of the drug could be hypothesized which is dependent on surface-modified properties of ASGP-R. Whatsoever,

ASGP-R protein could be expressed by both cancerous and normal hepatocytes and it is important to consider the targeted carrier-based ADDS that should be evaluated for both the HCC tissues and normal liver cells. For the development of ADDS for HCC, ASGP-R is a potential target for receptor-oriented delivery of the drug to hepatocarcinomatous cells as it has the potential to internalize macromolecules to the cell membranes with high affinity.

Another protein glypican 3 (GPC3) does not get expressed in hepatocytes but is expressed specifically on the cell surface of hepatocarcinoma cells. This leads to another potential approach to design DDS to target the protein in HCC. Depending upon the characteristics of the protein, many studies have been designed recently, which includes Wnt-β-catenin-associated and antibody-based approach for therapy and the results indicated that the antibody-based formulations have a significant targeting and bonding efficacy toward GPC3 and exhibited a significant therapeutic potential. Nanoparticle-based formulations are also shown to be suitable for targeting GPC3 where negatively charged nanoparticles with 50-100 nm size range could be suitable to target tumor cells expressing GPC3 [66].

In early 1991, it was discovered that hepatocyte membranes contain a definite site for glycyrrhetinic acid and, subsequently, it was found out that a carrier-mediated pathway was responsible for the hepatocytic uptake process for glycyrrhizin. Glycyrrhetinic acid and glycyrrhizin differ structurally by a hydroxyl group and glycosyl group in the C3 position, respectively. Research has indicated that glycyrrhetinic acid has an antihepatoma activity and it targets two important proteins (proteasome and peroxisome) playing a major role in HCC. The experimental results indicated that the glycyrrhetinic acid has a higher affinity toward the hepatocyte binding receptor, indicating the importance of the hydroxyl group in the structure of glycyrrhetinic acid [67, 68]. Results from research conducted in our labs (unpublished) have also indicated that the efficacy of liver targeting of a thiolated-based nanocomposite could be enhanced by functionalization of the formulation with glycyrrhetinic acid. Studies were conducted with nanoparticles prepared with carboxymethyl chitosan combined with glycyrrhizin to target HCC [69]. The results of the study indicated superior tumor-targeting effect of the developed nanoparticles. Liposomes containing plasmid DNA modified with glycyrrhetinic acid has shown its potential in hepatoma-targeting therapy [70]. The liposomes could potentially deliver plasmid DNA to HepG2 cells with the help of glycyrrhetinic acid.

Another target for the treatment of HCC is the transferrin receptor. The receptor has been formerly used as a specific ligand for targeting HCC as it has the efficiency to transport drugs to the intracellular regions by receptor-mediated endocytosis [71]. For the development of a superior targeted carrier, advancement in the designing of drug delivery capable of the transferrin receptor needs to be done. Liposome formulation conjugated with transferrin containing endostatin gene has indicated potential activity against tumor growth and angiogenesis in the liver. Also, a DDS consisting of sorafenib and doxorubicin encapsulated was observed to facilitate cellular uptake of the drugs as well as synergistic cytotoxicity toward HepG2 cells [72, 73]. The receptor targeting in the treatment of HCC has shown promising outcomes but in vivo evaluation of transferrin receptor drug delivery still needs to be explored.

Another broadly expressed cell surface adhesion molecule is CD44. The standard isoform of CD44 will not be expressed by normal liver cells but expressed by HCC tumor cells, especially the overriding form of CD44 mRNA (isoform) is expressed by HCC cells. When compared the expression of CD44 and isoform of CD44 in cancerous tissue and normal hepatocytes, it was found out that CD44 overexpression could be identified in HCC and poorly distinguished histology, and reduced survival was due to the upregulation of CD44 isoform. The overexpression of CD44 was specified to be a significant indication of tumor growth and metastasis and directs that inhibiting CD44 could be a potential approach to treat HCC [74–76]. Drug delivery systems mediated with CD44 have been developed by approach related to the conjugation of hyaluronic acid with nanodelivery systems. Results of the study conducted indicated that liposome delivery mediated with CD44 antibodies could potentially target the CD44 antigens on the cell surfaces of HCC.

8.3 Design and development of DDS/ADDS based on physicochemical properties of HCC

To increase the specificity of DDS, significant roles are played by physicochemical interactions. Designing a DDS to target the tumors, hypoxia is considered a significant microenvironment feature of tumor. Research studies have reported conjugating PEG groups with hypoxia-sensitive agents to facilitate PEG detachment in the microenvironment of hypoxia. A study demonstrated the development of nanoparticles with azobenzene used as a hypoxia-responsive bioreductive linker for transport of siRNA to tumor. The study exhibited the nanoparticle delivery to have superior downregulation of proteins expressing tumors as the degradation of azobenzene linker occurs there was a detachment of PEG groups from the conjugate leading to the uptake of the remaining complex by the cell [77, 78].

The effective delivery of anticancerous drugs could also depend upon the extracellular acidity of tumor. The chemotherapeutic agents which are weakly basic (doxorubicin, anthracyclines, mitoxantrone, etc.) will amass in acidic compartments of tumor cells and the acidic pH of the tumor will eventually hinder the internalization of weakly basic drugs leading to decreased therapeutic efficacy. So, one of the drug delivery approaches for the weakly basic drugs could be to encapsulate them in a pH-sensitive carrier to improve their uptake cytotoxic nature toward tumor cells. Drugs such as

5-fluorouracil, cyclophosphamide, chlorambucil, etc., which are weakly acidic, are projected to internalize more readily into the tumor tissues being acidic [79].

Interstitial fluid pressure (IFP) is a significant physicochemical property related to HCC to be considered for the design and development of DDS. IFP acts as an obstacle for the delivery of drugs inside tumor by reducing transcapillary fluid flow. So, reducing the tumor IFP with precise signal transduction antagonists could be a beneficial tactic to increase the potential activity of the DDS [80, 81].

8.4 Role of polymer hybrids/scaffolds with a specific target-oriented treatment of HCC

The design and development of suitable scaffolds for tissue engineering is one of the most desirable areas in the arena of regenerative medicine. In the field of biomedical sciences, the development of scaffolds with suitable physicochemical properties helps in the restoration of physiological functions of the damaged tissues. Studies performed utilizing inverted colloid crystals to develop liver organoids indicated that the functional properties of the normal liver might be engineered to functionalize the extracellular functions recognized by hepatic progenitors throughout human growth. The developed organoids were similar to human tissues concerning gene expression, protein exudation, morphology, and drug metabolism and could function like normal liver cells [82]. The bioengineered platform might be suitable for clinical organoid-related applications including treatment of HCC.

In another study, drug delivery systems based on hydrogels have been explored for the treatment of cancer [83]. The study described a facile method for the development of hydrogel scaffolds. A mixture of concentrated alginate inks and polydopamine was used as a shell layer and a drug-loaded core part was coinjected and 3D printed into core-shell hydrogel scaffolds. The scaffold can be conjugated with a targeted moiety and might be delivered to liver cancer cells and might be a significant therapeutic tactic for HCC.

Different proteins responsible for control and therapy of HCC might be recognized and their binding affinity with biodegradable polymers could be investigated by computer-aided drug design (docking). Based on the results, polymer hybrids and scaffolds could be designed and developed with a suitable targeting agent having an affinity toward tumorous cells and this could be an efficacious approach toward the control and treatment of HCC.

8.5 Miscellaneous DDS developmental strategies

Molecular docking studies could be associated with the development of ADDS strategies to investigate the new targets on HCC cells and therapeutically active agents that have anticancerous activity [84]. These agents could be functionalized onto the carrier surfaces to target the cancerous cells to be a potential substitute to synergize the therapeutic activity. For example, octreotide. For example, octreotide, an antiangiogenic agent that target and inhibit the growth factors (vascular endothelial growth factor) release might be a potential candidate for co-delivery with a carrier-based DDS like nanoparticles, liposomes, etc. to yield a synergistic effect [2]. Codelivery systems could also help overcome the multiple drug-oriented resistance associated with treatment of HCC such as the combinations of chemotherapeutics with efflux pump inhibitors, proapoptotic compounds, and siRNA (to silence the multidrug resistance genes) could be helpful in the therapeutic approach of HCC.

Alternative approaches of DDS related to intratumoral drug penetration have been explored like conjugation of cell-penetrating peptides to the surface of DDS where the peptides having a cationic structure can effectively attach to the cancer cell membrane and allow the carrier to achieve deep tumor penetration and show significant antitumor effect [85].

9 Conclusion and future perspectives

ADDSs oriented with carrier-mediated drug delivery for targeting cells and tissues have been broadly explored in the field of cancer therapy. Nanomedicines (nanoparticles and liposomes) in the treatment of HCC have been widely explored because of its superior targeting properties with the help of targeting moieties which could be channelized toward the membrane proteins and receptors expressed by HCC. However, a specific target site for tumor cells in HCC does not exist but the ones articulated should be exploited. For designing target-mediated DDS, receptor-oriented cellular actions arbitrated through endogenous ligands might be subjugated. Thus, ligand interceded drug targeting has become an innovative tactic for the treatment of HCC.

Future perspectives could be based on the development of polymer hybrid or scaffold-mediated drug delivery, where the scaffolds might be developed with biodegradable polymers having inherent anticancerous activity and could be an innovative way of treating HCC without the side effects or adverse effects of chemotherapeutic agents.

References

[1] Yin X, Xiao Y, Han L, Zhang B, Wang T, Su Z, et al. Ceramide-fabricated co-loaded liposomes for the synergistic treatment of hepatocellular carcinoma. AAPS PharmSciTech 2018;19(5):2133–43.
[2] Zhang X, Ng HLH, Lu A, Lin C, Zhou L, Lin G, et al. Drug delivery system targeting advanced hepatocellular carcinoma: current and future. Nanomedicine 2016;12(4):853–69.
[3] Ho EY, Cozen ML, Shen H, Lerrigo R, Trimble E, Ryan JC, et al. Expanded use of aggressive therapies improves survival in early and intermediate hepatocellular carcinoma. HPB 2014;16(8):758–67.

[4] Bae YH, Park K. Advanced drug delivery 2020 and beyond: perspectives on the future. Adv Drug Deliv Rev 2020.
[5] AAPS. Update. Pharm Res 2007;24(4):816–7.
[6] Laffleur F, Keckeis V. Advances in drug delivery systems: work in progress still needed? Int J Pharm X 2020;2:100050.
[7] Abdel-Misih SR, Bloomston M. Liver anatomy. Surg Clin North Am 2010;90(4):643–53.
[8] Vernon H, Wehrle CJ, Kasi A. Anatomy, abdomen and pelvis, liver. Treasure Island (FL): StatPearls; 2020. StatPearls Publishing Copyright © 2020, StatPearls Publishing LLC.
[9] Dixon LJ, Barnes M, Tang H, Pritchard MT, Nagy LE. Kupffer cells in the liver. Compreh Physiol 2013;3(2):785–97.
[10] Garza AZ, Park SB, Kocz R. Drug elimination. Treasure Island (FL): StatPearls; 2020. StatPearls Publishing Copyright © 2020, StatPearls Publishing LLC.
[11] Boyer JL. Bile formation and secretion. Compreh Physiol 2013;3(3):1035–78.
[12] Adeva-Andany MM, González-Lucán M, Donapetry-García C, Fernández-Fernández C, Ameneiros-Rodríguez E. Glycogen metabolism in humans. BBA Clin 2016;5:85–100.
[13] Kalra A, Yetiskul E, Wehrle CJ, Tuma F. Physiology, liver. Treasure Island (FL): StatPearls; 2020. StatPearls Publishing Copyright © 2020, StatPearls Publishing LLC.
[14] El-Serag HB. Epidemiology of viral hepatitis and hepatocellular carcinoma. Gastroenterology 2012;142(6). 1264-73.e1.
[15] Ghouri YA, Mian I, Rowe JH. Review of hepatocellular carcinoma: epidemiology, etiology, and carcinogenesis. J Carcinogen 2017;16:1.
[16] Tornesello ML, Buonaguro L, Izzo F, Buonaguro FM. Molecular alterations in hepatocellular carcinoma associated with hepatitis B and hepatitis C infections. Oncotarget 2016;7(18):25087–102.
[17] Arneth B. Tumor microenvironment. Medicina (Kaunas, Lithuania) 2019;56(1).
[18] Arriazu E, Ruiz de Galarreta M, Cubero FJ, Varela-Rey M, Pérez de Obanos MP, Leung TM, et al. Extracellular matrix and liver disease. Antioxid Redox Signal 2014;21(7):1078–97.
[19] LaGory EL, Giaccia AJ. The ever-expanding role of HIF in tumour and stromal biology. Nat Cell Biol 2016;18(4):356–65.
[20] Gonzalez H, Hagerling C, Werb Z. Roles of the immune system in cancer: from tumor initiation to metastatic progression. Genes Dev 2018;32(19–20):1267–84.
[21] Fujimura T, Kambayashi Y, Aiba S. Crosstalk between regulatory T cells (Tregs) and myeloid derived suppressor cells (MDSCs) during melanoma growth. Oncoimmunology 2012;1(8):1433–4.
[22] Seo HR. Roles of tumor microenvironment in hepatocelluar carcinoma. Curr Med Chem 2015;11(2):82–93.
[23] Niu G, Chen X. Vascular endothelial growth factor as an anti-angiogenic target for cancer therapy. Curr Drug Targets 2010;11(8):1000–17.
[24] Lo Presti E, Pizzolato G, Corsale AM, Caccamo N, Sireci G, Dieli F, et al. γδ T cells and tumor microenvironment: from immunosurveillance to tumor evasion. Front Immunol 2018;9:1395.
[25] Zamarron BF, Chen W. Dual roles of immune cells and their factors in cancer development and progression. Int J Biol Sci 2011;7(5):651–8.
[26] Tu E, Chia PZ, Chen W. TGFβ in T cell biology and tumor immunity: angel or devil? Cytokine Growth Factor Rev 2014;25(4):423–35.
[27] O'Neill RE, Cao X. Co-stimulatory and co-inhibitory pathways in cancer immunotherapy. Adv Cancer Res 2019;143:145–94.
[28] Ito Y, Takeda T, Sakon M, Tsujimoto M, Higashiyama S, Noda K, et al. Expression and clinical significance of erb-B receptor family in hepatocellular carcinoma. Br J Cancer 2001;84(10):1377–83.
[29] López-Luque J, Bertran E, Crosas-Molist E, Maiques O, Malfettone A, Caja L, et al. Downregulation of epidermal growth factor receptor in hepatocellular carcinoma facilitates transforming growth factor-β-induced epithelial to amoeboid transition. Cancer Lett 2019;464:15–24.
[30] Song P, Yang J, Li X, Huang H, Guo X, Zhou G, et al. Hepatocellular carcinoma treated with anti-epidermal growth factor receptor antibody nimotuzumab: A case report. Medicine 2017;96(39), e8122.
[31] Peng WT, Sun WY, Li XR, Sun JC, Du JJ, Wei W. Emerging roles of G protein-coupled receptors in hepatocellular carcinoma. Int J Mol Sci 2018;19(5).
[32] Chan KK, Lo RC. Deregulation of frizzled receptors in hepatocellular carcinoma. Int J Mol Sci 2018;19(1).
[33] Kassahun WT, Günl B, Jonas S, Ungemach FR, Abraham G. Altered liver α1-adrenoceptor density and phospholipase C activity in the human hepatocellular carcinoma. Eur J Pharmacol 2011;670(1):92–5.
[34] Bengochea A, de Souza MM, Lefrançois L, Le Roux E, Galy O, Chemin I, et al. Common dysregulation of Wnt/Frizzled receptor elements in human hepatocellular carcinoma. Br J Cancer 2008;99(1):143–50.
[35] Li Y, Wu J, Zhang P. CCL15/CCR1 axis is involved in hepatocellular carcinoma cells migration and invasion. Tumour Biol 2016;37(4):4501–7.
[36] Fantuzzi L, Tagliamonte M, Gauzzi MC, Lopalco L. Dual CCR5/CCR2 targeting: opportunities for the cure of complex disorders. Cell Mol Life Sci: CMLS 2019;76(24):4869–86.
[37] Zhuang H, Cao G, Kou C, Liu T. CCL2/CCR2 axis induces hepatocellular carcinoma invasion and epithelial-mesenchymal transition in vitro through activation of the Hedgehog pathway. Oncol Rep 2018;39(1):21–30.
[38] Xia S, Ma J, Bai X, Zhang H, Cheng S, Zhang M, et al. Prostaglandin E2 promotes the cell growth and invasive ability of hepatocellular carcinoma cells by upregulating c-Myc expression via EP4 receptor and the PKA signaling pathway. Oncol Rep 2014;32(4):1521–30.
[39] Zhang H, Cheng S, Zhang M, Ma X, Zhang L, Wang Y, et al. Prostaglandin E2 promotes hepatocellular carcinoma cell invasion through upregulation of YB-1 protein expression. Int J Oncol 2014;44(3):769–80.
[40] Flanagan BF, Dalchau R, Allen AK, Daar AS, Fabre JW. Chemical composition and tissue distribution of the human CDw44 glycoprotein. Immunology 1989;67(2):167–75.
[41] He G, Dhar D, Nakagawa H, Font-Burgada J, Ogata H, Jiang Y, et al. Identification of liver cancer progenitors whose malignant progression depends on autocrine IL-6 signaling. Cell 2013;155(2):384–96.
[42] Kasprzak A, Adamek A. Role of endoglin (CD105) in the progression of hepatocellular carcinoma and anti-angiogenic therapy. Int J Mol Sci 2018;19(12).
[43] Sökeland G, Schumacher U. The functional role of integrins during intra- and extravasation within the metastatic cascade. Mol Cancer 2019;18(1):12.
[44] Tian T, Li C-L, Fu X, Wang S-H, Lu J, Guo H, et al. β1 integrin-mediated multicellular resistance in hepatocellular carcinoma through activation of the FAK/Akt pathway. J Int Med Res 2018;46(4):1311–25.
[45] Siracusano G, Tagliamonte M, Buonaguro L, Lopalco L. Cell surface proteins in hepatocellular carcinoma: from bench to bedside. Vaccines 2018;8(1).
[46] Bruix J, Qin S, Merle P, Granito A, Huang Y-H, Bodoky G, et al. Regorafenib for patients with hepatocellular carcinoma who progressed on sorafenib treatment (RESORCE): a randomised, double-blind, placebo-controlled, phase 3 trial. Lancet 2017;389(10064):56–66.

[47] Dal Bo M, De Mattia E, Baboci L, Mezzalira S, Cecchin E, Assaraf YG, et al. New insights into the pharmacological, immunological, and CAR-T-cell approaches in the treatment of hepatocellular carcinoma. Drug Resist Updat 2020;51:100702.
[48] Kudo M, Finn RS, Qin S, Han K-H, Ikeda K, Piscaglia F, et al. Lenvatinib versus sorafenib in first-line treatment of patients with unresectable hepatocellular carcinoma: a randomised phase 3 non-inferiority trial. Lancet 2018;391(10126):1163–73.
[49] Abou-Alfa GK, Meyer T, Cheng A-L, El-Khoueiry AB, Rimassa L, Ryoo B-Y, et al. Cabozantinib in patients with advanced and progressing hepatocellular carcinoma. N Engl J Med 2018;379(1):54–63.
[50] Brizzi MP, Pignataro D, Tampellini M, Scagliotti GV, Di Maio M. Systemic treatment of hepatocellular carcinoma: why so many failures in the development of new drugs? Expert Rev Anticancer Ther 2016;16(10):1053–62.
[51] Siu EH-L, Chan AW-H, Chong CC-N, Chan SL, Lo K-W, Cheung ST. Treatment of advanced hepatocellular carcinoma: immunotherapy from checkpoint blockade to potential of cellular treatment. Transl Gastroenterol Hepatol 2018;3.
[52] Tawani A, Amanullah A, Mishra A, Kumar A. Evidences for piperine inhibiting cancer by targeting human G-quadruplex DNA sequences. Sci Rep 2016;6:39239.
[53] Anwanwan D, Singh SK, Singh S, Saikam V, Singh R. Challenges in liver cancer and possible treatment approaches. Biochim Biophys Acta (BBA) 1873;2020(1):188314.
[54] Sasaki H, Sunagawa Y, Takahashi K, Imaizumi A, Fukuda H, Hashimoto T, et al. Innovative preparation of curcumin for improved oral bioavailability. Biol Pharm Bull 2011;34(5):660–5.
[55] Di Sandro S, Benuzzi L, Lauterio A, Botta F, De Carlis R, Najjar M, et al. Single hepatocellular carcinoma approached by curative-intent treatment: a propensity score analysis comparing radiofrequency ablation and liver resection. Eur J Surg Oncol 2019;45(9):1691–9.
[56] Di Sandro S, Bagnardi V, Najjar M, Buscemi V, Lauterio A, De Carlis R, et al. Minor laparoscopic liver resection for Hepatocellular Carcinoma is safer than minor open resection, especially for less compensated cirrhotic patients: Propensity score analysis. Surg Oncol 2018;27(4):722–9.
[57] Mishra N, Yadav NP, Rai VK, Sinha P, Yadav KS, Jain S, et al. Efficient hepatic delivery of drugs: novel strategies and their significance. Biomed Res Int 2013;2013:382184.
[58] Bassyouni F, ElHalwany N, Abdel-Rehim M, Munir N. Advances and new technologies applied in controlled drug delivery system. Res Chem Intermed 2015;41(4):2165–200.
[59] Delehanty JB, Boeneman K, Bradburne CE, Robertson K, Bongard JE, Medintz IL. Peptides for specific intracellular delivery and targeting of nanoparticles: implications for developing nanoparticle-mediated drug delivery. Ther Deliv 2010;1(3):411–33.
[60] Torchilin VP. Drug targeting. Eur J Pharm Sci 2000;11(Suppl 2):S81–91.
[61] Maherani B, Arab-Tehrany E, Kheirolomoom A, Geny D, Linder M. Calcein release behavior from liposomal bilayer; influence of physicochemical/mechanical/structural properties of lipids. Biochimie 2013;95(11):2018–33.
[62] Yuan F, Dellian M, Fukumura D, Leunig M, Berk DA, Torchilin VP, et al. Vascular permeability in a human tumor xenograft: molecular size dependence and cutoff size. Cancer Res 1995;55(17):3752–6.
[63] Greish K, Fang J, Inutsuka T, Nagamitsu A, Maeda H. Macromolecular therapeutics: advantages and prospects with special emphasis on solid tumour targeting. Clin Pharmacokinet 2003;42(13):1089–105.
[64] Narang AS, Varia S. Role of tumor vascular architecture in drug delivery. Adv Drug Deliv Rev 2011;63(8):640–58.
[65] Li X, Ding L, Xu Y, Wang Y, Ping Q. Targeted delivery of doxorubicin using stealth liposomes modified with transferrin. Int J Pharm 2009;373(1-2):116–23.
[66] Park JO, Stephen Z, Sun C, Veiseh O, Kievit FM, Fang C, et al. Glypican-3 targeting of liver cancer cells using multifunctional nanoparticles. Mol Imaging 2011;10(1):69–77.
[67] Dong J, Olaleye OE, Jiang R, Li J, Lu C, Du F, et al. Glycyrrhizin has a high likelihood to be a victim of drug-drug interactions mediated by hepatic organic anion-transporting polypeptide 1B1/1B3. Br J Pharmacol 2018;175(17):3486–503.
[68] Tian Q, Wang X, Wang W, Zhang C, Liu Y, Yuan Z. Insight into glycyrrhetinic acid: the role of the hydroxyl group on liver targeting. Int J Pharm 2010;400(1-2):153–7.
[69] Shi L, Tang C, Yin C. Glycyrrhizin-modified O-carboxymethyl chitosan nanoparticles as drug vehicles targeting hepatocellular carcinoma. Biomaterials 2012;33(30):7594–604.
[70] He ZY, Zheng X, Wu XH, Song XR, He G, Wu WF, et al. Development of glycyrrhetinic acid-modified stealth cationic liposomes for gene delivery. Int J Pharm 2010;397(1–2):147–54.
[71] Ippoliti R, Lendaro E, D'Agostino I, Fiani ML, Guidarini D, Vestri S, et al. A chimeric saporin-transferrin conjugate compared to ricin toxin: role of the carrier in intracellular transport and toxicity. FASEB J 1995;9(12):1220–5.
[72] Li X, Fu GF, Fan YR, Shi CF, Liu XJ, Xu GX, et al. Potent inhibition of angiogenesis and liver tumor growth by administration of an aerosol containing a transferrin-liposome-endostatin complex. World J Gastroenterol 2003;9(2):262–6.
[73] Malarvizhi GL, Retnakumari AP, Nair S, Koyakutty M. Transferrin targeted core-shell nanomedicine for combinatorial delivery of doxorubicin and sorafenib against hepatocellular carcinoma. Nanomed Nanotechnol Biol Med 2014;10(8):1649–59.
[74] Mima K, Okabe H, Ishimoto T, Hayashi H, Nakagawa S, Kuroki H, et al. CD44s regulates the TGF-β-mediated mesenchymal phenotype and is associated with poor prognosis in patients with hepatocellular carcinoma. Cancer Res 2012;72(13):3414–23.
[75] Endo K, Terada T. Protein expression of CD44 (standard and variant isoforms) in hepatocellular carcinoma: relationships with tumor grade, clinicopathologic parameters, p53 expression, and patient survival. J Hepatol 2000;32(1):78–84.
[76] Xie Z, Choong PF, Poon LF, Zhou J, Khng J, Jasinghe VJ, et al. Inhibition of CD44 expression in hepatocellular carcinoma cells enhances apoptosis, chemosensitivity, and reduces tumorigenesis and invasion. Cancer Chemother Pharmacol 2008;62(6):949–57.
[77] Mishra S, Webster P, Davis ME. PEGylation significantly affects cellular uptake and intracellular trafficking of non-viral gene delivery particles. Eur J Cell Biol 2004;83(3):97–111.
[78] Perche F, Biswas S, Wang T, Zhu L, Torchilin VP. Hypoxia-targeted siRNA delivery. Angew Chem Int Ed Eng 2014;53(13):3362–6.
[79] Mahoney BP, Raghunand N, Baggett B, Gillies RJ. Tumor acidity, ion trapping and chemotherapeutics. I. Acid pH affects the distribution of chemotherapeutic agents in vitro. Biochem Pharmacol 2003;66(7):1207–18.

[80] Cairns R, Papandreou I, Denko N. Overcoming physiologic barriers to cancer treatment by molecularly targeting the tumor microenvironment. Mol Cancer Res: MCR 2006;4(2):61–70.
[81] Heldin CH, Rubin K, Pietras K, Ostman A. High interstitial fluid pressure—an obstacle in cancer therapy. Nat Rev Cancer 2004;4(10):806–13.
[82] Ng SS, Saeb-Parsy K, Blackford SJI, Segal JM, Serra MP, Horcas-Lopez M, et al. Human iPS derived progenitors bioengineered into liver organoids using an inverted colloidal crystal poly (ethylene glycol) scaffold. Biomaterials 2018;182:299–311.
[83] Wei X, Liu C, Wang Z, Luo Y. 3D printed core-shell hydrogel fiber scaffolds with NIR-triggered drug release for localized therapy of breast cancer. Int J Pharm 2020;580:119219.
[84] Santhosh S, Mukherjee D, Anbu J, Murahari M, Teja BV. Improved treatment efficacy of risedronate functionalized chitosan nanoparticles in osteoporosis: formulation development, in vivo, and molecular modelling studies. J Microencapsul 2019;36(4):338–55.
[85] Zhu Y, Yu F, Tan Y, Yuan H, Hu F. Strategies of targeting pathological stroma for enhanced antitumor therapies. Pharmacol Res 2019;148:104401.

Chapter 19

Advanced drug delivery systems in oral cancer

Subha Manoharan[a], Lakshmi Thangavelu[b], Kamal Dua[c,d], and Dinesh Kumar Chellappan[e]

[a]Department of Oral Medicine and Radiology, Saveetha Dental College & Hospital, Saveetha Institute of Medical and Technical Science, Chennai, India, [b]Department of Pharmacology, Saveetha Dental College & Hospital, Saveetha Institute of Medical and Technical Science, Chennai, India, [c]Discipline of Pharmacy, Graduate School of Health, University of Technology Sydney, Ultimo, NSW, Australia, [d]Priority Research Centre for Healthy Lungs, Hunter Medical Research Institute (HMRI) & School of Biomedical Sciences and Pharmacy, University of Newcastle, Callaghan, NSW, Australia, [e]Department of Life Sciences, School of Pharmacy, International Medical University, (IMU), Kuala Lumpur, Malaysia

1 Introduction

Oral cancer is one of the most prevalent oral diseases. It stands the 6th common disease of the world with a high mortality rate. It accounts for 4% of all the cancers. It was estimated that around 657,000 patients were diagnosed with oral cancer every year. Of this, 90%–95% of the malignancy is oral squamous cell carcinoma (OSCC). Males are affected twice as common as females [1].

The etiology includes lifestyle habits and genetic variations. Consumption of any form of tobacco and alcohol is an established etiological factor [1, 2]. However, infectious agents like Human Papilloma Virus, Treponema Pallidum (Syphilis), and Candida do contribute as etiological factors [1–3]. Compromised immune system can also be a contributing factor [4].

Oral squamous cell carcinoma can affect any part of the oral cavity. It can occur as a lump, ulcer, and white or red lesion persisting, nonhealing tooth extraction socket. Treating the malignancy is a challenge for the physician due to the complex anatomy of the head and neck region. Management of oral malignancies includes radiotherapy, chemotherapy, surgery, or combinations of any of them. Immunotherapy is yet another recent modality that is used when the other modalities fail.

Surgical management results in huge disfigurement and functional deformities while radio and chemotherapy causes generalized toxicity in the body. The adverse effects of radiation therapy include oral mucositis, xerostomia, dysgeusia, radiation caries, trismus, lymphedema, and osteoradionecrosis, which is a long-term complication [5, 6]. Chemotherapy delivers drugs to the whole body and causes nausea, vomiting, nerve damage, dehydration, malnutrition, and hair [7].

Although there are numerous adverse effects, chemotherapy is the most preferred form of combination therapy as well as supportive therapy. The targeted therapies considerably increases the bioavailability and distribution of drug to the primary tumor, however fails in sustained release of the drug. Advanced drug delivery system (DDS) is yet another milestone in management of oral cancer. DDS delivers bioactive molecules at a specific site with a specific delivery rate [8], thereby reducing the adverse effect on the surrounding normal tissues and also other systemic adverse effects.

2 Advanced drug delivery system in oral cancer

The advance in drug delivery is the control in drug delivery. As the drug is delivered to a specific site in a controlled manner, there will be a considerable reduction in the adverse effects. The time of drug administration also increases the efficacy of the drug and decreases the adverse effect too. World Health Organization had stated that chrono-chemotherapy [9] (time-specific drug delivery) can be utilized as there are target circadian clock genes (Bcl2, Top2a, and Tyms) which enhance the drug delivery based on circadian clock rhythm. They have recognized few drugs with better efficacy when employed as chrono chemotherapy, which include docetaxel, doxorubicin, fluorouracil, and paclitaxel. In OSCC, a combination of docetaxel, cisplatin, and fluorouracil when administered as chrono-chemotherapy has lesser severity. Controlled drug delivery systems are the different ways established to overcome the setbacks in the tradition or conventional chemotherapy.

The advanced drug delivery systems include:

1. Local drug delivery
2. Intratumoral injections
3. Phototherapy drug delivery
4. Ultra sonoporation with microbubbles

Advanced Drug Delivery Systems in the Management of Cancer. https://doi.org/10.1016/B978-0-323-85503-7.00022-5

2.1 Local drug delivery

Local drug delivery delivers the drug at the proximity of the tumor, thereby targeting the tumor cells in specific [10]. Nanodrug delivery is the application used for drug delivery for the tumor cells. The nanoparticles are smaller in size with larger surface area. Capillaries in the tumor have enhanced permeability and retention property (EPR) [11]. This property enhances the drug delivery by the nanoparticles. The three main advantages of nanodrug delivery are that the drugs have an increased stability, bioavailability, and target delivery.

2.1.1 Types of nano drug delivery systems

a. Passive targeting
b. Active targeting
c. Immune targeting
d. Magnetic targeting

2.1.1.1 Passive targeting

Passive targeting uses the EPR to diffuse into the neovascular capillaries in the tumor. In normal cells, the capillaries are tightly arranged, and the permeability is less. The EPR of the tumor capillaries allows passive diffusion of the nanoparticles. As they diffuse, they enter the cytoplasm by endocytosis to achieve the targeted therapy [12, 13].

2.1.1.2 Active targeting

Active targeting has ligands specific to a receptor in the tumor cells. The ligand can be a peptide, antibody, or aptamer. This ligand receptor interaction enhances the drug delivery in specific to the tumor cells with lesser toxicity to the surrounding cells or normal cells [13].

2.1.1.3 Immune targeting

Failure of immune surveillance is one of the theories in cancer pathogenesis. The immune target therapy focuses on enhancing the immune system of the body. The enhanced T-cell-mediated antitumor effect enhances the immune surveillance and destroys the tumor cells. Studies have been reported that encapsulated ellagic acid inside a chitosan nanoparticle has higher bioavailability and therapeutic effect in OSCC [14].

2.1.1.4 Magnetic targeting

Magnetic nanoparticles tagged with chemotherapeutic drugs prove to have increased antitumor effect. The superparamagnetism exhibited by ferric oxide locates and destroys tumor cells in specific. Paclitaxel combined with ferric oxide had improved efficacy on tumor cells and no damage to normal cells as reported by Oliveira et al. [15].

2.1.2 Target site in OSCC

The nanoparticles can target-specific sites of the tumor. It includes blood vessels, intracellular fluid and extracellular matrix, stromal cells, and dendritic cells. Blood vessels play a vital role in oncogenesis. Neoangiogenesis takes place to transport nutrients to the cells, enhance growth and multiplication of the tumor cells, and also invasion and metastasis of the tumor cells. There are numerous factors involved in this process and a few to mention include fibroblast growth factor (FGF), VEGF family, platelet-derived growth factor (PDGF), and angiogenin (ANG) [16]. Inhibiting these factors inhibits the neovascularization of the tumor thereby blocking the nutrition supplement to the tumor, which would cease the tumor progression [17]. Complex proteins secreted by the cells into the extracellular spaces form an extracellular matrix. They are dense and restrict transport across the cells. The nanoparticles have the advantage of smaller size hence diffuse through these matrices thereby improving the antitumor efficacy of the drug [17, 18]. Stromal cells are cells that are essential for adhesion of the cells. They are more stable than the tumor cells itself. Hence, targeting these cells with the nanoinfused drugs would cause the destruction of the tumor cell, resulting in better antitumor efficacy. Tumor-associated dendritic cells can be stimulated by the nanodrugs; they have better antitumor effect than the normal dendritic cells [19]. There are micelles loaded with nanodrugs that enhance the maturation and differentiation of the tumor cells thereby enhancing drug efficacy.

2.2 Intratumoral injection

Intratumoral chemotherapy (ITC) means injecting the chemotherapeutic drug directly into the tumor volume itself or in the surrounding region. It is commonly used as an adjuvant before surgical resurrection to shrink the tumor volume. It also helps to preserve the adjacent vital structures if any. Though there is little evidence on this procedure, it is hypothesized that it has higher intratumoral drug concentration and lesser systemic toxicity. Yet another advantage of ITC is that there is a reduction in volume of the distant metastasis also. The major drawbacks of ITC include rapid clearance of the drug, nonhomogenous drug distribution, and local adverse effects like necrosis of the surrounding tissue [19, 20]. Cisplastin, gemcitabine, and paclitaxel were drugs administered through ITC at different concentration from 20 to 200 mg/L. Paclitaxel and gemcitabine never had any adverse effect but cisplatin was safe at a dosage of 20 mg/L and more than that caused necrosis of surrounding normal tissue [21]. Li et al. did a study on ITC and stated that hydrogel is a promising agent in ITC.

2.3 Phototherapy drug delivery system

Phototherapy is a minimally invasive drug delivery system. It is of two types: photodynamic therapy (PDT) and photothermal therapy (PTT).

2.3.1 Photodynamic therapy

Photodynamic therapy involves administration of a photosensitizing agent, followed by irradiation. The photosensitizing agent absorbs a specific wavelength (near infrared light) and generates reactive oxygen species (ROS) resulting in a programmed cell death (apoptosis) of the cancer cells. The disadvantage is that the photosensitizers are absorbed in a relatively lesser quantity by the tumor cells. PDT in combination with chemotherapeutic drugs are quite promising especially in management of resistant head and neck tumors. Nanoparticles loaded with cisplatin and a photosensitizing agent show a remarkable tumor size reduction. The promising results are due to the synergistic effect, increased bioavailability of the drug due to smaller particles, and longer retention time of the drug and specific targeted site of delivery [22, 23].

2.3.2 Photothermal therapy

Photothermal therapy generates heat by utilizing light-absorbing agents. The generated heat destroys the cancer cells. However, they are not used in clinical applications as the heat generated destroys the normal cells too [23, 24].

2.4 Microbubble-mediated ultrasound in drug delivery

Microbubbles can be used as a contrast agent in ultrasonography. They are micro-sized bubbles (1–2 μm). When microbubbles are injected into the bloodstream, they enhance the contrast of the image. These microbubbles can be tagged to a ligand or a monoclonal antibody which is specific to a receptor in the tumor. They target the receptor and deliver the chemotherapeutic agent followed by ultrasound radiation thereby enhancing the drug availability with relatively lesser side effects [25]. This process is similar to sonoporation. Sonoporation is a type of drug delivery where intracellular delivery of drugs occurs, which is usually difficult to achieve. Sonoporation is also used in gene therapy in oral cancer [26]. The microbubbles carry the gene vectors and on application of ultrasound waves to the tumor site these bubbles enter the cells as its permeability is increased and deliver the vector [27–29]. Similarly, monoclonal antibodies can also be injected and made available for tumor cell destruction.

3 Drug vehicles

Drug vehicles also known as drug carriers are substances that control the drug release. It can prolong the duration of drug release or can release on specific targeted sites. It utilizes the temperature change, gets activated by light, or can respond to alteration in pH [30].

3.1 Types of drug vehicles

a. Nanoparticles
b. Nanolipids
c. Hydrogel
d. Exosomes

3.1.1 Nanoparticles

Nanoparticles are particles of smaller size ranging from 10 to 1000 μm. They have a large surface-to-volume ratio that enhances their bioavailability. These are vehicles which carry the chemotherapeutic drugs to specific sites, release them in a controlled, sustained manner thereby killing the cancer cells. This enhances the duration of drug delivery to specific sites and also minimizes the adverse effects [31].

Types of nanoparticles:

a. Polymeric nanoparticles
b. Inorganic nanoparticles

3.1.1.1 Polymeric nanoparticles

Polymeric nanoparticles are substances with polysaccharides, proteins, and biodegradable polymers like polyethylene gly-*col* (PEG), poly(γ-benzyl l-glutamate) (PBLG), poly(D,L-lactide), poly(lactic acid) (PLA), poly(D,L-glycolide), poly(lactide-co-glycolide), polycyanoacrylate, chitosan, gelatin, and sodium alginate. They are prepared by precipitation, polymerization, emulsification, and coacervation [32].

There is substantial evidence that nanoparticles provide better drug delivery, bioavailability, and efficacy. Cisplastin loaded in poly(ethylene glycol)-poly(glutamic acid) polymeric nanoparticles has been proved to have better efficacy with reduced toxicity [33]. Curcumin loaded in polysaccharide chitosan by nanoprecipitation produces polycaprolactone (PCL) nanoparticle that is used in OSCC and has proved to be effective [34]. This compound also has better mucoadhesive property that makes it available for a longer duration [35]. Paclitaxel loaded in human albumin (albumin NPs) NPs, used as induction chemotherapy before definitive advanced tongue cancer treatment proved to be effective [36]. Naringenin (NAR)-loaded nanoparticles (NARNPs) prepared in a NAR:aminoalkyl methacrylate copolymers, E:poly vinyl alcohol by a nanoprecipitation method proved to be promising in buccal tumors [37]. Anticancer effects of herpes simplex virus thymidine kinase (HSV-TK)-loaded PEG–PBLG nanoparticles and PEG–PBLG nanoparticle-mediated HSV-TK/ganciclovir nanoparticles toward OSCC also seemed to be promising [38].

Thus polymeric nanoparticles have better biocompatibility, sustained drug release, and increased efficacy with decreased toxicity. However, they show particle aggregation, cannot release proteins and antibodies, and can have immune response and local reactions. Biopolymeric nanoparticles may overcome these setbacks. One such study is chitosan nanoparticle with ellagic acid that proved to be effective in management of OSCC [14].

3.1.1.2 Inorganic nanoparticles

Noble metal (Gold) is used as a tag in inorganic nanoparticles. They have high optical properties used in diagnostic purposes. They also work as photothermal agents with high therapeutic efficacy. OSCC epithelial growth was controlled by using the epithelial growth factor receptor (EGFR) antibody to gold nanoparticle of size 40nm [39]. The anti-EGFR/Au nanoparticles cause EGFR overexpressing cancer cell destruction using continuous wave (CW) argon ion laser [39]. This photothermal destruction required lesser energy proving the nanoparticle to be effective. Near infrared (NIR) radiation also proves to have deeper penetration of this tag. Nanoparticles with PEgylated titanium dioxide (TiO_2) when used with NIR had deeper penetration into the tissue. These when combined with anti-EGFR forms anti-EGFR-PEG-TiO_2 which attacks the EGFR on cancer cell surface and inhibits the tumor progression [40]. This combination utilizes transmission electron microscopy (TEM). Also the above-mentioned nanoparticles had no toxic effects on healthy tissue. The destroyed cancer cells by apoptosis and tumor growth inhibition. The stability of the vehicle was the drawback of this carrier.

Magnetic nanoparticles using iron oxide and oleic acid in combination with doxorubicin and paclitaxel showed improved loading efficacy. Iron nanoparticles with size of 200nm on sonication for 30min and suspended in saline are used to deliver anticancer agents [41]. Alternating magnetic field of the site causes hyperthermia which enhances cancer cell destruction. Mesoporous silica nanoparticles (MSNP) have higher porosity thereby accommodating more drugs to be delivered to the cancer site [42]. The advantage of inorganic nanoparticles includes site-specific delivery and high photostability. However, they cannot deliver macromolecules, less penetration depth, and toxicity to normal cells.

3.1.2 Nanolipids

These are lipid components that act as carriers. They are of two types, solid lipid-based nanoparticles (SLNs) and nanostructured lipid carriers (NLCs). SLNs are physiological lipids of size 50–1000nm, colloidal in nature with crystalline structure. They are stable at body temperature and room temperature. They have better penetration than the nanoparticles. Their high stability enhances controlled drug delivery. It also prevents chemical degradation of the drug [43]. NLCs are prepared by mixing solid and liquid lipids. This distorts the crystalline structure and provides room for drug accumulation. Hence, the drug loading and the duration of drug release is better compared to SLN [44, 45]. Curcumin, docetaxel, and etoposide loaded into nanolipid particles had shown to be promising in management of OSCC [46].

In general, nanolipid vehicles are highly stable, have controlled drug release, penetrate deeper, have low solubility, and prevent drug destruction. However, the limitations are the crystalline structure that reduces the room space for the drug; SNLs would have an initial burst releasing most of the drug while NLCs would cluster during storage and most importantly being physiological component, it may induce immune reactions [47].

3.1.3 Hydrogel

Hydrogels, as the name suggests, is hydrophilic in nature. It has a mesh-like structure with water and biological fluids. Since they have fibers, they appear similar to the soft tissues of the body. Hydrogel is of two types based on the method of gelation, physical and chemical. Chemical hydrogels are more stable than the physical hydrogel [48]. Permeability, mechanical resistance, and modifiable surface structure are the properties that make them a drug vehicle [49].

Hydrogels can carry drugs, proteins, and genetic materials. Cisplatin is the commonly delivered drug with this vehicle. The mesh swells up as the drugs get released by dissolution and disintegration between the spaces. They act as localized targeted drug vehicles. When added to nanoparticles, the disadvantages of nanoparticles like early elimination from circulation, renal clearance, decreased efficacy due to increased interstitial fluid, and low penetration into the tumor can be overcome. Also the polymeric chain in their structure enhances adhesion with salivary glycoprotein, thereby enabling them to be mucoadhesive components too [50].

Hydrogels are advantageous as they are injectable at specific sites, stable at physiological temperature and pH, have high drug-loading capacity, extended-release property, can deliver both hydrophilic and hydrophobic drugs, and can carry multiple synergistic drugs together. The limitations include, poor mechanical property and reduced efficacy due to release of most of the drug in the initial burst itself [51].

3.1.4 Exosomes

Exosomes are small membranous vehicles with varying size between 40 and 120nm. They are secreted during both physiological and pathological processes in the body. They are secreted when the plasma membrane fuses with multivesicular bodies. Epithelial cells, endothelial cells, mesenchymal cells, macrophages, and dendritic cells secrete them into the intercellular spaces. Based on the type of cell they originate from, their characteristics also changes. The exosomal surface holds cell-specific antigens, fusion proteins, adhesion molecules, and integrins [52].

They serve as communicators between the cells. They adhere to the cell surface and enter the cells through endocytosis. They are more active in acidic environments. Hence, they are functionally active in tumor sites which adds to the advantage as a drug carrier. Evidence is available for using exosome as a drug vehicle with curcumin, DOX, and PTX in oral cancer therapy [53].

They also can transfer the biological molecules. This property is used for genetic alteration where they carry genetic vectors. The major drawback with exosomes is that purifying and analyzing them takes a long process, delivering high doses of drugs is not possible with them, and it might stimulate immune reaction in the patient. Also, the tumor-derived exosomes have immunosuppressive action which portrays a major challenge to overcome [54].

4 Oral cancer drugs formulation for DDS

In conventional methods, drugs were delivered intravenously. With the advent of newer drug delivery systems, FDA had approved certain drugs used in oral cancer for oral administration through carriers. Administering the drugs as pills and gels is much more convenient for the patient. It does not require inpatient admission, which is desirable in situations where the drug needs to be taken for a longer duration. However, the poor solubility and destruction of drugs by digestive acids is a problem. This hindrance was ruled out by the DDS discussed so far. The drugs, which are proved to be effective with the DDS, are discussed in the following section.

4.1 Paclitaxel (PTX)

Taxol (Paclitaxel) is an anticancer drug. It acts by blocking the cells in the G2 or M phase of the cell cycle, thereby preventing cell growth. When administered orally, it has low solubility and permeability across the mucosa, this reduces drug absorption resulting in decreased efficacy. When administered intravenously, it expresses severe systemic adverse effects including liver failure [55, 56]. To overcome these setbacks, it was combined with a vehicle. Paclitaxel with chitosan, nanoemulsion delivery of paclitaxel increases water solubility and bioavailability of the drug [57, 58]. Nanoparticle of PTX formulated by adding montmorillonite to PLGA showed enhanced uptake and retention of the drug thereby increasing the efficacy [55].

4.2 Cisplatin (DDP)

Cisplatin is an antineoplastic drug commonly used in OSCC. As the drug enters the cancer cells it releases reactive oxygen species and decreases mitochondrial glutathione, which results in cross-linking of purine base to DNA [59]. Cisplastin, when encapsulated with porous hollow nanoparticles of Fe_3O_4 has increased bioavailability, sustained release, and lesser destruction of the drug [60]. Cisplastin encapsulated with poly(acrylic acid-*co*-methyl methacrylate) microparticles also has a controlled, prolonged release, which increases the efficacy of the drug [61].

4.3 Doxorubicin (DOX)

Doxorubicin is an antineoplastic drug. The mechanism of action involves intercalation of DNA base pairs, leading to breakage of DNA strands thereby inhibiting DNA and RNA synthesis. It also acts by inhibiting topoisomerase II, which damages the DNA and initiates apoptosis. It destroys any rapidly replicating cell. Hence it tends to destroy normal cells too leading to cell death in major organs like heart, brain, kidney, and liver [62, 63]. To overcome this disadvantage, DOX is encapsulated with dextran nanoparticles to target the cancer cells specifically and they hit the cancer cell nuclei directly and destroy it sparing the normal cells [62]. It has been proved to show reduced toxicity on healthy cells and increases the efficacy on neoplastic cells.

4.4 Docetaxel (DTX)

Docetaxel is hydrophobic with low oral bioavailability hence administered intravenously. Docetaxel is associated with the phosphorylation and inactivation of the bcl-2 protein and the occurrence of apoptosis [64, 65]. Chitosan scaffold with folic acid and thiol, which help in targeting the cancer cells. Sohail et al. formulated silver nanoclusters with nanocapsules holding DTX embedded in a chitosan shell, which showed up to have better bioavailability to the tumor cells [64].

4.5 Methotrexate (MTX)

Methotrexate, a folate antagonist inhibits DHFR resulting in deficiency of the cellular pools of thymidylate and purines and thus in a decrease in nucleic acid synthesis. Therefore it interferes with DNA synthesis, repair, and cellular replication [66]. It is one of the common drugs used in management of OSCC. It has significantly low bioavailability and severe gastrointestinal adverse effects. This led to the advent of vehicles for this drug. Proteinoid microspheres were used as vehicles, which could effectively deliver MTX and other drugs also, which get degraded in gastric disorders [67]. MTX with SLNs also shows increased bioavailability and effectiveness under gastric disorders [68].

4.6 Fluoropyrimidine 5-fluorouracil (5-FU)

Fluoropyrimidine 5-fluorouracil is a thymidylate synthase (TS) inhibitor. It blocks the enzyme by preventing the synthesis of pyrimidine thymidylate. Pyrimidine thymidylate is a nucleotide necessary for DNA replication. 5-FU in combination with PLGA nanoparticle proves to be effective in oral cancer treatment with little or no gastric adverse effects. 5-FU in combination with polyvinyl alcohol-*co*-poly(methacrylic acid) hydrogel enabled a longer drug delivery [66, 69].

5 Conclusion

Oral squamous cell carcinoma is increasing in number day-by-day. The complex head and neck anatomy and the presence of vital structures around is a challenge for the physician. Be it, surgery, radiotherapy, or chemotherapy, the adverse effects of all these are immense and sometimes the hazards override the benefits. Treating a patient does not mean treating or curing the disease alone but to enhance the quality of life of the patient too. The current treatment modalities lead to huge compromise in the quality of life.

Drug vehicles to carry drugs to a specific site in the oral cavity are a promising therapeutic development. This enables us to overcome the drawbacks of the existing therapies to certain extent. However, most of these research studies are very expensive and require huge trials to establish their outcome. Though most of it are still under experimental stages, few of which are employed in practice and prove to be effective. In future there could be tailor-made drugs for every patient with the availability of drug vehicles, which would cure the disease and improve the quality of life.

References

[1] WHO. Oral cancer; 2018.

[2] Mimoza C. Tobacco smoke and alcohol consumption in relation to Oral Cancer in Albania. Int J Clin Oral Maxillofac Surg 2018;4:30. https://doi.org/10.11648/j.ijcoms.20180401.16.

[3] Thavaraj S, Stokes A, Odell EW. Is head and neck cancer infectious? Human papilloma virus in oral and pharyngeal cancer. Fac Dent J 2010;1:108–13. https://doi.org/10.1308/204268510x12804095837951.

[4] Irvine A. The immune system versus cancer: can the immune system win? Mol Med Today 2000;6:7–9. https://doi.org/10.1016/s1357-4310(99)01634-2.

[5] Sourati A, Ameri A, Malekzadeh M. Oral mucositis. Acute Side Eff Radiat Ther 2017;53–78. https://doi.org/10.1007/978-3-319-55950-6_6.

[6] Institute NC, National Cancer Institute. Oral complication of radiation therapy definitions; 2020. https://doi.org/10.32388/qrbk4u.

[7] Institute NC, National Cancer Institute. Nausea and vomiting during the first 24 hours after chemotherapy. Definitions; 2020. https://doi.org/10.32388/oo6su6.

[8] Pieter S, Morteza M. Drug delivery systems. World Scientific; 2017.

[9] Tsuchiya Y, Ushijima K, Noguchi T, Okada N, Hayasaka J-I, Jinbu Y, et al. Influence of a dosing-time on toxicities induced by docetaxel, cisplatin and 5-fluorouracil in patients with oral squamous cell carcinoma; a cross-over pilot study. Chronobiol Int 2018;35:289–94.

[10] Kaur L, Sohal HS, Kaur M, Malhi DS, Garg S. A mini-review on nano technology in the tumour targeting strategies: drug delivery to cancer cells. Anticancer Agents Med Chem 2020;20. https://doi.org/10.2174/1871520620666200804103714.

[11] Maeda H. The enhanced permeability and retention (EPR) effect in tumor vasculature: the key role of tumor-selective macromolecular drug targeting. Adv Enzyme Regul 2001;41:189–207. https://doi.org/10.1016/s0065-2571(00)00013-3.

[12] Allen TM. Passive targeting of anthracyclines entrapped in long-circulating(stealth) liposomes in the treatment of Cancer. Targeting Drugs 1994;4:119–28. https://doi.org/10.1007/978-1-4899-1207-7_10.

[13] Wu T-T, Zhou S-H. Nanoparticle-based targeted therapeutics in head-and-neck cancer. Int J Med Sci 2015;12:187–200.

[14] Arulmozhi V, Pandian K, Mirunalini S. Ellagic acid encapsulated chitosan nanoparticles for drug delivery system in human oral cancer cell line (KB). Colloids Surf B Biointerfaces 2013;110:313–20. https://doi.org/10.1016/j.colsurfb.2013.03.039.

[15] Oliveira FGL. Inflation targeting: análise de experiências e lições aproveitáveis n.d. doi:10.17771/pucrio.acad.11229.

[16] Laschke MW, Elitzsch A, Vollmar B, Vajkoczy P, Menger MD. Combined inhibition of vascular endothelial growth factor (VEGF), fibroblast growth factor and platelet-derived growth factor, but not inhibition of VEGF alone, effectively suppresses angiogenesis and vessel maturation in endometriotic lesions. Hum Reprod 2006;21:262–8. https://doi.org/10.1093/humrep/dei308.

[17] Jögi A. Tumour hypoxia and the hypoxia-inducible transcription factors: key players in cancer progression and metastasis. Tumor Cell Metab 2015;65–98. https://doi.org/10.1007/978-3-7091-1824-5_4.

[18] Reddy RS, Sudhakara Reddy R, Dathar S. Nano drug delivery in oral cancer therapy: an emerging avenue to unveil. J Med Radiol Pathol Surg 2015;1:17–22. https://doi.org/10.15713/ins.jmrps.31.

[19] Hirohashi Y, Sato N. Tumor-associated dendritic cells: molecular mechanisms to suppress antitumor immunity. Immunotherapy 2011;3:945–7. https://doi.org/10.2217/imt.11.94.

[20] Li J, Gong C, Feng X, Zhou X, Xu X, Xie L, et al. Biodegradable thermosensitive hydrogel for SAHA and DDP delivery: therapeutic effects on Oral squamous cell carcinoma xenografts. PLoS One 2012;7:e33860. https://doi.org/10.1371/journal.pone.0033860.

[21] Tegeder I. Cisplatin tumor concentrations after intra-arterial cisplatin infusion or embolization in patients with oral cancer. Clin Pharmacol Ther 2003;73:417–26. https://doi.org/10.1016/s0009-9236(03)00008-0.

[22] Agostinis P, Berg K, Cengel KA, Foster TH, Girotti AW, Gollnick SO, et al. Photodynamic therapy of cancer: an update. CA Cancer J Clin 2011;61:250–81. https://doi.org/10.3322/caac.20114.

[23] Guo R, Peng H, Tian Y, Shen S, Yang W. Mitochondria-targeting magnetic composite nanoparticles for enhanced phototherapy of Cancer. Small 2016;12:4541–52. https://doi.org/10.1002/smll.201601094.

[24] Huang X, Jain PK, El-Sayed IH, El-Sayed MA. Plasmonic photothermal therapy (PPTT) using gold nanoparticles. Lasers Med Sci 2008;23:217–28. https://doi.org/10.1007/s10103-007-0470-x.

[25] Ketabat K, Pundir M, Lobanova K, et al. Controlled drug delivery systems for oral cancer treatment—current status and future perspectives. Pharmaceutics 2019;11:302. https://doi.org/10.3390/pharmaceutics11070302.

[26] Hirabayashi F, Iwanaga K, Okinaga T, Takahashi O, Ariyoshi W, Suzuki R, et al. Epidermal growth factor receptor-targeted sonoporation with microbubbles enhances therapeutic efficacy in a squamous

cell carcinoma model. PLOS One 2017;12:e0185293. https://doi.org/10.1371/journal.pone.0185293.

[27] Carson AR, McTiernan CF, Lavery L, Grata M, Leng X, Wang J, et al. Ultrasound-targeted microbubble destruction to deliver siRNA cancer therapy. Cancer Res 2012;72:6191–9. https://doi.org/10.1158/0008-5472.can-11-4079.

[28] Carson AR, McTiernan CF, Lavery L, Hodnick A, Grata M, Leng X, et al. Gene therapy of carcinoma using ultrasound-targeted microbubble destruction. Ultrasound Med Biol 2011;37:393–402. https://doi.org/10.1016/j.ultrasmedbio.2010.11.011.

[29] Carson AR, Mctiernan CF, Lavery L, Schwartz A, Grata M, Leng X, et al. Abstract 585: treatment of squamous cell carcinoma with ultrasound and microbubble mediated gene therapy. Exp Mol Ther 2010. https://doi.org/10.1158/1538-7445.am10-585.

[30] Zhang X, Huang G, Huang H. The glyconanoparticle as carrier for drug delivery. Drug Deliv 2018;25:1840–5. https://doi.org/10.1080/10717544.2018.1519001.

[31] Calixto G, Bernegossi J, Fonseca-Santos B, Chorilli M. Nanotechnology-based drug delivery systems for treatment of oral cancer: a review. Int J Nanomedicine 2014;9:3719–35.

[32] Du F, Meng H, Xu K, Xu Y, Luo P, Luo Y, et al. CPT loaded nanoparticles based on beta-cyclodextrin-grafted poly(ethylene glycol)/poly (L-glutamic acid) diblock copolymer and their inclusion complexes with CPT. Colloids Surf B Biointerfaces 2014;113. https://doi.org/10.1016/j.colsurfb.2013.09.015.

[33] Endo K, Ueno T, Kondo S, Wakisaka N, Murono S, Ito M, et al. Tumor-targeted chemotherapy with the nanopolymer-based drug NC-6004 for oral squamous cell carcinoma. Cancer Sci 2013;104:369–74. https://doi.org/10.1111/cas.12079.

[34] Mazzarino L, Loch-Neckel G, Dos Santos BL, Mazzucco S, Santos-Silva MC, Borsali R, et al. Curcumin-loaded chitosan-coated nanoparticles as a new approach for the local treatment of oral cavity Cancer. J Nanosci Nanotechnol 2015;15:781–91. https://doi.org/10.1166/jnn.2015.9189.

[35] Mazzarino L, Borsali R, Lemos-Senna E. Mucoadhesive films containing chitosan-coated nanoparticles: a new strategy for buccal curcumin release. J Pharm Sci 2014;103:3764–71. https://doi.org/10.1002/jps.24142.

[36] Damascelli B, Patelli GL, Lanocita R, Di Tolla G, Frigerio LF, Marchianò A, et al. A novel Intraarterial chemotherapy using paclitaxel in albumin nanoparticles to treat advanced squamous cell carcinoma of the tongue: preliminary findings. Am J Roentgenol 2003;181:253–60. https://doi.org/10.2214/ajr.181.1.1810253.

[37] Sulfikkarali N, Krishnakumar N, Manoharan S, Nirmal RM. Chemopreventive efficacy of naringenin-loaded nanoparticles in 7,12-dimethylbenz(a)anthracene induced experimental oral carcinogenesis. Pathol Oncol Res 2013;19. https://doi.org/10.1007/s12253-012-9581-1.

[38] Yu D, Wang A, Huang H, Chen Y. PEG-PBLG nanoparticle-mediated HSV-TK/GCV gene therapy for oral squamous cell carcinoma. Nanomedicine 2008;3. https://doi.org/10.2217/17435889.3.6.813.

[39] Elsayed I, Huang X, Elsayed M. Selective laser photo-thermal therapy of epithelial carcinoma using anti-EGFR antibody conjugated gold nanoparticles. Cancer Lett 2006;239:129–35. https://doi.org/10.1016/j.canlet.2005.07.035.

[40] Lucky SS, Idris NM, Huang K, Kim J, Li Z, Thong PSP, et al. In vivo biocompatibility, biodistribution and therapeutic efficiency of Titania coated upconversion nanoparticles for photodynamic therapy of solid oral cancers. Theranostics 2016;6:1844–65. https://doi.org/10.7150/thno.15088.

[41] Eguchi H, Umemura M, Kurotani R, Fukumura H, Sato I, Kim J-H, et al. A magnetic anti-cancer compound for magnet-guided delivery and magnetic resonance imaging. Sci Rep 2015;5. https://doi.org/10.1038/srep09194.

[42] Wang D, Xu X, Zhang K, Sun B, Wang L, Meng L, et al. Codelivery of doxorubicin and MDR1-siRNA by mesoporous silica nanoparticles-polymerpolyethylenimine to improve oral squamous carcinoma treatment. Int J Nanomedicine 2017;13:187–98. https://doi.org/10.2147/ijn.s150610.

[43] Calixto G, Fonseca-Santos B, Chorilli M, Bernegossi J. Nanotechnology-based drug delivery systems for treatment of oral cancer: a review. Int J Nanomedicine 2014;3719. https://doi.org/10.2147/ijn.s61670.

[44] Sun M, Su X, Ding B, He X, Liu X, Yu A, et al. Advances in nanotechnology-based delivery systems for curcumin. Nanomedicine 2012;7. https://doi.org/10.2217/nnm.12.80.

[45] Beloqui A, Solinís MÁ, Rodríguez-Gascón A, Almeida AJ, Préat V. Nanostructured lipid carriers: promising drug delivery systems for future clinics. Nanomed: Nanotechnol Biol Med 2016;12:143–61. https://doi.org/10.1016/j.nano.2015.09.004.

[46] Fang C-L, Al-Suwayeh SA, Fang J-Y. Nanostructured lipid carriers (NLCs) for drug delivery and targeting. Recent Pat Nanotechnol 2012;7:41–55. https://doi.org/10.2174/187221051130701004 1.

[47] Chanburee S, Tiyaboonchai W. Mucoadhesive nanostructured lipid carriers (NLCs) as potential carriers for improving oral delivery of curcumin. Drug Dev Ind Pharm 2017;43:432–40. https://doi.org/10.1080/03639045.2016.1257020.

[48] Li J, Mooney DJ. Designing hydrogels for controlled drug delivery. Nat Rev Mater 2016;1. https://doi.org/10.1038/natrevmats.2016.71.

[49] Sharma S, Vijay P, Pathak PK, Singh P, Kant S. Role of nanotechnology in oral cancer diagnosis and therapeutics. Med Res Chronicles 2019;6:192–9.

[50] Sepantafar M, Maheronnaghsh R, Mohammadi H, Radmanesh F, Hasani-Sadrabadi MM, Ebrahimi M, et al. Engineered hydrogels in cancer therapy and diagnosis. Trends Biotechnol 2017;35:1074–87.

[51] Koutsopoulos S, Unsworth LD, Nagai Y, Zhang S. Controlled release of functional proteins through designer self-assembling peptide nanofiber hydrogel scaffold. Proc Natl Acad Sci U S A 2009;106. https://doi.org/10.1073/pnas.0807506106.

[52] Arrighetti N, Corbo C, Evangelopoulos M, Pastò A, Zuco V, Tasciotti E. Exosome-like Nanovectors for drug delivery in Cancer. Curr Med Chem 2019;26:6132–48. https://doi.org/10.2174/0929867325666180831150259.

[53] Tian Y, Li S, Song J, Ji T, Zhu M, Anderson GJ, et al. A doxorubicin delivery platform using engineered natural membrane vesicle exosomes for targeted tumor therapy. Biomaterials 2014;35:2383–90. https://doi.org/10.1016/j.biomaterials.2013.11.083.

[54] Ha D, Yang N, Nadithe V. Exosomes as therapeutic drug carriers and delivery vehicles across biological membranes: current perspectives and future challenges. Acta Pharm Sin B 2016;6:287–96.

[55] Dong Y, Feng S-S. Poly(d,l-lactide-co-glycolide)/montmorillonite nanoparticles for oral delivery of anticancer drugs. Biomaterials 2005;26:6068–76. https://doi.org/10.1016/j.biomaterials.2005.03.021.

[56] Horwitz SB. Taxol (paclitaxel): mechanisms of action. Ann Oncol 1994;5(Suppl 6).

[57] Tiwari SB, Amiji MM. Improved oral delivery of paclitaxel following administration in nanoemulsion formulations. J Nanosci Nanotechnol 2006;6:3215–21. https://doi.org/10.1166/jnn.2006.440.
[58] Lee E, Lee J, Lee I-H, Yu M, Kim H, Chae SY, et al. Conjugated chitosan as a novel platform for oral delivery of paclitaxel. J Med Chem 2008;51:6442–9. https://doi.org/10.1021/jm800767c.
[59] Aldossary SA. Review on pharmacology of cisplatin: clinical use, toxicity and mechanism of resistance of cisplatin. Biomed Pharmacol J 2019;12:7–15.
[60] Cheng K, Peng S, Xu C, Sun S. Porous hollow Fe3O4Nanoparticles for targeted delivery and controlled release of cisplatin. J Am Chem Soc 2009;131:10637–44. https://doi.org/10.1021/ja903300f.
[61] Yan X, Gemeinhart RA. Cisplatin delivery from poly(acrylic acid-co-methyl methacrylate) microparticles. J Control Release 2005;106:198–208. https://doi.org/10.1016/j.jconrel.2005.05.005.
[62] Li Y-L, Zhu L, Liu Z, Cheng R, Meng F, Cui J-H, et al. Reversibly stabilized multifunctional dextran nanoparticles efficiently deliver doxorubicin into the nuclei of cancer cells. Angew Chem 2009;121:10098–102. https://doi.org/10.1002/ange.200904260.
[63] Thorn CF, Oshiro C, Marsh S, Hernandez-Boussard T, McLeod H, Klein TE, et al. Doxorubicin pathways: pharmacodynamics and adverse effects. Pharmacogenet Genomics 2011;21:440.
[64] Sohail MF, Hussain SZ, Saeed H, Javed I, Sarwar HS, Nadhman A, et al. Polymeric nanocapsules embedded with ultra-small silver nanoclusters for synergistic pharmacology and improved oral delivery of docetaxel. Sci Rep 2018;8:13304.
[65] Herbst RS, Khuri FR. Mode of action of docetaxel - a basis for combination with novel anticancer agents. Cancer Treat Rev 2003;29. https://doi.org/10.1016/s0305-7372(03)00097-5.
[66] Cronstein BN, Aune TM. Methotrexate and its mechanisms of action in inflammatory arthritis. Nat Rev Rheumatol 2020;16:145–54.
[67] Kumar ABM, Madhan Kumar AB, Panduranga RK. Preparation and characterization of pH-sensitive proteinoid microspheres for the oral delivery of methotrexate. Biomaterials 1998;19:725–32. https://doi.org/10.1016/s0142-9612(97)00188-9.
[68] Paliwal R, Rai S, Vaidya B, Khatri K, Goyal AK, Mishra N, et al. Effect of lipid core material on characteristics of solid lipid nanoparticles designed for oral lymphatic delivery. Nanomed: Nanotechnol Biol Med 2009;5:184–91. https://doi.org/10.1016/j.nano.2008.08.003.
[69] Longley DB, Paul Harkin D, Johnston PG. 5-fluorouracil: mechanisms of action and clinical strategies. Nat Rev Cancer 2003;3:330–8.

Chapter 20

Advanced drug delivery system in pancreatic cancer

Vimal Arora[a], Dinesh Kumar Chellappan[b], Krishnan Anand[c], and Harish Dureja[d]

[a]*University Institute of Pharma Sciences, Chandigarh University, Mohali, Punjab, India,* [b]*Department of Life Sciences, School of Pharmacy, International Medical University, (IMU), Kuala Lumpur, Malaysia,* [c]*Department of Chemical Pathology, School of Pathology, Faculty of Health Sciences and National Health Laboratory Service, University of the Free State, Bloemfontein, South Africa,* [d]*Department of Pharmaceutical Sciences, Maharshi Dayanand University, Rohtak, Haryana, India*

1 Introduction

Cancer can be defined as the ailment or the state of the body when a group of conditions lead to uncontrolled growth of cells, initially confined to any given area or organ of the body followed by its extended expression. The cancer can be classified either on the basis of organ to which it is confined like lung cancer, blood cancer, etc., or on the basis of metastatic stage of the cancer, i.e., Stage I, II, III, and IV. One of the most vulnerable type of cancer is pancreatic cancer, which is considered to be the 14th most commonly occurring cancer globally and is 7th highest mortality causing type of cancer in the world and 4th in the western world. The general trend in the spread of this disease is commonly observed to be higher in developed countries like America, Europe, etc., whereas the spread of this ailment is comparatively seen less in the case of developing countries like India, Pakistan, Sri Lanka, etc. Therefore, the higher rate of spreading of pancreatic cancer in the western world irrespective of the gender is pointing out toward the risk factors associated with environmental changes along with the effects of ethnicity, genetic expression, microbial flora of gastrointestinal tract (GIT), other diseases state like diabetes, dietary factors, alcoholism, and smoking [1].

There are several theories and concepts explained from time to time regarding the prognosis and diagnosis of pancreatic cancer, but an interesting fact about this disease is that till date no clear prognosis or diagnosis is available. One reason for the delayed diagnosis can be the insignificant contribution of general screening methods employed for cancer scanning like cancer tumor and ultrasound scanning, which in turn lead to the higher mortality rate of this disease and its detection at a very late stage with high degree of malignancy as well as metastasis and this contributes to a low survival rate of 2–8 months after diagnosis. Though the present-day scenario reflects that the treatment of pancreatic cancer is a challenge due to its unknown etiology and pathogenesis, there can be certain preventive approaches toward the control of this ailment or decreasing the associated risk factors like management of associated diseases, namely diabetes, obesity, etc., routine examinations, and early diagnosis [2]. Other approaches toward the early diagnosis and treatment of pancreatic cancer with local tumor progression can be the molecular targeting, pharmacogenomics, and estimation of specific biomarkers associated with pancreatic cancer like transforming growth factor-β, phospho-glycogen synthase kinase, phosphoglycerate kinase, and ubiquitin-specific protease [3–6]. With the expanding horizons of global networking, digitalization, and drug targeting technologies, it is estimated that a promising early diagnosis and treatment can be developed for the fast-growing pancreatic cancer.

2 Etiology, pathogenesis, and structure

To understand completely the mode of treatment for successful management of any disease, it is always better to have complete insight about the pathogenesis and etiology of the disease. In this section, we will discuss about the pathogenesis and etiology of pancreatic cancer followed by the present-day scenario of its treatment.

2.1 Etiology

There are many factors that lead to this disease state and aggravation of it; undoubtedly age is the crucial one, as it is more commonly seen in elderly people as compared to that of young individuals. The exact cause of pancreatic cancer is complex and is not clearly known till date, but in various researches it has been found that there are

Advanced Drug Delivery Systems in the Management of Cancer. https://doi.org/10.1016/B978-0-323-85503-7.00030-4

many factors that lead to this disease [7]. The pancreatic carcinomas are majorly comprising of adenocarcinoma and variants of it and the associated precursor lesions [8]. Commonly it has been observed that about 60%–70% of pancreatic adenocarcinomas arise in the pancreatic heads and very less about 15% of it arises in body and tail region of the pancreas. Second common observation in pancreatic cancer is that, at the time of diagnosis, it is commonly found to be spread in the surrounding area and organs near by the pancreas; the major reason of this advanced stage diagnosis is the unavailability of diagnostic techniques at early stages [9].

The pancreatic adenocarcinomas is developed via a series of stepwise mutations, starting from the normal mucosa, leading to specific precursor lesion, and finally ending at invasive malignancy. The three well explained precursors of adenocarcinomas (malignant) having distinct molecular, pathological, and clinical characteristics are pancreatic intraepithelial neoplasia (PanIN), intraductal papillary mucinous neoplasm (IPMN), and mucinous cystic neoplasm (MCN) [10]. The various stages of development of pancreatic cancer are illustrated in Fig. 1.

There can be a number of factors influencing the onset of this disease and survival of the patients and these factors majorly cover the lifestyle of an individual or patient, these are [1]:

- Age: Majorly diagnosed in the patients having an age of more than 55 years; rarely diagnosed earlier.
- Sex: Higher number of incidences in males as compared to females.
- Ethnicity/race: Genomic factors play a very important role.
- Blood group: People with blood group A, B, and AB are at more risk.
- GIT microflora: Lower levels of Neisseria elongate along with *Streptococcus mitis* and higher levels of *Porphyromonas gingivalis* as well as *Granulicatella adiacens* and *Helicobacter pylori* are associated with higher risk to this disease state.
- Family history: The individuals with family history of this disease are more susceptible to it.
- Diabetes: It is a well-established reason; the patients with type-I diabetes are at more risk, as compared to the ones having type-II diabetes.

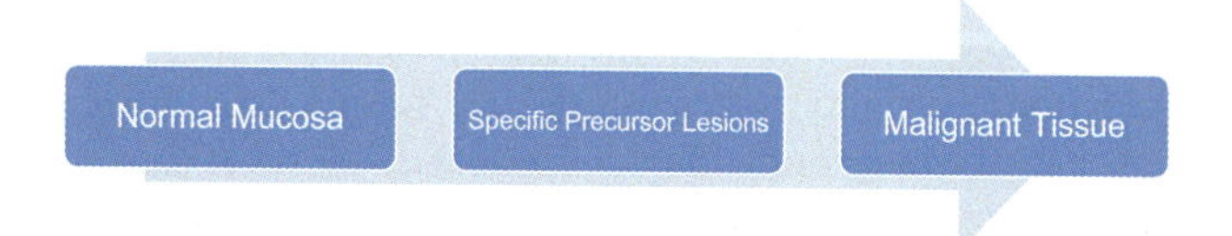

FIG. 1 Stages of development of pancreatic cancer.

- Chronic pancreatitis: The patients with chronic pancreatitis are at more risk.
- Obesity: The individuals with higher body mass index (BMI) are at more risk.
- Smoking: The smokers are more prone to this ailment.
- Diet: Vast dietary epidemiological studies have not been made till date, but it has been reported that few dietary components like red meat and high saturated fat diet are closely associated with the occurrence of pancreatic cancer.
- Infection: Patients with *H. pylori* and hepatitis-C are at more risk.

2.2 Pathogenesis

The malignant pancreas (neoplasm) is commonly categorized on the basis of cellular differentiation of neoplastic cells and is named as ductal, acinar, and neuroendocrinic pancreatic neoplasm, and secondly on the basis of the macroscopic appearance of the tumor they are classified as the solid and cystic neoplasm [11]. Generally, the exocrinic pancreatic cancer covers almost 95% of the pancreatic malignancy, out of which 90% of the pancreatic carcinomas are pancreatic adenocarcinomas or its variants and the precursor lesions [12, 13]. Moreover, 60%–70% of these pancreatic adenocarcinomas ascend in the pancreatic heads, followed by their occurrence in the pancreatic body (about 15%) and remaining 15% commonly exist in the pancreatic tail. Furthermore, it has also been observed that due to lack of early diagnosis the moment it is diagnosed, it is in a metastatic stage and spread in the surroundings [14, 15]. The most commonly reported pathogenic pathway in the case of pancreatic involves the following three stages/mutations:

- Normal mucosa
- Precursor lesions
- Invasive malignancy [16]

Three well-characterized precursors of malignancy are:

- Pancreatic intraepithelial neoplasms (PanIN)
- Intraductal papillary mucinous neoplasm (IPMN)
- Mucinous cystic neoplasm (MCN) [17]

Another important element in the diagnosis and treatment of pancreatic cancer can be the biomarkers available at the cancer sites and one such marker is epidermal growth factor receptor (EGFR). It is reported to be involved in the metastasis and growth of tumor along with its overexpression in the pancreatic cancer tissue, indicating the aggressive growth of cancer [18]. The pathogenesis of pancreatic cancer has not been defined very clearly till date and thus is treated as one of the major challenges in its diagnosis and treatment.

2.3 Structure (pancreatic cancer tissue)

The location of pancreas in the body is such that the availability of various other tissue in its vicinity permits an invasive metastasis of pancreatic cancer. There are some distinct parts or structures available in pancreatic cancer tissue, few of them are:

- Pancreatic cancer cells (PCC)
- Pancreatic cancer stem cells (PCSC)
- Tumor stroma [19]

Commonly it has been observed that the tumor stroma is very dense (desmoplastic reaction) comprising of blend of various cellular and noncellular components interacting with each other including pancreatic stellate cells (PSC), adipocytes, immune and inflammatory cells, cancer-associated fibroblasts (CAF), and endothelial cells. The CAFs are generally responsible for the chemotherapy resistance along with poor vascular density and epithelial mesenchymal transition (EMT), whereas the immunosuppressive response is commonly shown by the different types of immune and inflammatory cells [20].

3 Treatment and challenges

There are variety of treatments or combinational treatments that are available for the treatment of the pancreatic cancer like radiotherapy, chemotherapy, complimentary medicine along with the radiotherapy. In the segment of chemotherapy, gemcitabine and 5-fluorourocil are considered to be the first-line treatment for the pancreatic cancer. Besides all these, the last and the most effective treatment of pancreatic cancer is surgical resection in combination with the chemoradiation, along with the immunotherapy [21, 22].

The two major reasons for the need of advancement and exhaustive research in the development of new drug regimen and delivery system are drug resistance and failure in delivering the drug at the right site due to the lower vascularity at the cancer site. Therefore, these limitations of existing and under trial drugs for pancreatic cancer can pave the way to design and develop newer generation therapies for the pancreatic cancer, focusing on the pathogenesis, biological markers, nanocarriers, programmed drug delivery, etc.

4 Nanotechnology in drug delivery in pancreatic cancer

The drug delivery specifically to the pancreatic cancer cells is the biggest challenge with the existing drug delivery systems. The microenvironment of the cancerous cells (dense desmoplastic stroma) and hindered or inadequate perfusion prevents the drugs to penetrate the tumor cells effectively. The systemic drug delivery uses the vessels of tumor to reach the cancerous cells and the various physiological as well as chemical limitations lead to poor pharmacokinetic behavior of anticancer drugs in pancreatic cancer. The hypoxic conditions of the stroma lead to failure of accumulation of drug at the tumor site and thus making it difficult to respond due to faster clearance. Second most important limitation of anticancer therapy is the hydrophobic nature of the drugs used in the cancer treatment which had led to its limitation of administration via intravenous route. These limitations and challenges need effective and efficient approaches to make targeted delivery with intended output. To overcome this, polymer-based drug delivery, nanocarriers, and micelle have lead a new direction in the effective delivery of pancreatic cancer. These approaches can be effectively used for passive as well as active targeting of the drugs to the target site.

The nanotechnology has given a new direction to the cancer therapy and has proven to be a very efficient chemotherapeutic approach for treating different types of cancer. The major characteristics namely nanometric size range and surface engineering in nanocarriers have led to numerous advantages of using nanocarriers like small particle size, higher surface area, better target binding, narrow size range distribution, better stability of drug, and many more [23]. The present-day scenario in the drug delivery of anticancer drugs for the treatment of pancreatic cancer involves various types of nanocarriers as summarized in Fig. 2.

An optimized selection of drug and corresponding nanocarriers led to an effective drug delivery encompassing subcellular and intracellular drug delivery leading to a fairly effective anticancer therapy. These nanocarriers have also helped a lot in overcoming the issues related to the poor water solubility of chemotherapeutic agents and hence provided a better therapeutic value of such drugs [24, 25]. The nanoparticles are having a particle size of less than 100 nm and thus having distinct physicochemical properties due to this reduced size. Furthermore, the changes in the composition of these nanocarriers help in exploring new options like active targeting and passive vehicle designing, leading to water solubility, better bio-distribution, and increased half-life [26]. Another approach for targeting cancer tumor utilizing nanocarriers can be achieved via conjugating the drugs with molecules that can effectively bind to receptors (antigens) on cancer tumors [27, 28]. The newer concept of nanocarriers under trial is the smart nanocarriers based upon the drug delivery mediated via stimulus (pH, enzyme, temperature, etc.). Different types of nanocarriers are:

4.1 Liposomes

Liposomes are defined as vesicular structures comprising of lipid bilayer (concentric) covering the aqueous center core (Fig. 3). These are self-assembled lipid bilayers made up of either phospholipids, cholesterol, or any other synthetic amphiphilic component. Therefore, this drug delivery carrier

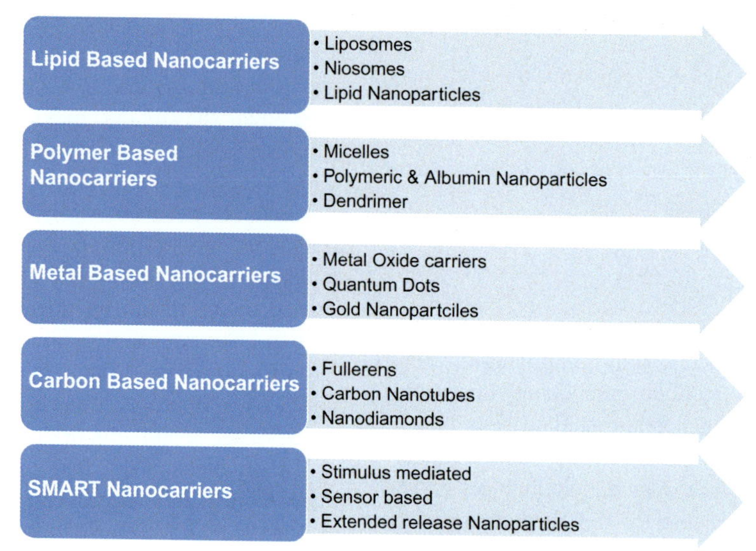

FIG. 2 Types of nanocarriers.

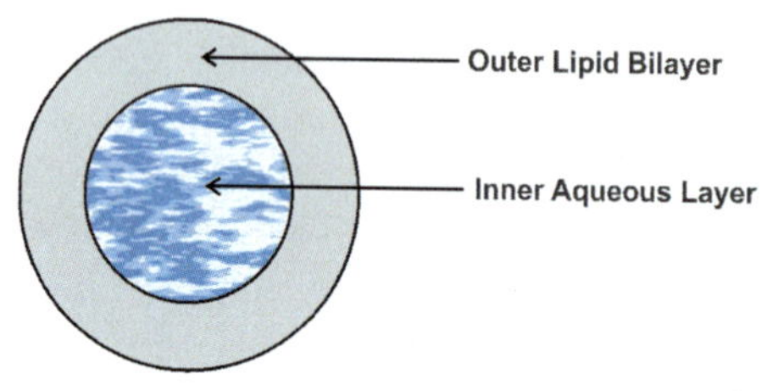

FIG. 3 Basic structure of liposome.

can be used to deliver both water soluble and lipid soluble bioactives; the hydrophilic drugs are dissolved/entrapped in the central core and hydrophobic drugs are entrapped in the lipid bilayer [29, 30]. These liposomes can be categorized in two ways, one based up on their size and another is based upon their nature or mechanism involved, as enlisted hereunder:

(1) Based upon size:
- Small unilamellar vesicles (20–100 nm)
- Large unilamellar vesicles (100–400 nm)
- Multilamellar vesicles (400–5000 nm)

(2) Based upon their nature or mechanism:
- Conventional liposomes
- pH-sensitive liposomes
- Cationic liposomes
- Immuno liposomes
- Long circulating liposomes

There are a number of advantages of liposomal drug delivery as compared to that of the conventional drug delivery, but the major ones are better pharmacokinetic profile, better stability, lesser toxicity, better half-life, and many more [31]. The first liposomal nanocarriers drug delivery system was designed (1995) for the doxorubicin hydrochloride, as liposomal injection and used in the treatment of AIDS.

4.2 Niosomes

Niosomes are the vesicular carrier systems comprising of nonionic surfactants and cholesterol, having a bilayer structure. The presence of surfactants gives an ease to modify and versatility to these vesicular structures with an extra advantage of being economical as compared to that of phospholipids [32, 33]. The surfactant system has an additional advantage to increase the bioavailability of some bioactives due to their p-glycoprotein inhibitory nature. These niosomal preparations have represented better inhibitory potential against tumors as compared to the conventional drug delivery system (solution form), when administered orally or intravenously [34]. This group of nanocarriers has also given remarkable outcomes for their transdermal application, e.g., the 5-fluorouracil transdermal preparation has shown better penetration as compared to the aqueous solution [35]. The niosomes are found to be more advantageous than the liposomal drug delivery system, due to two major limitations of liposomes—lower shelf life due to rancidification of lipids and higher cost of processing.

These nonionic surfactant-based nanocarriers are also found to be an efficient drug delivery system for targeting bioactives at the cancer sites, e.g., the niosomal preparation of methotrexate is found to be more effective than the normal drug solution. Similarly, the niosomes of doxorubicin hydrochloride designed with the help of C16 monoalkyl glycerol ether has exhibited better stability and drug release as compared to that of conventional drug delivery system [36]. Kong and his coworkers also reported effective hyaluronic acid-based niosomes for drug targeting using the transdermal route of administration [37]. Various advancements in designing niosomal drug delivery system utilize various approaches, like elastic cationic niosomes, magnetic niosomes, nanohybrid niosomes with nonionic surfactants, pH-sensitive niosomes, and encapsulated niosomes. The pH-sensitive niosomes have led to more effective and promising targeted drug delivery in cancer treatment; generally, drug release is carried out at ambient to slightly acidic pH value.

4.3 Polymeric nanoparticles

The polymeric nanocarriers or nanoparticles are simply defined as the submicron colloidal particle, having size range of 10–1000 nm. The nanoparticles may have a characteristic feature of having size range of less than 200 nm comprising of a single polymer [38]. There are various polymers available for designing polymeric nanoparticles when used individually or in combination, commonly used synthetic polymers are poly (D,L-lactide-coglycolide) (PLGA), poly (L-lactic acid) (PLL), poly (epsilon caprolactone) (PCL), poly (alkyl cyanoacrylate) (PACA), poly (isohexyl cyanoacrylate) (PIHCA), and poly (isobutyl cyanoacrylate) (PIBCA).

This drug delivery system also utilizes natural polymers as one of its components or simply in place of synthetic polymers which includes gelatin, chitosan, and hyaluronic acid.

This drug delivery system facilitates the encapsulation and thus the delivery of hydrophobic drugs, biological macromolecules, RNA, DNA, and proteins leading to their protection from systemic challenges, which in turn results in better tissue distribution and hence effective tumor targeting. The other approaches like the linking of inert polymers like poly ethylene glycol (PEG) in conjugation with either pH-sensitive or hypothermic polymers on the surface of nanoparticles has shown a remarkable outcome in terms of targeting the cancerous tumor. Furthermore, some other approaches can be utilized to engineer the surface of these nanocarriers to achieve effective target delivery via conjugation of their surfaces using aptamers, monoclonal antibodies, mannose, and folic acid [39].

The important factors while designing any polymeric nanoparticles are the biocompatibility among different sets of polymers used, particle size or magnitude of the monomer, its physic-chemical properties, and the far most important factor, i.e., biocompatibility between the drug and polymer duo used along with the method of manufacturing or processing. This system of drug delivery overcomes various limitations of liposomes, niosomes, and other nanocarriers namely drug stability in the systemic circulation, drug accumulation at the tumor site, drug loading efficiency, and intracellular drug concentration [40]. One of the most important advantages with this system is that it overcomes the drug resistance due to the presence of *p*-glycoproteins at the cancer site, as observed in other nanocarriers or delivery systems. The other major advantage of this delivery system is the use of multiple mechanism for controlled release used alone or in combination, viz., diffusion, dissolution, and erosion [41].

In the recent advancements, various other materials and processes are being used by various researchers to explore improvised targeted drug delivery via using albumin, cyclodextrin, and some polysaccharides, resulting in promising intended results in animal studies [42]. Another approach dealing with technique of developing hydrophobic core for encapsulation of drug using PLA and PEG block polymer with a hydrophilic surface for prolonged circulation has shown remarkable results during preclinical studies [43].

4.4 Dendrimers

The dendrimers are considered to be among novel approaches for targeted drug delivery with better pharmacokinetic profile and are designed using highly branched polymers in globular confirmation; arranged via attaching a large number of monomeric subunits to form large nanometric structures [44, 45]. In 1980s, the first dendrimer was designed comprising of poly(amidoamine), since then various other dendrimers are designed and studied for their utilization in drug targeting in cancer treatment [46]. These dendrimers are having three sites for varying the drug target potential namely, core, branching zone, and the branch edges. These dendrimeric structures have various advantages over other nanometric carriers in terms of its structural engineering capabilities, surface functionalities, chemical composition and simplified synthesis along with better pharmacokinetics, as well as biocompatibility [46, 47]. The major limitation of this delivery systems is that the commercially available are used successfully for imaging purposes and not for targeted drug delivery.

4.5 Lipid nanoparticles

The lipid nanoparticles can be simply considered as an advanced version of submicron size colloidal emulsions in which the liquid lipid component is replaced with the solid lipid phase. This structural change has led to change in his behavior in terms of drug entrapment and systemic stability. The composition of these structures displays a major advantage of accommodating both types of bioactives namely hydrophobic and hydrophilic drugs, along with its potential to design carriers for topical, oral, and parenteral administration of anticancer drugs [48–50]. The limitations associated with the design of lipid nanoparticles are:

- Differentiation between various types of lipid nanoparticles is not clear.
- Particle growth and polymorphic transition may lead to drug expulsion on storage.
- The crystalline nature may lead to lower drug entrapment efficiency [51].

4.6 Gold nanoparticles

The metal nanoparticles are having a different place of significance in the field of biomedical sciences, especially targeted drug delivery. These involve the various metal ions which are inert in nature, and gold is the most abundantly and successfully used metal. The gold metal-based nanoparticles are having two basic advantages over any other metal-based nanocarriers that are its lower toxicological profile and their standardization, because the oxidation of it does not interfere with the interaction of nanoparticles with various body fluids. Another advantage of dealing with gold nanocarriers is ease and flexibility in surface alterations at the time of association with bioactives using different processes. Thus gold nanoparticles being nonimmunogenic and nontoxic are having vast areas of applications like diagnosis, drug targeting, and radiation therapy because of its potential to convert the absorbed light into heat [52–54]. The gold nanoparticles are also used after due conjugation with various other components like antibodies, which is effectively used in drug targeting especially through the mechanism of its binding to the receptors available at the surface of cancer cells [55]. In various research studies, it has been observed that these nanocarriers can be efficiently used as delivery carriers for injectable nanomedicines comprising of anticancer drugs, protein, and DNA. The major drawback of this delivery system is the accumulation of gold in liver and same can be effectively overcome via its conjugation with PEG, i.e., PEG-coated gold nanoparticles show no or insignificant accumulation in liver [56]. In another research, it has been reported that the folic acid conjugated gold nanoparticles have shown a high degree of selective accumulation in folate responsive cancerous tumors [57]. The gold nanoparticle of bombesin has showcased the high degree of specificity and selectivity in uptake in case of gastrin-releasing peptide receptor in pancreatic acne [58]. These gold particles are found to be very effective in drug targeting to cancer cells overexpressing the EGFR. Moreover, they have shown remarkable adaptability in conjugating with a wider variety of compounds containing amine and/or thiol groups [59].

4.7 Micelles

The formulation of micelles is considered to be the most distinguished and basic approach of increasing the drug solubility. The formulation of micelles comprising of cytotoxic drugs was first carried in the early 1980s and 1990s [60]. The structural composition of self-assembled nanocolloidal micelles is such that they have hydrophobic core and hydrophilic outer covering, with a minimum particle size of less than 20 nm in diameter, which is of utmost importance as compared to any other nanocarrier system; as it allows better circulation time through escape from renal infiltration. Secondly, these micelles can be flexibly designed using either lipids or other polymer-based amphiphilic molecules like PEG, methacrylate, poly (amidoamine), poly (L-aspartic acid), 10-undecenoicacid, dimethyl acrylamide, 2-hydroxy ethyl methacrylate, methyl methacrylate, PGA, and many more. The hydrophobic core of the nanocarriers allows the loading of hydrophobic drugs and the outer hydrophilic shell facilitates the higher circulation time and stability to the drug carriers. Further, this outer hydrophilic covering can be conjugated with some specific ligands that can led to active targeting or binding to the malignant tumor sites via specific cellular structures and signaling biomarkers. This can further lead to lower systemic toxicity and drug-induced side effects because of lower drug penetration and/or binding to the normal tissue or organ [61].

4.8 Carbon nanotubes

The carbon nanotubes are the nano structures comprising of different allotropes of carbon having a cylindrical structure, arranged together either as graphite sheets or rolled as the tubular structures [62]. The carbon nanotubes are considered as one of the most remarkable outcome of nanotechnology having a variety of applications in the biomedical sciences. Due to its characteristic dark black color, they do have a tendency to absorb the infrared radiations and convert them to heat energy, therefore they have an application in photo thermal or photo acoustic therapy in cancer treatment [63]. The following characteristics of carbon nanotubes make it a suitable nanocarriers for the chemotherapeutic agents:

- Higher surface area
- Light weight
- Hollow monolithic structure
- High mechanical strength
- Distinct physical, biological, and chemical properties

The one major distinguishing feature of carbon nanotubes is its ability to diffuse through the biological membrane or its attachment to the cell membrane and subsequent endocytosis that makes it the most favorable tool for drug delivery and to increase the solubility of organic compounds [64, 65]. These carbon nanotubes have also reported to be very effective in drug targeting via their conjugation with metal atoms like platinum and polymers like PLA-PEG, TPGS, hyaluronic acid, and chitosan that has led to better drug efficacy via drug accumulation at the cancer tumor site.

4.9 Fullerenes

The fullerenes are defined as the symmetrical, spheroidal, nano-dimensional structures of carbon and the most commonly available fullerene is Buckmin-Sterfullerene (C60) containing 60 carbon atoms in a spherical shape. The application of fullerenes in biomedical sciences subject to its derivatization to make it water soluble and on the other hand its physical, chemical, and physicochemical properties. Similar to that of carbon nanotubes, the fullerenes are also used in drug targeting and photo acoustic therapy; however, its use is limited due to its toxicity. Fullerenes offer a unique property of covalent linkage of chemotherapeutic agents, even more than one agent/drug can be linked for a combination drug therapy.

4.10 Nanodiamonds

This nano-technological advancement belongs to the family of nano-carbons having a distinct set of physical, chemical, physic-chemical, mechanical, and biological properties due to their unique structural design [66, 67]. The properties of these nanodiamonds depend upon their method of production, like if these are obtained via breakdown of large or bulky diamond structures then their properties are similar to the parent structure. On the contrary, if these are obtained via the processes like detonation, then their properties are different and are depending up on the process conditions, because the nature of functional groups present on the particle surface depends upon the environmental conditions. The nanodiamonds are categorized as diamondiods (1–2 nm), ultra nanocrystalline particles (2–4 nm), and nanocrystalline particles (1–150 nm) on the basis of their particle size. All these structures are nontoxic, inert, having unique surface chemistry and tendency to cross cell membrane; therefore, these are considered to be an efficient candidate to be used as a drug carrier for targeted drug delivery in pancreatic cancer [68]. These nanodiamonds can be explored for their utilization in a wide variety of fields like bioactive linked drug delivery, embedded thin films, and device-based implants for targeted and controlled drug delivery. In some recent research studies, these nanodiamonds are conjugated with secondary plant metabolites like quercetin and ciproten that were successful in penetrating the cytoplasm and got embedded in the nuclear membrane [69]. In another study, these particles were conjugated with azide-modified transferrin that resulted in efficient targeting to the tumor cells capable of overexpressing the transferrin receptors [70].

4.11 Smart nanocarriers

The existing drug delivery using nanoparticles has undoubtedly given a new direction to the anticancer therapy still there is poor patient compliance due to lesser stability, lower bioavailability, poor pharmacodynamic and pharmacokinetic parameters, cytotoxicity, aspecific distribution, and the chemoresistance [71]. These limitations have led to the development of new generation of nanoparticles capable of targeting chemotherapeutic agents in line with certain stimulus or overexpressing biomarker compounds, these are known as smart nanoparticles. The various smart nanocarriers are illustrated in Fig. 4. These nanocarriers are based upon multifunctional targeting comprising of all three classes viz. passive, active, and stimuli mediated drug targeting. These nanoparticles can be designed as smart extended-release nanoparticles (SER NPs) that may in turn led to a better multifunctional targeting along with other physiological as well as clinical advantages and sometimes pathophysiological characteristics. The major disadvantage of presently available nanocarriers is their tendency to get accumulated in the liver, poor therapeutic efficiency (toxicity toward tumor), and other factors causing hindrance in the entry of nanocarriers in the cellular structure.

The principle of working of the smart nanoparticles is their capability to release the chemotherapeutic agent in line with the response to some specific physiological marker or trigger at a suitable time and specific target (due to these three parameters they are considered to be smart). The passive targeting of chemotherapeutic agent using these advanced nanocarriers is based upon the increased permeability and retention effect. This permeability and retention depends upon the vascular architecture and lymphatic draining linked with the given cancerous tumor (different types of tumor and their site affects the vasculature and lymphatic drainage). To overcome the uncertainty of drug target related to the passive targeting, the concept of targeting the overexpressed biomarkers or enzymes can lead to

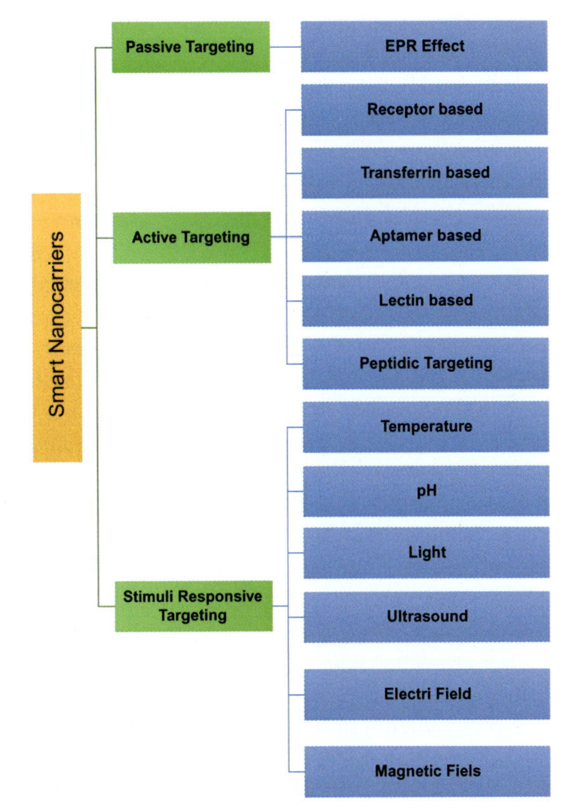

FIG. 4 Types of smart nanocarriers.

better targeting of therapeutic agents which is known as the active targeting via smart delivery. The major limitation of this drug delivery is the significant loss of dose due to lysosomal digestion followed by receptor-mediated endocytosis and immunogenicity leading to significant blood clearance.

The active targeting-based smart nanoparticles led to better therapeutic efficiency via reducing the drug content required for the given therapeutic effect and minimizing the unspecified targeted drug delivery. The nanoparticle is also reported to be effectively targeted delivery of siRNA. The most advanced category of smart nanoparticles is known as the stimuli responsive targeting of chemotherapeutic agents, which can be achieved via the change in physicochemical structure leading to site and time-specific delivery in the presence of external stimuli like pH, light, heat, electric field, magnetic field, and ultrasonic waves. This stimuli-based response is the resultant of some characteristic structural changes leading to the exploitation of basic characteristics of micro environment of the cancer tumor. The nanocarriers carrying pendant basic or acidic groups capable of releasing or accepting protons in line with the pH changes are called as pH-sensitive nanoparticles. This class of delivery or pH-sensitive targeting is beneficial to the delivery of thermolabile chemotherapeutic agents.

The smart nanoparticles can easily overcome the limitations of conventional nanotechnological drug delivery like hepatic degradation, renal clearance, higher intestinal fluid pressure, high tumor cellular density, and drug efflux pump [72, 73]. The stimuli-responsive smart nanoparticles are found to be more efficient as they led to faster drug release, better cellular binding, and a better drug perfusion in the cancerous tumor [74]. The controlled release drug delivery via stimuli responsive nanoparticles is also found to be better in performance and helps in maintaining a steady-state delivery of chemotherapeutic agent that can in turn lead to lower drug toxicity. The higher tumor density with highly dense tumor microenvironment supported by hypovasculature and lower perfusion lead to higher interstitial pressure that is considered to be the major barrier for effective targeting. All these obstacles can be effectively overcome by these stimuli responsive nanoparticles [75, 76].

Furthermore, due to administration of maximum tolerated dose of anticancer drug severe cytotoxic effects are observed with conventional and nanotechnology-based drug delivery systems which in turn may lead to discontinuation of therapy. The smart nanoparticles allow administration of lower dose and maintenance of effective drug concentrations at the target sites that widens their therapeutic window and hence are better than any other drug delivery system. The smart nanoparticles achieve these advantages via increased accumulation of drug at the tumor sites and specificity of drug targeting leading to decreased systemic side effects. The recent advancements made in the field of designing smart nanoparticles are now making use of natural polymers and proteins like albumin that helps in overcoming the limitation of toxicity shown by various other solvent systems. One of the major obstacles in designing an efficient drug delivery system for the treatment of pancreatic cancer is chemoresistance, which is supposed to be easily overcome by the newest advancement in designing smart nanoparticles, i.e., redox-sensitive smart nanoparticles. One another advanced smart nanoparticle which is significantly successful in overcoming the multidrug resistance is gene-silencing drugs [77, 78].

The drug delivery systems comprising of smart nanoparticles can be classified into three generations of delivery system on the basis of their specific targeting. These three generations are:

- First generation: lacking specific targeting.
- Second generation: based on active targeting.
- Third generation: multistage/multiple strategy drug delivery [79].

Undoubtedly there are numerous practical limitations of designing smart nanoparticles like the complexity of preparation process, regulatory barriers, and higher financial interventions. The advanced smart nanoparticles leading to improve pharmacokinetics and pharmacodynamics parameters of free drugs are leading to efficient clinical outcomes which are very important to circumvent the limitation associated with them. Therefore, it can be concluded that smart delivery systems are advantageous due to their noninvasive delivery, multi specificity, and multifunctional targeting [80]. One of the recently developed smart nanoparticle-based drug delivery systems is irinotecan (liposomal) stimuli-responsive drug delivery system based on photosensitive release for pancreatic cancer. Some of the newer advancements also include manifold functionalizing and activating stimuli-responsive drug delivery system like combination of pH-sensitive and reactive oxygen species layering.

4.12 Extended-release nanoparticles

The concept of extended-release nanoparticles is designed to enjoy two characteristic advantages, prolonged-release and steady-state drug release for achieving a better therapeutic window with minimal side effects [81]. Commonly the water insoluble polymeric substances (biodegradable) are used to encapsulate the active chemotherapeutic agents and their delivery at the tumor site. This technological advancement may also include the mechanism of changing the surface properties to attain extended circulation. One of such techniques is PEGylation of nanocarriers, which involves conjugation of polyethylene glycol (PEG) with the nanopolymer used [82]. This extended-release nanoparticle system is having a variety of advantages like extended plasma retention time, declined proteolysis, better hydrodynamicity, lower renal clearance, and shielded immune-detection.

The two keynote features of these extended-release nanoparticles are their prolonged circulation time of about 160 h and delayed elimination, which are assured in the size range of 40–200 nm [83, 84].

A variety of polymers can be utilized for designing the nanoparticles with distinct characteristic features and properties like nanocrystalline cellulose can be used to bind larger amount of drug due to its extensively charged surface area and these may be coated with cationic polymer like cetyltrimethylammonium bromide that can change its release profile [85]. Some other approaches to develop such extended-release system is use of amphiphilic polymers to achieve prolonged release and conjugation of micelles with biomarker specific ligands (overexpressed biomarkers by the cancer cells). These mechanisms led to high drug loading capacity and altered drug release profiles that make these nanoparticles the most favorable drug delivery system. The structures of various nanocarriers are shown in Figs. 5–13.

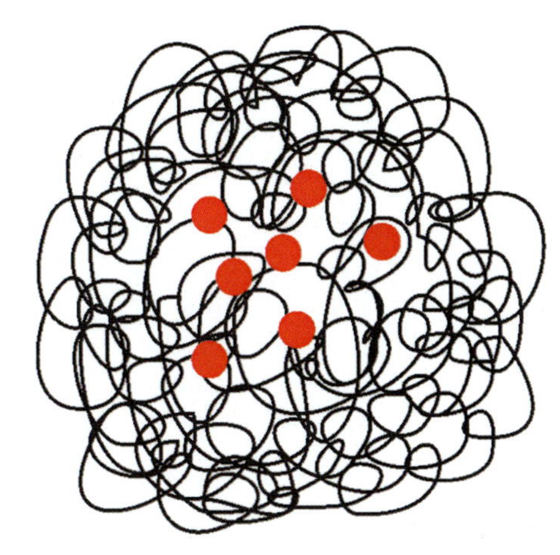

FIG. 7 Polymeric nanoparticle.

FIG. 5 Liposomes.

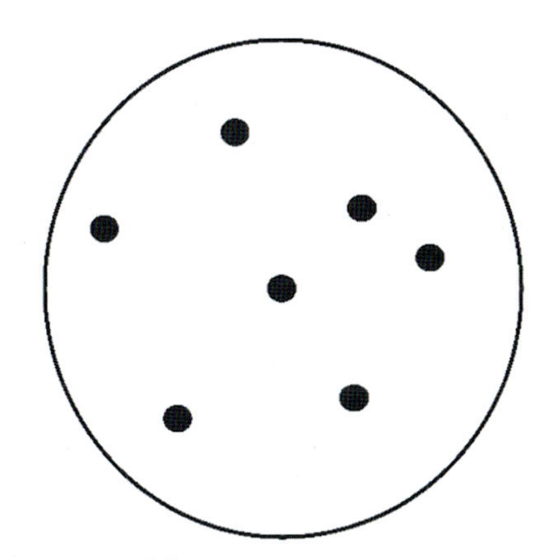

FIG. 8 Metal nanoparticles.

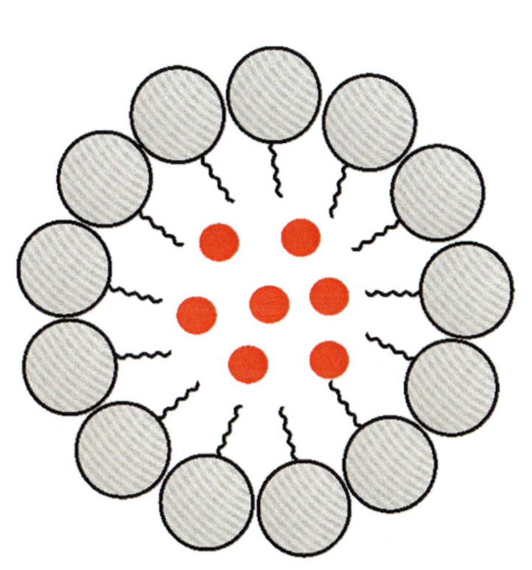

FIG. 6 Micelles.

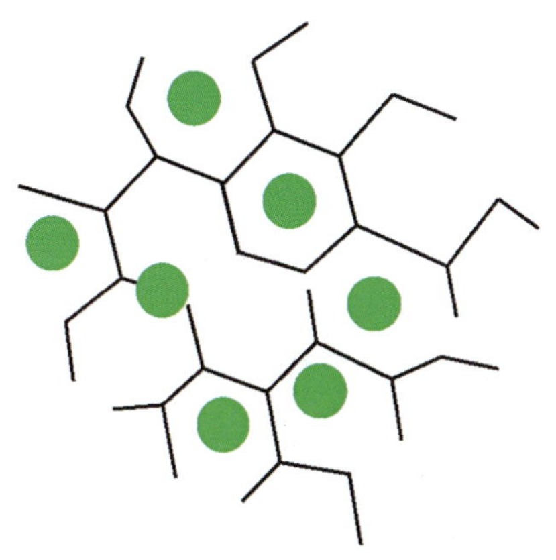

FIG. 9 Dendrimers.

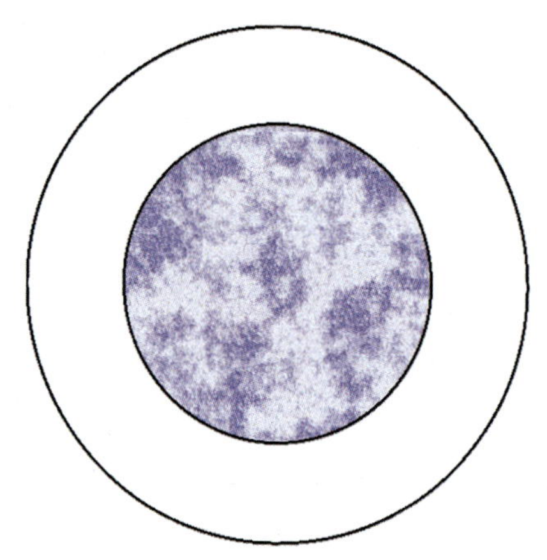

FIG. 10 Extended-release nanoparticles.

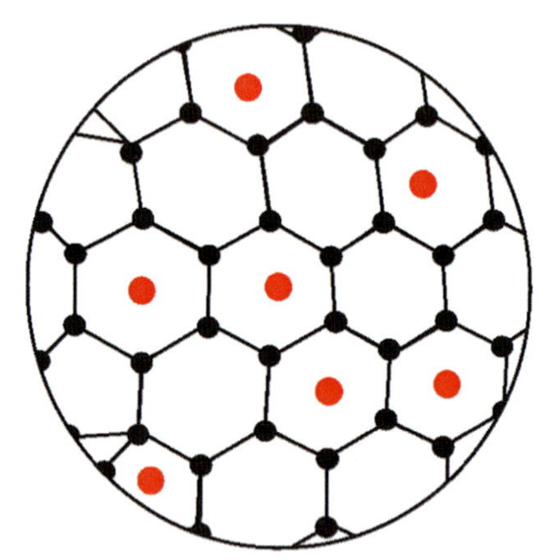

FIG. 11 Fullerene.

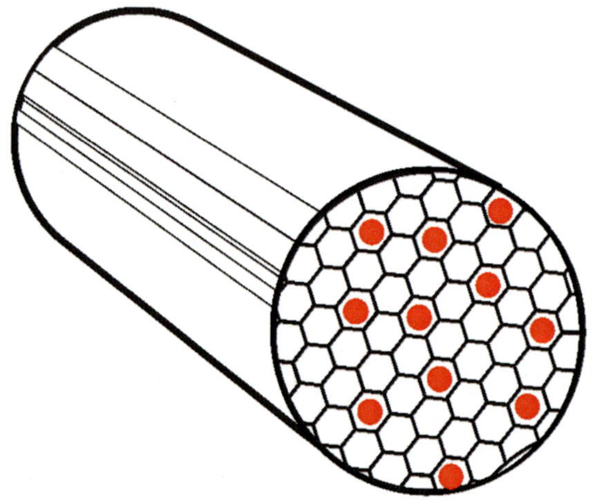

FIG. 12 Carbon nanotubes.

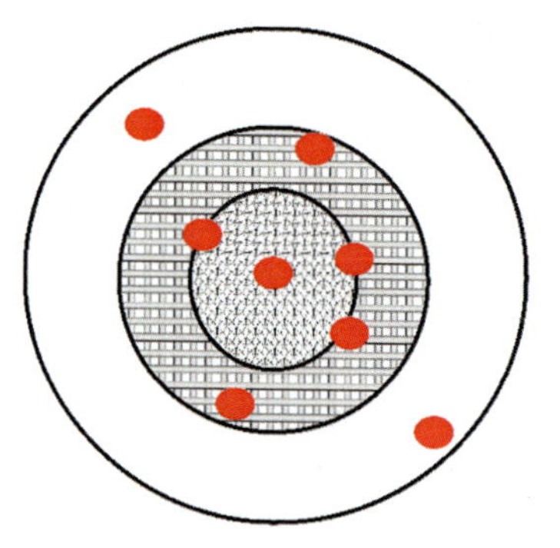

FIG. 13 Stimuli responsive nanocarrier.

4.13 Alternative treatment for pancreatic cancer

During the recent past, herbals are attracting many researchers toward developing treatment regimens to fight against deadly disease like cancer based on the traditional knowledge. The practice of plant-based natural remedies is used to combat with fibrotic-induced diseases as compared to other systems of complementary and alternative medicine; for example, Japanese herbal medicines, Ayurvedic medicines (traditional medicinal system of India), Chinese herbal medicine, etc., comprising of polyphenolic compounds (curcumin), antioxidants (vitamins), protease inhibitors, and other plant-derived pharmacologically active components are reported to prevent or treat pancreatic cancer. The numerous plant-derived compounds are under clinical trials for evaluation of their potential as antifibrotic agents. Another class of potentially active compounds is biosimilars or analogs of naturally occurring antifibrotic moieties like paricalcitol (a synthetic analog of vitamin D), undergoing clinical trials (Phase I and II), and are administered concomitantly with other well established chemotherapeutic drugs used for the treatment of pancreatic cancer [86, 87].

There are numerous phytoconstituents that have been studied in the past few decades for their potential to treat cancer, especially in the early stage of disease. Ayurveda treats the disease based upon the diseased state and its correlation with the metabolic state of the body. In case of cancer, the Ayurvedic treatment correlates the diseased state with metabolic syndrome and chronic inflammation, which are further considered to be linked with impaired lipid metabolism and cholesterol biosynthesis. This state of chronic inflammation is also considered to be promoting tumor cell growth and the inflammation microenvironment is treated as a reason for the spreading of cancer (metastasis). The pancreatic tumorigenesis is reported to be caused by oxidative stress in conjunction with inflammation which in turn leads to irreversible morphological changes and functional in the pancreas, that may lead to the chronic pancreatitis.

One of the amazing phytoconstituents is curcumin which has been considered to be the rejuvenating medicine in the field of Ayurveda, as it is reported to be the drug of choice against a number of diseases. The Indian herb *Curcuma longa* (Rhizome) is the source of curcumin, which is described as the universal antioxidant and antiinflammatory phytoconstituent. Curcumin is reported to be working through a dual mechanism comprising of increase in apoptosis and blocking the nuclear factor κB (NFκB) in the treatment of pancreatic cancer. It has been reported that curcumin suppresses cell proliferation, invasion, and angiogenesis [88]. Furthermore, it has also been reported that it escalates the islet content of glutathione, and insulin secretion, thus it protects them from oxidative stress [89]. Various clinical trials revealed that when curcumin is administered orally in combination with gemcitabine, it has been found to be very effective, regardless of its lower bioavailability. Recently, the nanoparticles of curcumin known as "nanocurcumin" is found to be very much effective against the pancreatic tumor and has also shown a synergistic effect with Gemcitabine; the clinical trials of nanocurcumin is expected to pave a new pathway for the treatment of pancreatic cancer. The physical and chemical properties of curcumin allow its conversion to the nanocurcumin. The efficiency of nanocurcumin is due to its promising physicochemical properties like surface area, particle size, surface charge, and hydrophobicity, leading to increased solubility and higher oral bioavailability which make it different from the naturally occurring curcumin. These improvised properties in turn lead to a better pharmacokinetic profile and thus makes it useful for active targeting. These characteristic properties diverge with change in particle size change in the given nanometric size range. The most commonly used size range of nanoparticles used during various clinical trials used for various medicinal applications is 10–100 nm. Nanocurcumin is reported to be having a higher intracellular absorption capacity and high systemic bioavailability [86, 90].

Another naturally occurring active anticancer agent is "Rhein" which is an anthraquinone derivative extracted from the rhizomes of *Rheum palmatum* and few other traditional medicinal plants. It shows its anticancer activity via inhibiting the cell growth of tumor cells via controlling the expression of hypoxia-inducible factor-1 alpha (HIF-1a) by decreasing the phosphorylation of phosphorylated protein kinases B (p-AKT) and ERK1/2 (p-ERK1/2). The antifibrotic and antitumorigenic effects of rhein are reported to be due to its mechanism of downregulation of the NF-kB and STAT3 signaling pathways [91].

Research studies also indicated that the consumption of cruciferous vegetables is very much effective against various malignancies including pancreatic cancer. Cruciferous vegetables are a rich source of cucurbitacins and glucosinolates. These glucosinolates are converted to isothiocyanates in the presence of enzyme myrosinase after chewing; moreover, the microflora of GIT is supposed to be synergizing this conversion. One of the phytoconstituents benzyl isothiocyanate is present in the vegetables (family cruciferae) such as cabbage, cauliflower, and mustard which are reported to inhibit initiation, growth, and metastasis in certain preclinical investigations. The second class of characteristic phytoconstituents is cucurbitacins; especially the cucurbitacin-B, which has been reported to be a very potential candidate to combat pancreatic cancer [86].

Camellia sinensi commonly known as Green tea and widely consumed in Asian countries for their antioxidant, antiinflammatory, antiatherosclerotic, and anticancer activities. It contains both polyphenolic and nonphenolic compounds as its major components mainly comprising of class of phenolic compounds called as catechins; consisting of epigallocatechin gallate, epigallocatechin epicatechin gallate, gallocatechin, and (þ)-catechin (C). Epigallocatechin gallate is known for its antioxidant activity, along with its potential to inhibit the PDGF-induced proliferation and migration of PSCs. It has been reported to inhibit the PDGF-induced tyrosine phosphorylation of the PDGF b-receptor, downstream activation of ERK, and phosphoinositide 3-kinase (PI3K)/AKT pathways [92, 93].

One another phytoconstituent having anticancer activity is "Resveratrol," which is a polyphenolic stilbene compound, commonly found in raspberries, blueberries, grapes, cocoa, and peanuts. It has been reported that it is generally synthesized in response to injury or attacks from pathogens in plants. A large number of in vitro and in vivo studies have demonstrated the anticancer activity, especially its use in treating pancreatic cancer, via inducing apoptosis in pancreatic cancer cells by encouraging caspase-3 activation. Likewise, resveratrol increases the chemo-sensitivity of pancreatic cancer cells by targeting the nutrient-deprived autophagy factor-1 (NAF-1) via using nuclear factor erythroid 2 (NRF2) signaling. It has also been reported that resveratrol increases the sensitivity of pancreatic tumor cells (MiaPaCa-2 and Panc-1) and reduces the overexpression of biomarkers of cancer stem cells [94, 95].

Emodin (1,3,8-trihydroxy-6-methylanthraquinone) is one another active anthraquinone derivative available in *Aloe vera* and plants (family Polygonaceae) such as *Palmatum rheum*. It is reported to exert antiinflammatory, antiangiogenic, and antidyslipidemic activities, along with anticancer potential. Emodin has shown potent anticancer activity via targeting cell proliferation and inducing apoptosis by utilizing various mechanisms. Additionally, the combinational therapy against pancreatic cancer, comprising of emodin and well-established anticancer agents revealed increased chemosensitivity of pancreatic cancer tumor cells [96, 97].

Ellagic acid is one more member among the various active phytoconstituents belonging to the category of nonflavonoid polyphenol that is mainly found in various nuts and fruits like walnuts, berries, pomegranates, and grapes. In recent years, various research studies have been conducted to explore the medicinal benefits of ellagic acid in the treatment of cancer. One more such phytoconstituent is embelin, which is a naturally occurring benzoquinone, extracted from the berries of the plant *Embelia ribes* Burm. (family Myrsinaceae). A number of its uses have been explained in the Indian traditional medicinal remedies. Embelin has been reported to show its anticancer activity via inhibiting cell migration, invasion, and induction of apoptosis in cancer cells. The antifibrotic activity of embelin has been indicated in pancreatic cancer cells in a dose-dependent manner. Furthermore, it has been reported that embelin in combination with ellagic acid (at lower concentrations) have shown an augmented apoptosis and decreased cellular proliferation [98–100].

Capsaicin, a derivative of homovanillic acid (*N*-vanillyl-8-methyl-nonenamide), is another phytoconstituent present in all chili pepper plants as the principal pungent constituent in the concentration range of 0.1%–1% (w/w). It has been successfully used in the treatment of pain, inflammation, and various other diseases, including anticancer potential expressed via suppression of caerulein-induced carcinogenesis in transgenic mice and overpowering pancreatic tumor growth. Capsaicin significantly reduced the phosphorylation of ERK, c-JUN, and Hedgehog/GLI1 activation and hence, it was found to be effective in inhibiting the carcinogenesis of pancreatic cancer. It has also been reported that capsaicin has shown dose-dependent induction of mitochondrial-dependent apoptosis in pancreatic cancer cells (ASPC-1 and BxPC-3). Capsaicin-induced apoptosis has also been reported to be via ROS-mediated JNK activation [101]. The various nanocarrier formulations are tabulated in Table 1.

TABLE 1 Nanocarrier formulations.

S. no.	Type of nanocarrier	Drug-polymer conjugate
1.	Micelle	Gemcitabine and PLGA
2.	Micelle	Gemcitabine and PEG conjugated hydrophobic stearic acid derivative bonded through acid-sensitive hydrazine
3.	Micelle	Docetaxel, PEG, and gemcitabine
4.	Metal oxide based	Gemcitabine, siRNA and iron-oxide nanoparticles
5.	Metal oxide based	Tetrasilylated porphyrin and ethylene periodic mesoporous organosilica nanoparticles
6.	Metal oxide based	Gold nanocluster bovine serum albumin cluster, mesoporous silica, 32% GEM and DOX combined with albumin, linked electrostatically to formulation
7.	Liposome	Docetexal liposomes
8.	Stimuli responsive	Irinotecan liposomes

5 Futuristic approaches to combat pancreatic cancer

Pancreatic cancer is supposed to be the second deadliest disease in the world till 2030; therefore, the challenges to circumvent this deadly disease are very crucial. In line with the present-day preclinical investigations and early-stage clinical trials, the approaches like phytoconstituent-based nanocarriers, magnetosome-based drug targeting, iRNA nanocarriers, biomarker-targeted drug delivery, and sensor-based (programmable) drug delivery can be some promising treatments against pancreatic cancer in the future. Additionally, the conjugated treatment (using complementary and integrative medicine) can also be beneficial in designing some promising treatments against pancreatic cancer.

References

[1] McGuigan A, et al. Pancreatic cancer: a review of clinical diagnosis, epidemiology, treatment and outcomes. World J Gastroenterol 2018;24(43):4846–61.

[2] Jin C, Bai L. Pancreatic cancer—current situation and challenges. Gastroenterol Hepatol Lett 2020;2(1):1–3.

[3] Javle M, Li Y, Tan D, et al. Biomarkers of TGF-β signaling pathway and prognosis of pancreatic cancer. PLoS One 2014;9(1), e85942.

[4] Naito S, Bilim V, Yuuki K, et al. Glycogen synthase kinase-3beta: a prognostic marker and a potential therapeutic target in human bladder cancer. Clin Cancer Res 2010;16(21):5124–32.

[5] Hwang TL, Liang Y, Chien KY, et al. Overexpression and elevated serum levels of phosphoglycerate kinase 1 in pancreatic ductal adenocarcinoma. Proteomics 2006;6(7):2259–72.

[6] Ning Z, Wang A, Liang J, et al. USP22 promotes the G1/S phase transition by upregulating FoxM1 expression via β-catenin nuclear localization and is associated with poor prognosis in stage II pancreatic ductal adenocarcinoma. Int J Oncol 2014;45(4):1594–608.

[7] Vincent A, Herman J, Schulick R, et al. Pancreatic cancer. Lancet 2011;378:607–20.

[8] Jung B, Hong S, Kim SC, et al. In vivo observation of endothelial cell-assisted vascularization in pancreatic cancer xenograft engineering. Tissue Eng Regen Med 2018;15:275–85.

[9] Lee WH, Loo CY, Leong CR, et al. The achievement of ligand-functionalized organic/polymeric nanoparticles for treating multidrug resistant cancer. Expert Opin Drug Deliv 2017;14:937–57.

[10] Yuen A, Diaz B. The impact of hypoxia in pancreatic cancer invasion and metastasis. Hypoxia (Auckl) 2014;2:91–106.

[11] Hackeng WM, Hruban RH, Offerhaus GJA, Brosens LAA. Surgical and molecular pathology of pancreatic neoplasms. Diagn Pathol 2016;11(47):1–17.

[12] Manzur A, Oluwasanmi A, Moss D, Curtis A, Hoskins C. Nanotechnologies in pancreatic cancer therapy. Pharmaceutics 2017;9(4):39.

[13] Zhou J, Lindsey E, Stojadinovic A, et al. Incidence rates of exocrine and endocrine pancreatic cancers in the United States. Cancer Causes Control 2010;21(6):853–61.

[14] Feldmann G, Beaty R, Hruban RH, Maitra A. Molecular genetics of pancreatic intraepithelial neoplasia. J Hepato-Biliary-Pancreat Surg 2007;14:224–32.

[15] Luchini C, Capelli P, Scarpa A. Pancreatic ductal adenocarcinoma and its variants. Surg Pathol Clin 2016;9:547–60.

[16] Mohammed S, Van Buren 2nd G, Fisher WE. Pancreatic cancer: advances in treatment. World J Gastroenterol 2014;20:9354–60.

[17] Esposito I, Konukiewitz B, Schlitter AM, Kloppel G. Pathology of pancreatic ductal adenocarcinoma: facts, challenges and future developments. World J Gastroenterol 2014;20:13833–41.

[18] Zhang Y, Satoh K, Li M. Novel therapeutic modalities and drug delivery in pancreatic cancer—an ongoing search for improved efficacy. Drugs Context 2012;212244. https://doi.org/10.7573/dic.212244.

[19] Hidalgo M. Pancreatic cancer. N Engl J Med 2010;362:1605–17.

[20] Tianqi S, Yang B, Gao T, Liu T, Li J. Polymer nanoparticle assisted chemotherapy of pancreatic cancer. Ther Adv Med Oncol 2020;12:1–33.

[21] Shaib Y, Davila J, Naumann C, et al. The impact of curative intent surgery on the survival of pancreatic cancer patients: a U.S. population based study. Am J Gastroenterol 2007;102:1377–82.

[22] Thind K, Padrnos LJ, Ramanathan RK, et al. Immunotherapy in pancreatic cancer treatment: a new frontier. Ther Adv Gastroenterol 2017;10:168–94.

[23] Jabr-Milane LS, Van Vlerken LE, Yadav S, Amiji MM. Multifunctional nanocarriers to overcome tumor drug resistance. Cancer Treat Rev 2008;34(7):579–602.

[24] Wagner V, Dullaart A, Bock AK, Zweck A. The emerging nanomedicine landscape. Nat Biotechnol 2006;24(10):1211–7.

[25] Khlebtsov N, Dykman L. Biodistribution and toxicity of engineered gold nanoparticles: a review of in-vitro and in-vivo studies. Chem Soc Rev 2011;40(3):1647–71.

[26] Prencipe G, Tabakman SM, Welsher K, Liu Z, Goodwin AP, Zhang L. PEG branched polymer for functionalization of nanomaterials with ultralong blood circulation. J Am Chem Soc 2009;131:4783–7.

[27] Polakis P. Arming antibodies for cancer therapy. Curr Opin Pharmacol 2005;5:382–7.

[28] Wu AM, Senter PD. Arming antibodies: prospects and challenges for immunoconjugates. Nat Biotechnol 2005;23:1137–46.

[29] Khan DR, Rezler EM, Lauer-Fields J, Fields GB. Effects of drug hydrophobicity on liposomal stability. Chem Biol Drug Des 2008;71(1):3–7.

[30] Slingerland M, Guchelaar H-J, Gelderblom H. Liposomal drug formulations in cancer therapy: 15 years along the road. Drug Discov Today 2012;17(3–4):160–6.

[31] Marianecci C, Marzio LD, Rinaldi F, et al. Niosomes from 80s to present: the state of the art. Adv Colloid Interf Sci 2014;205:187–206.

[32] Mahale NB, Thakkar RG, Mali DR, et al. Niosomes: novel sustained release nonionic stable vesicular systems—an overview. Adv Colloid Interf Sci 2012;183–184:46–54.

[33] Azmin MN, Florence AT, Handjani-Vila RM, Stuart JF, Vanlerberghe G, Whittaker JS. The effect of non-ionic surfactant vesicle (niosome) entrapment on the absorption and distribution of methotrexate in mice. J Pharm Pharmacol 1985;37(4):237–42.

[34] Paolino D, Cosco D, Muzzalupo R, Trapasso E, Picci N, Fresta M. Innovative bola-surfactant niosomes as topical delivery systems of 5-fluorouracil for the treatment of skin cancer. Int J Pharm 2008;353(1–2):233–42.

[35] Kumar GP, Rajeshwarrao P. Nonionic surfactant vesicular systems for effective drug delivery—an overview. Acta Pharm Sin B 2011;1:208–19.

[36] Uchegbu IF, Double JA, Kelland LR, Turton JA, Florence AT. The activity of doxorubicin niosomes against an ovarian cancer cell line and three in vivo mouse tumour models. J Drug Target 1996;3(5):399–409.

[37] Kong M, Park H, Feng C, Hou L, Cheng X, Chen X. Construction of hyaluronic acid noisome as functional transdermal nanocarrier for tumor therapy. Carbohydr Polym 2013;94(1):634–41.

[38] Kreuter J. Nanoparticles. In: Swarbrick J, Boylan JC, editors. Encyclopedia of pharma. New York: Tech. Marcel Dekker; 1994. p. 165–90.

[39] Moghimi SM, Hunter AC, Murray JC. Long-circulating and targetspecific nanoparticles: theory to practice. Pharmacol Rev 2001;53(2):283–318.

[40] Roney C, Kulkarni P, Arora V, et al. Targeted nanoparticles for drug delivery through the blood-brain barrier for Alzheimer's disease. J Control Release 2005;108(2–3):193–214.

[41] Veer LJ, Bernards R. Enabling personalized cancer medicine through analysis of gene-expression patterns. Nature 2000;452(7187):564–70.

[42] Prabaharan M. Chitosan-based nanoparticles for tumortargeted drug delivery. Int J Biol Macromol 2015;72:1313–22.

[43] Raj R, Mongia P, Sahu SK, Ram A. Nanocarriers based anticancer drugs: current scenario and future perceptions. Curr Drug Targets 2016;17(2016):206–28.

[44] Bulte JWM, Douglas T, Witwer B, et al. Magnetodendrimers allow endosomal magnetic labeling and in vivo tracking of stem cells. Nat Biotechnol 2001;19(12):1141–7.

[45] Kobayashi H, Kawamoto S, Brechbiel MW, et al. Detection of lymph node involvement in hematologic malignancies using micromagnetic resonance lymphangiography with a gadolinium labeled dendrimer nanoparticle. Neoplasia 2005;7(11):984–91.

[46] Svenson S, Tomalia DA. Dendrimers in biomedical applications: reflections on the field. Adv Drug Deliv Rev 2005;57(15):2106–29.

[47] Lee CC, MacKay JA, Fréchet JM, Szoka FC. Designing dendrimers for biological applications. Nat Biotechnol 2005;23(12):1517–26.

[48] Muller RH, Schwarz C, Mehnert W, Lucks JS. Production of solid lipid nanoparticles (SLN) for controlled drug delivery. Proc Int Symp Control Rel Bioact Mater 1993;20:480–1.

[49] Liedtke S, Wissing S, Muller RH, Mader K. Influence of high pressure homogenisation equipment on nanodispersions characteristics. Int J Pharm 2000;196(2):183–5.

[50] Wissing SA, Kayser O, Muller RH. Solid lipid nanoparticles for parenteral drug delivery. Adv Drug Deliv Rev 2004;56(9):1257–72.

[51] Wong HL, Bendayan R, Rauth A, Li Y, Wu XY. Chemotherapy with anticancer drugs encapsulated in solid lipid nanoparticles. Adv Drug Deliv Rev 2007;59(6):491–504.

[52] Chanda N, Shukla R, Zambre A, et al. An effective strategy for the synthesis of biocompatible gold nanoparticles using cinnamon phytochemicals for phantom CT imaging and photoacoustic detection of cancerous cells. Pharm Res 2011;28(2):279–91.

[53] Popovtzer R, Agrawal A, Kotov NA, et al. Targeted gold nanoparticles enable molecular CT imaging of cancer. Nano Lett 2008;8(12):4593–6.

[54] Huang X, Peng X, Wang Y, et al. A reexamination of active and passive tumor targeting by using rod-shaped gold nanocrystals and covalently conjugated peptide ligands. ACS Nano 2010;4(10):5887–96.

[55] Kumar S, Aaron J, Sokolov K. Directional conjugation of antibodies to nanoparticles for synthesis of multiplexed optical contrast agents with both delivery and targeting moieties. Nat Protoc 2008;3(2):314–20.

[56] Visaria RK, Griffin RJ, Williams BW, et al. Enhancement of tumor thermal therapy using gold nanoparticle-assisted tumor necrosis factor alpha delivery. Mol Cancer Ther 2006;5(4):1014–20.

[57] Dixit V, Bossche J, Sherman DM, Thompson DH, Andres RP. Synthesis and grafting of thioctic acid-PEG-folate conjugates onto Au nanoparticles for selective targeting of folate receptor-positive tumor cells. Bioconjug Chem 2006;17(3):603–9.

[58] Chanda N, Kattumuri V, Shukla R, Zambre A, Katti K, Upendran A. Bombesin functionalized gold nanoparticles show in vitro and in vivo cancer receptor specificity. Proc Natl Acad Sci U S A 2010;107(19):8760–5.

[59] Rebelo A, Molpeceres J, Rijo P, Reis CP. Pancreatic cancer therapy review: from classic therapeutic agents to modern nanotechnologies. Curr Drug Metab 2017;18(4):346–59.

[60] Yokoyama M, Okano T, Sakurai Y, Ekimoto H, Shibazaki C, Kataoka K. Toxicity and antitumor activity against solid tumors of micelle-forming polymeric anticancer drug and its extremely long circulation in blood. Cancer Res 1991;51:3229–36.

[61] Felber AE, Dufresne MH, Leroux JC. pH-sensitive vesicles, polymeric micelles, and nanospheres prepared with polycarboxylates. Adv Drug Deliv Rev 2012;64(12):979–92.

[62] Mehra NK, Jain AK, Lodhi N, et al. Challenges in the use of carbon nanotubes for biomedical applications. Crit Rev Ther Drug Carrier Syst 2008;25(2):169–206.

[63] Robertson CA, Evans DH, Abrahams H. Photodynamic therapy (PDT): a short review on cellular mechanisms and cancer research applications for PDT. J Photochem Photobiol B: Biol 2009;96(1):1–8.

[64] Sahoo NG, Bao H, Pan Y, et al. Functionalized carbon nanomaterials as nanocarriers for loading and delivery of a poorly water soluble anticancer drug: a comparative study. Chem Commun (Camb) 2011;47(18):5235–7.

[65] Sinha N, Yeow JT. Carbon nanotubes for biomedical applications. IEEE Trans Nanobioscience 2005;4(2):180–95.

[66] Kazi S. A review article on nanodiamonds discussing their properties and applications. Int J Pharm Sci Invent 2014;3(7):40–5.

[67] Kaur R, Badea I. Nanodiamond as novel nanomaterials for biomedical applications: drug delivery and imaging systems. Int J Nanomedicine 2013;8:203–20.

[68] Kong XL, Huang LCL, Hsu CM, Chen WH, Han CC, Chang HC. High-affinity capture of proteins by diamond nanoparticles for mass spectrometric analysis. Anal Chem 2005;77(1):259–65.

[69] Gismondi A, Reina G, Orlanducci S, et al. Nanodiamonds coupled with plant bioactive metabolites: a nanotech approach for cancer therapy. Biomaterials 2015;38:22–35.

[70] Rehor I, Lee KL, Chen K, et al. Plasmonic ND: targeted core-shell type nanoparticles for cancer cell thermoablation. Adv Healthc Mater 2015;4(3):460–8.

[71] Yang M, Yu L, Guo R, Dong A, Lin C, Zhang J. A modular coassembly approach to all-in-one multifunctional nanoplatform for synergistic codelivery of doxorubicin and curcumin. Nanomaterials 2018;8(3). https://doi.org/10.3390/nano8030167, E167.

[72] Kalaydina, et al. Recent advances in smart delivery systems for extended drug release in cancer therapy. Int J Nanomedicine 2018;13:4727–45.

[73] Lammers T. Improving the efficacy of combined modality anticancer therapy using HPMA copolymer-based nanomedicine formulations. Adv Drug Deliv Rev 2010;62(2):203–30.

[74] Du J, Lane LA, Nie S. Stimuli-responsive nanoparticles for targeting the tumor microenvironment. J Control Release 2015;219:205–14.

[75] Adiseshaiah PP, Crist RM, Hook SS, Mcneil SE. Nanomedicine strategies to overcome the pathophysiological barriers of pancreatic cancer. Nat Rev Clin Oncol 2016;13(12):750–65.

[76] Nielsen MF, Mortensen MB, Detlefsen S. Key players in pancreatic cancer-stroma interaction: cancer-associated fibroblasts, endothelial and inflammatory cells. World J Gastroenterol 2016;22(9):2678–700.

[77] Madhavan S, Gusev Y, Harris M, et al. G-DOC: a systems medicine platform for personalized oncology. Neoplasia 2011;13(9):771–83.

[78] Liu J, Li J, Liu N, et al. In vitro studies of phospholipid-modified PAMAM-siMDR1 complexes for the reversal of multidrug resistance in human breast cancer cells. Int J Pharm 2017;530(1–2):291–9.

[79] Park K. Controlled drug delivery systems: past forward and future back. J Control Release 2014;190:3–8.

[80] Cheng Z, Al Zaki A, Hui JZ, Muzykantov VR, Tsourkas A. Multifunctional nanoparticles: cost versus benefit of adding targeting and imaging capabilities. Science 2012;338(6109):903–10.

[81] Ramasamy T, Tran TH, Choi JY, et al. Layer-by-layer coated lipid–polymer hybrid nanoparticles designed for use in anticancer drug delivery. Carbohydr Polym 2014;102:653–61.

[82] Bobo D, Robinson KJ, Islam J, Thurecht KJ, Corrie SR. Nanoparticle-based medicines: a review of FDA-approved materials and clinical trials to date. Pharm Res 2016;33(10):2373–87.

[83] Sykes EA, Chen J, Zheng G, Chan WC. Investigating the impact of nanoparticle size on active and passive tumor targeting efficiency. ACS Nano 2014;8(6):5696–706.

[84] Wang B, Jiang W, Yan H, et al. Novel PEG-graft-PLA nanoparticles with the potential for encapsulation and controlled release of hydrophobic and hydrophilic medications in aqueous medium. Int J Nanomedicine 2011;6:1443–51.

[85] Jackson JK, Letchford K, Wasserman BZ, Ye L, Hamad WY, Burt HM. The use of nanocrystalline cellulose for the binding and controlled release of drugs. Int J Nanomedicine 2011;6:321–30.

[86] Long J, et al. Overcoming drug resistance in pancreatic cancer. Expert Opin Ther Targets 2011;15(7):817–28.

[87] Ramakrishnan P, et al. Selective phytochemicals targeting pancreatic stellate cells as new anti-fibrotic agents for chronic pancreatitis and pancreatic cancer. Acta Pharm Sin B 2020;10(3):399–413.

[88] Azimi H, Khakshur AA, Abdollahi M, Rahimi R. Potential new pharmacological agents derived from medicinal plants for the treatment of pancreatic cancer. Pancreas 2015;44:11–5.

[89] Kocaadam B, Sanlier N. Curcumin, an active component of turmeric (*Curcuma longa*), and its effects on health. Crit Rev Food Sci Nutr 2017;57:2889–95.

[90] Karthikeyan A, Senthil N, Min T. Nanocurcumin: a promising candidate for therapeutic applications. Front Pharmacol 2020;11:487. https://doi.org/10.3389/fphar.2020.00487.

[91] Chang CY, Chan HL, Lin HY, Way TD, Kao MC, Song MZ, et al. Rhein induces apoptosis in human breast cancer cells. Evid Based Complement Alternat Med 2012;2012:952504. https://doi.org/10.1155/2012/952504.

[92] Chacko SM, Thambi PT, Kuttan R, Nishigaki I. Beneficial effects of green tea: a literature review. Chin Med 2010;5:13. https://doi.org/10.1186/1749-8546-5-13.

[93] Hosseini A, Ghorbani A. Cancer therapy with phytochemicals: evidence from clinical studies. Avicenna J Phytomed 2015;5:84–97.

[94] Aggarwal BB, Bhardwaj A, Aggarwal RS, Seeram NP, Shishodia S, Takada Y. Role of resveratrol in prevention and therapy of cancer: preclinical and clinical studies. Anticancer Res 2004;24:2783–840.

[95] Zhou C, Qian W, Ma J, Cheng L, Jiang Z, Yan B, et al. Resveratrol enhances the chemotherapeutic response and reverses the stemness induced by gemcitabine in pancreatic cancer cells via targeting SREBP1. Cell Prolif 2019;52. https://doi.org/10.1111/cpr.12514, e12514.

[96] Guo HC, Bu HQ, Luo J, Wei WT, Liu DL, Chen H, et al. Emodin potentiates the antitumor effects of gemcitabine in PANC-1 pancreatic cancer xenograft model in vivo via inhibition of inhibitors of apoptosis. Int J Oncol 2012;40. https://doi.org/10.1371/journal.pone.0042146, 1849e57.

[97] Liu A, Chen H, Tong H, Ye S, Qiu M, Wang Z, et al. Emodin potentiates the antitumor effects of gemcitabine in pancreatic cancer cells via inhibition of nuclear factor κB. Mol Med Rep 2011;4(2):221–7.

[98] Saldanha E, Joseph N, Ravi R, Kumar A, Shetty V, Fayad R, et al. Polyphenols in the prevention of acute pancreatitis: preclinical observations. In: Watson RR, Preedy VR, Zibadi S, editors. Polyphenols in human health and disease. San Diago, CA: Academic Press; 2014. p. 427–33.

[99] Ceci C, Tentori L, Atzori MG, Lacal PM, Bonanno E, Scimeca M, et al. Ellagic acid inhibits bladder cancer invasiveness and in vivo tumor growth. Nutrients 2016;8. https://doi.org/10.3390/nu8110744, E744.

[100] Huang Y, Lu J, Gao X, Li J, Zhao W, Sun M, et al. PEG-derivatized embelin as a dual functional carrier for the delivery of paclitaxel. Bioconjug Chem 2012;23:1443–51.

[101] Boreddy SR, Srivastava SK. Pancreatic cancer chemoprevention by phytochemicals. Cancer Lett 2013;334(1):86–94.

Chapter 21

Advanced drug delivery system in colorectal cancer

Nitin Sharma[a], Ritu Karwasra[b], and Gaurav Kumar Jain[c]
[a]Department of Pharmaceutical Technology, Meerut Institute of Engineering and Technology, Meerut, Uttar Pradesh, India, [b]Department of Pathology, ICMR-National Institute of Pathology, New Delhi, India, [c]Delhi Pharmaceutical Science and Research University, New Delhi, India

1 Colorectal cancer

Carcinoma or tumors are the leading cause of mortality worldwide and among them, colorectal cancer (CRC) accounts to be the 3rd fatal and 4th most widespread carcinoma with high prevalence rate [1]. CRC is commonly termed as bowel carcinoma or either rectal carcinoma, or colorectal adrenocarcinoma and generally emerges from epithelial and glandular cells of large intestine [2]. CRC arises due to abnormal heightened replication and survival of epithelial cells and these hyperproliferative cells give rise to benign adenoma that further evolves to carcinoma. It usually arises from three different mechanisms i.e., CpG island methylator phenotype, microsatellite instability, and chromosomal instability. Genetic alterations results in impaired cellular pathways [3]. APC gene, known as adenomatous polyposis coli is responsible for producing APC protein that prevents accumulation of beta-catenin, and in absence of APC, this protein got accumulated and translocates to nucleus, where it binds to DNA thereby activating the process of proto-oncogenes. Mutations in APC gene resulted in formation of CRC. Generally, APC genes are responsible for stem cell renewal; however, if they are expressed inappropriately at high levels, it results in CRC [4, 5].

It is diagnosed in both males and females with yearly deaths recorded to be tens of thousands of people with 10% cancer-related mortality. Incidence of CRC, reported by GLOBOCAN 2018 data, states that it is among the 11% cancer diagnoses and the most commonly diagnosed cancer globally [6]. CRC-related mortality and incidence rate vary within the globe, with 1.8 million new cases every year. It is also seen that these rates vary between male and female, with substantially higher rate in males than females. Approximately, 521,000 women and 576,000 men were diagnosed with colon cancer 274,000 women and 430,000 men were diagnosed with rectum cancer. Incidence represents 1.51% increasing risk of colon carcinoma in men and 1.12% in women [1, 3]. Certain reports also stated that developed countries are at higher risk for CRC. In case of colon cancer, Northern and Southern Europe, New Zealand, Austria, and North America are at higher incidence, whereas in case of rectum cancer, Eastern Europe, Eastern Asia, and New Zealand are at higher risk [4, 7].

1.1 Cause and symptoms

Numerous risk factors are involved in development of CRC i.e., sedentary lifestyle, environmental, obesity, increase in age, high intake of meat and related products, lack of physical activity, and smoking or alcohol consumption. The survival rate noted by many physicians was found to be ~ 65%. Bacteria such as *Streptococcus gallolyticus* and *Streptococcus bovis/Streptococcus equinus* are associated with colorectal tumors. Polyps present in colon or rectum, which eventually become cancerous, are also a possible risk factor involved. Symptoms observed in CRC include weight loss, presence of RBC in stools, pain, discomfort in abdomen, alteration in bowel movements, appetite loss, exacerbate constipation, nausea, and vomiting [4]. Anemia and rectal bleeding are majorly high-risk symptoms in advanced age people (over 50 years). Etiology of CRC initiated with proliferation of noncancerous mucosal epithelial cells and these growths known as polyps progressively became cancerous after 10–20 years [8, 9].

1.2 Classification

CRC is classified into four stages on basis of histopathological and molecular studies. Stage 0 represents as earliest stage where cancer lies within the mucosa or in the internal epithelia of colon and rectum. Stage 1 refers to the initial stage in which the cancer has progressed in the internal layers of colon or rectum; however, it has not extended beyond the outer membrane of colon or rectum. Stage 2 refers to the

Advanced Drug Delivery Systems in the Management of Cancer. https://doi.org/10.1016/B978-0-323-85503-7.00012-2

invading stage in which cancer has progressed within the wall of colon or rectum; however, it has not extended on to lymph nodes. Stage 3 refers to the invasion of carcinoma to lymph nodes, but it has not extended or affects other parts of body. Stage 4 refers to the invasion of carcinoma to other body parts like as to lungs, liver, abdominal cavity, and ovaries. Lastly, recurrent phase comprises of reversal of carcinoma after treatment and affects colon, rectum, as well as to other parts of body [10,11].

It is categorized to be one of deadliest diseases worldwide, as there is lack of detection methods and proper drug delivery approaches. The review focuses on the advanced drug delivery methods for the treatment of CRC.

2 Drug delivery approaches in colorectal cancer

Along with traditional surgical treatment, various site-specific approaches are available for ensuring maximum availability of chemotherapeutics at cancer site. These approaches include oral delivery of colon-specific product such as colon tablet, colon capsules, and microspheres. These oral products prevent the drug release at upper gastrointestinal tract (GIT) while they allow at colonic area. Furthermore, numerous receptor-based targeting approaches are gaining popularity. Both active and passive targeting approaches are discussed here.

2.1 Oral colon-specific drug delivery for colorectal cancer

As discussed, these are the oral products that release maximum amount of chemotherapeutics at colon site and avoid drug release at upper GIT. There are several factors which should be taken under consideration while designing colon-specific oral formulations.

2.1.1 Factors affecting colon availability of therapeutics through oral route

Oral product intended to use for colon-specific delivery has to transit through the various parts of GIT. Whole GIT of an adult human can be up to 30 ft length and every part behaves entirely different in physiology to each other. Therefore these physiological parameters should be taken under consideration while designing oral product for colon-specific delivery [12]. Description of physiological parameters are as follows:

pH: The pH of GIT varies from 1.5 to 7.5 at different area. It increases from stomach to small intestine from 1.5 to 5 (depend on some other factors like feed condition and disease). At duodenum and jejunum, pH may rise up to 6 and 7.5, respectively. Now at ileocecal junction, this falls to 6.4 and remains in the range from 6.4 to 7.0 throughout the colon [13]. This wide change in pH should always be considered, which may affect the performance of oral product for colon delivery.

Transit time: Time taken by the oral product to reach into colon after transit through small intestine is a highly variable factor. The GI transit time may be influenced by several factors like feed condition, feed quantity, individual physiological difference, working/resting state, and size. Moreover, the normal transit time for oral product may vary from 4 to 12 h [14,15]. A therapeutic substance intended to deliver at colon should be designed in such a manner that delivery system should allow the release of therapeutic substance at desire site only.

Enzymatic degradation: Exposure of therapeutic substance to variety of enzymes from the mouth to the colon may cause serious damage to their therapeutic value. The majority of digestive enzymes such as amylase, maltase, pepsin, and trypsin involved in the digestion of food can also affect the susceptible drugs and dosage forms prepared for colon delivery [16].

Colonic Microflora: Approximately, 400 strains of different bacteria are present in human GIT for metabolic reactions such as hydrolysis, reduction, dehydroxylation, decarboxylation, and dealkylation [17,18]. Metabolic activities may also alter the stability and therapeutic value of active substance. Therefore, it is necessary to protect the colon-specific therapeutic product from the metabolic reactions by microflora.

Colonic absorption: The absorption of selected therapeutics may be delayed due to less availability of villi at colonic mucosa as compared to other parts of GIT and at the same time therapeutic substance are exposed to less absorption surface area into colon [12]. The lipid bilayer of colonocytes also play major role during drug absorption [19]. It is also reported that elevated viscosity of colonic content may reduce the diffusion of therapeutics substance, which may lead to poor drug absorption [20].

2.1.2 Approaches for oral colon delivery in colorectal cancer

Oral colon-targeted systems consisting of encapsulating drugs are developed with an objective to elude drug release in upper GIT and to target the release overtly in the colorectal area either to facilitate local delivery, particularly for CRC, or for systemic delivery. While developing the oral colon-targeted systems, the formulation scientist has to consider the physiological factors [12]. Cancer therapy involves delivery of high doses of anticancer drugs to achieve maximum efficacy; however, such high doses are liable to cause toxicity to normal cells or organs. Thus advanced delivery systems, capable of liberating drugs specifically to the target site are highly recommended. Utilizing the unique physiology of colon and advances in polymeric science, several colon targeted delivery systems have been designed as presented below:

2.1.2.1 pH-dependent systems

These systems are composed of pH responsive polymers, which solubilizes at pH of the colon (6.4–7), which is different from pH of the stomach and small intestine. The pH responsive polymers are insoluble at acidic pH of the stomach but solubilizes at colonic pH. Methacrylic resins also called as Eudragit® are the widely explored polymers for same purpose. Eudragit L, soluble at pH above than 6 is used for cecum targeting whereas Eudragit FS30D (soluble at pH above 7) and Eudragit S (soluble at pH above 7) are used for targeting of large intestine [15,21,22]. Using pH-responsive polymers, several dosage forms such as tablets, pellets, capsules, multiparticulates, microspheres, and nanoparticles have been designed to facilitate colon targeting. Alibolandi et al. [23] have demonstrated targeting of CRC by capecitabine-loaded microspheres composed of 1:2 ratio of Eudragit L100 and Eudragit S100 [24]. Microspheres prepared by mixture of EudragitP-4135F and Eudragit RS 100 loaded with 5-FU has shown promising results for the treatment of CRC [25]. In another study, Eudragit S100 coated, calcium pectinate beads demonstrated maximum delivery of 5-FU to colonic site, particularly to cecum and colon [26]. Although pH-dependent systems are widely used for colon targeting, their use for treatment of CRC is limited due to poor site specificity owing to similar pH, environment of colon, and lower small intestine. Further, the presence of gastrointestinal contents such as bile acid residues, carbon dioxide, fermentation products, and short chain fatty acids, reduces colonic pH and affect the performance of pH-dependent polymers.

2.1.2.2 Bacterially triggered systems

This system exploits the metabolic activity of occupant microflora in the colon to specifically release the drug to the colon. The two most important metabolic reactions of colonic bacteria are reduction and hydrolysis and based on these, different strategies for achieving colon targeting are designed and described in the following section:

2.1.2.3 Prodrugs

Colon-targeted prodrugs are designed in such a way that they are inactivated in the upper GIT and eventually undergo enzymatic transformation in the colon to release the drug [27]. Ability of colon microflora to metabolize azo compounds is widely utilized for developing prodrugs for colon targeting. Azo-based anticancer prodrugs of methotrexate, gemcitabine, and oxaliplatin were found to be effective for colon-targeted delivery [28,29]. Sauraj et al. has developed xylan-5-FU acetic acid conjugates as a prodrug for colon cancer treatment. The conjugates showed a potential increase in drug release with reduced cytotoxicity [30]. Inspired by the pH and enzymatic activity of colon, Gd3 +-doped mesoporous hydroxyapatite nanoparticles anchored with polyacrylic acid and chitosan were developed. Chitosan is an enzyme-sensitive polymer, degraded by β-glycosidase in the colon whereas polyacrylic acid is a pH-responsive polymer, which controls the drug release in the colon [31]. Variety of products such as glucose, galactose, and glucuronic acids have also been studied for the same purpose [27].

2.1.2.4 Hydrogels

The hydrogels protect the drug release in stomach due to a low degree of swelling in the acidic pH. As the hydrogels transit to lower GIT, the entrapped drug release due to increased degree of swelling hydrogel, and in colon the cross-links becomes accessible to enzymes. Recently, several types of hydrogels such as azo-containing dextran hydrogels, methacrylamide-based hydrogels [32], glycopolymer hydrogels [33], poly(vinylalcohol) hydrogels [34], chitosan hydrogels [35], and HPMC hydrogels [36] have been investigated for colon targeting.

2.1.2.5 Polysaccharides carriers

Natural polysaccharides such as xylan and pectin, due to their physicochemical characteristics are not digested in upper GIT, but are fermented in colon by resident bacteria. Thus polysaccharides became the primary choice of excipient for colon-specific targeting. Chitosan, pectin, guar gum, triglyceride esters, and mesoporous silica in form of microparticles or nanoparticles are used for delivery of chemotherapeutic agents for treatment of CRC [37,38].

2.1.2.6 Pressure-controlled systems

Pressure-controlled systems allow the release of drug from a system by utilizing enhanced intraluminal pressure of colon [39]. The newly designed pressure-controlled colon delivery capsules consist of capsular suppositories coated with a water-insoluble polymer. After administration the suppository melts at body temperature, however, ethyl cellulose coating prevents drug release and as a result the dosage form behaves like a balloon in stomach and small intestine. Increased intestinal pressure of colon causes rupturing of dosage form and release of drug in colon [40].

2.2 Nanocarriers-based targeting approaches for colorectal cancer

The development of the efficacious nanocarrier-based therapeutic strategies for CRC has gained popularity in last four decades. Most of the anticancer drugs cannot distinguish between normal and cancer cells, which cause poor efficacy and associated side effects. In this context, nanocarrier-based targeted drug delivery could be a choice of delivery system for improved efficacy and reduced side effects through passive or active mechanism. Several types of nanocarriers (e.g., polymeric nanoparticles, liposomes, micelles, quantum dotes, and dendrimers) have been reported for the therapeutic and diagnostic purpose. Several research publications have reported successful delivery of cancer therapeutics, including Si RNA [41], phytoconstituents [42], and chemotherapeutics [43]. Advantages of nanocarriers over the conventional treatment are due to the targeted delivery of therapeutics at CRC cell. Targeting may be achieved by the following ways.

2.2.1 *Passive targeting*

Passive targeting of loaded therapeutic content to the CRC cells may be achieved by exploiting tumor-associated pathophysiological changes. Enhanced permeation and further retention of nanoparticles at cancer cells are possible due to defective architecture of tumor vessels. It has been reported that endothelium space between blood vessels associated with cancer tissues are higher (100 nm to 2 μm) than the normal, thus it allows better permeation of size-specific drug-loaded nanocarriers to the cancer cells as compared to the conventional approach. Similarly, one more ground of passive targeting of nanocarriers is that cancer tissues may be associated with the poor lymphatic system, which leads to enhanced interstitial pressure at the center of cancer cells compared to their peripheries. Thus drug-loaded nanocarriers are retained for longer period of time at interstitial space [44].

Nanocarriers possess prolonged circulation in the systemic circulation by avoiding reticular endothelial system as compared to conventional delivery systems; thus it improves the therapeutic efficacy of loaded drug. Since loaded carrier remains available for prolonged period of time, they are ultimately taken up by the specific cancer cells. Prolonged circulation of nanocarriers can be achieved by opsonization (a process of making stealth nanocarriers), which is possible by coating or ligation of nanocarriers with polyethylene glycol [45].

2.2.2 *Active targeting*

Active targeting of nanocarriers involve molecular recognition of carriers specifically by the cancer cells thus having more chance of availability of therapeutic agents at cancer site only. Usually targeting molecules are ligated on the nanocarrier surface for molecular recognition and further uptake by cancer cells. Till date number of receptors and biomarkers have been identified, which are overexpressed on the CRC cells. A recent immunohistochemistry study revealed the overexpression of following receptors on CRC cell as compared to other cancer or normal cells [46]:

Carcinoembryonic antigen (CEA) receptor: These are most consistently overexpressed receptors in case of CRC. It has been reported that expressions of CEA in case of CRC reach up to 98.0%.

Tumor-associated glycoprotein-72 (TAG-72) receptor: These receptors are overexpressed on CRC cell up to 79%.

Folate receptor-α (FA): This receptor could also be a choice of targeting approach as FA is overexpressed up to 37.1% on CRC cells.

Epidermal growth factor receptor (EGFR): This receptor belongs to ErbB (erythroblastosis oncogene B)/HER (human epidermal growth factor receptor) family. It has been observed that overexpression of EGFR is 32.8% in case of CRC (Fig. 1).

2.2.3 *Novel carriers used in colorectal cancer treatment*

2.2.3.1 Nanoparticles

Over the past 50 years, many nanoparticles formulation loaded with anticancer drugs have come into consideration for the effective management of CRC. Initially, first generation nanoparticles were designed to target the disease passively, by taking advantage of physiological changes (enhanced permeation and retention effect offered by the vascular and lymphatic drainage of CRC) occurring during the disease. In later stage, next generation receptor-based active targeting came into consideration with selective targeting of drug-loaded nanoparticles into cancer cells only. Movement of nanoparticles was guided by the various ligands (monoclonal antibodies, peptides, and aptamers) attached on the nanoparticles surfaces for the purpose of cancer cell recognition.

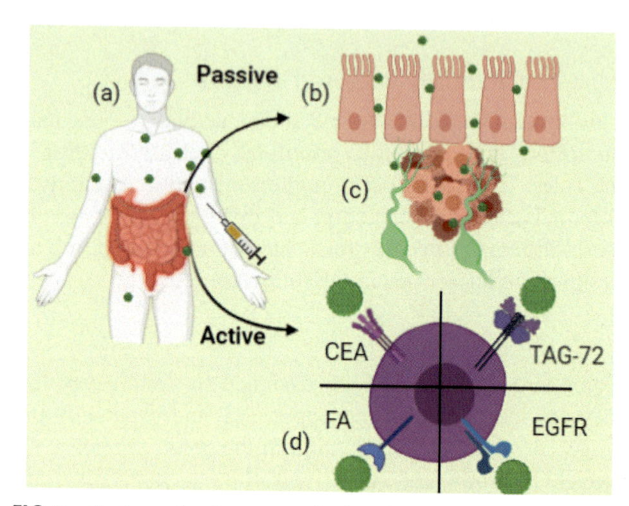

FIG. 1 Passive and active nanocarrier-based targeting approaches for effective treatment of colorectal cancer. (A) Nanocarriers ensure prolonged circulation of chemotherapeutics in systemic circulation ensuring more chance of drug targeting at desired site. (B) Enhanced permeation through space between blood vessels associated with cancer tissues. (C) Poor lymphatic system at cancer site allows prolonged retention of nanocarriers. (D) Receptor-based active targeting.

Fay et al. has demonstrated better uptake of Conatumumab (AMG 655)-coated nanoparticles by colorectal HCT116 cancer cells. They demonstrate that an encapsulated drug (camptothecin) was selectively delivered to HCT116 cancer cells for induction of apoptosis [47]. In a similar study by Abdelghany et al. hydrogel-based chitosan/alginate nanoparticles for the encapsulation of meso-Tetra (*N*-methyl-4-pyridyl) porphine tetra tosylate was prepared. Antibodies targeting to death receptor 5 were ligated on the prepared nanoparticles for selective uptake by the HCT116 [48]. Moreover, some more research related to the targeted nanoparticles encapsulated with anticancer drugs are also summarized in Table 1.

2.2.3.2 Liposome

These are nontoxic, biodegradable, or at least biocompatible spherical vesicles having one or more phospholipid bilayer with alternate internal and/or external aqueous phase where drugs are dissolved in any one of the phases. Liposomal drug delivery in cancer treatment has advantages over the nanoparticles as ligation of cancer-specific moiety over the surface of liposome is easy. Additionally, it also helps to increase drug internalization process with cancer cells. Various liposomal-based anticancer products (CPX-1, LE-SN38, and Thermodox) are under clinical trials [12]. So far, many studies have proved the capability of liposome for CRC cells targeting. One study focused on improved efficacy of SN-38 (poorly soluble active metabolite of CPT-11) through liposomal formulation. Therapeutic efficacy was evaluated against human colon cell lines (HT-29) and it was observed that there was significant improvement in growth inhibition of cancer cells in case of liposome [63]. In another study by Zalba et al., oxaliplatin-loaded liposome, which was ligated with Cetuximab, a monoclonal antibody for specific uptake

TABLE 1 Various nanocarriers used as targeting tools in colorectal cancer.

Formulation	Targeting approach	Therapeutics	Remarks	Reference
Poly(lactide-*co*-glycolide) nanoparticles	CD95/Apo-1 receptor antibody	Camptothecin	Enhanced cytotoxic activity in HCT116	[49]
Chitosan nanoparticles	Hyaluronic acid	Oxaliplatin	Higher uptake of drug at cancer site	[50]
Chitosan nanoparticles	Hyaluronic acid	5-Flurouracil	Higher uptake of nanoparticles in HT-29 colon cancer cells	[51]
Silica-coated iron oxide nanoparticles	Cetuximab conjugation	–	Enhanced uptake by HCT 116 cells and can be choice for diagnostic purpose	[52]
Guar Gum nanoparticles	Folic acid conjugation	Methotrexate	Better inhibition of Caco 2 cells	[53]
Silica nanoparticles for targeted drug delivery	Hyaluronic acid	Doxorubicin hydrochloride	Higher uptake of formulation by CD44-overexpressing HCT-116 cells	[54]
Liposome	Passive	Curcumin	Enhanced apoptosis in LoVo and Colo205 cells	[55]
Liposome	Humanized anti-EGFR monoclonal antibody	Doxorubicin	Improved cytotoxic effect on HCT116 cells	[56]
Liposome	EGFR monoclonal antibody	Oxaliplatin	Improved drug uptake	[57]
Liposome	Folic acid conjugation	5-Fluorouracil	Trigger necrosis in HT-29 cells and apoptotic pathway in HeLa cells. In vivo study revealed drug-loaded liposome reduce tumor volume	[58]
Liposome	Monoclonal antibodies against CEA	Adriamycin	Improved antitumor effect of conjugated liposome	[59]
Liposome	–	Antitumor-associated glycoprotein (TAG)-72	Effective localization of targeted liposome in LS174 T tumor tissues	[60]
Dendrimers	AS1411 aptamers	–	Inhibition of DNA replication and tumor cell growth	[61]
PAMAM dendrimer	Folic acid	Oxaliplatin	Decreased cellular viability and increased apoptosis	[62]

by the EGFR receptors overexpressed on CRC cells was developed. As result, increased uptake of targeted liposome was observed in CRC cells and further increased therapeutic efficacy in in vivo xenograft model [57].

2.2.3.3 Quantum dotes

These are tiny semiconductor particles having size range in nanometers and optical and electronic properties, which turns fluorescent from visible to infrared wavelengths upon excitation. Due to this property these particles are very useful for the diagnostic and imaging purpose in a variety of disease including CRC. Large number of studies have been done so far indicating the application of quantum dots as diagnostic tools in the CRC, which are guided by the certain ligands specifically taken up by the CRC cells. Carbary-Ganz et al. has prepared quantum dots targeted to vascular endothelial growth factor receptor 2 (VEGFR2). Formulations were conjugated to anti-VEGFR2 primary antibodies. Results of immunofluorescence/immunohistochemistry revealed that the prepared quantum dots were specifically accumulated at colon cancer in mice [64]. Similarly, Gazouli et al. also developed quantum dots conjugated with bevasizumab (for active targeting through vascular endothelial growth factor) for molecular imaging in CRC cell lines (DLD-1). The results of present study also revealed, simple, convenient, and noninvasive detection of overexpressed VEGF receptors on CRC cells [65].

In addition to the diagnostic application, quantum dots have proven their role as potential carrier of therapeutics in CRC. In a very recent study, curcumin-loaded quantum dots were prepared using Eudragit RS 100. Researchers found that prepared formulation show better inhibitory action on bacteria and CRC cells [66]. However, till date, only limited research have been conducted in this context, but it would be logical to say that similar to diagnostic application, receptor-based targeting of drug-loaded quantum dots may be a good alternative in the treatment of CRC.

2.2.3.4 Dendrimers

Unlike the polymeric nanoparticles, dendrimers have unique structure comprised of globular molecules made out of branched layers. Outer surface of dendrimers can be easily decorated with variety of functional groups, monoclonal antibodies, and many more structures. This property of dendrimers make them flexible for active targeting approaches [67].

Xie et al. demonstrate the application of polyamidoamine (PAMAM) dendrimers and multiple Sialyl Lewis X antibodies (aSlex)-conjugates for the specific targeting of CRC. Result indicates that the conjugates showed improved uptake by the HT29 cells up to 77.88%. In a similar study, Alibolandi et al. has developed camptothecin-loaded PAMAM dendrimer conjugated with S1411 antinucleolin aptamers for site-specific targeting to HT29 and C26 CRC cells. Further, the system was evaluated in C26 tumor-bearing mice and results of the study revealed better efficacy of therapeutic substance [23].

3 Conclusion

Due to inadequate early detection, CRC is a most life-threatening malignant disorders. Treatment of CRC with increased efficacy and reduced side effects are still challenging areas for researchers. Novel targeting approaches helps to ensure maximum drug availability at desired tissues. Some of the marketed available oral formulations are claiming maximum drug release at colon. Nanoformulations decorated with cell-specific ligands have shown next generation modalities providing cell-specific uptake of therapeutics. Some of them have reached advanced clinical trials phase and many more are in the pipelines. However, further research is required to encounter nanoformulations-related issues like stability, reproducibility, and large-scale production.

References

[1] Rawla P, Sunkara T, Barsouk A. Epidemiology of colorectal cancer: incidence, mortality, survival, and risk factors. Gastroenterol Rev 2019;14:89–103. https://doi.org/10.5114/pg.2018.81072.

[2] Ewing I, Hurley JJ, Josephides E, Millar A. The molecular genetics of colorectal cancer. Front Gastroenterol 2014;5:26–30. https://doi.org/10.1136/flgastro-2013-100329.

[3] Tariq K, Ghias K. Colorectal cancer carcinogenesis: a review of mechanisms. Cancer Biol Med 2016;13:120–35. https://doi.org/10.28092/j.issn.2095-3941.2015.0103.

[4] Dekker E, Tanis PJ, Vleugels JLA, Kasi PM, Wallace MB. Colorectal cancer. Lancet Lond Engl 2019;394:1467–80. https://doi.org/10.1016/S0140-6736(19)32319-0.

[5] Mármol I, Sánchez-de-Diego C, Pradilla Dieste A, Cerrada E, Rodriguez Yoldi M. Colorectal carcinoma: a general overview and future perspectives in colorectal Cancer. Int J Mol Sci 2017;18:197. https://doi.org/10.3390/ijms18010197.

[6] Bray F, Ferlay J, Soerjomataram I, Siegel RL, Torre LA, Jemal A. Global cancer statistics 2018: GLOBOCAN estimates of incidence and mortality worldwide for 36 cancers in 185 countries. CA Cancer J Clin 2018;68:394–424. https://doi.org/10.3322/caac.21492.

[7] Belali N, Wathoni N, Muchtaridi M. Advances in orally targeted drug delivery to colon. J Adv Pharm Technol Res 2019;10:100. https://doi.org/10.4103/japtr.JAPTR_26_19.

[8] Kolligs FT. Diagnostics and epidemiology of colorectal Cancer. Visc Med 2016;32:158–64. https://doi.org/10.1159/000446488.

[9] Nguyen HT, Duong H-Q. The molecular characteristics of colorectal cancer: implications for diagnosis and therapy. Oncol Lett 2018;16:9–18. https://doi.org/10.3892/ol.2018.8679.

[10] Kuipers EJ, Grady WV. Colorectal cancer. Nat Rev Dis Primers 2015;1:15065.

[11] Sachse C, Smith G, Wilkie MJ. A pharmacogenetic study to investigate the role of dietary carcinogenesis in the etiology of colorectal cancer. Carcinogenesis 2002;23(11):1839–50.

[12] Krishnaiah YSR, Khan MA. Strategies of targeting oral drug delivery systems to the colon and their potential use for the treatment of colorectal cancer. Pharm Dev Technol 2012;17:521–40. https://doi.org/10.3109/10837450.2012.696268.
[13] Friend DR. Colon-specific drug delivery. Adv Drug Deliv Rev 1991;7:149–99. https://doi.org/10.1016/0169-409X(91)90051-D. Gastrointestinal Tract as a Site for Drug Delivery.
[14] Reinholz J, Landfester K, Mailänder V. The challenges of oral drug delivery via nanocarriers. Drug Deliv 2018;25:1694–705. https://doi.org/10.1080/10717544.2018.1501119.
[15] Varum FJO, Merchant HA, Basit AW. Oral modified-release formulations in motion: the relationship between gastrointestinal transit and drug absorption. Int J Pharm 2010;395:26–36. https://doi.org/10.1016/j.ijpharm.2010.04.046.
[16] El Aidy S, van den Bogert B, Kleerebezem M. The small intestine microbiota, nutritional modulation and relevance for health. Curr Opin Biotechnol 2015;32:14–20. https://doi.org/10.1016/j.copbio.2014.09.005.
[17] Chourasia MK, Jain SK. Polysaccharides for colon targeted drug delivery. Drug Deliv 2004;11:129–48. https://doi.org/10.1080/10717540490280778.
[18] Linskens RK, Huijsdens XW, Savelkoul PH, Vandenbroucke-Grauls CM, Meuwissen SG. The bacterial flora in inflammatory bowel disease: current insights in pathogenesis and the influence of antibiotics and probiotics. Scand J Gastroenterol Suppl 2001;29–40. https://doi.org/10.1080/003655201753265082.
[19] Powell DW. Barrier function of epithelia. Am J Phys 1981;241:G275–88. https://doi.org/10.1152/ajpgi.1981.241.4.G275.
[20] Reppas C, Eleftheriou G, Macheras P, Symillides M, Dressman JB. Effect of elevated viscosity in the upper gastrointestinal tract on drug absorption in dogs. Eur J Pharm Sci Off J Eur Fed Pharm Sci 1998;6:131–9. https://doi.org/10.1016/s0928-0987(97)00077-8.
[21] Bak A, Ashford M, Brayden DJ. Local delivery of macromolecules to treat diseases associated with the colon. Adv Drug Deliv Rev 2018;136–137:2–27. https://doi.org/10.1016/j.addr.2018.10.009.
[22] Zhang T, Zhu G, Lu B, Peng Q. Oral nano-delivery systems for colon targeting therapy. Pharm Nanotechnol 2017;5. https://doi.org/10.2174/2211738505666170424122722.
[23] Alibolandi M, Taghdisi SM, Ramezani P, Hosseini Shamili F, Farzad SA, Abnous K, Ramezani M. Smart AS1411-aptamer conjugated pegylated PAMAM dendrimer for the superior delivery of camptothecin to colon adenocarcinoma in vitro and in vivo. Int J Pharm 2017;519:352–64. https://doi.org/10.1016/j.ijpharm.2017.01.044.
[24] Agarwal D, Ranawat MS, Chauhan CS, Kamble R. Formulation and charecterisation of colon targeted pH dependent microspheres of capecitabine for colorectal cancer. J Drug Deliv Ther 2013;3:215–22. https://doi.org/10.22270/jddt.v3i6.747.
[25] Lamprecht A, Yamamoto H, Takeuchi H, Kawashima Y. Microsphere design for the colonic delivery of 5-fluorouracil. J Control Release 2003;90:313–22. https://doi.org/10.1016/S0168-3659(03)00195-0.
[26] Jain A, Gupta Y, Jain SK. Potential of calcium pectinate beads for target specific drug release to colon. J Drug Target 2007;15:285–94. https://doi.org/10.1080/10611860601146134.
[27] Sinha VR, Kumria R. Colonic drug delivery: prodrug approach. Pharm Res 2001;18:557–64. https://doi.org/10.1023/a:1011033121528.
[28] Sharma R, Rawal RK, Gaba T, Singla N, Malhotra M, Matharoo S, Bhardwaj TR. Design, synthesis and ex vivo evaluation of colon-specific azo based prodrugs of anticancer agents. Bioorg Med Chem Lett 2013;23:5332–8. https://doi.org/10.1016/j.bmcl.2013.07.059.
[29] Van den Mooter G, Samyn C, Kinget R. Azo polymers for colon-specific drug delivery. II: Influence of the type of azo polymer on the degradation by intestinal microflora. Int J Pharm 1993;97:133–9. https://doi.org/10.1016/0378-5173(93)90133-Z.
[30] Sauraj, Kumar SU, Gopinath P, Negi YS. Synthesis and bio-evaluation of xylan-5-fluorouracil-1-acetic acid conjugates as prodrugs for colon cancer treatment. Carbohydr Polym 2017;157:1442–50. https://doi.org/10.1016/j.carbpol.2016.09.096.
[31] Song Q, Jia J, Niu X, Zheng C, Zhao H, Sun L, Zhang H, Wang L, Zhang Z, Zhang Y. An oral drug delivery system with programmed drug release and imaging properties for orthotopic colon cancer therapy. Nanoscale 2019;11:15958–70. https://doi.org/10.1039/c9nr03802g.
[32] Singh B, Sharma N, Chauhan N. Synthesis, characterization and swelling studies of pH responsive psyllium and methacrylamide based hydrogels for the use in colon specific drug delivery. Carbohydr Polym 2007;69:631–43. https://doi.org/10.1016/j.carbpol.2007.01.020.
[33] Mahkam M. New pH-sensitive glycopolymers for colon-specific drug delivery. Drug Deliv 2007;14:147–53. https://doi.org/10.1080/10717540601067745.
[34] Basak P, Adhikari B. Poly (vinyl alcohol) hydrogels for pH dependent colon targeted drug delivery. J Mater Sci Mater Med 2008;20:137. https://doi.org/10.1007/s10856-008-3496-0.
[35] Saboktakin MR, Tabatabaie RM, Maharramov A, Ramazanov MA. Synthesis and characterization of chitosan hydrogels containing 5-aminosalicylic acid Nanopendents for colon: specific drug delivery. J Pharm Sci 2010;99:4955–61. https://doi.org/10.1002/jps.22218.
[36] Davaran S, Rashidi MR, Khani A. Synthesis of chemically cross-linked hydroxypropyl methyl cellulose hydrogels and their application in controlled release of 5-amino salicylic acid. Drug Dev Ind Pharm 2007;33:881–7. https://doi.org/10.1080/03639040601150278.
[37] Basit AW, Short MD, McConnell EL. Microbiota-triggered colonic delivery: robustness of the polysaccharide approach in the fed state in man. J Drug Target 2009;17:64–71. https://doi.org/10.1080/10611860802455805.
[38] Jain A, Jain SK. Optimization of chitosan nanoparticles for colon tumors using experimental design methodology. Artif Cells Nanomed Biotechnol 2016;44:1917–26. https://doi.org/10.3109/21691401.2015.1111236.
[39] Barakat NS, Al-Suwayeh SA, Taha EI, Bakry Yassin AE. A new pressure-controlled colon delivery capsule for chronotherapeutic treatment of nocturnal asthma. J Drug Target 2011;19:365–72. https://doi.org/10.3109/1061186X.2010.504264.
[40] Hu Z, Kimura G, Mawatari S, Shimokawa T, Yoshikawa Y, Takada K. New preparation method of intestinal pressure-controlled colon delivery capsules by coating machine and evaluation in beagle dogs. J Control Release Off J Control Release Soc 1998;56:293–302. https://doi.org/10.1016/s0168-3659(98)00090-x.
[41] Rudzinski WE, Palacios A, Ahmed A, Lane MA, Aminabhavi TM. Targeted delivery of small interfering RNA to colon cancer cells using chitosan and PEGylated chitosan nanoparticles. Carbohydr Polym 2016;147:323–32. https://doi.org/10.1016/j.carbpol.2016.04.041.
[42] Khan T, Gurav P. PhytoNanotechnology: enhancing delivery of plant based anti-cancer drugs. Front Pharmacol 2018;8. https://doi.org/10.3389/fphar.2017.01002.
[43] Acharya S, Sahoo SK. PLGA nanoparticles containing various anticancer agents and tumour delivery by EPR effect. Adv Drug Deliv Rev 2011;63:170–83. https://doi.org/10.1016/j.addr.2010.10.008. EPR Effect Based Drug Design and Clinical Outlook for Enhanced Cancer Chemotherapy.

[44] Shah MR, Imran M, Ullah S. Nanocarriers for cancer diagnosis and targeted chemotherapy. Elsevier; 2019.
[45] Tila D, Ghasemi S, Yazdani-Arazi SN, Ghanbarzadeh S. Functional liposomes in the cancer-targeted drug delivery. J Biomater Appl 2015;30:3–16. https://doi.org/10.1177/0885328215578111.
[46] Tiernan JP, Perry SL, Verghese ET, West NP, Yeluri S, Jayne DG, Hughes TA. Carcinoembryonic antigen is the preferred biomarker for in vivo colorectal cancer targeting. Br J Cancer 2013;108:662–7. https://doi.org/10.1038/bjc.2012.605.
[47] Fay F, McLaughlin KM, Small DM, Fennell DA, Johnston PG, Longley DB, Scott CJ. Conatumumab (AMG 655) coated nanoparticles for targeted pro-apoptotic drug delivery. Biomaterials 2011;32:8645–53. https://doi.org/10.1016/j.biomaterials.2011.07.065.
[48] Abdelghany SM, Schmid D, Deacon J, Jaworski J, Fay F, McLaughlin KM, Gormley JA, Burrows JF, Longley DB, Donnelly RF, Scott CJ. Enhanced antitumor activity of the photosensitizer *meso* -tetra(*N*-methyl-4-pyridyl) Porphine tetra Tosylate through encapsulation in antibody-targeted chitosan/alginate nanoparticles. Biomacromolecules 2013;14:302–10. https://doi.org/10.1021/bm301858a.
[49] McCarron PA, Marouf WM, Quinn DJ, Fay F, Burden RE, Olwill SA, Scott CJ. Antibody targeting of Camptothecin-loaded PLGA nanoparticles to tumor cells. Bioconjug Chem 2008;19:1561–9. https://doi.org/10.1021/bc800057g.
[50] Jain A, Jain SK, Ganesh N, Barve J, Beg AM. Design and development of ligand-appended polysaccharidic nanoparticles for the delivery of oxaliplatin in colorectal cancer. Nanomed Nanotechnol Biol Med 2010;6:179–90. https://doi.org/10.1016/j.nano.2009.03.002.
[51] Jain A, Jain SK. In vitro and cell uptake studies for targeting of ligand anchored nanoparticles for colon tumors. Eur J Pharm Sci 2008;35:404–16. https://doi.org/10.1016/j.ejps.2008.08.008.
[52] Cho Y-S, Yoon T-J, Jang E-S, Soo Hong K, Young Lee S, Ran Kim O, Park C, Kim Y-J, Yi G-C, Chang K. Cetuximab-conjugated magneto-fluorescent silica nanoparticles for in vivo colon cancer targeting and imaging. Cancer Lett 2010;299:63–71. https://doi.org/10.1016/j.canlet.2010.08.004.
[53] Sharma M, Malik R, Verma A, Dwivedi P, Banoth GS, Pandey N, Sarkar J, Mishra PR, Dwivedi AK. Folic acid conjugated guar gum nanoparticles for targeting methotrexate to Colon Cancer. J Biomed Nanotechnol 2013;9:96–106. https://doi.org/10.1166/jbn.2013.1474.
[54] Yu M, Jambhrunkar S, Thorn P, Chen J, Gu W, Yu C. Hyaluronic acid modified mesoporous silica nanoparticles for targeted drug delivery to CD44-overexpressing cancer cells. Nanoscale 2013;5:178–83. https://doi.org/10.1039/C2NR32145A.
[55] Li L, Ahmed B, Mehta K, Kurzrock R. Liposomal curcumin with and without oxaliplatin: effects on cell growth, apoptosis, and angiogenesis in colorectal cancer. Mol Cancer Ther 2007;6:1276–82. https://doi.org/10.1158/1535-7163.MCT-06-0556.
[56] Mamot C, Ritschard R, Küng W, Park JW, Herrmann R, Rochlitz CF. EGFR-targeted immunoliposomes derived from the monoclonal antibody EMD72000 mediate specific and efficient drug delivery to a variety of colorectal cancer cells. J Drug Target 2006;14:215–23. https://doi.org/10.1080/10611860600691049.
[57] Zalba S, Contreras AM, Haeri A, ten Hagen TLM, Navarro I, Koning G, Garrido MJ. Cetuximab-oxaliplatin-liposomes for epidermal growth factor receptor targeted chemotherapy of colorectal cancer. J Control Release 2015;210:26–38. https://doi.org/10.1016/j.jconrel.2015.05.271.
[58] Handali S, Moghimipour E, Rezaei M, Ramezani Z, Kouchak M, Amini M, Angali KA, Saremy S, Dorkoosh FA. A novel 5-fluorouracil targeted delivery to colon cancer using folic acid conjugated liposomes. Biomed Pharmacother 2018;108:1259–73. https://doi.org/10.1016/j.biopha.2018.09.128.
[59] Konno H. Antitumor effect of adriamycin entrapped in liposomes conjugated with anti-human alpha-fetoprotein or anti-carcinoembryonic antigen monoclonal antibodies. Nihon Geka Gakkai Zasshi 1986;87:365–74.
[60] Kim KS, Lee YK, Kim JS, Koo KH, Hong HJ, Park YS. Targeted gene therapy of LS174 T human colon carcinoma by anti-TAG-72 immunoliposomes. Cancer Gene Ther 2008;15:331–40. https://doi.org/10.1038/cgt.2008.11.
[61] Fan X, Sun L, Wu Y, Zhang L, Yang Z. Bioactivity of 2′-deoxyinosine-incorporated aptamer AS 1411. Sci Rep 2016;6:25799. https://doi.org/10.1038/srep25799.
[62] Narmani A, Kamali M, Amini B, Salimi A, Panahi Y. Targeting delivery of oxaliplatin with smart PEG-modified PAMAM G4 to colorectal cell line: in vitro studies. Process Biochem 2018;69:178–87. https://doi.org/10.1016/j.procbio.2018.01.014.
[63] Lei S, Chien P-Y, Sheikh S, Zhang A, Ali S, Ahmad I. Enhanced therapeutic efficacy of a novel liposome-based formulation of SN-38 against human tumor models in SCID mice: anticancer. Drugs 2004;15:773–8. https://doi.org/10.1097/00001813-200409000-00006.
[64] Carbary-Ganz JL, Barton JK, Utzinger U. Quantum dots targeted to vascular endothelial growth factor receptor 2 as a contrast agent for the detection of colorectal cancer. J Biomed Opt 2014;19:086003. https://doi.org/10.1117/1.JBO.19.8.086003.
[65] Gazouli M, Bouziotis P, Lyberopoulou A, Ikonomopoulos J, Papalois A, Anagnou NP, Efstathopoulos EP. Quantum dots-bevacizumab complexes for in vivo imaging of tumors. Vivo Athens Greece 2014;28:1091–5.
[66] Khan FA, Lammari N, Muhammad Siar AS, Alkhater KM, Asiri S, Akhtar S, Almansour I, Alamoudi W, Haroun W, Louaer W, Meniai AH, Elaissari A. Quantum dots encapsulated with curcumin inhibit the growth of colon cancer, breast cancer and bacterial cells. Nanomedicine 2020;15:969–80. https://doi.org/10.2217/nnm-2019-0429.
[67] Carvalho MR, Reis RL, Oliveira JM. Dendrimer nanoparticles for colorectal cancer applications. J Mat Chem 2020;8:1128–38.

Chapter 22

Current strategies in targeted anticancer drug delivery systems to brain

Ratnali Bania[a], Pobitra Borah[a,k], Satyendra Deka[a], Lina A. Dahabiyeh[b], Vinayak Singh[c,d], Nizar A. Al-Shar'i[e], Anroop B. Nair[f], Manoj Goyal[g], Katharigatta N. Venugopala[f,h], Rakesh Kumar Tekade[i], and Pran Kishore Deb[j]

[a]*Pratiksha Institute of Pharmaceutical Sciences, Guwahati, Assam, India,* [b]*Department of Pharmaceutical Sciences, School of Pharmacy, The University of Jordan, Amman, Jordan,* [c]*Drug Discovery and Development Centre (H3D), University of Cape Town, Rondebosch, South Africa,* [d]*South African Medical Research Council Drug Discovery and Development Research Unit, Department of Chemistry and Institute of Infectious Disease and Molecular Medicine, University of Cape Town, Rondebosch, South Africa,* [e]*Department of Medicinal Chemistry and Pharmacognosy, Faculty of Pharmacy, Jordan University of Science and Technology, Irbid, Jordan,* [f]*Department of Pharmaceutical Sciences, College of Clinical Pharmacy, King Faisal University, Al-Ahsa, Saudi Arabia,* [g]*Department of Anesthesia Technology, College of Applied Medical Sciences in Jubail, Imam Abdulrahman bin Faisal University, Dammam, Saudi Arabia,* [h]*Department of Biotechnology and Food Technology, Durban University of Technology, Durban, South Africa,* [i]*National Institute of Pharmaceutical Education and Research–Ahmedabad (NIPER-A), Gandhinagar, Gujarat, India,* [j]*Department of Pharmaceutical Sciences, Faculty of Pharmacy, Philadelphia University, Amman, Jordan,* [k]*School of Pharmacy, Graphic Era Hill University, Dehradun, Uttarakhand, India*

1 Introduction

In 1873, brain cancer was first observed by Gupta Longati (a Russian scientist) in a deceased person who purportedly died of a benign tumor [1]. Brain cancer is a malignant form of tumor that either originated within the brain (referred to as primary) or is resulting from the brain tissue invasion as a consequence of metastasis (referred to as secondary). Generally, primary brain tumors are meningiomas, gliomas, vestibular schwannomas, pituitary adenomas, and medulloblastomas, a prototype of primitive neuroectodermal tumors. The term glioma refers to oligodendrogliomas, glioblastomas, astrocytomas, and ependymomas. Gliomas originate from neural progenitor cells, stem cells, or from dedifferentiated mature neural cells that transforms into malignant stem cells. Secondary or metastatic brain tumors are the consequence of the metastatic spreading of cancerous cells in other parts of the body [2]. Secondary tumors are more common in adults, whereas primary tumors are common in children [3].

Currently, the most commonly used methods for the treatment of brain cancer comprise either of surgery, radiation and chemotherapy, or a combination of these therapies. Although, surgery is recommended for the treatment of brain tumor patients, most tumor cells spread into the normal brain tissue, making it quite impossible for complete removal. Radiation therapy destroys the cancer cell using high-frequency X-rays. On the other hand, chemotherapy includes various drugs that kill tumor cells [4]. Nevertheless, diagnosis and treatment of brain cancer are very difficult due to the presence of different barriers like blood-brain barrier (BBB) and blood-cerebrospinal fluid barrier (BCSFB). Effective treatment of brain tumor greatly depends on the ability to sustain the necessary amount of drugs at the required site of tumor, which in turn can stop the spreading to other tissues [5]. Both BBB and BCSFB not only provide protection from foreign materials, toxins, and pathogens but also maintain the brain homeostasis by facilitating the transport of essential nutrients, ions, proteins, and metabolites. BCSFB is a barrier between cerebrospinal fluid (CSF) and blood circulation. Choroid plexus is the main component of BCSFB and acts as an immunological, physical, and enzymatic barrier that aids in the transport, metabolism, and signaling functions of a drug. The gap junction present in the choroid plexus limits the permeation through BCSFB, but to a lower extent than the tight junction of BBB. Thus, BCSFB is comparatively more permeable to the drugs and other substances when compared with BBB [6]. The BBB comprises of pericytes, astrocytes, basement membrane, end feet and a continuous endothelial layer linked through tight junctions. This complex CNS barrier regulates and limits the systemic delivery of desired therapeutics to the brain. More than 98% of small molecules and almost all large molecules are not able to penetrate the BBB [7].

Over the years, different methods are being investigated to ensure effective drug delivery to the brain. These include

Advanced Drug Delivery Systems in the Management of Cancer. https://doi.org/10.1016/B978-0-323-85503-7.00038-9

temporary BBB disruption, transport using endogenous transcytosis systems, modification of delivery agents, and receptor-mediated transcytosis. Exploration of nanoparticles is another approach to overcome the difficulties in brain cancer treatment [3]. The smaller size (1–100 nm) and the surface modifications of nanoparticles with different polymers, lipid and target specific ligands enhance drug transport through BBB [6, 8, 9]. This chapter discusses the rationale and the challenges associated with drug delivery across the brain, with an emphasis on the current remarkable advances in the targeted-drug delivery systems designed for brain cancer treatment.

2 Physiology of blood-brain barrier (BBB)

BBB is a unique, selective permeability barrier mainly formed by the endothelial cells, astrocytes, pericytes, and basement membrane surrounding the capillaries (Fig. 1). The difference between the cerebral endothelial cells from other endothelial cells are attributed to the lack of fenestrate, tight intracellular junctions, reduced vesicular transport, and excess mitochondrial number. The endothelial layer is entirely covered by a basement membrane consisting of fibronectin, type IV collagen, and laminin. The basement membrane embeds the pericytes covering approximately 20%–30% of endothelial lining [10], while basal lamina is surrounded by the astrocytic end-feet. The astrocyte end-feet process forms about 99% of the CNS capillaries, still they do not necessarily contribute to the physiology of the BBB [11–13]. They are very important for BBB integrity and blood flow to the brain. That is why if any damage occurs, they are readily renewed. Pericytes are known to mediate the inflammatory processes, regulate the physiology of the brain endothelium, and stimulate capillary-like structure formation. Pericytes also regulate the expression pattern of BBB-specific genes in endothelial cells, thus reducing the endothelial transcytosis process [13, 14].

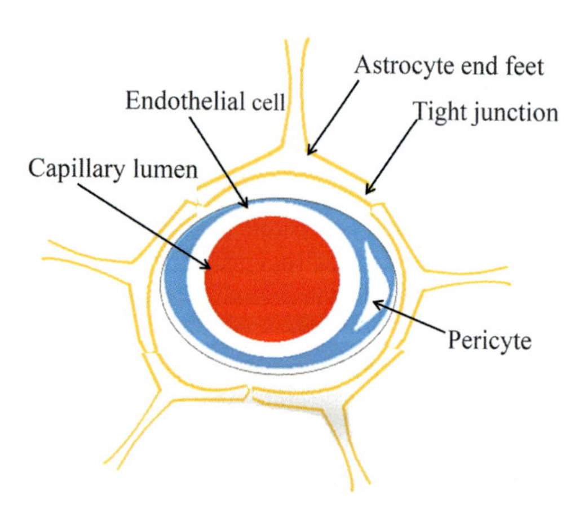

FIG. 1 The structure of blood brain barrier.

Astrocytes cover the outside of the pericytes and basement membrane [10]. The endings of the astrocyte appendices appear as cap-like structures, called as end-feet (Fig. 1). Astrocytes are tightly attached to the neurons and capillaries opposite to each other. They have a crucial role in the regulation and expression of a range of proteins within BBB. Astrocytes can produce some cytokines and growth factors. These factors contribute to the phenotyping of endothelial cells of the BBB, leading to tight junctions [15].

The tight junction is formed mainly due to the interaction of transmembrane proteins like occludins, claudins and junctional adhesion molecules (JAMs). They effectively enclose the paracellular transport pathway by dividing the endothelial membranes into luminal (blood) and abluminal (brain) sides, respectively [11]. Claudins are the integral part of tight junctions that forms the primary fusion by binding with the other claudins on adjacent endothelial linings. Occludin is the second component of the tight junctions having four transmembrane regions, and the C-terminal in the cytoplasm interacts directly with the zonula occludens (ZO). JAMs are the third group of protein that is involved in the tight junction formation. Apart from the tight junctions, capillary endothelial linings of the brain are also connected by adherents junctions situated near the basolateral surface of the endothelium and comprises cadherins [10, 14, 16]. Generally, the nonbrain capillaries possess fenestrations within the endothelial linings that allow the solutes to move readily by passive diffusion. On the other hand, the brain capillaries contain tight junctions rather than fenestrations, offering a high electrical resistance ($2000\,\Omega\,cm^2$) across intraparenchymal endothelial cells [10, 14].

Basement membrane covers the capillary tube and pericytes, and connects with astrocyte end-feet. It is about 30 to 40 nm thick and is composed of collagen type IV, fibronectin, heparin sulfate proteoglycans, and laminins [15]. The basement membrane can be divided into a vascular and parenchymal membrane. The vascular membrane is composed of α 4 and α 5 laminins, and is secreted by vascular pericytes and endothelial cells. The parenchymal membrane contains α 1 and α 2 laminins and is secreted by astrocytes [10].

3 Rationale of a brain-targeted anticancer drug delivery system

A targeted drug delivery system carries the therapeutic agents and delivers them to the site of action at the required amount. It prevents distribution of drugs to other tissues and thereby, avoids unnecessary toxicity. As brain tumor cells spread to the surrounding tissues quickly, a potential target-based drug delivery system is required to minimize the tumor spreading and reduce the toxic effects by delivering the desired drug to the site. To assure effective targeted drug delivery, various factors must be taken into consideration including the barriers and the microenvironment of the tumor cell [17]. For brain targeted drug delivery, carriers have to bypass the BBB, BCSFB and blood-brain tumor barrier (BBTB). BBTB is a barrier between the capillary vessels and brain tumor that restricts the release of hydrophilic drugs and antitumor agents into the brain-tumor site. BBTB is formed when the tumor cells grow up to a certain level damaging the BBB. In the initial stage of the tumor, the function of BBTB is quite similar to that of BBB. With tumor growth, the blood-brain tumor barrier (BBTB) becomes permeable and demonstrates different vascular functions. Although BBTB becomes permeable, still, it remains a barrier for the targeted delivery of required amount of therapeutics into the brain since the permeation is limited across the BBTB [18]. Passive and active targeting are the two main approaches used to deliver therapeutics to the required targeted site. Passive targeting depends on the properties of carriers (e.g., size and circulation time) and the tumor biology (vascularity and leakiness). Nanoparticles and modified nanoparticles are utilized for passive targeting. Doxil and Caelyx are examples of marketed nanocarriers used as passive tumor targeted systems [19].

Active targeting increases the efficiency of passive targeting to deliver the carriers at a more specific site. Usually, active targeting is achieved by surface alteration of nanoparticles by adding ligands capable of conjugating to the receptors located in the tumor cells. This approach increases the attraction of the nanoparticles toward the surface of tumor cells, and thus enhances drug penetration. Among the different receptors, transferrin receptor with nanoparticles or other carriers can easily reach the environment of brain tumor. Transferrin conjugates with the nanoparticle to target tumor cells that demonstrates transferrin receptor-mediated endocytosis on the membrane. This type of targeting enhances the uptake of drugs as compared with nonconjugated nanoparticles [19].

4 Transport across the BBB

Paracellular and transcellular pathways are the two primary transport mechanisms for crossing the BBB. In paracellular mechanism, substances must pass through the endothelial cells, while transcellular transport utilizes the luminal side of the endothelial cell. After that, substances cross the cytoplasm and pass through the abluminal surface of the endothelium into the brain interstitium [7]. Generally, hydrophilic drugs pass through the BBB by paracellular pathway, mainly through the diffusion process, while lipophilic substances follow the transcellular pathway. Along with these two pathways, some active transport mechanisms can also assist in BBB permeation, such as carrier-mediated transport, receptor-mediated transport, and adsorptive transcytosis (Fig. 2) [6, 20].

4.1 Paracellular pathway

In the paracellular pathway, the transient opening of the BBB can enhance the hydrophilic drug delivery into the brain tissue either by the cleft or the tight junction opening.

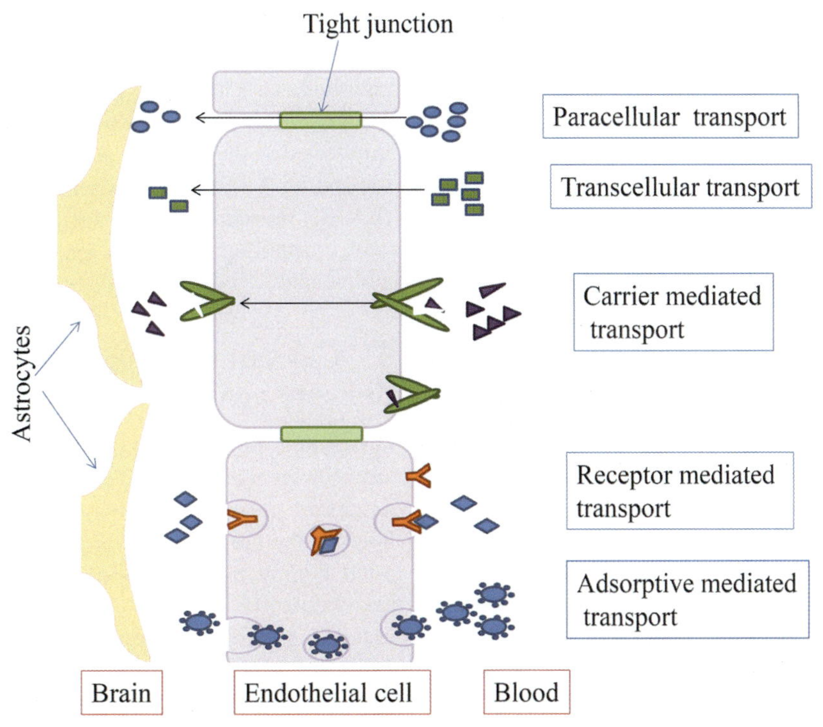

FIG. 2 Drug transport through BBB by different pathways.

Intracarotid administration of hyperosmotic mannitol provides an osmotic shock that leads to the shrinkage of endothelial cells and opens the tight junctions. It can also increase the concentration of subsequently administered drugs to a therapeutic level in the brain. Physical methods including utilization of electric as well as magnetic fields can enhance the drug uptake into the brain. Bradykinin derivative can also open the tight junction of BBB and facilitate drug delivery to the brain. However, those approaches are considered nonpatient-friendly and relatively expensive. Moreover, following BBB disruption and neuron damage, unwanted blood components might be transported into the brain, which may increase the chance of tumor dissemination [21].

4.2 Lipophilic diffusion pathway

Agents like nicotine, benzodiazepine and alcohol can easily enter the brain via the lipophilic diffusion pathway. Depending on lipophilicity, molecular weight, and charge, the molecules can have the ability to pass through the BBB passively. Studies suggest that modification of drugs or drug incorporation into the lipid carrier can increase the drug uptake into the brain. Antioxidants with pyrrolopyrimidines modification can also increase their capacity to access the target cells in the central nervous system (CNS). Although, these types of modifications boost the brain drug delivery, however, they often alter the therapeutic functions of the drug. Covalent attachment of 1-methyl-1,4-dihydronicotinate with a hydroxymethyl group can also increase the ganciclovir (an antiviral) delivery to the brain. In addition, higher drug lipophilicity tends to increase the susceptibility for P-glycoprotein (P-gp) efflux pump [21].

4.3 Carrier-mediated transport

The brain requires a specific amount of several nutrients and substances such as insulin, glucose, hormones, and low-density lipoprotein (LDL) for its survival and proper functioning. With the help of some specific receptors or transporters at the BBB, these substances can be transported into the brain. Modified drugs with carriers have the advantage of crossing the BBB. These physiological methodologies are considered as successful drug delivery approaches to the brain. Specific carriers expressed on both luminal and basolateral surfaces of the endothelium have been used to allow small molecules and peptides to cross into the brain. Structurally similar nutrients can be transported by different types of nutrient transporters. Until now, almost eight different carriers or transporters have been identified. Modified drugs can pass through the specific carrier-mediated transcytosis by mimicking the endogenous substrates of these transporters [22]. Specific essential factors like the kinetics, structural transporter binding, and therapeutic drug manipulation must be considered while utilizing the transporter for brain delivery [21].

4.4 Receptor-mediated transport

Some important molecules can be delivered into the brain through specific receptors, instead of transporters. The receptor-mediated transport includes three different steps, i.e., receptor-mediated endocytosis of the molecules at the luminal surface of the endothelium, followed by transport across the cytoplasm, and finally exocytosis at the abluminal surface of the brain endothelium [22]. A ligand can form a ligand-receptor complex by binding with a particular receptor. Through the receptor-mediated endocytosis, the complex that enters the endothelial cytoplasm releases the ligand-bound drug to the abluminal surface through exocytosis [23]. Ligands can be either natural or synthetic compounds. Generally, peptides and antibodies can bind with specific receptors at BBB. Examples of some receptors are insulin receptor, transferrin receptor, and LDL receptor 1/2 (LRP1/LRP 2) [21, 23].

4.5 Adsorptive mediated transport (AMT)

Adsorptive mediated transport (AMT) is the only pathway for the macromolecules like proteins, antibodies, and nanoparticles to cross the brain. The primary principle of AMT is the electrostatic interaction between the positive ligands (albumin, bovine serum albumin) and the negative glycoprotein at the endothelial cells of the brain. At a normal pH condition, the cerebral endothelium luminal surface and the surrounding basement membrane offers a suitable environment with a negative charge to deliver positively charged drugs and drug carriers. Different transporters are implicated in the AMT to the brain. These include glucose transporter 1 (GLUT1), excitatory amino acid transporter (EAAT), monocarboxylate transporter 1 (MCT1), L-amino acid transporter 1 (LAT1), organic cations, and cationic amino acid [21, 23].

5 Current strategies to enhance drug delivery against brain cancer

Different types of invasive and noninvasive techniques are utilized for drug delivery to the brain. Invasive techniques are performed by disrupting the BBB or by making holes in the brain. Various noninvasive techniques are applied to allow the distribution of drugs into the brain using the circulatory system. Some strategies implemented for brain drug delivery undergoing clinical trials are given in Table 1. Different invasive and noninvasive strategies are shown in Fig. 3 and a few important techniques are described below.

TABLE 1 Strategies for brain cancer drug delivery under clinical trials.

Strategies	Drug	Product	Study	Trial phase	CT identifier code
Intraventricular infusion	Methotrexate, etoposide	–	• Determination of the safety of methotrexate and etoposide infusions into the fourth ventricle of the brain in patients with recurring malignant posterior fossa brain tumors	Phase I	NCT02905110
	Topotecan hydrochloride	–	• Determination of maximum tolerated dose and toxicity of intrathecal or intraventricular topotecan in refractory leptomeningeal metastatic cancer patient • Study of pharmacokinetics and antitumor activity		NCT00025311
Convection-enhanced delivery	Radioactive iodine-labeled monoclonal antibody omburtamab	–	• Determination of maximum tolerated dose and toxicity profile	Phase I	NCT01502917
	D2C7-IT	–	• Determination of maximum tolerated dose and/or suggested dose of D2C7-IT (D2C7 Immunotoxin) when administered by CED to regular grade III and IV malignant glioma patients	Phase I	NCT02303678
	Topotecan	–	• Evaluation of topotecan distribution in brain tissue and determination of safety, toxicity and tolerance after administration by CED	Phase I	NCT03927274
Focused ultrasound	ExAblate BBBD	–	• Evaluation of safety and feasibility of utilizing the ExAblate Model 4000 Type 2 as a device to disrupt the BBB in brain and breast cancer patients	–	NCT03714243
	NaviFUS system	–	• Evaluation of safety and determination of tolerance ultrasound dose for temporary disruption of BBB by applying the NaviFUS System in Glioblastoma Multiforme patients	–	NCT03626896
Liposome	Doxorubicin HCl	Lipo-Dox	• Evaluation of the doxorubicin liposome efficacy in treatment of refractory solid tumors in children • Determination of tolerance and toxicity	Phase I	NCT00019630
	Vincristine	Marqibo	• Determination of maximum tolerated dose, toxicity profile, dose-limiting toxicities, and pharmacokinetics profile of Marqibo in children and adult patient hematologic malignancies receiving	Phase I	NCT01222780

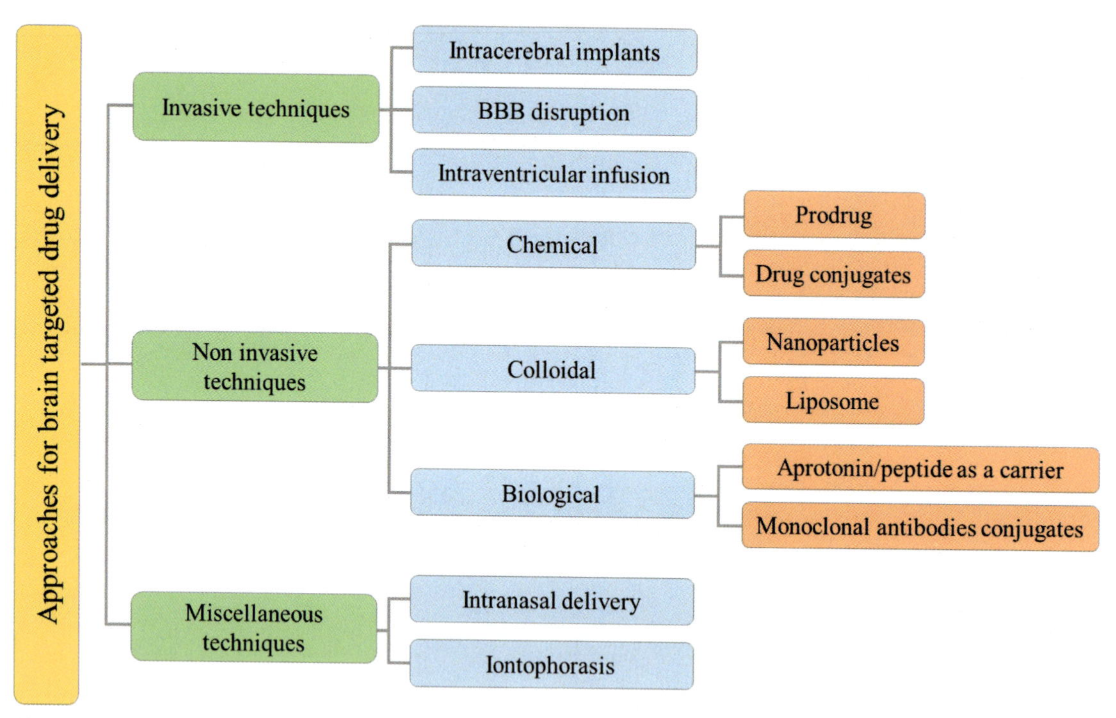

FIG. 3 Different approaches for drug delivery to brain.

5.1 Intracerebral implants

These are reservoirs of drugs loaded in a polymeric matrix. Polymers allow controlled release of the drug for a specific duration, prevent drug degradation, and decrease systemic side-effect. Different biodegradable and nonbiodegradable polymers are used for the development of the implant. Nonbiodegradable polymers, such as cellulose acetate, ethylene-vinyl acetate and polymethyl methacrylate are used for the preparation of solid implants. They follow zero-order kinetics and control drug release for a prolonged duration. However, they suffer from certain disadvantages like pain, discomfort, chances of infection, and higher cost during the surgical implantation and explantation of the implant. Various biodegradable polymers like polycarbonates, polyphosphate esters, polyanhydrides, polyphosphazenes, natural, and synthetic polyamides are also utilized for the preparation of implants for transporting different drugs with peptides and proteins [24].

The drug loaded polymeric implants administered intracranially can bypass the BBB and deliver a drug to the required site in a controlled rate. In this delivery system, the mechanism of drug distribution in brain is diffusion, which is altered with the changes of implantation distance. Therefore, implants are placed in required site only to obtain better effectiveness. Intracerebral implants are mainly used in case of malignant gliomas [25].

5.2 BBB disruption

BBB disruption results in the leakage of the tight junction, which permits the release of different anticancer drug in patients with cerebral lymphoma, malignant glioma, and disseminated CNS germ cell tumors. This method is associated with decreased mortality and morbidity rate as compared with systemic drug delivery [25]. The detailed description of the various techniques implemented for BBB disruption is given below.

5.2.1 Osmotic disruption

Temporary osmotic BBB disruption can be done by using some osmotic agents. Mannitol is the most commonly utilized osmotic agent. Mannitol causes shrinkage of endothelial cell by losing water molecules from endothelial matrix. Due to the shrinkage of endothelial cells tight junctions are opened resulting in passive diffusion of larger molecules through the BBB. The tight junction becomes porous and permeable to the exogenous substances [25]. However, several adverse events have been linked to the use of mannitol such as brain damage, alteration in glucose uptake, abnormal neural function, appearance of heat shock proteins, microembolisms, and passage of plasma proteins. Some other osmotic agents that can also be used include borneol, polydixylitol, and lyophosphatic acid [26].

5.2.2 Chemical disruption

A variety of chemical agents or vasoactive derivatives have been utilized to disrupt the BBB e.g., leukotriene C_4, interleukin-2 and bradykinin. Bradykinin is a common agent for the disruption of BBB because of its low concentration requirement. Bradykinin is given intraarterially which increases the BBB permeation in the brain tumor area. On endothelial cells, bradykinin binds to the B_2 receptor which temporarily increases cytosolic Ca^{2+}. Increased Ca^{2+} level stimulates nitric oxide synthase enzyme that is highly abundant in tumor vasculature. Nitric oxide causes vasodilation which leads to enhanced vascular permeation. RMP-7, a synthetic bradykinin analogue showed similar property with higher potency and specificity toward B_2 receptor. When compared with bradykinin, RMP-7 can be administered intravenously and shows a longer half-life and resistance to degradation [7].

5.3 Focused ultrasound (FUS)

Focused ultrasound (FUS) is a significant noninvasive technique, in which ultrasound is applied to induce temporary disruption of BBB. FUS enhances the cerebrovascular permeation by generating shear stress in cells or by activation of the signal pathways implicated in the regulation of permeation. Microbubbles, a contrast agent, can be included in FUS to decrease the ultrasound energy needed for the disruption of the BBB. Microbubbles oscillate when it passes through the ultrasound frequency. The oscillation of the microbubbles causes automatic stimulation of the blood vessels that lead to the temporary BBB disruption and increased permeation of the BBB with minimal side effects [27]. Microbubbles FUS have been utilized for the delivery of different therapeutic molecules e.g., temozolomide, doxorubicin, and methotrexate, small interfering RNA, and stem cells [25, 28]. The safety and efficacy of FUS depend on various parameters such as sonication time, the concentration of microbubble, pulse length and frequency, emission and pressure [26]. FUS combined with other targeting approaches tends to increase the accumulation of drugs and nanoparticles. Additionally, FUS modified with targeting ligands enhances the targeted delivery in the brain [29].

5.4 Intraventricular infusion

In intraventricular infusion method, therapeutic agents can be infused intraventricularly utilizing an Ommaya reservoir. Ommaya comprises a plastic reservoir that is implanted subcutaneously in the scalp and linked by an outlet catheter to the ventricles in the brain. The solution of therapeutic agents can be subcutaneously injected into the implanted reservoir, which is delivered by compressing the reservoir manually through the scalp. Although intraventricular infusion is considered as one of the important methods for the treatment of brain tumor, this method is associated with infection in brain cells, catheter obstruction, and insufficient drug delivery [30, 31].

5.5 Convection-enhanced delivery (CED)

CED is a new technique that enables continuous delivery of the infused drugs via intra-tumoral or intraparenchymal catheters from hours to days. Drug release depends on the hydrostatic pressure gradient to reach the tumor cells by convective flow through the parenchyma [32]. The CED method is utilized mainly for large molecular weight therapeutic agents which have lower penetration across the BBB and cause systemic toxicity. Some parameters like infusate characteristics (surface properties, molecular weight and tissue affinity), infusion parameters (cannula size, volume, rate of infusion, and time duration), and tissue properties (interstitial fluid pressure, tissue density, vascularity, extracellular space, and interstitial fluid pressure) can affect the CED volume of distribution [25]. CED is utilized for the delivery of different agents like antisense oligonucleotides, monoclonal antibodies, and viral vectors. CED is also used in antibody-mediated therapies and immunotherapies. CED has some advantages like delivery of a high amount of drug into the brain, targeted effect, less systemic side-effects and steady drug concentration. Whereas, disadvantages include the need for longer infusion time, irregular drug distribution, high intracranial pressure and local toxicity [30, 32].

5.6 Peptide-based drug delivery

The peptide can easily cross the BBB by adsorption or receptor-based transcytosis processes. Therapeutic agents are linked to a peptide that is occupied by a particular receptor to undergo endocytosis. Doxorubicin conjugated with small peptide showed increased brain delivery as compared with doxorubicin alone. Similarly, small peptides like AngioPep-2 showed increased transport and release of small molecules through the BBB by LRP1 [33, 34].

5.7 Intranasal delivery to the brain

In this delivery method, the drug is transported through the nasal cavity to the CSF and olfactory bulb via the olfactory neuron. The olfactory bulb offers fast nose-to-brain delivery, where therapeutic agents paracellularly cross the olfactory epithelium into the perineural space. Later on, the therapeutic agent goes to the subarachnoid space and CSF from the perineural space. From the CSF, the drug is directly absorbed by the brain tissue. The drug uptake in brain tissues and CSF depends on the molecular weight as well as the drug lipophilicity. Intranasal delivery has various advantages e.g., larger surface area for drug absorption, highly vascular submucosa and lymphatic system, noninvasiveness, reduced risk of infection and disease transmission, faster absorption, rapid onset of action, and ease of self-administration. However, it suffers from high mucociliary clearance, limited to potent drugs only, minimum volumes of drug can be administered, lower permeation of hydrophilic drugs, and decreased CNS delivery efficiency of proteins [25, 35]. To prevent enzymatic drug degradation and improve the efficacy of drug delivery, nanoparticles are utilized in intranasal delivery [29].

5.8 Colloidal drug-carrier systems

Colloidal carriers are mainly utilized to enhance the efficacy of specific brain tissue targeting by either increasing the bioavailability (due to easy permeation) or by protecting against enzymatic degradation. In a colloidal system, some therapeutic agents are encapsulated with lipid or polymers, so that they can easily penetrate the BBB [36]. Liposomes, polymeric nanoparticles, lipid nanoparticles, nanoemulsions, micelles, dendrimers, and quantum dots are the colloidal nanosized drug delivery system. Nanoparticles can be delivered to organs through the blood supply because of their smaller size. Additionally, surface modifications can also be done using different targeted ligands to assist easy permeation through the BBB to achieve better efficacy [37].

6 Nanoparticle as an emerging strategy for delivery across the BBB

The use of nanoparticles is an advanced technique for the delivery of anticancer drug molecules directly into the brain because of their desired properties like small size, increased bioavailability, enhanced drug solubility, and controlled release of drug at a specific site. The nanoparticle's surfaces are modified with different chemicals or ligands to enhance the BBB permeability. Different nanoparticles are available depending on the method of preparation, excipients used, release properties, and drug loading [5–7]. Some of the important nanoparticles are discussed below and are shown in Fig. 4.

Transportation of nanoparticles across the BBB involve different mechanisms, which include, binding of nanoparticles to the blood vessels on brain, surfactants used to fluidize the BBB endothelium, the opening of endothelium tight junctions, adsorptive-mediated and receptor-mediated transcytosis through the brain endothelial cells, endocytosis of endothelial cells in brain blood vessel, and glycoprotein blockage in the brain endothelial cells [38].

Numerous receptors in the BBB have also been considered for brain targeted drug delivery, e.g.; transferrin, insulin, glutathione, LDL, LRP-1/2 and angiopep-2 peptide [38, 39]. Therapeutic agents conjugated with the ligands for such receptors, e.g., peptide mimetic monoclonal antibodies (mAbs), have been broadly used to target the nanoparticles. The nanoparticles diffuse into the extracellular space of the brain after transcytosis. Nanoparticles can be eliminated from the brain by the cellular P-gp efflux pumps and an

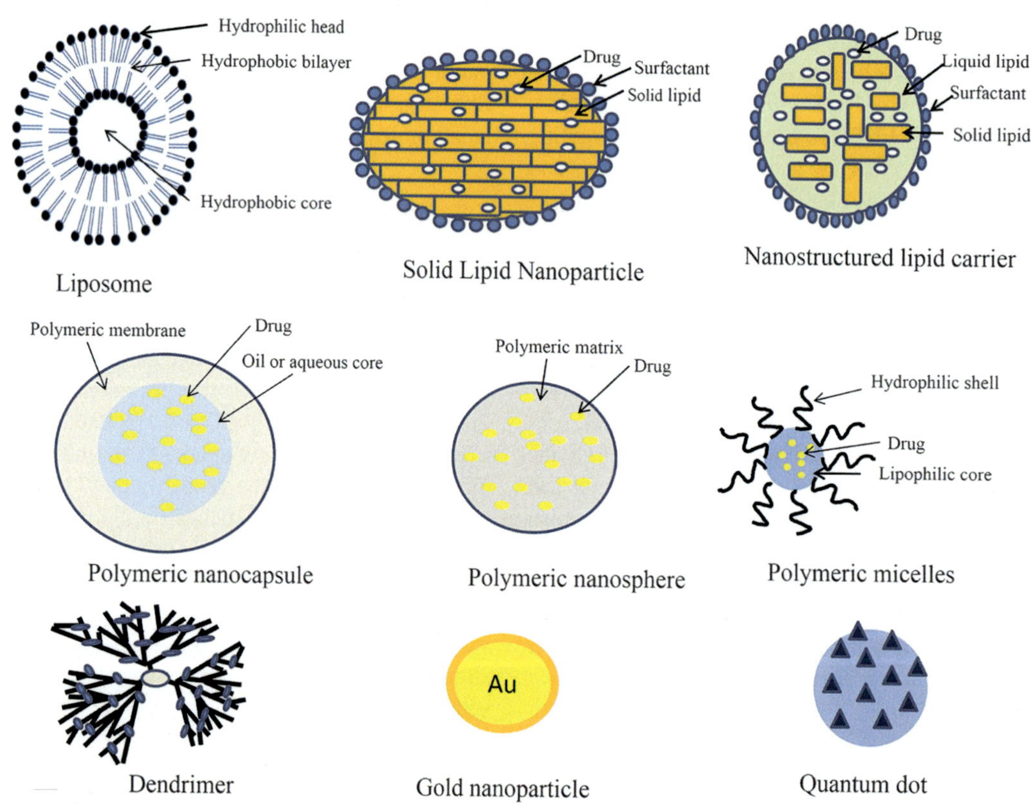

FIG. 4 Structures of some nanoparticles.

TABLE 2 Advantages and disadvantages of different nanoparticles.

Types of nanoparticle	Advantages	Disadvantages
Liposomes	• Low toxicity • Low antigenicity	• Low stability • Fast RES clearance
SLNs/NLCs	• Small size • High Stability • Easy surface modification • High-scale production • Controlled and sustained release • Delivery of hydrophobic and hydrophilic drugs • No toxicity • Biodegradation	• Low loading capacity • Drug expulsion during storage (only for SLNs) • Fast clearance from systemic circulation
Polymeric nanoparticles	• High encapsulation efficiency and loading capacity • Reactive to physical (magnetic fields, light, electric current, ultrasounds) and/or biological stimuli(ROS, pH, enzymes)	• Low encapsulation efficiency for hydrophilic drug • Some polymers have low solubility in water and their degradation leads to acidic by-products (e.g., PLGA)
Polymeric micelles	• Versatility in synthesis and modifications • Controllable size and shape	• Inherent toxicity
Dendrimer	• Controlled and sustained release • Long-term stability • Inherent ability to penetrate the BBB	• Difficulties in scaling up • Use of organic solvents during production
Metallic nanoparticles	• Small size • Multifunctionality	• Low biocompatibility • Fast RES clearance if they • Not modified
Quantum dots	• Large surface area and can encapsulate a wide variety of therapeutic agents	• Low drug release profile • Nonbiodegradability • High toxicity

elevated CSF turnover rate [40]. Nanoparticles have many advantages and limitations which are specified and summarized in Table 2. Some of the important nanoparticles are described below:

6.1 Liposomes

Liposomes are spherical-shaped vesicular systems having one or more lamellar phospholipid bilayers. Generally, the lamellar lipid bilayers are composed of biocompatible and biodegradable lipids like naturally produced phosphatidylcholine, sphingomyelin, or glycerophospholipids. Liposomes have the unique property of being amphiphilic, which makes them suitable for the delivery of both hydrophobic and hydrophilic drugs. The hydrophilic drug is enclosed in the aqueous compartment, whereas hydrophobic drug is encapsulated in the lipid bilayers. Liposomes can be useful in protecting the entrapped drug against enzymatic degradation and enhancing the intracellular uptake of the drug [41].

Several evidences are available on the effectiveness of liposomes to deliver drugs into the brain. One study reported the development of a dual-ligand liposome delivery system loaded with paclitaxel and doxorubicin for the treatment of brain glioma. The liposome was modified with cell-penetrating peptides, trans-activating transcriptional activator (TAT) and transferrin (TF). The effectiveness of paclitaxel and doxorubicin loaded TF-TAT conjugated liposomes were evaluated. In vitro studies confirmed that liposomes conjugated with TAT and TF efficiently target the endothelial and cancerous cells of the brain. In vivo studies have confirmed the maximum distribution properties of liposome conjugated with TAT and TF in tumor cells of the brain. The study revealed that paclitaxel and doxorubicin loaded TF-TAT conjugated liposome showed excellent antiglioma property as compared with doxorubicin and paclitaxel when used independently [42]. Another study reported the development of liposome with surface modification by (D) CDX, a D-peptide ligand of nicotinic acetylcholine receptors located in the BBB, and c(RGDyK), a ligand of integrin, highly expressed on the BBTB and glioma cells. From the in-vitro and in-vivo experiments, it was suggested that the surface-modified liposomes could effectively target the tumor cells and endothelial cells of the brain, and could easily cross the BBTB and BBB. Furthermore, the study confirmed that both modified liposomes (D)CDX and c(RGDyK) were associated with a better antiglioma effect as compared with the unmodified liposomes [37].

6.2 Solid lipid nanoparticles and nanostructured lipid carriers

Solid lipid nanoparticles (SLNs) and nanostructured lipid carriers (NLCs) have been utilized for the management of different brain diseases including brain cancer, due to their desired properties. SLNs are made up of stable lipid-based nanocarriers consisting of a hydrophobic lipid core, where the drug can be dissolved or dispersed [43]. The lipid matrix remains solid at both room and body temperature. Mostly these solid lipids belong to the physiological lipids such as monoglycerides, diglycerides, triglycerides, monostearin, fatty acids, stearyl alcohol, compritol, glycerol monostearate, stearic acid, precirol, and acetyl palmitate. Surfactants, e.g., tween 80, poloxamer, and dimethyl dioctadecyl ammonium bromide (DDAB), are commonly used to stabilize the product. SLNs are small in size which facilitates the sustained and controlled release of therapeutic agents. SLNs reduce drug toxicity, allow the drug to cross the tight brain endothelial cells, and protect the drug from the reticuloendothelial system (RES) [43]. SLNs have been found to have lower encapsulation efficiency, encapsulated therapeutic expulsion, and gelating tendency.

NLCs are usually colloidal carriers containing a solid-lipid core with a mixture of both solid and liquid lipids. NLCs are a modified versions of SLNs with better loading efficiencies, enhanced stability and the ability to prevent drug expulsion [44]. NLCs comprise of a mixture of several lipid molecules, i.e., long-chain solid lipid is blended with short-chain liquid, preferably in a 70:30 ratio to a 99.9:0.1 ratio. It has been reported that permeation, release, and pharmacodynamic properties of NLCs can be modulated by changing the ratio of these solid and liquid lipid. Similar to SLNs, NLCs use the same lipids in addition to liquid lipids (cetiol, almond oil, corn oil, oleic acid, mygliol, olive oil, sesame oil, peanut oil, tegosoft, soybean oil, phosphatidylcholine (PC), and soy lecithin). Surfactants used in the development of NLCs are eumulgin SML, cremophor, Pluronic F68, Span-85, Lutrol F68, Tween-20, Tween-80, and *N*-[1-(2,3-dioleyloxy)propyl]-*N*,*N*,*N*-trimethyl-ammonium chloride (DOTMA) [2].

Docetaxel-loaded SLNs with lactoferin conjugation has also been reported for targeting drug delivery to the brain. The cytotoxicity study of this developed system confirmed an increased apoptotic activity of lactoferin-conjugated SLNs as compared with the unconjugated SLNs and the pure drug. Additionally, the drug concentration in the brain was notably higher with the lactoferin conjugated SLNs compared with the marketed formulation [45].

Similarly, one comparison study reported the effectiveness of temozolomide- and vincristine-loaded SLNs/NLCs for the glioma treatment. The tumor inhibition activity was determined on U87 malignant glioma cells (U87MG) and malignant glioma mice model. The study confirmed that vincristine and temozolomide-loaded NLCs showed better tumor inhibition activity than the SLNs. The inhibition effectiveness of dual drug loaded NLCs showed higher efficacy than the single-loaded drug. A combination of drug-loaded NLCs might be considered as excellent drug delivery for the treatment of glioma [46].

6.3 Polymeric nanoparticles

Nowadays, polymeric nanoparticles are considered as one of the most promising approaches for brain drug delivery. Polymeric nanoparticles are composed of different natural or synthetic polymers. The drug molecules can be loaded by dissolving in solution or solid-state, or adsorbed, or chemically linked to the surface of the polymers. Depending upon the method of preparation, the polymeric nanoparticles can be categorized as, nanospheres, and nanocapsules. In polymeric nanocapsules, the drug is encapsulated by a polymeric membrane, whereas nanospheres exhibit uniform drug dispersion in the matrix. Generally, biodegradable polymers are used in polymeric nanoparticles, which are degraded within the body. Some of these polymers are polyglycolides (PGA), polylactides (PLA), polylactide-*co*-glycolides (PLGA), polycyanoacrylates, polyanhydrides, polycaprolactone, dextran, polypropylene oxide, propylene glycol, polyethylene glycol(PEG), a copolymer of poly (butylcyanoacrylate), poly (oxyethylene)-poly (oxypropylene), and poly (*N*-isopropylacrylamide) [7, 47]. Sometimes, drug-loaded polymeric nanoparticles are coated with different stabilizers, e.g., polysorbate-80 and PEG, to prolong the distribution time and decrease clearance.

Recently, biodegradable polymeric nanoparticles are gaining increased attention as a potential drug delivery system for drug targeting, gene therapy, and delivery of proteins as well as peptides. Some examples of therapeutics delivered to the brain using polymeric nanoparticles are doxorubicin, paclitaxel, methotraxate, and 5-fluorouracil. A recent study has reported the encapsulation of carmustine, etoposide and doxorubicin into folic acid, and wheat germ antibody conjugated methoxy PEG-PCL nanoparticles to enhance the delivery to the brain glioblastoma cells. The methoxy PEG and PCL maintained the sustained drug release property, and wheat germ antibody and folic acid support the targeting activity to the brain [48]. Similarly, another study has used lactoferrin conjugated on the surface of PEG-PLGA nanoparticles of shikonin for brain targeting against glioma. The lactoferrin conjugated PEG-PLGA nanoparticle confirmed enhanced brain targeting [49]. Another study has been performed on the development of methotrexate transferrin conjugate polymeric nanoparticles against brain tumor. The surface of PLGA polymeric nanoparticles was modified with polysorbate 80 and showed better antitumor activity as compared to methotrexate transferrin conjugate nanoparticle and methotrexate alone. Polysorbate 80 influenced the delivery of nanoparticles through BBB and the permeability to tumor cell by endocytosis. Transferrin increased the targeting and the distribution of methotrexate transferrin into the tumor cell [50].

6.4 Polymeric micelles

Polymeric micelles are spheroidal in shape having a hydrophobic core and a hydrophilic shell that are formed by aggregation of amphiphilic copolymers in aqueous solvent. Stability can be enhanced by crosslinking between the shell and the core chains. Polymeric micelles have additional features like the ability to become responsive to external stimulus (light, pH, ultrasound, temperature, etc.) eliciting a controllable release of encapsulated drugs. One of the commonly used polymers is pluronic type, a block copolymer based on propylene oxide and ethylene oxide [43]. A study has demonstrated the development of quercetin loaded lyophilized polymeric micelles for glioma treatment. The study revealed that lyophilized polymeric micelles showed effective sustained-release property, low toxicity, and effective BBB diffusion. In vivo studies exhibited that the quercetin-loaded lyophilized polymeric micelles have significant antitumor activity in the brain [51]. Another study has reported the development of myricetin encapsulated pluronic P123/F68 micelles modified with chitosan for glioblastoma cancer treatment. This study showed that drug-loaded polymeric micelles increased cellular uptake, antitumor activity, and were associated with effective BBB transport as compared with free myricetin [52].

6.5 Dendrimers

The term dendrimer is derived from the Greek word "dendron" which indicates the structure of a tree or branching. Dendrimers are strong bases having a number of positively charged surface groups. Dendrimers are usually 10–100 nm, particularly those used in targeted delivery [53]. Dendrimers consists of core molecule with at least 3 chemically reactive arms attached to multiple branches. As the branch density increases, the external branches modify themselves to a sphere form covering a lower density core. At higher generations, dendrimers form a sphere with numerous cavities within its branches which hold both drug and diagnostic agents. Dendrimers with hydrophobic groups at the surfaces are lipophilic in nature. Dendrimers also allow a wide range of contrast agents to be incorporated onto a single molecule, ultimately enhancing the imaging sensitivity of the imaging techniques. They also allow transportation of molecules across cell membranes by endocytosis via caveolae- and clathrin-mediated cellular internalization. Dendrimers are either polycationic or polyanionic in nature based on the charge of functional groups present at the surface of the dendrimers. Polycationic dendrimers can also construct polyplexes with nucleic acids at the normal physiological pH, whereas, polyanionic dendrimers bind to cationic agents or to coordinate into platinum complexes [40, 54]. Currently, researchers are focused on modifying the surface of dendrimers with some specific BBB or tumor cell ligands to facilitate the targeting effectiveness in brain [55]. A different study has reported the development of arsenic trioxide loaded PEG-modified PAMAM (polyamidoamine) to target the brain glioblastoma cells. The study revealed that PEG-modified PAMAM minimized the toxicity when compared with the unmodified PAMAM. Also, modified PAMAM sustained the drug release and enhanced the pharmacokinetic profile and the therapeutic activity of the drug [55].

6.6 Metallic nanoparticles

Metallic nanoparticles are composed of either some inorganic materials, e.g., gold, silver and iron oxide, or metallic allotropes of nonmetals, like carbon fullerenes [56]. Generally, these nanoparticles are smaller than the polymeric as well as lipid nanoparticles, which give them the advantage with regard to bypassing the BBB. Metallic nanoparticles are delivered to brain via passive diffusion, carrier-mediated transporter and trans-synaptic transport mechanism. Therapeutic agents cannot be encapsulated within the nanoparticle because of the solid and thick structure; it can only be conjugated to the surfaces of these nanoparticles. A number of in vivo studies have shown that metallic nanoparticles like gold, silver, and iron oxide can cross the BBB [7, 57].

Gold nanoparticles are widely employed in the delivery of active compounds targeting the brain as well as in various cargo-delivery strategies, because of their high biocompatibility and easy functionalization of their surface. Imaging of gold nanoparticles can be done using label-free, noninvasive X-ray computed tomography (CT). Typically, gold nanoparticles are composed of a gold core covered with an imaging tag. Then the gold particles are usually covered by polyethylene glycol (PEG) and silica shell. Ultimately, on the surface of gold nanoparticle, many molecules can be loaded to serve as internal enhancers for BBB crossing or as chemotherapy agents. Gold particles are available in different shapes and sizes depending on their method of preparation. Gold nanoparticles with rod, shell, sphere and cage shapes (referred to as nanorods, nonoshells, nanospheres and nanocages, respectively) possess different physiochemical properties, encapsulation efficiency, surface charge and polydispersity. One study has confirmed that modified gold nanoparticles with trans-activator of transcription peptide could cross the BBB and deliver both doxorubicin and gadolinium to a murine intracranial glioma xenograft [58].

Iron oxide nanoparticles depend on the magnetite particles embedded in synthetic polymers, polysaccharide, or monomer coatings. Iron particles included in these nanoparticles are either stored or fused by erythrocytes as a component of hemoglobin. Iron oxide nanoparticles demonstrated high biocompatibility and biodegradability in vivo. Kong et al., have reported that systemic administration of magnetic iron oxide nanoparticles with an external magnetic field could cross the BBB by transcellular pathway [59].

Carbon nanotubes (CNTs) are hollow sphere tubes of the graphene sheets [9, 60, 61]. They can be easily utilized in the drug delivery to the brain due to their needle-like flexible structures. Two types of CNTs are available i.e., single walled CNT (SWCNT) and multiwalled CNT (MWCNT) [62]. CNTs can be modified to increase the efficacy of the drug. PEGylation of CNTs increases the solubility and lowers the toxicity. CNTs can carry various agents like plasmid DNA, antisense oligonucleotide, and small-interfering ribonucleic acid to the tumor site [63]. A study has shown that modified multiwalled carbon nanotubes (MWNTs) administered systemically could cross the BBB and accumulate in the brain [64].

6.7 Quantum dots

Quantum dots are semiconductor nanocrystalline substances composed of metallic core and shell. The metallic shells cover the metalloid crystal core. These nanoparticles have different applications in brain drug delivery, e.g., diagnostic agent, gene delivery, and different therapeutic agents. Quantum dots have a large surface area and can encapsulate a wide variety of therapeutic agents. Due to the high photostability, brightness and size-tunable narrow emission spectra of quantum dots, they are used as diagnostic agents. The therapeutic agents are incorporated in the metalloid core of the quantum dots and the surface can be modified with different targeting ligands which increase the brain targeting. Nevertheless, quantum dots have several limitations including low drug release profile, nonbiodegradability and high toxicity profile [41]. One study has reported that PEGylated quantum dot nanoprobe conjugated with aptamer 32 showed the capability to bind with the glioma cells and could be used as fluorescent imaging for diagnosis, investigation and surgical intervention of brain tumor [65].

7 Conclusion

Currently, chemotherapeutics are exceptionally successful in cancer treatment. However, in brain cancer, drug delivery to the desired site is restricted by protective barriers like the BBB. The discovery of new drug methodologies to expand drug delivery into the brain is of outmost importance. Nowadays, brain drug delivery aims to bypass, penetrate, and temporarily disrupt the BBB. Biodegradable implants, FUS, osmotic solution, and CED are widely applied for safe and effective brain delivery. Nanocarriers are found to be very effective and functional for targeted brain cancer treatment. But the efficacy and the safety of targeted nanocarriers are hindered by the toxicity, bioavailability, and drug loading and release profile in the system. Much more scientific investigation is still required to decrease the complication rate of the disease state, and to discover an efficient delivery system for targeted delivery of the desired drugs in brain cancer.

References

[1] Shah V, Kochar P. Brain cancer: implication to disease, therapeutic strategies and tumor targeted drug delivery approaches. Recent Pat Anticancer Drug Discov 2018;13:70–85. https://doi.org/10.2174/1574892812666171129142023.

[2] Tapeinos C, Battaglini M, Ciofani G. Advances in the design of solid lipid nanoparticles and nanostructured lipid carriers for targeting brain diseases. J Control Release 2017;264:306–32. https://doi.org/10.1016/j.jconrel.2017.08.033.

[3] Cerna T, Stiborova M, Adam V, Kizek R, Eckschlager T. Nanocarrier drugs in the treatment of brain tumors. J Cancer Metastasis Treat 2016;2:407–16. https://doi.org/10.20517/2394-4722.2015.95.

[4] Tang W, Fan W, Lau J, Deng L, Shen Z, Chen X. Emerging blood–brain-barrier-crossing nanotechnology for brain cancer theranostics. Chem Soc Rev 2019;48:2967–3014. https://doi.org/10.1039/C8CS00805A.

[5] Sonali, Singh RP, Singh N, Sharma G, Vijayakumar MR, Koch B, Singh S, Singh U, Dash D, Pandey BL, Muthu MS. Transferrin liposomes of docetaxel for brain-targeted cancer applications: formulation and brain theranostics. Drug Deliv 2016;23:1261–71. https://doi.org/10.3109/10717544.2016.1162878.

[6] Alexander A, Agrawal M, Uddin A, Siddique S, Shehata AM, Shaker MA, Ata Ur Rahman S, Abdul MIM, Shaker MA. Recent expansions of novel strategies towards the drug targeting into the brain. Int J Nanomedicine 2019;14:5895–909. https://doi.org/10.2147/IJN.S210876.

[7] Hersh DS, Wadajkar AS, Roberts NB, Perez JG, Connolly NP, Frenkel V, Winkles JA, Woodworth GF, Kim AJ. Evolving drug delivery strategies to overcome the blood brain barrier. Curr Pharm Des 2016;22:1177–93.

[8] Agrawal M, Saraf S, Saraf S, Antimisiaris SG, Hamano N, Li S-D, Chougule M, Shoyele SA, Gupta U, Ajazuddin, Alexander A. Recent advancements in the field of nanotechnology for the delivery of anti-Alzheimer drug in the brain region. Expert Opin Drug Deliv 2018;15:589–617. https://doi.org/10.1080/17425247.2018.1471058.

[9] Mahajan S, Patharkar A, Kuche K, Maheshwari R, Deb PK, Kalia K, Tekade RK. Functionalized carbon nanotubes as emerging delivery system for the treatment of cancer. Int J Pharm 2018;548:540–58. https://doi.org/10.1016/j.ijpharm.2018.07.027.

[10] Sharif Y, Jumah F, Coplan L, Krosser A, Sharif K, Tubbs RS. Blood brain barrier: a review of its anatomy and physiology in health and disease. Clin Anat 2018;31:812–23. https://doi.org/10.1002/ca.23083.

[11] Bors LA, Erdő F. Overcoming the blood–brain barrier challenges and tricks for CNS drug delivery. Sci Pharm 2019;87:6. https://doi.org/10.3390/scipharm87010006.
[12] Dong X. Current strategies for brain drug delivery. Theranostics 2018;8:1481–93. https://doi.org/10.7150/thno.21254.
[13] Patel M, Souto EB, Singh KK. Advances in brain drug targeting and delivery: limitations and challenges of solid lipid nanoparticles. Expert Opin Drug Deliv 2013;10:889–905. https://doi.org/10.1517/17425247.2013.784742.
[14] Papademetriou IT, Porter T. Promising approaches to circumvent the blood–brain barrier: progress, pitfalls and clinical prospects in brain cancer. Ther Deliv 2015;6:989–1016. https://doi.org/10.4155/tde.15.48.
[15] Serlin Y, Shelef I, Knyazer B, Friedman A. Anatomy and physiology of the blood–brain barrier. Semin Cell Dev Biol 2015;38:2–6. https://doi.org/10.1016/j.semcdb.2015.01.002. Blood-Brain Barrier in Health and Disease & Lymphatic vessels in health and disease.
[16] Pulgar VM. Transcytosis to cross the blood brain barrier, new advancements and challenges. Front Neurosci 2019;12. https://doi.org/10.3389/fnins.2018.01019.
[17] Wei X, Chen X, Ying M, Lu W. Brain tumor-targeted drug delivery strategies. Acta Pharm Sin B 2014;4:193–201. https://doi.org/10.1016/j.apsb.2014.03.001.
[18] Meng J, Agrahari V, Youm I. Advances in targeted drug delivery approaches for the central nervous system Tumors: the inspiration of nanobiotechnology. J Neuroimmune Pharmacol 2017;12:84–98. https://doi.org/10.1007/s11481-016-9698-1.
[19] Attia MF, Anton N, Wallyn J, Omran Z, Vandamme TF. An overview of active and passive targeting strategies to improve the nanocarriers efficiency to tumour sites. J Pharm Pharmacol 2019;71:1185–98. https://doi.org/10.1111/jphp.13098.
[20] Chen Y, Liu L. Modern methods for delivery of drugs across the blood–brain barrier. Adv Drug Deliv Rev 2012;64:640–65. https://doi.org/10.1016/j.addr.2011.11.010. Delivery of Therapeutics to the Central Nervous System.
[21] Fu BM. Transport across the blood-brain barrier. In: Fu BM, Wright NT, editors. Molecular, cellular, and tissue engineering of the vascular system, advances in experimental medicine and biology. Cham: Springer International Publishing; 2018. p. 235–59. https://doi.org/10.1007/978-3-319-96445-4_13.
[22] Pardridge WM. Delivery of biologics across the blood-brain barrier with molecular Trojan horse technology. Bio Drugs Clin Immunother Biopharm Gene Ther 2017;31:503–19. https://doi.org/10.1007/s40259-017-0248-z.
[23] Garg T, Bhandari S, Rath G, Goyal AK. Current strategies for targeted delivery of bio-active drug molecules in the treatment of brain tumor. J Drug Target 2015;23:865–87. https://doi.org/10.3109/1061186X.2015.1029930.
[24] Kaurav H, Kapoor DN. Implantable systems for drug delivery to the brain. Ther Deliv 2017;8:1097–107. https://doi.org/10.4155/tde-2017-0082.
[25] Sharma K, Harikumar S. Recent advancement in drug delivery system for brain: an overview. World J Pharm Pharm Sci 2017;6:292–305. https://doi.org/10.20959/wjpps20177-9508.
[26] Patel MM, Patel BM. Crossing the blood–brain barrier: recent advances in drug delivery to the brain. CNS Drugs 2017;31:109–33. https://doi.org/10.1007/s40263-016-0405-9.
[27] Burgess A, Shah K, Hough O, Hynynen K. Focused ultrasound-mediated drug delivery through the blood-brain barrier. Expert Rev Neurother 2015;15:477–91. https://doi.org/10.1586/14737175.2015.1028369.
[28] Zhang F, Xu C-L, Liu C-M. Drug delivery strategies to enhance the permeability of the blood–brain barrier for treatment of glioma. Drug Des Dev Ther 2015;9:2089–100. https://doi.org/10.2147/DDDT.S79592.
[29] Gao H. Progress and perspectives on targeting nanoparticles for brain drug delivery. Acta Pharm Sin B 2016;6:268–86. https://doi.org/10.1016/j.apsb.2016.05.013. Functional Materials, Nanocarriers, and Formulations for Targeted Therapy.
[30] Lu C-T, Zhao Y-Z, Wong HL, Cai J, Peng L, Tian X-Q. Current approaches to enhance CNS delivery of drugs across the brain barriers. Int J Nanomed 2014;9:2241–57. https://doi.org/10.2147/IJN.S61288.
[31] Soni V, Pandey V, Asati S, Jain P, Tekade RK. Chapter 14—design and fabrication of brain-targeted drug delivery. In: Tekade RK, editor. Basic fundamentals of drug delivery, advances in pharmaceutical product development and research. Academic Press; 2019. p. 539–93. https://doi.org/10.1016/B978-0-12-817909-3.00014-5.
[32] Parrish KE, Sarkaria JN, Elmquist WF. Improving drug delivery to primary and metastatic brain tumors: strategies to overcome the blood–brain barrier. Clin Pharmacol Ther 2015;97:336–46. https://doi.org/10.1002/cpt.71.
[33] Deb PK, Al-Attraqchi O, Chandrasekaran B, Paradkar A, Tekade RK. Chapter 16 - protein/peptide drug delivery systems: practical considerations in pharmaceutical product development. In: Tekade RK, editor. Basic fundamentals of drug delivery, advances in pharmaceutical product development and research. Academic Press; 2019. p. 651–84. https://doi.org/10.1016/B978-0-12-817909-3.00016-9.
[34] Obcroi RK, Parrish KE, Sio TT, Mittapalli RK, Elmquist WF, Sarkaria JN. Strategies to improve delivery of anticancer drugs across the blood–brain barrier to treat glioblastoma. Neuro-Oncology 2016;18:27–36. https://doi.org/10.1093/neuonc/nov164.
[35] Singh K, Ahmad Z, Shakya P, Ansari VA, Kumar A, Arif M. Nano formulation: a novel approach for nose to brain drug delivery. J Chem Pharm Res 2016;8:208–15.
[36] Deeksha, Malviya R, Sharma PK. Brain targeted drug delivery: factors, approaches and patents. Recent Pat Nanomed 2014;4:2–14.
[37] Wei X, Gao J, Zhan C, Xie C, Chai Z, Ran D, Ying M, Zheng P, Lu W. Liposome-based glioma targeted drug delivery enabled by stable peptide ligands. J Control Release 2015;218:13–21. https://doi.org/10.1016/j.jconrel.2015.09.059.
[38] Mehmood Y, Tariq A, Siddiqui FA. Brain targeting drug delivery system: a review. Int J Basic Med Sci Pharm 2015;5:10.
[39] Sharma U, Badyal PN, Gupta S. Polymeric nanoparticles drug delivery to brain: a review. Int J Pharmacol Pharm Sci 2015;2(5):60–9.
[40] Posadas I, Monteagudo S, Ceña V. Nanoparticles for brain-specific drug and genetic material delivery, imaging and diagnosis. Nanomedicine 2016;11:833–49. https://doi.org/10.2217/nnm.16.15.
[41] Li X, Tsibouklis J, Weng T, Zhang B, Yin G, Feng G, Cui Y, Savina IN, Mikhalovska LI, Sandeman SR, Howel CA, Mikhalovsky SV. Nano carriers for drug transport across the blood–brain barrier. J Drug Target 2017;25:17–28. https://doi.org/10.1080/1061186X.2016.1184272.
[42] Chen X, Yuan M, Zhang Q, Ting Yang Y, Gao H, He Q. Synergistic combination of doxorubicin and paclitaxel delivered by blood brain barrier and Glioma cells dual targeting liposomes for chemotherapy of brain Glioma. Curr Pharm Biotechnol 2016;17:636–50.
[43] Masserini M. Nanoparticles for brain drug delivery [WWW document]. ISRN Biochem 2013. https://doi.org/10.1155/2013/238428.
[44] Attama AA, Momoh MA, Builders PF. Lipid nanoparticulate drug delivery systems: a revolution in dosage form design and development. In: Recent advances in novel drug carrier systems; 2012. https://doi.org/10.5772/50486.

[45] Singh I, Swami R, Pooja D, Jeengar MK, Khan W, Sistla R. Lactoferrin bioconjugated solid lipid nanoparticles: a new drug delivery system for potential brain targeting. J Drug Target 2016;24:212–23. https://doi.org/10.3109/1061186X.2015.1068320.
[46] Wu M, Fan Y, Lv S, Xiao B, Ye M, Zhu X. Vincristine and temozolomide combined chemotherapy for the treatment of glioma: a comparison of solid lipid nanoparticles and nanostructured lipid carriers for dual drugs delivery. Drug Deliv 2016;23:2720–5. https://doi.org/10.3109/10717544.2015.1058434.
[47] Deb PK, Kokaz SF, Abed SN, Paradkar A, Tekade RK. Chapter 6 - pharmaceutical and biomedical applications of polymers. In: Tekade RK, editor. Basic fundamentals of drug delivery, advances in pharmaceutical product development and research. Academic Press; 2019. p. 203–67. https://doi.org/10.1016/B978-0-12-817909-3.00006-6.
[48] Kuo Y-C, Chang Y-H, Rajesh R. Targeted delivery of etoposide, carmustine and doxorubicin to human glioblastoma cells using methoxy poly(ethylene glycol)-poly(ε-caprolactone) nanoparticles conjugated with wheat germ agglutinin and folic acid. Mater Sci Eng C 2019;96:114–28. https://doi.org/10.1016/j.msec.2018.10.094.
[49] Li H, Tong Y, Bai L, Ye L, Zhong L, Duan X, Zhu Y. Lactoferrin functionalized PEG-PLGA nanoparticles of shikonin for brain targeting therapy of glioma. Int J Biol Macromol 2018;107:204–11. https://doi.org/10.1016/j.ijbiomac.2017.08.155.
[50] Jain A, Jain A, Garg NK, Tyagi RK, Singh B, Katare OP, Webster TJ, Soni V. Surface engineered polymeric nanocarriers mediate the delivery of transferrin–methotrexate conjugates for an improved understanding of brain cancer. Acta Biomater 2015;24:140–51. https://doi.org/10.1016/j.actbio.2015.06.027.
[51] Wang G, Wang J-J, Chen X-L, Du L, Li F. Quercetin-loaded freeze-dried nanomicelles: improving absorption and anti-glioma efficiency in vitro and in vivo. J Control Release 2016;235:276–90. https://doi.org/10.1016/j.jconrel.2016.05.045.
[52] Wang G, Wang J-J, Tang X-J, Du L, Li F. In vitro and in vivo evaluation of functionalized chitosan–Pluronic micelles loaded with myricetin on glioblastoma cancer. Nanomed Nanotechnol Biol Med 2016;12:1263–78. https://doi.org/10.1016/j.nano.2016.02.004.
[53] Gupte AH, Kathpalia HT. Recent advances in brain targeted drug delivery systems: a review. Int J Pharm Pharm Sci 2014;6:51–7.
[54] Gorain B, Choudhury H, Pandey M, Nair AB, Iqbal Mohd Amin MC, Molugulu N, Deb PK, Tripathi PK, Khurana S, Shukla R, Kohli K, Kesharwani P. Chapter 7-Dendrimer-based nanocarriers in lung cancer therapy. In: Kesharwani P, editor. Nanotechnology-based targeted drug delivery systems for lung cancer. Academic Press; 2019. p. 161–92. https://doi.org/10.1016/B978-0-12-815720-6.00007-1.
[55] Lu Y, Han S, Zheng H, Ma R, Ping Y, Zou J, Tang H, Zhang Y, Xu X, Li F. A novel RGDyC/PEG co-modified PAMAM dendrimer-loaded arsenic trioxide of glioma targeting delivery system [WWW document]. Int J Nanomedicine 2018. https://doi.org/10.2147/IJN.S175418.
[56] Shnoudeh AJ, Hamad I, Abdo RW, Qadumii L, Jaber AY, Surchi HS, Alkelany SZ. Chapter 15—synthesis, characterization, and applications of metal nanoparticles. In: Tekade RK, editor. Biomaterials and bionanotechnology, advances in pharmaceutical product development and research. Academic Press; 2019. p. 527–612. https://doi.org/10.1016/B978-0-12-814427-5.00015-9.
[57] Chawla S, Kalyane D, Tambe V, Deb PK, Kalia K, Tekade RK. Evolving nanoformulation strategies for diagnosis and clinical interventions for Parkinson's disease. Drug Discov Today 2020;25:392–405. https://doi.org/10.1016/j.drudis.2019.12.005.
[58] Cheng Y, Dai Q, Morshed RA, Fan X, Wegscheid ML, Wainwright DA, Han Y, Zhang L, Auffinger B, Tobias AL, Rincón E, Thaci B, Ahmed AU, Warnke PC, He C, Lesniak MS. Blood-brain barrier permeable gold nanoparticles: an efficient delivery platform for enhanced malignant Glioma therapy and imaging. Small 2014;10:5137–50. https://doi.org/10.1002/smll.201400654.
[59] Kong SD, Lee J, Ramachandran S, Eliceiri BP, Shubayev VI, Lal R, Jin S. Magnetic targeting of nanoparticles across the intact blood–brain barrier. J Control Release 2012;164:49–57. https://doi.org/10.1016/j.jconrel.2012.09.021.
[60] Al-Qattan MN, Deb PK, Tekade RK. Molecular dynamics simulation strategies for designing carbon-nanotube-based targeted drug delivery. Drug Discov Today 2018;23:235–50. https://doi.org/10.1016/j.drudis.2017.10.002.
[61] Maheshwari N, Tekade M, Soni N, Ghode P, Sharma MC, Deb PK, Tekade RK. Chapter 16—Functionalized carbon nanotubes for protein, peptide, and gene delivery. In: Tekade RK, editor. Biomaterials and bionanotechnology, advances in pharmaceutical product development and research. Academic Press; 2019. p. 613–37. https://doi.org/10.1016/B978-0-12-814427-5.00016-0.
[62] Wang JT-W, Al-Jamal KT. Functionalized carbon nanotubes: revolution in brain delivery. Nanomedicine 2015;10:2639–42. https://doi.org/10.2217/nnm.15.114.
[63] Hossen S, Hossain MK, Basher MK, Mia MNH, Rahman MT, Uddin MJ. Smart nanocarrier-based drug delivery systems for cancer therapy and toxicity studies: a review. J Adv Res 2019;15:1–18. https://doi.org/10.1016/j.jare.2018.06.005.
[64] Kafa H, Wang JT-W, Rubio N, Venner K, Anderson G, Pach E, Ballesteros B, Preston JE, Abbott NJ, Al-Jamal KT. The interaction of carbon nanotubes with an in vitro blood-brain barrier model and mouse brain in vivo. Biomaterials 2015;53:437–52. https://doi.org/10.1016/j.biomaterials.2015.02.083.
[65] Tang J, Huang N, Zhang X, Zhou T, Tan Y, Pi J, Pi L, Cheng S, Zheng H, Cheng Y. Aptamer-conjugated PEGylated quantum dots targeting epidermal growth factor receptor variant III for fluorescence imaging of glioma. Int J Nanomedicine 2017;12:3899–911. https://doi.org/10.2147/IJN.S133166.

Chapter 23

Neuroblastoma: Current advancements and future therapeutics

Sin Wi Ng[a,b,*], Yinghan Chan[a,c,*], Xin Yi Ng[d,*], Kamal Dua[e], and Dinesh Kumar Chellappan[f]

[a]*School of Pharmacy, International Medical University (IMU), Bukit Jalil, Kuala Lumpur, Malaysia,* [b]*Head and Neck Cancer Research Team, Cancer Research Malaysia, Subang Jaya, Selangor, Malaysia,* [c]*Department of Pharmacology, Faculty of Medicine, University of Malaya, Kuala Lumpur, Malaysia,* [d]*Program in Neuroscience and Behavioural Disorders, Duke-NUS Medical School, National University of Singapore, Singapore, Singapore,* [e]*Discipline of Pharmacy, Graduate School of Health, University of Technology Sydney, Ultimo, NSW, Australia,* [f]*Department of Life Sciences, School of Pharmacy, International Medical University (IMU), Kuala Lumpur, Malaysia*

1 Introduction

Neuroblastoma is an extracranial solid tumor and embryonal malignancy that is the most common cancer in infants and accounts for 7% of all cancers in the pediatric population [1, 2]. According to the American Cancer Society, it affects roughly 800 children annually in the United States and represents approximately 15% of all pediatric cancer–related deaths. Up to 90% of neuroblastoma cases are diagnosed prior to the age of 5 with a median age of diagnosis of 18 months, and its incidence is minimal in adolescents and adults [3]. As neuroblastoma is an embryonal tumor that stems from neural crest cells in the autonomic nervous system, it is known to cause abruption to the development of the adrenal medulla and paravertebral sympathetic ganglia in early childhood, often resulting in mass lesions in the abdomen, neck, chest, or pelvis. The presence of metastasis at diagnosis is also often observed in 50% of patients, with involvement of the bone marrow, bone, and regional lymph nodes. The clinical behavior in neuroblastoma can be attributed to a variety of biological factors such as cytogenetic features, histological features, and molecular changes, especially the amplification of MYCN oncogene [2, 4, 5].

In the recent years, considerable amount of research has been conducted to improve the outcomes of children with solid tumors, with an overall improvement from <50% to 80%, however, this is not seen in neuroblastoma. While prognosis in infants remains favorable, the 5-year survival rates in high-risk children persist below 70%, attributed to the heterogeneity of neuroblastoma and factors such as late diagnosis, spontaneous recurrences, and relapses [6, 7]. Current conventional therapeutic modalities for neuroblastoma include surgery, radiotherapy, and chemotherapy, with recent advancements in targeted therapies and gene/stem cell therapies. Although these therapies have significantly improved outcome in patients, they are also associated with various drawbacks. For example, neuroblastoma treatment is particularly aggressive, resulting in increased systemic toxicities due to prolonged exposure and poor targeting, as well as treatment resistance [8, 9]. These limitations denote the need to develop novel therapeutic strategies capable of combating the disease, meanwhile improving the efficacy of conventional therapies. In this context, delivery systems for both drugs and genes have shown major advantages through improving bioavailability, minimizing toxicity, and improving targeting [4, 6, 10].

One such delivery system is the use of nanocarriers for drug and gene delivery. Nanomedicine is an emerging field whereby materials or structures in the nanoscale range, with sizes of less than 100 nm, are utilized in the diagnosis and delivery of therapeutic moieties. They are characterized by their high stability and large surface area, making them good potential drug carriers in a variety of diseases. Moreover, their unique biological, physical, and chemical properties allow interaction with a range of biological substances and intracellular environment, thereby improving tolerance and limiting toxicities in the body [11, 12]. Aside from nanocarriers which are synthetic, the use of viral delivery systems also provides a range of opportunities in enhancing the efficacies of conventional treatment [13, 14]. This chapter sheds light on the current treatment options and their limitations in neuroblastoma, which tie into the rationales for the use of delivery system. We will also focus on the different systems employed in the delivery of drugs and genes, highlighting some of the most recent studies along with pros and cons of their use.

* Authors with equal contributions.

Advanced Drug Delivery Systems in the Management of Cancer. https://doi.org/10.1016/B978-0-323-85503-7.00001-8

2 Current advancements in the management of neuroblastoma and their limitations

According to the International Neuroblastoma Risk Group (INRG) Task Force, neuroblastoma can be classified into four different stages: L1, L2, M, and MS, based on image-defined risk factors (IDRFs) before treatment. These risk factors foresee poor prognosis with low event-free and overall survival, including tumor extension into other organs or body cavities, and encasement of blood vessels [15]. L1 encompasses patients with localized tumors which do not meet any IDRFs; L2 consists of locoregional tumors with at least one IDRF, whereas both M and MS involve patients with metastatic tumors, with the only difference being that MS includes only patients below the age of 1.5 years with metastasis to the skin, bone marrow, or liver. Aside from cancer staging, INRG also stratifies risk of neuroblastoma into very low, low, intermediate, and high. Based on Cohn et al., these risks are comprised of INRG staging, age, histological category, ploidy, MYCN amplification, grade of tumor differentiation, and 11q aberration [16]. Typically, very low risk has an event-free survival (EFS) of more than 85%, whereas low risk, intermediate risk, and high risk have EFS of 75%–85%, 50%–75%, and less than 50%, respectively. The Children's Oncology Group (COG) and the International Society of Pediatric Oncology European Neuroblastoma (SIOPEN) have issued a guideline on treatment options in line with the risk classification (Fig. 1) [2].

The prognosis for low- and intermediate-risk neuroblastoma patients is very promising, as these patients can achieve an overall survival of greater than 90% with minimal therapy [17, 18]. On the contrary, overall survival is still under 50% for high-risk patients, in spite of attempts to enhance treatment efficiency with targeted therapies [2]. Typically, low-risk patients are treated by either observation alone or surgical resection, as surgery can cure low-risk neuroblastoma in most patients, including those without a complete resection [19]. Studies have also shown that observation is suitable for low-risk neuroblastoma, especially among infants, as these tumors commonly regress without the need for any interventions [20, 21]. Patients on observation are usually scheduled with physical examinations, imaging via MRI or ultrasound, and urine catecholamines every 6 weeks [2]. For intermediate-risk neuroblastoma patients, they can be treated with either surgery or chemotherapy. Patients are prescribed with two to eight cycles of chemotherapy, depending on their response toward the treatment. This mild and reduced form of treatment has displayed excellent survival rates, particularly in infants with unresectable neuroblastoma without MYCN amplification [17, 22]. Surgery is recommended if the primary tumor can be removed without any complications. However, if the patient is above the age of 1.5 years with presence of L2 tumors which are irremovable by surgery, a more intense therapy such as radiation is encouraged to improve overall survival [23].

On the other hand, treatment regimens for high-risk neuroblastoma are more complex, involving three different phases: induction, consolidation, and maintenance [2, 24].

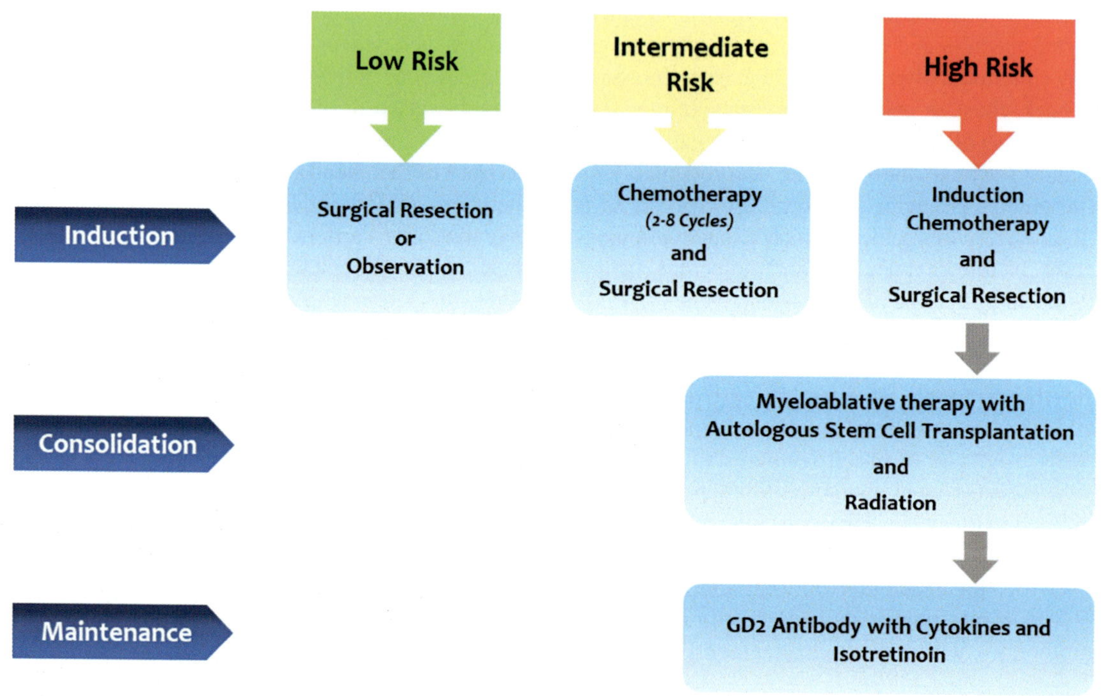

FIG. 1 Overview of treatment regimens according to risk stratification for neuroblastoma [2].

During the induction phase, multiple cycles of high-dose chemotherapy are administered to reduce tumor burden, thereby ameliorating surgical resection rate. Patients' response to induction chemotherapy serves as an important prognostic criterion, as those with either complete or very good partial response usually have a superior outcome with significantly higher EFS [25]. Additionally, myeloablative therapies such as chemoimmunotherapy or ^{131}I-MIBG radiotherapy are often prescribed to attain complete removal of detectable cancer before performing autologous stem cell transplantation [26–28]. Despite all these efforts, about 50% of high-risk patients experience a relapse after completion of induction and consolidation therapy due to therapy-resistant residual cancer cells. Hence, maintenance therapy with targeted therapies such as anti-GD2 antibody immunotherapy and isotretinoin are recommended to treat and suppress residual disease, improving EFS and reducing relapse probability [29, 30].

2.1 Surgery

Over the decades, major advances were achieved in this field to enhance the efficacy of tumor resection in neuroblastoma patients. Apart from being a therapeutic option, surgery is also required for biopsy to obtain tissue sample for histological and biological studies to confirm the diagnosis, which is paramount for oncologists to decide the best treatment regimens [31]. Open surgical biopsy is the major method used, however, it has been shown to possess high risks for complications, such as hemorrhage and increased morbidity, compared to the less invasive needle core biopsy [32]. Similar to needle core biopsy, image-guided biopsy is a minimally invasive procedure accompanied with a lower complication rate and maintains high efficiency in diagnosis [31]. Surgical resection is the most common form of therapy among neuroblastoma patients of all stages, with the goal of accomplishing maximal macroscopic tumor resection with minimal disruption to organ and neurologic function [33]. However, complete resection is difficult in intermediate- and high-risk patients, especially if the tumor is encasing major arteries, veins, or nerve roots. Moreover, metastasis is often present in high-risk patients which cannot be removed by primary tumor resection. Hence, there is a high incidence of relapse and negligible improvement in OS and EFS among high-risk patients [34]. Although tumor resection is relatively safe and can cure most low-risk neuroblastomas, that is not the case for high- and intermediate-risk patients due to the involvement of IDRFs, increasing the challenge for complete resection as well as higher complications' risk [31]. Although operative mortality is rare, complications such as massive hemorrhage, respiratory failure, and major vascular injury can be very severe. Depending on the site of tumor, complications can range from chronic diarrhea to nerve injury (Table 1) [31].

TABLE 1 Complications associated with surgical resection of tumor at the primary site.

Organ system	Potential complications
Vascular	Arterial or venous laceration
	Lymphatic ascites
	Renovascular hypertension
Nervous	Paralysis
	Recurrent nerve injury
	Horner syndrome
	Brachial or lumbosacral plexus injury
	Sensory loss
Genitourinary	Nephrectomy
	Renal infarction
	Ureteral transection or fibrosis
	Neurogenic bladder
	Bladder perforation
Gastrointestinal	Intussusception
	Chronic diarrhea
	Gastric atony

For example, tumor resection at the retroperitoneum can cause nephrectomy and renal infarction. Moreover, surgical resection of pelvic tumors could injure the pelvic nerve roots, leading to foot drop, as well as paralysis for tumor removal at the spinal foramina or epidural space [35].

2.2 Chemotherapy

As shown in Fig. 1, chemotherapy plays a crucial role in neuroblastoma therapies, particularly for high-risk patients. As these patients are positive for various IDRFs such as encasement of large vessels and vital nerve roots as well as invasion of spinal cord, the risk for surgical resection complications is greatly increased. As such, patients are often prescribed with induction chemotherapy to reduce the tumor size for safer removal. Aside from that, chemotherapy is also effective in eradicating microscopic tumor residue from a surgery and distant metastases.

Canonically, neuroblastoma chemotherapy regimen is comprised of melphalan, vincristine, cyclophosphamide, cisplatin, doxorubicin, etoposide (VP16), and teniposide (VP26) [36]. These drugs exert their effect through disruption of nucleotide DNA replication, cell division, and RNA/protein synthesis (Table 2) [37]. For instance, both melphalan and cyclophosphamide are alkylating agents which

TABLE 2 Acute and late toxicities associated with various neuroblastoma chemotherapeutic drugs.

Class	Drugs	Mechanism of action	Toxicity		References
Platinum-based chemotherapeutics	• Cisplatin • Carboplatin	• Cross-link with purine bases and interferes with DNA replication and repair • Induces mitochondrial DNA damage • Carboplatin is less potent but safer with fewer side effects compared to cisplatin	Short-term effects	Hearing loss, stomatitis, vomiting, myelosuppression, hemolytic anemia, and alopecia	[36,37,41]
			Long-term effects	Renal function impairment and infertility (cisplatin)	
Alkylating agents	• Cyclophosphamide • Melphalan	• Both are members of the nitrogen mustard family • Melphalan alkylates guanine • Alkylation results in inter-/intrastrand DNA cross-linking, inhibiting DNA and RNA synthesis	Short-term effects	Myelosuppression, vomiting, alopecia, hemorrhagic cystitis, and pulmonary fibrosis	[36, 37]
			Long-term effects	Infertility	
Microtubule inhibitors	Vinca alkaloids (e.g., vincristine)	Inhibits tubulin polymerization and microtubule formation, thus disrupting mitotic spindle assembly and mitosis	Short-term effects	Sensory, motor, and autonomic neuropathies which encompass a broad range of dysfunctions such as paresthesia, impaired balance, constipation, urinary retention, and orthostatic hypotension	[36, 37, 42]
			Long-term effects	Unknown. More studies required	
Antitumor antibiotics	Anthracyclines (e.g., doxorubicin)	• Intercalated into DNA double helix, inhibiting DNA replication, RNA synthesis, and transcription • Disrupts topoisomerase-II-mediated DNA repair • Generation of free radicals, promoting DNA damage	Short-term effects	Stomatitis, typhlitis, erythema nausea and vomiting, and alopecia	[36,37,43]
			Long-term effects	Cardiotoxicity and dyspigmentation	
Topoisomerase inhibitors	• Etoposide • Teniposide	Blocks topoisomerase II relegation function leading to permanent double-strand DNA breaks	Short-term effects	Myelosuppression, stomatitis, hypotension, vomiting, hypersensitivity reactions, and alopecia	[36–38]
			Long-term effects	Acute myelogenous leukemia	

trigger cross-linking between DNA strands, resulting in the inhibition of DNA and RNA synthesis. On the other hand, etoposide and teniposide are topoisomerase inhibitors which cause extensive DNA double-strand breakage, promoting caspase activation and apoptosis [38]. Currently, SIOPEN has recommended a rapid COJEC regimen which yields 5-year and 10-year EFS outcomes of 30.2% and 27.1%, respectively, with reduced toxicity for high-risk neuroblastoma (1.5-fold of the standard regimen). This is made up of cisplatin (C), vincristine (O), carboplatin (J), etoposide (E), and cyclophosphamide (C), whereas standard treatment can either be OPEC or OJEC regimen. The rapid COJEC regimen is administered with a shorter interval between each treatment (10 days instead of 21 days), regardless of hematological recovery, with a higher dosage intensity [39, 40].

Chemotherapeutic drugs act on cancer cells by targeting various pathways involved in cell division such as DNA replication. As a result, chemotherapy comes with side effects, as these drugs can also exert their functions on actively dividing cells such as hair follicle cells, hematopoietic stem cells in the bone marrow as well as epithelial cells lining the mouth and digestive tract. Hence, some of the most common short-term side effects of chemotherapy are stomatitis, alopecia (hair loss), vomiting, and myelosuppression. Occasionally, these drugs also lead to severe long-term side effects, one of it being secondary cancer, primarily acute myeloid leukemia which is often implicated after administration of topoisomerase inhibitors and alkylating agents [44]. Microtubule inhibitors such as vincristine induce neuropathies (e.g., ototoxicity, cranial, sensory, and motor neuropathy) as microtubules are one of the major components for axonal stabilization and transport [42]. Table 2 summarizes both short- and long-term side effects of chemotherapy for neuroblastoma.

2.3 Autologous hemopoietic stem cell transplantation

Autologous stem cell transplantation (AHSCT) is incorporated into the multimodal treatment regimen for high-risk neuroblastoma patients, as it has been shown to greatly elevate EFS [45]. This involves the reinfusion of HSC harvested from the bone marrow or blood of the patients after high-dose chemotherapy. At present, conventional autologous bone marrow HSC has been replaced with peripheral blood stem cells (PBSCs) for transplantation [46]. This is attributed to the fact that PBSCs accelerate hematopoietic recovery, leading to lower risk of infection and earlier discharge from the hospital. PBSCs are collected via apheresis after they have been mobilized by chemotherapeutic drugs (i.e.: cyclophosphamide) or hematopoietic growth factors (i.e., granulocyte-colony-stimulating factor) [47]. Studies have demonstrated that adequate amount of PBSCs can be collected after two cycles of induction chemotherapy with negligible tumor contamination [48]. The collected PBSCs are then processed and purged to remove unwanted malignant cells which can be done via CD34 selection of HSC or removal of neuroblastoma cells using antitumor monoclonal antibodies [46]. Prior to performing AHSCT, chemotherapy is prescribed to eliminate residual metastatic cancer cells. In the beginning, myeloablative chemotherapy was used to achieve total body irradiation, however, this was later replaced by high-dose chemotherapy due to the adverse effects such as infertility, secondary malignancy, and growth failure correlated with total body irradiation [28].

Majority of the complications of AHSCT arise from the high-dose, myeloablative chemotherapy. These include common side effects associated with chemotherapy, such as stomatitis, alopecia, and nausea. Moreover, there is also an increased risk of infection following AHSCT treatment due to the low number of neutrophils and immune cells in the patients, which often lead to complications such as febrile neutropenia or even sepsis and death. Other toxicities include hepatic toxicity which often leads to sinusoidal obstruction syndrome and nephrotoxicity [49, 50]. Furthermore, growth and puberty failure are also common long-term side effects among children, and these children were observed to have a lower median height score and primary hypogonadism in males [51].

2.4 Radiotherapy

Following AHSCT, patients are usually administered with radiotherapy for control of local disease and prevention of relapse as well as to ablate any persistent metastatic sites [52–54]. Conventionally, three-dimensional conformal radiotherapy (3D-CRT) has been used to perform radiotherapy. In this technique, the radiation beam will be shaped to match the tumor, reducing unwanted irradiation of surrounding tissues. Over the years, various techniques have been developed to improve dose distribution and reduce radiation-related side effects. Intensity-modulated radiation therapy (IMRT) has been demonstrated to improve dose distribution and provide better protection for organs at risk than 3D-CRT [55]. Like 3D-CRT, the photon beams are also shaped to outline the tumor, however, instead of using a single beam, the beam is separated into beamlets. With the help of specialized software, these beamlets can be programmed to deliver sufficient dose to the tumor while avoiding healthy tissues. Another new technique is proton therapy (PT) which utilizes a proton beam and has a better normal tissue sparing capacity compared to IMRT [56, 57]. Radiotherapy can also be performed in two different forms for refractory neuroblastoma and metastatic patients. For patients with refractory neuroblastoma, intraoperative radiotherapy (IORT) will be considered as during the operation, surgeons can move healthy tissues out of the radiation beams and protect them with lead shielding. This allows for a much higher radiation dosage with compromising toxicity. In the case of metastatic neuroblastoma, palliative radiotherapy is prescribed to relieve symptoms and improve quality of life [58, 59].

Another form of radiotherapy for neuroblastoma is the use of ^{131}I-*meta*-iodobenzylguanidine (mIBG). mIBG is an aralkylguanidine analog of norepinephrine (Fig. 2). It can bind to human norepinephrine transporter (hNET) which is expressed in 90% of neuroblastomas and taken up by these cancer cells. Based on this property, mIBG has been exploited for both diagnostic and therapeutic options for neuroblastoma, as it enables specific targeting of the radiation into the tumor cells. This specificity reduces side effects caused by conventional radiotherapy, where mIBG radiotherapy only causes myelosuppression [60]. Two different iodine isotopes have been used in mIBG: I-123 for diagnosis and I-131 for therapy. I-123 isotope has a shorter half-life of only 13h, as compared to 8days for I-131 isotope. This makes it safer for diagnosis, whereas the long half-life of I-131 enables an efficient killing of cancer cells with lower dosage [61]. Generally, radiotherapy causes a range of long-term effects such as growth and developmental failure, gastrointestinal dysfunction, hypothyroidism, pulmonary and cardiac abnormalities, neurocognitive defects, sensorineural hearing loss, dental caries, infertility, and secondary cancers. The most frequent side effect of radiation therapy is musculoskeletal abnormalities. As there is a high percentage of neuroblastoma patients presenting with paravertebral primary tumor, radiotherapy of these lesions often results in growth impairment of bones such as bone hypoplasia, kyphosis, scoliosis, and short stature [58, 61].

2.5 Targeted therapies

Ensuing induction and consolidation therapy, maintenance therapy is prescribed for high-risk neuroblastoma patients for treating minimal residual disease and to reduce risk of relapse. Targeted therapies such as GD2-targeted immunotherapy and small molecule inhibitors have been explored to specifically interfere with genetic and molecular pathways associated with neuroblastoma pathogenesis as well as for relapsed and metastatic neuroblastoma.

GD2 is a disialoganglioside that is highly abundant in neuroectodermal tumors including neuroblastoma ($>98\%$) with an expression level of about 10^7 molecules per cell [62]. Moreover, as GD2 levels in the blood and cerebrospinal fluid are insufficient to sequester GD2-specific monoclonal antibodies, this makes it an ideal target for immunotherapy. Dinutuximab, also known as ch14.18, is a chimeric anti-GD2 IgG1 which has been approved by FDA in 2015 for neuroblastoma treatment [63]. ch14.18 is created by fusing the variable regions of murine 14G2a GD2-specific mAb with the constant regions of human IgG1, and this can improve its half-life as well as efficacy. Dinutuximab treatment functions on the concept of passive cancer immunotherapy where it binds to GD2 to kill cancer via antibody-dependent cell-mediated cytotoxicity (ADCC) by activating neutrophils, natural killer (NK), and lymphokine-activated killer cells and is often accompanied with granulocyte-macrophage colony-stimulating factor (GM-CSF) or interleukin-2 (IL-2) [30, 62, 63]. Both cytokines enhance the ADCC of effector cells. Additionally, GM-CSF also stimulates the production of monocytes, eosinophils, and neutrophils. GM-CSF also stimulates neutrophils to increase phagocytic activity as well as production of reactive oxygen/nitrogen species. Common toxicities of anti-GD2 immunotherapy include pain, fever, nausea, diarrhea, urticaria, mild elevation of hepatic transaminases, capillary leak syndrome, and hypotension [64].

Besides that, isotretinoin, also known as 13-*cis*-retinoic acid, which is primarily used for treatment of cystic acne is also capable of triggering differentiation of neuroblastoma cells, thus decreasing their proliferative property. It has been reported that the administration of isotretinoin significantly increased EFS from 29% to 46% in neuroblastoma patients [29]. However, the incorporation of anti-GD2 monoclonal antibody, ch14.18, coupled with GM-CSF or IL-2 to the standard isotretinoin therapy further enhances 2-year EFS from 46% to 66% [30]. Some of the most prevalent short-term side effects associated with isotretinoin involve musculoskeletal, ophthalmic, and mucocutaneous systems, and they resolve quickly after discontinuation of the treatment. Other studies have also reported that pediatric patients treated with isotretinoin experienced poor bone growth, hypertriglyceridemia, hypercalcemia, pseudotumor cerebri, and relapse-like hyperplastic bone lesions [65, 66]. Moreover, isotretinoin treatment has also been associated with inflammatory bowel disease, cardiovascular disorders, and depression/anxiety [64].

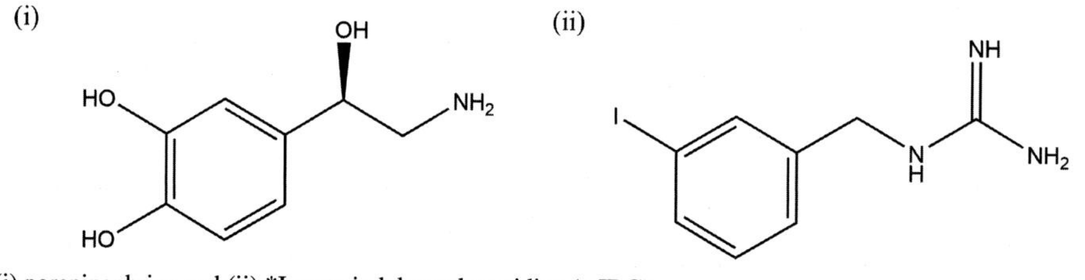

(i) OH HO NH$_2$ HO

(ii) NH I N H NH$_2$

FIG. 2 Structures of (i) norepinephrine and (ii) *I-*meta*-iodobenzylguanidine (mIBG).

Anaplastic lymphoma kinase (ALK) is a receptor tyrosine kinase and part of the insulin receptor superfamily. ALK acts upstream of both the RAS-MAPK pathway which plays a role in cellular growth, survival, differentiation, and the PI3K/Akt/mTOR pathway which is crucial for cell survival. Hence, inhibition of ALK could disrupt these two important pathways for malignancy, coupled with the fact that around 14% of high-risk neuroblastoma has a gain-of-function mutation in their *ALK* gene [67]. Mutations are the most prevalent in these three positions: R1275, F1174, and F1245, accounting for 85% of ALK mutations in neuroblastoma [68]. Crizotinib is the first identified ALK inhibitor and has been approved by FDA for use against nonsmall-cell lung cancer with ALK translocations [69]. Clinical trials on ALK-positive neuroblastoma have revealed promising results where one patient was reported to have made a complete recovery, whereas two other patients have reached a stable disease stage. Aside from crizotinib, second- (ceritinib and brigatinib) and third- (lorlatinib) generation ALK inhibitors with improved potency have also been developed. The ALK F1174L mutations have been reported to exhibit resistance against crizotinib. Despite their improved potency, the second-generation inhibitors are still only partially active against ALK mutations with the exception of brigatinib [67, 70].

MYCN amplification is the best-characterized genetic marker of risk in neuroblastoma. Twenty-five percent of neuroblastoma patients exhibit amplified *MYCN* and this often results in high-risk neuroblastoma with poor prognosis. *MYCN* is part of the *MYC* family of proto-oncogenes, which encodes for a group of proteins functioning as transcription factors. Like MYC protein, MYCN protein also heterodimerizes with MAX at consensus *E*-box sequences (CANNTG) for transcription regulation. This drives the expression of tumor-promoting genes which are involved in angiogenesis, proliferation, self-renewal, pluripotency, and metastasis while inhibiting the expression of tumor suppressor genes such as those associated with differentiation, immune surveillance, and cell cycle arrest [71]. The bromodomain and extra-terminal domain (BET) family regulates gene expression via recognition and binding of acetylation site on histones in the active promoter and enhancer regions via two conserved N-terminal regions, BD1 and BD2. This binding enables the recruitment of other transcriptional regulatory proteins such as mediator, positive transcription elongation factor b (P-TEFb), and jumonji domain containing 6 (Jmjd6). These proteins come together to form a complex to promote RNA polymerase II (RNAPII) elongation, hence inducing transcription. BET acts upstream of *MYCN* and can regulate *MYCN* expression levels. As a result, efforts have been made to discover small molecule inhibitors which can bind to histone-binding proteins such as BET. One such inhibitor is JQ1, a cell-permeable small molecule which binds to the BD1 and BD2 domains, blocking it from binding with acetylation sites on histones [72, 73]. Neuroblastoma with *MYCN* amplification has been shown to be sensitive to JQ1 treatment. Apart from downregulation of MYCN gene expression and protein levels, JQ1 treatment also induces apoptosis and cell cycle arrest in tumor cells. Hence, various clinical trials involving BET inhibitors in neuroblastoma are currently still ongoing [69, 74].

Although these small molecule–based targeted therapies have been shown to be effective in controlling tumor growth in neuroblastoma patients, similar to chemotherapy, various challenges such as potential systemic/tissue toxicity and resistance to treatment limit the capacity of these therapeutic options. Moreover, studies have also reported that around 1.7% to 11.7% of neuroblastoma can metastasize to the brain and central nervous system after treatment. In these cases, the solubility and ability of these drugs to cross the blood-brain barrier (BBB) needs to be considered. This increases the complexity of drug development as drugs which are able to cross the BBB might induce severe neurotoxicity [74, 75].

2.6 Gene therapy

Despite the advancements in treatments throughout the years, long-term life expectancy and survival remain low for neuroblastoma patients. As such, gene therapy has recently been explored as a potential therapeutic option for neuroblastoma. Both MYCN amplification and ALK mutations predispose children to neuroblastoma, hence, these genes, or their downstream pathways can serve as prospective targets for gene therapy in neuroblastoma patients. Gene therapy utilizes vectors to carry transgene into tumor cells to initiate apoptosis and halt tumor growth. Firstly, these transgenes can code for cytokines to activate T-cells and NK cells for elevation of immune response against cancer cells [76]. Transgenes can also code for miRNA to block transcription of oncogene, thus converting cancer cells back into normal cells. These transgenes can be delivered into tumor cells via viral or nonviral methods. Adenovirus and retrovirus are the most prevalent form of viral vectors [77, 78]. Retroviruses can ensure a permanent gene transfer due to their stable integration into host genome. Moreover, infection of tumor cells by viruses can trigger the lytic or lysogenic cycle to initiate cell death. Adenovirus is an oncolytic virus which can recognize and kill cancer cells. However, the usage of viral vectors often triggers high immune response, resulting in low efficacy in eradicating tumor cells from patients [79]. Systemic introduction of viral vectors can also induce tissue toxicity. These challenges arising from viral delivery methods have called for the usage of nonviral vectors. For example, studies on polyethylene glycol–grafted polyethylenimine-coated siRNAs packaged in superparamagnetic iron oxide nanoparticle were conducted to silence expression of antiapoptotic *Bcl2* gene [80]. The use of viral and nonviral vectors for gene therapy will be further explored in the following sections.

3 Nanomedicine-based approach in the management of neuroblastoma

Overall, the prognosis of patients with high risk or relapse neuroblastoma remained poor, despite the standard therapies including chemotherapy, surgery, and radiotherapy, as presented in the above section. Despite advancing medical and biological knowledge, many patients are still suffering from long-term complications associated with aggressive neuroblastoma therapy [81]. Therefore, it is of utmost importance for researchers to elucidate next-generation strategies with the aim of effectively combating the disease, as well as minimizing the occurrence of chemotherapy- and radiotherapy-associated toxicities. For this instance, although the discovery of novel drug compounds is one straightforward strategy to improve therapeutic outcomes in neuroblastoma patients, a more convincing strategy is to improve the therapeutic profiles of existing agents that are already known to exert beneficial activity on neuroblastoma patients. We believe that nanomedicine is an important platform to facilitate this second strategy, and, depending on the novel compounds being discovered, nanomedicine may also be useful in the development of novel therapeutics identified in the first strategy [82].

Nanomedicine is a rapidly developing area of science which utilizes colloid systems at the nanoscale size for the diagnosis, treatment, and prevention of diseases. These can be constructed on lipids, polymers, metal, or other nanomaterials that allow the encapsulation of diverse chemical moieties [83–85]. Over the years, nanomedicine has been an alternative to conventional therapeutic approaches in various diseases attributed to their advantageous capabilities as demonstrated in numerous studies, including improved bioavailability and biodistribution of therapeutic agents due to increased solubility of hydrophobic moieties, reduced clearance, and minimal degradation of unstable compounds [6, 81–83]. Hence, nanomedicine-based approach may prove useful in neuroblastoma management with respect to their uniquely appealing properties, in particular, their ability to differentiate and selectively eradicate malignant cancer cells [6, 86].

3.1 Enhanced permeability and retention effect

Generally, the basis of tumor-targeting delivery systems is the ability of such vehicles to accumulate in targeted cells and tissues. One method of achieving tumor targeting is via the enhanced permeability and retention (EPR) effect, known as passive targeting, whereby molecules of 100–1000nm preferentially accumulate in tumor tissues, instead of in normal tissues [86, 87]. As tumor cells are usually not responsive to cell signaling processes that regulate vasculogenesis, the endothelial gap of angiogenic tumor vasculature is larger as compared to that of normal vasculature, thus facilitating the entrance of macromolecules into the tumor via its leaky vasculature. Besides, the lack of lymphatic vessels in tumor tissues also assists in the accumulation of macromolecular substances due to reduced clearance [83, 87, 88]. Hence, via the EPR effect, researchers can design nanovehicles to carry imaging and/or chemotherapeutic agents to target tumor tissues, in which a high concentration of such agents can be released and retained at specific sites, thus achieving promising therapeutic outcome while minimizing toxicity to other normal tissues.

3.2 Active tumor targeting

Nanomedicine-based approach allows the design of nanovehicles that can attach to specific biological structures in tumor tissues via the recognition of surface-bound ligands. Such attachment can facilitate the localization of drug-loaded nanovehicles to malignant cells rather than nonmalignant cells, at the same time, it reduces nonspecific interactions of nanovehicles with the cell membranes [86]. The advent of nanomedicine provides a platform for surface functionalization with ligands such as transferrin, epidermal growth factor receptor, or monoclonal antibodies to deliver drugs selectively to tumor tissues, known as active targeting. Active targeting is commonly utilized as a complementary strategy to EPR-based passive targeting to strengthen nanovehicle retention in tumor tissues [86, 88]. In addition, hydrophilic molecules such as poly(ethylene glycol) (PEG) can also be functionalized on the surface of nanovehicles to improve their biodistribution time by avoiding clearance by the reticuloendothelial system [6, 85, 89]. Nonetheless, the targeting efficiency of ligands is highly dependent on their molecular weight, density, targeting affinity, as well as their biocompatibility to the nanovehicle. Hence, it is crucial that ligand-conjugated nanovehicles are designed based on the specificity of overexpressed antigen on tumor cell surface, while considering the biocompatibility between nanomaterial and ligand [86, 90]. This ensures that such nanovehicles can effectively target tumor cells and exert their intended therapeutic effects.

3.3 Tumor microenvironment

The physiological state of tumor microenvironment (TME) greatly contributes to the pathogenesis of cancer. Throughout the years, TME has been shown to modulate the functions of aberrant tissues which lead to the evolution of cell malignancies. Therefore, medical researchers are focused on developing novel therapeutics that effectively targets TME in the management of cancer [91]. Apart from EPR-based passive targeting and active targeting, TME-targeted nanovehicles can be designed to be responsive to various physiological features of tumor tissues, such as TME-associated pH abnormalities, enzyme redox environment, hypoxia, and reactive oxygen species [91–93]. Hence, nanomedicine appears to be a promising tool to provide a universal approach for the effective management of cancer.

For example, the acidic condition of TME, which results from the abnormal metabolism and protein regulation of tumor tissues, can be exploited to modulate chemical bonds in the structure of nanovehicles, which leads to the dissociation or charge reversal for the targeted release of therapeutic moieties in tumor tissues [83, 86, 94]. On the other hand, the higher temperature of tumor tissues can also be exploited for temperature-responsive delivery by nanovehicles. Such temperature difference between normal tissues and tumor tissues may trigger targeted release of drugs from the nanovehicles. Moreover, the rate of drug release can be modified by simply adjusting external temperature [83, 92]. Another example is the enzyme-responsiveness of nanovehicles. Upregulation of enzymes, namely, matrix metalloproteinase (MMP), is commonly observed in TME, whereby MMP plays a central role in the development and proliferation of tumor cells. As such, nanovehicles can be designed with specific substrates to induce binding with MMP enzymes, producing an intelligent targeted response. This will result in enhancement in therapeutic efficacy, as drugs can be delivered and retained in tumor tissues at a high concentration [86, 92]. In a nutshell, nanomedicine-based approach has demonstrated great potential for the targeted management of cancer, attributed to the tunable and customizable properties of nanomaterials which allowed the design of multifunctional drug delivery nanovehicles.

4 Drug and gene delivery systems in neuroblastoma

A variety of delivery systems have been engineered and studied preclinically in the management of neuroblastoma. These can be categorized into viral and nonviral, with nonviral delivery systems further divided into natural compounds and synthetic compounds. Viral delivery systems are more commonly used for gene therapy, whereas nonviral delivery systems are mostly used in the delivery of drugs [76,95,96]. On the circumstances of drug delivery, the use of synthetic delivery systems such as polymeric and nonpolymeric nanoparticles has garnered recent attention due to their multitude of physicochemical properties [84–86]. All these delivery systems can deliver therapeutic moieties to the targeted sites at reduced dosage frequency and in a controlled manner, thereby reducing occurrence of adverse events associated with conventional therapeutic strategies [97]. Here, we offer some insights into the utilization of viral and nonviral delivery systems in neuroblastoma and their observed anticancer effects in several studies.

4.1 Viral delivery systems

Viral vectors are the most used delivery system in gene therapy for neuroblastoma as they are able to effectively target, bind, and infect host cells, transferring their genetic material into the host cell and activating killing mechanisms. They can be further subdivided into two categories: nonreplicating viruses (NRVs) and replication-competent/oncolytic viruses (OVs) [98]. NRVs are believed to have a better safety profile, as they consist of viruses whereby most or all of its genome has been removed and thereby cannot undergo replication to form infectious progenies, limiting toxicity and antivector immune responses. Predominantly, adenoviruses, retroviruses, and lentiviruses have been used as NRVs for neuroblastoma gene therapy. On the other hand, OVs are believed to possess higher efficiency due to their ability to replicate in tumor cells, forming progenies that can spread through the tumor mass. However, they are highly cytotoxic and often activate antitumor immune responses, hence, they are often used in combination with drugs that reduce these responses. Examples of OVs include herpes simplex virus (HSV) and measles virus [13, 14].

Retroviruses are RNA-based vectors that have long been used since the 1990s. They have been pivotal to the rise of gene therapy, as they possess the capability to integrate stably into the host genome, allowing the permanent transfer of genes. Moreover, they are able to work concurrently with the immune system by means of immunotherapy, through the expression of cytokines such as CD4+ and CD8+ T-cells [99, 100]. In an isolated study by Cho et al., no tumor growth was detected in mice when injected with neuro-2a cells transduced with IL-2 expressing engineered retrovirus, in contrast to the appearance of tumors in the group injected with IL-2 gene–nontransduced cells. Besides that, a decrease in the growth of preexisting tumors was observed in the mice when immunized with IL-2 transfected neuro-2a cells. This is attributed to the secretion of IFN-γ by infiltrating T-cells, thereby increasing the susceptibility to cytotoxic T-cell mediated lysis. In the immunization model, the appearance of tumor-specific cytotoxic T-cells was observed, due to the development of specific immune response, which thereby elicits an antitumor effect [101]. In another study, Prapa et al. evaluated the efficacy of an antidisialoganglioside (GD2) chimeric antigen receptor (CAR), which is expressed in T lymphocytes via retroviral transduction, in neuroblastoma. The T-cells are stably transduced by the retroviral vectors, thereby expressing remarkably high levels of anti-GD2 CAR and exhibited a robust and specific anticancer activity in cultures of neuroblastoma cells, most likely attributed to the release of proapoptotic factors such as IFN-γ and TRAIL. In the xenograft model, the retroviral vectors facilitated the infiltration of T-cells into tumor cells, which then strongly induced neuroblastoma cell apoptosis to a great extent and eventual ablation of tumor growth [102].

Adenovirus is another type of vector that has gained increasing popularity over the recent years, primarily due to its nonpathogenicity. In terms of neuroblastoma treatment, adenoviruses are able to transduce both dividing and nondividing cells; besides, they have broad tropism whereby their capsids can undergo genetic modifications to modulate such tropism, therefore, they can be a viable candidate as a gene therapy vector. Besides, as adenoviral vectors contain large capacity to hold large segments of DNA, they can be easily modified using recombinant DNA technology to produce high titers [76,103]. Generally, replication-defective adenoviral vectors are utilized as gene delivery vehicles or vaccines, whereas replication-competent adenoviral vectors, or oncolytic vectors, are mostly used for cancer gene therapy [76]. The role of adenoviral vectors as gene delivery vehicle systems can be seen in an early study by Maerken et al. which utilized adenovirus to mediate transfer of human polynucleotide phosphorylase (hPNPase) in neuroblastoma cells. Profound tumor growth suppression and apoptosis of neuroblastoma cells were documented, supporting future applications of adenovirus for gene delivery in neuroblastoma [104]. In terms of gene therapy, a recent study by Tanimoto et al. attempted to eliminate MYCN oncogene–amplified neuroblastoma cells using human telomerase reverse transcriptase (hTERT)–targeted oncolytic adenoviruses. The oncolytic adenoviruses demonstrated tumor-specific replication along with a remarkable antitumor effect that is associated with autophagy in neuroblastoma cells, which then significantly inhibited tumor growth. The results proved that oncolytic adenoviruses are promising gene therapy candidates that suppressed the expression of MYCN oncogene in neuroblastoma cells [105]. However, studies have reported the lack of targeting capability of oncolytic adenoviruses, which reduces the efficacy of gene therapy. Mesenchymal stem cells (MSCs) are widely regarded as the cellular vehicles for transporting oncolytic adenoviruses to neuroblastoma sites, attributed to the natural tumor stroma engraftment property of human MSCs [76,106]. The effectiveness of MSCs has already been proven by García-Castro et al., whereby autologous MSCs infected with oncolytic adenovirus were used to treat metastatic neuroblastoma. In the study, it was demonstrated that MSCs allowed the replication of adenoviruses and protected them from neutralizing antibodies and the human innate immune responses. Great tolerance was also observed in children receiving the virotherapy with minimal systemic toxicity, indicating that such formulation can produce great therapeutic outcomes in treating neuroblastoma [106, 107].

Apart from that, adeno-associated virus is another example of gene therapy vector, also known for its nonpathogenicity in humans. One major advantage of adeno-associated viruses is their unique ability for sustained expression, thereby establishing a practical solution for long-term delivery of therapeutic genes in the management of neuroblastoma. Furthermore, stable transgene expression can also be achieved, and the normal angiogenesis processes will not be disrupted, which facilitate the targeting of aggressive metastatic tumors [76,108]. Streck et al. in their study evaluated the antitumor efficacy of human IFN-β-encoded adeno-associated viral vector (AAV hIFN-β) in murine neuroblastoma models. Significant antiangiogenic effect was observed, which prevented the continual growth and dissemination of neuroblastoma. In addition, when AAV hIFN-β was used in conjunction with cyclophosphamide, a conventional chemotherapy, complete tumor regression was observed [109]. Similarly, the ability of adeno-associated viral vector–mediated delivery of pigment epithelium–derived factor (AAV PEDF) to inhibit neuroblastoma was evaluated by Streck et al. Significant impact on neuroblastoma growth was observed as a result of angiogenesis inhibition by AAV PEDF and subsequent intratumoral apoptosis, suggesting that adeno-associated virus is a promising delivery vector for the treatment of neuroblastoma [110].

HSVs are neurotropic, double-stranded DNA viruses that have been employed in the development of gene transfer vectors. As HSVs exist in their dormant form in neuronal cells, they can be exploited to induce a site-specific cytotoxic activity. Like adenoviruses, HSVs also show tropism for various cell types with high infectivity for both nondividing and dividing cells [76,111]. In a study, Gillory et al. reported that in murine neuroblastoma xenografts, treatment with neuroattenuated oncolytic HSV-1 demonstrated remarkable tumor growth suppression. This is due to the high expression of CD111 in neuroblastoma cells, which is the primary entry protein for oncolytic HSV-1. As a result, oncolytic HSV-1 can effectively target and replicate in neuroblastoma cells, thereby acting as a potential therapy in children with unresponsive or relapsed neuroblastoma [112]. Oncolytic HSVs can also be used in combination with dendritic cells to amplify the antitumor effect, as presented by Farrell et al. The study showed that the combination therapy exhibited an approximately fourfold increase in IFN-γ producing cells, resulting in an enhanced systemic antitumor immune response in a murine neuroblastoma model [113].

4.2 Nonviral delivery systems

While viral delivery systems have shown promising effects in gene therapy for neuroblastoma, it is known that viral systems demonstrated lower efficacy due to body immune response that prevents complete tumor regression [76]. It is also impossible to neglect the potential immunogenicity, pathogenicity, and carcinogenicity of viral-based approaches, whereby such issues have been the center of research attention in the recent years [114]. As such, considerable attention has been diverted to the use of nonviral vectors. Generally, the advantages of nonviral vectors include lower immunogenicity and better stability, ease of modification, and large-scale engineering which can be done at a lower cost [115]. These characteristics make nonviral vectors attractive for gene therapy. Many compounds have been employed in the fabrication of nonviral delivery systems, which include lipids, peptides, and polymers. To achieve gene transfer, nonviral vectors must overcome three barriers: (i) the vector must be able to cross the cell membrane effectively; (ii) the vector must be able to protect the loaded content from various degrading factors; and (iii) the vector must penetrate the nuclear membrane, prior to releasing its content inside the nucleus [76,97,114,116]. Nonetheless, utilization of nonviral vectors for gene therapy is relatively uncommon. This is most likely due to their significantly lower gene transfer efficiency in contrast to that of viral vectors. As a result, nonviral vectors are more widely used as delivery systems for drug compounds [114, 117, 118].

Liposomal delivery systems are lipid vesicles that consist of phospholipid bilayers encapsulating an aqueous solution. As they are amphiphilic in nature, liposomes can bind to both hydrophilic and lipophilic molecules in their core and bilayers, respectively, therefore, they have garnered much attention as a drug delivery vehicle throughout many years. There have been multiple studies conducted using liposomal nanoparticles to deliver tumor suppressor miRNA and/or siRNA [119, 120]. A study by Paolo et al. has developed antidisialoganglioside (GD_2)–targeted liposomes encapsulating siRNA targeting anaplastic lymphoma kinase (ALK) in GD_2-expressing neuroblastoma cells. ALK is a receptor tyrosine kinase that has been shown to contribute tumor growth and progression in human neuroblastoma. Such nanoformulation demonstrated minimal plasma clearance along with improved siRNA stability, uptake, cell death induction, as well as specificity to neuroblastoma cells. This study suggested that liposomal nanoparticles are suitable candidates as carriers for siRNA, as they are resistant to degradation and can be surface modified to produce specific targeting to tumor cell surfaces [121]. Apart from biomolecules, liposomes have also been utilized to deliver natural medicinal compound for targeted neuroblastoma therapy. Orr et al. in a study fabricated curcumin-encapsulated liposome to inhibit tumor cell proliferation and tumor angiogenesis via the nuclear factor-κB pathway. In vitro results demonstrated that liposomal curcumin inhibited the proliferation of all neuroblastoma cell lines (NB1691, CHLA-20, SK-N-AS), while in vivo results showed a significant reduction in disseminated tumor burden. As curcumin is a natural compound with extremely poor bioavailability, it was thought that the observed effects on neuroblastoma are due to the enhancement of bioavailability by liposomes via increased dissolution and subsequent absorption of the compound [122]. Bhunia et al. developed CDC20siRNA and paclitaxel coloaded liposomes and studied their effects on xenografted neuroblastoma. The study showed that liposomes are able to protect the encapsulated siRNAs from RNase attack, with high liposomal entry in human neuroblastoma IMR-32 cells via the $GABA_A$ receptor. Upon treatment with CDC20siRNA and paclitaxel coloaded liposomes, 80% of neuroblastoma cell death was observed, which is assumed to originate from enhanced cell apoptosis. The growth of xenografted human neuroblastoma was also significantly suppressed. These findings indicate that liposomal nanoparticles may be a viable option for drug and gene delivery in neuroblastoma management, owing to their advantageous physicochemical properties [123].

In gene delivery research, the potential advantages of polymeric carriers have been evaluated by many researchers. Cationic polymers are one of the most utilized nonviral vectors for the delivery of genes. Polyethylenimine (PEI) is an example of cationic polymer that has the ability to condense DNA, and to facilitate cell internalization and endosomal escape [95, 124]. It is an attractive carrier because it has extremely high cation charge density and it can rupture cellular endosomes via the proton-sponge effect, which then protects the encapsulated genes from lysosomal degradation. Reports have also shown that PEI exhibited improved cellular uptake, resulting in more effective delivery into immune cells, as compared to other nonviral vectors [95, 124, 125]. Stegantseva et al. investigated the immunogenicity and antitumor effect of a DNA vaccine loaded as a PEI carrier in a mouse neuroblastoma model. It was shown that the DNA-PEI vaccine exhibited the highest cytotoxicity at $37.3 \pm 6.9\%$, as compared to unconjugated DNA vaccine at $26.2 \pm 4.0\%$ and placebo control at $21.9 \pm 3.7\%$. Mice vaccinated with the DNA-PEI vaccine also exhibited a remarkably improved survival in contrast to unconjugated DNA and control groups. In short, this study proved that PEI can effectively facilitate in the retardation of neuroblastoma cell growth [126]. Similarly, poly(L-lysine) (PLL) is another example of cationic polymer widely utilized in therapeutic applications. It contains intrinsic affinity for negatively charged oligonucleotides and cell surfaces. Although both PEI and PLL can effectively condense DNA and shield it from enzymatic degradation, PLL has a lower transfection efficiency as compared to PEI. On the contrary, unlike PLL, PEI is nonbiodegradable and has varying cytotoxicity depending on its molecular weight and degree of polymerization [127–129]. A study by Askarian et al. has conjugated low molecular weight PEI to a PLL core to incorporate both the properties of PEI and PLL for effective gene therapy in neuroblastoma. They demonstrated that the PEI-PLL conjugate can form polyplexes that are capable of condensing DNA at a low concentration. Besides, the transfection efficiency also improved between 2.8- and 4-fold in comparison to unmodified PEI. In a nutshell, this study showed that chemical modifications of PEI using PLL can lead to remarkable improvement in gene transfection efficiency, which suggested that PEI-PLL conjugate can be developed as a next-generation gene delivery system in the management of neuroblastoma [127].

On the other hand, polyamidoamine (PAMAM) is also a cationic polymer which has been widely utilized in the synthesis of dendrimers. As dendrimers are usually engineered in a tightly controlled manner, they have well-defined size and favorable physicochemical properties [95, 130]. As such, they are versatile nanovehicles whose surface can be modified to attach target-specific ligands for recognizing specific overexpressed receptors on tumor cells. In addition, PAMAM dendrimers can also display great electrostatic interactions with biomolecules such as RNAs and DNAs due to their high density of cationic charges, forming dendriplexes that can shield the nucleic acids from enzymatic degradation [131]. Despite that, the main drawback concerning the use of PAMAM dendrimers is their potential toxicity due to their high positive charge interacting with negatively charged cell membranes. For this instance, PEGylation is an established method to improve the biocompatibility of PAMAM dendrimers, to extend their biocirculation time, as well as to enhance tumor accumulation via the EPR effect [132, 133]. A recent study by Dąbkowska et al. has designed a PEGylated PAMAM-based dendrimer to deliver brain-derived neurotrophic factor (BDNF) to neuroblastoma SH-SY5Y cells. It was found that PAMAM dendrimer–treated group exhibited significantly increased BDNF concentration in comparison to the control group. It was also observed that PEGylated PAMAM dendrimer–treated cells exhibited remarkably higher level of BDNF, as compared with non-PEGylated PAMAM dendrimers. This is attributed to the rapid cellular interactions between non-PEGylated PAMAM dendrimers with cell membranes that impacted the long-term release of BDNF. Thus, it was proven that PEGylation can enhance therapeutic efficacy in neuroblastoma and allow the design of stable dendrimer-based nanovehicles [134].

Inorganic nanoparticles are another class of nonviral delivery systems commonly investigated as potential candidates for therapeutic and imaging treatments attributed to their multitude of advantages, including high drug-loading capacity, greater bioavailability, large surface area, controlled drug release profile, as well as good tolerance toward many organic solvents [135]. One study by Yang et al. investigated the effects of silica nanoparticles on human neuroblastoma cell line, SH-SY5Y. The results showed that the nanoparticles induced cytotoxicity and apoptosis in SH-SY5Y cells. Prolonged exposure of silica nanoparticles also caused the loss of mitochondrial membrane potential and mitochondrial damage, which led to elevated Apafl and CytC expressions, and eventually resulted in neural apoptosis. Thus, it was suggested that silica nanoparticles can induce apoptotic cell death in neuroblastoma via intrinsic apoptosis [136]. Inorganic nanoparticles can also be modified using cationic polymers to achieve greater therapeutic efficacy as both their intrinsic advantages can be exploited to produce a synergistic effect. Babaei et al. have formulated a gene delivery system using PEI-modified silica nanoparticles. Investigation showed a high level of transfection activity, attributed to the ability of the PEI-conjugated silica carrier to form nanosized complexes with DNA. The presence of high positive charge density of PEI contributed to effective DNA condensation and electrostatic interaction with the cell membranes via the proton-sponge mechanism; while the intrinsic ability of silica nanoparticles to increase DNA concentration on cell surface enhanced cellular uptake and transfection efficiencies. The same study also reported that such PEI-silica nanoparticles significantly improved gene delivery up to 38-fold, as compared to PEI alone in a murine neuroblastoma model [137].

5 Conclusion and future directions

Currently, the management of neuroblastoma is based on the administration of chemotherapy to immunotherapy regimens. Nevertheless, the prognosis of neuroblastoma remained poor and high-risk neuroblastoma is still incurable in most of the patients. Efforts have been made by researchers in the last few years to develop novel strategies for replacing or complementing current management strategies. The emerging field of nanomedicine has been said to provide solutions to longstanding issues in medical research, ranging from poor drug bioavailability to lack of targeting specificity for therapeutic agents. Moreover, nanomedicine also demonstrated great potential as a noninvasive tool for diagnostic imaging, detection of tumors, as well as drug delivery due to their unique and advantageous physicochemical properties that other tools do not possess. In terms of neuroblastoma, nanomedicine presents novel opportunities to improve the efficacy and safety profile of conventional therapeutics. Various types of delivery vehicles have been exploited for their benefits in combating neuroblastoma, which can be categorized into viral vectors and nonviral vectors. These delivery systems have been investigated by many researchers and promising outcomes have been obtained in many preclinical studies. Despite that, the complex pathophysiology of neuroblastoma remained as the main obstacle for this novel strategy, as some aspects such as the safety profile, biodistribution, and underlying molecular mechanisms are yet to be fully elucidated. Thus, further research must be continued with emphasis on clinical studies, as certain adverse events may not be observed in preclinical models. Interdisciplinary collaboration between biologists, medical scientists, and pediatric oncologists is also crucial to ensure research success and eventual translation of this new, innovative strategy as the future therapeutics for neuroblastoma.

References

[1] Colon NC, Chung DH. Neuroblastoma. Adv Pediatr 2011;58:297–311. https://doi.org/10.1016/j.yapd.2011.03.011.

[2] Tolbert VP, Matthay KK. Neuroblastoma: clinical and biological approach to risk stratification and treatment. Cell Tissue Res 2018;372:195–209. https://doi.org/10.1007/s00441-018-2821-2.

[3] American Cancer Society. Key statistics of neuroblastoma. American Cancer Society; 2020. https://www.cancer.org/cancer/neuroblastoma/about/key-statistics.html.

[4] Bhatnagar SN, Sarin YK. Neuroblastoma: a review of management and outcome. Indian J Pediatr 2012;79:787–92. https://doi.org/10.1007/s12098-012-0748-2.

[5] Filbin M, Monje M. Developmental origins and emerging therapeutic opportunities for childhood cancer. Nat Med 2019;25:367–76. https://doi.org/10.1038/s41591-019-0383-9.

[6] Rodríguez-Nogales C, Noguera R, Couvreur P, Blanco-Prieto MJ. Therapeutic opportunities in neuroblastoma using nanotechnology. J Pharmacol Exp Ther 2019;370:625–35. https://doi.org/10.1124/jpet.118.255067.

[7] Tsubota S, Kadomatsu K. Origin and initiation mechanisms of neuroblastoma. Cell Tissue Res 2018;372:211–21. https://doi.org/10.1007/s00441-018-2796-z.

[8] Brodeur GM. Spontaneous regression of neuroblastoma. Cell Tissue Res 2018;372:277–86. https://doi.org/10.1007/s00441-017-2761-2.

[9] Whittle SB, Smith V, Doherty E, Zhao S, McCarty S, Zage PE. Overview and recent advances in the treatment of neuroblastoma. Expert Rev Anticancer Ther 2017;17:369–86. https://doi.org/10.1080/14737140.2017.1285230.

[10] Maris JM. Recent advances in neuroblastoma. N Engl J Med 2010;362:2202. https://doi.org/10.1056/nejmra0804577.

[11] Li Z, Tan S, Li S, Shen Q, Wang K. Cancer drug delivery in the nano era: an overview and perspectives. Oncol Rep 2017;38:611–24. https://doi.org/10.3892/or.2017.5718.

[12] Sutradhar KB, Amin ML. Nanotechnology in cancer drug delivery and selective targeting. ISRN Nanotechnol 2014;2014:1–12. https://doi.org/10.1155/2014/939378.

[13] Kim JW, Chang AL, Kane JR, Young JS, Qiao J, Lesniak MS. Gene therapy and virotherapy of gliomas. Prog Neurol Surg 2018;32:112–23. https://doi.org/10.1159/000469685.

[14] Caffery B, Lee JS, Alexander-Bryant AA. Vectors for glioblastoma gene therapy: viral & non-viral delivery strategies. Nanomaterials (Basel) 2019;9:105. https://doi.org/10.3390/nano9010105.

[15] Monclair T, Brodeur GM, Ambros PF, Brisse HJ, Cecchetto G, Holmes K, Kaneko M, London WB, Matthay KK, Nuchtern JG, Von Schweinitz D, Simon T, Cohn SL, Pearson ADJ. The International Neuroblastoma Risk Group (INRG) staging system: an INRG task force report. J Clin Oncol 2009;27:298–303. https://doi.org/10.1200/JCO.2008.16.6876.

[16] Cohn SL, Pearson ADJ, London WB, Monclair T, Ambros PF, Brodeur GM, Faldum A, Hero B, Iehara T, Machin D, Mosseri V, Simon T, Garaventa A, Castel V, Matthay KK. The International Neuroblastoma Risk Group (INRG) classification system: an INRG task force report. J Clin Oncol 2009;27:289–97. https://doi.org/10.1200/JCO.2008.16.6785.

[17] Baker DL, Schmidt ML, Cohn SL, Maris JM, London WB, Buxton A, Stram D, Castleberry RP, Shimada H, Sandler A, Shamberger RC, Look AT, Reynolds CP, Seeger RC, Matthay KK. Outcome after reduced chemotherapy for intermediate-risk neuroblastoma. N Engl J Med 2010;363:1313–23. https://doi.org/10.1056/NEJMoa1001527.

[18] Strother DR, London WB, Lou Schmidt M, Brodeur GM, Shimada H, Thorner P, Collins MH, Tagge E, Adkins S, Reynolds CP, Murray K, Lavey RS, Matthay KK, Castleberry R, Maris JM, Cohn SL. Outcome after surgery alone or with restricted use of chemotherapy for patients with low-risk neuroblastoma: results of children's Oncology Group study P9641. J Clin Oncol 2012;30:1842–8. https://doi.org/10.1200/JCO.2011.37.9990.

[19] De Bernardi B, Mosseri V, Rubie H, Castel V, Foot A, Ladenstein R, Laureys G, Beck-Popovic M, De Lacerda AF, Pearson ADJ, De Kraker J, Ambros PF, De Rycke Y, Conte M, Bruzzi P, Michon J. Treatment of localised resectable neuroblastoma. Results of the LNESG1 study by the SIOP Europe Neuroblastoma Group. Br J Cancer 2008;99:1027–33. https://doi.org/10.1038/sj.bjc.6604640.

[20] Nuchtern JG, London WB, Barnewolt CE, Naranjo A, McGrady PW, Geiger JD, Diller L, Lou Schmidt M, Maris JM, Cohn SL, Shamberger RC. A prospective study of expectant observation as primary therapy for neuroblastoma in young infants: a children's oncology group study. Ann Surg 2012;256:573–80. https://doi.org/10.1097/SLA.0b013e31826cbbbd.

[21] Hero B, Simon T, Spitz R, Ernestus K, Gnekow AK, Scheel-Walter HG, Schwabe D, Schilling FH, Benz-Bohm G, Berthold F. Localized infant neuroblastomas often show spontaneous regression: results of the prospective trials NB95-S and NB97. J Clin Oncol 2008;26:1504–10. https://doi.org/10.1200/JCO.2007.12.3349.

[22] Rubie H, De Bernardi B, Gerrard M, Canete A, Ladenstein R, Couturier J, Ambros P, Munzer C, Pearson ADJ, Garaventa A, Brock P, Castel V, Valteau-Couanet D, Holmes K, Di Cataldo A, Brichard B, Mosseri V, Marquez C, Plantaz D, Boni L, Michon J. Excellent outcome with reduced treatment in infants with nonmetastatic and unresectable neuroblastoma without MYCN amplification: results of the prospective INES 99.1. J Clin Oncol 2011;29:449–55. https://doi.org/10.1200/JCO.2010.29.5196.

[23] Kohler JA, Rubie H, Castel V, Beiske K, Holmes K, Gambini C, Casale F, Munzer C, Erminio G, Parodi S, Navarro S, Marquez C, Peuchmaur M, Cullinane C, Brock P, Valteau-Couanet D, Garaventa A, Haupt R. Treatment of children over the age of one year with unresectable localised neuroblastoma without MYCN amplification: results of the SIOPEN study. Eur J Cancer 2013;49:3671–9. https://doi.org/10.1016/j.ejca.2013.07.002.

[24] Pinto NR, Applebaum MA, Volchenboum SL, Matthay KK, London WB, Ambros PF, Nakagawara A, Berthold F, Schleiermacher G, Park JR, Valteau-Couanet D, Pearson ADJ, Cohn SL. Advances in risk classification and treatment strategies for neuroblastoma. J Clin Oncol 2015;33:3008–17. https://doi.org/10.1200/JCO.2014.59.4648.

[25] Yanik GA, Parisi MT, Naranjo A, Nadel H, Gelfand MJ, Park JR, Ladenstein RL, Poetschger U, Boubaker A, Valteau-Couanet D, Lambert B, Castellani MR, Bar-Sever Z, Oudoux A, Kaminska A, Kreissman SG, Shulkin BL, Matthay KK. Validation of postinduction curie scores in high-risk neuroblastoma: a children's oncology group and SIOPEN group report on SIOPEN/HR-NBL1. J Nucl Med 2018;59:502–8. https://doi.org/10.2967/jnumed.117.195883.

[26] Modak S, Kushner BH, Basu E, Roberts SS, Cheung NKV. Combination of bevacizumab, irinotecan, and temozolomide for refractory or relapsed neuroblastoma: results of a phase II study. Pediatr Blood Cancer 2017;64. https://doi.org/10.1002/pbc.26448.

[27] Mody R, Naranjo A, Van Ryn C, Yu AL, London WB, Shulkin BL, Parisi MT, Servaes SEN, Diccianni MB, Sondel PM, Bender JG, Maris JM, Park JR, Bagatell R. Irinotecan–temozolomide with temsirolimus or dinutuximab in children with refractory or relapsed neuroblastoma (COG ANBL1221): an open-label, randomised, phase 2 trial. Lancet Oncol 2017;18:946–57. https://doi.org/10.1016/S1470-2045(17)30355-8.

[28] Matthay KK, Maris JM, Schleiermacher G, Nakagawara A, Mackall CL, Diller L, Weiss WA. Neuroblastoma. Nat Rev Dis Primers 2016;2:16078. https://doi.org/10.1038/nrdp.2016.78.

[29] Matthay KK, Villablanca JG, Seeger RC, Stram DO, Harris RE, Ramsay NK, Swift P, Shimada H, Black CT, Brodeur GM, Gerbing RB, Reynolds CP. Treatment of high-risk neuroblastoma with intensive chemotherapy, radiotherapy, autologous bone marrow transplantation, and 13-cis-retinoic acid. N Engl J Med 1999;341:1165–73. https://doi.org/10.1056/NEJM199910143411601.

[30] Yu AL, Gilman AL, Ozkaynak MF, London WB, Kreissman SG, Chen HX, Smith M, Anderson B, Villablanca JG, Matthay KK, Shimada H, Grupp SA, Seeger R, Reynolds CP, Buxton A, Reisfeld RA, Gillies SD, Cohn SL, Maris JM, Sondel PM. Anti-GD2 antibody with GM-CSF, interleukin-2, and isotretinoin for neuroblastoma. N Engl J Med 2010;363:1324–34. https://doi.org/10.1056/NEJMoa0911123.

[31] Croteau NJ, Saltsman JA, La Quaglia MP. Advances in the surgical treatment of neuroblastoma. In: Ray SK, editor. Neuroblastoma: molecular mechanisms and therapeutic interventions. Academic Press; 2019. p. 175–86.

[32] Hassan SF, Mathur S, Magliaro TJ, Larimer EL, Ferrell LB, Vasudevan SA, Patterson DM, Louis CU, Russell HV, Nuchtern JG, Kim ES. Needle core vs open biopsy for diagnosis of intermediate- and high-risk neuroblastoma in children. J Pediatr Surg 2012;47:1261–6. https://doi.org/10.1016/j.jpedsurg.2012.03.040.

[33] La Quaglia MP. The role of primary tumor resection in neuroblastoma: when and how much? Pediatr Blood Cancer 2015;62:1516–7. https://doi.org/10.1002/pbc.25585.

[34] Yeung F, Chung PHY, Tam PKH, Wong KKY. Is complete resection of high-risk stage IV neuroblastoma associated with better survival? J Pediatr Surg 2015;50:2107–11. https://doi.org/10.1016/j.jpedsurg.2015.08.038.

[35] Cruccetti A, Kiely EM, Spitz L, Drake DP, Pritchard J, Pierro A. Pelvic neuroblastoma: low mortality and high morbidity. J Pediatr Surg 2000;35:724–8. https://doi.org/10.1053/jpsu.2000.6076.

[36] Schor NF. Current pharmacotherapy for neuroblastoma. In: Ray SK, editor. Neuroblastoma mol. mech. ther. interv. Academic Press; 2019. p. 203–11. https://doi.org/10.1016/B978-0-12-812005-7.00012-6.

[37] Sprangers B, Cosmai L, Porta C. Conventional chemotherapy. In: Finkel KW, Perazella MA, Cohen EP, editors. Onco-nephrology. Elsevier; 2020. p. 127–53.

[38] Hande KR. Topoisomerase II inhibitors. Update Cancer Ther 2008;3:13–26. https://doi.org/10.1016/j.uct.2008.02.001.

[39] Garaventa A, Poetschger U, Valteau-Couanet D, Castel V, Elliott M, Ash S, Chan GC-F, Laureys G, Beck Popovic M, Vettenranta K, Balwierz W, Schroeder H, Owens C, Cesen M, Papadakis V, Trahair T, Luksch R, Schleiermacher G, Ambros PF, Ladenstein RL. The randomised induction for high-risk neuroblastoma comparing COJEC and N5-MSKCC regimens: early results from the HR-NBL1.5/SIOPEN trial. J Clin Oncol 2018;36:10507. https://doi.org/10.1200/jco.2018.36.15_suppl.10507.

[40] Pearson AD, Pinkerton CR, Lewis IJ, Imeson J, Ellershaw C, Machin D. High-dose rapid and standard induction chemotherapy for patients aged over 1 year with stage 4 neuroblastoma: a randomised trial. Lancet Oncol 2008;9:247–56. https://doi.org/10.1016/S1470-2045(08)70069-X.

[41] Dasari S, Bernard Tchounwou P. Cisplatin in cancer therapy: molecular mechanisms of action. Eur J Pharmacol 2014;740:364–78. https://doi.org/10.1016/j.ejphar.2014.07.025.

[42] Madsen ML, Due H, Ejskjær N, Jensen P, Madsen J, Dybkær K. Aspects of vincristine-induced neuropathy in hematologic malignancies: a systematic review. Cancer Chemother Pharmacol 2019;84:471–85. https://doi.org/10.1007/s00280-019-03884-5.

[43] Thorn CF, Oshiro C, Marsh S, Hernandez-Boussard T, McLeod H, Klein TE, Altman RB. Doxorubicin pathways: pharmacodynamics and adverse effects. Pharmacogenet Genomics 2011;21:440–6. https://doi.org/10.1097/FPC.0b013e32833ffb56.

[44] Applebaum MA, Vaksman Z, Lee SM, Hungate EA, Henderson TO, London WB, Pinto N, Volchenboum SL, Park JR, Naranjo A, Hero B, Pearson AD, Stranger BE, Cohn SL, Diskin SJ. Neuroblastoma survivors are at increased risk for second malignancies: a report from the International Neuroblastoma Risk Group Project. Eur J Cancer 2017;72:177–85. https://doi.org/10.1016/j.ejca.2016.11.022.

[45] Berthold F, Boos J, Burdach S, Erttmann R, Henze G, Hermann J, Klingebiel T, Kremens B, Schilling FH, Schrappe M, Simon T, Hero B. Myeloablative megatherapy with autologous stem-cell rescue versus oral maintenance chemotherapy as consolidation treatment in patients with high-risk neuroblastoma: a randomised controlled trial. Lancet Oncol 2005;6:649–58. https://doi.org/10.1016/S1470-2045(05)70291-6.

[46] Fish JD, Grupp SA. Stem cell transplantation for neuroblastoma. Bone Marrow Transplant 2008;41:159–65. https://doi.org/10.1038/sj.bmt.1705929.

[47] Alexander ET, Towery JA, Miller AN, Kramer C, Hogan KR, Squires JE, Stuart RK, Costa LJ. Beyond CD34+ cell dose: impact of method of peripheral blood hematopoietic stem cell mobilization (granulocyte-colony-stimulating factor [G-CSF], G-CSF plus plerixafor, or cyclophosphamide G-CSF/granulocyte-macrophage [GM]-CSF) on number of colony-forming. Transfusion 2011;51:1995–2000. https://doi.org/10.1111/j.1537-2995.2011.03085.x.

[48] Kreissman SG, Seeger RC, Matthay KK, London WB, Sposto R, Grupp SA, Haas-Kogan DA, LaQuaglia MP, Yu AL, Diller L, Buxton A, Park JR, Cohn SL, Maris JM, Reynolds CP, Villablanca JG. Purged versus non-purged peripheral blood stem-cell transplantation for high-risk neuroblastoma (COG A3973): a randomised phase 3 trial. Lancet Oncol 2013;14:999–1008. https://doi.org/10.1016/S1470-2045(13)70309-7.

[49] Elzembely MM, Park JR, Riad KF, Sayed HA, Pinto N, Carpenter PA, El-Haddad A, Scott Baker K. Acute complications after high-dose chemotherapy and stem-cell rescue in pediatric patients with high-risk neuroblastoma treated in countries with different resources. J Glob Oncol 2018;2018:1–12. https://doi.org/10.1200/JGO.17.00118.

[50] Park JR, Kreissman SG, London WB, Naranjo A, Cohn SL, Hogarty MD, Tenney SC, Haas-Kogan D, Shaw PJ, Kraveka JM, Roberts SS, Geiger JD, Doski JJ, Voss SD, Maris JM, Grupp SA, Diller L. Effect of tandem autologous stem cell transplant vs single transplant on event-free survival in patients with high-risk neuroblastoma: a randomized clinical trial. J Am Med Assoc 2019;322:746–55. https://doi.org/10.1001/jama.2019.11642.

[51] Trahair TN, Vowels MR, Johnston K, Cohn RJ, Russell SJ, Neville KA, Carroll S, Marshall GM. Long-term outcomes in children with high-risk neuroblastoma treated with autologous stem cell transplantation. Bone Marrow Transplant 2007;40:741–6. https://doi.org/10.1038/sj.bmt.1705809.

[52] Bagley AF, Grosshans DR, Philip NV, Foster J, McAleer MF, McGovern SL, Lassen-Ramshad Y, Mahajan A, Paulino AC. Efficacy of proton therapy in children with high-risk and locally recurrent neuroblastoma. Pediatr Blood Cancer 2019;66. https://doi.org/10.1002/pbc.27786, e27786.

[53] Casey DL, Kushner BH, Cheung NKV, Modak S, LaQuaglia MP, Wolden SL. Local control with 21-Gy radiation therapy for high-risk neuroblastoma. Int J Radiat Oncol Biol Phys 2016;96:393–400. https://doi.org/10.1016/j.ijrobp.2016.05.020.

[54] Hill-Kayser CE, Tochner Z, Li Y, Kurtz G, Lustig RA, James P, Balamuth N, Womer R, Mattei P, Grupp S, Mosse YP, Maris JM, Bagatell R. Outcomes after proton therapy for treatment of pediatric high-risk neuroblastoma. Int J Radiat Oncol Biol Phys 2019;104:401–8. https://doi.org/10.1016/j.ijrobp.2019.01.095.

[55] Beneyton V, Niederst C, Vigneron C, Meyer P, Becmeur F, Marcellin L, Lutz P, Noel G. Comparison of the dosimetries of 3-dimensions radiotherapy (3D-RT) with linear accelerator and intensity modulated radiotherapy (IMRT) with helical tomotherapy in children irradiated for neuroblastoma. BMC Med Phys 2012;12:2. https://doi.org/10.1186/1756-6649-12-2.

[56] Cho B. Intensity-modulated radiation therapy: a review with a physics perspective. Radiat Oncol J 2018;36:1–10. https://doi.org/10.3857/roj.2018.00122.

[57] Hattangadi JA, Rombi B, Yock TI, Broussard G, Friedmann AM, Huang M, Chen YLE, Lu HM, Kooy H, MacDonald SM. Proton radiotherapy for high-risk pediatric neuroblastoma: early outcomes and dose comparison. Int J Radiat Oncol Biol Phys 2012;83:1015–22. https://doi.org/10.1016/j.ijrobp.2011.08.035.

[58] Zhao Q, Liu Y, Zhang Y, Meng L, Wei J, Wang B, Wang H, Xin Y, Dong L, Jiang X. Role and toxicity of radiation therapy in neuroblastoma patients: a literature review. Crit Rev Oncol Hematol 2020;149:102924. https://doi.org/10.1016/j.critrevonc.2020.102924.

[59] Caussa L, Hijal T, Michon J, Helfre S. Role of palliative radiotherapy in the management of metastatic pediatric neuroblastoma: a retrospective single-institution study. Int J Radiat Oncol Biol Phys 2011;79:214–9. https://doi.org/10.1016/j.ijrobp.2009.10.031.

[60] Matthay KK, George RE, Yu AL. Promising therapeutic targets in neuroblastoma. Clin Cancer Res 2012;18:2740–53. https://doi.org/10.1158/1078-0432.CCR-11-1939.

[61] Rubio PM, Galán V, Rodado S, Plaza D, Martínez L. MIBG therapy for neuroblastoma: precision achieved with dosimetry, and concern for false responders. Front Med 2020;7. https://doi.org/10.3389/fmed.2020.00173.

[62] Hung JT, Yu AL. GD2-targeted immunotherapy of neuroblastoma. In: Ray SK, editor. Neuroblastoma mol. mech. ther. interv. Academic Press; 2019. p. 63–78. https://doi.org/10.1016/B978-0-12-812005-7.00004-7.

[63] Mora J. Dinutuximab for the treatment of pediatric patients with high-risk neuroblastoma. Expert Rev Clin Pharmacol 2016;9:647–53. https://doi.org/10.1586/17512433.2016.1160775.

[64] Zaenglein AL, Pathy AL, Schlosser BJ, Alikhan A, Baldwin HE, Berson DS, et al. Guidelines of care for the management of acne vulgaris. J Am Acad Dermatol 2016;74:945–73. https://doi.org/10.1016/j.jaad.2015.12.037.

[65] Lowenstein EB, Lowenstein EJ. Isotretinoin systemic therapy and the shadow cast upon dermatology's downtrodden hero. Clin Dermatol 2011;29:652–61. https://doi.org/10.1016/j.clindermatol.2011.08.026.

[66] Prevost N, English JC. Isotretinoin: update on controversial issues. J Pediatr Adolesc Gynecol 2013;26:290–3. https://doi.org/10.1016/j.jpag.2013.05.007.

[67] Trigg RM, Turner SD. ALK in neuroblastoma: biological and therapeutic implications. Cancers (Basel) 2018;10:113. https://doi.org/10.3390/cancers10040113.

[68] Bresler SC, Weiser DA, Huwe PJ, Park JH, Krytska K, Ryles H, Laudenslager M, Rappaport EF, Wood AC, McGrady PW, Hogarty MD, London WB, Radhakrishnan R, Lemmon MA, Mossé YP. ALK mutations confer differential oncogenic activation and sensitivity to ALK inhibition therapy in neuroblastoma. Cancer Cell 2014;26:682–94. https://doi.org/10.1016/j.ccell.2014.09.019.

[69] Applebaum MA, Desai AV, Glade Bender JL, Cohn SL. Emerging and investigational therapies for neuroblastoma. Expert Opin Orphan Drugs 2017;5:355–68. https://doi.org/10.1080/21678707.2017.1304212.

[70] Mossé YP, Lim MS, Voss SD, Wilner K, Ruffner K, Laliberte J, Rolland D, Balis FM, Maris JM, Weigel BJ, Ingle AM, Ahern C, Adamson PC, Blaney SM. Safety and activity of crizotinib for paediatric patients with refractory solid tumours or anaplastic large-cell lymphoma: a children's Oncology Group phase 1 consortium study. Lancet Oncol 2013;14:472–80. https://doi.org/10.1016/S1470-2045(13)70095-0.

[71] Huang M, Weiss WA. Neuroblastoma and MYCN. Cold Spring Harb Perspect Med 2013;3:a014415. https://doi.org/10.1101/cshperspect.a014415.

[72] Jiang G, Deng W, Liu Y, Wang C. General mechanism of JQ1 in inhibiting various types of cancer. Mol Med Rep 2020;21:1021–34. https://doi.org/10.3892/mmr.2020.10927.

[73] Filippakopoulos P, Qi J, Picaud S, Shen Y, Smith WB, Fedorov O, Morse EM, Keates T, Hickman TT, Felletar I, Philpott M, Munro S, McKeown MR, Wang Y, Christie AL, West N, Cameron MJ, Schwartz B, Heightman TD, La Thangue N, French CA, Wiest O, Kung AL, Knapp S, Bradner JE. Selective inhibition of BET bromodomains. Nature 2010;468:1067–73. https://doi.org/10.1038/nature09504.

[74] Puissant A, Frumm SM, Alexe G, Bassil CF, Qi J, Chanthery YH, Nekritz EA, Zeid R, Gustafson WC, Greninger P, Garnett MJ, Mcdermott U, Benes CH, Kung AL, Weiss WA, Bradner JE, Stegmaier K. Targeting MYCN in neuroblastoma by BET bromodomain inhibition. Cancer Discov 2013;3:309–23. https://doi.org/10.1158/2159-8290.CD-12-0418.

[75] Zhu J, Wang J, Zhen ZJ, Lu SY, Zhang F, Sun FF, Li PF, Huang JT, Cai RQ, Sun XF. Brain metastasis in children with stage 4 neuroblastoma after multidisciplinary treatment. Chin J Cancer 2015;34:531–7. https://doi.org/10.1186/s40880-015-0038-2.

[76] Kumar MD, Dravid A, Kumar A, Sen D. Gene therapy as a potential tool for treating neuroblastoma—a focused review. Cancer Gene Ther 2016;23:115–24. https://doi.org/10.1038/cgt.2016.16.

[77] Wold W, Toth K. Adenovirus vectors for gene therapy, vaccination and cancer gene therapy. Curr Gene Ther 2014;13:421–33. https://doi.org/10.2174/1566523213666131125095046.

[78] Solly SK, Trajcevski S, Frisén C, Holzer GW, Nelson E, Clerc B, Abordo-Adesida E, Castro M, Lowenstein P, Klatzmann D. Replicative retroviral vectors for cancer gene therapy. Cancer Gene Ther 2003;10:30–9. https://doi.org/10.1038/sj.cgt.7700521.

[79] Ogris M, Wagner E. Targeting tumors with non-viral gene delivery systems. Drug Discov Today 2002;7:479–85. https://doi.org/10.1016/S1359-6446(02)02243-2.

[80] Shen M, Gong F, Pang P, Zhu K, Meng X, Wu C, Wang J, Shan H, Shuai X. An MRI-visible non-viral vector for targeted Bcl-2 siRNA delivery to neuroblastoma. Int J Nanomedicine 2012;7:3319–32. https://doi.org/10.2147/IJN.S32900.

[81] Morandi F, Frassoni F, Ponzoni M, Brignole C. Novel immunotherapeutic approaches for neuroblastoma and malignant melanoma. J Immunol Res 2018;2018. https://doi.org/10.1155/2018/8097398.

[82] Gilabert-Oriol R, Chernov L, Deyell RJ, Bally MB. Developing liposomal nanomedicines for treatment of patients with neuroblastoma. In: Grumezescu AM, editor. Lipid Nanocarriers Drug Target. Elsevier; 2018. p. 361–411.

[83] Xin Y, Yin M, Zhao L, Meng F, Luo L. Recent progress on nanoparticle-based drug delivery systems for cancer therapy. Cancer Biol Med 2017;14:228–41. https://doi.org/10.20892/j.issn.2095-3941.2017.0052.

[84] Chan Y, Ng SW, Chellappan DK, Madheswaran T, Zeeshan F, Kumar P, et al. Celastrol-loaded liquid crystalline nanoparticles as an anti-inflammatory intervention for the treatment of asthma. Int J Polym Mater Polym Biomater 2020;1–10. https://doi.org/10.1080/00914037.2020.1765350. In press.

[85] Chan Y, Ng SW, Mehta M, Anand K, Kumar Singh S, Gupta G, Chellappan DK, Dua K. Advanced drug delivery systems can assist in managing influenza virus infection: a hypothesis. Med Hypotheses 2020;144:110298. https://doi.org/10.1016/j.mehy.2020.110298.

[86] Tan YY, Yap PK, Xin Lim GL, Mehta M, Chan Y, Ng SW, Kapoor DN, Negi P, Anand K, Singh SK, Jha NK, Lim LC, Madheswaran T, Satija S, Gupta G, Dua K, Chellappan DK. Perspectives and advancements in the design of nanomaterials for targeted cancer theranostics. Chem Biol Interact 2020;329:109221. https://doi.org/10.1016/j.cbi.2020.109221.

[87] Golombek SK, May JN, Theek B, Appold L, Drude N, Kiessling F, Lammers T. Tumor targeting via EPR: strategies to enhance patient responses. Adv Drug Deliv Rev 2018;130:17–38. https://doi.org/10.1016/j.addr.2018.07.007.

[88] Shi Y, van der Meel R, Chen X, Lammers T. The EPR effect and beyond: strategies to improve tumor targeting and cancer nanomedicine treatment efficacy. Theranostics 2020;10:7921–4. https://doi.org/10.7150/thno.49577.

[89] Suk JS, Xu Q, Kim N, Hanes J, Ensign LM. PEGylation as a strategy for improving nanoparticle-based drug and gene delivery. Adv Drug Deliv Rev 2016;99:28–51. https://doi.org/10.1016/j.addr.2015.09.012.

[90] Yoo J, Park C, Yi G, Lee D, Koo H. Active targeting strategies using biological ligands for nanoparticle drug delivery systems. Cancers (Basel) 2019;11(5). https://doi.org/10.3390/cancers11050640, 640.

[91] Roma-Rodrigues C, Mendes R, Baptista PV, Fernandes AR. Targeting tumor microenvironment for cancer therapy. Int J Mol Sci 2019;20(4). https://doi.org/10.3390/ijms20040840, 840.

[92] Uthaman S, Huh KM, Park IK. Tumor microenvironment-responsive nanoparticles for cancer theragnostic applications. Biomater Res 2018;22. https://doi.org/10.1186/s40824-018-0132-z, 22.

[93] Zhong S, Jeong JH, Chen Z, Chen Z, Luo JL. Targeting tumor microenvironment by small-molecule inhibitors. Transl Oncol 2020;13:57–69. https://doi.org/10.1016/j.tranon.2019.10.001.

[94] Feng L, Dong Z, Tao D, Zhang Y, Liu Z. The acidic tumor microenvironment: a target for smart cancer nano-theranostics. Natl Sci Rev 2018;5:269–86. https://doi.org/10.1093/nsr/nwx062.

[95] Kullberg M, McCarthy R, Anchordoquy TJ. Systemic tumor-specific gene delivery. J Control Release 2013;172:730–6. https://doi.org/10.1016/j.jconrel.2013.08.300.

[96] Sung YK, Kim SW. Recent advances in the development of gene delivery systems. Biomater Res 2019;23:1–7. https://doi.org/10.1186/s40824-019-0156-z.

[97] Hossen S, Hossain MK, Basher MK, Mia MNH, Rahman MT, Uddin MJ. Smart nanocarrier-based drug delivery systems for cancer therapy and toxicity studies: a review. J Adv Res 2019;15:1–18. https://doi.org/10.1016/j.jare.2018.06.005.

[98] Ning J, Rabkin SD. Current status of gene therapy for brain tumors. In: Laurence J, Franklin M, editors. Translating gene therapy to the clinic: techniques and approaches. Elsevier; 2015. p. 305–23.

[99] Worgall S, Crystal RG. Gene therapy. In: Lanza R, Langer R, Vacanti J, editors. Principles of tissue engineering. 4th ed. Elsevier Inc.; 2013. p. 657–86.

[100] Listopad JJ, Kammertoens T, Anders K, Silkenstedt B, Willimsky G, Schmidt K, Kuehl AA, Loddenkemper C, Blankenstein T. Fas expression by tumor stroma is required for cancer eradication. Proc Natl Acad Sci USA 2013;110:2276–81. https://doi.org/10.1073/pnas.1218295110.

[101] Cho HS, Song JY, Park CY, Lyu CJ, Kim BS, Kim KY. Retroviral-mediated IL-2 gene transfer into murine neuroblastoma. Yonsei Med J 2000;41:76–81. https://doi.org/10.3349/ymj.2000.41.1.76.

[102] Prapa M, Caldrer S, Spano C, Bestagno M, Golinelli G, Grisendi G, et al. A novel anti-GD2/4-1BB chimeric antigen receptor triggers neuroblastoma cell killing. Oncotarget 2015;6:24884–94. https://doi.org/10.18632/oncotarget.4670.

[103] Walsh MP, Seto J, Liu EB, Dehghan S, Hudson NR, Lukashev AN, Ivanova O, Chodosh J, Dyer DW, Jones MS, Seto D. Computational analysis of two species C human adenoviruses provides evidence of a novel virus. J Clin Microbiol 2011;49:3482–90. https://doi.org/10.1128/JCM.00156-11.

[104] Van Maerken T, Sarkar D, Speleman F, Dent P, Weiss WA, Fisher PB. Adenovirus-mediated hPNPaseold-35 gene transfer as a therapeutic strategy for neuroblastoma. J Cell Physiol 2009;219:707–15. https://doi.org/10.1002/jcp.21719.

[105] Tanimoto T, Tazawa H, Ieda T, Nouso H, Tani M, Oyama T, Urata Y, Kagawa S, Noda T, Fujiwara T. Elimination of MYCN-amplified neuroblastoma cells by telomerase-targeted oncolytic virus via MYCN suppression. Mol Ther Oncolytics 2020;18:14–23. https://doi.org/10.1016/j.omto.2020.05.015.

[106] García-Castro J, Alemany R, Cascalló M, Martínez-Quintanilla J, Del Mar Arriero M, Lassaletta Á, Madero L, Ramírez M. Treatment of metastatic neuroblastoma with systemic oncolytic virotherapy delivered by autologous mesenchymal stem cells: an exploratory study. Cancer Gene Ther 2010;17:476–83. https://doi.org/10.1038/cgt.2010.4.

[107] Franco-Luzón L, González-Murillo Á, Alcántara-Sánchez C, García-García L, Tabasi M, Huertas AL, et al. Systemic oncolytic adenovirus delivered in mesenchymal carrier cells modulate tumor infiltrating immune cells and tumor microenvironment in mice with neuroblastoma. Oncotarget 2020;11:347–61. https://doi.org/10.18632/oncotarget.27401.

[108] Wang D, Tai PWL, Gao G. Adeno-associated virus vector as a platform for gene therapy delivery. Nat Rev Drug Discov 2019;18:358–78. https://doi.org/10.1038/s41573-019-0012-9.

[109] Streck CJ, Dickson PV, Ng CYC, Zhou J, Gray JT, Nathwani AC, Davidoff AM. Adeno-associated virus vector-mediated systemic delivery of IFN-β combined with low-dose cyclophosphamide affects tumor regression in murine neuroblastoma models. Clin Cancer Res 2005;11:6020–9. https://doi.org/10.1158/1078-0432.CCR-05-0502.

[110] Streck CJ, Zhang Y, Zhou J, Ng C, Nathwani AC, Davidoff AM. Adeno-associated virus vector-mediated delivery of pigment epithelium-derived factor restricts neuroblastoma angiogenesis and growth. J Pediatr Surg 2005;40:236–43. https://doi.org/10.1016/j.jpedsurg.2004.09.049.

[111] Artusi S, Miyagawa Y, Goins W, Cohen J, Glorioso J. Herpes simplex virus vectors for gene transfer to the central nervous system. Diseases 2018;6:74. https://doi.org/10.3390/diseases6030074.

[112] Gillory LA, Megison ML, Stewart JE, Mroczek-Musulman E, Nabers HC, Waters AM, Kelly V, Coleman JM, Markert JM, Gillespie GY, Friedman GK, Beierle EA. Preclinical evaluation of engineered oncolytic herpes simplex virus for the treatment of neuroblastoma. PLoS One 2013;8. https://doi.org/10.1371/journal.pone.0077753, e77753.

[113] Tang Y, Akubulut H, Maynard J, Ped-Ersen L, Deisseroth A, 1 Farrell C, Zaupa CM, Martuza RL, Rabkin SD, Curry WT. 169. Combination oncolytic herpes simplex virus and dendritic cell immunotherapy for the treatment of established murine neuroblastomas. Mol Ther 2007;15:S64–5. https://doi.org/10.1016/s1525-0016(16)44375-3.

[114] Hidai C, Kitano H. Nonviral gene therapy for cancer: a review. Diseases 2018;6:57. https://doi.org/10.3390/diseases6030057.

[115] Zhang XX, McIntosh TJ, Grinstaff MW. Functional lipids and lipoplexes for improved gene delivery. Biochimie 2012;94:42–58. https://doi.org/10.1016/j.biochi.2011.05.005.

[116] Mohammadinejad R, Dehshahri A, Sagar Madamsetty V, Zahmatkeshan M, Tavakol S, Makvandi P, Khorsandi D, Pardakhty A, Ashrafizadeh M, Ghasemipour Afshar E, Zarrabi A. In vivo gene delivery mediated by non-viral vectors for cancer therapy. J Control Release 2020;325:249–75. https://doi.org/10.1016/j.jconrel.2020.06.038.

[117] Ramamoorth M, Narvekar A. Non viral vectors in gene therapy—an overview. J Clin Diagn Res 2015;9:GE01–6. https://doi.org/10.7860/JCDR/2015/10443.5394.

[118] Walker GF, Wagner E. Nonviral vector systems for cancer gene therapy. In: Curiel DT, Douglas JT, editors. Cancer gene therapy. Humana Press; 2007. p. 367–78.

[119] Mobasheri T, Rayzan E, Shabani M, Hosseini M, Mahmoodi Chalbatani G, Rezaei N. Neuroblastoma-targeted nanoparticles and novel nanotechnology-based treatment methods. J Cell Physiol 2020. https://doi.org/10.1002/jcp.29979. jcp.29979.

[120] O'Neill CP, Dwyer RM. Nanoparticle-based delivery of tumor suppressor microRNA for cancer therapy. Cell 2020;9:521. https://doi.org/10.3390/cells9020521.

[121] Di Paolo D, Brignole C, Pastorino F, Carosio R, Zorzoli A, Rossi M, Loi M, Pagnan G, Emionite L, Cilli M, Bruno S, Chiarle R, Allen TM, Ponzoni M, Perri P. Neuroblastoma-targeted nanoparticles entrapping siRNA specifically knockdown ALK. Mol Ther 2011;19:1131–40. https://doi.org/10.1038/mt.2011.54.

[122] Orr WS, Denbo JW, Saab KR, Myers AL, Ng CY, Zhou J, Morton CL, Pfeffer LM, Davidoff AM. Liposome-encapsulated curcumin suppresses neuroblastoma growth through nuclear factor-kappa B inhibition. Surgery 2012;151:736–44. https://doi.org/10.1016/j.surg.2011.12.014.

[123] Bhunia S, Radha V, Chaudhuri A. CDC20siRNA and paclitaxel co-loaded nanometric liposomes of a nipecotic acid-derived cationic amphiphile inhibit xenografted neuroblastoma. Nanoscale 2017;9:1201–12. https://doi.org/10.1039/c6nr07532k.

[124] Bono N, Ponti F, Mantovani D, Candiani G. Non-viral in vitro gene delivery: it is now time to set the bar! Pharmaceutics 2020;12:183. https://doi.org/10.3390/pharmaceutics12020183.

[125] Zakeri A, Kouhbanani MAJ, Beheshtkhoo N, Beigi V, Mousavi SM, Hashemi SAR, Karimi Zade A, Amani AM, Savardashtaki A, Mirzaei E, Jahandideh S, Movahedpour A. Polyethylenimine-based nanocarriers in co-delivery of drug and gene: a developing horizon. Nano Rev Exp 2018;9:1488497. https://doi.org/10.1080/20022727.2018.1488497.

[126] Stegantseva MV, Shinkevich VA, Tumar EM, Meleshko AN. Conjugation of new DNA vaccine with polyethylenimine induces cellular immune response and tumor regression in neuroblastoma mouse model. Exp Oncol 2020;42:120–5. https://doi.org/10.32471/exp-oncology.2312-8852.vol-42-no-2.14473.

[127] Askarian S, Abnous K, Darroudi M, Oskuee RK, Ramezani M. Gene delivery to neuroblastoma cells by poly (L-lysine)-grafted low molecular weight polyethylenimine copolymers. Biologicals 2016;44:212–8. https://doi.org/10.1016/j.biologicals.2016.03.007.

[128] Ke L, Cai P, Wu Y, Chen X. Polymeric nonviral gene delivery systems for cancer immunotherapy. Adv Ther 2020;3:1900213. https://doi.org/10.1002/adtp.201900213.

[129] Kim SW. Polylysine copolymers for gene delivery. Cold Spring Harb Protoc 2012;7:433–8. https://doi.org/10.1101/pdb.ip068619.

[130] Fana M, Gallien J, Srinageshwar B, Dunbar GL, Rossignol J. PAMAM dendrimer nanomolecules utilized as drug delivery systems for potential treatment of glioblastoma: a systematic review. Int J Nanomedicine 2020;15:2789–808. https://doi.org/10.2147/IJN.S243155.

[131] Abedi-Gaballu F, Dehghan G, Ghaffari M, Yekta R, Abbaspour-Ravasjani S, Baradaran B, Dolatabadi JEN, Hamblin MR. PAMAM dendrimers as efficient drug and gene delivery nanosystems for cancer therapy. Appl Mater Today 2018;12:177–90. https://doi.org/10.1016/j.apmt.2018.05.002.

[132] Castro RI, Forero-Doria O, Guzmán L. Perspectives of dendrimer-based nanoparticles in cancer therapy. An Acad Bras Cienc 2018;90:2331–46. https://doi.org/10.1590/0001-3765201820170387.

[133] Xiong Z, Shen M, Shi X. Dendrimer-based strategies for cancer therapy: recent advances and future perspectives. Sci China Mater 2018;61:1387–403. https://doi.org/10.1007/s40843-018-9271-4.

[134] Dąbkowska M, Łuczkowska K, Rogińska D, Sobuś A, Wasilewska M, Ulańczyk Z, Machaliński B. Novel design of (PEG-ylated)PAMAM-based nanoparticles for sustained delivery of BDNF to neurotoxin-injured differentiated neuroblastoma cells. J Nanobiotechnol 2020;18:120. https://doi.org/10.1186/s12951-020-00673-8.

[135] Senapati S, Mahanta AK, Kumar S, Maiti P. Controlled drug delivery vehicles for cancer treatment and their performance. Signal Transduct Target Ther 2018;3:1–19. https://doi.org/10.1038/s41392-017-0004-3.

[136] Yang Y, Yu Y, Wang J, Li Y, Li Y, Wei J, Zheng T, Jin M, Sun Z. Silica nanoparticles induced intrinsic apoptosis in neuroblastoma SH-SY5Y cells via CytC/Apaf-1 pathway. Environ Toxicol Pharmacol 2017;52:161–9. https://doi.org/10.1016/j.etap.2017.01.010.

[137] Babaei M, Eshghi H, Abnous K, Rahimizadeh M, Ramezani M. Promising gene delivery system based on polyethylenimine-modified silica nanoparticles. Cancer Gene Ther 2017;24:156–64. https://doi.org/10.1038/cgt.2016.73.

Chapter 24

Nanoparticulate systems and their translation potential for breast cancer therapeutics

Shashank Chaturvedi[a] and Kamla Pathak[b]
[a]Institute of Pharmaceutical Research, GLA University, Mathura, Uttar Pradesh, India, [b]Pharmacy College Saifai, Uttar Pradesh University of Medical Sciences, Etawah, Uttar Pradesh, India

1 Introduction

Cancer is regarded as a complex medical condition arising due to the uncontrolled proliferation of cells. The rapidly growing cells tend to migrate from their origin to the distant parts of the body. This condition is regarded as metastasis. In the past, metastasis has been recognized as the principal cause (>90%) for cancer-related morbidities and mortalities [1]. Breast cancer (BC) is the most commonly occurring type of cancer among all the cancers in the female population, and the number of cases increased from 870.2 thousand to 1937.6 thousand from 1990 to 2017. Similar trends have been observed in the male population and the number of BC cases has shown a steep rise from 8.5 thousand in 1990 to 23.1 thousand during the same duration [2]. The metastatic spread in the BC has been associated prominently with the involvement of the lymphatic system [3]. The proliferating cells after gaining access into the lymph (the carrier) interact with the primary and/or the sentinel lymph nodes. The cancer cells thereafter invade the lymph nodes and generate secondary tumors [4].

The strategies for BC treatment include surgical removal of the malignant tumor, accompanied by radiation and/or chemotherapy [5]. But surgical intervention is of limited use in the advanced stages of BC emerging due to recurrence or metastasis. Additionally, nonselective delivery of chemotherapeutic agents precipitates serious side effects to normal cells [6]. Furthermore, multidrug resistance (MDR) due to presystemic first-pass effects by cytochrome P450 3A4 (CYP3A4), para-glycoprotein pump (P-gp)-mediated efflux, low aqueous solubility (tamoxifen, rubitecan, sorafenib, gefitinib) and/or permeability (doxorubicin, anastrozole, cyclophosphamide, letrozole, methotrexate) cumulatively restrict the therapeutic benefits of anticancer drugs [7].

2 BC identifiers and the role of P-gp and CYP3A4 in limiting cancer therapeutics

Recent progress in the field of tumor biology and nanoparticulate delivery systems has opened new frontiers for effective and safe delivery of anticancer drugs. Therefore, identifying the clinically measurable biomarkers and developing targeted nanoparticulate cargo according to them offers multiple benefits like low dose, site-specific delivery and enhanced bioavailability. The most frequently identified biological markers associated with BC are the progesterone receptor (PR), estrogen receptor (ER) and human epidermal growth factor receptor (HER2/ERBB2) [8]. The ER overexpression has been ranked number one, while HER2 overexpression accounts for 25% of the population diagnosed with BC. However, there have been few BC cases that do not express any of the above biomarkers and are difficult to treat. These are considered triple-negative BC (TNBC) [9,10].

There are multiple transportation systems in the body like P-gp, cytoplasmic transport, breast cancer resistant proteins (BCRP), multidrug resistance-associated proteins (MRP) which limit the access of administered anticancer drugs at the target site [11,12]. The P-gp encrypted alongside the multidrug resistance-1 (MDR1) gene is present at the enterocytes and has been regarded as the major drug efflux pump. It is a membrane-linked protein and has a relationship with the super-family of ATP binding cassette (ABC) transporter. This limits access to drugs into the intestinal lumen [13]. Multiple models have been developed for explaining the mechanism of P-gp including flippase, pore, and hydrophobic vacuum cleaner (HVC) model. The HVC model has been accepted widely [14]. A drug molecule is effluxed out owing to ATP binding and subsequent hydrolysis.

Advanced Drug Delivery Systems in the Management of Cancer. https://doi.org/10.1016/B978-0-323-85503-7.00021-3

The net outcome is the expense of two ATP molecules [15]. Additionally, if a drug is a substrate to CYP3A4 this tandem situation becomes even more critical. The CYP3A4 is available in the enterocytes, causing the metabolism of substrate drugs at the luminal wall. Even if the drug absorbs into the systemic circulation, the portal circulation induces the metabolism of anticancer drugs. The net outcome is reduced bioavailability and poor therapeutics.

3 Nanoparticulate systems in BC therapeutics

These systems comprise particles in the range of 1–100 nm. Furthermore, lipid-based drug delivery systems (LBDDS) encompass P-gp and CYP3A4 modulators like Tween 80, Solutol HS-15, TPGS, Cremophor EL, thereby overcoming their associated effects [16,17]. Additionally, recent trends have depicted the lymphatic targeting ability of LBDDS [18] thereby imparting a cumulative increment in the cancer therapeutics. Furthermore, the nanoparticulate systems can be employed in passive targeting utilizing the EPR effects. The EPR effect is related to the enhanced angiogenesis by proliferating cells to make up their higher demand for nutrients and oxygen. Angiogenesis facilitates irregularities in the blood vessels and heterogeneous epithelium. This heterogeneous make-up of the epithelium promotes the permeation of nanoparticulate systems (through the interstitial spaces) compared to that of the normal tissues. Furthermore, the lymphatic system is less functional in the tumors, thereby leading to enhanced accumulation [19]. However, these strategies sometimes fail during the clinical translation. Therefore, active targeting strategies have emerged in recent years employing ligands that are specific to receptor overexpressed on the tumor. This facilitates bypass of the reticuloendothelial system, enhanced uptake by the tumor cells, and an overall enhancement in the bioavailability of anticancer targeting [20–22]. A schematic representation of passive and/or active targeting strategies employing nanoparticulate systems have been depicted in Fig. 1.

The nanoparticulate systems explored to tackle the metastatic spread and overcoming the MDR in BC with passive and/or active targeting abilities are liposomes [23, 24], solid lipid nanoparticle (SLN) [25–27], nanostructured lipid carriers (NLCs) [28–30], nanoemulsions [31], self-emulsifying drug delivery systems (SEDDS) [32, 33], lipid nanocapsules (LNC) [34, 35], lipoproteins [36, 37], polymeric micelles (PMs) [38, 39], mixed micelles (MC) [40–42], niosomes [43, 44], carbon nanotubes (CNT) [45, 46], and dendrimers [47, 48]. Further, there are some other nanoparticulate systems explored in recent years like metallic nanoparticles [49], mesoporous silica nanoparticles [50], and viral nanoparticles [51] for the said purpose, these have also been covered in this chapter. Fig. 2 depicts the types of nanoparticulate systems explored in BC therapeutics.

4 Liposomes

Liposomes have been a major thrust area in the past with major objectives to enhance the therapeutic efficacy of anticancer drugs [52]. These were the first LBDDS developed and successfully approved for clinical use in the treatment of different cancers. The first liposomal anticancer formulation approved by FDA for cancer therapeutics employing passive targeting abilities was Doxil in 1995 [50]. The main components of liposomes are phospholipids and cholesterol, which together constitute a spherical lipid bilayer structure of liposomes. The vesicle size can be in the range of 25 nm to 2.5 μm with single or multiple bilayer membranes [53]. The liposomes have been classified on the basis of their particle size and the number of bilayers into two categories namely unilamellar vesicles (ULVs) and multilamellar vesicles (MLVs). The ULVs comprise of single bilipid layer whereas the MLVs have more than one bilipid layer. The latter resembles like an onion. The ULVs have further been classified into viz. large ULVs and small ULVs [54]. These vesicles can be tailored for the specific delivery (active targeting) of the cytotoxic drugs to the proliferating cells employing specific antibodies [55], or ligands like transferrin [56], estrone [57], and folate [58]. These tailored vesicles bind selectively to receptors overexpressed on the tumor cells. Additionally, numerous attempts have been made to target the HER-2 positive and TNBC. In agreement to which Zahmatkeshan et al. (2016) reported an enhancement in the HER2 specific targeting (overexpressed in BC) and therapeutic outcomes from PEGylated liposomal formulation of doxorubicin as compared to free drug. The PEGylated liposomes were engineered with a peptide ligand AHNP (Anti-HER2/neu peptide). The efficacy of the system was evaluated in TUBO BC tumor model in mice [59]. Recently, Burande et al. (2020) reported selective targeting to TNBC with cetuximab conjugated TPGS-based liposomes of paclitaxel with or without piperine. The cellular uptake studies and cell viability assay studies on MDA-MB-231 cells exhibited enhanced tumor uptake and cytotoxic effects compared to nonconjugated formulations. The concomitant use of paclitaxel and piperine provided synergistic anticancer effects [60]. The preclinical studies on different liposomal anticancer formulations have been presented in Table 1.

5 Solid lipid nanoparticles and nanostructured lipid carriers

The SLNs are particles in the submicron size range (50–1000 nm), that have emerged as a safer alternative in comparison to the inorganic and polymeric nanoparticulate systems for anticancer drug delivery [65]. These systems are generally developed employing biodegradable and nontoxic lipid constituents, surfactants with occasional incorporation of the

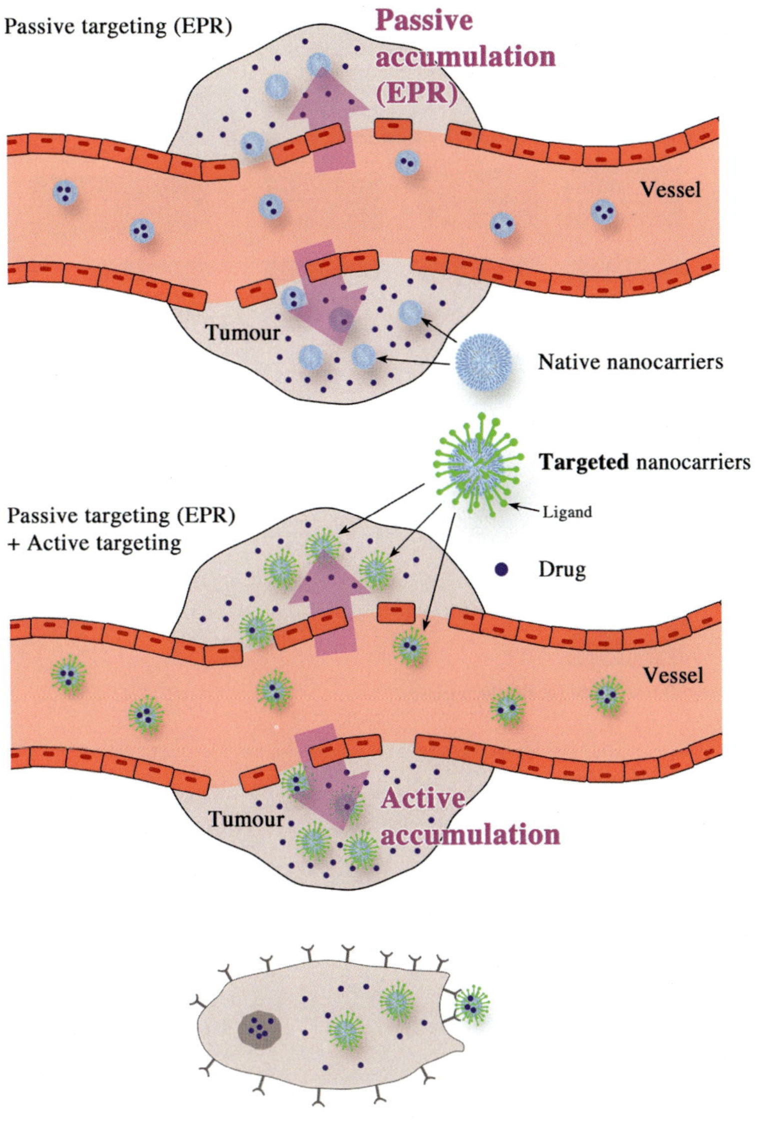

FIG. 1 Scheme illustrating the passive targeting (EPR) and the active targeting into a tumor. *Reproduced with permission from Attia MF, Anton N, Wallyn J, Omran Z, Vandamme TF. An overview of active and passive targeting strategies to improve the nanocarriers efficiency to tumour sites. J Pharm Pharmacol 2019;71(8):1185–98. Available from: https://onlinelibrary.wiley.com/doi/abs/10.1111/jphp.13098. Copyright 2019 John Wiley and sons.*

cosurfactants and are solid at physiological temperature. The lipid constituents employed for the preparation of SLN/NLC are approved by Food Drug and Administration (FDA) and are also enlisted in the generally regarded as safe (GRAS) list [66]. These systems offer multiple advantages like enhanced entrapment efficiency for hydrophilic and hydrophobic drugs, higher surface area, controlled release, and are easy to scale-up [67]. SLNs can be prepared in a large scale by employing technologies like high-pressure homogenization (HPH) [67] with hot/cold processing and solvent emulsification/evaporation process [68]. The preparation of SLN by HPH requires an initial dispersion/dissolution of the drug in the molten lipid. The drug-lipid molten mixture is then passed through a tapered passage in a homogenizer maintained under high pressure (up to 200 bars). The mixture under the influence of enormous forces transforms into a nanoparticulate system [69]. NLCs encompass the characteristics of both the solid and liquid lipids and are derived from SLNs. These systems are designed to reduce the limitations associated with SLNs viz. poor drug loading efficiency, alteration in the lipid matrix structures during storage which often leads to drug loss [70]. Hence, these systems offer stability to the encapsulated drug molecules during storage. NLCs can be prepared with similar techniques to those addressed for the SLN.

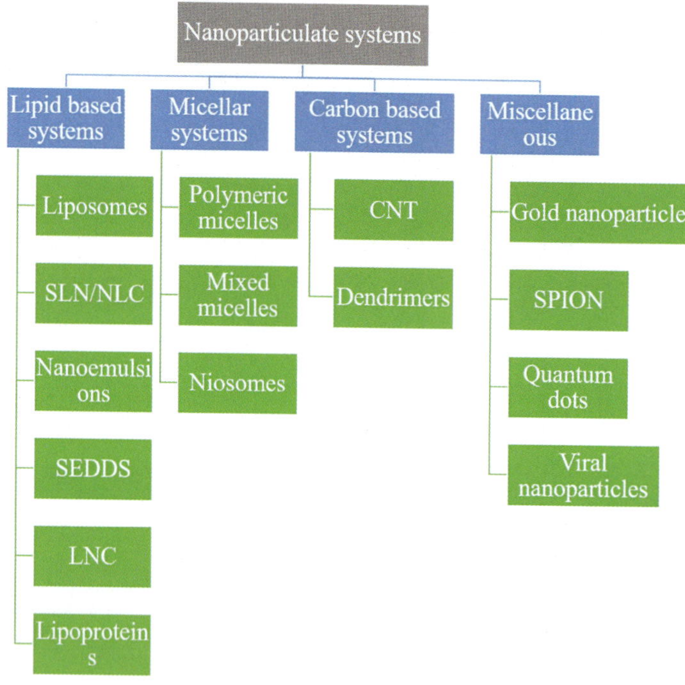

FIG. 2 Various nanoparticulate systems researched for BC therapeutics.

TABLE 1 Recent preclinical studies depicting the applicability of liposomes in the BC therapeutics.

Active moiety	Formulation/modification	Breast cancer target	Outcome	Reference
Docetaxel-trastuzumab	Stealth immunoliposomes (antibody nanoconjugate-1 and artificial lipids)	HER-2 + breast cancer	In vitro antiproliferative studies on the negative (MDA-MB-231), positive (MDA-MB-453), and overexpressing (SKBR3) HER2 human breast cancer cell models revealed that the formulations exhibited the highest uptake in SKBR3	[61]
Doxorubicin	Bispecific PEG-binding antibodies (PEG engagers) were used for selective targeting of PEGylated nanomedicines	$EGFR^{+}$ TNBC	The use of PEG engagers provided enhanced antiproliferative effects up to 100 times for PEGylated liposomal formulations of doxorubicin in TNBC human xenograft model	[62]
Doxorubicin	Aptamer-labeled liposomal nanoparticles	HER-2 + breast cancer	Liposomal formulation for docetaxel exhibited enhanced cytotoxicity in MCF-7 and SKBR-3 cell lines. Furthermore, aptamer-labeled formulation provided greater than 60% cellular uptake in MCF-7 and SKBR-3 cell lines as compared with nontargeted formulations	[63]
Epirubicin and paclitaxel	PEGylated epirubicin and paclitaxel loaded liposomal formulations linked with estrone ligand	ER	In vitro cellular uptake, studies revealed enhanced uptake of the developed formulation in MCF-7 cell lines in contrast with the nontargeted counterparts. In vivo targeting studies in a tumor-bearing mouse model (MCF-7 derived) proved enhanced accumulation in the tumor cells and suppressed tumor growth	[64]

5.1 SLNs and NLCs in BC targeting

SLN/NLC-incorporated anticancer drugs have been frequently explored with passive and/or active targeting capabilities for BC therapeutics in the recent past [71–73]. In agreement with which doxorubicin-incorporated SLNs exhibited enhanced apoptotic cell death of P-gp overexpressed MCF-7/ADR cell lines (a doxorubicin-resistant cell lines). The probable mechanism for this enhanced anticancer effect was attributed to an increased accumulation (EPR effect) of doxorubicin in the MCF-7/ADR cell lines. Hence, proving the efficacy of developed SLN in doxorubicin-resistant cancer cells [74]. Chen et al. (2015) developed a pH-sensitive cholesterol-PEG-coated trilaurin-based SLN formulation of doxorubicin and investigated in vivo antiproliferative activity on MCF-7/MDR xenografted tumor Balb/c nude mice. The results revealed an enhanced accumulation of doxorubicin in the tumor through EPR effects. Furthermore, cholesterol-PEG coating on SLN enabled P-gp inhibitory effect on the tumor cells with reduced side effects compared to the control. Hence, this study exhibited enhanced efficacy on MDR BC cells [75].

Similar results were obtained by delivering curcumin through folate-functionalized NLCs. The in vitro studies on MCF-7 cell lines and in vivo investigations on the tumor-bearing mouse revealed an enhanced cytotoxic potential for the developed functionalized formulations as compared to that of control, respectively [76]. Campos et al. (2017) reported an enhanced antiproliferative activity for paclitaxel employing surface-modified SLNs. The system comprises of SLNs in the core with layer-by-layer deposition of the chitosan and hyaluronic acid on the outermost layer for active targeting on the over expressed hyaluronic (CD44 receptors). The cytotoxic assay (MTT) and cellular uptake studies on MCF-7 cell lines revealed selective cytotoxicity and cellular uptake for the developed formulation vis-à-vis paclitaxel suspension and nonfunctionalized paclitaxel-incorporated SLNs [77]. Additionally, codelivering two anticancer drugs for synergistic anticancer effects can further address the condition of MDR. These include examples like curcumin and paclitaxel [78], curcumin and piperine [79] baicalein and doxorubicin [80], lapachone and doxorubicin [81], cisplatin and doxorubicin [82]. Recent preclinical reports on the passive and active targeting potential of SLNs and NLCs have been presented in Table 2.

6 Nanoemulsions

These nanoparticulate systems essentially comprise two immiscible phases, viz. oil and water. If the oil constitutes the internal phase and water is in the external phase or vice versa, it is regarded as an oil-in-water (o/w) or water-in-oil (w/o) type of nanoemulsion, respectively [87]. Normally emulsifier (hydrophilic and/or lipophilic) and often a coemulsifier is used to enhance their physical stability [88]. The nanosize of the internal phase (mean radii < 100 nm) offers multiple benefits like enhanced visibility, kinetic stability (up to 6 months), and greater solubilization potential as compared to that of coarse emulsions, and micellar dispersions [89]. The internal phase of coarse emulsions has a mean radius between 100 nm and 100 μm [90]. Hence, the nanoparticulate systems are free from destabilization processes that are associated with conventional emulsions like coalescence, sedimentation, and creaming. Furthermore, as the particle size of the internal phase is very small, therefore, it overcomes the effect of gravity and viscosity-mediated instabilities [91]. These multiparticulates when in the vicinity of a large volume of aqueous environment in vivo exhibit drug release through a series of events, namely (a) initial partitioning of the drug from the core into the emulsifier layer; (b) subsequent partitioning from the emulsifier layer into the aqueous phase; (c) nanoprecipitation of the partitioned drug. This nanoprecipitation step increases the surface area of the drug substantially, which in turn improves the dissolution rate of the drug (*Noyes-Whitney equation*) [88]. Hence, nanoemulsions have been investigated for positive alteration in the bioavailability of poorly water-soluble anticancer drugs like tamoxifen [92], doxorubicin [93,94], paclitaxel [95].

The nanoparticulate size (20–100 nm) of these systems can be exploited for enhanced accumulation of the anticancer drugs in the BC tissues exploiting the EPR effects [96]. However, nanoemulsions with a size range below 100 nm may undergo opsonization by mononuclear phagocytic system (MPS), thereby reducing their circulation time in vivo [97]. This shortcoming has been addressed by employing surface modifications on the nanoemulsion by hydrophilic polymers like poloxamers and polyethylene glycols (PEG) [98,99]. It reduces the risk of adsorption with the blood proteins and identification by the MPS cells. The net outcome is prolonged in vivo circulation time and enhanced efficacy [100]. In agreement with this, Afzal et al. (2016) reported enhanced tumor specificity (4.81- and 2.08-fold) for docetaxel-loaded folate-PEG-functionalized nanoemulsions as compared to that of control [101]. Recently, Tripathi et al. (2018) developed doxorubicin-incorporated folate-decorated nanoemulsions. They also employed α-linolenic acid as the oily component to further enhance the anticancer activity of doxorubicin. The developed system exhibited enhanced in vitro cytotoxicity in MCF-7 cell lines by reactive oxygen species/mitochondrial membrane-assisted apoptosis as compared to that of the pure drug or a commercial formulation. Additionally, in vivo studies in DMBA-mammary gland tumor model in female Albino Wistar rats revealed an enhancement in the tumor specificity and safety for doxorubicin in relation with the pure drug or a commercial formulation [102].

TABLE 2 Applications of SLNs and NLCs in BC therapeutics.

Active moiety	Formulation/ modification	BC target	Outcome	Reference
BMN 673 (Talazoparib)	SLN	BRCA1 Mutant TNBC	In vitro cell line studies in HCC1937 and HCC1937-R cells revealed enhanced cytotoxicity with SLN-based formulations as compared with BMN 673. The developed formulation was less lethal to MCF-10A healthy human BC cell line (epithelial type)	[83]
Docetaxel-curcumin (codelivery)	Folic acid functionalized SLN	Folate receptor	In vitro cytotoxic and cellular uptake studies in MCF-7 and MDA-MB-231 cell lines revealed enhanced cellular uptake and cytotoxic effects. Furthermore, in vivo biodistribution studies exhibited reduced accumulation in the other vital organs (heart and kidney) suggesting selective targeting to the breast cancer	[71]
Epigallocatechin-gallate (EGCG)	Bombesin-functionalized SLN	Gastrin-releasing peptide receptors (GRPR)	The bombesin conjugated formulation of EGCG exhibited enhanced cytotoxicity, higher cellular uptake, reduced tumor growth and an overall survival of the tumor-bearing animals as compared to plain EGCG/nonconjugated EGCG-SLN formulation	[84]
Letrozole	Folic-acid functionalized SLN	Estrogen receptor-positive	In vitro cytotoxicity evaluation revealed enhanced cytotoxicity on MCF-7 cell line and minimal cytotoxicity on MCF-10 A cell lines for functionalized formulations as compared to nonfunctionalized SLN and lone drug	[85]
Resveratrol	Folate-modified NLC	Folate receptors	In vitro cytotoxicity studies on MCF-7 cell lines revealed enhanced targeting specificity for functionalized NLC. Additionally, in vivo pharmacokinetic studies depicted a ninefold increment in the AUC values for folate modified NLC as compared with nonfunctionalized NLC	[73]
Docetaxel	Trastuzumab (Herceptin)-functionalized NLC	HER2 +	In vitro cytotoxicity studies with functionalized NLC exhibited enhanced cytotoxicity on BT-474 (HER2 positive) cells. Further, flow cytometry studies revealed enhanced cellular uptake on HER2-positive cell lines as compared with nonfunctionalized NLC	[86]

7 SEDDS

SEDDS are LBDDS that comprise of lipid-surfactant with or without the incorporation of a cosurfactant. In recent years, these systems have exhibited their potential in overcoming challenges associated with anticancer drug delivery like poor bioavailability, high first-pass effects, extensive P-gp-mediated efflux [103–105]. SEDDS are further classified as self-micro emulsifying (SMEDDs) and self-nanoemulsifying (SNEDDs). SMEDDs are thermodynamically stable formulations and have a globule size between 100 and 250nm. Whereas, SNEDDs have low thermodynamic but higher pharmacokinetic stability and their particle size is normally <100nm [106, 107]. Therefore, SNEDDs formulations resist any change under stress conditions like change in the temperature or dilution effects. However, the SMEDDs are sensitive to changes in temperature and/or dilution. The percentage of the oil phase varies from 40% to 80% in the SEDDS, while in SMEDDs/SNEDDs it is $>20\%$. Similarly, the percentage of surfactant varies from 30% to 40% in the SEDDS, and it is on the higher side (40%–80%) in the SMEDDs/SNEDDs. The surfactants with HLB value <12 are employed in the preparation of SEDDS, while hydrophilic surfactant with HLB value >12 is the essential part in the SMEDDs/SNEDDs [108].

7.1 SEDDS in BC

The poor aqueous solubility of anticancer drug resists their dissolution into the physiological systems; additionally, their activity is further restricted if the drug is poorly permeable. However, in the LBDDS the drug is in the solubilized state, thereby avoiding the requirement of dissolution step (a rate-limiting step). The lipid preconcentrate under the influence of gastric lipase activity transforms into a crude emulsion. These are further converted (pancreatic lipase) into a mixture of micelles and mixed micelles thereby potentiating the process of intestinal drug absorption through chylomicron formation and subsequent lymphatic transport [109]. The availability of CYP3A4 and P-gp modulators in the formulation further promotes their luminal absorption [110]. Hence, promoting overall bioavailability of anticancer drugs like resveratrol and coenzyme Q10 [111], paclitaxel [112, 113], docetaxel [114], curcumin [115]. Recent preclinical reports depicting the applicability of SMEDDs/SNEDDs in BC have been presented in Table 3.

TABLE 3 Preclinical reports on SMEDDs/SNEDDs for BC therapeutics.

Active moiety	Objective	Formulation/Modification	Outcome	Reference
Docetaxel	Improving biopharmaceutical attributes of docetaxel	Long-chain polyunsaturated fatty acids-based SNEDDs	In vitro permeation studies in Caco-2 cell line (bidirectional) exhibited enhanced permeation with less P-gp-mediated efflux. Moreover, ex vivo permeation and in situ single-pass intestinal perfusion studies further confirmed enhanced permeation by PUFA enriched SNEDDs	[116]
Docetaxel	Enhancing BC therapeutics	SNEDDs (ethyl oleate, Tween 80 and PEG 600 as lipid, emulsifier, and coemulsifier, respectively)	The in vitro cytotoxicity studies in MCF-7 cell lines exhibited IC_{50} value of $0.98 \pm 0.05\,\mu g\,mL^{-1}$. Furthermore, in vivo pharmacokinetic studies revealed enhancement in the C_{max} and AUC by 2- and 1.5-fold respectively, for docetaxel-loaded SNEDDs versus docetaxel solution	[117]
Docetaxel-cyclo-sporine A (codelivery)	Enhanced bioavailability by overcoming the P-gp, CYP3A4-mediated efflux, and metabolism, respectively	SNEDDS (Capryol 90, Cremophor EL, and Transcutol HP as lipid, emulsifier, and coemulsifier, respectively)	Codelivery approach employing SNEDDs-based formulation promoted intestinal permeation (intestinal permeation studies). Furthermore, bioavailability was enhanced by 3.4 and 9.2 times with codelivered SNEDDs formulations vis-à-vis docetaxel-loaded SNEDDs and docetaxel solution	[118]
Sunitinib	SNEDDs-based formulation for enhanced BC therapeutics	SNEDDs (ethyl oleate, Tween 80, and PEG 600 as lipid, emulsifier, and coemulsifier, respectively)	In vitro cytotoxicity studies revealed enhanced antiproliferative activity in MCF-7 and 4T1 cell lines. The values for C_{max} and AUC were increased by 1.45 and 1.24 folds, respectively, for SNEDDS as compared to that of drug suspension	[119]
LyP-1 bearing doxorubicin coloaded SMEDDs	Lymphatic targeting of BC	LyP-1 (for active binding to p-32 receptors on BC), SMEDDs containing maisine/peceol (as oil), PEG 300 (coemulsifier) and TPGS:Gelucire 44/14/TPGS: Labrasol or Labrasol:Tween 80 (emulsifier)	LyP-1 incorporation provided selective targeting to p-32 expressing 4T1 and MDA-MB-231 BC cell lines. Furthermore, these formulations provided a significant reduction in cancer metastasis through lymphatic voyage	[120]

8 Lipid nanocapsules

The LNCs are colloidal drug delivery systems with the ability to mimic lipoproteins. A patent (WO02688000) was granted to Heurtault et al. (2000) for the manufacturing process of LNCs [121]. LNCs offer advantages over liposomes specifically superior physical stabilization (up to 18 months), and are unaffected by biological fluids, and confer enhanced biodegradability [122]. LNCs possess a lipophilic core which is made up of medium-chain triglycerides and protected by a covering composed of PEGylated emulsifier and lecithin [123]. The LNCs have been successfully employed in the passive and active targeting of anticancer drugs, both from the synthetic and natural origin [124–126]. Gaber et al. (2019) reported the formulation of albumin-coated oily nanocapsules for the codelivery of exemestane and hesperetin. The outer covering of the albumin shell (albondin and SPARC) was additionally coupled with phenylboronic acid for active targeting in BC. The codelivery approach provided synergistic cytotoxic efficacy in MCF-7 cell lines and the combination index (0.662) and dose reduction index values were found to be 8.22 for exemestane and 1.84 for hesperetin, respectively. Furthermore, the in vivo studies revealed that the developed nanocapsules significantly reduced the tumor volume and Ki67 expression by 7 and 3 times in tumor-bearing mice, respectively, suggesting selective targeting ability of the developed nanocapsules compared to the nonselective formulations [127].

Additionally, the codelivery approach has also been explored in the recent past for enhanced BC therapeutics. These include delivering two anticancer drugs simultaneously, or drug(s) that can potentiate the activity of anticancer drugs like COX-2 inhibitors (celecoxib) [128]. Recent preclinical reports on LNCs with passive/active targeting abilities have been summarized in Table 4.

9 Polymeric micelles

The PMs are block-copolymers based nanoparticulate systems. These systems self-assemble when in contact with the aqueous environment, into a hydrophobic core and an outer hydrophilic shell. Their make-up can be of A-B and A-B-A type backed by the diblock and triblock copolymeric arrangement respectively. Additionally, grafted polymers are also employed having characteristic hydrophilic support with multiple lipophilic side chains [135, 136]. The particle size normally ranges in between 10 and 100 nm, therefore promotes passive targeting in BC through EPR effects. Furthermore, their hydrophobic core offers encapsulation of hydrophobic anticancer drugs and the hydrophilic shell provides a prolonged systemic availability due to their enhanced colloidal stability. Moreover, the hydrophilic shell can be functionalized with tumor receptor-guided ligands for active BC targeting. The drug release profile, stability and drug entrapment efficiency can be altered by modifying the constitution of the PMs [137, 138]. Hence, these systems have been exploited tremendously for active and/or passive targeting of anticancer drugs in BC including examples like epirubicin [139], dasatinib [140], and quercetin [141]. Recently, codelivery approaches have also been explored in overcoming the MDR BC, for example, resveratrol and docetaxel from mPEG-b-PLA copolymer-based PM system. Multiple ratios were tested for best synergistic in vitro cytotoxic effect for resveratrol and docetaxel on MCF-7 cell lines and the best efficacy was obtained with 1:1 ratio. Further, concomitant delivery of these two drugs provided sustained cytotoxic effects with enhanced $AUC_{(0\rightarrow t)}$ for resveratrol and docetaxel (i.v. injection) and it was found to be 1.6-times and 3.0-times higher as compared with the individual i.v. injection for both the drugs [142]. Similarly, Wan et al. (2019) developed a stable PM system [poly(2-methyl-2-oxazoline-block-2-butyl-2-oxazoline-block-2-methyl-2-oxazoline—(P(MeOx-b-BuOx-b-MeOx))] for the codelivery of paclitaxel and alkylated cisplatin prodrug combination. The system exhibited enhanced antiproliferative efficacy in LCC-6-MDR BC cell line (orthotopic tumor model). Additionally, it was also efficacious against A2780/CisR xenograft tumor model (a cisplatin-resistant ovarian cell line) [143].

However, PMs possess some limitations that restrict their clinical translations like post i.v. administration of the PM systems and their subsequent dilution with the blood often reduces the polymeric concentration below the critical micellar concentration. This reduces the tumor-specific drug release specificity. Additionally, the monomeric units of the polymeric system get entangled with the plasma proteins like albumin and apolipoproteins thereby leading to the destruction of the PM structure [144, 145]. Therefore, recent studies have shown a remarkable improvement in the PM systems employing stimuli-responsive approaches either alone or in combination like pH-sensitive [146, 147], thermoregulated [148], redox-sensitive [149, 150], ultrasonic waves-assisted [151, 152] and mixed micellar systems [153]. Recent research reports on PM systems for BC therapeutics have been presented in Table 5.

10 Lipoproteins

The applicability of lipoprotein-assisted delivery for anticancer drugs was first recognized by Gal et al. in (1981) [160]. They have a core-shell-like structure. The shell is made up of a phospholipid monolayer containing cholesterol and some proteins (apolipoproteins); whereas, the core is made up of highly lipophilic moieties like cholesteryl esters and triglycerides. Their main function is to transport the lipids and various other endogenous lipophilic molecules in the circulatory system [161]. Their biocompatible nature offers prolonged availability in the circulatory system due

TABLE 4 Preclinical reports on LNC with passive/active targeting abilities in BC.

Active moiety	Objective	Formulation/ Modification	Outcome	Reference
Doxorubicin	Selective targeting on αvβ3 integrin overexpressed tumor cells	Arginylglycylaspartic acid-surface-functionalized-doxorubicin-loaded LNCs	In vitro cellular uptake and cytotoxic studies on MCF-7 and U87MG cell lines with varying levels of αvβ3 integrin expression revealed that the functionalized LNCs were specifically targeted to U87MG cell	[129]
Docetaxel-thymoquinone (codelivery)	Enhanced cytotoxic, and antiproliferative efficacy (TNBC) with reduced side effects	PEGylated lipid nanocapsules (mPEG-DSPE-Vitamin E TPGS-Lipid nanocapsules)	The in vivo studies revealed enhanced cytotoxic and antiproliferative efficacy of the developed LNCs. Furthermore, the formulations were minimally cytotoxic on kidney and liver cells as compared to drug suspension	[130]
Doxorubicin	Antiproliferative and cellular uptake ability in MCF-7 cells lines	LNC (poly(ε-caprolactone, sorbitan monostearate, capric/caprylic triglyceride)	The developed formulation exhibited sustained cytotoxic effect with IC_{50} values 4.49 and 1.60 micromolar at 24 and 72 h respectively. Furthermore, the cellular uptake studies revealed that the LNCs internalized by caveolin and fluid phase endocytosis pathways	[131]
Curcumin	Rapid internalization in BC cells	Albumin-covered olive oil-based nanocapsules	The developed formulation exhibited fast internalization into the MCF-7 cells with IC_{50} values comparable to that of the free curcumin	[132]
Doxorubicin and tanespimycin (codelivery)	Synergistic cytotoxic efficacy in BC	Folate receptor directed hybrid lipid-core nanocapsules. Tanespimycin (lipid core) and doxorubicin (polymeric corona)	In vitro studies on MCF-7 cell lines revealed enhancement in the cellular uptake, cytotoxicity and apoptotic activity. Additionally, in vivo studies exhibited increased bioavailability as compared to that of free drug or nonfunctionalized formulations	[133]
Tretinoin alone or in combination with 5-FU or doxorubicin	Antiproliferative activity assessment in TNBC for LNC-loaded Tretinoin alone or in combination with 5-FU or doxorubicin	LNC employing poly(ε-caprolactone, sorbitan monostearate, and caprylic/capric triglyceride mixture	Tretinoin-incorporated LNCs with or without 5-FU/doxorubicin enhanced the cytotoxic effects on the TNBC MDA-MB-231 cell lines. Moreover, codelivery of 5-FU or doxorubicin with tretinoin resulted in a synergistic cytotoxic activity	[134]

to reduced clearance by liver and spleen. They have been classified on the basis of their buoyancy, size, chemical make-up, and electrophoretic mobility into the low-density lipoproteins (LDL), high-density lipoproteins (HDL), very low-density lipoproteins, and the chylomicrons [162]. Thus, in the past years, numerous research reports depicted the applicability of lipoproteins, and their synthetic/reconstituted alternatives in BC therapeutics [163].

10.1 Lipoprotein-based drug delivery system for BC therapeutics

Recently LDL and HDL as nanoparticulate system have been explored for enhancing the BC therapeutics with examples like paclitaxel [164] and curcumin [165]. Further, HDL-based drug delivery systems are easy to modify as compared to the LDL for targeting specificity due to the availability of hydrophilic small-sized apolipoproteins (Apo-A1) in the former as compared to apo-B100 in the latter [166]. Zhang et al. (2016) reported an improvement in the cytotoxicity potential of paclitaxel from reconstituted HDL-assisted codelivery of paclitaxel with N-cyano-1-[(3,4-dimethoxyphenyl) methyl]-3,4-dihydro-6,7-dimethoxy-*N′*-octyl-2(1H)-isoquinoline carboximidamide (HZ08 an MDR reversal agent) in BC resistant and sensitive MCF-7 cell lines [37]. Additionally, the synthetic HDL-based systems have also shown potential in cancer theranostics

TABLE 5 PM/Mixed micelles in BC therapeutics.

Active moiety	Objective/Stimuli employed	Formulation/ modification	Outcome	Reference
Codelivery of $CGKRK_D$ $(KLAKLAK)_2$ a proapoptotic peptide and docetaxel (anticancer drug)	pH-sensitive PM system for subcellular delivery to MCF-7 cell lines for enhanced anticancer efficacy	Doxorubicin core-encapsulated PM system comprising poly (β-amino esters)-poly (ethylene glycol) as amphiphilic copolymer in conjugation with $GKRK_D(KLAKLAK)_2$	The pH-sensitive PM system provided a synergistic and selective targeting on MCF-7 cell lines	[154]
Paclitaxel-apatinib (codelivery)	MDR reversal through redox-sensitive PM system	Paclitaxel-apatinib codelivery through hyaluronic acid-g-cystamine dihydrochloride-poly-ε-(benzyloxy-carbonyl)-L-lysine PM systems	The system exhibited selective targeting on MCF-7 and ADR cell lines. Furthermore, this proved effective P-gp-mediated efflux of the codelivered drugs	[155]
Doxorubicin	pH-responsive PM system for controlled release of anticancer drug	Poly(ethylene glycol)-b-poly(L-lysine) (PEG-PLL) with coumarin and imidazole grafting to enhance the lipophilic drug loading	The developed grafted (coumarin-imidazole) PM system provide site-specific in vitro antiproliferative activity	[156]
Doxorubicin	Thermo and reactive oxygen species sensitive delivery for enhanced antiproliferative efficacy	Diblock copolymer with PPS-PNIPAm for thermo- and oxidant-sensitive doxorubicin release	The dual-responsive system provided an enhanced antiproliferative activity in in vitro studies on MCF-7 cell lines with enhanced accumulation in the cell cytoplasm	[157]
Doxorubicin	Enhancement of antiproliferative property of doxorubicin against breast and ovarian cancer cell line (MDA-MB-231 and SKOV-3 respectively)	Mixed micellar systems composed of T1107: TPGS (1:3, weight ratio)	The developed mixed micellar system of doxorubicin exhibited enhancement in the in vitro antiproliferative activity against MDA-MB-231 and SKOV-3 cell lines respectively as compared to that of Doxil. Furthermore, the cellular uptake studies revealed an enhancement of doxorubicin in both the cell lines	[158]
Paclitaxel-naringin (coencapsulated)	Reversal of the MDR employing mixed micellar approach	Paclitaxel-naringin coencapsulated amphiphilic block copolymer based mixed micellar system	The developed formulation exhibited synergistic improvement in the cellular uptake and a marked cytotoxic enhancement (65%) on the breast cancer cell lines with a low dose of 15 μg/mL	[159]
Paclitaxel	Glucose membrane transporter targeted mixed micellar system for BC therapeutics	Soluplus, Soluplus (Glu) and TPGS-based glycosylated mixed micellar formulation. (3:1:1)	The developed formulation exhibited enhanced cytotoxicity against MCF-7 (1.6 times) and MDA-MB-231 (14.1times) as compared to that of Genexol. Moreover, these formulations also provided an enhanced cellular uptake (after 6 h) in both the cell lines viz. MCF-7 and MDA-MB-231 (30.5 times and 5 times, respectively)	[153]

(therapy with diagnosis) [167,168]. The techniques employed for loading anticancer drugs to the core of the lipoproteins include solvent evaporation, lyophilization, and reconstitution procedures [169,170].

11 Carbon nanotubes

CNT have immense potential in both, cancer therapeutics and diagnosis amongst all nanocarriers [171,172]. The fact that CNT have a large surface area readily enables their surface engineering, thus rendering them biocompatible for harnessing large benefits from them [173]. Their ability to stack interactions, encapsulate small molecules and undergo conjugation can significantly improve the pharmacokinetic and pharmacodynamic profile of anticancer drugs like cisplatin [174], paclitaxel [175,176]. The ability to absorb near-infrared radiation also facilitates photoacoustic and photothermal therapy using CNT [177]. The CNT can be synthesized employing any of the three techniques viz. electric arc discharge, laser ablation, and chemical vapor deposition and among them chemical vapor deposition offers CNT with high quality and yield [178,179].

11.1 Selective cancer cell destruction

CNT can be functionalized either covalently (single/double/triple group) or noncovalently (by surfactant/polymer adsorption or wrapping on the CNT) for improving their solubilization and in vivo processing characteristics [180, 181]. In noncovalent functionalization, the electric and optical properties of the CNT remain intact and advantage over covalent functionalization [179]. CNT can selectively destruct cancer cells and are also employed for diagnostic applications [182]. Owing to the fact that the in vivo biological conditions are highly transparent to 700–1100nm near-infrared radiation, strong optical absorbance of single-walled carbon nanotubes (SWCNT) in this spectral region has been employed for optical sensitization of CNT intracellularly to provide multifunctional biological transporters [183]. Furthermore, selective demolition of cancerous cell was achievable by functionalization on single walled/multiwalled carbon nanotubes with a folate moiety [184]. Tavakolifard et al. (2016) reported that the folate functionalization afforded a selective internalization of SWCNT (incorporated with paclitaxel) inside the cells labelled with FR tumor markers, without damaging the folate receptor-free noncancerous (normal) cells [185]. In a different study by Suo et al. (2018) reported a selective destruction of the P-gp mediated MDR cells using MWCNT with anti-PGP antibody functionalization and photothermal therapy. This approach provided enhanced P-gp specific cellular uptake along with marked phototoxicity in tumor spheroids of MDR cells [186]. Thus, the conveying potential of CNT amalgamated with appropriate surface modifications, and their inherited optical attributes have opened new frontiers in the BC therapy and diagnosis [187].

12 Dendrimers

Dendrimers are elaborately branched (analogous with the branch of trees), monodisperse, three-dimensional molecules with extensively controlled architectures [188]. The maiden research work on dendrimer was performed by Vogtle et al. and for which US patents 4289872 and 4410688 were granted [189]. These nanoparticulate systems consisted of dendrons originating from the central core and essentially consisting of three parts, namely (a) core, (b) repeating units (homocentric layers) attached to the core and regarded as the generations, and (c) terminal functional groups which decides the pharmacokinetic and biocompatibility of the system [190]. Their water solubility, encapsulation ability, monodispersity, and a large number of peripheral functional groups, render them suitable candidates for drug delivery system [191–193].

12.1 Dendrimers in BC therapeutics

Of late, dendrimers have been researched for drug delivery system of anticancer agents, combination therapies (dendrimer-antibody conjugates) with examples such as chlorambucil [194], doxorubicin [195,196], paclitaxel [197], docetaxel [198], and camptothecin [199]. Further, the exquisite attribute of pH-sensitive drug release by poly(amido amine) PAMAM and poly(propylene imine) PPI dendrimers have been employed in the BC therapeutics [200]. The tertiary amine moieties of these systems at physiological pH conditions lose H^+, which in turn envelopes the core. This property restricts the drug release from the core in normal pH conditions, whereas relatively acidic condition in the vicinity of proliferating cells induces protonation of the amine groups. This facilitates selective drug release at the target site [201]. Furthermore, the extensively tailorable surface of the dendrimer facilitates attachment of various ligands onto it; this property can be explored for active targeting in BC. In agreement to which Kulhari et al. (2016) reported selective and enhanced cytotoxic efficacy of docetaxel from Trastuzumab-grafted PAMAM dendrimers in HER2-positive MDA-MB-453 human BC cells in contrast with their HER2-negative cell lines [48]. Similarly, Gu et al. (2017) exploited graphene oxide functionalized PAMAM dendrimer for effective and targeted concomitant administration of doxorubicin (drug) and MMP-9 shRNA plasmid (gene). The system exhibited pH-sensitive drug release. Further, MMP-9 protein expression in MCF-7 cells was significantly inhibited. Thus, these results suggested that functionalized PAMAM dendrimeric system can be employed for efficient gene and drug codelivery [202].

13 Miscellaneous nanoparticulate systems for BC therapeutics

Besides the nanoparticulate systems discussed above numerous other systems have been exploited in the recent past for passive and/or active targeting in BC. These are metallic nanoparticles, mesoporous silica nanoparticles, niosomes, viral nanoparticles for therapeutics and imaging applications. Metallic nanoparticles are categorized as gold nanoparticles [203, 204], superparamagnetic iron oxide nanoparticle [205,206], and quantum dots [207, 208]. Gold nanoparticles provide selective drug targeting and enhanced cytotoxic efficacy through surface functionalization for transferrin/EGFR/VEGFR-2 receptors. These receptors are overexpressed in BC and associated metastasis [209]. Whereas, superparamagnetic iron oxide nanoparticles and quantum dots find elaborate applications in BC theranostics [210,211]. Similarly, mesoporous silica nanoparticles have also gained valuable recognition among researchers owing to their properties like enormous surface area and pore volume for controlled and targeted drug delivery in BC [212]. However, they are less penetrable in the tumor mass which limits their applications [213]. Additionally, niosomes (nonionic surfactant-dependent nanovesicles) have evolved since their inception in the 1970s and are successfully employed in BC therapeutics [214]. In agreement to which Salem et al. (2018) developed pH-sensitive niosomes (in situ gelling system) for site-specific and sustained delivery of tamoxifen [215]. Recently, another class of nanoparticles have also been explored that are closely related to the protein envelope/ viral capsids and are referred to as viral nanoparticles. These systems are devoid of the viral genome and hence are not infectious [216]. Shukla et al. (2017) reported successful development of plant viral nanoparticle-assisted [icosahedral cowpea mosaic virus (CPMV) and filamentous potato virus X (PVX)] vaccine for delivering HER 2 receptor epitopes in the HER2$^+$ BC [217].

14 Challenges associated with the clinical translation of nanoparticulate systems

The nanoparticulate system discussed above has immense potential in improving the BC therapeutics; however, their clinical translation is a costlier affair both in terms of money and time. The challenges encountered are addressing large-scale manufacturing issues, biological challenges, biocompatibility and safety, regulatory aspects, and overall higher cost involved compared to the conventional formulations [218]. All these can inflict significant hurdles, thereby limiting the commercialization of nanocarriers. For lab to clinical transition, it is essential to use a manufacturing method that allows large-scale production of nanocarriers with minimum batch-to-batch variability and high quality [219]. The industrially scalable methods for liposomes have been successfully developed with minimal manufacturing steps [220]. The challenges escalate when the nanocarrier system is decorated with ligands and/or coating for active targeting, codelivery approach, or employing more than one targeting ligands [221]. The codelivery approach for a synergistic effect in BC will further complex the manufacturing process, posing hurdles in their bulk scale-up and good manufacturing (cGMP) production, thereby increasing the overall cost of manufacturing [222]. The quality control and quality assurance evaluation of such products is an additional difficult terrain owing to numerous characterization parameters to be catered.

Batch-to-batch variabilities during the production of nanocarriers often alter their physicochemical properties, pharmacokinetic parameters, and/or pharmacodynamic interactions [218,221]. Postmanufacturing, the nanocarriers should possess sufficient shelf stability, additionally, these systems should release the drug in a controlled fashion to avoid any possibility of accumulation before reaching the target site [219]. Furthermore, a detailed toxicological investigation is mandatory for ascertaining the overall safety and efficacy and to expedite the clinical translation of nanocarriers [223].

15 Conclusion

The research on BC nanotherapeutics is consistently progressing and the nanoformulations have the potential to overcome the limitations of conventional therapies used for the diagnosis and treatment of BC. Various nanoparticle-based formulations are in the differential preclinical and clinical stages of development. Liposomes, polymeric drug micelles, lipid nanoparticles, nanoemulsions, dendrimers, lipidic systems and CNT are the commonly explored. Functionalization of these systems can evade uptake by macrophages and be used for efficient targeting. Many nanoformulations are already in clinical practice and some in the later stages of clinical trials. Perennial efforts by pharmaceutical researchers will consistently produce a new platform for nanoparticles that would find greater oncological applications.

References

[1] Karaman S, Detmar M. Review series mechanisms of lymphatic metastasis. J Clin Invest 2014;124(3):922–8. Available from: http://www.ncbi.nlm.nih.gov/pubmed/24590277.

[2] Chen Z, Xu L, Shi W, Zeng F, Zhuo R, Hao X, et al. Trends of female and male breast cancer incidence at the global, regional, and national levels, 1990–2017. Breast Cancer Res Treat 2020;180(2):481–90. Available from: https://link.springer.com/article/10.1007/s10549-020-05561-1.

[3] Ran S, Volk L, Hall K, Flister MJ. Lymphangiogenesis and lymphatic metastasis in breast cancer. Pathophysiology 2010;17(4):229–51. Available from: /pmc/articles/PMC2891887/?report=abstract.

[4] Sleeman JP. The lymph node as a bridgehead in the metastatic dissemination of tumors. Recent Results Cancer Res 2000;157:55–81.
[5] Nounou MI, Elamrawy F, Ahmed N, Abdelraouf K, Goda S, Syed-Sha-Qhattal H. Breast cancer: Conventional diagnosis and treatment modalities and recent patents and technologies supplementary issue: targeted therapies in breast cancer treatment. Breast Cancer Basic Clin Res 2015;9(Suppl 2):17–34. Available from: /pmc/articles/PMC4589089/?report=abstract.
[6] Alqaraghuli HGJ, Kashanian S, Rafipour R. A review on targeting nanoparticles for breast cancer. Curr Pharm Biotechnol 2019;20(13):1087–107.
[7] Šemeláková M, Jendželovský R, Fedoročko P. Drug membrane transporters and CYP3A4 are affected by hypericin, hyperforin or aristoforin in colon adenocarcinoma cells. Biomed Pharmacother 2016;81:38–47.
[8] Liyanage PY, Hettiarachchi SD, Zhou Y, Ouhtit A, Seven ES, Oztan CY, et al. Nanoparticle-mediated targeted drug delivery for breast cancer treatment. Biochim Biophys Acta—Rev Cancer 2019;1871(2):419–33.
[9] Jin S, Ye K. Targeted drug delivery for breast cancer treatment. Recent Pat Anticancer Drug Discov 2013;8(2):143–53.
[10] Tai W, Mahato R, Cheng K. The role of HER2 in cancer therapy and targeted drug delivery. J Control Release 2010;146(3):264–75.
[11] Hunter J, Hirst BH, Simmons NL. Drug absorption limited by p-glycoprotein-mediated secretory drug transport in human intestinal epithelial Caco-2 cell layers. Pharm Res An Off J Am Assoc Pharm Sci 1993;10(5):743–9. Available from: https://link.springer.com/article/10.1023/A:1018972102702.
[12] Breedveld P, Beijnen JH, Schellens JHM. Use of P-glycoprotein and BCRP inhibitors to improve oral bioavailability and CNS penetration of anticancer drugs. Trends Pharmacol Sci 2006;27(1):17–24. Available from: https://pubmed.ncbi.nlm.nih.gov/16337012/.
[13] Ambudkar SV, Dey S, Hrycyna CA, Ramachandra M, Pastan I, Gottesman MM. Biochemical, cellular, and pharmacological aspects of the multidrug transporter. Annu Rev Pharmacol Toxicol 1999;39(1):361–98. Available from: http://www.annualreviews.org/doi/10.1146/annurev.pharmtox.39.1.361.
[14] Varma MVS, Ashokraj Y, Dey CS, Panchagnula R. P-glycoprotein inhibitors and their screening: a perspective from bioavailability enhancement. Pharmacol Res 2003;48(4):347–59. Available from: https://pubmed.ncbi.nlm.nih.gov/12902205/.
[15] Schinkel AH, Jonker JW. Mammalian drug efflux transporters of the ATP binding cassette (ABC) family: an overview. Adv Drug Deliv Rev 2012;64(Suppl):138–53.
[16] Aboulfotouh K, Allam AA, El-Badry M, El-Sayed AM. Self-emulsifying drug-delivery systems modulate P-glycoprotein activity: role of excipients and formulation aspects. Nanomedicine 2018;13(14):1813–34. Available from: https://www.futuremedicine.com/doi/abs/10.2217/nnm-2017-0354.
[17] Talegaonkar S, Bhattacharyya A. Potential of lipid nanoparticles (SLNs and NLCs) in enhancing oral bioavailability of drugs with poor intestinal permeability. AAPS PharmSciTech 2019;20(3):1–15. Available from: https://link.springer.com/article/10.1208/s12249-019-1337-8.
[18] Chaturvedi S, Garg A, Verma A. Nano lipid based carriers for lymphatic voyage of anti-cancer drugs: an insight into the in-vitro, ex-vivo, in-situ and in-vivo study models. J Drug Deliv Sci Technol 2020;59:101899.
[19] Sun T, Zhang YS, Pang B, Hyun DC, Yang M, Xia Y. Engineered nanoparticles for drug delivery in cancer therapy. Angew Chem Int Ed 2014;53(46):12320–64. Available from: https://onlinelibrary.wiley.com/doi/full/10.1002/anie.201403036.
[20] Attia MF, Anton N, Wallyn J, Omran Z, Vandamme TF. An overview of active and passive targeting strategies to improve the nanocarriers efficiency to tumour sites. J Pharm Pharmacol 2019;71(8):1185–98. Available from: https://onlinelibrary.wiley.com/doi/abs/10.1111/jphp.13098.
[21] Fang J, Nakamura H, Maeda H. The EPR effect: unique features of tumor blood vessels for drug delivery, factors involved, and limitations and augmentation of the effect. Adv Drug Deliv Rev 2011;63(3):136–51.
[22] Kalyane D, Raval N, Maheshwari R, Tambe V, Kalia K, Tekade RK. Employment of enhanced permeability and retention effect (EPR): nanoparticle-based precision tools for targeting of therapeutic and diagnostic agent in cancer. Mater Sci Eng C 2019;98:1252–76.
[23] Cao H, Dan Z, He X, Zhang Z, Yu H, Yin Q, et al. Liposomes coated with isolated macrophage membrane can target lung metastasis of breast cancer. ACS Nano 2016;10(8):7738–48. Available from: https://pubs.acs.org/doi/abs/10.1021/acsnano.6b03148.
[24] Franco YL, Vaidya TR, Ait-Oudhia S. Anticancer and cardio-protective effects of liposomal doxorubicin in the treatment of breast cancer. Breast Cancer Targets Ther 2018;10:131–41. Available from: /pmc/articles/PMC6138971/?report=abstract.
[25] Wang W, Zhang L, Chen T, Guo W, Bao X, Wang D, et al. Anticancer effects of resveratrol-loaded solid lipid nanoparticles on human breast cancer cells. Molecules 2017;22(11):1814. Available from: http://www.mdpi.com/1420-3049/22/11/1814.
[26] Abd-Rabou AA, Bharali DJ, Mousa SA. Taribavirin and 5-fluorouracil-loaded pegylated-lipid nanoparticle synthesis, p38 docking, and antiproliferative effects on MCF-7 breast cancer. Pharm Res 2018;35(4):1–10. Available from: https://link.springer.com/article/10.1007/s11095-017-2283-3.
[27] Guney Eskiler G, Cecener G, Dikmen G, Egeli U, Tunca B. Solid lipid nanoparticles: reversal of tamoxifen resistance in breast cancer. Eur J Pharm Sci 2018;120:73–88.
[28] Sabzichi M, Mohammadian J, Mohammadi M, Jahanfar F, Movassagh Pour AA, Hamishehkar H, et al. Vitamin D-loaded nanostructured lipid carrier (NLC): a new strategy for enhancing efficacy of doxorubicin in breast cancer treatment. Nutr Cancer 2017;69(6):840–8. Available from: https://www.tandfonline.com/doi/abs/10.1080/01635581.2017.1339820.
[29] Nordin N, Yeap SK, Rahman HS, Zamberi NR, Mohamad NE, Abu N, et al. Antitumor and anti-metastatic effects of citral-loaded nanostructured lipid carrier in 4T1-induced breast cancer mouse model. Molecules 2020;25(11):2670. Available from: https://www.mdpi.com/1420-3049/25/11/2670.
[30] Fernandes RS, Silva JO, Seabra HA, Oliveira MS, Carregal VM, Vilela JMC, et al. α-Tocopherol succinate loaded nano-structed lipid carriers improves antitumor activity of doxorubicin in breast cancer models in vivo. Biomed Pharmacother 2018;103:1348–54.
[31] Cao X, Luo J, Gong T, Zhang ZR, Sun X, Fu Y. Coencapsulated doxorubicin and bromotetrandrine lipid nanoemulsions in reversing multidrug resistance in breast cancer in vitro and in vivo. Mol Pharm 2015;12(1):274–86. Available from: https://pubs.acs.org/doi/abs/10.1021/mp500637b.
[32] Kamal MM, Nazzal S. Novel sulforaphane-enabled self-microemulsifying delivery systems (SFN-SMEDDS) of taxanes: Formulation development and in vitro cytotoxicity against breast cancer cells. Int J Pharm 2018;536(1):187–98.

[33] Sandhu PS, Kumar R, Beg S, Jain S, Kushwah V, Katare OP, et al. Natural lipids enriched self-nano-emulsifying systems for effective co-delivery of tamoxifen and naringenin: systematic approach for improved breast cancer therapeutics. Nanomed Nanotechnol Biol Med 2017;13(5):1703–13.

[34] Oliveira CP, Prado WA, Lavayen V, Büttenbender SL, Beckenkamp A, Martins BS, et al. Bromelain-functionalized multiple-wall lipid-core nanocapsules: formulation, chemical structure and antiproliferative effect against human breast cancer cells (MCF-7). Pharm Res 2017;34(2):438–52. Available from: https://link.springer.com/article/10.1007/s11095-016-2074-2.

[35] Zafar S, Akhter S, Ahmad I, Hafeez Z, Alam Rizvi MM, Jain GK, et al. Improved chemotherapeutic efficacy against resistant human breast cancer cells with co-delivery of Docetaxel and Thymoquinone by Chitosan grafted lipid nanocapsules: formulation optimization, in vitro and in vivo studies. Colloids Surf B Biointerfaces 2020;186:110603.

[36] Gong M, Zhang Q, Zhao Q, Zheng J, Li Y, Wang S, et al. Development of synthetic high-density lipoprotein-based ApoA-I mimetic peptide-loaded docetaxel as a drug delivery nanocarrier for breast cancer chemotherapy. Drug Deliv 2019;26(1):708–16. Available from: https://www.tandfonline.com/doi/full/10.1080/10717544.2019.1618420.

[37] Zhang F, Wang X, Xu X, Li M, Zhou J, Wang W. Reconstituted high density lipoprotein mediated targeted co-delivery of HZ08 and paclitaxel enhances the efficacy of paclitaxel in multidrug-resistant MCF-7 breast cancer cells. Eur J Pharm Sci 2016;92:11–21.

[38] Wang Z, Li X, Wang D, Zou Y, Qu X, He C, et al. Concurrently suppressing multidrug resistance and metastasis of breast cancer by co-delivery of paclitaxel and honokiol with pH-sensitive polymeric micelles. Acta Biomater 2017;62:144–56.

[39] Logie J, Ganesh AN, Aman AM, Al-awar RS, Shoichet MS. Preclinical evaluation of taxane-binding peptide-modified polymeric micelles loaded with docetaxel in an orthotopic breast cancer mouse model. Biomaterials 2017;123:39–47.

[40] Hou J, Sun E, Sun C, Wang J, Yang L, Jia X. bin, et al. Improved oral bioavailability and anticancer efficacy on breast cancer of paclitaxel via Novel Soluplus®—Solutol® HS15 binary mixed micelles system. Int J Pharm 2016;512(1):186–93.

[41] Song J, Huang H, Xia Z, Wei Y, Yao N, Zhang L, et al. TPGS/phospholipids mixed micelles for delivery of icariside II to multidrug-resistant breast cancer. Integr Cancer Ther 2016;15(3):390–9. Available from: http://journals.sagepub.com/doi/10.1177/1534735415596571.

[42] Cagel M, Moretton MA, Bernabeu E, Zubillaga M, Lagomarsino E, Vanzulli S, et al. Antitumor efficacy and cardiotoxic effect of doxorubicin-loaded mixed micelles in 4T1 murine breast cancer model. Comparative studies using Doxil® and free doxorubicin. J Drug Deliv Sci Technol 2020;56:101506.

[43] Shah HS, Khalid F, Bashir S, Asad MHB, KUR K, Usman F, et al. Emulsion-templated synthesis and in vitro characterizations of niosomes for improved therapeutic potential of hydrophobic anti-cancer drug: tamoxifen. J Nanoparticle Res 2019;21(2):1–10. Available from: https://link.springer.com/article/10.1007/s11051-019-4464-y.

[44] Mirzaei-Parsa MJ, Najafabadi MRH, Haeri A, Zahmatkeshan M, Ebrahimi SA, Pazoki-Toroudi H, et al. Preparation, characterization, and evaluation of the anticancer activity of artemether-loaded nano-niosomes against breast cancer. Breast Cancer 2020;27(2):243–51. Available from: https://link.springer.com/article/10.1007/s12282-019-01014-w.

[45] Taghavi S, HashemNia A, Mosaffa F, Askarian S, Abnous K, Ramezani M. Preparation and evaluation of polyethylenimine-functionalized carbon nanotubes tagged with 5TR1 aptamer for targeted delivery of Bcl-xL shRNA into breast cancer cells. Colloids Surf B Biointerfaces 2016;140:28–39.

[46] Badea M, Prodana M, Dinischiotu A, Crihana C, Ionita D, Balas M. Cisplatin loaded multiwalled carbon nanotubes induce resistance in triple negative breast cancer cells. Pharmaceutics 2018;10(4):228. Available from: http://www.mdpi.com/1999-4923/10/4/228.

[47] Chittasupho C, Anuchapreeda S, Sarisuta N. CXCR4 targeted dendrimer for anti-cancer drug delivery and breast cancer cell migration inhibition. Eur J Pharm Biopharm 2017;119:310–21.

[48] Kulhari H, Pooja D, Shrivastava S, Kuncha M, Naidu VGM, Bansal V, et al. Trastuzumab-grafted PAMAM dendrimers for the selective delivery of anticancer drugs to HER2-positive breast cancer. Sci Rep 2016;6(1):1–13. Available from: www.nature.com/scientificreports/.

[49] Loutfy SA, Al-Ansary NA, Abdel-Ghani NT, Hamed AR, Mohamed MB, Craik JD, et al. Anti-proliferative activities of metallic nanoparticles in an in vitro breast cancer model. Asian Pacific J Cancer Prev 2015;16(14):6039–46. Available from: http://dx.doi.org/10.7314/.

[50] Meng H, Mai WX, Zhang H, Xue M, Xia T, Lin S, et al. Codelivery of an optimal drug/siRNA combination using mesoporous silica nanoparticles to overcome drug resistance in breast cancer in vitro and in vivo. ACS Nano 2013;7(2):994–1005. Available from: https://pubs.acs.org/doi/abs/10.1021/nn3044066.

[51] Bruckman MA, Czapar AE, VanMeter A, Randolph LN, Steinmetz NF. Tobacco mosaic virus-based protein nanoparticles and nanorods for chemotherapy delivery targeting breast cancer. J Control Release 2016;231:103–13.

[52] Bangham AD, Standish MM, Watkins JC. Diffusion of univalent ions across the lamellae of swollen phospholipids. J Mol Biol 1965;13(1):238–52.

[53] Akbarzadeh A, Rezaei-Sadabady R, Davaran S, Joo SW, Zarghami N, Hanifehpour Y, et al. Liposome: classification, preparation, and applications. Nanoscale Res Lett 2013;8(1):102. Available from: https://nanoscalereslett.springeropen.com/articles/10.1186/1556-276X-8-102.

[54] Sharma A, Sharma US. Liposomes in drug delivery: progress and limitations. Int J Pharm 1997;154(2):123–40.

[55] Munster P, Krop IE, LoRusso P, Ma C, Siegel BA, Shields AF, et al. Safety and pharmacokinetics of MM-302, a HER2-targeted antibody-liposomal doxorubicin conjugate, in patients with advanced HER2-positive breast cancer: a phase 1 dose-escalation study. Br J Cancer 2018;119(9):1086–93. https://doi.org/10.1038/s41416-018-0235-2.

[56] Gandhi R, Khatri N, Baradia D, Vhora I, Misra A. Surface-modified Epirubicin-HCl liposomes and its *in vitro* assessment in breast cancer cell-line: MCF-7. Drug Deliv 2016;23(4):1152–62. Available from: https://www.tandfonline.com/doi/full/10.3109/10717544.2014.999960.

[57] Salkho NM, Paul V, Kawak P, Vitor RF, Martins AM, Al Sayah M, et al. Ultrasonically controlled estrone-modified liposomes for estrogen-positive breast cancer therapy. Artif Cells Nanomed Biotechnol 2018;46(Suppl 2):462–72. Available from: https://www.tandfonline.com/doi/full/10.1080/21691401.2018.1459634.

[58] Lohade AA, Jain RR, Iyer K, Roy SK, Shimpi HH, Pawar Y, et al. A novel folate-targeted nanoliposomal system of doxorubicin for cancer targeting. AAPS PharmSciTech 2016;17(6):1298–311. Available from: https://link.springer.com/article/10.1208/s12249-015-0462-2.

[59] Zahmatkeshan M, Gheybi F, Rezayat SM, Jaafari MR. Improved drug delivery and therapeutic efficacy of PEgylated liposomal doxorubicin by targeting anti-HER2 peptide in murine breast tumor model. Eur J Pharm Sci 2016;86:125–35.
[60] Burande AS, Viswanadh MK, Jha A, Mehata AK, Shaik A, Agrawal N, et al. EGFR targeted paclitaxel and piperine Co-loaded liposomes for the treatment of triple negative breast cancer. AAPS PharmSciTech 2020;21(5):1–12. Available from: https://link.springer.com/article/10.1208/s12249-020-01671-7.
[61] Rodallec A, Brunel JM, Giacometti S, Maccario H, Correard F, Mas E, et al. Docetaxel-trastuzumab stealth immunoliposome: development and in vitro proof of concept studies in breast cancer. Int J Nanomed 2018;13:3451–65. Available from: /pmc/articles/PMC6014390/?report=abstract.
[62] Su YC, Burnouf PA, Chuang KH, Chen BM, Cheng TL, Roffler SR. Conditional internalization of PEGylated nanomedicines by PEG engagers for triple negative breast cancer therapy. Nat Commun 2017;8(1):1–12. Available from: www.nature.com/naturecommunications.
[63] Chowdhury N, Chaudhry S, Hall N, Olverson G, Zhang QJ, Mandal T, et al. Targeted delivery of doxorubicin liposomes for Her-2+ breast cancer treatment. AAPS PharmSciTech 2020;21(6):1–12. Available from: https://link.springer.com/article/10.1208/s12249-020-01743-8.
[64] Tang H, Chen J, Wang L, Li Q, Yang Y, Lv Z, et al. Co-delivery of epirubicin and paclitaxel using an estrone-targeted PEGylated liposomal nanoparticle for breast cancer. Int J Pharm 2020;573:118806.
[65] Talluri SV, Kuppusamy G, Karri VVSR, Tummala S, Madhunapantula SV. Lipid-based nanocarriers for breast cancer treatment—comprehensive review. Drug Deliv 2016;23(4):1291–305. Available from: https://www.tandfonline.com/doi/full/10.3109/10717544.2015.1092183.
[66] Kumar S, Randhawa JK. High melting lipid based approach for drug delivery: solid lipid nanoparticles. Mater Sci Eng C 2013;33(4):1842–52.
[67] Müller RH, Mäder K, Gohla S. Solid lipid nanoparticles (SLN) for controlled drug delivery—a review of the state of the art. Eur J Pharm Biopharm 2000;50(1):161–77.
[68] Marengo E, Cavalli R, Caputo O, Rodriguez L, Gasco MR. Scale-up of the preparation process of solid lipid nanospheres. Part I. Int J Pharm 2000;205(1–2):3–13.
[69] Mehnert W, Mäder K. Solid lipid nanoparticles: Production, characterization and applications. Adv Drug Deliv Rev 2012;64(Suppl):83–101.
[70] Salvi VR, Pawar P. Nanostructured lipid carriers (NLC) system: a novel drug targeting carrier. J Drug Deliv Sci Technol 2019;51:255–67.
[71] Pawar H, Surapaneni SK, Tikoo K, Singh C, Burman R, Gill MS, et al. Folic acid functionalized long-circulating co-encapsulated docetaxel and curcumin solid lipid nanoparticles: in vitro evaluation, pharmacokinetic and biodistribution in rats. Drug Deliv 2016;23(4):1453–68. Available from: https://www.tandfonline.com/doi/full/10.3109/10717544.2016.1138339.
[72] Yang XY, Li YX, Li M, Zhang L, Feng LX, Zhang N. Hyaluronic acid-coated nanostructured lipid carriers for targeting paclitaxel to cancer. Cancer Lett 2013;334(2):338–45.
[73] Poonia N, Kaur Narang J, Lather V, Beg S, Sharma T, Singh B, et al. Resveratrol loaded functionalized nanostructured lipid carriers for breast cancer targeting: systematic development, characterization and pharmacokinetic evaluation. Colloids Surf B Biointerfaces 2019;181:756–66.
[74] Kang KW, Chun MK, Kim O, Subedi RK, Ahn SG, Yoon JH, et al. Doxorubicin-loaded solid lipid nanoparticles to overcome multidrug resistance in cancer therapy. Nanomed Nanotechnol Biol Med 2010;6(2):210–3.
[75] Chen HH, Huang WC, Chiang WH, Liu TI, Shen MY, Hsu YH, et al. Ph-responsive therapeutic solid lipid nanoparticles for reducing P-glycoprotein-mediated drug efflux of multidrug resistant cancer cells. Int J Nanomed 2015;10:5035–48. Available from: /pmc/articles/PMC4531030/?report=abstract.
[76] Lin M, Teng L, Wang Y, Zhang J, Sun X. Curcumin-guided nanotherapy: a lipid-based nanomedicine for targeted drug delivery in breast cancer therapy. Drug Deliv 2016;23(4):1420–5. Available from: https://www.tandfonline.com/doi/full/10.3109/10717544.2015.1066902.
[77] Campos J, Varas-Godoy M, Haidar ZS. Physicochemical characterization of chitosan-hyaluronan-coated solid lipid nanoparticles for the targeted delivery of paclitaxel: a proof-of-concept study in breast cancer cells. Nanomedicine 2017;12(5):473–90. Available from: https://www.futuremedicine.com/doi/abs/10.2217/nnm-2016-0371.
[78] Baek JS, Cho CW. A multifunctional lipid nanoparticle for co-delivery of paclitaxel and curcumin for targeted delivery and enhanced cytotoxicity in multidrug resistant breast cancer cells. Oncotarget 2017;8(18):30369–82. Available from: /pmc/articles/PMC5444749/?report=abstract.
[79] Tang J, Ji H, Ren J, Li M, Zheng N, Wu L. Solid lipid nanoparticles with TPGS and brij 78: a co-delivery vehicle of curcumin and piperine for reversing P-glycoprotein-mediated multidrug resistance in vitro. Oncol Lett 2017;13(1):389–95. Available from: http://www.spandidos-publications.com/10.3892/ol.2016.5421/abstract.
[80] Liu Q, Li J, Pu G, Zhang F, Liu H, Zhang Y. Co-delivery of baicalein and doxorubicin by hyaluronic acid decorated nanostructured lipid carriers for breast cancer therapy. Drug Deliv 2016;23(4):1364–8. Available from: https://www.tandfonline.com/doi/full/10.3109/10717544.2015.1031295.
[81] Li X, Jia X, Niu H. Nanostructured lipid carriers co-delivering lapachone and doxorubicin for overcoming multidrug resistance in breast cancer therapy. Int J Nanomed 2018;13:4107–19. Available from: /pmc/articles/PMC6047616/?report=abstract.
[82] Di H, Wu H, Gao Y, Li W, Zou D, Dong C. Doxorubicin- and cisplatin-loaded nanostructured lipid carriers for breast cancer combination chemotherapy. Drug Dev Ind Pharm 2016;42(12):2038–43. Available from: https://www.tandfonline.com/doi/abs/10.1080/03639045.2016.1190743.
[83] Guney Eskiler G, Cecener G, Egeli U, Tunca B. Synthetically lethal BMN 673 (talazoparib) loaded solid lipid nanoparticles for BRCA1 mutant triple negative breast cancer. Pharm Res 2018;35(11):1–20. Available from: https://link.springer.com/article/10.1007/s11095-018-2502-6.
[84] Radhakrishnan R, Pooja D, Kulhari H, Gudem S, Ravuri HG, Bhargava S, et al. Bombesin conjugated solid lipid nanoparticles for improved delivery of epigallocatechin gallate for breast cancer treatment. Chem Phys Lipids 2019;224:104770.
[85] Yassemi A, Kashanian S, Zhaleh H. Folic acid receptor-targeted solid lipid nanoparticles to enhance cytotoxicity of letrozole through induction of caspase-3 dependent-apoptosis for breast cancer treatment. Pharm Dev Technol 2020;25(4):397–407. Available from: https://www.tandfonline.com/doi/abs/10.1080/10837450.2019.1703739.

[86] Varshosaz J, Davoudi MA, Rasoul-Amini S. Docetaxel-loaded nanostructured lipid carriers functionalized with trastuzumab (Herceptin) for HER2-positive breast cancer cells. J Liposome Res 2018;28(4):285–95. Available from: https://www.tandfonline.com/doi/abs/10.1080/08982104.2017.1370471.

[87] Mason TG, Wilking JN, Meleson K, Chang CB, Graves SM. Nanoemulsions: formation, structure, and physical properties. J Phys Condens Matter 2006;18(41):R635. Available from: https://iopscience.iop.org/article/10.1088/0953-8984/18/41/R01.

[88] Singh Y, Meher JG, Raval K, Khan FA, Chaurasia M, Jain NK, et al. Nanoemulsion: Concepts, development and applications in drug delivery. J Control Release 2017;252:28–49.

[89] Montes de Oca-Ávalos JM, Candal RJ, Herrera ML. Nanoemulsions: stability and physical properties. Curr Opin Food Sci 2017;16:1–6.

[90] McClements DJ. Edible delivery systems for nutraceuticals: designing functional foods for improved health. Ther Deliv 2012;3(7):801–3. Available from: www.future-science.com.

[91] Ganta S, Talekar M, Singh A, Coleman TP, Amiji MM. Nanoemulsions in translational research—opportunities and challenges in targeted cancer therapy. AAPS PharmSciTech 2014;15(3):694–708. Available from: https://link.springer.com/article/10.1208/s12249-014-0088-9.

[92] Tagne JB, Kakumanu S, Ortiz D, Shea T, Nicolosi RJ. A nanoemulsion formulation of tamoxifen increases its efficacy in a breast cancer cell line. Mol Pharm 2008;5(2):280–6. Available from: https://pubs.acs.org/doi/abs/10.1021/mp700091j.

[93] Dos Santos Câmara AL, Nagel G, Tschiche HR, Cardador CM, Muehlmann LA, De Oliveira DM, et al. Acid-sensitive lipidated doxorubicin prodrug entrapped in nanoemulsion impairs lung tumor metastasis in a breast cancer model. Nanomedicine 2017;12(15):1751–65. Available from: https://www.futuremedicine.com/doi/abs/10.2217/nnm-2017-0091.

[94] Alkhatib MH, Albishi HM. In vitro evaluation of antitumor activity of doxorubicin-loaded nanoemulsion in MCF-7 human breast cancer cells. J Nanoparticle Res 2013;15(3):1–15. Available from: https://link.springer.com/article/10.1007/s11051-013-1489-5.

[95] Meng L, Xia X, Yang Y, Ye J, Dong W, Ma P, et al. Co-encapsulation of paclitaxel and baicalein in nanoemulsions to overcome multidrug resistance via oxidative stress augmentation and P-glycoprotein inhibition. Int J Pharm 2016;513(1–2):8–16.

[96] Pawar VK, Panchal SB, Singh Y, Meher JG, Sharma K, Singh P, et al. Immunotherapeutic vitamin e nanoemulsion synergies the antiproliferative activity of paclitaxel in breast cancer cells via modulating Th1 and Th2 immune response. J Control Release 2014;196:295–306.

[97] Mahato R. Nanoemulsion as targeted drug delivery system for cancer therapeutics. J Pharm Sci Pharmacol 2017;3(2):83–97.

[98] Alayoubi A, Alqahtani S, Kaddoumi A, Nazzal S. Effect of PEG surface conformation on anticancer activity and blood circulation of nanoemulsions loaded with tocotrienol-rich fraction of palm oil. AAPS J 2013;15(4):1168–79. Available from: https://link.springer.com/article/10.1208/s12248-013-9525-z.

[99] Parmar K, Patel JK. Surface modification of nanoparticles to oppose uptake by the mononuclear phagocyte system. In: Surface modification of nanoparticles for targeted drug delivery. Springer International Publishing; 2019. p. 221–36. Available from: https://link.springer.com/chapter/10.1007/978-3-030-06115-9_12.

[100] Attia MF, Dieng SM, Collot M, Klymchenko AS, Bouillot C, Serra CA, et al. Functionalizing nanoemulsions with carboxylates: impact on the biodistribution and pharmacokinetics in mice. Macromol Biosci 2017;17(7):1600471. Available from: http://doi.wiley.com/10.1002/mabi.201600471.

[101] Afzal SM, Shareef MZ, Dinesh T, Kishan V. Folate-PEG-decorated docetaxel lipid nanoemulsion for improved antitumor activity. Nanomedicine 2016;11(16):2171–84. Available from: https://www.futuremedicine.com/doi/abs/10.2217/nnm-2016-0120.

[102] Tripathi CB, Parashar P, Arya M, Singh M, Kanoujia J, Kaithwas G, et al. QbD-based development of α-linolenic acid potentiated nanoemulsion for targeted delivery of doxorubicin in DMBA-induced mammary gland carcinoma: in vitro and in vivo evaluation. Drug Deliv Transl Res 2018;8(5):1313–34. Available from: https://link.springer.com/article/10.1007/s13346-018-0525-5.

[103] Dokania S, Joshi AK. Self-microemulsifying drug delivery system (SMEDDS)-challenges and road ahead. Drug Deliv 2015;22(6):675–90. Available from: https://pubmed.ncbi.nlm.nih.gov/24670091/.

[104] Singh AK, Chaurasiya A, Awasthi A, Mishra G, Asati D, Khar RK, et al. Oral bioavailability enhancement of exemestane from self-microemulsifying drug delivery system (SMEDDS). AAPS PharmSciTech 2009;10(3):906–16. Available from: https://link.springer.com/article/10.1208/s12249-009-9281-7.

[105] Wang M, You S-K, Lee H-K, Han M-G, Lee H-M, Pham TMA, et al. Development and evaluation of docetaxel-phospholipid complex loaded self-microemulsifying drug delivery system: optimization and in vitro/ex vivo studies. Pharmaceutics 2020;12(6):544. Available from: https://www.mdpi.com/1999-4923/12/6/544.

[106] Solans C, Izquierdo P, Nolla J, Azemar N, Garcia-Celma MJ. Nano-emulsions. Curr Opin Colloid Interface Sci 2005;10(3–4):102–10.

[107] Rao J, McClements DJ. Food-grade microemulsions and nanoemulsions: role of oil phase composition on formation and stability. Food Hydrocoll 2012;29(2):326–34.

[108] Rahman MA, Hussain A, Hussain MS, Mirza MA, Iqbal Z. Role of excipients in successful development of self-emulsifying/microemulsifying drug delivery system (SEDDS/SMEDDS). Drug Dev Ind Pharm 2013;39(1):1–19. Available from: https://pubmed.ncbi.nlm.nih.gov/22372916/.

[109] Porter CJH, Trevaskis NL, Charman WN. Lipids and lipid-based formulations: optimizing the oral delivery of lipophilic drugs. Nat Rev Drug Discov 2007;6(3):231–48.

[110] Yáñez JA, Wang SWJ, Knemeyer IW, Wirth MA, Alton KB. Intestinal lymphatic transport for drug delivery. Adv Drug Deliv Rev 2011;63(10–11):923–42. https://doi.org/10.1016/j.addr.2011.05.019.

[111] Jain S, Garg T, Kushwah V, Thanki K, Agrawal AK, Dora CP. α-Tocopherol as functional excipient for resveratrol and coenzyme Q10-loaded SNEDDS for improved bioavailability and prophylaxis of breast cancer. J Drug Target 2017;25(6):554–65. Available from: https://www.tandfonline.com/doi/abs/10.1080/1061186X.2017.1298603.

[112] Meher JG, Dixit S, Pathan DK, Singh Y, Chandasana H, Pawar VK, et al. Paclitaxel-loaded TPGS enriched self-emulsifying carrier causes apoptosis by modulating survivin expression and inhibits tumour growth in syngeneic mammary tumours. Artif Cells Nanomed Biotechnol 2018;46(Suppl 3):S344–58. Available from: https://www.tandfonline.com/doi/full/10.1080/21691401.2018.1492933.

[113] Cho HY, Kang JH, Ngo L, Tran P, Lee YB. Preparation and evaluation of solid-self-emulsifying drug delivery system containing paclitaxel for lymphatic delivery. J Nanomater 2016;2016.

[114] Valicherla GR, Dave KM, Syed AA, Riyazuddin M, Gupta AP, Singh A, et al. Formulation optimization of Docetaxel loaded self-emulsifying drug delivery system to enhance bioavailability and anti-tumor activity. Sci Rep 2016;6(1):1–11. Available from: www.nature.com/scientificreports.

[115] Shukla M, Jaiswal S, Sharma A, Srivastava PK, Arya A, Dwivedi AK, et al. A combination of complexation and self-nanoemulsifying drug delivery system for enhancing oral bioavailability and anticancer efficacy of curcumin. Drug Dev Ind Pharm 2017;43(5):847–61. Available from: https://www.tandfonline.com/doi/abs/10.1080/03639045.2016.1239732.

[116] Khurana RK, Beg S, Burrow AJ, Vashishta RK, Katare OP, Kaur S, et al. Enhancing biopharmaceutical performance of an anticancer drug by long chain PUFA based self-nanoemulsifying lipidic nanomicellar systems. Eur J Pharm Biopharm 2017;121:42–60. https://doi.org/10.1016/j.ejpb.2017.09.001.

[117] Akhtartavan S, Karimi M, Karimian K, Azarpira N, Khatami M, Heli H. Evaluation of a self-nanoemulsifying docetaxel delivery system. Biomed Pharmacother 2019;109:2427–33.

[118] Cui W, Zhao H, Wang C, Chen Y, Luo C, Zhang S, et al. Co-encapsulation of docetaxel and cyclosporin A into SNEDDS to promote oral cancer chemotherapy. Drug Deliv 2019;26(1):542–50. Available from: https://www.tandfonline.com/doi/full/10.1080/10717544.2019.1616237.

[119] Nazari-Vanani R, Azarpira N, Heli H, Karimian K, Sattarahmady N. A novel self-nanoemulsifying formulation for sunitinib: evaluation of anticancer efficacy. Colloids Surfaces B Biointerfaces 2017;160:65–72.

[120] Timur SS, Yöyen-Ermiş D, Esendağlı G, Yonat S, Horzum U, Esendağlı G, et al. Efficacy of a novel LyP-1-containing self-microemulsifying drug delivery system (SMEDDS) for active targeting to breast cancer. Eur J Pharm Biopharm 2019;136:138–46.

[121] Heurtault B, Saulnier P, Pech B, Proust JE, Richard J, Benoit JP. Lipidic nanocapsules: preparation process and use as Drug Delivery Systems. WO02688000; 2000.

[122] Huynh NT, Passirani C, Saulnier P, Benoit JP. Lipid nanocapsules: a new platform for nanomedicine. Int J Pharm 2009;379(2):201–9.

[123] Heurtault B, Saulnier P, Pech B, Proust JE, Benoit JP. A novel phase inversion-based process for the preparation of lipid nanocarriers. Pharm Res 2002;19(6):875–80. Available from: https://link.springer.com/article/10.1023/A:1016121319668.

[124] Safwat S, Hathout RM, Ishak RA, Mortada ND. Augmented simvastatin cytotoxicity using optimized lipid nanocapsules: a potential for breast cancer treatment. J Liposome Res 2017;27(1):1–10. Available from: https://www.tandfonline.com/doi/abs/10.3109/08982104.2015.1137313.

[125] Lainé AL, Adriaenssens E, Vessières A, Jaouen G, Corbet C, Desruelles E, et al. The invivo performance of ferrocenyl tamoxifen lipid nanocapsules in xenografted triple negative breast cancer. Biomaterials 2013;34(28):6949–56.

[126] Montigaud Y, Ucakar B, Krishnamachary B, Bhujwalla ZM, Feron O, Préat V, et al. Optimized acriflavine-loaded lipid nanocapsules as a safe and effective delivery system to treat breast cancer. Int J Pharm 2018;551(1–2):322–8.

[127] Gaber M, Hany M, Mokhtar S, Helmy MW, Elkodairy KA, Elzoghby AO. Boronic-targeted albumin-shell oily-core nanocapsules for synergistic aromatase inhibitor/herbal breast cancer therapy. Mater Sci Eng C 2019;105:110099.

[128] Elzoghby AO, Mostafa SK, Helmy MW, ElDemellawy MA, Sheweita SA. Superiority of aromatase inhibitor and cyclooxygenase-2 inhibitor combined delivery: hyaluronate-targeted versus PEGylated protamine nanocapsules for breast cancer therapy. Int J Pharm 2017;529(1–2):178–92.

[129] Antonow M, Franco C, Prado W, Beckenkamp A, Silveira G, Buffon A, et al. Arginylglycylaspartic acid-surface-functionalized doxorubicin-loaded lipid-core nanocapsules as a strategy to target Alpha(V) Beta(3) integrin expressed on tumor cells. Nanomaterials 2017;8(1):2. Available from: http://www.mdpi.com/2079-4991/8/1/2.

[130] Zafar S, Akhter S, Garg N, Selvapandiyan A, Kumar Jain G, Ahmad FJ. Co-encapsulation of docetaxel and thymoquinone in mPEG-DSPE-vitamin E TPGS-lipid nanocapsules for breast cancer therapy: formulation optimization and implications on cellular and in vivo toxicity. Eur J Pharm Biopharm 2020;148:10–26.

[131] Antonow MB, Asbahr ACC, Raddatz P, Beckenkamp A, Buffon A, Guterres SS, et al. Liquid formulation containing doxorubicin-loaded lipid-core nanocapsules: cytotoxicity in human breast cancer cell line and in vitro uptake mechanism. Mater Sci Eng C 2017;76:374–82.

[132] Galisteo-González F, Molina-Bolívar JA, Navarro SA, Boulaiz H, Aguilera-Garrido A, Ramírez A, et al. Albumin-covered lipid nanocapsules exhibit enhanced uptake performance by breast-tumor cells. Colloids Surf B Biointerfaces 2018;165:103–10.

[133] de Oliveira C, Büttenbender S, Prado W, Beckenkamp A, Asbahr A, Buffon A, et al. Enhanced and selective antiproliferative activity of methotrexate-functionalized-nanocapsules to human breast cancer cells (MCF-7). Nanomaterials 2018;8(1):24. Available from: http://www.mdpi.com/2079-4991/8/1/24.

[134] Schultze E, Buss J, Coradini K, Begnini KR, Guterres SS, Collares T, et al. Tretinoin-loaded lipid-core nanocapsules overcome the triple-negative breast cancer cell resistance to tretinoin and show synergistic effect on cytotoxicity induced by doxorubicin and 5-fluororacil. Biomed Pharmacother 2017;96:404–9.

[135] Torchilin VP. Micellar nanocarriers: Pharmaceutical perspectives. Pharm Res 2007;24(1):1–16. Available from: https://link.springer.com/article/10.1007/s11095-006-9132-0.

[136] Jones MC, Leroux JC. Polymeric micelles—a new generation of colloidal drug carriers. Eur J Pharm Biopharm 1999;48(2):101–11. Available from: https://pubmed.ncbi.nlm.nih.gov/10469928/.

[137] Shi Y, Lammers T, Storm G, Hennink WE. Physico-chemical strategies to enhance stability and drug retention of polymeric micelles for tumor-targeted drug delivery. Macromol Biosci 2017;17(1):1600160. Available from: http://doi.wiley.com/10.1002/mabi.201600160.

[138] Zhang Y, Huang Y, Li S. Polymeric micelles: nanocarriers for cancer-targeted drug delivery. AAPS PharmSciTech 2014;15(4):862–71. Available from: https://link.springer.com/article/10.1208/s12249-014-0113-z.

[139] Chida T, Miura Y, Cabral H, Nomoto T, Kataoka K, Nishiyama N. Epirubicin-loaded polymeric micelles effectively treat axillary lymph nodes metastasis of breast cancer through selective accumulation and pH-triggered drug release. J Control Release 2018;292:130–40.

[140] Yao Q, Choi JH, Dai Z, Wang J, Kim D, Tang X, et al. Improving tumor specificity and anticancer activity of dasatinib by dual-targeted polymeric micelles. ACS Appl Mater Interfaces 2017;9(42):36642–54. Available from: https://pubs.acs.org/doi/abs/10.1021/acsami.7b12233.

[141] Chen LC, Chen YC, Su CY, Hong CS, Ho HO, Sheu MT. Development and characterization of self-assembling lecithin-based mixed polymeric micelles containing quercetin in cancer treatment and an in vivo pharmacokinetic study. Int J Nanomed 2016;11:1557–66. Available from: /pmc/articles/PMC4841422/?report=abstract.

[142] Guo X, Zhao Z, Chen D, Qiao M, Wan F, Cun D, et al. Co-delivery of resveratrol and docetaxel via polymeric micelles to improve the treatment of drug-resistant tumors. Asian J Pharm Sci 2019;14(1):78–85.

[143] Wan X, Beaudoin JJ, Vinod N, Min Y, Makita N, Bludau H, et al. Co-delivery of paclitaxel and cisplatin in poly(2-oxazoline) polymeric micelles: implications for drug loading, release, pharmacokinetics and outcome of ovarian and breast cancer treatments. Biomaterials 2019;192:1–14.

[144] Talelli M, Barz M, Rijcken CJF, Kiessling F, Hennink WE, Lammers T. Core-crosslinked polymeric micelles: principles, preparation, biomedical applications and clinical translation. Nano Today 2015;10(1):93–117. Available from: https://pubmed.ncbi.nlm.nih.gov/25893004/.

[145] Kim S, Shi Y, Kim JY, Park K, Cheng JX. Overcoming the barriers in micellar drug delivery: loading efficiency, in vivo stability, and micelle-cell interaction. Expert Opin Drug Deliv 2010;7(1):49–62. Available from: https://pubmed.ncbi.nlm.nih.gov/20017660/.

[146] Zhou H, Qi Z, Xue X, Wang C. Novel pH-sensitive urushiol-loaded polymeric micelles for enhanced anticancer activity. Int J Nanomed 2020;15:3851–68. Available from: https://www.dovepress.com/novel-ph-sensitive-urushiol-loaded-polymeric-micelles-for-enhanced-ant-peer-reviewed-article-IJN.

[147] Wang C, Qi P, Lu Y, Liu L, Zhang Y, Sheng Q, et al. Bicomponent polymeric micelles for pH-controlled delivery of doxorubicin. Drug Deliv 2020;27(1):344–57. Available from: https://www.tandfonline.com/doi/full/10.1080/10717544.2020.1726526.

[148] Soltantabar P, Calubaquib EL, Mostafavi E, Biewer MC, Stefan MC. Enhancement of loading efficiency by coloading of doxorubicin and quercetin in thermoresponsive polymeric micelles. Biomacromolecules 2020;21(4):1427–36. Available from: https://pubs.acs.org/doi/abs/10.1021/acs.biomac.9b01742.

[149] Yang Y, Li Y, Chen K, Zhang L, Qiao S, Tan G, et al. Dual receptor-targeted and redox-sensitive polymeric micelles self-assembled from a folic acid-hyaluronic acid-SS-vitamin E succinate polymer for precise cancer therapy. Int J Nanomed 2020;15:2885–902. Available from: /pmc/articles/PMC7188338/?report=abstract.

[150] Teo JY, Chin W, Ke X, Gao S, Liu S, Cheng W, et al. pH and redox dual-responsive biodegradable polymeric micelles with high drug loading for effective anticancer drug delivery. Nanomed Nanotechnol Biol Med 2017;13(2):431–42.

[151] Wu P, Jia Y, Qu F, Sun Y, Wang P, Zhang K, et al. Ultrasound-responsive polymeric micelles for sonoporation-assisted site-specific therapeutic action. ACS Appl Mater Interfaces 2017;9(31):25706–16. Available from: https://pubs.acs.org/doi/abs/10.1021/acsami.7b05469.

[152] Takemae K, Okamoto J, Horise Y, Masamune K, Muragaki Y. Function of epirubicin-conjugated polymeric micelles in sonodynamic therapy. Front Pharmacol 2019;10(May):546. Available from: https://www.frontiersin.org/article/10.3389/fphar.2019.00546/full.

[153] Moretton MA, Bernabeu E, Grotz E, Gonzalez L, Zubillaga M, Chiappetta DA. A glucose-targeted mixed micellar formulation outperforms Genexol in breast cancer cells. Eur J Pharm Biopharm 2017;114:305–16.

[154] Mozhi A, Ahmad I, Okeke CI, Li C, Liang XJ. pH-sensitive polymeric micelles for the Co-delivery of proapoptotic peptide and anticancer drug for synergistic cancer therapy. RSC Adv 2017;7(21):12886–96. Available from: https://pubs.rsc.org/en/content/articlehtml/2017/ra/c6ra27054a.

[155] Zhang X, Ren X, Tang J, Wang J, Zhang X, He P, et al. Hyaluronic acid reduction-sensitive polymeric micelles achieving co-delivery of tumor-targeting paclitaxel/apatinib effectively reverse cancer multidrug resistance. Drug Deliv 2020;27(1):825–35. Available from: https://www.tandfonline.com/doi/full/10.1080/10717544.2020.1770373.

[156] Zhang L, Pu Y, Li J, Yan J, Gu Z, Gao W, et al. pH responsive coumarin and imidazole grafted polymeric micelles for cancer therapy. J Drug Deliv Sci Technol 2020;58:101789.

[157] Tang M, Hu P, Zheng Q, Tirelli N, Yang X, Wang Z, et al. Polymeric micelles with dual thermal and reactive oxygen species (ROS)-responsiveness for inflammatory cancer cell delivery. J Nanobiotechnol 2017;15(1):39. Available from: http://jnanobiotechnology.biomedcentral.com/articles/10.1186/s12951-017-0275-4.

[158] Cagel M, Bernabeu E, Gonzalez L, Lagomarsino E, Zubillaga M, Moretton MA, et al. Mixed micelles for encapsulation of doxorubicin with enhanced in vitro cytotoxicity on breast and ovarian cancer cell lines versus Doxil®. Biomed Pharmacother 2017;95:894–903.

[159] Jabri T, Imran M, Aziz A, Rao K, Kawish M, Irfan M, et al. Design and synthesis of mixed micellar system for enhanced anticancer efficacy of Paclitaxel through its co-delivery with Naringin. Drug Dev Ind Pharm 2019;45(5):703–14. Available from: https://www.tandfonline.com/doi/abs/10.1080/03639045.2018.1550091.

[160] Gal D, Ohashi M, MacDonald PC, Buchsbaum HJ, Simpson ER. Low-density lipoprotein as a potential vehicle for chemotherapeutic agents and radionucleotides in the management of gynecologic neoplasms. Am J Obstet Gynecol 1981;139(8):877–85. Available from: https://pubmed.ncbi.nlm.nih.gov/7223790/.

[161] Mahmoudian M, Salatin S, Khosroushahi AY. Natural low- and high-density lipoproteins as mighty bio-nanocarriers for anticancer drug delivery. Cancer Chemother Pharmacol 2018;82(3):371–82. Available from: https://link.springer.com/article/10.1007/s00280-018-3626-4.

[162] Almer G, Mangge H, Zimmer A, Prassl R. Lipoprotein-related and apolipoprotein-mediated delivery systems for drug targeting and imaging. Curr Med Chem 2015;22(31):3631–51. Available from: https://pubmed.ncbi.nlm.nih.gov/26180001/.

[163] Sabnis N, Lacko AG. Drug delivery via lipoprotein-based carriers: Answering the challenges in systemic therapeutics. Ther Deliv 2012;3(5):599–608. Available from: https://www.future-science.com/doi/abs/10.4155/tde.12.41.

[164] Mooberry LK, Nair M, Paranjape S, McConathy WJ, Lacko AG. Receptor mediated uptake of paclitaxel from a synthetic high density lipoprotein nanocarrier. J Drug Target 2010;18(1):53–8. Available from: https://www.tandfonline.com/doi/abs/10.3109/10611860903156419.

[165] Jutkova A, Chorvat D, Miskovsky P, Jancura D, Datta S. Encapsulation of anticancer drug curcumin and co-loading with photosensitizer hypericin into lipoproteins investigated by fluorescence resonance energy transfer. Int J Pharm 2019;564:369–78.

[166] Zhang Z, Chen J, Ding L, Jin H, Lovell JF, Corbin IR, et al. HDL-mimicking peptide-lipid nanoparticles with improved tumor targeting. Small 2010;6(3):430–7. Available from: https://onlinelibrary.wiley.com/doi/abs/10.1002/smll.200901515.

[167] Ng KK, Lovell JF, Zheng G. Lipoprotein-inspired nanoparticles for cancer theranostics. Acc Chem Res 2011;44(10):1105–13. Available from: /pmc/articles/PMC3196219/?report=abstract.

[168] Mirzaei M, Akbari ME, Mohagheghi MA, Ziaee SAM, Mohseni M. A novel biocompatible nanoprobe based on lipoproteins for breast cancer cell imaging. Nanomed J 2020;7(1):73–9. Available from: http://nmj.mums.ac.ir/article_13980.html.

[169] Hammel M, Laggner P, Prassl R. Structural characterisation of nucleoside loaded low density lipoprotein as a main criterion for the applicability as drug delivery system. Chem Phys Lipids 2003;123(2):193–207.

[170] Zhu C, Xia Y. Biomimetics: reconstitution of low-density lipoprotein for targeted drug delivery and related theranostic applications. Chem Soc Rev 2017;46(24):7668–82. Available from: /pmc/articles/PMC5725233/?report=abstract.

[171] Panchapakesan B, Lu S, Sivakumar K, Teker K, Cesarone G, Wickstrom E. Single-wall carbon nanotube nanobomb agents for killing breast cancer cells. Nanobiotechnology 2005;1(2):133–9. Available from: https://link.springer.com/article/10.1385/NBT:1:2:133.

[172] Ogbodu RO, Limson JL, Prinsloo E, Nyokong T. Photophysical properties and photodynamic therapy effect of zinc phthalocyanine-spermine-single walled carbon nanotube conjugate on MCF-7 breast cancer cell line. Synth Met 2015;204:122–32.

[173] Arora S, Kumar R, Kaur H, Rayat CS, Kaur I, Arora SK, et al. Translocation and toxicity of docetaxel multi-walled carbon nanotube conjugates in mammalian breast cancer cells. J Biomed Nanotechnol 2014;10(12):3601–9.

[174] Guven A, Villares GJ, Hilsenbeck SG, Lewis A, Landua JD, Dobrolecki LE, et al. Carbon nanotube capsules enhance the in vivo efficacy of cisplatin. Acta Biomater 2017;58:466–78.

[175] Hashemzadeh H, Raissi H. The functionalization of carbon nanotubes to enhance the efficacy of the anticancer drug paclitaxel: a molecular dynamics simulation study. J Mol Model 2017;23(8):1–10. Available from: https://link.springer.com/article/10.1007/s00894-017-3391-z.

[176] Shao W, Paul A, Zhao B, Lee C, Rodes L, Prakash S. Carbon nanotube lipid drug approach for targeted delivery of a chemotherapy drug in a human breast cancer xenograft animal model. Biomaterials 2013;34(38):10109–19.

[177] Pramanik M, Song KH, Swierczewska M, Green D, Sitharaman B, Wang LV. In vivo carbon nanotube-enhanced non-invasive photoacoustic mapping of the sentinel lymph node. Phys Med Biol 2009;54(11):3291–301. Available from: https://iopscience.iop.org/article/10.1088/0031-9155/54/11/001.

[178] Mubarak NM, Abdullah EC, Jayakumar NS, Sahu JN. An overview on methods for the production of carbon nanotubes. J Ind Eng Chem 2014;20(4):1186–97.

[179] Ravi Kiran AVVV, Kusuma Kumari G, Krishnamurthy PT. Carbon nanotubes in drug delivery: focus on anticancer therapies. J Drug Deliv Sci Technol 2020;59:101892.

[180] Tagmatarchis N, Prato M. Functionalization of carbon nanotubes via 1,3-dipolar cycloadditions. J Mater Chem 2004;14(4):437–9. Available from: https://pubs.rsc.org/en/content/articlehtml/2004/jm/b314039c.

[181] Tasis D, Tagmatarchis N, Bianco A, Prato M. Chemistry of carbon nanotubes. Chem Rev 2006;106(3):1105–36. Available from: https://pubmed.ncbi.nlm.nih.gov/16522018/.

[182] Moon HK, Lee SH, Choi HC. In vivo near-infrared mediated tumor destruction by photothermal effect of carbon nanotubes. ACS Nano 2009;3(11):3707–13. Available from: https://pubs.acs.org/doi/abs/10.1021/nn900904h.

[183] Tang S, Chen M, Zheng N. Multifunctional ultrasmall Pd nanosheets for enhanced near-infrared photothermal therapy and chemotherapy of cancer. Nano Res 2014;8(1):165–74. Available from: https://link.springer.com/article/10.1007/s12274-014-0605-x.

[184] Shi X, Su HW, Shen M, Antwerp ME, Chen X, Li C, et al. Multifunctional dendrimer-modified multiwalled carbon nanotubes: synthesis, characterization, and in vitro cancer cell targeting and imaging. Biomacromolecules 2009;10(7):1744–50. Available from: https://pubs.acs.org/doi/abs/10.1021/bm9001624.

[185] Tavakolifard S, Biazar E, Pourshamsian K, Moslemin MH. Synthesis and evaluation of single-wall carbon nanotube-paclitaxel-folic acid conjugate as an anti-cancer targeting agent. Artif Cells Nanomed Biotechnol 2016;44(5):1247–53. Available from: https://www.tandfonline.com/doi/full/10.3109/21691401.2015.1019670.

[186] Suo X, Eldridge BN, Zhang H, Mao C, Min Y, Sun Y, et al. P-glycoprotein-targeted photothermal therapy of drug-resistant cancer cells using antibody-conjugated carbon nanotubes. ACS Appl Mater Interfaces 2018;10(39):33464–73. Available from: https://pubs.acs.org/doi/abs/10.1021/acsami.8b11974.

[187] Tian Z, Tian Z, Shi Y, Yin M, Shen H, Jia N. Nano Biomed Eng OPEN ACCESS article functionalized multiwalled carbon nanotubes-anticancer drug carriers: synthesis, target-ing ability and antitumor activity. Nano Biomed Eng 2011;3(3):157–62. Available from: http://nanobe.org.

[188] Markowicz-Piasecka M, Mikiciuk-Olasik E. Dendrimers in drug delivery. In: Nanobiomaterials in drug delivery: applications of nanobiomaterials. Elsevier Inc.; 2016. p. 39–74.

[189] Wehner BWVF. Cascade and nonskid-chain like synthesis of molecule cavity topologies. In: Vogtle synthesis; 1978. p. 155. US Patents 4289872 (1981) and 4410688 (1985).

[190] Chen CZ, Cooper SL. Recent advances in antimicrobial dendrimers. Adv Mater 2000;12(11):843–6. Available from: https://onlinelibrary.wiley.com/doi/full/10.1002/%28SICI%291521-4095%28200006%2912%3A11%3C843%3A%3AAID-ADMA843%3E3.0.CO%3B2-T.

[191] Caminade AM, Turrin CO. Dendrimers for drug delivery. J Mater Chem B 2014;2(26):4055–66. Available from: https://pubs.rsc.org/en/content/articlehtml/2014/tb/c4tb00171k.

[192] Mandal AK. Dendrimers in targeted drug delivery applications: a review of diseases and cancer. Int J Polym Mater Polym Biomater 2020. Available from: https://www.tandfonline.com/doi/abs/10.1080/00914037.2020.1713780.

[193] Wolinsky JB, Grinstaff MW. Therapeutic and diagnostic applications of dendrimers for cancer treatment. Adv Drug Deliv Rev 2008;60(9):1037–55.

[194] Bielawski K, Bielawska A, Muszyńska A, Popławska B, Czarnomysy R. Cytotoxic activity of G3 PAMAM-NH 2 dendrimer-chlorambucil conjugate in human breast cancer cells. Environ Toxicol Pharmacol 2011;32(3):364–72.

[195] Wang M, Li Y, Huangfu M, Xiao Y, Zhang T, Han M, et al. Pluronic-attached polyamidoamine dendrimer conjugates overcome drug resistance in breast cancer. Nanomedicine 2016;11(22):2917–34. Available from: https://www.futuremedicine.com/doi/abs/10.2217/nnm-2016-0252.

[196] Zhang C, Pan D, Luo K, She W, Guo C, Yang Y, et al. Peptide dendrimer-doxorubicin conjugate-based nanoparticles as an enzyme-responsive drug delivery system for cancer therapy. Adv

Healthc Mater 2014;3(8):1299–308. Available from: http://doi.wiley.com/10.1002/adhm.201300601.
[197] Li N, Cai H, Jiang L, Hu J, Bains A, Hu J, et al. Enzyme-sensitive and amphiphilic PEGylated dendrimer-paclitaxel prodrug-based nanoparticles for enhanced stability and anticancer efficacy. ACS Appl Mater Interfaces 2017;9(8):6865–77. Available from: https://pubs.acs.org/doi/abs/10.1021/acsami.6b15505.
[198] Marcinkowska M, Stanczyk M, Janaszewska A, Sobierajska E, Chworos A, Klajnert-Maculewicz B. Multicomponent conjugates of anticancer drugs and monoclonal antibody with PAMAM dendrimers to increase efficacy of HER-2 positive breast cancer therapy. Pharm Res 2019;36(11):1–17. https://doi.org/10.1007/s11095-019-2683-7.
[199] Morgan MT, Nakanishi Y, Kroll DJ, Griset AP, Carnahan MA, Wathier M, et al. Dendrimer-encapsulated camptothecins: increased solubility, cellular uptake, and cellular retention affords enhanced anticancer activity in vitro. Cancer Res 2006;66(24):11913–21. Available from: www.aacrjournals.org.
[200] Kaur D, Jain K, Mehra NK, Kesharwani P, Jain NK. A review on comparative study of PPI and PAMAM dendrimers. J Nanopart Res 2016;18(6):1–14. Available from: https://link.springer.com/article/10.1007/s11051-016-3423-0.
[201] Lai PS, Lou PJ, Peng CL, Pai CL, Yen WN, Huang MY, et al. Doxorubicin delivery by polyamidoamine dendrimer conjugation and photochemical internalization for cancer therapy. J Control Release 2007;122(1):39–46.
[202] Gu Y, Guo Y, Wang C, Xu J, Wu J, Kirk TB, et al. A polyamidoamne dendrimer functionalized graphene oxide for DOX and MMP-9 shRNA plasmid co-delivery. Mater Sci Eng C 2017;70:572–85.
[203] Suganya USU, Govindaraju K, Kumar GG, Prabhu D, Arulvasu C, Dhas SS, et al. Anti-proliferative effect of biogenic gold nanoparticles against breast cancer cell lines (MDA-MB-231 & MCF-7). Appl Surf Sci 2016;371:415–24.
[204] Kefayat A, Ghahremani F, Motaghi H, Mehrgardi MA. Investigation of different targeting decorations effect on the radiosensitizing efficacy of albumin-stabilized gold nanoparticles for breast cancer radiation therapy. Eur J Pharm Sci 2019;130:225–33.
[205] Klein S, Sommer A, Distel LVR, Neuhuber W, Kryschi C. Superparamagnetic iron oxide nanoparticles as radiosensitizer via enhanced reactive oxygen species formation. Biochem Biophys Res Commun 2012;425(2):393–7.
[206] Aliakbari M, Mohammadian E, Esmaeili A, Pahlevanneshan Z. Differential effect of polyvinylpyrrolidone-coated superparamagnetic iron oxide nanoparticles on BT-474 human breast cancer cell viability. Toxicol Vitr 2019;54:114–22.
[207] Chen C, Sun SR, Gong YP, Qi CB, Peng CW, Yang XQ, et al. Quantum dots-based molecular classification of breast cancer by quantitative spectroanalysis of hormone receptors and HER2. Biomaterials 2011;32(30):7592–9.
[208] Abdelhamid AS, Zayed DG, Helmy MW, Ebrahim SM, Bahey-El-Din M, Zein-El-Dein EA, et al. Lactoferrin-tagged quantum dots-based theranostic nanocapsules for combined COX-2 inhibitor/herbal therapy of breast cancer. Nanomedicine 2018;13(20). Available from: https://www.futuremedicine.com/doi/abs/10.2217/nnm-2018-0196.
[209] Balakrishnan S, Bhat FA, Raja Singh P, Mukherjee S, Elumalai P, Das S, et al. Gold nanoparticle-conjugated quercetin inhibits epithelial-mesenchymal transition, angiogenesis and invasiveness *via* EGFR/VEGFR-2-mediated pathway in breast cancer. Cell Prolif 2016;49(6):678–97. Available from: http://doi.wiley.com/10.1111/cpr.12296.
[210] Yan L, Amirshaghaghi A, Huang D, Miller J, Stein JM, Busch TM, et al. Protoporphyrin IX (PpIX)-coated superparamagnetic iron oxide nanoparticle (SPION) nanoclusters for magnetic resonance imaging and photodynamic therapy. Adv Funct Mater 2018;28(16):1707030. Available from: http://doi.wiley.com/10.1002/adfm.201707030.
[211] Freitas M, Neves MMPS, Nouws HPA, Delerue-Matos C. Quantum dots as nanolabels for breast cancer biomarker HER2-ECD analysis in human serum. Talanta 2020;208:120430.
[212] Lu J, Liong M, Zink JI, Tamanoi F. Mesoporous silica nanoparticles as a delivery system for hydrophobic anticancer drugs. Small 2007;3(8):1341–6. Available from: https://pubmed.ncbi.nlm.nih.gov/17566138/.
[213] Vallet-Regí M, Colilla M, Izquierdo-Barba I, Manzano M. Mesoporous silica nanoparticles for drug delivery: current insights. Molecules 2018;23(1). Available from: /pmc/articles/PMC5943960/?report=abstract.
[214] Uchegbu IF, Vyas SP. Non-ionic surfactant based vesicles (niosomes) in drug delivery. Int J Pharm 1998;172(1–2):33–70.
[215] Salem HF, Kharshoum RM, El-Ela FIA, Gamal AF, Abdellatif KRA. Evaluation and optimization of pH-responsive niosomes as a carrier for efficient treatment of breast cancer. Drug Deliv Transl Res 2018;8(3):633–44. Available from: https://link.springer.com/article/10.1007/s13346-018-0499-3.
[216] Brunel FM, Lewis JD, Destito G, Steinmetz NF, Manchester M, Stuhlmann H, et al. Hydrazone ligation strategy to assemble multifunctional viral nanoparticles for cell imaging and tumor targeting. Nano Lett 2010;10(3):1093–7. Available from: https://pubs.acs.org/doi/abs/10.1021/nl1002526.
[217] Shukla S, Myers JT, Woods SE, Gong X, Czapar AE, Commandeur U, et al. Plant viral nanoparticles-based HER2 vaccine: immune response influenced by differential transport, localization and cellular interactions of particulate carriers. Biomaterials 2017;121:15–27.
[218] Kumar Teli M, Mutalik S, Rajanikant GK. Nanotechnology and nanomedicine: going small means aiming big. Curr Pharm Des 2010;16(16):1882–92. Available from: https://pubmed.ncbi.nlm.nih.gov/20222866/.
[219] Barz M, Luxenhofer R, Schillmeier M. Quo vadis nanomedicine? Nanomedicine 2015;10(20):3089–91. Available from: https://pubmed.ncbi.nlm.nih.gov/26465063/.
[220] Kraft JC, Freeling JP, Wang Z, Ho RJY. Emerging research and clinical development trends of liposome and lipid nanoparticle drug delivery systems. J Pharm Sci 2014;103(1):29–52. Available from: https://pubmed.ncbi.nlm.nih.gov/24338748/.
[221] Tinkle S, Mcneil SE, Mühlebach S, Bawa R, Borchard G, Barenholz YC, et al. Nanomedicines: addressing the scientific and regulatory gap. Ann N Y Acad Sci 2014;1313(1):35–56. Available from: https://pubmed.ncbi.nlm.nih.gov/24673240/.
[222] Svenson S. Clinical translation of nanomedicines. Curr Opin Solid State Mater Sci 2012;16(6):287–94.
[223] Nyström AM, Fadeel B. Safety assessment of nanomaterials: implications for nanomedicine. J Control Release 2012;161(2):403–8. Available from: https://pubmed.ncbi.nlm.nih.gov/22306428/.

Chapter 25

Advancement on nanoparticle-based drug delivery systems for cancer therapy

Brahmeshwar Mishra and Mansi Upadhyay
Department of Pharmaceutical Engineering & Technology, Indian Institute of Technology (BHU), Varanasi, Uttar Pradesh, India

1 Introduction

The term "cancer" is not new to the society; it is a highly lethal health issue worldwide however, with respect to detection of probability of cancer from cellular to the molecular level, through research efforts and early diagnosis via advancement in technology seems that there is a tremendous scope of nanotechnology to explore more and more about this deadly problem [1]. Within past few decades nanoparticles (NPs), a gift from nanotechnology, have gained remarkable attention in cancer therapy, screening, and diagnosis [2]. Nano has always been in demand in the broad area of research field, for instance, according to the data available at PubMed.gov, the number of articles published alone on NPs in 2004 was 1840 and has remained constantly increased to 27,072 till 2019. Nano has always been in demand in broad area of research field, for instance according to the data available at PubMed.gov, the number of articles published alone on NPs in 2004 were 1840 that increased to 27,072 in 2019. Whereas, the articles published alone on role of NPs in cancer, were also in the increasing range, i.e., 233 (2004) to 5902 (2019) [3]. The reason behind such progressive publications remains in NP special features. Nanoparticles (NPs) are one of the smartest creations of nanotechnology. Maximum drugs available in the market as well as the molecules in the pipelines are lipophilic in nature. Further, the drugs to treat life-threatening and serious diseases, for example, cancer, asthma, Alzheimer's, have poor bioavailability and water solubility problem thus requires frequent exposure of drugs that can lead to serious toxic effects and can cause death too. In surmounting the aforesaid problem, NPs are sufficient [4]. They can be tailored in any form to meet the therapeutic requirements for instance can be directly administered or targeted at the disease site either coating with particular polymer or by decorating its surface by specific ligands [5]. Taking into account the above-mentioned special features of NPs, several nanodrug products, for instance, Doxil (the first approved nanodrug by FDA, 1995), Marquibo, Auroshell, and many more have been approved by FDA for cancer therapy [6]. Herein, the aim of this chapter is to focus on the key role and recent developments in cancer therapy through nanotechnology in conquering the malignancy via different ways such as by in vivo imaging or study of molecular change during cancer cell growth or by therapeutic organic NPs (liposomes, dendrimers, solid lipid NPs, etc.), targeted anticancer therapeutics drug delivery, polymeric NPs or surface-coated drug delivery (PLGA, PEG/PEGylation) or inorganic NPs (for example, gold, silica, and magnetic NPs) formulation.

2 Structure and importance of nanoparticles in cancer therapy

2.1 Structure of nanoparticle

In general, NPs are small colloidal solid nanostructure whose size ranges between 10 and 100 nm. Though the size nano sounds tiny, however, the structure that exhibits its advantage as well as the property is somewhat complex. The "surface" layer is said to be organized with a series of metallic ions, small molecules, surfactants, or polymers. Surface of any nanoparticle is an important component because usually in many cases it is the surface of the NPs, decorated with the desired functional group or short or long peptide chains is formed to interact with the biological system [7]. Next layer is called as the "core" present in the middle of the NP. This layer is often referred as NPs as the specific property of any NP is generally related to the composition of the core. Lastly, the "shell," which is the entirely different chemical material from the layer core [8]. An image of structure of NPs containing its layer is shown in Fig. 1.

2.2 Importance of NPs in cancer therapy: Anatomy based

There are several reasons and facts to why prefer NPs in cancer. Unlike normal tissue, the anatomy of cancerous

Advanced Drug Delivery Systems in the Management of Cancer. https://doi.org/10.1016/B978-0-323-85503-7.00026-2

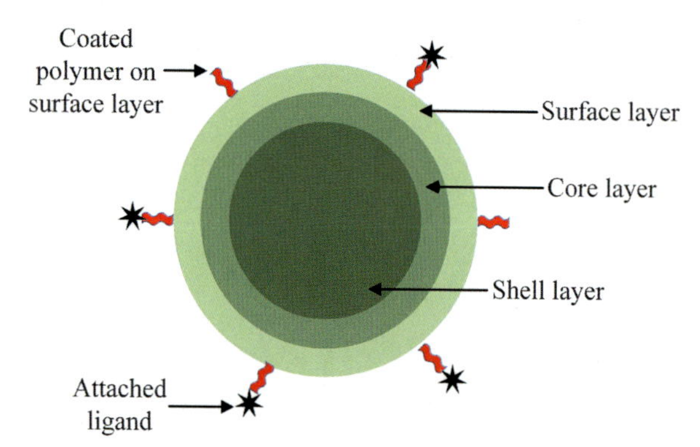

FIG. 1 Layers of nanoparticle.

tissue changes dramatically [9]. Tumor tissues become heterogenous, highly dense, and strongly strives for ample supply of nutrients and oxygen for fast growth leading to rapid uncontrolled cell differentiation and slower angiogenesis that consequently forms an immature, defective, leaky, and hyperpermeable vasculatures whose pore size ranges approximately 380–780 nm [10–12]. Further, the interstitial compartment of tumor tissue also loses its structured network (composed of collagen and elastic fiber) and rigidity, resulting into the outward movement of interstitial fluid and nonfunctional lymphatic network. Now, here NPs take the advantage of the above aberrant vasculature. Due to dysfunctional leakage, drainage, lack of lymphatic circulation, and presence of nanopores in the tight junction of endothelial cells of tumor blood vessels allow NPs or nanoformulations to retain within those fenestrae. This mechanism of retention of the particles within nanopores of abnormal cancer tissue is called as enhanced permeation and retention effect (EPR), where the concentrated NPs loaded with therapeutics release the drug locally to the tumor tissues [13, 14]. The EPR effect of particles has also been proved in studies, for instance, Unezaki et al. using in vivo inflorescence have reported the extravasation of PEG liposomes of size 400 nm into tumor vasculature [15].

3 Advantage of nanoformulations over other formulations

Different nanoformulations such as liposomes, dendrimers, solid lipid NPs, and micelles have marked their potential in the treatment of cancer [16, 17]. The success of nanocarriers over other formulation are as follows:

3.1 Nanosize

NPs are said to be 100–1000 times smaller in comparison to a single malignant cell, therefore, exhibit good cellular uptake over micro formulations [18]. For instance, it is reported that NPs of an average size of 100 nm show two and half times of greater cellular uptake when compared to a 1 μm diameter while the same particle exhibits six times greater cellular uptake for a 10 μm microparticle [19]. Moreover, the NPs of size range between 100 and 200 nm can easily reach to the tumor tissue by surpassing the sinusoids of reticuloendothelial system (RES). Studies have shown that often, microparticles of size resembling to a microbe can be easily taken up by the macrophages (resting in the sinusoidal gaps of RES organs like liver and spleen (~ 150–200 nm)) and are distributed throughout the body resulting in rapid clearance from blood circulation. Therefore, it is suggested to prepare NPs of size range between 100 and 200 nm as they can easily reach to the tumor tissue by surpassing the sinusoids of RES [20–22].

3.2 Nanosurface property

In comparison to small micro/macromolecule, NPs exhibit large surface area to volume ratio, thus exhibits huge reactive interface between nanomaterial and the local environment [23]. This enormous ratio of NPs enables its surface to attach and facilitate the binding of several biological ligands, for instance, proteins (antibodies/monoclonal antibodies), aptamers, peptides, polysaccharides (hyaluronic acid), and small molecules like folate [24].

3.3 Nanosolubility and bioavailability

Drugs falling into BCS (Biopharmaceutical Classification System) class II (low solubility and high permeability) and IV (low solubility and low permeability) are generally called as "brick dust." Nanonization of these drugs not only increase the surface area (thus, promotes solubility with enhanced bioavailability) but also improves pharmacokinetics. The best suggested formulation for these class of drugs are "nanosuspensions." This drug delivery vehicle is regarded as the most stable formulation due to few important reasons such as presence of stabilizing agent, its preparation method, and suitability to carry and solubilize both water soluble as well as insoluble drug as nanosuspensions are composed of aqueous and oil media thus could be an ideal formulation to enhance the bioavailability of most of the drugs [25].

4 Application of NPs in cancer therapy

Over the past decades, a variety of approaches of NPs have been tried and prepared. The NPs on their nanoscale have always been formulated with reference to the rationale, objective, type, and pathological site of cancer. Regarding these above-mentioned points here under this heading, we will summarize numerous traditional nanodelivery vehicles, their mechanism to reach to the cancer site as well as the advancement done on nano-based systems for cancer therapy.

4.1 Nanomedicines

Nanomedicine (NMD) is an interdisciplinary field of "nanotechnology" and "medicine." The field involves the application of nanomaterials to enhance and improve the medications [26]. Functionalization of NPs in different several regards has enabled the NMDs for early detection of cancer and cancer treatment. Functionalization means alterations in the NPs property by physical or chemical modification to achieve the desired effect, for instance, prolong drug effect, local or direct delivery, detection of location of tumor or site of infection [27]. There are number of ways to functionalize NPs and they are as follows:

4.1.1 Functionalization via targeting

Due to nano in size NPs are the most preferred carrier to understand the tumor biology as well as accessible to target the tumor tissues. This targeting is achieved by two methods, i.e., active or passive targeting (Fig. 2) [28]. *Active targeting* is based on molecular recognition. In this method, NPs (composed of polymers or lipids) carrying the active chemotherapeutic drugs interact with the cancerous cell directly. Cancerous cells are said to be different from healthy cells as they possess some specific receptors on their surface. Decorating the complementary ligands on the surface of NPs loaded with therapeutic helps the nanomaterial to recognize only the infected or tumorous cell. After reaching to the main site, NPs interact to the overexpressed receptor by ligand-receptor interaction or by antigen-antibody interaction. Afterward these NPs are uptake by phagocytes followed by cell internalization of the entrapped dug and finally, rapid cleared off by RES system [29]. In contrast, the principle of passive targeting is based on tumor biology that involves the concentration of NPs to the leaky and perforated vasculatures through the EPR effect. As a result of hypoxia, inflammation, and stopped apoptosis, tumor cells start sucking the nutritious materials from the existing blood vessels and exhibit defective angiogenesis, i.e., formation of new leaky blood vessels which are devoid of basement membranes and pericytes. Both these components surround the endothelium wall. Due to their absence, the permeability of molecules (100–780 nm) to pass into the interstitium of tumor blood vessel is increased. Therefore, NPs can easily diffuse and retain within these perforated vessels and allows the release of drug into the locality of tumor [30, 31].

4.1.2 Functionalization via conjugation of nanoparticle surface

Surface conjugation of NPs is meant only for specific targeting to the specific organ so that off-target effects can be minimized to the other healthy tissues. Conjugation is done by adding several wide varieties of ligands to the surface of the particle such as aptamers, and antibody protein domains [32]. In case of cancer therapy, this surface conjugation is very effective, for instance, anticancer drugs such as paclitaxel, tamoxifen, and doxorubicin are majorly used in the treatment of different types of cancer and cancer therapies. Besides their active application these drugs in combination or alone cause serious side effects too. To avoid their adverse effects, efforts toward preparation of targeted nanoformulations, for example, dendrimers, liposomes, and micelles are prepared by attaching ligands so that the drug loaded within these formulations can recognize and target only its specific receptor present on the surface of cancer and can release the drug to that site only without harming the healthy tissues. Functionalization via conjugation has the advantage over nonconjugated surface NPs as it increases the NPs biological half-life; therefore, these functionalized drug-loaded NPs can accumulate over a long time at the cancer site [33]. Conjugation of NPs also helps in screening and diagnosis of cancer and cancerous site. It is possible to attach fluorescent dyes for imaging purpose, cell penetrating peptides

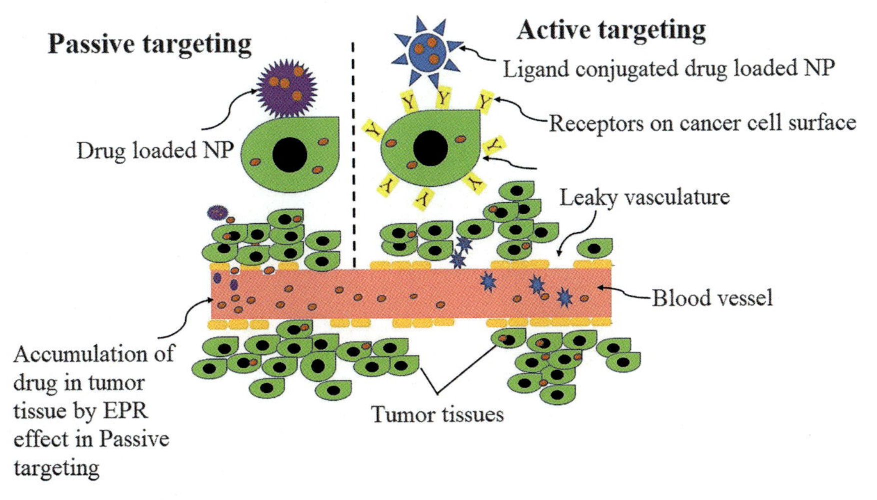

FIG. 2 Targeting mechanism of nanoparticles.

TABLE 1 Nanoparticles with their conjugated ligands and receptors to be targeted.

S. no.	Ligand	Receptor/target	Nanoparticle	Type of cancer	Reference
1	siRNA	EZH2	Iron NP	Ovarian	[34]
2	mAb	PDL-1	PEG-PCL	Gastric	[35]
3	Transferrin	Folate	PEG	Breast	[36]
4	Translocator protein	Benzodiazepine	PLGA	Glioma	[37]
5	SP204 peptide	Androgen	QDs, SPIONs	Prostate	[38]
6	L-Carnitine	OCTN2, $ATB^{0,+}$ transporter	PLGA	Colon	[39]
7	Galactosamine	HepG2	Poly (γ-glutamic acid) and poly (lactide)	Liver	[40]
8	Herceptin	HER2	PLGA-PHis-PEG	Breast	[41]
9	Polygalacturonic acid	EphA2	Magnetic cobalt spinel ferrite	Ovarian	[42]
10	Thiol PEGylated	Estrogen	Gold	Breast	[43]

such as small inhibitory RNA (siRNA) [26]. Few examples of nanoparticles conjugated with different types of ligands for therapeutic applications are given in Table 1.

4.1.3 Functionalization via stimuli to trigger the control release

Release of therapeutics from NPs can be controlled temporally and spatially by functionalizing with several stimuli such as pH, temperature, light, heat, magnetic, and enzyme (Table 2). Preparing the NPs as bio-responsive and target only pathological and infected site so that minimal or no side effects occur to the healthy tissues is a better approach to cure the cancer.

4.2 Nanomedicines and related nanopharmaceuticals

Nanomedicines and related nanopharmaceuticals are rapidly developing and establishing their importance in cancer therapy. Since 1980, continuously, they are being developed and entering in the clinical trial for the marketed purpose [54]. The first-ever nanomedicine approved by USFDA

TABLE 2 Smart nanoparticles triggered with stimuli.

S. no.	Stimuli	Drug	Type of NP	Type of cancer	Reference
1	pH	Doxorubicin	Liposome	Lung	[44]
2	pH	Dopamine	Mesoporous silicon	Tumor in blood vessel	[45]
3	Magnetic field-induced	Peptide	Mesoporous silicon	Pancreatic	[46]
4	Enzyme	Methotrexate	Iron	Breast	[47]
5	Photosensitive	Doxorubicin	Liposome	All type tumor cells	[48]
6	pH	Doxorubicin	QDs	Lung	[49]
7	pH	Curcumin	Polymeric	Colon	[50]
8	pH	Doxorubicin and Curcumin	PEG	All type tumor cells	[51]
9	pH, redox and enzyme	Doxorubicin	Iron	All type tumor cells	[52]
10	Magnetic field-induced	Doxorubicin	Iron	All type tumor cells	[53]

TABLE 3 Marketed and approved nanomedicines and their nanoformulations [55–57].

S. no.	Type of NMDs	Brand/bioactive	Cancer	Make
1	PEGylated liposome	Doxil/doxorubicin-HCl	Ovarian	Ortho Biotech
2	Non-PEGylated liposome	Myocet/doxorubicin-HCl	Metastatic breast	Teva Pharma B.V.
3	Nanosphere	Abraxane/albumuin bound	Various cancers	Celgene, Abraxis, Bioscience
4	Micelle	Apealea/paclitaxel	Ovarian, fallopian tube, peritoneal	Osamia Pharmaceutical AB
5	Micelle	Genoxol-PM/paclitaxel	Ovarian, gastric	Samyang Biopharm
6	Liposome	DaunoXome/daunorubicin	Breast, Kaposi's	Galen Ltd.
7	Liposome	Lipusu/paclitaxel	Breast, nonsmall-cell lung	Luye Pharma
8	Liposome	DepoCyt/cytarabine	Lymphoma/leukemia	Sigma-Tau Pharmaceuticals
9	Liposome	Marquibo/vincristine sulfate	Acute lymphoblastic leukemia	Spectrum Pharmaceuticals
10	Liposome	Mepact/mifamurtide	Brain tumor, osteogenic sarcoma	Takeda
11	Liposome	ThermoDox/doxorubicin	Breast, liver	Celsion
12	Oncaspar	PEGasparaginase	Acute lymphocytic leukemia	Enzon
13	Rexin-G	Targeting protein-tagged phospholipid/microRNA122	Pancreatic, sarcoma, osteosarcoma	Epeius Biotechnologies Corp
14	Resovist	Iron oxide NP coated with carboxydextran	Liver/spleen lesion imaging	Bayer Schering Pharma AG
15	Endorem	Iron oxide NP coated with dextran	Liver/spleen lesion imaging	Guerbet
16	Feridex	Iron oxide NP coated with dextran	Liver/spleen lesion imaging	Berlex Laboratories

in 1995 to treat the variety of cancer was a PEGylated liposomal Doxorubicin under name of "Doxil" and since then different nanopharmaceuticals or nanomedicine-based nanoformulations such as nanocrystals, liposomes, micelles, dendrimers, protein, and metallic-based NP products were launched and were marketed successfully. Few of the marketed nanomedicines are summarized in Table 3.

4.3 Role of NPs in combined drug therapy for cancer

Codelivery of therapeutics through nanotechnology is another strategy toward standard clinical practice for the treatment of cancer [58]. Progression of tumor follows multiple pathways that led the cancer cells to mutate continuously. This makes the disease even more complex when the mutated tumor cells show acquired resistance to the chemotherapeutics. Therefore, the use of a single chemotherapeutic during chemotherapy or cancer therapy would not be sufficient to overcome the successive mutation [59]. In combined drug therapy it is believed that the application of two or more drugs will exhibit synergistic effect in targeting different cancerous pathways or receptors or genes overexpressed on cancerous cells or any stage in the cell cycle [60]. This therapy has potential advantages over mono therapy such as reduction in drug resistance, reduction in metastatic and tumor growth, reduction in dose, and reduction in cancer stem cell population [61]. Through this approach, the delivery of small interfering RNA (siRNA) and micro RNA (miR) could be a promising candidate for cancer therapy; however, their low stability, immunogenicity, and inability to cross the cell membrane make their application limited [62]. Use of nanocarriers or nanodrug delivery systems will be efficient in delivering these biologics to their active site. In addition, various nanodrug delivery systems, for example, liposomes, nanocrystals, nanotubes, and micelles, have been utilized for the delivery of active drug in combination to target the respective sites. Another, concept associated with combination therapy is that, it is not only restricted to drug-drug combination. Now, the combination therapy is not only restricted to drug-drug combination, however, other combinations such as, drug-peptide combination, drug-protein (RNAi, i.e., inhibitory RNA, mRNA) combination, RNAi-RNAi combination, etc., have also shown positive results when delivered at nanoplatform. These kinds of combinations have also shown positive results when delivered at nanoplatform [63]. Few examples of these combinations delivered through different nanodelivery system are tabulated in Table 4.

With respect to this approach, FDA has also approved various combination of therapeutics to be used during chemotherapy (Table 5). In practice, combination of chemotherapeutics could show a better response and survival rate than single therapy, for instance, combination of cisplatin (Platinol) and vinorelbine (Navelbine) for treatment of nonsmall-cell lung cancer, and use of Taxol, Carboplatin, and Herceptin (TCH) for breast cancer [81].

4.4 Nanodevices in cancer screening and diagnosis

Conventional and current screening methods, for example, X-ray, magnetic resonance imaging (MRI), endoscopy, computed tomography (CT), and ultrasound can detect only a visual change in the tumor tissues and till their detection, there is a possibility of further progression in the tumor development. Another drawback with these detection techniques is that they cannot differentiate between malignant and benign tumor [82]. Cancer therapies such as phototherapy, radiotherapy, surgery, and chemotherapy, are also considered to treat cancer clinically; however, their nonspecificity to the other healthy tissues can exhibit additional serious problem during cancer treatment [83]. Conclusively, all these traditional approaches are not enough to diagnose and cure cancer at the initial stage due to their low efficiency and working at tissue level. Advancement in nanotechnology by employing NPs for diagnosis purpose in

TABLE 4 Codelivery of therapeutics with different combinations.

S. no.	Nanodrug delivery system	Therapeutic agent 1	Therapeutic agent 2	Reference
Drug-drug combination				
1	Polymeric core shell NP	Doxurubicin	Docetaxel	[64]
2	Polymeric micelle	Doxurubicin	6-Mercaptopurine	[65]
3	Liposome	Paclitaxel	Resveratrol	[66]
4	Liposome	Doxurubicin	Dihydroartemisinin	[67]
5	Liposome	Doxurubicin	Paclitaxel	[68]
6	Dendrimer	Cisplatin	Doxurubicin	[69]
Drug-gene combination				
1	Liposome	Doxurubicin	SATB1shRNA	[70]
2	Dendrimer	Curcumin	Bcl2-siRNA	[71]
3	Dendrimer	Paclitaxel	TR3 siRNA	[72]
4	Micelle	Paclitaxel	shRNA	[73]
5	Micelle	Doxurubicin	P-glycoprotein siRNA	[74]
6	Liposome	Doxurubicin	Vimentin siRNA	[75]
Drug-plasmid combination				
1	Liposome	Doxurubicin	Msurvivin T34A plasmid	[76]
2	Dendrimer	Doxurubicin	pORF-hTRAIL	[77]
3	Dendrimer	Doxurubicin	TRAIL	[78]
4	Micelle	Doxurubicin	IL-36γ	[79]
5	Solid lipid NP	Doxurubicin	DNA	[70]

TABLE 5 FDA-approved therapeutics [80].

S. no.	Therapeutic 1	Therapeutic 2	Cancer	Approval Date
1	Carfilzomib (anticancer)	Daratumumab (anticancer)	Relapsed or refractory multiple myeloma	August 2020
2	Tafasitamab-cxix (antibody)	Lenalidomide (anticancer)	B-cell lymphoma	July 2020
3	Decitabine (nucleic acid synthesis inhibitor)	Cedazuridine (anticancer)	Myelodysplastic syndromes	July 2020
4	Ramucirumab (antibody)	Erlotinib (anticancer)	Metastatic nonsmall cell lung cancer	May 2020
5	Atezolizumab (anticancer)	Bevacizumab (anticancer)	Metastatic hepatocellular carcinoma	May 2020
6	Nivolumab (anticancer)	Ipilimumab (monoclonal antibody)	Metastatic nonsmall cell lung cancer	May 2020
7	Olaparib (anticancer)	Bevacizumab (anticancer)	Advanced epithelial ovarian, fallopian tube, or primary peritoneal cancer	May 2020
8	Daratumumab (anticancer)	Hyaluronidase-fihj (enzyme)	Relapsed/refractory multiple myeloma	April 2020
9	Ibrutinib (anticancer)	Rituximab (monoclonal antibody)	Small lymphocytic lymphoma	April 2020
10	Nivolumab (anticancer)	Ipilimumab (monoclonal antibody)	Hepatocellular carcinoma	March 2020
11	Neratinib (anticancer)	Capecitabine (anticancer)	Metastatic breast cancer	February 2020
12	Pembrolizumab (anticancer)	Lenvatinib (anticancer)	Advanced endometrial carcinoma	September 2019
13	Lenalidomide (anticancer)	Rituximab (monoclonal antibody)	Follicular lymphoma	May 2019
14	Pembrolizumab (anticancer)	Paclitaxel (anticancer)	Metastatic squamous nonsmall cell lung cancer	October 2018
15	Abemaciclib (anticancer)	Fulvestrant (antiestrogen)	Metastatic breast cancer	September 2017
16	Lenvatinib capsule (cytotoxic)	Everolimus (cytotoxic)	Advanced renal cell carcinoma	May 2016
17	Palbociclib (anticancer)	Fulvestrant (antiestrogen)	Metastatic breast cancer	February 2016
18	Obinutuzumab (monoclonal antibody)	Bendamustine (anticancer)	Follicular lymphoma	February 2016
19	Panobinostat (anticancer)	Bortezomib (anticancer)	Multiple myeloma	February 2015

cancer, recognized as nanodiagnostic that allows the use of different nanomaterials for cancer detection at an early stage [84]. The commonly used nanomaterials involved in nanodiagnosis are quantum dots, nanotubes, nanoshells, nanowires, nanodiamonds, nanosponges, metallic NPs, etc.

4.4.1 Quantum dots (QDs)

QDs are semiconductor fluorescent nanocrystals ranging normally within 2–10nm in size. These nanocrystals are composed of approximately 100–100,000 atom similar to a large protein molecule [85]. These semiconducting QDs when absorb photonic energy more than their bandgap causes excitation of electrons from lower level to higher level. When the excited electrons return back to their lower and stable state an additional light energy is emitted equivalent to a specific frequency. However, the emitted light frequency is due to the quantum confined properties [86, 87]. QDs are very useful and important tool in cancer detection.

The quantum confinement effects involving excitation of multiple fluorescence color, photostability, flexibility in size and composition make it a perfect device for in vivo imaging, in vitro diagnosis, and multicolor imaging for detection of multiple targets in cancer [88]. Targeted QDs have great potential in cancer imaging. For example, Li et al. developed DNA aptamer-conjugated dendrimer modified QD to specifically target U251 human glioblastoma cells. The prepared modified QD showed potential results in cancer targeting and imaging [89]. Chen et al. developed Ag_2S-based QDs and conjugated it covalently with a cyclic tri-peptide RGD (cRGD) and the most widely used chemotherapeutic drug doxorubicin (DOX). The formed Ag_2S-DOX-cRGD showed a promising, selective, and favorable tumor inhibition [70].

4.4.2 Gold nanoparticles (AuNPs)

AuNPs have been widely used in a variety of cancer detection. AuNPs are widely used in variety of cancer detection due to their unique optical property and their ability to exhibit size dependent absorption in ultraviolet visible range. Size and thickness of material is crucial for a metal surface which is related to surface plasmon resonance (SPR) [87]. SPR is a phenomenon where the beam of light when incident on the metallic surface such as silver or gold causes oscillation of free electrons followed by fast electronic relaxation resulting into strong absorption of light [90]. Because of their ability to absorb strong light they are well preferred for labeling technique in cancer imaging. For instance, Aydogan et al. utilized 2-deoxy-D-glucose labeled AuNP as a CT contrast agent to observe the in vitro cellular uptake of the agent by human alveolar epithelial cancer cell line A-549 [91]. In labeling method, the labeled AuNPs are targeted and accumulate at their desirable site, and based on their unique optical scattering property they enable the visualization of the cancerous/tumor site by microscopic techniques, for example, photoacoustic imaging, photothermal imaging, phase contrast optical microscopy, or dark field microscopy [92]. Being photostable they can easily be conjugated with antibodies in a form of biosensor and can act as an excellent probe for cancer imaging [93]. For example, Emami et al. prepared doxorubicin and anti-PD-L1 antibody-conjugated targeted gold nanoparticles against overexpressed PD-L1 for colorectal cancer. The study was a combination of chemo and photothermal therapy [94]. AuNPs due to their characteristic SPR effect can absorb efficiently in the near infra-red light, therefore, suitable for cancer photothermal therapy [95].

4.4.3 Carbon nanotubes (CNTs)

CNTs belong to the fullerenes family of carbon allotropes rolled up in the shape of single or multiple layers of graphene sheets into seamless hollow cylindrical structure [96]. CNTs have sp^2 hybridized hexagonal arrangement of carbon atoms. Based on their cylindrical structures they are classified as single-walled CNTs (SWCNTs) made up of single layer of cylindrical graphene sheet and multiwalled CNTs (MWCNTs) made up of multiple concentric layers of graphene sheet. Special and unique properties such as easy surface manipulation, excellent mechanical strength, electrical conductivity, light weight and good thermal conductivity making CNTs an efficient tool for both cancer diagnosis and therapy [97]. In comparison to other nanodevices, CNTs are more versatile with respect to their biological action. For instance, QDs are best used in cancer cell imaging only whereas CNTs are not only used for cancer cell imaging but also for thermal ablation and in drug delivery [98]. Lancu et al. investigated a novel method for the treatment of human hepatocellular carcinoma by functionalizing human serum albumin to MWCNT (HSA-MWCNT) and performed selective laser thermal ablation therapy against human hepatocellular liver carcinoma cell line HepG2. The experiment was based on comparative concentration uptake of HSA between normal hepatocyte and HepG2 cancerous cell line. Finally, through various instrumental studies such as transmission electron, confocal, and phase contrast microscopy it was observed that maximum internalization of HSA by HepG2 cell line was found to be maximum in comparison to normal liver cell. The maximum concentration in HepG2 was due to the overexpression of Gp60 receptor, common in case of liver cancer. The results of the study showed that MWCNT functionalized with HSA were suitable for selective targeting of Gp60 receptors present on HepG2 cancer cell line [99]. Another study by Ghosh et al. also exhibited the selective thermal ablation of cancerous tissue mass in vivo via DNA encased MWNTs [100]. Labeling of CNTs with QDs is also a step toward advancement in cancer screening and diagnosis. For example, detection of prostate-specific antigen (PSA) cancer biomarker by using QDs functionalized to CNTs and this entire assembly was further labeled to an immunosensor that acted as a probe. The primary anti-PSA antibody was also immobilized on CNT. The immunosensor equipped with QD labeled CNT showed potential results in patient serum samples and proved that could be effective for cancer clinical diagnosis [101]. Therefore, it can be said that there are many exciting prospects with CNTs to make it more extra functional in the field of cancer therapy.

4.4.4 Nanofibers (NF)

NFs are unique one-dimensional system where the diameter of fibers measures below 1 μm to a minimum 500 nm. NFs possess very high aspect ratios in an order of magnitude of longer in one direction than their breadth, i.e., fiber could be 1 cm in length; however, may have a diameter of 100 nm. The nano size of these fibers exhibits an excellent increase in the surface area [83]. NFs can be prepared by several techniques, for example, electrospinning (most widely used method), melt blowing, flash spinning, and force spinning. The NFs produced by electrospinning techniques possess characteristics such as large surface area, consistent size with controllable pores and are mostly used in cancer diagnosis. Eradication of tumor cell through surgery and fear of its recurrence is common. Treating the cells by chemotherapy is also a traditional way that destroys healthy cells too along with cancerous cells. Therefore, NFs loaded with anticancer drug can be directly inserted to the targeted area and can show its action in a sustained form thus, risk to the recurrence of tumor after surgery can be reduced NFs are superior to NPs in a sense that NPs encased within polymeric shell can be uptake by spleen or liver while NFs can be directly implanted to the tumor site [102]. Another benefit with NFs that it can track the circulating tumor cells also, for example, Zhao et al. for early cancer diagnosis prepared an electrospun polyvinylalcohol/polyethyleneimine (PVA/PEI) NFs and conjugate it with hyaluronic acid (HA) for capturing and tracking the CD44 overexpressed receptors on cancer cells. The NFs produced had a mean diameter of 459.57 nm and exhibited good capability in capturing the circulating cancer cells [103]. As discussed above polymeric NPs when given alone can undergo cell uptake mechanism, however when impregnated within NFs can deliver therapeutics posing low bioavailability and physical instability easily. Balan et al. developed chitosan NPs embedded in polycaprolactone NFs for successful delivery of resveratrol and ferulic acid against A431 epidermoid human carcinoma cells [104].

4.4.5 Upconversion nanoparticles (UCNPs)

UCNPs are rare earth element nanocrystals that emit a photon when excited by NIR. They are generally doped with elements such as lanthanum (Ln^{3+}), ytterbium (Yb^{3+}), thulium (Tm^{3+}), and erbium (Er^{3+}) in varying compositions [105]. These photoluminescent materials exhibit excellent tissue penetration and high photochemical stability; this property assures it as a good nanodevice in cancer imaging with higher sensitivity [106]. UCNPs can also be conjugated as a contrast agent for imaging purpose. Han et al. reported gadolinium ion-doped UCNPs-based antihuman CD326 grafted micelles with magnetic resource imaging in pancreatic cancer [107]. UCNPs can also be conjugated with photosensitizers so useful in photothermal therapy (PTT) and photodynamic therapy (PDT) [106]. Wang et al. loaded photosensitizer Chlorin e6 (Ce6) on polymer-coated UCNPs. The complex UCNPs-Ce6 injected intratumorally in tumor bearing mice showed excellent NIR-induced PDT [108].

4.4.6 Polymer dots (PDs)

PDs are conductive nanoprobes and are well suited for biological analysis, imaging, and diagnosis purpose. These probes are highly emissive, show potential quantum yields, photostable, and nontoxic [109]. Semiconducting PDs, emerging as fluorescent probe, are also utilized in labeling, tracking, imaging, and in in vivo imaging for cancer cells and pathway [72]. Zhou et al., based on PDT, studied intracellular hypoxia detection by conjugating porphyrin as a photosensitizer to polymer dots [110]. PDs in imaging of cancer cells have also been utilized, for instance, Zhang et al. prepared folic acid and horseradish peroxidase biofunctionalized PDs for targeted PDT and imaging of tumor cells [111].

5 Conclusion

Advancement in nanotechnology in the area of cancer has made remarkable achievements in the past decade. Compared with the present traditional treatment to cancer, the novel and innovative nanodevices have exhibited potential results in cancer imaging and diagnosis at early stage. Definitely these advances and progress will be effective in improving the mortality rate of cancer. The three basic properties of NPs, i.e., high specificity, sensitivity, and selectivity are showing that still there are many opportunities in improving nanotechnology to bring this multidisciplinary field to the clinical platform.

References

[1] Parvanian S, Mostafavi SM, Aghashiri M. Multifunctional nanoparticle developments in cancer diagnosis and treatment. Sens Biosensing Res 2017;13:81–7.
[2] Jain KK. Current status and future prospects of drug delivery systems. Methods Mol Biol 2014;1–56.
[3] N.L.O. Medicine. PubMed; 2019.
[4] Jeevanandam J, Barhoum A, Chan YS, Dufresne A, Danquah MK. Review on nanoparticles and nanostructured materials: history, sources, toxicity and regulations. Beilstein J Nanotechnol 2018;9(1):1050–74.
[5] Hoshyar N, Gray S, Han H, Bao G. The effect of nanoparticle size on in vivo pharmacokinetics and cellular interaction. Nanomedicine 2016;11(6):673–92.
[6] Blagosklonny MV. Analysis of FDA approved anticancer drugs reveals the future of cancer therapy. Cell Cycle 2004;3(8):1033–40.
[7] Christian P, Von der Kammer F, Baalousha M, Hofmann T. Nanoparticles: structure, properties, preparation and behaviour in environmental media. Ecotoxicology 2008;17(5):326–43.
[8] Malik MA, O'Brien P, Revaprasadu N. A simple route to the synthesis of core/shell nanoparticles of chalcogenides. Chem Mater 2002;14(5):2004–10.
[9] Jain RK. Transport of molecules in the tumor interstitium: a review. Cancer Res 1987;47(12):3039–51.
[10] Hobbs SK, Monsky WL, Yuan F, Roberts WG, Griffith L, Torchilin VP, Jain RK. Regulation of transport pathways in tumor vessels: role of tumor type and microenvironment. Proc Natl Acad Sci U S A 1998;95(8):4607–12.
[11] Yuan F, Dellian M, Fukumura D, Leunig M, Berk DA, Torchilin VP, Jain RK. Vascular permeability in a human tumor xenograft: molecular size dependence and cutoff size. Cancer Res 1995;55(17):3752–6.

[12] Nalwa HS, Webster TJ, editors. Cancer nanotechnology: nanomaterials for cancer diagnosis and therapy. American Scientific Publishers; 2007.
[13] Brigger I, Dubernet C, Couvreur P. Nanoparticles in cancer therapy and diagnosis. Adv Drug Deliv Rev 2012;64:24–36.
[14] Matsumura Y, Maeda H. A new concept for macromolecular therapeutics in cancer chemotherapy: mechanism of tumoritropic accumulation of proteins and the antitumor agent smancs. Cancer Res 1986;46(12 Pt 1):6387–92.
[15] Unezaki S, Maruyama K, Hosoda JI, Nagae I, Koyanagi Y, Nakata M, Ishida O, Iwatsuru M, Tsuchiya S. Direct measurement of the extravasation of polyethyleneglycol-coated liposomes into solid tumor tissue by in vivo fluorescence microscopy. Int J Pharm 1996;144(1):11–7.
[16] Torchilin VP. Recent advances with liposomes as pharmaceutical carriers. Nat Rev Drug Discov 2005;4(2):145–60.
[17] Wagner V, Dullaart A, Bock AK, Zweck A. The emerging nanomedicine landscape. Nat Biotechnol 2006;24(10):1211–7.
[18] Goldberg M, Langer R, Jia X. Nanostructured materials for applications in drug delivery and tissue engineering. J Biomater Sci Polym Ed 2007;18(3):241–68.
[19] Rizvi SA, Saleh AM. Applications of nanoparticle systems in drug delivery technology. Saudi Pharm J 2018;26(1):64–70.
[20] Wisse E, Braet F, Luo D, De Zanger R, Jans D, Crabbe E, Vermoesen AN. Structure and function of sinusoidal lining cells in the liver. Toxicol Pathol 1996;24(1):100–11.
[21] Peer D, Karp JM, Hong S, Farokhzad OC, Margalit R, Langer R. Nanocarriers as an emerging platform for cancer therapy. Nat Nanotechnol 2007;2(12):751–60.
[22] Mok H, Park TG. Self-crosslinked and reducible fusogenic peptides for intracellular delivery of siRNA. Biopolymers 2008;89(10):881–8.
[23] Bantz C, Koshkina O, Lang T, Galla HJ, Kirkpatrick CJ, Stauber RH, Maskos M. The surface properties of nanoparticles determine the agglomeration state and the size of the particles under physiological conditions. Beilstein J Nanotechnol 2014;5(1):1774–86.
[24] Yoo J, Park C, Yi G, Lee D, Koo H. Active targeting strategies using biological ligands for nanoparticle drug delivery systems. Cancer 2019;11(5):640.
[25] Ravichandran R. Nanoparticles in drug delivery: potential green nanobiomedicine applications. Int J Green Nanotechnol Biomed 2009;1(2):B108–30.
[26] Ventola CL. The nanomedicine revolution: part 1: emerging concepts. Pharm Ther 2012;37(9):512.
[27] Wang R, Billone PS, Mullett WM. Nanomedicine in action: an overview of cancer nanomedicine on the market and in clinical trials. J Nanomater 2013;1:2013.
[28] Clemons TD, Singh R, Sorolla A, Chaudhari N, Hubbard A, Iyer KS. Distinction between active and passive targeting of nanoparticles dictate their overall therapeutic efficacy. Langmuir 2018;34(50):15343–9.
[29] Sutradhar KB, Amin M. Nanotechnology in cancer drug delivery and selective targeting. Int Sch Res Notices 2014;2014.
[30] Dadwal A, Baldi A, Kumar Narang R. Nanoparticles as carriers for drug delivery in cancer. Artif Cells Nanomed Biotechnol 2018;46(Suppl. 2):295–305.
[31] Attia MF, Anton N, Wallyn J, Omran Z, Vandamme TF. An overview of active and passive targeting strategies to improve the nanocarriers efficiency to tumour sites. J Pharm Pharmacol 2019;71(8):1185–98.
[32] Friedman AD, Claypool SE, Liu R. The smart targeting of nanoparticles. Curr Pharm Des 2013;19(35):6315–29.
[33] Yahyaei B, Pourali P. One step conjugation of some chemotherapeutic drugs to the biologically produced gold nanoparticles and assessment of their anticancer effects. Sci Rep 2019;9(1):1–15.
[34] Yu C, Ding B, Zhang X, Deng X, Deng K, Cheng Z, Xing B, Jin D, Lin J. Targeted iron nanoparticles with platinum-(IV) prodrugs and anti-EZH2 siRNA show great synergy in combating drug resistance in vitro and in vivo. Biomaterials 2018;155:112–23.
[35] Xu S, Cui F, Huang D, Zhang D, Zhu A, Sun X, Cao Y, Ding S, Wang Y, Gao E, Zhang F. PD-L1 monoclonal antibody-conjugated nanoparticles enhance drug delivery level and chemotherapy efficacy in gastric cancer cells. Int J Nanomedicine 2019;14:17.
[36] Cui T, Zhang S, Sun H. Co-delivery of doxorubicin and pH-sensitive curcumin prodrug by transferrin-targeted nanoparticles for breast cancer treatment. Oncol Rep 2017;37(2):1253–60.
[37] Laquintana V, Denora N, Lopalco A, Lopedota A, Cutrignelli A, Lasorsa FM, Agostino G, Franco M. Translocator protein ligand–PLGA conjugated nanoparticles for 5-fluorouracil delivery to glioma cancer cells. Mol Pharm 2014;11(3):859–71.
[38] Yeh CY, Hsiao JK, Wang YP, Lan CH, Wu HC. Peptide-conjugated nanoparticles for targeted imaging and therapy of prostate cancer. Biomaterials 2016;99:1–15.
[39] Kou L, Yao Q, Sivaprakasam S, Luo Q, Sun Y, Fu Q, He Z, Sun J, Ganapathy V. Dual targeting of l-carnitine-conjugated nanoparticles to OCTN2 and $ATB^{0,+}$ to deliver chemotherapeutic agents for colon cancer therapy. Drug Deliv 2017;24(1):1338–49.
[40] Liang HF, Chen CT, Chen SC, Kulkarni AR, Chiu YL, Chen MC, Sung HW. Paclitaxel-loaded poly (γ-glutamic acid)-poly (lactide) nanoparticles as a targeted drug delivery system for the treatment of liver cancer. Biomaterials 2006;27(9):2051–9.
[41] Zhou Z, Badkas A, Stevenson M, Lee JY, Leung YK. Herceptin conjugated PLGA-PHis-PEG pH sensitive nanoparticles for targeted and controlled drug delivery. Int J Pharm 2015;487(1–2):81–90.
[42] Scarberry KE, Dickerson EB, McDonald JF, Zhang ZJ. Magnetic nanoparticle-peptide conjugates for in vitro and in vivo targeting and extraction of cancer cells. J Am Chem Soc 2008;130(31):10258–62.
[43] Dreaden EC, Mwakwari SC, Sodji QH, Oyelere AK, El-Sayed MA. Tamoxifen-poly (ethylene glycol)-thiol gold nanoparticle conjugates: enhanced potency and selective delivery for breast cancer treatment. Bioconjug Chem 2009;20(12):2247–53.
[44] Men W, Zhu P, Dong S, Liu W, Zhou K, Bai Y, Liu X, Gong S, Zhang S. Layer-by-layer pH-sensitive nanoparticles for drug delivery and controlled release with improved therapeutic efficacy in vivo. Drug Deliv 2020;27(1):180–90.
[45] Taleb M, Ding Y, Wang B, Yang N, Han X, Du C, Qi Y, Zhang Y, Sabet ZF, Alanagh HR, Mujeeb A. Dopamine delivery via pH-sensitive nanoparticles for tumor blood vessel normalization and an improved effect of cancer chemotherapeutic drugs. Adv Healthc Mater 2019;8(18):1900283.
[46] Ruan L, Chen W, Wang R, Lu J, Zink JI. Magnetically stimulated drug release using nanoparticles capped by self-assembling peptides. ACS Appl Mater Interfaces 2019;11(47):43835–42.
[47] Nosrati H, Mojtahedi A, Danafar H, Kheiri MH. Enzymatic stimuli-responsive methotrexate-conjugated magnetic nanoparticles for target delivery to breast cancer cells and release study in lysosomal condition. J Biomed Mater Res A 2018;106(6):1646–54.

[48] Shah SA, Khan MA, Arshad M, Awan SU, Hashmi MU, Ahmad N. Doxorubicin-loaded photosensitive magnetic liposomes for multi-modal cancer therapy. Colloids Surf B: Biointerfaces 2016;148:157–64.
[49] Cai X, Luo Y, Zhang W, Du D, Lin Y. pH-sensitive ZnO quantum dots–doxorubicin nanoparticles for lung cancer targeted drug delivery. ACS Appl Mater Interfaces 2016;8(34):22442–50.
[50] Prajakta D, Ratnesh J, Chandan K, Suresh S, Grace S, Meera V, Vandana P. Curcumin loaded pH-sensitive nanoparticles for the treatment of colon cancer. J Biomed Nanotechnol 2009;5(5):445–55.
[51] Zhang Y, Yang C, Wang W, Liu J, Liu Q, Huang F, Chu L, Gao H, Li C, Kong D, Liu Q. Co-delivery of doxorubicin and curcumin by pH-sensitive prodrug nanoparticle for combination therapy of cancer. Sci Rep 2016;6(1):1–12.
[52] Qi A, Deng L, Liu X, Wang S, Zhang X, Wang B, Li L. Gelatin-encapsulated magnetic nanoparticles for pH, redox, and enzyme multiple stimuli-responsive drug delivery and magnetic resonance imaging. J Biomed Nanotechnol 2017;13(11):1386–97.
[53] Hayashi K, Nakamura M, Miki H, Ozaki S, Abe M, Matsumoto T, Sakamoto W, Yogo T, Ishimura K. Magnetically responsive smart nanoparticles for cancer treatment with a combination of magnetic hyperthermia and remote-control drug release. Theranostics 2014;4(8):834.
[54] Soares S, Sousa J, Pais A, Vitorino C. Nanomedicine: principles, properties, and regulatory issues. Front Chem 2018;6:360.
[55] Pillai G. Nanomedicines for cancer therapy: an update of FDA approved and those under various stages of development. SOJ Pharm Pharm Sci 2014;1(2):13.
[56] Salvioni L, Rizzuto MA, Bertolini JA, Pandolfi L, Colombo M, Prosperi D. Thirty years of cancer nanomedicine: success, frustration, and hope. Cancer 2019;11(12):1855.
[57] Hartshorn CM, Bradbury MS, Lanza GM, Nel AE, Rao J, Wang AZ, Wiesner UB, Yang L, Grodzinski P. Nanotechnology strategies to advance outcomes in clinical cancer care. ACS Nano 2018;12(1):24–43.
[58] Woodcock J, Griffin JP, Behrman RE. Development of novel combination therapies. N Engl J Med 2011;364(11):985–7.
[59] Iyer AK, Singh A, Ganta S, Amiji MM. Role of integrated cancer nanomedicine in overcoming drug resistance. Adv Drug Deliv Rev 2013;65(13–14):1784–802.
[60] Xu X, Ho W, Zhang X, Bertrand N, Farokhzad O. Cancer nanomedicine: from targeted delivery to combination therapy. Trends Mol Med 2015;21(4):223–32.
[61] Mokhtari RB, Homayouni TS, Baluch N, Morgatskaya E, Kumar S, Das B, Yeger H. Combination therapy in combating cancer. Oncotarget 2017;8(23):38022.
[62] Tiram G, Scomparin A, Ofek P, Satchi-Fainaro R. Interfering cancer with polymeric siRNA nanocarriers. Nanotechnology 2013;9:1–17.
[63] Eldar-Boock A, Polyak D, Scomparin A, Satchi-Fainaro R. Nano-sized polymers and liposomes designed to deliver combination therapy for cancer. Curr Opin Biotechnol 2013;24(4):682–9.
[64] Shanavas A, Jain NK, Kaur N, Thummuri D, Prasanna M, Prasad R, Naidu VG, Bahadur D, Srivastava R. Polymeric core–shell combinatorial nanomedicine for synergistic anticancer therapy. ACS Omega 2019;4(22):19614–22.
[65] Ghorbani M, Mahmoodzadeh F, Nezhad-Mokhtari P, Hamishehkar H. A novel polymeric micelle-decorated Fe_3O_4/Au core–shell nanoparticle for pH and reduction-responsive intracellular co-delivery of doxorubicin and 6-mercaptopurine. New J Chem 2018;42(22):18038–49.
[66] Meng J, Guo F, Xu H, Liang W, Wang C, Yang XD. Combination therapy using co-encapsulated resveratrol and paclitaxel in liposomes for drug resistance reversal in breast cancer cells in vivo. Sci Rep 2016;6(1):1–11.
[67] Kang XJ, Wang HY, Peng HG, Chen BF, Zhang WY, Wu AH, Xu Q, Huang YZ. Codelivery of dihydroartemisinin and doxorubicin in mannosylated liposomes for drug-resistant colon cancer therapy. Acta Pharmacol Sin 2017;38(6):885–96.
[68] Liu Y, Fang J, Kim YJ, Wong MK, Wang P. Codelivery of doxorubicin and paclitaxel by cross-linked multilamellar liposome enables synergistic antitumor activity. Mol Pharm 2014;11(5):1651–61.
[69] Guo XL, Kang XX, Wang YQ, Zhang XJ, Li CJ, Liu Y, Du LB. Co-delivery of cisplatin and doxorubicin by covalently conjugating with polyamidoamine dendrimer for enhanced synergistic cancer therapy. Acta Biomater 2019;84:367–77.
[70] Chen H, Li B, Zhang M, Sun K, Wang Y, Peng K, Ao M, Guo Y, Gu Y. Characterization of tumor-targeting Ag_2S quantum dots for cancer imaging and therapy in vivo. Nanoscale 2014;6(21):12580–90.
[71] Ghaffari M, Dehghan G, Baradaran B, Zarebkohan A, Mansoori B, Soleymani J, Dolatabadi JE, Hamblin MR. Co-delivery of curcumin and Bcl-2 siRNA by PAMAM dendrimers for enhancement of the therapeutic efficacy in HeLa cancer cells. Colloids Surf B: Biointerfaces 2020;188:110762.
[72] Chen D, Wu IC, Liu Z, Tang Y, Chen H, Yu J, Wu C, Chiu DT. Semiconducting polymer dots with bright narrow-band emission at 800nm for biological applications. Chem Sci 2017;8(5):3390–8.
[73] Hu Q, Li W, Hu X, Hu Q, Shen J, Jin X, Zhou J, Tang G, Chu PK. Synergistic treatment of ovarian cancer by co-delivery of survivin shRNA and paclitaxel via supramolecular micellar assembly. Biomaterials 2012;33(27):6580–91.
[74] Zhang CG, Zhu WJ, Liu Y, Yuan ZQ, Chen WL, Li JZ, Zhou XF, Liu C, Zhang XN. Novel polymer micelle mediated co-delivery of doxorubicin and P-glycoprotein siRNA for reversal of multidrug resistance and synergistic tumor therapy. Sci Rep 2016;6(1):1–12.
[75] Oh HR, Jo HY, Park JS, Kim DE, Cho JY, Kim PH, Kim KS. Galactosylated liposomes for targeted co-delivery of doxorubicin/vimentin siRNA to hepatocellular carcinoma. Nanomaterials 2016;6(8):141.
[76] Xiao W, Chen X, Yang L, Mao Y, Wei Y, Chen L. Co-delivery of doxorubicin and plasmid by a novel FGFR-mediated cationic liposome. Int J Pharm 2010;393(1–2):120–7.
[77] Han L, Huang R, Li J, Liu S, Huang S, Jiang C. Plasmid pORF-hTRAIL and doxorubicin co-delivery targeting to tumor using peptide-conjugated polyamidoamine dendrimer. Biomaterials 2011;32(4):1242–52.
[78] Pishavar E, Ramezani M, Hashemi M. Co-delivery of doxorubicin and TRAIL plasmid by modified PAMAM dendrimer in colon cancer cells, in vitro and in vivo evaluation. Drug Dev Ind Pharm 2019;45(12):1931–9.
[79] Chen Y, Sun J, Huang Y, Liu Y, Liang L, Yang D, Lu B, Li S. Targeted codelivery of doxorubicin and IL-36γ expression plasmid for an optimal chemo-gene combination therapy against cancer lung metastasis. Nanomedicine 2019;15(1):129–41.
[80] Hilal T, Gonzalez-Velez M, Prasad V. Limitations in clinical trials leading to anticancer drug approvals by the US Food and Drug Administration. JAMA Intern Med 2020;180(8):1108–15.
[81] Douillard JY, Rosell R, De Lena M, Carpagnano F, Ramlau R, Gonzáles-Larriba JL, Grodzki T, Pereira JR, Le Groumellec A, Lorusso V, Clary C. Adjuvant vinorelbine plus cisplatin versus observation in patients with completely resected stage IB–IIIA non-small-cell lung cancer (Adjuvant Navelbine International Trialist Association [ANITA]): a randomised controlled trial. Lancet Oncol 2006;7(9):719–27.

[82] Choi YE, Kwak JW, Park JW. Nanotechnology for early cancer detection. Sensors 2010;10(1):428–55.
[83] Horne J, McLoughlin L, Bridgers B, Wujcik EK. Recent developments in nanofiber-based sensors for disease detection, immunosensing, and monitoring. Sens Actuators Rep 2020;21:100005.
[84] Alexis F, Rhee JW, Richie JP, Radovic-Moreno AF, Langer R, Farokhzad OC. New frontiers in nanotechnology for cancer treatment. In: Urologic oncology: seminars and original investigations, vol. 26(1). Elsevier; 2008. p. 74–85.
[85] Zhang H, Yee D, Wang C. Quantum dots for cancer diagnosis and therapy: biological and clinical perspectives. Nanomedicine (Lond) 2008;3(1):83–91.
[86] Jaishree V, Gupta PD. Nanotechnology: a revolution in cancer diagnosis. Indian J Clin Biochem 2012;27(3):214–20.
[87] Chinen AB, Guan CM, Ferrer JR, Barnaby SN, Merkel TJ, Mirkin CA. Nanoparticle probes for the detection of cancer biomarkers, cells, and tissues by fluorescence. Chem Rev 2015;115(19):10530–74.
[88] Peng CW, Li Y. Application of quantum dots-based biotechnology in cancer diagnosis: current status and future perspectives. J Nanomater 2010;1:2010.
[89] Li Z, Huang P, He R, Lin J, Yang S, Zhang X, Ren Q, Cui D. Aptamer-conjugated dendrimer-modified quantum dots for cancer cell targeting and imaging. Mater Lett 2010;64(3):375–8.
[90] Lyberopoulou A, Efstathopoulos EP, Gazouli M. Nanotechnology-based rapid diagnostic tests. In: Proof and concepts in rapid diagnostic tests and technologies, vol. 7. IntechOpen; 2016. p. 89–105.
[91] Aydogan B, Li J, Rajh T, Chaudhary A, Chmura SJ, Pelizzari C, Wietholt C, Kurtoglu M, Redmond P. AuNP-DG: deoxyglucose-labeled gold nanoparticles as X-ray computed tomography contrast agents for cancer imaging. Mol Imaging Biol 2010;12(5):463–7.
[92] Lim ZZ, Li JE, Ng CT, Yung LY, Bay BH. Gold nanoparticles in cancer therapy. Acta Pharmacol Sin 2011;32(8):983–90.
[93] Ahmad T, Sarwar R, Iqbal A, Bashir U, Farooq U, Halim SA, Khan A, Al-Harrasi A. Recent advances in combinatorial cancer therapy via multifunctionalized gold nanoparticles. Nanomedicine 2020;15(12):1221–37.
[94] Emami F, Banstola A, Vatanara A, Lee S, Kim JO, Jeong JH, Yook S. Doxorubicin and anti-PD-L1 antibody conjugated gold nanoparticles for colorectal cancer photochemotherapy. Mol Pharm 2019;16(3):1184–99.
[95] Bucharskaya AB, Maslyakova GN, Terentyuk GS, Navolokin NA, Bashkatov AN, Genina EA, Khlebtsov BN, Khlebtsov NG, Tuchin VV. Gold nanoparticle-based technologies in photothermal/photodynamic treatment: the challenges and prospects. In: Nanotechnology and biosensors. Elsevier; 2018. p. 151–73.
[96] Thakare VS, Das M, Jain AK, Patil S, Jain S. Carbon nanotubes in cancer theragnosis. Nanomedicine 2010;5(8):1277–301.
[97] Ji SR, Liu C, Zhang B, Yang F, Xu J, Long J, Jin C, Fu DL, Ni QX, Yu XJ. Carbon nanotubes in cancer diagnosis and therapy. Biochim Biophys Acta Rev Cancer 2010;1806(1):29–35.
[98] Madani SY, Naderi N, Dissanayake O, Tan A, Seifalian AM. A new era of cancer treatment: carbon nanotubes as drug delivery tools. Int J Nanomedicine 2011;6:2963.
[99] Iancu C, Mocan L, Bele C, Orza AI, Tabaran FA, Catoi C, Stiufiuc R, Stir A, Matea C, Iancu D, Agoston-Coldea L. Enhanced laser thermal ablation for the in vitro treatment of liver cancer by specific delivery of multiwalled carbon nanotubes functionalized with human serum albumin. Int J Nanomedicine 2011;6:129.
[100] Ghosh S, Dutta S, Gomes E, Carroll D, D'Agostino Jr R, Olson J, Guthold M, Gmeiner WH. Increased heating efficiency and selective thermal ablation of malignant tissue with DNA-encased multiwalled carbon nanotubes. ACS Nano 2009;3(9):2667–73.
[101] Yang M, Javadi A, Gong S. Sensitive electrochemical immunosensor for the detection of cancer biomarker using quantum dot functionalized graphene sheets as labels. Sens Actuators B: Chem 2011;155(1):357–60.
[102] Abid S, Hussain T, Raza ZA, Nazir A. Current applications of electrospun polymeric nanofibers in cancer therapy. Mater Sci Eng C 2019;97:966–77.
[103] Zhao Y, Fan Z, Shen M, Shi X. Hyaluronic acid-functionalized electrospun polyvinyl alcohol/polyethyleneimine nanofibers for cancer cell capture applications. Adv Mater Interfaces 2015;2(15):1500256.
[104] Balan P, Indrakumar J, Murali P, Korrapati PS. Bi-faceted delivery of phytochemicals through chitosan nanoparticles impregnated nanofibers for cancer therapeutics. Int J Biol Macromol 2020;142:201–11.
[105] Wang F, Banerjee D, Liu Y, Chen X, Liu X. Upconversion nanoparticles in biological labeling, imaging, and therapy. Analyst 2010;135(8):1839–54.
[106] Cheng L, Wang C, Liu Z. Upconversion nanoparticles and their composite nanostructures for biomedical imaging and cancer therapy. Nanoscale 2013;5:23–37.
[107] Han Y, An Y, Jia G, Wang X, He C, Ding Y, Tang Q. Facile assembly of upconversion nanoparticle-based micelles for active targeted dual-mode imaging in pancreatic cancer. J Nanobiotechnol 2018;16(1):1–13.
[108] Wang C, Tao H, Cheng L, Liu Z. Near-infrared light induced in vivo photodynamic therapy of cancer based on upconversion nanoparticles. Biomaterials 2011;32(26):6145–54.
[109] Yu J, Rong Y, Kuo CT, Zhou XH, Chiu DT. Recent advances in the development of highly luminescent semiconducting polymer dots and nanoparticles for biological imaging and medicine. Anal Chem 2017;89(1):42–56.
[110] Zhou X, Liang H, Jiang P, Zhang KY, Liu S, Yang T, Zhao Q, Yang L, Lv W, Yu Q, Huang W. Multifunctional phosphorescent conjugated polymer dots for hypoxia imaging and photodynamic therapy of cancer cells. Adv Sci 2016;3(2):1500155.
[111] Zhang Y, Pang L, Ma C, Tu Q, Zhang R, Saeed E, Mahmoud AE, Wang J. Small molecule-initiated light-activated semiconducting polymer dots: an integrated nanoplatform for targeted photodynamic therapy and imaging of cancer cells. Anal Chem 2014;86(6):3092–9.

Chapter 26

Advances in polymeric nanoparticles for drug delivery systems in cancer: Production and characterization

Valker Araujo Feitosa[a,b], Terezinha de Jesus Andreoli Pinto[b], Kamal Dua[c], and Natalia Neto Pereira Cerize[a]

[a]*Bionanomanufacturing Center, Institute for Technological Research, São Paulo, Brazil,* [b]*School of Pharmaceutical Sciences, University of São Paulo, São Paulo, Brazil,* [c]*Discipline of Pharmacy, Graduate School of Health, University of Technology Sydney, Ultimo, NSW, Australia*

1 Introduction

Cancer is, by far, one of the main challenging chronic diseases and a public healthy challenge. Although chemotherapy-loaded nanoparticles (NPs) have been improving the preclinical and clinical results of cancer therapies, however, therapeutic efficacy still remains limited to the tumors heterogeneity and complexity. By modifying the chemical and/or physical properties of polymeric NPs, it is possible to overcome the challenges of drugs related to solubility, chemical degradation, and incompatibility [1].

Efforts have been made to minimize the drug toxicity during the cancer therapy, among them: (i) prevent damage on nearby healthy cells, tissues, and organs; (ii) increase the accumulation and effectiveness of the drug in the local organ or tissue; (iii) develop new pharmacokinetic profiles for drug regarding absorption, distribution, metabolism, and clearance [1].

Several biological applications have been reported for nanoscale particles, for examples, targeted and controlled release, as well as enhanced bioavailability of drugs [2]. In this way, use of polymeric NPs has increased drug safety, providing a higher bioavailability exploring the controlled and targeted release [1].

NPs are now present in commercial products, but manufacturing challenges still limit their production. These challenges have become more urgent, as initial efforts have led to the second era of nanotechnology, in which the initiative is shifting from research to commerce [3].

In this chapter, we will brief introduction to polymeric NPs, the incorporation of drugs in these NPs, and their application in the treatment of cancer. Besides, we will show NPs production methods and characterization techniques. This chapter updates the work related to production and characterization of NPs and points out the promising advances in this field.

2 Polymeric NPs for drug delivery systems

At the moment, there is a growing apprehension of the pharmaceutical industry regarding the improvement of drug nanoformulations designed in a size and structure suitable for pharmaceutical applications, including drug delivery systems (DDSs) to address improved bioavailability and therapeutic effects with a decrease in the side effects of active (bio)molecules [4].

Generally, traditional chemotherapy drugs have poor biodistribution, and due to their low molecular weight, they have a short bloodstream time and low accumulation in the affected cancer tissues. To improve these obstacles, nano/microscale DDSs have emerged for modern anticancer therapy [5].

Thus, through the DDSs, the anticancer drugs should be preferentially released in the tumor in a controllable and sustainable way. Such progress in the development of drug formulations, personalized medicine strategies, as well as gene- and cell-based therapies, can be understood by nanobiotechnology tools and techniques [4].

Although several new nanostructures are being presented, polymeric NPs remain a good choice for drug encapsulation. NPs are also extensively explored in pharmaceutical research because of their favorable physical and chemical properties, such as reduced particle size and surface charge flexibility, as well as mechanical rigidity, biodegradability, and biocompatibility of their polymers [1].

In last decades, polymers have been extensively explored as raw materials because their excellent properties, regarding to biocompatibility, friendly design and synthesis, diversity of morphologies, and biomimetic properties. Additionally, related to smart DDS development, polymers have been an important material allowing to modulate, orientate, and deliver therapeutic actives directly to the target

Advanced Drug Delivery Systems in the Management of Cancer. https://doi.org/10.1016/B978-0-323-85503-7.00029-8

organ or tissue, compared to free drugs [2]. Thus, the research interest is concentrated mainly in the design of new therapies based on polymeric NPs for cancer treatment [1].

Indeed, polymeric NPs offer a promising strategy by encapsulating chemotherapeutic drugs and have been shown to reduce side effects by providing a protection compartment for the drug limiting its interaction with healthy tissues [6].

Efforts have been made to develop nano-drugs; consequently, the nanoformulations have developed significantly. There are several DDSs, including those with the drug nanoencapsulated in a carrier, for example, dendrimers, liposomes, micelles, and NPs [4]. In this context, polymeric NPs have been granted as potential and interesting smart DDSs due to singular biological, physical, and chemical characteristics in the nano/microsize.

The surface modification of polymeric NPs has also proved useful as multifunctional tools. It can optimize the modified NPs distribution and accumulation of tumor tissues and promote cell absorption or reduce interactions between loaded-drug and healthy cell [1]. Several studies have also shown that once in contact with biological fluids, biomolecules (e.g., plasma proteins) are adsorbed onto the surface of NPs, forming a surrounding corona. The biomolecular corona formation interfere with cell interactions and internalization mechanisms, thereby controlling their fate in vivo, as well as modulating the distribution, accumulation, and elimination of NPs from organisms [7].

Thus, NPs can enter cells by endocytosis, phagocytosis, and pinocytosis. In contrast, bulky/agglomerated NPs are quickly opsonized/eliminated from the circulation by macrophages and reticuloendothelial system. In this way, it is essential to reduce opsonization and increase NPs' circulation for clinical application by modifying the NPs surface through different methods, including polymers with hydrophilic properties, surfactants, or copolymers with hydrophilic/hydrophobic chains (e.g., poloxamers/Pluronic and poloxamines/Tetronics) [4].

Also, polymeric NPs can be conveniently produced with cell-recognizable portions attached to the particle surface, for example, by using transferrin, growth factors, and hormones (e.g., EGF, VEGF), RGD peptide (i.e., arginine, glycine, aspartic acid), and antibodies (e.g., anti-EGF, anti-EGFR, anti-HER2, anti-CD20, anti-CD30, anti-CD52). Besides, folate, hyaluronic acid, and some specific carbohydrates have also been used as functionalizing agents [1].

3 Preparation of polymeric NPs

This part of the chapter summarizes different preparation methods that have been proposed to produce polymer NPs. The exploitation of NPs for DDSs has intensified in recent years. The main advantages of NPs that can be highlighted are regarding to chemical and physical stability, target specificity, capacity of sustain, retard or prolong release, easiness to be incorporated in different formulations, possibility to address distinct routes of administration, and increased capacity to transport hydrophilic and hydrophobic drugs with the ability to transport substances compared to conventional pharmaceutical formulations [4]. Quality control is a critical issue in the manufacture of NPs; few studies have systematically investigated the spread of product heterogeneity through processes, beginning from synthesis, stabilization, functionalization, purification, and characterization to the integration [3].

The different methods used for NP synthesis are an important part in this field because they allow the construction of different structures with adequate properties to guarantee the adequate drug delivery and targeting. Through progress in the polymer synthesis, it is possible to obtain polymeric materials with a large range of compositions, structures, and physicochemical properties under well-controlled conditions and with a low degree of variability and thus can prepare NPs with the desired properties well adjusted [8].

NPs made of biodegradable polymers still seen as first option over nonbiodegradable ones because they degrade, gradually release drugs by diffusion, swelling, erosion, and degradation and are eliminated from the body [6]. In this sense, polymeric NPs are produced from a natural, semisynthetic, and synthetic polymers in scale from the nanosize to microsize [2].

Polymeric nanocarriers include micelles, colloids, spheres, capsules, dendrimers, nanogels, among others. Generally, polymeric NPs are collectively referred to as spheres and capsules [2]. The particle synthesis process can be designed to encapsulate the drugs in internal compartment (into the sphere and capsules) onto the surface NP or in both compartments [4].

Nanocapsules have a core-shell structure. Usually, the drugs are dissolved or precipitated internally in the nucleus (i.e., core); this inner region is covered by a polymeric coating, forming the outer layer (i.e., shell). In nanocapsules, the nucleus can be lipidic, allowing the encapsulation of fat-soluble drugs. In contrast, nanocapsules with an aqueous core are capable of encapsulating hydrophilic drugs. Various techniques were developed for the production of NPs using emulsification process to obtain mixture of different phases (internal and external), using different surfactants to stabilize the systems. In general, the polymers are precipitated on droplets surface by polymerization reaction from monomers [8]. In some cases, the formation of shell is due to polymer precipitation in a solvent evaporation or diffusion process. These methods are known as synthesis by preformed polymers.

In contrast, nanospheres are matrix particles, that is, they are integrally in the solid state. These NPs are generally spherical, but they can also assume other geometric shapes [8].

The processes used for the preparation of polymeric NPs are mainly based on techniques known as "top-down" and "bottom-up." The methods can be chosen according to the physical and chemical properties of the polymer and the drugs. It should be noted that most of the formulation processes involve some aggressive mechanisms, using, for example, organic solvents, higher temperatures, application of mechanical agitation and ultrasound, which can degrade the most sensitive drugs. Thus, the nanoparticulate system must be developed considering a methodology that does not affect the stability and activity of drugs [2].

As already mentioned, the active ingredient can stay trapped in internal compartment of NPs, and in other cases adsorbed on particle surface. Commonly, encapsulating instable drugs into nanocarrier is a promising strategy to prevent chemical degradation, by embodying the drug in phases of NP synthesis. However, the adsorption onto particle surface is an alternative for solvent and thermal sensitive drugs [8].

Generally, methods for preparing NPs start with an emulsion and finalize with proper formation of NPs. The emulsified systems formed can be conventional emulsions, nanoemulsions, and microemulsions [8].

NPs, similar to all colloidal systems, to remain well dispersed in liquids, need to be stabilized using surface-active agents (i.e., surfactants) to ensure the reduced tension interfaces, droplets stability, and consequently maintenance of NPs suspension. These surfactants are small amphiphilic molecules or (co)polymers that stabilize NPs preventing its agglomeration and precipitation [2]. Emulsions are related as first-step process in lipid nanocarriers, such as solid-lipid nanoparticles (SLN) and lipid nanocapsules (LNC) [8]. Some stabilizers most used in colloidal systems are listed in Table 1.

In fact, the properties of the surfactant strongly influence the formation of emulsion. The very common Tweens or Spans are now replaced by polymeric surfactants, or even several polysaccharides. Surfactant copolymers (e.g., Pluronics) have been explored as stabilizing agents for emulsions. These molecules anchor onto surface of droplets dissolving partially on hydrophilic/hydrophobic phase, accordingly the solvent system and emulsion properties. It is still possible to carry out the conjugation of the hydrophobic polymer used to form the NPs, e.g., PLGA, PCL, PLA, with hydrophilic polymers, especially poly(ethylene glycol), PEG. An advantage of synthesizing such copolymers is that their hydrophobic portion is the same used to make the nucleus of the NPs [8].

The methods of preparing emulsified systems have evolved significantly, in general, requiring two immiscible phases and a stabilizer. The physical and chemical methods used to obtain the emulsification between immiscible phases are diverse [8]. Generally, polymeric NPs are prepared initially by dissolving both (the desired drug and polymer) in solvents that have a minimum water miscibility (e.g., chloroform, methanol, or ethyl acetate) and then by emulsion through (i) sonication probe method, (ii) high-pressure homogenization, (iii) high-speed mixer, and (iv) spontaneous emulsification, as shown in Table 2. Then, the organic solvent needs to be removed using reduced pressure and increased temperature, as an example in rotary evaporation. Finally, the NPs are washed and can be collected by centrifugation [2].

Other different emulsion processes employed in formulation of NPs for sensitive drugs is spontaneous emulsification, since to obtain the emulsion, procedures with low energy dissipation are applied. Polymeric nanocapsules are also produced by spontaneous emulsification using a low-energy dissipation process based on the displacement of water-soluble solvents (e.g., ethanol and acetone). This phenomenon is also called "ouzo effect" or "pastis effect" [8].

Generally, the stage of NP formation gives the method its name, as shown in Table 3.

TABLE 1 Common colloid stabilizers applied to produce nanoparticles.

Stabilizer	Abbreviation or examples
Dextran	
Poly(vinyl alcohol)	PVA
Polysorbate	Tween 20 and Tween 80
Poloxamer	Pluronics, F68, F127
Sodium dodecyl sulphate	SDS
Poly(ethyleneoxide) lauryl ester	Brij 35

TABLE 2 Preparation of the emulsified systems. According to Vauthier and Bouchemal [8] and Bennet and Kim [2].

Process	Method
Mechanical	Sonication probe method
	High-pressure homogenization
	High-speed mixer (colloidal mill with a rotor stator device)
	Extrusion (microfiltration device)
Spontaneous emulsification	"ouzo or pastis effect"

TABLE 3 Two-steps-based bottom-up processes applied to NPs production from an emulsion system. According to Vauthier and Bouchemal [8] and Bennet and Kim [2].

Procedure	Method	Comments
Solvent extraction (inducing polymer precipitation)	Emulsification-solvent evaporation	Emulsions are formulated with polymer solutions prepared in volatile solvents (e.g., dichloromethane, chloroform, ethyl acetate). Conversion of the emulsion into a suspension of NPs occurs by the evaporation of polymer-solvent that is allowed to diffuse through the emulsion's continuous phase. This is a slow vacuum process
	Emulsification-solvent diffusion (or displacement)	The polymer-solvent used to prepare the emulsion must be partially soluble in water. The emulsion is then prepared with water saturated with the polymer-solvent that makes up the oil phase and with an oil phase saturated with water as a continuous phase. Since the oil-in-water emulsion is obtained using the previously saturated solvents, it is diluted with a large amount of pure water. As a result of this dilution, the additional organic solvent from the organic phase contained in the dispersed droplets can diffuse out of the droplets leading to the precipitation of the polymer. The diameter of the NPs produced is around 150 nm
	Emulsification-reverse salting-out	The polymer-solvent used to prepare the emulsion is completely miscible in water (e.g., acetone). The artefact used to emulsify the polymer solution in the aqueous phase consists of dissolving a high concentration of salt or sucrose chosen for its strong salting-out effect in the aqueous phase. These components retain water molecules for their solubilization; therefore, modify the miscibility properties of water with other solvents. Precipitation of the polymer dissolved in the emulsion droplets is induced through a reverse salting-out effect that is obtained by diluting the emulsion with a large excess of water
Gelation		This method involves gelling the polymer dissolved in the droplets of the emulsion. It can be applied with polymers that exhibit gelling properties (e.g., agarose, alginate, pectin)
"In situ" polymerization	Polymerization of alkylcyanoacrylates	A monomer is added to the emulsion instead of a polymer solution, and the polymer is formed by polymerization to give rise to the nanoparticles. Most poly(alkylcyanoacrylate) NPs are obtained by anionic polymerization of the corresponding monomer that is spontaneously initiated by hydroxyl groups of water or any type of nucleophilic groups found in molecules dissolved in the polymerization system. In general, emulsion polymerizations used to produce NPs is performed in acidic conditions to slow down the rate of the anionic polymerization and allow nanospheres to form instead of polymer aggregates. Contrarily, the capacity of monomers to develop extremely rapid polymerizations is advantageous to achieve successful encapsulation of drugs within nanocapsules
	Interfacial polycondensation	This method leads to forming a polymer film at the interface between two nonmiscible phases from two monomers' reaction, each dissolved in one phase. The reaction is carried out at the interface of oil droplets dispersed in the aqueous phase of an emulsion that is obtained by a spontaneous emulsification technique. Usually, the organic phase comprises a water-miscible solvent, the lipophilic monomer, oil, and a lipophilic surfactant, while the aqueous phase contains water, the hydrophilic monomer, and a hydrophilic surfactant. As monomers are included in the composition of both phases of the emulsion, polycondensation can occur readily at the interface of the oil droplets that form during the mixing of the two phases leading to the spontaneous formation of the oil-in-water emulsion. Examples of monomers: polyamides, polyurea, polyurethanes, and poly(urethane ether)

3.1 Bottom-up processes

Synthesis techniques include salting-out, solvent displacement, phase separation, evaporation, (antisolvent) precipitation, and spraying methods [2]. Thus, polymeric NPs are commonly prepared by following methods: (i) solvent emulsification-evaporation, (ii) solvent displacement, (iii) salting-out method, (iv) emulsification-diffusion of solvent, and (v) evaporation of double solvent emulsion [2].

In fact, the first technique described to prepare NPs from preformed polymers, dates back 1979, is emulsification-solvent evaporation (US Patent 4,177,177—Polymer emulsification process). Even though polymer chemists originally proposed this method, pharmaceutical technology made it possible by the pioneering work of Gurny et al. [9] who had proposed the use of biodegradable polymer to prepare controlled release system from NPs. In this purpose, polymers are dissolved in volatile solvent, this phase is used to prepare the emulsion and the emulsion is transformed into a suspension of NPs after solvent removal by diffusion and evaporation. For this purpose, the appropriated polymer is dissolved in a volatile solvent (e.g., chloroform and ethyl acetate); this organic phase is used to prepare an oil-in-water (O/W) emulsion, where, after diffusion and solvent extraction, each drop becomes NPs that remain in aqueous solution as a dispersion. The drop size reduction is accomplished by homogenization with high speed and high shear mixing. Then, the washing step is applied to remove excess surfactants, or other additives used [5].

Additionally, some one-step-based bottom-up processes applied to NPs production, such as self-assembling polymers and ionic gelation, are presented in Table 4. Then, spraying methods are presented below.

3.2 Spray methods

The researchers used various drying techniques to obtain formulations of dry polymeric NPs, such as spray-drying and freeze-drying. However, these NPs quickly tend to aggregate. Usually, polydispersity of the NPs obtained by spray-dryer is lower than those obtained by freeze-dryer. One advantage of spray-drying technique to NPs formation is facility of transformation of the polymeric solution into dry NPs. In these techniques, NPs could be obtained continuously, and in general they result in amorphous structures and the content of the drug present in the NP matrix is high [2].

3.3 Top-down processes

In addition, many top-down approaches have been explored to particle size reduction to the nanometric size range with a significant increase in surface area. Equipment such as high-pressure homogenizers, high shear mixers, and wet grinders are often employed to produce small-sized NPs. High-pressure homogenization is an effective method to reduce particle size in emulsions, suspensions, and complex fluids. Also, sophisticated equipment must withstand the increase in pressure and temperature [2].

TABLE 4 One-step-based bottom-up processes applied to NPs production. According to Vauthier and Bouchemal [8] and Bennet and Kim [2].

Procedure	Method	Comments
Self-assembling polymers	Polyelectrolyte complexes	Typical examples of NPs obtained from polyelectrolyte complexes are formed between polycations and polyanios (e.g., nucleic acids) thanks to complementary charge annealing. According to the nature of the polyelectrolyte used, either positively or negatively charged nanoparticles can be synthesized. Polycations such as polyamines, poly(ethylenimine) (PEI), poly(lysine) (PLL), poly(ornithine) or chitosan usually are used to prepare polyplexes. The resulting nanospheres were also named nanoplexes. These NPs are widely developed to delivery of nucleic acids including plasmid genes, siRNA and antisense oligodeoxynucleotides
	Neutral nanogels	This method uses polymers that can associate together to form supramolecular nanoassemblies of spherical shape. The NPs formed are composed of a hydrogel in which a large amount of water is entrapped in the polymer network formed by the assemblies of the polymers such as modified dextran and poly(beta-cyclodextrin). The resulting nanospheres were also named nanogels
	Thin-film hydration	According to amphiphilic employed this method can leads to forming polymeric micelles or polymeric vesicles (i.e., polymersomes)
Ionic gelation		Ionic nanogels can be obtained from aqueous solutions of charged polysaccharides (e.g., alginate) which gel in the presence of small ions of opposite charges (e.g., calcium). The gelation of the polysaccharide should be performed in very dilute solution using concentrations of the gelling agent below the gel point. Chitosan is the second gelling polysaccharide which was used to produce nanoparticles through a gelling process
Spray	Spray-dried	These techniques are employed to obtain dry polymeric NPs
	Nano-spray-dried	
	Freeze-drying	
Supercritical fluid		This method can obtain a dry product without any solution, even without additional drying steps. Still, the supercritical fluid can swell some of the polymers and act as a softener, extender, and lubricant, which lead to aggregation. Furthermore, this method cannot obtain monodisperse particles due to the different kinetics

3.4 Next-generation production systems

Microelectromechanical and nanoelectromechanical systems offer an innovative possibility to produce micromanufactured devices, including wearables, implantable systems, lab-on-a-chip. Microfabricated systems for controlled drug delivery have several benefits. Drugs can be stored in NPs inside microdevices, from an internal or external signal, and through an electronic interface; they promote the continuous and prolonged release of the loaded drugs. Thus, the start of drug release can be triggered by the microfabricated reservoir device through chemical or biochemical, physical, mechanical, or electrical stimuli [2].

4 Polymers used in the NPs process

Polymeric NPs have been considered efficient carriers for drug delivery because their excellent pharmacokinetics, regarding drug loading, drug release, structure stability, and degradability [1]. Generally, drug-loaded NPs consist of the drug, and nontoxic, biocompatible polymer (biodegradable or not) with stabilizing agents [2].

Several biodegradable and biocompatible polymers, including natural or semisynthetic or synthetic, with different physical-chemical characters are offered to prepare NPs [2]. Of these polymers, polyethylene glycol (PEG), poly-ε-caprolactone (PCL), polylactic acid (PLA), poly(D,L-lactic-co-glycolic acid) (PLGA), and polyglycolic acid (PGA) are mainly studied because they are biocompatible and biodegradable [6]. Besides, the combination of polymers and their copolymers (e.g., di-block, tri-block, multiblock, and radial block copolymer) has generally been used in the composition of NPs [2].

In fact, even if the diversity of polymers offered by chemical industry nowadays, merely a few polymers are used as NP constituents for in vivo application. In general, a suitable polymer must meet several requirements to be used clinically, as follows:

- **(i)** The polymer must have adequate physicochemical properties in relation to the intended drug delivery.
- **(ii)** Preferentially property of biodegradability or quickly body clearance, allowing repetitive administration reducing risk of accumulates in organism.
- **(iii)** Does not produce toxic degradation products or derivatives that trigger immunogenic reactions.

In this sense, the regulatory approval is also an important aspect, several polymers are approved by European Medicines Agency (EMEA), National Agency of Sanitaria Vigilance (ANVISA) and Food and Drug Administration (FDA), and other regulatory agencies for human use [2].

Table 5 provides a list of polymers mainly explored in nanocarrier preparation. Currently, a few polymers are approved by regulatory agencies for injectable administration, and some can be used for oral or topical use. In the past decade, many copolymers, including PEG or polysaccharides, have been studied. The necessity of NPs with flexible surface properties to modulate their interactions with biological macromolecules present mainly in the mucosa and in the bloodstream required for the development of new copolymers. In addition, these molecules showed interesting stabilizing properties by replacing surfactants with toxic characteristics.

5 API loading and release

A significant obstacle to traditional chemotherapy is the difficulty of administering adequate doses of medication in the areas affected by the tumor. The rapid elimination of blood circulation requires large amounts to be effective; however, the systemic toxicity of these drugs also limits the dose increase [6].

In addition, polymeric particles have proven their effectiveness in protecting and stabilizing peptides, proteins, and gene therapies. Therefore, these polymers provide the potential for biopharmaceutical production and gene delivery [2]. NPs are also promising for targeting molecules such as oligonucleotides and interfering RNA [8].

Generally, DDSs that explore nanometric scale and nanostructures are formulated to be administered by injection, transdermally or orally. Accordingly, the smart nanostructured DDSs must accelerate pharmaceutical development and improve the therapeutic potential of encapsulated drugs [4]. As a result of the drug's nanoencapsulation, the drug's pharmacokinetic properties are now modulated by the NP's pharmacokinetic properties, since the drug remains in the carrier until the effective release [6].

The potential benefits of NPs in pharmacokinetics also include increased drug absorption, long-term controlled release rates, and prolonged half-life. From the therapeutic point of view, there is a reduction in side effects, greater patient compliance due to the possibility of reducing doses and frequency, in addition to the ability to associate more than one drug simultaneously [6]. Thus, the delivery of multiple drugs together would improve the action on different signaling pathways related to tumor growth, enabling a synergistic therapy effect at the same location. However, these approaches are still in the initial stages due to technical limitations, difficulties in determining the ideal dose of each drug and also the possible toxic effects of the associations [1].

TABLE 5 Common synthetic and natural (co)polymers applied to produce nanoparticles.

Material	Full name	Abbreviation or examples	Comments
Synthetic homopolymers	Poly(ethylene glycol)	PEG	It is a hydrophilic polymer widely used in biomedicine due to its excellent water solubility, chain mobility, nontoxicity, and nonimmunogenicity. Enable to be furtive and blind to the immune systems being used to vehicle different chemical structures
	Poly(lactide)	PLA	It is an FDA-approved biodegradable polymer, and fumaric acid existed in the Kreb's cycle. It can be entirely excreted by metabolism after degradation
	Poly(lactide-co-glycolide)	PLGA	It is one of the few polymers approved by the FDA for clinical applications due to its excellent properties, such as biodegradability and biocompatibility. So far, it is the most widely used synthetic polymer in developing drug-loaded NPs for cancer therapy
	Poly(epsilon-caprolactone)	PCL	It is biodegradable polyester used as a matrix to carrier mainly hydrophobic drugs. PCL can be mixed with starch to lower its cost and increase biodegradability. Another application that can be mentioned and can be added as a polymeric plasticizer to polyvinyl chloride
	Poly(isobutylcyanoacrylate)	PICBA	
	Poly(isohexylcyanoacrylate)	PIHCA	
	Poly(*n*-butylcyanoacrylate)	PBCA	
	Poly(acrylate) and poly(methacrylate)	Eudragits	
Synthetic copolymers	Poly(lactide)-poly(ethylene glycol)	PLA-PEG	
	Poly(lactide-co-glycolide)-poly(ethylene glycol)	PLGA-PEG	
	Poly(epsilon-caprolactone)-poly(ethylene glycol)	PCL-PEG	
	Poly(hexadecylcyanoacrylate-co-poly(ethylene glycol) cyanoacrylate	Poly(HDCA-PEGCA)	
	Poly(ethyleneglycol)-poly(propyleneglycol)-poly(ethyleneglycol)	Pluronics	It is a block copolymer biodegradable material approved for pharmaceutical use. The chemical structure provides surfactant properties allowing micellar self-assembled nanostructures besides be used as matrix polymers or blended with other polymers to confer improving dissolution characteristics
Natural polymers	Chitosan		
	Alginate		
	Gelatin		
	Albumin		
	Starch		
	Maltodextrin		

6 Characterization of polymeric NPs

The ideal requirements that must be met during the production of NPs are (i) particle size and surface characteristics properly controlled; (ii) increased solubility, absorption, and distribution of drugs in the intended target; and (iii) therapeutic activity [2]. Table 6 provides some general techniques to characterize NPs.

TABLE 6 General techniques applied to characterize nanoparticles.

Technique	Comments
Dynamic light scattering	Dynamic light scattering (DLS) is a method conventionally used as size analysis of NPs. The fine particles are 100–2500nm, and the ultrafine particles are 1–100nm in size
Microscopy	The microscopies provide information about the real size and shape of particles as drugs and polymeric particles. The scanning electronic microscopy (SEM) allows visualizing particles' shape, external and surface appearance, while transition electronic microscopy (TEM) shows the internal structure and morphology
Spectroscopy	Fourier-transform infrared spectroscopy (FTIR) provides information about chemical structure and main functional groups that characterize drugs and polymers used in drug delivery systems
Diffraction	X-ray diffraction (XRD) technique allows the evaluation of the drug crystallography profile. This characterization share information about physical structures from drugs, and crystals could be related to polymorphism changes during the manufacturing process. These properties are relevant to pharmaceutical development, implying in changed therapeutic effects in some cases
Thermal analysis	Differential scanning calorimetry (DSC) and thermogravimetric analysis (TGA) are both techniques that provide information about degradation profile in controlled programming of temperature. It is possible to infer transition phases, for example, melting point from a drug and compatibility with physical mixture polymers. We can measure the loss mass with the increase of temperature and correlate with drug or complex formulations' chemical stability

7 Applications of polymer NPs

For an anticancer drug carrier to be effective, prolonged blood circulation and controlled release at the tumor site are necessary [6]. In fact, polymeric NPs can prevent drug degradation, increase systemic circulation half-life, minimize contact with healthy tissues. Controlled DDSs can be categorized into different profiles of active ingredient release, as described (i) controlled release, by drug diffusion through NPs, (ii) release induced by different triggers as chemical or biochemical (i.e., enzymatically), physical or electrical factors, (iii) release activated by feedback, and (iv) localized release [2].

Due to the small size of NPs, targeting several cancers has shown promise [2]. DDSs that exploit polymeric NPs generally comprise approaches related to target in passive, active, or responsive to stimuli [4]. In general, the most widely used mechanisms of DDSs for targeting are based on passive targeting. Passive targeting is achieved by reduced size of NPs, increasing the residence time through the permeability and enhanced retention (RPE) effect on tumors [6].

The EPR effect is characterized by tumor vessels hyperpermeable to circulating macromolecules, such as NPs, which, together with the lack of a lymphatic drainage system, result in their localized accumulation [6]. The improved EPR effect allows NPs of less than 500nm to better penetrate tumor tissue. Larger NPs would be prevented from getting there because they are easily removed from the organs by the reticuloendothelial system. NPs are generally designed to achieve a neutral behavior (regarding the charge of surface), for example, by coating with PEG to prevent aggregation of NPs and its removal via RES [1]. Thus, PEG carriage on NPs surface reduces the binding of plasma proteins and recognition by RES. This gives NPs "stealth" properties, significantly increasing their blood circulation time [6].

8 Patent research: A briefly overview

A patent research was carried out using Questel/Orbit patent database [10], a business intelligence tool that allows a powerful patent searching and analysis about the main players and their portfolios. The research strategy has comprised the key words: NPs or nanostructures and cancer or neoplasia or oncology (and other key elements to support and complete the investigation).

Fig. 1 shows the size of the applicants' portfolios in the patent pool analyzed, involving NPs and cancer. This data is a good indicator of the level of inventiveness of the active players. Clearly, the University of California, the Abraxis Bioscience, the Korea Institute of Science & Technology, and the Massachusetts Institute of Technology—MIT are the top-5 principal applicants. The applicants' portfolios

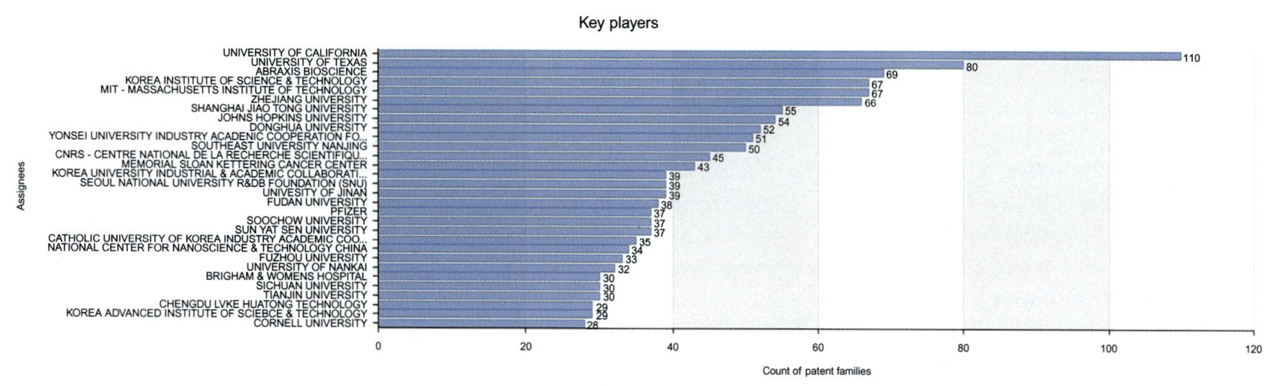

FIG. 1 Major patent assignees (key players) on the field of nanoparticles and cancer, according to Questel/Orbit patent database. *(This figure was created on the Questel's IP Business Intelligence application Orbit Intelligence. Credits to Mr. Yuri Tukoff regarding the patent research in September, 2020.)*

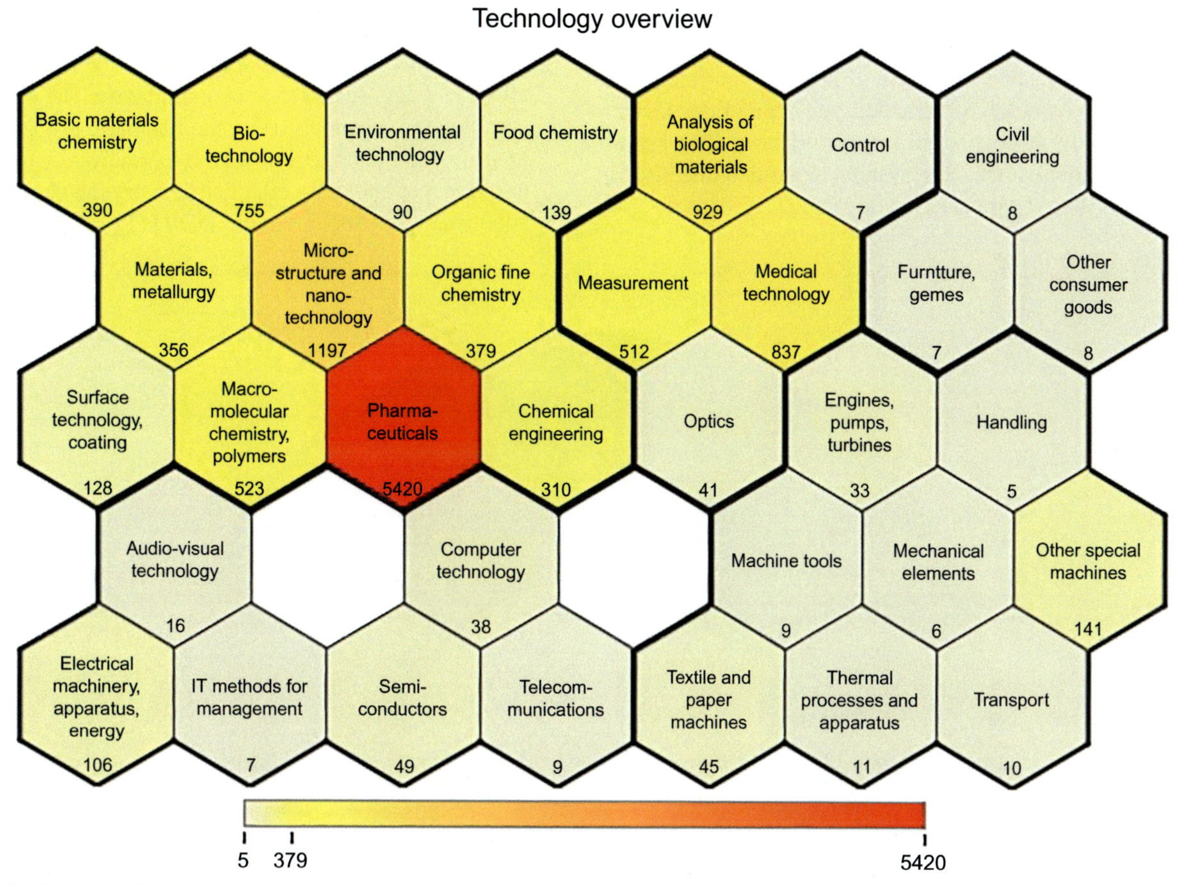

FIG. 2 Overview of technologies according to International Patent Classification (IPC) codes, related to the field of nanoparticles and cancer, according to the Questel/Orbit patent database. *(This figure was created on the Questel's IP Business Intelligence application Orbit Intelligence. Credits to Mr. Yuri Tukoff regarding the patent research in September, 2020.)*

represent a good indicator of the candidate's propensity to collaborate and identify their preferred partners. This graph presents the top applicants by volume of the topic studied. This represents the applicants who have the largest number of patents in their portfolios in the subject area analyzed.

Fig. 2 is based on the International Patent Classification (IPC) codes contained in a patent set being analyzed. The IPC codes have been grouped in 35 technology fields, which are represented here. This illustration enables users to very quickly identify the core business of the player being studied. The least represented categories also serve as a means to identify other potential applications of this actor's patents. Finally, this graph is useful in identifying patents in a domain and in a field that may have multiple uses. It can be a good way to identify new uses for patents already filed. Note: Categorizations by technology domain are based on

IPC code groupings, so patents can appear in several different categories.

Fig. 3 illustrates the distribution of the main concepts contained in the analyzed portfolio. This illustration makes it possible to very quickly identify the concepts most used by the applicants analyzed, and this graph provides the most-used concepts in the study area. This can be a source of ideation for new developments or identification of protected technologies in a new field. It is evident cancer treatment is the most promisor area for investors.

9 Future perspectives of polymeric NPs in oncology pathologies

New developments in DDSs are needed, coupled with improvements to preclinical and clinical models for cancer treatment [1]. The more important progress that can be mentioned are related to the use of NPs for controlled and targeted delivery, microfabricated self-regulating devices with the ability to retain and release various active agents on demand, biorecognizable systems and microneedles for delivery of drugs in transdermic layers [2].

Current research, therefore, has focused on advancing these "smart" technologies that respond to internal and external stimuli. In this sense, the systems can be grouped into two categories: (i) targeting the site of action, in which NPs actively seek and bind to specific tumor cells through molecules such as ligands, antibodies, and aptamers and (ii) local triggering, where biological, physical, chemical changes in the microenvironment activate the release of the drug, e.g., enzyme activity, pH, and temperature variation [6].

We would like to end this chapter by providing some multifunctional NP trends that were developed by our research group:

- Thermoresponsive NPs (hyperthermia): new multimodal therapies have been proposed to address the aggressive oncological pathologies as glioblastomas. The combination of superparamagnetic particles with genetic material has been reported as theranostics agents since its allow to diagnose and treat concomitantly. The aim is to combine the diagnosis using the magnetic properties to identify the tumor region with hyperthermal capacities to cause a cytotoxicity effect or delivery the chemotherapeutical agent anchored into DDS [11].

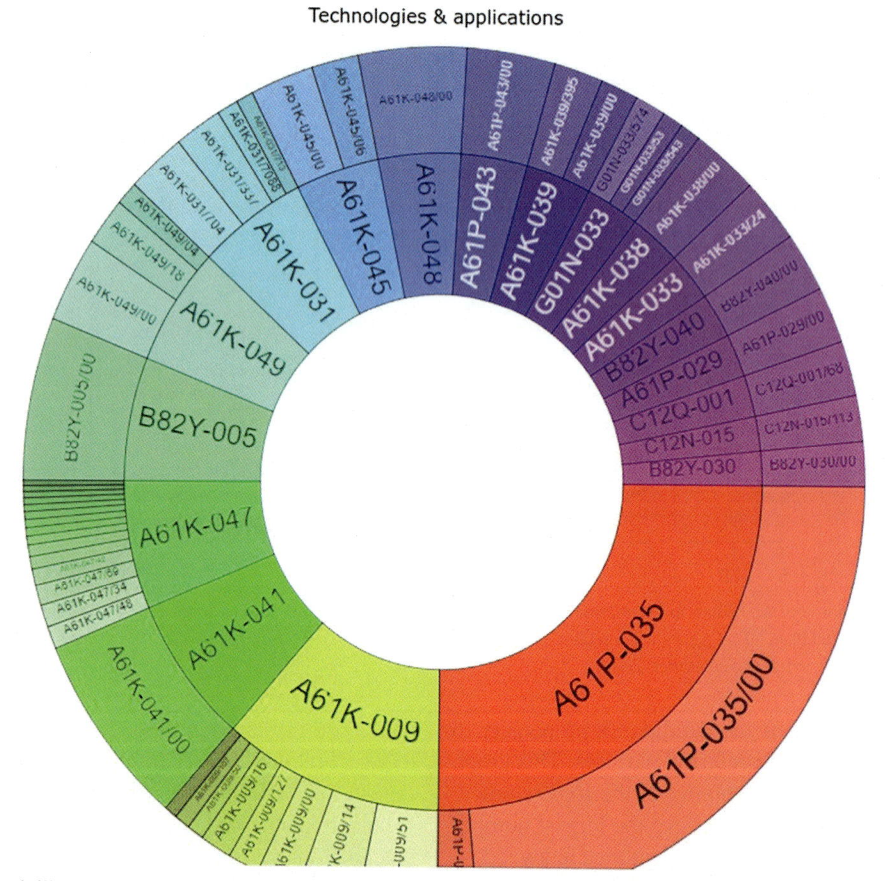

FIG. 3 Schematic graph illustration of the distribution of the main concepts contained in the analyzed portfolio related to nanoparticles and cancer, according to the Questel/Orbit patent database. *(This figure was created on the Questel's IP Business Intelligence application Orbit Intelligence. Credits to Mr. Yuri Tukoff regarding the patent research in September, 2020.)*

- Development and characterization of antitumor miltefosine encapsulated into polymeric micelles of Pluronic F127 for cancer therapy. Miltefosine presents antineoplastic potential despite high hemolytic effect limiting its application to cutaneous metastasis therapy derived of breast neoplasias. Thus, incorporation of miltefosine into polymeric micelles provides hemolytic protection. By lowering hemolytic potential, a promising application is on the view, bringing the parenteral use of miltefosine and other amphiphilic drugs [12, 13].
- pH-responsive NPs: particles responsive to acid-resistant and selective to neutral and basic gastrointestinal portion have been proposed to carrier drugs as antiretroviral (e.g., azithromycin) and chemotherapeutics (e.g., idarrubicin) [14].
- Polymeric NPs applied for immunotherapy, using nanospray dryer technology for vectorizing α-galactosylceramide (αGC). A single one-step process, free of organic solvent, was developed to encapsulate αGC into a polymeric matrix using a cationic copolymer to increase invariant natural killer T-lymphocyte responses. The production of polymeric NPs to vectorized αGC with an increased biological effect than the free agonist was reported by authors [15].
- Sugar-based NPs are recommended for developing DDSs, as they are biocompatible and biodegradable and can modulate meglumine antimoniate effect, increasing their bioavailability and reducing their toxicity [16].

References

[1] Li B, et al. Drug-Loaded Polymeric Nanoparticles for cancer stem cell targeting. Front Pharmacol 2017;14. https://doi.org/10.3389/fphar.2017.00051.

[2] Bennet D, Kim S. Polymer nanoparticles for smart drug delivery, application of nanotechnology in drug delivery, Ali Demir Sezer. IntechOpen; 2014. https://doi.org/10.5772/58422. Available from: https://www.intechopen.com/books/application-of-nanotechnology-in-drug-delivery/polymer-NPs-for-smart-drug-delivery.

[3] Stavis SM, et al. Nanoparticle manufacturing—heterogeneity through processes to products. ACS Appl Nano Mater 2018;1(9). https://doi.org/10.1021/acsanm.8b01239.

[4] Singh AK, et al. Engineering nanomaterials for smart drug release: recent advances and challenges. Elsevier; 2019. https://doi.org/10.1016/B978-0-12-814029-1.00015-6. Available from: https://www.sciencedirect.com/science/article/pii/B9780128140291000156?via%3Dihub.

[5] Pantapasis K, et al. Bioengineered nanomaterials for chemotherapy. In: Nanostructures for cancer therapy. Elsevier; 2017. Available from: https://www.elsevier.com/books/nanostructures-for-cancer-therapy/grumezescu/978-0-323-46144-3.

[6] Brewer E, Coleman J, Lowman A. Emerging technologies of polymeric nanoparticles in cancer drug delivery. J Nanomater 2011. https://doi.org/10.1155/2011/408675.

[7] Francia V, et al. Corona composition can affect the mechanisms cells use to internalize nanoparticles. ACS Nano 2019;13:10. https://doi.org/10.1021/acsnano.9b03824.

[8] Vauthier C, Bouchemal K. Methods for the preparation and manufacture of polymeric nanoparticles. Pharm Res 2009;26:5. https://doi.org/10.1007/s11095-008-9800-3.

[9] Gurny R, Peppas NA, Harrington DD, Banker GS. Development of biodegradable and injectable lattices for controlled release potent drugs. Drug Dev Ind Pharm 1981;7:1–25. https://doi.org/10.3109/03639048109055684.

[10] ORBIT. Questel's IP Business Intelligence application Orbit Intelligence—Patent research report. Available from: https://www.orbit.com/reportviewer/5CDBB472-470A-4EF2-B604-A9A6319EDC12/#12; 2020.

[11] Perecin C, Cerize N, Chitta V, Gratens X, Léo P, Oliveira A, Yoshioka S. Magnetite nanoparticles encapsulated with PCL and poloxamer by nano spray drying technique. Nanosci Nanotechnol 2016;6(4):68–73.

[12] Feitosa VA, et al. Polymeric Micelles of Pluronic F127 reduce hemolytic potential of amphiphilic drugs. Coll Surf B: Biointerfaces 2019;180:177–85. https://doi.org/10.1016/j.colsurfb.2019.04.045.

[13] Valenzuela-Oses JK, et al. Development and characterization of miltefosine-loaded polymeric micelles for cancer treatment. Mater Sci Eng C Biomimetic Mater Sens Syst 2017;81:327.

[14] Braga MS, Carvalho G, Dua K, Cerize N, Pinto T. Idarubicin loaded nanoparticles for breast cancer. In: Conference: 3rd world congress on recent advances in nanotechnology; 2018.

[15] Gonzatti MB, et al. Nano spray dryer for vectorizing α-Galactosylceramide in polymeric nanoparticles: a single step process to enhance invariant Natural Killer T lymphocyte responses. Int J Pharm 2019;565. https://doi.org/10.1016/j.ijpharm.2019.05.013.

[16] Horoiwa TA, et al. Sugar-based colloidal nanocarriers for topical meglumine antimoniate application to cutaneous leishmaniasis treatment: ex vivo cutaneous retention and in vivo evaluation. Eur J Pharm Sci 2020;147.

Chapter 27

Lipid-polymer hybrid nanocarriers for delivering cancer therapeutics

Viney Chawla[a], Pooja A. Chawla[b], and Anju Dhiman[c]

[a]University Institute of Pharmaceutical Sciences and Research, Baba Farid University of Health Sciences, Faridkot, Punjab, India, [b]Department of Pharmaceutical Chemistry and Analysis, ISF College of Pharmacy, Moga, Punjab, India, [c]Department of Pharmaceutical Sciences, M.D. University, Rohtak, Haryana, India

1 Introduction

Cancer is one of the leading causes of death in humans, and its prevalence is increasing with each passing year. As per an estimate, 30 million people are likely to be affected by cancer by the year 2030 [1]. Targeting drugs to the cancer cells, development of resistance and cancer relapse are some of the limitations that further make this disease more dreadful. Drug delivery system has an important role to play in the ADME of a drug, rate of its release and its duration of action, site of drug action, and finally the therapeutic and side effects [2]. Conventional drug delivery systems exhibit nonspecific biodistribution and thus are not only limited in efficacy but also possess high toxic potential. Novel drug delivery systems are capable of delivering a variety of chemotherapeutic agents and offer advantages of increased biological half-life, improved targeting of cancer cells, reduced toxicity, and sustained release of active molecules. Colloidal drug delivery systems exemplified by microspheres, nanoparticles, nanosponges, micelles, and liposomes are gaining popularity as a delivery vehicle for cancer chemotherapeutics. Use of nanoparticles has met with overwhelming success in targeting tumors of varied etiology and reducing toxicity. This is evident from the quantum of research and review papers published in this field [3–7]. Nanoparticles can either be polymer based or lipid based. The former include polymer nanoparticles, micelles based on polymer based, and conjugates of drug with polymer, while the latter are exemplified by liposomes, solid lipid nanoparticles, and nanostructured lipid carriers. An instant comparison of these two systems reveals that lipid-based systems offer improved entrapment of drugs coupled with low cost and their GRAS (generally regarded as safe) status [8]. However, at the same time they are plagued by problems of poor size distribution, burst release of entrapped drugs, poor stability, and cumbersome sterilization procedures. It is difficult to chemically modify such particles or exercise control over their surface properties in an attempt to improve targeting or entrapment efficiency. Polymer-based systems on the other hand render themselves suitable for chemical modification, thereby making it easy to design nanoparticles suited to a particular requirement. Further their small size, favorable polydispersity index, ease of scale up and manufacture, and improved stability add to their suitability. At the same time, the use of organic solvents in their manufacture, toxic degradation products and poor loading capacities limit their use [9].

Formulation scientists have devised lipid-polymer hybrid (LPH) nanoparticles, which offer the advantages of both lipid-based systems such as higher drug loading and those of polymeric systems like ease of obtaining desired release profile and drug targeting. Essentially, an LPH nanocarrier consists of a central polymer core that is surrounded by single or multiple lipid(s). The outer lipid layers may or may not be coated with PEG to provide stealth effect. It is possible to produce LPH nanocarriers in different morphology like core shell or the matrix type [10]. Some hydrophobic drugs that are otherwise poorly bioavailable and cannot be used in their free from can be easily encapsulated in the core of LPH. Further, it is possible to load both lipid soluble and water-soluble drugs within the same particle. The release of drugs from such nanoparticles can be tailor made either by controlling the polymer properties or through the use of stimuli responsive polymers. Moreover, the outer stealth coating not only acts as a barrier toward diffusion but also increases the mean residence time of PLH in systemic circulation [11, 12]. The structure of a typical polymer lipid hybrid nanoparticle is shown in Fig. 1.

1.1 Advantages of LPH

(1) It is easy to synthesize lipid-polymer hybrid nanoparticles.
(2) Ease of scale up through fluidic mixing devices.

Advanced Drug Delivery Systems in the Management of Cancer. https://doi.org/10.1016/B978-0-323-85503-7.00031-6

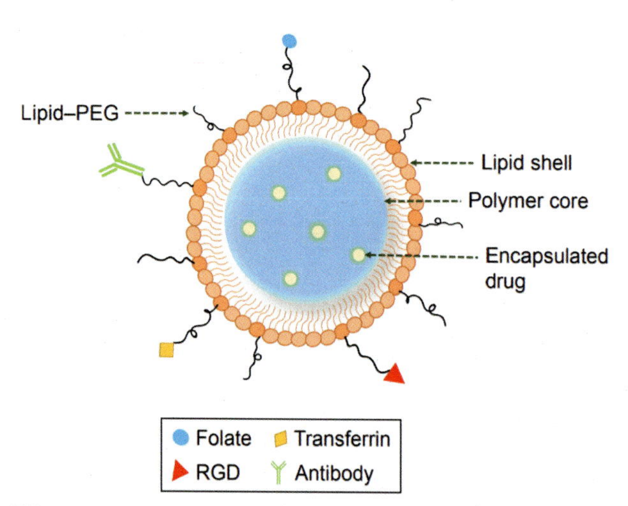

FIG. 1 Schematic diagram of a polymer lipid hybrid nanoparticle. *(Reproduced from Mukherjee A, Waters AK, Kalyan P, Achrol AS, Kesari S, Yenugonda VM. Lipid–polymer hybrid nanoparticles as a next-generation drug delivery platform: state of the art, emerging technologies, and perspectives. Int J Nanomedicine 2019;14:1937.)*

(3) Cancer chemotherapeutics, nucleic acids, and other materials can be easily incorporated.
(4) It is possible to load both lipophilic and hydrophilic drugs in the same particle.
(5) By coating with PEG, such particles remain in blood stream for a long period of time.
(6) It is possible to exercise control over release rate of drugs from such particles by suitable choice of polymers.
(7) Targeting of drugs to tumor sites is possible.

2 Challenges in the delivery of cancer chemotherapeutics

2.1 Physiology of tumors

The penetration of PLH nanocarriers into the microenvironment of the tumor often gets affected by tumor physiology at various levels, viz., molecular level, cellular level, microscopic and macroscopic level. When these nanocarriers interact with RES, they get eaten up by the macrophages and phagocytes leading to their short circulation half-life. Therefore, these nanocarriers get less time to accumulate in the tumor region [13]. Large size of liver accounts for off-target accumulation of nanocarriers. An impaired vasculature often results from angiogenesis in tumor brought about through activation of factors like PDGF and VEGF. The resulting high permeability of tumor tissues is beneficial on one hand by allowing better penetration of nanocarriers and detrimental on the other hand because it blocks extravasation into other tumor tissues arising out of fluid loss. As a consequence, some tumor tissues get less than required dose of chemotherapeutic agent and thus develop multidrug resistance (MDR). Cancer cells also possess efflux transporters like PGP, which shunt the drugs out of tissues, thereby posing another challenge in the delivery of anticancer drugs [14, 15]. It is difficult on part of nanocarriers to gain entry into tumor cells owing to elevated interstitial fluid pressure (which is an outcome of frequent entry and build-up of fluids in the interstitial space) and leaky vasculature [16]. Another barrier toward delivery of cancer chemotherapeutics is the stromal density, which is known to be comprised of fibroblasts, immune cells, extracellular matrix, and the basement membrane. Interstitial fluid pressure can be increased by the fibroblasts through the induction of contractile forces. The negatively charged components, namely, glycocalyx (found in cell membrane) and hyaluronic acid (present in interstitial stroma), interact with nanocarriers leading to their inactivation and thus affecting their transport [17].

Mesenchymal cells in some types of advanced cancers like prostate cancer, thyroid cancer, and ovarian cancer change to epithelial cells and thus form barriers called tight junctions, which act as a deterrent for drug delivery. There is a documented evidence of association of these junctions with occluding and claudins—two proteins involved in the formation of a paracellular barrier. Therefore, it becomes difficult for the drug(s) to cross this barrier. Malignant cells often exhibit elevated levels of another junction protein called desmoglein-2 (DSG-2) and this can be exploited for targeted delivery of cancer chemotherapeutics. Proteins that reverse the transition of epithelial cells back to mesenchymal cells, thereby compromising the efficacy of tight junctions have been discovered and are often given with cancer chemotherapeutics. Such proteins, called junction openers, often improve the penetration of cancer chemotherapeutics into tumor tissues [18, 19].

2.2 Multiple drug resistance

It is one of the main challenges in cancer chemotherapeutics. Different mechanisms are involved in this. One is pumping out of drugs by transporters like ATP-binding cassettes (ABC), it can also arise out of those cells that survive chemotherapeutic agents. Further, cell defense mechanism, mutation of oncogenes might also be responsible for development of resistance. Sometimes, the biological environment of the cancer cells undergoes adaptation leading to development of resistance [20]. Moreover, the overexpression of some transporters, like ABCB1, known as P-glycoprotein, P-gp, ABCG2, known as breast cancer resistant protein, BCRP, and ABCC1, known as multidrug resistance-associated protein 1 (MRP1), contributes to MDR [20]. It leads to expulsion of drugs away from the cell, leading to reduced intracellular accumulation. By formulating the drugs in lipid polymer hybrid nanoparticles along with PGP-1 inhibitor can help overcome MDR [21]. However, the overall ability of LPH to overcome drug resistance depends on the extent and ability of LPH to cleave from the nanoparticles [22].

2.3 Rapid clearance

Another challenge in the delivery of cancer chemotherapeutics to tumor cells is the short stay of drugs at the desired site of action because of its short circulation half-life. When formulated as LPH, the drug release can be sustained either through controlled diffusion or erosion of the polymer matrix. Circulation half-life can also be improved by controlling properties of vehicles such as their size, charge, surface hydrophobicity, ligand type, and density of ligands on the surface [23]. PEG coating can also help to improve the half-life of chemotherapeutic agents. In general pegylated nanoparticles that bear a slight negative surface charge and possess a size between 10 and 150nm are known to possess higher circulation times and are able to permeate into tumor cells either by passive diffusion or active targeting [24, 25].

3 Methods of preparation of lipid polymer hybrid nanocarriers

There are different methods for preparing lipid-polymer hybrid nanoparticles such as solvent-evaporation, emulsion method, nanoprecipitation, followed by self-assembly, and a sonication method. They can either be single-step processes, multistep processes, or a nonconventional process. However, multistep processes have been reported to provide better results in terms of providing a better control over the ratio of lipid and polymer component in the system.

3.1 Multistep process

Here, the lipid vesicles and polymeric nanoparticles are prepared separately and then the aqueous suspension of these nanoparticles is added to preformed lipid vesicles (Fig. 2). Alternatively the suspension may be added to a dried film of lipid. In both cases, an outer coat of lipids is formed around the polymeric nanoparticles. Scientists have recommended the use of extrusion (frequent passage of formulation through mesh like membrane) and homogenization methods (formulation is passed through a high-speed homogenizer where it experiences high shear) to achieve a uniform size distribution and favorable polydispersity index. Extrusion method has been employed by Hu et al. to fabricate PLGA-based lipid hybrid nanoparticles using membrane lipids. The authors mixed preformed nanoparticles of PLGA with vesicles prepared from RBC membrane. The resulting hybrid nanoparticles (size 80nm) were obtained by extrusion from a 100-nm polycarbonate porous membrane [26].

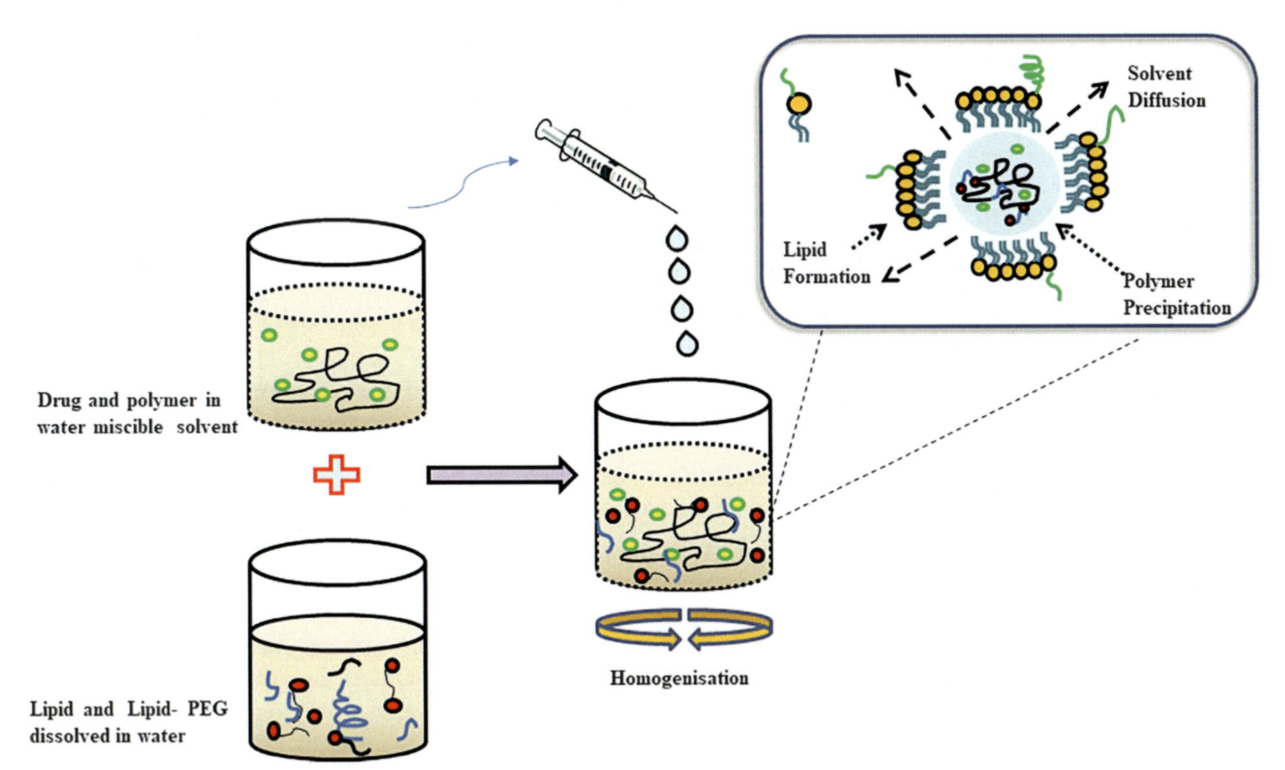

FIG. 2 Two variants of the two-step-process in the preparation of LPH. *(Reproduced from Mukherjee A, Waters AK, Kalyan P, Achrol AS, Kesari S, Yenugonda VM. Lipid–polymer hybrid nanoparticles as a next-generation drug delivery platform: state of the art, emerging technologies, and perspectives. Int J Nanomedicine 2019;14:1937.)*

3.2 Single-step nanoprecipitation method

Unlike the multistep process described above which is very labor intensive, this method is simple and easy. Here, there is no need to prepare lipid vesicles or polymeric nanoparticles separately.

In this method, the aqueous lipid dispersion is mixed with polymer solution (inorganic phase) giving self-assembly-based core-shell LPH formulation. Both nanoprecipitation method and emulsification solvent evaporation method have been employed for self-assembly of LPH systems. Sometimes nanoprecipitation is combined with the use of high-energy sonication to have a better control over particle size of resulting nanocarriers.

In this method, polymer is first added to water miscible organic solvent like acetonitrile or acetone along with drug. This solution is then added drop by drop with constant stirring into hot aqueous dispersion of lipid and/or lipid-PEG components, which results in the precipitation of polymer and drug component [27, 28].

4 Applications of polymer lipid hybrid nanoparticles

4.1 Use of PLH nanoparticles in diagnosis of cancer

Using nanocarriers for in vitro diagnostics helps to improve sensitivity, reduce cost, and consumes less of the sample. LPH find applications as the carriers for contrast or imaging agents. Some examples of imaging agents are magnetic nanocrystals (for MRI and magnetic targeting), quantum dots (for optical imaging), and carbon dots (for photoablation therapy). Imaging helps to detect cancers at an early stage and thus treatment options can be swiftly explored.

Kandel et al. developed an LPH system containing a core of polymers with high intrinsic fluorescence {e.g., poly-(fluorene-*alt*-benzothidiazole)-PFBT} surrounded by lipid-PEG (DMPE-PEG) layer. These LPH exhibited bright fluorescence owing to the insertion of lipid tail in the polymer core leading to increased spacing and hence intrachain quenching, giving higher quantum yields [28]. Wu et al. used folate-receptor-targeted LPH coated with gadolinium for imaging and targeted delivery of doxorubicin [29]. Mieszawska et al. employed nanoprecipitation method to prepare PLH based on gold nanoparticles and quantum dots. The latter were placed in the center of the lipid and polymer and attached through esterification process yielding PLGA conjugates. The authors reported successful bioimaging of mice macrophages by using the prepared nanoparticles [30].

4.2 Use of LPH nanoparticles in the treatment of cancer

LPH nanoparticles have varied applications in the management of cancers. Some of these are summarized below.

Dehaini et al. employed modified nanoprecipitation method to design ultrasmall LPH containing docetaxel in the size range of 25 nm. The prepared formulation was found to exhibit a superior tumor inhibition when compared to taxotere. The LPH-treated mice exhibited a survival time of 64 days vis-à-vis 44 days observed in taxotere-treated mice. This could be attributed to superior permeation and biodistribution profile of LPH [27].

Joshi et al. prepared LPH of paclitaxel in the size range of 200–300 nm designed for oral use. The authors aimed to formulate a formulation that could withstand gut conditions with improved bioavailability. When compared with taxol, the bioavailability improved one and a half times with concomitant 5.5-fold increase in elimination half-life. Further due to stealth effect of the coating, RES uptake was significantly lower [31].

Curcumin-loaded long circulating LPH nanoparticles based on PLGA were designed by Palanage et al. to stop cancer metastasis. There was a marked reduction in the adhesion of cancerous cells to endothelial cells (70%) with an associated reduction in vascular deposition (50%). Curcumin-loaded LPH successfully prevented metastasis and reduced the number of tumor cells in circulation [32].

Mandal et al. prepared core shell LPH of erlotinib employing sonication method. The fabricated LPH exhibited an entrapment efficiency of 66% and a particle size of 170 nm. By using human lung adenocarcinoma cell line (A 549 cells), the authors carried out studies to assess in vitro cellular uptake, luminescent cell viability, and colony-forming assay. The authors reported an increased and efficient uptake of prepared LPH by the cancerous cells, thereby emphasizing the suitability of this carrier medium for tumor uptake [33].

The development of resistance against sorafenib, which is an antiangiogenic drug active against highly vascular hepatocellular carcinoma, was mainly attributed to the activation of CXC receptor type 4 (CXCR4). LPH nanoparticles based on PLGA and loaded with sorafenib were fabricated by Gao et al. using single-step precipitation method and tested in vitro and in vivo against tumors. When compared to control group, the sorafenib-loaded LPH delayed tumor progression, improved antiangiogenic effect, and an improvement in the survival rate of animals in HCC model [34].

Park et al. designed LPH based on SB505124-entrapped cyclodextrins and cytokine-encapsulating PLGA within a PC-Chol-DSPE-PEG shell capable of delivering both agents simultaneously to the microenvironment of the tumor. The authors reported significant increase in the survival of tumor bearing mice, which was attributed to sustained release of TGF-β receptor-I inhibitor by LPH [35].

The use of a nanoburr system was put forward by Chan et al. for the delivery of paclitexal meant to treat injured vasculature. This system essentially consisted of a drug-conjugated

PLA core and a lipoidal shell comprising lecithin/DSPE-PEG, and the shell was further modified through the use of a basement membrane targeting peptide. When tested in a rat model, the prepared LPH were found to accumulate specifically in the injured vasculature where they were reported to release paclitexal for more than 2 weeks [36].

A step-by-step approach to fight tumors was propagated by Sengupta et al. through the use of a nanocell system. The authors designed a system that carried doxorubicin in a PLGA core and an outer lipid shell bearing combretastatin—an antiangiogenic drug. The designed nanocell system was able to increase the survival rate of tumor-bearing mice compared to those groups of animals that received either a single drug or a mixture of two drugs. Both the drugs were reported to act synergistically. First, combretastatin was released from the outer shell and caused vascular shutdown followed by release of doxorubicin from the inner PLGA core, thereby causing tumor reduction [37].

To improve delivery of anticancer agents, Aryal et al. recommended the use of LPH based on combinatorial drug conjugation method. The prepared LPH were able to incorporate both hydrophilic and lipophilic anticancer agents [38].

Zhang et al. reported preparing novel PLH capable of encapsulating large amounts of drug and providing sustained release. The serum stability of these particles prepared by nanoprecipitation method was quite high, and they were amenable to targeting. Docetaxel was selected as a model hydrophobic drug. The authors were able to exercise control over size of LPH in a highly reproducible manner with minimal effect on their surface charge [11].

Zhu et al., in an attempt to explore the potential of prohibitin 1 (PHB1) in the treatment of NSCLC, developed LPH containing si-RNA by employing a self-assembly approach where the polymer core consisted of PLGA, whereas the lipid shell comprised of lecithin. The authors prepared small-sized (< 100 nm) PLH capable of providing longer circulation times and higher tumor accumulation. These particles were effective in gene silencing with fewer side effects [39].

Gao et al. prepared hybrid vesicles containing isoliquiritigenin (ISL) by using PLGA as the core material that was surrounded with lipid layers of DSPE-PEG-Mal and lecithin. To achieve targeting of the tumor, iRGD peptides were bound to vesicle surface. The results of in vitro study were encouraging and displayed superior inhibitory and apoptotic effects of the targeted vesicles vis-à-vis the free drug or the nontargeted form. The findings of the in vivo studies corroborated the results indicating that the targeted vesicles exhibited higher anticancer potential in 4T1 tumor-bearing mice models [40].

Ma et al. developed nanohybrid vesicles using cholesteryl succinyl silane (CSS) capable of exerting selective antitumor effects on leukemia cells while protecting normal blood cells. The prepared vesicles were able to encapsulate both lipophilic and hydrophilic drugs. The authors reported enhanced anticancer effects of doxorubicin using this system with fewer side effects [41].

Jadon and Sharma developed LPH of docetaxel in an attempt to combat breast cancer. The authors used nanoprecipitation method where the core was made of PLGA and the lipid layer consisted of DSPE-PEG2000-NH2. The authors tested the prepared formulations in MDA-MB-231 cells lines to evaluate cell viability and cellular uptake. Further apoptosis, ADME studies with special emphasis on biodistribution in important organs, percent changes in tumor volume, and rates of animal survival were also compared to free drug. The findings of in vitro cell studies revealed superior cytotoxicity and enhanced uptake of drug at lower IC50 in MDA-MB-231 cell lines. FACS analysis exhibited improved apoptosis for LPH nanoparticles. The pharmacokinetic profile and ability to target the tumors was found to be better for LPH vis-à-vis the free drug. In addition a noticeable reduction in tumor burden was observed for LPH in comparison to free docetaxel [42].

In an attempt to improve the aqueous solubility and bioavailability of psoralen, Du et al. developed PLH using soy lecithin as the lipid and PLGA as the polymer by nanoprecipitation method. The developed formulations were characterized and tested against MCF-7 breast tumor model. The authors reported equal efficiency to doxorubicin with less toxicity [5].

Jain et al. prepared LPH of beta-carotene and methotrexate by using nanoprecipitation technique. The authors intended to study the protective effects of beta-carotene in methotrexate-induced toxicity in breast cancer models. The prepared LPH exhibited an apoptosis index of 0.89 against MCF-7 breast cancer cells. Further, beta-carotene was found to ameliorate toxicity induced by methotrexate in kidney and liver cells [43].

Yu et al. developed a charge reversal strategy to achieve in vivo mitochondrial targeting of doxorubicin. PLGA polymeric core was used to prepare PLH, whereas C18-PEG2000-TPP formed the lipid component. Acidic pH of the triggered reversal of charge in TPP-containing LPH and made it possible to achieve mitochondrial targeting for the delivery of selected drug into the mitochondria of cancer cells while overcoming drug resistance of selected MCF-7 cell lines [44].

Epithelial cell adhesion molecule (EpCAM) is a protein that is frequently overexpressed in cells of colorectal adenocarcinoma. Li et al. prepared LPH-containing curcumin conjugated with a synthetic aptamer to target EpCAM. Two types of formulations were designed, namely, those bearing curcumin alone and those bearing aptamer-conjugated curcumin in the size range of less than 100 nm. The circulation half-life of curcumin was increased six times and mean retention time was enhanced three times following its encapsulation in LPH nanoparticles [45].

Zhao et al. advocated the synergistic use of curcumin and doxorubicin in combating hepatocellular carcinoma of mice induced by diethylnitrosamine. Both these drugs were formulated together as PLH, and the results were encouraging in terms of increased anticancer and antiangiogenic effects of both drugs in terms of apoptosis index and inhibition of cell proliferation. Additionally, the antitumor effects of the two-drug combination in PLH were superior to nanoparticles containing doxorubicin alone [46]. Table 1 highlights a comparative account of different polymer lipid hybrid nanoparticles.

In a similar study, Zhang et al. [48] successfully bound iRGD peptide to PLH nanoparticles in an attempt to increase the antitumor effect of doxorubicin (DOX) and sorafenib (SOR). The tumor inhibition of prepared LPH was tested in a mouse model of liver cancer. When compared with the free drugs and nanoparticles that did not have iRGD peptide modification, DOX + SOR/iRGD-NPs had higher synergistic cytotoxic effects. Not only did they prolong the cycle time but also improved the biocompatibility. Further, endocytosis was also enhanced along with higher drug accumulation in the tumor.

Tahir et al. compared doxorubicin and doxorubicin hydrochloride by formulating these two molecules in separate PLH through modified single-step precipitation method. The PLH exhibited superior encapsulation of DOX base (59.8 ± 1.4) compared to DOX·HCl (43.8 ± 4.4) owing to its lipophilic character. The drug release over a 24-h period was slower in lipophilic base (32.9%) vis-à-vis the hydrochloride salt (56.7%). The authors advocated the use of PLH owing to their improved therapeutic effects, which they attributed to the better safety profile, improved antitumor effect, and cellular uptake of lipid polymer hybrid nanoparticles [49].

Yang et al. described a complete ablation of the tumor growth through their PLGA-based PLH formulation containing si-RNA and a cationic lipid. The prepared formulation was tested in MDA-MB-435s murine xenograft model. The authors employed solvent evaporation technique to prepare the nanoparticles in the range of 170–200 nm. Further, a downregulation of gene expression was reported when the same formulation was tested in a mice model of liver cancer [47].

Zeng et al. prepared PLH nanoparticles for delivery of paclitexal and celecoxib to overcome multiple drug resistance and to reduce PGP efflux. The polymer core consisted of PLA, whereas lecithin and DSPE-PEG2000 with a molar ratio of 7:3 constituted the lipid phase. In vitro cell inhibition potential of the drug-loaded PLH was evaluated in both drug-resistant cells (MCF-7/ADR) and nonresistant cells (HeLa). When the nanoparticles loaded with a single drug were compared with dual drug-loaded PLH, the latter exhibited superior tumor inhibition owing to downregulation of P-gp expression by celecoxib [50].

Tahir et al. reported preparing methotrexate-loaded PLH nanoparticles by modified nanoprecipitation method employing PLGA as polymer and lipoid S100 as lipid material. The cytotoxicity of the prepared nanoparticles was tested in MDA-MB-231, PC3, and HT 29 cell lines after 24 h incubation and plotted against different concentrations. The authors reported remarkable cellular uptake of prepared PLH with better antiproliferative effects in a time- and dose-dependent manner vis-à-vis a plain solution of the drug [51] [LPH35].

Overall, PLH nanoparticles have proved their worth in diagnosis and treatment of a variety of cancers and can be potential delivery vehicle in future.

TABLE 1 A comparative account of different polymer lipid hybrid nanoparticles.

S. no	Lipid/polymer used	Method employed	Application	References
1	Soy lecithin/PLGA	Nanoprecipitation method	Liver cancer	[5]
2	HSPC/polycaprolactone	Sonication method	Lung adenocarcinoma	[33]
3	PLGA	Nanoprecipitation method	Hepatocellular carcinoma	[34]
4	PC-Chol-DSPE-PEG/PLGA	Two-step method	Metastatic melanoma	[35]
5	Lecithin/DSPE-PEG/PLA	Nanoprecipitation method	Treatment of injured vasculature	[36]
6	Lecithin/PLGA	Self-assembly approach	NSCLC	[39]
7	DSPE-PEG-Mal and lecithin/PLGA	Two-step method	4T1 tumor-bearing mice models	[40]
8	DSPE-PEG2000-NH2/PLGA	Nanoprecipitation method	Breast cancer	[42]
9	Lecithin/PLGA	Nanoprecipitation method	Breast cancer	[43]
10	Cationic lipid/PLGA	Solvent evaporation technique	Liver cancer	[47]

5 Conclusion

Polymer lipid hybrid nanoparticles can provide a potential platform for delivery of drugs in various types of cancer. The advantages of these carriers over other carriers or free drugs have been frequently documented. The PEG coating on their outer surface provides them with an innate ability to overcome uptake by phagocytes and provide higher amounts of the drugs at the desired site for prolonged periods of time. Their ability to manipulate microenvironment of the tumor provides them with an additional advantage in delivering cancer chemotherapeutics.

6 Future prospects

Preparations based on polymer lipid hybrid nanoparticles like Genexol-PM that contains paclitaxel (PCX) and poly(D,L-lactide)-*b*-polyethylene glycol-methoxy (PLGA-mPEG) has been approved for metastatic breast cancer treatment in European Union and Korea. However, a lot of research needs to be carried out in this field. Biomarker-guided delivery of anticancer agents can be attempted by encapsulating both in the same nanoparticles. The prospects of using more and more natural lipids and polymers do exist and can be taken as a challenge by formulation scientists. This may pave a way for the discovery of more biocompatible and biodegradable lipids.

References

[1] Jadon RS, Sharma M. Lipid-polymer hybrid nanoparticles (Lipobrid) based targeted delivery of Docetaxel to breast cancer. In: Proc. 2018 2nd international conference and exhibition on nanomedicne and drug delivery; 2018. https://doi.org/10.4172/2325-9604-C2-030.

[2] Siepmann J, Siegel RA, Rathbone MJ, editors. Fundamentals and applications of controlled release drug delivery. Spring Street, NY: Springer Science & Business Media; 2011.

[3] Pukale SS, Sharma S, Dalela M, Kumar Singh A, Mohanty S, Mittal A, Chitkara D. Multi-component clobetasol-loaded monolithic lipid-polymer hybrid nanoparticles ameliorate imiquimod-induced psoriasis-like skin inflammation in Swiss albino mice. Acta Biomater 2020;115:393–409.

[4] Yalcin TE, Ilbasmis-Tamer S, Takka S. Antitumor activity of gemcitabine hydrochloride loaded lipid polymer hybrid nanoparticles (LPHNS) in vitro and in vivo. Int J Pharm 2020;580, 119246.

[5] Du M, Ouyang Y, Meng F, Zhang X, Ma Q, Zhuang Y, Liu H, Pang M, Cai T, Cai Y. Polymer-lipid hybrid nanoparticles: a novel drug delivery system for enhancing the activity of Psoralen against breast cancer. Int J Pharm 2019;561:274–82.

[6] Mukherjee A, Waters AK, Kalyan P, Achrol AS, Kesari S, Yenugonda VM. Lipid–polymer hybrid nanoparticles as a next-generation drug delivery platform: state of the art, emerging technologies, and perspectives. Int J Nanomedicine 2019;14:1937.

[7] Date T, Nimbalkar V, Kamat J, Mittal A, Mahato RI, Chitkara D. Lipid-polymer hybrid nanocarriers for delivering cancer therapeutics. J Control Release 2018;271:60–73.

[8] Anon, https://www.fda.gov/food/food-ingredients-packaging/generally-recognized-safe-gras. [Accessed 13 October 2020].

[9] Dave V, Tak K, Sohgaura A, Gupta A, Sadhu V, Reddy KR. Lipid-polymer hybrid nanoparticles: synthesis strategies and biomedical applications. J Microbiol Methods 2019;160:130–42.

[10] Tahir N, Haseeb MT, Madni A, Parveen F, Khan MM, Khan S, Jan N, Khan A. Lipid polymer hybrid nanoparticles: a novel approach for drug delivery. In: Role of Novel Drug Delivery Vehicles in Nanobiomedicine. IntechOpen; 2019.

[11] Zhang L, Chan JM, Gu FX, Rhee JW, Wang AZ, Radovic-Moreno AF, Alexis F, Langer R, Farokhzad OC. Self-assembled lipid-polymer hybrid nanoparticles: a robust drug delivery platform. ACS Nano 2008;2(8):1696–702.

[12] Gref R, Domb A, Quellec P, Blunk T, Müller RH, Verbavatz JM, Langer R. The controlled intravenous delivery of drugs using PEG-coated sterically stabilized nanospheres. Adv Drug Deliv Rev 1995;16:215–33.

[13] Heimann IW, Ratge D, Wisser H, Frölich JC. Influence of prostaglandin inhibition on dihydralazine induced acute effects in patients with essential hypertension. Clin Exp Pharmacol Physiol 1985;12(1):79–89.

[14] Stylianopoulos T. Intelligent drug delivery systems for the treatment of solid tumors. Eur J Nanomed 2016;8(1):9–16.

[15] Zhang RX, Ahmed T, Li LY, Li J, Abbasi AZ, Wu XY. Design of nanocarriers for nanoscale drug delivery to enhance cancer treatment using hybrid polymer and lipid building blocks. Nanoscale 2017;9(4):1334–55.

[16] Carlier C, Mathys A, De Jaeghere E, Steuperaert M, De Wever O, Ceelen W. Tumour tissue transport after intraperitoneal anticancer drug delivery. Int J Hyperth 2017;33(5):534–42.

[17] Beyer I, van Rensburg R, Lieber A. Overcoming physical barriers in cancer therapy. Tissue Barriers 2013;1(1):3340–51.

[18] Wang H, Li ZY, Liu Y, Persson J, Beyer I, Möller T, Koyuncu D, Drescher MR, Strauss R, Zhang XB, Wahl JK. Desmoglein 2 is a receptor for adenovirus serotypes 3, 7, 11 and 14. Nat Med 2011;17(1):96–104.

[19] Beyer I, Cao H, Persson J, Song H, Richter M, Feng Q, Yumul R, van Rensburg R, Li Z, Berenson R, Carter D. Coadministration of epithelial junction opener JO-1 improves the efficacy and safety of chemotherapeutic drugs. Clin Cancer Res 2012;18(12):3340–51.

[20] Ma P, Dong X, Swadley CL, Gupte A, Leggas M, Ledebur HC, Mumper RJ. Development of idarubicin and doxorubicin solid lipid nanoparticles to overcome Pgp-mediated multiple drug resistance in leukemia. J Biomed Nanotechnol 2009;5(2):151–61.

[21] Jain RK. Delivery of molecular medicine to solid tumors: lessons from in vivo imaging of gene expression and function. J Control Release 2001;74(1–3):7–25.

[22] Siegfried JM. Biology and chemoprevention of lung cancer. Chest 1998;113(1):40S–5S.

[23] Zhou C, Wu YL, Chen G, Feng J, Liu XQ, Wang C, Zhang S, Wang J, Zhou S, Ren S, Lu S. Erlotinib versus chemotherapy as first-line treatment for patients with advanced EGFR mutation-positive non-small-cell lung cancer (OPTIMAL, CTONG-0802): a multicentre, open-label, randomised, phase 3 study. Lancet Oncol 2011;12(8):735–42.

[24] Fang C, Shi B, Pei YY, Hong MH, Wu J, Chen HZ. In vivo tumor targeting of tumor necrosis factor-α-loaded stealth nanoparticles: effect of MePEG molecular weight and particle size. Eur J Pharm Sci 2006;27(1):27–36.

[25] Perrault SD, Walkey C, Jennings T, Fischer HC, Chan WC. Mediating tumor targeting efficiency of nanoparticles through design. Nano Lett 2009;9(5):1909–15.

[26] Hu CM, Zhang L, Aryal S, Cheung C, Fang RH, Zhang L. Erythrocyte membrane-camouflaged polymeric nanoparticles as a biomimetic delivery platform. Proc Natl Acad Sci 2011;108(27):10980–5.

[27] Dehaini D, Fang RH, Luk BT, Pang Z, Hu CM, Kroll AV, Yu CL, Gao W, Zhang L. Ultra-small lipid–polymer hybrid nanoparticles for tumor-penetrating drug delivery. Nanoscale 2016;8(30):14411–9.

[28] Kandel PK, Fernando LP, Ackroyd PC, Christensen KA. Incorporating functionalized polyethylene glycol lipids into reprecipitated conjugated polymer nanoparticles for bioconjugation and targeted labeling of cells. Nanoscale 2011;3(3):1037–45.

[29] Wu B, Cui C, Liu L, Yu P, Zhang Y, Wu M, Zhang LJ, Zhuo RX, Huang SW. Co-delivery of doxorubicin and amphiphilic derivative of Gd-DTPA with lipid-polymer hybrid nanoparticles for simultaneous imaging and targeted therapy of cancer. J Control Release 2015;213:e13–4.

[30] Mieszawska AJ, Gianella A, Cormode DP, Zhao Y, Meijerink A, Langer R, Farokhzad OC, Fayad ZA, Mulder WJ. Engineering of lipid-coated PLGA nanoparticles with a tunable payload of diagnostically active nanocrystals for medical imaging. Chem Commun 2012;48(47):5835–7.

[31] Joshi N, Saha R, Shanmugam T, Balakrishnan B, More P, Banerjee R. Carboxymethyl-chitosan-tethered lipid vesicles: hybrid nanoblanket for oral delivery of paclitaxel. Biomacromolecules 2013;14(7):2272–82.

[32] Palange AL, Di Mascolo D, Carallo C, Gnasso A, Decuzzi P. Lipid–polymer nanoparticles encapsulating curcumin for modulating the vascular deposition of breast cancer cells. Nanomedicine 2014;10(5):e991–1002.

[33] Mandal B, Mittal NK, Balabathula P, Thoma LA, Wood GC. Development and in vitro evaluation of core–shell type lipid–polymer hybrid nanoparticles for the delivery of erlotinib in non-small cell lung cancer. Eur J Pharm Sci 2016;81:162–71.

[34] Gao DY, Lin TT, Sung YC, Liu YC, Chiang WH, Chang CC, Liu JY, Chen Y. CXCR4-targeted lipid-coated PLGA nanoparticles deliver sorafenib and overcome acquired drug resistance in liver cancer. Biomaterials 2015;67:194–203.

[35] Park J, Wrzesinski SH, Stern E, Look M, Criscione J, Ragheb R, Jay SM, Demento SL, Agawu A, Limon PL, Ferrandino AF. Combination delivery of TGF-β inhibitor and IL-2 by nanoscale liposomal polymeric gels enhances tumour immunotherapy. Nat Mater 2012;11(10):895–905.

[36] Chan JM, Zhang L, Tong R, Ghosh D, Gao W, Liao G, et al. Spatiotemporal controlled delivery of nanoparticles to injured vasculature. Proc Natl Acad Sci 2010;107(5):2213–8.

[37] Sengupta S, Eavarone D, Capila I, Zhao G, Watson N, Kiziltepe T, Sasisekharan R. Temporal targeting of tumour cells and neovasculature with a nanoscale delivery system. Nature 2005;436(7050):568–72.

[38] Aryal S, Hu CM, Zhang L. Combinatorial drug conjugation enables nanoparticle dual-drug delivery. Small 2010;6(13):1442–8.

[39] Zhu X, Xu Y, Solis LM, Tao W, Wang L, Behrens C, Xu X, Zhao L, Liu D, Wu J, Zhang N. Long-circulating siRNA nanoparticles for validating Prohibitin1-targeted non-small cell lung cancer treatment. Proc Natl Acad Sci 2015;112(25):7779–84.

[40] Gao F, Zhang J, Fu C, Xie X, Peng F, You J, Tang H, Wang Z, Li P, Chen J. iRGD-modified lipid–polymer hybrid nanoparticles loaded with isoliquiritigenin to enhance anti-breast cancer effect and tumor-targeting ability. Int J Nanomedicine 2017;12:4147.

[41] Ma Y, Dai Z, Zha Z, Gao Y, Yue X. Selective antileukemia effect of stabilized nanohybrid vesicles based on cholesteryl succinyl silane. Biomaterials 2011;32(35):9300–7.

[42] Jadon RS, Sharma M. Docetaxel-loaded lipid-polymer hybrid nanoparticles for breast cancer therapeutics. J Drug Delivery Sci Technol 2019;51:475–84.

[43] Jain A, Sharma G, Kushwah V, Garg NK, Kesharwani P, Ghoshal G, Singh B, Shivhare US, Jain S, Katare OP. Methotrexate and beta-carotene loaded-lipid polymer hybrid nanoparticles: a preclinical study for breast cancer. Nanomedicine 2017;12(15):1851–72.

[44] Yu H, Li JM, Deng K, Zhou W, Wang CX, Wang Q, Li KH, Zhao HY, Huang SW. Tumor acidity activated triphenylphosphonium-based mitochondrial targeting nanocarriers for overcoming drug resistance of cancer therapy. Theranostics 2019;9(23):7033.

[45] Khalil NM, do Nascimento TCF, Casa DM, Dalmolin LF, de Mattos AC, Hoss I, Romano MA, Mainardes RM. Pharmacokinetics of curcumin-loaded PLGA and PLGA–PEG blend nanoparticles after oral administration in rats. Colloids Surf B: Biointerfaces 2013;101:353–60.

[46] Zhao X, Chen Q, Li Y, Tang H, Liu W, Yang X. Doxorubicin and curcumin co-delivery by lipid nanoparticles for enhanced treatment of diethylnitrosamine-induced hepatocellular carcinoma in mice. Eur J Pharm Biopharm 2015;93:27–36.

[47] Yang XZ, Dou S, Sun TM, Mao CQ, Wang HX, Wang J. Systemic delivery of siRNA with cationic lipid assisted PEG-PLA nanoparticles for cancer therapy. J Control Release 2011;156(2):203–11.

[48] Zhang J, Hu J, Chan HF, Skibba M, Liang G, Chen M. iRGD decorated lipid-polymer hybrid nanoparticles for targeted co-delivery of doxorubicin and sorafenib to enhance anti-hepatocellular carcinoma efficacy. Nanomedicine 2016;12(5):1303–11.

[49] Tahir N, Madni A, Correia A, Rehman M, Balasubramanian V, Khan MM, Santos HA. Lipid-polymer hybrid nanoparticles for controlled delivery of hydrophilic and lipophilic doxorubicin for breast cancer therapy. Int J Nanomedicine 2019;14:4961.

[50] Zeng SQ, Chen YZ, Chen Y, Liu H. Lipid–polymer hybrid nanoparticles for synergistic drug delivery to overcome cancer drug resistance. New J Chem 2017;41(4):1518–25.

[51] Tahir N, Madni A, Balasubramanian V, Rehman M, Correia A, Kashif PM, Mäkilä E, Salonen J, Santos HA. Development and optimization of methotrexate-loaded lipid-polymer hybrid nanoparticles for controlled drug delivery applications. Int J Pharm 2017;533(1):156–68.

Chapter 28

Advancements on microparticles-based drug delivery systems for cancer therapy

Dhriti Verma[a], Amit Bhatia[b], Shruti Chopra[c], Kamal Dua[a,d], Parteek Prasher[e], Gaurav Gupta[f], Murtaza M. Tambuwala[g], Dinesh Kumar Chellappan[h], Alaa A.A. Aljabali[i], Mousmee Sharma[j], and Deepak N. Kapoor[a]

[a]*School of Pharmaceutical Sciences, Shoolini University of Biotechnology and Management Sciences, Solan, Himachal Pradesh, India,* [b]*Department of Pharmaceutical Sciences and Technology, Maharaja Ranjit Singh Punjab Technical University, Punjab, India,* [c]*Amity Institute of Pharmacy, Amity University, Noida, Uttar Pradesh, India,* [d]*Discipline of Pharmacy, Graduate School of Health, University of Technology Sydney, Ultimo, NSW, Australia,* [e]*Department of Chemistry, University of Petroleum & Energy Studies, Dehradun, India,* [f]*School of Pharmacy, Suresh Gyan Vihar University, Jaipur, India,* [g]*SAAD Centre for Pharmacy and Diabetes, School of Pharmacy and Pharmaceutical Sciences, Ulster University, Coleraine, Northern Ireland, United Kingdom,* [h]*Department of Life Sciences, School of Pharmacy, International Medical University (IMU), Kuala Lumpur, Malaysia,* [i]*Department of Pharmaceutics and Pharmaceutical Technology, Yarmouk University, Irbid, Jordan,* [j]*Department of Chemistry, Uttaranchal University, Dehradun, India*

1 Introduction

Microparticles are solid particles consisting of synthetic or natural polymers and are generally in the micrometer size range (1–1000 μm). Microparticles can be prepared by dispersing, entrapping, encapsulating, or suspending the active drug within a polymer matrix. Depending on the drug to be encapsulated and the polymer chosen, the type of microparticle can be formulated [1]. Microparticles are categorized as microcapsules and microspheres as shown in Fig. 1. Microspheres are microparticles with a broad surface-to-volume ratio, spherical shape, which homogeneously entraps the drug in the matrix. They are also categorized as solid and hollow types. Because of their enhanced tumor targeting, magnetic microspheres have also been examined for targeted drug delivery, in particular magnetic-targeted chemotherapy [2]. In the case of microcapsules, the drug is present as a reservoir or core, which is surrounded by a polymeric shell [3–5].

Microparticles are prepared from different materials both natural and synthetic, including polymers, ceramics, and glass, which are also commercially available [6].

Microparticles as a drug delivery system can increase the effectiveness of several treatments due to their well-defined pharmacokinetic and physiological benefits, such as protection of the bioactive agents, ability to achieve a controlled release rate, and ability to transport drugs to the targeted site [7]. Microparticles when used as a drug delivery system have enhanced the effectiveness of various medical treatments and have been widely investigated for various applications owing to their superior properties [8–12]. Various advantages of microparticles are:

(1) Microparticles can improve the relative bioavailability of drugs.
(2) The composition of microparticles also offers the drug-targeting approach to particular locations and has significant potential to minimize the level of dosage and toxicity of different medications.
(3) Microcapsules can also be used as carrier for vaccines and drugs as screening agents and in surgical procedures.
(4) It provides protection against degradation of bioactive agents, small molecules, peptides, and proteins.
(5) When compared to nanoparticles they are able to release an active substance for longer periods of time (up to 6 months).
(6) It is possible to achieve site-specific targeting by adding targeting ligands to the particles' surface or by using magnetic guidance.
(7) It helps to modify and strengthen the pharmacokinetics and pharmacodynamic properties of drug molecule.
(8) Solid lipid microparticles have a great potential for topical drug targeting and may be used as a feasible cargo carrier for the cutaneous delivery of various drugs.

Drug release from the microparticles occurs by polymer degradation, diffusion, and hydrolysis/erosion. Microspheres follow first-order diffusion release rate kinetics and microcapsule follows zero-order kinetics [5]. There is clinical importance for the structural nature of microparticles as they may be used to assess therapeutic response and direct individualized treatment strategies in cancer patients [13].

Advanced Drug Delivery Systems in the Management of Cancer. https://doi.org/10.1016/B978-0-323-85503-7.00003-1

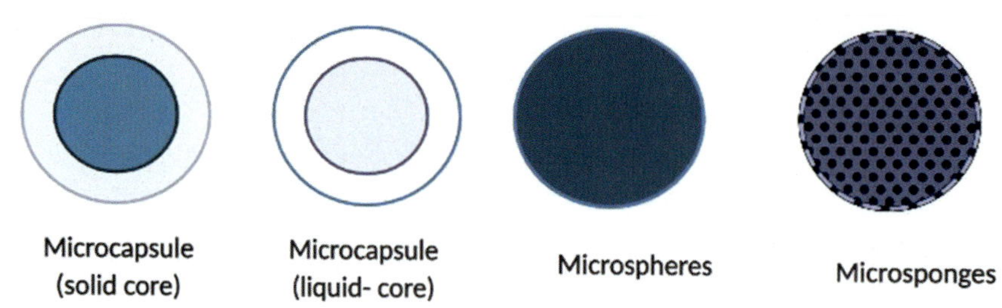

FIG. 1 Schematic representation of types of microparticles.

2 Materials used for the preparation of microparticles

For preparation of microparticles, the selection of biomaterial depends on the application, type of bioactive agent for drug delivery applications, type of the polymer-expected treatment duration, and any other requirement to modify or functionalize the delivery system [14, 15]. Both biodegradable and nonbiodegradable materials have been evaluated for drug delivery using microparticles [16]. Current focus of research on microparticles is reduction in side effects, controlled release, and drug targeting. Biodegradable polymers used in the preparation of microparticles are obtained either from renewable or nonrenewable resources [17]. These polymers can be further classified as follows:

(1) Natural polymers:
- **(a)** Polymers from animal sources, including collagen, gelatin, elastin, albumin, fibrin, alginates, chitin, chitosan, casein, chondroitin sulfate, etc.
- **(b)** Polymers from plant sources such as starch, cellulose derivatives, and dextrin.
- **(c)** Polymers from other sources such as dextran and hyaluronic acid.

(2) Synthetic biodegradable polymers:

Polyglycolic acid (PGA), polylactic acid (PLA), polycaprolactone (PCL), poly(lactide-*co*-glycolide) (PLGA), polyvinyl alcohol (PVA), poly acrylamide (PAA), etc. [17–20].

Some of the important polymers used for the preparation of microparticles are discussed here.

2.1 Natural polymers

2.1.1 Chitosan

Chitosan [poly(β-(10-/4)-2-amino-2-deoxy D-glucose)] is a natural cationic polysaccharide derived from chitin, which is copolymer where a glucosamine and an *N*-acetyl glucosamine unit are joined together [21], which makes chitosan a biocompatible and biodegradable polymer having many biomedical applications. Because of the presence of free amino groups, chitosan is insoluble in neutral or basic environment and require acidic pH environment to dissolve completely for complete solubilization. The protonation of amino group in acidic environment leads to the swelling of the polymer due to which, maximum drug from a chitosan-based drug delivery system is released in the stomach [22]. Drug delivery systems based on chitosan and its derivatives enhance drug absorption and in addition to drug release improvement, stabilize drug constituents for drug targeting. Studies have shown that antitumor agents when conjugated with chitosan or its derivatives, demonstrate stronger anticancer effects than the pure drug with decreasing adverse effects [23]. Ko et al. prepared chitosan beads with TPP (Tripolyphosphate). Chitosan gel was prepared and cross-linked with GA (Glutaraldehyde) for formation of chitosan microparticles. These microparticles improved the loading efficiency of drug and prolonged the drug release period [24]. Microspheres of chitosan have been found to be an effective delivery system for long-term drug delivery [25] (Fig. 2).

2.1.2 Alginate

Alginates are a block copolymer with 1,4 linked p-D-mannuronic acid and L-guluronic acid residues in different ratios and configurations. It is a natural, biocompatible, biodegradable, anionic, and unbranched polysaccharide obtained from brown seaweed. It forms gels with divalent ions like calcium. The gel formation is also related to the amount and length of L-guluronic units [26]. Ca-alginate beads are one of the most commonly used systems for the controlled release of drugs [27]. Alginate microparticles, especially with chitosan,

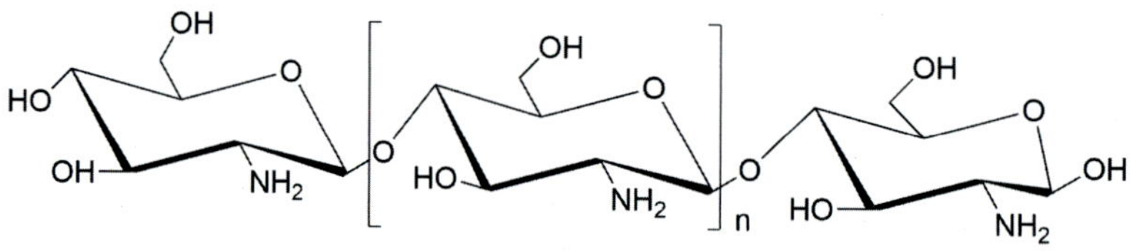

FIG. 2 Chemical structure of chitosan.

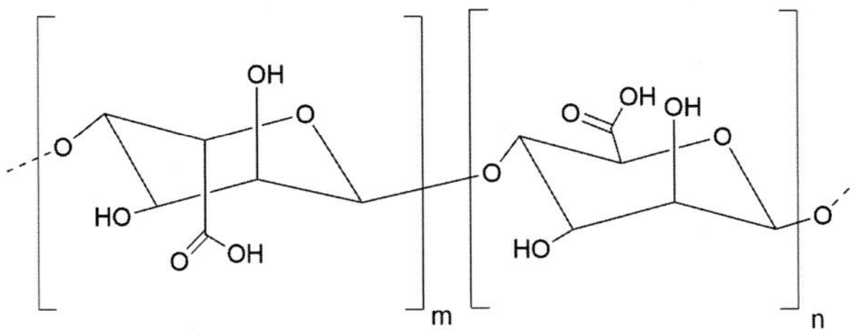

FIG. 3 Chemical structure of alginate.

have demonstrated excellent bioadhesive properties that had a strong affinity to the gastric mucosa. Alginate microparticles have also been extensively evaluated as systems for the immobilization of cells, proteins, and enzymes [28] (Fig. 3).

If used like a coating material, the structure of the semipermeable gel serves as a contact shield to the unfavorable environment and allows the release of embedded materials in a controlled manner. Alginate is a natural polymer with good biodegradability, nonantigenicity, and biocompatibility [29]. The loading capacity of alginate microparticles can be controlled by its porosity. Porous alginate microparticles achieve a greater drug loading and a quicker delivery rate as compared to nonporous microparticles. Drug release of poorly water-soluble drugs can be delayed by making enteric bypass delivery system using alginate microparticles [28].

2.1.3 Gelatin

Gelatin is a biodegradable, denatured protein formed by the acid and alkaline treatment of collagen. It's a versatile, natural, and commonly used structural protein in pharmaceutics. The key desirable features are its biocompatibility and degradation to nontoxic materials. Gelatin, being a hydrophilic substance, has a wall-forming potential that must be altered to prepare drug delivery systems [30, 31]. It's a promising candidate in biomedical applications for the development of microspheres and microcapsules with controlled release properties [32]. Microparticles of the starch/gelatin mixture can be used as controlled release systems for water-soluble drugs [33]. One study reported development of type A and type B gelatin microparticles that were lyophilized with PEG. The microparticles were prepared by S/O/O/W emulsification method [34] (Fig. 4).

2.1.4 Dextran

Dextran and its derivatives are nontoxic polysaccharides, composed of α-D-glucans with anhydro-D-glucopyranose units as part of their key molecular chain [35]. Dextran or its derivatives are easy to process, biocompatible, and pH sensitive due to which, these have versatile applications in drug delivery [36]. The dextran pendant hydroxyl is acetalized. Acetalization converts hydrophilic dextran into a hydrophobic polymer and when exposed to acid, acetals degrade to make dextran hydrophilic. This acetalized polymer (Ac-DEX) allows entrapment of hydrophobic and other usually hard-to-deliver drugs. Since this polymer degrades faster at lower pH as it happens in the endosome of phagocytic cells, tumors, or inflammatory regions, this allows for site-specific polymer degradation and drug delivery [37]. Hydrogels based on dextran can be regarded as biocompatible materials, making hydrogels suitable for drug delivery purposes [38] (Fig. 5).

2.2 Synthetic polymers

2.2.1 Polylactic acid/poly (lactic acid-*co*-glycolic acid) (PLA/PLGA)

Lactic acid is the essential building block of PLA, which was first isolated from sour milk in 1780 [39]. It is used as a model biodegradable and biocompatible polymer for preparing sustained release particles [40]. It exhibits a broad range of erosion time and has tunable mechanical properties and can be used for the formulation of controlled delivery systems of therapeutic molecules, including vaccines, proteins, genes, and anticancer drugs [41]. PLA and PLGA microspheres-containing bioactive agents are biocompatible and do not react unfavorably when used in in vivo therapeutic applications locally or systemically [42]. Thus PLA/PLGA microparticles are one of the most promising drug delivery systems [43]. PLA-PEG-PLA microspheres are also important candidates for the therapeutic delivery of extremely hydrophobic anticancer medications such as paclitaxel [44]. Entrapment and release of budenoside from the matrix of PLA microparticles

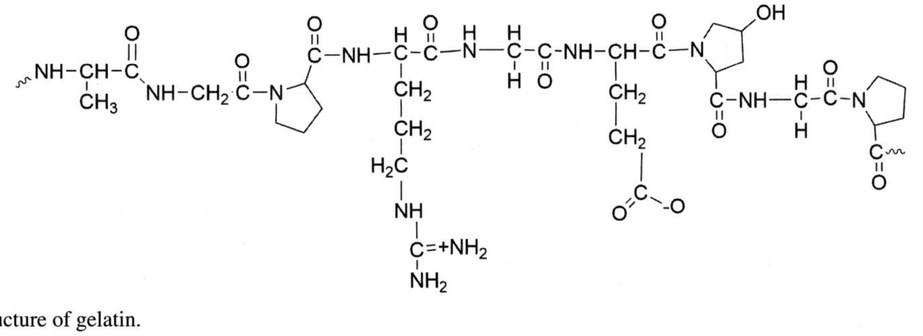

FIG. 4 Chemical structure of gelatin.

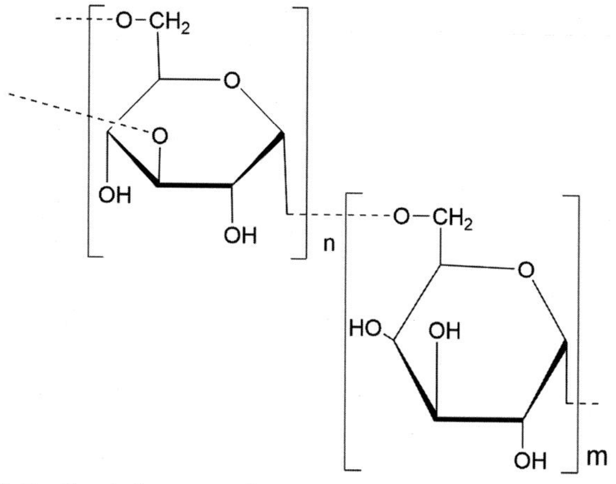

FIG. 5 Chemical structure of dextran.

showed extended release for up to 4 weeks [45]. Enhanced cytotoxicity of doxorubicin against Glioma C6 cancer cells was observed when doxorubicin was delivered from a PLGA-based polymeric system [46] (Fig. 6).

2.2.2 *Poly(ϵ-caprolactone) (PCL)*

Polycaprolactone is a widely used biodegradable, semicrystalline, and hydrophobic polymer having low transition temperature, which is commonly used in polymeric drug delivery systems [47, 48]. Moreover, since PCL can easily blend with other polymers, it has been reported to entrap various drugs both in its native or blended form. It is an interesting candidate for the preparation of long-term implants due to its slower degradation compared to PLA [49]. PCL microparticles/microspheres have been prepared using different techniques and there are several applications of PCL fields for long-term drug delivery [50]. PHBV/PCL microparticles were prepared showing high drug-loading efficiencies, controlled release of resveratrol, and can be used as an alternative in chronic diseases prevention [51] (Fig. 7).

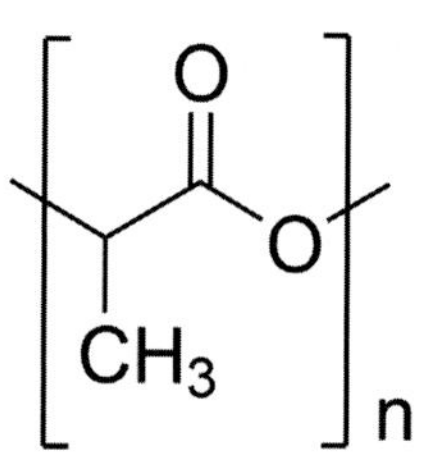

FIG. 6 Chemical structure of polylactic acid.

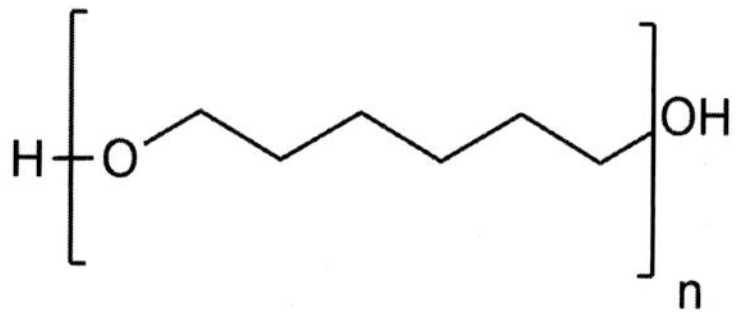

FIG. 7 Chemical structure of poly(ϵ-caprolactone).

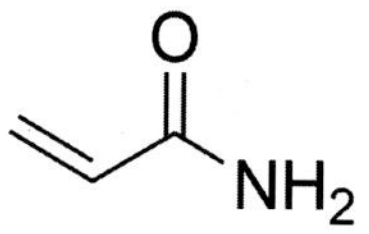

FIG. 8 Chemical structure of polyacrylamide.

2.2.3 *Polyacrylamide (PAA)*

Polyacrylamide is a hydrogel produced by cross-linking acrylamide monomers. It is a biocompatible polymer that has porous structure, which allows drug loading [52]. PAA provides excellent bioadhesive properties and therefore is ideal for entrapment of peptides and other water-soluble drugs [53]. PAA is a suitable polymer for drug delivery applications (Fig. 8).

2.2.4 *Polyvinyl alcohol (PVA)*

Polyvinyl alcohol (PVA) is neutral, water-soluble, and hydrogel-forming synthetic polymer having good mechanical properties and biocompatibility. It has been used to develop different drug delivery systems, such as microspheres, floating microspheres, nanoparticles, and mucoadhesive-targeted drug delivery systems [54]. Applications of PVA nanocomposites were reported in bone tissue engineering and drug delivery systems due to its biocompatibility, nontoxicity, noncarcinogenicity, smoothness, and durability [55]. PVA microparticles with size ranging from 150 to more than 1400 μm have been prepared [19]. Spontaneous adsorption of PVA results from aqueous suspensions onto the surface of hydrophobic drugs such as budesonide and betamethasone valerate (BMV) [56] (Fig. 9).

3 Microparticles for the treatment or diagnosis of cancer

Delivery of anticancer drugs entrapped in polymeric microparticles offer a stable medium for the sustained release of drugs into the cancer tissue, reducing the use of large doses of medicines and their adverse effects [57]. Microparticles, as biomarkers of cancer, provide an important and noninvasive method of cancer identification, diagnosis, and tracking to customize and personalize the treatment [58]. Microparticles also have a potential role in patients with asthma, diffused parenchymal lung disease, lung cancer, thromboembolism, and pulmonary arterial hypertension [59]. They can also act as a marker for early detection of endometrial cancer [60]. Microsphere technology

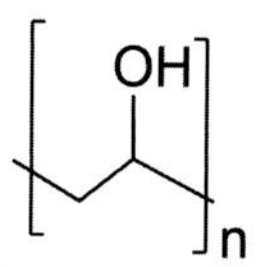

FIG. 9 Chemical structure of polyvinyl alcohol.

is one of the promising approaches that can be used for site-specific intervention without producing major side effects to normal cells [61]. Activation of the Stimulator Interferon Gene (STING) pathway within the microenvironment of the tumor induces a powerful antitumor response that could directly support current STING agonist therapy by reducing the number of doses, decreasing the possibility of metastases, and increasing its applicability to cancers that are difficult to reach [62]. For oral delivery of drugs, hollow microspheres loaded with 5-Florouracil can be a promising chemotherapeutic agent for sustained and controlled drug delivery. It showed high drug loading, excellent floating characteristics, increased absorption, and oral bioavailability with extended $t_{1/2}$, MRT, and improved tissue distribution [63].

Layer-by-layer magnetic platforms have also been developed using a magnetic field to develop fast, simple, and nondestructive methods for early cancer detection [64]. Extracellular vesicles (EVs) can also constitute novel diagnostic targets for particular tumors, where their contents are diagnostic targets indicating the existence (or absence) of a tumor. The EVs themselves or the EV cargo molecules, may be used to observe cancer risk, the survival of cancer, and cancer cure or relapse as an analyst of treatment outcomes. In this age of personalized medicine, the production of EVs in cancer offers more precise, less toxic, and high-efficiency cancer therapies [65].

Chen et al. reported the quantitative analysis of microparticles isolated from different biological materials and the ability of human urinary microparticles has been identified in the noninvasive diagnosis of bladder cancer [66].

Bi_2Se_3 nanodots and doxorubicin hydrochloride (DOX) coembedded tumor cell-derived microparticles (Bi_2Se_3/DOX@MPs) were successfully prepared, which showed great potential against small hemolytic activity, considerable metabolizability, and low systemic toxicity for tumor therostics [67].

3.1 Lung cancer

Microparticles formulated by w/o/w double emulsion-solvent evaporation method showed results that make microparticles a promising candidate to treat lung cancer. According to Li et al., porous poly(cyclohexane-1,4-diyl acetone dimethylene ketal)(PCADK)/poly(D,L-lactide-*co*-glycolide) (PLGA) mixed-matrix porous microspheres containing doxorubicin can be a promising system for lung cancer treatment by pulmonary administration. The porous structure showed excellent lung deposition and also a stronger antiproliferative effect [68]. Kim et al. reported that doxorubicin-loaded PLGA microparticles had appropriate aerosolization, strong characteristics of encapsulation, and also avoided phagocytosis. These microparticles were found to be a potential long-term prolonged-release inhalation agent for the therapy of lung tumors [69]. Novel, inhalable, bioresponsive, large porous stealth lipid MPs were formed by Abdelaziz et al., as DPI microparticles for deep lung deposition. However, these microparticles suffered from opsonization by alveolar macrophages and cohesiveness with impaired aerosolization [70]. Single-photo emission computed tomography (SPECT) can be used for lung cancer imaging examinations where etoposide microparticles labeled with 99mTc were used as micro-radiopharmaceutical for early detection of lung cancer. Particularly, SPECT may be used for lung cancer imaging examinations [71].

3.2 Breast cancer

Microspheres synthesized via emulsion/solvent extraction method by Defail et al. were used for controlled delivery of DOX to maintain local levels of drug. A cytotoxic effect on 4T1 tumor cells was shown by both microspheres alone and microspheres introduced into gelatin [72]. When microsphere was prepared via suspension cross-linking method, the intratumoral mitoxantrone significantly enhanced the survival rate and reduced systemic toxicity. This makes intratumoral chemotherapy using microspheres a less toxic therapy for breast cancer and in particular a modality for neoadjuvant (preoperative) therapy [73]. Sanchez et al. reported that when cannabidiol-in-solution (CBD sol) encapsulated in polymeric microparticles with PTX and DOX using O/W emulsion-solvent evaporation technique in breast cancer treatment, in vitro and in vivo findings demonstrated potential use of CBD-loaded microparticles in conjunction with PTX or DOX in the treatment of both estrogen-receptor-positive and triple-negative breast cancer [74].

3.3 Colon cancer

Hydrogel beads were synthesized via ionotropic gelation technique and the tissue distribution tests showed a high amount of 5-FU in colon, a high degree of drug entrapment efficiency, and an improved anticancer activity of the neoplastic drug. Prolonged release of 5-FU up to 16h in the colonic area and IC50 11.50 μg/mL against HT-29 colon cancer cell line was reported. This makes hydrogel beads an appropriate carrier system for targeting drugs to the colon [75]. Enteric-coated HPMC capsules sealed with 5-FU loaded microsponges and beads of calcium pectinate have been shown to be a successful technique for the delivery of colon-targeted drugs for colorectal cancer treatment [76]. The guar gum microspheres of methotrexate are a possible colon delivery system for colorectal cancer chemotherapy as these particles delay the release of MTX efficiently until they reach the colon [77]. Zhao et al. reported that for the prevention of clotting in patients with colon cancer, the hypercoagulability of phosphatidylserine-positive platelets and microparticles (MPs) can be a possible therapeutic target [78]. Alhakamy et al. prepared simvastatin microparticles (STVMPs) with significant release at pH7.4 and muco-adhesion to the colonic tissues, which enhanced the cytotoxicity and proapoptotic activity against colon cancer cells and improved targeting [79]. Another study reports that in comparison to raw ellagic acid, chitosan loaded with ellagic acid and treated with Eudragit S100 demonstrated better colon targeting and enhanced cytotoxic and proapoptotic activity against HCT 116 colon cancer cell line. Valdecoxib microspheres formulated with chitosan as a core and coated with Eudragit S100 could also be used for colon targeting of drugs [80, 81].

3.4 Brain cancer

It is possible to target drugs in the CNS through the implantation of biodegradable microspheres that provide numerous advantages as discussed by Menei et al. Methotrexate-loaded magnetic microspheres were synthesized via a spacer, amino-hexanol, by covalently linking magnetic microspheres with methotrexate. The formulation resulted in high concentrations of drug in brain tumor [82]. In solid cancers, including meningioma, chordoma, and numerous noncentral nervous system-derived carcinomas, verteporfin can be used when enclosed in PLGA-based microparticles. These methods serve as effective therapeutic interventions for different cancers of brain and ultimately leading to improved patient outcomes [83].

3.5 Ovarian cancer

The biomarker used predominantly to diagnose and predict ovarian cancer is CA125. Isolation and analysis of circulating microvesicles (MVs) released into body fluids exposed to primary tumors (e.g., blood, urine, saliva, ascites, pleural effusion, and spinal fluid) can provide an opportunity to evaluate biological information related to pathology and cancer [84]. PTX-SLMPs have an ability with higher toxicity to SKOV-3 ovarian cancer cell lines. The core-shell structure of PTX-SLMPs will improve the passive tumor targeting of PTX not only in the therapy of ovarian cancer, but also in another highly encapsulatory and sustained release profiles of peritoneal cancers [85]. Microbubbles, when prepared by modified emulsification process can dramatically increase local oxygen release and may effectively increase antiproliferative activity and induce cell apoptosis in hypoxic ovarian cancer cells [86]. The toxicity and compliance-related issues that have restricted the effectiveness of intraperitoneal treatment can be solved by tumor-penetrating microparticles (TPM). It is a valuable technique for ovarian cancer as it has demonstrated higher yield, better tumor targeting, sustained release, lower host toxicity, and improved therapeutic efficacy [87, 88].

4 Conclusion

Microparticles play a very important role for the treatment and diagnosis of various cancers. Microparticles have shown promising results in various types of cancers, which include brain cancer, lung cancer, colon cancer, breast cancer, and ovarian cancer, etc. Microparticles in the form of DPI can be used for deep lung deposition. Microparticles as drug delivery system can also help in reducing dose of chemotherapeutic drugs, which will reduce the side effects. Thus it may be concluded that microparticles are promising drug delivery systems, which can be further explored for the treatment and diagnosis of cancer.

References

[1] Campos E, et al. Designing polymeric microparticles for biomedical and industrial applications. Eur Polym J 2013;49(8):2005–21.
[2] Chandna A, et al. A review on target drug delivery: magnetic microspheres. J Acute Dis 2013;2(3):189–95.
[3] Li SP, et al. Recent advances in microencapsulation technology and equipment. Drug Dev Ind Pharm 1988;14(2–3):353–76.
[4] Aggarwal A, Chhajer P, Maheshwari S. Magnetic drug delivery in therapeutics. Int J Pharm Sci Res 2012;3(12):4670.
[5] Bale S, et al. Overview on therapeutic applications of microparticulate drug delivery systems. Crit Rev Ther Drug Carrier Syst 2016;33(4).
[6] Beyatricks KJ, et al. Recent trends in microsphere drug delivery system and its therapeutic application—a review. Crit Rev Pharm Sci 2013;2(1).
[7] Lengyel M, et al. Microparticles, microspheres, and microcapsules for advanced drug delivery. Sci Pharm 2019;87(3):20.
[8] Madhav NVS, Kala S. Review on microparticulate drug delivery system. Int J PharmTech Res 2011;3(3):1242–4.
[9] Stack M, et al. Electrospun nanofibers for drug delivery. In: Electrospinning: Nanofabrication and applications. Elsevier; 2019. p. 735–64.
[10] Padalkar AN, Shahi SR, Thube MW. Microparticles: an approach for betterment of drug delivery system. Int J Pharm Res Dev 2011;1:99–115.
[11] Shivani Sujitha H. Review article on microparticles. Int J Pharm Anal Res 2015;4(3):302–9.
[12] Rahimpour Y, Javadzadeh Y, Hamishehkar H. Solid lipid microparticles for enhanced dermal delivery of tetracycline HCl. Colloids Surf B: Biointerfaces 2016;145:14–20.
[13] Gong J, et al. Microparticles and their emerging role in cancer multidrug resistance. Cancer Treat Rev 2012;38(3):226–34.
[14] Serrano-Ruiz D, et al. Hybrid microparticles for drug delivery and magnetic resonance imaging. J Biomed Mater Res B Appl Biomater 2013;101(4):498–505.
[15] Suri S, et al. Microparticles and nanoparticles. In: Biomaterials science. Elsevier; 2013. p. 360–88.
[16] Birnbaum DT, Brannon-Peppas L. Microparticle drug delivery systems. In: Drug delivery systems in cancer therapy. Springer; 2004. p. 117–35.
[17] Vroman I, Tighzert L. Biodegradable polymers. Materials 2009;2:307–44.
[18] Lehr C-M, et al. In vitro evaluation of mucoadhesive properties of chitosan and some other natural polymers. Int J Pharm 1992;78(1–3):43–8.

[19] Ficek BJ, Peppas NA. Novel preparation of poly (vinyl alcohol) microparticles without crosslinking agent for controlled drug delivery of proteins. J Control Release 1993;27(3):259–64.
[20] Fathima A, Vedha Hari BN, Ramya Devi D. Micro particulate drug delivery system for anti-retroviral drugs: a review. Int J Pharm Clin Res 2011;4(2).
[21] Lee O-S, et al. Studies on the pH-dependent swelling properties and morphologies of chitosan/calcium-alginate complexed beads. Macromol Chem Phys 1997;198(9):2971–6.
[22] Agnihotri SA, Mallikarjuna NN, Aminabhavi TM. Recent advances on chitosan-based micro-and nanoparticles in drug delivery. J Control Release 2004;100(1):5–28.
[23] Cheung RCF, et al. Chitosan: an update on potential biomedical and pharmaceutical applications. Mar Drugs 2015;13(8):5156–86.
[24] Ko JA, et al. Preparation and characterization of chitosan microparticles intended for controlled drug delivery. Int J Pharm 2002;249(1–2):165–74.
[25] Mi F-L, Sung H-W, Shyu S-S. Release of indomethacin from a novel chitosan microsphere prepared by a naturally occurring crosslinker: examination of crosslinking and polycation-anionic drug interaction. J Appl Polym Sci 2001;81(7):1700–11.
[26] Martinsen A, Skjak-Braek G, Smidsrod O. Alginate as immobilization material: I. Correlation between chemical and physical properties of alginate gel beads. Biotechnol Bioeng 1989;33(1):79–89.
[27] Fundueanu G, et al. Physico-chemical characterization of ca-alginate microparticles produced with different methods. Biomaterials 1999;20(15):1427–35.
[28] Tu J, et al. Alginate microparticles prepared by spray-coagulation method: preparation, drug loading and release characterization. Int J Pharm 2005;303(1–2):171–81.
[29] Murtaza G, Waseem A, Hussain I. Alginate microparticles for biodelivery: a review. Afr J Pharm Pharmacol 2011;5(25):2726–37.
[30] Tabata Y, Ikada Y. Protein release from gelatin matrices. Adv Drug Deliv Rev 1998;31(3):287–301.
[31] Bruschi ML, et al. Gelatin microparticles containing propolis obtained by spray-drying technique: preparation and characterization. Int J Pharm 2003;264(1–2):45–55.
[32] Akin H, Hasirci N. Preparation and characterization of crosslinked gelatin microspheres. J Appl Polym Sci 1995;58(1):95–100.
[33] Phromsopha T, Baimark Y. Preparation of starch/gelatin blend microparticles by a water-in-oil emulsion method for controlled release drug delivery. Int J Biomater 2014;2014, 829490.
[34] Morita T, et al. Preparation of gelatin microparticles by co-lyophilization with poly (ethylene glycol): characterization and application to entrapment into biodegradable microspheres. Int J Pharm 2001;219(1–2):127–37.
[35] Dhaneshwar SS, et al. Dextran: a promising macromolecular drug carrier. Indian J Pharm Sci 2006;68(6):705.
[36] Bachelder EM, Pino EN, Ainslie KM. Acetalated dextran: a tunable and acid-labile biopolymer with facile synthesis and a range of applications. Chem Rev 2017;117(3):1915–26.
[37] Bachelder EM, et al. Acetal-derivatized dextran: an acid-responsive biodegradable material for therapeutic applications. J Am Chem Soc 2008;130(32):10494–5.
[38] Cadee JA, et al. In vivo biocompatibility of dextran-based hydrogels. J Biomed Mater Res 2000;50(3):397–404.
[39] Garlotta D. A literature review of poly (lactic acid). J Polym Environ 2001;9(2):63–84.
[40] Lee BK, Yun Y, Park K. PLA micro-and nano-particles. Adv Drug Deliv Rev 2016;107:176–91.
[41] Makadia HK, Siegel SJ. Poly lactic-co-glycolic acid (PLGA) as biodegradable controlled drug delivery carrier. Polymers 2011;3(3):1377–97.
[42] Anderson JM, Shive MS. Biodegradation and biocompatibility of PLA and PLGA microspheres. Adv Drug Deliv Rev 2012;64:72–82.
[43] Blasi P. Poly (lactic acid)/poly (lactic-co-glycolic acid)-based microparticles: an overview. J Pharm Investig 2019;1–10.
[44] Ruan G, Feng S-S. Preparation and characterization of poly (lactic acid)-poly (ethylene glycol)-poly (lactic acid)(PLA-PEG-PLA) microspheres for controlled release of paclitaxel. Biomaterials 2003;24(27):5037–44.
[45] Martin TM, et al. Preparation of budesonide and budesonide-PLA microparticles using supercritical fluid precipitation technology. AAPS PharmSciTech 2002;3(3):16–26.
[46] Lin R, Ng LS, Wang C-H. In vitro study of anticancer drug doxorubicin in PLGA-based microparticles. Biomaterials 2005;26(21):4476–85.
[47] Sinha VR, et al. Poly-ϵ-caprolactone microspheres and nanospheres: an overview. Int J Pharm 2004;278(1):1–23.
[48] Park J, Ye M, Park K. Biodegradable polymers for microencapsulation of drugs. Molecules 2005;10(1):146–61.
[49] Kumari A, Yadav SK, Yadav SC. Biodegradable polymeric nanoparticles based drug delivery systems. Colloids Surf B: Biointerfaces 2010;75(1):1–18.
[50] Azimi B, et al. Poly (ϵ-caprolactone) fiber: an overview. J Eng Fibers Fabr 2014;9(3). https://doi.org/10.1177/155892501400900309.
[51] Mendes JBE, et al. PHBV/PCL microparticles for controlled release of resveratrol: physicochemical characterization, antioxidant potential, and effect on hemolysis of human erythrocytes. Sci World J 2012;2012, 542937.
[52] Risbud MV, Bhonde RR. Polyacrylamide-chitosan hydrogels: in vitro biocompatibility and sustained antibiotic release studies. Drug Deliv 2000;7(2):69–75.
[53] Kriwet B, Walter E, Kissel T. Synthesis of bioadhesive poly (acrylic acid) nano-and microparticles using an inverse emulsion polymerization method for the entrapment of hydrophilic drug candidates. J Control Release 1998;56(1–3):149–58.
[54] Gajra B, et al. Poly vinyl alcohol hydrogel and its pharmaceutical and biomedical applications: a review. Int J Pharm Res 2012;4(2):20–6.
[55] Gaaz TS, et al. Properties and applications of polyvinyl alcohol, halloysite nanotubes and their nanocomposites. Molecules 2015;20(12):22833–47.
[56] Buttini F, et al. Multilayer PVA adsorption onto hydrophobic drug substrates to engineer drug-rich microparticles. Eur J Pharm Sci 2008;33(1):20–8.
[57] Sawyer AJ, Piepmeier JM, Saltzman WM. Cancer issue: new methods for direct delivery of chemotherapy for treating brain tumors. Yale J Biol Med 2006;79(3–4):141.
[58] Gong J, et al. Microparticles in cancer: a review of recent developments and the potential for clinical application. In: Seminars in cell & developmental biology. Elsevier; 2015.
[59] Nieri D, et al. Cell-derived microparticles and the lung. Eur Respir Rev 2016;25(141):266–77.
[60] Dziechciowski M, et al. Diagnostic and prognostic relevance of microparticles in peripheral and uterine blood of patients with endometrial cancer. Ginekol Pol 2018;89(12):682–7.
[61] Rajput MS, Agrawal P. Microspheres in cancer therapy. Indian J Cancer 2010;47(4):458.

[62] Lu X, et al. Engineered PLGA microparticles for long-term, pulsatile release of STING agonist for cancer immunotherapy. Sci Transl Med 2020;12(556).
[63] Huang Y, et al. A 5-fluorouracil-loaded floating gastroretentive hollow microsphere: development, pharmacokinetic in rabbits, and biodistribution in tumor-bearing mice. Drug Des Devel Ther 2016;10:997.
[64] Liu XQ, Picart C. Layer- by-layer assemblies for cancer treatment and diagnosis. Adv Mater 2016;28(6):1295–301.
[65] Jaiswal R, Sedger LM. Intercellular vesicular transfer by exosomes, microparticles and oncosomes-implications for cancer biology and treatments. Front Oncol 2019;9:125.
[66] Chen C-L, et al. Comparative and targeted proteomic analyses of urinary microparticles from bladder cancer and hernia patients. J Proteome Res 2012;11(12):5611–29.
[67] Wang D, et al. Engineered cell-derived microparticles Bi2Se3/DOX@ MPs for imaging guided synergistic photothermal/low-dose chemotherapy of cancer. Adv Sci 2020;7(3):1901293.
[68] Li W, et al. Inhalable functional mixed-polymer microspheres to enhance doxorubicin release behavior for lung cancer treatment. Colloids Surf B: Biointerfaces 2020;196:111350.
[69] Kim I, et al. Doxorubicin-loaded highly porous large PLGA microparticles as a sustained-release inhalation system for the treatment of metastatic lung cancer. Biomaterials 2012;33(22):5574–83.
[70] Abdelaziz HM, et al. Inhalable particulate drug delivery systems for lung cancer therapy: nanoparticles, microparticles, nanocomposites and nanoaggregates. J Control Release 2018;269:374–92.
[71] Salvi R, et al. Diagnosing lung cancer using etoposide microparticles labeled with 99mTc. Artif Cells Nanomed Biotechnol 2018;46(2):341–5.
[72] DeFail AJ, et al. Controlled release of bioactive doxorubicin from microspheres embedded within gelatin scaffolds. J Biomed Mater Res A 2006;79(4):954–62.
[73] Almond BA, et al. Efficacy of mitoxantrone-loaded albumin microspheres for intratumoral chemotherapy of breast cancer. J Control Release 2003;91(1–2):147–55.
[74] Fraguas-Sanchez AI, et al. CBD loaded microparticles as a potential formulation to improve paclitaxel and doxorubicin-based chemotherapy in breast cancer. Int J Pharm 2020;574:118916.
[75] Asnani GP, Kokare CR. In vitro and in vivo evaluation of colon cancer targeted epichlorohydrin crosslinked Portulaca-alginate beads. Biomol Concepts 2018;9(1):190–9.
[76] Gupta A, et al. Enteric coated HPMC capsules plugged with 5-FU loaded microsponges: a potential approach for treatment of colon cancer. Braz J Pharm Sci 2015;51(3):591–605.
[77] Chaurasia M, et al. Cross-linked guar gum microspheres: a viable approach for improved delivery of anticancer drugs for the treatment of colorectal cancer. AAPS PharmSciTech 2006;7(3), E143.
[78] Zhao L, et al. Phosphatidylserine exposing-platelets and microparticles promote procoagulant activity in colon cancer patients. J Exp Clin Cancer Res 2016;35(1):1–12.
[79] Alhakamy NA, et al. Chitosan coated microparticles enhance simvastatin colon targeting and pro-apoptotic activity. Mar Drugs 2020;18(4):226.
[80] Thakral NK, Ray AR, Majumdar DK. Eudragit S-100 entrapped chitosan microspheres of valdecoxib for colon cancer. J Mater Sci Mater Med 2010;21(9):2691–9.
[81] Alhakamy NA, et al. Chitosan-based microparticles enhance ellagic acid's colon targeting and proapoptotic activity. Pharmaceutics 2020;12(7):652.
[82] Devineni D, Klein-Szanto A, Gallo JM. Tissue distribution of methotrexate following administration as a solution and as a magnetic microsphere conjugate in rats bearing brain tumors. J Neuro-Oncol 1995;24(2):143–52.
[83] Shah SR, et al. Verteporfin-loaded polymeric microparticles for intratumoral treatment of brain cancer. Mol Pharm 2019;16(4):1433–43.
[84] Lu Z, et al. Tumor-penetrating microparticles for intraperitoneal therapy of ovarian cancer. J Pharmacol Exp Ther 2008;327(3):673–82.
[85] Han S, et al. Sustained release paclitaxel-loaded core-shell-structured solid lipid microparticles for intraperitoneal chemotherapy of ovarian cancer. Artif Cells Nanomed Biotechnol 2019;47(1):957–67.
[86] Sun J, et al. Ultrasound-mediated destruction of oxygen and paclitaxel loaded lipid microbubbles for combination therapy in hypoxic ovarian cancer cells. Ultrason Sonochem 2016;28:319–26.
[87] Giusti I, D'Ascenzo S, Dolo V. Microvesicles as potential ovarian cancer biomarkers. Biomed Res Int 2012;2013, 703048.
[88] Van Doormaal FF, et al. Cell-derived microvesicles and cancer. Neth J Med 2009;67(7):266–73.

Chapter 29

Microparticles for cancer therapy

Varun Kumar[a,*], Nitesh Kumar[a,*], Akansha Mehra[a,*], Priya Shrivastava[a,*], and Pawan Kumar Maurya[b]
[a]*Amity Institute for Advanced Research and Studies (Materials & Devices), Amity University, Noida, India,* [b]*Department of Biochemistry, Central University of Haryana, Mahendergarh, Haryana, India*

1 Introduction

Multiparticulate drug delivery systems provides us with both therapeutic and technological benefits because of the some useful elements like microparticles (MPs), microspheres, and microcapsules. The size of MPs usually lies in 1–1000 μm range [1]. They have distinct physiological and pharmacokinetic advantages ameliorating the potency, appropriateness, and patient conformity.

On the basis of formulation, they can be blended to form various pharmaceutical dosage, including solids, semisolids, or liquids. Solids comprise of capsules, tablets, sachets, semisolids include gels, creams, pastes and liquid form covers solutions, suspensions, and parenterals. Microcarrier offers an advantage over nanoparticles by acting locally because of their huge size, which restricts them to not overpass interstitium. Perhaps, encapsulation can be used to carry toxic substances and solid MPs can be used to handle liquids. Multiparticulates offer an exquisite superiority for the distribution of dose in various tiny separate particles that release a portion of the dose without leading to breakdown of the entire dosage if at any point single subunit gets collapsed.

1.1 Advantages of Microparticulate drug delivery system [1]

- Alternative of dosage form for the preferred route of drug delivery
- Improved and site-specific drug release
- More uniform administration in anatomical conditions
- Sturdy and fastened dose fusion of drugs
- Dose titration and minimized dose-dumping
- Better and stabilized medicinal preparations
- Constituent separation ensuring stronger compatibility
- Innovative products with an enhanced shelf life

1.2 Characterization of microparticles

The essential characterization parameters for MPs are as follows:

1.2.1 Mean particle size analysis and size distribution

The size of MPs can be easily tailored and adjusted to achieve drug targeting. The mean size of MPs and size distribution are the important characteristic features since they direct saturation solubility and biological performance of MPs. Photon correlation spectroscopy can be employed to measure average size of MPs [2].

1.2.2 Morphology and crystalline state of the microparticles

Crystalline state and particle morphology help us to understand better the morphological changes associated with drug when it is subjected to microsizing. Besides, amorphous particle generation in drug also takes place while preparing MPs, which can be further determined using X-ray diffraction analysis. Scanning electron microscopy is preferred for actual determination of particle morphology [2].

* Equal contributions.

Advanced Drug Delivery Systems in the Management of Cancer. https://doi.org/10.1016/B978-0-323-85503-7.00019-5

1.2.3 Saturation solubility and dissolution velocity of the microparticles

These parameters are important to study to assess any change in in vivo performance of the drug. The saturation solubility can be checked in various physiological buffers using different methods of assessment available. The dissolution velocity helps in understanding the benefits of latest methods of released dosage forms over conventional methods of formulations [2].

1.3 Morphology and structure

Size of MPs ranges from 1 to 1000 μm and they exist in a familiar structure having different forms. Apart from the excipients used, structure as well as shape of MPs affect their function. These systems of drug delivery catch eye because of broad spectrum of beneficial technological distinctiveness. On the basis of formulation and involved steps of processing, characterization of MPs may be done as either a homogenous or heterogenous assembly. Since the spheroid shape makes the further processing easier, it is usually chosen.

1.3.1 Microspheres and microcapsules

Microspheres can be characterized as matrix systems in which the drug is homogeneously dispersed, either dissolved or suspended. An ideal microsphere contains slid or liquid API mixed in a matrix.

Microcapsules come under the category of heterogeneous structure where a membrane shell surrounds the core forming a reservoir to carry drug particles. They are reservoirs of microscopic size by a wall that is able to control the release from the reservoir.

1.3.2 Liposomes

They are basically lipid vesicles containing one or more phospholipid bilayers and their configuration is comprised of small unilamellar vesicles (SUV: 20–100 nm), large unilamellar vesicles (LUV: > 100 nm), multilamellar vesicles (MLV: > 500 nm), oligolamellar (OLV: 0.1–1 μm), giant unilamellar liposomes (GUV: > 1 μm), and multivesicular vesicles (MVV: > 1 μm). They perform a cell membrane-like structure acting as an appropriate drug delivery carriers for most of the APIs (hydrophilic and hydrophobic). Several underlying mechanisms are there via which liposomes can deliver drug into the cells:

- liposomes interaction with the components of cell surface,
- fuse with the cell membrane,
- through endocytic pathway via phagocytic cells, and
- entry via bilayer components of cells.

1.3.3 Colloidosomes

They are microcapsules containing hollow or hydrogel core with self-assembled colloidal particles embodied in their wall. Their sizes range between 10 and 20 nm to micrometers. Selective permeability is one of the main features of colloidosomes, where drugs diffusion occurred via size exclusion.

1.4 Various approaches in the fabrication of microparticles

Biodegradability, biocompatibility, and target drug delivery should always be the characteristics of efficient MPs. Depending upon the type of polymer used and method of preparation, the characteristics might change. The various approaches to fabricate the MPs are mentioned in the following section [3].

1.4.1 Spray-drying technique

In the spraying-drying method, APIs are dissolved in a polymer solution (either organic or aqueous solvent, as per the polymer used). The suspension/solution is allowed to put through nozzle into apparatus used for spray-drying. Mixing of solution takes place in the presence of air. Droplets formed due to nebulization at nozzle, gets evaporated under high pressure and can be collected later on. This technique of MPs fabrication offers high encapsulation efficiencies, leaving no residual surfactant on the surface of the MPs [4].

1.4.2 Coacervation

This method of preparation of microspheres and microcapsules is considered as one of the traditional methods and is related to the preparation of calcium alginate. Calcium ions form cross-links between the α-L-glucuronic acid and β-D-mannuronic acid units of alginate, thus organizing the polymer chains into an "eggbox-structure." This chemical bonding contributes to the possibility of establishing either a stable microspheres or core-shelled microcapsules [1].

1.4.3 Solvent evaporation/solvent extraction

In this method of MPs preparation, the polymer is usually a hydrophobic polyester. Polymers are allowed to dissolve in a water-insoluble volatile organic solvent, into which the central material is also dispersed. The prepared solution is added dropwise to a stirring aqueous having stabilizer like polyvinyl alcohol or polyvinylpyrrolidone leading to formation of polymer droplets containing encapsulated material. By this time, the microdroplets are hardened to prepare the resultant polymer microcapsules. This hardening step is achieved by the solvent removal from the polymer droplets either by solvent evaporation (by heat or reduced pressure) or by solvent extraction (with a third liquid which is a precipitant for the polymer and soluble in both aqueous and nonaqueous solvent). The process of solvent extraction forms microcapsules with higher porosities than those formed from solvent evaporation of microencapsulation via solvent evaporation process. Solvent evaporation/extraction is the most appropriate processes for the fabrication of drug-loaded microcapsules made up with the biodegradable polyesters like polylactide, lactidecoglycolide, and polyhydroxybutyrate [5].

1.4.4 Polymerization techniques

The polymerization techniques are generally classified as:

(i) normal polymerization
(ii) interfacial polymerization

Normal polymerization: Techniques such as suspension, emulsion, and bulk are used to carry out normal polymerization. A continuous aqueous phase is employed to carry out microdroplets heating in suspension technique. This approach is also termed as bead or pearl polymerization. The main example of emulsion polymerization is nylon microcapsules.

Bulk polymerization: Bulk polymerization involves heating of one or more monomer units under presence of catalyst for the initiation of polymerization. This form of polymerization is most suitable in the formation of pure polymers [6].

2 Microparticles: An efficient tool in drug delivery

The targeted drug mechanism aims at focusing the majority of active agents in inflamed tissues by using specific delivery to achieve therapeutic potential while limiting adverse effects at the same time. Additionally, such site specific delivery system must be capable enough to cause complete biodegradation and high biocompatibility without evoking any proinflammatory responses.

Drug delivery system for controlled release includes carriers like polymer-based, MPs, nanoparticles, etc. Drug encapsulated pellets that release at controlled rates for prolonged period is the most efficient approach to deal with the difficulties associated with other approaches of drug administration.

2.1 Advantages associated with controlled drug delivery system

2.1.1 Adjustable drug release

Rate of drug release can be adjusted according to the needs of a specific application, for example, whether an application demands drug delivery at uniform rate or at pulsatile rate.

2.1.2 No drug degradation

Controlled drug release system also protects drugs, especially proteins, from degradation that are otherwise rapidly destroyed by the body.

2.1.3 Comfort of patient

Daily drug administration can be substituted by giving once per month injection, which eventually favors patient comfort and compliance.

Selection of carrier is very important for particular drug, i.e., either hydrophilic or lipophilic. This is based on the diseased conditions as well as on the physicochemical properties of administered drug [3].

2.2 Mechanism involved in drug release from microparticulates

Various mechanisms and phenomena influence the drug release profile from microparticulates that involves dissolution/diffusion, osmotic release, and erosion (Fig. 1). These mechanisms have stronger impact and greater roles during drug release.

In the microparticulate system, when API is encapsulated in a polymer core, the features of the polymer system are important for dissolution, but influenced by many factors, including properties of drug, formulation, release environment, etc.

Whereas in polymer matrix, the drug diffusion can be through the intact polymer system or through the pores filled with water. Water-soluble drugs may also dissolve in the aqueous pore networks. Water uptake causes polymer chains to swell, indicating the development of new pores and/or osmotic pressure. During the process of swelling, volume increases, increase in the effective diffusion coefficient of the drug and other pharmacon molecules comes into the aqueous part. The polymer matrix (bulk/surface) may also erode.

Further, the film-forming polymer-coated MPs may dissolve in the medium or act as a water-immiscible, permeable, or semipermeable membrane. In the former case, the diffusion is primarily due to the release of the drug. In semipermeable coating, the osmotic process should be taken into consideration. It is also possible to use water-soluble pore formers, which, by creating pores, accelerate the dissolution profile.

In the case of smart drug delivery MP systems, the release of the drug occurs via a stimulus. It can be possible that one or multiple stimuli are needed for dissolution of polymer system. The stimulus for drug release from MPs could be internal or external and can be categorized as physical, chemical, or even microbiological. Opening and closing signals of these polymer systems may also possible for generating feedback [1]. Some more mechanisms are explained in the following section [3].

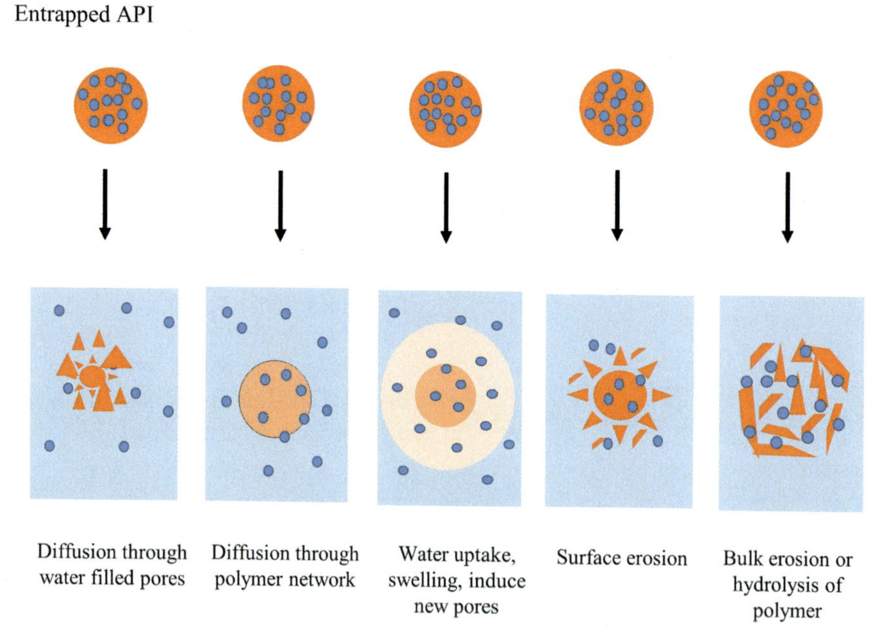

FIG. 1 Schematic representation of mechanism involved in drug release from microparticles.

2.2.1 Diffusion-controlled reservoir system

This system involve a controlling membrane that is present around the drug, through which drug get released out. Here, matrix degradation doesn't affect with drug release rate. After the complete diffusion, membrane gets eroded.

2.2.2 Diffusion-controlled monolithic system

In this system, drug is dispersed in polymer matrix. The drug starts diffusing out along with the degradation of polymer core. Here, matrix degradation highly affects the release rate of drug.

2.2.3 Degradation-controlled monolithic system

This system involves uniform dispersion of drug into the polymer core and diffusion rate depends upon degradation of the polymer matrix. Generally, diffusion rate is slow as compared to the degradation of polymer matrix.

3 Advances in microparticle technology for cancer therapy

The MP technology has experienced a series of changes to deal with the current issues associated with cancer therapy, and some of the advances related to fabrication, encapsulation, and coating of MPs are discussed in this section. Complex MPs with unique characteristics, including various morphologies, desired sizes, and multicompartments have been reported by many scientists in the past decades. Conversely, established methods for the fabrication of MPs are lyophilization, freeze-drying, spray-drying, ionotropic gelation, emulsion, and coacervation, which have an inadequacy in achieving the MPs with desired characteristics (Fig. 2). Recently, several methods such as the flow lithography and microdroplet (microfluidics-assisted), electrohydrodynamics (EHDs), centrifugation, and template-based methods have been employed in the preparation of MPs [7]. The unique characteristics of the complex MPs prepared through these suggested approaches lead to the enormous applications, including drug delivery vehicles, microsensors, catalyst substrate, and cell carriers [8].

To enhance the functionality of engineered MPs in multiplexing assay, cell or drug delivery, as well as self-assembling building blocks, various fabrication approaches have been recommended Table 1. The fabrication approaches are broadly classified into two main categories (Fig. 2):

1. Microfluidics-assisted methods
   - Droplet-based
   - Flow lithography-based
2. Nonmicrofluidics-assisted methods
   - Centrifugation-based
   - EHD-based
   - Template-based

3.1 Advances in the MPs engineering through Microfluidics-assisted approach

3.1.1 Droplet-based

In general, the droplet-based method is a simple and precise method that controls the shape, size, and substance of the fabricated MPs. The droplet-based method involves shear

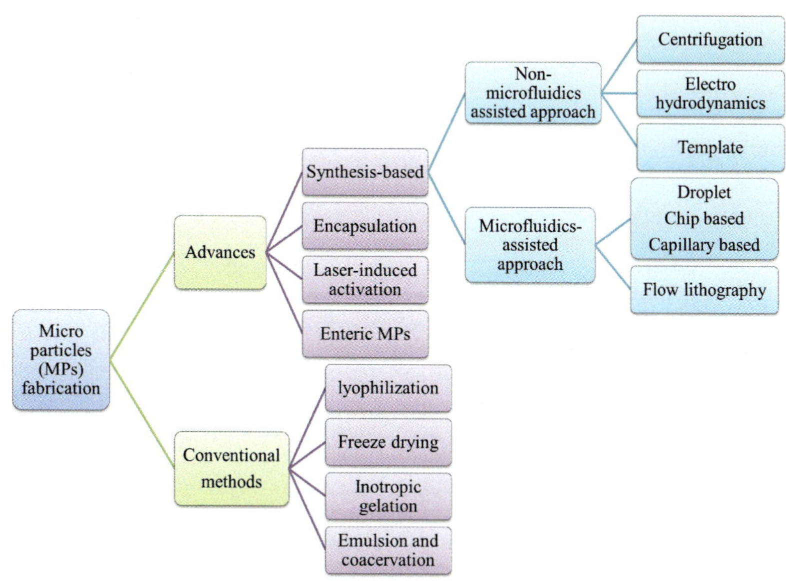

FIG. 2 Advances in the fabrication, encapsulation, and activation of MPs for the cancer therapy.

TABLE 1 Diverse advances in the engineering of microparticles (MPs) for cancer therapy.

Approach	Methods	Pros and Cons	Morphology	Size (μm)
Microfluidics-assisted	Droplet-based	Pros: simple, escape complicated systems, Uniform products Cons: production rate is very low, morphological flexibility is restricted	Spherical, Spherical (core-shell) and Janus	2.5–500
	Flow lithography -based	Pros: unrestricted morphological complexity Cons: need for intricate systems	Complex 3D geometries and Barcode shaped	70–400
Nonmicrofluidics-assisted	Centrifugation-based	Pros: comparatively short operational time, easy systems, MPs monodispersity rate is high Con: limited control over morphology	Spherical, ellipses, fibers, and Janus	15–300
	Electro hydrodynamics-based	Pro: limited morphological complexity, target specific control Cons: less control over morphological flexibility	Spherical and cylindrical	0.1–20
	Template-based	Pros: unrestricted morphological complexity Cons: need for template	Complex 3D geometries, spherical and cylindrical	0.1–300

force persuaded inside the microfluidic device by immiscible fluids to generate microdroplets and hardened into the MPs. In this method, two basic types of microfluidic devices that have been employed [8] are as follows:

- a chip-based-microfluidic device made up of T-/Ψ-shaped microchannels inside the chip and
- a capillary-based-microfluidic device in which microcapillaries are aligned in a coaxillary fashion.

The basic principle behind the generation of microdroplets in this method is pinching of the flowing dispersed phase in the microchannel by an immiscible phase under the influence of shear force (Fig. 3). Furthermore, several modified microfluidic devices for the synthesis of MPs have been proposed so far by integrating two or more basic microfluidic devices. For example, study conducted by Nisisako and coworkers synthesized core-shell structured MPs by cascading more than one T-shaped microfluidic device [8].

3.1.2 Flow lithography-based

Droplet-based microfluidics can only synthesize spherical-shaped MPs due to the minimization of surface energy. To deal with the issue associated with the synthesis of MPs through the droplet-based method, a predefined geometry is introduced within the microfluidic channel in the initial stage to synthesize rod and disk-shaped MPs [9]. Moreover,

Droplet method

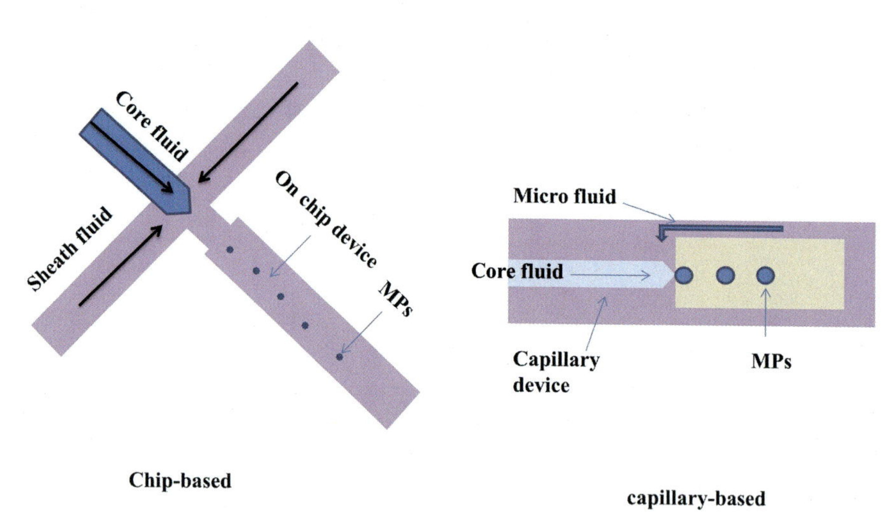

FIG. 3 MPs synthesis via droplet-based method.

Flow lithography

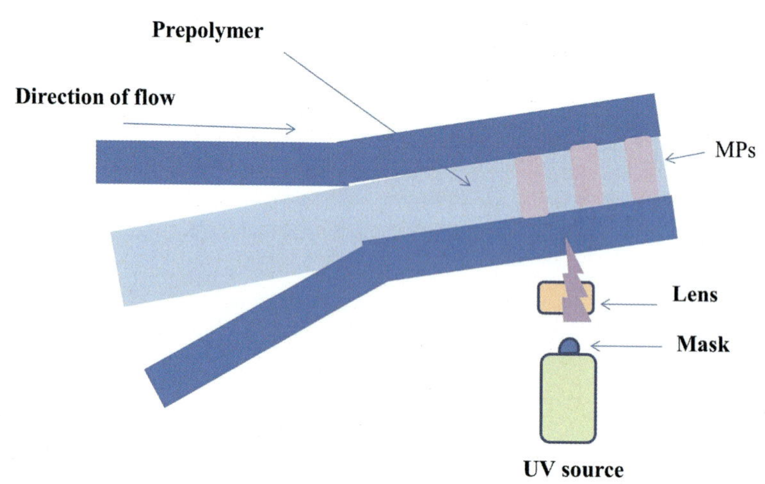

FIG. 4 MPs synthesis via flow lithography-based method.

multicompartment MPs are fabricated with the aid of convoluted microfluidic devices containing multiple inlets though it is difficult to handle and find an ideal platform for the fabrication of complex MPs (Fig. 4). To overcome the limitations, Doyle's group [10] and Kwon's [11] simultaneously introduced the concept of flow lithography-based microfluidics. This was the only method that provides a degree of freedom to the shape of MPs but restricted up to the fabrication of 2D MPs due to the height limits of microfluidic channels. Therefore several studies have been conducted with the use of innovative channels-based flow lithographic fabrication of the 3D MPs. Paulsen et al. recommended optofluidic fabrication to develop engineered 3D MPs comprising of two combined processes, UV light polymerization and inertial low shaping [12].

3.1.3 *Parallelization (scaling-up)*

Microfluidics-assisted fabrication of the complex MPs at a laboratory scale can only produce 100 g/day. For instance, the fabrication of 100 μm sized MPs with a breakup rate of 103 micro drops/s^{-1} resembles an approximate output of 50 g/day [13]. Thus fabrication of the complex MPs using microfluidics-assisted devices is an exclusively challenging task in comparison with the advanced technology like emulsion polymerization. Hence, the fabrication process needs to be scaled up for commercial production of the complex MPs. There is the requirement of kg/h production of the MPs to meet the industrial demands. The fabrication of MPs by a single microfluidic device is restricted due to its upper limit; therefore, the parallelization of a single microfluidic device is imperative in scaling up the MPs.

3.2 Advances in MPs engineering through Nonmicrofluidics-assisted approach

3.2.1 *Centrifugation-based*

In the centrifugation-based method, centrifugal force is the chief driving force in the generation of MPs. In 2007, Haeberle et al. first introduced the method for fabrication and manipulation of mono-dispersed droplets [14]. Note that the fabrication of multicompartmental MPs is obtained through the immediate solidification of microdroplets in the two miscible and unstable phases of liquids. In brief, two fluorescent dyes along with Na-alginate solutions were introduced into the two barrels of the capillary, and to fix the capillary, an acrylic holder was placed at the bottom of the microtube containing $CaCl_2$ solution. Solutions inside the microtube were centrifuged and streamed by centrifugal force, and finally drained off at the barrel tip (Fig. 5). Thereafter, the synthesized Na-alginate droplets were hardened in the $CaCl_2$ containing a solution. In this system, barrel configuration can be easily tuned up to get the desired compartmentalization and size of the fabricated MPs. However, this centrifugation-based approach can only synthesize spherical-shaped MPs. Recently, Hayakawa et al. introduced the concept of nonequilibrium-prompted microflows for the fabrication of MPs with high complexity and diversity [15]. The two microflows named, diffusional flow and the Marangoni flow, were produced by the differences in the surface tension between the compartments and concentration of monomer tend to manipulate microdrop before the solidification process. The complexity of the morphology can also be increased through the incomplete dissolution of the fabricated MPs in the specific compartment. The short operating time and the simplicity of the centrifugation-based approach showed immense potential in the fabrication of MPs.

3.2.2 *Electrohydrodynamic-based*

The most advanced approach in the fabrication of MPs is the electrohydrodynamic cojetting system. The concept of EHD was introduced by Lahman et al. in support of the economical synthesis of anisotropic MPs concerning the size of various orders of magnitude from 0.1 to 100 μm [8]. This system often comprises of two needles pumping different solutions (polymer) through capillaries with a laminar flow rate. The fluid gladly stretches near the grounded electrode after the application of an external electric field to the system. The solvent evaporates and leaves the fabricated MPs in the solution during the stretching process (Fig. 6). The EHD processes are generally being carried out in a normal atmosphere, increased capillary numbers and controlled process parameters affect the exclusive properties such as surface tension, electrical force, and the speed of solvent

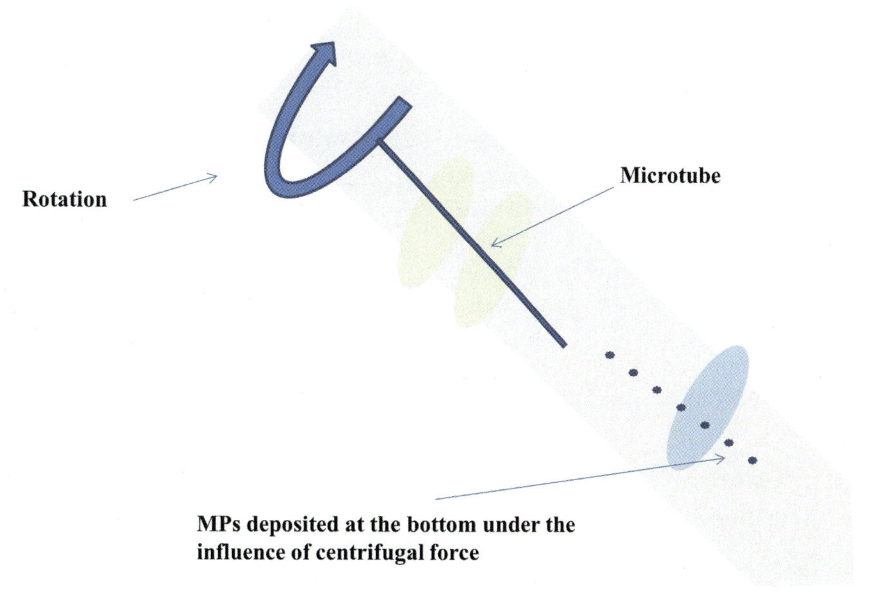

FIG. 5 MPs synthesis via centrifugation-based method.

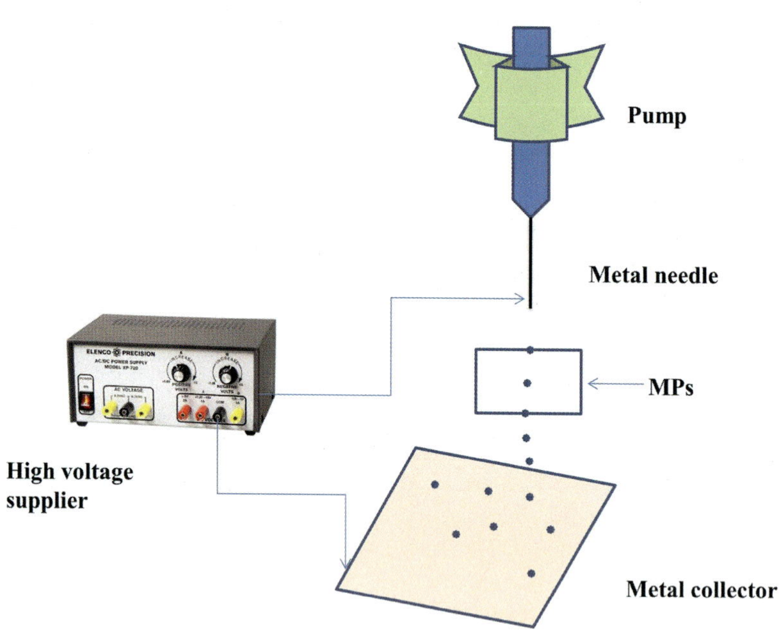

FIG. 6 MPs synthesis via EHD-based method.

evaporation from the system facilitates the synthesis of diverse complex MPs [16]. Lee et al. and Yoon et al. have succeeded in the fabrication of MPs with controlled shape and size by varying different parameters of the EHD process, like polymer viscosity, flow rate, and the polymer concentration.

3.2.3 *Template-based*

This is the most advantageous technique in the fabrication of complex MPs. The template-based method can reproduce the same shape and sized MPs by utilizing a micropatterned master mold. The shape and size of the fabricated MPs are determined by the predefined master mold. The master mold is prepared from the materials such as silicon and glass, a polymer precursor is cast on the master mold for the fabrication of MPs (Fig. 7). The template-based approach can synthesize mono-dispersed 2D MPs with outstanding control over the size range However, the external morphology of the fabricated MPs through template-based technique is limited due to a predefined geometry of the master mold [17]. To deal with this limitation, Lee's et al. has suggested a series of an improved template-based method for the précised fabrication of the MPs in terms of shape and size [8].

3.3 Laser-induced activation of microparticles for image-guided cancer therapy

The laser-induced activation is an emerging trend in the area of MPs and several laser-inducible lipid MPs have been exploited for their applications in cancer therapy, drug targeting, and imaging studies. For instance, poly-lactide-co-glycolic acid (PLGA) MPs are biodegradable and biocompatible and can be employed in drug delivery of many synthetic drugs, targeting moieties, and imaging studies [8]. PLGA MPs were fabricated with the use of gold nanoparticles and dye (DiI), and a core containing perfluorohexane liquid (PFH). The liquid core is activated with the absorption of laser energy by PLGA shell and converting liquid into gas. The converted gas is released from the particle, resulting in the formation of microbubbles that can kill cells and tissue. In vitro studies with the cell culture showed that PLGA MPs are phagocytized by a single cell and are vaporized by $90\,MJ\,cm^{-2}$ laser energies, resulting in cell death [8].

3.4 Advances in MPs encapsulation technology

The encapsulation process is used in the surface modification of the MPs to enhance their overall properties. For instance, multifunctional polymeric PLGA MPs with quantum dots and camptothecin encapsulation have been fabricated as a model for imaging and anticancer therapy. The PLGA-QD MPs are designed in such a way that the absorption and accumulation of the MPs occurred in cancerous cells only; and such difference was unnoticed in the efficient absorption of nanoparticles into the cancerous and noncancerous cells [18]. The PLGA-QD MPs were designed to achieve fluorescent core-shell structure by incorporating QDs into the shell. Such an arrangement renders clear identification of individual MPs in the cancerous cells to analyze their selective and size-based absorption with the aid of three-color-nonoverlapped-fluorescent imaging [18].

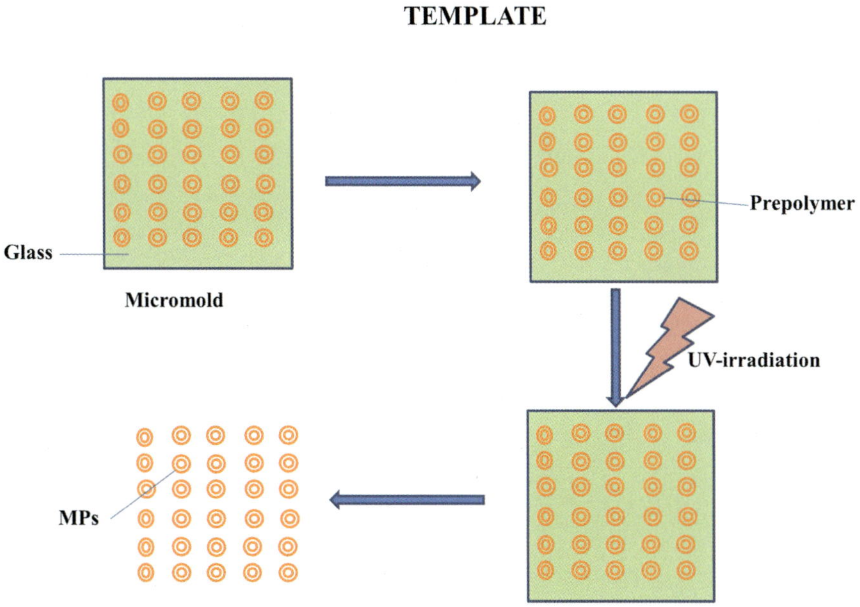

TEMPLATE

Glass
Micromold
Prepolymer
UV-irradiation
MPs

FIG. 7 MPs synthesis via template-based method.

3.4.1 Enteric microparticles

Enteric MPs contain a thin single or multilayer of polymer formulation that provide complete coverage to the pharmaceutical ingredients. The polyelectrolyte-coated MPs may serve the following purposes [19, 20]:

- improving stability,
- masking undesirable taste or odor,
- enhancing appearance,
- protecting product from humidity,
- easing consumption, and
- modifying drug release properties.

In 2012, Patil et al. reported the improved oral dosage form for the treatment of IBD by the enteric coating of a bioactive agent; this not only prevents its degradation in the stomach but also allows the release of the drug in the intestinal fluid [21]. Another motive of the enteric coating is to prevent stomach lining from the adverse reactions of nonsteroidal antiinflammatory drugs which may result in ulcer formation [22]. The ionotropic gelation method followed by the coating of poly-cation using polyelectrolyte complexation was first utilized in the preparation of enteric alginate MPs encapsulating bioactive ingredients. This method has been extended to the further process named coacervation or blending to enhance the synthesis of enteric MPs [7].

3.5 Miscellaneous advances in MPs technology

Recently, the supercritical fluids (SCFs) can be utilized in the synthesis of a powdered formulation of MPs either as antisolvent or solvent, which confers various advantages depending upon their properties such as dryness, moderate working temperature, and purity of the synthesized MPs [23]. The most commonly used SCF is supercritical carbon dioxide (SC-CO_2) based on its mechanical and chemical properties working as a spray enhancer. In another approach, dry powder of inhalable magnetically active thermosensitive lipid-based MPs have been explored with the aid of the freeze-drying process followed by jet milling [24]. The transformation of lipid-based MPs from dispersion to redispersible powder was enabled during the extremely low temperature in freeze-drying [25]. For the codelivery of budesonide and SPIONs, glyceryl behenate-based MPs were prepared with the aid of emulsification (o/w) followed by freeze-drying [24]. Then the size was reduced through jet milling to achieve appropriate NMAD (number median aerodynamic diameter) of 2–3 μm. The increased particle-wall and particle-particle impaction because of the high stirring rate in the emulsification process resulted in the breakdown of lipid droplets into MPs. Therefore disaggregation of the MPs was done by the jet-milling process without affecting the state of lipid droplets. Under the influence of the external magnetic field, SPIONs (superparamagnetic iron oxide nanoparticles) can be directed toward the tumor cells for targeted drug delivery and specifically kills tumor cells by the process called magnetic-induced hyperthermia. Moreover, the hyperthermia at the tumor site significantly boosts drug release from the lipid MPs. Jinturkar et al. synthesized liposomal etoposide (ETP) and DTX (pretreated) and administered in lung cancer cells along with the p53 tumor suppressor gene [26, 27].

4 Challenges and limitations of microparticle-based cancer therapy

However, the challenges encountered during the synthesis, encapsulation, burst, and stability of multilayered particles

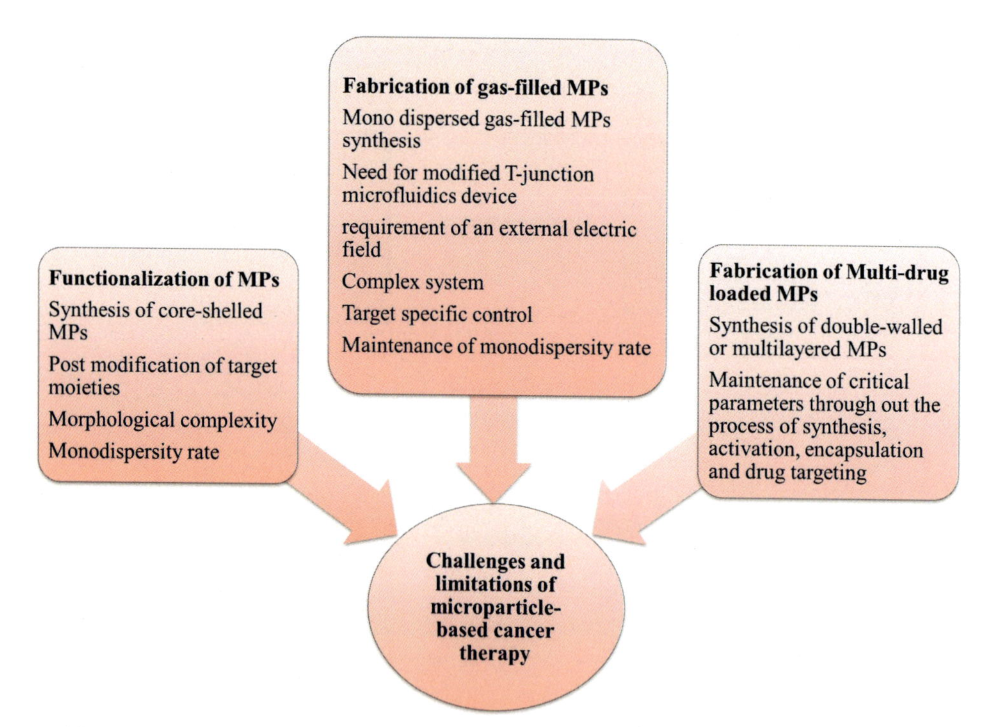

FIG. 8 Challenges and limitations of microparticle-based cancer therapy.

are discussed in this section (Fig. 8). The technical advancement of EDH in the synthesis of double-walled MPs involves core-shell structure, optimum size, and varieties of available polymer system and pharmaceutically active ingredients for encapsulation. The limitations associated with loading, release, and instability of drugs could have been solved with the idea of controlled size and well-designed structure of the MPs. In the EDHA method, the controlled droplet size was produced from the source and resulted in the formation of particle size ranging from nanometer to micrometers in diameter [28, 29]. Another big challenge for the synthesis of MPs is the generation of mono-dispersed particles; they have good economic and technological benefits over polydispersed particles. It is very difficult to obtain monodispersity in the MPs using conventional methods. To produce MPs of the desired size in the micrometric scale, the three most key parameters need to be critically analyzed and optimized:

- the concentration of the polymer solution,
- a flow rate of the polymer solution, and
- external electric field.

However, selected material for the synthesis of MPs can restrict the optimum value for these parameters.

4.1 Challenges in the synthesis of functionalized MPs

A new era of drug delivery applications demands for an advanced multifunctional micro-carrier loaded with drug and functionalized with drug targeting moieties [30, 31]. However, the preparation of core-shell MPs along with functional modification in both core and shell phases by using the coaxial electro-spraying technique is a very challenging task. While the drug can be incorporated into the central core [32–34], the targeting moieties are often protruded on the shell that may need postmodification of the synthesized MPs [35]. Duong et al. proposed a new particle synthesis approach that brings together a top-down technology, electrospray, continuous MPs fabrication technology, and bottom-up self-assembly of micelles to form a semi-continuous micellar technology, i.e., micellar electrospray [36]. Since the fabricated composites are water-soluble, the method should be readily acceptable to any hydrophobic drug and could form a large array of composites including those having targeted moieties on the shell. Gun et al., fabricated PLGA MPs encapsulating iron oxide (Fe_3O_4) nanoparticles by a coaxial electro-spraying technique [37]. The study showed a convenient method for the fabrication of MPs, which shows great potential in magnetically induced clinical imaging and directed drug delivery.

4.2 Challenges in the fabrication of gas-filled microparticles

Gas-filled MPs are recognized as ultrasound contrast agents for ultrasound imaging because of their potential to reflect and scatter ultrasound waves. Gas-filled MPs with drug incorporation into their shell can be used as a drug delivery vehicle and can be monitored by using low-intensity ultrasound waves. Coated gas-filled MPs have become more popular in the area of targeted drug delivery [38, 39]. Because of the limitations associated with sonication-, agitation-, and microfluidic-based methods [40], such as a wide range of MPs size distribution or microchannel blockage, the coaxial EHDA-based approach is most suitable for the fabrication of relatively mono-dispersed gas-filled MPs for the targeted delivery of drugs. More recently, Parhizkar et al. introduced a fabrication method by combining a microfluidic set-up with the EHDA process to create mono-dispersed gas-filled MPs [41]. The external electric field across the channel output was applied to modify the T-junction microfluidic device. The modified T-junction-based microchannels could synthesize gas-filled MPs with a much smaller diameter than that of the microchannel with a polydispersity index ~1% [7].

4.3 Challenges in the fabrication of multidrug loaded microparticles

Multidrug loading into the fabricated MPs is a highly challenging task. Earlier studies have reported the loading of multidrug in double-walled MPs synthesized from coaxial electrospraying that could render parallel or successive release of drug at the targeted site. In general, the characteristics of drug delivery are affected by diffusion of a drug, degradation of the polymer, and water penetration. The drug release rate can be altered and fine-tuned by separate loading of drug in each layer of the fabricated MPs, accomplishing optimal multidrug release kinetics and profiles [42–44]. However, the interfacial tension and material phase separation of the solution in each layer limits the maximum number of layers presented in the MPs. This carries more opportunities and limitations for the EHDA method to fabricate multilayered MPs [7, 45].

5 Conclusion

There is a huge progress that has been made in the field of diagnosis for cancer therapy using MP. The MPs were employed for several drug targeting therapies using rodent tumor models. MPs blend to form pharmaceutical dosage as solids, liquids, and semisolids. Here relation between the microspheres, MPs, and microcarrier has been described and the requirement of site-specific activities for therapies using MPs-based system is known to be successful. These particles are site-specific, shows maximum therapeutic efficacy rather than undesirable adverse effects. The limitations associated with conventional methods including biodegradability, stability, and sustained release of drug can be overcome by using MPs based drug delivery system. MPs based drug delivery system provides more uniform administration in anatomical conditions with sturdy and fastened dose fusing response. The main characteristics of these MPs were its particle size, morphology, solubility, and dissolution velocity, which provide the researchers to use these method over conventional methods of formulations. Several techniques used to develop these MPs were discussed earlier in detail, including spray-drying technique, coacervation, solvent extraction/solvent evaporation, and polymerization technique. These site-specific deliveries were found to be highly biocompatible without causing proinflammatory response. These methods became the most efficient method to overcome most of the difficulties associated with other methods of administration. The MP-based technology has undergone numerous changes and thus to cope with the current issues associated with cancer therapy, some of the advanced methods related to fabrication, encapsulation, and coating of MPs were discussed in this chapter. Several challenges and limitations were faced during the MP-based cancer therapy, which includes its synthesis, fabrication, encapsulation, and stability of these MPs that were also discussed. The limitations associated with loading, release, and instability of drugs could have been solved with the idea of controlled size and well-designed structure of the MPs.

6 Future perspectives

MPs were found to be highly advantageous as compared to other traditional drug delivery system. For wider application in clinical traits, the MPs structure, degradability, properties, and formation need to be studied. The toxicity and biodegradability with long-term biological behavior of these MPs make them very widely used system. The chapter basically focuses on the types of MPs, that is microsphere and microcapsules, along with their properties, preparation methods, and the mechanism how these MP helps in drug delivery were discussed in detail. The various approaches followed in MP technology were also briefed. The advanced technology based on MPs for cancer therapy related to fabrication, encapsulation, and coating were given under the chapter. Several recent methods were employed in the preparation of MPs that is flow lithographic and microfluidic-assisted methods. In multiplexing task, cell and drug delivery the MP properties were enhanced using various approaches like fabrication, which include microfluidic and nonmicrofluidic-assisted methods. In microfluidic approach, some other methods employed such as droplet-based, flow lithography, and scaling up methods from which the droplet method was the easiest and simple and can synthesize only spherical-shaped MPs due to less surface energy; thus to overcome this flow lithography technique came, which provided greater freedom to the shape of MPs. But the method is applicable for only 2Ds MP because of the height of microfluidic channels. Some other ways were followed to develop 3Ds MP using UV-light polymerization and inertial low-shaping methods. Other nonmicrofluidic-assisted approaches came in

MP fabrication. The centrifugation-based method showed immense potential while the EHD cojetting method was found to be the most advanced method. For a precise fabrication of MPs, another improved template-based method was employed in terms of shape and size. Other advanced technologies followed for image-guided cancer therapy based on MPs-induced activation method that emerged at a high rate. For this, polylactide-co-glycolic acid was used as they are biocompatible and biodegradable for drug delivery process. Image-guided drug delivery incorporates magnetic resonance imaging (MRI) with drug delivery nanoparticles to monitor the biodistribution, circulation, and targeting behavior of nanoparticles. The overall property of MPs can be improved by surface encapsulation for imaging and anticancer therapy. Some miscellaneous advances in MP-based drug delivery were developed where SCFs, which used powdered formulation, became an advantageous method. With the development of technology, advanced multifunctional microcarrier demands increase with the loading of drugs and functionalization with drug targeting moieties. Technique like electro-spraying method became a challenging task but can be used for delivery of drug in the central core. Each of the MPs that has been used has its own advantages and disadvantages. In clinical traits, MPs need to be tested, which are more appropriate and practical for the concerned usage. The MPs size, mean diameter, drug encapsulation, fabrication challenges, and other loading criteria need to be solved. Thus in future, all these issues will be eradicated and are expected to promote the use of MPs in cancer therapy for better results with great efficiency.

References

[1] Lengyel M, Kállai-Szabó N, Antal V, Laki AJ, Antal I. Microparticles, microspheres, and microcapsules for advanced drug delivery. Sci Pharm 2019;87(3):20.

[2] Shivani SH. Review article on microparticles. IJPAR 2015;4(3):302–9.

[3] Rashid M, Kaur V, Hallan SS, Sharma S, Mishra N. Microparticles as controlled drug delivery carrier for the treatment of ulcerative colitis: a brief review. Saudi Pharm J 2016;24(4):458–72.

[4] Padalkar AN, Shahi SR, Thube M. Microparticles: an approach for betterment of drug delivery system. Int J Pharm Res Dev 2011;1:99–115.

[5] Hwisa N, Katakam P, Chandu B, Adiki S. Solvent evaporation techniques as promising advancement in microencapsulation. Vedic Res Int Biol Medicinal Chem 2013;1(1):8–22.

[6] Singh M, Hemant K, Ram M, Shivakumar H. Microencapsulation: a promising technique for controlled drug delivery. Res Pharm Sci 2010;5(2):65.

[7] Agüero L, Zaldivar-Silva D, Peña L, Dias ML. Alginate microparticles as oral colon drug delivery device: a review. Carbohydr Polym 2017;168:32–43.

[8] Choi A, Seo KD, Kim BC, Kim DS. Recent advances in engineering microparticles and their nascent utilization in biomedical delivery and diagnostic applications. Lab Chip 2017;17(4):591–613.

[9] Xu S, Nie Z, Seo M, Lewis P, Kumacheva E, Stone HA, et al. Generation of monodisperse particles by using microfluidics: control over size, shape, and composition. Angew Chem 2005;117(5):734–8.

[10] Pregibon DC, Toner M, Doyle PS. Multifunctional encoded particles for high-throughput biomolecule analysis. Science 2007;315(5817):1393–6.

[11] Chung SE, Park W, Park H, Yu K, Park N, Kwon S. Optofluidic maskless lithography system for real-time synthesis of photopolymerized microstructures in microfluidic channels. Appl Phys Lett 2007;91(4), 041106.

[12] Paulsen KS, Di Carlo D, Chung AJ. Optofluidic fabrication for 3D-shaped particles. Nat Commun 2015;6(1):1–9.

[13] Nisisako T, Torii T, Takahashi T, Takizawa Y. Synthesis of monodisperse bicolored janus particles with electrical anisotropy using a microfluidic Co-Flow system. Adv Mater 2006;18(9):1152–6.

[14] Haeberle S, Zengerle R, Ducrée J. Centrifugal generation and manipulation of droplet emulsions. Microfluid Nanofluid 2007;3(1):65–75.

[15] Hayakawa M, Onoe H, Nagai KH, Takinoue M. Complex-shaped three-dimensional multi-compartmental microparticles generated by diffusional and Marangoni microflows in centrifugally discharged droplets. Sci Rep 2016;6:20793.

[16] Bhaskar S, Pollock KM, Yoshida M, Lahann J. Towards designer microparticles: simultaneous control of anisotropy, shape, and size. Small 2010;6(3):404–11.

[17] Choi CH, Lee J, Yoon K, Tripathi A, Stone HA, Weitz DA, et al. Surface-tension-induced synthesis of complex particles using confined polymeric fluids. Angew Chem 2010;122(42):7914–8.

[18] Win KY, Ye E, Teng CP, Jiang S, Han MY. Engineering polymeric microparticles as theranostic carriers for selective delivery and cancer therapy. Adv Healthcare Mater 2013;2(12):1571–5.

[19] Kan S, Lu J, Liu J, Wang J, Zhao Y. A quality by design (QbD) case study on enteric-coated pellets: screening of critical variables and establishment of design space at laboratory scale. Asian J Pharm Sci 2014;9(5):268–78.

[20] Mounika A, Sirisha B, Rao VUM. Pharmaceutical mini tablets, its advantages and different enteric coating processes. World J Pharm Pharm Sci 2015;4(8):523–41.

[21] Patil AT, Khobragade DS, Chafle SA, Ujjainkar AP, Umathe SN, Lakhotia CL. Development and evaluation of a hot-melt coating technique for enteric coating. Braz J Pharm Sci 2012;48(1):69–77.

[22] Hashmat D, Shoaib MH, Mehmood ZA, Bushra R, Yousuf RI, Lakhani F. Development of enteric coated flurbiprofen tablets using Opadry/Acryl-Eze System—a technical note. AAPS PharmSciTech 2008;9(1):116.

[23] Okuda T, Kito D, Oiwa A, Fukushima M, Hira D, Okamoto H. Gene silencing in a mouse lung metastasis model by an inhalable dry small interfering RNA powder prepared using the supercritical carbon dioxide technique. Biol Pharm Bull 2013;36(7):1183–91.

[24] Upadhyay D, Scalia S, Vogel R, Wheate N, Salama RO, Young PM, et al. Magnetised thermo responsive lipid vehicles for targeted and controlled lung drug delivery. Pharm Res 2012;29(9):2456–67.

[25] Dixit M, Kulkarni P. Lyophilization monophase solution technique for improvement of the solubility and dissolution of piroxicam. Res Pharm Sci 2012;7(1):13.

[26] Jinturkar KA, Anish C, Kumar MK, Bagchi T, Panda AK, Misra AR. Liposomal formulations of Etoposide and Docetaxel for p53 mediated enhanced cytotoxicity in lung cancer cell lines. Biomaterials 2012;33(8):2492–507.

[27] Abdelaziz HM, Gaber M, Abd-Elwakil MM, Mabrouk MT, Elgohary MM, Kamel NM, et al. Inhalable particulate drug delivery systems for lung cancer therapy: nanoparticles, microparticles, nanocomposites and nanoaggregates. J Control Release 2018;269:374–92.
[28] Zamani M, Prabhakaran MP, Ramakrishna S. Advances in drug delivery via electrospun and electrosprayed nanomaterials. Int J Nanomedicine 2013;8:2997.
[29] Paine MD, Alexander MS, Stark JP. Nozzle and liquid effects on the spray modes in nanoelectrospray. J Colloid Interface Sci 2007;305(1):111–23.
[30] Shapira A, Livney YD, Broxterman HJ, Assaraf YG. Nanomedicine for targeted cancer therapy: towards the overcoming of drug resistance. Drug Resist Updat 2011;14(3):150–63.
[31] Felice B, Prabhakaran MP, Rodriguez AP, Ramakrishna S. Drug delivery vehicles on a nano-engineering perspective. Mater Sci Eng C 2014;41:178–95.
[32] Bungay PM, Morrison PF, Dedrick RL. Steady-state theory for quantitative microdialysis of solutes and water in vivo and in vitro. Life Sci 1990;46(2):105–19.
[33] Tan WHK, Wang F, Lee T, Wang CH. Computer simulation of the delivery of etanidazole to brain tumor from PLGA wafers: comparison between linear and double burst release systems. Biotechnol Bioeng 2003;82(3):278–88.
[34] Wang C-H, Li J, Teo CS, Lee T. The delivery of BCNU to brain tumors. J Control Release 1999;61(1-2):21–41.
[35] Arifin DY, Lee KYT, Wang C-H. Chemotherapeutic drug transport to brain tumor. J Control Release 2009;137(3):203–10.
[36] Duong AD, Ruan G, Mahajan K, Winter JO, Wyslouzil BE. Scalable, semicontinuous production of micelles encapsulating nanoparticles via electrospray. Langmuir 2014;30(14):3939–48.
[37] Gun S, Edirisinghe M, Stride E. Encapsulation of superparamagnetic iron oxide nanoparticles in poly-(lactide-co-glycolic acid) microspheres for biomedical applications. Mater Sci Eng C 2013;33(6):3129–37.
[38] Unger EC, Matsunaga TO, McCreery T, Schumann P, Sweitzer R, Quigley R. Therapeutic applications of microbubbles. Eur J Radiol 2002;42(2):160–8.
[39] Dijkmans P, Juffermans L, Musters R, van Wamel A, Ten Cate F, van Gilst W, et al. Microbubbles and ultrasound: from diagnosis to therapy. Eur J Echocardiogr 2004;5(4):245–6.
[40] Wyss HM, Blair DL, Morris JF, Stone HA, Weitz DA. Mechanism for clogging of microchannels. Phys Rev E 2006;74(6), 061402.
[41] Parhizkar M, Stride E, Edirisinghe M. Preparation of monodisperse microbubbles using an integrated embedded capillary T-junction with electrohydrodynamic focusing. Lab Chip 2014;14(14):2437–46.
[42] Labbaf S, Deb S, Cama G, Stride E, Edirisinghe M. Preparation of multicompartment sub-micron particles using a triple-needle electrohydrodynamic device. J Colloid Interface Sci 2013;409:245–54.
[43] Misra AC, Bhaskar S, Clay N, Lahann J. Multicompartmental particles for combined imaging and siRNA delivery. Adv Mater 2012;24(28):3850–6.
[44] Kim W, Kim SS. Multishell encapsulation using a triple coaxial electrospray system. Anal Chem 2010;82(11):4644–7.
[45] Davoodi P, Feng F, Xu Q, Yan W-C, Tong YW, Srinivasan M, et al. Coaxial electrohydrodynamic atomization: Microparticles for drug delivery applications. J Control Release 2015;205:70–82.

Chapter 30

Biosynthetic exosome nanoparticles isolation, characterization, and their diagnostic and therapeutic applications

Krishnan Anand[a], Balakumar Chandrasekaran[b,c], Gaurav Gupta[d], Harish Dureja[e], Sachin Kumar Singh[f], Monica Gulati[f], Dinesh Kumar Chellappan[g], Balamuralikrishnan Balasubramanian[h], Ireen Femeela[i], Vijaya Anand Arumugam[i], and Kamal Dua[j,k,l]

[a]*Department of Chemical Pathology, School of Pathology, Faculty of Health Sciences and National Health Laboratory Service, University of the Free State, Bloemfontein, South Africa,* [b]*Faculty of Pharmacy, Philadelphia University, Amman, Jordan,* [c]*Faculty of Pharmaceutical Sciences, Block-G, UCSI University, Kuala Lumpur, Malaysia,* [d]*School of Pharmacy, Suresh Gyan Vihar University, Jagatpura, Jaipur, India,* [e]*Department of Pharmaceutical Sciences, Maharshi Dayanand University, Rohtak, Haryana, India,* [f]*School of Pharmaceutical Sciences, Lovely Professional University, Phagwara, Punjab, India,* [g]*Department of Life Sciences, School of Pharmacy, International Medical University (IMU), Kuala Lumpur, Malaysia,* [h]*Department of Food Science and Biotechnology, College of Life Science, Sejong University, Seoul, South Korea,* [i]*Department of Human Genetics and Molecular Biology, Bharathiar University, Coimbatore, TN, India,* [j]*Discipline of Pharmacy, Graduate School of Health, University of Technology Sydney, Ultimo, NSW, Australia,* [k]*Priority Research Centre for Healthy Lungs, Hunter Medical Research Institute (HMRI) & School of Biomedical Sciences and Pharmacy, University of Newcastle, Callaghan, NSW, Australia,* [l]*School of Pharmaceutical Sciences, Shoolini University of Biotechnology and Management Sciences, Solan, Himachal Pradesh, India*

1 Introduction

Exosomes are of emerging interest in diverse pathological conditions due to their key role in cellular physiology. These entities can signal and change the phenotype of target cells and hence they have quantitative and qualitative impacts on disease. Exosomes are organically found nanoparticles released endogenously by the cellular structure of mammals. Due to limitations like instability in donor cells, lower production, and incapability to target the required cells, exosomes are not utilized clinically. According to Pillay et al. [1], exosomes are the basic components of the syncytiotrophoblast extracellular vesicles (STBEVs) ranging from 20 to 130 nm in size with the ability to migrate endothelial cells. They also play a major latent role in immune tolerance between fetus and mother. The end products of the lysosomal pathway and secretions by cells undergoing fusion between multi-vesicular bodies (MVBs) and their plasma membranes results in the production of the exosomes. These bi-lipid vesicle membranes are packed with a wide range of molecules involved in signaling along with molecules for cellular adhesion and receptors for growth factors. Furthermore, the presence of RNA (mRNA and miRNA) aides in the function of providing immunity. The basic function of exosomes is to transfer signals to cells in adjacent or distal positions to reboot their function both at phenotypic and regulatory levels.

The tiny membranous extracellular vesicles mediating communication between the cells are categorized as exosomes, the bodies causing apoptosis, and MVBs based on their length and origin. Being the smallest of extracellular vesicles (30–150 nm), exosomes have a key function in mediating cellular interaction with the potency to alter phenotype via nucleic acid transport or lipid and protein transport from their progenitor cells to remote or adjacent cells. Every cell in the mammalian body can produce exosomes in urine, blood, saliva, and breast milk. Regionally and systemically, exosomes appear to be vital mediators of intercellular conversation, which regulates an extensive variety of organic processes. Moreover, exosomes are membranous proteins with the ability to suggest the pathological and physiological states of parental cells and specify their favored purpose. First-rate varieties of micro-vesicles, exosomes among them, are secreted through cells that encompass micro- and macro-molecules [1].

Exosomes are nano-sized vesicles that might switch lipids, DNAs, non-coding RNAs, and microRNAs (miRNAs) with or without contacting the cells, thus providing capability for intracellular conversation. Exosomes can be used as providers among precise locations of the frame; their maximum critical position is to hold facts and ship several signaling molecules and effectors among specific cells. Direct or changed membrane protein expression of exosomes may be useful in therapeutic platforms and drug transport

Advanced Drug Delivery Systems in the Management of Cancer. https://doi.org/10.1016/B978-0-323-85503-7.00037-7

systems, including centered remedy strategies, due to their natural transportation houses, exquisite biocompatibility, and prolonged intrinsic motion functionality. These properties make exosomes suitable for turning in a good-sized range of proteins, nucleic acids, chemical materials, and therapeutic markers [2]. Of the few endogenous non-coding RNAs, round RNA (circRNA) is a unique member that is broadly distributed and has copious cellular functions. Recently, spherical RNA was determined to be strong and enriched in exosomes. The exosomal circRNAs can be effective for improving novel healing desires and biomarkers. Long-time exosome research has focused on their switch of miRNAs and mRNAs to adjacent cells and their nucleic acid composition [3].

2 Types of exosomes

2.1 Plant-derived exosomes

According to Zhang et al. [4] the smaller-size molecules with miRNA, the active biomolecules (lipids and proteins) could communicate between the cells in plants and can play a vital role of messengers resembling the exosomes released by mammalian cells. Depending on their source and delivery, these consumable plant-based nanoparticles carry out signaling between molecules and species, thus they can cure a wide range of pathologies naturally. Efficient drug delivery was exhibited for specific drugs made of lipid nanoparticles from plants. These palatable particles can be more easily generated on a large scale than synthetic ones. Our review focuses on the current advancements in obtaining edible plant nanoparticles and gaining deeper understanding of how to produce these nanoparticles in a sustainable and bio-renewable way. We opted to study the following four plant parts: grape, carrot, grapefruit, and ginger. These plant parts are consumed and imported as delicacies worldwide. The rhizome of *Zingiber officinale* is ginger. The modified root of *Daucus carota* (carrot) is economically important for its import and export throughout the world. Similarly, the phytochemicals and nutrients of grapefruits and grapes act as inhibitors of pathologies like viral infections and even diseases of the heart, nerve degeneration, Alzheimer's, and cancer [5] (Fig. 1).

2.2 Animal-derived exosomes

The small vesicles (50–200nm) secreted by mammalian cells are known as MVs or exosomes and they are present in bodily fluids like blood with stable nucleic acids like various types of RNA (mRNA, miRNA, and other types), proteins, and DNA in special packages with the capability to communicate with surrounding or distinct cells. Their role can be more prominent in detecting and monitoring a specific disease. Tumor profiling can be done using miRNA profiling of exosomes in the absence of tissue (Fig. 2).

3 Isolation and purification of exosomes

Théry et al. [6] described an efficient protocol to isolate exosomes from the biopsies of natural liquids like milk, fruit extract, plasma, urine, and others by diluting the liquids with equal volumes of phosphate-buffered saline (PBS; pH 7.4). The obtained exosomes were purified using differential ultra-centrifugation with a cushion of 30% sucrose. In brief, initial centrifugation was carried out at 4°C and 2000 RCFin 30min, and the next centrifugation was carried out at 4°C and 12000 RCF for 45min. The obtained pellet was suspended with PBS and filtered through a 0.22-mm mesh. The obtained filtrate was re-centrifuged at 4°C and 110,000 RCF for 70min to obtain a pellet, which was re-suspend in PBS (pH 7.4) and again centrifuged at 4°C and 110000 RCF for 70min. The obtained exosome pellet was re-suspended in PBS followed by purification through a sucrose cushion of 30%. The resultant pellet was resuspended with PBS (100mL) and stored at 80°C. Goetzl et al. [7] used RC DC protein assay to obtain the concentration of protein in an exosome. A filtration process was used to remove debris and cells and their membranes. The obtained supernatant was centrifuged again at. 4°C and 100,000RCF for about 70min to obtain a pellet, which was then resuspended with 60mL of cold PBS and centrifuged again at 4°C and 100,000RCF for 70min. The obtained pellet was mixed with 0.0.95M of 2mL sucrose solution and then poured into a step gradient column of sucrose (six.2mL steps starting from 2.0M sucrose up to 0.25M sucrose in 0.35M increments, with the 0.95M sucrose step containing the exosomes) and centrifuged at 4°C and 200,000RCF for 16h. After centrifugation, 1mL was collected from the topmost fraction of the gradient, and the fractions at the interphase between the two neighboring sucrose layers was pooled for total fractions. The obtained fraction was infused with cold PBS and recentrifuged at 4°C and 100,000RCF for 70min. The in-between fraction of sucrose pellet was mixed with cold PBS (100µL) with 2µL to assess the activity of the enzyme and another 2µL for electron microscopy (EM). Analysis of total exosomal protein was carried out by combining the remaining solution (16.µL) with an equal volume of buffer containing a cocktail of protease inhibitors. The overall protein composition of exosomes from various cellular sources has been investigated by immunoblotting (Fig. 3).

4 Characterization of exosomes

4.1 Nanoparticle tracking analysis (NTA)

Pillay *et al.* (2016) used a 405-nm NS500 laser and sC-MOS camera (NanoSight.NTA 3.0) nanoparticle tracking to ascertain exosome quantification and distribution based on size. The samples were primarily diluted with PBS to obtain distribution of particles of 10–100 per image

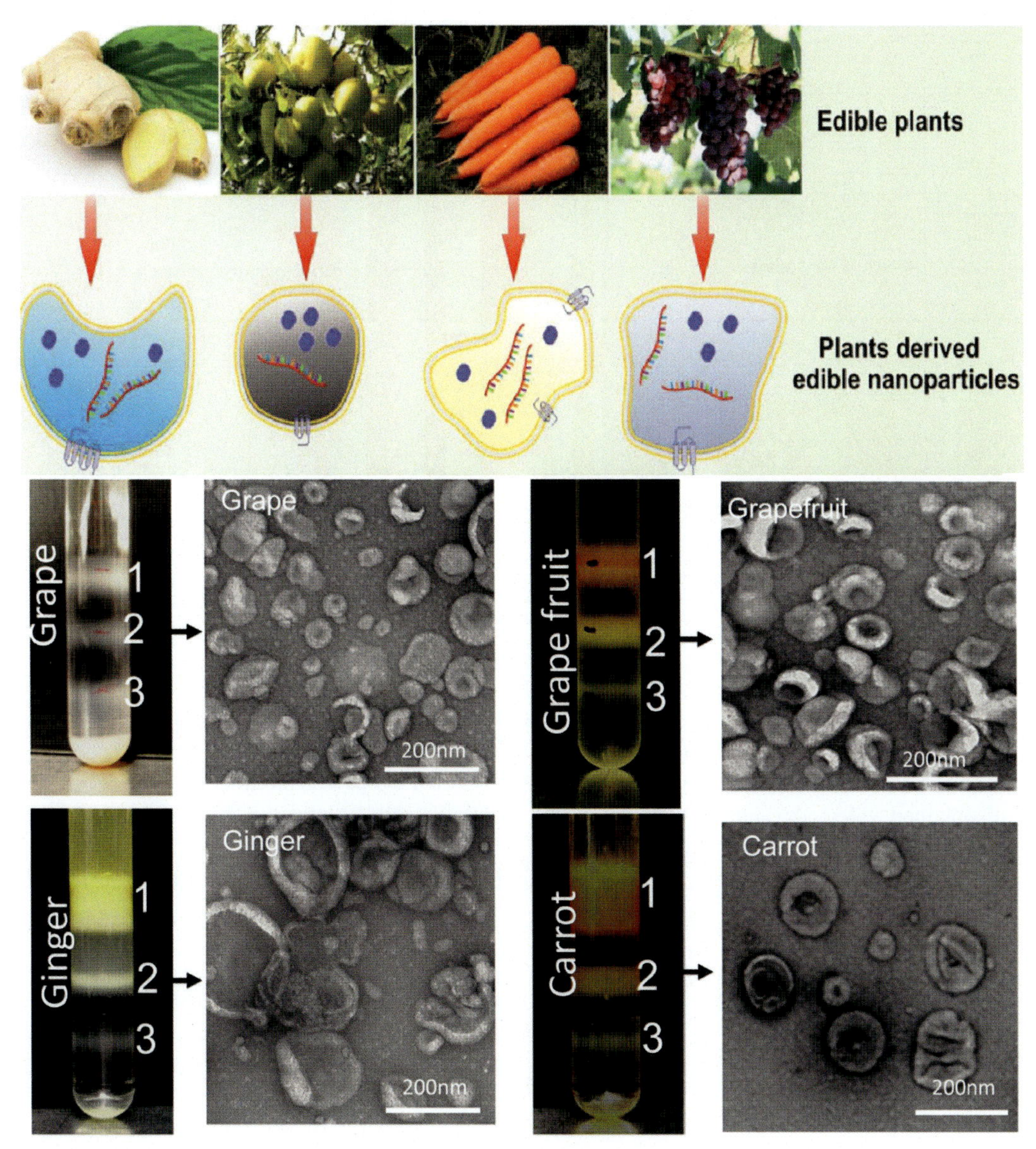

FIG. 1 Plants derived edible nanoparticles (PDEN) and drug loaded PDEN nanoparticles.

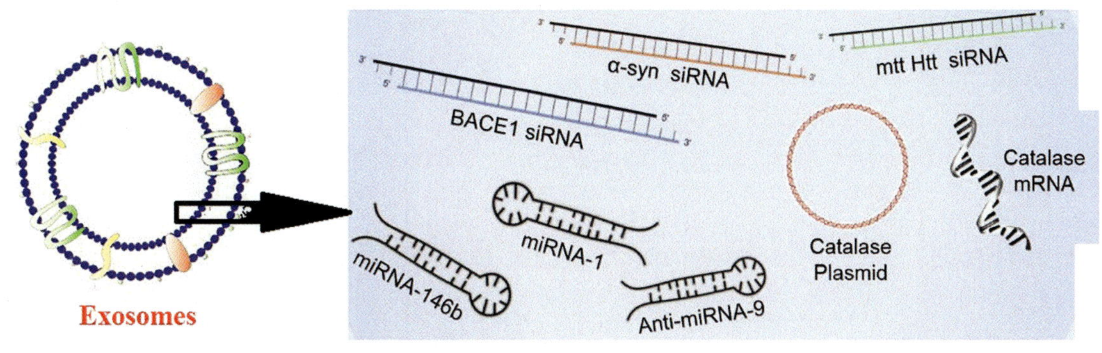

FIG. 2 The figure shows the composition and RNAs in exosome.

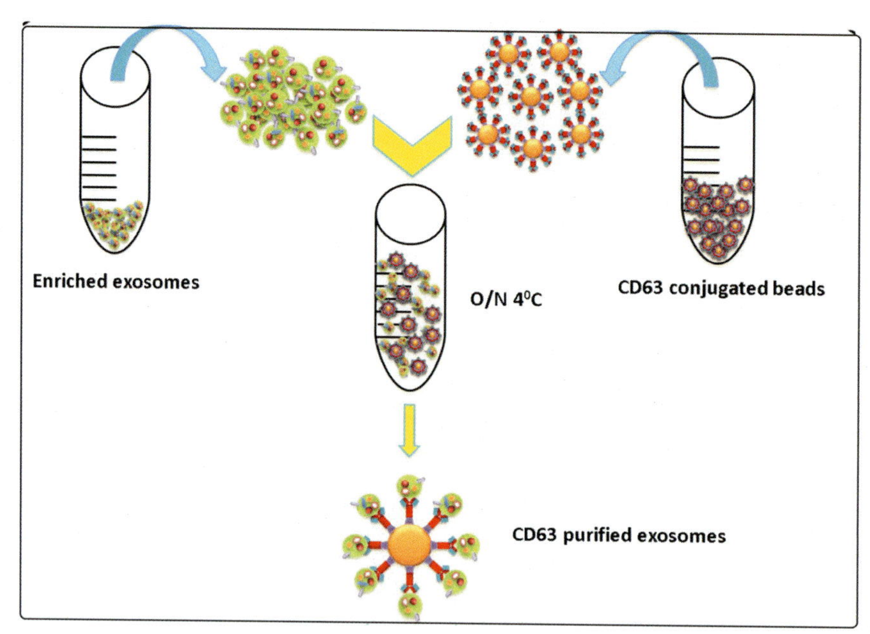

FIG. 3 General centrifugation protocol.

(50.particles per image being optimal) prior to using the NTA software for analysis. The camera was leveled at 10 with a shutter speed of 20ms and gain of 600 to record consistent video with same consistency for samples. Following analysis of all obtained video, the authors calculated the mean of particle size and concentration of particles, represented in ±SD (Fig. 4).

4.1.1 Exosome identification by hyperspectral electron microscopy

Kalluri et al. [8] reported transmission electron microscopy (TEM) images from a cell culture supernatant embedded with epon and sectioned exosomes (NCI-H292 cells, which are human lung mucoepidermoid cells) (Fig. 5).

A continuous carbon grid with negative stain of 2% uranyl acetate was coated on the exosomes to examine their morphology and dimension using TEM (Pillay *et al.*, 2016). To avoid and remove any cross contamination with proteins, the pellet rich in exomes was resuspended in PBS and then centrifuged at 4°C and at.120000*g for.70min. A small amount of sample was collected to assess the whole protein and isolate it. Only one-fourth of the sample was lysed to extract the protein, while the remaining sample was kept safely with exosomes (intact) on ice at −80°C in PBS to proceed with the experiment. To perform EM, a drop (10μg) of intact exosome was taken on parafilm and placed on a carbon-coated nickel grid with forceps for about an hour. The grid was positioned with the carbon coating on the drop to trap the exosomes. No immunostaining was required to identify the exosomes, as their peculiar morphology and shape differentiate them from other cellular components.

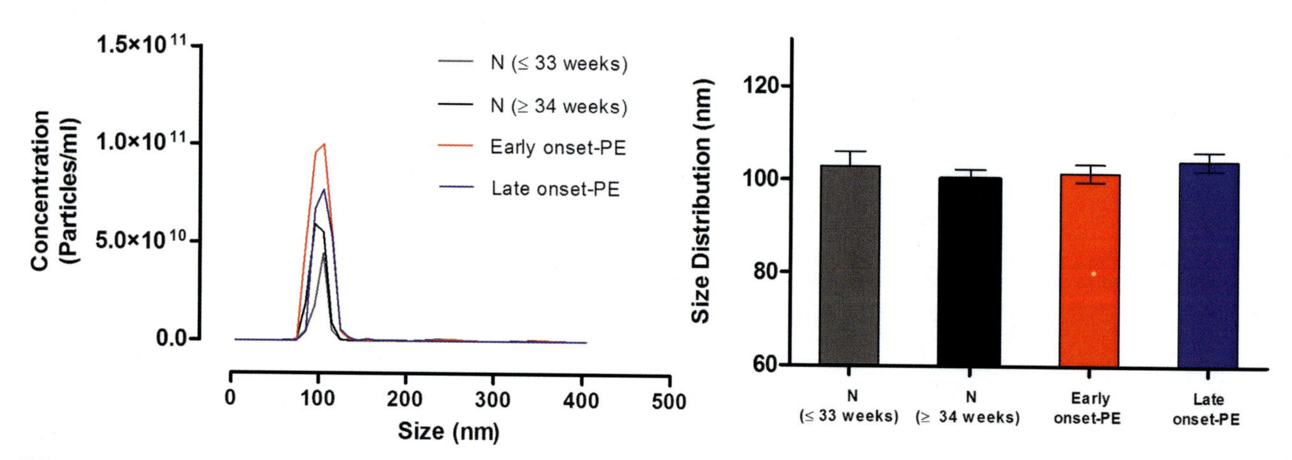

FIG. 4 Nanoparticle tracking analysis (NTA): Representative vesicle size distribution (nm) and vesicle size distribution.

4.1.2 Western blot analysis

The extraction of proteins from the exosomes of brain tissue is carried out in an amended SDS-PAGE sample buffer comprised of protease inhibitor tablets that were incubated in the sample for 5 min at.100 °C. Centrifuging for 10 min at 10000*g removed the insoluble matters, and the proteins were estimated via commercial protein assay kits. The next steps included boiling the extracted proteins at 100 °C for 10 min with Laemmli sample buffer (2% *v*/v β-mercaptoethanol, 0.2% *w*/*v* SDS, and.10% w/v sucrose in 0.1875 M Tris, pH 6.8, with bromophenol blue) and resolving them in SDS-PAGE polyacrylamide gels followed by transferring on the membrane of nitrocellulose. After successful transfer, the membrane was blocked and washed with probes corresponding to the antibodies for kinases. Detection was carried out by developing bands using chemiluminescent assay. The immunoblotting analyses were done in triplicate with their respective blots. Mizutani *et al.* [9] purified exosomes from a prostate cancer cell line maintained in conditioned media by centrifuging sequentially the total cell lysate (5 μg) of both LNCaP and PC-3 cells with antibodies, indicated in the right panel. The obtained fraction of whole exosome (10 μg protein) was removed from the conditioned media of the cell lines through sequential centrifugation followed by western blotting to analyze the protein, as shown in the left panels (Fig. 6).

5 Exosome nanoparticles: Biomarkers and drug delivery applications

Exosomes are capable of transporting drugs due to their biocompatibility, lengthy circulatory functionality, and herbal material transportation properties. They are suitable for turning in an expansion of chemicals, nucleic acids, proteins, and gene healing markers. . The absorption by target cells is nevertheless lacking in this technique. The utility ability of exosomes is still constrained [10]. Cells continuously secrete a wide range of micro-vesicles, including

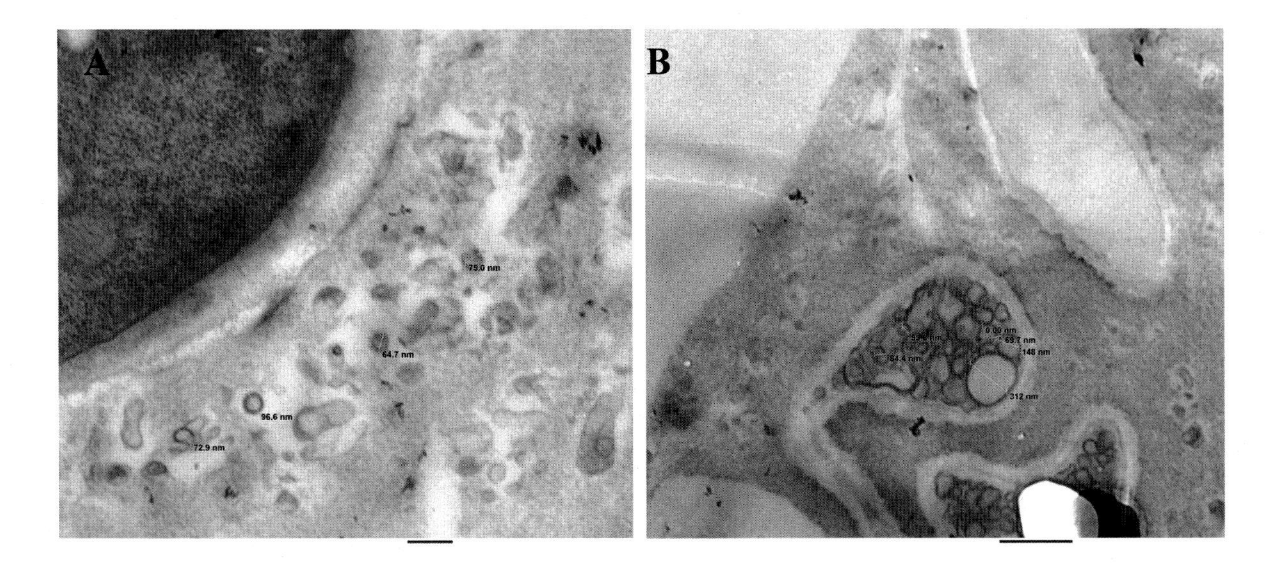

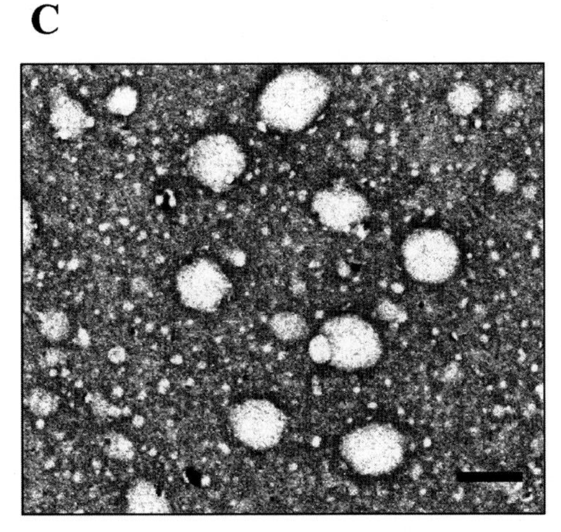

FIG. 5 TEM images of epon-embedded and sectioned exosomes (A and B) and placental exosome (C).

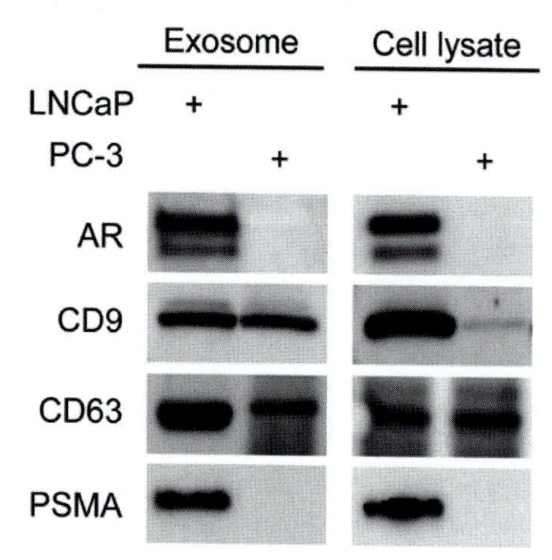

FIG. 6 Exosome protein western blot analysis.

micro- and macromolecules, into extracellular fluids. Exosomes are nano-sized vesicles able to transfer lipids, DNAs, non-coding RNAs, and miRNAs with or without direct cell-to-cell.contact, representing the manner of intracellular verbal exchange. Exosomes are utilized as transporters between diverse areas of the frame. Exosomes are secreted by all types of cells and are abundant in saliva, breast milk, urine, and blood (Fig. 7). The features, biological, therapeutic, and emerging exosomes from patient-specific provide a platform for personalized therapy [5]. Fig. 8 highlights the diagnostic and therapeutic targets of exosomes.

6 Exosomes and cardiovascular disease

Exosomes are being used as biomarkers for cardiovascular ailments, particularly atherosclerosis. Atherosclerosis is the most common cause of cardiovascular deaths worldwide. Therefore, it is important to determine prognostic factors like biomarkers to manage and treat this disease effectively [11]. Exosomes have a prime function in communication between cells and between cells and bilipid membranous vesicles packed with nucleic acids, proteins, or lipids, which may be maximally released from residing cells in atherosclerosis. Exosomes have a vital role in transferring proteins, mRNA, and miRNA in cellular conversation, migration, proliferation, differentiation, and specific bioactive molecules among cells, and can alter gene expression inside recipient cells. In the cardiovascular system, exosomes are related to endothelial cells, cardiac myocytes, vascular cells, and stem and progenitor cells.

Recent evidence implicates exosomes in the development of cardiovascular disorders. Exosomal miRNA play roles in cardiac regeneration and protection as well as in non-cardiac cells, stem cells, and progenitor cells. The miRNAs may be isolated from blood and used for diagnosing cardiovascular sicknesses [12]. Exosomes are 30–100 nm membrane-bound vesicles secreted by all cells and found in all body fluids. The mechanism of exosomes on recipient cells is unknown, but it is hypothesized that exosomes can be genetically engineered to hold ligands or homing peptides on their surfaces, so one can help in exosome concentrated on to various receptors. With the invention of peptides homing to diseased tissues or organs through phage show and in vivo biopanning technology, there is the possibility of discovering the capability of exosomes for centered gene remedy [13].

Myocardial hypertrophy is a common cardiac condition in reaction to hemodynamic and neurohormonal modifications. Pathological hypertrophic increase in the heart triggers the decline of cardiac skills and eventually develops into congestive coronary heart failure. Recent research uncovered that the exosomes from cardiac fibroblasts and

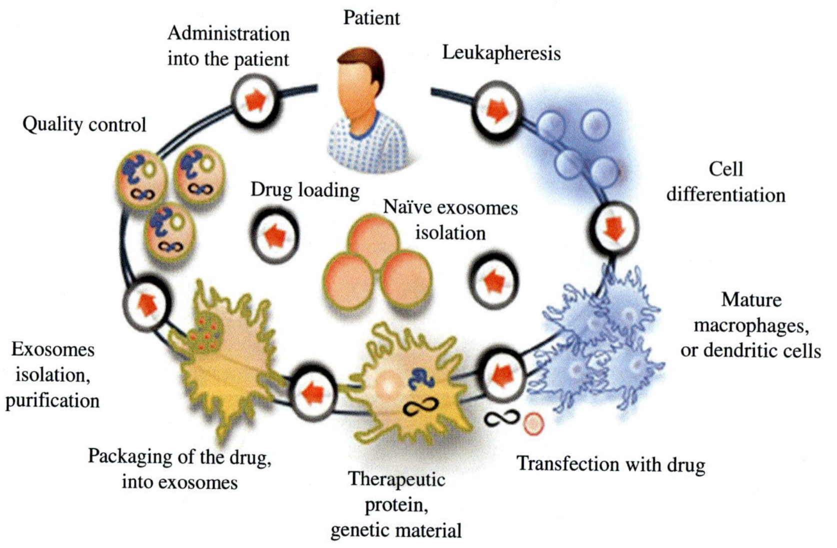

FIG. 7 Exosomal production and drug delivery in patients. Replace image requested.

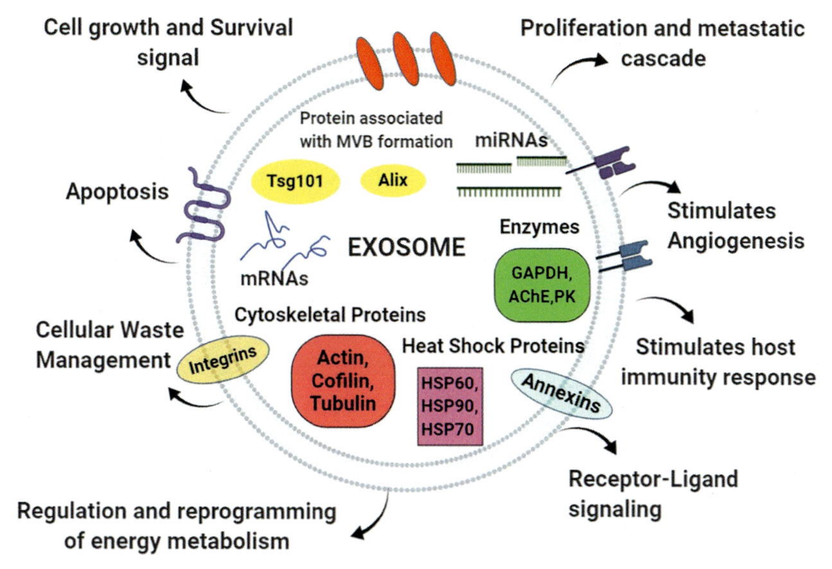

FIG. 8 Diagnostic and therapeutic targets of exosome.

other tissues take part in the improvement of myocardial hypertrophy. Cardiac progenitor cells and exosomes from mesenchymal stem cells (MSCs) protect against myocardial hypertrophy. The exosomes function as new intercellular mediators among cardiomyocytes and special cells, which display software capability in the diagnosis and treatment of cardiomyocyte hypertrophy.

Although the function of exosomes in arrhythmia has not been established, an acceptable case for exosome involvement can be made. Exosomes secreted by ischemic cardiomyocytes from coronary artery disease patients and ischemic rat heart patients are rich in MR-1 and MR-133 [14], which focus on the effects of Ca^{2+}/calmodulin-structured protein kinase II on action energy and cardiac circulation [15,16]. Specific progeserogenic MIRs that regulate exosomes during ischemia include MIR-328, which is abundant in platelet-derived exosomes [17] and round L-type calcium channels [18,19].

Exosomes play a pre-inflammatory role in the heart at the onset of various cardiomyopathies. Sepsis-triggered cardiomyopathy occurs when a systemic bacterial infection eventually leads to a breakdown of heart function, which can be observed through more than one organ failure. Platelet-derived exosomes contribute to inflammation in lipospecies (LPS), triggering sepsis in a mouse model. Exosomes taken from injured cardiovascular tissue promote progression of disease through a variety of pathological signals [20]. Fig. 9 highlights the role of exosomes in cardiovascular diseases.

6.1 Roles of exosomes in HIV

Patients with HIV infection may develop a variety of musculoskeletal syndromes, such as soft tissue rheumatic syndromes, and a wide spectrum of articular manifestations including reactive arthritis, psoriatic arthritis, septic arthritis, osteonecrosis, and severe inflammatory oligo- or

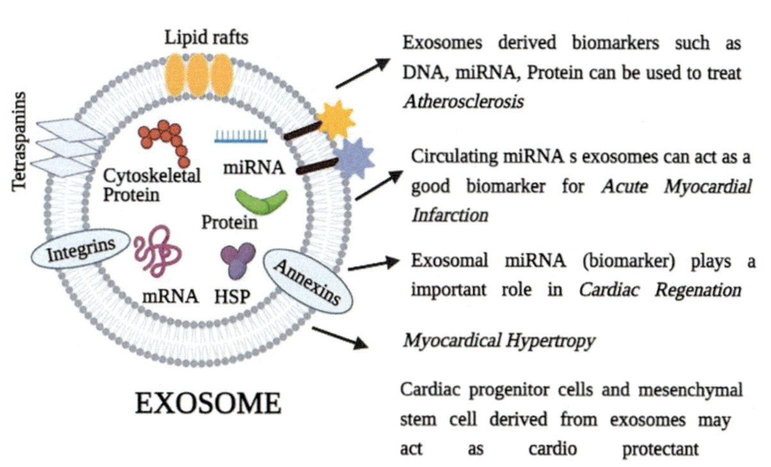

FIG. 9 Cardiovascular disease and exosomes.

polyarthritis. These comorbidities may present at any stage of the disease with the most common manifestation being arthralgia, which is seen in approximately 45% of patients and usually resolves with the commencement of antiretroviral therapy. Rheumatic manifestations have been associated with significant morbidity by causing persistent pain, functional disability, and impaired quality of life. In Europe and the United States, articular syndromes in HIV-infected patients were initially reported among intravenous drug users and homosexuals. The pattern of arthritis in these individuals was similar to that in HIV-negative patients with arthritis. In Africans with HIV infection, the patterns of arthritis were reactive arthritis, psoriatic arthritis, and undifferentiated spondyloarthritis, African-American which were previously uncommon in this population due to the relatively low prevalence of HLA-B27. Additionally, the incidence of polyarticular joint involvement in Western communities has significantly escalated compared to earlier reports. Since exosomes may have a key regulatory role in HIV, we hypothesize that exosomes, being nanoparticulates, are altered in HIV-influenced pathological mechanisms. This may be related to the functional modification of HIV-derived exosomes in circulation and tissues. In addition, the therapeutic potential of exosomes for viral diseases necessitates an understanding of how they cross the lymphocytes, optimization of methods for isolation and characterization, and targeting to blood cells. Hence, in addition we would like to investigate these particles in HIV-associated diseases.

6.2 Roles of exosomes in cancer

The exosomes have a pivotal role in intercellular communications and package deal delivery in addition to diagnostic biomarkers. They are also viable drug delivery vehicles in cancer therapy because of their precise biocompatibility, excessive balance, good tumor targeting, and adjustability.

Prostate and lung cancer are the most common cancers worldwide. Together they account for more than 250,000 deaths globally; they predominate in developing countries like Africa. In South Africa, both cancer types are number one in men followed by lung, esophageal, colorectal, and bladder cancers. In women, breast cancer is the most prevalent followed by cervical, uterus, colorectal, and esophageal cancers. A recent study predicts the incidence of cancer cases will increase to.78% by 2030 in South Africa. From a global point of view, an increase of 75% is expected, with total numbers of new cancer patients increasing from 12.7 million in 2008 to 22.2 million in 2030. The development of certain types of cancers is more likely in aging persons and those suffering from HIV/AIDS. This may be due to the correlation between immune competence and certain cancers.

Several types of extracellular vesicles, including exosomes, are released from cells. As discussed, exosomes are 40–120 nm in diameter and released at the time of fusion of multi-vesicular endosomes with the cell membrane.

Exosomes are naturally occurring nanoparticles that are endogenously secreted by the mammalian cells. Clinical application of exosomes is challenging due to their low scalability, unsuitable donors, and insufficient targeting ability. It is a key component of cell to cell communications, which have a smaller size range of 20–130 nm and have a latent role in fetal maternal immune tolerance and migration of endothelial cells.

In the tumor microenvironment, exosomes play an important role in intermolecular communication and progression of cancer, particularly in drug resistance, tumor immune tolerance, metastasis, and angiogenesis. Prostate-specific membrane antigen (PSMA) is a specific and well-known prostate tumor biomarker that is enriched in cancer exosomes by PSMA-positive prostate cancer cells, including LNCaP cells. Recent studies support the development of a PSMA-based affinity capture stage for exosome isolation. This will provide distinctive information on the disease stage and guide treatment.

Cancer cells produce exosomes, which contain molecules to induce a plethora of organic techniques like metastasis formation, angiogenesis, epigenetic/stemness (re) programming, epithelial–mesenchymal transition, and healing resistance. Exosome proteins probable replicate their cellular foundation and may additionally useful resource to detect most of the cancers. Intercellular switch off oncoproteins by using exosomes may also play a function in facilitating tumorigenesis [21,22]. Cancer research has suggested that exosome-mediated cell-to-cell communication has led to drug resistance of most cancers cells. Exosome-based drug transport strategies have proven effective for inhibiting the secretion of exosomes and thus inhibiting tumor growth. The non-coding RNAs (ncRNAs) present in exosomes are found to inhibit tumor progression and drug resistance. Consequently, tumor-suppressed exosomal ncRNAs may be used in the treatment of cancers. The exosomes related to cancers are diagnosed from the fluids of patients and the detection of exosomes within cancer patients is a potential strategy for obtaining the pathogenic records of most cancers. Usage of exosomes as delivery vehicles for siRNAs and therapeutic tablets brings out new principles consisting of biometrics in cancer remedy.

Tumor heterogeneity involves genomic heterogeneity between neoplastic cells within a tumor, in addition to non-cancer cell tumor microenvironmental heterogeneity. The various cellular structures, MSCs, and cellular components contribute to the active heterogeneity of tumors. Tumor nano-environment (TNE) heterogeneity increases as another layer of molecular weight. TNE incorporates extracellular vesicles of various types as well as apoptotic bodies. Exosomes extracted from multiple cancer cells can contribute to the formation of a protumorigenic or antitumorigenic milieu through the host stromal responses. The substances released from the stem cells can also increase and further limit the growth of cancer in a context-dependent manner [8].

Resistance to chemotherapy, radiation, and targeted treatment remains the primary barrier to cancer treatment [23]. Resistance enhancement is multifactorial, involving the conversion of multiple cancer cells into secondary salvage pathways while closing the true sign [24], epigenetic suppression of protein initiation into activation using miRNAs, the presence of anti-cancer cells epithelial-to-mesenchymal transition (EMT) phenotype [25], low drug intake (due to desmoplastic reaction (DR) within tumor microenvironment), and more [26]. Cancer and anti-drug radiotherapy is a recent field of intensive research. Fig. 10 highlights exosomal miRNAs for diagnosis and prognosis of non-small-cell lung cancer (NSCLC).

The usage of nanoscale debris like exosomes may also improve cancer vaccines. Antigen-imparting cell technology can be used to apprehend and kill most cancerous cells. Exosomes blended with nanotechnology may prove useful for developing cancer vaccines. Present-day vaccines for cancers precipitated by viral infections include Cervarix, Gardasil, and hepatitis B vaccines. However, these are not "proper" vaccines, because even though they act effectively against the viruses that cause the cancer, they do not destroy the cancerous cells. The primary approved therapeutic vaccine for cancer is Provenge (sipuleucel-T), which was approved by the FDA in April 2010 for the treatment of prostate cancer. Fig. 11 highlights some important mechanisms of exosomes in supporting cancer-resistance networks.

Rapid precipitation and ultracentrifugation are used to isolate exosomes, but are not effective for isolating tumor-derived exosomes from those obtained from normal tissue,

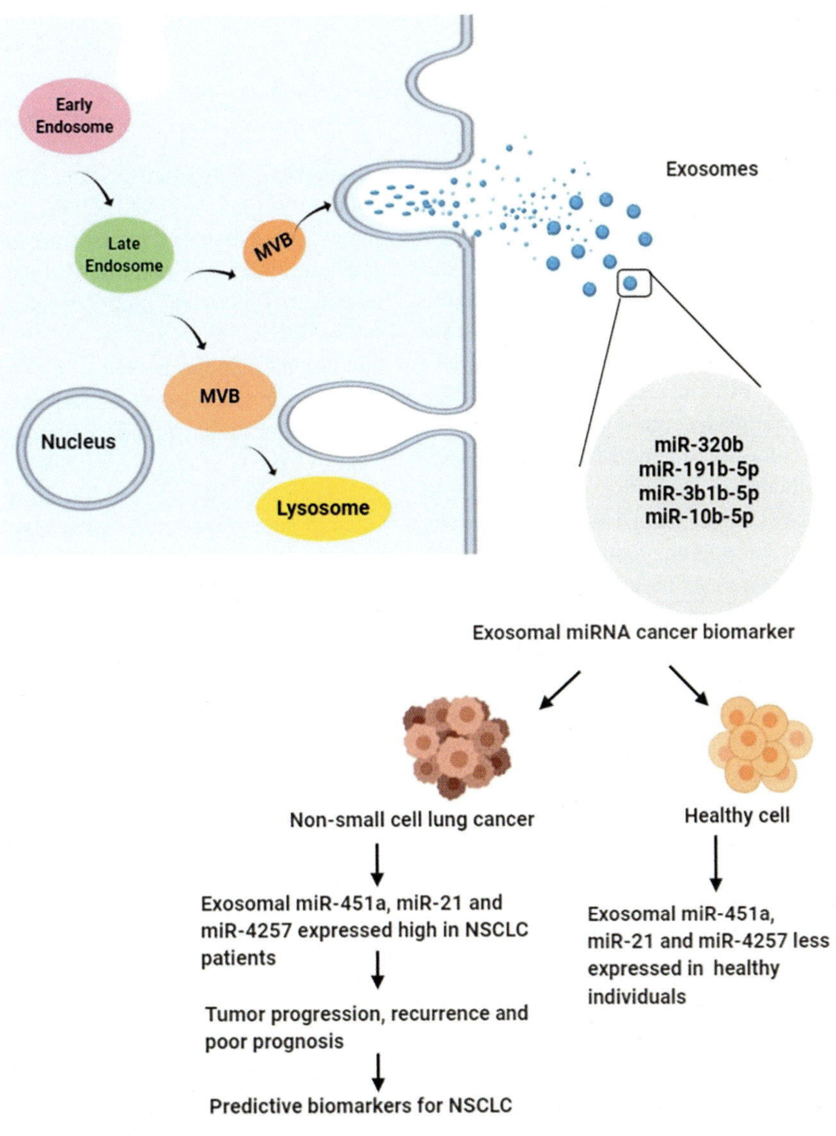

FIG. 10 Exosomal miRNAs for NSCLC cancer diagnosis and prognosis.

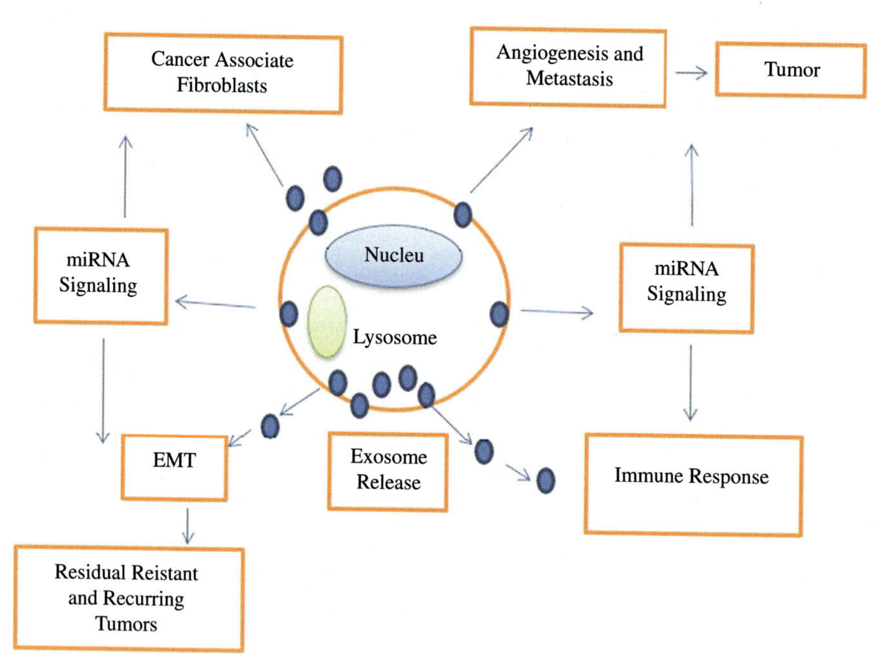

FIG. 11 Some important mechanisms of exosomes in supporting cancer-resistance networks.

due to low yield recovery and the extensive time needed (8 h) for the process. To overcome the disadvantages of available methods, several microfluidic techniques have been developed to extract micro-vesicles and exosomes in an easy and rapid manner. Microfluidic approaches are selected based on physical characteristics (e.g., density, size, and surface charge). Recently, Ghosh et al. described a method for isolation using Heat Shock Protein (HSP)-binding peptide Vn96 to comprehensive HSP-decorated extracellular vesicles that may perform at the small peptide-mediated "miniprep" scale. These methods are susceptible to clogging and normally exhibit low exosome recovery. A nanoporous membrane is used for the entrapping of exosomes on porous structures. The benefits of microfluidic-based immune affinity approaches compared to physical-based purification are the capability to isolate a precise subpopulation of exosomes relying on the expression of a specific surface marker and improved exosome recovery of the blood plasma. An additional benefit of immunological separation is that it can be carried out in low volumes over short times.

6.3 Roles of exosomes in diabetes-linked diseases

Type 2 diabetes mellitus (T2DM) and Alzheimer's disease (AD) are interconnected; cognitive impairment has been associated with diabetes mellitus (hyperglycemia and hypoglycemia), abnormalities in insulin (insulin deficiency or resistance), vascular abnormalities, and oxidative stress in the central nervous system (CNS). Cognitive impairment in diabetic individuals is two-to three-times greater than in non-diabetic individuals. Since exosomes may have a key regulatory role in AD and T2DM, we hypothesize that exosomes, being nanoparticulates, are altered in T2DM and AD and could influence pathological mechanisms. This may be related to functional modification of neural-derived exosomes (NDE) in circulation and at the tissue level. In addition, the therapeutic potential of exosomes for neurological diseases necessitates an understanding of how they cross the blood–brain barrier (BBB), optimization of methods for isolation and characterization, and targeting to neural cells. Hence, in addition we would like to investigate these particles in neurodegenerative diseases, and the more recently described type 3 diabetes mellitus.

Metabolic disorders are clinically complex and include diseases and conditions like obesity, metabolic syndrome, and diabetes mellitus. The cellular functions in these conditions may change due to chronic inflammation, oxidative stress, and fat toxicity. Exosomes play an important role in the pathogenesis of metabolic disorders. Exosomes are nanoscale vesicles secreted by cells that contain nucleic acids, lipids, proteins, and membrane receptors. They mediate material transport and signal transduction to neighboring as well as distant cells. Exosomes participate in multiple pathological processes, including tumor metastasis, chronic inflammation, atherosclerosis, and insulin resistance. Exosome-associated miRNAs regulate the pathological function and physiological progression of metabolic disorders. The application of exosome-associated miRNAs is useful in novel therapeutics and diagnostics of metabolic disorders. They exhibit special features of non-immunogenicity and quick extraction [27]. In normal and diseased states, various membrane vesicles such as exosomes are constantly secreted in different cell types. Exosome in extracellular space can isolate various biological fluids and cell culture supernatants. Exosomes are potential biomarkers for autoimmune and infectious disorders. The utilization of exosomes as drug delivery tools for treating autoimmune diseases and diabetes requires further research [28].

6.4 Autoimmune diseases and exosomes

In autoimmune diseases, the body attacks its own tissues by releasing autoantibodies that target healthy cells. Therapeutic drugs for these diseases are linked with adverse side effects and their beneficial effects are limited to specific populations. Exosomes are being investigated for use as transporters between various areas of the body. They play an important role in immune tolerance and stimulation because they are involved in several processes, including inflammation, angiogenesis, and immune signaling. Exosomes are promising tools for treatment due to their intrinsic features of stability and biocompatibility. Exosomes have possible therapeutic effects in autoimmune diseases such as type 1 diabetes mellitus, Sjogren's syndrome, systemic lupus erythematosus, multiple sclerosis, and rheumatoid arthritis [29] (Fig. 12).

6.5 Biomedical field

The concept of gene therapy is that several disorders can be cured through gene transfer, although this is still up for debate. The challenges and difficulties associated with evolving therapies and adverse effects in patients undergoing gene therapy raise concerns about the safety of this approach in molecular medicine. The critical problems in the field of gene therapy are discovering and developing good bio-vectors with low toxicity and high efficiency. The uses of exosomes as vectors for the transmission of genes or gene delivery is an opportunity for gene-based therapy [30]. The effective exchange of cellular components through exosomes will inform their application in the design of exosome-based therapies [31].

6.6 Wound healing and exosomes

Wound healing may be complicated, especially in individuals suffering from method underlying conditions. In addition, the costs of wound treatment can be an enormous. The advantages of MSCs feature in the release of secretory factors, matrix proteins and recruitment of host cells growth wound recuperation. Those secrete bioactive trophic elements from MSCs additionally encompass exosomes or extracellular vesicles, latter mediating cell-to-cell communication in wound recovery. Use of extracellular vesicles in wound recuperation is promising as a substitute for cell-based remedy [32].

Extracellular vesicles obtained from stem cells may be the next "off-the-shelf" products for treating wounds. Extracellular vesicles have been shown to contribute to cataractous wound healing, including proangiogenic effects to heal wounds by promoting the secretion of growth factor and synthesis of collagen. Extracellular vesicles produce nanovesicles and are appreciated for their immune response ability with low toxicity that may penetrate cell membranes. Extracellular vesicles can be used and modified for targeted treatment. In addition to RNA therapy, they can also be used to transfer drugs to toxin-reducing cells [32].

6.7 Nanomedicine and exosome

Exosomes have good innate structure and biocompatible nanocarrier homes. Exosomes are extracellular nanovesicles

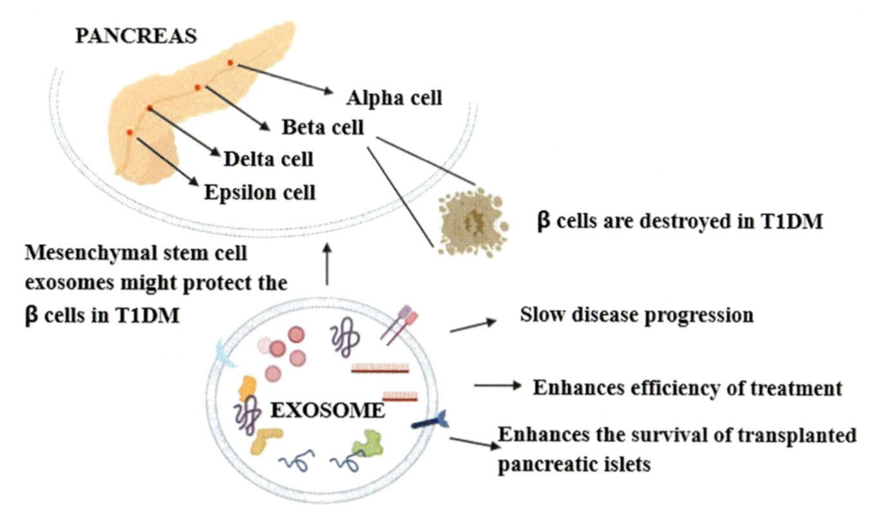

FIG. 12 Therapeutic effects of exosomes in autoimmune disease (clinical phase) – Type 1 diabetes.

and their components may be engineered at the cell level. Its miles imitative from a increase of biological origins may be modified submit isolation and exosome imaging and monitoring by *in vivo* may performed by means of change of exosome surface systems. To enhance drug stability and efficacy, hydrophobic therapeutics are loaded with exosome membranes. Hydrophilic therapeutics such as RNA may be encapsulated in exosomes to increasethe cellular delivery [33].

6.8 Anti-viral infection and exosome

Extracellular vesicles, includes microvesicles and exosomes, can be mediated the viral spread by non-enveloped and enveloped viruses. Viruses can enter host cells, either avoiding or enhancing the immune reaction by exosomes. However, the complex interplay between extracellular vesicles and viruses gives rise to opposed biological tasks advantage to benefits the viruses, enhancing infection to benefit the host or, interacting with the immunity by means of mediating anti-viral responses. Herpes simplex type 1 virus with cells infected from exosomes may additionally help to transfer the virus and host transcripts and innate immune machine [34].

Exosomes are excretory nano-vesicles shed from the surface of almost all types of cells and that are formed by the cell's endocytic system. These tiny extracellular vesicles, once considered to be "garbage bags for cells." It carries varieties of molecules of cellular origin, such as RNAs, lipids, and proteins, which might be integrated selectively throughout the formation of exosomes. There are many numbers of animal viruses that make use of the pathways of exosomes through interacting viral elements inside the exosomes or particular cells. The components of the virus, which appears to be selectively integrated into exosomes, and the capability of these exosomes in viral pathogenesis have been studied. The identity of viral signatures in exosomes and their mechanisms of action are key for designing diagnostic and treatment strategies for viral infections [35]. This intercellular translocation of exosomes inhibits viruses and induced the expression of interferon of genes encoding antiviral substances, which may lead to the restoration of antiviral status in virus-infected cells [36].

6.9 Tuberculosis and exosomes

Host–pathogen interaction is highly important for identifying the new therapeutics target for treating tuberculosis. Exosomes are rich in lipids, proteins, and nucleic acids, which may additionally act as messengers. Proteins are concerned with plasma heme scavenging, neutrophil degranulation, lipid metabolism, and display deregulation in patients with tuberculosis. The protein panel that circulates small extracellular vesicles has a crucial position in tuberculosis, identifying the objectives of and developing host-directed healing interventions [37].

Exosomal RNA, which mediates protection against rod RNA deficiency due to axial-product stability, appears to be an ideal platform for the development of clinical biomarkers. In tuberculosis diagnostics, there are no sensitive or specific diagnostic tests to detect the various manifestations of tuberculosis infection. Therefore, the application of exosome based biomarkers for clinical and therapeutic purposes or for vaccine development may be promising. Future studies need to address the benefits of exosome ohm-based biomarkers in tuberculosis diagnosis and vaccination [38–40].

7 Summary and perspectives

Bio-nanotechnology is an important field in biochemistry, protein science, and nanochemistry. Many scientific fields are engaged in producing exosomes and building integrated devices to take advantage of the optimal properties of exosomal proteins. Scientists are continuously seeking to identify and characterize precise groups of exosome nanoparticles that have intrinsic targeting abilities. These efforts are part of a new era of nanomaterials with multiple applications in biomedical fields including disease diagnosis and biomarkers, sensing devices, therapeutics, and nanotechnology. Exosomes are ideal materials to work with because they are easily purified, can decrease the complexity of the biofluid source, and improve the scientific interest of RNA. Exosomes are dynamic and can be derived from a wide variety of biofluids. They can be used for vaccine delivery, as their numbers increase during inflammatory processes during infection. They possess several properties including biocompatibility, stability, biological barrier permeability, low immunogenicity, and low toxicity, all of which make them important vehicles for therapeutic delivery. This chapter examined exosomes and their potential applications for drug delivery and biomarker identification. Further research is needed to improve next-generation nanomedicines.

References

[1] Pillay P, Maharaj N, Moodley J, Hu TY. Placental exosomes and pre-eclampsia: maternal circulating levels in normal pregnancies and, early and late onset pre-eclamptic pregnancies. Placenta 2016;46(1):18–25. https://doi.org/10.1016/j.placenta.2016.08.078.

[2] Wang Y, Liu J, Ma J, Sun T, Zhou Q, Wang W, Wang G, Wu P, Wang H, Jiang L, Yuan W, Sun Z, Ming L. Exosomal circRNAs: biogenesis, effect and application in human diseases. Mol Cancer 2019;18(1):116.

[3] Xu R, Greening DW, Zhu HJ, Takahashi N, Simpson RJ. Extracellular vesicle isolation and characterization: toward clinical application. J Clin Invest 2016;126(4):1152–62.

[4] Zhang M, Viennois E, Xu C, Merlin D. Plant derived edible nanoparticles as a new therapeutic approach against diseases. Tissue Barriers 2016;4(2):e1134415.

[5] Qin J, Qing X. Functions and application of exosomes. Acta Pol Pharm 2014;71(4):537–43.

[6] Théry C, Amigorena S, Raposo G, Clayton A. Isolation and characterization of exosomes from cell culture supernatants and biological fluids. Curr Protoc Cell Biol 2006. https://doi.org/10.1002/0471143030.cb0322s30.

[7] Goetzl EJ, Ledreux A, Granholm A-C, Elahi FM, Goetzl L, Hiramoto J, Kapogiannis D. Neuron-derived exosome proteins may contribute to progression from repetitive mild traumatic brain injuries to chronic traumatic encephalopathy. Front Neurosci 2019. https://doi.org/10.3389/fnins.2019.00452.

[8] Kalluri R. The biology and function of exosomes in cancer. J Clin Invest 2016;126(4):1208–15.

[9] Mizutani K, Terazawa R, Kameyama K, Kato T, et al. Isolation of prostate cancer-related exosomes. Anticancer Res 2014;34:3419–23.

[10] Liu C, Changqing S. Design strategies and application progress of therapeutic exosomes. Theranostics 2019;9(4):1015–28.

[11] Lu M, Yuan S, Li S, Li L, Liu M, Wan S. The exosome-derived biomarker in atherosclerosis and its clinical application. J Cardiovasc Transl Res 2019;12(1):68–74.

[12] Zamani P, Fereydouni N, Butler AE, Navashenaq JG, Sahebkar A. The therapeutic and diagnostic role of exosomes in cardiovascular diseases. Trends Cardiovasc Med 2019;29(6):313–23.

[13] Xitong D, Xiaorong Z. Targeted therapeutic delivery using engineered exosomes and its applications in cardiovascular diseases. Gene 2016;575(2 Pt 2):377–84.

[14] Xu C, Lu Y, Pan Z, Chu W, Luo X, et al. The muscle-specific microRNAs miR-1 and miR-133 produce opposing effects on apoptosis by targeting HSP60, HSP70 and caspase-9 in cardiomyocytes. J Cell Sci 2007;120:3045–52.

[15] Belevych AE, Sansom SE, Terentyeva R, Ho HT, Nishijima Y, et al. Microrna-1 and -133 increase arrhythmogenesis in heart failure by dissociating phosphatase activity from RyR2 complex. PLoS One 2011;6:e28324.

[16] Elton TS, Martin MM, Sansom SE, Belevych AE, Gyorke S, Terentyev D. MiRNAs got rhythm. Life Sci 2011;88:373–83.

[17] Dempsey E, Feidhlim D, Maguire PB. Platelet derived exosomes are enriched for specific microRNAs which regulate Wnt signalling in endothelial cells. In: Presented at American Society of Hematology Annual Meeting, San Francisco, CA, December 6–9; 2014.

[18] Lu Y, Zhang Y, Wang N, Pan Z, Gao X, et al. MicroRNA-328 contributes to adverse electrical remodeling in atrial fibrillation. Circulation 2010;122:2378–87.

[19] Neppl RL, Wang DZ. The myriad essential roles of microRNAs in cardiovascular homeostasis and disease. Genes Dis 2014;1:18–39.

[20] Ibrahim A, Marbán E. Exosomes: fundamental biology and roles in cardiovascular physiology. Annu Rev Physiol 2016;78(1):67–83.

[21] Al-Nedawi K, et al. Intercellular transfer of the oncogenic receptor EGFRvIII by microvesicles derived from tumour cells. Nat Cell Biol 2008;10(5):619–24.

[22] DemoryBeckler M, et al. Proteomic analysis of exosomes from mutant KRAS colon cancer cells identifies intercellular transfer of mutant KRAS. Mol Cell Proteomics 2013;12(2):343–55.

[23] Meads MB, Gatenby RA, Dalton WS. Environment-mediated drug resistance: a major contributor to minimal residual disease. Nat Rev Cancer 2009;9:665–74.

[24] Shain KH, Landowski TH, Dalton WS. The tumour microenvironment as a determinant of cancer cell survival: a possible mechanism for *de novo* drug resistance. Curr Opin Oncol 2000;12:557–63.

[25] Holzel M, Bovier A, andTuting, T. Plasticity of tumour and immune cells: a source of heterogeneity and a cause for therapy resistance? Nature reviews. Cancer 2013;13:365–76.

[26] McMillin DW, Negri JM, andMitsiades, C. S. The role of tumour-stromal interactions in modifying drug response: challenges and opportunities. Nat Rev Drug Discov 2013;12:217–28.

[27] Yao ZY, Chen WB, Shao SS, Ma SZ, Yang CB, Li MZ, Zhao JJ, Gao L. Role of exosome-associated microRNA in diagnostic and therapeutic applications to metabolic disorders. J Zhejiang Univ Sci B 2018;19(3):183–98.

[28] Wang Q, Ding X, Zhen F, Ma J, Meng F. Remedial applications of exosomes in cancer, infections and diabetes. Acta Theriol 2017;74(2):313–20.

[29] Xu H, Jia S, Xu H. Potential therapeutic applications of exosomes in different autoimmune diseases. Clin Immunol (Orlando, Fla) 2019;205:116–24.

[30] Rezaie J, Ajezi S, Avci ÇB, Karimipour M, Geranmayeh MH, Nourazarian A, Sokullu E, Rezabakhsh A, Rahbarghazi R. Exosomes and their application in biomedical field: difficulties and advantages. Mol Neurobiol 2018;55(4):3372–93.

[31] Kalluri R, LeBleu VS. The biology, function, and biomedical applications of exosomes. Science (New York, NY) 2020;367(6478):eaau6977.

[32] Rani S, Ritter T. The exosome - a naturally secreted nanoparticle and its application to wound healing. Adv Mater 2016;28(27):5542–52. https://doi.org/10.1002/adma.201504009.

[33] Hood JL. Post isolation modification of exosomes for nanomedicine applications. Nanomedicine (Lond) 2016;11(13):1745–56.

[34] Bello-Morales R, Ripa I, López-Guerrero JA. Extracellular vesicles in viral spread and antiviral response. Viruses 2020;12(6):E623.

[35] Khan G, Ahmed W, Philip PS. In: Wang J, editor. Exosomes and their role in viral infections, novel implications of exosomes in diagnosis and treatment of cancer and infectious diseases. IntechOpen; 2017.

[36] Li J, Liu K, Liu Y, Xu Y, Zhang F, Yang H, Yuan Z. Exosomes mediate the cell-to-cell transmission of IFN-α-induced antiviral activity. Nat Immunol 2013;14(8):793–803. https://doi.org/10.1038/ni.2647.

[37] Mateos J, Estévez O, González-Fernández Á, et al. Serum proteomics of active tuberculosis patients and contacts reveals unique processes activated during *Mycobacterium tuberculosis* infection. Sci Rep 2020;10:3844.

[38] Hadifar S, Fateh A, Yousefi MH, Siadat SD, Vaziri F. Exosomes in tuberculosis: still terra incognita? J Cell Physiol 2018;1–8.

[39] Hunter MP, Ismail N, Zhang X, Aguda BD, Lee EJ, Yu L, Marsh CB. Detection of microRNA expression in human peripheral blood microvesicles. PLoS One 2008;3(11):e3694.

[40] Mitchell PS, Parkin RK, Kroh EM, Fritz BR, Wyman SK, Pogosova-Agadjanyan EL, Tewari M. Circulating microRNAs as stable blood-based markers for cancer detection. Proc Natl Acad Sci 2008;105(30):10513–8.

Chapter 31

Advances in delivery of nanomedicines and theranostics for targeting breast cancer

Ajay Kumar Pal[a], Mukesh Nandave[a], and Rupesh K. Gautam[b]

[a]*Department of Pharmacology, Delhi Pharmaceutical Sciences and Research University (DPSRU), New Delhi, India,* [b]*Department of Pharmacology, MM School of Pharmacy, Maharishi Markandeshwar University, Sadopur-Ambala, Haryana, India*

1 Background

Breast cancer (BC) is defined as a phenotypically variable and heterogeneous disease caused by malignant lesions growing in the breast tissues [1, 2]. It is diagnosed as the presence of localized mass under/within breast tissues, nipple discharge, breast asymmetry, nipple inversion, erythema, and thickening of breast skin, and it is screened by mammography [3]. It is highly mortal disease among women due lack of early diagnosis and long duration of therapeutic regimen. The factors responsible for breast cancer development and occurrence have been detailed in Table 1.

BC is classified into many subtypes on the basis of the receptor that is overexpressed on their cell membrane. Common receptors that are overexpressed include estrogen receptor (ER), human epidermal growth factor receptor 2 (HER2), and progesterone receptor (PR). The type of treatment that patient should receive is depends on the diagnosis of its subtypes [2]. Hence, BC is subdivided into luminal A (ER^+, PR^+, $HER2^-$), luminal B (ER^+, PR^+, $HER2^-$), HER2 positive (ER/PR^-, $HER2^+$), and basal like (ER^-, PR^-, $HER2^-$) [1]. Triple negative breast cancer (TNBC) is also known as basal-like breast cancer.

TNBC is the most malignant and aggressive form of breast cancer [4], which accounts for 15%–20% of all cases of breast cancer [5–8]. Because of the unresponsiveness of the hormonal and/or targeted therapies due to the absence of any hormonal receptors on breast tumors of TNBC subtype [5, 9], chemotherapy is the only alternative to treat TNBC. Systemically administered chemotherapy with clinically approved drugs also shows poor efficacy, high toxicity, and development of multidrug resistance in TNBC tumors. Additionally, high chances of metastasis to secondary organs (lungs, bones, and brain), molecular heterogeneity, high relapse rate, and mutations (BRCA*) contribute a challenging hurdle in its management and thus poor prognosis [10, 11]. Also, TNBC tumors often metastasize to secondary nearby organs, such as lungs, bones, and brain [5, 12]. TNBC is subdivided into six subgroups: (a) basal-like 1 (BL1); (b) basal-like 2 (BL2); (c) mesenchymal (M); (d) mesenchymal stem like (MSL); (e) immunomodulatory (IM); and (f) luminal androgen receptor (LAR) [4].

Globally breast cancer (BC) is a major health issue among women, which gains 1st rank in both the mortality (6.6%) and incidence (11.6%) of all cancer-related cases globally according to cancer statistics 2019 [13]. A report released in 2018 from American Cancer Society states that approximately 2.6 lakhs new cases of BC (invasive) were found, which contributes as 30% of total cancer cases. In addition, approximately 50,000 cases were found fatal [14] and at least 1 in every 10 women found to have BC [15, 16]. However, there is significant variation in survival rate of BC patients among developed and developing countries. The developed countries estimate 80% (high) 5-year survival rate of BC patients compared with 40% (low) in developing countries [17]. It is a very alarming that incidence of BC shows an upward trend. Its incidence is predicted to rise by 85 cases per 1 lakh women by the end of 2021 [18]. Among men, BC accounts for 1% of the total BC cases as per the data revealed in 2011 [19].

With the emerging advancements in the nanobiotechnology field, the therapy for breast cancer has been in continuous upgradation in parallel with the advancing molecular stratification of breast cancer, rising chemoresistance, undesirable harm to other organs/tissues, simultaneous alterations in tumor microenvironment and metastasis which have pushed the research community to develop more precise alternatives/approaches to defeat cancer. The incorporation of nanomedicines and theranostics in every

Advanced Drug Delivery Systems in the Management of Cancer. https://doi.org/10.1016/B978-0-323-85503-7.00014-6

TABLE 1 Factors associated with occurrence and development of breast cancer.

Category	Example falling under the category
Diet and lifestyle-associated factors	Obesity and factors such as smoking and drinking are related to the development of BC.
Genetic factors	Study results reveal that if mother is breast cancer patient then her daughter will have two times risk of having breast cancer. Mutation of gene BRCA1 or 2.
Reproductive factors	Women who do not breastfeed or with late menopause, premature menarche, and nulliparity are more susceptible to develop BC.
Environmental factors	Chronic exposure of exogenous estrogen increases the risk of BC. High-dose radiation exposure to chest at young age.

BC, breast cancer; *BRCA1 or 2*, tumor suppressor oncogene responsible for breast cancer.

module of cancer therapy and diagnostics with precise targeted delivery of these led to the advancements to defeat biology of BC. Therefore, this chapter details the advancements and challenges underway in the development of targeted delivery systems for nanomedicines and theranostics in BC.

2 Current trends and challenges in diagnosis as well as treatment of breast cancer

The current treatment direction available for cancers includes chemotherapy, radiation, and surgery [20]. Considering the aggressiveness and unresponsiveness of distinct molecular strata of breast cancers to the conventional chemotherapy, the advanced approach of target delivery systems aims to selectively and efficiently target chemotherapeutic agents onto all forms of breast tumors [21]. The conventional drug delivery system includes direct oral and intravenous administration of chemotherapeutic drugs [21]. Table 2 details the currently approved standard chemotherapy for breast cancer. Compared to conventional cancer chemotherapy, paradigm shift occurred in current cancer therapeutics. Therefore, the emergence of approach for targeting specific proteins in cancer cell by exploiting the cytotoxic potential of traditional drugs had opened new doors for adjuvant chemotherapeutic [22]. To keep up with the pace and for safe and effective drug administration, parallel advancements in drug delivery systems are also need of the hour [23]. Thus, the recent developments in tumor-targeted drug delivery systems (TTDDs) have led to improved targeting of drugs onto the tumors, controlled drug release, longer retention, and higher efficacy accompanied with reduction in drug-associated toxicities, nontarget effects, and side-effects.

Perhaps, with the advancing discoveries in the molecular stratifications of breast cancer, their precise diagnosis is also crucial for choosing the selective therapeutic intervention. Till now, palpation, mammography, ultrasonography, magnetic resonance imaging (MRI), and immunohistochemistry (IHC) are the best available diagnostics methods for early diagnosis for breast cancer. Of them, mammography is the most widely employed radiological examination which utilizes X-rays to detect any early pathological alteration in breast tissues. In case of TNBC tumors, the diagnosis becomes inaccurate due to absences of abnormal pathological features [24]. Such limitation can be overcome by ultrasonography which possesses higher sensitivity (greater than 90%) [25], with accuracy for benign tumors which again limit its application for TNBC diagnosis. In TNBC diagnosis, MRI has high sensitivity; however, it also has false-positive predictive values. This can lead to avoidable painful biopsies [26].

The accurate detection for TNBC through above radiological examinations needs expertise and further clinical experience for designing of new diagnostic modalities considering either benign or early stage of cancer. Furthermore, the evolution of immunohistochemistry (IHC) and the role of oncopathologist become crucial in clinical diagnosis of TNBC. The hallmark behind the immunohistochemical diagnosis of TNBC relies on the absence of hormonal receptors (ER, PR) and HER-2 in patient's biopsy tissues and regarded as best TNBC diagnostic technique [27]. Upon proper TNBC diagnosis, other factors need to be considered for designing selective pharmacological intervention. Such factors include metastatic nature, drug resistance, relapse, and inappropriate prognosis. Breast sustentation treatment is the first and most important goal of TNBC treatment. However, the high chances of relapse even post radiation treatment (RT) require the mastectomy along with radiotherapy [28].

Hormonal therapy gained success in hormonal-dependent breast cancer, which is not effective to TNBC with due absence of HER2, ER, and PR hormone receptors. Therefore, it requires the chemotherapy for systematic treatment of hormone-independent subtypes of breast cancer [29]. Anthracyclines and taxanes are commonly used chemotherapeutic drug for TNBC tumor treatment [30]. However, their chemotherapeutic efficacy could be potentiated via utilization of novel drug delivery target technologies involving nanomedicine field. The continuous chemotherapy cycles kill not only the cancer cells but also healthy cells in close vicinity. So, for avoiding the nonspecific targeting and associated side effects, nanotechnology-

TABLE 2 FDA-approved therapeutics for breast cancer.

Category	Drug	Marketed name	Mechanism of action (MOA)
Chemotherapeutic agents			
Alkylating agents	Cyclophosphamide Thiotepa	Cytoxan	Cause alkylation of DNA structure and prevents cancer cells division.
Anthracyclines	Doxorubicin Liposomal DOX Epirubicin Mitoxantrone	Adriamycin Doxil Epirubicin Novantrone	These attack cancer cells by disrupting the re-ligation of DNA strands through its intercalation with topoisomerase-II, thus preventing cancer cell division.
Platinum drugs	Carboplatin Cisplatin	Paraplatin Platinol	These drugs contain the metal platinum, which bind to DNA and interfere with the cell's ability to repair itself, eventually leading to cancer cell death.
Taxanes	PTX protein bound PTX Docetaxel	Abraxane Taxol Taxotere	These drugs inhibit cell growth by stopping cell division; they act by interfering with with β-tubulin and enhances its polymerization. The cellular microtubules get stabilized and their depolymerization gets retarded. Thus abnormal arrays of microtubule bundles are produced which are nonfunctional for cancer cells to divide. These are often used in combination with other chemotherapy agents.
Vinca agents	Vinorelbine	Navelbine	It inhibits cancer cell growth by stopping cell division which works by interfering with the formation of cellular microtubules by binding with protein tubulin.
Hormonal therapy agents			
Antiestrogen drugs	Raloxifene Fareston Fulvestrant Tamoxifen	Evista Toremifene Faslodex Nolvadex	These drugs block the effects of estrogen in the breast by binding to estrogen receptors on tumor cells; thus prevent endogenous estrogen to bind to its receptor and retard tumor growth. Raloxifene; an FDA approved antiestrogen drug for the prevention of BC.
Aromatase inhibitors	Anastrozole Exemestane Letrozole	Arimidex Aromasin Femara	These drugs block aromatase enzyme due to which aromatization of androgens to estrogen does not take place. These drugs are only effective in postmenopausal women whose ovaries are no longer producing estrogen.
Ovarian suppression	Leuprolide Abarelix Buserelin Goserelin	Lupron Plenaxis Suprefact Zoladex	These drugs suppresses the ability of the ovaries to produce estrogen. It is a viable treatment for estrogen-sensitive breast tumors in premenopausal women in combination with chemotherapeutic drugs.
Anticancer drugs (brand name)			
Everolimus (Afinitor®)	Afinitor® is a targeted therapy that acts by inhibiting an enzyme involved in cancer cell growth; it specifically targets mTOR.		
Bevacizumab (Avastin®)	Avastin® functions by inhibiting angiogenesis by cutting off the blood supply to tumors and thus limits growth of tumor.		
Trastuzumab (Herceptin®)	Herceptin® is Mab which specifically attaches to domain IV of HER2/neu receptor protein (induce tumor growth by its binding to growth-promoting agent); thus prevents the tumor growth signaling.		
Adotrastuzumab emtansine (Kadcyla®)	Kadcyla® delivers chemotherapy drugs Herceptin® and DM1 directly to HER2-positive cells and limits exposure of the rest of the body to the chemotherapy.		
Lapatinib (Tykerb®)	Tykerb® (lapatinib) targets tumor cells that express the HER2/neu protein; in contrast to Herceptin®. It blocks the HER2/neu protein inside the cell rather than at the cell surface.		
Pertuzumab (Perjeta®)	It works by targeting the subdomain II of HER2 protein, and retard the growth of HER2-positive BC cells.		

BC, breast cancer; *DM1*, maytansinoid (a cytotoxic component); *mTOR*, a protein kinase enzyme mammalian target of rapamycin; *HER2*, human epidermal growth factor receptor 2.

based drug formulations are promising tools. The recent advancements in nanomedicines and emergence of theranostics as a co-delivery approach would not only target the cancer selectively but also cumbersome the cytotoxicity of therapeutic agent to other organs as well as making the diagnostics more precise and accurate.

Thus, with the advancing tumor biology and genetics, the heterogeneity of breast cancer had come into existence. Thus, to conquer the advancing subtypes and different stages of breast cancer, it becomes necessary to upgrade diagnostic techniques along with pharmacological intervention. The prior diagnosis of tumor subtype and its heterogeneity could help us to understand and establish new therapeutic interventions with considering the challenges like metastasis, multidrug chemo-resistance, associated tumor markers, etc.

Henceforth, it becomes necessary to design novel approaches that could inculcate both diagnostics and therapeutics domains of breast cancer/tumor together. One such approach is known as "theranostic approach" which becomes possible with nano-technological advancements in contrast agents and drug delivery carriers. So, the term "theranostics" can be defined as precise, smart, and targeted co-delivery of both diagnosis and treatment drug. These smart formulations carry nanocarriers made of polymers, proteins, lipids, nucleic acid, carbon, and metals, including dendrimers, micelles, liposomes, nanoparticles, DNA tetrahedral/pyramids, and nanotubes [21, 31–33]. These smart nanocarriers encapsulate antitumor agents with coated surface ligands which bind to receptor/target on BC cells and destroy the cells with release of pharmacological agents along with imaging via tracer agents assuring molecular imaging of the tumor for excellent diagnosis to simultaneously evaluate the efficacy of regimen. Such approach is known as *theranostic approach*, an advancing regimen in cancer diagnosis and therapeutics. It becomes more successful in recent years with the parallel advancement in the development of efficient drug delivery system which could cross the biological barriers with delivery of rationale as well as controlled drug release at desired location, which minimizes the side effects and improves therapeutic efficacy [34]. With the limited pharmacological intervention available for TNBC, the cancer immunotherapy has also played a successful role in treatment of various malignancies. The first FDA-approved immunotherapy for treating TNBC is atezolizumab. Therefore, exploring immunotherapies for treating TNBC at clinical levels has been a worth gaining step [35, 36].

3 Application of nanomedicine in theranostics for breast cancer nanomedicine

The science of nanosystems/carriers has come into existence from the discovery of nano-biomaterials, and it is one of the applications of nanotechnology. The clinically approved nanosystem for BC treatment is classified into lipid, polymer, inorganic, viral, and drug-conjugated nanoparticles (NPs). These vary from each other in terms of their structure, shape, size, and charge. Each carry their unique properties such as drug release, drug loading capacity, targeting at cellular level, and stability [37]. Table 3 enlists the nanomedicines used clinically for BC treatment. Out of all these formulations, liposomal, protein, and polymeric nanoformulations are approved BC nanotherapeutics [52]. They mainly minimize toxicity rather than improving the efficacy compared with the conventional drug formulations [53]. Table 4 enlists the nanomedicines for breast cancer which are under clinical development.

4 Nanotechnology advancements for targeted theranostics

In nanoscience, developing a promising nanoformulation entails numerous physiochemical, functional, and biological properties for biomedicinal uses. The desired size (1–200 nm) and conformation are the deciding factors for the trajectory dynamics of the nanoparticles and are the decisive elements for nanomedicine formulation. The conjugation capacity, encapsulation, and surface modification of the nanoparticles are the key factors behind the designing of precise targeted drug delivery using conjugated ligand against ligands/target receptor on cancer cells. Fig. 1 details the implication of nanomedicines in the delivery of therapeutics for BC.

4.1 Liposomal-mediated advancements in theranostics for breast cancer

Liposomes are the spherical-shaped vesicular nanocarriers of size up to 400 nm composed of hydrophilic core inside with surrounding lipid bilayer. It is extensively employed nanocarrier in designing and development of targeted delivery systems and theranostics for breast cancer especially TNBC due to their various characteristics. These include low toxicity, high biocompatibility, and high biodegradability and could encapsulate both lipophilic and hydrophilic drug agents [54].

Liposome has been developed with conjugation or encapsulation of alone or both the drug and photosensitizers. For instances, indocyanine green (LPICG), a photosensitizer, has been formulated into liposomal carrier, and the release of which is triggered by NIR-based photo dermal therapy (PDT). Its efficacy on mice model of TNBC shows enhanced circulation time, tumor targeting, and tumor growth regression [38]. The loading of gambogic acid (GA) with PEGylated liposomal carriers has been designed for TNBC treatment [39], which enhances the permeation and retention effect through accumulation and prolonged

TABLE 3 Clinically approved nanomedicines for treatment of breast cancer.

Drug name	Type of nanosystem	Trade name	Company	Indication(s)	Year of approval	Reference
Doxorubicin	PEGylated liposome	Doxil	Janssen products	Metastatic breast cancer; Kaposi's sarcoma; multiple myeloma; ovarian cancer	1995	[38]
	Non-PEGylated liposome	Myocet	Sopherion Therapeutics	Metastatic breast cancer	2000 EU and Canada in combination with cyclophosphamide	[38]
	PEGylated liposome	Lipodox	Sun Pharma Global FZE	Breast and ovarian cancer	2013 (as generic of Doxil); preclinical studies	[38]
Daunorubicin	Heat-activated liposome	DaunoXome	NeXstar Pharmaceuticals	AIDS-related Kaposi's Sarcoma	1996	[39, 40]
Paclitaxel	Liposome	Lipusu	Sike Pharmaceutical Co. Ltd.	Breast cancer and NSCLC	2006 (China)	[41]
	Protein NPs: albumin-bound paclitaxel (nabpaclitaxel)	Abraxane	Celgene	Breast cancer	2005	[42, 43]
	PEG-PLA micelles	Genexol-PM	Samyang Biopharm	Breast, ovarian and advanced lung cancer	2007 (South Korea)	[44, 45]
	mPEG-PDLLA	Nanoxel	Fresenius Kabi	Breast cancer and gastroesophageal cancer	2006 (India); Phase I	[46, 47]
SPIONs	Inorganic NPs (iron oxide NPs)	NanoTherm	MagForce AG	Glioblastoma and other brain tumors, prostate cancer	2010 (EU); late clinical trials in USA	[48, 49]
Ado-trastuzumab	Others: Ado-trastuzumab emtansine (MCC-DM1 complex)	Kadcyla	Hoffmann-La Roche	$HER2^{+}$ breast cancer (metastatic)	2018 (Canada) and 2019 (USA)	[50, 51]

TABLE 4 Breast cancer nanomedicines and theranostics in clinical trial.

Nanomedicine formulation	Type of nanoformulation system	Indications	Status in clinical trial
Liposomal NPs			
ThermoDox®	Lysolipid thermally sensitive liposome/doxorubicin	Solid tumors of liver and metastatic BC	Phase III
EndoTAG™-1	DOTAP/paclitaxel	BC and pancreatic cancer	Phase II
MM-302	HER2-targeting antibody liposomal-doxorubicin conjugate	HER2-positive advanced BC	Phase II
LEP-ETU®	NeoLipid technology liposomes/ paclitaxel	BC and ovarian cancer	Approved orphan drug (2015)
2B3-101	GSH PEG-liposome/doxorubicin	Metastatic BC and brain cancer	Phase II
LEM-ETU	NeoLipid technology liposomes/ Mitoxantrone	BC, leukemia, stomach, liver, ovarian cancer	Phase I
Lipolatin®	PEGylated liposome/cisplatin	Pancreatic cancer, nonsmall-cell lung cancer (NSCLC), metastatic BC	Approved in 2015, metastatic pancreatic adenocarcinoma; Phase III for metastatic BC
Onivyde®	PEGylated liposome/irinotecan sucrosofate salt	Metastatic pancreatic adenocarcinoma, small-cell lung cancer (SCLC), metastatic BC	Approved in 2005; Phases II and III; Phase I
Polymeric NPs			
Accurins®	PEG-PLGA/docetaxel	Solid tumors (BC, prostate, endometrial cancer, head and neck cancer, melanoma)	Phase II
NK105®	PEG polyaspartate micelle/ paclitaxel	Metastatic or recurrent BC	Phase III
Protein NPs			
ABI-008 (Abraxis BioScience)	Albumin-bound docetaxel	Prostate cancer, NSCLC, metastatic BC	Phase I/II, preclinical studies
Inorganic NPs			
Aurimune™	PEGylated colloidal GNPs/CYT-6091 THF-targeting ligand	Metastatic BC, adenocarcinoma, colorectal cancer	Phase I/II
AuroLase®	Silica core coated with gold shell and PEG/PTA with gold NPs	Refractory and/or recurrent head and neck tumors, lung and prostate cancers	Phase I for prostate cancer
Others			
Rexin-G®	Tumor-targeted retroviral expression vector/micro-RNA-122	Recurrent or metastatic BC	Approved in Philippines in 2007, in Phase II in USA

BC, breast cancer; *DOTAP*, 1,2-dioleoyl-sn-glycero-3-phosphocholine; *PTA*, photothermal ablation.

exposure at target site. In vitro cytotoxicity assay conducted on MDA-MB-231 cell line showed no significant difference between free-GA and GAL, but GAL-treated cells show decrease in apoptotic marker (BCL-2), cyclin D1, survivin, and rise in BAX expression when compared with control and free-GA treated cells. Thus, GAL induces apoptosis in TNBC. While, GAL treatment on MDA-MB-231 xenograft model enhances the antiangiogenic potential of GA along with enhanced circulation time and decrease in tumor growth and tumor volume [39]. Another instance that folate when conjugated with PEGylated liposomes enhances specific targeting to overexpressed folate receptors on TNBC cell surfaces [40]. When these liposomes encapsulated with photosensitive agent, benzoporphyrin derivative monoacid (BPD) utilizes photodynamic therapy for irradiation to get activated and triggers tumor cell death [40]. Furthermore,

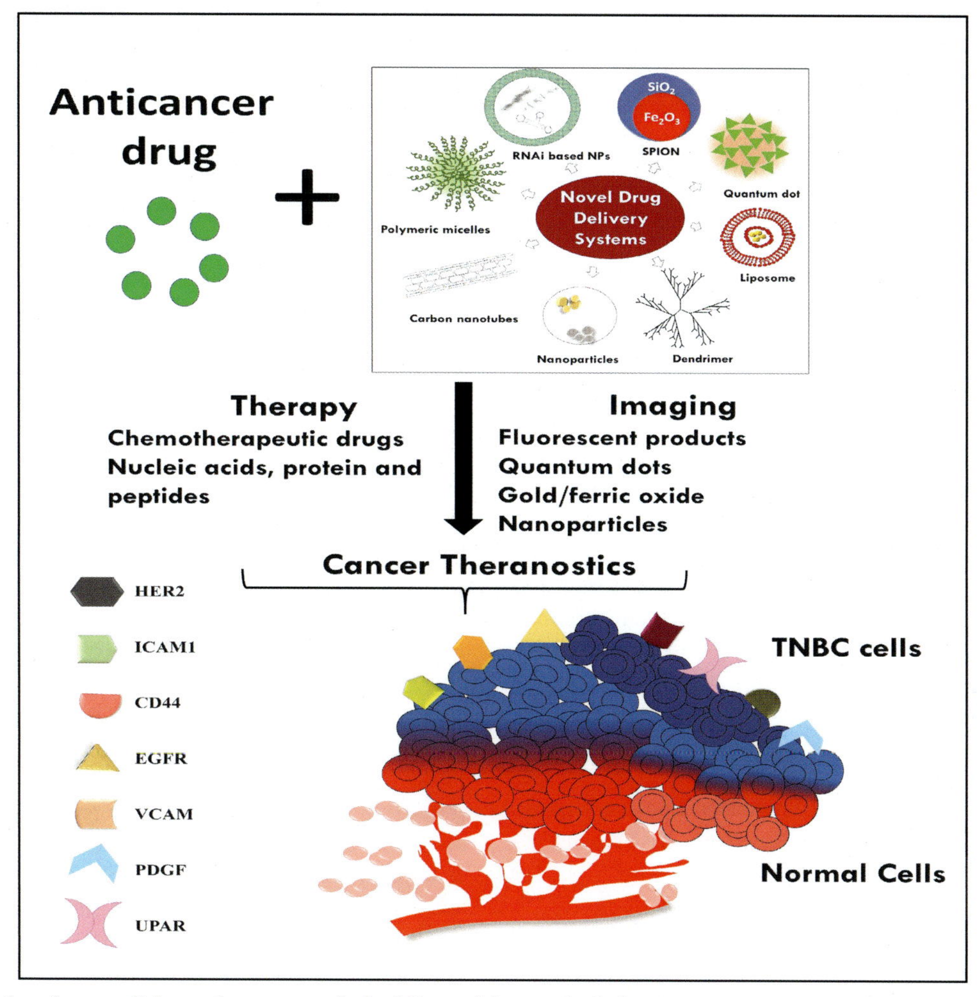

FIG. 1 Implication of nanomedicines and nanosystems in the delivery of therapeutics for breast cancer.

encapsulation of 6BrCaQ (analogue of novobiocin) into PEGylated liposomal carrier inhibits tumor growth at low dose in MDA-MB-231 tumor mice model [41]. Dual drug encapsulation of PEGylated liposomes with doxorubicin (DOX) and hexadecyl oxaliplatin carboxylic acid showed significant antitumor efficacy [42].

The surface-modified liposomes were designed with a cyclic peptide, c(RGDyK), in conjugation with DSPE-PEG2000-NHS for effective targeting and cellular uptake by TNBC cells [43]. These peptides possess high affinity for integrin-α3 receptor overexpressed on malignant breast cancer cells. These surface-modified liposomes are utilized for encapsulating dual agents, namely vincristine and dasatinib, to prevent cancer relapse by inhibiting formation of vasculogenic mimicry channels [43]. Liposomal carriers with prodrug-modified surfaces for co-delivery of docetaxel (DTX) and gemcitabine for targeting CD44 overexpressed TNBC cells have been formulated and possess high drug loading capacity and dual stimulus-responsive drug release [44]. The in vitro studies showed synergistic effect of drug on apoptosis, cytotoxicity, and wound healing. Further, in vivo studies demonstrated better accumulation in tumor cells and enhanced antitumor and antiproliferative activity with no systemic side effects [44]. The encapsulation of ruthenium coordination complexes in hydrophobic bilayer of liposomal carriers (Lipo-Ru) was done, and this induced double-stranded breaks (DSBs) and apoptosis on MDA-MB-231 cells. The in vivo studies in TNBC mice models showed tumor growth inhibition [45].

Furthermore, TNBC overexpressed ICAM1 has been targeted by engineering of lipocalin-2 SiRNA into encapsulation with liposomal drug delivery system [46]. The liposomes targeting ICAM1 possess more binding affinity to MDA-MB-231 cells than MCF-10A cells (nonneoplastic cells) [46]. Perhaps, strategies are on the way to design dual complementary liposome (DCLs) with encapsulating DOX for targeting both EGFRs and ICAM1 overexpressed on TNBC cells [47]. The DCLs possess following advantages: (i) increase in cellular binding due to multivalent surface interactions, (ii) increased endocytic internalization, and (iii) increased antitumor efficacy with blocking of ICAM1 and EGFR pathways [47]. Perhaps, with the inherent limitation of liposomes (poor stability, low drug loading capacity, drug leakage, and difficult sterilization), liposomes are most studied targeted delivery system not only for delivery of pharmacological agent but also for theranostics [48, 49]. For instance, TNBC overexpressing integrin-α3 with cyclic octapeptide LXY (Cys-Asp-Gly-Phe (3,5-DiF)-Gly-Hyp-Asn-Cys conjugated liposome with dual drug encapsulation (DOX and rapamycin) [50]. However, DOX and sorafenib-loaded liposome express better anticancer activity in TNBC xenograft mice model [51].

Commercially available DOX liposomal formulations are associated with cardiotoxicity, whereas a novel micelle-encapsulated DOC formulation (NK911) that is in clinical trial has improved tumor penetration and reduced in vivo toxicity [55]. Perhaps, liposomal delivery system found effective in successful delivery of antimetastatic agent (antagomir-10b) with co-delivery of anticancer agent paclitaxel (PTX) to delay 4T1 tumor growth and reduce lung metastasis of breast cancer [56]. In addition to this, PEGylated PTX nanocrystals cause significant inhibition and 82% reduction in tumor growth against MDA-MB-231 nude mice and lung tumor metastasis model [57]. However, lipid-conjugated estrogenic (47.03% bioactive) NPs when combined with cisplatin show up to 87% inhibition of tumor growth against MDA-MB-231 xenograft mice model [58]. So far, irinotecan and PTX-loaded liposomes named as MM-398 and EndoTAG-1, respectively, have reached the clinical trial for TNBC patients [59].

4.2 Polymeric micelles (PMs)

Polymeric micelles are the promising self-assembled copolymers composed of hydrophobic core which is formed from stabilized hydrophilic shell and Van-der Waals bonds [60] which imparts amphiphilic properties. Owing to their amphiphilic properties, micelle can deliver both water soluble and hydrophobic anticancer drugs. PMs offer inherent advantage over other nanomedicines, that is, avoidance of the reticuloendothelial system (RES) [61]. PMs enhance both water solubility and permeability of anticancer drugs with least nonspecific toxicity. The recent advancement made in PMs for TNBC therapeutics is their incorporation into quantum dots (quantum dot-based micelles) and magnetic micelles. The incorporation of aminoflavone into quantum dot-based micelle demonstrates better tumor targeting and significant tumor regression in MDA-MB-23 cells and requires minimum dose to inhibit tumor growth with reduced systemic toxicity [62]. Magnetic nanomicelles have also been developed with diverse application in tumor-targeted drug delivery, MRI, and induce hyperthermia in localized tumor regions. Magnetic micelle loaded with dasatinib (multitarget kinase inhibitor) showed 1.35-fold increase in cytotoxicity against MDA-MB-231 cells. These magnetic-based nano-medicines possess applications in tumor-targeted drug delivery [63]. In addition to this, combination drug nano-micelles have also been designed such as cetuximab-conjugated DTX-loaded micelle [31] and SAHA (suberoylanilide hydroxamic acid)-conjugated PTX-loaded micelle [64]. All these combinations were prepared to enhance therapeutic potential of drugs by several folds. These attempts were made to lipophilize bortezomib (proteosomal inhibitor) with broad range of effectiveness against various cancer types [65]. The lipophilized-micellar nanoformulation of bortezomib elicits its enhanced potential through prolonged circulation time, enhanced tolerability, and elective tumor targeting [65]. This micellar system containing styrene-*co*-maleic acid (SMA)-conjugated hydrophobic derivative of curcumin (RL71) for treatment of TNBC showed high cytotoxicity due to endocytosis mediated faster cellular uptake and slower drug release profile [66]. This approach enhanced the drug uptake; however, it still lacks the specificity. Therefore, it is still a serious challenge in the treatment of metastatic TNBC.

In vitro study of transferrin conjugated to vitamin-E-D-alpha-tocopheryl polyethylene glycol succinate (TPGS) conjugated with cetuximab on MDA-MB-231 showed higher IC_{50} (0.1715 μg/mL) than TPSG without cetuximab (1.12 μg/mL) and free drug (35.26 μg/mL), respectively. Furthermore, such approaches for nano-medicine have promising capability in breast cancer to further clinical trials could be explored as theranostics [67]. The classical example of theranostics is TPSG micelle-conjugated DTX and diagnostic nanocluster AUNc for simultaneous detection and treatment against MDA-MB-231-Luc breast cancer in vitro model [67]. In xenograft model, the real-time imaging and tumor inhibition were imaged.

The copolymeric micelles, that is, poly(acrylic acid)-g-PEG loaded DOX, was developed, which showed significant reduction in 4T1 mouse breast tumor growth and lung cancer metastasis [56]. However, the only miracle micelle which enters the clinical phase II successfully for TNBC patients is NK012 micelle (composed of PAG-polyGlu-conjugated irinotecan) and requires validation in other clinical trial phases [68].

In comparison with other DDSs, the polymeric micelles possess more stability, high loading capacity, and effectively deliver hydrophobic drugs with high therapeutic potential. But challenges like improving the drug-loading efficiency, increasing thermodynamic and kinetic stability, and enhancing transport across the cell membrane need to be resolved before fully utilizing the potential of micelles as a TTDDs [49].

4.3 Dendrimers

Dendrimers are the synthetic highly branched polymeric macromolecule (10–100 nm) with central core surrounded by branched polymeric chains arranged either in convergent or divergent manner [69]. They are classified by their form as polymers, hyperbranched polymers, or brush polymers and can also be classified by their molecular weight as low or high. These are categorized into generations depending on the number of polymeric branches, that is, G1, G2, G3, G4, and G5. Dendrimers were first described in 1978 by Vogtle, and different types of dendrimers were developed, namely, poly(amidoamine) (PAMAM) dendrimer [70–74], polylysine (PLL) dendrimer [75], poly(propylenimine) (PPI) dendrimer [76, 77], and phosphorus dendrimer [78]. Due to its special three-dimensional structure, excellent water solubility, easiness of functionalization, and low immunogenicity, the dendrimer-based therapeutics as well as theranostics had become the most promising and efficient platforms for the treatment of breast [79].

4.3.1 Dendrimer-based cancer therapy

There are numerous nanomedicinal applications involving platforms like physical loading of drugs and therapeutics via electrostatic interactions or by hydrophobic interaction, covalently linking drug molecules onto dendrimer surface, compress genetic materials (noncoding DNA, SiRNA, miRNA, etc.), via electrostatic interaction, entrap inorganic NPs within the dendrimer interior, or stabilize inorganic NPs. These approaches allow various dendrimer-based nanoplatforms for treatment, diagnosis, or both using chemotherapy, radiotherapy, photo-thermal therapy (PTT), photodynamic therapy (PDT), gene therapy, or combination therapy [80–82]. The types of dendrimers utilized for breast cancer therapy have been tabulated in Table 5.

Dendrimers are the suitable candidate for delivery of siRNA to tumor cells because of its cavity-enriched spherical shape with a hydrophilic polymeric chain and hydrophobic core [32, 112]. In TNBC xenograft mouse model, antisense oligo (AODNs)-conjugated PAMAM (polyamidoamine) dendrimer demonstrates the decline in tumor vascularization [113]. The AODN-conjugated PAMAM dendrimers utilize overexpressed VEGF receptors as their therapeutic target. Another approach made utilizing siRNA-conjugated PAMAM dendrimers had shown promising downregulation of TWISTI transcriptors, a potential

TABLE 5 Types of dendrimer studied for cancer therapy.

Type of cancer therapy	Type of dendrimer	Functional group	Reference
Chemotherapy	PAMAM	DOX α-TOS 2-ME GEM	[83] [84, 85] [86] [87]
	PPI	DOX	[88]
	PLL	MTX	[89]
Radiotherapy	PAMAM	^{131}I ^{177}Lu	[90, 91] [92, 93]
PTT	PAMAM	GNRs ICG MoS2 CuS AuNSs	[94] [95] [96] [97] [98]
PDT	PAMAM	Ce6	[88]
	PPI	Phthalocyanines Methylene blue	[99] [100]
	Phosphorus dendrimers	Rose bengal	[100, 101]
Gene therapy	PAMAM	Luciferase-targeted siRNA or Bcl-2 siRNA Bcl-2 siRNA siBcl-xl, siBcl-2, siMcl-1 siRNAs and a siScrambled sequence	[102] [103–106] [107]
	PPI	Bcl-2 siRNA Luciferase-targeted siRNA pDNA	[108] [109] [110, 111]
	Phosphorus dendrimers	siBcl-xl, siBcl-2, siMcl-1 siRNAs and a siScrambled sequence	[107]

TNBC target [114]. Dendrimers had gained immense light not only in delivery of therapeutics but also in diagnostics. For instance, GdDOTA and DL680 (NIR dye) conjugated G4PAMAM were prepared and injected subcutaneously in TNBC mouse model for imaging and drug delivery. MRI scan and near-infrared (NIR) fluorescence imaging show homing of NPs, and high-frequency signal in TNBC model, respectively, demonstrated targeted diagnostic potential/utilization of (GdDOTA) 42-G4PAMAM-DL680 dendrimeric agent. Moreover, dodecylated G2, G3, and G4 PAMAM dendrimers were developed to deliver luciferase (Luc)-targeted siRNA or Bcl-2 siRNA to MDA-MB231 cell line for gene silencing [102]. The dodecylated G2, G3, and G4 dendrimers showed significant high rate of gene silencing than unmodified G2, G3, and G4 PAMAM dendrimers. Moreover, 84.6% Luc gene was knocked down by dodecylated G4 PAMAM dendrimers with more than 90% cell survival implicating best drug delivery [102].

4.4 Polymeric nanoparticles

Polymeric nanoparticle size ranges from 50 nm to 10 μm and when these nanoparticle sizes range up to 10 μm, it is classified as misnomer nanoparticle. Besides drug and protein incorporation without chemical modification, these nanoparticles offer special advantage that these are prepared from drugs and proteins as well. Drug and/or protein molecule could be effectively encapsulated by electrospray, emulsification, and nanoprecipitation. Moreover, in concern to improve biocompatibility and reduce toxicity properties, biodegradable polymeric particles, that is, poly(lactic) and copolymer like poly(lactide-*co*-glycolide), had been under application for synthesis of polymeric nanoparticles [115].

The development of PRINT (particle replication in nonwetting templates) method for synthesis of polymeric NPs of uniform size provides the scope for customization of properties for building effective cancer therapy [116]. The nontargeted delivery of Pt(IV) mitaplatin through PLGA-PEG NPs showed higher rate of tumor inhibition in TNBC xenograft model (nude mice with MDA MB 468 cell line) [117]. Researchers identify a novel peptide (Gle-Ile-Arg-Leu-Arg-Gly) which recognizes glucose-regulated protein (GRP78) [118]. In TNBC xenograft mice model, induction of apoptosis in GRP78 expressing tumors via targeting through specific GIALAG-conjugated PTX encapsulated polyester NPs. A recent clinical study performed on pretreated metastatic breast cancer patients expressing high-protein Trop-2 reported 30% response rate when treated with anti-Trop-2-SN-38 antibody (IMMU-1322) [119]. The combination regimen of succinobucol and P188 (poloxamer) is emerging as best oral breast cancer treatment as evidenced by 13-fold improvement in bioavailability of succinobucol NPs enhancing inhibition of VCAM-1 invasion and migration of tumor cells [119]. Polymeric NPs are well known for delivery of gene therapy (miRNA and siRNA) with therapeutic drug for reducing tumor volume and growth. This is evidenced by the findings when PLGA-b-PEG polymeric NPs delivered both antisense-miR-21 and antisense-miR-10b with 0.15 mg/kg of drug loading; of them, co-loaded siRNA-DOX-NP induces eightfold reduction in tumor volume and growth [120, 121]. Polymeric NPs also successfully deliver ligand, Arg-Gly-Asp (RGD), which either target drug delivery or inhibit caner invasion in metastatic breast cancer tumor models. For example, cyclic RGD functionalized solid lipid NP (RGD-SLN) inhibited both adhesion and invasion of alpha-v beta 3 (αvβ-3) integrin receptor in an invasive metastatic breast cancer tumors [122]. This study provides the excellent example of ligand targeting and delivery in breast cancer cells through polymeric NPs. Zhang et al. carried out the synthesis of hybrid shealth polymer lipid nanoparticle (PLN) conjugated to peptide ligand RGD loaded DOX and mitomycin C (RGD-PMPLN). The synthesized RGD-PMPLN which was evaluated for therapeutic efficacy in metastatic breast cancer mouse model showed the enhanced cytotoxicity with due synergism of DOX-MMC with further enhancement by target RGD-PMPLN [123]. Such targeted delivery synergistic drugs enhance the overall efficacy and open doors for wide applications in breast cancer treatment.

4.5 Metal nanoparticles (MNPs)

MNPs are nanosized metal particles of size ranging between 1 and 100 nm. MNPs include gold (Au), platinum (Pt), silver (Ag), zinc (ZnO), titanium dioxide (TiO_2), etc., which had shown potential in cancer therapy. These nanoparticles may owe their wide applicability in therapeutic and diagnostic fields due to their unique optical, magnetic, thermal, and electrical properties, as well as their chemical composition. The size of metal plays an important role in interaction with membrane protein, enzymes, and other components of the cell. The surface modification of MNPs via conjugation with different groups expands their acceptance in clinical utility for cancer therapeutics. The different MNPs work by different molecular mechanism, likely via generation of reactive oxygen species (ROS) and increase in cellular oxidative stress, DNA intercalation, and specific apoptotic tumor cell death [124]. MNPs radiate electromagnetic radiations into heat to induce hyperthermia to kill the tumor cells. Because of their unique physiochemical properties, some MNPs possess inherent potential as anticancer activity.

Gold nanoparticle (AuNP) is the most studied and promising MNP known to deliver PTX. Au NPs have been designed in various shapes and configurations like Au-nano shells (AuNS), Au-nanorods (AuNR), and Au-nanocages (AuNC). These are an emerging versatile nanovehicle for breast cancer therapy. AUNP coated with PEG provides higher survival rate in breast cancer mice model beside the ionizing radiation [125]. The serum-coated AuNR possesses inherent capability of downregulating the gene expressions responsible for energy kinetics of tumor cells. Thus, with low energy, the invasion and migration of breast cancer cells are inhibited in both in vitro and in vivo models. However, the introduction of combinational approach involving cisplatin-loaded AuNR and NIR laser showed promising suppression of TNBC tumors metastasis [58]. Silver nanoparticles (AgNPs) are also very well-explored NPs for their antiangiogenic, antiproliferative, and proapoptotic effects on breast cancer cells. The AuNPs are a potent radiosensitizing agent which reacts with acidic environment of cancer cells and raise oxidative stress with due production of ROS which induce damage and cell death/apoptosis. AgNP treatment after radiotherapy studied on gliomas showed promising results. These NPs also cause inhibition of VEGF on breast cancer cells, thus limiting the metastasis [126]. Additionally, zinc oxide nanoparticles (ZnO NPs) function as genotoxic for breast cancer treatment. It forms micronucleus in tumor cells, which increases mitotic and interphase apoptotic cell death [127]. Asparaginase is an enzyme which hydrolyzes the asparagine into nonfunctional aspartic acid and ammonia leading to deprivation of required amino acid and blocks tumor cell proliferation by interrupting the asparagine-dependent protein synthesis. Therefore, asparagine is well known for anticancer potential. It was formulated with ZnO NPs, which improves stability and specificity when given in combination with PTA and DOX [128]. Even, PTX and cisplatin when combined with ZnO NPs show improved efficacy and reduced toxicity in breast cancer cells [129]. Similarly, other MNPs like copper oxide NPs (CuO NPs), iron oxide (Fe_2O_3), silica, titanium oxide, and cerium oxide are also under developmental pipeline for exploration of their potential in breast cancer treatment and diagnosis. CuO NPs have been described as green NPs as they are synthesized from *Acalypha indic*a and *Ficus religioss*. These NPs treat metastatic lung tumor cells (B16-F10 cell line) by inducing apoptosis and generating ROS [130]. The advancements in breast cancer subtypes and continuous pathological alterations occurring in cancerous cells have led the introduction of dual-modal theranostic therapy, which utilizes the photo-dermal and radiotherapy with Cu-64 labeled copper sulfide NP (CuS NP). Such therapy resulted the suppression of tumor growth in subcutaneous B747 breast cancer model and prolonged the survival of orthotopic 4T1 breast tumor bearing mice [131]. The anti-HER2 antibody-conjugated silica-gold nanoshells in photothermal therapy specifically target HER2 overexpressed human breast cancer cells (in vitro). Cerium oxide NPs (CNPs) work as radio-sensitizing agent by increasing oxidative stress and induce DNA damage leading to apoptosis of tumor cells [132]. It supplements the conventional chemotherapy by delivering chemotherapeutic drugs like DOX, thus providing smart approach for breast cancer therapy. Additionally, platinum- and titanium-based NPs are also a promising nanocarriers and therapeutic candidate in cancer photodynamic therapy, respectively. The iron-oxide nanoparticles (Fe_2O_3 NPs) utilize the magnetic property for accurate diagnoses and targeted treatment in mouse model of squamous carcinoma [133]. The conjugation of multivalent pseudopeptide (N6L) and DOX with Fe_2O_3 NPs (MF66) forms a multifunctionalized Fe_2O_3 NPs known as MF66-N6L-DOX. It involved combination of hyperthermia and drug delivery performs better specificity and tumor killing potential in breast cancer model [134]. The cyclic RGD peptide conjugation with Fe_2O_3 NPs and Mab conjugated SPIONs had improved the diagnosis of micrometastasis and metastatic breast cancer in transgenic mice, respectively [135, 136]. Herceptin®, a trastuzumab conjugated modified magnetic polymerosomes, had reached clinical platform to target bone metastasis in HER2 overexpressed breast cancer model (BT474).

4.6 DNA nanostructures

DNA nanostructures had become the emerging nanomaterial for various application in nanotechnology due to their various advantages, namely specific self-assembly ability through complementary (Watson-Crick) base pairing, biodegradability, biocompatibility, precise control over size and shape, and functional modifications with other molecules. The variety of DNA nanostructure has been produced in 2D and 3D conformations, namely cube [137], tube [138, 139], tetrahedron [140], truncated octahedron [141], icosahedron [142], 2D-DNA arrays [143, 144], and DNA origami nanostructures [145, 146]. Among these reported DNA nanostructures, tetrahedral DNA nanostructures (TDN) have been widely utilized for biomedical purposes. Such DNA nanostructures can entrap ligands and/or small functional compounds for site-specific attachment and/or for bioimaging. Various studies have been conducted by researchers in relation to designing and their testing against metastatic breast cancer. Likely, Kutty et al. designed a novel self-assembled DNA nanopyramid, and it was designed with tagging glutathione (red-emissive) protected gold nanoclusters (GSH-Au NCs) at the base and incorporated dactinomycin in the DNA groove [147]. Such nanoclusters were designed for detection and killing of *E. coli*, which warranted the evaluation for cancer treatment [147]. The major hurdle behind their utilization for breast cancer is to escape them from endosomal degradation in mammalian breast cancer cells. Another nanostructure, DNA tetrahedron (TH), was introduced for bioimaging and antibody-mediated targeted drug delivery. TH self-assembled into four vertices. Cetuximab-conjugated TH (THC3) with intercalated DOX (THDC3) resulted in killing EGFR overexpressed TNBC cells with low IC_{50} of THDC3 (0.91 μM) than IC_{50} of free DOX

(3.06 μM) [33]. This formulation was modified carrying three cetuximab and one Cy3 probe, that is, Cy3-THC3, which showed high-signaling intensity with enhanced uptake by MDA-MB-68 cells. Both THDC3 and Cy3-THC3 with slight modification show improved targeting and more cidal on cancer cells, which could be an excellent candidate as nanomedicine for TNBC theranostic.

4.7 Carbon nanotubes (CNTs)

CNTs are the cylindrical geometric structures carrying graphene ring knitted flat sheets molded either single-walled nanotubes (SWNTs) or multilayer-walled nanotubes (MWNTs). CNTs have their vast applications in nano-biotechnology science as a carrier for proteins and vaccines, sensor for protein and DNA detection, and diagnostic tool for various types of serum protein estimation [148]. The chemical modification in CNTs offers various therapeutic advantages with huge possibility in cancer therapy. They possess high electrical and thermal conductance, high chemical stability, surface area, and mechanical strength [149]. As CNTs are completely insoluble in all solvents, toxicity and health-related issues may occur due to this limitation. However, chemical modification improves their interaction with active molecules, which include proteins, peptides, nucleic acids, and other therapeutic agents and solubility [150]. SWNTs (1–2 nm in diameter) possess the ability to penetrate inside the cells imparting prolonged distribution which improved localized effects. Oxidized MWNTs (o-MWNTs) carry novel approach for cancer therapy through reducing macrophages and vascular density in the tumor [60, 151]. CNTs induced hyperthermia promoting cell membrane permeability leading to destruction in tumor mass [152]. Therefore, photo-thermal-induced ablation using MWNTs has been proposed for metastatic breast cancer therapy. DOX-complexed nanodiamond had been designed and tested on breast cancer mice model, which showed improved drug loading, minimized drug efflux, increased apoptosis [153], and lung metastasis of breast cancer [154]. Hyaluronic acid (HA)-modified SWNTs conjugated with DOX was developed in 2019 for enhancement of its breast cancer therapeutic efficacy [155]. HA binds specifically to CD44 receptor and was noncovalently coated with simply electrostatic adsorption. The prepared formulation showed significant improvement in intracellular delivery of DOX in CD44 overexpressing TNBC cell line, along with inhibiting proliferation and induced apoptosis. Additionally, it also blocked MDA-MB-231 cell migration and also inhibits growth of cancer cell spheroids [155]. PTX-loaded riboflavin and thiamine-conjugated MWNTs have also been synthesized and studied on MCF-7 breast cancer cell lines [156].

For selective and specific treatment of tumors, molecular targeting and photodynamic therapy have high potential. For targeting SWCNT on breast cancer, HER2- and IGF1-specific monoclonal antibodies can bind to SWCNT cells for specific near-infrared phototherapy. Because of this approach, these nanotubes selectively attach to the breast cancer cells.

Mashal et al. demonstrated that SWCNTs are a probable dielectric contrast enhancer between normal and malignant breast tumor microwave detection and induced microwave hyperthermia in malignant breast cancer diagnosis and treatment [157]. Collagen-coated MWNTs were designed, and seeding of breast cancer cells demonstrated the rise in cell adhesion and decreased in cell migration which indicates up-regulation of E-cadherin. This study provides an alternative approach for metastatic breast cancer [158]. With simultaneous investigation of two unique optical properties of SWNTs, that is, very strong NIR absorbance and a very strong Raman signal [159], SWNTs have made wide application in cancer therapy. HER2 IgY-SWNT complex was synthesized by covalently functionalizing SWNTs with anti-HER2 IgY antibody to produce new SWNTs with high sensitivity and specificity for the IgY antibody. When this complex was applied in vitro, it detected and selectively destroyed HER2 receptor expressing breast cancer cells. The Raman signal obtained from tumor cells was identified at unicellular level. The novelty of this dual-function agent is that there is no internalization by tumor cells for achieving selective photothermal ablation and it offers advantage of its easy transmission to other tumor types. Therefore, more research is needed to justify its applicability in the clinical settings. The recent studies on PEG and estradiol were used for modification of CNTs along with loading of lobaplatin for treatment of breast cancer. After 72 h of incubation, significant inhibition of tumorigenic activity was observed at 200 μg/mL of PEG-E2-CNT-lobaplatin without any toxic effects [160].

4.8 Ligands for targeted TNBC therapy

The ligands for targeting breast cancer are the small sequence/pair of nucleotides, peptide, or small molecules which bind specifically to its receptor through ligand-receptor interaction. These include antibodies, aptamers, peptides, and others like carbon, and quantum dots are widely utilized ligands for targeted or probe-based diagnostic in breast cancer nanomedicine.

4.8.1 Aptamers: Nucleic acid-based ligands

Aptamers are short oligonucleotides which stretch of single-stranded DNA/RNA. With its unique 3D confirmation, aptamers specifically bind to the target molecule with high affinity and strength. The only limitation is the degradation by nucleases; however, its high stability gained attention for the development of molecular probes. A newly LXL-1 aptamer was identified by cell-SELEX method which specifically targets surface protein on TNBC tumors [161].

PDGF-aptamer conjugated to gold nanoparticles were used to detect the differential overexpression of platelet-derived growth factor (PDGF) receptor in TNBC cells [162]. MCF7 and MDA-MB-415 breast cancer cells are known to overexpress mammaglobin B1 and the mammaglobin A2. Hassann et al. [163] used MAMA2 and MAMB1 aptamers to detect the metastatic breast cancer by using highly sensitive terahertz (THz) chemical microscopy (TCM) using THz radiations. Additionally, 26-mer G-rich DNA aptamer was used specifically to target the nucleolin receptor breast cancer cells [164]. However, aptamer-based diagnostic approach still needs a lot of improvement and should be combined with drug delivery for TNBC theranostic use.

4.8.2 Antibodies as theranostics for breast cancer

Antibodies are regarded as best ligand for targeting any receptor/target as antigen due to its high selectivity and affinity for its target receptor. The antibodies are utilized in diagnosis and immunotherapy of cancer. To conceptualize the differential upregulation in expression level of tissue factor (TF) receptor and urokinase plasminogen activator receptor (uPAR) in TNBC, researchers validated the utility of anti-TF antibody labeled with Cu-64 (anti-TF-antibody-64Cu) using PET imaging in TNBC in vitro model [165]. LeBeau et al. detected NIR fluorophore and Indium-111 (^{111}In) labeled uPAR antibodies using optical and SPECT imaging, respectively [166]. Similarly, anti-EGFR and anti-VEGFR antibodies were conjugated with fluorescent NP, and ultrasound contrast agents are detected using fluorescence microscopy and ultrasonography. The in vivo study performed on TNBC xenograft mice [167] using Iodine-124(^{124}I)-labeled B-B4 antibody (targeting syndecan-1; CD138 antigen) and I-131 (^{131}I)-radiolabeled B-B4 antibody demonstrates good visualization of TNBC tumor and excellent treatment, respectively [167].

4.8.3 Peptides: Cell-penetrating ligands

Peptides are the low-molecular-weight peptide sequence with ability to specifically target intracellular molecules [168]. These target binding peptide sequences are fused with bacterial coat proteins and can be expressed using genetic engineering which are finally screened by phage display library technique [169]. There are few peptides which have been employed for targeting metastatic BC-P-selectin, arginyl-glycyl-aspartic acid (RGD), tumor metastasis targeting (TMT), and chlorotoxin. Peptide sequence known as CK3 (Cys-Leu-Lys-Ala-asp-Lys-Ala-Lys-Cys) was prepared and observed for its binding to neuropilin-1 (NRP-1) protein using NIR fluorescence imaging in TNBC mice models [170]. Activatable cell-penetrating peptides (ACPPs) target the enzyme matrix metalloproteinase-2 (MMP-2) was covalently bounded to cyclic-RGD peptide and evaluated for its potential in TNBC mice model showed enhanced uptake by tumors and improved contrast imaging [171]. The modified Fe_2O_3 NPs when linked to cRGD peptide showed better targeting of $\alpha v\beta 3$ integrin receptors [172]. Additionally, P-selectin and RGD peptide when linked together to liposomal NP showed efficient targeting at their respective overexpressed receptor at breast tumor sites [173]. The pH-responsive low insertion peptide (pHLIP) was designed in conjugation with MRI-NP (pH-responsive MRI nano-probe) which showed specific internalization and accumulation in TNBC cells [174].

4.8.4 Other small molecules

These are the low-molecular-weight (< 500 Da) ligands having targeting potential for cancer imaging. The clinically utilized molecule is 2-deoxy-2-[fluorine-18]fluoro-D-glucose (^{18}F-FDG), with PET imaging providing information on the basis of glucose uptake and glycolysis by cancer cells [175], while other molecule, folate, has potential as direct imaging agents. Studies performed show that folate molecule aid in the reaching the SPIO contrast agent (P1133) onto the folate receptors and get internalized in growing TNBC [176]. Even folic acid (FA)-conjugated gold nanoparticle (FA-AuNP) also targets the folate receptors and showed high uptake in 4T1 metastatic breast cancer cells [177]. Similarly, carbon dots (CDots) and quantum dots (QDs) are useful tool in cancer imaging [178] and are utilized for detection of early-stage TNBC. Not only this, but chemokine receptor type also 4 (CXCR4) is a cellular target responsible for tumor growth and metastasis of breast cancer.

By improving the cellular uptake into MDA-MB-231 cells with Plerixafor or AMD3100 (CXCR4 ligand) conjugated poly(lactide-*co*-glycolide) NPs found to enhances the siRNA mediated gene silencing [179]. Moreover, AMD3100-loaded human serum albumin-encapsulated NPs found to target the CXCR4 on metastatic model of breast cancer [180]. Hyaluronic acid (HA) found to possess high affinity to CD44 receptor. Therefore, an ultra-small (~5 kDa) HA-PTX nanoconjugate is taken up via CD44 receptor-mediated endocytosis into metastatic breast cancer (MDA-MB-231Br) cells [181]. Urokinase plasminogen activator receptor (uPAR) in conjugation with poly(lactic-*co*-glycolic acid)-*b*-PEG polymers loaded with two antisense miRNA showed potent tumor inhibitory activity through [120] functionalized fullerenes and have been regarded as novel contrast agents in MRI. Other small carbon molecules like nanodiamonds and nanocarbons with distinctive physical and chemical properties are also emerging very well as biomedicines [182, 183]; however, they need to be extensively studied.

4.9 Virus-like particles (VLPs)

VLPs are the self-assembled nanostructures of size ranging from 20 to 200 nm, which closely resemble viral protein complexes [89,202]. They are called virus like because they do not contain viral genetic material making them multipurpose nanovehicle for delivery of therapeutics. There are peptides whose origin could be from plant, microbial, or mammalian origin, which are assembled into filamentous or spherical form [184]. VLPs carry highly organized and repetitive structures which are recognized as potent geometric pathogen-associated structural patterns (PASPs) [185]. Such characteristic does help in cross-linking the B-cell receptors as well as recruits members of the innate humoral immune system such as natural antibodies and complements, further enhancing innate and adaptive immune responses [186]. The functional group modifications in the capsid proteins of functional groups helps in designing target-mediated therapy. The uniqueness of VLPs is their particle size, which is quite small to pass through blood vessels and lymph nodes, while functional viral proteins of their surface aid penetration into the cell. VLPs function by encapsulating the small drug molecule/radiotagged material, and thus targeting the tumor cells here required by specifically entering tumor cells of interest using receptor-mediated endocytosis and finally release the free drug into the tumor microenvironment or cancer cells. Their promising potential is rewarded by their ability to escape from endosomal lysosomal degradation, favoring availability of therapeutics and in the blood stream as unchanged. However, the only one disadvantage which restricts their utilization is that VLPs produce innate immune response due to viral proteinaceous particle and their fast uptake by dendritic cells [187].

VLPs gained popularity, which is attributable to its versatility, cell-specific targeting, efficient cell entry, lack of endosomal sequestering, multivalency, biocompatibility, large encapsulation, and safe delivery system. However, VLPs are at their infancy for drug targeting and delivery systems. VLPs offer a promising platform of vaccine development for breast cancer. There are several task performed in which VLP-based vaccine has been tested for their potential on breast cancer at preclinical level. The HER2 receptor offers a suitable candidate to target via VLP-based vaccine [188]. A VLP-based vaccine, AX09-0M6, was derived from MS2 bacteriophage, which can be efficiently used in aggressive forms of breast cancer, especially HER2-overexpressed breast cancer [189]. A similar study was designed in which cysteine-glutamate exchanging transporter (xCT), a member of heterodimeric sodium-independent transporter system, is highly specific for transport of cysteine and glutamate across breast tumor cells [190]. Thus, its expression is highly observed in breast cancer stem cells [191]. The AX09-OM6 vaccine that was tested on 4T1 subcutaneous breast tumor mice model showed significant decrease in growth of breast tumors and also prevented pulmonary metastasis [189].

Researchers Palladini and colleagues [192] designed VLP-based vaccine system in 2018 which utilizes AP205-VLPs conjugated with SpyTag/SpyCatcher allowing bidirectional and high-density fusion of HER-2 antigen with exposed SpyTag on Ap205-VLPs. The study demonstrated that vaccine system has B-cell tolerance and induced potent anti-HER-2 IgG titers. Its therapeutic potential was evaluated on wild mice with HER2 mammary gland cancer. Thus, induced Ab titers had easily retarded tumor progression. Such results were also obtained when evaluated on transgenic mice with spontaneous mammary gland tumors expressing HER-2 [192].

5 Nanomedicine for gene therapy

Gene silencing is currently developing as tailored approach for treating deadly disease like cancer by means of RNA interference (RNAi). The microRNAs and siRNAs come in the category of effector molecules that are employed to mute particular oncogenes [193]. Now, the main limitation to siRNA- and miRNA-based therapies is their clinical lack into effective delivery system, which can inhibit nuclease degradation of RNA molecules, deliver them within cancer cells, and release them into target organelles in smart delivery system which can inhibit degradation. Several formulations of nanoparticles are fabricated by formulation scientists with the aim to minimize above-mentioned drawbacks for efficient gene-targeted therapy as well as verified through cell line studies and in vivo results [194].

The nanoparticle formulation of chitosan and HA was formed by Deng and coworkers for co-encapsulated of DOX and miR-34a; a combined approach for TNBC treatment. The tumor suppressant activity is demonstrated by miR-34a gene. Results of in vivo investigation displayed that miR-34a and DOX entered tumor cells of breast with substantial antitumor action of DOX through suppressing nonpump resistance expression and antiapoptosis proto-oncogene Bcl-2. Furthermore, by targeting Notch-1 signaling, miR-34a prevented migration of breast cancer cells. Both agents possess a good synergistic effect for breast cancer cell suppression [195, 196]. Moreover, Juneja and his team fabricated nanoparticle for treatment of TNBC using approach, i.e., simultaneous delivery of RNA, protoporphyrin IX, and curcumin. For the combined phototherapy and chemotherapy of TNBC, this multidelivery method has a major synergistic impact and can be employed for incorporating nucleic acids carrying therapeutic activity [197, 198]. Jafari et al. conducted research for TNBC treatment with the aptamer-hybrid chitosan NPs using simultaneous targeted delivery of gene (DTX and IGF-1R siRNA) and chemotherapeutic agent [199]. The findings presented an increase in cell uptake and a major reduction in the cell viability of SKBR3 cells. In addition, Apt-conjugated NPs substantially decreased the genetic expression of IGF-1R, transcription 3 signal transducers and activators, vascular growth factor, and matrix metalloproteinases [200, 201].

The advancing heterogeneity of BC and raising resistance to the established therapeutic regimens has led the genetics to enter for BC treatment. The gene therapy, RNAi (miRNA and siRNA), has evolved with theranostics for fast and excellent treatment of BC.

6 Conclusion

With due marked advancements in understanding the biology of BC and hurdles that overcome in defeating BC, the flourishing footholds of nanotechnology and biomedical sciences in cancer biology have led to understand the BC biology better in all aspects, either pathology, diagnosis, or treatment. The path of BC regimen, conventional noncytoselective anticancer drugs, chemotherapy, radiation therapy, adjuvant/neoadjuvant therapy, immunotherapy, and gene therapy has been an astonishing to state that nanomedicines had occupied all these aspects strongly. The development of nanomedicine had open new doors for defeating challenges that came on the path with conventional anticancer therapies. The drug resistance, noncytoselectivity, oncogene mutations, advancing heterogeneity of BC, and metastasis to distant organ/tissues (lung and brain) have been a great challenge so far concerned behind the BC treatment. The nanomedicine incorporation even in theranostics is promising to state that nanomedicine is backbone for BC therapeutics and diagnosis. The introduction of theranostics has changed the scenario of cancer biology with simultaneous smart diagnosis and targeted drug delivery to specific breast tumors/tissues. The content discussed earlier had shown how nanomedicines and theranostics are an essential component for managing BC. Now, it is better to conclude that researchers of polymer sciences, medical sciences, and biotechnology background must collaborate in setting these disclosed nanomedicines for their incorporation at clinical level. The gene therapy holds a potential ahead for defeating all forms of BC even resistant types. Gene therapy can also be personalized from individual to individual based on the genetic makeup of the caner that the patient is suffering from.

Acknowledgment

Authors are thankful to Mr. Prateek Sharma, doctoral student, Department of Pharmacology, Delhi Institute of Pharmaceutical Sciences and Research, New Delhi, India for his continuous dedicated support in drafting the manuscript.

Funding

Authors also express sincere thanks to Indian Council of Medical Research (ICMR) for providing financial support as senior research fellowship to Mr. Ajay Kumar Pal.

Conflict of interest

The authors have no conflict of interest to declare.

References

[1] Fragomeni SM, Sciallis A, Jeruss JS. Molecular subtypes and local-regional control of breast cancer. Surg Oncol Clin 2018;27(1):95–120.

[2] Prado-Vázquez G, Gámez-Pozo A, Trilla-Fuertes L, Arevalillo JM, Zapater-Moros A, Ferrer-Gómez M, Díaz-Almirón M, López-Vacas R, Navarro H, Maín P, Feliú J. A novel approach to triple-negative breast cancer molecular classification reveals a luminal immune-positive subgroup with good prognoses. Sci Rep 2019;9(1):1–2.

[3] Bradbury AR, Olopade OI. Genetic susceptibility to breast cancer. Rev Endocr Metab Disord 2007;8(3):255–67.

[4] Wang DY, Jiang Z, Ben-David Y, Woodgett JR, Zacksenhaus E. Molecular stratification within triple-negative breast cancer subtypes. Sci Rep 2019;9(1):1–10.

[5] Al-Mahmood S, Sapiezynski J, Garbuzenko OB, Minko T. Metastatic and triple-negative breast cancer: challenges and treatment options. Drug Deliv Transl Res 2018;8(5):1483–507.

[6] Jiang YZ, Ma D, Suo C, Shi J, Xue M, Hu X, Xiao Y, Yu KD, Liu YR, Yu Y, Zheng Y. Genomic and transcriptomic landscape of triple-negative breast cancers: subtypes and treatment strategies. Cancer Cell 2019;35(3):428–40.

[7] Dent R, Trudeau M, Pritchard KI, Hanna WM, Kahn HK, Sawka CA, Lickley LA, Rawlinson E, Sun P, Narod SA. Triple-negative breast cancer: clinical features and patterns of recurrence. Clin Cancer Res 2007;13(15):4429–34.

[8] Verma P, Mittal P, Singh A, Singh IK. New entrants into clinical trials for targeted therapy of breast cancer: an insight. Anticancer Agents Med Chem 2019;19(18):2156–76.

[9] He Y, Jiang Z, Chen C, Wang X. Classification of triple-negative breast cancers based on immunogenomic profiling. J Exp Clin Cancer Res 2018;37(1):327.

[10] Dai X, Li T, Bai Z, Yang Y, Liu X, Zhan J, Shi B. Breast cancer intrinsic subtype classification, clinical use and future trends. Am J Cancer Res 2015;5(10):2929.

[11] Lord CJ, Ashworth A. BRCAness revisited. Nat Rev Cancer 2016;16(2):110.

[12] Neophytou C, Boutsikos P, Papageorgis P. Molecular mechanisms and emerging therapeutic targets of triple-negative breast cancer metastasis. Front Oncol 2018;8:31.

[13] Siegel RL, Miller KD, Jemal A. Cancer statistics, 2019. CA Cancer J Clin 2019;69(1):7–34.

[14] Staff A. Cancer facts & figures 2018. Atlanta: American Cancer Society, Cancer; 2018. p. 19–20.

[15] Tao Z, Shi A, Lu C, Song T, Zhang Z, Zhao J. Breast cancer: epidemiology and etiology. Cell Biochem Biophys 2015;72(2):333–8.

[16] Haghighat S, Akbari ME, Ghaffari S, Yavari P. Standardized breast cancer mortality rate compared to the general female population of Iran. Asian Pac J Cancer Prev 2012;13(11):5525–8.

[17] Coleman MP, Quaresma M, Berrino F, Lutz JM, De Angelis R, Capocaccia R, Baili P, Rachet B, Gatta G, Hakulinen T, Micheli A. Cancer survival in five continents: a worldwide population-based study (CONCORD). Lancet Oncol 2008;9(8):730–56.

[18] Han SJ, Guo QQ, Wang T, Wang YX, Zhang YX, Liu F, Luo YX, Zhang J, Wang YL, Yan YX, Peng XX. Prognostic significance of interactions between ER alpha and ER beta and lymph node status in breast cancer cases. Asian Pac J Cancer Prev 2013;14(10):6081–4.
[19] de la Mare JA, Contu L, Hunter MC, Moyo B, Sterrenberg JN, KCH D, Mutsvunguma LZ, Edkins AL. Breast cancer: current developments in molecular approaches to diagnosis and treatment. Recent Pat Anticancer Drug Discov 2014;9(2):153–75.
[20] Khodabandehloo H, Zahednasab H, Hafez AA. Nanocarriers usage for drug delivery in cancer therapy. Iran J Cancer Prev 2016;9(2).
[21] Kumari P, Ghosh B, Biswas S. Nanocarriers for cancer-targeted drug delivery. J Drug Target 2016;24(3):179–91.
[22] Chessum N, Jones K, Pasqua E, Tucker M. Recent advances in cancer therapeutics. In: Progress in medicinal chemistry, vol. 54. Elsevier; 2015. p. 1–63.
[23] Tibbitt MW, Dahlman JE, Langer R. Emerging frontiers in drug delivery. J Am Chem Soc 2016;138(3):704–17.
[24] Schmadeka R, Harmon BE, Singh M. Triple-negative breast carcinoma: current and emerging concepts. Am J Clin Pathol 2014;141(4):462–77.
[25] Herranz M, Ruibal A. Optical imaging in breast cancer diagnosis: the next evolution. J Oncol 2012;1:2012.
[26] Dogan BE, Turnbull LW. Imaging of triple-negative breast cancer. Ann Oncol 2012;23(Suppl_6):vi23–9.
[27] Kreike B, van Kouwenhove M, Horlings H, Weigelt B, Peterse H, Bartelink H, van de Vijver MJ. Gene expression profiling and histopathological characterization of triple-negative/basal-like breast carcinomas. Breast Cancer Res 2007;9(5):R65.
[28] Kyndi M, Sørensen FB, Knudsen H, Overgaard M, Nielsen HM, Overgaard J. Estrogen receptor, progesterone receptor, HER-2, and response to postmastectomy radiotherapy in high-risk breast cancer: the Danish Breast Cancer Cooperative Group. J Clin Oncol 2008;26(9):1419–26.
[29] Bayraktar S, Glück S. Molecularly targeted therapies for metastatic triple-negative breast cancer. Breast Cancer Res Treat 2013;138(1):21–35.
[30] Shi Y, Jin J, Ji W, Guan X. Therapeutic landscape in mutational triple negative breast cancer. Mol Cancer 2018;17(1):99.
[31] Kutty RV, Feng SS. Cetuximab conjugated vitamin E TPGS micelles for targeted delivery of docetaxel for treatment of triple negative breast cancers. Biomaterials 2013;34(38):10160–71.
[32] Svenson S, Tomalia DA. Dendrimers in biomedical applications—reflections on the field. Adv Drug Deliv Rev 2012;64:102–15.
[33] Setyawati MI, Kutty RV, Leong DT. DNA nanostructures carrying stoichiometrically definable antibodies. Small 2016;12(40):5601–11.
[34] Mu Q, Wang H, Zhang M. Nanoparticles for imaging and treatment of metastatic breast cancer. Expert Opin Drug Deliv 2017;14(1):123–36.
[35] Schmid P, Adams S, Rugo HS, Schneeweiss A, Barrios CH, Iwata H, Diéras V, Hegg R, Im SA, Shaw Wright G, Henschel V. Atezolizumab and nab-paclitaxel in advanced triple-negative breast cancer. N Engl J Med 2018;379(22):2108–21.
[36] Del Paggio JC. Cancer immunotherapy and the value of cure. Nat Rev Clin Oncol 2018;15(5):268–70.
[37] Lammers T, Aime S, Hennink WE, Storm G, Kiessling F. Theranostic nanomedicine. Acc Chem Res 2011;44(10):1029–38.
[38] Shemesh CS, Moshkelani D, Zhang H. Thermosensitive liposome formulated indocyanine green for near-infrared triggered photodynamic therapy: in vivo evaluation for triple-negative breast cancer. Pharm Res 2015;32(5):1604–14.
[39] Doddapaneni R, Patel K, Owaid IH, Singh M. Tumor neovasculature-targeted cationic PEGylated liposomes of gambogic acid for the treatment of triple-negative breast cancer. Drug Deliv 2016;23(4):1232–41.
[40] Sneider A, Jadia R, Piel B, VanDyke D, Tsiros C, Rai P. Engineering remotely triggered liposomes to target triple negative breast cancer. Oncomedicine 2017;2:1.
[41] Sauvage F, Fattal E, Al-Shaer W, Denis S, Brotin E, Denoyelle C, Blanc-Fournier C, Toussaint B, Messaoudi S, Alami M, Barratt G. Antitumor activity of nanoliposomes encapsulating the novobiocin analog 6BrCaQ in a triple-negative breast cancer model in mice. Cancer Lett 2018;432:103–11.
[42] Zhou F, Feng B, Wang T, Wang D, Meng Q, Zeng J, Zhang Z, Wang S, Yu H, Li Y. Programmed multiresponsive vesicles for enhanced tumor penetration and combination therapy of triple-negative breast cancer. Adv Funct Mater 2017;27(20):1606530.
[43] Zeng F, Ju RJ, Liu L, Xie HJ, Mu LM, Zhao Y, Yan Y, Hu YJ, Wu JS, Lu WL. Application of functional vincristine plus dasatinib liposomes to deletion of vasculogenic mimicry channels in triple-negative breast cancer. Oncotarget 2015;6(34):36625.
[44] Fan Y, Wang Q, Lin G, Shi Y, Gu Z, Ding T. Combination of using prodrug-modified cationic liposome nanocomplexes and a potentiating strategy via targeted co-delivery of gemcitabine and docetaxel for CD44-overexpressed triple negative breast cancer therapy. Acta Biomater 2017;62:257–72.
[45] Shen J, Kim HC, Wolfram J, Mu C, Zhang W, Liu H, Xie Y, Mai J, Zhang H, Li Z, Guevara M. A liposome encapsulated ruthenium polypyridine complex as a theranostic platform for triple-negative breast cancer. Nano Lett 2017;17(5):2913–20.
[46] Guo P, Yang J, Di Jia MA, Auguste DT. ICAM-1-targeted, Lcn2 siRNA-encapsulating liposomes are potent anti-angiogenic agents for triple negative breast cancer. Theranostics 2016;6(1):1.
[47] Guo P, Yang J, Liu D, Huang L, Fell G, Huang J, Moses MA, Auguste DT. Dual complementary liposomes inhibit triple-negative breast tumor progression and metastasis. Sci Adv 2019;5(3), eaav5010.
[48] Pawar A, Prabhu P. Nanosoldiers: a promising strategy to combat triple negative breast cancer. Biomed Pharmacother 2019;110:319–41.
[49] Ahmad Z, Shah A, Siddiq M, Kraatz HB. Polymeric micelles as drug delivery vehicles. RSC Adv 2014;4(33):17028–38.
[50] Dai W, Yang F, Ma L, Fan Y, He B, He Q, Wang X, Zhang H, Zhang Q. Combined mTOR inhibitor rapamycin and doxorubicin-loaded cyclic octapeptide modified liposomes for targeting integrin α3 in triple-negative breast cancer. Biomaterials 2014;35(20):5347–58.
[51] Lee SM, Ahn RW, Chen F, Fought AJ, O'halloran TV, Cryns VL, Nguyen ST. Biological evaluation of pH-responsive polymer-caged nanobins for breast cancer therapy. ACS Nano 2010;4(9):4971–8.
[52] Sainz V, Conniot J, Matos AI, Peres C, Zupanǒiǒ E, Moura L, Silva LC, Florindo HF, Gaspar RS. Regulatory aspects on nanomedicines. Biochem Biophys Res Commun 2015;468(3):504–10.
[53] Caster JM, Patel AN, Zhang T, Wang A. Investigational nanomedicines in 2016: a review of nanotherapeutics currently undergoing clinical trials. Wiley Interdiscip Rev Nanomed Nanobiotechnol 2017;9(1), e1416.
[54] Deshpande PP, Biswas S, Torchilin VP. Current trends in the use of liposomes for tumor targeting. Nanomedicine 2013;8(9):1509–28.
[55] Matsumura Y, Hamaguchi T, Ura T, Muro K, Yamada Y, Shimada Y, Shirao K, Okusaka T, Ueno H, Ikeda M, Watanabe N. Phase I clinical trial and pharmacokinetic evaluation of NK911, a micelle-encapsulated doxorubicin. Br J Cancer 2004;91(10):1775–81.

[56] Sun Y, Zou W, Bian S, Huang Y, Tan Y, Liang J, Fan Y, Zhang X. Bioreducible PAA-g-PEG graft micelles with high doxorubicin loading for targeted antitumor effect against mouse breast carcinoma. Biomaterials 2013;34(28):6818–28.
[57] Zhang H, Hu H, Zhang H, Dai W, Wang X, Wang X, Zhang Q. Effects of PEGylated paclitaxel nanocrystals on breast cancer and its lung metastasis. Nanoscale 2015;7(24):10790–800.
[58] Andey T, Sudhakar G, Marepally S, Patel A, Banerjee R, Singh M. Lipid nanocarriers of a lipid-conjugated estrogenic derivative inhibit tumor growth and enhance cisplatin activity against triple-negative breast cancer: pharmacokinetic and efficacy evaluation. Mol Pharm 2015;12(4):1105–20.
[59] Awada A, Bondarenko IN, Bonneterre J, Nowara E, Ferrero JM, Bakshi AV, Wilke C, Piccart M, CT4002 Study Group. A randomized controlled phase II trial of a novel composition of paclitaxel embedded into neutral and cationic lipids targeting tumor endothelial cells in advanced triple-negative breast cancer (TNBC). Ann Oncol 2014;25(4):824–31.
[60] Sharma A, Jain N, Sareen R. Nanocarriers for diagnosis and targeting of breast cancer. Biomed Res Int 2013;2013.
[61] Torchilin VP. Micellar nanocarriers: pharmaceutical perspectives. Pharm Res 2007;24(1):1.
[62] Wang Y, Wang Y, Chen G, Li Y, Xu W, Gong S. Quantum-dot-based theranostic micelles conjugated with an anti-EGFR nanobody for triple-negative breast cancer therapy. ACS Appl Mater Interfaces 2017;9(36):30297–305.
[63] Sabra SA, Sheweita SA, Haroun M, Ragab D, Eldemellawy MA, Xia Y, Goodale D, Allan AL, Elzoghby AO, Rohani S. Magnetically guided self-assembled protein micelles for enhanced delivery of dasatinib to human triple-negative breast cancer cells. J Pharm Sci 2019;108(5):1713–25.
[64] Kutty RV, Tay CY, Lim CS, Feng SS, Leong DT. Anti-migratory and increased cytotoxic effects of novel dual drug-loaded complex hybrid micelles in triple negative breast cancer cells. Nano Res 2015;8(8):2533–47.
[65] Wu K, Cheng R, Zhang J, Meng F, Deng C, Zhong Z. Micellar nanoformulation of lipophilized bortezomib: high drug loading, improved tolerability and targeted treatment of triple negative breast cancer. J Mater Chem B 2017;5(28):5658–67.
[66] Taurin S, Nehoff H, Diong J, Larsen L, Rosengren RJ, Greish K. Curcumin-derivative nanomicelles for the treatment of triple negative breast cancer. J Drug Target 2013;21(7):675–83.
[67] Muthu MS, Kutty RV, Luo Z, Xie J, Feng SS. Theranostic vitamin E TPGS micelles of transferrin conjugation for targeted co-delivery of docetaxel and ultra bright gold nanoclusters. Biomaterials 2015;39:234–48.
[68] Matsumura Y. Preclinical and clinical studies of NK012, an SN-38-incorporating polymeric micelles, which is designed based on EPR effect. Adv Drug Deliv Rev 2011;63(3):184–92.
[69] Morikawa A. Comparison of properties among dendritic and hyperbranched poly (ether ether ketone) s and linear poly (ether ketone) s. Molecules 2016;21(2):219.
[70] Brothers II HM, Piehler LT, Tomalia DA. Slab-gel and capillary electrophoretic characterization of polyamidoamine dendrimers. J Chromatogr A 1998;814(1–2):233–46.
[71] Hecht S, Fréchet JM. Dendritic encapsulation of function: applying nature's site isolation principle from biomimetics to materials science. Angew Chem Int Ed 2001;40(1):74–91.
[72] Dandliker PJ, Diederich F, Zingg A, Gisselbrecht JP, Gross M, Louati A, Sanford E. Dendrimers with porphyrin cores: synthetic models for globular heme proteins. Helv Chim Acta 1997;80(6):1773–801.
[73] Jiang DL, Aida T. A dendritic iron porphyrin as a novel haemoprotein mimic: effects of the dendrimer cage on dioxygen-binding activity. Chem Commun 1996;13:1523–4.
[74] Weyermann P, Gisselbrecht JP, Boudon C, Diederich F, Gross M. Dendritic iron porphyrins with tethered axial ligands: new model compounds for cytochromes. Angew Chem Int Ed 1999;38(21):3215–9.
[75] Sadler K, Tam JP. Peptide dendrimers: applications and synthesis. Rev Mol Biotechnol 2002;90(3–4):195–229.
[76] Richter-Egger DL, Tesfai A, Tucker SA. Spectroscopic investigations of poly (propyleneimine) dendrimers using the solvatochromic probe phenol blue and comparisons to poly (amidoamine) dendrimers. Anal Chem 2001;73(23):5743–51.
[77] Shao N, Su Y, Hu J, Zhang J, Zhang H, Cheng Y. Comparison of generation 3 polyamidoamine dendrimer and generation 4 polypropylenimine dendrimer on drug loading, complex structure, release behavior, and cytotoxicity. Int J Nanomedicine 2011;6:3361.
[78] Ionov M, Wróbel D, Gardikis K, Hatziantoniou S, Demetzos C, Majoral JP, Klajnert B, Bryszewska M. Effect of phosphorus dendrimers on DMPC lipid membranes. Chem Phys Lipids 2012;165(4):408–13.
[79] Esfand R, Tomalia DA. Poly (amidoamine)(PAMAM) dendrimers: from biomimicry to drug delivery and biomedical applications. Drug Discov Today 2001;6(8):427–36.
[80] Cheng Y, Zhao L, Li Y, Xu T. Design of biocompatible dendrimers for cancer diagnosis and therapy: current status and future perspectives. Chem Soc Rev 2011;40(5):2673–703.
[81] Zhu J, Shi X. Dendrimer-based nanodevices for targeted drug delivery applications. J Mater Chem B 2013;1(34):4199–211.
[82] Shen M, Shi X. Dendrimer-based organic/inorganic hybrid nanoparticles in biomedical applications. Nanoscale 2010;2(9):1596–610.
[83] Li X, Takashima M, Yuba E, Harada A, Kono K. PEGylated PAMAM dendrimer–doxorubicin conjugate-hybridized gold nanorod for combined photothermal-chemotherapy. Biomaterials 2014;35(24):6576–84.
[84] Zhu J, Fu F, Xiong Z, Shen M, Shi X. Dendrimer-entrapped gold nanoparticles modified with RGD peptide and alpha-tocopheryl succinate enable targeted theranostics of cancer cells. Colloids Surf B: Biointerfaces 2015;133:36–42.
[85] Zhu J, Zheng L, Wen S, Tang Y, Shen M, Zhang G, Shi X. Targeted cancer theranostics using alpha-tocopheryl succinate-conjugated multifunctional dendrimer-entrapped gold nanoparticles. Biomaterials 2014;35(26):7635–46.
[86] Shi X, Lee I, Chen X, Shen M, Xiao S, Zhu M, Baker JR, Wang SH. Influence of dendrimer surface charge on the bioactivity of 2-methoxyestradiol complexed with dendrimers. Soft Matter 2010;6(11):2539–45.
[87] Zhang C, Pan D, Li J, Hu J, Bains A, Guys N, Zhu H, Li X, Luo K, Gong Q, Gu Z. Enzyme-responsive peptide dendrimer-gemcitabine conjugate as a controlled-release drug delivery vehicle with enhanced antitumor efficacy. Acta Biomater 2017;55:153–62.
[88] Bastien E, Schneider R, Hackbarth S, Dumas D, Jasniewski J, Röder B, Bezdetnaya L, Lassalle HP. PAMAM G4.5-chlorin e6 dendrimeric nanoparticles for enhanced photodynamic effects. Photochem Photobiol Sci 2015;14(12):2203–12.
[89] Thakur S, Tekade RK, Kesharwani P, Jain NK. The effect of polyethylene glycol spacer chain length on the tumor-targeting potential of folate-modified PPI dendrimers. J Nanopart Res 2013;15(5):1625.
[90] Zhu J, Zhao L, Cheng Y, Xiong Z, Tang Y, Shen M, Zhao J, Shi X. Radionuclide 131 I-labeled multifunctional dendrimers for targeted SPECT imaging and radiotherapy of tumors. Nanoscale 2015;7(43):18169–78.

[91] Cheng Y, Zhu J, Zhao L, Xiong Z, Tang Y, Liu C, Guo L, Qiao W, Shi X, Zhao J. ^{131}I-labeled multifunctional dendrimers modified with BmK CT for targeted SPECT imaging and radiotherapy of gliomas. Nanomedicine 2016;11(10):1253–66.
[92] Mendoza-Nava H, Ferro-Flores G, Ramírez FD, Ocampo-García B, Santos-Cuevas C, Aranda-Lara L, Azorín-Vega E, Morales-Avila E, Isaac-Olivé K. ^{177}Lu-dendrimer conjugated to folate and bombesin with gold nanoparticles in the dendritic cavity: a potential theranostic radiopharmaceutical. J Nanomater 2016;2016.
[93] Mendoza-Nava H, Ferro-Flores G, Ramírez FD, Ocampo-García B, Santos-Cuevas C, Azorín-Vega E, Jiménez-Mancilla N, Luna-Gutiérrez M, Isaac-Olivé K. Fluorescent, plasmonic, and radiotherapeutic properties of the ^{177}Lu–dendrimer-AuNP–folate–bombesin nanoprobe located inside cancer cells. Mol Imaging 2017;16, 1536012117704768.
[94] Li X, Takeda K, Yuba E, Harada A, Kono K. Preparation of PEG-modified PAMAM dendrimers having a gold nanorod core and their application to photothermal therapy. J Mater Chem B 2014;2(26):4167–76.
[95] Zan M, Li J, Huang M, Lin S, Luo D, Luo S, Ge Z. Near-infrared light-triggered drug release nanogels for combined photothermal-chemotherapy of cancer. Biomater Sci 2015;3(7):1147–56.
[96] Kong L, Xing L, Zhou B, Du L, Shi X. Dendrimer-modified MoS2 nanoflakes as a platform for combinational gene silencing and photothermal therapy of tumors. ACS Appl Mater Interfaces 2017;9(19):15995–6005.
[97] Zhou Z, Wang Y, Yan Y, Zhang Q, Cheng Y. Dendrimer-templated ultrasmall and multifunctional photothermal agents for efficient tumor ablation. ACS Nano 2016;10(4):4863–72.
[98] Wei P, Chen J, Hu Y, Li X, Wang H, Shen M, Shi X. Dendrimer-stabilized gold nanostars as a multifunctional theranostic nanoplatform for CT imaging, photothermal therapy, and gene silencing of tumors. Adv Healthc Mater 2016;5(24):3203–13.
[99] Taratula O, Schumann C, Naleway MA, Pang AJ, Chon KJ, Taratula O. A multifunctional theranostic platform based on phthalocyanine-loaded dendrimer for image-guided drug delivery and photodynamic therapy. Mol Pharm 2013;10(10):3946–58.
[100] Dabrzalska M, Zablocka M, Mignani S, Majoral JP, Klajnert-Maculewicz B. Phosphorus dendrimers and photodynamic therapy. Spectroscopic studies on two dendrimer-photosensitizer complexes: cationic phosphorus dendrimer with rose bengal and anionic phosphorus dendrimer with methylene blue. Int J Pharm 2015;492(1–2):266–74.
[101] Dabrzalska M, Janaszewska A, Zablocka M, Mignani S, Majoral JP, Klajnert-Maculewicz B. Cationic phosphorus dendrimer enhances photodynamic activity of rose bengal against basal cell carcinoma cell lines. Mol Pharm 2017;14(5):1821–30.
[102] Chang H, Zhang Y, Li L, Cheng Y. Efficient delivery of small interfering RNA into cancer cells using dodecylated dendrimers. J Mater Chem B 2015;3(41):8197–202.
[103] Shan Y, Luo T, Peng C, Sheng R, Cao A, Cao X, Shen M, Guo R, Tomás H, Shi X. Gene delivery using dendrimer-entrapped gold nanoparticles as nonviral vectors. Biomaterials 2012;33(10):3025–35.
[104] Qiu J, Kong L, Cao X, Li A, Tan H, Shi X. Dendrimer-entrapped gold nanoparticles modified with β-cyclodextrin for enhanced gene delivery applications. RSC Adv 2016;6(31):25633–40.
[105] Hou W, Wen S, Guo R, Wang S, Shi X. Partially acetylated dendrimer-entrapped gold nanoparticles with reduced cytotoxicity for gene delivery applications. J Nanosci Nanotechnol 2015;15(6):4094–105.
[106] Hou W, Wei P, Kong L, Guo R, Wang S, Shi X. Partially PEGylated dendrimer-entrapped gold nanoparticles: a promising nanoplatform for highly efficient DNA and siRNA delivery. J Mater Chem B 2016;4(17):2933–43.
[107] Ionov M, Lazniewska J, Dzmitruk V, Halets I, Loznikova S, Novopashina D, Apartsin E, Krasheninina O, Venyaminova A, Milowska K, Nowacka O. Anticancer siRNA cocktails as a novel tool to treat cancer cells. Part (A). Mechanisms of interaction. Int J Pharm 2015;485(1–2):261–9.
[108] Liu H, Wang Y, Wang M, Xiao J, Cheng Y. Fluorinated poly (propylenimine) dendrimers as gene vectors. Biomaterials 2014;35(20):5407–13.
[109] Tietze S, Schau I, Michen S, Ennen F, Janke A, Schackert G, Aigner A, Appelhans D, Temme A. A poly (propyleneimine) dendrimer-based polyplex-system for single-chain antibody-mediated targeted delivery and cellular uptake of SiRNA. Small 2017;13(27):1700072.
[110] Hashemi M, Ayatollahi S, Parhiz H, Mokhtarzadeh A, Javidi S, Ramezani M. PEGylation of polypropylenimine dendrimer with alkylcarboxylate chain linkage to improve DNA delivery and cytotoxicity. Appl Biochem Biotechnol 2015;177(1):1–7.
[111] Hashemi M, Tabatabai SM, Parhiz H, Milanizadeh S, Farzad SA, Abnous K, Ramezani M. Gene delivery efficiency and cytotoxicity of heterocyclic amine-modified PAMAM and PPI dendrimers. Mater Sci Eng C 2016;61:791–800.
[112] Bawarski WE, Chidlowsky E, Bharali DJ, Mousa SA. Emerging nanopharmaceuticals. Nanomedicine 2008;4(4):273–82.
[113] Wang P, Zhao XH, Wang ZY, Meng M, Li X, Ning Q. Generation 4 polyamidoamine dendrimers is a novel candidate of nano-carrier for gene delivery agents in breast cancer treatment. Cancer Lett 2010;298(1):34–49.
[114] Finlay J, Roberts CM, Lowe G, Loeza J, Rossi JJ, Glackin CA. RNA-based TWIST1 inhibition via dendrimer complex to reduce breast cancer cell metastasis. Biomed Res Int 2015;11:2015.
[115] Elsabahy M, Wooley KL. Design of polymeric nanoparticles for biomedical delivery applications. Chem Soc Rev 2012;41(7):2545–61.
[116] Xu J, Luft JC, Yi X, Tian S, Owens G, Wang J, Johnson A, Berglund P, Smith J, Napier ME, DeSimone JM. RNA replicon delivery via lipid-complexed PRINT protein particles. Mol Pharm 2013;10(9):3366–74.
[117] Johnstone TC, Kulak N, Pridgen EM, Farokhzad OC, Langer R, Lippard SJ. Nanoparticle encapsulation of mitaplatin and the effect thereof on in vivo properties. ACS Nano 2013;7(7):5675–83.
[118] Passarella RJ, Spratt DE, Van Der Ende AE, Phillips JG, Wu H, Sathiyakumar V, Zhou L, Hallahan DE, Harth E, Diaz R. Targeted nanoparticles that deliver a sustained, specific release of Paclitaxel to irradiated tumors. Cancer Res 2010;70(11):4550–9.
[119] Cardillo TM, Govindan SV, Sharkey RM, Trisal P, Arrojo R, Liu D, Rossi EA, Chang CH, Goldenberg DM. Sacituzumab govitecan (IMMU-132), an anti-Trop-2/SN-38 antibody–drug conjugate: characterization and efficacy in pancreatic, gastric, and other cancers. Bioconjug Chem 2015;26(5):919–31.
[120] Devulapally R, Sekar NM, Sekar TV, Foygel K, Massoud TF, Willmann JK, Paulmurugan R. Polymer nanoparticles mediated codelivery of antimiR-10b and antimiR-21 for achieving triple negative breast cancer therapy. ACS Nano 2015;9(3):2290–302.

[121] Deng ZJ, Morton SW, Ben-Akiva E, Dreaden EC, Shopsowitz KE, Hammond PT. Layer-by-layer nanoparticles for systemic codelivery of an anticancer drug and siRNA for potential triple-negative breast cancer treatment. ACS Nano 2013;7(11):9571–84.

[122] Shan D, Li J, Cai P, Prasad P, Liu F, Rauth AM, Wu XY. RGD-conjugated solid lipid nanoparticles inhibit adhesion and invasion of αvβ3 integrin-overexpressing breast cancer cells. Drug Deliv Transl Res 2015;5(1):15–26.

[123] Zhang T, Prasad P, Cai P, He C, Shan D, Rauth AM, Wu XY. Dual-targeted hybrid nanoparticles of synergistic drugs for treating lung metastases of triple negative breast cancer in mice. Acta Pharmacol Sin 2017;38(6):835–47.

[124] Su XY, Liu PD, Wu H, Gu N. Enhancement of radiosensitization by metal-based nanoparticles in cancer radiation therapy. Cancer Biol Med 2014;11(2):86.

[125] Kong T, Zeng J, Wang X, Yang X, Yang J, McQuarrie S, McEwan A, Roa W, Chen J, Xing JZ. Enhancement of radiation cytotoxicity in breast-cancer cells by localized attachment of gold nanoparticles. Small 2008;4(9):1537–43.

[126] Liu P, Huang Z, Chen Z, Xu R, Wu H, Zang F, Wang C, Gu N. Silver nanoparticles: a novel radiation sensitizer for glioma? Nanoscale 2013;5(23):11829–36.

[127] Wahab R, Siddiqui MA, Saquib Q, Dwivedi S, Ahmad J, Musarrat J, Al-Khedhairy AA, Shin HS. ZnO nanoparticles induced oxidative stress and apoptosis in HepG2 and MCF-7 cancer cells and their antibacterial activity. Colloids Surf B: Biointerfaces 2014;117:267–76.

[128] Baskar G, Chandhuru J, Fahad KS, Praveen AS, Chamundeeswari M, Muthukumar T. Anticancer activity of fungal L-asparaginase conjugated with zinc oxide nanoparticles. J Mater Sci Mater Med 2015;26(1):43.

[129] Hackenberg S, Scherzed A, Harnisch W, Froelich K, Ginzkey C, Koehler C, Hagen R, Kleinsasser N. Antitumor activity of photo-stimulated zinc oxide nanoparticles combined with paclitaxel or cisplatin in HNSCC cell lines. J Photochem Photobiol B: Biol 2012;114:87–93.

[130] Wang Y, Yang F, Zhang HX, Zi XY, Pan XH, Chen F, Luo WD, Li JX, Zhu HY, Hu YP. Cuprous oxide nanoparticles inhibit the growth and metastasis of melanoma by targeting mitochondria. Cell Death Dis 2013;4(8), e783.

[131] Zhou M, Zhao J, Tian M, Song S, Zhang R, Gupta S, Tan D, Shen H, Ferrari M, Li C. Radio-photothermal therapy mediated by a single compartment nanoplatform depletes tumor initiating cells and reduces lung metastasis in the orthotopic 4T1 breast tumor model. Nanoscale 2015;7(46):19438–47.

[132] Pešić M, Podolski-Renić A, Stojković S, Matović B, Zmejkoski D, Kojić V, Bogdanović G, Pavićević A, Mojović M, Savić A, Milenković I. Anti-cancer effects of cerium oxide nanoparticles and its intracellular redox activity. Chem Biol Interact 2015;232:85–93.

[133] Huang HS, Hainfeld JF. Intravenous magnetic nanoparticle cancer hyperthermia. Int J Nanomedicine 2013;8:2521.

[134] Kossatz S, Grandke J, Couleaud P, Latorre A, Aires A, Crosbie-Staunton K, Ludwig R, Dähring H, Ettelt V, Lazaro-Carrillo A, Calero M. Efficient treatment of breast cancer xenografts with multifunctionalized iron oxide nanoparticles combining magnetic hyperthermia and anti-cancer drug delivery. Breast Cancer Res 2015;17(1):66.

[135] Pourtau L, Oliveira H, Thevenot J, Wan Y, Brisson AR, Sandre O, Miraux S, Thiaudiere E, Lecommandoux S. Antibody-functionalized magnetic polymersomes: in vivo targeting and imaging of bone metastases using high resolution MRI. Adv Healthc Mater 2013;2(11):1420–4.

[136] Kievit FM, Stephen ZR, Veiseh O, Arami H, Wang T, Lai VP, Park JO, Ellenbogen RG, Disis ML, Zhang M. Targeting of primary breast cancers and metastases in a transgenic mouse model using rationally designed multifunctional SPIONs. ACS Nano 2012;6(3):2591–601.

[137] Chen J, Seeman NC. Synthesis from DNA of a molecule with the connectivity of a cube. Nature 1991;350(6319):631–3.

[138] Rothemund PW, Ekani-Nkodo A, Papadakis N, Kumar A, Fygenson DK, Winfree E. Design and characterization of programmable DNA nanotubes. J Am Chem Soc 2004;126(50):16344–52.

[139] Mathieu F, Liao S, Kopatsch J, Wang T, Mao C, Seeman NC. Six-helix bundles designed from DNA. Nano Lett 2005;5(4):661–5.

[140] Goodman RP, Berry RM, Turberfield AJ. The single-step synthesis of a DNA tetrahedron. Chem Commun 2004;12:1372–3.

[141] Zhang Y, Seeman NC. Construction of a DNA-truncated octahedron. J Am Chem Soc 1994;116(5):1661–9.

[142] Zhang C, Su M, He Y, Zhao X, Fang PA, Ribbe AE, Jiang W, Mao C. Conformational flexibility facilitates self-assembly of complex DNA nanostructures. Proc Natl Acad Sci U S A 2008;105(31):10665–9.

[143] Winfree E, Liu F, Wenzler LA, Seeman NC. Design and self-assembly of two-dimensional DNA crystals. Nature 1998;394(6693):539–44.

[144] Majumder U, Rangnekar A, Gothelf KV, Reif JH, LaBean TH. Design and construction of double-decker tile as a route to three-dimensional periodic assembly of DNA. J Am Chem Soc 2011;133(11):3843–5.

[145] Rothemund PW. Folding DNA to create nanoscale shapes and patterns. Nature 2006;440(7082):297–302.

[146] Han D, Pal S, Nangreave J, Deng Z, Liu Y, Yan H. DNA origami with complex curvatures in three-dimensional space. Science 2011;332(6027):342–6.

[147] Setyawati MI, Kutty RV, Tay CY, Yuan X, Xie J, Leong DT. Novel theranostic DNA nanoscaffolds for the simultaneous detection and killing of *Escherichia coli* and *Staphylococcus aureus*. ACS Appl Mater Interfaces 2014;6(24):21822–31.

[148] Roldo M, Fatouros DG. Biomedical applications of carbon nanotubes. Ann Rep C Phys Chem 2013;109:10–35.

[149] Casais-Molina ML, Cab C, Canto G, Medina J, Tapia A. Carbon nanomaterials for breast cancer treatment. J Nanomater 2018;2018.

[150] Bianco A, Kostarelos K, Prato M. Applications of carbon nanotubes in drug delivery. Curr Opin Chem Biol 2005;9(6):674–9.

[151] Yang M, Meng J, Cheng X, Lei J, Guo H, Zhang W, Kong H, Xu H. Multiwalled carbon nanotubes interact with macrophages and influence tumor progression and metastasis. Theranostics 2012;2(3):258.

[152] Burke AR, Singh RN, Carroll DL, Wood JC, D'Agostino Jr RB, Ajayan PM, Torti FM, Torti SV. The resistance of breast cancer stem cells to conventional hyperthermia and their sensitivity to nanoparticle-mediated photothermal therapy. Biomaterials 2012;33(10):2961–70.

[153] Chow EK, Zhang XQ, Chen M, Lam R, Robinson E, Huang H, Schaffer D, Osawa E, Goga A, Ho D. Nanodiamond therapeutic delivery agents mediate enhanced chemoresistant tumor treatment. Sci Transl Med 2011;3(73), 73ra21.

[154] Xiao J, Duan X, Yin Q, Zhang Z, Yu H, Li Y. Nanodiamonds-mediated doxorubicin nuclear delivery to inhibit lung metastasis of breast cancer. Biomaterials 2013;34(37):9648–56.

[155] Liu D, Zhang Q, Wang J, Fan L, Zhu W, Cai D. Hyaluronic acid-coated single-walled carbon nanotubes loaded with doxorubicin for the treatment of breast cancer. Die Pharmazie 2019;74(2):83–90.
[156] Singh S, Mehra NK, Jain NK. Development and characterization of the paclitaxel loaded riboflavin and thiamine conjugated carbon nanotubes for cancer treatment. Pharm Res 2016;33(7):1769–81.
[157] Mashal A, Sitharaman B, Li X, Avti PK, Sahakian AV, Booske JH, Hagness SC. Toward carbon-nanotube-based theranostic agents for microwave detection and treatment of breast cancer: enhanced dielectric and heating response of tissue-mimicking materials. IEEE Trans Biomed Eng 2010;57(8):1831–4.
[158] Dineshkumar B, Krishnakumar K, Bhatt AR, Paul D, Cherian J, John A, Suresh S. Single-walled and multi-walled carbon nanotubes based drug delivery system: cancer therapy: a review. Indian J Cancer 2015;52(3):262.
[159] Xiao Y, Gao X, Taratula O, Treado S, Urbas A, Holbrook RD, Cavicchi RE, Avedisian CT, Mitra S, Savla R, Wagner PD. Anti-HER2 IgY antibody-functionalized single-walled carbon nanotubes for detection and selective destruction of breast cancer cells. BMC Cancer 2009;9(1):1.
[160] Yu S, Zhang Y, Chen L, Li Q, Du J, Gao Y, Zhang L, Yang Y. Antitumor effects of carbon nanotube drug complex against human breast cancer cells. Exp Ther Med 2018;16(2):1103–10.
[161] Li X, Zhang W, Liu L, Zhu Z, Ouyang G, An Y, Zhao C, Yang CJ. In vitro selection of DNA aptamers for metastatic breast cancer cell recognition and tissue imaging. Anal Chem 2014;86(13):6596–603.
[162] Huang YF, Lin YW, Lin ZH, Chang HT. Aptamer-modified gold nanoparticles for targeting breast cancer cells through light scattering. J Nanopart Res 2009;11(4):775–83.
[163] Yang Y, Kang M, Fang S, Wang M, He L, Zhao J, et al. Electrochemical biosensor based on three-dimensional reduced graphene oxide and polyaniline nanocomposite for selective detection of mercury ions. Sens Actuators B Chem 2015;214:63–9.
[164] Tang L, Yang X, Dobrucki LW, Chaudhury I, Yin Q, Yao C, Lezmi S, Helferich WG, Fan TM, Cheng J. Aptamer-functionalized, ultra-small, monodisperse silica nanoconjugates for targeted dual-modal imaging of lymph nodes with metastatic tumors. Angew Chem 2012;124(51):12893–8.
[165] Shi S, Hong H, Orbay H, Graves SA, Yang Y, Ohman JD, Liu B, Nickles RJ, Wong HC, Cai W. ImmunoPET of tissue factor expression in triple-negative breast cancer with a radiolabeled antibody Fab fragment. Eur J Nucl Med Mol Imaging 2015;42(8):1295–303.
[166] LeBeau AM, Sevillano N, King ML, Duriseti S, Murphy ST, Craik CS, Murphy LL, VanBrocklin HF. Imaging the urokinase plasminongen activator receptor in preclinical breast cancer models of acquired drug resistance. Theranostics 2014;4(3):267.
[167] Rousseau C, Ruellan AL, Bernardeau K, Kraeber-Bodéré F, Gouard S, Loussouarn D, Saï-Maurel C, Faivre-Chauvet A, Wijdenes J, Barbet J, Gaschet J. Syndecan-1 antigen, a promising new target for triple-negative breast cancer immuno-PET and radioimmunotherapy. A preclinical study on MDA-MB-468 xenograft tumors. EJNMMI Res 2011;1(1):1.
[168] Reubi JC, Maecke HR. Peptide-based probes for cancer imaging. J Nucl Med 2008;49(11):1735–8.
[169] Yu MK, Park J, Jon S. Targeting strategies for multifunctional nanoparticles in cancer imaging and therapy. Theranostics 2012;2(1):3.
[170] Feng GK, Liu RB, Zhang MQ, Ye XX, Zhong Q, Xia YF, Li MZ, Wang J, Song EW, Zhang X, Wu ZZ. SPECT and near-infrared fluorescence imaging of breast cancer with a neuropilin-1-targeting peptide. J Control Release 2014;192:236–42.
[171] Crisp JL, Savariar EN, Glasgow HL, Ellies LG, Whitney MA, Tsien RY. Dual targeting of integrin αvβ3 and matrix metalloproteinase-2 for optical imaging of tumors and chemotherapeutic delivery. Mol Cancer Ther 2014;13(6):1514–25.
[172] Peiris PM, Toy R, Doolittle E, Pansky J, Abramowski A, Tam M, Vicente P, Tran E, Hayden E, Camann A, Mayer A. Imaging metastasis using an integrin-targeting chain-shaped nanoparticle. ACS Nano 2012;6(10):8783–95.
[173] Doolittle E, Peiris PM, Doron G, Goldberg A, Tucci S, Rao S, Shah S, Sylvestre M, Govender P, Turan O, Lee Z. Spatiotemporal targeting of a dual-ligand nanoparticle to cancer metastasis. ACS Nano 2015;9(8):8012–21.
[174] Costello JF. Abstract IA09: imaging to guide genomics and epigenomics of glioma. 75. Cancer Research: AACR; 2015, IA09.
[175] Chung ZY, Kim MW. Biomd Res Int 2014;2014(81932).
[176] Meier R, Henning TD, Boddington S, Tavri S, Arora S, Piontek G, Rudelius M, Corot C, Daldrup-Link HE. Breast cancers: MR imaging of folate-receptor expression with the folate-specific nanoparticle P1133. Radiology 2010;255(2):527–35.
[177] Feng B, Xu Z, Zhou F, Yu H, Sun Q, Wang D, Tang Z, Yu H, Yin Q, Zhang Z, Li Y. Near infrared light-actuated gold nanorods with cisplatin–polypeptide wrapping for targeted therapy of triple negative breast cancer. Nanoscale 2015;7(36):14854–64.
[178] Cao L, Yang ST, Wang X, Luo PG, Liu JH, Sahu S, Liu Y, Sun YP. Competitive performance of carbon "quantum" dots in optical bioimaging. Theranostics 2012;2(3):295.
[179] Misra AC, Luker KE, Durmaz H, Luker GD, Lahann J. CXCR4-targeted nanocarriers for triple negative breast cancers. Biomacromolecules 2015;16(8):2412–7.
[180] Zevon M, Ganapathy V, Kantamneni H, Mingozzi M, Kim P, Adler D, Sheng Y, Tan MC, Pierce M, Riman RE, Roth CM. CXCR-4 targeted, short wave infrared (SWIR) emitting nanoprobes for enhanced deep tissue imaging and micrometastatic cancer lesion detection. Small 2015;11(47):6347–57.
[181] Wu Z, Zheng J. Nanoparticles for taxanes delivery in cancer treatment. J Nanosci Nanotechnol 2016;16(7):6634–47.
[182] Liu Z, Liang XJ. Nano-carbons as theranostics. Theranostics 2012;2(3):235.
[183] Chen Z, Ma L, Liu Y, Chen C. Applications of functionalized fullerenes in tumor theranostics. Theranostics 2012;2(3):238.
[184] Zeltins A. Construction and characterization of virus-like particles: a review. Mol Biotechnol 2013;53(1):92–107.
[185] Bachmann MF, Jennings GT. Vaccine delivery: a matter of size, geometry, kinetics and molecular patterns. Nat Rev Immunol 2010;10(11):787–96.
[186] Rynda-Apple A, Patterson DP, Douglas T. Virus-like particles as antigenic nanomaterials for inducing protective immune responses in the lung. Nanomedicine 2014;9(12):1857–68.
[187] Grasso S, Santi L. Viral nanoparticles as macromolecular devices for new therapeutic and pharmaceutical approaches. Int J Physiol Pathophysiol Pharmacol 2010;2(2):161–78.
[188] Mitri Z, Constantine T, O'Regan R. The HER2 receptor in breast cancer: pathophysiology, clinical use, and new advances in therapy. Chemother Res Pract 2012;2012.

[189] Bolli E, O'Rourke JP, Conti L, Lanzardo S, Rolih V, Christen JM, Barutello G, Forni M, Pericle F, Cavallo F. A virus-like-particle immunotherapy targeting epitope-specific anti-xCT expressed on cancer stem cell inhibits the progression of metastatic cancer in vivo. Oncoimmunology 2018;7(3), e1408746.

[190] Lewerenz J, Hewett SJ, Huang Y, Lambros M, Gout PW, Kalivas PW, Massie A, Smolders I, Methner A, Pergande M, Smith SB. The cystine/glutamate antiporter system xc− in health and disease: from molecular mechanisms to novel therapeutic opportunities. Antioxid Redox Signal 2013;18(5):522–55.

[191] Yoshida GY, Saya H. The novel anti-tumor therapy targeting the "functional" cancer stem cell markers. Clin Exp Pharmacol 2014;4(147). 2161-1459.

[192] Palladini A, Thrane S, Janitzek CM, Pihl J, Clemmensen SB, de Jongh WA, Clausen TM, Nicoletti G, Landuzzi L, Penichet ML, Balboni T. Virus-like particle display of HER2 induces potent anticancer responses. Oncoimmunology 2018;7(3), e1408749.

[193] Zafar S, Beg S, Panda SK, Rahman M, Alharbi KS, Jain GK, Ahmad FJ. Novel therapeutic interventions in cancer treatment using protein and peptide-based targeted smart systems. In: Seminars in cancer biology. Academic Press; 2019.

[194] Ahmadzada T, Reid G, McKenzie DR. Fundamentals of siRNA and miRNA therapeutics and a review of targeted nanoparticle delivery systems in breast cancer. Biophys Rev 2018;10(1):69–86.

[195] Rahman M, Beg S, Verma A, Kazmi I, Kumar Patel D, Anwar F, Al Abbasi FA, Kumar V. Therapeutic applications of liposomal based drug delivery and drug targeting for immune linked inflammatory maladies: a contemporary view point. Curr Drug Targets 2017;18(13):1558–71.

[196] Deng X, Cao M, Zhang J, Hu K, Yin Z, Zhou Z, Xiao X, Yang Y, Sheng W, Wu Y, Zeng Y. Hyaluronic acid-chitosan nanoparticles for co-delivery of MiR-34a and doxorubicin in therapy against triple negative breast cancer. Biomaterials 2014;35(14):4333–44.

[197] Rahman M, Al-Ghamdi SA, Alharbi KS, Beg S, Sharma K, Anwar F, Al-Abbasi FA, Kumar V. Ganoderic acid loaded nano-lipidic carriers improvise treatment of hepatocellular carcinoma. Drug Deliv 2019;26(1):782–93.

[198] Juneja R, Lyles Z, Vadarevu H, Afonin KA, Vivero-Escoto JL. Multimodal polysilsesquioxane nanoparticles for combinatorial therapy and gene delivery in triple-negative breast cancer. ACS Appl Mater Interfaces 2019;11(13):12308–20.

[199] Pandey P, Rahman M, Bhatt PC, Beg S, Paul B, Hafeez A, Al-Abbasi FA, Nadeem MS, Baothman O, Anwar F, Kumar V. Implication of nano-antioxidant therapy for treatment of hepatocellular carcinoma using PLGA nanoparticles of rutin. Nanomedicine 2018;13(8):849–70.

[200] Jafari R, Zolbanin NM, Majidi J, Atyabi F, Yousefi M, Jadidi-Niaragh F, Aghebati-Maleki L, Shanehbandi D, Zangbar MS, Rafatpanah H. Anti-Mucin1 aptamer-conjugated chitosan nanoparticles for targeted co-delivery of docetaxel and IGF-1R siRNA to SKBR3 metastatic breast cancer cells. Iran Biomed J 2019;23(1):21.

[201] Beg S, Rahman M, Jain A, Saini S, Midoux P, Pichon C, Ahmad FJ, Akhter S. Nanoporous metal organic frameworks as hybrid polymer–metal composites for drug delivery and biomedical applications. Drug Discov Today 2017;22(4):625–37.

[202] Gomes AC, Mohsen M, Bachmann MF. Harnessing nanoparticles for immunomodulation and vaccines. Vaccines 2017;5(1):6.

Chapter 32

Bioresponsive nano-theranostic approaches for cancer targeting

Debarshi Kar Mahapatra[a], Dileep Kumar[b], Atmaram Pawar[b], Gopal Kumar Rai[c], and Sanjay Kumar Bharti[d]

[a]*Department of Pharmaceutical Chemistry, Dadasaheb Balpande College of Pharmacy, Nagpur, MH, India,* [b]*Department of Pharmaceutical Chemistry, Poona College of Pharmacy, Bharati Vidyapeeth (Deemed to be University), Pune, MH, India,* [c]*S. P. College of Pharmaceutical Sciences and Research, Bijnor, UP, India,* [d]*Institute of Pharmaceutical Sciences, Guru Ghasidas Vishwavidyalaya (A Central University), Bilaspur, CG, India*

1 Introduction

Cancer is a gene-mutation-induced disease that leads to unbridled cell division. These atypical components can produce malignant abrasions and even relocate (metastasize) to other human body organs, which significantly threatens the health of the patient. In 2018, there were 18.5 million cases of cancer worldwide and 10 million cancer deaths. Lung, prostate, and breast cancers are the most lethal carcinomas among all cancer types, responsible for 18.4%, 7.1%, and 11.6% of all cancer-associated deaths, respectively [1]. Magnetic resonance imaging (MRI), computed tomography (CT), positron emission tomography (PET), surgery, hormone therapy, radiotherapy, and chemotherapy are commonly used to diagnose and treat cancer patients [2].

Most techniques, however, fail to effectively identify and remove cancer cells. Moreover, it is difficult to fully prevent treatment-associated adverse side effects, such as vomiting and asthenia. While immunotherapy may be the best alternative to circumvent these conditions, only a minute section of patients may be able to afford the high cost [3]. The convenient nanomaterials could be consistently fabricated at the nanoscale with the advancement of nanotechnology, showing an enormous assurance in biomedical applications such as nanotheranostics. These developed nanomaterials might serve as delivery systems for monoclonal antibodies or anti-cancer agents by hauling effectual products and straightforwardly treating them with therapeutic outcomes like photothermal therapy, photodynamic therapy, and so on [4]. The effectiveness of nanopharmacotherapeutics (such as extensive circulation of drugs, augmented loading compacity, united therapy, etc.) help reduce side effects (specific targeting of the tumors and encapsulated form of anti-cancer drugs) dramatically compared to general therapies. Up to now, only a few FDA-approved cancer therapies, such as paclitaxel albumin-bound nanoparticles and liposomal irinotecan, are being used clinically. Although these nano-agents display anti-cancer properties, their continuously active cytotoxicity may cause harm to normal tissues [5].

Novel nanomaterials have been known to react successfully with biological factors surrounding tumor regions, such as temperature, concentration of copper ion, enzyme levels, redox potential, acidity, and others, have been recruited through imaging (MRI, near infrared, and fluorescence) and therapy (photothermal, chemotherapy, chemodynamic therapy, and anti-angiogenesis) to battle fatal forms of cancer [6]. Therefore, the current analysis highlights these latest bioresponsive nanotheranostics and their use in therapy for lung, prostate, and breast carcinomas, along with associated challenges [7].

2 Current situation of cancer

The lungs are closely correlated with other structures of the body, which makes it extremely susceptible to cancer. Lung cancer is a life-threatening condition, particularly lung cancer that migrates from the lungs to other body parts, which is a typical sign of advanced metastatic disease [8]. The five-year survival rate for lung cancer patients is around 5%. Breast cancer is the most common cancer in women. It is anticipated that 400,000 females will be diagnosed with either invasive or non-invasive forms of breast cancer by 2022. In men, prostate cancer causes nearly 300,000 thousand deaths and two million new cases annually worldwide. While cancer typically affects individuals with elevated risks (elderly, those with a family history of cancer, etc.), young patients are also being affected by prostate and lung cancers [9].

3 General therapeutic strategies

Generally known therapies like chemotherapies, surgeries, and radiotherapies have been employed as routine techniques for years to combat lethal forms of cancer. While

Advanced Drug Delivery Systems in the Management of Cancer. https://doi.org/10.1016/B978-0-323-85503-7.00036-5

contemporary clinical practice may encourage early-stage outcomes of these treatments, the efficacy of most advanced chemotherapies is restricted. The human genetic information of precision-based medicine has provided successful management options to patients via targeted therapy, such as blocking certain growth factor receptors [10]. While these precision-based chemotherapies may work in most general-target cancers, their effectiveness for triple-negative breast cancer (~ 11.2%) or epidermal growth factor receptor (EGFR) mutation (~ 29.3%) in mutated forms of non-small cell lung cancer (NSCLC) is greatly restricted by the accessibility of monoclonal antibodies or small molecular entities [11]. Because of this, new pharmacotherapeutic options for treating developed and mutated cancers are desperately needed (Table 1).

4 Present status of nanomedicine for cancer therapy

As a superior method, nanomedicine has been steadily functionalized to combat cancer (Fig. 1), particularly lung, breast, and prostate carcinomas. Since 2018, nano-sized components have dominated the field, though little components are practically intended for chemotherapy. At this time, around 27 clinical trials associated with breast carcinoma therapy are focusing on nanoparticles [12]. These studies are investigating nanoparticles in imaging and treatment of assorted breast carcinomas, ranging from the triple-negative form to metastatic breast cancer. In clinical trials, Imx-110, a curcumin/doxorubicin encapsulated nanoparticle system; mRNA-2752, a lipid nanoparticle system; quantum dots; paclitaxel albumin-bound nanoparticles; and silica nanoparticle systems are being investigated in the treatment of breast carcinoma [13]. In particular, CRLX101, a cyclodextrin-based polymer, or Abraxane polymer, are used in clinical practice as functional materials that cause potential harm to normal human cells. Therefore, new nano-delivery systems smartly enabled by particular conditions, such as bioresponsive nanomedicine, could provide more powerful and safer cancer therapy [14].

5 Bioresponsive nanotheranostics

In the tumor microenvironment (TME), some biological aspects such as elevated redox, cooper ions, unique enzymes, acidic extracellular environment, and others are well known. TME-activated functionalized advanced nanotheranostics have recently been developed to treat deadly cancers based on these TME factors. The pharmacotherapeutic potential of these bioresponsive nanocomponents could be elegantly activated in TME as opposed to existing nanomedicines that typically exhibit continuously activating functions (cytotoxic effects of anti-cancer molecules), effectively preventing many of the unfavorable effects induced because of mistargeting [15].

6 pH-responsive nanotheranostics

The extracellular region surrounding the cancerous foci has high lactic acid levels with unwarranted aerobic glycolysis, suggesting acidic surroundings with varying pH. Different nanomaterials, including polymers, silicas, and upconver-

TABLE 1 Recent reports of bioresponsive nanotheranostics.

Bioresponsive factor	Nanomaterial	Theranostic system	Specific therapy	Carcinoma form
Acidic environment	Polymeric-based system	PWMs	siRNA delivery	Breast cancer with lung metastasis
Acidic environment	Polymeric-based system	NP15	siRNA delivery	Breast cancer
Acidic environment	Silica-based system	TPZ@HHSN-C/P-mAb	USI and MRI	Prostate cancer
Redox-based system	Nanozyme	Lipo-OGzyme-AIE	PDT	Breast cancer
Enzymatic system	Polymeric-based system	HACE	NIRF/PAI-PDT	Lung cancer
Enzymatic system	Polymeric-based system	Self-assembled polymer	Chemo-based system	Lung cancer
Copper ion-based system	Silica-based system	Imi-OSi	Antiangiogenic	Breast cancer
Thermal and acidic environment	Polymeric-based system	mPEG-PAAV	NIRF/PAI-PTT	Breast cancer and metastasis
Copper ion-based system and acidic environment	Polymeric-based system	RPTDH	Anti-angiogenic	Metastatic breast cancer

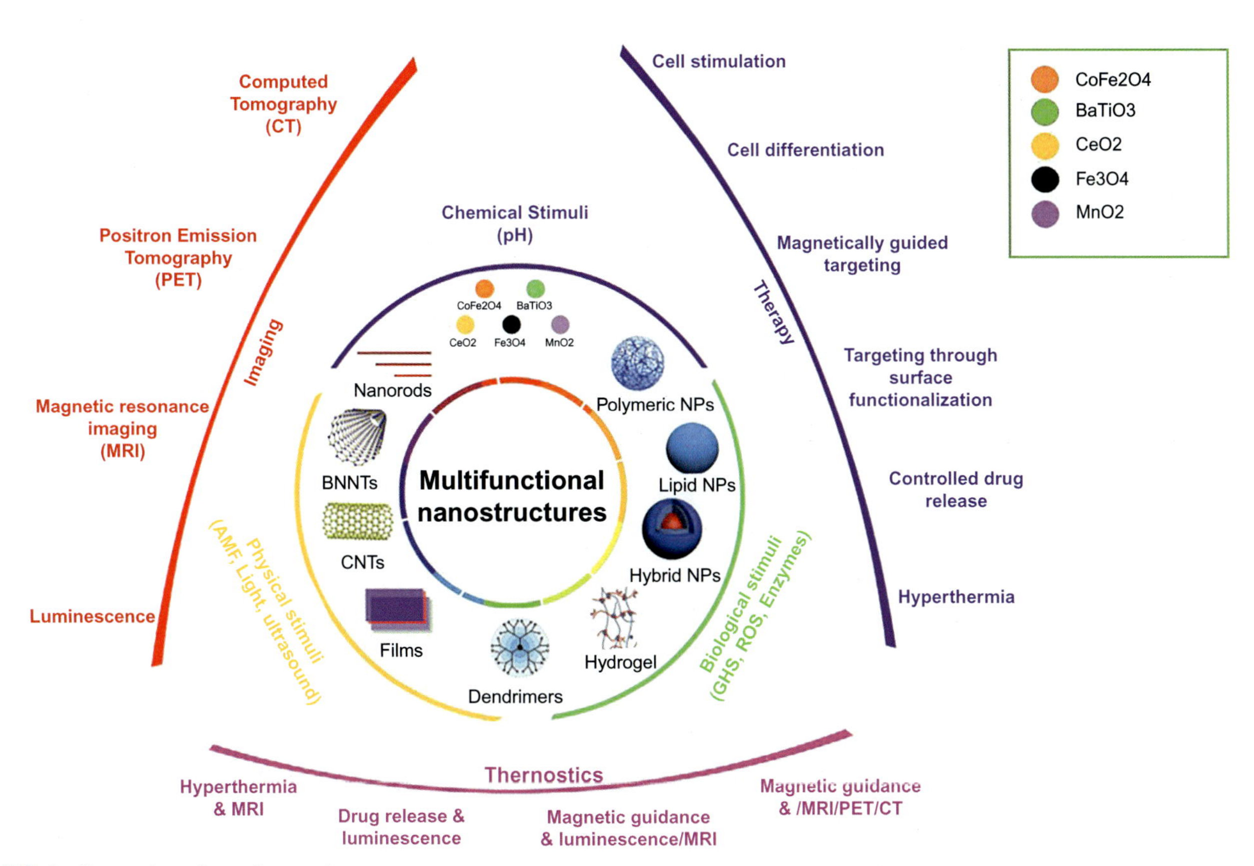

FIG. 1 Perspectives of nanotheranostics.

sion nanoparticles, were created for pH response-based smart drug delivery systems for tumor targeting. Polymer-based nanoplatforms show an immense benefit in pH-triggered drug release with exceptional pH-responsive characteristics like solubility or structural alteration [16]. Upon coating with pH sensitive element, the inhibitor succinobucol, can effectively escape from the micelles at TME, thereby inhibiting breast cancer and lung tumor metastasis by 4.5- and 6.25-fold, respectively, as compared to the saline group and SCB group [17]. In addition, Saw et al. productively developed < 100 nm-sized N-15 polymeric nanomaterial and a pH-responsive polyethylene glycol (PEG) shell by combining enzyme-induced features (esterase). Polo-like kinase 1 (PLK1) siRNA (> 90%) would only be liberated after a two-step decomposition triggered by tumor-area esterase and acid pH, which effectively inhibited about 70% PLK1 expression and two-fold tumor growth in around 18 days of MDA-MB-231 [18]. Meanwhile, silica-based, multi-module nanotheranostic components for imaging and treating prostate cancer (sonodynamic therapy and bioreductive therapy) have been developed. The acid-degraded silica nanotheranostics were capable of targeting PC3 tumors (via customized prostate stem cell antigen monoclonal antibody) and elegantly releasing tirapazamine in the TME, ultimately restraining ultrasound irradiation tumor growth by > 91.5% [19].

7 Redox-responsive nanotheranostics

The concentrations of glutathione (GSH) and reactive oxygen species (ROS) in TMEs that allow various nanocomponents to be utilized for treating hostile forms of cancers like erlotinib-resistant, EFGR-mutated non-small-cell lung cancer, taxane-resistant prostate cancer, and triple-negative breast cancer via redox-induced therapeutic functions, are also drastically superior [20]. A new P-RUB micelle was developed for co-delivery of rubone and docetaxel in taxane-resistant prostate cancer in amalgamation with acidic and GSH-responsive characteristics [21]. When GSH was accessible, the prepared nanomedicine could easily release both drugs at acidic pH of 5 [21]. The PC3-TXR proliferation was effectively repressed by about 50% with two-step activated therapy as compared to the other groups. Correspondingly, ECMI silica nanoparticles for treating lung carcinoma, disulfide bond-linked chitosan components, and pH-responsive zinc oxide quantum dots were used as the main initiators [22]. Precise allocation and successful reticence of EGFR-mutated NSCLC tumors with NIR irradiation were exhibited by ECMI carrying ICG and erlotinib. Gao et al. developed hypoxia-tropic nanozymes as nanotheranostics to suppress the growth of lung metastasis and orthotopic breast cancer [23]. Appreciably, the MnO_2 component within the ferritin nanocages (FTn) catalyze hydrogen ion and hydrogen peroxide into oxygen, which could be produced under irradiation by AIE modules into toxic ROS. These responses could successfully adapt TME (e.g., hypoxic environment neutralization, with a hypoxic tissue reduction of more than 14.5%) and forbid the uncontrolled growth and further migration of breast carcinoma with a migration of only ~5% relative to the phosphate buffer saline (PBS) community.

8 Enzyme-responsive nanotheranostics

Different enzymes are heavily connected with cancerous cells. Excess hyaluronidase has been situated in malignant tissues in breast, prostate, and lung carcinomas [24]. More specifically, hyaluronidase presentation may expedite the extension of cancer development. Li et al. fabricated chlorine-e6 theranostic micelles and hyaluronidase-activated hyaluronic acid. This polymer was capable of creating both A549 tumor imaging systems (photoacoustic imaging and near-infrared (NIR) fluorescence) and hyaluronidase-released photodynamic therapy [25]. Similarly, HA-responsive and ROS-responsive nanocomponents have been successfully employed through PDT and checkpoint blockade in the management of breast carcinoma and lung metastasis [26]. A specific category of tumor-specific enzyme is matrix metalloproteinase (MMP). Notably, in context of MMP expression, a considerable disparity is found between normal lung tissues and cancerous lung tissues, suggesting immense aspects of MMP as the bioresponsive feature for theranosis of lung cancer [27]. For treatment of lung cancer, more than a few MMP2/9-responsive nanotheranostics have recently been developed. For example, by MMP-2-cleavable peptide conjugation, a team of researchers produced nano-self-assembled nanomedicine that fosters delivery of trans-retinoic acid and camptothecin antitumor drugs selectively to cancer cells with elevated MMP-2 response [28]. Investigators engineered an innovative nanovehicle for unique extracellular delivery in comparison with traditional intracellular drug delivery. This system could effectively defend and discharge nimotuzumab loaded to the extracellular region of the lung tumor (LN-229 and PC-9) with MMP-2-receptive peptides and regulate the surface filling ratio of phosphorylcholine (PC), showing the maximum antitumor effect as compared to the free form of nimotuzumab [29].

9 Miscellaneous bio-responsive nanomedicine

More studies are showing that several cancers are related with high serum copper ions, especially breast carcinoma. In tumor angiogenesis, copper ions mediate decisive functions as capable tumor stimuli. For treating lung and breast cancer, two copper-chelator-based nanotheranostic systems were created. These smart nanoparticles produce anti-angiogenesis effects via copper-mediated chelation, which triggers better anti-tumor activity through tumor vessel obstruction (nanochelator aggregation) and TLR-mediated immune cell activation (TLR8 and TLR7 agonists). Meanwhile, for breast and prostate carcinoma therapy, numerous temperature-responsive nanomedicines have been created. These nanotheranostics are receptive to temperature transformation and release the loads at the temperature of the lower or upper critical solution, gradually activating anti-proliferative activity by pharmaco-chemotherapeutics [30].

10 Conclusion

Cancers of the lungs, breasts, and prostate cause hundreds of thousands of deaths annually, while the effectiveness of widespread approaches to treatment are often restricted, particularly for those individuals with drug resistance or genetic mutations. In contrast, bioresponsive nanomedicine has immense potential for treating cancers through intelligently activated functions (temperature, concentration of copper ion, enzyme levels, redox potential, acidity, etc.). A variety of nanotheranostics have since been productively created and demonstrated hopeful results. Incontrovertibly, the subsequent theragnostic steps for these lethal cancers will be bioresponsive nanomedicine. Basic models of bioresponsive nanotheranostics are strongly recommended. For example, MMP-activated nanotheranostics are promising against lung cancer. For treating TME, novel directions such as extracellular distribution are planned. For function activation, plural activation (e.g., pH–enzyme as well as pH–redox) integrating efficient nanotheranostics like polymers could provide more specific therapy. In the meantime, several functions are needed for future bioresponsive nanomedicine (therapy and image-guided surgery). While bioresponsive nanomedicine is smarter and more precise compared to general therapy, specialized training may be needed for treating this form of novel medicine (pH-responsive nanomedicine and redox-responsive) during the production, storage, processing, and therapeutic phases, which can potentially escalate the value of care. The redesigning of these nanotheranostics as bioresponsive components though partnership between research organizations and industry may be an option to decrease the price of quality management and additional development.

References

[1] Ventola CL. Progress in nanomedicine: approved and investigational nanodrugs. Pharm Ther 2017;42:742.

[2] Oun R, Moussa YE, Wheate NJ. The side effects of platinum based chemotherapy drugs: a review for chemists. Dalton Trans 2018;47:6645–53. https://doi.org/10.1039/C8DT00838H.

[3] Flühmann B, Ntai I, Borchard G, Simoens S, Mühlebach S. Nanomedicines: the magic bullets reaching their target? Eur J Pharm Sci 2019;128:73–80. https://doi.org/10.1016/j.ejps.2018.11.019.

[4] Kocak G, Tuncer C, Bütün V. pH-responsive polymers. Polym Chem 2017;8:144–76. https://doi.org/10.1039/C6PY01872F.

[5] Cook JA, Gius D, Wink DA, Krishna MC, Russo A, Mitchell JB. Oxidative stress, redox, and the tumor microenvironment. Semin Radiat Oncol 2004;14:259–66. https://doi.org/10.1016/j.semradonc.2004.04.001 [Elsevier].

[6] Battistella C, Callmann CE, Thompson MP, Yao S, Yeldandi AV, Hayashi T, et al. Delivery of immunotherapeutic nanoparticles to tumors via enzyme-directed assembly. Adv Healthc Mater 2019;8:1901105. https://doi.org/10.1002/adhm.201901105.

[7] Dienstmann R, Vermeulen L, Guinney J, Kopetz S, Tejpar S, Tabernero J. Consensus molecular subtypes and the evolution of precision medicine in colorectal cancer. Nat Rev Cancer 2017;17:79. https://doi.org/10.1038/nrc.2016.126.

[8] Torre LA, Siegel RL, Jemal A. Lung cancer statistics. In: Lung cancer and personalized medicine, vol. 893. Cham: Springer; 2016. p. 1–19. https://doi.org/10.1007/978-3-319-24223-1_1.

[9] Bray F, Ferlay J, Soerjomataram I, Siegel RL, Torre LA, Jemal A. Global cancer statistics 2018: GLOBOCAN estimates of incidence and mortality worldwide for 36 cancers in 185 countries. CA Cancer J Clin 2018;68:394–424. https://doi.org/10.3322/caac.21492.

[10] Shen S, Wu Y, Li K, Wang Y, Wu J, Zeng Y, et al. Versatile hyaluronic acid modified AQ4N-cu (II)-gossypol infinite coordination polymer nanoparticles: multiple tumor targeting, highly efficient synergistic chemotherapy, and real-time self-monitoring. Biomaterials 2018;154:197–212. https://doi.org/10.1016/j.biomaterials.2017.11.001.

[11] Tan T, Dent R. Triple-negative breast cancer: clinical features. In: Triple-Negative Breast cancer. Cham: Springer; 2018. p. 23–32. https://doi.org/10.1007/978-3-319-69980-6_2.

[12] Mcatee CO, Barycki JJ, Simpson MA. Emerging roles for hyaluronidase in cancer metastasis and therapy. Adv Cancer Res 2014;123:1–34. https://doi.org/10.1016/B978-0-12-800092-2.00001-0 [Elsevier].

[13] Qiao H, Cui Z, Yang S, Ji D, Wang Y, Yang Y, et al. Targeting osteocytes to attenuate early breast cancer bone metastasis by theranostic upconversion nanoparticles with responsive plumbagin release. ACS Nano 2017;11:7259–73. https://doi.org/10.1021/acsnano.7b03197.

[14] Dai L, Li X, Duan X, Li M, Niu P, Xu H, et al. A pH/ROS cascade-responsive charge-reversal nanosystem with self-amplified drug release for synergistic oxidation-chemotherapy. Adv Sci 2019;6:1801807. https://doi.org/10.1002/advs.201801807.

[15] He X, Cai K, Zhang Y, Lu Y, Guo Q, Zhang Y, et al. Dimeric prodrug self-delivery nanoparticles with enhanced drug loading and bioreduction responsiveness for targeted Cancer therapy. ACS Appl Mater Interfaces 2018;10:39455–67. https://doi.org/10.1021/acsami.8b09730.

[16] Kato Y, Ozawa S, Miyamoto C, Maehata Y, Suzuki A, Maeda T, et al. Acidic extracellular microenvironment and cancer. Cancer Cell Int 2013;13:89. https://doi.org/10.1186/1475-2867-13-89.

[17] Midha A, Dearden S, McCormack R. EGFR mutation incidence in non-small-cell lung cancer of adenocarcinoma histology: a systematic review and global map by ethnicity (mutMapII). Am J Cancer Res 2015;5:2892.

[18] Saw PE, Yao H, Lin C, Tao W, Farokhzad OC, Xu X. Stimuli-responsive polymer–prodrug hybrid nanoplatform for multistage siRNA delivery and combination cancer therapy. Nano Lett 2019;19:5967–74. https://doi.org/10.1021/acs.nanolett.9b01660.

[19] Wang Y, Liu X, Deng G, Sun J, Yuan H, Li Q, et al. Se@ SiO2-FA–CuS nanocomposites for targeted delivery of DOX and nano selenium in synergistic combination of chemo-photothermal therapy. Nanoscale 2018;10:2866–75. https://doi.org/10.1039/C7NR09237G.

[20] He X, Yu H, Bao X, Cao H, Yin Q, Zhang Z, et al. pH responsive wormlike micelles with sequential metastasis targeting inhibit lung metastasis of breast cancer. Adv Healthc Mater 2016;5:439–48. https://doi.org/10.1002/adhm.201500626.

[21] Hu D, Zhong L, Wang M, Li H, Qu Y, Liu Q, et al. Perfluorocarbon-loaded and redox-activatable photosensitizing agent with oxygen supply for enhancement of fluorescence/photoacoustic imaging guided tumor photodynamic therapy. Adv Funct Mater 2019;29:1806199. https://doi.org/10.1002/adfm.201806199.

[22] Van Rijt SH, Bölükbas DA, Argyo C, Datz S, Lindner M, Eickelberg O, et al. Protease-mediated release of chemotherapeutics from mesoporous silica nanoparticles to ex vivo human and mouse lung tumors. ACS Nano 2015;9:2377–89. https://doi.org/10.1021/nn5070343.

[23] Gao F, Wu J, Gao H, Hu X, Liu L, Midgley AC, et al. Hypoxia-tropic nanozymes as oxygen generators for tumor-favoring theranostics. Biomaterials 2020;230:119635. https://doi.org/10.1016/j.biomaterials.2019.119635.

[24] Li W, Zheng C, Pan Z, Chen C, Hu D, Gao G, et al. Smart hyaluronidase-actived theranostic micelles for dual-modal imaging guided photodynamic therapy. Biomaterials 2016;101:10–9. https://doi.org/10.1016/j.biomaterials.2016.05.019.

[25] Li S, Chen L, Huang K, Chen N, Zhan Q, Yi K, et al. Tumor microenvironment-tailored weakly cell-interacted extracellular delivery platform enables precise antibody release and function. Adv Funct Mater 2019;29:1903296. https://doi.org/10.1002/adfm.201970301.

[26] Liu C, Wang D, Zhang S, Cheng Y, Yang F, Xing Y, et al. Biodegradable biomimic copper/manganese silicate nanospheres for chemodynamic/photodynamic synergistic therapy with simultaneous glutathione depletion and hypoxia relief. ACS Nano 2019;13:4267–77. https://doi.org/10.1021/acsnano.8b09387.

[27] Merchant N, Nagaraju GP, Rajitha B, Lammata S, Jella KK, Buchwald ZS, et al. Matrix metalloproteinases: their functional role in lung cancer. Carcinogenesis 2017;38:766–80. https://doi.org/10.1093/carcin/bgx063.

[28] Yang Y, Tang J, Zhang M, Gu Z, Song H, Yang Y, et al. Responsively aggregatable sub-6 nm nanochelators induce simultaneous antiangiogenesis and vascular obstruction for enhanced tumor vasculature targeted therapy. Nano Lett 2019;19:7750–9. https://doi.org/10.1021/acs.nanolett.9b02691.

[29] Lin F, Wen D, Wang X, Mahato RI. Dual responsive micelles capable of modulating miRNA-34a to combat taxane resistance in prostate cancer. Biomaterials 2019;192:95–108. https://doi.org/10.1016/j.biomaterials.2018.10.036.

[30] Denoyer D, Masaldan S, La Fontaine S, Cater MA. Targeting copper in cancer therapy:'copper that cancer'. Metallomics 2015;7:1459–76. https://doi.org/10.1039/C5MT00149H.

Chapter 33

Two-dimensional materials-based nanoplatforms for lung cancer management: Synthesis, properties, and targeted therapy

Bharath Singh Padya, Abhijeet Pandey, Ajinkya NIkam (Nitin), Sanjay Kulkarni, Gasper Fernandes, and Srinivas Mutalik

Department of Pharmaceutics, Manipal College of Pharmaceutical Sciences, Manipal Academy of Higher Education, Manipal, KA, India

1 Introduction

Cancer, after cardiovascular diseases, represents the second-most reason of death around the world. Conferring to a current report provided by WHO (World Health Organization), every sixth death worldwide is caused by cancer [1]. In particular, among the cancers, lung cancer is the second-most common type. Major lifestyle factors like smoking, alcohol consumption, and the diet habits have worsened the condition. Exposure to pollutants such as asbestos, toxins, arsenic, and others have increased the risk-inducing lung cancer [2]. Based on the stage of cancer, first-line treatment options like radiation therapy, surgery, and chemotherapy are currently available. However, these treatment options fail to control the metastasis of tumors to different organs. In addition, the use of chemotherapeutic agent causes several undesirable effects on the healthy cells like bone marrow, skin, gastrointestinal tract cells, and hair follicles resulting in toxicity. Other limitations of the conventional therapeutic strategies include improper biodistribution, chances of recurrence, lack of targetability, and multidrug resistance phenomenon [3, 4]. In this context, different type of nanomaterials has been explored for targeting the cancer; active targeting by functionalizing the carriers and passive targeting by enhanced permeability and retention effect [2].

2 Two-dimensional (2D) nanoplatforms for lung cancer targeting

Two-dimensional (2D) nanomaterials constitute an emerging novel class of nanomaterials with sheet-like structures having larger lateral size (up to a few microns), but the thickness is much smaller like approximately less than 5 nm [5]. The 2D feature is exceptional and crucial for access to unparalleled electronic, chemical, and physical properties because of the electron confinement in 2D [6]. Owing to these properties and their ultrathin planar nanostructure, 2D nanomaterials have been proved to exhibit notable advantages in the field of oncotherapy and diagnosis. Furthermore, these planar nanostructures provide several anchoring sites for the loading of therapeutics as they have also shown prolonged blood circulation and efficient tumor targetability. Graphene, due to its properties like ultrahigh carrier mobility [7], specific surface area [8], optical transparency [9], quantum hall effect [10], and thermal [11] and electrical [7] conductivities, stands as an exemplary model of 2D nanomaterials. The study of other 2D nanomaterials like black phosphorous, transition metal dichalcogenides, copper, and tin is also emerging [12]. However, there are few reports describing the role of 2D metals targeting lung cancer. Keeping this in mind, the chapter discusses about the role of advanced materials involving 2D materials for effective cancer therapy. Various 2D materials such as graphene, transition metal dichalcogenides, black phosphorous, titanium dioxide, and copper have been focused. This chapter deals in detail with the synthesis as well as biomedical application of these materials in lung cancer. Emphasis has been given to the role of these nanostructures in chemotherapy, immunotherapy, and gene therapy along with alternative therapeutic strategy such as photodynamic therapy (PDT) and photothermal therapy (PTT) of lung cancer as described in Fig. 1.

Advanced Drug Delivery Systems in the Management of Cancer. https://doi.org/10.1016/B978-0-323-85503-7.00016-X

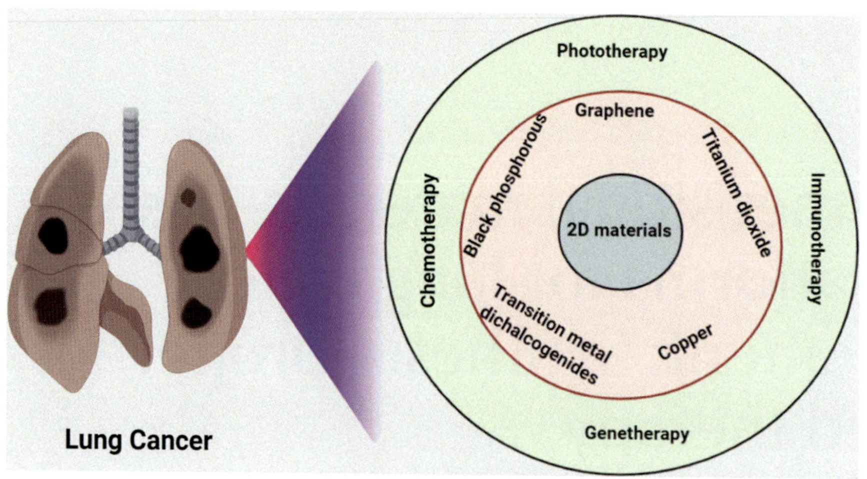

FIG. 1 Schematic representation of the role of various 2D nanomaterials in different therapeutic strategies for lung cancer.

2.1 Graphene-based 2D nanoplatforms

The scientist around the world are exploring graphene-based 2D nanoplatforms for the past many years as cancer theranostic agent. Herein, we focus on graphene-based 2D nanoplatforms targeting lung cancer as shown in Fig. 2 and their method of synthesis.

2.1.1 Graphite oxide synthesis

The Brodie [13], Staudenmaier [14], or Hummers and Offeman [15] methods or variants of the latter, namely, the Modified Hummers method or the Improved Hummers method [16], have been used to synthesize graphite oxide. Table 1 summarizes the key differences between the above approaches, with specific focus on the toxicity, nature of oxidant, and the main advantages or drawbacks of each strategy.

2.1.2 Graphene-based 2D nanoplatforms for lung cancer targeting

Nie et al. produced theranostic GO-based DOX nanocarriers for the drug delivery and PAI (photoacoustic imaging) in lung cancer (H1975 cells) in vitro and in vivo. To enable fluorescence imaging, the framework was also attached to the Cy5.5 dye. Furthermore, the material was also able to persuade efficient PAI-monitored chemotherapy in mice owing to its high loading capacity [17]. Khatun and colleagues also examined the delivery of DOX in accordance with PTT. Their graphene built nanosystem coupled with a nanogel of hyaluronic acid for photothermal imaging was capable of inducing an efficient destruction of A549 cells (lung cancer cells), while displaying only mild toxicity in the healthy MDCK cells [18]. In recent studies, porphyrin derivatives were paired with aptamer-decorated GQDs for the delivery of miRNA and subsequent PTT/PDT,

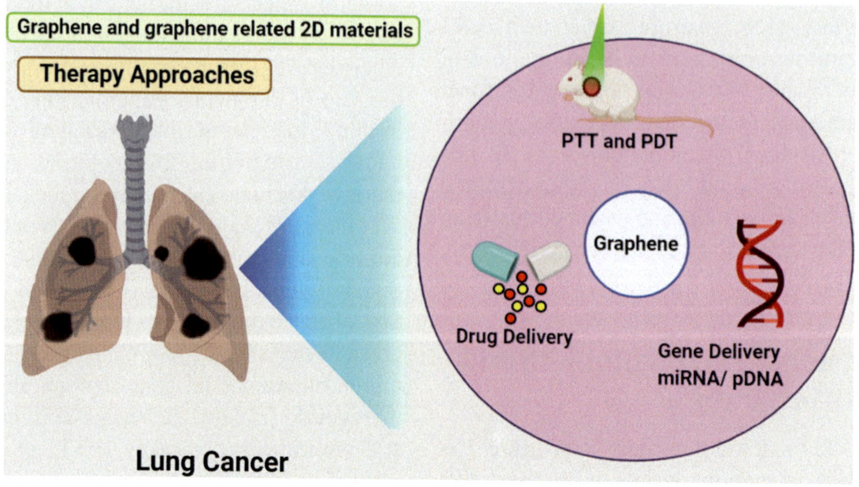

FIG. 2 Therapeutic approaches of graphene and graphene-related 2D materials.

TABLE 1 Fabrication approaches of graphite oxide.

Synthesis approaches	Brodie method	Staudenmaier method	Hummers method	Modified Hummers method	Improved Hummers method
Oxidants used	HNO_3, $KClO_3$	($NaClO_3$), $KClO_3$, H_2SO_4, HNO_3	$KMnO_4$, $NaNO_3$, H_2SO_4	$KMnO_4$, H_2SO_4, $NaNO_3$	$KMnO_4$, H_2SO_4, H_3PO_4
Toxicity	Yes	Yes	No	No	No
Advantages	–	–	• Elevated oxidation degree than that of in Brodie or Staudenmaier approaches	• Higher level of oxidation and, consequently, better performance of product	• Flaws in the basal plane are condensed • Higher quantity of oxidized graphite is obtained • The degree of reduction provides an equal level of conductivity when related to other approaches • Toxic gases are not produced during preparations, which is why its environmentally safe • The method has much organized framework
Drawbacks	• Soft dispersibility in basic solutions • Small size, limiting thickness, and providing an imperfect structure	• Time-consuming and dangerous method • The incorporation of $KClO_3$ normally takes longer than a week and CO_2 is evolved, thus making it essential to eliminate an inert gas • Danger of explosion is a persistent risk	• The oxidation is also known to be incomplete.	• Procedures of isolation and purification are lengthy • Highly time consuming	–

facilitating breast and lung cancer cell CLSM [19]. A lot of researches have reported on the use of graphene for targeting lung cancer and are summarized in Table 2.

2.2 Black phosphorous-based 2D nanoplatforms for lung cancer targeting

Phosphorous is one of Earth's most ample elements and can be found in numerous allotropic forms like white, black, violet, and red phosphorous. Among these, the black phosphorous (BP) is the most stable form and similar to graphite, it has a layered structure wherein the layers are held together by van der Waals forces and can be exfoliated to the few/single-layer structure [26]. It is a 2D material having semiconductor properties with a layer-dependent tunable bandgap which helps the phosphorous in actively interacting with the incident light ranging from UV to NIR. Several studies have shown that the BP is biocompatible and poses minimum toxicity. Due to these properties, it has a very good applicability in the field of drug delivery and is one among the monoelemental materials which has demonstrated remarkable ability in the biomedical field [27]. For instance, BP nanosheets display a good photothermal conversion efficiency.

Like other 2D materials, BP nanosheets (BPNS) can also be prepared by the bottom-up and top-down strategies. But for the fabrication of single-layer phosphorous-phosphorene, top-down approach stands out to be the most convenient one [26]. The first method for its preparation was reported in 2014 via the micromechanical cleavage and then transferred onto the SiO_2/Si substrate [28], and till date it is the most commonly used method in the development of BP-based optoelectronic devices. For the large-scale use, mechanical exfoliation associated with the high-power ultrasonication is used which yields nanosized flakes of few layer BP [29]. Further, to develop the phosphorene, liquid exfoliation involving the dispersion of bulk BP in polar aprotic solvents like DMSO and DMF has proved to be advantageous [30]. Another study synthesized the BP nanosheets through ultrasonication-assisted liquid exfoliation of BP dispersed in water for about 30 min and the resulting suspension of several layered BP was further subjected to ultrasonication for 10 min to produce BP nanosheets [31]. There are studies which have reported the exfoliation of BP by employing the low-power radiations like Ar plasma (plasma etching) which also produce the monolayered BP [32]. However, these top-down methods have a limitation in providing high-quality sheets of BP necessary for the field effect transistor (FET) and related applications in field of biosensing. Further, the chemical stability also remains as a crucial unanswered issue. Few studies have countered this problem and have developed surface-modified BP nanosheets to enhance its chemical stability. For instance, Sun et al. developed the PEGylated BP to improvise the stability and biocompatibility. The in vivo studies proved the accumulation of BP in the tumor and it induced the photothermal ablation of tumor in the presence of NIR radiation [33].

In particular to lung cancer, Latiff et al. [34] addressed the toxicological aspects of the BPNS on A549 (human lung carcinoma) cells. Plate-like BP sheets were developed by treating the red phosphorous with Au/Sn alloy like solvent and the SnI_4 as vapor transport medium and subjected it to 400–600 °C in a muffle furnace. Furthermore, the dye-based cell viability assays like MTT and WST-8 were performed by incubating the A549 cells to various concentrations of BP for a period of 24 h. Results demonstrated the dose-dependent toxicity of BP and it had an intermediate toxicity between dichalcogenides and graphene. The control experiments proved that interference of the nanomaterial above concentration of 100 mg/mL in the assay via the reduction of the assay reagents. Nevertheless, the study demonstrated the cytotoxicity of this 2D nanomaterial toward the lung cancer. In another research, gold nanobipyramids (GNBP)-decorated BP nanosheets were evaluated against the orthotropic lung tumor-bearing mice [35]. Herein, the nanosheets were fabricated using the liquid exfoliation method by dispersing the bulk BPs in NMP (*N*-methyl-2-pyrrolidone) and probe sonicated for 6 h in an ice bath followed by sonication at 0 °C for 12 h. The resulting suspension was centrifuged and the precipitate was washed with water to collect the BP nanosheets. The GNBP was gradually mixed with BPNS under stirring to obtain the BPNS-GNBP nanocomposite. Later, the in vitro cytotoxicity of this nanocomposite was assessed in the HeLa cell line using MTT assay and confirmed the in vivo synergistic antitumor efficacy of photodynamic (PDT) and photothermic (PTT) therapies in orthotropic human lung tumor models established by inducing the A549 cells in the lungs of nude mice. The nanocomposite demonstrated the inhibition of the tumor growth when subjected to the combined PDT-PTT therapy upon exposure to 808 nm laser radiation. Results of both in vitro as well as the in vivo studies established the enhanced antitumor efficacy of the BPNS-GNBP nanocomposite when compared to the individual effects of BPNS and GNBP. Other cell viability studies and toxicity assessments are the need of the hour to understand the potential effects of 2D BP nanoplatforms in providing the enhanced anticancer efficacy.

Similarly, NIR-controlled drug delivery platform of BPNS was developed and evaluated for its efficacy against several cancer cell lines [36]. The BPNSs were developed by a modified liquid exfoliation method using isopropanol as a solvent and cup ultrasound sonication to exfoliate the bulk BP. Though the tip/probe sonication yields high-quality BPNSs in large quantity, it contaminates the sample when in direct contact with the tip. This contamination was taken care by employing cup sonication method to

TABLE 2 Graphene-related materials in lung cancer theranostics.

Graphene-related materials		Application		Coating/functionalization			Target			
Material	Name	Therapy	Imaging	Drug	Gene	Targeting moieties and other molecules	Cell line	Cancer	Model	References
GQDs	AS1411-GQDs	Drug delivery	Confocal laser scanning microscopy (CLSM)	Aptamer	–	–	A549, HeLa, MCF-7, and HepG-2	Lung, cervical, breast, and liver cancer	In vitro	[20]
GO	GO-Cy5.5-Dox	Drug delivery	Fluorescence imaging PAI	DOX	–	Cy5.5	H1975	Lung cancer	In vivo	[17]
GO	DOX-CMG-GFP-DNA	Drug and gene delivery	Fluorescence imaging	DOX	pDNA	–	A549, LLC1, PC3, and C42b	Lung and prostate cancer	In vivo and in vitro	[21]
NGO	NGO-808	PTT and PDT	CLSM, NIR fluorescence, and thermal imaging PAI	–	–	–	A549 and Lewis lung cancer cells	Lung cancer	In vivo and in vitro	[22]
GO	GO-PEG-DVDMS	PTT and PDT	PAI, Fluorescence imaging	–	–	DVDMS	PC9	Lung cancer	In vivo and in vitro	[23]
GO	GO-HA-Ce6	PTT and PDT	NIR fluorescence imaging	–	–	Ce6	A549	Lung cancer	In vitro	[24]
GQDs	GQD-PEG-P	Gene delivery, drug delivery, PTT, and PDT	CLSM	Aptamer	miRNA	–	A549 and MCF-7	Lung and breast cancer	In vitro	[19]
rGO	Anti-EGFR-PEG-rGO@CPSSAu-R6G	Drug delivery and PTT	CLSM, fluorescence imaging, optical imaging, SERS imaging, and Raman imaging	Anti-EGFR SERS probes	–	–	A549	Lung cancer	In vitro	[25]
Graphene	GDH	Drug delivery and PTT	Optical and photothermal imaging	DOX	–	–	A549	Lung cancer	In vivo and in vitro	[18]

synthesize BPNSs and further, surface modified with PEG to improve the biocompatibility. The PEGylated BPNSs were mixed with agarose solution and loaded with doxorubicin (DOX) to form the DOX-loaded BP@Hydrogel. The hydrogel system exhibited NIR-controlled drug release. The cytotoxicity studies performed on the A549 cells showed the potential biosafety of the material due to the absence of cytotoxicity at higher concentrations of 200 mg/mL. In another study, a novel p–p stacking interaction of aromatic 1-pyrenylbutyric acid (p) was employed in the fabrication of BPNSs [37]. This strategy enhances the stability of BPNSs and also paves a way to conveniently anchor the bioactive moieties to enable the efficient targeting and thus extending the biomedical applicability of BPNSs. The bulk BP together with 1-pyrenylbutyric acid was added in dimethylformamide (DMF) and was sonicated in ice bath for 16 h. The resultant suspension was centrifuged and a stable brown supernatant was further centrifuged to obtain the p-BPNSs. The p-BPNS was conjugated with arginine-glycine-aspartic acid (RGD) peptides via carbodiimide chemistry and evaluated for its efficacy against the A549 cell line through cytotoxicity assay. The in vitro cytotoxicity assay proved the biocompatibility of the p-BPNSs and RGD-p-BPNS before therapy. Upon irradiation of 808 nm NIR laser, the cell viability of A549 cells was reduced owing to the higher uptake of nanosheets by the cells. The in vivo studies were performed on the Balb/c nude mice induced with A549 cells. The nanosheets exhibited good hemocompatibility enabling further in vivo studies. The photoacoustic imaging confirmed the efficient tumor accumulation ability of the RGD-p-BPNSs at the tumor site. After 24 h of I.V. injection of RGD-p-BPNSs, the entire tumor area was exposed to 808 nm NIR laser for 10 min. Upon laser irradiation, the temperature increase in the treated mice increased by 2.9 °C in 10 min and had reached 56.4 °C, high enough to induce hyperthermia and cause tumor ablation. These results proved the PTT efficacy of RGD-modified BPNSs for tumor therapy.

2.3 Copper-based 2D nanoplatforms

Copper (Cu) is the most widely available transition metal that exists freely in nature and plays a vital role in the variety of metalloproteins in biological systems, such as hemocyanins, copper blue proteins, cytochrome oxidase, and superoxide dismutase [38]. The electronic configuration of Cu with totally loaded "d" levels displays very similar levels of energy among both the bulk valence and conduction bands [39]. Because of their biocompatibility, nontoxic nature, and low cost, Cu has attracted great interest. The threshold energy for Cu interband transitions is 2.0 eV, which is close to gold (Au) and slightly very lower than silver (Ag) (4.0 eV), hence, Cu displays a strong reduction and expansion of copper nanoclusters in the surface plasmon resonance (SPR) [40]. In optoelectronic devices like luminescence [41], sensors [42], catalysis [43], etc., noble metal nanoclusters like $CuMoS_2$ of different sizes ranging from ~ 1.1 to ~ 2.2 nm recently received tremendous attention. Owing to their unique characteristics which are not seen in bulk corresponding item, two-dimensional (2D) constituents have drawn wide interest [44, 45]. Because of their high availability, excellent physicochemical properties, and simple preparation methods, copper sulfide nanosheets were being extensively studied [9–12]. Very recently, β-Cu2S 2D sheet of 1.8 nm thickness was synthesized and proved to exist as a mixture of solid–liquid phase [46–48]. For high chalcocite to low chalcocite transition, the temperature is significantly lower than the bulk materials, making 2D β-Cu2S sheets easily accessible for ambient temperature applications [47]. Also, with decreasing the nanoparticle size, transition temperature from β-Cu2S to γ-Cu2S decreases as well. Furthermore, Cu2S sheets possess greater photoactivity than their bulk counterparts for conversion of solar energy [49]. A recent theoretical analysis predicted that a thin bilayer of β-Cu2S is more stable because of its close bound between the two layers and there by exists a high binding energy. In addition, owing to its high direct band gap of 0.9 eV, bilayer β-Cu2S could be a potential candidate for its optical and electronic applications [49].

An unexpected 2D copper oxide (CuO) atomic geometry leads to unexpected magnetic and electronic properties. It is estimated that 2D copper oxide will have semiconducting properties with a 2.7 eV band gap, whereas bulk CuO has 1.5 eV [50]. In addition, 2D CuO has ground state of an antiferromagnetic, while CuO and Cu are paramagnetic in bulk. 2D CuO has high rigidity because of binding energy of 4.07 eV due to strong covalent bonds that are formed by O and Cu atoms in the 2D sheet. Furthermore, by considering the total energy of flat 2D CuO with CuO nanotubes, the 2D cluster's mechanical stability against bending was examined and studied. Calculations of result obtained concluded that the finite radius of the CuO nanotube is dynamically highly stable than the 2D sheet, suggesting the need for graphene as a 2D CuO supporting material [50]. Finally, owing to the strong affinity of Cu for sulfur, radioactive 64Cu was fully absorbed to MoS_2 that makes it suitable for positron emission tomography (PET) chelating agent-free radiolabeling [51]. Cu-based metal ions and clusters with controllable quantum yields (QLs) contribute considerably to their photochemical properties. However, the synthetic obstacle of stabilizing the ultrafine size and airborne oxidation stability tends to leave major challenges in the synthesis of Cu nanoclusters (CuNCs) [52].

2.3.1 Copper synthesis routes for 2D materials

Due to the greater lateral size, distinct physicochemical properties, atomically organized networks, strong in-plane, and weak our-plane bonding, 2D materials have gained a lot of interest in biomedical applications. The rapid development of many reliable synthesis techniques for 2D material synthesis was already explored due to such persuasive properties and limited applications. Two separate approaches have been made to develop 2D materials from their precursor, like a top-down approach and a bottom-up approach depending on the number of layers or sheet exfoliation. These various synthetic techniques have different chemical, surface or electronic, and physical characteristics that can be used to interpret the relation between functional and structural characteristics. To know the relativity among structural and functional properties for different applications, the synthesis of 2D materials with desirable physiochemical parameters like size, composition, thickness, surface, and defect characteristics is of great importance [53].

2.3.2 Top-down approach

By disassembling or splitting them down while maintaining the original integrity, the top-down process is used to perform reproducible production of thin-layer sheets using parent bulk materials. This is often done by methods of liquid and mechanical exfoliation. It is specifically concerned with the elimination of the contact among the fixed layers of parent bulk constituents mainly because of Van der Waals forces. Top-down approach is advantageous owing to its flexibility in large-scale manufacturing and reduced cost, and hence, potentially desirable in various research applications. Nevertheless, complications like surface structure defects and remarkable crystal disruption to the final structures frequently occur [53].

2.3.2.1 Bottom-up approach

Many reactions-based chemical methods are being used to generate monolayer nanosheets (NSs) from molecules or atoms based on structural changes in the reaction mixture, like metal–organic decay, chemical vapor deposition, solvothermal synthesis, molecular or atomic condensation, etc. The existence of appropriate "metal–organic" compounds as substrates for the direct fabrication of the ultrathin and discrete 2D NSs depends on most of these chemical reactions. Such synthesized 2D materials are mostly utilized in biological or chemical methods due to large-scale processing, effective development of the smallest nanomaterials (~10nm), and also cost effective. To achieve atomic to nanoscale 2D materials, two different approaches are used commonly, namely, wet-chemical synthesis and chemical vapor deposition [53].

2.3.3 Cu-based 2D nanoplatforms for lung cancer targeting

Tumor tissues have an elevated copper content that is used in cellular processes like angiogenesis to facilitate the growth of cancer. Organic compounds binding to Cu offer useful methods for transforming cancer-promoting copper into cancer-fighting agents. To inhibit tumor progression, these copper complexes may serve as antiangiogenic agents or ROS generators. These compounds may also target cell proteins like proteasomes to prevent the proliferation of cancer cells. Taken together, to design and develop more efficient and selective antitumor copper ligands, combinatorial chemistry, computation modeling, and medicinal chemistry can provide effective tools to understand structure–activity and chemical space relationships [54]. CuO-doped substances have shown significant anticancer effects and are strongly suggested for use in cancer treatment [55].

Copper radionuclides decay characteristics make them ideal for various medical applications, including radioimmunological tracing, positron emission tomography (PET) imaging, and cancer radiotherapy. Two features are necessary for the widespread use of any radioisotope in medicine: the accessibility of the isotope and potential binding modes with a suitable chemical carrier. Over the past 20–30 years, the efficient development of copper isotopes has been extensively studied, and many prospective chelators have also been implemented during that time. Production methods, nuclear medicine applications, and copper radioisotope chelating agents have been reviewed recently by Szymanski et al. [56]

Compared to other conventional nanomaterials, the exceptional physical properties of 2D materials, especially high ultralight and heat transfer, make them highly attractive in PTT. In vitro studies using A549 cell line and human RBCs have being carried out for the toxicity of CuNCs. CuNCs had strong hemocompatibility till 16 μg/mL and biocompatibility of up to 20 μg/mL as synthesized, which is consistent with recorded approaches. CuNCs were protected by the MPS staples and, thus, aided in an efficient bioimaging of A549 cells [53]. The flexibility of copper and its derivatives have given it a significant position to generate appropriate pharmaceutical products. Although there are currently only a few implementations of Cu in medicine, in the future, multiple emerging researches will most likely lead to new uses. Radiopharmaceuticals made from copper are likely to be the first to be approved for clinical use. Green-synthesized CuO NPs suppressed HDAC and thus proved anticancer activity mediated by apoptosis in lung cancer cell line A549 [57]. Since there are reports only on Cu-based nanoparticle and nanocluster for treatment of lung cancer, 2D material-based nanosystem can be implemented soon for treating lung cancer in the near future. Cu-containing complexes and nanomaterials are also highly promising and many other applications in different biomedical fields should be found soon [56].

2.4 Transition metal dichalcogenides (TMDs)-based 2D nanoplatforms for lung cancer targeting

Amidst the discovery of well-defined pyrolytic graphene and its derivatives, research in the field of 2D layered materials is rapidly growing [7, 58]. Such layered materials consist of atomically thin stacked sheets held together by weak bonding forces (Van-der-Waals). On exfoliation to monolayers, 2D layered materials exhibited unique properties wherein 2D transition metal dichalcogenides (TMDs) possessed broken inversion symmetry and quantum confinement effect unlike graphene and bulk TMDs and thus received substantial research interest. The 2D TMDs are generally represented by MX_2, where M denotes transition metals belonging from the group IVB to VIIB of the periodic table and X denotes the chalcogenide elements (S, Se, and Te). The transition metals are sandwiched between two chalcogen layers, assuming an octahedral, trigonal, or prismatic morphology [59]. The higher surface area and low cytotoxicity of ultrathin 2D nanolayer render 2D TMDs as attractive metals for application in the biomedical arena like drug delivery (loading multiple drug molecules, genes, or organic molecules) [60], biosensing [61], bioimaging [62], photothermal, and photoacoustic therapy (due to intrinsic NIR absorption and high photoacoustic conversion coefficient) [63]. Among the several types of TMDs, molybdenum disulfide is also represented as MoS_2 [64], tungsten selenide (WS_2) [61], bismuth selenide (Bi_2S_3) [65], and titanium dichalcogenides (TiS_2) [66] are the most studied and well-described TMDs so far. The 2D TMDs exhibit poor stability due to high surface area and abundant atoms exposed leading to higher surface free energy [59, 64]; hence, to improve its stability, dispersibility, and biocompatibility, biodegradable or synthetic polymers such as chitosan, bovine serum albumin, polyethylene glycol, polyethylenimine, and polyvinylpyrrolidone are used to modify its surface by either physical adsorption (electrostatic interaction and hydrophobic interactions) or chemical bonding (covalent or coordination bonds) [67, 68].

Mechanical exfoliation, solvothermal, chemical exfoliation, chemical vapor deposition, and liquid phase exfoliation are the most commonly used techniques for synthesizing 2D TMDs. The synthetic schemes can be categorized into the bottom-up approach and top-down approach, where the top-down method is the favored one for fabrication of 2D TMDs specifically for biomedical applications while the bottom-up approach is favored for optoelectronic and catalysis field which is depicted in Table 3 [80].

As cancer is an imminent risk to human health, lung cancer and metastatic lung cancer are the key determinants of cancer-related deaths [81]. Nonsmall-cell lung cancer (NSCL) accounts for greater than 85% of the mortality rate worldwide [82]. Therefore, lung targeting is a long-term approach to lung cancer treatment, where nanotechnology is unquestionably integrating ground-breaking science and technology to achieve this aim [83]. Since graphene oxide showed the potential for localization in the lung after administration through several routes, other 2D metals, including MoS_2, were explored and gained increasing popularity due to its high drug surface, resulting in high drug loading. As discussed in the earlier sections, the functionalization of MoS_2 is mandatory to enhance its stability in the physiological environment. Drugs with multiple hydroxyl and amino groups can form coordinate bonds with Mo resulting in higher loading of drugs. Hence, a combination of GO and MoS_2 would be beneficial in targeting tumors specific to lung due to GO specificity in the lungs, whereas higher drug loading would be possible due to the unique properties exhibited by MoS_2. Hence, utilizing a similar concept in a study, anticancer drug doxorubicin was loaded in MoS_2/GO nanocomposites. The prepared composites showed good dispersibility in aqueous solutions, as well as the tendency of the composites to localize in the lung, which was higher than GO. Additionally, MoS_2/GO composites substantially suppressed B16 murine melanoma cancer cell tumor growth in the lungs of the mice [84]. Similarly, Zhang et al. synthesized MoS_2 nanosheets using chemical exfoliation and modified its surface using HA which not only improves physiological stability but also has specific activity in tumor cells that are overexpressed with CD44. The MoS_2 was fabricated to target CD44 cells and release the drug in a controlled manner on NIR irradiation as represented in Fig. 3 [63]. Chen et al. fabricated MoS_2 nanoflakes using a hydrothermal process and coated it with PAA and PEG to improve its biocompatibility. The bioaccumulation studies showed a high concentration of Mo in the lung without any inflammatory response or toxicity [85]. Kumar et al. designed MoS_2 nanosheets by one-step hydrothermal synthesis and prepared PEGylated microspheres of MoS_2 nanosheets to test its toxicity on lung (A549) and breast cancer cell lines. They reported that the PEGylated MoS_2 nanosheets cytotoxicity depends on their ability to release metal ions, solubility, cellular absorption, and specific surface area [86].

2.5 TiO_2-based 2D nanoplatforms for lung cancer targeting

Due to its low cost, multifunctionality, and biomedical applications, titanium dioxide nanosheets (TiO2) have been among the most researched material system over the past few years [87, 88]. The controlled facile synthesis of TiO_2 nanosheets can be achieved by controlling the growth of TiO_2 nanoparticles in some direction to achieve 2D morphology [89, 90]. TiO_2 nanomaterials with varied morphology, such as zero-dimensional nanoparticles, one-dimensional nanowires or nanotubes, two-dimensional nanosheets, as well as three-dimensional nanostructures, have been synthesized [91–93].

A standard approach for synthesizing TiO2 nanosheets is via the intercalation-exfoliation technique. In this technique, in terms of crystallinity, structure, size, and functionality, the initial layered titanates greatly influence the "offspring" nanosheets. Another traditional approach is

TABLE 3 Summary of synthesis of TMDs nanosheets.

Methods		Description	Advantages	Disadvantages	Variables/chemical reagents/equipment	References
Top-down synthesis	Mechanical exfoliation	The TMDs molecular layers are bound to a sheet of Scotch tape that would then be placed against a solid substrate, such as an oxidized Si wafer.	High yield	The thickness and size of the nanosheets are difficult to control	Scotch tape	[69, 70]
	Liquid exfoliation	Ultrasonication in a specific solvent	Large-scale preparation	Toxic solvents, and solvent limits few applications	Ultrasonic time and intensity/ dimethylformamide (DMF), and *N*-methyl pyrrolidone (NMP)	[71, 72]
	Chemical exfoliation	Repeated process of introducing intercalators between interlayers of bulk crystals followed by ultrasonication	No use of toxic solvents	Additional purification step	Butyl lithium and naphthyl sodium	[73, 74]
	Electrochemical exfoliation	Controlled rate of intercalation of lithium ion into MoS_2 powder (lithium intercalation process)	No decomposition due to extreme intercalation, and enhanced quality and safety	–	Lithium foil as anode and TMDS dispersed in PVDF or NMP as cathode	[75]
Bottom-up approach	Chemical vapor deposition	Exposure of the reaction precursor (alternate transition metals and chalcogen) to the substrate under high pressure and temperature	Exceptional electronic property with tunable size and high quality of crystal	High temperature and vacuum required	Ex. of substrates SiO_2	[76, 77]
	Solvothermal process	Reacting the precursor with surfactant at low temperature under high pressure	Controlled size and surface chemistry, and good colloidal stability	–	Teflon-coated autoclave	[78, 79]

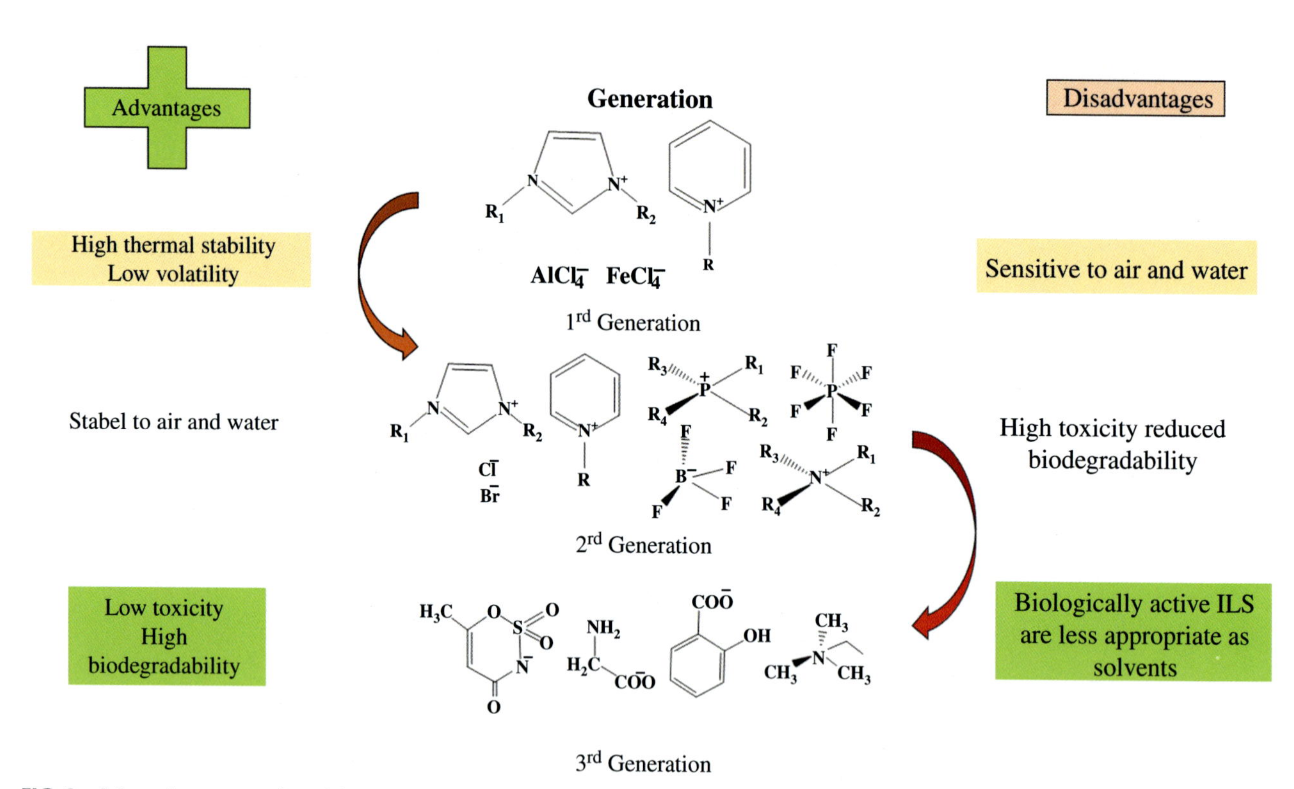

FIG. 3 Schematic representation of the construction of nanocomposites based on MoS_2-SS-HA for synergistic chemo-photothermal therapy.

solid-state calcination, in which TiO_2 and alkali metal are calcinated at high temperatures to produce micrometer-sized TiO_2 nanosheets. The polymorphic forms of TiO_2, namely brookite, rutile, and anatase, are nonlayered crystals owing to the strong electrostatic association between host layers and counter ions, i.e., metal cations and anionic oxygen, thereby making it difficult to mechanically exfoliate into 2D nanosheets using conventional exfoliation techniques [94]. To address this issue, a study utilizes a novel technique similar to synthesizing TiO_2 nanotubes, wherein anatase TiO_2 nanoparticles were treated with an alkaline solution (NaOH) for 24 h at elevated temperature (130 °C) with the incorporation of a cationic surfactant, namely tetrabutylammonium hydroxide (TBAOH), as shown in Fig. 4. Utilizing this technique, large-scale synthesis of single-layered TiO_2 nanosheets was possible [95]. A bottom-up technique can be used to fabricate single nanosheets by either the process of wet-chemical one-pot synthesis [96] or vapor deposition [97].

Lung cancer is considered a serious life-threatening cancer with a high mortality rate, several strategies have been undertaken, namely, radiotherapy, surgery, chemotherapy, and surgery, depending on the cancer subtype and stage [98]. While medical research has made major progress in recent years, the mortality rate for patients with lung cancer is extremely high, accounting for less than 10% for advanced lung cancer [99, 100]. Exfoliated nanosheets can be a suitable host material for immobilizing protein or drug molecules due to their incredibly high surface area and charge-bearing nature [101]. In fascinating research, TiO_2 nanosheets were fabricated by a combination of solid-state synthesis followed by proton exchange and exfoliation. The precursor selected for the study was potassium zinc titanate which was prepared by calcinating a mixture of potassium carbonate, zinc oxide, and titanium dioxide at 800 °C for 1 h followed by reacting at 900 °C for an additional 20 h. TBAOH was utilized for exfoliation by shaking mechanically for 14 days at 180 rpm. The prepared TiO_2 nanosheets suppressed cancer stem cells by superoxide anion generation in the lungs and showed no toxicity toward the noncancerous cells which were demonstrated in lung cancer cell lines. In addition, TiO_2 nanosheets also suppressed alpha and β1 integrins that are potential markers of metastasis [102]. TiO_2 nanosheets can have potential as a drug carrier in pharmaceutical applications similar to nanoporous capsules. For the synergistic treatment of lung cancer, Dai et al. synthesized pH-sensitive dual stimulus-responsive TiO peroxide nanosheets loaded with DOX. The TiO_2 nanosheets were fabricated using a hydrothermal process and using polymer 123 to control the shape. The fabricated nanosheets were oxidized to form TiO nanosheets and loaded with DOX by conjugating it to the surface. It was observed that on X-ray irradiation ROS were generated for effective cancer radiotherapy along with the release of DOX in acidic conditions. Therefore, they concluded that a combination of radiotherapy and chemotherapy could be a good cancer

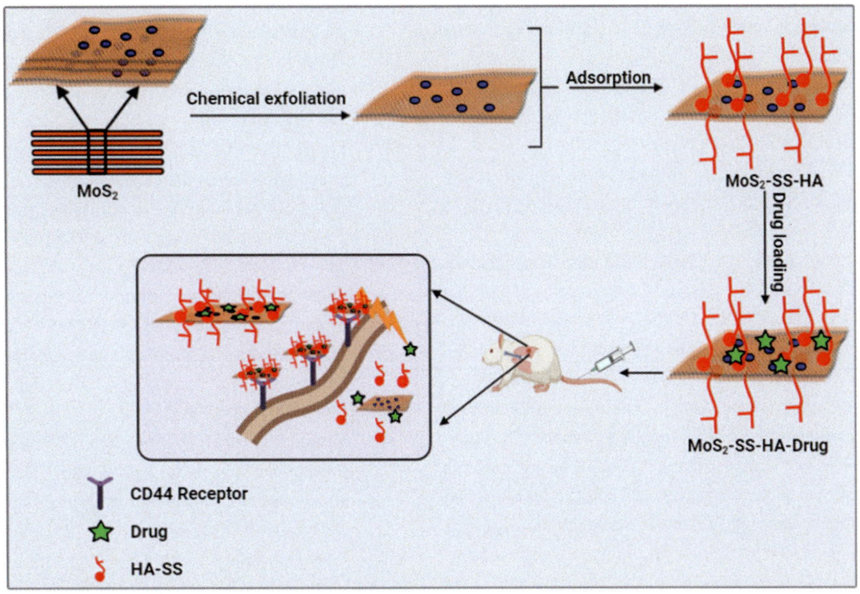

FIG. 4 Synthesis scheme of TiO_2 nanosheets in presence of a surfactant.

treatment option, demonstrating a new strategy for future cancer treatment [103].

3 Conclusion

Nanotechnology has partaken the newly fostered strategies for the therapy of lung cancer. Specifically, the 2D materials have gained a greater attention due to their properties like flexibility, higher surface area, photothermal conversion efficiency, specific planar structure, and bioactivity. The introduction of 2D materials has revolutionized the importance of dimensionality in determining the intrinsic properties and the potential applicability of the nanomaterials. The anchoring sites provided by these 2D materials not only increase the loading of drug but also helps in the surface modification to enable the targeting of overexpressed receptors. Extensive exploration of the 2D nanomaterials is also associated with newer challenges. From the point of synthesis, the quantity, production yield, and the quality of 2D nanomaterials being produced from the current synthetic approaches are lagging far behind from the requirements to be commercialized in an industrial scale. Hence, the major challenge would be to improve the yield quality and quantity to go on par with the industry requirements. Despite great advances, current 2D nanomaterial research is still far from mature, particularly beyond graphene. Concerning its applicability in the lung cancer, there is a need of detailed investigation on understanding the interaction of these nanomaterials with hematological factors as well as with the immune factors to ensure the safety and toxicity. Further, it has also been reported that the physicochemical factors of these nanomaterials play an important part in enhancing the therapeutic efficacy by improving the internalization efficiency. It is therefore firmly believed that in the scientific communities, these 2D nanomaterials would ignite wider interests in addressing these issues and finding effective ways of clinical applications to battle cancer.

References

[1] Ritchie H, Roser M. Causes of death, our world data, https://ourworldindata.org/causes-of-death; 2018. [Accessed 13 October 2020].

[2] Vanza JD, Patel RB, Patel MR. Nanocarrier centered therapeutic approaches: recent developments with insight towards the future in the management of lung cancer. J Drug Deliv Sci Technol 2020;102070.

[3] Shewach DS, Kuchta RD. Introduction to cancer chemotherapeutics. ACS Publications; 2009.

[4] Peer D, Karp JM, Hong S, Farokhzad OC, Margalit R, Langer R. Nanocarriers as an emerging platform for cancer therapy. Nat Nanotechnol 2007;2:751.

[5] Zhang H. Ultrathin two-dimensional nanomaterials. ACS Nano 2015;9:9451–69.

[6] Geim AK, Novoselov KS. The rise of graphene, in: Nanosci. Technol. Collect. Rev. Nat. J: World Scientific; 2010. p. 11–9.

[7] Novoselov KS, Geim AK, Morozov SV, Jiang D, Zhang Y, Dubonos SV, Grigorieva IV, Firsov AA. Electric field effect in atomically thin carbon films. Science 2004;306:666–9.

[8] Stoller MD, Park S, Zhu Y, An J, Ruoff RS. Graphene-based ultracapacitors. Nano Lett 2008;8:3498–502.

[9] Nair RR, Blake P, Grigorenko AN, Novoselov KS, Booth TJ, Stauber T, Peres NM, Geim AK. Fine structure constant defines visual transparency of graphene. Science 2008;320:1308.

[10] Zhang Y, Tan Y-W, Stormer HL, Kim P. Experimental observation of the quantum hall effect and Berry's phase in graphene. Nature 2005;438:201–4.

[11] Balandin AA, Ghosh S, Bao W, Calizo I, Teweldebrhan D, Miao F, Lau CN. Superior thermal conductivity of single-layer graphene. Nano Lett 2008;8:902–7.

[12] Tan C, Cao X, Wu X-J, He Q, Yang J, Zhang X, Chen J, Zhao W, Han S, Nam G-H. Recent advances in ultrathin two-dimensional nanomaterials. Chem Rev 2017;117:6225–331.

[13] Brodie BC. XXIII.—Researches on the atomic weight of graphite. Q J Chem Soc Lond 1860;12:261–8. https://doi.org/10.1039/QJ8601200261.

[14] Staudenmaier L. Verfahren zur Darstellung der Graphitsäure. Berichte Dtsch Chem Ges 1898;31:1481–7. https://doi.org/10.1002/cber.18980310237.

[15] Hummers WS, Offeman RE. Preparation of graphitic oxide. J Am Chem Soc 1958;80:1339. https://doi.org/10.1021/ja01539a017.

[16] Improved synthesis of graphene oxide - PubMed, (n.d.). https://pubmed.ncbi.nlm.nih.gov/20731455/ (accessed October 12, 2020).

[17] Nie L, Huang P, Li W, Yan X, Jin A, Wang Z, Tang Y, Wang S, Zhang X, Niu G, Chen X. Early-stage imaging of Nanocarrier-enhanced chemotherapy response in living subjects by scalable photoacoustic microscopy. ACS Nano 2014;8:12141–50. https://doi.org/10.1021/nn505989e.

[18] Khatun Z, Nurunnabi M, Nafiujjaman M, Reeck GR, Khan HA, Cho KJ, Lee Y. A hyaluronic acid nanogel for photo-chemo theranostics of lung cancer with simultaneous light-responsive controlled release of doxorubicin. Nanoscale 2015;7:10680–9. https://doi.org/10.1039/c5nr01075f.

[19] Cao Y, Dong H, Yang Z, Zhong X, Chen Y, Dai W, Zhang X. Aptamer-conjugated graphene quantum dots/porphyrin derivative theranostic agent for intracellular cancer-related microRNA detection and fluorescence-guided photothermal/photodynamic synergetic therapy. ACS Appl Mater Interfaces 2017;9:159–66. https://doi.org/10.1021/acsami.6b13150.

[20] Wang X, Sun X, He H, Yang H, Lao J, Song Y, Xia Y, Xu H, Zhang X, Huang F. A two-component active targeting theranostic agent based on graphene quantum dots. J Mater Chem B 2015;3:3583–90. https://doi.org/10.1039/C5TB00211G.

[21] Wang C, Ravi S, Garapati US, Das M, Howell M, MallelaMallela J, Alwarapan S, Mohapatra SS, Mohapatra S. Multifunctional chitosan magnetic-graphene (CMG) nanoparticles: a Theranostic platform for tumor-targeted co-delivery of drugs, genes and MRI contrast agents. J Mater Chem B 2013;1:4396–405. https://doi.org/10.1039/C3TB20452A.

[22] Luo S, Yang Z, Tan X, Wang Y, Zeng Y, Wang Y, Li C, Li R, Shi C. Multifunctional photosensitizer grafted on polyethylene glycol and Polyethylenimine dual-functionalized Nanographene oxide for Cancer-targeted near-infrared imaging and synergistic phototherapy. ACS Appl Mater Interfaces 2016;8:17176–86. https://doi.org/10.1021/acsami.6b05383.

[23] Yan X, Hu H, Lin J, Jin A, Niu G, Zhang S, et al. Optical and photoacoustic dual-modality imaging guided synergistic photodynamic/photothermal therapies. Nanoscale 2015;7:2520–6. https://doi.org/10.1039/c4nr06868h.

[24] Cho Y, Kim H, Choi Y. A graphene oxide–photosensitizer complex as an enzyme-activatable theranostic agent. Chem Commun 2013;49:1202–4. https://doi.org/10.1039/C2CC36297J.

[25] Chen Y-W, Liu T-Y, Chen P-J, Chang P-H, Chen S-Y. A high-sensitivity and low-power Theranostic Nanosystem for cell SERS imaging and selectively Photothermal therapy using anti-EGFR-conjugated reduced graphene oxide/mesoporous silica/AuNPs Nanosheets. Small Weinh Bergstr Ger 2016;12:1458–68. https://doi.org/10.1002/smll.201502917.

[26] Gusmao R, Sofer Z, Pumera M. Black phosphorus rediscovered: from bulk material to monolayers. Angew Chem Int Ed 2017;56:8052–72.

[27] Liu G, Tsai H-I, Zeng X, Qi J, Luo M, Wang X, Mei L, Deng W. Black phosphorus nanosheets-based stable drug delivery system via drug-self-stabilization for combined photothermal and chemo cancer therapy. Chem Eng J 2019;375:121917.

[28] Liu H, Neal AT, Zhu Z, Luo Z, Xu X, Tománek D, Ye PD. Phosphorene: an unexplored 2D semiconductor with a high hole mobility. ACS Nano 2014;8:4033–41.

[29] Mu Y, Si MS. The mechanical exfoliation mechanism of black phosphorus to phosphorene: a first-principles study. EPL Europhys Lett 2015;112:37003.

[30] Yasaei P, Kumar B, Foroozan T, Wang C, Asadi M, Tuschel D, Indacochea JE, Klie RF, Salehi-Khojin A. High-quality black phosphorus atomic layers by liquid-phase exfoliation. Adv Mater 2015;27:1887–92.

[31] Song S-J, Raja IS, Lee YB, Kang MS, Seo HJ, Lee HU, Han D-W. Comparison of cytotoxicity of black phosphorus nanosheets in different types of fibroblasts. Biomater Res 2019;23:1–7.

[32] Lu W, Nan H, Hong J, Chen Y, Zhu C, Liang Z, Ma X, Ni Z, Jin C, Zhang Z. Plasma-assisted fabrication of monolayer phosphorene and its Raman characterization. Nano Res 2014;7:853–9.

[33] Sun C, Wen L, Zeng J, Wang Y, Sun Q, Deng L, Zhao C, Li Z. One-pot solventless preparation of PEGylated black phosphorus nanoparticles for photoacoustic imaging and photothermal therapy of cancer. Biomaterials 2016;91:81–9.

[34] Latiff NM, Teo WZ, Sofer Z, Fisher AC, Pumera M. The cytotoxicity of layered black phosphorus. Chem A Eur J 2015;21:13991–5.

[35] Wang J, Zhang H, Xiao X, Liang D, Liang X, Mi L, Wang J, Liu J. Gold nanobipyramid-loaded black phosphorus nanosheets for plasmon-enhanced photodynamic and photothermal therapy of deep-seated orthotopic lung tumors, Acta biomater; 2020.

[36] Qiu M, Wang D, Liang W, Liu L, Zhang Y, Chen X, Sang DK, Xing C, Li Z, Dong B. Novel concept of the smart NIR-light–controlled drug release of black phosphorus nanostructure for cancer therapy. Proc Natl Acad Sci 2018;115:501–6.

[37] Li Z, Guo T, Hu Y, Qiu Y, Liu Y, Wang H, Li Y, Chen X, Song J, Yang H. A highly effective π–π stacking strategy to modify black phosphorus with aromatic molecules for cancer theranostics. ACS Appl Mater Interfaces 2019;11:9860–71.

[38] Liu Y, Yao D, Zhang H. Self-assembly driven aggregation-induced emission of copper nanoclusters: a novel Technology for Lighting. ACS Appl Mater Interfaces 2018;10:12071–80. https://doi.org/10.1021/acsami.7b13940.

[39] Häkkinen H, Moseler M, Landman U. Bonding in Cu, Ag, and Au clusters: relativistic effects, trends, and surprises. Phys Rev Lett 2002;89. https://doi.org/10.1103/PhysRevLett.89.033401.

[40] Cottancin E, Celep G, Lermé J, Pellarin M, Huntzinger JR, Vialle JL, Broyer M. Optical properties of Noble metal clusters as a function of the size: comparison between experiments and a semi-quantal theory. Theor Chem Acc 2006;116:514–23. https://doi.org/10.1007/s00214-006-0089-1.

[41] Pyo K, Thanthirige VD, Kwak K, Pandurangan P, Ramakrishna G, Lee D. Ultrabright luminescence from gold nanoclusters: rigidifying the au(I)–thiolate Shell. J Am Chem Soc 2015;137:8244–50. https://doi.org/10.1021/jacs.5b04210.

[42] Kwak K, Kumar SS, Pyo K, Lee D. Ionic liquid of a gold nanocluster: a versatile matrix for electrochemical biosensors. ACS Nano 2014;8:671–9. https://doi.org/10.1021/nn4053217.

[43] Zhu Y, Qian H, Jin R. Catalysis opportunities of atomically precise gold nanoclusters. J Mater Chem 2011;21:6793. https://doi.org/10.1039/c1jm10082c.

[44] Xu M, Liang T, Shi M, Chen H. Graphene-like two-dimensional materials. Chem Rev 2013;113:3766–98. https://doi.org/10.1021/cr300263a.

[45] Novoselov KS. Electric field effect in atomically thin carbon films. Science 2004;306:666–9. https://doi.org/10.1126/science.1102896.

[46] Romdhane FB, Cretu O, Debbichi L, Eriksson O, Lebègue S, Banhart F. Quasi-2D cu $_2$ S crystals on graphene: in-situ growth and ab-initio calculations. Small 2015;11:1253–7. https://doi.org/10.1002/smll.201400444.

[47] Li B, Huang L, Zhao G, Wei Z, Dong H, Hu W, Wang L-W, Li J. Large-size 2D β-cu $_2$ S Nanosheets with Giant phase transition temperature lowering (120*K*) synthesized by a novel method of supercooling chemical-vapor-deposition. Adv Mater 2016;28:8271–6. https://doi.org/10.1002/adma.201602701.

[48] Shahzad R, Kim T, Mun J, Kang S-W. Observation of photoluminescence from large-scale layer-controlled 2D ß-cu $_2$ S synthesized by the vapor-phase sulfurization of copper thin films. Nanotechnology 2017;28:505601. https://doi.org/10.1088/1361-6528/aa972b.

[49] Guo Y, Wu Q, Li Y, Lu N, Mao K, Bai Y, Zhao J, Wang J, Zeng XC. Copper(I) sulfide: a two-dimensional semiconductor with superior oxidation resistance and high carrier mobility. Nanoscale Horiz 2019;4:223–30. https://doi.org/10.1039/C8NH00216A.

[50] Kano E, Kvashnin DG, Sakai S, Chernozatonskii LA, Sorokin PB, Hashimoto A, Takeguchi M. One-atom-thick 2D copper oxide clusters on graphene. Nanoscale 2017;9:3980–5. https://doi.org/10.1039/C6NR06874J.

[51] Kurapati R, Kostarelos K, Prato M, Bianco A. Biomedical uses for 2D materials beyond graphene: current advances and challenges ahead. Adv Mater 2016;28:6052–74. https://doi.org/10.1002/adma.201506306.

[52] Jia X, Yang X, Li J, Li D, Wang E. Stable cu nanoclusters: from an aggregation-induced emission mechanism to biosensing and catalytic applications. Chem Commun 2014;50:237–9. https://doi.org/10.1039/C3CC47771A.

[53] Murugan C, Sharma V, Murugan RK, Malaimegu G, Sundaramurthy A. Two-dimensional cancer theranostic nanomaterials: synthesis, surface functionalization and applications in photothermal therapy. J Control Release 2019;299:1–20. https://doi.org/10.1016/j.jconrel.2019.02.015.

[54] Wang F, Jiao P, Qi M, Frezza M, Dou QP, Yan B. Turning tumor-promoting copper into an anti-cancer weapon via high-throughput chemistry. Curr Med Chem 2010;17:2685–98. https://doi.org/10.2174/092986710791859315.

[55] Mabrouk M, Kenawy SH, El-Bassyouni GET, Ibrahim Soliman AAE-F, Aly Hamzawy EM. Cancer cells treated by clusters of copper oxide doped calcium silicate. Adv Pharm Bull 2019;9:102–9. https://doi.org/10.15171/apb.2019.013.

[56] Szymański P, Frączek T, Markowicz M, Mikiciuk-Olasik E. Development of copper based drugs, radiopharmaceuticals and medical materials. Biometals 2012;25:1089–112. https://doi.org/10.1007/s10534-012-9578-y.

[57] Kalaiarasi A, Sankar R, Anusha C, Saravanan K, Aarthy K, Karthic S. T. Lemuel Mathuram, V. Ravikumar, copper oxide nanoparticles induce anticancer activity in A549 lung cancer cells by inhibition of histone deacetylase. Biotechnol Lett 2018;40:249–56. https://doi.org/10.1007/s10529-017-2463-6.

[58] Xu M, Liang T, Shi M, Chen H. Graphene-like two-dimensional materials. Chem Rev 2013;113:3766–98. https://doi.org/10.1021/cr300263a.

[59] Chhowalla M, Shin HS, Eda G, Li L-J, Loh KP, Zhang H. The chemistry of two-dimensional layered transition metal dichalcogenide nanosheets. Nat Chem 2013;5:263–75. https://doi.org/10.1038/nchem.1589.

[60] Yang X, Li J, Liang T, Ma C, Zhang Y, Chen H, Hanagata N, Su H, Xu M. Antibacterial activity of two-dimensional MoS2 sheets. Nanoscale 2014;6:10126–33. https://doi.org/10.1039/C4NR01965B.

[61] Yuan Y, Li R, Liu Z. Establishing water-soluble layered WS2 Nanosheet as a platform for biosensing. Anal Chem 2014;86:3610–5. https://doi.org/10.1021/ac5002096.

[62] Ma Y-H, Dou W-T, Pan Y-F, Dong L-W, Tan Y-X, He X-P, Tian H, Wang H-Y. Fluorogenic 2D Peptidosheet unravels CD47 as a potential biomarker for profiling hepatocellular carcinoma and cholangiocarcinoma tissues. Adv Mater 2017;29:1604253. https://doi.org/10.1002/adma.201604253.

[63] Zhang C, Zhang D, Liu J, Wang J, Lu Y, Zheng J, Li B, Jia L. Functionalized MoS2-erlotinib produces hyperthermia under NIR. J Nanobiotechnology 2019;17:76. https://doi.org/10.1186/s12951-019-0508-9.

[64] Liu T, Wang C, Gu X, Gong H, Cheng L, Shi X, Feng L, Sun B, Liu Z. Drug delivery with PEGylated MoS2 Nano-sheets for combined Photothermal and chemotherapy of Cancer. Adv Mater 2014;26:3433–40. https://doi.org/10.1002/adma.201305256.

[65] Cheng L, Shen S, Shi S, Yi Y, Wang X, Song G, Yang K, Liu G, Barnhart TE, Cai W, Liu Z. FeSe2-decorated Bi2Se3 Nanosheets fabricated via cation exchange for chelator-free 64Cu-labeling and multimodal image-guided Photothermal-radiation therapy. Adv Funct Mater 2016;26:2185–97. https://doi.org/10.1002/adfm.201504810.

[66] Qian X, Shen S, Liu T, Cheng L, Liu Z. Two-dimensional TiS2 nanosheets for in vivo photoacoustic imaging and photothermal cancer therapy. Nanoscale 2015;7:6380–7. https://doi.org/10.1039/C5NR00893J.

[67] Kou Z, Wang X, Yuan R, Chen H, Zhi Q, Gao L, Wang B, Guo Z, Xue X, Cao W, Guo L. A promising gene delivery system developed from PEGylated MoS2 nanosheets for gene therapy. Nanoscale Res Lett 2014;9:587. https://doi.org/10.1186/1556-276X-9-587.

[68] Afaneh T, Sahoo PK, Nobrega IAP, Xin Y, Gutierrez HR. Laser-assisted chemical modification of monolayer transition metal Dichalcogenides. Adv Funct Mater 2018;28:1802949. https://doi.org/10.1002/adfm.201802949.

[69] Chen J, Liu C, Hu D, Wang F, Wu H, Gong X, Liu X, Song L, Sheng Z, Zheng H. Single-layer MoS2 Nanosheets with amplified photoacoustic effect for highly sensitive photoacoustic imaging of Orthotopic brain tumors. Adv Funct Mater 2016;26:8715–25. https://doi.org/10.1002/adfm.201603758.

[70] Novoselov KS, Jiang D, Schedin F, Booth TJ, Khotkevich VV, Morozov SV, Geim AK. Two-dimensional atomic crystals. Proc Natl Acad Sci 2005;102:10451–3. https://doi.org/10.1073/pnas.0502848102.

[71] Jawaid A, Nepal D, Park K, Jespersen M, Qualley A, Mirau P, Drummy LF, Vaia RA. Mechanism for liquid phase exfoliation of MoS2. Chem Mater 2016;28:337–48. https://doi.org/10.1021/acs.chemmater.5b04224.

[72] Yong Y, Zhou L, Gu Z, Yan L, Tian G, Zheng X, Liu X, Zhang X, Shi J, Cong W, Yin W, Zhao Y. WS2 nanosheet as a new photosensitizer carrier for combined photodynamic and photothermal therapy of cancer cells. Nanoscale 2014;6:10394–403. https://doi.org/10.1039/C4NR02453B.

[73] Eda G, Yamaguchi H, Voiry D, Fujita T, Chen M, Chhowalla M. Photoluminescence from chemically exfoliated MoS2. Nano Lett 2011;11:5111–6. https://doi.org/10.1021/nl201874w.

[74] Lukowski MA, Daniel AS, Meng F, Forticaux A, Li L, Jin S. Enhanced hydrogen evolution catalysis from chemically exfoliated metallic MoS2 Nanosheets. J Am Chem Soc 2013;135:10274–7. https://doi.org/10.1021/ja404523s.

[75] Zeng Z, Sun T, Zhu J, Huang X, Yin Z, Lu G, Fan Z, Yan Q, Hng HH, Zhang H. An effective method for the fabrication of few-layer-thick inorganic Nanosheets. Angew Chem Int Ed 2012;51:9052–6. https://doi.org/10.1002/anie.201204208.

[76] Ling X, Lee Y-H, Lin Y, Fang W, Yu L, Dresselhaus MS, Kong J. Role of the seeding promoter in MoS2 growth by chemical vapor deposition. Nano Lett 2014;14:464–72. https://doi.org/10.1021/nl4033704.

[77] Liu K-K, Zhang W, Lee Y-H, Lin Y-C, Chang M-T, Su C-Y, Chang C-S, Li H, Shi Y, Zhang H, Lai C-S, Li L-J. Growth of large-area and highly crystalline MoS2 thin layers on insulating substrates. Nano Lett 2012;12:1538–44. https://doi.org/10.1021/nl2043612.

[78] Wang S, Li K, Chen Y, Chen H, Ma M, Feng J, Zhao Q, Shi J. Biocompatible PEGylated MoS2 nanosheets: controllable bottom-up synthesis and highly efficient photothermal regression of tumor. Biomaterials 2015;39:206–17. https://doi.org/10.1016/j.biomaterials.2014.11.009.

[79] Cao S, Liu T, Hussain S, Zeng W, Peng X, Pan F. Hydrothermal synthesis of variety low dimensional WS2 nanostructures. Mater Lett 2014;129:205–8. https://doi.org/10.1016/j.matlet.2014.05.013.

[80] Liu T, Liu Z. 2D MoS2 nanostructures for biomedical applications. Adv Healthc Mater 2018;7:1701158. https://doi.org/10.1002/adhm.201701158.

[81] Miller KD, Siegel RL, Lin CC, Mariotto AB, Kramer JL, Rowland JH, Stein KD, Alteri R, Jemal A. Cancer treatment and survivorship statistics, 2016. CA Cancer J Clin 2016;66:271–89. https://doi.org/10.3322/caac.21349.

[82] Zhang Y, Xu W, Guo H, Zhang Y, He Y, Lee SH, Song X, Li X, Guo Y, Zhao Y, Ding C, Ning F, Ma Y, Lei Q-Y, Hu X, Li S, Guo W. NOTCH1 signaling regulates self-renewal and platinum Chemoresistance of Cancer stem–like cells in human non–small cell lung Cancer. Cancer Res 2017;77:3082–91. https://doi.org/10.1158/0008-5472.CAN-16-1633.

[83] Chen Y, Wu Y, Sun B, Liu S, Liu H. Two-dimensional nanomaterials for Cancer Nanotheranostics. Small 2017;13:1603446. https://doi.org/10.1002/smll.201603446.

[84] Liu Y, Peng J, Wang S, Xu M, Gao M, Xia T, Weng J, Xu A, Liu S. Molybdenum disulfide/graphene oxide nanocomposites show favorable lung targeting and enhanced drug loading/tumor-killing efficacy with improved biocompatibility. NPG Asia Mater 2018;10:e458. https://doi.org/10.1038/am.2017.225.

[85] Chen L, Feng Y, Zhou X, Zhang Q, Nie W, Wang W, Zhang Y, He C. One-pot synthesis of MoS2 Nanoflakes with desirable degradability for Photothermal Cancer therapy. ACS Appl Mater Interfaces 2017;9:17347–58. https://doi.org/10.1021/acsami.7b02657.

[86] Kumar N, George BPA, Abrahamse H, Parashar V, Ngila JC. Sustainable one-step synthesis of hierarchical microspheres of PEGylated MoS2 nanosheets and MoO3 nanorods: their cytotoxicity towards lung and breast cancer cells. Appl Surf Sci 2017;396:8–18. https://doi.org/10.1016/j.apsusc.2016.11.027.

[87] Chen JS, David Lou XW. SnO2 and TiO2 nanosheets for lithium-ion batteries. Mater Today 2012;15:246–54. https://doi.org/10.1016/S1369-7021(12)70115-3.

[88] Fujishima A, Zhang X. Titanium dioxide photocatalysis: present situation and future approaches. Comptes Rendus Chim 2006;9:750–60. https://doi.org/10.1016/j.crci.2005.02.055.

[89] Liu G, Yang HG, Wang X, Cheng L, Lu H, Wang L, Lu GQ, Cheng H-M. Enhanced photoactivity of oxygen-deficient anatase TiO2 sheets with dominant {001} facets. J Phys Chem C 2009;113:21784–8. https://doi.org/10.1021/jp907749r.

[90] Liu S, Yu J, Jaroniec M. Tunable photocatalytic selectivity of hollow TiO2 microspheres composed of Anatase Polyhedra with exposed {001} facets. J Am Chem Soc 2010;132:11914–6. https://doi.org/10.1021/ja105283s.

[91] Fröschl T, Hörmann U, Kubiak P, Kučerová G, Pfanzelt M, Weiss CK, Behm RJ, Hüsing N, Kaiser U, Landfester K, Wohlfahrt-Mehrens M. High surface area crystalline titanium dioxide: potential and limits in electrochemical energy storage and catalysis. Chem Soc Rev 2012;41:5313–60. https://doi.org/10.1039/C2CS35013K.

[92] Jiang C, Zhang J. Nanoengineering Titania for high rate Lithium storage: a review. J Mater Sci Technol 2013;29:97–122. https://doi.org/10.1016/j.jmst.2012.11.017.

[93] Chen X, Mao SS. Titanium dioxide nanomaterials: synthesis, properties, modifications, and applications. Chem Rev 2007;107:2891–959. https://doi.org/10.1021/cr0500535.

[94] Wu B, Guo C, Zheng N, Xie Z, Stucky GD. Nonaqueous production of nanostructured Anatase with high-energy facets. J Am Chem Soc 2008;130:17563–7. https://doi.org/10.1021/ja8069715.

[95] Leng M, Chen Y, Xue J. Synthesis of TiO2 nanosheets via an exfoliation route assisted by a surfactant. Nanoscale 2014;6:8531–4. https://doi.org/10.1039/C4NR00946K.

[96] Tae EL, Lee KE, Jeong JS, Yoon KB. Synthesis of diamond-shape Titanate molecular sheets with different sizes and realization of quantum confinement effect during dimensionality reduction from two to zero. J Am Chem Soc 2008;130:6534–43. https://doi.org/10.1021/ja711467g.

[97] Orzali T, Casarin M, Granozzi G, Sambi M, Vittadini A. Bottom-up assembly of single-domain titania nanosheets on (1 × 2)-Pt(110). Phys Rev Lett 2006;97:156101. https://doi.org/10.1103/PhysRevLett.97.156101.

[98] Molina JR, Yang P, Cassivi SD, Schild SE, Adjei AA. Non-small cell lung Cancer: epidemiology, risk factors, treatment, and survivorship. Mayo Clin Proc 2008;83:584–94. https://doi.org/10.4065/83.5.584.

[99] Sarris EG, Saif MW, Syrigos KN. The biological role of PI3K pathway in lung Cancer. Pharmaceuticals 2012;5:1236–64. https://doi.org/10.3390/ph5111236.

[100] Whitsett TG, Inge LJ, Dhruv HD, Cheung PY, Weiss GJ, Bremner RM, Winkles JA, Tran NL. Molecular determinants of lung cancer metastasis to the central nervous system. Transl Lung Cancer Res 2013;2:273–83. https://doi.org/10.3978/j.issn.2218-6751.2013.03.12.

[101] Han Z-P, Fu J, Ye P, Dong X-P. A general strategy for protein immobilization in layered titanates: polyelectrolyte-assisted self-assembly. Enzyme Microb Technol 2013;53:79–84. https://doi.org/10.1016/j.enzmictec.2013.04.011.

[102] Petpiroon N, Bhummaphan N, Soonnarong R, Chantarawong W, Maluangnont T, Pongrakhananon V, Chanvorachote P. Ti0.8O2 Nanosheets inhibit lung Cancer stem cells by inducing production of superoxide anion. Mol Pharmacol 2019;95:418–32. https://doi.org/10.1124/mol.118.114447.

[103] Dai Z, Song X-Z, Cao J, He Y, Wen W, Xu X, Tan Z. Dual-stimuli-responsive TiO x DOX nanodrug system for lung cancer synergistic therapy. RSC Adv 2018;8:21975–84. https://doi.org/10.1039/C8RA02899K.

Chapter 34

Cell and gene therapies—Emerging technologies and drug delivery systems for treating brain cancer

Lakshmi Pallavi Ganipineni[a,*], **Yinghan Chan**[b,c,*], **Sin Wi Ng**[b,d,*], **Saikrishna Kandalam**[a,*], **and Kiran Kumar Chereddy**[e,*]

[a]*Advanced Drug Delivery and Biomaterials, Louvain Drug Research Institute (LDRI), Université catholique de Louvain, Brussels, Belgium,* [b]*School of Pharmacy, International Medical University (IMU), Bukit Jalil, Kuala Lumpur, Malaysia,* [c]*Department of Pharmacology, Faculty of Medicine, University of Malaya, Kuala Lumpur, Malaysia,* [d]*Head and Neck Cancer Research Team, Cancer Research Malaysia, Subang Jaya, Selangor, Malaysia,* [e]*Cell and Gene Therapy Development and Manufacturing, Lonza Pharma and Biotech, Houston, TX, United States*

1 Introduction

Cancer is a result of genetic alteration on a cellular and molecular level, otherwise known as carcinogenesis. It remains a major leading cause of mortality worldwide, with the World Health Organization (WHO) reporting approximately 9.6 million deaths in 2018 alone [1]. Among the various types of cancer, brain cancer remains one of the most lethal and fast-growing cancer due to treatment resistance, continual neurological degradation, and low survival rates [2, 3]. Brain cancer has been identified as the most common cancer among those aged between 0 and 19 years, and in recent years, it has been recognized to be the primary cause of cancer-related deaths among children [4, 5]. According to the Global Cancer Observatory (GLOBOCAN) by the International Agency for Research on Cancer (IARC), there was an estimate of 296,861 new brain and central nervous system (CNS) cancer cases diagnosed in 2018, accounting for 1.7% of all cancer cases. Besides that, the 2018 GLOBOCAN database also predicted the incidence of brain and CNS cancer worldwide to continue to rise by 46.7%, leading to an estimate of 435,554 cases in the year 2040 (Fig. 1) [6]. The 5-year survival rate for brain and CNS cancer patients was reported to be as low as 36%, with an average survival duration of between 15 and 22 months, and this number is still decreasing as opposed to other types of malignancies [3].

Brain cancer encompasses malignancies that are either primary or metastatic. Primary brain cancers are malignancies that originate from the brain itself, such as the glia or its precursors, whereas metastatic brain cancers are derived from systemic neoplasms, which continue to grow into the brain parenchyma [7]. The types of primary brain cancer, or brain tumors, are classified according to the type of cells and/or the area of the brain in which they develop from. Moreover, the WHO further classifies each brain tumor into subtypes according to their malignancy grade, which are ranked from grade I to grade IV based on their histological findings [8, 9]. Generally, gliomas, such as astrocytoma, oligodendroglioma, and oligo-astrocytoma, account for up to 80% of primary malignant brain tumors. Based on the malignancy grade by WHO, oligodendroglioma, and oligo-astrocytoma falls under grade II and grade III respectively, whereas astrocytoma are subcategorized as follows: pilocytic astrocytoma (grade I), low-grade diffuse astrocytoma (grade II), anaplastic astrocytoma (grade III), and glioblastoma multiforme (GBM). GBM is the commonest form of brain tumor, accounting for up to 70% of all gliomas, and remains the most aggressive, particularly due to its resistance to therapeutic interventions [10, 11].

Genetic and environmental factors have both been identified as implications in the pathogenesis of brain cancers. Primary glioblastomas are usually seen in individuals aged >50, presented with genetic characteristics such as the phosphatase and tensin homologue gene deletion, mutations, and amplification in endothelial growth factor receptor (EGFR) and chromosomes with a loss of heterozygosity. As for secondary glioblastomas, they are more commonly seen in younger individuals, manifesting as low-grade tumors and progressing to glioblastomas over the years. They can be defined by genetic mutations and abnormalities in tumor suppressor genes, retinoblastoma pathways, and

* Equal contribution.

Advanced Drug Delivery Systems in the Management of Cancer. https://doi.org/10.1016/B978-0-323-85503-7.00017-1

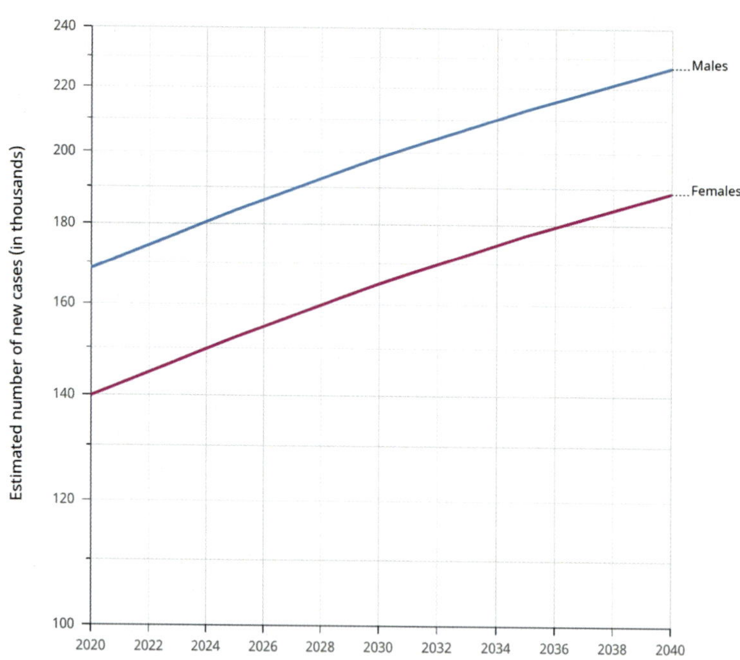

FIG. 1 The estimated increase in brain and CNS cancer incidences worldwide for both sexes of all ages, up to 2040. *Reproduced from Cancer Tomorrow, 2020 (https://gco.iarc.fr/tomorrow/en/dataviz/trends?types=0&sexes=1_2&mode=cancer&group_populations=1&multiple_populations=1&multiple_cancers=1&cancers=31&populations=903_904_90 5_908_909_935&apc=cat_ca20v1.5_ca23v-1.5), with permission from the International Agency for Research on Cancer, World Health Organization.*

aberration in the copy number of DNA [11, 12]. Moreover, brain tumor affects Caucasians more frequently compared to people of Asian or African descent, and this difference is also seen in children [13]. With regards to environmental factors, tobacco use, a high nitrite diet, prolonged radiation exposure, occupational exposure, and environmental carcinogens are examples that could be linked to the development of brain tumors [14, 15].

The following sections elaborate on the currently approved treatments for GBM and several active preclinical and clinical studies.

2 Current approved medications for brain cancers

Patients diagnosed with GBM undergo maximum surgical resection, followed by temozolomide (TMZ) and radiation therapy [16] . Elimination of all the tumor cells during the surgical resection is rare due to the invasive nature of the GBM. Moreover, prevention of recurrence requires post-surgical treatment. Based on the age and disease stage, the treatment options vary for GBM patients. Even for clinical trials, patient enrollment will be on their disease status and overall health.

For GBM treatment, the Food and Drug Administration (FDA) approved approximately 1600 drugs. Among them, 33% target G-protein-coupled receptors (GPCRs), 18% target ion channels, 16% target nuclear receptors, and only 3% target kinases [17]. For the remaining 30% of drugs, the target and mechanism of action were not fully established. This shows a new proteomic space identified to focus on new drug targets. For better patient outcomes, the researcher's efforts are in place in new treatment areas like immunotherapy, antibody-drug conjugates, and alternating electric field therapy (tumor-treating fields). Reaching the therapeutic concentration at the target site is the greatest challenge in developing therapeutics to GBM. Factors such as systemic toxicity, resistance to treatment, lack of specificity, and presence of blood-brain barrier (BBB) limit the effectiveness of conventional treatment; hence, there is an incipient need for new therapeutic strategies to overcome these factors [18]. Only a few drug molecules can cross the BBB, and those were exported through efflux pumps, which are described in the following paragraphs.

Temozolomide (TMZ), a DNA-alkylating agent, was first discovered in the 1970s, received FDA approval in 2005 as the standard treatment option for patients diagnosed with brain tumors. In 1993, temozolomide first clinical trials were conducted. Among 10 patients who received temozolomide as the adjuvant, 5 patients showed significant clinical and radiographic improvement [19]. The accomplishments of this trial incited more successful studies on temozolomide treatment in GBM patients. Compared to other therapies, temozolomide-treated patients were more responsive in these studies. Responsive patients who showed higher survival rates expressed methyl-guanine-methyltransferase (MGMT) genes with methylated promoters than the patients with hypomethylated MGMT genes

[20]. Being a DNA repair enzyme, MGMT repairs the N7 and O6 positions of guanine base alkylated by temozolomide. However, MGMT reduction does not seem to be an effective treatment plan [21, 22]. For GBM diagnosis, MGMT gene methylation status remains an important biomarker. Even though temozolomide is part of the standard chemotherapeutic regimen for GBM, it holds unwanted toxicity and does not eliminate the disease. As an alternative approach, targeted treatments may limit undesirable toxicity and more effectively block tumor proliferation.

Bevacizumab, an antivascular endothelial growth factor (VEGF) monoclonal antibody, is a promising targeted treatment option in cancer patients. In 2004, the FDA first approved this antibody for metastatic colorectal cancer treatment. Later it has been approved for different cancer treatments, including GBM. Angiogenesis is an essential feature for tumors to get nutrients from the vasculature for proliferation. VEGF mediates the neovascularization in tumors, and GBM tumorigenicity is linked to VEGF expression [23]. In 2004, bevacizumab was first tested in malignant glioma patients ($n=21$), at a dose of 5 mg/kg and irinotecan at 125 mg/m^2 every 2 weeks, bevacizumab shows a 43% of significant response rate in the treated patients [24]. On the contrary, Chinot OL et al. (2014) observed in their Phase III "Avaglio" trial that there is no overall survival benefit in bevacizumab-treated (16.8 months) vs placebo-treated (16.7 months) patients ($n=921$) [25]. Likewise, in RTOG 0825 trial (Phase III trial, $n=637$), patients receiving either 10 mg/kg bevacizumab every 2 weeks or placebo, the overall survival benefits between the two groups have no significant difference (15.7 and 16.1 months, respectively) [26]. Hence, bevacizumab treatment is a choice kept for patients with recurrent GBM.

After first-line therapy, there is a possibility for GBM tumor recurrence, and there is no standard approach for treating the recurring GBM. Second-line treatment depends on different factors like tumor size and location, age, time from initial diagnosis, and tumor location. Various treatment options for recurring tumors include reirradiation, nitrosoureas, surgical resection, temozolomide re-challenge, bevacizumab, or tyrosine kinase inhibitors. Even after these treatments for recurring tumors, the survival benefit is 6.2 months [27]. In a Phase II study directed to provisional FDA approval, the most prolonged median progression-free survival (5.6 months) was seen with a combination of bevacizumab and irinotecan [28], whereas the most prolonged overall survival (12 months) resulted from lomustine + bevacizumab [29].

3 Preclinical studies on cell and gene therapies

Recently, molecular medicine has proven to be beneficial in presenting novel therapeutic approaches, with multiple clinical trials underway to develop alternate strategies and provide different perspectives in treating brain tumors, especially with cell and gene therapies. Cell and gene therapies take on a direct approach to treatment by targeting the underlying cause of a disease rather than just reducing the symptoms, which is unique compared to most other forms of treatment. Generally, this is done by introducing a specific gene for the replacement of mutated or deleted genes, or as a means of introducing a therapeutic gene that aims to improve the disease environment, and can involve loss-of-function, as well as gain-of-function alterations [30, 31]. Gene therapy remains a viable option for the treatment of brain cancer, as most brain neoplasms, such as GBM, are confined within the brain and rarely form metastatic deposits. Besides that, the physiological isolation of the brain and the body reduces the occurrence of systemic toxicities [32]. The considerations that should be accounted for in developing a gene therapy strategy for brain cancers are the gene/sequence that should be delivered and the method of delivery/vector to the targeted tumor tissues [18].

There are several different types of genes used in treating brain cancers, and they can be grouped as follows: (i) oncolytic viruses that replicate in and destroy targeted tumor cells (i.e., herpes simplex viruses (HSV-1), AdV, PV); (ii) suicide genes that activate nontoxic prodrugs into metabolites that cause tumor cell death (i.e., HSV-tk); (iii) immunomodulatory genes that induce a T-cell mediated immune response (i.e., IL-2, IFN-γ); (iv) antisense/interfering RNAs that silences gene expression at the translational level (i.e., IGF-1, TGF-β2); and (v) tumor suppressor genes that elicits effects such as inhibiting angiogenesis (i.e., TP53) [18, 31, 33, 34]. Similar to a majority of cancer therapies, a combination approach involving multiple genes will improve potency and is comparatively easier than developing new drugs. Following this, the development of a reliable and specific gene delivery system is of high importance in gene therapy, which involves the development and utilization of multiple types of vectors, which include viral, nonviral, cellular, and synthetic [35–37].

A study utilizing adeno-associated viruses (AdV) mediated CRISPR screen for GBM has managed to establish concurrent driver combinations that act as functional suppressors of GBM, such as Mll2, B2m-Nf1, Mll3-Nf1, Zc3h13-Rb1, in conditional Cas9 mice [38]. Potential therapeutic targets such as EphA2, IL-13Rα2, and survivin were also discovered and could be utilized to develop vaccines for pediatric glioma [39].

Currently, gene therapy has limited clinical use and is under analysis for treating mostly hereditary disorders, notably hemophilia and cystic fibrosis [30]. This section discusses the numerous applications of gene therapy in the therapeutics and management of various brain neoplasms in clinical and preclinical settings.

3.1 Gene delivery systems

3.1.1 Brain tropic viral vectors

3.1.1.1 Adeno-associated viruses (AAV)

AAV is a single-stranded DNA (ssDNA) parvovirus with multiple serotypes demonstrating brain tropism, such as AAV9 having great tropism for neurons that are predominantly located in brain tissue compared with the other serotypes [40]. Multiple studies have demonstrated that AAV was capable of ependymal cell transduction and crossing of the BBB in neuronal and glial cell types present in rodents and nonhuman primates, which suggests that AAV could be a highly specific vector able to carry therapeutic particles across the BBB [40]. This is reflected in AAV9 being suitable to deliver gene-altering oligonucleotides, such as microRNA (miRNA) and antisense oligonucleotides (ASOs), with multiple studies validating the cost-effectiveness of using in gene therapy for a multitude of CNS diseases [41–43].

The AAV9-sTRAIL study utilizes AAV9 for the delivery of sTRAIL with active chicken β-actin (CBA) and neuron-specific enolase (NSE) promoter genes for the regulation of sTRAIL expression to reduce the toxicity levels. Reduction of tumor expansion alongside increased survival rates was observed when AAV9-CBA-sTRAIL and AAV9-NSE-sTRAIL were administered intravenously in mice models.

Next-generation AAV vectors are also being developed through genetic modification of the viral capsid to enhance delivery through tumor cell barriers, as seen in the RGD4C-AAV vector [44]. RGD4C-AAV was the result of fusing a tumor-specific double-cyclic ligand RGD4C into the minor coat protein of fUSE5, and the ability of tumor cell transduction was enhanced by host immune response against the virus through repeated RGDDC-AAV intravenous injections [44].

It was found that different routes of administrations can influence AAV9 tropism in brain tissue. For example, the brain region where AAV9 replicates after an intraparenchymal administration depends on the efferent and afferent neuronal projections located at the injection site, whereas AAV9 intravascular administrations will prompt neuronal and astrocyte transduction in the brain; alongside the transduction of motor neurons within the spinal cord of animals as seen in rodent and nonhuman primates [40]. This finding suggests that the route of administration needs to take into account the target brain region or cell types to ensure greater specificity.

3.1.1.2 Herpes simplex viruses (HSV)

The herpes simplex virus 1-derived thymidine kinase/ganciclovir (HSV-TK/GCV) system was developed for an efficient and reliable system for GBM cell research that uses HSV-TK carrying the prodrug ganciclovir and inducing apoptosis in selected TK-expressing cells through the phosphorylation of GCV [45–48]. A modified trial using human mesenchymal stem cells (MSC) along with the HSV-TK/GCV system (MSCTK/GCV) shows that human MSCs are able to exert a bystander effect that enhances the therapeutic efficiency of the treatment [49].

These trials suggest that the system was conducive and safe in the setting of high-grade glioma treatment when tested using U-87-derived GBM cells, which prompts further research in combination therapies, especially with conventional radiotherapy or chemotherapy, to increase the median survival rate of GBM [46–49]. However, a 2019 study reported that the viability of prodrug-based suicide gene therapies such as HSV-TK/GCV in GBM treatment is lower than the standard conventional therapy, due to the dependency of HSV-TK on the prodrug concentration and distribution [50]. Instead, the same study suggests using the apceth's Good Manufacturing Practice (GMP) process to produce hMSC for tumor growth reduction and increase survivability in intracerebral U-87 glioma xenografts [50]. HSV-TK was combined with yeast cytosine deaminase (CD) in another study as part of a double-suicide gene transfer mediated by a retroviral vector [51].

3.1.1.3 Retroviral replicating vectors (RRV)

Currently, the RRV known as vocimagene amiretrorepvec (Toca 511), which is derived from a gammaretrovirus with an amphotropic envelope that allows for high selectivity toward human tumor cells, is a subject of interest in a Phase I trial for the treatment of high-grade glioma, which aims to improve 5-FU chemotherapeutic efficacy, as 5-FU encounters major difficulties in bypassing the BBB [52]. Injections of Toca 511 into tumor cavity shows successful transportation of the cytosine deaminase gene (CD) into tumor cells, and since the conversion of prodrug 5-fluorocytosine (5-FC) into 5-FU was facilitated by CD intracellularly, this results in cytotoxicity and heightened local immune response, leading to direct tumor cell lysis [52–54]. The inherent short half-life of 5-FU when formed intracellularly indicates there is limited myelotoxicity and immunosuppression effects of 5-FU when used with Toca 511 [52, 53]. Toca 511 and Toca-FC, which is the investigational extended release form of 5-FC, have demonstrated favorable results when compared to lomustine control in a Phase I trial, with a Phase II/III trial underway [52–54].

3.1.2 Nanoparticles (NPs)

NPs have been well studied and documented as a mode of delivering therapeutics that are innately repelled by the BBB, and as a diagnostic aid for various CNS conditions. Generally, nanovehicles could be categorized into two types, organic and inorganic NPs, where their impact on CNS tumors will be discussed below.

3.1.2.1 Liposomes

Liposomes are vesicles of singular or multiple phospholipid bilayers arranged in a spherical manner, which possess biocompatibility, high safety profile, and specific targeting of tumor tissues, which makes it a viable candidate for delivering therapeutics into brain tumors [55, 56]. To better overcome the BBB obstacle, a 2019 study utilizing liposomes with a modified surface transferrin (Tf) receptor in the combination of cell-penetrating peptide PFVYLI (PFV) to create a liposomal delivery system of two modus operandi, known as Tf-PFV for increased tumor receptor targeting and uptake [57]. The study reported significantly higher BBB liposomal translocation resulting in 62% of GBM tumor cell death when using modified liposomes that carry erlotinib and doxorubicin using in vitro brain tumor models [57]. In addition, liposomal NPs could also carry oligonucleotides such as siRNA and miRNA for GBM delivery, especially with the novel gold-liposome enhancing the delivery of miRNA in U87 models [58, 59].

A separate study has developed liposome-templated hydrogel NPs (LHNPs) using 1,2-dioleoyl-3-trimethylammonium-propane chloride salt (DOTAP) liposomes as the template through the autocatalytic brain tumor-targeting (ABTT) mechanism to deliver CRISPR/Cas9 into U87 glioma-bearing mice models, which reported 33.1% uptake of LHNPs in mice, which demonstrated significant tumor-tissue targeting [60].

3.1.2.2 Polymeric NPs

Polymeric NPs are a suitable vector in nanomedicine due to their ability of controlled drug release and high specificity [61–63]. Biodegradable poly(beta-amino ester)s (PBAE) carrying the tumoricidal gene HSV-TK has demonstrated improved efficacy and median overall survival in treating atypical teratoid or rhabdoid tumors (AT/RT) and medulloblastoma (MB) in mice when compared to previous studies, which suggests that the delivery of genes and oligonucleotides such as HSV-TK, siRNA, and miRNA to AT/RT and MB is viable with PBAE [62, 64–66]. A 2018 study utilizing a novel nano-miRNA system with PBAE R646 and miR-148a and miR-296-5p on GBM mice models reported diminishing tumor load after 28 days [66]. The same study also suggests that the combination of multiple miRNAs into the same nano-miRNA system could potentially has led to a significant increase in cellular uptake, such as using both Cy3-labeled miRNA and Cy5-labeled miRNA [66]. The use of poly(ethylene imine) coupled with L-glutathione (GSH) also produces favorable results for gene delivery across BBB [67].

3.1.2.3 Dendritic polymers

Dendritic polymers are a hyperbranched macromolecule that is stable in the blood circulation system and could be modified for various uses, including encapsulating therapeutics for delivery into a target tissue [68, 69]. Liu et al. introduced angiopep-2 (Ang2) into a dendritic polymer alongside epidermal growth factor receptor (EGFR)-targeting peptide (EP-1), creating a dual-targeting poly(amidoamine) (PAMAM) dendritic carrier for enhanced specificity, and showed promising results for in vitro and in vivo BBB translocation in mice-based specimens [68].

3.1.2.4 Inorganic NPs

The main difference of inorganic NPs such as metallic NPs and organic NPs are the lack of biodegradability and are therefore used mostly for diagnostic or imaging purposes [61]. Gold-, silver- and other metal-based NPs could be used for radiosensitization since they have high X-ray absorption capacities and aids the diagnosis of GBM utilizing computed tomography (CT) scanning in virtue of the enhanced permeability and retention (EPR) effect [70, 71]. Inorganic NPs could be used to improve or circumvent certain drawbacks of current treatment modalities, which would be discussed below [70].

Photodynamic therapy (PDT) was shown to be effective at treating tumor tissues, but the nanomaterial-mediated photothermal (NmPTT) effect seen in PDT is highly dependent on the choice of photosensitizers, which are used to absorb the energy of the light wavelengths and causing increased oxidative stress in the target tissue, leading to cell death [72]. Gold NPs was shown to be an efficient photosensitizer due to its biocompatibility, selectivity, and malleability, which was observed when the temperature of melanoma tumors in mice models changed significantly under light irradiation when treated with gold NPs, with 12.6 °C caused by the NmPTT effect and 10.5 °C caused by the combination of NmPTT effect and nano-material-mediated photodynamic therapeutic (NmPDT) effects [70, 72–74].

Gold NPs, especially gold nanorods, possess good biocompatibility and selectivity for brain tumors, that it is utilized as a radiosensitizer or a photosensitizer in multiple applications of oncological therapy, most notably in radiotherapy, laser-induced hyperthermia, in PDT, and imaging [70]. Gold NPs also tend to accumulate in tumor tissue with a 19:1 tumor-to-healthy parenchyma ratio and a 1.5% tumor uptake, which provides high-resolution computed tomography (HRCT) brain tumor imaging [70].

3.2 Cytotoxic approach

Cytotoxic therapeutic methods are centralized on two main concepts, either by suicide gene therapy, which involves transferring conditional or direct cytotoxic transgenes into tumor cells; or by virotherapy, with a modus operandi of delivering engineered viral vectors that selectively targets and replicates within tumor cells and subsequently kills them [45, 75–81]. Gene therapy for GBM involves a targeted toxin approach, with toxic protein transgene delivery through viral vectors, such as the *Pseudomonas* exotoxin (PE), that selectively targets brain tumor cells to disrupt protein translation and synthesis [81].

The hIL-13-PE protein formulation known as cintredekin besudotox was developed based on the overexpression of human IL-13 variant receptor hIL3Rα2, which is present in 50%–80% of human GBM cells, that indicates Cintredekin Besudotox may have a high selectivity for human GBM cells [39, 82]. However, multiple studies concluded that the survival rates of convection-enhanced delivery (CED) of cintredekin besudotox was not significant when compared to standard gliadel wafers (GW) in the treatment of adult patients with GBM, mainly due to hIL-13-PE protein having a short half-life and low specificity, leading to dose-limiting toxicity and the drug requiring multiple or continued administrations to achieve adequate therapeutic effect [81, 83–86]. A follow-up preclinical trial using hIL-13.E13K, a mutated variant of hIL-13, that is fused to the mutant human IL-4 gene (mhIL-4) that limits binding of the compound to native healthy brain cells, PE and a TRE promoter, delivered by an adenoviral vector (AdV) intracranially, generating the protein formulation Ad.mhIL-4.TRE. mhIL-13-PE, which can exhibit powerful antitumor effects with relatively negligible neurotoxicity, longer half-life, and can be regulated through the withdrawal of the inducer doxycycline. This results in a more robust and efficient protein compound compared to Cintredekin Besudotox and was shown to induce tumor regression and yields a survival rate of over 70% in mice models [86].

3.3 Virotherapy

Virotherapy is an approach using an oncolytic conditionally replicating viral vectors to directly cause tumor cell lysis and death by specific tumor cell targeting [81]. The specificity of the oncolytic viral vector is paramount to achieve effective antitumor activity with limited toxicity to the normal native brain tissues, and these viral vectors are obtained from a diverse pool of preexisting viruses, such as HSV, adenoviruses, retroviruses, reoviruses, and measles virus [81].

3.3.1 Conditionally replicating HSV vectors

G207 is a modified HSV-1 with a biallelic deletion of γ34.5 gene with a lack of U_L39 gene due to the lacZ gene insertion at the loci, which results in a viral vector that selectively replicates within rapidly dividing cells [81]. The intratumoral and tumor cavity injection of G207 in six patients in a Phase Ib clinical trial involving recurrent GBM that is resectable was shown to be safe, even with multiple direct injections. Despite the original aim being the demonstration of G207 safety profile, there was observable viral replication and limited oncolytic activity, with no signs of encephalitis in the patients, which shows the potential therapeutic efficacy of G207 in treating GBM [81]. Based on this principle, multiple studies are able to proceed with using G207, which prompts the development of the HSVQuik system, which could conjugate transgenes of interest with an oncolytic HSV vector within 10 days to facilitate clinical studies using HSV [81, 87].

Besides exerting a direct antitumoral effect, G207 is able to deliver therapeutics such as tumor necrosis factor (TNFα) directly into tumor cells along with radiosensitive US11 HSV promoter gene to regulate the extent of viral replication, and G207 can potentially disrupt tumor cell vasculature through an expression of antiangiogenic factors such as Platelet Factor 4 [88, 89]. However, the presence of the γ34.5 gene plays an important role regarding the efficacy of viral replication and specificity of the virus [90,91]. Although mutant HSV lacking γ34.5 was more specific, the cytotoxic effects are not as marked when compared to mutants with γ34.5 intact [90,91]. To overcome this predicament, two separate preclinical studies were shown to have different approaches, creating glioma-selective HSV-1 mutant Myb34.5 and rQNestin34.5, respectively. The study involving Myb34.5 reintroduced the γ34 gene to the HSV virus but a B-myb promoter gene served as regulation, demonstrated that in mouse models, Myb34.5 was able to remain as oncolytic as the wild type variant with γ34.5 gene intact, yet had a relatively favorable toxicity profile, with its 50% lethal dose (LD_{50}) to remain at $>10^7$ PFU, which is a value comparable to the γ34.5 deleted HSV variant [90]. The latter study involving rQNestin34.5 was based on the concept of targeting a tumor-specific promoter and utilizing it to drive the expression of the viral gene, which in this case is based on the glioma molecular marker nestin [91]. The study was able to demonstrate better oncolytic activity in rQNestin34.5 in glioma cells expressing high concentrations of nestin when compared to Myb34.5 in animal models [91].

3.3.2 Conditionally replicating adenovirus vectors

Most AdV used in gene therapy have a deletion in the early region 1A (E1A), which results in a virus with poor replication abilities and are mainly used to contain large inserts of up to 37 kb in high capacity vectors [30]. Recent studies have managed to create genetically engineered adenoviruses to have a high specificity for tumor cells and lyse them, by manipulating the E1 region in the adenoviral genome [79, 92, 93]. ONYX-015 has a deletion in the E1B region encoding for a 55 kDa protein in infected cells that deactivates cellular p53, hence it is thought that ONYX-015 would selectively target and lyse tumor cells, which normally lacks p53 [94]. A Phase I trial had shown that ONYX-015 has a high safety profile and is relatively well tolerated in patients with recurrent or refractory head and neck squamous cell carcinoma (HNSCC) [95–97]. Although mainly targeting p53-defective tumor cells, a study has found that the tumor selectivity of ONYX-015 is determined by late viral RNA export [98].

4 Current cell and gene clinical trials for brain cancers

For the last two decades, several investigational studies have been performed to target the neuro-onco-immune environment. However, immunotherapy in brain cancer had a limited success rate and led to specific challenges [99, 100]. For instance, drug delivery to the brain has been restricted by a healthy BBB. In contrast, the brain tumor environment enables drug access due to leaky vasculature, and the raised interstitial pressure of fluid can hinder therapeutics infiltration. Moreover, the unproductive immunotherapeutic delivery to tumor cells and homogeneous immunological response in the brain is potentially due to a poor understanding of the brain lymphatic system. The brain has its own native immune cells to contribute to the immunological function, and it was done not only by resident cells but also by the peripheral response. Furthermore, there are not many immune cells in the brain during cancer, along with a few antigen-presenting cells (APCs). Lastly, although there is a clear desired anticancer immune response due to neuropil sensitivity, a serious concern with immunotherapeutics is neurotoxicity and autoimmunity [101]. There were several studies that reviewed immunotherapy for a brain tumor. This paper focuses on the most recent and late-stage clinical trials on brain tumors and the challenges that immunotherapies have in the brain tumor environment. The relevant search parameter on the National Institute of Health Clinical Trial database [102] found 439 clinical trials related to brain tumor immunotherapy. Among them, 106 trials started in January 2016 and later continued to Phase II. Only 7% of the analyzed trials were either active or completed Phase III or IV trials, which clearly indicates a challenge in product translation to later-stage clinical trials [103, 104].

4.1 Gene therapies

The cancer treatment with gene therapy includes the insertion of growth-regulating or tumor-suppressing genes. Recently, to prevent oncogenes' action triggering tumorigenesis or proliferation, RNA interference (RNAi) has been introduced. For nontoxic prodrugs converting to lethal active compounds, the suicide gene therapy (a common viral gene therapy) is one approach that has been used. Oncolytic and immunomodulatory gene therapy is the other approach for this. To successfully deliver these therapies into tumor tissues, nucleic acid vectors or carriers must be used. In the following sections, delivery vectors like viral vectors, polymeric NPs, and nonpolymeric NPs used in GBM gene therapy were discussed [105].

4.1.1 Viral vector

In glioma clinical trials, viral vectors were first used as the vehicles for gene therapy. Neurotropic retroviruses and adenoviruses that can infect neurons and glial cells were commonly used viral vectors in the GBM gene therapy. During glioma preclinical studies, AAV are shown as a promising vehicle for gene therapy. However, these vectors were not tested in clinical trials [106–108]. Table 1 presents the ongoing and completed clinical trials that used different viral vectors for gene therapy in GBM treatment.

4.1.1.1 Retroviral

During the clinical trials for glioma therapy, retroviral vectors were first evaluated as the delivery systems. In 1992, NCT00001328 clinical trials first assessed the combination of ganciclovir (Cytovene) with modified murine cells comprising retroviral herpes simplex virus-thymidine kinase (HSV-tk) in glioma therapy. Inside HSV-tk-transfected cells, HSV-tk acts as a suicide gene and translates ganciclovir (which is a prodrug) to ganciclovir-triphosphate (active form), which in turn inhibits the DNA replication and cell division [110]. This study proved that retroviral vectors upon intratumoral implantation through cells facilitated HSV-tk transfection had antitumor activity only in small tumors. This indicates that retroviral vectors have partial transfection efficiency [111].

Toca511, another retroviral vector delivering cytosine deaminase (CD) suicide gene associated with Toca FC (oral prodrug), converts 5-fluorocytosine into the active 5-fluorouracil, which is an antineoplastic drug. The CD enzyme has facilitated this conversion. During the Phase-1 clinical trials, patients with recurrent high-grade glioma, infusion of Toca 511 and Toca FC, was well tolerated and promoted tumor regression. At present, for GBM and anaplastic astrocytoma treatment, Toca 511 and Toca FC retroviral vectors were considered for phase 2/3 clinical trials [112–114].

4.1.1.2 Adenoviral

AdV as the delivery system was extensively assessed in clinical trials. In a phase-1 trial, intratumoral injection of the AdV carrying the wild-type p53 gene (Ad-p53) during pre- and postresection of the glioma tumor effectively transfected astrocytic tumor cells with minimal toxicity. However, the transfected cells were only spotted within 5 mm of the injection site, representing the therapeutics' restricted ability to enter the tumor tissue [115]. Combination therapy exploiting ganciclovir and intratumoral injection of HSV-tk administered either by retrovirus-packaging cells or adenoviruses was studied in patients with late-phase glioma. The results demonstrated that 3 out of 7 patients treated

TABLE 1 Different ongoing and completed clinical trials that used different viral vectors for gene therapy in GBM treatment.

Vector	Gene therapy agent and mechanism	Combination therapy	Clinical trial phase	Clinical trial number
Retrovirus	HSV-tk: HSV-tk transforms ganciclovir to antiviral drug ganciclovir triphosphate	Ganciclovir	Phase I	NCT00001328
	Toca 511: CD transforms prodrug 5-FC to antineoplastic 5-FU	Oral 5-FC	Phase II/III	NCT02414165
Adenovirus	SCH-58500: transfects p53 gene	–	Phase I	NCT00004080
	Ad-p53: transfects p53 gene	–	Phase I	NCT00004041
Retro or adenovirus	HSV-tk: HSV-tk transforms ganciclovir to antiviral drug ganciclovir triphosphate	Ganciclovir	Phase I	[109]
Adenovirus	AdV-tk: HSV-tk transforms valacyclovir to antiviral drug acyclovir	Valacyclovir	Phase I	NCT00751270
	AdV-tk: HSV-tk transforms valacyclovir to antiviral drug acyclovir	Valacyclovir and radiation therapy	Phase IIa	NCT00589875
Liposome	SGT-53: transfects p53 gene	TMZ	Phase II	NCT02340156
Spherical nucleic acid gold NP	NU-0129: transfects siRNAs targeting oncogene Bcl2L12	–	Phase I	NCT03020017

with the adenovirus has a stable tumor, while treatment with the retrovirus, all patients have tumor progression 3 months posttreatment. Moreover, patients treated with adenovirus had doubled survival time than treatment with the retrovirus, with an average of 15 months and 7.4 months, respectively. Low transfection and low availability at brain tumor tissues may have resulted in poor efficacy [109].

Many clinical studies assessed the delivery of an AdV containing HSV-tk (AdV-tk) along with valacyclovir (an antiherpetic prodrug). This combination induced both conventional antitumoral effects of valacyclovir and activation of immune responses triggered by AdV-tk. Here, thymidine kinase facilitated the translation of the inactive prodrug into toxic nucleotide analogs. A Phase 1B study was conducted in malignant/late-phase glioma patients using AdV-tk in combination with valacyclovir and radiation therapy, followed by TMZ. Postresection, injecting the AdV-tk into the tumor bed, with subsequent radiation and chemotherapy, led to 33% subjects' survival rate increased to 2 years and about 25% to 3 years. Even it was a minor increase over the existing standard of care, still a major improvement in the realm of brain tumor management. Furthermore, posttreatment tumor analysis found CD3+ T-cells, which indicates immune system activation [116]. In phase 2 clinical study, patients treated with AdV-tk combined with valacyclovir and standard of care showed a significant increase in survival time compared to the patients received only standard of care treatment (17.1 months vs 13.5 months, respectively) [46]. In addition, a Phase I study was conducted in pediatric patients diagnosed with malignant glioma and recurrent ependymoma using AdV-tk combined with valacyclovir and radiation therapy. Posttreatment, 50% of the pediatric subjects survived up to 16 months with no dose-limited toxicities. However, a grade 3 lymphopenia was commonly observed in all patients [117].

Even though viral vectors have been studied extensively, there was a borderline increase in overall survival time and have yet to attain clinical translation via FDA approval to treat GBM patients. The major challenge that viral vectors are facing in glioma therapy was inefficient tumor penetration. However, in some cancer treatments, viral vectors were clinically successful. For the treatment of melanoma, FDA approved talimogene laherparepvec virotherapy, which consists of a genetically modified HSV-1 gene containing the human granulocyte-macrophage colony-stimulating factor [118].

4.1.2 Nonviral

Along with viral vectors, nonpolymeric and polymeric delivery systems are available to deliver genes for glioma therapy. These delivery systems showed promising results as gene vectors during preclinical and clinical studies. In 2018, FDA approved patisiran (RNAi therapeutic), a lipid NP-containing siRNAs for the treatment of transthyretin-mediated amyloidosis (a neurodegenerative disease). Being a nonviral vector-based nucleic acid therapy, patisiran showed promising results [119]. For the treatment of GBM, the FDA has not approved any of these vectors yet, but few nonpolymeric vectors like liposomes, gold NPs, polymeric NPs, and RNA NPs were still under clinical evaluation.

4.1.2.1 Gold NPs

In a preclinical study, it was observed that SGT-53 (a transferrin receptor-targeted liposomal vector) encapsulating wild-type p53 plasmid DNA able to cross the BBB and target GBM cells. The accumulation of sufficient NPs resulted in the decrease of MGMT and apoptosis in GBM xenografts. When SGT-53 and TMZ were administered in combination, it enhanced the antitumor efficacy than TMZ alone treatment. SGT-53 shows the capability to improve chemosensitivity [120]. For the treatment of recurrent glioblastoma, SGT-53, combined with TMZ, is currently under Phase-II clinical trials. Also, NU-0129 (siRNAs containing gold NPs) aiming Bcl-2-like protein 12 (Bcl2L12) is under phase-1 clinical trials. NU-0129 promotes inhibition of tumor progression and resistance to apoptosis. Confirming the hypothesis, a systemic administration of NU-0129 in mice xenograft models, enhanced apoptosis of tumor cells, and decreased tumor progression were observed [121].

4.1.2.2 RNA NPs

Additional to clinical studies, RNA-containing polymeric NPs (new vectors) have been under preclinical studies for glioma gene therapy. In a xenograft GBM mice models, Croce et al. used RNA NPs, delivering anti-miR-21-locked nucleic acid sequences to prevent oncogenic miR-21. Compared to the control group, the RNA NPs group demonstrated tumor regression and increased survival [122].

4.1.2.3 Polymeric NPs

For better therapeutic outcomes in cancer treatment, polymeric nanoparticle delivering genes is an emerging approach. To deliver the genes and/or drugs systematically or locally, micro- and NPs attained popularity in current research. For gene therapy, NPs are advantageous because they allow conjugation of nucleic acids, be highly tailorable, homing peptides, or target ligands. In gene therapy for glioma treatment, the use of several polymeric delivery systems has been studied. However, nothing has reached the clinical trial stage yet.

4.2 Cell therapies

In adoptive cell therapy (ACT), T-cells and numerous other tumor-infiltrating lymphocytes (TILs) are activated precisely without relying on dendritic cell activation. During ACT, TILS were isolated from biopsies and expanded, assuming that these were tumor-specific. TILS were ultimately homing back to the tumor before being injected into the patient. These autologous approaches depend on the immune response of the patient and the tissue's lymphatic architecture where the tumor resides [123]. In the ACT, the most used TILs were cytokine-induced killer (CIK) cells and chimeric antigen receptor (CAR) T-cells. When peripheral blood lymphocytes are induced in vitro with IL2, CD3 monoclonal antibodies, and IFNγ, a series of bioreactions occur and result in CIK cells [124].

CAR T-cells are engineered to express an intracellular activation specially designed to an extracellular domain-specific receptors. These CARs bind to a single or multiple costimulatory tumor-associated antigens [123]. During the clinical trials for GBM, both CIK cell and CAR T-cells were identified as useful because both cells give MHC-unrestrictive antigen recognition and are thus beneficial for use in the ACT. Intravenous injection of autologous CIK cells associated with temozolomide treatment during GBM Phase III randomized trials showed 5.6 months increased median survival time than the standard of care [124]. In a Phase I trial with CAR T-cells engineered to target the GBM tumor antigen IL13Rα2, preliminary results demonstrated that the CAR T-cells were detected in the tumor cyst fluid or cerebrospinal fluid for minimum 7 days, and also had safe intracranial delivery. A fraction of these patients showed a reduction in IL13Rα2 antigen expression, and in one patient, >79% drop of recurrent tumor mass was detected [125].

5 Challenges in call and gene therapy for brain cancer

5.1 Blood-brain barrier

Among the many challenges faced by medical professionals in treating CNS diseases would be the BBB, which consists of specialized endothelial cells forming a series of tight junctions and astrocyte processes terminating in an overlapping manner on capillary walls that creates a high resistance barrier that only allows lipid-soluble molecules of 400–600 Da to pass through, which maintains a stable tumor microenvironment (TME) and limits the amount and type of medication from being delivered into the brain effectively [126, 127]. BBB maintains TME by isolating the brain from the blood, besides supporting tumor progression and chemotherapeutic resistance by creating tumor-favorable conditions such as hypoxia and acidosis, which is often seen in drug-resistant glioblastoma [127].

5.2 Routes of viral delivery

Typically, vectors are delivered into the tumor bed via direct injection of the virus after the surgical removal of tumors. However, the interstitial pressure from the skull due to the presence of normal and tumor tissues limited the intracerebral distribution of vectors [32]. One method to improve the intratumoral spread of the virus is by applying a pressure gradient to create convection flow (CED), resulting in distal diffusion of injected substances through tumor tissues [128]. Liposomes have been established as effective carriers for the delivery of toxins to tumor sites, attributed to their ability to circumvent the BBB. Nevertheless, further investigation of the relative toxicity and efficacy of such an approach will be needed, as some studies have shown that gene delivery by liposomes via CED may not be therapeutically advantageous [32, 128]. As an alternative, the intraventricular injection has been considered a route of viral delivery due to its noninvasive nature and the circulating cerebrospinal fluid, which may assist in viral distribution. However, several preclinical studies have demonstrated high toxicity due to the presence of inflammatory responses [32]. Conversely, the intravascular route of delivery allows the targeting of tumor neovascularization regions representing migrating areas of gliomas. Despite this being proven safe in animal studies, the clinical efficacy may be reduced due to inhibition of viral passage through the BBB, as well as virus neutralization by host immunity [32, 129].

5.3 Tumor heterogeneity

Generally, cancer is a complex disease that becomes more heterogeneous during the course of the disease. As a result of this heterogeneity, efficient treatment of brain tumors is hampered, attributed to the multiple genetic alterations that may be present in the neoplastic cells forming such tumors. Besides, not all types of tumors can be susceptible to vector interaction as a result of the heterogeneous presentation of vector receptors used in gene therapy. Furthermore, heterogeneity offers the opportunity for treatment resistance and eventually, the relapse of disease. Hence, the engineering of viral vectors for gene therapy in brain cancer must be adapted and improvised from the lessons learned from preclinical and patient models allow us to predict the evolutionary trajectory of brain cancer, thereby allowing the design of safe, effective, and durable therapies [32, 130].

5.4 Host immune responses

Gene therapy vectors generally contain components of viruses, bacteria, as well as some other microorganisms. For this instance, viruses are appealing as vectors for gene therapy as they are naturally capable of incorporating foreign genetic material within the genome of the host cell. Nevertheless, as the body immune system functions primarily to combat infections by bacteria and viruses, when such pathogen or a component thereof penetrates within the body, it leads to the development of innate immune response, which leads to the production of inflammatory cytokines and influx of nonspecific inflammatory cells [131, 132]. In numerous preclinical and clinical studies, infiltration of immune cells was widely documented, which results in reduced therapeutic outcomes [32, 131]. To increase the efficacy of gene therapy in brain cancer, it is essential to elucidate the role of preexisting immunity to viral vectors and whether such immunity would be concerning when viral vectors are delivered in situ. It is also important to determine whether these innate immune responses are derived from the periphery or primarily modulated by intracerebral defense mechanisms. These knowledges are of utmost importance to formulate immunomodulating strategies in the design of viral vectors, inducing vector tolerance, which ensures that gene transfer can be done safely without inducing severe pathogenic effects. Nonetheless, the ultimate satisfying approach is to develop viral vectors with little to negligible potential for inducing an immune response [32, 131, 133].

6 Conclusions

The brain is a critical organ and has exclusive and natural anatomical, physiological, and immunological barriers for a successful treatment. These constraints have led to inherent engineering challenges for developing therapies, which need to show pharmacological action either directly in the brain cells and/or within the brain tumor immune microenvironment. Specifically, targeting and regulating the oncogenes and tumor suppressor gene is the way of cell and gene therapy in glioma. Additionally, chemotherapy resistance barriers can be overcome using cell- and gene-based therapies through downregulating resistance genes or using suicide gene therapy. Regardless of these clues at a successful therapeutic avenue, there can be severe side-effects due to their direct involvement in brain physiology and immune pathways. For now, it is clear that the completed clinical trials were not successful and extensive studies should be performed to understand both the short-term and long-term effects of these advanced medical products on brain activities. Further assessment of the safety and efficacy of different types of stem cells, immune cells, gene vectors, and their combinations in clinical trials in the future may deliver advanced treatment options for cell- and gene-based therapies in GBM.

References

[1] Cancer, World Heal. Organ. (2018). https://www.who.int/news-room/fact-sheets/detail/cancer (accessed November 30, 2019).

[2] Ghorbani M, Bigdeli B, Jalili-baleh L, Baharifar H, Akrami M, Dehghani S, Goliaei B, Amani A, Lotfabadi A, Rashedi H, Haririan I, Alam NR, Hamedani MP, Khoobi M. Curcumin-lipoic acid conjugate as a promising anticancer agent on the surface of gold iron oxide nanocomposites: a pH-sensitive targeted drug delivery system for brain cancer theranostics. Eur J Pharm Sci 2018;114:175–88. https://doi.org/10.1016/j.ejps.2017.12.008.

[3] Sonali, Viswanadh MK, Singh RP, Agrawal P, Mehata AK, Pawde DM, Narendra, Sonkar R, Muthu MS. Nanotheranostics: Emerging strategies for early diagnosis and therapy of brain cancer. Nanotheranostics 2018;2:70–86. https://doi.org/10.7150/ntno.21638.

[4] Abou-Antoun TJ, Hale JS, Lathia JD, Dombrowski SM. Brain cancer stem cells in adults and children: cell biology and therapeutic implications. Neurotherapeutics 2017;14:372–84. https://doi.org/10.1007/s13311-017-0524-0.

[5] Ostrom QT, Gittleman H, Farah P, Ondracek A, Chen Y, Wolinsky Y, Stroup NE, Kruchko C, Barnholtz-Sloan JS. CBTRUS statistical report: Primary brain and central nervous system tumors diagnosed in the United States in 2006–2010. Neuro Oncol 2013;15. ii1 https://doi.org/10.1093/neuonc/not151.

[6] Bray F, Ferlay J, Soerjomataram I, Siegel RL, Torre LA, Jemal A. Global cancer statistics 2018: GLOBOCAN estimates of incidence and mortality worldwide for 36 cancers in 185 countries. CA Cancer J Clin 2018;68:394–424. https://doi.org/10.3322/caac.21492.

[7] Kabitha K, Rajan MS, Hegde K, Koshy S, Shenoy A. A comprehensive review on brain tumor. Int J Pharm Chem Biol Sci 2013;3:1165–71.

[8] Gupta A, Dwivedi T. A simplified overview of World Health Organization classification update of central nervous system tumors 2016. J Neurosci Rural Pract 2017;8:629–41. https://doi.org/10.4103/jnrp.jnrp_168_17.

[9] Shai RM, Reichardt JKV, Chen TC. Pharmacogenomics of brain cancer and personalized medicine in malignant gliomas. Future Oncol 2008;4:525–34. https://doi.org/10.2217/14796694.4.4.525.

[10] de Melo MT, Piva HL, Tedesco AC. Design of new protein drug delivery system (PDDS) with photoactive compounds as a potential application in the treatment of glioblastoma brain cancer. Mater Sci Eng C 2020;110:110638. https://doi.org/10.1016/j.msec.2020.110638.

[11] Jovčevska I, Kočevar N, Komel R. Glioma and glioblastoma—how much do we (not) know? Mol Clin Oncol 2013;1:935–41. https://doi.org/10.3892/mco.2013.172.

[12] Wen PY, Kesari S. Malignant gliomas in adults. N Engl J Med 2008;359:492–507. https://doi.org/10.1056/NEJMra0708126.

[13] Ohgaki H, Kleihues P. Epidemiology and etiology of gliomas. Acta Neuropathol 2005;109:93–108. https://doi.org/10.1007/s00401-005-0991-y.

[14] Bondy ML, Scheurer ME, Malmer B, Barnholtz-Sloan JS, Davis FG, Il'yasova D, Kruchko C, McCarthy BJ, Rajaraman P, Schwartzbaum JA, Sadetzki S, Schlehofer B, Tihan T, Wiemels JL, Wrensch M, Buffler PA. Brain tumor epidemiology: consensus from the brain tumor epidemiology consortium. Cancer 2008;113:1953–68. https://doi.org/10.1002/cncr.23741.

[15] Vienne-Jumeau A, Tafani C, Ricard D. Environmental risk factors of primary brain tumors: a review. Rev Neurol (Paris) 2019;175:664–78. https://doi.org/10.1016/j.neurol.2019.08.004.

[16] Stupp R, Mason WP, Van Den Bent MJ, Weller M, Fisher B, Taphoorn MJB, Belanger K, Brandes AA, Marosi C, Bogdahn U, Curschmann J, Janzer RC, Ludwin SK, Gorlia T, Allgeier A, Lacombe D, Cairncross JG, Eisenhauer E, Mirimanoff RO. Radiotherapy plus concomitant and adjuvant temozolomide for glioblastoma. N Engl J Med 2005;352:987–96. https://doi.org/10.1056/NEJMoa043330.

[17] Santos R, Ursu O, Gaulton A, Bento AP, Donadi RS, Bologa CG, Karlsson A, Al-Lazikani B, Hersey A, Oprea TI, Overington JP. A comprehensive map of molecular drug targets. Nat Rev Drug Discov 2016;16:19–34. https://doi.org/10.1038/nrd.2016.230.

[18] Ning J, Rabkin SD. Current status of gene therapy for brain tumors. Transl Gene Ther Clin Tech Approaches 2015;305–23. https://doi.org/10.1016/B978-0-12-800563-7.00019-1.

[19] O'Reilly SM, Newlands ES, Brampton M, Glaser MG, Rice-Edwards JM, Illingworth RD, Richards PG, Kennard C, Colquhoun IR, Lewis P, Stevens MFG. Temozolomide: a new oral cytotoxic chemotherapeutic agent with promising activity against primary brain tumours. Eur J Cancer 1993;29:940–2. https://doi.org/10.1016/S0959-8049(05)80198-4.

[20] Hegi ME, Diserens AC, Gorlia T, Hamou MF, De Tribolet N, Weller M, Kros JM, Hainfellner JA, Mason W, Mariani L, Bromberg JEC, Hau P, Mirimanoff RO, Cairncross JG, Janzer RC, Stupp R. MGMT gene silencing and benefit from temozolomide in glioblastoma. N Engl J Med 2005;352:997–1003. https://doi.org/10.1056/NEJMoa043331.

[21] Quinn JA, Jiang SX, Reardon DA, Desjardins A, Vredenburgh JJ, Rich JN, Gururangan S, Friedman AH, Signer DD, Sampson JH, McLendon RE, Herndon JE, Walker A, Friedman HS. Phase II trial of Temozolomide plus O6-Benzylguanine in adults with recurrent, Temozolomide-resistant malignant glioma. J Clin Oncol 2009;27:1262–7. https://doi.org/10.1200/JCO.2008.18.8417.

[22] Robinson CG, Palomo JM, Rahmathulla G, McGraw M, Donze J, Liu L, Vogelbaum MA. Effect of alternative temozolomide schedules on glioblastoma O 6-methylguanine-DNA methyltransferase activity and survival. Br J Cancer 2010;103:498–504. https://doi.org/10.1038/sj.bjc.6605792.

[23] Cheng SY, Huang HJS, Nagane M, Ji XD, Wang D, Shih CCY, Arap W, Huang CM, Cavenee WK. Suppression of glioblastoma angiogenicity and tumorigenicity by inhibition of endogenous expression of vascular endothelial growth factor. Proc Natl Acad Sci U S A 1996;93:8502–7. https://doi.org/10.1073/pnas.93.16.8502.

[24] Ruiz-Sánchez D, Calero MA, Sastre-Heres AJ, García MTI, Hernandez MAC, Martinez FM, Peña JD. Effectiveness of the bevacizumab-irinotecan regimen in the treatment of recurrent glioblastoma multiforme: comparison with other second-line treatments without this regimen. Oncol Lett 2012;4:1114–8. https://doi.org/10.3892/ol.2012.861.

[25] Chinot OL, Wick W, Mason W, Henriksson R, Saran F, Nishikawa R, Carpentier AF, Hoang-Xuan K, Kavan P, Cernea D, Brandes AA, Hilton M, Abrey L, Cloughesy T. Bevacizumab plus radiotherapy-temozolomide for newly diagnosed glioblastoma. N Engl J Med 2014;370:709–22. https://doi.org/10.1056/NEJMoa1308345.

[26] Gilbert MR, Dignam JJ, Armstrong TS, Wefel JS, Blumenthal DT, Vogelbaum MA, Colman H, Chakravarti A, Pugh S, Won M, Jeraj R, Brown PD, Jaeckle KA, Schiff D, Stieber VW, Brachman DG, Werner-Wasik M, Tremont-Lukats IW, Sulman EP, Aldape KD, Curran WJ, Mehta MP. A randomized trial of bevacizumab for newly diagnosed glioblastoma. N Engl J Med 2014;370:699–708. https://doi.org/10.1056/NEJMoa1308573.

[27] Gorlia T, Stupp R, Brandes AA, Rampling RR, Fumoleau P, Dittrich C, Campone MM, Twelves CC, Raymond E, Hegi ME, Lacombe D, Van Den Bent MJ. New prognostic factors and calculators for outcome prediction in patients with recurrent glioblastoma: a pooled analysis of EORTC brain tumour group phase i and II clinical trials. Eur J Cancer 2012;48:1176–84. https://doi.org/10.1016/j.ejca.2012.02.004.

[28] Friedman HS, Prados MD, Wen PY, Mikkelsen T, Schiff D, Abrey LE, Yung WKA, Paleologos N, Nicholas MK, Jensen R, Vredenburgh J, Huang J, Zheng M, Cloughesy T. Bevacizumab alone and in combination with irinotecan in recurrent glioblastoma. J Clin Oncol 2009;27:4733–40. https://doi.org/10.1200/JCO.2008.19.8721.

[29] Taal W, Oosterkamp HM, Walenkamp AME, Dubbink HJ, Beerepoot LV, Hanse MCJ, Buter J, Honkoop AH, Boerman D, de Vos FYF, Dinjens WNM, Enting RH, Taphoorn MJB, van den Berkmortel FWPJ, Jansen RLH, Brandsma D, Bromberg JEC, van Heuvel I, Vernhout RM, van der Holt B, Van Den Bent MJ. Single-agent bevacizumab or lomustine versus a combination of bevacizumab plus lomustine in patients with recurrent glioblastoma (BELOB trial): A randomised controlled phase 2 trial. Lancet Oncol 2014;15:943–53. https://doi.org/10.1016/S1470-2045(14)70314-6.

[30] Flomenberg P, Daniel R. In: Tirnauer JS, editor. Overview of gene therapy, gene editing, and gene silencing; 2019. p. 1–38.

[31] Kim JW, Chang AL, Kane JR, Young JS, Qiao J, Lesniak MS. Gene therapy and Virotherapy of gliomas. Prog Neurol Surg 2018;32:112–23. https://doi.org/10.1159/000469685.

[32] Fulci G, Chiocca EAA. The status of gene therapy for brain tumors. Expert Opin Biol Ther 2007;7:197–208. https://doi.org/10.1517/14712598.7.2.197.

[33] Natsume A, Yoshida J. Gene therapy for high-grade glioma. Neurol Med Chir 2017;6918. https://doi.org/10.4161/cam.2.3.6278.

[34] Asadi-Moghaddam K, Chiocca EA. Gene- and Viral-based therapies for brain tumors. Neurotherapeutics 2009;6:547–57. https://doi.org/10.1016/j.nurt.2009.04.007.

[35] O'Connor DM. Introduction to Gene and Stem-Cell Therapy. Mol Cell Ther Mot Neuron Dis 2017;141–65. https://doi.org/10.1016/B978-0-12-802257-3.00007-9.

[36] Caffery B, Lee JS, Alexander-Bryant AA. Vectors for glioblastoma gene therapy: Viral & non-viral delivery strategies. Nanomaterials 2019;9. https://doi.org/10.3390/nano9010105.

[37] Tan YY, Yap PK, Xin Lim GL, Mehta M, Chan Y, Ng SW, Kapoor DN, Negi P, Anand K, Singh SK, Jha NK, Lim LC, Madheswaran T, Satija S, Gupta G, Dua K, Chellappan DK. Perspectives and advancements in the design of nanomaterials for targeted cancer theranostics. Chem Biol Interact 2020;329:109221. https://doi.org/10.1016/j.cbi.2020.109221.

[38] Chow RD, Guzman CD, Wang G, Schmidt F, Youngblood MW, Ye L, Errami Y, Dong MB, Martinez MA, Zhang S, Renauer P, Bilguvar K, Gunel M, Sharp PA, Zhang F, Platt RJ, Chen S. AAV-mediated direct in vivo CRISPR screen identifies functional suppressors in glioblastoma. Nat Neurosci 2017;20:1329–41. https://doi.org/10.1038/nn.4620.

[39] Okada H, Low KL, Kohanbash G, McDonald HA, Hamilton RL, Pollack IF. Expression of glioma-associated antigens in pediatric brain stem and non-brain stem gliomas. J Neurooncol 2008;88:245–50. https://doi.org/10.1007/s11060-008-9566-9.

[40] Stanimirovic DB, Sandhu JK, Costain WJ. Emerging technologies for delivery of biotherapeutics and gene therapy across the blood–brain barrier. BioDrugs 2018;32:547–59. https://doi.org/10.1007/s40259-018-0309-y.

[41] Tasfaout H, Lionello VM, Kretz C, Koebel P, Messaddeq N, Bitz D, Laporte J, Cowling BS. Single intramuscular injection of AAV-shRNA reduces DNM2 and prevents Myotubular myopathy in mice. Mol Ther 2018;26:1082–92. https://doi.org/10.1016/j.ymthe.2018.02.008.

[42] Pfister EL, Chase KO, Sun H, Kennington LA, Conroy F, Johnson E, Miller R, Borel F, Aronin N, Mueller C. Safe and efficient silencing with a pol II, but not a pol III, promoter expressing an artificial miRNA targeting human huntingtin. Mol Ther - Nucleic Acids 2017;7:324–34. https://doi.org/10.1016/j.omtn.2017.04.011.

[43] Biferi MG, Cohen-Tannoudji M, Cappelletto A, Giroux B, Roda M, Astord S, Marais T, Bos C, Voit T, Ferry A, Barkats M. A New AAV10-U7-mediated gene therapy prolongs survival and restores function in an ALS mouse model. Mol Ther 2017;25:2038–52. https://doi.org/10.1016/j.ymthe.2017.05.017.

[44] Suwan K, Yata T, Waramit S, Przystal JM, Stoneham CA, Bentayebi K, Asavarut P, Chongchai A, Pothachareon P, Lee KY, Topanurak S, Smith TL, Gelovani JG, Sidman RL, Pasqualini R, Arap W, Hajitou A. Next-generation of targeted AAVP vectors for systemic transgene delivery against cancer. Proc Natl Acad Sci U S A 2019;116:18571–7. https://doi.org/10.1073/pnas.1906653116.

[45] Rangel-Sosa MM, Aguilar-Córdova E, Rojas-Martínez A. Immunotherapy and gene therapy as novel treatments for cancer. Colomb Medica (Cali, Colomb) 2017;48:138–47. https://doi.org/10.25100/cm.v48i3.2997.

[46] Wheeler LA, Manzanera AG, Bell SD, Cavaliere R, McGregor JM, Grecula JC, Newton HB, Lo SS, Badie B, Portnow J, Teh BS, Trask TW, Baskin DS, New PZ, Aguilar LK, Aguilar-Cordova E, Chiocca EA. Phase II multicenter study of gene-mediated cytotoxic immunotherapy as adjuvant to surgical resection for newly diagnosed malignant glioma. Neuro Oncol 2016;18:1137–45. https://doi.org/10.1093/neuonc/now002.

[47] Zhao F, Tian J, An L, Yang K. Prognostic utility of gene therapy with herpes simplex virus thymidine kinase for patients with high-grade malignant gliomas: a systematic review and meta analysis. J Neurooncol 2014;118:239–46. https://doi.org/10.1007/s11060-014-1444-z.

[48] De Melo SM, Bittencourt S, Ferrazoli EG, Da Silva CS, Da Cunha FF, Da Silva FH, Stilhano RS, Denapoli PMA, Zanetti BF, Martin PKM, Silva LM, Dos Santos AA, Baptista LS, Longo BM, Han SW. The anti-tumor effects of adipose tissue mesenchymal stem cell transduced with HSV-Tk gene on U-87-Driven brain tumor. PLoS One 2015;10. https://doi.org/10.1371/journal.pone.0128922.

[49] Wei D, Hou J, Zheng K, Jin X, Xie Q, Cheng L, Sun X. Suicide gene therapy against malignant gliomas by the local delivery of

genetically engineered umbilical cord mesenchymal stem cells as cellular vehicles. Curr Gene Ther 2019;19:330–41. https://doi.org/10.2174/1566523219666191028103703.

[50] Dührsen L, Hartfuß S, Hirsch D, Geiger S, Mair CL, Sedlacik J, Guenther C, Westphal M, Lamszus K, Hermann FG, Schmidt NO. Preclinical analysis of human mesenchymal stem cells: tumor tropism and therapeutic efficiency of local HSV-TK suicide gene therapy in glioblastoma. Oncotarget 2019;10:6049–61. https://doi.org/10.18632/oncotarget.27071.

[51] Lee M, Kim YS, Lee K, Kang M, Shin H, Oh JW, Koo H, Kim D, Kim Y, Kong DS, Nam DH, Lee HW. Novel semi-replicative retroviral vector mediated double suicide gene transfer enhances antitumor effects in patient-derived glioblastoma models. Cancers (Basel) 2019;11. https://doi.org/10.3390/cancers11081090.

[52] Strebe JK, Lubin JA, Kuo JS. "Tag team" glioblastoma therapy. Neurosurgery 2016;79:N18–20. https://doi.org/10.1227/01.neu.0000508605.38694.fd.

[53] Cloughesy TF, Landolfi J, Hogan DJ, Bloomfield S, Carter B, Chen CC, Elder JB, Kalkanis SN, Kesari S, Lai A, Lee IY, Liau LM, Mikkelsen T, Nghiemphu PL, Piccioni D, Walbert T, Chu A, Das A, Diago OR, Gammon D, Gruber HE, Hanna M, Jolly DJ, Kasahara N, McCarthy D, Mitchell L, Ostertag D, Robbins JM, Rodriguez-Aguirre M, Vogelbaum MA. Phase 1 trial of vocimagene amiretrorepvec and 5-fluorocytosine for recurrent high-grade glioma. Sci Transl Med 2016;8. 341ra75-341ra75 https://doi.org/10.1126/scitranslmed.aad9784.

[54] Yagiz K, Huang TT, Lopez Espinoza F, Mendoza D, Ibañez CE, Gruber HE, Jolly DJ, Robbins JM. Toca 511 plus 5-fluorocytosine in combination with lomustine shows chemotoxic and immunotherapeutic activity with no additive toxicity in rodent glioblastoma models. Neuro Oncol 2016;18:1390–401. https://doi.org/10.1093/neuonc/now089.

[55] Akbarzadeh A, Rezaei-Sadabady R, Davaran S, Joo SW, Zarghami N, Hanifehpour Y, Samiei M, Kouhi M, Nejati-Koshki K. Liposome: classification, preparation, and applications. Nanoscale Res Lett 2013;8:102. https://doi.org/10.1186/1556-276X-8-102.

[56] Hofheinz RD, Gnad-Vogt SU, Beyer U, Hochhaus A. Liposomal encapsulated anti-cancer drugs. Anticancer Drugs 2005;16:691–707. https://doi.org/10.1097/01.cad.0000167902.53039.5a.

[57] Lakkadwala S, Singh J. Co-delivery of doxorubicin and erlotinib through liposomal nanoparticles for glioblastoma tumor regression using an in vitro brain tumor model. Colloids Surfaces B Biointerfaces 2019;173:27–35. https://doi.org/10.1016/j.colsurfb.2018.09.047.

[58] Saw PE, Zhang A, Nie Y, Zhang L, Xu Y, Xu X. Tumor-associated fibronectin targeted liposomal nanoplatform for cyclophilin A siRNA delivery and targeted malignant glioblastoma therapy. Front Pharmacol 2018;9. https://doi.org/10.3389/fphar.2018.01194.

[59] Grafals-Ruiz N, Rios-Vicil CI, Lozada-Delgado EL, Quiñones-Díaz BI, Noriega-Rivera RA, Martínez-Zayas G, Santana-Rivera Y, Santiago-Sánchez GS, Valiyeva F, Vivas-Mejía PE. Brain targeted gold liposomes improve rnai delivery for glioblastoma. Int J Nanomedicine 2020;15:2809–28. https://doi.org/10.2147/IJN.S241055.

[60] Chen Z, Liu F, Chen Y, Liu J, Wang X, Chen AT, Deng G, Zhang H, Liu J, Hong Z, Zhou J. Targeted delivery of CRISPR/Cas9-mediated cancer gene therapy via liposome-templated hydrogel nanoparticles. Adv Funct Mater 2017;27. https://doi.org/10.1002/adfm.201703036.

[61] Teleanu DM, Chircov C, Grumezescu AM, Volceanov A, Teleanu RI. Blood-brain delivery methods using nanotechnology. Pharmaceutics 2018;10. https://doi.org/10.3390/pharmaceutics10040269.

[62] Choi J, Rui Y, Kim J, Gorelick N, Wilson DR, Kozielski K, Mangraviti A, Sankey E, Brem H, Tyler B, Green JJ, Jackson EM. Nonviral polymeric nanoparticles for gene therapy in pediatric CNS malignancies, nanomedicine nanotechnology. Biol Med 2020;23:102115. https://doi.org/10.1016/j.nano.2019.102115.

[63] Chan Y, Ng SW, Mehta M, Anand K, Kumar Singh S, Gupta G, Chellappan DK, Dua K. Advanced drug delivery systems can assist in managing influenza virus infection: a hypothesis. Med Hypotheses 2020;144:110298. https://doi.org/10.1016/j.mehy.2020.110298.

[64] Kozielski KL, Ruiz-Valls A, Tzeng SY, Guerrero-Cázares H, Rui Y, Li Y, Vaughan HJ, Gionet-Gonzales M, Vantucci C, Kim J, Schiapparelli P, Al-Kharboosh R, Quiñones-Hinojosa A, Green JJ. Cancer-selective nanoparticles for combinatorial siRNA delivery to primary human GBM in vitro and in vivo. Biomaterials 2019;209:79–87. https://doi.org/10.1016/j.biomaterials.2019.04.020.

[65] Kozielski KL, Tzeng SY, Hurtado De Mendoza BA, Green JJ. Bioreducible cationic polymer-based nanoparticles for efficient and environmentally triggered cytoplasmic siRNA delivery to primary human brain cancer cells. ACS Nano 2014;8:3232–41. https://doi.org/10.1021/nn500704t.

[66] Lopez-Bertoni H, Kozielski KL, Rui Y, Lal B, Vaughan H, Wilson DR, Mihelson N, Eberhart CG, Laterra J, Green JJ. Bioreducible polymeric nanoparticles containing multiplexed cancer stem cell regulating miRNAs inhibit glioblastoma growth and prolong survival. Nano Lett 2018;18:4086–94. https://doi.org/10.1021/acs.nanolett.8b00390.

[67] Englert C, Trützschler AK, Raasch M, Bus T, Borchers P, Mosig AS, Traeger A, Schubert US. Crossing the blood-brain barrier: glutathione-conjugated poly(ethylene imine) for gene delivery. J Control Release 2016;241:1–14. https://doi.org/10.1016/j.jconrel.2016.08.039.

[68] Liu C, Zhao Z, Gao H, Rostami I, You Q, Jia X, Wang C, Zhu L, Yang Y. Enhanced blood-brain-barrier penetrability and tumor-targeting efficiency by peptide-functionalized poly(Amidoamine) dendrimer for the therapy of gliomas. Nanotheranostics 2019;3:311–30. https://doi.org/10.7150/ntno.38954.

[69] He H, Li Y, Jia XR, Du J, Ying X, Lu WL, Lou JN, Wei Y. PEGylated poly(amidoamine) dendrimer-based dual-targeting carrier for treating brain tumors. Biomaterials 2011;32:478–87. https://doi.org/10.1016/j.biomaterials.2010.09.002.

[70] Pinel S, Thomas N, Boura C, Barberi-Heyob M. Approaches to physical stimulation of metallic nanoparticles for glioblastoma treatment. Adv Drug Deliv Rev 2019;138:344–57. https://doi.org/10.1016/j.addr.2018.10.013.

[71] Hainfeld JF, Smilowitz HM, O'connor MJ, Dilmanian FA, Slatkin DN. Gold nanoparticle imaging and radiotherapy of brain tumors in mice. Nanomedicine 2013;8:1601–9. https://doi.org/10.2217/nnm.12.165.

[72] Sun J, Kormakov S, Liu Y, Huang Y, Wu D, Yang Z. Recent progress in metal-based nanoparticles mediated photodynamic therapy. Molecules 2018;23:1704. https://doi.org/10.3390/molecules23071704.

[73] Srivatsan A, Jenkins SV, Jeon M, Wu Z, Kim C, Chen J, Pandey RK. Gold nanocage-photosensitizer conjugates for dual-modal image-guided enhanced photodynamic therapy. Theranostics 2014;4:163–74. https://doi.org/10.7150/thno.7064.

[74] Vankayala R, Lin CC, Kalluru P, Chiang CS, Hwang KC. Gold nanoshells-mediated bimodal photodynamic and photothermal cancer treatment using ultra-low doses of near infra-red light. Biomaterials 2014;35:5527–38. https://doi.org/10.1016/j.biomaterials.2014.03.065.

[75] Aboody KS, Najbauer J, Danks MK. Stem and progenitor cell-mediated tumor selective gene therapy. Gene Ther 2008;15:739–52. https://doi.org/10.1038/gt.2008.41.

[76] Aghi M, Chiocca EA. Gene therapy for glioblastoma. Neurosurg Focus 2006;20.

[77] Candolfi M, Kroeger K, Muhammad A, Yagiz K, Farrokhi C, Pechnick R, Lowenstein P, Castro M. Gene therapy for brain cancer: combination therapies provide enhanced efficacy and safety. Curr Gene Ther 2009;9:409–21. https://doi.org/10.2174/156652309789753301.

[78] Ferguson SD, Ahmed AU, Thaci B, Mercer RW, Lesniak MS. Crossing the boundaries: stem cells and gene therapy. Discov Med 2010;9:192–6.

[79] Jiang H, Gomez-Manzano C, Lang F, Alemany R, Fueyo J. Oncolytic adenovirus: preclinical and clinical studies in patients with human malignant gliomas. Curr Gene Ther 2009;9:422–7. https://doi.org/10.2174/156652309789753356.

[80] Markert JM, Parker JN, Buchsbaum DJ, Grizzle WE, Gillespie GY, Whitley RJ. Oncolytic HSV-1 for the treatment of brain tumours. Herpes 2006;13:66–71.

[81] Kroeger KM, Muhammad AKMG, Baker GJ, Assi H, Wibowo MK, Xiong W, Yagiz K, Candolfi M, Lowenstein PR, Castro MG. Gene therapy and virotherapy: novel therapeutic approaches for brain tumors. Discov Med 2010;10:293–304.

[82] Wykosky J, Gibo DM, Stanton C, Debinski W. Interleukin-13 receptor α2, EphA2, and Fos-related antigen 1 as molecular denominators of high-grade astrocytomas and specific targets for combinatorial therapy. Clin Cancer Res 2008;14:199–208. https://doi.org/10.1158/1078-0432.CCR-07-1990.

[83] Kawakami K, Kawakami M, Kioi M, Husain SR, Puri RK. Distribution kinetics of targeted cytotoxin in glioma by bolus or convection-enhanced delivery in a murine model. J Neurosurg 2004;101:1004–11. https://doi.org/10.3171/jns.2004.101.6.1004.

[84] Vogelbaum MA, Sampson JH, Kunwar S, Chang SM, Shaffrey M, Asher AL, Lang FF, Croteau D, Parker K, Grahn AY, Sherman JW, Husain SR, Puri RK. Convection-enhanced delivery of cintredekin besudotox (interleukin-13- PE38QQR) followed by radiation therapy with and without temozolomide in newly diagnosed malignant gliomas: phase 1 study of final safety results. Neurosurgery 2007;61:1031–7. https://doi.org/10.1227/01.neu.0000303199.77370.9e.

[85] Kunwar S, Prados MD, Chang SM, Berger MS, Lang FF, Piepmeier JM, Sampson JH, Ram Z, Gutin PH, Gibbons RD, Aldape KD, Croteau DJ, Sherman JW, Puri RK. Direct intracerebral delivery of cintredekin besudotox (IL13-PE38QQR) in recurrent malignant glioma: a report by the cintredekin besudotox intraparenchymal study group. J Clin Oncol 2007;25:837–44. https://doi.org/10.1200/JCO.2006.08.1117.

[86] Candolfi M, Xiong W, Yagiz K, Liu C, Muhammad AKMG, Puntel M, Foulad D, Zadmehr A, Ahlzadeh GE, Kroeger KM, Tesarfreund M, Lee S, Debinski W, Sareen D, Svendsen CN, Rodriguez R, Lowenstein PR, Castro MG. Gene therapy-mediated delivery of targeted cytotoxins for glioma therapeutics. Proc Natl Acad Sci U S A 2010;107:20021–6. https://doi.org/10.1073/pnas.1008261107.

[87] Terada K, Wakimoto H, Tyminski E, Chiocca EA, Saeki Y. Development of a rapid method to generate multiple oncolytic HSV vectors and their in vivo evaluation using syngeneic mouse tumor models. Gene Ther 2006;13:705–14. https://doi.org/10.1038/sj.gt.3302717.

[88] Han ZQ, Assenberg M, Liu BL, Wang YB, Simpson G, Thomas S, Coffin RS. Development of a second-generation oncolytic herpes simplex virus expressing TNFα for cancer therapy. J Gene Med 2007;9:99–106. https://doi.org/10.1002/jgm.999.

[89] Liu TC, Zhang T, Fukuhara H, Kuroda T, Todo T, Martuza RL, Rabkin SD, Kurtz A. Oncolytic HSV armed with platelet factor 4, an antiangiogenic agent, shows enhanced efficacy. Mol Ther 2006;14:789–97. https://doi.org/10.1016/j.ymthe.2006.07.011.

[90] Chung RY, Saeki Y, Chiocca EA. B-myb promoter retargeting of Herpes simplex virus γ34.5 gene-mediated virulence toward tumor and cycling cells. J Virol 1999;73:7556–64. https://doi.org/10.1128/jvi.73.9.7556-7564.1999.

[91] Kambara H, Okano H, Chiocca EA, Saeki Y. An oncolytic HSV-1 mutant expressing ICP34.5 under control of a nestin promoter increases survival of animals even when symptomatic from a brain tumor. Cancer Res 2005;65:2832–9. https://doi.org/10.1158/0008-5472.CAN-04-3227.

[92] Chiocca EA, Aghi M, Fulci G. Viral therapy for glioblastoma. Cancer J 2003;9:167–79. https://doi.org/10.1097/00130404-200305000-00005.

[93] Fueyo J, Alemany R, Gomez-Manzano C, Fuller GN, Khan A, Conrad CA, Liu TJ, Jiang H, Lemoine MG, Suzuki K, Sawaya R, Curiel DT, Yung WKA, Lang FF. Preclinical characterization of the antiglioma activity of a tropism-enhanced adenovirus targeted to the retinoblastoma pathway. J Natl Cancer Inst 2003;95:652–60. https://doi.org/10.1093/jnci/95.9.652.

[94] Birgit G, Jacques G, Paule O, Jackie M, Genevieve A, Marie-Jose T-L, Bressac D-PB, Michel B, Jean F, Kirn David H, Gilles V. Oncolytic activity of the E1B-55 kDa-deleted adenovirus ONYX-015 is independent of cellular p53 status in human malignant glioma xenografts. Cancer Res 2002;62:764–72.

[95] Ganly I, Kirn D, Eckhardt SG, Rodriguez GI, Soutar DS, Otto R, Robertson AG, Park O, Gulley ML, Heise C, Von Hoff DD, Kaye SB. A phase I study of Onyx-015, an E1B attenuated adenovirus, administered intratumorally to patients with recurrent head and neck cancer. Clin Cancer Res 2000;6:798–806.

[96] Khuri FR, Nemunaitis J, Ganly I, Arseneau J, Tannock IF, Romel L, Gore M, Ironside J, MacDougall RH, Heise C, Randlev B, Gillenwater AM, Bruso P, Kaye SB, Hong WK, Kirn DH. A controlled trial of intratumoral ONYX-015, a selectively-replicating adenovirus, in combination with cisplatin and 5-fluorouracil in patients with recurrent head and neck cancer. Nat Med 2000;6:879–85. https://doi.org/10.1038/78638.

[97] Nemunaitis J, Ganly I, Khuri F, Arseneau J, Kuhn J, McCarty T, Landers S, Maples P, Romel L, Randlev B, Reid T, Kaye S, Kirn D. Selective replication and oncolysis in p53 mutant tumors with ONYX-015, an E1B-55kD gene-deleted adenovirus, in patients with advanced head and neck cancer: a phase II trial. Cancer Res 2000;60:6359–66.

[98] O'Shea CC, Johnson L, Bagus B, Choi S, Nicholas C, Shen A, Boyle L, Pandey K, Soria C, Kunich J, Shen Y, Habets G, Ginzinger D, McCormick F. Late viral RNA export, rather than p53 inactivation, determines ONYX-015 tumor selectivity. Cancer Cell 2004;6:611–23. https://doi.org/10.1016/j.ccr.2004.11.012.

[99] Fecci PE, Heimberger AB, Sampson JH. Immunotherapy for primary brain tumors: no longer a matter of privilege. Clin Cancer Res 2014;20:5620–9. https://doi.org/10.1158/1078-0432.CCR-14-0832.
[100] Tivnan A, Heilinger T, Lavelle EC, Prehn JHM. Advances in immunotherapy for the treatment of glioblastoma. J Neurooncol 2017;131. https://doi.org/10.1007/s11060-016-2299-2.
[101] Goel S, Duda DG, Xu L, Munn LL, Boucher Y, Fukumura D, Jain RK. Normalization of the vasculature for treatment of cancer and other diseases. Physiol Rev 2011;91:1071–121. https://doi.org/10.1152/physrev.00038.2010.
[102] Home - ClinicalTrials.gov, (n.d.).
[103] Huang B, Zhang H, Gu L, Ye B, Jian Z, Stary C, Xiong X. Advances in immunotherapy for glioblastoma multiforme. J Immunol Res 2017;2017. https://doi.org/10.1155/2017/3597613.
[104] Dunn-Pirio AM, Vlahovic G. Immunotherapy approaches in the treatment of malignant brain tumors. Cancer 2017;123:734–50. https://doi.org/10.1002/cncr.30371.
[105] Natsume A, Yoshida J. Gene therapy for malignant glioma. Biotherapy 2006;20:300–4. https://doi.org/10.1186/2052-8426-2-21.
[106] Guhasarkar D, Su Q, Gao G, Sena-Esteves M. Systemic AAV9-IFNβ gene delivery treats highly invasive glioblastoma. Neuro Oncol 2016;18:1508–18. https://doi.org/10.1093/neuonc/now097.
[107] Crommentuijn MHW, Maguire CA, Niers JM, Vandertop WP, Badr CE, Würdinger T, Tannous BA. Intracranial AAV-sTRAIL combined with lanatoside C prolongs survival in an orthotopic xenograft mouse model of invasive glioblastoma. Mol Oncol 2016;10:625–34. https://doi.org/10.1016/j.molonc.2015.11.011.
[108] Watanabe R, Takase-Yoden S. Gene expression of neurotropic retrovirus in the CNS. Prog Brain Res 1995;105:255–62. https://doi.org/10.1016/S0079-6123(08)63302-6.
[109] Sandmair AM, Vanninen R, Lehtolainen P, Paljarvi L, Johansson R, Vapalahti M, Yla-Herttuala S, Loimas S, Puranen P, Immonen A, Kossila M, Puranen M, Hurskainen H, Tyynela K, Turunen M. Thymidine kinase gene therapy for human malignant glioma, using replication-deficient retroviruses or adenoviruses. Hum Gene Ther 2000;11:2197–205. https://doi.org/10.1089/104303400750035726.
[110] Rainov NG. A phase III clinical evaluation of herpes simplex virus type 1 thymidine kinase and ganciclovir gene therapy as an adjuvant to surgical resection and radiation in adults with previously untreated glioblastoma multiforme. Hum Gene Ther 2000;11:2389–401. https://doi.org/10.1089/104303400750038499.
[111] Ram Z, Culver KW, Oshiro EM, Viola JJ, Devroom HL, Otto E, Long Z, Chiang Y, Mcgarrity GJ, Muul LM, Katz D, Blaese RM, Oldfield EH. Therapy of malignant brain tumors by intratumoral implantation of retroviral vector-producing cells. Nat Med 1997;3:1354–61. https://doi.org/10.1038/nm1297-1354.
[112] Aghi M, Vogelbaum MA, Kesari S, Chen CC, Liau LM, Piccioni D, Portnow J, Chang S, Robbins JM, Boyce T, Huang TT, Pertschuk D, Ostertag D, Cloughesy TF. At-02 * Intratumoral delivery of the retroviral replicating vector (Rrv) Toca 511 in subjects with recurrent high grade glioma: interim report of phase 1 study (Nct 01156584). Neuro Oncol 2014;16:v8–. https://doi.org/10.1093/neuonc/nou237.2.
[113] Takahashi M, Valdes G, Hiraoka K, Inagaki A, Kamijima S, Micewicz E, Gruber HE, Robbins JM, Jolly DJ, McBride WH, Iwamoto KS, Kasahara N. Radiosensitization of gliomas by intracellular generation of 5-fluorouracil potentiates prodrug activator gene therapy with a retroviral replicating vector. Cancer Gene Ther 2014;21:405–10. https://doi.org/10.1038/cgt.2014.38.
[114] Huang TT, Hlavaty J, Ostertag D, Espinoza FL, Martin B, Petznek H, Rodriguez-Aguirre M, Ibañez CE, Kasahara N, Gunzburg W, Gruber HE, Pertschuk D, Jolly DJ, Robbins JM. Toca 511 gene transfer and 5-fluorocytosine in combination with temozolomide demonstrates synergistic therapeutic efficacy in a temozolomide-sensitive glioblastoma model. Cancer Gene Ther 2013;20:544–51. https://doi.org/10.1038/cgt.2013.51.
[115] Lang FF, Bruner JM, Fuller GN, Aldape K, Prados MD, Chang S, Berger MS, McDermoff MW, Kunwar SM, Junck LR, Chandler W, Zwiebel JA, Kaplan RS, Yung WKA. Phase I trial of adenovirus-mediated p53 gene therapy for recurrent glioma: biological and clinical results. J Clin Oncol 2003;21:2508–18. https://doi.org/10.1200/JCO.2003.21.13.2508.
[116] Chiocca EA, Aguilar LK, Bell SD, Kaur B, Hardcastle J, Cavaliere R, McGregor J, Lo S, Ray-Chaudhuri A, Chakravarti A, Grecula J, Newton H, Harris KS, Grossman RG, Trask TW, Baskin DS, Monterroso C, Manzanera AG, Aguilar-Cordova E, New PZ. Phase IB study of gene-mediated cytotoxic immunotherapy adjuvant to up-front surgery and intensive timing radiation for malignant glioma. J Clin Oncol 2011;29:3611–9. https://doi.org/10.1200/JCO.2011.35.5222.
[117] Kieran MW, Goumnerova L, Manley P, Chi SN, Marcus KJ, Manzanera AG, Polanco MLS, Guzik BW, Aguilar-Cordova E, Diaz-Montero CM, Dipatri AJ, Tomita T, Lulla R, Greenspan L, Aguilar LK, Goldman S. Phase I study of gene-mediated cytotoxic immunotherapy with AdV tk as adjuvant to surgery and radiation for pediatric malignant glioma and recurrent ependymoma. Neuro Oncol 2019;21:537–46. https://doi.org/10.1093/neuonc/noy202.
[118] Conry RM, Westbrook B, McKee S, Norwood TG. Talimogene laherparepvec: first in class oncolytic virotherapy. Hum Vaccines Immunother 2018;14:839–46. https://doi.org/10.1080/21645515.2017.1412896.
[119] Adams D, Gonzalez-Duarte A, O'Riordan WD, Yang CC, Ueda M, Kristen AV, Tournev I, Schmidt HH, Coelho T, Berk JL, Lin KP, Vita G, Attarian S, Planté-Bordeneuve V, Mezei MM, Campistol JM, Buades J, Brannagan TH, Kim BJ, Oh J, Parman Y, Sekijima Y, Hawkins PN, Solomon SD, Polydefkis M, Dyck PJ, Gandhi PJ, Goyal S, Chen J, Strahs AL, Nochur SV, Sweetser MT, Garg PP, Vaishnaw AK, Gollob JA, Suhr OB. Patisiran, an RNAi therapeutic, for hereditary transthyretin amyloidosis. N Engl J Med 2018;379:11–21. https://doi.org/10.1056/NEJMoa1716153.
[120] Kim SS, Rait A, Kim E, Pirollo KF, Nishida M, Farkas N, Dagata JA, Chang EH. A nanoparticle carrying the p53 gene targets tumors including cancer stem cells, sensitizes glioblastoma to chemotherapy and improves survival. ACS Nano 2014;8:5494–514. https://doi.org/10.1021/nn5014484.
[121] Jensen SA, Day ES, Ko CH, Hurley LA, Luciano JP, Kouri FM, Merkel TJ, Luthi AC, Patel PC, Cutler JI, Daniel WL, Scott AW, Rotz MW, Meade TJ, Giljohann DA, Mirkin CA, Stegh AH. Spherical nucleic acid nanoparticle conjugates as an RNAi-based therapy for glioblastoma. Sci Transl Med 2013;5, 209ra152. https://doi.org/10.1126/scitranslmed.3006839.
[122] Shu Y, Pi F, Sharma A, Rajabi M, Haque F, Shu D, Leggas M, Evers BM, Guo P. Stable RNA nanoparticles as potential new generation drugs for cancer therapy. Adv Drug Deliv Rev 2014;66:74–89.
[123] Fesnak AD, June CH, Levine BL. Engineered T cells: the promise and challenges of cancer immunotherapy. Nat Rev Cancer 2016;16:566–81. https://doi.org/10.1038/nrc.2016.97.

[124] Kong DS, Nam DH, Kang SH, Lee JW, Chang JH, Kim JH, Lim YJ, Koh YC, Chung YG, Kim JM, Kim CH. Phase III randomized trial of autologous cytokine-induced killer cell immunotherapy for newly diagnosed glioblastoma in Korea. Oncotarget 2017;8:7003–13. https://doi.org/10.18632/oncotarget.12273.

[125] Brown C, Alizadeh D, Starr R, Weng L, Wagner J, Naranjo A, Blanchard S, Kilpatrick J, Simpson J, Ressler JA, Jensen M, Portnow J, D'Apuzzo M, Barish M, Forman S, Badie B. Atim-13. Phase Istudy of chimeric antigen receptor-engineered T cells targeting Il13Rα2 for the treatment of Glioblastoma. Neuro Oncol 2016;18. vi20–vi20 https://doi.org/10.1093/neuonc/now212.078.

[126] D.J. Sexton, Cerebrospinal fluid: Physiology and utility of an examination in disease ... http://www.uptodate.com/contents/cerebrospinal-fluid-physiology-and- ... Cerebrospinal fluid: Physiology and utility of an examination in disease ... http://www.uptodate.com/co, 20 (2013) 1–12.

[127] Da Ros M, De Gregorio V, Iorio AL, Giunti L, Guidi M, de Martino M, Genitori L, Sardi I. Glioblastoma chemoresistance: the double play by microenvironment and blood-brain barrier. Int J Mol Sci 2018;19. https://doi.org/10.3390/ijms19102879.

[128] Fiandaca MS, Berger MS, Bankiewicz KS. The use of convection-enhanced delivery with liposomal toxins in neurooncology. Toxins (Basel) 2011;3:369–97. https://doi.org/10.3390/toxins3040369.

[129] Medina-Kauwe LK, MacVeigh M, Chen X, Maguire M, Kedes L, Hamm-Alvarez S, Dirven CMF, Lamfers MLM, Molenaar B, van Beusechem VW, Curiel DT, Gerritsen WR, Vandertop WP, Grill J, Toietta G, Koehler DR, Finegold MJ, Lee B, Hu J, Beaudet AL. Opening of the blood-brain barrier for improved adenovirus delivery to the brain also increases delivery to the liver. Mol Ther 2002;5:S196. https://doi.org/10.1016/s1525-0016(16)43430-1.

[130] Dagogo-Jack I, Shaw AT. Tumour heterogeneity and resistance to cancer therapies. Nat Rev Clin Oncol 2018;15:81–94. https://doi.org/10.1038/nrclinonc.2017.166.

[131] Bessis N, GarciaCozar FJ, Boissier MC. Immune responses to gene therapy vectors: influence on vector function and effector mechanisms. Gene Ther 2004;11:S10–7. https://doi.org/10.1038/sj.gt.3302364.

[132] Nayak S, Herzog RW. Progress and prospects: immune responses to viral vectors. Gene Ther 2010;17:295–304. https://doi.org/10.1038/gt.2009.148.

[133] De Haan P, Van Diemen FR, Toscano MG. Viral gene delivery vectors: the next generation medicines for immune-related diseases. Hum Vaccines Immunother 2020;1–8. https://doi.org/10.1080/21645515.2020.1757989.

Chapter 35

Targeting siRNAs in cancer drug delivery

Mohammad A. Obeid[a], Alaa A.A. Aljabali[a], Walhan Alshaer[b], Nitin Bharat Charbe[c], Dinesh Kumar Chellappan[d], Kamal Dua[e,f], Saurabh Satija[e,g], and Murtaza M. Tambuwala[h]

[a]*Faculty of Pharmacy, Department of Pharmaceutics and Pharmaceutical Technology, Yarmouk University, Irbid, Jordan,* [b]*Cell Therapy Center, The University of Jordan, Amman, Jordan,* [c]*Departamento de Quimica Orgánica, Facultad de Química y de Farmacia, Pontificia Universidad Católica de Chile, Santiago, Chile,* [d]*Department of Life Sciences, School of Pharmacy, International Medical University (IMU), Kuala Lumpur, Malaysia,* [e]*Discipline of Pharmacy, Graduate School of Health, University of Technology Sydney, Ultimo, Sydney, NSW, Australia,* [f]*Priority Research Centre for Healthy Lungs, Hunter Medical Research Institute (HMRI) & School of Biomedical Sciences and Pharmacy, University of Newcastle, Callaghan, NSW, Australia,* [g]*School of Pharmaceutical Sciences, Lovely Professional University, Phagwara, Punjab, India,* [h]*SAAD Centre for Pharmacy and Diabetes, School of Pharmacy and Pharmaceutical Sciences, Ulster University, Coleraine, Northern Ireland, United Kingdom*

1 Introduction

RNA inhibitory (RNAi) is a natural posttranscriptional gene regulatory mechanism discovered by Fire et al. in 1998. This discovery led to a Nobel prize as it represents a promising therapeutic pathway for several diseases. In this mechanism, double-stranded RNA (dsRNA) will result in the selective inhibition for gene expression nematode *Caenorhabditis elegans* [1]. This mechanism is illustrated in Fig. 1.

The concept of the RNAi is that noncoding, double-stranded RNA molecules will result in the inhibition of certain gene expressions and the subsequent knockdown of the protein expression encoded in that gene [3]. The siRNA represents an example of a double-stranded RNA molecule that is involved in this regulatory mechanism. The structure of siRNA is composed of 21–23 base pairs with highly specific sense and antisense strands and is usually generated in human cells through the activity of the RNase III endonuclease dicer on longer RNA molecules [4, 5].

Following the generation of specific siRNA molecules, it will be integrated into a complex called the RNA-induced silencing complex (RISC). This integration will result in the cleavage of these double-stranded molecules into two strands. The antisense strand will remain incorporated into the RISC, and the sense strand will be eliminated out of the complex.

The next step in this mechanism will be the incorporation of specific messenger RNA (mRNA) molecules that are complementary to the antisense strand of the siRNA in the RISC. This binding will result in the cleavage of these complementary mRNA molecules by the action of the RNA endonuclease (Argonaut 2) enzyme, and this cleavage will end up with the shutdown of these mRNAs and the prevention of the protein expression that is originally encoded through these mRNA molecules [6]. After the mRNA cleavage, the RISC will regenerate for another cycle of mRNA molecules cleavage by the complementary siRNA molecules [7, 8].

As there is a growing need to develop new therapeutic agents for the treatment of many diseases, this discovery led to a new research area to apply this mechanism to treat many diseases, especially cancer. This was followed by the conclusion by many researchers that the use of synthetic siRNA targeting a specific protein in the body will result in one selective knockdown of that protein [2]. This means that in certain diseases, the knockdown of a particular protein crucial for that disease might lead to an effective treatment option.

In cancer treatment, siRNA therapeutics have been extensively studied to develop alternative therapeutic options from current anticancer medications. Theoretically, cancer treatment with siRNA is based on determining the protein that is upregulated or mutated in the cancer tissues and is crucial for cancer growth, survival, metastasis, cell cycle regulation, or cell resistance to chemotherapy radiotherapy. Then based on the structure of this protein, synthetic siRNA molecules targeting that protein will be delivered to the cancer cells to inhibit that crucial protein and then subsequently resulting in cancer cell death [3, 7].

Compared to current therapeutic options for cancer, the siRNA approach has good therapeutic advantages as it will result in functional interference with endogenous gene expression within the target cells. This interference with the subsequent protein synthesis inhibition is highly specific as the siRNA will result in the degradation of its complementary mRNA [9].

Advanced Drug Delivery Systems in the Management of Cancer. https://doi.org/10.1016/B978-0-323-85503-7.00027-4

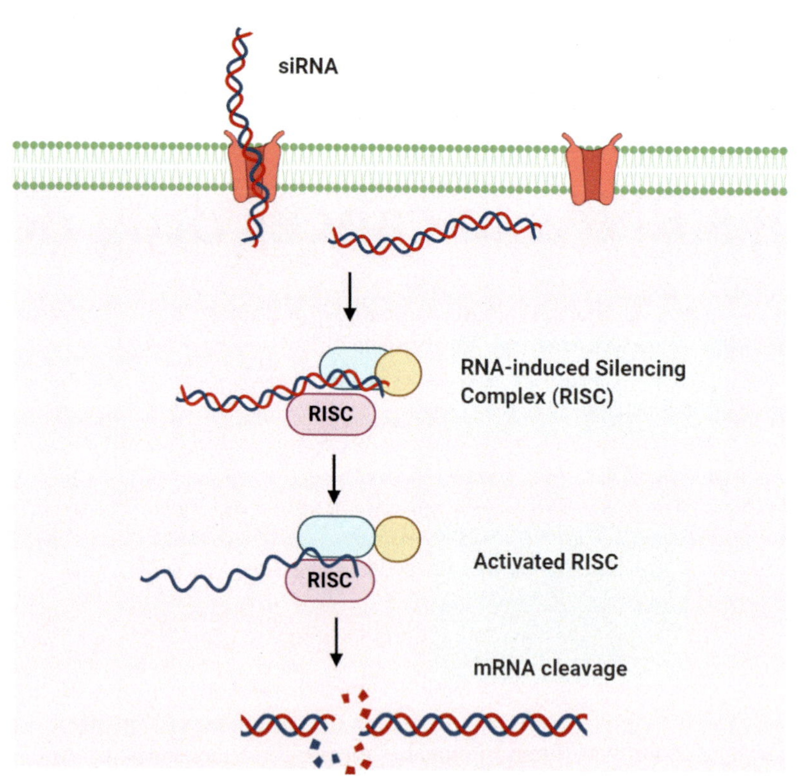

FIG. 1 Schematic diagram of the mRNA degradation through siRNA. *Adapted from Ref. [2].*

Moreover, the currently used chemotherapeutic agents for cancer treatment will act by blocking or activating their targets, which means they are not specific in their action, and this can result in various side effects in comparison with siRNA therapeutics that are considered as safe with low side effects [10, 11].

1.1 The critical challenge of siRNA translation in therapy

The concept of RNAi mechanism for inhibiting the expression of target protein genes in the targeted cells with a known sequence of that protein [12]. Nucleic acid therapies revolutionized the research of cancer treatment development using this strategy and ended up with thousands of publications in this field. This increase in this research area revealed many challenges that stand against developing an effective siRNA-based therapeutics for cancer treatment [13].

The primary challenge is the determination of the target gene that is crucial for cancer tissue growth as the success of the siRNA therapeutics is mainly dependent on the structure of that protein. Moreover, even after determining the sequence of the target protein and designing siRNA molecules to inhibit this protein, simple administration of this siRNA will not generate any therapeutic effect. This is related to the structure of siRNA as the administration of naked siRNA into cellular plasma will be followed by its rapid degradation and elimination resulting in a very short half-life ranging from minutes to around an hour [14–17]. This rapid degradation and elimination of the administered siRNA will be due to nucleases which are abundant in the plasma. This short half-life of the siRNA molecules will not be enough to get distributed to the target site and exert their therapeutic effect.

Moreover, even if the siRNA molecules were able to accumulate at the target tissues, their cellular uptake by the target cells will be minimum. This is because of the considerable molecular weight of the siRNA molecules, which is around 13 kDa, and the high negative charge of the siRNA molecules due to the presence of the phosphate groups, which will result in electrostatic repulsion with the cell membrane proteins that have the same negative charge [18]. These properties of siRNA molecules are quite challenging to across the cellular membrane naturally [19].

Even if the siRNA molecules have been uptaken by the target cells, they need to escape the endosomes and being released into the cytoplasm where the RNAi mechanism occurs. This endosome escape represents a significant challenge for the siRNA therapeutics. If the siRNA molecules remain trapped into the endosomal for an extended period, they will be degraded and eliminated out of the cells after their uptake without exerting any protein expression inhibition [20, 21].

Besides the expansion in the RNAi mechanism, research revealed other limitations such as the possible saturation of this mechanism in the cytoplasm of the target cells, the development of certain immunogenic reaction after the siRNA administration, and the unwanted off-target effects of these siRNA molecules inside the body by inhibiting other types of proteins other than the target protein [22–24].

The abovementioned limitations of therapeutic siRNA restricts the early applications of these molecules in the clinical trials including intranasal and intraocular or intravitreal administration [25].

1.2 The use of nanoparticles as drug delivery systems for siRNA

Based on the limitations of the use of free therapeutic siRNA for disease treatment, different solutions have been proposed to utilize the RNAi mechanism in the development of new therapeutic options. Nanotechnology is one of the methods, which has been used for improving therapeutic siRNA characteristics. It is based on using specific nanoparticles as a carrier for the therapeutic siRNA to deliver it to its target site, as shown in Fig. 2. The use of nanoparticles as a delivery system for siRNA will increase the stability and decrease the degradation rate of the siRNA by the nucleases as the siRNA is encapsulated and protected inside these nanocarriers.

Moreover, these nanocarriers will enhance the accumulation of the therapeutic siRNA at its target site through passive and active targeting, which will result in facilitating the targeting and avoiding the unwanted distribution and reducing the chance of off-target effect of the loaded siRNA [26]. The encapsulation of siRNA into the nanocarriers will also enhance their cellular uptake. This encapsulation will mask the negative charge of siRNA, which will reduce the repulsion with the target cell membrane [18, 27].

Concerning the endosomal escape, currently, different types of nanoparticles can be used with a composition that will reduce the stability of the endosomes and result in their rupture, which eventually leads to the release of the loaded siRNA into the cytoplasm where it induces its action on its complementary mRNA [28]. Therefore, the design of a specific nanocarrier for siRNA delivery should be based on improving the characteristics and avoiding the siRNA obstacles to increase the half-life of the therapeutic siRNA and overcome any intracellular or extracellular barriers that will prevent the siRNA from reaching the RISC and induce its action on its complementary mRNA [24].

Here, we will describe the most studied types of nanoparticles in siRNA delivery and the results of the most significant research that has been done in this field.

2 Types of nanoparticles used for siRNA delivery

2.1 Liposomes

Lipid-based nanoparticles are among the most studied nanoparticle drug delivery systems used for siRNA delivery. The studies that have been done with these nanoparticles revealed that they are the most promising nanoparticles for systemic siRNA delivery. This is because of the ease of engineering these nanoparticles to enhance their accumulation at specific sites and avoid any early clearance and degradation of the loaded siRNA from the body, as depicted in Fig. 3. Lipid-based nanoparticles include different types of nanoparticles such as liposomes, niosomes, solid lipid

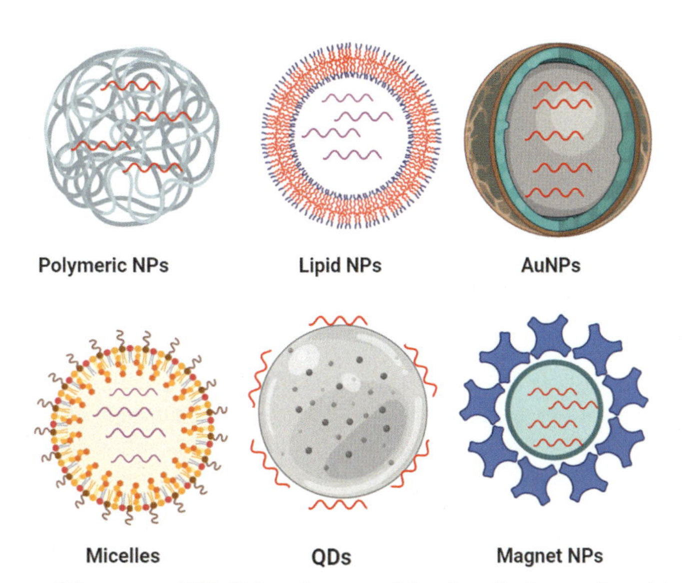

FIG. 2 Schematic illustration of some of the most used NPs. Polymeric nanoparticles where the therapeutic nucleic acid can be incorporated within the polymer, within the lipid-based NPs, covalently attached on quantum dots (QDs) and incorporated with magnetic shells.

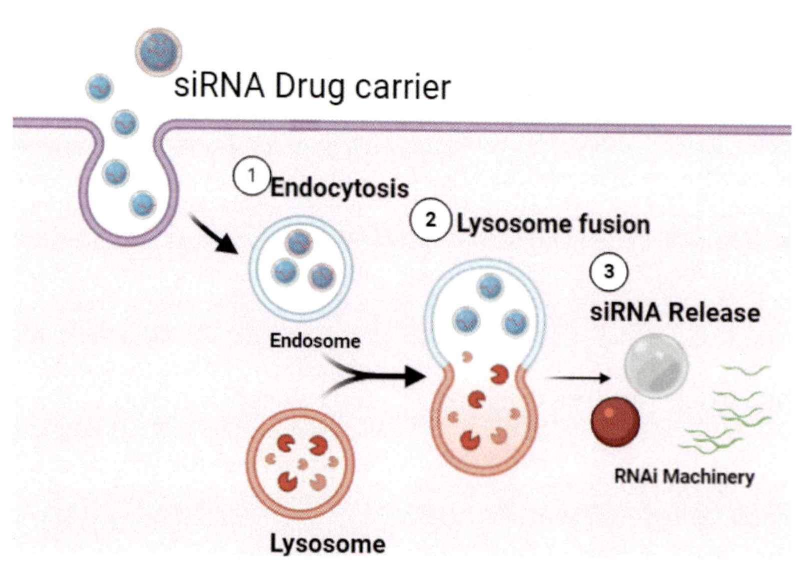

FIG. 3 Schematic illustration of siRNA NPs intake by the targeted cells by the endocytosis process followed the fusion of lysosomes and peroxisome with the drug carrier resulting in releasing the therapeutic siRNA moieties to their targets.

nanoparticles, and micelles. All are composed of either natural or synthetic lipid components.

Liposomes are the most advanced and most studied drug delivery systems. They are composed of a spherical bilayer structure encapsulating an aqueous moiety. Liposomes are mainly prepared using phospholipids, which are amphiphilic lipids that will self-assemble into the bilayer structure upon contact with the aqueous compartment [22, 29]. This unique structure of liposomes allows for the encapsulation and delivery of various types of therapeutic agents. The hydrophilic drugs will be encapsulated into the aqueous core of liposomes, whereas the lipophilic drugs will be incorporated into the lipid membrane bilayer of the liposomes [30]. These nanoparticles were first investigated for drug delivery by Gregoriadis et al. to deliver various antimicrobial and anticancer agents [31, 32]. Following this application, many therapeutic formulations have been developed using liposomes, and these efforts have been translated into the first two liposomes-based product in the market; AmBisome (liposomal amphotericin B for the treatment of fungal infections) and Doxil (liposomal doxorubicin) [33, 34].

Liposomes are prepared using various types of natural or synthetic phospholipids along with other components such as cholesterol and different types of polymers. The ability of phospholipids to form a bilayer structure is based on the presence of a hydrophilic head and hydrophobic tail, which will spontaneously orient themselves into a bilayer structure upon contact with water. The use of cholesterol in the liposome formulations will increase the rigidity of the membrane bilayer, which eventually reduces the leakage of the encapsulated drug from these nanoparticles [30, 35].

Since liposomes are mainly composed of phospholipids and cholesterol, which are also the components of the cell membranes, liposomes are characterized by their high safety profile with less toxic effects. Also, because of their ability to encapsulate both hydrophilic and hydrophobic drugs, they have been investigated to deliver various agents such as DNA, siRNA, anticancer agents, vaccines, and many other small-molecular-weight drugs. The encapsulation of these agents into the liposomes will protect them from the external environment and increase their stability [36].

The encapsulation of the hydrophilic siRNA molecules in the aqueous core of the liposome nanoparticles will protect them from degradation by nuclease activity, which will increase the stability and the half-life of the loaded siRNA molecules. Moreover, this encapsulation will facilitate the cellular uptake of the loaded siRNA molecules [23, 37].

With increasing research and attention toward liposomes, several manufacturing techniques have been developed, such as the thin-film hydration method, the heating method, the ether injection method, and the microfluidic mixing method. The feasibility of preparing the liposomes using different composition and different preparation technique makes the control over the primary liposomes features such as the particles size and the surface charge relatively easy which enhance the ability of these nanoparticles to deliver their therapeutic payloads [23, 30, 33]. The surface charge of the liposomes will affect their distribution along with their interaction with the biological surfaces. The liposomes can be prepared to have cationic, anionic, or neutral surface charge. The final net surface charge will be based on phospholipids used in the preparation of liposomes.

The use of cationic phospholipids such as dioleoyl phosphatidylethanolamine, 1,2-dioleoyl-3-trimethylammonium-propane (DOTAP), dioleoyl phosphatidylethanolamine (DOPE), oleic acid (OA), and dimethyl-dioctadecyl ammonium bromide (DDAB) will end up with the preparation of cationic liposomes. These liposomes will be efficient when the loaded drug has a negative charge, resulting in the formation of stable complexes. These complexes between the cationic liposomes and the negatively charged drugs are referred to as lipoplexes [7, 25, 29].

This would be of great interest in siRNA delivery as the siRNA molecules bear a high negative charge, which will be complexed with the cationic liposomes and thus the delivery of the siRNA is enhanced unchanged to the target site. Moreover, the presence of a cationic charge on the surface of the liposome nanoparticles will improve their uptake by the target cells as this will increase their interaction with the negatively charged cell membrane [25].

However, individual toxicity concerns are present and need to be investigated before using these cationic liposomes in drug delivery. Several reports indicated that cationic liposomes would have various dose-dependent toxicities such as hepatotoxicity and pulmonary inflammation [7, 23, 29]. Besides, the cationic liposomes will be recognized by the components of the reticuloendothelial system after their interaction with the serum proteins such opsonins, which will result in the rapid elimination of these nanoparticles by the macrophages [7, 23].

In terms of siRNA delivery, cationic liposomes can form lipoplexes with the negatively charged siRNA, which will result in high encapsulation and transfection efficiencies of these nanoparticles. This has been reported in several studies in the literature. For example, in the work of Chae et al., antisphingosine-1-phosphate receptor-1-siRNA were prepared and complexed with DOTAP-based cationic liposomes. The transfection of these lipoplexes into mouse lung cancer cells resulted in a significant knockdown of the target protein [38]. P-selectin targeted cationic liposomes were also used to form lipoplexes with anti-GADPH siRNA designed to act as a targeted formulation to P-selectin, a cell adhesion molecule. The formed lipoplexes were able to protect the siRNA molecules and successfully deliver siRNA molecules to the target cells with successful GADPH gene silencing [39]. Hattori et al. prepared various cationic liposomes for siRNA encapsulation and transfection using multiple types of cationic lipids. They have reported that the prepared cationic liposomes were encapsulated siRNA molecules targeting the protein kinase N3 (PKN3) and successfully delivered to the lung cancer cells after intravenous administration in a mice model with efficient tumor growth suppression [40]. Tekmira Pharmaceuticals Corporation developed one type of cationic liposomes called stabilized nucleic acid-lipid particles (SNALPs) for systemic siRNA delivery using different kinds of ionizable cationic lipid components. These nanoparticles were reported to have excellent protection of the loaded siRNA from the exogenous conditions and efficient delivery to the target cells [24, 41].

Phospholipids with anionic head groups can be used to prepare liposomes with anionic surface charge. The naturally occurring types of anionic phospholipids are phosphatidylglycerol, phosphatidylinositol, phosphatidic acid, and phosphatidylserine [20]. The anionic liposomes were also significantly investigated as a possible vector for siRNA delivery as they offer a safer alternative than cationic liposomes [10]. Yu et al. prepared phosphatidylethanolamine-based anionic liposomes for the delivery of siRNA targeting VEGF protein expression. The flow cytometry and Western blotting results indicate that the prepared, loaded anionic liposomes can successfully carry and deliver siRNA into the cytoplasm of the target cells to suppress the VEGF protein [42].

However, since the siRNA molecules have a negative charge, this might result in low encapsulation efficiency of the siRNA into anionic liposomes due to charge repulsion. This issue was resolved by many researchers by the use of divalent cationic bridging ions such as calcium ions [23]. This was reported in Kapoor et al. where siRNA was complexed with anionic liposome using calcium ion bridges and resulted in about 99% encapsulation efficiency [43, 44].

Neutral liposomes can also be prepared using neutral phospholipids such as dioleoyl-sn-glycerol-phosphatidylcholine (DOPC), which are proposed to overcome some of the limitations that have been reported for charged liposomes such as toxicity concerns [23, 37]. Several studies reported that the use of neutral liposomes as a drug delivery system would have a high safety profile and less interaction with the opsonin, which will result in increasing the circulation time of these nanoparticles [7, 23, 37]. Because of these advantages, neutral liposomes were preferred by many researchers for the delivery of various therapeutic agents. This was translated into some products in the market, such as the introduction of Doxil, which is an FDA-approved medication composed of neutral liposomes encapsulating the anticancer agent doxorubicin. DaunoXome is another example of an FDA-approved drug for the delivery of daunorubicin using di-stearoyl phosphatidylcholine-based neutral liposomes [45].

Merritt et al. reported using neutral liposome formulations to deliver siRNA to downregulate the expressions of various target proteins such as the EphA2 oncoprotein and cytokine interleukin 8 (IL-8). They have tested this system in vivo on the ovarian cancer-bearing mouse and reported an effective target gene knockdown [46].

2.2 Niosomes

Niosomes or sometimes referred to as nonionic surfactant vesicles (NISV) are another type of nanoparticle drug delivery system that has been studied extensively to deliver various therapeutic agents, including siRNA. The basic structure of niosomes is similar to liposomes in having a membrane bilayer structure encapsulating an aqueous compartment. However, instead of using phospholipids in the bilayer moiety, niosomes are prepared using various types of nonionic surfactants such as sorbitan fatty acid esters (Spans) and polyoxyethylene fatty acid esters (Tweens) [47, 48]. These surfactants are amphiphilic molecules with a hydrophilic head and hydrophobic tails. This structure favors the formation of a membrane bilayer following exposure to an aqueous moiety in the same way as with the liposome formulation. Cholesterol will also be incorporated in the niosomes formulation to add rigidity to the bilayer membrane. Other components, such as charging lipids, can be used to form niosomes to induce a specific charge on the surface of these nanoparticles [49]. These prepared niosomes using nonionic surfactants have superiority over liposomes in terms of cost and stability. The niosome nanoparticles were first used in the cosmetic industry in the 1970s, and then these vesicles gained more attention for the delivery of therapeutic agents [50].

Like liposomes, niosomes are also considered versatile nanoparticles owing to their ability to encapsulate hydrophilic drugs in their aqueous core as well as lipophilic drugs in their membrane bilayer. These favorable characteristics of niosomes were the reason for their investigation by many researchers to deliver various agents such as anticancer agents, antibiotics, nonsteroidal antiinflammatory drugs, and vaccines [45].

In terms of siRNA delivery, there were limited reports for the use of niosomes in these fields as they were used in the cosmetics industry only. For example, in the work of Yang et al. niosome nanoparticles were prepared using Span 80 and DOTAP. They were efficiently used in delivering siRNA targeting the negative regulator of osteoblast differentiation, miR-138, to human mesenchymal stem cells (hMSCs) with successful inhibition of the target protein [51]. However, this used system is not a pure niosomes as it combines between the nonionic surfactant Span 80 and the phospholipid DOTAP. Some researchers referred to this system as spanosomes as it comes in between the liposomes and niosomes.

Obeid et al. developed a useful cationic niosome nanoparticle by microfluidic mixing to deliver siRNA both in vitro and in vivo. Their prepared niosomes were composed of Tween 85 as a nonionic surfactant, cholesterol, and a cationic lipid to induce a positive charge on the surface of these niosomes [52]. The prepared niosomes were first tested for their stability, cytotoxicity, and encapsulation efficiency using fluorescently labeled negative siRNA control. These niosomes were then used to encapsulate and deliver antiluciferase siRNA molecules into mouse melanoma cells in vitro, which are genetically modified to express luciferase enzyme. The prepared complexes between the cationic niosomes and siRNA resulted in a significant cellular uptake and a significant reduction in the luciferase expression. The intratumoral administration of cationic niosomes encapsulating antiluciferase siRNA significantly suppresses the luciferase expression in a nude mice model compared to mice injected with naked siRNA [53].

2.3 Micelles

Micelles are another type of nanoparticles that are prepared using amphiphilic molecules. However, in contrast to liposomes and niosomes, the amphiphilic molecules in the structure of the micelles will arrange as a monolayer with a hydrophobic core consisting of the lipophilic tails and a hydrophilic surface composed of the hydrophilic heads exposed to the aqueous environment. These nanoparticles have a spherical, rod, or ellipsoidal shape depending on their composition. To prepare a specific type of micelles, the concentration of the amphiphilic molecules must be above the critical micelle concentration (CMC) over which the micelles will start to form [54, 55].

Micelles were also investigated as a potential carrier for siRNA delivery into different targets, as shown in Fig. 4. The use of micelles in this field will be based on the formation of complexes between the micelles and the siRNA molecules as these hydrophilic molecules cannot be impeded in

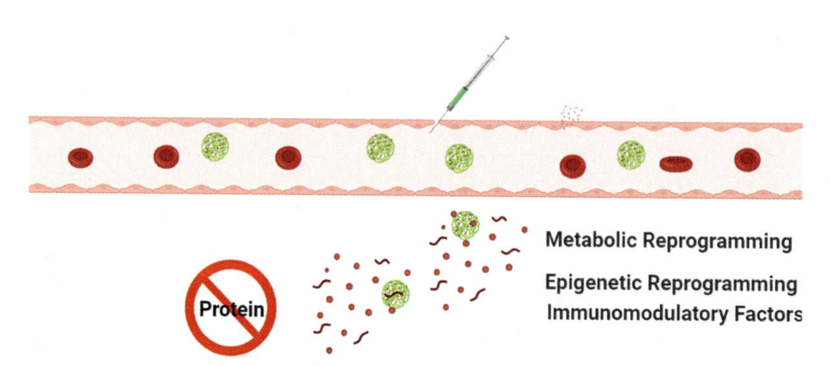

FIG. 4 Illustration of the direct delivery of siRNA molecules through the injection to the circulatory system and their potential mode of action from metabolic reprogramming of the cancerous cell, cellular epigenetic, and immunomodulatory factors target; alternatively, siRNA can target protein synthesis production machinery.

the hydrophobic core of the micelles. This means a cationic surface charge of the micelles is needed to aid in the complex formation with siRNA.

Hoang et al. designed polydiacetylene micelles using different amine-terminated amphiphiles, and the prepared micelles were investigated for their ability to complex with siRNA and deliver its payload to the target cells. They have reported that the efficacy of micelles in siRNA delivery varies depending on the cationic head of the micelles [56].

Zhong et al. prepared a type of micelle called DMP hybrid micelle by modifying the MPEG-PCL copolymer using DOTAP to create cationic micelles. The prepared micelles were then complexed with siRNA targeting the C26 colon cancer cells. In vitro transfection indicated that the DMP cationic micelles were able to successfully deliver siRNA into the target cells and inhibit the growth of C26 colon cancer cells. Moreover, these complexes were administered in vivo through intertumoral administration and successfully suppress the subcutaneous tumor in the mice model [57].

2.4 Dendrimers

Dendrimers are branched three-dimensional and polymeric-based macromolecules that have been employed extensively as a drug delivery system for therapeutic agents, including nucleic acids such as siRNA. These molecules are composed of highly branched polymers that originate from a center core molecule. Depending on the type of polymer and the level of branching, dendrimers can have specific unique properties in terms of having controlled size and low particle size distribution. Moreover, the chemistry of dendrimers allows for the preparation of various types of nanoparticles with different functional groups. This represents an advantage when using dendrimers for drug and nucleic acid delivery, especially for active targeting [58].

In terms of siRNA delivery, the unique chemical and physical properties of dendrimers allow for the successful delivery of various siRNA molecules into specific targets with promising outcomes. Dendrimers can carry the siRNA molecules through complexation where the cationic functional groups on the surface of the dendrimer will form an electrostatic complex with the opposing phosphorus groups of the siRNA to protect from nuclease activities and the subsequent degradation [59].

Several researchers have investigated these nanoparticles for siRNA delivery to provide an alternative to the commonly used transfection nanoparticles such as liposomes. For example, Dong et al. developed a targeted amphiphilic dendrimer for siRNA delivery to knock down the expression of Hsp27, which is one of the proteins that are involved in the proliferation and drug resistance by the prostate cancer cells in both in vitro and in vivo models. The formed targeted siRNA-dendrimers complexes resulted in a significant reduction (˃90%) in the Hsp27 expression. This was associated with a significant reduction in tumor growth in the in vivo model [60].

2.5 Gold nanoparticles

Gold nanoparticles (from now on AuNPs) are shown to be quite essential for the delivery of nucleic acid [61]. Conversely, there is anything to draw from the staggering amount of literature in this area; there is no universal siRNA delivery using AuNPs. In comparison, one of the key benefits of using AuNPs would be that the specification of the most suitable size, structure, external surface, and functioning method within each task may be planned precisely for the desired application and route of administration. Those are just the essential explanations for siRNA facilitated AuNP distribution, which have been addressed in the following paragraphs. These considerations must be considered for the development of AuNPs as siRNA-mediated delivery. A range of synthetic routes would be used in the production of AuNPs with wide variety of sizes, and morphologies are also essential variables in the assessment of their very own optical characteristics, molecular associations, and biological distribution characteristics. That prevention of toxic substances throughout the AuNP synthesis is another critical factor to remember since the rest remnants of the AuNP product are always correlated with the exterior: impurities that are limiting to their applicability in biological research may also be undermined. While it can appear to be a straightforward variable to specify the proper form of AuNP, key factors, such as yield and flexible nanoparticles, have a crucial impact on the actual clinical environment. Many traditional synthetic methods can accomplish nearly any form of AuNP structure and the AuNP scale can easily be managed in many of these cases by adjusting one of several critical experimental synthesis parameters, for example, reagent level, reaction times, pH-resolution, potential reductions, aqueous ionic resistance, so on and so forth [62, 63].

Moreover, such processes are significantly modified by the AuNP exterior protective coating. In particular, AuNPs are internalizing in a size-dependent manner through multiple cellular pathways [63, 64]. Thus, strictly speaking, fewer than the larger counterparts (pieces of the same structure and different size) are more likely to be possibly internalized and generally cause massive toxicity [65]. Once AuNPs are subcutaneously delivered, they would continue in the circulation to accomplish their target location of intervention and overcome the immune response elimination and renal clearance. Glomerulus (in the kidneys) pore size is roughly 8 nm, so to prevent kidney clearance for drug delivery, nanomaterials tend to be greater than 20 nm in diameter [66]. A wide variety of AuNP sizes and shapes is already documented to be available for siRNA delivery. However, the AuNP shape must not be chosen casually, since features such as cellular internalization and cellular distribution are not to be taken lightly, siRNA power loading and optical characteristics can differ significantly. While AuNP structure and size are important factors that affect tissue and cellular interactions, the functionality of siRNA can alter the two factors. This is why studies have compared various types of AuNPs that must be conducted out after siRNA functionalization like Yue et al. [67]. The cells in U87 for the gold 50nm nanospheres and 40nm nanostars were shown to attain much greater cellular uptake efficiency than the 13nm spheres. Fortunately, after 24h, more massive particles were released from endosomes (40nm stars and 50nm spheres) and spread worldwide. Cytoplasm stayed in the endosomes, while the 13nm spheres indicated that bigger nanomaterials were more vital targets for transmission of siRNA. As already stated, the unique optical properties are among the key advantages of AuNPs for siRNA distribution. These optical properties are a function of consistent oscillations of conducting electrons on the metal surface attributable to electromagnetic radiation enthusiasm on the metal-dielectric interface, called SPR [68].

Even though literature currently records the synthesis in AuNPs in different forms, their use for siRNA distribution is amazingly sparse. Gold nanocrystals or nanorods are the most widely used for in vivo siRNA dissemination though other essential experiments have been carried out lately using gold nanostructures and gold nanoshells in mice animal models [69–71]. Gold nanospheres, nanorods, and nanoshells are the most widely examined in vitro delivery studies; nevertheless, gold nanoclusters, dendritic nanomaterials, and nanoflower were also documented [72–74].

After the first production of siRNA facilitated by AuNP in 2006 [75], the "functionality" (i.e., combine or conjugate) AuNPs with siRNA has been used for an extensive list of strategies. However, the inference of the most-effective siRNA functionalization technology for any required loading presents considerable challenges, owing to the wide variety of AuNP varieties, practical methods, siRNA-targeted genes, cell lines, and the analytical techniques used to detect siRNA operation.

Covalent or electrostatic surface modification techniques for integrating siRNAs mostly on the gold surface can be used as anchoring points as shown in Fig. 3. In such alterations to gold nanoparticles, available thiol groups could be used to bind selected molecules or stimulation-responsive compounds to regulated siRNA release. Giljohann et al. added multipurpose siRNA complexes to the AuNP interface by thiol groups, which improved the half-life and durability over free siRNA many folds. Such complexity of siRNA-AuNP culminated in an in vitro prototype that significantly decomposes the desired gene [76]. Attaching PEIs to the AuNP nucleus would cationized nanomaterials, which will make them an appealing delivery method for siRNA, Song et al. [77] report electrostatic interactions tightly connect AuNPs surrounded by PEIs to siRNA. The GFP production in MDA-MB-435 cells was downregulated by siRNA provided by this method. Besides, siRNA-PEI-capped AuNPs were substantially knock down and improved cell proliferation and apoptosis of the cells by inducing the polo-like kinase 1 (PLK1) gene. Such results revealed that AuNPs with cationic surface coatings have improved properties to be utilized as a unique gene delivery platform, avoiding cellular cytotoxicity entirely (Fig. 5).

In order to achieve the recombination of the supplementary nuclear base-pairing system, Son et al. [78] created an exquisite RNAi-AuNP nanoconstruct of different geometrical arrangements, which displayed the exact recombination and isolation of a defined set of therapeutically assigned siRNA into AuNP. Moreover, the RGD modulation to dendrimer-trapped AuNPs was an exceptionally reliable and efficient gene delivery to stem cells by using the feature of mesenchymal stem cells (MSC) as a high affinity for the relation of arginine-glycine-aspartic (RGD) ligand. This confirmed that even a dendritic vector with a remarkable ability to connect to the cell surface by integral receptors and to improve dendritic density confirmation, enabling effective and defined stem cell gene delivery applications, was developed following RGD ligand and AuNPs. The results of this project are available [79]. In addition, poly(thymine)-functional AuNP (AuNP-p(T)-DNA) is yet another gene technique. Each gene of interest was injected individually into the BGH (P(A)) signal plasmid vector pcDNA6; the poly(A) tail conjugated to the AuNPs can be hybridized with poly(T) oligonucleotide, and thus enhancing the corresponding protein synthesis [80, 81].

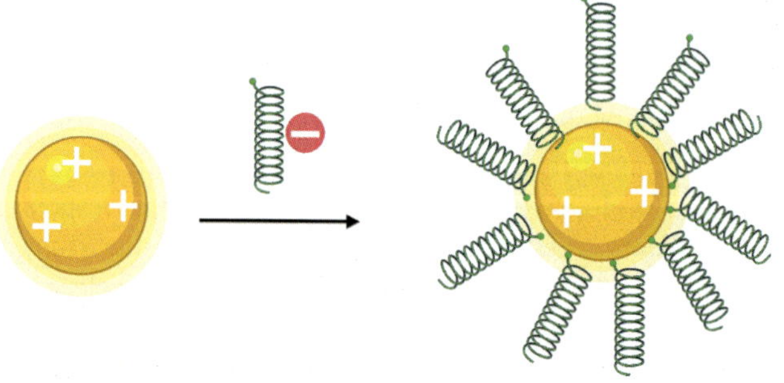

FIG. 5 Schematic illustration of siRNA electrostatically adsorbed to the cationic gold nanoparticles.

Furthermore, the integration of siRNA has been dramatically improved by Moller et al. when modifying MSNs with block copolymer comprising positively charged amino acids and oleic acids. The framework of nanomaterials demonstrated better siRNA deliveries and successful endosomal escape from positive amino acids and oil acid chains. Even with low MSN-carry concentrations of siRNA (2.5 μg MSN comprising 32 pM siRNA) per usual, transfection efficiencies in KB cells were 90%. Many of these methods are being tested with MSNs to achieve improved clinical effectiveness for siRNA. These methods can be translated for the treatment of cancer [82].

AuNPs may also be used to promote the effective distribution of siRNA. As just an illustrative example, the effective distribution of siRNA in the cytoplasm to a glutathione reduction condition was demonstrated by the functionality of the amine AuNPs complexed by SiRNA-PEG moieties that include disulfide interconnection. The released siRNA had demonstrated an effective knockdown of the target gene, while the transporter had low cytotoxicity [83]. AuNPs have reportedly used their photothermal capabilities in the supply of siRNA, as defined in many published research [84, 85]. A close infrared (NIR) laser irradiance is performed near the resonant frequency of the gold nanoshell (800 nm) by the sustained delivery of siRNA trapped into poly-L-lysine (PLL) epilayer. They reported sufficient green fluorescent protein (GFP) gene silencing in H1299 cells, whether either single-stranded antisense DNA or GFP-specific siRNA is delivered by the gold nanoshell-PLL framework [85]. Using NIR radiation in a particular capacity tends to heat the endosomal cavitation locally, allowing siRNA to migrate in the cytoplasm. In melanoma, gene therapy applications, such as structural, stimulatory, and photothermal features of gold-based NPs carriers, are highly beneficial.

3 Clinical trials of siRNA by nanoparticles

The significant advancements of RNAi-based therapy into clinical trials has been rapid. Besides, a variety of disorders, including cancer, respiratory ailments, age-related macular degeneration (ADM), glaucoma, and hypercholesterolemia, have been researched. Utilizing siRNA- and miRNA-related therapies, at least 20 clinical trials have been conducted over the past few years. Though RNAi can be used in many instances as summarizes in Table 1, most clinical trials currently include siRNA technology [95–97].

TABLE 1 siRNA-based clinical trials that are currently under different stages of evaluation for cancer therapy.

siRNA	Construction	Cellular target	Disease	Clinical trial phase	References
Atu027	Cationic lipid-based NPs combining siRNA with 2′-O-Me	Kinase N3 protein (PKN3)	Late-stage of metastatic pancreatic cancer	Phase-I	[86]
CLAA-01	Rondel® NPs	Ribonucleotide reductase subunit M2 (RPM2)	Solid cancer	Phase-I	[87]
ALN-RSV01	Naked siRNA	RSV nucleocapsid	RSV infection	Phase-II	[88]
siG12D LODER	Biodegradable polymer matrix	KRAS	Pancreatic ductal adenocarcinoma, solid cancers	Phase-I	[89]
ALN-VSPO2	Lipid nanoparticles	Liver-related solid cancers	KSP and VEGF	Phase-I	[90]
15NP	Naked siRNA	p53	Acute kidney injury	Phase-I and -II	
DCR-MYC	Lipid-based nanoparticles	MYC gene	Hepatocellular carcinoma; myeloma, solid tumors, non-Hodgkin lymphoma, and pancreatic, neuroendocrine tumors	Phase-I and -II	[91]
TKM-080301	Lipid-based nanoparticles	PLK1	Liver cancer (primary and secondary), neuroendocrine tumors, hepatocellular carcinoma	Phase-I	[92]
siRNA-EphA2-DOPC	Lipid-based nanoparticles	EphA2	Solid cancer	Recruiting stage	[93]
SNS01-T	Polyethyleneimine	eIF5A	Multiple myeloma	Phase-I and -II	[94]

A huge proportion of NP modifications will obstruct RISC connections to siRNA, thereby avoiding the process of interference with RNA. Sometimes these improvements occur between the 3- and 5-in passage and guidance strands, where the following properties have been shown to have significant repercussions: (i) suppress exonuclease degradation [98]; (ii) improve RISC affinity [99]; (iii) help avoid the loading of the payload strand [100]; (iv) suppress innate immunostimulant reaction [101]; and (v) stop miRNA-like responses [101].

However, many clinical studies were investigated the systemic delivery of siRNA particles utilizing various lipid compositions include liposomes, SNALPs, and other lipid substances containing siRNA residues to develop siRNA clusters. Two siRNA compounds developed in the lipid nanoparticles (LNP) were used for the first human clinical cancer studies in treating colon cancer and liver metastases. The liver actively took up these drugs, but cancer treatment of other tissues is not reported [86]. Carbohydrate-siRNA complexes are often used for in vivo hepatocyte targets, particularly *N*-acetyl-galactosamine (GalNAc)-siRNA analogs. A significant effect of siRNAs with triantennary GalNAc was observed on cellular uptake. Transfection studies have reported elevated levels of liver specificities and gene silencing. A potential cellular uptake and improve endosomal escape has also been identified as a vital substitute to anticancer siRNA distribution for peptides-siRNA targeting ligands, in particular cell-penetrating peptides (CPPs) and fusogenic peptides [86, 102].

Davis et al. [103] have generated the very first clinical proof that RNAi could be attained by delivering siRNA with M2 subunit ribonucleotide reductase (RRM2) in such a latest study from phase I trials in patients who suffer from relapsed or recovered solid tumor. Furthermore, in treating cancer for combat inaccessible cancerous cells through systematic management, Atu027 [104] and CALAA-01 [87], produced by Silence Therapeutics and Calando Pharmaceuticals, simultaneously, both have demonstrated the therapeutic ability of RNAi as a novel new class of medicines. CALAA-01 was included in phase I clinical study in 2008 to treat the first patient. In this study, siRNA nanoparticles were intravenously administered for the treatment of solid malignant tumor, targeting protein N3 (PKN3), or RRM2 kinases. Lately, Atu027, siRNA targeted on protein kinase N3 (PKN3), alongside gemcitabine, has been used in a clinical study to combat advanced or metastatic pancreatic cancer [87] (registered at clinicaltrials.gov with the identifier: NCT01808638). Tabernero et al. [86] showed a fascinating clinical approach. Two distinct siRNAs were designed to target two genes [VAGF and KSP (Kinesin spindle Protein)], involved in angiogenesis, which were conceived in this analysis in a related lipid formulation. The study indicated the siRNAs could encourage mRNA cleavage and antitumor activities in mammalian cells. Such a lipid (ALN-RSV) formulation developed by Alnylam Pharmaceuticals has been shown in some cases to regress entirely from tumor progression and endometrial disease.

For the silencing of Myc oncoprotein, related lipid formulation is added and deregulated in more than half of malignant tumors. DCR-MYC has been tested in individuals with advanced cancer types, lymphoma, multiple myeloma, with strong clinical and biochemical outcomes at different dosage levels during a successful multicenter clinical work [105]. Furthermore, A phase 1 clinical trial to suppress the ephrin type-A (EphA2) gene has recently evaluated the use of 1,2-dioleoyl-sn-glycerol-3-phosphatidylcholine liposomes for systemic siRNA distribution. The liposomal formulation has been well accepted in currently ongoing clinical trials [106]. In a different approach, silenced used to incorporate the tumor tissue kernel of a highly specialized biopolymer as a nanoscaffold containing the siRNA moieties. In such a clinical study of the silence of a defectively mutated K-RAS oncogene for pancreatic ductal adenocarcinoma, the LODER polymer matrix has been developed [107]. While in highly developed clinical trials, there are not many siRNAs, and RNAi has become a unique technique for innovative new therapies, as siRNAs are substantially decreased in cancerous cells in the functioning and production of high expression disrupting genes. However, after the RNAi strategies have been applied in vivo, it is therapeutic siRNA agents that are the most impediment to obtaining gene silencing. There are also many problems facing the therapeutic implementation of siRNA therapy in order to achieve adequate doses of healthy target cells and to preserve circulating oligonucleotide stability [105].

It is essential to establish approaches for controlling the delivery including management features and to improve cellular uptake of siRNA delivery approaches. Initial clinical tests showed promising outcomes for both in vivo and in vitro studies. The use of chemotherapy and siRNA combinations and the combined effect of many siRNAs targeting elevated proteins in different metabolic pathways, for example, angiogenesis, telomerase, nucleotide pool, etc., are fascinating and may hold great potential soon.

4 Conclusions

RNA interference has been used as an approach to cancer treatment as it is less harmful than traditional chemotherapeutics. For the treatment of various ailments, such as genetic abnormalities, skin conditions, cardiovascular problems, infectious infections, or tumors, siRNAs have also been documented [108]. If protein-based therapies or comparatively molecules could not adequately be used, the ability to target practically any gene(s) is one of the best features of siRNA for cancer therapy. siRNA provides significant benefits, including excellent safety, higher efficiency, limitless selection of targets, and precision, which is among the most successful therapies for the treatment of cancer. Numerous delivery methodologies have been proposed to resolve the delivery problems of siRNA. The architecture, scale, and composition of such advantageous delivery systems differ tremendously, yet guidelines also exist regarding the features of optimized, targeted delivery. The distribution systems of nanoparticulate should be between 20 and 200 nm particle size, i.e., broad enough but for preventing renal filtration, but small enough just to prevent phagocytosis clearance.

Furthermore, exogenous or endogenous trigger-binding affinities are often effective toward cancerous cells siRNA penetration. Whereas various studies show that siRNA is highly probable and useful in cancer therapy, the maximum capabilities of siRNA remain challenging, yet most siRNA drug delivery is still in preclinical studies. While a great deal of outstanding experimental studies has been established, future research is required to focus on the in vivo safety and efficacy of nanoparticular delivery methods, particularly undesirable cytotoxicity and immune activation, such as composites, cationic lipids, dendrimers, and inorganic NPs. The progress, in conjunction with the development of viable and straightforward procedures for both the development of a load of regulations and clinical trial evaluation, of biocompatible, biodegradable, and healthy environmentally friendly NPs, is still crucial for the clinical use of siRNA-based cancer therapy. To summarize, the cornerstone to siRNA drug development is a successful delivery system. After a substantial advance has been achieved in siRNA drug delivery systems, siRNA is well placed mostly on the drug market, particularly on the market for chemotherapeutic agents.

References

[1] Fire A, Xu S, Montgomery MK, Kostas SA, Driver SE, Mello CC. Potent and specific genetic interference by double-stranded RNA in *Caenorhabditis elegans*. Nature **1998**;*391*(6669):806–11.

[2] Elbashir SM, Harborth J, Lendeckel W, Yalcin A, Weber K, Tuschl T. Duplexes of 21-nucleotide RNAs mediate RNA interference in cultured mammalian cells. Nature **2001**;*411*(6836):494–8.

[3] Miele E, Spinelli GP, Miele E, Di Fabrizio E, Ferretti E, Tomao S, Gulino A. Nanoparticle-based delivery of small interfering RNA: challenges for cancer therapy. Int J Nanomedicine 2012;7:3637–57.

[4] Hammond SM. Dicing and slicing. FEBS Lett 2005;579(26):5822–9.

[5] Obeid MA, Aljabali AA, Amawi H, Abdeljaber SN. Carbon nanotubes in nucleic acids delivery. 6. Whites Science Innovation Ltd; 2020. p. 26–30.

[6] Ameres SL, Martinez J, Schroeder R. Molecular basis for target RNA recognition and cleavage by human RISC. Cell 2007;130(1):101–12.

[7] Ozpolat B, Sood AK, Lopez-Berestein G. Liposomal siRNA nanocarriers for cancer therapy. Adv Drug Deliv Rev 2014;66:110–6.

[8] Matranga C, Tomari Y, Shin C, Bartel DP, Zamore PD. Passenger-strand cleavage facilitates assembly of siRNA into Ago2-containing RNAi enzyme complexes. Cell 2005;123(4):607–20.

[9] Obeid MA, Teeravatcharoenchai T, Connell D, Niwasabutra K, Hussain M, Carter K, Ferro VA. Examination of the effect of niosome preparation methods in encapsulating model antigens on the vesicle characteristics and their ability to induce immune responses. J Liposome Res 2020;1–30.

[10] Tam YYC, Chen S, Cullis PR. Advances in lipid nanoparticles for siRNA delivery. Pharmaceutics 2013;5(3):498–507.

[11] Zhou Y, Zhang C, Liang W. Development of RNAi technology for targeted therapy—a track of siRNA based agents to RNAi therapeutics. J Control Release 2014;193:270–81.

[12] Resnier P, Montier T, Mathieu V, Benoit J-P, Passirani C. A review of the current status of siRNA nanomedicines in the treatment of cancer. Biomaterials 2013;34(27):6429–43.

[13] Aljabali AA, Al Zoubi MS, Alzoubi L, Al-Batanyeh KM, Obeid MA, Tambwala MM. Chemical engineering of protein cages and nanoparticles for pharmaceutical applications. In: Nanofabrication for smart nanosensor applications. Elsevier; 2020. p. 415–33.

[14] Morrissey DV, Blanchard K, Shaw L, Jensen K, Lockridge JA, Dickinson B, McSwiggen JA, Vargeese C, Bowman K, Shaffer CS. Activity of stabilized short interfering RNA in a mouse model of hepatitis B virus replication. Hepatology 2005;41(6):1349–56.

[15] Zimmermann TS, Lee AC, Akinc A, Bramlage B, Bumcrot D, Fedoruk MN, Harborth J, Heyes JA, Jeffs LB, John M. RNAi-mediated gene silencing in non-human primates. Nature 2006;441(7089):111–4.

[16] Dykxhoorn D, Palliser D, Lieberman J. The silent treatment: siRNAs as small molecule drugs. Gene Ther 2006;13(6):541–52.

[17] Soutschek J, Akinc A, Bramlage B, Charisse K, Constien R, Donoghue M, Elbashir S, Geick A, Hadwiger P, Harborth J. Therapeutic silencing of an endogenous gene by systemic administration of modified siRNAs. Nature 2004;432(7014):173–8.

[18] Kim WJ, Kim SW. Efficient siRNA delivery with non-viral polymeric vehicles. Pharm Res 2009;26(3):657–66.

[19] Aljabali AA, Obeid MA, Amawi HA, Rezigue MM, Hamzat Y, Satija S, Tambwala MM. Application of nanomaterials in the diagnosis and treatment of genetic disorders. In: Applications of nanomaterials in human health. Springer; 2020. p. 125–46.

[20] Draz MS, Fang BA, Zhang P, Hu Z, Gu S, Weng KC, Gray JW, Chen FF. Nanoparticle-mediated systemic delivery of siRNA for treatment of cancers and viral infections. Theranostics 2014;4(9):872.

[21] Tagalakis AD, Do Hyang DL, Bienemann AS, Zhou H, Munye MM, Saraiva L, McCarthy D, Du Z, Vink CA, Maeshima R. Multifunctional, self-assembling anionic peptide-lipid nanocomplexes for targeted siRNA delivery. Biomaterials 2014;35(29):8406–15.

[22] Díaz MR, Vivas-Mejia PE. Nanoparticles as drug delivery systems in cancer medicine: emphasis on RNAi-containing nanoliposomes. Pharmaceuticals 2013;6(11):1361–80.

[23] Zhang J, Li X, Huang L. Non-viral nanocarriers for siRNA delivery in breast cancer. J Control Release 2014.

[24] Gomes-da-Silva LGC, Fonseca NA, Moura V, Pedroso de Lima MC, Simões SR, Moreira JON. Lipid-based nanoparticles for siRNA delivery in cancer therapy: paradigms and challenges. Acc Chem Res 2012;45(7):1163–71.

[25] Laouini A, Jaafar-Maalej C, Limayem-Blouza I, Sfar S, Charcosset C, Fessi H. Preparation, characterization and applications of liposomes: state of the art. J Colloid Sci Biotechnol 2012;1(2):147–68.

[26] Obeid MA, Khadra I, Albaloushi A, Mullin M, Alyamani H, Ferro VA. Microfluidic manufacturing of different niosomes nanoparticles for curcumin encapsulation: physical characteristics, encapsulation efficacy, and drug release. Beilstein J Nanotechnol 2019;10(1):1826–32.

[27] David S, Pitard B, Benoît J-P, Passirani C. Non-viral nanosystems for systemic siRNA delivery. Pharmacol Res 2010;62(2):100–14.
[28] Al Qaraghuli MM, Obeid MA, Aldulaimi O, Ferro VA. Control of malaria by bio-therapeutics and drug delivery systems. J Med Microbiol Diagn 2017;6(3):1–8.
[29] Lee J-M, Yoon T-J, Cho Y-S. Recent developments in nanoparticle-based siRNA delivery for cancer therapy. Biomed Res Int 2013;2013.
[30] Yingchoncharoen P, Kalinowski DS, Richardson DR. Lipid-based drug delivery systems in cancer therapy: what is available and what is yet to come. Pharmacol Rev 2016;68(3):701–87.
[31] Gregoriadis G. The carrier potential of liposomes in biology and medicine. N Engl J Med 1976;295(14):765–70.
[32] Gregoriadis G. Engineering liposomes for drug delivery: progress and problems. Trends Biotechnol 1995;13(12):527–37.
[33] Allen TM, Cullis PR. Liposomal drug delivery systems: from concept to clinical applications. Adv Drug Deliv Rev 2013;65(1):36–48.
[34] Al Qaraghuli MM, Alzahrani AR, Niwasabutra K, Obeid MA, Ferro VA. Where traditional drug discovery meets modern technology in the quest for new drugs. Ann Pharmacol Pharm 2017;2(11):1–5.
[35] Sharma AK, Prasher P, Aljabali AA, Mishra V, Gandhi H, Kumar S, Mutalik S, Chellappan DK, Tambuwala MM, Dua KJDD, Research T. Emerging era of "Somes": polymersomes as versatile drug delivery carrier for cancer diagnostics and therapy; 2020.
[36] Aljabali AA, Al Zoubi MS, Al-Batanyeh KM, Al-Radaideh A, Obeid MA, Al Sharabi A, Alshaer W, AbuFares B, Al-Zanati T, Tambuwala MM. Gold-coated plant virus as computed tomography imaging contrast agent. Beilstein J Nanotechnol 2019;10(1):1983–93.
[37] Buyens K, De Smedt SC, Braeckmans K, Demeester J, Peeters L, van Grunsven LA, de Mollerat du Jeu X, Sawant R, Torchilin V, Farkasova K. Liposome based systems for systemic siRNA delivery: stability in blood sets the requirements for optimal carrier design. J Control Release 2012;158(3):362–70.
[38] Chae S-S, Paik J-H, Furneaux H, Hla T. Requirement for sphingosine 1–phosphate receptor-1 in tumor angiogenesis demonstrated by in vivo RNA interference. J Clin Invest 2004;114(8):1082–9.
[39] Constantinescu CA, Fuior EV, Rebleanu D, Deleanu M, Simion V, Voicu G, Escriou V, Manduteanu I, Simionescu M, Calin M. Targeted transfection using PEGylated cationic liposomes directed towards P-selectin increases siRNA delivery into activated endothelial cells. Pharmaceutics 2019;11(1):47.
[40] Hattori Y, Nakamura M, Takeuchi N, Tamaki K, Shimizu S, Yoshiike Y, Taguchi M, Ohno H, Ozaki K-I, Onishi H. Effect of cationic lipid in cationic liposomes on siRNA delivery into the lung by intravenous injection of cationic lipoplex. J Drug Target 2019;27(2):217–27.
[41] Shrivastava G, Bakshi HA, Aljabali AA, Mishra V, Hakkim FL, Charbe NB, Kesharwani P, Chellappan DK, Dua K, Tambuwala MMJCDD. Nucleic acid aptamers as a potential nucleus targeted drug delivery system. Curr Drug Deliv 2020;17(2):101–11.
[42] Yu Q, Zhang B, Zhou Y, Ge Q, Chang J, Chen Y, Zhang K, Peng D, Chen W. Co-delivery of gambogenic acid and VEGF-siRNA with anionic liposome and polyethylenimine complexes to HepG2 cells. J Liposome Res 2019;29(4):322–31.
[43] Kapoor M, Burgess DJ. Efficient and safe delivery of siRNA using anionic lipids: formulation optimization studies. Int J Pharm 2012;432(1):80–90.
[44] Alyamani H, Obeid MA, Tate RJ, Ferro VA. Exosomes: fighting cancer with cancer. Ther Deliv 2019;10(1):37–61.
[45] Obeid MA, Tate RJ, Mullen AB, Ferro VA. Lipid-based nanoparticles for cancer treatment. In: Lipid nanocarriers for drug targeting. Elsevier; 2018. p. 313–59.
[46] Merritt WM, Lin YG, Spannuth WA, Fletcher MS, Kamat AA, Han LY, Landen CN, Jennings N, De Geest K, Langley RR. Effect of interleukin-8 gene silencing with liposome-encapsulated small interfering RNA on ovarian cancer cell growth. J Natl Cancer Inst 2008;100(5):359–72.
[47] Obeid MA, Khadra I, Mullen AB, Tate RJ, Ferro VA. The effects of hydration media on the characteristics of non-ionic surfactant vesicles (NISV) prepared by microfluidics. Int J Pharm 2017;516(1–2):52–60.
[48] Obeid MA, Gebril AM, Tate RJ, Mullen AB, Ferro VA. Comparison of the physical characteristics of monodisperse non-ionic surfactant vesicles (NISV) prepared using different manufacturing methods. Int J Pharm 2017;521(1–2):54–60.
[49] Marianecci C, Di Marzio L, Rinaldi F, Celia C, Paolino D, Alhaique F, Esposito S, Carafa M. Niosomes from 80s to present: the state of the art. Adv Colloid Interface Sci 2014;205:187–206.
[50] Uchegbu IF, Vyas SP. Non-ionic surfactant based vesicles (niosomes) in drug delivery. Int J Pharm 1998;172(1):33–70.
[51] Yang C, Gao S, Song P, Dagnæs-Hansen F, Jakobsen M, Kjems J. Theranostic niosomes for efficient siRNA/MicroRNA delivery and activatable near-infrared fluorescent tracking of stem cells. ACS Appl Mater Interfaces 2018;10(23):19494–503.
[52] Obeid MA, Elburi A, Young LC, Mullen AB, Tate RJ, Ferro VA. Formulation of nonionic surfactant vesicles (NISV) prepared by microfluidics for therapeutic delivery of siRNA into cancer cells. Mol Pharm 2017;14(7):2450–8.
[53] Obeid MA, Dufès C, Somani S, Mullen AB, Tate RJ, Ferro VA. Proof of concept studies for siRNA delivery by nonionic surfactant vesicles: in vitro and in vivo evaluation of protein knockdown. J Liposome Res 2018;1–10.
[54] Arleth L, Ashok B, Onyuksel H, Thiyagarajan P, Jacob J, Hjelm RP. Detailed structure of hairy mixed micelles formed by phosphatidylcholine and PEGylated phospholipids in aqueous media. Langmuir 2005;21(8):3279–90.
[55] Feng L, Mumper RJ. A critical review of lipid-based nanoparticles for taxane delivery. Cancer Lett 2013;334(2):157–75.
[56] Hoang M-D, Vandamme M, Kratassiouk G, Pinna G, Gravel E, Doris E. Tuning the cationic interface of simple polydiacetylene micelles to improve siRNA delivery at the cellular level. Nanoscale Adv 2019;1(11):4331–8.
[57] Lu Y, Zhong L, Jiang Z, Pan H, Zhang Y, Zhu G, Bai L, Tong R, Shi J, Duan X. Cationic micelle-based siRNA delivery for efficient colon cancer gene therapy. Nanoscale Res Lett 2019;14(1):193.
[58] Dufes C, Uchegbu IF, Schätzlein AG. Dendrimers in gene delivery. Adv Drug Deliv Rev 2005;57(15):2177–202.
[59] Lamberti G, Barba AA. Drug delivery of siRNA therapeutics. Multidisciplinary Digital Publishing Institute; 2020.
[60] Dong Y, Yu T, Ding L, Laurini E, Huang Y, Zhang M, Weng Y, Lin S, Chen P, Marson D. A dual targeting dendrimer-mediated siRNA delivery system for effective gene silencing in cancer therapy. J Am Chem Soc 2018;140(47):16264–74.
[61] Ding Y, Jiang Z, Saha K, Kim CS, Kim ST, Landis RF, Rotello VM. Gold nanoparticles for nucleic acid delivery. Mol Ther 2014;22(6):1075–83.
[62] Piella J, Bastús NG, Puntes V. Size-controlled synthesis of sub-10-nanometer citrate-stabilized gold nanoparticles and related optical properties. Chem Mater 2016;28(4):1066–75.

[63] Vigderman L, Zubarev ER. High-yield synthesis of gold nanorods with longitudinal SPR peak greater than 1200 nm using hydroquinone as a reducing agent. Chem Mater 2013;25(8):1450–7.
[64] Cheng X, Tian X, Wu A, Li J, Tian J, Chong Y, Chai Z, Zhao Y, Chen C, Ge C. Protein corona influences cellular uptake of gold nanoparticles by phagocytic and nonphagocytic cells in a size-dependent manner. ACS Appl Mater Interfaces 2015;7(37):20568–75.
[65] Shang L, Nienhaus K, Nienhaus G. Engineered nanoparticles interacting with cells: size matters. J Nanobiotechnol 2014;12(1):5.
[66] Kanasty R, Dorkin JR, Vegas A, Anderson D. Delivery materials for siRNA therapeutics. Nat Mater 2013;12(11):967–77.
[67] Wang Y, Black KC, Luehmann H, Li W, Zhang Y, Cai X, Wan D, Liu S-Y, Li M, Kim P. Comparison study of gold nanohexapods, nanorods, and nanocages for photothermal cancer treatment. ACS Nano 2013;7(3):2068–77.
[68] Liz-Marzán LM. Tailoring surface plasmons through the morphology and assembly of metal nanoparticles. Langmuir **2006**;*22*(1):32–41.
[69] Wang B-K, Yu X-F, Wang J-H, Li Z-B, Li P-H, Wang H, Song L, Chu PK, Li C. Gold-nanorods-siRNA nanoplex for improved photothermal therapy by gene silencing. Biomaterials **2016**;*78*:27–39.
[70] Wang S, Tian Y, Tian W, Sun J, Zhao S, Liu Y, Wang C, Tang Y, Ma X, Teng Z. Selectively sensitizing malignant cells to photothermal therapy using a CD44-targeting heat shock protein 72 depletion nanosystem. ACS Nano 2016;10(9):8578–90.
[71] Wang Z, Li S, Zhang M, Ma Y, Liu Y, Gao W, Zhang J, Gu Y. Laser-triggered small interfering rna releasing gold nanoshells against heat shock protein for sensitized photothermal therapy. Adv Sci 2017;4(2):1600327.
[72] Rahme K, Nolan MT, Doody T, McGlacken GP, Morris MA, O'Driscoll C, Holmes JD. Highly stable PEGylated gold nanoparticles in water: applications in biology and catalysis. RSC Adv 2013;3(43):21016–24.
[73] Huang X, Pallaoro A, Braun GB, Morales DP, Ogunyankin MO, Zasadzinski J, Reich NO. Modular plasmonic nanocarriers for efficient and targeted delivery of cancer-therapeutic siRNA. Nano Lett 2014;14(4):2046–51.
[74] Wang Z, Wu H, Shi H, Wang M, Huang C, Jia N. A novel multifunctional biomimetic Au@BSA nanocarrier as a potential siRNA theranostic nanoplatform. J Mater Chem B 2016;4(14):2519–26.
[75] Oishi M, Nakaogami J, Ishii T, Nagasaki Y. Smart PEGylated gold nanoparticles for the cytoplasmic delivery of siRNA to induce enhanced gene silencing. Chem Lett 2006;35(9):1046–7.
[76] Giljohann DA, Seferos DS, Prigodich AE, Patel PC, Mirkin CA. Gene regulation with polyvalent siRNA–nanoparticle conjugates. J Am Chem Soc 2009;131(6):2072–3.
[77] Song WJ, Du JZ, Sun TM, Zhang PZ, Wang J. Gold nanoparticles capped with polyethyleneimine for enhanced siRNA delivery. Small 2010;6(2):239–46.
[78] Son S, Kim N, You DG, Yoon HY, Yhee JY, Kim K, Kwon IC, Kim SH. Antitumor therapeutic application of self-assembled RNAi-AuNP nanoconstructs: combination of VEGF-RNAi and photothermal ablation. Theranostics 2017;7(1):9.
[79] Kong L, Alves CS, Hou W, Qiu J, Möhwald H, Tomás H, Shi X. RGD peptide-modified dendrimer-entrapped gold nanoparticles enable highly efficient and specific gene delivery to stem cells. ACS Appl Mater Interfaces 2015;7(8):4833–43.
[80] Chan KP, Chao S-H, Kah JCY. Universal mRNA translation enhancement with gold nanoparticles conjugated to oligonucleotides with a poly (T) sequence. ACS Appl Mater Interfaces 2018;10(6):5203–12.
[81] Mehta M, Dhanjal DS, Paudel KR, Singh B, Gupta G, Rajeshkumar S, Thangavelu L, Tambuwala MM, Bakshi HA, Chellappan DK. Cellular signalling pathways mediating the pathogenesis of chronic inflammatory respiratory diseases: an update. J Inflamm 2020;1–23.
[82] Möller K, Müller K, Engelke H, Bräuchle C, Wagner E, Bein T. Highly efficient siRNA delivery from core–shell mesoporous silica nanoparticles with multifunctional polymer caps. Nano Adv 2016;8(7):4007–19.
[83] Lee SH, Bae KH, Kim SH, Lee KR, Park TG. Amine-functionalized gold nanoparticles as non-cytotoxic and efficient intracellular siRNA delivery carriers. Int J Pharm 2008;364(1):94–101.
[84] Braun GB, Pallaoro A, Wu G, Missirlis D, Zasadzinski JA, Tirrell M, Reich NO. Laser-activated gene silencing via gold nanoshell–siRNA conjugates. ACS Nano 2009;3(7):2007–15.
[85] Huschka R, Barhoumi A, Liu Q, Roth JA, Ji L, Halas N. Gene silencing by gold nanoshell-mediated delivery and laser-triggered release of antisense oligonucleotide and siRNA. ACS Nano 2012;6(9):7681–91.
[86] Tabernero J, Shapiro GI, LoRusso PM, Cervantes A, Schwartz GK, Weiss GJ, Paz-Ares L, Cho DC, Infante JR, Alsina M. First-in-humans trial of an RNA interference therapeutic targeting VEGF and KSP in cancer patients with liver involvement. Cancer Discov 2013;3(4):406–17.
[87] Davis ME, Zuckerman JE, Choi CHJ, Seligson D, Tolcher A, Alabi CA, Yen Y, Heidel JD, Ribas A. Evidence of RNAi in humans from systemically administered siRNA via targeted nanoparticles. Nature 2010;464(7291):1067–70.
[88] Gottlieb J, Zamora MR, Hodges T, Musk A, Sommerwerk U, Dilling D, Arcasoy S, DeVincenzo J, Karsten V, Shah S. ALN-RSV01 for prevention of bronchiolitis obliterans syndrome after respiratory syncytial virus infection in lung transplant recipients. J Heart Lung Transplant 2016;35(2):213–21.
[89] Varghese AM, Ang C, Dimaio CJ, Javle MM, Gutierrez M, Yarom N, et al. A phase II study of siG12D-LODER in combination with chemotherapy in patients with locally advanced pancreatic cancer (PROTACT). Proc Am Soc Clin Oncol 2020;38(15).
[90] Large DE, Soucy JR, Hebert J, Auguste DT. Advances in receptor-mediated, tumor-targeted drug delivery. Adv Ther 2019;2(1):1800091.
[91] Miller AJ, Chang A, Cunningham PN. Chronic microangiopathy due to DCR-MYC, a myc-targeted short interfering RNA. Am J Kidney Dis 2020;75(4):513–6.
[92] El Dika I, Lim HY, Yong WP, Lin CC, Yoon JH, Modiano M, Freilich B, Choi HJ, Chao TY, Kelley RK. An open-label, multicenter, phase I, dose escalation study with phase II expansion cohort to determine the safety, pharmacokinetics, and preliminary antitumor activity of intravenous TKM-080301 in subjects with advanced hepatocellular carcinoma. Oncologist 2019;24(6):747.
[93] Ozcan G, Ozpolat B, Coleman RL, Sood AK, Lopez-Berestein G. Preclinical and clinical development of siRNA-based therapeutics. Adv Drug Deliv Rev 2015;87:108–19.
[94] Chou S-T, Mixson AJ. siRNA nanoparticles: the future of RNAi therapeutics for oncology? Nanomedicine 2014;9(15):2251–4.
[95] Class, I.; USPC, A. K. F. Delivery, engineering and optimization of systems, methods and compositions for sequence manipulation and therapeutic applications. Inventors: Feng Zhang (Cambridge, MA, US) Feng Zhang (Cambridge, MA, US) Randall Jeffrey Platt (Cambridge, MA, US) Guoping Feng (Cambridge, MA, US) Yang Zhou (Cambridge, MA, US); 2015.

[96] Xin Y, Huang M, Guo WW, Huang Q, Zhen Zhang L, Jiang G. Nano-based delivery of RNAi in cancer therapy. Mol Cancer 2017;16(1):1–9.

[97] Chakraborty C, Sharma AR, Sharma G, Doss CGP, Lee S-S. Therapeutic miRNA and siRNA: moving from bench to clinic as next generation medicine. Mol Ther Nucleic Acids 2017;8:132–43.

[98] Bramsen JB, Pakula MM, Hansen TB, Bus C, Langkjær N, Odadzic D, Smicius R, Wengel SL, Chattopadhyaya J, Engels JW. A screen of chemical modifications identifies position-specific modification by UNA to most potently reduce siRNA off-target effects. Nucleic Acids Res 2010;38(17):5761–73.

[99] Terrazas M, Ocampo SM, Perales JC, Marquez VE, Eritja R. Effect of north bicyclo [3.1. 0] hexane 2′-deoxy-pseudosugars on RNA interference: a novel class of siRNA modification. ChemBioChem 2011;12(7):1056–65.

[100] Alagia A, Jorge AF, Aviñó A, Cova TF, Crehuet R, Grijalvo S, Pais AA, Eritja R. Exploring PAZ/3′-overhang interaction to improve siRNA specificity. A combined experimental and modeling study. Chem Sci 2018;9(8):2074–86.

[101] Eberle F, Gießler K, Deck C, Heeg K, Peter M, Richert C, Dalpke AH. Modifications in small interfering RNA that separate immunostimulation from RNA interference. J Immunol 2008;180(5):3229–37.

[102] Grijalvo S, Alagia A, Jorge AF, Eritja R. Covalent strategies for targeting messenger and non-coding RNAs: an updated review on siRNA, miRNA and antimiR conjugates. Genes 2018;9(2):74.

[103] Davis ME. The first targeted delivery of siRNA in humans via a self-assembling, cyclodextrin polymer-based nanoparticle: from concept to clinic. Mol Pharm 2009;6(3):659–68.

[104] Schultheis B, Strumberg D, Santel A, Gebhardt F, Khan M, Keil O, et al. First-in-human phase I study of the liposomal RNAi therapeutic Atu027 in patients with advanced cancer. Proc Am Soc Clin Oncol 2013;32(36):4141–8.

[105] Tolcher AW, Papadopoulos KP, Patnaik A, Rasco DW, Martinez D, Wood DL, et al. Safety and activity of DCR-MYC, a first-in-class dicer-substrate small interfering RNA (DsiRNA) targeting MYC, in a phase I study in patients with advanced solid tumors. Proc Am Soc Clin Oncol 2015;33(15). 11006-11006.

[106] Wagner MJ, Mitra R, McArthur MJ, Baze W, Barnhart K, Wu SY, Rodriguez-Aguayo C, Zhang X, Coleman RL, Lopez-Berestein G. Preclinical mammalian safety studies of EPHARNA (DOPC nanoliposomal EphA2-targeted siRNA). Mol Cancer Ther 2017;16(6):1114–23.

[107] Golan T, Khvalevsky EZ, Hubert A, Gabai RM, Hen N, Segal A, et al. RNAi therapy targeting KRAS in combination with chemotherapy for locally advanced pancreatic cancer patients. Oncotarget 2015;6(27):24560.

[108] Aljabali AA, Obeid MA, Amawi HA, Rezigue MM, Hamzat Y, Satija S, Tambuwala MM. Correction to: Application of nanomaterials in the diagnosis and treatment of genetic disorders. In: Applications of nanomaterials in human health. Springer; 2020. p. C1.

Chapter 36

Targeting micro-ribonucleic acid (miRNA) in cancer using advanced drug delivery systems

Farrukh Zeeshan
Department of Pharmaceutical Technology, School of Pharmacy, International Medical University (IMU), Kuala Lumpur, Malaysia

1 Introduction

Cancer is one of the most challenging pathologic conditions owing to the involvement of several cellular abnormalities and genetic disorders [1]. Cancer has always been a major public health issue globally and the emergence of cancer cases is increasing [2]. According to the World Health Organization (WHO), the new cancer patients can be up to 15 million per year with expected fatalities of approximately 10 million by the year 2020 [1, 3]. The available cure options for cancer include surgical intervention, radiation, and chemotherapy which though can mitigate the progression of cancer, but relapse frequently occurs [3]. Despite the phenomenal advancements in the prevention, diagnosis, and management of cancer along with remarkable developments in comprehending the signaling pathways involved in the progression of cancer and the diversity of the tumor microenvironment, metastasis remains a primitive obstacle [1]. Conventional chemotherapy involves non-targeted treatments which frequently lead to adverse and toxic effects primarily attributed to the destruction of normal organs and healthy tissues. The course of cancer chemotherapy is significantly halted by these adverse effects [3]. Consequently, there is a growing interest in the development of novel techniques and approaches to cure cancer; for instance, one strategy is the co-administration of multiple pharmacological agents belonging to different classes of therapeutics [3]. Another strategy namely RNA interference (RNAi) is the identification and alteration of the genes responsible for the pathophysiology of cancer [1, 3].

RNAi is employed to identify the distinctive roles of genes in the modulation of various cellular processes and ailments. In addition, RNAi is also involved in the augmentation of RNA-based therapies to regulate disease-inductive genes in cancer, viral infection, and hepatitis. RNAi is a potential gene maneuvering approach capable of inducing a therapeutic response alone or in combination with other pharmacological agents [1, 3]. The RNA-based therapy involves antisense oligonucleotides (ASOs), aptamers, microRNA (miRNA), small-interfering RNA (siRNA), synthetic mRNA, and CRISPR-Cas9 [4]. The RNA-based novel therapies are auspicious in the management of cancer by alteration of mRNA splicing, prevention of gene expression, genome editing, and targeting gene upregulation [5]. In contrast to conventional drug therapy, RNA-based therapeutic approaches selectively target a particular single gene and thus avoid off-target effects, which renders better therapeutic efficacy with enhanced tolerance [6]. Hitherto, micro-RNAs (miRNAs), small interfering RNAs (siRNAs), and short hairpin RNA (shRNA) are emerging RNAi effectors. The miRNA is single stranded possessing 19–24 nucleotides which generally triggers functions in an endogenous pathway. Primary miRNA is firstly transcribed in the cellular nucleus followed by processing into both precursors and mature forms through cascade of biogenesis mechanisms involving enzymes Drosha (class 2 RNase III enzyme) and Dicer (type class 3 RNase III enzyme). The miRNAs are attached to the cytosolic RISC and modulate the particular genes involved in protein-coding, primarily via degradation of mRNA translational suppression [7]. Argonaute (Ago) protein complexes identify and readily combine with passenger strand of miRNA and search for the reciprocal target sites located in miRNAs to induce endonucleolytic breakdown of mRNA leading to ultimate mRNA degradation (Fig. 1). Apart from endonucleolytic degradation of mRNA, some ago protein-containing complexes induce translation suppression using mRNA sequestration in the processing bodies (p-bodies) [8].

In this chapter, the targeted use of miRNA using advanced drug delivery systems such as liposome, solid lipid nanoparticles, and nanostructured lipid carriers in the treatment of cancer would be discussed.

Advanced Drug Delivery Systems in the Management of Cancer. https://doi.org/10.1016/B978-0-323-85503-7.00004-3

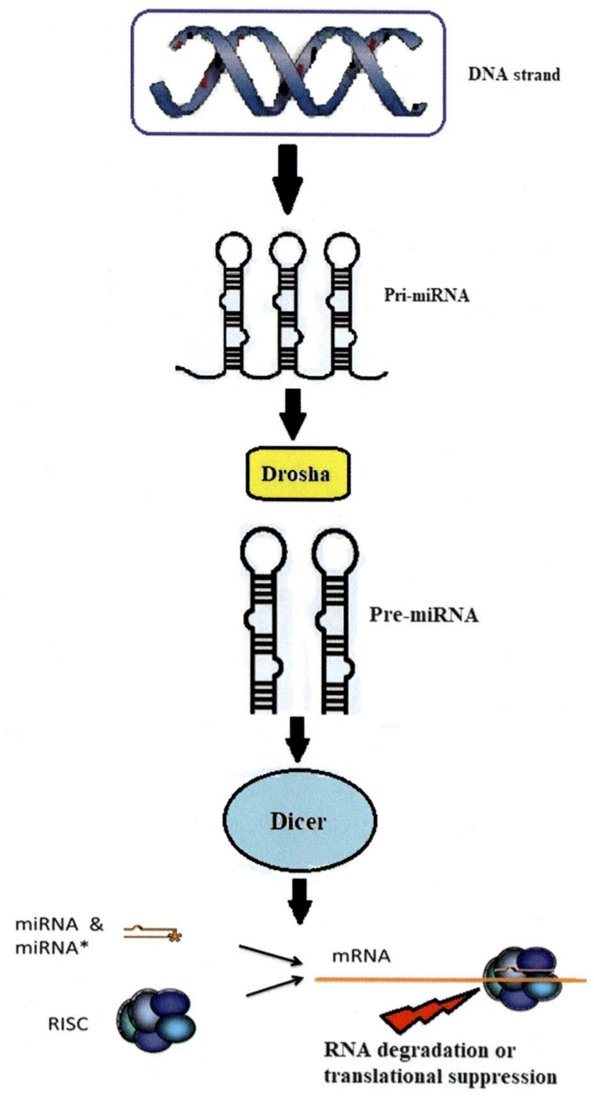

FIG. 1 miRNA maturation. The RISC containing the mature miRNA binds a target mRNA to inhibit translation either by repression of translation or by mRNA degradation. *Adapted from Motameny, S., Wolters, S., Nürnberg, P., & Schumacher, B. (2010). Next generation sequencing of miRNAs—strategies, resources and methods. Genes, 1(1), 70–84.*

2 Advanced drug delivery systems

Numerous potential therapeutic agents are insoluble in aqueous environment, biologically and structurally frail and induce serious untoward effects. Lipid-based nanocarriers (LBNCs) are accounted as one of the most established colloidal carrier systems for the delivery of such drug molecules. Their ongoing use in oncology has transformed the cancer therapy primarily by enhancing the tumor suppression activity of numerous chemotherapeutic drugs. LBNCs offer several merits such as immense physiochemical and thermic stability, better drug loading efficiency, comfort fabrication at reasonable cost and scale-up potential owing to the involvement of natural materials [9]. Nanocarriers possess a size range between 1 and 1000 nm and they enhance the bioavailability and specificity of the administered drug [10]. Although several nanocarriers have been employed in oncology, this chapter describes lipid-based lipid nanocarrier systems used in the delivery of combination therapy comprising of micro-RNA and an anticancer drug. Lipid-based nanocarriers such as liposomes, solid lipid nanoparticles (SLN), and nanostructured lipid carriers (NLC) have attracted immense focus in cancer management [11] and are capable to carry both hydrophilic as well as hydrophobic drug molecules with almost zero toxicity with enhanced duration of drug therapeutic response by extending the biological half-life and controlling the rate of drug release [12]. In addition, lipid-based nanocarrier systems may uptake chemical alterations to evade human immune system and polyethylene glycol (PEG) or gangliosides can be added to enhance the miscibility of the drug. Moreover, they could be fabricated with a pH-responsive design to induce the drug release solely in an acidic media and could also be bound with specific antibodies to facilitate the recognition of tumor tissues (such as folic acid (FoA)) [13]. There are two types of nano-scaled advanced drug delivery systems namely lipid-based and polymer-based.

2.1 Lipid-based systems

2.1.1 Liposomes

Liposomes are the most extensively investigated lipid nanocarriers owing to their better biocompatible and biodegradable characteristics. The primary constituents of liposomes are phospholipids, arranged in a bilayer scaffold owing to amphipathic attributes. Upon contact with aqueous media, liposomes forge sacs which enhances the dissolution and stability of the incorporated drug molecules. Being amphipathic, liposomes are suited for the encapsulation of both hydrophilic and hydrophobic drugs [14]. Besides phospholipids, other excipients such as cholesterol can also be included in liposomes to mitigate the hydrophilicity and to enhance the penetration of hydrophobic drugs via the bilayer membrane [14]. Cholesterol-altered liposomes may offer a multiple bilayer namely multilaminar vesicles (size range between 0.5 and 10 nm), intermediate sizes (10–100 nm), a single bilayer unilamellar vesicles (size more than 100 nm), small unilamellar vesicles, and large unilamellar vesicles [14]. There are two approaches to achieve the efficacy of liposomes namely passive targeting and active targeting [15]. In passive targeting, liposomes penetrate the tumor cell using molecular influx through the cellular membrane. On the contrary, in active targeting, liposomes target the intended tumor tissues with the help of antibodies capable to identify a specific tumor tissue [15]. Nevertheless, a third mechanism is also emanating namely using stimulus-sensitive structures involving pH, temperature or magnetic fields to achieve the targeted delivery of entrapped drug using liposomes [16].

2.1.2 Solid lipid nanoparticles (SLN)

SLNs are colloidal nanoparticulate drug delivery system comprised of lipids that ensure their solid state both at body as well as room temperature. SLNs exhibit a size range between 50 and 1000 nm and the lipid materials employed provide a matrix scaffold for the encapsulation of drug molecules. The solid lipid matrix could be fatty acids, mono-, di- or triglycerides or complex glyceride mixtures and is stabilized using a blend of polymers surfactants or polymers. SLNs offer crucial advantages such as long-term physical stability, cushioning of instable drugs, site-specific targeting, probability of controlled drug release, cost-effectiveness, smooth fabrication, and safety with remarkably low noxious activity on human granulocytes [17]. Nonetheless, SLNs possess some drawbacks as well such as mediocre drug-entrapment efficiency and drug discharge owing to crystallization upon storage [18].

2.1.3 Nanostructured lipid carriers (NLC)

NLCs are second generation of lipid-based nanocarrier systems, evolved from SLN and are consisted of a blend of solid and liquid lipid materials. Basically, NLCs were designed to rectify the problems associated with SLNs such as limited drug entrapment efficiency, drug discharge during long-term storage primarily by reducing the lipid crystallization because of the presence of liquid lipids in the formula. NLCs are prepared from lipid materials such as glyceryl dioleate, glyceryl tricaprylate, ethyl oleate, and isopropyl myristate [9]. The average particle size range of NLCs is closely similar to SLNs usually in the range between 10 and 1000 nm and is influenced by the type of lipid material employed and the method of preparation. Despite their prominent merits they still present some pitfalls such as drug discharge following the polymorphic transformation of the lipid from the nanocarrier matrix amid the storage time and relatively average drug loading capability [19].

2.2 Polymer-based systems

Polymer-based advanced drug delivery systems involve natural polymers such as gelatin, collagen and chitosan, synthetic biodegradable polymers, and copolymers. The advantages of polymer-based nanoparticle systems are ease of substitution, high stability, flexibility, and the incorporation of functional groups [20]. The specific attributes of these systems are size, charge, morphology, and alteration simply by changing polymer composition, molecular weight, and design including branched, linear, dendrimer, and copolymer. A cationic polymer namely polyethyleneimine (PEI) is the most commonly employed polymer-based system for the therapeutic delivery of miRNA. Cationic polymers possess multiple amino groups and exploit the proton sponge effect to trigger flee from endosomal membranes [21].

3 Targeting miRNA in the management of cancer using advanced drug delivery systems

As aforementioned, RNAi is an inherent mechanism encompassing the regulation of gene expression using the double-stranded ribonucleic acids (dsRNAs) [22]. In addition, RNAi also promotes the modulation of numerous biological pathways protecting the living cell from several dangers [23, 24]. The RNAi-based therapeutic approaches involve the delivery of double-stranded miRNA or siRNA to the targeted area [22] leading to alterations in RNAi sequences to target the specific genes. Conventional therapeutic approaches cannot access some of the proteins or genes accountable for the emergence of a particular disease such as cancer. Nonetheless, RNAi-based therapeutics can access these targets easily using miRNA/siRNA [22]. To treat cancer, combination therapy is preferable due to better therapeutics efficacy resulting from synergism [25, 26]. However, in order to achieve a robust synergistic effect, accurate selection of the appropriate chemotherapeutic agent is mandatory which would not only mitigate the dose and drug resistance but also could improve patient compliance [27–29]. miRNA and siRNA-mediated RNAi is emerging as a meritorious novel approach in the treatment of cancer [30]. Being a nucleic acid bioactive, miRNA has enormous potential as an anticancer substance [31] and has been documented to be misexpressed amid horrific proliferation of cancerous cells and perform as tumor suppressors or oncogenes [32].

Being a relatively novel approach to treat cancer, reported studies on the miRNA therapeutics using lipid-based nano-carriers are scant [33, 34]. Liposome-polycation-hyaluronic acid (LPH) formulation modulated using single-chain variable fragment (scFv) and GC4 (phage identified internalizing) was prepared. The authors demonstrated the capability of targeting the tumor cells for the systemic administration of siRNA and miRNA into experimentally induced lung metastasis of murine B16F10 melanoma model [35]. The authors also reported the zeta potential and the size of the miRNA- and siRNA-loaded LPH nanoparticles to be 10.9 ± 4.8 mV and 170 nm, respectively. In addition, liposomes were found capable of effective cytosolic disposition of fluorescein isothiocyanate (FITC)-labeled siRNA in B16F10 tumor cells. The authors reported the suppression in the protein expression of c-Myc, MDM2, and VEGFR in the B16F10 lung metastasis, following the delivery of siRNA with GC4-targeted liposomes, implying the concurrent muzzling by siRNAs. Furthermore, the metastasis nodules proliferation was also halted following the combined administration of siRNA by the GC4-targeted liposomes with a depletion in the tumor load up to 30%. The PEGylated siRNA and miRNA GC4-targeted liposomes exhibited negligible toxicity as the pro-inflammatory markers [interleukin (IL)-6, IL-12, and interferon (IFN)-γ] were not activated and the liver toxicity marker (alanine aminotransferase and aspartate aminotransferase) concentrations were also found normal in the C57BL/6 mice [36]. In other research work, cationic liposomal-based delivery of miR-29 was found to emulate and miR-7-encoding plasmid effectively deposited at the targeted tumor area [34, 37]. The researcher also reported significant tumor reduction in a murine xenograft tumor model of pulmonary cancer [34, 37]. In another work, miR-34a and let-7b loading neutral liposomes were fabricated and promoted lung tumor suppression without increasing kidney and liver enzymes in serum and did not cause any immune response in the mice [33, 38]. In another study, ionizable liposomes (NOV340) were prepared to target miR-200c at lung cancer in vivo [39]. The authors reported enhanced cellular uptake of the administrated miR-200c leading to increased radiosensitivity of lung cancer in vivo which was attributed to pH-dependent ionization of NOV340.

Studies have also been reported on the delivery of miRNA using polymer-based advanced drug delivery systems. PEI-mediated miRNA mimics found to be effective in suppressing in glioblastoma by miR-145, hepatocellular carcinoma using miR-34a and in colon cancer using miR-145 and miR-33 [40–42]. Researchers have reported the effectiveness of polymer-based nanoparticles fabricated for the co-delivery of miRNA and chemotherapeutic agents such as temozolomide, doxorubicin, and gemcitabine in in vivo tumor models [43–45]. Despite the promising results, the toxicity of PEI toxicity contributed by excessive positive charge and slow biological degradation resulted from non-specific binding to proteins in vivo abstained them from entering the clinical phase. In another study, positively charged PEI complexed with PLGA (poly lactic-*co*-glycolic acid) and anionic hyaluronic acid was reported to effectively deliver a DNA plasmid-containing miR-145 to the colon cancer tumor in vivo with enhanced pro-apoptotic and anti-proliferative activities [46]. In another documented study, miR-34a mimics loaded in the PLGA-chitosan complexes reduced the tumor size and prolonged the survival time in a multiple myeloma xenograft model [47]. In yet another reported study, it was reported that let-7g miRNA mimics loaded in ester-based dendrimer carrier profoundly repressed the liver tumor growth with negligible dendrimer-related toxicity in vivo [48]. Shatsberg and co-workers fabricated polymer-based nanogel complex loaded with miR-34a mimics aimed for active cytoplasmic delivery [49]. The authors reported suppression of tumor progression in human U-87 MG GBM-bearing SCID mice.

4 Future perspectives and conclusion

Although advanced drug delivery systems such as lipid-based nanoparticles and polymer-based nanoparticles have been designed and employed for delivery of miRNA at tumor site, however, some serious issues are still present. For instance, the intricacy of advanced drug delivery system particularly for lipid-based nanocarriers and the concerns of safety of each substance incorporated in these systems. In addition to the aforementioned formulation challenges, commercialization of these advanced nanocarrier systems is cumbersome and time consuming. Moreover, the composition of these nanocarrier systems should have FDA-approved materials which is troublesome for a formulator. Although nanoparticles can be altered using targeting ligands and other materials to enhance their efficacy, yet these substances must be compatible with the targeted receptors overexpressed in the tumor tissues. Definitely, the therapeutic efficacy and safety of miRNA-based advanced drug delivery systems can be enhanced by ratifying these approaches in the upcoming future.

miRNA is a rapid, robust, potent, and adaptable approach to target and silence particular cancer-causing genes. Consequently, miRNA-based therapeutics have acquired immense attention for cancer treatment at the molecular level. Nonetheless, the use of miRNA technology intended for cancer management at the clinical level is a potential challenge which is ascribable to certain physiological barriers influencing their stability. The advanced drug delivery systems such as nanoparticles have emerged as potential novel carriers for targeting miRNA because nanoparticles can mitigate the aforementioned pitfalls associated with miRNA administration. The noticeable advantages of nanoparticles are capability of incorporation of both hydrophilic and hydrophobic substances, better protection against degradation, enhanced solubility and bioavailability, improved drug stability and retention time in the body, improved penetration of miRNA through biological barriers, capability to undergo chemical and surface modification and thus allowing ligand attachment to the nanoparticles surface which facilitates in active targeting. Consequently, these nanotechnology-based advanced drug delivery systems have successfully been exploited for the delivery of miRNA alone and in combination with other anticancer agents resulting in the enhancement of the therapeutic efficiency by instant manipulation of multiple pathways or regulatory proteins accountable for cancer cell proliferation and metastasis. Although the potential of both lipid-based and polymer-based nanocarriers for targeting miRNA in cancer treatment has been well advocated by documented scientific studies, there is a need of modulation methods to enhance their physicochemical, biological, and long-term shelf stabilities along with toxicity reduction in vivo. These modulation strategies would evaluate the appropriateness of advanced drug delivery system for the delivery of miRNA mimics to target and combat cancer which may result in their clinical or commercial use in the forthcoming future.

Acknowledgment

The author would like to acknowledge the School of Pharmacy, International Medical University (IMU), Kuala Lumpur, Malaysia for providing resources and support in completing this work.

Conflict of interest

The authors declare no conflict of interest.

References

[1] Chalbatani GM, Dana H, Gharagouzloo E, Grijalvo S, Eritja R, Logsdon CD, Memari F, Miri SR, Rad MR, Marmari V. Small interfering RNAs (siRNAs) in cancer therapy: a nano-based approach. Int J Nanomedicine 2019;14:3111.

[2] Xin Y, Huang M, Guo WW, Huang Q, Zhen Zhang L, Jiang G. Nano-based delivery of RNAi in cancer therapy. Mol Cancer 2017;16(1):134.

[3] Subhan MA, Torchilin VP. Efficient nanocarriers of siRNA therapeutics for cancer treatment. Transl Res 2019.

[4] Zhang P, An K, Duan X, Xu H, Li F, Xu F. Recent advances in siRNA delivery for cancer therapy using smart nanocarriers. Drug Discov Today 2018;23(4):900–11.

[5] Wittrup A, Lieberman J. Knocking down disease: a progress report on siRNA therapeutics. Nat Rev Genet 2015 Sep;16(9):543–52.

[6] Jackson AL, Linsley PS. Recognizing and avoiding siRNA off-target effects for target identification and therapeutic application. Nat Rev Drug Discov 2010;9(1):57–67.

[7] Ling H, Fabbri M, Calin GA. MicroRNAs and other non-coding RNAs as targets for anticancer drug development. Nat Rev Drug Discov 2013;12(11):847.

[8] Crocco P, Montesanto A, Passarino G, Rose G. Polymorphisms falling within putative miRNA target sites in the 3′ UTR region of SIRT2 and DRD2 genes are correlated with human longevity. J Gerontol A Biomed Sci Med Sci 2016;71(5):586–92.

[9] García-Pinel B, Porras-Alcalá C, Ortega-Rodríguez A, Sarabia F, Prados J, Melguizo C, López-Romero J. Lipid-based nanoparticles: application and recent advances in Cancer treatment. Nanomaterials (Basel) 2019;9(4):638.

[10] Miele E, Spinelli GP, Miele E, Di Fabrizio E, Ferretti E, Tomao S, Gulino A. Nanoparticle-based delivery of small interfering RNA: challenges for cancer therapy. Int J Nanomedicine 2012;7:3637–57.

[11] Cormode DP, Naha PC, Fayad ZA. Nanoparticle contrast agents for computed tomography: a focus on micelles. Contrast Media Mol Imaging 2014;9:37–52.

[12] Ozpolat B, Sood AK, Lopez-Berestein G. Liposomal siRNA nanocarriers for cancer therapy. Adv Drug Deliv Rev 2014;66:110–6.

[13] Rama AR, Jimenez-Lopez J, Cabeza L, Jimenez-Luna C, Leiva MC, Perazzoli G, Hernandez R, Zafra I, Ortiz R, Melguizo C, Prados J. Last advances in nanocarriers-based drug delivery systems for colorectal cancer. Curr Drug Deliv 2016;13(6):830–8.

[14] Yingchoncharoen P, Kalinowski DS, Richardson DR. Lipid-based drug delivery Systems in Cancer Therapy: what is available and what is yet to come. Pharmacol Rev 2016;68:701–87.

[15] Gogoi M, Kumar N, Patra S. Multifunctional magnetic liposomes for Cancer imaging and therapeutic applications. In: Holban AM, Grumezescu AM, editors. Nanoarchitectonics for smart delivery and drug targeting. New York, USA: William Andrew Publishing; 2016. p. 743–82.

[16] Kunjachan S, Ehling J, Storm G, Kiessling F, Lammers T. Noninvasive imaging of Nanomedicines and Nanotheranostics: principles, Progress, and prospects. Chem Rev 2015;115:10907–37.

[17] Mydin RBSMN, Moshawih S. Nanoparticles in Nanomedicine application: Lipid-based nanoparticles and their safety concerns. In: Siddiquee S, Melvin GJH, Rahman MM, editors. Nanotechnology: Applications in energy. Cham, Switzerland: Drug and Food. Springer International Publishing; 2007. p. 227–32.

[18] Rajabi M, Mousa SA. Lipid nanoparticles and their application in Nanomedicine. Curr Pharm Biotechnol 2016;17:662–72.

[19] Obeid MA, Tate RJ, Mullen AB, Ferro VA. Lipid-based nanoparticles for Cancer. Amsterdam, The Netherlands: Treatment. Elsevier Inc.; 2018.

[20] Fernandez CA, Rice KG. Engineered nanoscaled polyplex gene delivery systems. Mol Pharm 2009;6:1277–89.

[21] Forterre A, Komuro H, Aminova S, Harada M. A comprehensive review of Cancer MicroRNA therapeutic delivery strategies. Cancers (Basel) 2020;12(7):1852. Published 2020 Jul 9 https://doi.org/10.3390/cancers12071852.

[22] Daka A, Peer D. RNAi-based nanomedicines for targeted personalized therapy. Adv Drug Deliv Rev 2012;64(13):1508–21.

[23] Hutvágner G, Zamore PD. RNAi: nature abhors a double-strand. Curr Opin Genet Dev 2002;12(2):225–32.

[24] Sledz CA, Williams BR. RNA interference in biology and disease. Blood 2005;106(3):787–94.

[25] Kogan G, Šandula J, Korolenko TA, Falameeva OV, Poteryaeva ON, Zhanaeva SY, Levina OA, Filatova TG, Kaledin VI. Increased efficiency of Lewis lung carcinoma chemotherapy with a macrophage stimulator—yeast carboxymethyl glucan. Int Immunopharmacol 2002;2(6):775–81.

[26] Uckun FM, Ramakrishnan S, Houston LL. Increased efficiency in selective elimination of leukemia cells by a combination of a stable derivative of cyclophosphamide and a human B-cell-specific immunotoxin containing pokeweed antiviral protein. Cancer Res 1985;45(1):69–75.

[27] Hu CMJ, Zhang L. Nanoparticle-based combination therapy toward overcoming drug resistance in cancer. Biochem Pharmacol 2012;83(8):1104–11.

[28] Ihde DC. Chemotherapy of lung cancer. N Engl J Med 1992;327(20):1434–41.

[29] Myhr G, Moan J. Synergistic and tumour selective effects of chemotherapy and ultrasound treatment. Cancer Lett 2006;232(2):206–13.

[30] Filleur S, Courtin A, Ait-Si-Ali S, Guglielmi J, Merle C, Harel-Bellan A, Clézardin P, Cabon F. SiRNA-mediated inhibition of vascular endothelial growth factor severely limits tumor resistance to antiangiogenic thrombospondin-1 and slows tumor vascularization and growth. Cancer Res 2003;63(14):3919–22.

[31] Fabian MR, Sonenberg N. The mechanics of miRNA-mediated gene silencing: a look under the hood of miRISC. Nat Struct Mol Biol 2012;19(6):586.

[32] Trang P, Weidhaas JB, Slack FJ. MicroRNAs as potential cancer therapeutics. Oncogene 2008;27(2):S52–7.

[33] Trang P, Wiggins JF, Daige CL, Cho C, Omotola M, Brown D, Weidhaas JB, Bader AG, Slack FJ. Systemic delivery of tumor suppressor microRNA mimics using a neutral lipid emulsion inhibits lung tumors in mice. Mol Ther 2011;19(6):1116–22.

[34] Wu Y, Crawford M, Mao Y, Lee RJ, Davis IC, Elton TS, Lee LJ, Nana-Sinkam SP. Therapeutic delivery of microRNA-29b by cationic lipoplexes for lung cancer. Molecular Therapy-Nucleic Acids 2013;2:e84.

[35] Chen Y, Bathula SR, Li J, Huang L. Multifunctional nanoparticles delivering small interfering RNA and doxorubicin overcome drug resistance in cancer. J Biol Chem 2010;285(29):22639–50.

[36] Chen Y, Zhu X, Zhang X, Liu B, Huang L. Nanoparticles modified with tumor-targeting scFv deliver siRNA and miRNA for cancer therapy. Mol Ther 2010;18(9):1650–6.

[37] Rai K, Takigawa N, Ito S, Kashihara H, Ichihara E, Yasuda T, Shimizu K, Tanimoto M, Kiura K. Liposomal delivery of MicroRNA-7-expressing plasmid overcomes epidermal growth factor receptor tyrosine kinase inhibitor-resistance in lung cancer cells. Mol Cancer Ther 2011;10:1720–7.

[38] Wiggins JF, Ruffino L, Kelnar K, Omotola M, Patrawala L, Brown D, Bader AG. Development of a lung cancer therapeutic based on the tumor suppressor microRNA-34. Cancer Res 2010;70:5923–30.

[39] Cortez MA, Valdecanas D, Zhang X, Zhan Y, Bhardwaj V, Calin GA, Komaki R, Giri DK, Quini CC, Wolfe T, et al. Therapeutic delivery of miR-200c enhances radiosensitivity in lung cancer. Mol Ther 2014;22:1494–503.

[40] Hu Q, Wang K, Sun X, Li Y, Fu Q, Liang T, Tang G. A redox-sensitive, oligopeptide-guided, self-assembling, and efficiency-enhanced (ROSE) system for functional delivery of microRNA therapeutics for treatment of hepatocellular carcinoma. Biomaterials 2016;104:192–200.

[41] Ibrahim AF, Weirauch U, Thomas M, Grünweller A, Hartmann RK, Aigner A. MicroRNA replacement therapy for miR-145 and miR-33a is efficacious in a model of colon carcinoma. Cancer Res 2011;71:5214–24.

[42] Yang YP, Chien Y, Chiou GY, Cherng JY, Wang ML, Lo WL, Chang YL, Huang PI, Chen YW, Shih YH, et al. Inhibition of cancer stem cell-like properties and reduced chemoradioresistance of glioblastoma using microRNA145 with cationic polyurethane-short branch PEI. Biomaterials 2012;33:1462–76.

[43] Mittal A, Chitkara D, Behrman SW, Mahato RI. Efficacy of gemcitabine conjugated and miRNA-205 complexed micelles for treatment of advanced pancreatic cancer. Biomaterials 2014;35:7077–87.

[44] Qian X, Long L, Shi Z, Liu C, Qiu M, Sheng J, Pu P, Yuan X, Ren Y, Kang C. Star-branched amphiphilic PLA-b-PDMAEMA copolymers for co-delivery of miR-21 inhibitor and doxorubicin to treat glioma. Biomaterials 2014;35:2322–35.

[45] Seo YE, Suh HW, Bahal R, Josowitz A, Zhang J, Song E, Cui J, Noorbakhsh S, Jackson C, Bu T, et al. Nanoparticle-mediated intratumoral inhibition of miR-21 for improved survival in glioblastoma. Biomaterials 2019;201:87–98.

[46] Liang G, Zhu Y, Jing A, Wang J, Hu F, Feng W, Xiao Z, Chen B. Cationic microRNA-delivering nanocarriers for efficient treatment of colon carcinoma in xenograft model. Gene Ther 2016;23:829–38.

[47] Cosco D, Cilurzo F, Maiuolo J, Federico C, Di Martino MT, Cristiano MC, Tassone P, Fresta M, Paolino D. Delivery of miR-34a by chitosan/PLGA nanoplexes for the anticancer treatment of multiple myeloma. Sci Rep 2015;5:17579.

[48] Zhou K, Nguyen LH, Miller JB, Yan Y, Kos P, Xiong H, Li L, Hao J, Minnig JT, Zhu H, et al. Modular degradable dendrimers enable small RNAs to extend survival in an aggressive liver cancer model. Proc Natl Acad Sci U S A 2016;113:520–5.

[49] Shatsberg Z, Zhang X, Ofek P, Malhotra S, Krivitsky A, Scomparin A, Tiram G, Calderón M, Haag R, Satchi-Fainaro R. Functionalized nanogels carrying an anticancer microRNA for glioblastoma therapy. J Control Release 2016;239:159–68.

Chapter 37

Organic nanocarriers for targeted delivery of anticancer agents

Sunita Dahiya[a] and Rajiv Dahiya[b]

[a]Department of Pharmaceutical Sciences, School of Pharmacy, Medical Sciences Campus, University of Puerto Rico, San Juan, PR, United States,
[b]School of Pharmacy, Faculty of Medical Sciences, The University of the West Indies, St. Augustine, Trinidad and Tobago

1 Introduction

Cancer is a group of diseases that is characterized by abnormal and uncontrollable cell death in any organ or tissue of the body [1]. Being the second cause of deaths globally, the deadly cancer accounted for 9.6 million deaths in 2018, which is forecasted to grow up to 16.3 million deaths and 27.5 million new cases by 2040 [2]. Among the large number of cancers, lung, prostate, colorectal, stomach, and liver cancers are usual in men, while breast, colorectal, lung, cervical, and thyroid cancers are commonly found in women [1]. It is usual practice to use chemotherapy in cancer treatment employing the nonselective chemotherapeutic agents which kill fast-growing cancer cells by interfering with their cell DNA synthesis and mitosis. Being nonselective, chemotherapeutic agents damage healthy cells, resulting in severe undesirable side effects such as nausea, appetite loss, and other dreadful effects causing high fatality among cancer patients. In addition, conventional anticancer drug delivery exhibits poor bioaccessibility which requires higher doses that leads to multiple drug resistance (MDR). Delivery of anticancer agents via passive or active targeting improves bioaccessibility to tumor tissues, which results in reduced adverse effects and improved therapeutic efficacy [3].

Since past two decades, there has been continuous development in nanotechnological approaches for delivery of anticancer agents to provide safe and effective cancer therapy. Nanocarrier, a nanoscale particle, capable of transporting a therapeutic agent, possesses distinct physicochemical and biological characteristics which are advantageous in the delivery of anticancer agents [4]. In recent times, the increased understanding of tumor pathophysiology and availability of versatile materials such as lipids, polymers, and inorganic materials have led to advanced strategies for cancer treatments. These nanomaterials are capable of developing nanocarrier systems that can deliver anticancer agents to specific organ or cancer tissues in more controllable way to attain higher selective drug accumulation with substantially improved clinical efficacy [5, 6]. The targeted delivery of anticancer agents via nanocarrier-based systems demonstrates the potential to address challenges associated with conventional anticancer delivery. The most common approach to target and deliver the drug to a specific organ or tissue can be achieved via tumor-selective drug accumulation by the enhanced permeability and retention (EPR) effect, which also reduces normal cell damage [7]. The advantages of organic nanocarriers for cancer targeting are summarized in Table 1.

This chapter discusses how tumor vasculature and tumor microenvironment can be exploited for design of cancer targeting nanocarriers using different targeting strategies with focus on major categories of organic nanocarriers such as liposomes, solid lipid nanoparticles, polymeric nanoparticles, polymeric micelles, polymer drug conjugates, dendrimers, and lipid-polymer hybrid nanocarriers along with their properties and functionalization strategies desirable for cancer targeting. The state-of-the-art tumor-targeting approaches in cancer treatment with updates on their clinical development have been discussed. The challenges associated with clinical translation of organic nanocarriers for cancer treatment are also presented.

2 Exploiting tumor microvasculature and tumor microenvironment as natural cancer targets

The basic approach for efficient anticancer delivery of small molecules, antibodies, or drug-loaded nanocarrier system to a solid tumor relies on the relationship between the tumor cells and the blood vessels supporting their growth. Although different cancers have definite causes that lead to the abnormality in normal vasculature, all types of cancer require a suitable blood supply to facilitate the tumor growth, whereas a regular blood supply to abnormal vasculature is rare in solid tumors [8]. Cancer is masses of malignant cells that behaves as complex devil organs. During progressive tumor proliferation, tumor cells release various

Advanced Drug Delivery Systems in the Management of Cancer. https://doi.org/10.1016/B978-0-323-85503-7.00010-9

TABLE 1 Beneficial properties of organic nanocarrier in cancer targeting.
Targeting to specific organs, tissues, or cells via passive or active mechanism
Elimination from the body in biodegradable form
Protection of encapsulated drug for decreased drug degradation and enhanced stability
Decreased concentration of anticancer drug in normal healthy cells
Increased aqueous drug solubility for intravenous administration Surface functionalization abilities to meet clinical needs
Improved pharmacokinetics and pharmacodynamics
Controlled release of incorporated active agents
Enhanced cell internalization and uptake for intracellular delivery

cytokines and growth factors within their vicinity and either recruit, reprogram, or corrupt many other types of cells which in turn contribute to establish the tumor microenvironment, generating both malignant and nontransformed cell. The tumor microenvironment is created by interactions of malignant and nontransformed cells [9, 10]. In addition, the tumor microenvironment contributes to the dysfunction of dendritic cells, which contributes to induce the tumor-specific immune responses [11]. Likewise, the nonmalignant cells of the tumor microenvironment function as a dynamic, tumor-promoting factor at all stages of cancer. The roles of B7 molecules in the dynamic interactions between tumors and the host immune system during their expression, regulation, and function in the tumor microenvironment were demonstrated in a study. It was revealed that the overexpression of immune-inhibitory B7 molecules in the tumor microenvironment promoted the immune evasion of tumors emerging the possibility of inhibitory B7 molecules and their signaling pathways to become tumor targets in the treatment of various cancers [11].

When tumor reaches the particular size, it cannot depend on the normal adjacent vasculature to supply the necessary oxygen demand for its further growth. This initiates the death of the oxygen-deprived cells and the secretion of growth factors, which promotes the generation of new blood vessels from the adjacent capillaries. The process of generation of new blood vessels from the existing ones is known as angiogenesis [7], which in turn brings physiological changes to blood flow and transport characteristics of tumor vessels leading to abnormal structural changes to vascular tumor pathophysiology. These structural vascular abnormalities include high density of proliferating endothelial cells, deficient pericyte, and formation of aberrant basal membrane, creating a disconnected or gapping endothelium which causes an enhanced vascular permeability. This discontinuous endothelium could provide opportunities for the use of nanocarriers in the size range of 20–200 nm, since they can extravasate and accumulate inside the interstitial spaces of about 10–1000 nm which have been created between the endothelial cells in presence of tumors [12]. In metastasis stage, the tumor cells penetrate blood or lymphatic vessels, circulate through the intravascular stream, and then proliferate at another site necessitating further growth of lymphatic network for proliferation. The process of generation of new lymphatic vessels is known as lymphangiogenesis. The role of both angiogenesis and lymphangiogenesis is crucial for supplying nutrients, oxygen, and immune cells as well as eliminating the waste products. A typical tumor microenvironment is depicted in Fig. 1.

For solid tumors of 1–2 mm size, oxygen and nutrients can reach the center of the tumor by simple diffusion. On the contrary, in case of nonangiogenic tumors, the vasculature is still nonfunctional, making the nonangiogenic tumor dependent on its microenvironment for oxygen and nutrient supply. Progressive tumor enlargement leads to increased demand for nutrients, oxygen, and growth factors, and the need of eliminating increased metabolic waste by the cancer cells. Since the abnormal vasculature cannot satisfy these needs, there is creation of domains with low glucose pH and oxygen concentration (hypoxia), causing shrinkage and death of these cells [13]. To avoid this condition, the tumors must expand its blood supply network by initiating angiogenesis and creating its own microvasculature to provide constant supply of oxygen and nutrients for its continuous growth [14–16]. Thus, angiogenesis and tumor microvasculature have emerged as a potential natural target for cancer therapy by selective tumor targeting through the EPR effect [17–19].

The intracellular pH of normal healthy tissues and tumors is similar, whereas the extracellular pH of tumors is lower than normal tissues. For instance, an average extracellular tumor pH is between 6.0 and 7.0, but extracellular normal tissues and blood pH are about 7.4 [20, 21]. The reason behind low extracellular tumor pH is higher glycolysis rate in oxygen-deprived cancer cells, which requires conversion of pyruvate into lactate to generate nicotinamide adenine NAD^+, which is an essential factor for different glycolytic enzymes. However, removal of lactate from the cell is required to increase metabolic flux and avoid cytotoxicity. With each lactate molecule, one proton is exported by monocarboxylate transporter, whereas hypoxia-induced carbonic anhydrase causes subsequent exacerbation to raise pH gradient and acidification of the tumor extracellular space [22]. The pH differences between both intra- and extra-cellular environment as well as tumor cells and normal host tissues act as significant sources of differential drug partitioning and distribution. The acidic extracellular environment favors diffusion of the unionized fraction of

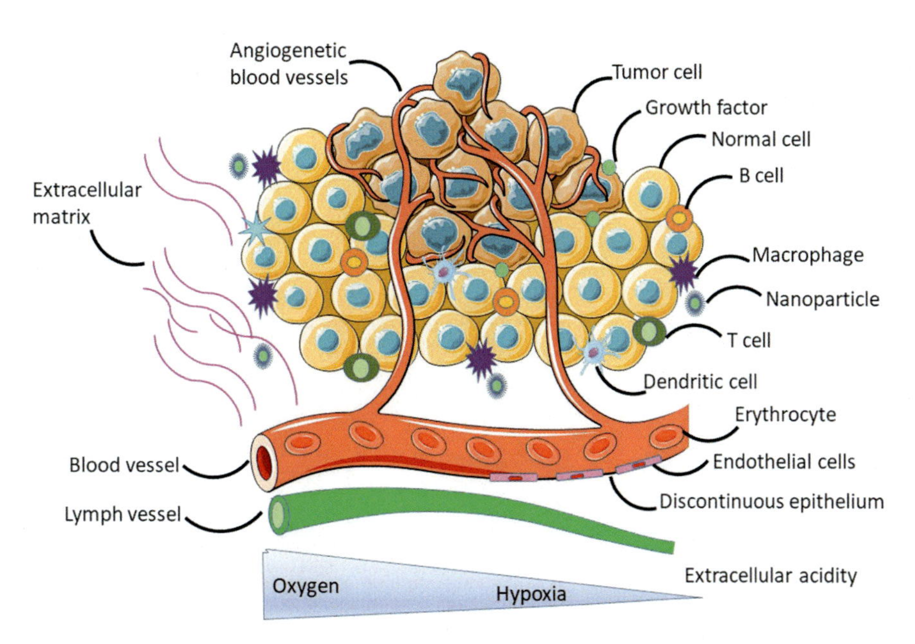

FIG. 1 A typical tumor microenvironment which allows nanocarriers in the size range of 20–200nm to extravasate and accumulate inside the interstitial space.

a weak acidic drug, whereas relatively alkaline intracellular compartment favors ionization of weakly acidic drug molecule, which in turn promotes the drug to accumulate in cytosol. If this process is modified, MDR phenomenon may occur [23].

3 Approaches for tumor targeting

Targeted delivery of anticancer agents is broadly divided into three approaches: (i) passive targeting, (ii) active targeting, and (iii) external stimuli triggering. In addition, current practice involves cancer immunotherapy as effective treatment options for various types of nanocarriers (Fig. 2). Each approach has merits and challenges which are discussed in subsequent section.

3.1 Passive targeting

Passive targeting is a major approach to achieve targeted drug delivery of anticancer agents. Passive targeting exploits the tumor pathophysiology, which exhibits anatomical and pathological differences to the normal tissues [24]. In nontumor condition, normal blood vessel walls possess intact endothelium as the essential component, as well as additional supportive layers in most organs. This intact structure enables the endothelium to act as a physical obstacle between blood and tissues. On the other hand, in presence of tumor, the blood vessel network is grown enormously to fulfill abnormal need of nutrients and oxygen necessary for tumor cells' proliferation. In addition, the tumor creates structurally defective irregular-shaped endothelial cells and defective support layers giving rise to disordered blood vessels. Moreover, the tumor microenvironment overexpresses the vascular permeability factor/vascular endothelial growth factor (VPF/VEGF) making the vasculature highly permeable to blood-circulating macromolecules. VEGF is a group of powerful vascular-permeabilizing substances that renders the tumor microvasculature highly permeable to plasma proteins, thereby promoting the tumor growth [25]. The morphological studies of leaky tumor vasculature in subcutaneously grown rodent tumor indicated the vascular pore cutoff size between 200 and 780nm [24]. On the other hand, the tight junctions' openings for normal endothelial cells were below 2nm. This difference in pore size between normal and tumor vasculature allows a drug delivery system with particles smaller than 200nm to cross tumor vasculature. However, the normal tissues of kidney glomerulus, liver sinusoid, and pulmonary region also exhibit disconnected endothelium, allowing the drug delivery systems to access these tissues. This similarity between diseased and normal tissues compromises the passive delivery of tumor-targeting systems [24]. Besides the tumor blood vessels, the tumor lymphatic system also contributes to enhance the passive tumor targeting. For instance, the normal lymphatic system collects and returns interstitial fluid to the blood to retain organ's fluid balance. The lymphatic system that exists at tumor exterior is functional, whereas the lymphatic systems that exist in the interior of tumor tissues are nonfunctional [26]. Hence, when the growing tumor cells create a pressure, the tumor lymphatic system collapses due to the absence of internal balancing pressure, causing retention of the high-molecular-weight molecules inside the tumors. Thus, the leaky vasculature and impaired lymphatic systems in tumor sites together describe the EPR effect as a

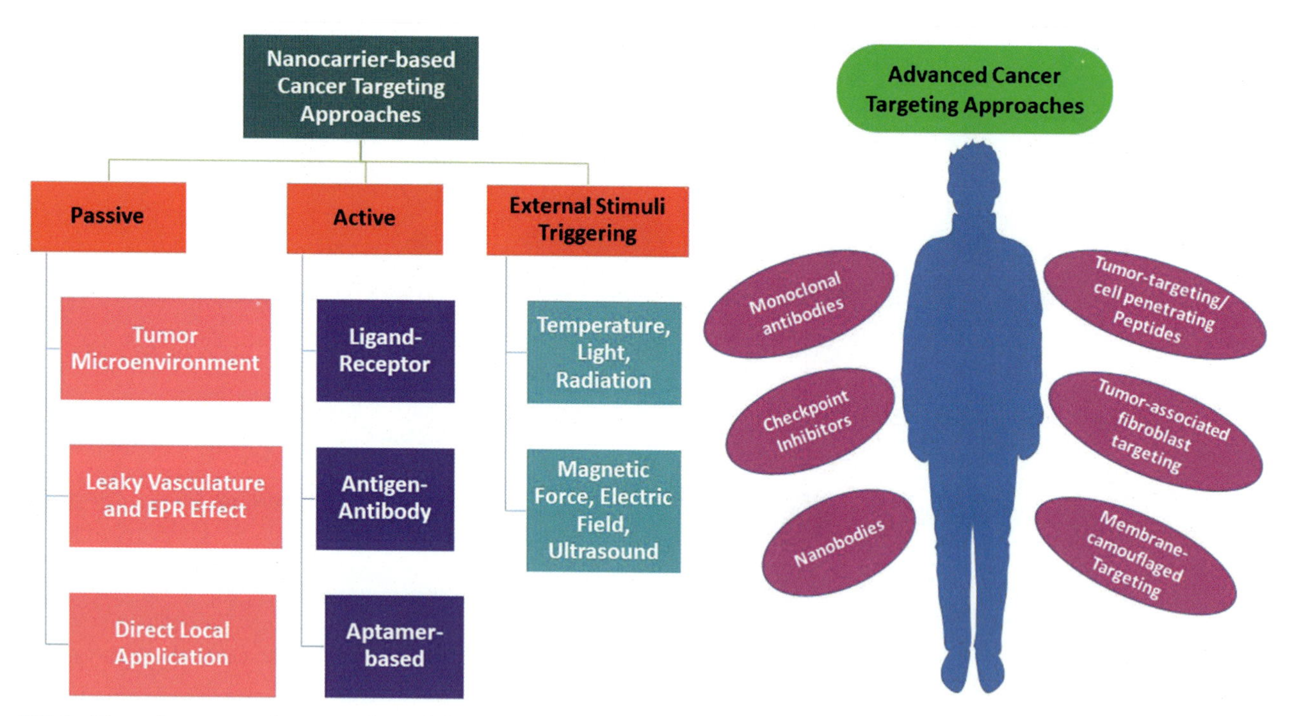

FIG. 2 Types of cancer targeting approaches to deliver anticancer agents. Current practice involves targeted nanocarriers that employ advanced targeting approaches including cancer immunotherapy.

well-recognized pathway to achieve passive tumor targeting [25]. Thus, passive targeting is a good strategy for effective cancer therapy of those delivery systems which can pass through the leaky tumor vasculature causing increased local accumulation of drugs [27] (Fig. 3A).

Nanocarriers, even without surface modification, can utilize the EPR effect to target tumor tissues, allowing nanoparticles to passively accumulate in the leaky tumor vasculature. On the other side, the passive targeting also leads to accumulation of nanoparticles in normally fenestrated tissues like liver or spleen. In addition, the tumor microenvironments vary depending on various stages of cancer and may act as barriers, limiting the passive targeting. Therefore, EPR-dependent passive targeting can be promoted by increasing the blood circulation time of a drug delivery system. In general, renal clearance and reticuloendothelial system (RES) uptake are responsible for the clearance of the blood-circulating substances [28]. Therefore, bypassing the clearance mechanism (renal clearance or RES uptake) can be a strategy to increase the blood circulation time of anticancer drug targeting system. In case of water-soluble polymeric drug carrier, renal clearance is controlled by the molecular weight and size. Therefore, polymer molecular weight higher than the renal clearance threshold imparts increased blood circulation [29]. To reduce RES uptake, particle size should be controlled below 100nm, whereas particle surfaces can be modified using nonimmunogenic water-soluble polymers. Although EPR effect-based passively targeting anticancer drug delivery has shown promise for nanocarrier-based delivery systems, it faces several barriers in effectively delivering the drugs to the tumor sites such as (i) interstitial fluid pressure that prevents the nanoparticle penetration inside the tumor tissues; (ii) increased interstitial fluid pressure with tumor growth cause leakage of plasma fluids and proteins from the capillaries, developing colloidal pressure due to increased protein content in the interstitial space and preventing the entry of any macromolecules from the blood; (iii) compression of lymphatic system with tumor growth causes a reduction of interstitial fluid drainage and a gain of fluid pressure; (iv) tumor heterogeneity factors such as presence of tumor stem cells in central part of the tumor show less accumulation of nanocarriers and poor EPR effect in center of tumor compared with other parts of the tumor [30]. Thus, the EPR-dependent passive targeting strategy shows the absence of selectivity which does not provide adequate nanoparticle accumulation inside the treated tumor area essential for the effective cancer therapy. However, modifications in physicochemical properties and surface functionalization of nanocarriers improve the passive targeted drug delivery.

3.2 Active targeting

Active targeting comprises of a delivery system which contains ligands/binding moieties that can attach to the specific receptors which are overexpressed in tumors [31]. An active tumor-targeted drug delivery system essentially contains ligands, whereas other components include anticancer agent,

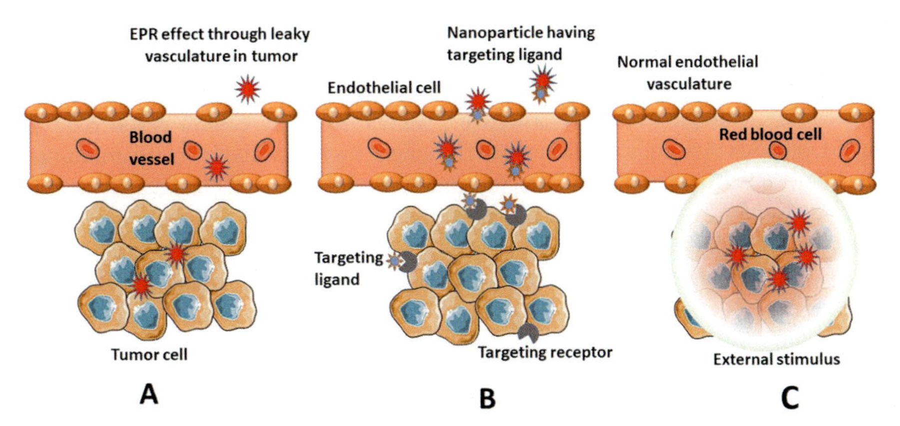

FIG. 3 Different drug targeting mechanisms for nanoparticle delivery to tumor tissue. (A) Passive targeting is achieved via leaky vasculature of tumors that allows nanocarriers to pass across the fenestrations and reach specific tumor tissues. (B) Active targeting is achieved using nanocarriers having ligands on their surfaces. The ligand can recognize and bind receptors that are expressed onto tumor cells. (C) Externally triggered release is achieved via external stimuli which allows nanoparticles to bunch up when exposed to an external stimulus such as temperature, magnetic field, and light.

drug carrier, or drug release enhancers. Some tumor-specific ligands are transferrin, folic acid, enzymes, engineered antibodies, as well as protein or carbohydrate macromolecules [3, 6, 27] (Fig. 3B). The optimized ligand density allows nonrecognition of nanoparticles by the RES and interaction with serum proteins causing enhanced blood residence time [3]. Anticancer agents are coupled with macromolecular or particulate drug carriers via physical interaction or chemical conjugation, where the drug release can be triggered by changes due to physiological tumor conditions such as low pH, high temperature, or presence of enzymes as compared to normal environment [32].

Active targeting poses three major challenges: (i) identification of tumor-specific ligands for a tumor-targeted delivery system since most tumors do not express unique antigens; (ii) the antigens overexpressed in tumors are also expressed in normal tissues; and (iii) the heterogeneous nature of antigens overexpressed among different types or stages of tumor growth. Tumor antigens are expressed on the cell surface or on the vasculature. Drug carriers that are bound to cellular receptors can release the drug to attack the tumor cells by cell-internalization via receptor-mediated endocytosis, whereas binding of drug carriers to noninternalizing receptors or vascular receptors prompts the local accumulation of drug-loaded carriers in the tumor interstitium. The extracellular drug release is taken up by cells via regular passive diffusion or active transport pathways [33].

Actively targeted drug delivery systems reduce the drug deposition in normal tissues, which minimizes toxic side effects. An adequate blood-circulation time is critical to achieve a considerable level of active targeting for a tumor-targeted system. Coupling targeting moieties with a drug carrier may result in long blood circulation due to an increased molecular weight and decreased tendency of renal clearance. However, when macromolecular targeting systems such as monoclonal antibodies are introduced, the acquired immunogenicity can be the major consequence. Injected immunoliposomes can be found at about 50% concentration in the liver after 2h of injection, whereas only 10% of the liposomes without the bound antibodies accumulate in the tissues. Use of humanized monoclonal antibodies or fragments reduces the immunogenicity of an actively targeted system [33]. Thus, unlike passive targeting, active targeting possesses higher specificity to receptors or other tumor-specific targets overexpressed on tumor cell surface. This ligand-mediated active targeting eliminates possibility of nonspecific uptake of nanocarriers, ensuring higher tumor accumulation, improved therapeutic efficacy, and lower incidents of adverse side effects during cancer treatment.

3.3 External stimuli triggering

External stimuli triggering is a promising approach to reduce nonspecific exposure of anticancer agents. These smart systems control and target the delivery of drug by using external input like temperature, magnetic force, electric field, ultrasound, light, or radiation which can activate the drug release. The release of the drug from such systems can be then modulated through some action or external input such as externally supplied energy (Fig.3C). Unlike internal stimuli-based systems that depend on internal stimuli-like pH, temperature, redox potential, or enzymes, the external stimuli-responsive systems show better control of the drug release by overcoming interpatient variability and using the external stimuli precisely [34]. Besides controlling the drug release, external stimuli can also control the intracellular drug fate and improve cell-internalization, resulting in an enhanced tumor targeting [35]. In designing external temperature-responsive systems, the role of thermo-responsive polymers is crucial. Such polymers respond to heat-generation causing an increase in temperature which triggers drug release through thermo-responsive polymer. For instance, hyperthermal temperature change from 37°C to 42°C may increase the vascular permeability for enhanced nanocarrier delivery. ThermoDox employs LTSL (lysolipid thermally sensitive liposome) technology to fabricate doxorubicin-encapsulated thermosensitive liposomes, which showed rapid structural changes when heated to 40–45°C. ThermoDox delivered by intravenous infusion with simultaneous heat-based treatment created pores in the liposome, releasing about 25 times more doxorubicin directly into and around the targeted tumor as compared with intravenous infusion alone [36]. Likewise, near-infrared wavelengths better penetrated into the body than that of superficially adsorbed UV light by the skin, blood, and tissues less than 10cm in depth [37]. Ultrasound systems aided cancer diagnosis by triggering the release of contrast agents [38], whereas magnetic and electric fields allowed nanoparticle aggregation in specific sites [39]. Biocompatible verteporfin, a benzoporphyrin derivative (BPD) in nanoparticles, allowed a light-triggered external stimulus which produced reactive oxygen species capable of damaging the cell DNA and causing the cell death upon exposure to infrared light. Konan-Kouakou et al. injected verteporfin-loaded Poly(d,L-lactide-*co*-glycolide) (PLGA) nanoparticles into tumor-bearing mice which indicated that verteporfin-incorporated small nanoparticles of up to 167nm diameter could be an effective controller of tumor growth in mice when coadministered with an early light irradiation prior to drug administration [40].

4 Organic nanocarriers in cancer targeting

4.1 Physicochemical properties

A successful nanocarrier must meet some design criteria to achieve drug loading; passive, active, or triggered release; and surface functionalization ability to tailor specific needs of cancer treatment. Physicochemical parameters like size, shape, surface chemistry, solubility, degradation, and clearance are significant in terms of nanocarriers' blood circulation kinetics in the body and their subsequent biodistribution, tumor tissue penetration, accumulation in target tissues, and cellular uptake. Thus, physicochemical properties play significant role in determining overall in vivo fate of nanocarriers.

Nanoparticle size impacts its binding capacity to the cell membrane due to its influence on the enthalpic and entropic properties that govern the adhesion strength between nanoparticles and cellular receptors, which is significant for its travel across the bloodstream, its cellular and macrophage uptake, its delivery to the cancer cells, as well as size-dependent nanotoxicity [41]. Moreover, nanoparticles tend to form aggregates in solution giving rise to larger-sized aggregates. In fact, smaller nanoparticles show much greater tendency to accumulate in the leaky tumor blood vessels as compared with the larger nanoparticles. In addition, smaller nanoparticles can extravasate into normal tissues, whereas larger particles cannot extravasate as easily, resulting in their highly variable biodistribution in the bloodstream. However, the renal clearance is very fast with nanoparticles of size less than 5nm. Studies on liposomes ranging between 40 and 700nm in size revealed that liposomes larger than 200nm did not extravasate into tumors [42]. In general, nanoparticle size greater than 500nm enables its cell internalization via phagocyte-mediated phagocytosis, whereas smaller particles can enter the cells by receptor-mediated endocytosis [43].

Although unusual particle shapes such as hollow tubes, rods, and stars are available, spherical geometry is the most common due to ease of manufacturing. Moreover, wide variety of sizes and shapes of nanoparticles can also be customized to gain maximum tumor accumulation [44]. In case of nonspherical nanoparticles, the angular orientation of nanoparticles relative to the cell membrane is influenced by hydrodynamic forces, shear rate, blood vessel geometry, and blood flow pattern, showing common tendency to marginate and escape the blood flow [45]. In the transport of flowing nanoparticles, near-the-wall margination is critical for a nanoparticle to show meaningful interactions with the tumor vascular bed in case of both cell targeting and vascular targeting [43]. Christian et al. demonstrated the potential of long cylinder filamentous micelles to achieve prolonged blood circulation, enhanced tumor accumulation, large maximum tolerated dose and high tumor inhibition in mouse xenograft tumors [46]. Nanoparticle shape can be tailored depending on ligand attachment to achieve enhanced tumor accumulation of nanoparticles and improved therapeutic efficacy.

Surface chemistry including charge, hydrophilicity, and hydrophobicity is another important parameter that affects stability and biodistribution of nanoparticles in blood. Many studies reported that the surface charge impacts tumor penetration. Campbell et al. reported significantly enhanced accumulation of positively charged liposomes in tumor tissue and tumor vasculature as compared with that exhibited by negatively charged and neutral liposomes [47]. In addition, positively charged liposomes exhibited preferential uptake in angiogenic tumor vessels, which indicated that the positively charged liposomes could be selectively delivered to blood vessels of solid tumors for diagnostic or therapeutic application. Likewise, negative and neutral liposomes can be utilized to deliver drug to the extravascular compartment since their extravasation tendency allows them to escape from blood vessels into the tissues [47, 48]. The surface coating of nanoparticle by hydrophilic polymer such as polyethylene glycol (PEG), known as PEGylation, is an effective approach to provide stealth properties to the nanoparticles enabling their longer blood circulation in vivo [49]. The benefits of PEGylation is elaborated in the subsequent sections. Moreover, poorly water-soluble drug may be cleared-off from the blood circulation before it reaches the tumor tissues. Therefore, the use of hydrophilic nanoparticles to encapsulate poorly water-soluble drugs can enhance their solubility and bioavailability for more effective delivery [50].

4.2 Surface functionalization

Surface properties such as hydrophilicity, hydrophobicity, and surface charge are important determinants of cell interactions and cellular uptake of nanoparticles. Moreover, the organic nanocarriers obtained through conventional preparation methods face challenges such as formulation instability, low blood circulation time, and unfavorable surfaces limiting their delivery potential. Therefore, the nanocarrier surfaces can be modified via addition of various functional groups for tailoring their properties and achieve enhanced tumor targeting to fulfill unmet clinical needs. Various strategies to achieve functionalized nanocarriers are described in subsequent section.

4.2.1 Functionalization strategies for organic nanocarriers

Various functionalization strategies were developed to impart certain benefits such as (i) protection of the nanocarrier against premature erosion to enhance formulation stability; (ii) control of nanocarrier surface charges and prolongation of their blood circulation time; (iii) attachment of ligands on the nanocarrier surfaces for active targeting of cancer tissues; and (iv) coating of organic nanocarrier surfaces using various biomimetic cell membranes to replicate their biological features to the nanocarrier surfaces [51].

Nanocarriers with positively charged surfaces can target the tumor vessels in the most effective way [47–49]. However, after extravasation, nanocarriers could shift to neutral charge and displayed rapid diffusion into the tumor tissues [42]. In addition, a hydrophobic molecule generally gets easily recognized by the RES causing their uptake from the blood via liver and spleen [51]. The charged molecule like DNA cannot be disrupted by uncharged hydrophilic molecule like PEG [52]. Grafting of PEG on nanocarrier surfaces provides hydrophilicity to the surfaces which mask their hydrophobicity and increases solubility and stability, which in turn leads to prolongation of their blood circulation time. Such PEGylated long-circulated nanoparticles show less clearance, and therefore, get sufficient time to reach tumor tissue. PEGylation can also prevent nanoparticle degradation before reaching the desired tissue site, and consequently improves the bioavailability [53].

Development of cell-membrane camouflaged organic nanocarriers is an emerging potential bioinspired strategy that employs biomimetic cell membrane-derived materials to achieve immune escape and improved tumor targeting [54]. Examples of biomimetic cell membrane-derived materials such as red blood cell membrane, cancer cell membrane, T-cell membrane, and activated fibroblast cell membranes have been used by the extrusion method or microfluidic electroporation technique to modify the organic nanocarrier properties [54]. A large number of biological ligands have been identified for decorating organic nanocarriers to target those substances which are overexpressed during tumor development. Some molecular ligands such as antibody, affibody (engineered, high-affinity, smaller size than normal antibodies), aptamer, and peptide displayed strong affinity to receptors, such as transferrin, epidermal growth factor, and folate, or tumor endothelium, such as $\alpha v\beta 3$ integrins and matrix metalloproteinases, which are overexpressed on cancer cell surfaces [55]. Various other biomacromolecules including monoclonal antibodies (mAbs), oligonucleotides, proteins, and siRNA can be used for surface functionalization of nanocarriers to enable receptor-specific binding of nanocarrier system for greater therapeutic effectiveness with minimal cytotoxicity, which is not easy to achieve with synthetic molecules [56]. Major strategies of nanocarrier functionalization that used ligand attachment [57–59], polymer coating [60, 61], biomimetic material coating [62–65], inorganic material coating [66–68], and biocompatible material coating [69, 70] are summarized in Table 2.

4.3 Types of organic nanocarriers for cancer targeting

Nanocarriers can be broadly classified as organic, inorganic, and hybrid nanocarriers. Organic nanocarriers can be fabricated employing wide variety of natural or synthetic polymers, including biocompatible and biodegradable polymers and lipids. The construction and tailorability of organic nanocarriers' properties involve covalent bonding and noncovalent interactions like hydrogen bonding; van der Waals forces; and electrostatic, hydrophobic, host-guest, or π-π interactions as driving forces which are influenced by composition of systems, type of solvents, conditions such as temperature and pH, etc., which can be utilized for designing stimuli-responsive smart nanocarrier systems [54]. Major types of organic nanocarriers studied for targeted delivery of anticancer agents such as liposomes, solid lipid nanocarriers, polymeric nanocarriers, polymeric micelles, polymer conjugates, dendrimers, and lipid-polymer hybrid nanocarriers are discussed in this chapter. Main advantages and disadvantages of these nanocarriers for anticancer drug targeting are summarized in Table 3.

TABLE 2 Major functionalization strategies for organic nanocarriers.

Strategy	Example	Benefits	References
Attachment of targeting ligands	Protein, peptide, polysaccharide, aptamer, small molecules	Improve targeting efficacy, cellular uptake, and internalization for intracellular delivery	[57–59]
Polymer coating	PEG	Improve stability, enhanced blood circulation and cancer targeting, reducing phagocytic capture	[60, 61]
Biomimetic material coating	Cell membrane (e.g., red blood cell membrane, cancer cell membrane, T lymphocyte cell membrane)	Ability of immune escape and reduced clearance, improved targeting capacity	[62–65]
Biomolecules attachment	Monoclonal antibodies, oligonucleotides, proteins, and siRNA	Improve targeting, minimal cytotoxicity	[56, 71]
Semiconducting polymeric nanocarriers coating with inorganic material	Silica	High fluorescence and photo acoustic intensities	[66, 67]
Semiconducting hybrid core-shell nanocarriers coating with inorganic material	MnO_2 nanosheets	Enhances performance in photodynamic therapy under hypoxic conditions	[68]
Coating nanocarriers by biocompatible material	Polydopamine, high-density lipoprotein	Impart good biocompatibility	[69, 70]

TABLE 3 Summary of advantages and disadvantages of various organic nanocarriers.

Organic nanocarrier	Advantages	Limitations
Liposomes	Amphiphilic and biocompatible Capacity to carry large drug payloads Capacity to combine and deliver multiple medications in single system Can be tuned to modify or control their biological characteristics, e.g., long circulating by PEGylation	Leakage of encapsulated agent Short half-life Possible triggering of immune response High production cost
Solid lipid nanoparticles	Excellent biocompatibility Improved stability Controlled and/or target drug release Incorporates both lipophilic and hydrophilic drugs Water-based technology that avoids organic solvents Easy to scale-up and sterilize Cost-effective, easier regulatory approval	Limited capacity for hydrophilic drug loading Drug discharge during storage Poor entrapment efficiency Lipid polymorphism Stability issues due to possible super cooled melts
Polymeric nanoparticles	Controlled/sustained drug release Incorporates both lipophilic and hydrophilic drugs Tunable physical and chemical properties Reproducible with synthetic polymers Several methods of preparation	Low drug-loading capacity Difficult scale-up Inadequate toxicological data Chances of toxic degradation
Polymeric micelles	Increasing solubility of hydrophobic drugs Protecting drug from environmental conditions Tailorability by tuning physicochemical properties Controlled drug release	Load only hydrophobic drugs Low drug-loading capacity Need of critical micelle concentration
Polymer-drug conjugates	High drug loading Sustained release of drug High stability compared with other nanocarriers which encapsulates the drug	Poor aqueous solubility Short circulation half-life Chances of embolism
Dendrimers	Increasing solubility of hydrophobic drugs Tailorability by tuning physicochemical properties High functionalization properties for targeting Covalent association for drug loading	High cost of synthesis Elimination and metabolism depend on the materials and generation Cellular toxicity
Lipid-polymer hybrid nanoparticles	High structural integrity Storage stability High biocompatibility and bioavailability Controlled drug release Easy fabrication process	Difficulty to load multiple drugs with different hydrophobicity Little attention paid to large-scale production that may affect its clinical translation

4.3.1 Liposomes

Lipid-based liposomes are the most studied nanocarrier for targeted anticancer drug delivery. Liposomes are spherical vesicular systems comprising an aqueous core which is surrounded by either single or multiple lipidic bilayer membrane. Liposome size ranges from 25 nm to several microns [72]. The lipids employed for liposome preparation can be of natural or synthetic origin. Liposomes are classified based on number of bilayer membranes as small unilamellar vesicles (one bilayer membrane) or multilamellar vesicles (more than one bilayer membranes), whereas different liposomal forms can be achieved via modification of liposomal composition, size, surface charge, and preparation method [72]. Besides being amphiphilic, biocompatible, and biodegradable, liposomes offer several other advantages including encapsulation ability for both hydrophilic (in aqueous core) and lipophilic (in lipid bilayer) drugs, cost-effectiveness, reduced systemic toxicity, and improved therapeutic performance of the product which are desirable in anticancer targeted drug delivery [73, 74]. Moreover, drugs loaded in liposomes displayed altered pharmacokinetic and improved biodistribution compared with free drugs in solution [75].

Liposomal system was clinically available for the first time in 1995 with USFDA-approved liposomal antifungal amphotericin B (Ambisome) and a liposomal doxorubicin (Myocet) [75]. With progressive advancements, liposomes are being preferably formulated as multifunctional nanocarriers by surface coating with hydrophilic PEG which imparts stealth properties to these PEGylated liposomes to provide longevity and prolonged blood circulation time, or can be surface coated using PLGA to enhance their stability and half-life, or can be attached to antibodies or ligands for enhanced specificity toward the cancer targets [75]. Liposomes can be designed as a multifunctional nanocarrier system that simultaneously target three molecular targets. One target can be the extracellular receptors or antigen expressed on the plasma membranes of the cancer cell surfaces. This sort of targeting directs the liposomal system to the specific tumor and limits their adverse side effects on normal tissues. The second target can be inhibition of drug efflux pumps like permeability glycoprotein (P-gp). This sort of targeting enhances the retention of drug in cancer cells and increases tumor cell accumulation of drug, resulting in reduced drug doses with lesser adverse drug effects or reduced risk of MDR. Liposomes can also target intracellular apoptosis-controlling mechanisms to conquer cellular antiapoptotic defense [76]. In addition, mAb or small proteins like folate, epidermal growth factor, or transferrin can be attached to lateral part of PEG chain to modify PEGylated liposomes for more target specificity. Liposomes localize in the tumor interstitial spaces and slowly release the anticancer agent within the tumor. The bioavailability and sustained-release ability of liposomes can be manipulated by altering liposome diameter [77]. The lipoplexes, which are cationic liposome-DNA complexes, were used for gene transfer into tumor cells; however, it showed plasma instability also [78, 79]. Various liposome nanocarriers in cancer targeting can be presented as depicted in Fig. 4. Recently, Pal et al. coadministered everolimus and vinorelbine using a proprietary tumor-targeting-peptide-conjugated lipopeptide which demonstrated significantly higher suppression of tumor growth compared to the single drug-loaded liposomes in renal cell carcinoma xenografts, and demonstrated excellent tumor-specific uptake [80].

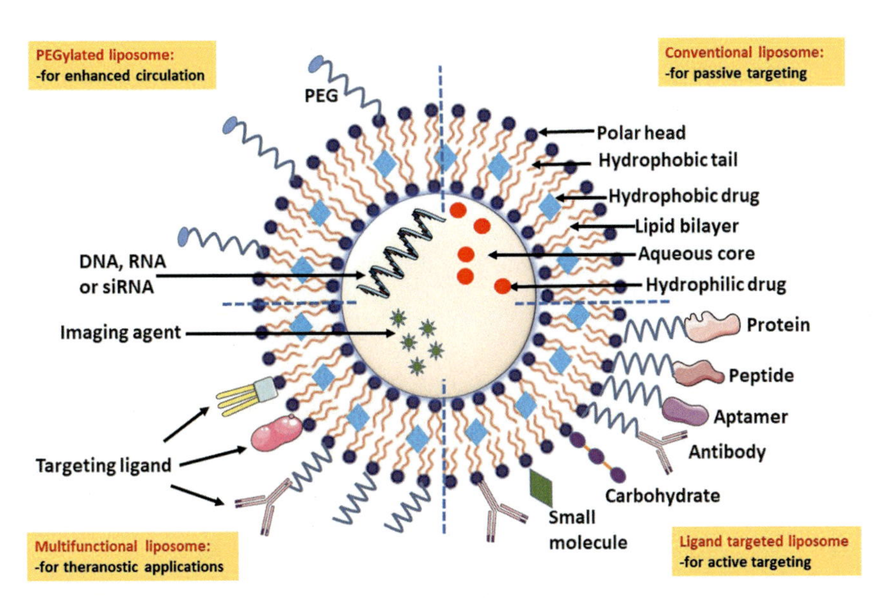

FIG. 4 Various types of liposomes that can be fabricated via surface functionalization strategies for cancer targeting applications.

4.3.2 Solid-lipid nanocarriers (SLNs)

SLNs are lipid-based colloidal carriers of size range 50–1000nm, which have been employed for delivery of lipophilic drugs since the early 1990s [81]. SLNs can be prepared by melting the solid lipid(s) and dispersing in emulsifier-containing water to stabilize the dispersion using high-pressure homogenization or microemulsification. Polymorphism of lipids also affects the properties of a SLN system. In addition, composition of lipid determines the overall hydrophobicity of SLN, which varies with the hydrophile–lipophile balance of the hydrophobic and hydrophilic functional groups of lipid molecules [82]. Commonly used solid lipids that achieve a highly lipophilic matrix to disperse or dissolve lipophilic drugs include mono-, di-, and triglycerides, free fatty acids and fatty alcohols, waxes, and steroids [83]. In addition to lipids, SLN surfaces can be modified using surfactants and cosurfactants (Fig. 5A). For example, hydrophilic polymers such as PEG, poloxamer, or poloxamine can be used for coating SLN carrier which provides stealth or prolonged circulation properties to the SLN to reduce their RES clearance and enhance systemic circulation time [84, 85]. The hydrophilic coating agents can be preconjugated to the hydrophobic lipid moieties on SLN surfaces, which leads to the formation of amphiphilic molecules, thereby securing attachment of hydrophilic coating agents onto the SLN particle surface. Surface modification technique by hydrophilic polymer grafting can minimize phagocytosis-mediated clearance of SLNs [86].

SLNs provide advantages including biosafety, controllable drug release, high drug loading, improved bioavailability of hydrophobic drugs, stability, and higher feasibility for manufacturing [81–83]. Although SLNs traditionally incorporate hydrophobic drugs, recent approaches such as lipid-polymer hybrid or lipid-drug conjugates enabled incorporation of hydrophilic and ionic drugs also to achieve their controlled release. Despite many advantages, SLNs eliminate quickly from the blood by RES, presenting a challenge for sustaining the drug release [87, 88]. Quian et al. developed biomimetic low-density lipoprotein nanocarrier in which the native apolipoprotein B-100 (apoB-100) was replaced with a lab-synthesized artificial amphipathic hybrid peptide (FPL), which consisted of a lipid-binding motif of apoB-100 via modified solvent emulsification method. The FPL-decorated paclitaxel-loaded lipoprotein-mimicking nanoparticles demonstrated markedly improved in vitro antitumor effect than Taxol formulation [89].

4.3.3 Polymeric nanocarriers (PNCs)

PNCs are solid nanosized carriers made up of biodegradable polymers within 10–1000nm size range [90, 91]. PNCs can be classified as matrix type (nanospheres) and reservoir type (nanocapsules). In nanospheres, polymer matrix dissolves or disperses the drug, whereas in nanocapsules, the drug is either dissolved or dispersed in liquid oily or aqueous core which is encapsulated by a solid polymeric membrane [90]. A nanospheres PNC with surface functionalization is depicted in Fig. 5B. Drug loading in nanospheres or nanocapsules can be done via adsorption or chemical conjugation to the nanoparticle surface [92]. Several methods have been developed to prepare PNCs based on their composition and the desired properties. The preparation methods of PNCs are based on either the use of the dispersion of preformed polymers or the use the direct polymerization of monomers. The methods such as solvent evaporation, salting out, nanoprecipitation, dialysis, and supercritical fluid technology involve the dispersion of preformed polymers, whereas methods such as emulsification polymerization, interfacial

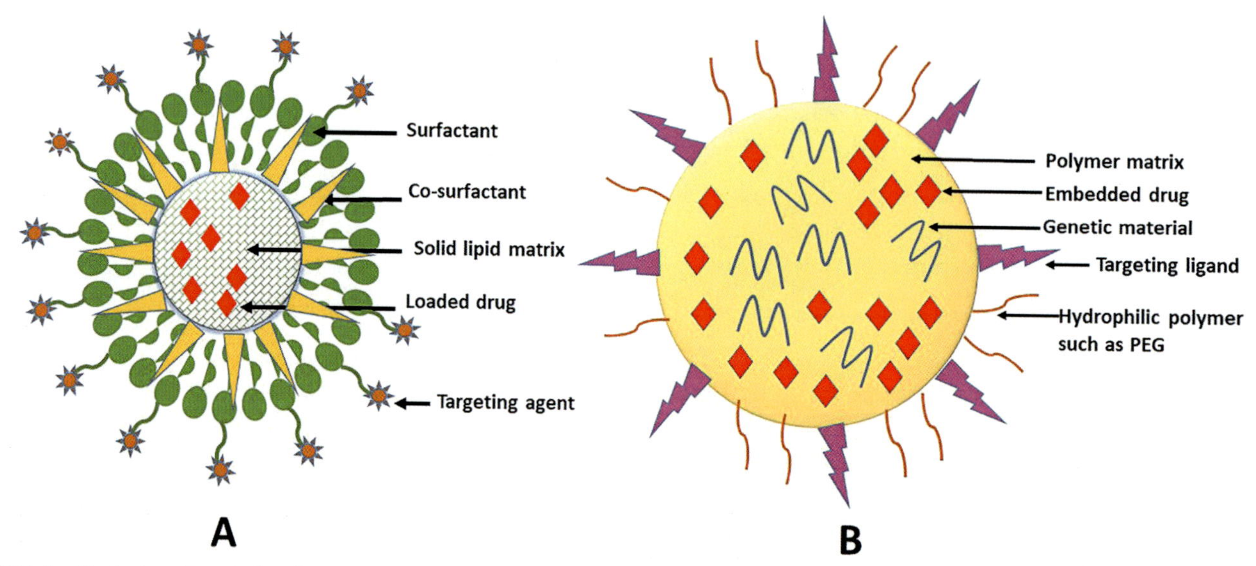

FIG. 5 Illustration of a typical and/or surface modified (A) solid lipid nanocarrier and (B) polymeric nanocarrier.

polymerization, and controlled/living radical polymerization involve direct polymerization of monomers [90].

PNCs employ natural and synthetic biocompatible and biodegradable polymers which degrade into individual monomer units in the body via normal metabolic pathway, and finally removed from the body [92]. Most commonly employed synthetic polymers in PNCs include PEG, polylactic acid (PLA), polyglycolic acid (PGA), PLGA, polycaprolactone (PCL), hydroxypropyl methacrylamide (HPMA) copolymer, and polyaspartic acid (PAA), whereas most commonly employed natural polymers include albumin, alginate, chitosan, collagen, dextran, gelatin, and heparin [93]. Advantages offered by PNCs include high payload, uniform particle size-distribution, controllable physicochemical properties, in vivo stability, and high blood circulation times, which are desirable characteristics for successful cancer therapy [94]. The controllable physicochemical properties of PNCs are achievable due to flexible polymer properties like molecular weight, dispersity index, hydrophobicity, crystallinity, and biodegradability. Modification of these polymer properties enables accurately controlled and desired drug release profiles from the PNCs [95]. In addition, surface modification of polymers is used in developing stimuli-sensitive polymer systems capable of altering their own characteristics in response to certain stimuli which prove to be more accurate drug delivery in cancer treatment [91, 96]. Zhu et al. developed docetaxel-loaded polydopamine-modified nanoparticles using D-α-tocopherol polyethylene glycol 1000 succinate-poly(lactide), which was conjugated to galactosamine for enhancing the docetaxel delivery via ligand-mediated endocytosis. The results suggested specific nanoparticle interactions with the hepatocellular carcinoma cells via ligand-receptor recognition demonstrating higher cellular uptake, enhanced tumor inhibition, and reduced tumor size on hepatoma-bearing nude mice [97].

4.3.4 *Polymeric micelles (PMs)*

Amphiphilic (molecule with both hydrophobic and hydrophilic groups) block copolymers can self-assemble in aqueous solution to form micelles [98]. To form PM, two copolymers are used in which one copolymer is soluble in solvent and forms hydrophilic shell, while the other copolymer is insoluble in solvent and forms hydrophobic core, resulting in the "core-shell structure" [98–100]. The hydrophilic shell of micelles reduces their clearance by the mononuclear phagocytic system (MPS), whereas the hydrophobic core functions as a hydrophobic drug reservoir. In addition, the hydrophobic core entraps hydrophobic drug and controls the drug release, while the hydrophilic shell ensures the solubility of micelles in aqueous media, stabilizes the core controls the pharmacokinetics, and prevents their recognition uptake by the RES, resulting in increased blood circulation time [101]. There are mainly three techniques to form drug-loaded block copolymer micelles (Fig. 6). In the first technique known as "direct dissolution method,"

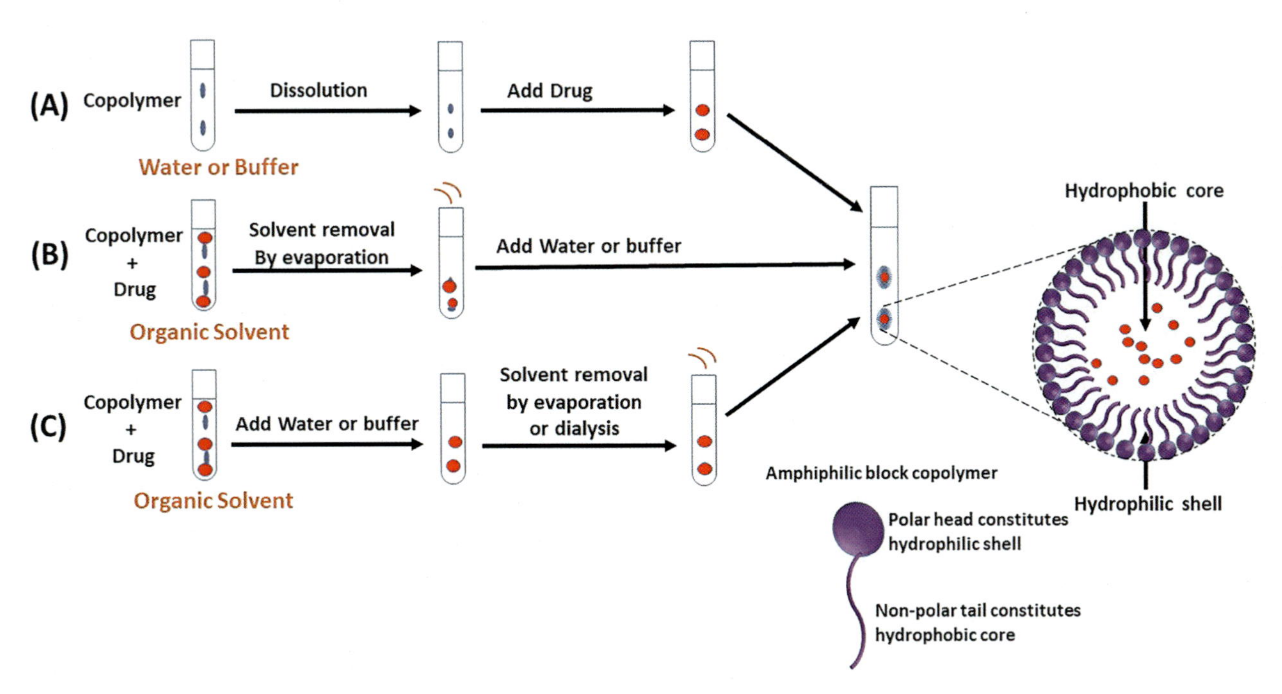

FIG. 6 Preparation technique for drug-loaded polymeric micelles. (A) In direct dissolution method, copolymers and drugs are dissolved in water, which spontaneously forms drug-loaded polymeric micelles. (B) In solvent evaporation method, copolymers and drugs are first dissolved in organic solvent, followed by evaporation of solvent and subsequent addition of water to form drug-loaded micelles. (C) In dialysis method, copolymers and drugs are dissolved in water-miscible organic solvent followed by addition of water to form micelles, followed by solvent removal via dialysis.

the copolymers and drugs are directly dissolved in water to spontaneously form the drug-loaded micelles. In the second technique known as "solvent evaporation," the copolymers and drugs are dissolved in an organic solvent, the solvent is evaporated, and water is added to form drug-loaded micelles. In the third technique known as "dialysis method," the copolymers and drugs are dissolved in a water-miscible organic solvent and water is added to initiate formation of micelles with swollen cores. The solvent is removed by dialysis to form drug-loaded micelles [98–100].

The functionality of PMs allows attachment of different tumor-targeting ligands like transferrin, folate, antibody fragments, and epidermal growth factors at the outer shell surface, which is used for active targeting of anticancer agent to the cancer tissues [101, 102]. The small size and increased retention time of PMs favor their accumulation in the cancer tissues for passive targeting [100]. Besides, PMs can be designed as smarter stimuli-responsive systems which are sensitive and responsive to changes in pH, temperature, heat, ultrasound, etc. to enhance drug accumulation in tumoral cells by active targeting [103]. Block copolymers for preparation of PMs include poloxamers, PEGylated polylactic acid (PEG–PLA), PEGylated polyaspartic acid (PEG-PAA), PEGylated polyglutamic acid (PEG-PGA), PEGylated polycaprolactone (PEG-PCL), and PEGylated poly(lactic-*co*-glycolic acid) (PEG-PLGA) [98–101]. Zhong et al. reported 71.6% tumor growth inhibition with 93.5% reduction in lung metastasis by cabazitaxel-loaded polymeric micelles in metastatic breast cancer model [104].

4.3.5 Polymer-drug conjugates (PDCs)

The concept of polymer-drug conjugation was first introduced in 1975 where a polymer carrier was used as a targetable drug carrier to enhance the drug's performance such as selectivity toward the target site, cellular uptake, solubility, and stability [105]. Since then, plethora of work has been done in this area resulting in several PDCs reaching clinical approval, for example, the approval of the PEGylated protein conjugate PEG-L-asparaginase (Oncaspar™) for treatment of leukemia [106]. The polymer conjugates can also use protein drugs to form polymer-protein conjugate with reduced protein immunogenicity and increased protein solubility and stability. In early 1990s, SMANCS was approved for the treatment of hepatocellular carcinoma. Since then, it is used routinely in cancer treatments as an adjunct to chemotherapy and combination therapies [107].

A PDC typically consists of a drug, a polymer carrier, and a linker, with a targeting group optionally, whereas the linker and targeting ligand possess desirable properties and functional groups responsible for conjugation (Fig. 7). In a PDC, the drug is covalently attached to the polymer via a biodegradable linker. This is absent in other nanocarrier systems such as liposomes and PNPs where the drug is loaded by physical entrapment within the carrier. The principle behind conjugation of a drug to a polymer via covalent bonding was based on the concept that increasing the molecular weight of a drug can significantly alter its biological behavior, which results in drug delivery benefits including long blood circulation time, restricted body distribution, and selective drug release [29]. Each constituent of a PDC

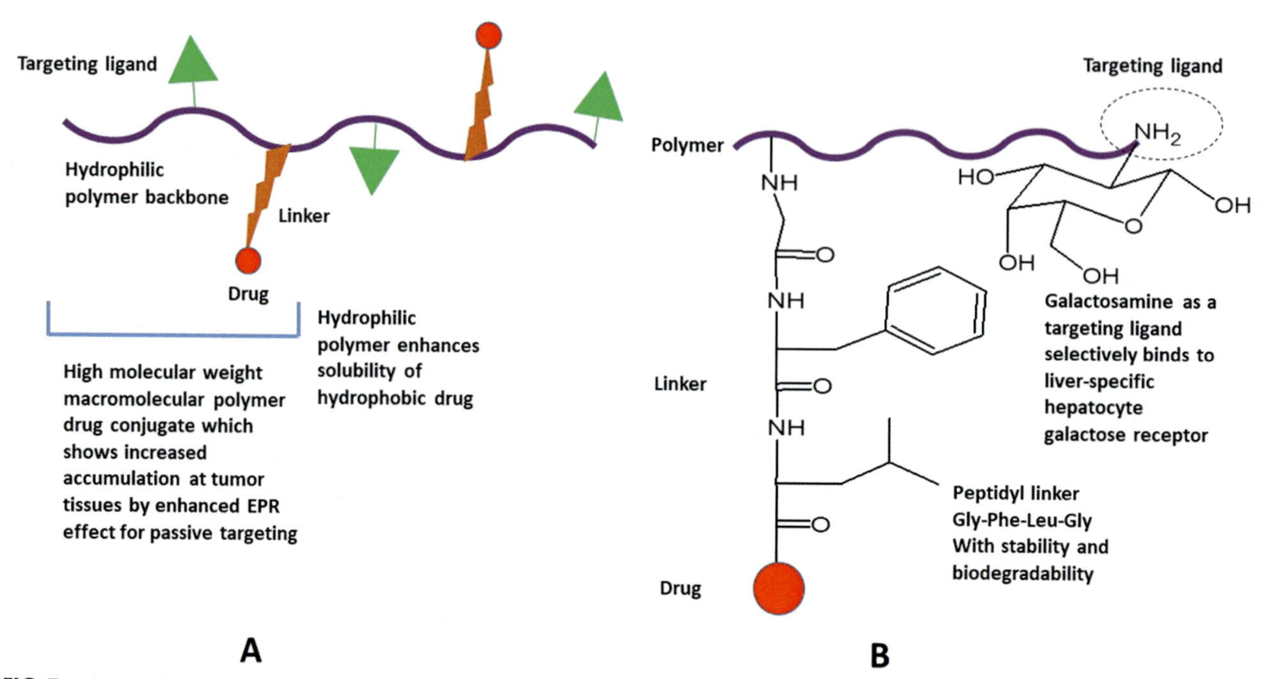

FIG. 7 (A) A typical polymer-drug conjugate for active tumor targeting. (B) Peptidyl Gly-Phe-Leu-Gly linker is stable in blood and can be degraded at the target site. Galactosamine is a targeting ligand which can be specifically linked to the polymer via primary amino functional group.

strongly affects the biological behavior as well as physicochemical properties like solubility of the PDC system. Polymeric carriers included for clinically tested PDCs include PEG, PGA, HPMA copolymers, and oxidized dextran [107]. The linker is used to connect the drug to the polymeric carrier. Linkers such as peptidyl linkers and pH-labile linkers have been suggested for use in PDCs [107]. A good linker needs to be biodegradable at target site and stable in the blood so that premature release of the drug can be prevented. Most of the PDCs use EPR effect for passive tumor targeting. However, the PDC can be prepared with a targeting group galactosamine (amino sugar), which selectively binds to the liver-specific hepatocyte galactose receptor for active targeting of the tumor [108].

The macromolecular size of a PDC restricts extravasation of the drug in regions where the vascular endothelium is intact, causing the drug access to only those tissues where the vascular endothelium exhibits gaps of size >20nm, which is generally seen in case of the defective tumor vasculature. It indicates that the PDC utilizes EPR effect for passive targeting of the tumor tissues [109]. Moreover, PDC can modify the cellular pharmacokinetics of the drug due to its macromolecular size that allows its uptake by endosomes, followed by lysosomes. The lysosomal compartment typically possesses acidic pH 4–5 and a proteolytic enzyme-rich environment that acts as drug release triggers [110]. In fact, PDCs have been designed with acidic pH-sensitive or proteolytic enzyme-sensitive biodegradable linkers, leading to exclusively intracellular release of anticancer drug to reach its biological target such as the nucleus and the DNA effectively [111]. In PDC, the conjugated drug is protected by polymer from its early inactivation during delivery to the site of action, whereas the macromolecular size prevents the early clearance of the drug through renal filtration. Moreover, PDC has been employed for solubility enhancement and delivery benefits of poorly water-soluble anticancer compound 1,5-diazaanthraquinones, using hydrophilic water-soluble HPMA polymer [112]. Yang et al. developed polymer conjugates by covalently attaching high, medium, and low contents of anticancer drug podophyllotoxin and PEG with acetylated carboxymethyl cellulose. It was found that the drug loading in PDC influenced the nanoparticle sizes where low drug content conjugates formed the smallest nanoparticles (20nm), showed faster drug release, and improved tumor penetration as compared with the high drug content conjugates (30–120nm) [113].

4.3.6 Dendrimers

Dendrimers are three-dimensional, exceedingly branched, star-shaped macromolecules with various arms originating from the central core. In addition to central core, dendrimer consists of interior branches and exterior surface. The surface consists of terminal functional surface groups [114] (Fig. 8). The dendrimers can be produced using polymers, carbohydrates, sugars, nucleotides, and amino acids using step-by-step synthesis techniques to yield highly regular hyperbranched molecules with distinct number of peripheral groups. The step-by-step synthesis processes for dendrimers are different as compared with routine polymerization processes in terms of highly arranged branching patterns and unique architecture compared with traditional polymerization processes which typically result in irregular branching patterns [114, 115]. Moreover, multivalency, monodispersity, distinct shape, nanosize (1.5–14.5nm diameter), and surface functionality further extends dendrimers' drug delivery applications [115]. During the dendrimer synthesis, branches can be added to the core at each level throughout the synthesis, and addition at each synthetic cycle is called a "generation" (G). The calculated properties of the primary amine surface polyamidoamine (PAMAM) dendrimers having ethylenediamine core with increasing generation are summarized in Table 4 [116].

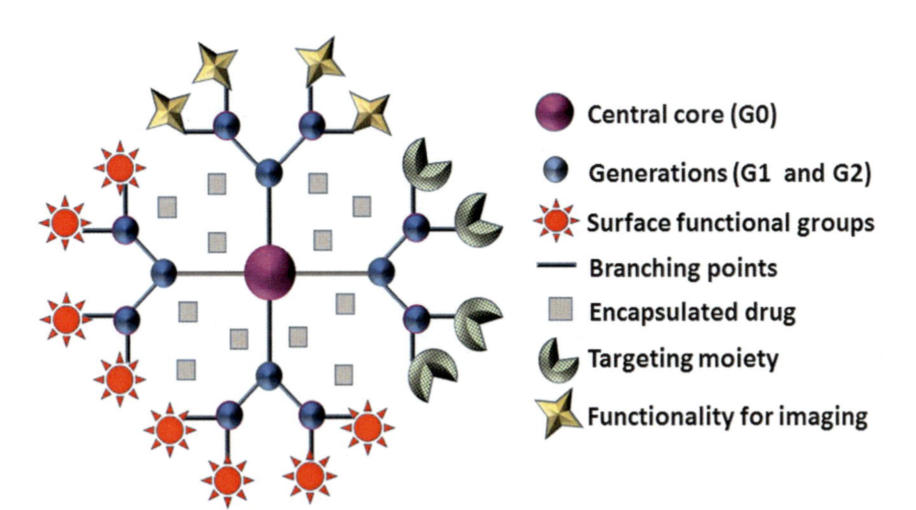

FIG. 8 A typical dendrimer structure with functional surfaces.

TABLE 4 The calculated properties of the primary amine surface PAMAM dendrimers by generation.

S. No.	Generation	Measured diameter (nm)	Molecular weight	Number of surface groups
1	G0	1.5	517	4
2	G1	2.2	1430	8
3	G2	2.9	3256	16
4	G3	3.6	6909	32
5	G4	4.5	14,215	64
6	G5	5.4	28,226	128
7	G6	6.7	58,048	256
8	G7	8.1	116,493	512
9	G8	9.7	233,383	1024
10	G9	11.4	467,162	2048
11	G10	13.5	934,720	4096

Recently, the pioneers of dendrimer research, the Tomalia group, reviewed the progress of over three decades of dendrimer chemistry and branch cell symmetry for advanced drug delivery applications. It was revealed that dendrimer family can be classified into two categories of symmetry: (i) symmetrical branch cell dendrimers comprise of interior hollowness/porosity and (ii) asymmetrical branch cell dendrimers comprise of no interior/void space. This information could be critical during preliminary studies, which confirms the effect of branch cell symmetry on interior packing properties [117].

The dendrimers can be functionalized by attachment of the drugs and targeting ligands to their surfaces, enabling the multifunctional dendrimers to contact precisely at active sites which can reduce the side effects as presented in Fig. 8. Besides anticancer drug delivery, dendrimers have been extensively used for gene delivery, immunology, magnetic resonance imaging, and vaccines delivery [118]. Drug loading in dendrimer cores can be achieved by hydrogen bonding, chemical linkages, or hydrophobic interactions; however, the drug is rarely loaded in the interior branches to prepare a dendrimer-drug conjugate. De Groot et al. investigated a "cascade release" dendrimer in which the dendritic boundary was used to attach the drug using acid-labile or disulfide linkers. These linkers created acidic or reduction-prone atmosphere close to the cancer cells, which led to release of all the attached drug molecules to the outside upon single triggering with masked 4-aminocinnamyl alcohol linker. This release utilized a reaction in which the nitro group underwent reduction to an amino group at slightly acidic cancer environment [119]. Different drug-dendrimer conjugates have been developed for improved anticancer activities including the cisplatin-dendrimer conjugation for enhanced antitumor activities as compared with cisplatin alone [120]; the 5-Flourouracil-dendrimer conjugates of G0.5–5.5 using sequential propagation technique for controlled release of anticancer drugs [121]; and PEGylated dendrimer-doxorubicin conjugation for enhanced retention and circulation time while reducing the toxicity [122]. Recently, Liu et al. screened and identified a high-affinity peptide ligand "EGFR-binding peptide 1 (EBP-1)" and conjugated this targeting ligand with PAMAM dendrimer to encapsulate doxorubicin. The bi-functional drug carrier substantially enhanced both tumor-targeting efficiency and antiproliferation effect, which resulted in remarkable tumor inhibition efficacy against human breast cancer [123].

4.3.7 Lipid-polymer hybrid nanocarriers (LPHNs)

The concept of hybrid nanocarriers emerged with a view of incorporating two nanocarriers together to attain the dual nature of both components, which might result in considerable enhancement in its properties [124]. In advanced cancer therapy, polymer and lipid materials have been widely employed due to certain advantages. For instance, liposomes offer excellent biocompatibility but lack structural integrity, which results in leakage of liposomal contents and instability during storage. On the other hand, polymeric carriers possess structural integrity, storage stability, high drug loading capacity, and controlled release ability. Moreover, liposome by itself readily cleared off by the RES and most often requires to be PEGylated to enhance in vivo circulation time. Therefore, researchers endeavored to combine their advantages and address their limitations by formulating new-generation therapeutic class of LPHN [125]. The hydrophilic drug has been delivered with a high drug-loading capacity as well as release control using this new variation of nanocarrier [124]. The schematic illustration of LPHN is presented in Fig. 9.

In LPHN, the small molecule drug and/or diagnostic agent is encapsulated in polymer core surrounded by an inner layer of lipid. Thus, the inner lipid layer envelopes the polymer core, rendering the polymer core biocompatible. An outer lipid-PEG layer serves as a stealth coating provides steric stabilization with simultaneous prolongation of blood circulation time of the LPHNs [126]. In addition, the inner lipid layer provides a molecular defense that reduces the discharge of the encapsulated content and also delays the rate of polymer degradation in LPHNs product by limiting ingoing water diffusion, enabling sustained release of the content. Therefore, the core-shell structure of the LPHNs provides structural integrity, storage stability, and sustained release [124]. The most preferred core in LPHN is PLGA due to its biocompatible and biodegradable properties as well as versatileness of drug loading [127]. Lipid-coated PLGA nanoparticles can be conjugated to transferrin for enhanced delivery of aromatase inhibitors [128]. Various lipids used in the shell of LPHN include phosphatidylcholine, 1,2-dilauroyl-sn-glycero-3-phosphocholine, 1,2-distearoylsn-glycero-3-phosphoethanolamine, cholesterol, myristic acid, stearic acid, and 1,2-dipalmitoyl-sn-glycero-3-phosphocholine [127]. Core-shell type PLHN exhibited potential of siRNA delivery for in vivo tumor targeting [129]. Wang et al. developed hyaluronic acid (ligand)-modified irinotecan and gene-loaded LPHNs using solvent evaporation method as a targeted therapy for treatment of colorectal cancer. The hybrid nanocarriers were able to encapsulate over 80% drug and over 90% of gene with superior efficacy against tumor growth inhibition and the distinguished transfection efficiency in vivo [130].

5 Clinical status of organic nanocarriers for cancer treatment

The growing interest in the use of nanotechnology for cancer treatment has led to several nanocarrier-based approved drug delivery systems in clinics, whereas many more are under clinical development [50]. Approved organic nanocarriers on market which employs passive targeting modality are summarized in Table 5. Representative clinical-stage products based on both passive and active targeting mechanisms are summarized in Table 6. In recent years, most nanoformulations approved by either the FDA or under clinical trials repurposed the previously approved technologies for treatment of breast, gynecological, lung, lymphoma, solid tumor, mesenchymal tissue, central nervous system, and genitourinary cancers.

6 Advanced cancer targeting approaches

6.1 Tumor-targeting antibodies

Tumor-targeting approaches based on cancer immunotherapy imply two strategies: the first strategy is to identify antigens specific to the cancer cells which is known as "tumor-specific antigens," whereas the second strategy implies to target specific antigens overexpressed by the cancer cells which is known as "tumor-associated antigens" [132]. When antibodies or ligands target specific antigens, it causes

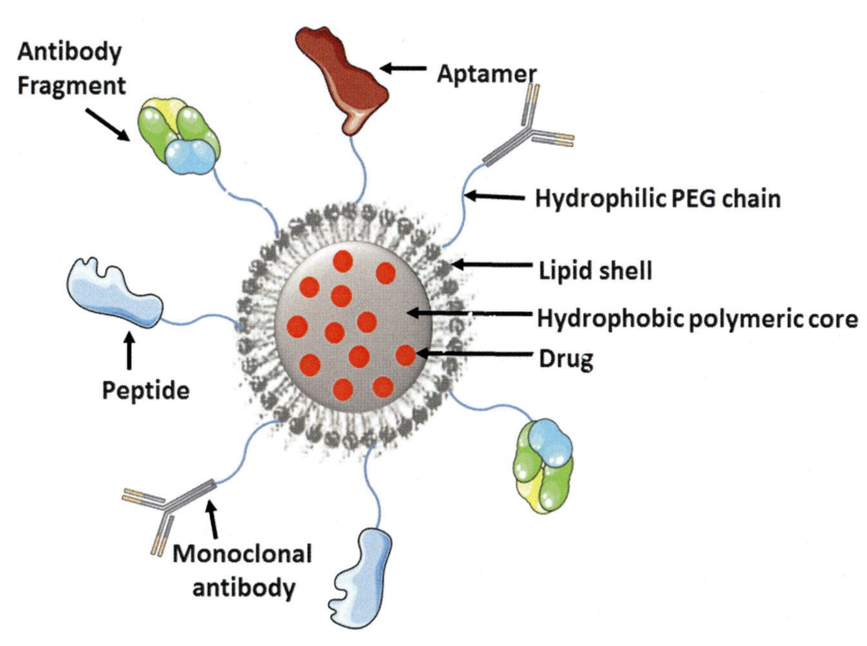

FIG. 9 A lipid-polymer hybrid nanocarrier with functionalization for various applications.

TABLE 5 Approved passive targeting organic nanocarriers for cancer treatment.

Generic/brand name	Type of nanocarrier	Therapeutic agent	Indication	Company (approval year)	References
Styrene maleic anhydride neocarzinostatin (SMANCS)	Polymer conjugate	Neocarzinostatin	Liver and renal cancer	Yamanouchi, Japan (1993)	[21]
PEG-L-asparaginase (Oncaspar)	Polymer conjugate	Pegaspargase	Leukemia	Baxter International, USA (1994)	[111]
Liposomal doxorubicin (Doxil)	PEGylated liposome	Doxorubicin	Sarcoma, ovarian cancer, multiple myeloma	Janssen, USA (1995)	[35]
Liposomal daunorubicin (DaunoXome)	Liposome	Daunorubicin	Sarcoma	NeXstar Pharmaceuticals, USA (1996)	[35]
Liposomal doxorubicin (Myocet)	Liposome	Doxorubicin	Breast cancer	Elan Corporation, Ireland (2000)	[34]
Nab-paclitaxel (Abraxane)	Albumin nanoparticle	Paclitaxel	Breast, lung, and pancreatic cancer	American Pharmaceutical Partners, Inc., USA (2005)	[34]
Polymeric micelle paclitaxel (Genexol®PM)	Polymeric micelle	Paclitaxel	Breast cancer, NSCLC[a]	Samyang Corporation, Korea (2007)	[35]
Mifamurtide (Mepact)	Liposome	Muramyl tripeptide Phosphatidylethanolamine	Osteosarcoma	Takeda Pharmaceutical Company Limited, Japan (2009)	[35]
Liposomal vincristine (Marqibo)	Liposome	Vincristine sulfate	Leukemia	Talon Therapeutics, USA (2012)	[55]
Liposomal irinotecan (Onivyde)	PEGylated liposome	Irinotecan	Pancreatic cancer	Merrimack Pharmaceuticals, Inc., USA (2015)	[55]

[a] *Non-small-cell lung cancer.*

TABLE 6 Representative nanocarriers under clinical development for cancer treatments.

Targeting mechanism	Proprietary name	Nanocarrier system	Therapeutic agent	Indication	Clinical stage	References
Passive targeting	Liposomal cisplatin (Lipoplatin)	PEGylated liposome	Cisplatin	NSCLC[a]	Phase III	[34]
	NK-105	Polymeric micelle	Paclitaxel	Breast cancer	Phase III	[34]
	Liposomal paclitaxel (EndoTAG-1)	Liposome	Paclitaxel	Pancreatic cancer, liver cancer, HER2-negative and triple-negative breast cancer	Phase II	[19]
	Nab-rapamycin (ABI-009)	Albumin nanoparticle	Rapamycin	Advanced malignant PEComa[b]; advanced cancer with mTOR[c] mutations	Phase II	[19]
	CRLX-101	Polymeric nanoparticle	Camptothecin	NSCLC; renal cell carcinoma; ovarian, tubal, or peritoneal cancer	Phase II	[71]
Active targeting	MM-302	HER2[d]-targeting liposome	Doxorubicin	HER2-positive breast cancer	Phase II/III	[21]
	Polyglutamate-paclitaxel (CT-2103; Xyotax)	Polymer-drug conjugate	Paclitaxel	Different cancers like NSCLC; ovarian cancer	Phase III	[131]
	HPMA-doxorubicin PK1 (FCE28068)	Polymer-drug conjugate	Doxorubicin	Different cancers like lung and breast cancers	Phase II	[19, 34]
	Polyglutamate-camptothecin (CT-2106)	Polymer-drug conjugate	Camptothecin	Different cancers	Phase I	[19, 21]
	BIND-014	PSMA[c]-targeting polymeric nanoparticle	Docetaxel	NSCLC, prostate cancer	Phase II	[19]
	MBP-426	Transferrin receptor-targeting liposome	Oxaliplatin	Gastric, esophageal, and gastroesophageal adenocarcinoma	Phase I/II	[21]
	Immunoliposomes loaded with doxorubicin	EGFR-targeting liposome	Doxorubicin	Solid tumors	Phase I	[19, 72]
	MM-302	HER2-targeting liposome	Doxorubicin	HER2-positive breast cancer	Phase II/III	[21]
External stimuli triggering	ThermoDox	Liposome	Doxorubicin	Hepatocellular carcinoma	Phase III	[19]

[a] *Non-small-cell lung cancer.*
[b] *Malignant PEComa is a rare and aggressive perivascular epithelioid cell tumor.*
[c] *Prostate-specific membrane antigen.*
[d] *Human epidermal growth factor receptor 2.*

the accumulation of anticancer agent at the specific cancer tissues. However, when targeting is achieved with nonantibody ligands such as arginine-glycine-aspartate (RGD), folate, and transferrin, it suffers from a disadvantage of being adequately available on normal cells also. Therefore, the anticancer agent targeted through nonantibody ligands can show binding to nontarget tissues as well [133]. Moreover, ligands such as folate are abundantly present as its free form in diet, which can compete for the ligand-modified targeted delivery. Due to these reasons, antibodies serve as much better target-specific and selective ligands compared with nonantibody ligands [132]. Various immunoglobulin (Ig) antibodies such as IgG, IgA, IgM, and IgE circulate in serum, among which IgG is mainly used for therapeutic and diagnostic applications. Igs are proteins consisting of two heavy (H) and two light (L) chains, and they can be functionally divided into two domains: variable (V) domains and constant (C) domains. The V domain binds to antigens, whereas C domain specifies effector functions via activation or binding to fragment crystallizable (Fc) receptors [134]. IgG has a Y-shaped structure with two arms containing antigen recognition sites, whereas the stem structure in lower part conducts effector functions. Advanced antibody engineering techniques have enabled development of various antibody-based cancer targeting systems for cancer immunotherapy [132].

6.1.1 Monoclonal antibodies (mAbs)

mAbs is a major treatment approach based on cancer immunotherapy. mAbs are man-made antibodies developed in vitro either to act as antibodies themselves or to strengthen body's natural antibodies to fight cancers. mABs act by recognizing, targeting, and blocking a specific protein antigen on cancer cells (Fig. 10); therefore, mAb-targeted immunotherapy-based cancer treatment is limited only to cancers in which antigens have been identified [131]. Georges Köhler and César Milstein in 1975 developed mAbs using mice hybridoma technique to create hybridoma using fusions of myeloma cell lines with B cells that produced immortalized antigen-specific antibodies [135]. In 1997, rituximab was the first clinically approved mAb for the treatment of B cell lymphoma [135]. Since then, the approach is being employed as a standard treatment for many hematological and solid tumors [131, 135]. Most of the targeted therapies are either small molecule compounds or mAbs; however, the difference being that the small molecules can target and enter inside the cell unlike mAbs usually cannot enter cells and only targets the cell surface [136].

Chemotherapeutic mAbs (CmAbs) target the tumor cells by binding to various cell surface antigens. These cell-surface antigens are categorized as (i) cluster of differentiation (CD), e.g., CD20, CD30, CD33, and CD52; (ii) epidermal growth factor receptor (EGFR), e.g., receptor activator of nuclear factor kappa-B ligand (RANKL) and human epidermal growth factor receptor 2 (HER2); (iii) vascular endothelial growth factor (VEGF), e.g., VEGF receptor (VEGFR); (iv) integrins, e.g., αVβ3 and α5β1; (v) carcinoembryonic antigen (CEA); (vi) fibroblast activation protein (FAP); and (vii) extracellular matrix metalloproteinase inducers (EMMPRIN) [137]. mAbs can be of two types: unconjugated or conjugated. The unconjugated type is the most common which works by themselves without attaching to the radioactive materials [136]. For instance, CD52 is an antigen found on lymphocytes, to which Alemtuzumab can

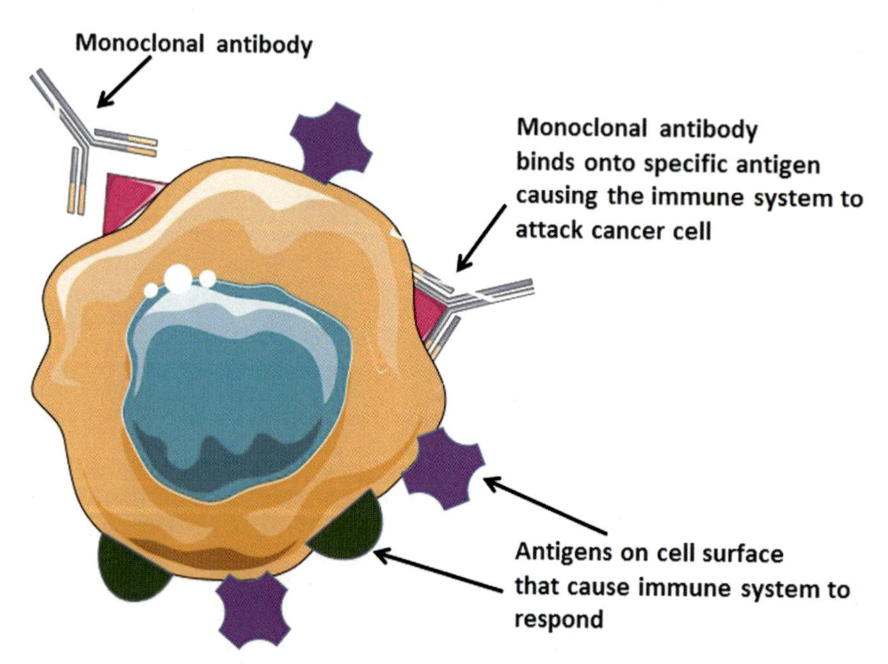

FIG. 10 An illustration of monoclonal antibody which can bind to specific antigen proteins present on surface of cancer cell.

bind. Once bind to the antigen, the antibody stimulates immune cells to kill these cells. The conjugated mAbs joined to chemotherapeutic agent or radioactive substance by working as homing device to take these substances directly to the cancer cells [138]. The mAb circulates all over the body, finds and attaches to the target antigen, and then delivers the therapeutic agent where it is most required, thereby minimizing the destruction to healthy cells [136, 138].

CmAb sticks to a specific protein antigen overexpressed in tumor and engages the immune system to target and kill antigen-containing cells. The cell destruction happens through three main mechanisms which are (i) direct tumor cell death, (ii) immune-mediated tumor cell killing, and (iii) vascular and stromal ablation [136, 138]. In direct tumor cell death, the cell death occurs by targeting and inhibiting cell survival signals via receptor blockade, apoptosis induction, anticancer drug delivery, or radiation modalities [139, 140]. Immune-mediated tumor cell killing occurs by occupying antibody-dependent-cell-mediated cytotoxicity (ADCC), T-cell function regulation, targeting gene-modified T cells, and specific effects on tumor vasculature, activating complement-mediated cytotoxicity, or cellular phagocytosis. Immunostimulatory CmAbs can activate T lymphocyte cells by inhibiting T-lymphocyte inhibitory receptors [131]. Moreover, CmAbs use effector mechanisms such as abrogation of tumor-cell signals (e.g., cetuximab and trastuzumab), ADCC (e.g., rituximab), and immune modulation of T-cell function (e.g., ipilimumab) as most successful cancer treatment approaches [131]. Since 1997 to present, several CmAbs gained FDA approval for treatment of different cancers [136–140], which are summarized in Table 7.

TABLE 7 FDA-approved chemotherapeutic monoclonal antibodies in clinic.

CmAb	Type	Target antigen	Approval year	Cancer-type indication
Rituximab	Chimeric	CD[a]20	1997	Non-Hodgkin's lymphoma
Trastuzumab	Humanized	HER[b]2	1998	HER2-positive metastatic/nonmetastatic breast cancer; gastric cancers
Alemtuzumab	Humanized	CD52	2001	B-cell chronic lymphocytic leukemia
Tositumomab	Murine	CD20	2003	CD20-positive non-Hodgkin's lymphoma
Cetuximab	Chimeric	EGFR[c]	2004	Metastatic colorectal cancer, metastatic non-small-cell lung cancer, and head and neck cancer
Bevacizumab	Humanized	VEGF[d]-A	2004	Colorectal cancer, NSCLC[e], HER2-negative breast cancer, RCC, cervical cancer, malignant brain tumor, ovarian, fallopian tube, or peritoneal cancer
Panitumumab	Human	EGFR	2006	Colorectal cancer
Catumaxomab*	Chimeric mouse-rat hybrid	EpCAM[f]/CD3 multitargeting	2009	Malignant ascites
Ofatumumab	Human	CD20	2009	Chronic lymphocytic leukemia
Ipilimumab	Human	CTLA-4[g]	2011	Late-stage metastatic melanoma and cutaneous melanoma
Cetuximab	Chimeric	EGFR	2011	Head and neck cancers
Pertuzumab	Humanized	HER2	2012	Breast cancer
Ado-trastuzumab	Humanized	HER2	2013	HER2-positive breast cancer
Elotuzumab	Humanized	SLAMF7[h]	2015	Multiple myeloma
Necitumumab	Human	EGFR	2015	Metastatic squamous NSCLC
Daratumumab	Human	CD38	2015	Multiple myeloma
Dinutuximab	Chimeric	GD2[i]	2015	Pediatric high-risk neuroblastoma
Ramucirumab	uman	VEGFR-2[j]	2015	Gastric cancers; NSCLC; colorectal cancer

Continued

TABLE 7 FDA-approved chemotherapeutic monoclonal antibodies in clinic.—cont'd

CmAb	Type	Target antigen	Approval year	Cancer-type indication
Nivolumab	Human	PD-1[k]	2015	Melanoma, NSCLC, RCC[l], recurrent or metastatic HNSCC[m]
Olaratumab	Human	PDGFR-α[n]	2016	Soft tissue sarcoma
Atezolizumab	Humanized	PD-L1[o]	2016	Urothelial carcinoma
Olaratumab	Human	PDGFR	2016	Soft tissue sarcoma
Atezolizumab	Humanized	PD-L1	2016	Urothelial carcinoma, metastatic NSCLC
Pertuzumab	Humanized	HER2	2017	HER2-positive breast cancer; approved for cutaneous squamous cell carcinoma in 2020
Gemtuzumab ozogamicin	Humanized	CD33	2017	Blood and bone marrow cancer
Ocrelizumab	Humanized	CD20	2017	Multiple sclerosis
Inotuzumab-ozogamicin	Humanized	CD22	2017	Precursor B-cell acute lymphoblastic leukemia
Avelumab	Human	PD-L1	2017	Trabecular cancer
Durvalumab	Human	PD-L1	2017	Urothelial carcinoma
Rituximab-abbs[p]	Humanized (90%–95% human)	CD20	2018	Non-Hodgkin lymphoma
Cemiplimab	Human	PD-1	2018	Squamous cell carcinoma
Mogamulizumab	Humanized	CCR4[q]	2018	Cutaneous T-cell lymphoma
Polatuzumab vedotin-piiq	Humanized	CD79b	2019	Diffuse large B-cell lymphoma that has progressed or returned after at least 2 prior therapies
Tafasitamab-cxix	Humanized	CD19	2020	Relapsed or refractory large B-cell lymphoma

[a] *Cluster of differentiation.*
[b] *Human epidermal growth factor receptor 2.*
[c] *Epidermal growth factor, receptor.*
[d] *Vascular endothelial growth factor.*
[e] *Non-small-cell lung cancer.*
[f] *Epithelial cell adhesion molecule.*
[g] *Cytotoxic T-lymphocyte-associated protein 4.*
[h] *Signaling lymphocytic activation molecule family member 7.*
[i] *Ganglioside.*
[j] *Vascular endothelial growth factor receptor-2.*
[k] *Programmed cell death 1 receptor.*
[l] *Renal cell carcinoma.*
[m] *Head and neck squamous cell carcinoma.*
[n] *Platelet-derived growth factor receptor α.*
[o] *Programmed death-ligand 1.*
[p] *First biosimilar of rituximab approved by FDA.*
[q] *Chemokine receptor.*

6.1.2 Immune checkpoint targeting

Immune checkpoint refers to several inhibitory pathways constructed into the immune system that are essential for regulating T-cell receptor recognition as antigen during the process of immune response [141]. Immune checkpoint comprises two types of signals: (i) costimulatory immune checkpoint, which is responsible for stimulating immune progress. The examples of costimulatory receptors include CD28 and CD137 (also known as agonist antibodies) and (ii) coinhibitory immune checkpoint, which is responsible for inhibiting immune progress. The examples of inhibitory receptors include programmed cell death protein 1 (PD1) or cytotoxic T-lymphocyte-associated antigen 4 (CTLA-4) (known as antagonist antibodies) [142]. The function of immune checkpoint molecules is to protect the normal tissues when the immune system attacks pathogens. However, the cancer cells smartly escape immune attack since they impair immune resistance mechanism by dysregulating expression of immune checkpoint-related proteins. In case of T cells, the immune responses initiated through antigen recognition by the T-cell receptor are regulated by a balance between costimulatory and inhibitory signaling pathways [141]. Since these immune checkpoints are stimulated by a combination of receptors and their respective ligands, the immune checkpoint therapy depends on simultaneous functioning of agonists or antagonist inhibitory signals of the immune system [143].

Thus, immune checkpoint inhibitor is a type of targeted treatment that uses one's own immune system in attempt to destroy cancer cells by blocking proteins called as checkpoints. Some lymphocytes such as T cells and B cells perform specific functions in immune system and contribute to make immune checkpoints. Since the checkpoint prevents the immune responses to become very strong and kill the cancer cells, blockade of these checkpoints frees the T cells to do its function by killing cancer cells [144]. Checkpoint inhibitor therapy utilizes this fact by targeting and blocking the immune checkpoints to restore the immune functions [143]. Antibodies (Abs) blocking the immune checkpoints is a successful type of cancer immunotherapy [145]. Unlike antibodies used for cancer therapy, antibodies developed as checkpoint inhibitor do not target tumor cells directly, but they target lymphocyte receptors or their ligands to enhance endogenous antitumor activity [143]. The two FDA-approved immune checkpoint receptors for clinical use are CTLA-4 and PD-1 or its ligand PD-L1. T cells express CTLA-4, which controls T-cell activation and competes with the costimulatory receptor CD28 for binding to ligands CD80 and CD86 [146]. During the T-cell activation process, the overexpression of CTLA-4 causes greater ligand affinity for the CD80/86 as compared with CD28 [147], which prevents or controls further increase of the initial T-cell response. PD-1 is expressed on activated T cells, B cells, and natural killer cells which builds a mechanism of tumor immune resistance in peripheral tissues which shows binding to PD-L1 [148, 149]. Fig. 11 illustrates PD-1/PD-L1 immune checkpoint with a cancer cell where the checkpoint inhibitor blocks the PD-1/PD-L1 "checkpoint" and activates T cells, allowing the T cells to recognize the cancer cells as "abnormal" and eventually destroy these cells. FDA-approved checkpoint inhibitors are summarized in Table 8.

6.1.3 Nanobodies

Despite huge clinical success, mAbs suffer from problems such as possible immunogenicity, large size (~150kDa), and incomplete tumor penetration, which are major limitations to their therapeutic efficacy [150]. This led to a research wave toward "third-generation" heavy-chain only antibodies (HcAbs), which consisted of only two heavy chains (HH) and named as "nanobodies" [151]. Nanobodies are a novel class of single-chain antigen-binding fragments derived from naturally occurring heavy-chain-only antibodies present in the camelids serum. Unlike other antibody fragments, huge assembly and molecular optimization are not essential to create nanobodies [151, 152]. Nanobodies are antibodies with a single variable domain (V) located on a heavy chain, also known as VHH antibodies, and lack a VL domain causing destructive effect to antigen binding. However, three peptide loops or complementarity determining regions (CDRs) as CDR1, CDR2, and CDR3 determine the antigen specificity of nanobodies [152]. Among these three loops, the nanobodies are rich with CDR3 loop. Also, the CDR3 is much longer than CDR1 and CDR2 and forms finger-like epitopes which contribute most significantly to their specificity and diversity, and increases their interactions with target antigens [153]. The CDR1 and CDR2 regions aid the nanobody binding to antigen, enabling diversity in antigen-binding sites as compared with mAbs [154]. The structural differences in mAb and HcAb are depicted in Fig. 12.

The compact-sized nanobodies combine the therapeutic advantages of mAbs and the targeting potential of

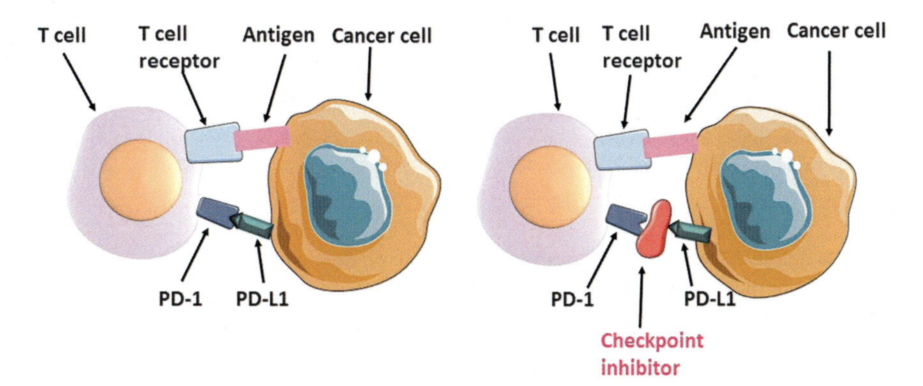

FIG. 11 The PD-1/PD-L1 immune checkpoint with a cancer cell. Checkpoint inhibitor blocks the PD-1/PD-L1 "checkpoint" and activates T cell allowing the T cells to recognize the cancer cells as "abnormal" and eventually destroy these cells.

TABLE 8 FDA-approved checkpoint inhibitors for cancer immunotherapy.

Name (trade name)	Target	Type and source	Cancer type	Company	Approval year
Ipilimumab (Yervoy)	CTLA-4[a]	Whole antibody, human	Metastatic melanoma	Bristol Myers Squibb, USA	2011
Nivolumab (Opdivo)	PD-1[b]	Whole antibody, human	Metastatic melanoma, NSCLC[c], RCC[d], Hodgkin's lymphoma, head and neck cancer, urothelial carcinoma	Ono Pharmaceuticals Co. Ltd., Japan	2014
Pembrolizumab (Keytruda)	PD-1	Whole antibody, humanized (from mouse)	Melanoma, NSCLC, head and neck cancer, Hodgkin's lymphoma	Merck & Co. Inc., USA	2014
Atezolizumab (Tecentriq)	PD-L1[e]	Whole antibody, humanized	NSCLC, bladder cancer	Genentech, USA	2016
Avelumab (Bavencio)	PD-L1	Whole antibody, human	Urothelial carcinoma, neuroendocrine carcinoma	Merck KGaA, Germany and Pfizer Inc., USA	2017
Durvalumab (Imfinzi)	PD-L1	Whole antibody, human	Urothelial carcinoma	AstraZeneca, UK	2017
Cemiplimab (Libtayo)	PD-1	Whole antibody, human	Cutaneous squamous cell carcinoma	Regeneron Pharmaceuticals, Inc., USA	2018

[a] *Cytotoxic T-lymphocyte-associated protein 4*
[b] *Programmed death-1.*
[c] *Non-small-cell lung cancer.*
[d] *Renal cell carcinoma.*
[e] *Programmed death-ligand 1.*

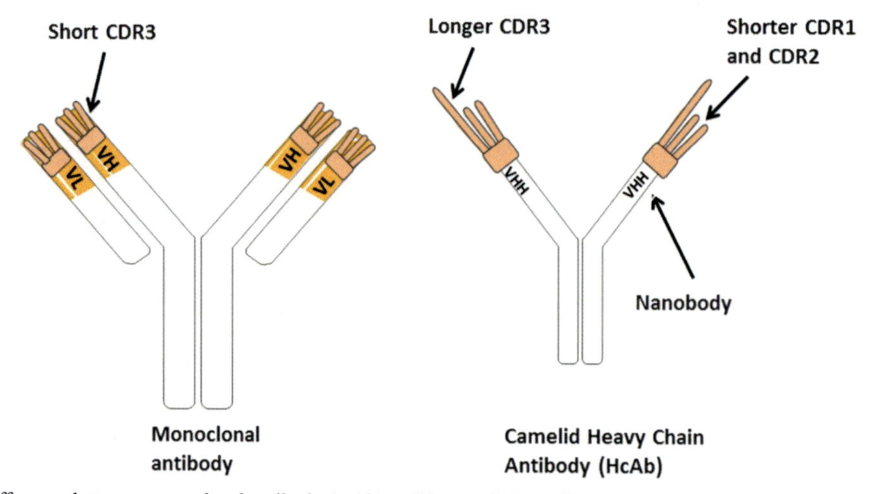

FIG. 12 Structural difference between monoclonal antibody (mAb) and heavy-chain antibody (HcAb) that is VHH nanobody. Compared to VH-VL domain in mAb, the VHH nanobody lacks the VL domain and possess a longer CDR3 loop as compared with that of mAb, enabling stronger antigen affinity by nanobody.

nanoscale delivery, resulting in enhanced tumor penetration [151]. In addition, nanobodies possess superior properties like natural origin, water-solubility, high stability, and strong antigen-binding affinity, which are suitable for next-generation biodrugs in advanced cancer treatments [151, 155]. Moreover, nanobodies possess relative ease of large-scale production than costly manufacture of mAbs [152]. However, the nanobodies can be rapidly eliminated by renal system due to their small size, which leads to shortening their therapeutic lifetime, or increasing its dose and frequency to attain clinical efficacy. Modification of nanobodies to extend their plasma half-life via PEGylation, fusion to antialbumin nanobodies or Fc domains, and multimerization can reduce their renal clearance [153, 154]. In spite of wide avenues of nanobodies' application in cancer therapy, very limited number of clinical trials completed till date. In 2019, the FDA approved Sanofi's caplacizumab (which targets von Willebrand factor, vWF) as the first drug for treatment of a rare disease called acquired thrombotic thrombocytopenic purpura (aTTP), which is characterized by excessive blood clotting in small blood vessels. The clinical status of camelid-based VHH domains is prominent with one product approval. In addition, eight VHH nanobodies are under clinical development for various diseases. Two VHH nanobodies for cancer treatment under clinical trials are LCAR-B38M for multiple myeloma under Phase II by Janssen and BI 836880 for solid tumor under Phase I by Boehringer Ingelheim [156].

6.2 Tumor-targeting and cell-penetrating peptides

Advancements in molecular biology has steered the discovery of several types of cancer targets. Peptides play significant roles in cellular functions and intercellular communication [157]. Peptide ligands are composed of about less than hundred amino acid monomers connected by amide bonds and are considered highly flexible in their chemical compositions. Although modifications such as cyclization, unnatural amino acids, or their linking with chemical linkers are possible, they might result in either diminution or total loss of peptide's target affinity and must be considered carefully. Moreover, they can be rapidly cleared off from the blood and nonspecific binding sites [158]. Cell-penetrating peptides (CPPs) refer to peptides that are internalized into cells and are considered efficient vectors for achieving cellular uptake. CPPs are a group of small, diverse polypeptides that typically comprised of 5–30 amino acids [158]. Among polypeptide chain, several amino acids are positively charged such as lysine or arginine, or the chain may also contain a pattern of alternating polar and nonpolar amino acids [159]. Two main types of CPP structures are polycationic and amphipathic; however, positively charged CPPs have been shown to penetrate inside cancer cells in a receptor-independent manner [160]. Cationic CPPs with at least eight positive charges show electrostatic interactions with cell surface glycosaminoglycans (GAGs) and enter the cell through endocytic pathways [159]. Although CPPs can cross the cellular membrane unlike common peptides, they still suffer other limitations of peptides such as short half-life due to enzymatic degradation and rapid renal elimination, as well as lack of oral bioavailability [159].

The concept of developing tumor-targeting peptides (TTPs) or homing peptides was to target the cancer tissues and accumulate the therapeutic agent inside these tissues using tumor-specific markers found on tumors [161]. The phage-displayed peptide libraries have successfully identified several TTPs which bind to tumor-expressed molecules to function in vivo [161, 162].

Among two types of TTPs, the first type of TTPs are linked to CPPs allowing the transportation of cargo through the cell membrane, thereby adding specificity to the CPPs, which is otherwise absent in CPPs on their own. Attachment of the trans-acting activator of transcription (TAT) sequence (GRKKRRQRRRPPQ) could achieve cargo-internalization from the surrounding media. In second type, TTPs bind directly to a tumor marker causing the internalization of the cargo [163]. Arap et al. demonstrated enhanced efficacy and reduced toxicity of doxorubicin coupled to two of peptides motifs in human breast cancer mice xenografts, indicating inherent tumor-targeting ability RGD (arginine/glycine/aspartic acid) or NGR (asparagine/glycine/arginine) sequenced peptides [164]. TTPs with RGD-like or TAT-like sequences can be linked with cytotoxic agent to form hybrid anticancer systems, which can efficiently deliver cargo to tumors [164]. Few CPPs for treatment of solid tumors are under Phase I trial [165–168].

6.3 Cell membrane-camouflaged biomimetic nanoparticles

Synthetic stealth particles obtained by coating nanoparticles with PEG do not guarantee their escape from RES uptakes, as nanoparticles are recognized as foreign materials by the immune system and renal or hepatic clearance systems [49]. This limitation can be resolved if the nanocarrier systems are more biomimetic and biocompatible to escape early RES uptake. Nanoparticles coating with natural cell membrane are a powerful strategy to make the nanoparticles biomimetic, thereby hiding their recognition by RES and avoiding early clearance. The cell membrane coating results in a core-shell nanostructure where the nanoparticle is the core that carries payload to be delivered at desired site, while the cell membrane is the shell [169]. This cell membrane-camouflaged nanoparticles find applications for targeted drug delivery, photothermal therapy, and combinational therapy [170]. Membranes can be obtained from different source cells which are isolated through ultracentrifugation techniques, while extrusion, sonication, or electroporation techniques are used for coating nanoparticle surfaces with cell membranes [169]. Once the membrane is coated onto the nanoparticle surface, the translocation of membrane proteins available on the cell membranes of source cells is initiated onto the newly nanoparticle surface. The cell membrane coating provides immune escape, longevity, and tumor-targeting abilities to the nanoparticles [171]. Membrane-camouflaged nanoparticles can be fabricated employing PLGA, gelatin, human serum albumin, silica, and liposome or magnetic nanoparticles [169]. Different source cells such as red blood cells, immune cells, cancer cells, stem cell, white blood cells, platelets, macrophages, natural killer cells, or hybrid cells can be employed for camouflaging [170]. Natural blood cell membrane-camouflaged polymeric nanoparticles exhibited 72-h retention in the blood when administered via particle injection, indicating the protein translocation of natural cellular membranes as an effective nanoparticle functionalization approach [172]. Artificial antigen-presenting cells (aAPCs) developed with biomimetic magnetosomes using azide-engineered leukocyte membrane coating and decorated with T-cell stimuli indicated that the persistent T cells were able to slow down tumor growth without marked toxicity in a murine lymphoma model [173].

6.4 Tumor-associated fibroblast targeting

The vastly heterogeneous and multicellular tumor microenvironment is characterized by the presence of tumor cells; adjacent dynamic nontumor cells such as fibroblasts, endothelial cells, and mesenchymal stem cells; and cells of hematopoietic origins such as immune cells in the extracellular matrix [174]. These stromal cells interact closely with tumor cells, contributing to tumorigenesis. Tumor-associated fibroblasts (TAFs) are very dominant constituent of tumor stroma and TAFs are found in all cancers. However, TAFs are abundant in breast, pancreatic, colorectal, prostate, and non-small-scale lung cancer, whereas less prevalent in brain, renal, ovarian, head, and neck cancers [175]. The presence of large number of spindle-shaped TAFs can construct or remodel the extracellular matrix and produce cytokines and growth factors that contribute to metastasis or tumor proliferation in breast, lung, and pancreatic cancer or can even lead to drug resistance [176–179]. β-Catenin signaling can deactivate stromal fibroblasts and reduce the generation of extracellular matrix proteins resulting in tumor-stroma interactions creating a tumor-suppressive microenvironment for improved skin cancer treatment [180]. The dynamic roles of TAFs in genesis, growth, and suppression of tumors led to emergence of TAF-targeted immunotherapies in multiple ways [181]. Although the clear role of TAFs in tumor microenvironment makes them promising targets for immunotherapy, drug delivery strategies to TAFs with high-level specificity are still limited. Nanotechnology-based approaches can be employed as a strategy to achieve enhanced targeting efficiencies in immunotherapy, chemotherapy, or radiation therapy [182]. Therapeutic targets of TAFs and their clinical development are shown in Table 9 [183–191].

7 Challenges and future perspectives of organic nanocarriers

The progress in nanocarriers' applications in cancer treatment is ample as evidenced by several approved nanoproducts on market, and increasing number of products under various stages of clinical trials as well. Myriad of research efforts in the field of passive, active, stimuli-responsive and immunotherapy-based advanced active targeting approaches are underway and several products based on these approaches are under clinical development. However, there are many challenges including biological challenges, manufacturing complexities, biocompatibility and safety, in vitro and in vivo screening, and availability of characterization models, as well as financial and regulatory issues, which pose obstacles in the pathway of clinical translation of nanocarriers for cancer-targeted therapies [192]. The challenges associated with nanoparticle transition from bench to bedside are depicted in Fig. 13.

In addition, current nanocarrier-based delivery systems suffer from limitations such as uptake and rapid clearance by the immune system, poor targeting efficiency, and issues in passing through biological barriers. To resolve these issues, it is essential to understand thoroughly the nanoparticle transport in complex biological network and develop technologies to manipulate the drug or carrier properties in

TABLE 9 Clinical development of therapeutic tumor-associated fibroblasts.

Therapeutic target	Function of therapeutic target	Therapeutic agent	Status	References
VEGF	Angiogenesis	IMC-IC11	Phase I	[183]
		RPI 4610	Phase II	[184]
		Bevacizumab	Phase III	[185]
JAK2[a]	Promotes progression	SAR302503	Phase III	[186, 187]
FAP[b]	Tumor angiogenesis	PT-100	Phase I	[188]
		Sibrotuzumab	Phase I	[189]
MMPs[c]	Cell proliferation	Marimastat	Phase III	[190]
		Tenomastat	Phase III	[191]

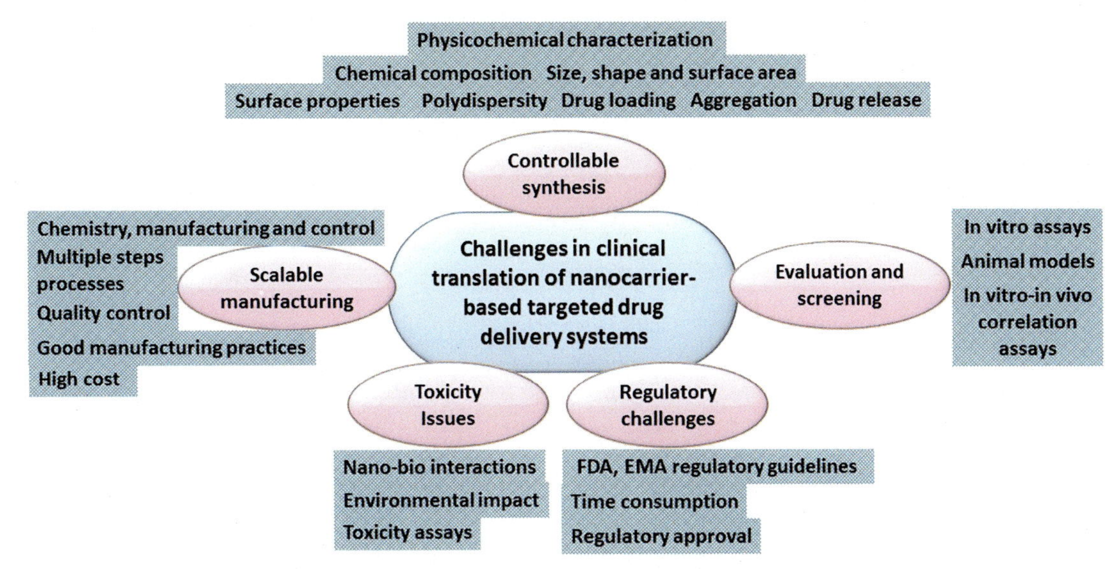

FIG. 13 Various challenges in clinical translation of nanocarrier-based targeted drug delivery systems.

the most effective way [29]. The chemistry, manufacturing, and control of nanoparticles become more challenging for complex nanocarrier system as it transitions from preclinical to clinical development followed by its regulatory requirements for marketing approval. Recent development in particle replication in nonwetting template (PRINT) technology has enabled synthesis of monodisperse nanoparticles with precise control over chemical composition, physicochemical properties (e.g., size, shape, and surface charge), and drug loading efficiencies to aid development of controlled and reproducible synthesis of nanocarriers; however, the technology is still under evaluation for large-scale manufacturing of nanocarriers [193]. In addition, it is mandatory to assess nanocarriers' in vivo performances to study their pharmacokinetics, biodistribution, efficacy, and safety using animal models. Although some dynamic cell line-specific orthotopic, patient-derived, and genetically engineered xenografts have been developed for this purpose, no single animal model is currently available, which can fully mimic human malignancy and EPR effect [193]. To promote clinical translation of nanocarriers, the R&D cost and cycle time should also be reduced [194].

In designing novel cancer targeting nanomedicine, it is crucial to consider cancer dynamics and complexity by conducting appropriate studies to understand tumor microenvironment [182]. In this milieu, it is critical to consider the complex tumor heterogeneity within and among human tumors to design effective personalized cancer treatments [195] that aim to individualize the molecular interventions of cancer disease to optimize and tailor the treatment based on the tumor grade and stage by chasing the therapeutic monitoring and response data [196]. In addition, molecular targeted therapy which uses specific parts of cancer cells to target and block cancer genes or proteins, such as signal transduction inhibitors, angiogenesis inhibitors, gene expression modulators, and hormone therapies, is an emerging clinical approach to treat many cancers. For example, in advanced non-small-cell lung cancer, EGFR transformations are common oncogenic drivers [197]. Accurate detection and selection of the right "druggable" targets are still limited due to the genetic diversity in cancer, limiting the development of new anticancer drug [197]. Despite extensive research efforts in nanomaterial field, only a few nanocarriers have reached the market approval or clinical evaluation, whereas the long-term toxicity, biocompatibility, and genotoxicity of nanomaterials are not adequately studied. Future development of peptidic ligands can be envisaged by developing methods to select and generate target-specific peptides and developing methods to improve their chemical/physical properties which make them stable in vivo [193]. In this context, cell-derived microvesicles which construct the cell-based nanosystems can be foreseen as emerging strategies to achieve enhanced drug delivery efficiency in cancer treatment. Cell-based nanosystems employing techniques including camouflaging nanoparticles using versatile cell membrane coatings or loading of nanoparticles in living carrier cells for targeted antimetastasis therapy will remain in focus [198]. Advances in understanding of cancer metastasis are expected to undertake great efforts to improve the functionality of cell-based nanosystems so that clinical translation of antimetastasis therapy becomes reality in future [198].

8 Conclusion

Organic nanocarriers are among effective and successful targeted therapies for treatment of various types of cancers by passive, active, or external stimuli triggering. Among other types, liposomes and polymer conjugates are the most successfully clinically translated organic nanocarriers based on passive targeting modality, whereas monoclonal antibodies and checkpoint inhibitors are currently the major treatment option in cancer immunotherapy. Several other modified organic nanoforms based on passive or active targeting mechanisms are under clinical development. Moreover, research advancements and increased understanding of tumor heterogeneity have led to more efficient target-specific nanosystems, including nanobodies, cell-membrane camouflaged nanoparticles, cell-penetrating and tumor-targeting peptides, tumor-associated fibroblast targeting, antimetastasis therapy, etc. to demonstrate potential therapeutic outcomes in cancer treatment as compared with traditional targeting strategies, which has enabled several advanced targeted nanosystems under various stages of clinical development. However, manufacturing complexities, biocompatibility and safety, cost of R&D, and regulatory requirements are among major challenges for translation of complex nanocarrier systems from bench to bedside. Emerging future strategies in search of new druggable targets based on genetic information will lead to more efficacious molecularly targeted therapy, antiangiogenesis therapy, immunotherapy, and signal-transduction therapy to develop "maximum-effect–minimal-side-effect" nanocarriers as precision medicine in future.

Conflict of interest

The authors declare no conflict of interest in this work.

Funding

Not applicable.

References

[1] World Health Organization. Health topics, cancer, https://www.who.int/health-topics/cancer#tab=tab_1.

[2] Ferlay J, Soerjomataram I, Dikshit R, Eser S, Mathers C, Rebelo M, et al. Cancer incidence and mortality worldwide: sources, methods and major patterns in GLOBOCAN 2012. Int J Cancer 2015;136:E359–86.

[3] Pérez-Herrero E, Fernández-Medarde A. Advanced targeted therapies in cancer: drug nanocarriers, the future of chemotherapy. Eur J Pharm Biopharm 2015;93:52–79.

[4] Din FU, Aman W, Ullah I, Qureshi OS, Mustapha O, Shafique S, et al. Effective use of nanocarriers as drug delivery systems for the treatment of selected tumors. Int J Nanomedicine 2017;12:7291–309.

[5] Estanqueiro M, Amaral MH, Conceição J, Sousa Lobo JM. Nanotechnological carriers for cancer chemotherapy: the state of the art. Colloids Surf B Biointerfaces 2015;126:631–48.

[6] Jin SE, Jin HE, Hong SS. Targeted delivery system of nanobiomaterials in anticancer therapy: from cells to clinics. Biomed Res Int 2014;2014:814208.

[7] Maeda H, Wu J, Sawa T, Matsumura Y, Hori K. Tumor vascular permeability and the EPR effect in macromolecular therapeutics: a review. J Control Release 2000;65:271–84.

[8] Khawar IA, Kim JH, Kuh HJ. Improving drug delivery to solid tumors: priming the tumor microenvironment. J Control Release 2015;201:78–89.

[9] Balkwill FR, Capasso M, Hagemann T. The tumor microenvironment at a glance. J Cell Sci 2012;125:5591–6.

[10] Hanahan D, Coussens LM. Accessories to the crime: functions of cells recruited to the tumor microenvironment. Cancer Cell 2012;21:309–22.

[11] Lee JH, Choi SY, Jung NC, Song JY, Seo HG, Lee HS, et al. The effect of the tumor microenvironment and tumor-derived metabolites on dendritic cell function. J Cancer 2020;11:769–75.

[12] Sheu BC, Chang WC, Cheng CY, Lin HH, Chang DY, Huang SC. Cytokine regulation networks in the cancer microenvironment. Front Biosci 2008;13:6255–68.

[13] Vaupel P, Kallinowski F, Okunieff P. Blood flow, oxygen and nutrient supply, and metabolic microenvironment of human tumors: a review. Cancer Res 1989;49:6449–65.

[14] Folkman J. Tumor angiogenesis: therapeutic implications. N Engl J Med 1971;285:1182–6.

[15] Rofstad EK. Microenvironment-induced cancer metastasis. Int J Radiat Biol 2000;76:589–605.

[16] Subarsky P, Hill RP. The hypoxic tumour microenvironment and metastatic progression. Clin Exp Metastasis 2003;20:237–50.

[17] Matsumura Y, Maeda H. A new concept for macromolecular therapeutics in cancer chemotherapy: mechanism of tumoritropic accumulation of proteins and the antitumor agent smancs. Cancer Res 1986;46:6387–92.

[18] Maeda H, Sawa T, Konno T. Mechanism of tumor-targeted delivery of macromolecular drugs, including the EPR effect in solid tumor and clinical overview of the prototype polymeric drug SMANCS. J Control Release 2001;74:47–61.

[19] Maeda H, Bharate GY, Daruwalla J. Polymeric drugs for efficient tumor targeted drug delivery based on EPR-effect. Eur J Pharm Biopharm 2009;71:409–19.

[20] van Sluis R, Bhujwalla ZM, Raghunand N, Ballesteros P, Alvarez J, Cerdán S, et al. In vivo imaging of extracellular pH using 1H MRSI. Magn Reson Med 1999;41:743–50.

[21] Feron O. Pyruvate into lactate and back: from the Warburg effect to symbiotic energy fuel exchange in cancer cells. Radiother Oncol 2009;92:329–33.

[22] Cardone RA, Casavola V, Reshkin SJ. The role of disturbed pH dynamics and the Na+/H+ exchanger in metastasis. Nat Rev Cancer 2005;5:786–95.

[23] Simon SM. Role of organelle pH in tumor cell biology and drug resistance. Drug Discov Today 1999;4(1):32–8.

[24] Jang SH, Wientjes MG, Lu D, Au JL. Drug delivery and transport to solid tumors. Pharm Res 2003;20:1337–50.

[25] Maeda H, Fang J, Inutsuka T, Kitamoto Y. Vascular permeability enhancement in solid tumor: various factors, mechanisms involved and its implications. Int Immunopharmacol 2003;3:319–28.

[26] Padera T, Kadambi A, Tomaso E, Carreira CM, Brown EB, Boucher Y, et al. Lymphatic metastasis in the absence of functional intratumor lymphatics. Science 2002;296:1883–6.

[27] Yu L, Lee C-P. Drug delivery: tumor-targeted systems. In: Swarbrick J, editor. Encyclopedia of pharmaceutical technology. 3rd ed, vol. 1. New York: Informa Healthcare; 2007. p. 1326–38.

[28] Yu M, Zheng J. Clearance pathways and tumor targeting of imaging nanoparticles. ACS Nano 2015;9:6655–74.

[29] Patra JK, Das G, Fraceto LF, et al. Nano based drug delivery systems: recent developments and future prospects. J Nanobiotechnol 2018;16:71.

[30] Nakamura Y, Mochida A, Choyke PL, Kobayashi H. Nanodrug delivery: is the enhanced permeability and retention effect sufficient for curing Cancer? Bioconjug Chem 2016;27:2225–38.

[31] Luo Y, Prestwich GD. Cancer-targeted polymeric drugs. Curr Cancer Drug Targets 2002;2:209–26.

[32] Liu D, Auguste DT. Cancer targeted therapeutics: from molecules to drug delivery vehicles. J Control Release 2015;219:632–43.

[33] Lin MZ, Teitell MA, Schiller GJ. The evaluation of antibodies into versatile tumor-targeting agents. Clin Cancer Res 2005;11:129–38.

[34] Das SS, Bharadwaj P, Bilal M, Barani M, Rahdar A, Taboada P, et al. Stimuli-responsive polymeric Nanocarriers for drug delivery, imaging, and Theragnosis. Polymers (Basel) 2020;12:1397.

[35] Zhu L, Torchilin VP. Stimulus-responsive nanopreparations for tumor targeting. Integr Biol (Camb) 2013;5:96–107.

[36] Chen K-J, Liang H-F, Chen H-L, Wang Y, Cheng P-Y, Liu H-L, et al. A thermoresponsive bubble-generating liposomal system for triggering localized extracellular drug delivery. ACS Nano 2013;7:438–46.

[37] Jhaveri A, Deshpande P, Torchilin V. Stimuli-sensitive nanopreparations for combination cancer therapy. J Control Release 2014;190:352–70.

[38] Rapoport N, Gao Z, Kennedy A. Multifunctional nanoparticles for combining ultrasonic tumor imaging and targeted chemotherapy. J Natl Cancer Inst 2007;99:1095–106.

[39] Guduru R, Liang P, Runowicz C, Nair M, Atluri V, Khizroev S. Magneto-electric nanoparticles to enable field-controlled high-specificity drug delivery to eradicate ovarian cancer cells. Sci Rep 2013;3:2953.

[40] Konan-Kouakou YN, Boch R, Gurny R, Allemann E. In vitro and in vivo activities of verteporfin-loaded nanoparticles. J Control Release 2005;103:83–91.

[41] Hoshyar N, Gray S, Han H, Bao G. The effect of nanoparticle size on in vivo pharmacokinetics and cellular interaction. Nanomedicine (Lond) 2016;11:673–92.

[42] Blanco E, Shen H, Ferrari M. Principles of nanoparticle design for overcoming biological barriers to drug delivery. Nat Biotechnol 2015;33:941–51.

[43] Toy R, Peiris PM, Ghaghada KB, Karathanasis E. Shaping cancer nanomedicine: the effect of particle shape on the in vivo journey of nanoparticles. Nanomedicine (Lond) 2014;9:121–34.

[44] Sun Q, Ojha T, Kiessling F, Lammers T, Shi Y. Enhancing tumor penetration of nanomedicines. Biomacromolecules 2017;18:1449–59.

[45] Truong NP, Whittaker MR, Mak CW, Davis TP. The importance of nanoparticle shape in cancer drug delivery. Expert Opin Drug Deliv 2015;12:129–42.

[46] Christian DA, Cai S, Garbuzenko OB, Harada T, Zajac AL, Minko T, et al. Flexible filaments for in vivo imaging and delivery: persistent circulation of filomicelles opens the dosage window for sustained tumor. Mol Pharm 2009;6:1343–52.

[47] Campbell RB, Fukumura D, Brown EB, Mazzola LM, Izumi Y, Jain RK, et al. Cancer Res 2002;62:6831–6.

[48] Krasnici S, Werner A, Eichhorn ME, Schmitt-Sody M, Pahernik SA, Sauer B, et al. Effect of the surface charge of liposomes on their uptake by angiogenic tumor vessels. Int J Cancer 2003;105:561–7.

[49] Suk JS, Xu Q, Kim N, Hanes J, Ensign LM. PEGylation as a strategy for improving nanoparticle-based drug and gene delivery. Adv Drug Deliv Rev 2016;99:28–51.

[50] Tran S, DeGiovanni PJ, Piel B, Rai P. Cancer nanomedicine: a review of recent success in drug delivery. Clin Transl Med 2017;6:44.

[51] Stylianopoulos T, Poh M-Z, Insin N, Bawendi MG, Fukumura D, Munn LL, et al. Diffusion of particles in the extracellular matrix: the effect of repulsive electrostatic interactions. Biophys J 2010;99:1342–9.

[52] Albanese A, Tang PS, Chan WC. The effect of nanoparticle size, shape, and surface chemistry on biological systems. Annu Rev Biomed Eng 2012;14:1–16.

[53] Locatelli E, Franchini MC. Biodegradable PLGA-b-PEG polymeric nanoparticles: synthesis, properties, and nanomedical applications as drug delivery system. J Nanopart Res 2012;14:1.

[54] Fang F, Li M, Zhang J, Lee CS. Different strategies for organic nanoparticle preparation in biomedicine. ACS Mater Lett 2020;2:531–49.

[55] Yoo J, Park C, Yi G, Lee D, Koo H. Active targeting strategies using biological ligands for nanoparticle drug delivery systems. Cancers (Basel) 2019;11:640.

[56] Verma A, Stellacci F. Effect of surface properties on nanoparticle-cell interactions. Small 2010;6:12–21.

[57] Dreaden EC, Mackey MA, Huang X, Kang B, El-Sayed MA. Beating cancer in multiple ways using nanogold. Chem Soc Rev 2011;40:3391–404.

[58] Kim B, Han G, Toley BJ, Kim CK, Rotello VM, Forbes NS. Tuning payload delivery in tumour cylindroids using gold nanoparticles. Nat Nanotechnol 2010;5:465–72.

[59] Shi X, Thomas TP, Myc LA, Kotlyar A, Baker Jr J. Synthesis, characterization, and intracellular uptake of carboxyl-terminated poly(amidoamine) dendrimer-stabilized iron oxide nanoparticles. Phys Chem Chem Phys 2007;9:5712–20.

[60] Cheng W, Zeng X, Chen H, Li Z, Zeng W, Mei L, et al. Versatile polydopamine platforms: synthesis and promising applications for surface modification and advanced nanomedicine. ACS Nano 2019;13:8537–65.

[61] Jia X, Zhang Y, Zou Y, Wang Y, Niu D, He Q, et al. Dual intratumoral redox/enzyme-responsive NO-releasing nanomedicine for the specific, high-efficacy, and low-toxic cancer therapy. Adv Mater 2018;30:1704490.

[62] Chai Z, Ran D, Lu L, Zhan C, Ruan H, Hu X, et al. Ligand-modified cell membrane enables the targeted delivery of drug nanocrystals to glioma. ACS Nano 2019;13:5591–601.

[63] Lv Y, Liu M, Zhang Y, Wang X, Zhang F, Li F, et al. Cancer cell membrane-biomimetic nanoprobes with two-photon excitation and near-infrared emission for intravital tumor fluorescence imaging. ACS Nano 2018;12:1350–8.

[64] Han Y, Pan H, Li W, Chen Z, Ma A, Yin T, et al. Cell membrane mimicking nanoparticles with bioorthogonal targeting and immune recognition for enhanced photothermal therapy. Adv Sci 2019;6:1900251.

[65] Li J, Zhen X, Lyu Y, Jiang Y, Huang J, Pu K. Cell membrane coated semiconducting polymer nanoparticles for enhanced multimodal cancer phototheranostics. ACS Nano 2018;12:8520–30.

[66] Zhen X, Feng X, Xie C, Zheng Y, Pu K. Surface engineering of semiconducting polymer nanoparticles for amplified photoacoustic imaging. Biomaterials 2017;127:97–106.

[67] Zhu H, Fang Y, Zhen X, Wei N, Gao Y, Luo KQ, et al. Multilayered semiconducting polymer nanoparticles with enhanced NIR fluorescence for molecular imaging in cells, zebrafish and mice. Chem Sci 2016;7:5118–25.
[68] Zhu H, Li J, Qi X, Chen P, Pu K. Oxygenic hybrid semiconducting nanoparticles for enhanced photodynamic therapy. Nano Lett 2018;18:586–94.
[69] Ding Y, Su S, Zhang R, Shao L, Zhang Y, Wang B, et al. Precision combination therapy for triple negative breast cancer via biomimetic polydopamine polymer core-shell nanostructures. Biomaterials 2017;113:243–52.
[70] Harmatys KM, Chen J, Charron DM, MacLaughlin CM, Zheng G. Multipronged biomimetic approach to create optically tunable nanoparticles. Angew Chem Int Ed 2018;57:8125–9.
[71] Jiang W, Kim BY, Rutka JT, Chan WC. Nanoparticle-mediated cellular response is size-dependent. Nat Nanotechnol 2008;3:145–50.
[72] Akbarzadeh A, Rezaei-Sadabady R, Davaran S, Joo SW, Zarghami N, Hanifehpour Y, Samiei M, Kouhi M, Nejati-Koshki K. Liposome: classification, preparation, and applications. Nanoscale Res Lett 2013;8:102.
[73] Bozzuto G, Molinari A. Liposomes as nanomedical devices. Int J Nanomedicine 2015;2(10):975–99.
[74] Abu Lila AS, Ishida T. Liposomal delivery systems: design optimization and current applications. Biol Pharm Bull 2017;40:1–10.
[75] Bulbake U, Doppalapudi S, Kommineni N, Khan W. Liposomal formulations in clinical use: An updated review. Pharmaceutics 2017;9:12.
[76] Minko T, Pakunlu RI, Wang Y, Khandare JJ, Saad M. New generation of liposomal drugs for cancer. Anticancer Agents Med Chem 2006;6:537–52.
[77] Allen TM, Cheng WW, Hare JI, Laginha KM. Pharmacokinetics and pharmacodynamics of lipidic nano-particles in cancer. Anticancer Agents Med Chem 2006;6:513–23.
[78] Wasungu L, Hoekstra D. Cationic lipids, lipoplexes and intracellular delivery of genes. J Control Release 2006;116:255–64.
[79] Ross PC, Hui SW. Lipoplex size is a major determinant of in vitro lipofection efficiency. Gene Ther 1999;6:651–9.
[80] Pal K, Madamsetty VS, Dutta SK, Mukhopadhyay D. Co-delivery of everolimus and vinorelbine via a tumor-targeted liposomal formulation inhibits tumor growth and metastasis in RCC. Int J Nanomedicine 2019;11:5109–23.
[81] Müller RH, Mäder K, Gohla S. Solid lipid nanoparticles (SLN) for controlled drug delivery - a review of the state of the art. Eur J Pharm Biopharm 2000;50:161–77.
[82] Malam Y, Loizidou M, Seifalian AM. Liposomes and nanoparticles: Nanosized vehicles for drug delivery in cancer. Trends Pharmacol Sci 2009;30:592–9.
[83] Mehnert W, Mäder K. Solid lipid nanoparticles: production, characterization and applications. Adv Drug Deliv Rev 2001;47:165–96.
[84] Gref R, Domb A, Quellec P, Blunk T, Müller RH, Verbavatz JM, et al. The controlled intravenous delivery of drugs using PEG-coated sterically stabilized nanoparticles. Adv Drug Deliv Rev 1995;16:215–33.
[85] Illum L, Davis SS, Müller RH, Mak E, West P. The organ distribution and circulation time of intravenous injected colloidal carriers sterically stabilized with a block copolymer-poloxamine 908. Life Sci 1987;40:367–74.
[86] Bocca C, Caputo O, Cavalli R, Gabriel L, Miglietta A, Gasco MR. Phagocytic uptake of fluorescent stealth and non-stealth solid lipid nanoparticles. Int J Pharm 1998;175:185–93.
[87] Wong HL, Bendayan R, Rauth AM, Li Y, Wu XY. Chemotherapy with anticancer drugs encapsulated in solid lipid nanoparticles. Adv Drug Deliv Rev 2007;59:491–504.
[88] Hallan SS, Kaur P, Kaur V, Mishra N, Vaidya B. Lipid polymer hybrid as emerging tool in nanocarriers for oral drug delivery. Artif Cells Nanomed Biotechnol 2016;44:334–49.
[89] Qian J, Xu N, Zhou X, Shi K, Du Q, Yin X, et al. Low density lipoprotein mimic nanoparticles composed of amphipathic hybrid peptides and lipids for tumor-targeted delivery of paclitaxel. Int J Nanomedicine 2019;14:7431–46.
[90] Rao JP, Geckeler KE. Polymer nanoparticles: preparation techniques and size-control parameters. Prog Polym Sci 2011;36:887–913.
[91] Bamrungsap S, Zhao Z, Chen T, Wang L, Li C, Fu T, et al. Nanotechnology in therapeutics: a focus on nanoparticles as a drug delivery system. Nanomedicine (Lond) 2012;7:1253–71.
[92] Prabhu RH, Patravale VB, Joshi MD. Polymeric nanoparticles for targeted treatment in oncology: current insights. Int J Nanomedicine 2015;10:1001–18.
[93] Wang X, Wang Y, Chen ZG, Shin DM. Advances of cancer therapy by nanotechnology. Cancer Res Treat 2009;41:1–11.
[94] Hu CM, Aryal S, Zhang L. Nanoparticle-assisted combination therapies for effective cancer treatment. Ther Deliv 2010;1:323–34.
[95] Alexis F, Pridgen EM, Langer R, Farokhzad OC. Nanoparticle technologies for cancer therapy. In: Schäfer-Korting M, editor. Drug delivery. Berlin, Heidelberg: Springer; 2010. p. 55–86.
[96] Zhu Y, Liao L. Applications of nanoparticles for anticancer drug delivery: a review. J Nanosci Nanotechnol 2015;15:4753–73.
[97] Zhu D, Tao W, Zhang H, Liu G, Wang T, Zhang L, et al. Docetaxel (DTX)-loaded polydopamine-modified TPGS-PLA nanoparticles as a targeted drug delivery system for the treatment of liver cancer. Acta Biomater 2016;30:14454.
[98] Dahiya S. Block copolymeric micelles: basic concept and preparation techniques. Bull Pharm Res 2019;9:166.
[99] Zhang J, Dubay MR, Houtman CJ, Severtson SJ. Sulfonated amphiphilic block copolymers: syntheses, self-assembly in water, and applications as stabilizer in emulsion polymerization. Macromolecules 2009;42:5080–90.
[100] Biswas S, Kumari P, Lakhani PM, Ghosh B. Recent advances in polymeric micelles for anti-cancer drug delivery. Eur J Pharm Sci 2016;83:184–202.
[101] Gothwal A, Khan I, Gupta U. Polymeric micelles: recent advancements in the delivery of anticancer drugs. Pharm Res 2016;33:18–39.
[102] Nakanishi T, Fukushima S, Okamoto K, Suzuki M, Matsumura Y, Yokoyama M, et al. Development of the polymer micelle carrier system for doxorubicin. J Control Release 2001;74:295–302.
[103] Rapoport N. Physical stimuli-responsive polymeric micelles for anti-cancer drug delivery. Prog Polym Sci 2007;32:962–90.
[104] Zhong T, He B, Cao HQ, Tan T, Hu HY, Li YP, et al. Treating breast cancer metastasis with cabazitaxel-loaded polymeric micelles. Acta Pharmacol Sin 2017;38:924–30.
[105] Ringsdorf H. Structure and properties of pharmacologically active polymers. J Polym Sci 1975;51:135–53.
[106] Feng Q, Tong R. Anticancer nanoparticulate polymer-drug conjugate. Bioeng Transl Med 2016;1:277–96.
[107] Duncan R. Polymer conjugates as anticancer nanomedicines. Nat Rev Cancer 2006;6:688–701.
[108] Bayram B, Ozgur A, Tutar L, Tutar Y. Tumor targeting of polymeric nanoparticles conjugated with peptides, saccharides, and small molecules for anticancer drugs. Curr Pharm Des 2017;23:5349–57.

[109] Nehoff H, Parayath NN, Domanovitch L, Taurin S, Greish K. Nanomedicine for drug targeting: strategies beyond the enhanced permeability and retention effect. Int J Nanomedicine 2014;9:2539–55.
[110] Duncan R. Designing polymer conjugates as lysosomotropic nanomedicines. Biochem Soc Trans 2007;35:56–60.
[111] Larson N, Ghandehari H. Polymeric conjugates for drug delivery. Chem Mater 2012;24:840–53.
[112] Vicent MJ, Manzanaro S, de la Fuente JA, Duncan R. HPMA copolymer-1,5-diazaanthraquinone conjugates as novel anticancer therapeutics. J Drug Target 2004;12:503–15.
[113] Yang Y, Roy A, Zhao Y, Undzys E, Li SD. Comparison of tumor penetration of podophyllotoxin-carboxymethylcellulose conjugates with various chemical compositions in tumor spheroid culture and in vivo solid tumor. Bioconjug Chem 2017;28:1505–18.
[114] Abbasi E, Aval SF, Akbarzadeh A, Milani M, Nasrabadi HT, Joo SW, et al. Dendrimers: synthesis, applications, and properties. Nanoscale Res Lett 2014;9:247.
[115] Bello M, Fragoso-Vázquez J, Correa-Basurto J. Theoretical studies for dendrimer-based drug delivery. Curr Pharm Des 2017;23:3048–61.
[116] Kharwade R, More S, Warokar A, Agrawal P, Mahajan N. Starburst pamam dendrimers: synthetic approaches, surface modifications, and biomedical applications. Arab J Chem 2020;13(7):6009–39.
[117] Tomalia DA, Nixon LS, Hedstrand DM. The role of branch cell symmetry and other critical nanoscale design parameters in the determination of dendrimer encapsulation properties. Biomolecules 2020;10:642.
[118] Stiriba SE, Frey H, Haag R. Dendritic polymers in biomedical applications: from potential to clinical use in diagnostics and therapy. Angew Chem Int Ed Engl 2002;41:1329–34.
[119] de Groot FM, Albrecht C, Koekkoek R, Beusker PH, Scheeren HW. Cascade-release dendrimers liberate all end groups upon a single triggering event in the dendritic core. Angew Chem Int Ed Engl 2003;42:4490–4.
[120] Malik N, Evagorou EG, Duncan R. Dendrimer-platinate: a novel approach to cancer chemotherapy. Anticancer Drugs 1999;10:767–76.
[121] Zhuo RX, Du B, Lu ZR. In vitro release of 5-fluorouracil with cyclic core dendritic polymer. J Control Release 1999;57:249–57.
[122] Lee CC, Gillies ER, Fox ME, Guillaudeu SJ, Fréchet JM, Dy EE, et al. A single dose of doxorubicin-functionalized bow-tie dendrimer cures mice bearing C-26 colon carcinomas. Proc Natl Acad Sci U S A 2006;103:16649–54.
[123] Liu C, Gao H, Zhao Z, Rostami I, Wang C, Zhu L, et al. Improved tumor targeting and penetration by a dual-functional poly(amidoamine) dendrimer for the therapy of triple-negative breast cancer. J Mater Chem B 2019;7:3724–36.
[124] Hadinoto K, Sundaresan A, Cheow WS. Lipid-polymer hybrid nanoparticles as a new generation therapeutic delivery platform: a review. Eur J Pharm Biopharm 2013;85:427–43.
[125] Zhang L, Chan JM, Gu FX, Rhee JW, Wang AZ, Radovic-Moreno AF, et al. Self-assembled lipid-polymer hybrid nanoparticles: a robust drug delivery platform. ACS Nano 2008;2:1696–702.
[126] Mukherjee A, Waters AK, Kalyan P, Achrol AS, Kesari S, Yenugonda VM. Lipid-polymer hybrid nanoparticles as a next-generation drug delivery platform: state of the art, emerging technologies, and perspectives. Int J Nanomedicine 2019;14:1937–52.
[127] Hasan W, Chu K, Gullapalli A, Dunn S, Enlow E, Luft JC, et al. Delivery of multiple siRNAs using lipid-coated PLGA nanoparticles for treatment of prostate cancer. J Nano Lett 2012;12:287–92.
[128] Zheng Y, Yu B, Weecharangsan W, Piao L, Darby M, Mao Y, et al. Transferrin-conjugated lipid-coated PLGA nanoparticles for targeted delivery of aromatase inhibitor 7alpha-APTADD to breast cancer cells. Int J Pharm 2010;390:234–41.
[129] Gao LY, Liu XY, Chen CJ, Wang JC, Feng Q, Yu MZ, et al. Core-shell type lipid/rPAA-Chol polymer hybrid nanoparticles for in vivo siRNA delivery. Biomaterials 2014;35:2066–78.
[130] Wang Z, Zang A, Wei Y, An L, Hong D, Shi Y, et al. Hyaluronic acid capped, irinotecan and gene co-loaded lipid-polymer hybrid nanocarrier-based combination therapy platform for colorectal cancer. Drug Des Devel Ther 2020;14:1095–105.
[131] Yang JC, Hughes M, Kammula U, Royal R, Sherry RM, Topalian SL, et al. Ipilimumab (anti-CTLA4 antibody) causes regression of metastatic renal cell cancer associated with enteritis and hypophysitis. J Immunother 2007;30:825.
[132] ·Attarwala H. Role of antibodies in cancer targeting. J Nat Sci Biol Med 2010;1:53–6.
[133] Bazak R, Houri M, El Achy S, Kamel S, Refaat T. Cancer active targeting by nanoparticles: a comprehensive review of literature. J Cancer Res Clin Oncol 2015;141:769–84.
[134] Schroeder Jr HW, Cavacini L. Structure and function of immunoglobulins. J Allergy Clin Immunol 2010;125:S41–52.
[135] Scott AM, Allison JP, Wolchok JD. Monoclonal antibodies in cancer therapy. Cancer Immun 2012;12:14.
[136] Coulson A, Levy A, Gossell-Williams M. Monoclonal antibodies in cancer therapy: mechanisms, successes and limitations. West Indian Med J 2014;63:650–4.
[137] Nessa MU, Rahman MA, Kabir Y. Plant-produced monoclonal antibody as immunotherapy for cancer. Biomed Res Int 2020;2020:3038564.
[138] Redman JM, Hill EM, AlDeghaither D, Weiner LM. Mechanisms of action of therapeutic antibodies for cancer. Mol Immunol 2015;67:28–45.
[139] Tsumoto K, Isozaki Y, Yagami H, Tomita M. Future perspectives of therapeutic monoclonal antibodies. Immunotherapy 2019;11:119–27.
[140] Chiavenna SM, Jaworski JP, Vendrell A. State of the art in anticancer mAbs. J Biomed Sci 2017;24:15.
[141] Pardoll DM. The blockade of immune checkpoints in cancer immunotherapy. Nat Rev Cancer 2012;12:252–64.
[142] Lu KL, Wu MY, Wang CH, Wang CW, Hung SI, Chung WH, et al. The role of immune checkpoint receptors in regulating immune reactivity in lupus. Cells 2019;8:1213.
[143] Marhelava K, Pilch Z, Bajor M, Graczyk-Jarzynka A, Zagozdzon R. Targeting negative and positive immune checkpoints with monoclonal antibodies in therapy of cancer. Cancers (Basel) 2019;11:1756.
[144] National Cancer Institute at the National Institutes of Health. Immune checkpoint inhibitor 2020. https://www.cancer.gov/publications/dictionaries/cancer-terms/def/immune-checkpoint-inhibitor.
[145] Zappasodi R, Merghoub T, Wolchok JD. Emerging concepts for immune checkpoint blockade-based combination therapies. Cancer Cell 2018;33:581–98.
[146] Callahan MK, Wolchok JD. At the bedside: CTLA-4- and PD-1-blocking antibodies in cancer immunotherapy. J Leukoc Biol 2013;94:41–53.
[147] Podojil JR, Miller SD. Molecular mechanisms of T-cell receptor and costimulatory molecule ligation/blockade in autoimmune disease therapy. Immunol Rev 2009;229:337–55.

[148] Sansom DM. CD28, CTLA-4 and their ligands: who does what and to whom? Immunology 2000;101:169–77.
[149] Boussiotis VA, Chatterjee P, Li L. Biochemical signaling of PD-1 on T cells and its functional implications. Cancer J 2014;20:265–71.
[150] Liu JKH. The history of monoclonal antibody development - progress, remaining challenges and future innovations. Ann Med Surg 2014;3:113–6.
[151] Yang EY, Shah K. Nanobodies: next generation of cancer diagnostics and therapeutics. Front Oncol 2020;10:1182.
[152] Bannas P, Hambach J, Koch-Nolte F. Nanobodies and nanobody-based human heavy chain antibodies as antitumor therapeutics. Front Immunol 2017;8:1603.
[153] De Genst E, Silence K, Decanniere K, Conrath K, Loris R, Kinne J, et al. Molecular basis for the preferential cleft recognition by dromedary heavy-chain antibodies. Proc Natl Acad Sci U S A 2006;103:4586–91.
[154] Mitchell LS, Colwell LJ. Analysis of nanobody paratopes reveals greater diversity than classical antibodies. Protein Eng Des Sel 2018;31:267–75.
[155] Jovčevska I, Muyldermans S. The therapeutic potential of nanobodies. BioDrugs 2020;34:11–26.
[156] Morrison C. Nanobody approval gives domain antibodies a boost. Nat Rev Drug Discov 2019;18(7):485–7. News.
[157] Sun X, Li Y, Liu T, Li Z, Zhang X, Chen X. Peptide-based imaging agents for cancer detection. Adv Drug Deliv Rev 2017;110–111:38–51.
[158] Ruseska I, Zimmer A. Internalization mechanisms of cell-penetrating peptides. Beilstein J Nanotechnol 2020;11:101–23.
[159] Milletti F. Cell-penetrating peptides: classes, origin, and current landscape. Drug Discov Today 2012;17:850–60.
[160] Zorko M, Langel U. Cell-penetrating peptides: mechanism and kinetics of cargo delivery. Adv Drug Deliv Rev 2005;57:529–45.
[161] Boohaker RJ, Lee MW, Vishnubhotla P, Perez JM, Khaled AR. The use of therapeutic peptides to target and to kill cancer cells. Curr Med Chem 2012;19:3794–804.
[162] Brown KC. Peptidic tumor targeting agents: the road from phage display peptide selections to clinical applications. Curr Pharm Des 2010;16:1040–54.
[163] Torchilin VP. Tat peptide-mediated intracellular delivery of pharmaceutical nanocarriers. Adv Drug Deliv Rev 2008;60:548–58.
[164] Arap W, Pasqualini R, Ruoslahti E. Cancer treatment by targeted drug delivery to tumor vasculature in a mouse model. Science 1998;279:377–80.
[165] Lulla RR, Goldman S, Yamada T, Beattie CW, Bressler L, Pacini M, et al. Phase I trial of p28 (NSC745104), a non-HDM2-mediated peptide inhibitor of p53 ubiquitination in pediatric patients with recurrent or progressive central nervous system tumors: a pediatric brain tumor consortium study. Neuro Oncol 2016;18:1319–25.
[166] Safety study of a cell penetrating peptide (p28) to treat solid tumors that resist standard methods of treatment. https://clinicaltrials.gov/ct2/show/NCT00914914.
[167] Coriat R, Faivre SJ, Mir O, Dreyer C, Ropert S, Bouattour M, et al. Pharmacokinetics and safety of DTS-108, a human oligopeptide bound to SN-38 with an esterase-sensitive cross-linker in patients with advanced malignancies: a phase I study. Int J Nanomedicine 2016;11:6207–16.
[168] Miampamba M, Liu J, Harootunian A, Gale AJ, Baird S, Chen SL, et al. Sensitive in vivo visualization of breast cancer using ratiometric protease-activatable fluorescent imaging agent, AVB-620. Theranostics 2017;7:3369–86.
[169] Vijayan V, Uthaman S, Park IK. Cell membrane-camouflaged nanoparticles: a promising biomimetic strategy for cancer theragnostics. Polymers (Basel) 2018;10:983.
[170] Wu M, Le W, Mei T, Wang Y, Chen B, Liu Z, et al. Cell membrane camouflaged nanoparticles: a new biomimetic platform for cancer photothermal therapy. Int J Nanomedicine 2019;14:4431–48.
[171] Harris JC, Scully MA, Day ES. Cancer cell membrane-coated nanoparticles for cancer management. Cancers (Basel) 2019;11:1836.
[172] Hu CM, Zhang L, Aryal S, Cheung C, Fang RH, Zhang L. Erythrocyte membrane-camouflaged polymeric nanoparticles as a biomimetic delivery platform. Proc Natl Acad Sci U S A 2011;108:10980–5.
[173] Zhang Q, Wei W, Wang P, Zuo L, Li F, Xu J, et al. Biomimetic magnetosomes as versatile artificial antigen-presenting cells to potentiate t-cell-based anticancer therapy. ACS Nano 2017;11:10724–32.
[174] Bussard KM, Mutkus L, Stumpf K, Gomez-Manzano C, Marini FC. Tumor-associated stromal cells as key contributors to the tumor microenvironment. Breast Cancer Res 2016;18:84.
[175] Smith NR, Baker D, Farren M, Pommier A, Swann R, Wang X, et al. Tumor stromal architecture can define the intrinsic tumor response to VEGF-targeted therapy. Clin Cancer Res 2013;19:6943–56.
[176] Kalluri R. The biology and function of fibroblasts in cancer. Nat Rev Cancer 2016;16:582–98.
[177] Micke P, Ostman A. Exploring the tumour environment: Cancer-associated fibroblasts as targets in cancer therapy. Expert Opin Ther Targets 2005;9:1217–33.
[178] Orimo A, Gupta PB, Sgroi DC, Arenzana-Seisdedos F, Delaunay T, Naeem R, et al. Stromal fibroblasts present in invasive human breast carcinomas promote tumor growth and angiogenesis through elevated SDF-1/CXCL12 secretion. Cell 2005;121:335–48.
[179] Rasanen K, Vaheri A. Activation of fibroblasts in cancer stroma. Exp Cell Res 2010;316:2713–22.
[180] Zhou L, Yang K, Andl T, Wickett RR, Zhang Y. Perspective of targeting cancer-associated fibroblasts in melanoma. J Cancer 2015;6:717–26.
[181] Spaw M, Anant S, Thomas SM. Stromal contributions to the carcinogenic process. Mol Carcinog 2017;56:1199–213.
[182] Fernandes C, Suares D, Yergeri MC. Tumor microenvironment targeted nanotherapy. Front Pharmacol 2018;9:1230.
[183] Posey JA, Ng TC, Yang BL, Khazaeli MB, Carpenter MD, Fox F, et al. A phase I study of anti-kinase insert domain-containing receptor antibody, IMC-1C11, in patients with liver metastases from colorectal carcinoma. Clin Cancer Res 2003;9:1323–32.
[184] Kobayashi H, Eckhardt SG, Lockridge JA, Rothenberg ML, Sandler AB, O'Bryant CL, et al. Safety and pharmacokinetic study of RPI.4610 (ANGIOZYME), an anti-VEGFR-1 ribozyme, in combination with carboplatin and paclitaxel in patients with advanced solid tumors. Cancer Chemother Pharmacol 2005;56:329–36.
[185] Hurwitz H, Fehrenbacher L, Novotny W, Cartwright T, Hainsworth J, Heim W, et al. Bevacizumab plus irinotecan, fluorouracil, and leucovorin for metastatic colorectal cancer. N Engl J Med 2004;350:2335–42.
[186] Greco R, Hurley R, Sun F, Yang L, Yu Q, Williams J, et al. Abstract No. 1796: SAR302503: a jak2 inhibitor with antitumor activity in solid tumor models. In: Proceedings of the 103rd Annual Meeting of the American Association for Cancer Research; 2012 Mar 31-Apr 4;

Chicago, IL. Philadelphia (PA): AACR; Cancer Research, vol. 72; 2012.

[187] Mascarenhas J, Hoffman R, Talpaz M, Gerds AT, Stein B, Gupta V, et al. Pacritinib vs best available therapy, including ruxolitinib, in patients with myelofibrosis: a randomized clinical trial. JAMA Oncol 2018;4:652–9.

[188] Adams S, Miller GT, Jesson MI, Watanabe T, Jones B, Wallner BP. PT-100, a small molecule dipeptidyl peptidase inhibitor, has potent antitumor effects and augments antibody-mediated cytotoxicity via a novel immune mechanism. Cancer Res 2004;64:5471–80.

[189] Wuest T, Moosmayer D, Pfizenmaier K. Construction of a bispecific single chain antibody for recruitment of cytotoxic T cells to the tumour stroma associated antigen fibroblast activation protein. J Biotechnol 2001;92:159–68.

[190] Bramhall SR, Hallissey MT, Whiting J, Scholefield J, Tierney G, Stuart RC, et al. Marimastat as maintenance therapy for patients with advanced gastric cancer: a randomised trial. Br J Cancer 2002;86:1864–70.

[191] Hirte H, Vergote IB, Jeffrey JR, Grimshaw RN, Coppieters S, Schwartz B, et al. A phase III randomized trial of BAY 12-9566 (tanomastat) as maintenance therapy in patients with advanced ovarian cancer responsive to primary surgery and paclitaxel/platinum containing chemotherapy: a National Cancer Institute of Canada clinical trials group study. Gynecol Oncol 2006;102:300–8.

[192] Hua S, de Matos MBC, Metselaar JM, Storm G. Current trends and challenges in the clinical translation of nanoparticulate nanomedicines: pathways for translational development and commercialization. Front Pharmacol 2018;9:790.

[193] Shi J, Kantoff PW, Wooster R, Farokhzad OC. Cancer nanomedicine: progress, challenges and opportunities. Nat Rev Cancer 2017;17:20–37.

[194] Metselaar JM, Lammers T. Challenges in nanomedicine clinical translation. Drug Deliv Transl Res 2020;10:721–5.

[195] Bedard PL, Hansen AR, Ratain MJ, Siu LL. Tumour heterogeneity in the clinic. Nature 2013;501:355–64.

[196] Matchett KB, Lynam-Lennon N, Watson RW, Brown JAL. Advances in precision medicine: tailoring individualized therapies. Cancers (Basel) 2017;9:146.

[197] Lheureux S, Denoyelle C, Ohashi PS, De Bono JS, Mottaghy FM. Molecularly targeted therapies in cancer: a guide for the nuclear medicine physician. Eur J Nucl Med Mol Imaging 2017;44:41–54.

[198] Gong X, Li J, Tan T, Wang Z, Wang H, Wang Y, et al. Emerging approaches of cell-based nanosystems to target cancer metastasis. Adv Funct Mater 2019;29:1903441.

Chapter 38

Targeting cancer using phytoconstituents-based drug delivery

Rati Yadav[a], Joydeep Das[b], H. Lalhlenmawia[c], Rajiv K. Tonk[d], Lubhan Singh[e], and Deepak Kumar[a]

[a]*Department of Pharmaceutical Chemistry, School of Pharmaceutical Sciences, Shoolini University, Solan, HP, India,* [b]*Advance School of Chemical Sciences, Shoolini University, Solan, HP, India,* [c]*Department of Pharmacy, Regional Institute of Paramedical and Nursing Sciences, Aizawl, MZ, India,* [d]*Department of Pharmaceutical Chemistry, Delhi Pharmaceutical Sciences & Research University, New Delhi, India,* [e]*Kharvel Subharti College of Pharmacy, Swami Vivekanand Subharti University, Meerut, India*

1 Introduction

The word cancer is derived from the Latin word "cancrum," which means crab, and resembles the crab limbs that never let go of the swollen veins around the tumor [1]. It is a cluster of diseases that arise due to abnormal growth of the cells and can spread from its origin (primary tumor) to other body parts (secondary tumors) [2]. Benign tumors are usually encapsulated or surrounded by a membrane that restrains their invasive capacity. Despite being slow growing, they can reach a considerable size (as big as a grapefruit) and if compress other tissues (e.g., blood vessels or the brain) can have serious effects that require surgical treatment. In addition, some benign tumors can have harmful, indirect effects if they occur in endocrine tissues and result in abnormal levels of hormone production, for example, adenomas in the thyroid, adrenocortex, or pituitary glands. Examples of benign tumors are fibroma, chondroma, choristoma, hamartoma, lipoma, and pappilioma [3]. Approximately 85% of the malignant tumors (adenocarcinomas) are of epithelial origins. Cancer advances due to gene mutations that change cellular processes including cell growth, division, transcription, mutation, and gene expressions. Changes in DNA sequence lead to rapid proliferation followed by successive mutations, resulting in malignancy and metastases [4,5].

Metastatic lesions lead to the molecular and biophysical barriers that would present insurmountable challenges to their nonmetastasized counterparts [6–8]. Progression of metastasis is not only a cell-autonomic occurrence but is also highly affected by complex microenvironmental tissues. Interactions among cancerous cells, endothelial cells, stromal fibroblasts, and immune cells elevated tissue oxygen stress and caused a significant influence on tumor progression. Maintenance of the transcription factor and hypoxia-inducible factor causes ameboid migration as oxygen voltage fluctuates in the tumor, stimulating reciprocal signaling of the mesenchymal stem cells and cancer cells which boost the metastatic phenotype [9,10].

Nowadays, cancer has drawn the major public health concern globally and is a leading cause of mortality. In between 2010 and 2020, cancer cases have increased by 24% in males (1 million cases) and about 21% in females (more than 900,000 cases) per year in the United States [11]. The type of cancers that are expected to be increased the most are melanoma, prostate, lung, liver, and bladder cancers in males and lung, breast, uterine, and thyroid cancers in females [12].

Natural compounds derived from numerous plant and animal sources are used for the treatment and prevention of various chronic ailments since many decades. Natural compounds application has shown considerable clinical effectiveness against cancers. Chemoprevention by ingestion of dietary phytoconstituents represents an important strategy in the battle against cancer [13]. Natural compounds obtained from plants and animals have traditionally been used especially as anticancer and antimicrobial agents [14]. The potential use of natural products as anticancer agents was approved by the National Cancer Institute, USA, since 1950, which resulted in the development of natural anticancer compounds. Later in 1980s, several novel screening methods helped the researchers to find new anticancer agents from plants and other species worldwide. Chemotherapy is in the forefront of cancer treatment as compared with the other treatment modalities such as surgery, radiotherapy, and phototherapy, but causes severe side effects with tumor recurrence and metastasis, whereas several plants- and animal-derived natural compounds have been used for cancer treatment without causing serious side effects since ancient times [15–20].

Several natural compounds are recognized as significant source of drugs used in the treatment of cancer, and 70% of these products are at different stages of clinical trials

Advanced Drug Delivery Systems in the Management of Cancer. https://doi.org/10.1016/B978-0-323-85503-7.00033-X

[21,22]. One such natural compound extracted from plants for anticancer activity is Paclitaxel (Taxol ®), which is obtained from the plant *Taxus brevifolia* Nutt's bark. Taxol prevents microtubule disassembly by linking polymerized microtubules [23,24].

The principal features of the compounds obtained from biological/natural sources used for the treatment of cancer are less side effects, lower costs, and prevention of drug resistance. Phytochemicals derived from diverse sources have the ability to stimulate a variety of biochemical mechanisms that are essential for curing different diseases, including cancer. Therefore, use of natural compounds signify a capable strategy for the treatment and control of cancers.

2 Traditional Indian and Chinese medicine

Over a decade, studies have shown that traditional Indian and Chinese herbs are used as medicines to treat cancer. The compounds have shown efficient anticancer activities by inhibiting the development, proliferation, angiogenesis, and metastasis of cancer. For example, curcumin, resveratrol, quercetin, and berberine are used for clinical trials in treating several types of cancer and its related health issues. In addition, traditional medicines are also taken as an additive therapy for the cure of cancer and related problems. The drugs including vincristine, vinblastine (vinca alkaloids), and campothecin derivatives, e.g., topotecan, irinotecan, and crocetin, are promising anticancer agents which are being tested in preclinical or clinical stages [25]. There are large number of plant-based active compounds that are used for treatment and regulation of cancer and cancer-related conditions from the ancient time in China and South-East Asia [26].

2.1 Curcuma longa

Curcumin (Fig. 1) is a diketone type of compound which has diverse biological activities like anticancer, anti-inflammatory, analgesic, and for the control of diabetes and cardiovascular disease [27]. It can prevent the spread and proliferation of cervical cancer cells and induce metastatic cell apoptosis without altering normal cervical epithelial cells [27]. In thyroid cancer, curcumin can downregulate the Bcl-xl, restrict the expression of Cyclin B1, and thereby inhibit the spread of cancer and promote apoptosis [28]. In an in vivo study, curcumin has been shown to reduce the amount of tumor marker molecules and nitric oxide in cervical cancer-bearing mice, thereby suggesting inhibition of cancer development [29]. In addition, curcumin dramatically decreases the oral squamous cancer cell by preventing the AKT/PI3K/mTOR signaling [29].

Curcumin can, by downregulating snail expression, reverse TGF-induced epithelial mesenchymal (EMT) in liver cancer cell. The proliferation and migration of the glioma cells have been reported to be inhibited by compound **3** and its mode of action may be due to prevention of oncogenic protein expression, NEDD4. Curcumin can also control the spread, apoptosis, and invasion of pancreatic cancer cells by NEDD4 inhibition [30–32].

2.2 Oldenlan diadiffusa

Oldenlan diadiffusa (OD) is an annual herb found mainly in China and Korea with a number of biological activities, viz. immune regulation, antimicrobial, antioxidant activities, and is well known for its therapeutic anticancer effect. Some of the bioactive compounds include 2-Hydroxy-3-methylanthraquinone (**1**), Vanillic acid (**2**), (Fig. 2) also contains anthraquinone and flavonoids.

The active components of OD can induce apoptosis in different cancer conditions such as prostate, colorectal, cervical, and gastric cancer. Studies of OD against breast cancer cells prevent the expression and invasion of MMP-9. Besides, its aqueous extract causes apoptosis of lung cancer cells by inhibiting the MAPK pathway [33]. OD limit the spread and metastases of liver carcinoma via obstructing chemokine receptors like CXCR1, CXCR2, and CXCR4 expression by inducing apoptosis through caspase 3 pathways [34]. In vivo studies report that OD flavonoids against the cervical cancer in mice may significantly inhibit the cancer growth. The process may be associated with the rise in TNF-α, IFN-γ, and IL2 serum levels. In addition, studies have demonstrated that the herb limits cancer by improving body immunity [35].

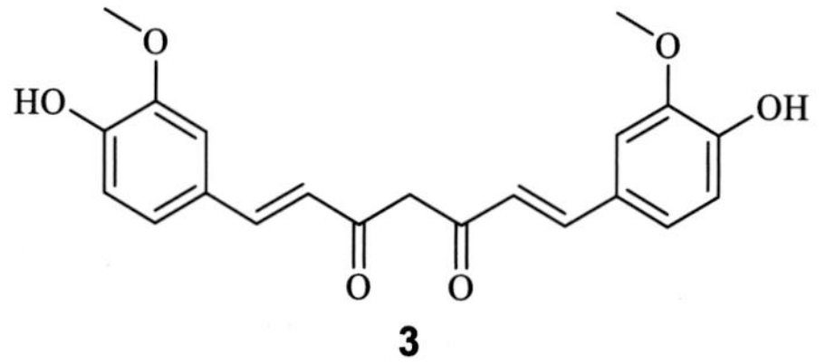

HO O O O OH O
3

FIG. 1 Chemical structure of curcumin.

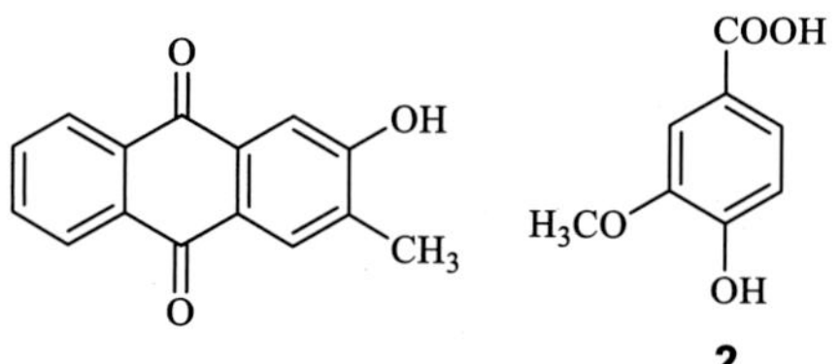

O OH CH_3 O COOH H_3CO OH
1 **2**

FIG. 2 Chemical structure of compounds.

2.3 Astragalus membranaceus

This herb is one of the most common Chinese medicine which belongs to the leguminous plant. The chief constituents of this herb are astragalus polysaccharide and astragaloside (Fig. 3) with antiviral, antiinflammatory, antioxidant, and immunoblotting activity. Astragaloside have the ability to downregulate the expression of PD-1 and PD-L1 (immune checkpoint proteins) and decrease the migration-associated proteins MMP-9 and MMP-2, thereby causing inhibition of cervical cancer cells [36].

Astragalus can control the growth and metastasis of ovarian carcinoma by limiting the expression of MMP-9 and MMP-2 proteins. It controls the apoptosis by upregulating caspase 3 and caspase 9 (apoptosis-related proteins) and elevating ratio of Bax-to-Bcl-2 proteins [37]. Astralagus may also potentially cause the macrophages polarization into M1 type which result in tumor suppression. However, combination with cisplatin further increases the anticancer activity. [38]

2.4 Ganoderma lucidum

Ganoderma lucidum is a plant that contains triterpenoids, amino acids, polysaccharides, alkaloids, proteins, and sterols. It controls diabetes and enhances cardiovascular and antiaging effects. Previous research has shown that Ganoderma triterpenoids (Fig. 4) prevent the spread of lung carcinoma and its action is linked to cell cycle control and Bax/Bcl protein ratio [39].

Triterpenoids inhibit the proliferation of prostate cancerous cells and suppress cancer activity [40]. In the breast cancer, GL decreases the quantity of breast cancer stem cells by lowering the STAT3 pathway and prevents cancer cell invasion [41]. The polysaccharides present in GL may have anticancer function of the standard drug paclitaxel (PTX) and suppress cancer metabolism [42]. A recent study showed that GL polysaccharide-gold nanocomposites stimulate dendritic cells that facilitate T-cell activation, and has a significant inhibitory impact on lung cancer cell proliferation and metastasis [43].

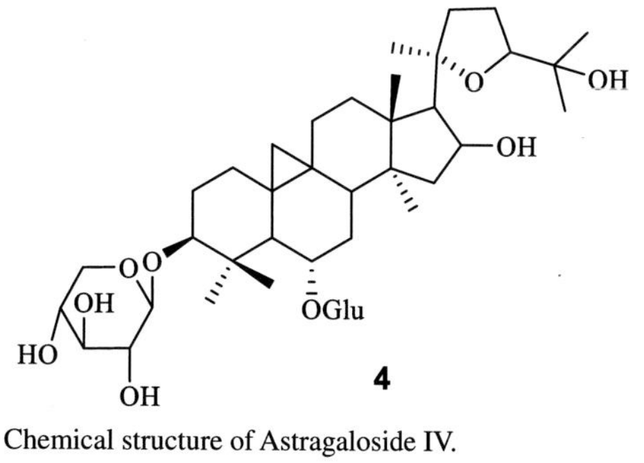

OH OH O O O OH HO OH ŌGlu 4

FIG. 3 Chemical structure of Astragaloside IV.

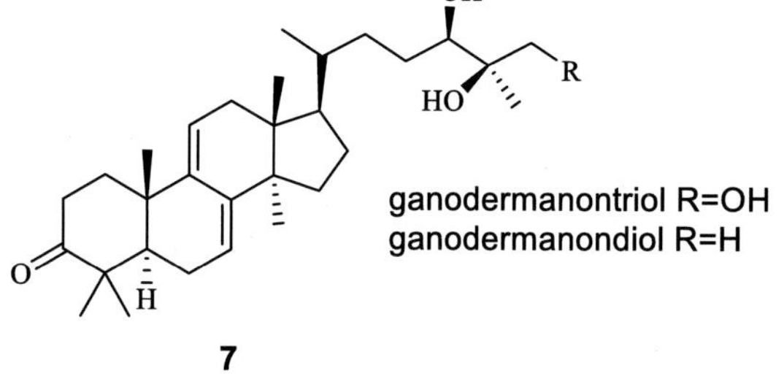

OH R HO O H ganodermanontriol R=OH ganodermanondiol R=H 7

FIG. 4 Chemical structure of Ganoderma triterpenoids.

2.5 Panax ginseng

Ginseng has drawn significant interest owing to its various pharmacological roles like cardiovascular defense, antitumor, and antiaging activities [44]. The main ingredients of Ginseng are ginsenosides (Fig. 5), which are triterpenoids that primarily contain Rg1, Rg2, Rg3, Rh1, and Rh2. Studies have shown that Rh2 ginsenosides may induce endometrial cancer cells apoptosis, and their action is correlated with increased caspase 3 and PARP protein levels [45].

2.6 Angelica sinensis

Angelica sinensis (AS) is a perennial herbaceous plant with analgesic, cardiovascular, and antiinflammatory properties. It contains organic acids, volatile oils, and polysaccharides [46]. The volatile component consists mainly *Z*-ligustilide (Fig. 6) that has an anticancer and immune-enhancing effect.

Angelica in acetone decreased vascular endothelial growth factor (VEGF) and hypoxia-inducible factor-1-alpha (HIF-1-α) expression in cancer cells of bladder cancer, thereby effectively decreased tumor microenvironment development. Meanwhile, PI3K/AKT /mTOR signaling pathway which is required for cancer cell survival is also inhibited; Angelica polysaccharides release cAMP-responsive element-binding protein (CREB), upregulated caspase 3/9, and cleaved PARP, and contribute to apoptosis in breast cancer cells [47,48]

3 Plant-based natural compounds for treatment of cancers

3.1 Skin cancer (melanoma)

Out of one million new cases reported annually, ~4.5% of cancer patients are suffering from melanoma. The morbidity rate is significantly high; therefore, research is required to identify effective formulations which have potential therapeutic effectiveness with low side effects [49]. Plant-based active ingredients enriched with phenolic compounds displayed important anticancer activities toward skin cancer [50,51]. The polyphenol extracts from *Euphorbia lagascae Spreng* plant caused cytotoxicity against SK-MEL-28 melanoma cells by cell cycle arrest at G2/M phase via downregulating cyclins A, E, and B1 proteins [50]. Procyanidins obtained from *Pinus pinaster*

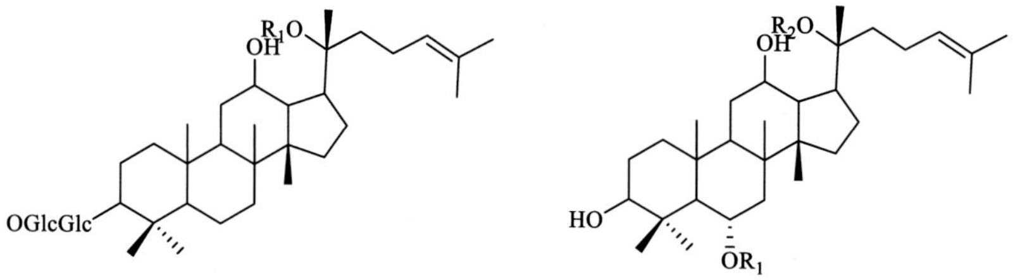

Ginsenosides	R_1	Ginsenosides	R_1	R_2
Rb1	Glc-^{6}Glc	Re	Glc-^{2}Rha	Glc
Rb2	Ara-6Glc	Rf	Glc-^{2}Glc	H
Rc	Ara-^{6}Glc	Rg1	Glc	Glc
Rd	Glc	Rg2	Glc-^{2}Rha	H

FIG. 5 Chemical structure of *Panax ginseng*.

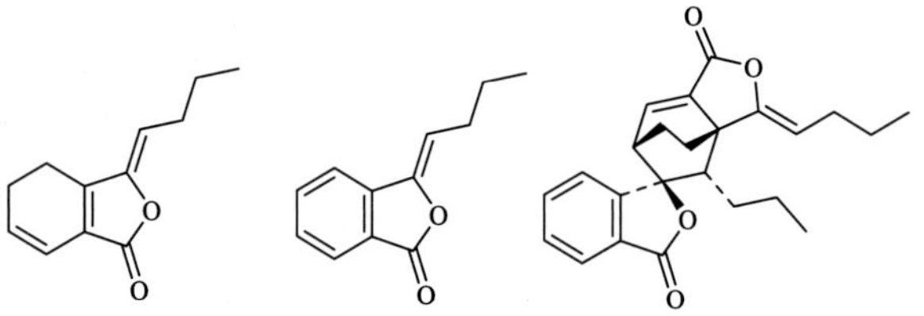

Z-ligustilide E-butylidenepthalide Z-ligustilide dimer E-232

FIG. 6 Chemical structure of identified phthalides found in *Angelica sinensis*.

Aiton also induced a cytotoxic effect in SK-MEL-28 cells [52]. 4-Nerolidylcatechol derived from *Piper umbellatum* L. acts by arresting G-1 phase, loss of membrane integrity with inhibition of matrix metalloproteinase MMP-2 and MMP-9 activity [53]. Galangin obtained from *Alpinia officinarum Hance* reduces the mitochondrial membrane potential in melanoma [53]. *Indigofera aspalathoides DC* and *Saturejathymbra L* possess cytotoxic activity against melanoma [53].

3.2 Glioblastoma or brain cancer

Brain cancer has the highest fatality rate if found in the brain or spinal cord [54]. There are numerous compounds obtained from plants such *as Bursera microphylla, Annona glabra L., Angelica sinensis (Oliv.) Diels, and Bupleurumscorzonerifolium Wild* which possess promising anticancer effects against brain tumor cells [55].

Carvone, a volatile component obtained from plants, showed outstanding anticancer activity against brain cancer. Its anticancer activity is linked to a rise in antioxidant levels in cancer cells [56]. β-Elemene extracted from Curcuma aromatic Salisb has also shown profound anticancer activity against brain cancer. Different phytoconstituents and their mechanism are given in Table 1 [56].

3.3 Breast cancer

It is the most prevalent cancer in women worldwide. According to an earlier survey in the United States, 255,180 breast cancer cases have arisen and 41,070 death reports were registered in 2017 [62, 63]. Breast cancer is categorized into two types: estrogen-positive (such as MCF-7 and T47D cells) and estrogen receptor-negative (such as MDA-MB-231, MDA-MB-468, SKBR3, and MDA-MB-453 cells) breast cancer. It can further be classified into many molecular subtypes, including luminal A, luminal B, basal, and HER 2-positive cancers, based on the screening of the biomarkers such as progesterone receptor (PR) and human epidermal growth factor receptor 2 (HER2) [64, 65]. Researchers have demonstrated that natural compounds such as sanguinarine isolated from *Sanguinaria canadensis L* induce apoptosis in human breast cancer MDA-MB-231 cells [66]. Studies have also reportedthe use of active components isolated from *Sophora flavescens Aiton* that could be used against metastatic breast cancer (MCF-7 and 4T1 cells) treatment. Piperine, an alkaloid isolated from *Piper nigrum*, suppresses 4T1 cell growth and causes apoptosis [67, 68]. Evodiamine phytoconstituent acts by decreasing the growth of NCI/ADR-RES cells of human breast cancer resistant to adriamycin. [67, 68] Neobractatin isolated from *Garcinia bracteata* prevents metastasis [67, 68].

TABLE 1 Anticancer activities of phytochemicals in brain cancer.

Plants	Constituents	Mechanism
Croton regelianus Müll.Arg.	Essential oil	Cytotoxic effects [55]
Angelica sinensis	Polysaccharides	Acts by changing the cell cycle distribution along with inducing programmed cell death, that is, apoptosis [57]
Croton zambesicus	Carvone	Acts by increasing antioxidant level [58]
Trifolium pratense L.	Isoflavones	Isoflavones combines with rapamycin, which results isoflavones inhibits phosphorylation of eIF4E and Akt proteins [59]
Camellia sinensis	Epigallocatechin gallate	Induced programmed cell death by downregulation of p-Akt and Bcl-2 [60]
Phellinus linteus	Hispolon	Activate G2/M cell cycle arrest and apoptosis [61]

4 An approach of using phytochemicals or natural products for reducing damage caused by chemotherapy

Phytochemicals are plant-derived compounds and have been used as herbal drugs for several decades to heal multiple disorders as well as cancer due to reduced side effects [69]. Therefore, the combination chemotherapy using the natural compounds can not only suppress the side effects of the chemotherapeutic drug but also synergistically improve the anticancer activity [70].

Hesperetin is a bioflavonoid found in citrus fruits and have potential to exhibit antiinflammatory activity. It can restore the normal renal function by decreasing cisplatin-mediated oxidative stress and inflammatory cytokine levels in rats [71]. Cisplatin supplementation with quercetin also inhibits cisplatin-induced nephrotoxicity [72]. Berberine isoquinoline alkaloids isolated from Berberis have nephroprotective role for cisplatin-mediated kidney damage via modulation of oxidative/nitrosive stress, inflammation, autophagy, and apoptosis [73]. Cannabidiol isolated from *Cannabis sativa* attenuates cisplatin-induced oxidative/nitrosive stress, inflammation, and apoptosis in the kidneys, thereby restoring the normal renal function. Besides, this natural compound also defends for paclitaxel-induced neurotoxicity in female C57Bl/6 mice [74]. Dioscin derived from *diosgenin* acts against doxorubicin (DOX)-induced farnesoid X receptor (FXR)-mediated oxidative stress and inflammation [75]. Lycopene carotenoid administration with cisplatin could reduce nephrotoxicity in cancer patients [76].

5 Natural products originated from marine sources

In addition to plant origins, marine-derived anticancer compounds are also used for the prevention of cancer and associated health hazards. Marine species also possess bioactive compounds with high therapeutic potential, but their production is limited due to difficulties in extraction of those compounds. For example, the first report of the isolation of anticancer agent ecteinascidia-743 (known as trabectedin) from *Ecteinascidia turbinata* was registered in 1984 [77]. Recent works suggest that marine species may play an important role in the future toward the development of anticancer drugs. The approved lists of marine-derived drugs are given in Table 2.

6 Molecular targets of phytochemicals

In broader sense, chemotherapeutics confers their protecting effects in two different manners; either by inhibiting the formation of carcinogens or their interaction with biological molecules, mostly with DNA, hence, the chemotherapeutics act as blocking agents. Irrespective of this, mutations may also direct cells to reproduce abnormal cells even after damage. At this point, chemopreventive agents inhibit further progression of the transformed cells into tumor [72, 78]. It is reported that the combination of curcumin and piperine causes decrease in breast stem cell markers, Wnt signaling, and mammosphere formation in breast cancer cells [78, 79]. *Musa paradisiaca* causes selective induction of apoptosis in jurkat T and prostate cancer cells (LNCaP) as compared with normal human fibroblasts [80]. Catechins act by inhibiting the activity of proteasome in human Jurkat T cells with rise in capase-3, calpain 1, and P-27 activities. Besides, it also cause prostate and breast cancer cell arrest in G1 phase [81]. Sugiol (terpenoid) inhibits STAT3 signaling, causes ROS generation, and G_0/G1 phase cell cycle arrests in human prostate cancer cells (LNCaP, DU145, PC3), breast cancer cells (MDA-MB648), and colon cancer cells (HCT-116), together with tumor growth inhibition of DU145 cells in SPF female Balb/c nude mice [78].

7 Novel approaches in the field of nanotechnologies with targeting strategies for delivery of bioactives phytoconstituents as a plant pharmaceuticals

Pharmaceutical nanotechnology is gaining a remarkable and marvelous scope and interest toward the prospective of plant-based active constituents and herbal extracts (Table 3).

TABLE 2 Marine-derived chemotherapeutic pipeline for cancer-related issues [77].

		Approved drugs		
Active compound	**Source**	**Target**	**Cancer-related condition**	
Cytarabine	Spongothymidine (from sponges)	DNA polymerase	Lymphomatous meningitis and leukemia	
Trabectedin	Trabectedin (from tunicate)	DNA minor groove	Ovarian cancer and soft tissue sarcoma	
Eribulin mesylate	Halichondrin (from sponge)	Microtubules	Advanced liposarcoma and metastatic breast cancer	
Brentuximab vedotin	Dolastatin (mollusk/ cyanobacterium)	CD30 and microtubules	Hodgkin lymphoma	
		Under phase 3 trial		
Salinosporamide A Marizomib	Salinosporamide A (bacterium)	20S proteasome	Newly diagnosed glioblastoma	
Plinabulin	Halimide (from fungus)	Microtubules	NSCLC	
Lurbinectedin (Zepsyre®)	Trabectedin	Alkaloid	DNA minor groove	Ovarian cancer and SCLC

TABLE 3 Various biologically active compounds and their nanotechnology-based novel delivery systems [82].

Delivery system	Features	Herbal extract
Phyto-phospholipid complex	It is one of the successful novel approach toward delivery of plant-based bioactive constituents This novel strategy has successful impact over on improvement of low oral bioavailability of molecules which can enhance the bioavailability by complex formation of phospholipid	Rutin [83], Gallic acid [84], *Ginkgo biloba* [85],catechin [86], and curcumin [87]
Liposomes	Liposomes, first developed by Alec D. Bangham in 1963, are microscopic vesicles, the potential carrier mechanism for transporting hydrophilic and hydrophobic drugs, and is the recent advanced drug delivery mechanisms. Amount of traditional chemotherapeutics containing herbal compounds has limited side effects during tumor targeting Liposomes demonstrate exceptional aggregation in highly vascularized and permeable tissues, such as tumors or cancerous tissues	Silymarin [88],quercetin [89], propolis [90], berberine and palmitine [91], and carotenoid [92]
Nanoparticles	A nanoparticle is ultrafine particle and have diameter in range between 1 and 100 nm (nm) Nanonization of phytochemicals leads to variety of advantages like significant improvement in absorption, solubility, and decreased dose as that of the crude drug extract	Silver nanoparticles of Aloe [93], gold nanoparticles of *Euphorbia hirta* [94],silver nanoparticles of Ajwain [95], and iron magnetic nanoparticles of *Argemone Mexicana L* [96]
Microspheres	These are small emulsion cells or solid particles, having range of diameter of typically 1 to 1000 μm This current approach remains an option of delivery systems that integrate herbal constituents. Among the different benefits of the microsphere is release and the drug targeting to a particular location	Campothecin loaded microspheres [97], rutin-alginate chitosan microcaps, and [98] essential oil microspheres [99]
Phytosomes	This novel system comprises of complex of natural bioactive ingredient and phospholipid chiefly lecithin Phytosomes are phospholipid complexes for the efficient delivery of plant bioactive. This novel strategy is helpful for better targeting and to enhance their bioavailability by phospholipid complexation	Silybin, green tea, ginseng, grape seed, and hawthorn [90, 100]

The growth of pharmaceutical nanotechnology techniques resolves the difficulties in delivering poorly soluble herbal extracts for increasing their therapeutic efficacy and bioavailability [78].

8 Conclusion

Efficient treatment of cancers is one of the biggest challenges in the world. Conventional cancer therapies have shown significant side effects which limit their uses. Besides single treatment, combined therapeutic approaches are highly anticipated. However, highly efficient therapeutics with no/reduced side effects is warranted. Natural products have traditionally been used for cancer treatment in most parts of the world since ancient times due to their remarkable pharmacological activity without causing much side effects.

Acknowledgment

The authors are thankful to School of Pharmaceutical Sciences, Shoolini University, Solan, Himachal Pradesh 173229, India.

Conflicts of interest

The authors declare no conflict of interest.

References

[1] Korgaonkar N, Yadav KS. Understanding the biology and advent of physics of cancer with perspicacity in current treatment therapy. Life Sci 2019;239:117060.
[2] Cooper GM. The development and causes of cancer. The cell: a molecular approach. Sinauer Associates; 2000. https://www.ncbi.nlm.nih.gov/books/NBK9963/.
[3] Hesketh R. Introduction to cancer biology. United Kingdom: Cambridge University Press; 2013.
[4] Hassanpour SH, Dehghani M. Review of cancer from perspective of molecular. J Cancer Res Pract 2017;4(4):127–9.
[5] Suhail Y, Cain MP, Vanaja K, Kurywchak PA, Levchenko A, Kalluri R. Systems biology of cancer metastasis. Cell Syst 2019;9(2):109–27.
[6] Aceto N, Bardia A, Miyamoto DT, Donaldson MC, Wittner BS, Spencer JA, Yu M, Pely A, Engstrom A, Zhu H, Brannigan BW. Circulating tumor cell clusters are oligoclonal precursors of breast cancer metastasis. Cell 2014;158(5):1110–22.
[7] Lambert AW, Pattabiraman DR, Weinberg RA. Emerging biological principles of metastasis. Cell 2017;168(4):670–91.
[8] McFarland J, Hussar B, De Brey C, Snyder T, Wang X, Wilkinson-Flicker S, Gebrekristos S, Zhang J, Rathbun A, Barmer A, Bullock MF. The condition of education 2017. NCES 2017-144. National Center for Education Statistics; 2017.
[9] Quail DF, Joyce JA. Microenvironmental regulation of tumor progression and metastasis. Nat Med 2013;19(11):1423–37. https://doi.org/10.1038/nm.3394.
[10] Moreno-Smith M, et al. Impact of stress on cancer metastasis. Future Oncol 2010;6(12):1863–81. https://doi.org/10.2217/fon.10.142.
[11] Siegel RL, Miller KD, Goding Sauer A, Fedewa SA, Butterly LF, Anderson JC, et al. Colorectal cancer statistics, 2020. CA Cancer J Clin 2020;70(3):145–64.
[12] https:/www.cdc.gov/cancer/dcpc/search/articles/cancer%202020.html.
[13] Bishayee A, Sethi G. Bioactive natural products in cancer prevention and therapy: progress and promise. In: Seminars in cancer biology, vol. 40. Academic Press; 2016. p. 1–3.
[14] Harvey AL, Edrada-Ebel R, Quinn RJ. The re-emergence of natural products for drug discovery in the genomics era. Nat Rev Drug Discov 2015;14(2):111–29.
[15] Ferreira PM, Farias DF, Viana MP, Souza TM, Vasconcelos IM, Soares BM, Pessoa C, Costa-Lotufo LV, Moraes MO, Carvalho AF. Study of the antiproliferative potential of seed extracts from northeastern Brazilian plants. An Acad Bras Cienc 2011;83(3):1045–58.
[16] Jiao L, Bi L, Lu Y, et al. Cancer chemoprevention and therapy using chinese herbal medicine. Biol Proced Online 2018;20:1.
[17] Ma X, Wang Z. Anticancer drug discovery in the future: an evolutionary perspective. Drug Discov Today 2009;14(23–24):1136–42.
[18] Widmer N, Bardin C, Chatelut E, Paci A, Beijnen J, Levêque D, Veal G, Astier A. Review of therapeutic drug monitoring of anticancer drugs part two–targeted therapies. Eur J Cancer 2014;50(12):2020–36.
[19] Kuczynski EA, Sargent DJ, Grothey A, Kerbel RS. Drug rechallenge and treatment beyond progression—implications for drug resistance. Nat Rev Clin Oncol 2013;10(10):571.
[20] Hayden EC. Cancer complexity slows quest for cure. Nature 2008;455(7210):148.
[21] Newman DJ, Cragg GM. Drugs and drug candidates from marine sources: an assessment of the current "state of play". Planta Med 2016;82(09/10):775–89.
[22] Newman DJ, Cragg GM. Natural products as sources of new drugs from 1981 to 2014. J Nat Prod 2016;79(3):629–61.
[23] Malik S, Cusidó RM, Mirjalili MH, Moyano E, Palazón J, Bonfill M. Production of the anticancer drug taxol in Taxus baccata suspension cultures: a review. Process Biochem 2011;46(1):23–34.
[24] Morales-Cano D, Calviño E, Rubio V, Herráez A, Sancho P, Tejedor MC, Diez JC. Apoptosis induced by paclitaxel via Bcl-2, Bax and caspases 3 and 9 activation in NB4 human leukaemia cells is not modulated by ERK inhibition. Exp Toxicol Pathol 2013;65(7–8):1101–8.
[25] McCubrey JA, Kvin L, Linda SS, Steve LA, Li VY, Ramiro MM, AS PLR, Luca MN, Lucio C, Stefano R, Alberto MM, Piotr L, Dulińska-Litewka J, Dariusz R, Agnieszka G, Paolo L, Ferdinando N, Saverio C, Massimo L, Giuseppe M, Cervello M. Effects of resveratrol, curcumin, berberine and other nutraceuticals on aging, cancer development, cancer stem cells and microRNAs. Aging (Albany NY) 2017;9(6):1477–536.
[26] Liu W, Yang B, Yang L, Kaur J, Jessop C, Fadhil R, Good D, Ni G, Liu X, Mosaiab T, Yi Z. Therapeutic effects of ten commonly used Chinese herbs and their bioactive compounds on cancers. Evid Based Complement Alternat Med 2019;2019:6057837.
[27] Perrone D, Ardito F, Giannatempo G, Dioguardi M, Troiano G, Lo Russo L, DE Lillo A, Laino L, Lo ML. Biological and therapeutic activities, and anticancer properties of curcumin. Exp Ther Med 2015;10(5):1615–23.

[28] Yan MD, Cen XY. Effects of curcumin on the proliferation and apoptosis of thyroid cancer cell line SW579. Chin Gen Med 2015;13:396–8.

[29] Feng RX, Cai RG, Xie DD. Effects of curcumin on proliferation and apoptosis of oral squamous cell carcinoma Tca8113 cells. J Chengdu Med College 2019;14:25–30.

[30] Wilken R, Veena MS, Wang MB, Srivatsan ES. Curcumin: A review of anti-cancer properties and therapeutic activity in head and neck squamous cell carcinoma. Mol Cancer 2011;10:12.

[31] Wang X, Deng J, Yuan J, Tang X, Wang Y, Chen H, Liu Y, Zhou L. Curcumin exerts its tumor suppressive function via inhibition of NEDD4 oncoprotein in glioma cancer cells. Int J Oncol 2017;51(2):467–77.

[32] Su J, Zhou X, Yin X, Wang L, Zhao Z, Hou Y, Zheng N, Xia J, Wang Z. The effects of curcumin on proliferation, apoptosis, invasion, and NEDD4 expression in pancreatic cancer. Biochem Pharmacol 2017;140:28–40.

[33] Chung TW, Choi H, Lee JM, Ha SH, Kwak CH, Abekura F, Park JY, Chang YC, Ha KT, Cho SH, Chang HW. Oldenlandiadiffusa suppresses metastatic potential through inhibiting matrix metalloproteinase-9 and intercellular adhesion molecule-1 expression via p38 and ERK1/2 MAPK pathways and induces apoptosis in human breast cancer MCF-7 cells. J Ethnopharmacol 2017;195:309–17.

[34] Sunwoo YY, Lee JH, Jung HY, Jung YJ, Park MS, Chung YA, Maeng LS, Han YM, Shin HS, Lee J, Park SI. Oldenlandiadiffusa promotes antiproliferative and apoptotic effects in a rat hepatocellular carcinoma with liver cirrhosis. Evid Based Complement Alternat Med 2015;2015.

[35] Zhao YL. Application of hedyotis diffusa in tumor treatment. Gansu Sci Technol 2018;34:141–2.

[36] Liu W, Yang B, Yang L, Kaur J, Jessop C, Fadhil R, Good D, Ni G, Liu X, Mosaiab T, Yi Z. Therapeutic effects of ten commonly used Chinese herbs and their bioactive compounds on cancers. Evid Based Complement Alternat Med 2019;15:2019.

[37] Zhang J, Liu L, Wang J, Ren B, Zhang L, Li W. Formononetin, an isoflavone from Astragalus membranaceus inhibits proliferation and metastasis of ovarian cancer cells. J Ethnopharmacol 2018;221:91–9.

[38] Wang LX, Wu WB, Xu ZH. Astragaloside a plays an anti-tumor role by inducing polarization of M1-type macrophages. Chin J Exp Formulae 2019;10:1–7.

[39] Zolj S, Smith MP, Goines JC, T'Shura SA, Huff MO, Robinson DL, Lau JM. Antiproliferative effects of a triterpene-enriched extract from lingzhi or reishi medicinal mushroom, ganodermalucidum (agaricomycetes), on human lung cancer cells. Int J Med Mushrooms 2018;20(12).

[40] Qu L, Li S, Zhuo Y, Chen J, Qin X, Guo G. Anticancer effect of triterpenes from Ganoderma lucidum in human prostate cancer cells. Oncol Lett 2017;14(6):7467–72.

[41] Rios-Fuller TJ, Ortiz-Soto G, Lacourt-Ventura M, Maldonado-Martinez G, Cubano LA, Schneider RJ, Martinez-Montemayor MM. Ganoderma lucidum extract (GLE) impairs breast cancer stem cells by targeting the STAT3 pathway. Oncotarget 2018;9(89):35907.

[42] Su J, Li D, Chen Q, Li M, Su L, Luo T, Liang D, Lai G, Shuai O, Jiao C, Wu Q. Anti-breast cancer enhancement of a polysaccharide from spore of ganoderma lucidum with paclitaxel: suppression on tumor metabolism with gut microbiota reshaping. Front Microbiol 2018;9:3099.

[43] Zhang S, Pang G, Chen C, Qin J, Yu H, Liu Y, Zhang X, Song Z, Zhao J, Wang F, Wang Y. Effective cancer immunotherapy by Ganoderma lucidum polysaccharide-gold nano-composites through dendritic cell activation and memory T cell response. Carbohydr Polym 2019;205:192–202.

[44] Kim JH. Pharmacological and medical applications of Panax ginseng and ginsenosides: a review for use in cardiovascular diseases. J Ginseng Res 2018;42(3):264–9.

[45] Wang M. Ginsenoside Rh2 promotes apoptosis of human endometrial cancer cells and its mechanism. Southwest Natl Def Med 2018;28:720–3.

[46] Xie G, Peng W, Li P, Xia Z, Zhong Y, He F, Tulake Y, Feng D, Wang Y, Xing Z. A network pharmacology analysis to explore the effect of astragali radix-radix angelica sinensis on traumatic brain injury. Biomed Res Int 2018;2018.

[47] Chen MC, Hsu WL, Chang WL, Chou TC. Antiangiogenic activity of phthalides-enriched Angelica Sinensis extract by suppressing WSB-1/pVHL/HIF-1α/VEGF signaling in bladder cancer. Sci Rep 2017;7(1):5376.

[48] Zhou WJ, Wang S, Hu Z, Zhou ZY, Song CJ. Angelica sinensis polysaccharides promotes apoptosis in human breast cancer cells via CREB-regulated caspase-3 activation. Biochem Biophys Res Commun 2015;467(3):562–9.

[49] Anastyuk SD, Shevchenko NM, Ermakova SP, Vishchuk OS, Nazarenko EL, Dmitrenok PS, Zvyagintseva TN. Anticancer activity in vitro of a fucoidan from the brown alga Fucusevanescens and its low-molecular fragments, structurally characterized by tandem mass-spectrometry. Carbohydr Polym 2012;87(1):186–94.

[50] Elmasri WA, Hegazy ME, Mechref Y, Paré PW. Cytotoxic saponin poliusaposide from Teucrium polium. RSC Adv 2015;5(34):27126–33.

[51] Touriño S, Selga A, Jiménez A, Juliá L, Lozano C, Lizárraga D, Cascante M, Torres JL. Procyanidin fractions from pine (Pinus pinaster) bark: radical scavenging power in solution, antioxidant activity in emulsion, and antiproliferative effect in melanoma cells. J Agric Food Chem 2005;53(12):4728–35.

[52] Tran QL, Tezuka Y, Banskota AH, Tran QK, Saiki I, Kadota S. New spirostanol steroids and steroidal Saponins from roots and rhizomes of dracaena a ngustifolia and their antiproliferative activity. J Nat Prod 2001;64(9):1127–32.

[53] Sharifi-Rad J, Ozleyen A, BoyunegmezTumer T, OluwaseunAdetunji C, El Omari N, Balahbib A, Taheri Y, Bouyahya A, Martorell M, Martins N, Cho WC. Natural products and synthetic analogs as a source of antitumor drugs. Biomolecules 2019;9(11):679.

[54] Tandel GS, Biswas M, Kakde OG, Tiwari A, Suri HS, Turk M, Laird JR, Asare CK, Ankrah AA, Khanna NN, Madhusudhan BK. A review on a deep learning perspective in brain cancer classification. Cancer 2019;11(1):111.

[55] Bezerra DP, Filho JD, Alves AP, Pessoa C, de Moraes MO, Pessoa OD, Torres MC, Silveira ER, Viana FA, Costa-Lotufo LV. Antitumor activity of the essential oil from the leaves of Croton regelianus and its component ascaridole. Chem Biodivers 2009;6(8):1224–31.

[56] Cai Y, Luo Q, Sun M, Corke H. Antioxidant activity and phenolic compounds of 112 traditional Chinese medicinal plants associated with anticancer. Life Sci 2004;74(17):2157–84.

[57] Tsai NM, Lin SZ, Lee CC, Chen SP, Su HC, Chang WL, Harn HJ. The antitumor effects of *Angelica sinensis* on malignant brain tumors in vitro and in vivo. Clin Cancer Res 2005;11(9):3475–84.

[58] Okokon JE, Dar A, Choudhary MI. Immunomodulatory, cytotoxic and antileishmanial activity of phytoconstituents of Croton zambesicus. Phytopharmacol J 2013;4(1):31–40.
[59] Puli S, Jain A, Lai JC, Bhushan A. Effect of combination treatment of rapamycin and isoflavones on mTOR pathway in human glioblastoma (U87) cells. Neurochem Res 2010;35(7):986–93.
[60] Zhang Y, Wang SX, Ma JW, Li HY, Ye JC, Xie SM, Du B, Zhong XY. EGCG inhibits properties of glioma stem- like cells and synergizes with temozolomide through downregulation of P-glycoprotein inhibition. J Neurooncol 2015;121(1):41–52.
[61] Arcella A, Oliva MA, Sanchez M, Staffieri S, Esposito V, Giangaspero F, Cantore G. Effects of hispolon on glioblastoma cell growth. Environ Toxicol 2017;32(9):2113–23.
[62] Torre LA, Bray F, Siegel RL, Ferlay J, Lortet-Tieulent J, Jemal A. Global cancer statistics, 2012. CA Cancer J Clin 2015;65(2):87–108.
[63] Siegel RL, Miller KD, Fedewa SA, Ahnen DJ, Meester RG, Barzi A, Jemal A. Colorectal cancer statistics, 2017. CA Cancer J Clin 2017;67(3):177–93.
[64] Reis-Filho JS, Pusztai L. Gene expression profiling in breast cancer: classification, prognostication, and prediction. Lancet 2011;378(9805):1812–23.
[65] Perou CM, Børresen-Dale AL. Systems biology and genomics of breast cancer. Cold Spring Harb Perspect Biol 2011;3(2):a003293.
[66] Kim S, Lee TJ, Leem J, Choi KS, Park JW, Kwon TK. Sanguinarine-induced apoptosis: generation of ROS, down-regulation of Bcl-2, c-FLIP, and synergy with TRAIL. J Cell Biochem 2008;104(3):895–907.
[67] Gautam N, Mantha AK, Mittal S. Essential oils and their constituents as anticancer agents: a mechanistic view. Biomed Res Int 2014;9:2014.
[68] Li H, Tan G, Jiang X, Qiao H, Pan S, Jiang H, Kanwar JR, Sun X. Therapeutic effects of matrine on primary and metastatic breast cancer. Am J Chin Med 2010;38(06):1115–30.
[69] Agbarya A, Ruimi N, Epelbaum R, Ben-Arye E, Mahajna J. Natural products as potential cancer therapy enhancers: A preclinical update. SAGE Open Med 2014;2. 2050312114546924.
[70] Prša P, Karademir B, Biçim G, Mahmoud H, Dahan I, Yalçın AS, Mahajna J, Milisav I. The potential use of natural products to negate hepatic, renal and neuronal toxicity induced by cancer therapeutics. Biochem Pharmacol 2020;173:113551.
[71] Bai X, Yang P, Zhou Q, Cai B, Buist-Homan M, Cheng H, Jiang J, Shen D, Li L, Luo X, Faber KN. The protective effect of the natural compound hesperetin against fulminant hepatitis in vivo and in vitro. Br J Pharmacol 2017;174(1):41–56.
[72] Waseem M, Bhardwaj M, Tabassum H, Raisuddin S, Parvez S. Cisplatin hepatotoxicity mediated by mitochondrial stress. Drug Chem Toxicol 2015;38(4):452–9.
[73] Domitrović R, Cvijanović O, Pernjak-Pugel E, Škoda M, Mikelić L, Crnčević-Orlić Ž. Berberine exerts nephroprotective effect against cisplatin-induced kidney damage through inhibition of oxidative/nitrosative stress, inflammation, autophagy and apoptosis. Food Chem Toxicol 2013;62:397–406.
[74] Pan H, Mukhopadhyay P, Rajesh M, Patel V, Mukhopadhyay B, Gao B, Haskó G, Pacher P. Cannabidiol attenuates cisplatin-induced nephrotoxicity by decreasing oxidative/nitrosative stress, inflammation, and cell death. J Pharmacol Exp Ther 2009;328(3):708–14.
[75] Zhang Y, Xu Y, Qi Y, Xu L, Song S, Yin L, Tao X, Zhen Y, Han X, Ma X, Liu K. Protective effects of dioscin against doxorubicin-induced nephrotoxicity via adjusting FXR-mediated oxidative stress and inflammation. Toxicology 2017;378:53–64.
[76] Cheng HM, Koutsidis G, Lodge JK, Ashor A, Siervo M, Lara J. Tomato and lycopene supplementation and cardiovascular risk factors: A systematic review and meta-analysis. Atherosclerosis 2017;257:100–8.
[77] Pereira RB, Evdokimov NM, Lefranc F, Valentão P, Kornienko A, Pereira DM, Andrade PB, Gomes NG. Marine-derived anticancer agents: clinical benefits, innovative mechanisms, and new targets. Mar Drugs 2019;17(6):329.
[78] Kaur V, Kumar M, Kumar A, Kaur K, Dhillon VS, Kaur S. Pharmacotherapeutic potential of phytochemicals: implications in cancer chemoprevention and future perspectives. Biomed Pharmacother 2018;97:564–86.
[79] Kakarala M, Brenner DE, Korkaya H, Cheng C, Tazi K, Ginestier C, Liu S, Dontu G, Wicha MS. Targeting breast stem cells with the cancer preventive compounds curcumin and piperine. Breast Cancer Res Treat 2010;122(3):777–85.
[80] Kazi A, Urbizu DA, Kuhn DJ, Acebo AL, Jackson ER, Greenfelder GP, Kumar NB, Dou QP. A natural musaceas plant extract inhibits proteasome activity and induces apoptosis selectively in human tumor and transformed, but not normal and non-transformed, cells. Int J Mol Med 2003;12(6):879–87.
[81] Jung SN, Shin DS, Kim HN, Jeon YJ, Yun J, Lee YJ, Kang JS, Han DC, Kwon BM. Sugiol inhibits STAT3 activity via regulation of transketolase and ROS-mediated ERK activation in DU145 prostate carcinoma cells. Biochem Pharmacol 2015;97(1):38–50.
[82] Alexander A, Patel RJ, Saraf S, Saraf S. Recent expansion of pharmaceutical nanotechnologies and targeting strategies in the field of phytopharmaceuticals for the delivery of herbal extracts and bioactives. J Control Release 2016;241:110–24.
[83] Singh D, Rawat MS, Semalty A, Semalty M. Rutin-phospholipid complex: an innovative technique in novel drug delivery system-NDDS. Curr Drug Deliv 2012;9(3):305–14.
[84] Kuamwat RS, Mruthunjaya K, Gupta MK. Hepatoprotective effect of Gallic acid and Gallic acid Phytosome against carbon tetrachloride induced damage in albino rats. Res J Pharm Technol 2012;5(5):677–81.
[85] Chen ZP, Sun J, Chen HX, Xiao YY, Liu D, Chen J, Cai H, Cai BC. Comparative pharmacokinetics and bioavailability studies of quercetin, kaempferol and isorhamnetin after oral administration of Ginkgo biloba extracts, Ginkgo biloba extract phospholipid complexes and Ginkgo biloba extract solid dispersions in rats. Fitoterapia 2010;81(8):1045–52.
[86] Semalty A, Semalty M, Singh D, Rawat MS. Phyto-phospholipid complex of catechin in value added herbal drug delivery. J Incl Phenom Macrocycl Chem 2012;73(1–4):377–86.
[87] Zhang J, Tang Q, Xu X, Li N. Development and evaluation of a novel phytosome-loaded chitosan microsphere system for curcumin delivery. Int J Pharm 2013;448(1):168–74.
[88] Elmowafy M, Viitala T, Ibrahim HM, Abu-Elyazid SK, Samy A, Kassem A, Yliperttula M. Silymarin loaded liposomes for hepatic targeting: in vitro evaluation and HepG2 drug uptake. Eur J Pharm Sci 2013;50(2):161–71.
[89] Liu D, Hu H, Lin Z, Chen D, Zhu Y, Hou S, Shi X. Quercetin deformable liposome: preparation and efficacy against ultraviolet B induced skin damages in vitro and in vivo. J Photochem Photobiol B Biol 2013;127:8–17.

[90] Yanyu X, Yunmei S, Zhipeng C, Qineng P. The preparation of silybin–phospholipid complex and the study on its pharmacokinetics in rats. Int J Pharm 2006;307(1):77–82.

[91] Pustovidko AV, Rokitskaya TI, Severina II, Simonyan RA, Trendeleva TA, Lyamzaev KG, Antonenko YN, Rogov AG, Zvyagilskaya RA, Skulachev VP, Chernyak BV. Derivatives of the cationic plant alkaloids berberine and palmatine amplify protonophorous activity of fatty acids in model membranes and mitochondria. Mitochondrion 2013;13(5):520–5.

[92] Nacke C, Schrader J. Liposome based solubilisation of carotenoid substrates for enzymatic conversion in aqueous media. J Mol Catal B: Enzym 2011;71(3–4):133–8.

[93] Zhang Y, Cheng X, Zhang Y, Xue X, Fu Y. Biosynthesis of silver nanoparticles at room temperature using aqueous aloe leaf extract and antibacterial properties. Colloids Surf A Physicochem Eng Asp 2013;423:63–8.

[94] Annamalai A, Christina VL, Sudha D, Kalpana M, Lakshmi PT. Green synthesis, characterization and antimicrobial activity of au NPs using Euphorbia hirta L. leaf extract. Colloids Surf B Biointerfaces 2013;108:60–5.

[95] Vijayaraghavan K, Nalini SK, Prakash NU, Madhankumar D. One step green synthesis of silver nano/microparticles using extracts of Trachyspermumammi and Papaver somniferum. Colloids Surf B Biointerfaces 2012;94:114–7.

[96] Arokiyaraj S, Saravanan M, Prakash NU, Arasu MV, Vijayakumar B, Vincent S. Enhanced antibacterial activity of iron oxide magnetic nanoparticles treated with Argemone mexicana L. leaf extract: an in vitro study. Mater Res Bull 2013;48(9):3323–7.

[97] Machida Y, Onishi H, Kurita A, Hata H, Morikawa A, Machida Y. Pharmacokinetics of prolonged-release CPT-11-loaded microspheres in rats. J Control Release 2000;66(2–3):159–75.

[98] Xiao L, Zhang YH, Xu JC. Rutin chitosan–alginate microcapsules floating research. Chin Tradit Herb Drug 2008;39:209–12.

[99] Glenn GM, Klamczynski AP, Woods DF, Chiou B, Orts WJ, Imam SH. Encapsulation of plant oils in porous starch microspheres. J Agric Food Chem 2010;58(7):4180–4.

[100] Bhattacharya S. Phytosomes emerging strategy in delivery of herbal drugs and nutraceuticals. Pharmatimes 2009;41:3.

Chapter 39

Clinical trials in drug delivery for the treatment of cancer

Nitesh Kumar[a,*], Tania Patwal[b,*], Varun Kumar[a,*], Priya Shrivastava[a,*], Akansha Mehra[a], and Pawan Kumar Maurya[c]

[a]*Amity Institute for Advanced Research and Studies (Materials & Devices), Amity University, Noida, India,* [b]*Department of Biotechnology, Thapar Institute of Engineering and Technology, Patiala, India,* [c]*Department of Biochemistry, Central University of Haryana, Mahendergarh, Haryana, India*

1 Introduction

Research and development work for drug delivery systems, devices, biologics, and therapeutics has been going on since the early 1900s. The 2007 Food and Drug Amendments Act mandated that all drugs, therapeutics, biologics, and devices in clinical trials be registered at ClinicalTrials.gov, which has shown a rapid growth in total number of registrations of clinical trials [1]. The database contains all the registered clinical trials for different diseases: 89% are investigating cancer and cardiac diseases in the elderly, 64% are investigating mental and behavioral disorders in the elderly, 33% are investigating AIDS in young subjects, and 15% are investigating cardiovascular diseases in young subjects [1]. Cancer is the top-ranked research area with the highest percentage of clinical trials (87%), mostly enrolling elderly patients older than 65 years. Cancer, although more common among older adults, is now the most diagnosed disease in patients aged 30–45 years. This is likely due to several factors such as pollution, diet, cigarette smoking, alcohol, hormones, dust and processed foods [2]. One study projected 600,920 deaths and 1,688,780 new cases of cancer by 2017 with an estimated increase of 70% over the next 20 years [3]. Rapidly advancing technology for the treatment of cancer includes chemotherapy, monoclonal antibodies, proteins, and nanotechnologies.

Clinical trials of different formulated drugs for cancer therapy are conducted in four phases. If a drug exhibits efficacy and safety at the fourth stage, it is likely to be approved by either the United States Food and Drug Administration (FDA) or European Medicines Association (EMA). The EMA was established in 1995 for the evaluation of novel cancer drugs [4]. Phase I studies include safety trials, in which a sponsor submits documentation for an investigational new drug (IND) to the FDA. Phase I trials determine the safety, toxicity, and dosage regimen of the drug [5]. Phase II clinical trials begin after the safety of the drug is determined. It includes testing on 30–300 subjects, usually with a specific type of cancer [6]. Phase II trials determine the effectiveness of the drug on the subjects [7] and collect data on time-tumor progression (TTP), time-treatment failure (TTF), and progression-free survival (PFS). At the completion of Phase II trials, the FDA and the sponsoring team meet to discuss if the drug has met the safety hypothesis and to determine how many subjects are needed to conduct the Phase III trials. The Phase III trials are comparative or approval trials [5], which are usually conducted on more than 3000 subjects to review the drug for its effectiveness and safety. Their purpose is to determine whether the new therapy has potential effectiveness from small to moderate to the advanced level for the treatment of cancer [5]. Phase IV clinical trials require the FDA and sponsors to have a post approval for the drug to be released in the market. Phase IV trials typically include thousands of subjects and results determine either a drug's rejection or approval [8] (Fig. 1).

Many different antibodies/protein drugs for treating different types of cancers have been approved by the FDA and EMA and are currently available on the market [9]. Some of these drugs include brentuximab vedotin (SGN-35), inotuzumab, pertuzumab, rituximab, and cetuximab for Hodgkin and non-Hodgkin lymphoma; trastuzumab and paclitaxel for breast cancer; gefitinib (ZD1839) for nonsmall cell lung cancer [10], anti-Flk-1, MAb-NP-90Y, and IA-NP-90Y for a variety of tumors [11], catumaxomab for ovarian and epithelial cancer [12], ipilimumab for metastatic-resistant prostate cancer; tremelimumab for solid malignancies; and anti-CTLA-4 therapy for malignancies [13]. Other drugs include liposomal-encapsulated drugs, nanoparticle drugs, and antibody drug conjugates (ADC), for example, anti-HER2, anti-TNF-α (antitumor necrosis factor-α), infliximab (Remicade), anti-RhD, alemtuzumab, ofatumumab, veltuzumab, anti-AME-133,

* Equal contributions.

Advanced Drug Delivery Systems in the Management of Cancer. https://doi.org/10.1016/B978-0-323-85503-7.00002-X

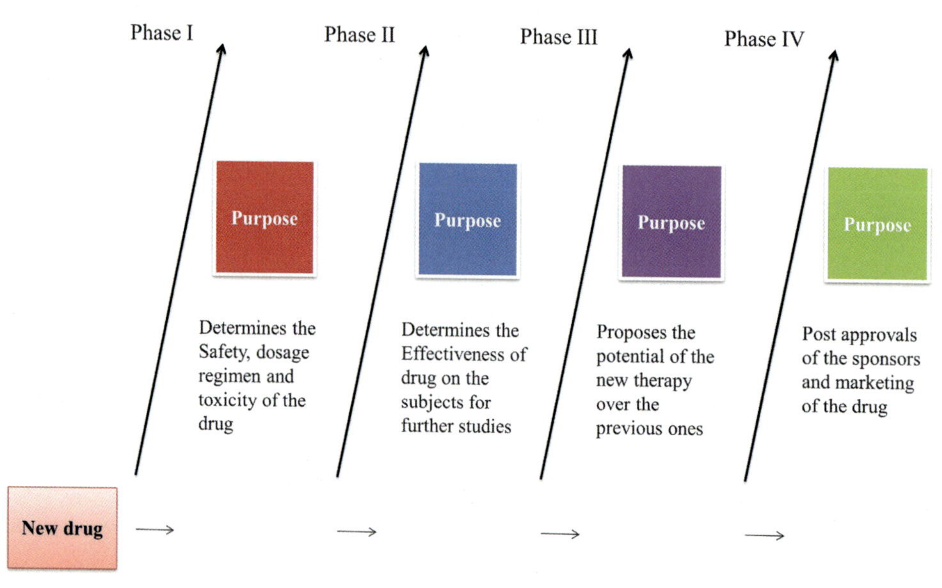

FIG. 1 Clinical trials sequence for new drug formulation.

and bevacizumab [14]. These drugs are delivered to patients through various new technologies.

1.1 Drug delivery technologies

Since the formulation of new cancer drugs, delivery technology has advanced significantly. Initially, chemotherapy (chemicals to kill cancer cells) and photothermal (fluorescent light) treatments were administered to cancer patients. One study showed that the combination of these two therapies, as "photochemotherapy" (PCT), had promising effects for cancer therapy [15]. The technology includes usage of different hybrid nanomaterials integrating PCT with near-infrared (NIR) dye-based nanostructures [16]. By the time grows scientists drawn out the immune therapies and transcriptional control based therapies to target cancer treatments [17]. Immunotherapy has proven to be a powerful additional cancer treatment that targets the immune system. It merges drug delivery to the system with adoptive cellular therapies and targets selective immune cells for destruction of cancer cells [18]. One study demonstrated the use of Chinese medicines for cancer treatment [19]. New technology is necessary in the treatment of cancer due to the rapid proliferation, spreading, and mutation of cancer cells in the body. The most promising new technologies include nanotherapy [20], liposomal-encapsulated drug delivery systems, and usage of clinically approved drugs, antibodies, biologics, and proteins [21] (Fig. 2).

2 Clinical trial advances in drug delivery for cancer treatment

Technological and computational advances in clinical trials have upgraded the process for selecting new treatment approaches with favorable safety and efficacy pharmacokinetics profiles of anticancer drugs, as shown in Fig. 3 [22]. The advances in the field of clinical trials for cancer therapy are categorized as follows [22, 23].

2.1 Advances in research methods

Advanced research methodologies employed during clinical trials of cancer therapy include:

- randomized controlled trials
- nonrandomized trials
- standard and placebo treatments
- crossover studies
- blinded studies

2.2 Randomized controlled trials

Randomized controlled trials are the most appropriate model to determine treatment efficacy. These types of investigations avoid biases that may result from human influence, anticipations, or various other factors. Most Phase II and Phase III clinical trials are randomized. This means that populations in these trial phases are randomly divided into two or more groups, and the consequences of the different groups are compared [24].

- Test group: this group is treated with the potential drug during trial
- Reference group: this group is treated with the current standard drug for the disease

When a comparison is made among the randomly allocated groups, it is easy to determine the most effective treatment for a particular disease.

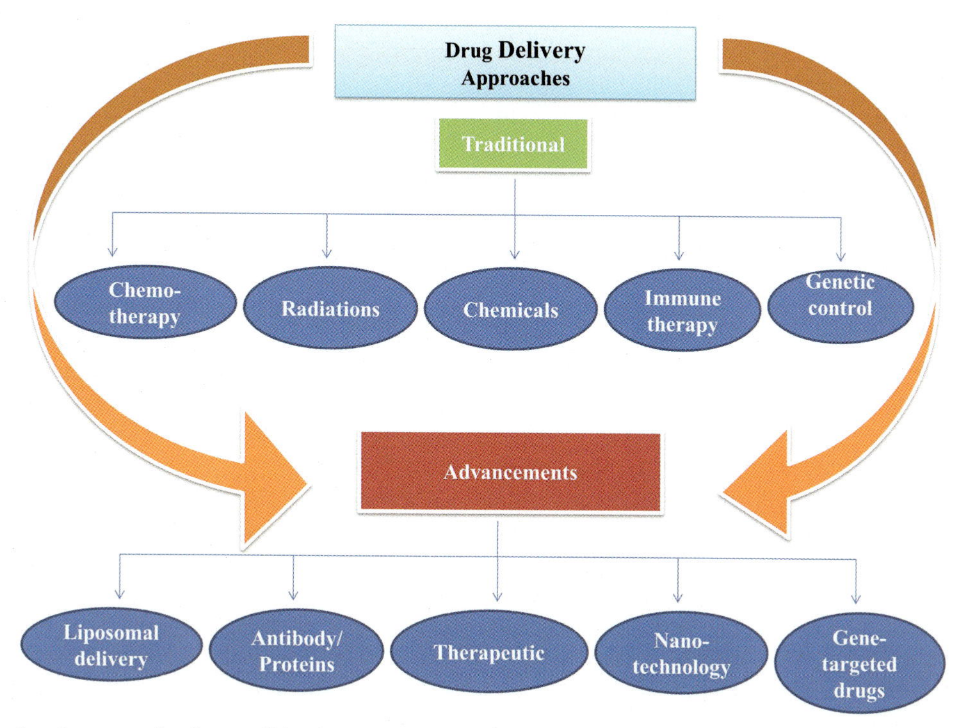

FIG. 2 Different changing approaches from traditional to advancements of drug delivery for cancer treatment.

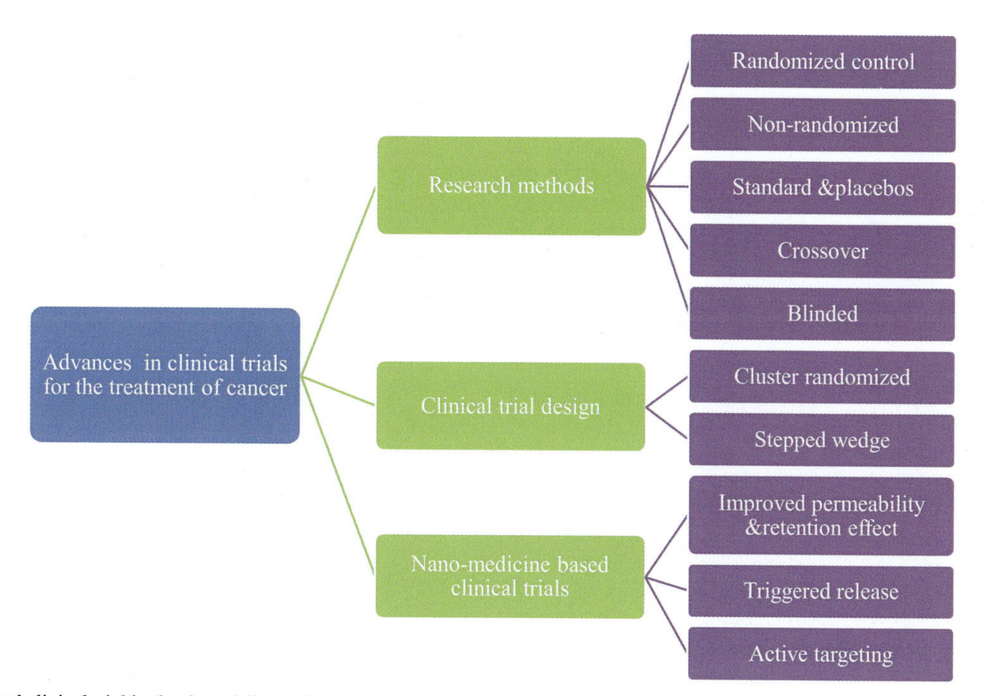

FIG. 3 Advanced clinical trial in the drug delivery for cancer treatment.

2.3 Nonrandomized trials

A nonrandomized trial is a study in a single group, in which the entire population receives the same kind of treatment. Generally, this method may be used during Phase I and Phase II trials, or where it is impossible to perform a randomized trial.

2.4 Standard treatment and placebos

Standard treatment: This is the most advanced and effective treatment given to the population for the specific disease. For instance, tumor surgery is recommended for newly diagnosed breast cancer, and followed by other therapies like hormone, radiation, and chemotherapy treatments.

Placebo: This is a mock treatment performed to look, taste, or feel like the treatment being checked, but the treatment does not involve any active drug. Saline injections or sugar pills are examples of placebo treatments. This type of treatment helps in concluding whether the results obtained from patients are due to the actual treatment or may be the result of various other factors in trials. If improvements are noticed within experimental groups receiving the actual drug, this provides evidence of the drug's efficacy over the placebo [25].

In clinical trials for cancer treatment, one group may be treated with reference therapy plus experimental therapy, and the second group may receive reference plus a placebo treatment.

Crossover studies: In these studies, the assigned treatment to a group of participants is given for a defined time before switching to another treatment. These studies enable all group members to experience all treatments and confirm the most efficient treatment method.

Blinded studies: In these studies, participants are unaware of the treatment group they belong to. Double-blind studies are randomized controlled clinical trials in which neither the researchers nor the study cohort know who is getting the control or experimental treatment. A blinded study is useful only when participants are unable to differentiate between two types of treatment methods. It is not possible to conduct a blinded study when the control and experimental treatment are different. The main objective of this type of study is to avoid biases in the reporting of good and bad effects of the trial method. The results of the trials are completely unaffected by the participant's and doctor's thoughts if they are unaware of the treatment method. For decades, randomized trials have been providing supreme results, but there are limitations associated with this traditional study design. At the 2018 ASCO Quality Care Symposium, George J. debated the need for modernization of clinical trials to provide high-value arbitration in quality enhancement [26].

3 Modernization in clinical trial design

The main challenge in quality enhancement for cancer treatment is to implement the findings of randomized controlled trials into clinical practice [27]. Traditional randomized clinical trials are performed in strict and idealized conditions that may not accurately reflect patients and situations that clinicians will be faced with in real life. The main goal of modern trial designing is pragmatism. Pragmatic trials are studies that involve the effectiveness of treatment under real conditions. The modern trial depends on the existing research infrastructure, principally a data infrastructure, so the collected data and results can be monitored without the need for complex research infrastructure. This is completely clinician-focused trials. The two pragmatic designs for clinical trials are:

- cluster randomized design
- stepped wedge design

3.1 Cluster randomized design

These trials are specifically useful for quality-enhancement interventions because the interventions are focused on the investigators, hospitals, and clinics, and not the patients. These designs can randomize social entities or groups (e.g., clinics, practices, even whole groups) and in so doing reduce the contamination of interventions from control to treated group [27].

With these trials, quality enhancement is not confined at the degree of a solitary point in continuous care. "It's not just centered on patients or investigators however require regular commitment at various levels." Furthermore, the unit of investigation probably will not be equivalent to the unit of randomization. The collected data could be examined at the patient or investigator level or various levels and may contrast from the unit of speculation as well. There is a ton of adaptability in the trial plan.

In another cluster randomized trial design, investigators looked for approaches to diminish the danger of falling among older patients in nursing homes depend on preliminary proof that enhances the navigation among the nursing home providers can improve the efficacy of the study. When nursing home representatives were randomized to receive detailing for enhancing navigation or not, though, no progress was noticed in either fall risk or counseling about fall risk.

3.2 Stepped wedge design

The second most used pragmatic design is the stepped wedge design, which accommodates the requirement for analysis within logistics limitations. These trial designs, which are being used with more frequency, are perfect for circumstances in which all associations ought to have a chance to utilize another intervention. "For complex interventions where the quick rollout isn't achievable, the stepped-wedge design is—compelling because all clusters will, in the end, be presented to the intervention, yet they likewise fill as in their controls."

3.3 Other approaches to quality improvement

Another tactic for quality enhancement includes incident learning schemes. In these schemes, bad outcomes of treatment are identified to determine at which point(s) the treatment went wrong. A significant risk of biases is acknowledged due to voluntary reporting. There is poor involvement of physicians in many of these schemes, and because this provides only the numerator, it is rarely possible for us to have understanding into the actual effect on the organization. Finally, the approach to quality improvement is quality measurements, but they also limited to some challenges, involving large resource utilization. Eventually, clinical trials for cancer therapy are all about data-driven quality enhancement [26].

4 Advances in clinical trials of nanomedicine-based cancer therapy

Nanomedicines (NNMs) are frequently utilized for targeted drug delivery, as they mitigate local adverse effects in non-targeted tissues [28]. Most NNMs designed to target various cancers and tumors are in the preclinical and clinical development phases [29].

4.1 Enhanced permeability and retention effect of NNMs

The preferential internalization of NNMs in cancerous tissue is due to improved permeability of the vasculature in such tissues (tumors and cancerous tissues). The factor-like activation of vascular permeability increased expression and/or deregulated angiogenesis predominates at these targeted sites, which may permit passage of NNMs [30]. Furthermore, the enhanced permeability and retention (EPR) effect is also linked to the lack of functional lymphatic drainage tendency in solid tumors, which is responsible to limit burping of NNMs from tumor cells [31, 32]. Accumulations of NNMs at the targeted site as the action of pathological properties are referred to as passive targeting. To attain this, prolonged circulation kinetics of NNMs with drug cargo in the bloodstream is important. This can be accomplished by attaching polyethylene glycol (PEG) on the surface of NNMs.

Thus, the therapeutic effect of NNMs is expected to improve through the coating of concerning small molecules, those having lesser pharmacokinetic properties [31, 33–35].

4.2 Active targeting through NNMs

Receptor-mediated targeting or ligand-mediated targeting, also known as active targeting, involves the chemical or physical attachment of ligands to the surface of NNMs to facilitate proper localization/uptake by tumor cells [32, 35]. Targeted NNMs have extreme affinity for site-specific delivery of the drug to the targeted cells in vivo, which selectively express targeted receptor or adhesion molecule at the site of the tumor [36, 37].

For instance, the following cellular targets are usually considered for ligand-based targeting in cancer:

- receptors-based targeting of cancer cells through the overexpression of receptors such as transferrin, folate, epidermal growth factor, or glycoproteins,
- endothelium-based targeting of tumor cells by the overexpression of vascular endothelial growth factors, integrins, vascular cell adhesion molecule-1, or matrix metalloproteinase, and
- targeting of stromal cells (e.g., macrophages, fibroblasts) that are responsible for tumor cell survival in response to cytokines [32, 38–40].

4.3 Triggered release through NNMs

Endogenous or exogenous stimuli were employed in this class to improve drug release in the tumor environment. NNMs responding to endogenous stimuli exploit associated factors within the local environment of targeted sites. For example, redox gradient presence or enzymes, low pH in the tumor micro-environment. Exogenous-responding NNMs respond to external stimuli such as increased temperature, magnetic field, or light to trigger drug release. A hyperthermic trigger is the most promising strategy used as external stimuli in the release of drug from NNMs (e.g., thermosensitive liposomal doxorubicin) [41].

Several NNM-derived products are available in the market and more are in the clinical development phase. Most NNMs in the developmental phase encapsulate FDA-approved drugs and rely on various drug delivery carriers involving polymeric micelles, dendrimers, liposomes, and other inorganic nanoparticles [42–45].

5 Challenges and limitations of clinical trials of drug delivery in cancer

Beyond the expense of developing and promoting a novel cancer drug (approximately $1 billion in 2006), pharmaceutical and biotech research in clinical cancer trials face many exceptional challenges and limitations that are frequently difficult to deal with without acquiring outside facilities [29]. Fig. 4 lists some of these challenges and limitations.

5.1 Deficient scientific knowledge

New treatments can't seem to uncover the full extent of their capacity (or scarcity) over a wide range of human variables, making it hard to determine the safety and adequacy of new cancer drugs [46]. This absence of preclinical information and postadvertising reports can drastically drive the expense of the program up and increase failure risk in Phase III clinical trials [23].

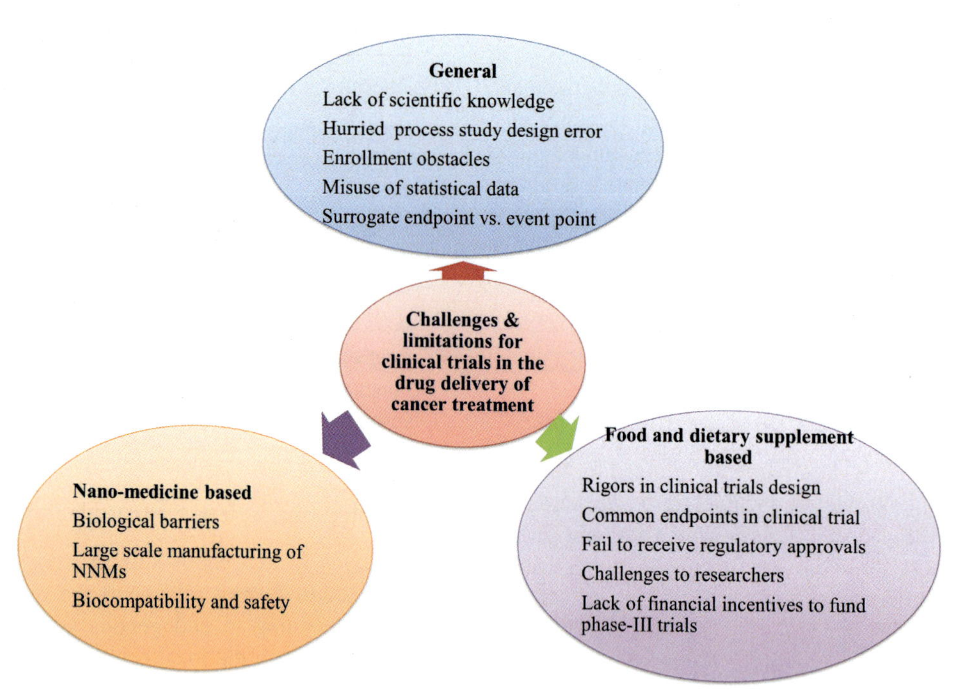

FIG. 4 Challenges and limitations for clinical trial in the drug delivery for cancer treatment.

The positive or negative impacts of chemotherapies are regularly noticed during clinical treatment, and thus it can take a long time to build up the efficacy and toxicity levels of MTAs. This can importantly affect enrollment, as patients might be hesitant to try out a program that bears an excessive amount of vulnerability and insufficient data.

5.2 Time-consuming

Advertising a novel cancer formulation is a time-consuming matter. The need to keep moving can put a strain on groups and lead to miscues that come from deficiently staffed programs.

5.3 Errors in study design

Effectively planning and arranging a clinical trial is critical in any given space. There are numerous phases to consider and incorporate to guarantee the objectives. Clinical trials for cancer have many limitations, mostly because of the heterogeneity of the disease.

5.4 Enrollment obstacles

Successful clinical trials in cancer rely on the commitment among its members, including oncologists, patients, and clinical staff. Tragically, several disputes emerge within these groups for both strategic and psychic reasons. As a result, fewer than 5% of patients (adults) diagnosed with cancer register for oncology trials. This is likely due to the following reasons [27]:

- Availability of trials: Is there a trial for the patient's type of cancer? Is the trial adequately staffed and equipped?
- Site accessibility: Is the trial location easy to get to? Will it be expensive for the patient to travel there?
- Oncologist's approval: Is the patient's oncologist on board with the patient enrolling in the trial?
- Commitment from clinical staff: Is the staff on site adequately skilled and ready to take on the additional managerial responsibilities related to cancer clinical trial enrollment?
- Patient's worries: Is the patient uneasy about joining the trial? Is the patient fully aware of the risks of participating in the trial?

5.5 Misuse of statistical data

On the off chance that your data are superfluous or do not convey logical approval, they ought not to be utilized. In any event, the data will affect the patient enrollment procedure and potentially keep qualified patients from enrolling.

Reports and scholastic publications progressively depend on statistics to affirm speculation. Nonetheless, you should continue with an alert, particularly if your group does exclude biostatisticians:

- interruption of poor *p*-value
- must include confidence intervals
- the intent-to-treat principle
- data doss
- diversity within data
- analyses of subgroup
- association vs. causation
- recording
- possibility and Bayesian statistics
- collaboration and communication between clinician and statistician

5.6 Surrogate endpoints vs. event endpoints

Anticancer drugs are expected to build patients' general endurance or/and enhance their quality of life. The utilization of surrogate endpoints in cancer trials is not exceptional, due typically to the fact that they can be estimated sooner, whereas concentrating on event endpoints frequently implies a more drawn-out investigation [37]. The other clear advantage is quickened FDA authorization.

Pharmaceutical industries, in any case, ought to consider surrogate endpoints for clinical trials in cancer treatment. The surrogate endpoints may not be treated as true markers for medication's adequacy, particularly if no further analysis can build up the quality of the surrogate-survival relationship.

In a 2017 publication, Kim and Prasad questioned the extensive utilization of surrogate endpoints to quicken authorization of cancer drugs. They stated "Of 55 regulatory authorization made by the FDA based on enhancements in surrogates somewhere in between 2009 and 2014, 65% had no consideration for preliminary level approval. Of the 35% that were considered, just 16% associated exceptionally with survival" [47].

Removal of weak surrogate endpoints can be accomplished through:

- coherent strategy for custom study design
- data-backed selection of endpoints at every phase
- using epidemiological evidence to validate selection of surrogates
- advances in global data management
- demonstration of clinical benefits through confirmatory trials

However, numerous things can go wrong at any stage further delaying patients' opportunity to live more. From poor study structure to execution boundaries or inadequate dominance of information sciences, clinicians face one-of-a-kind difficulties and cancer trial crash. Organizations that effectively take an interest in the journey to combat cancer have the charge to take all measures to execute fruitful clinical trials and put up sheltered and proficient treatments for sale to the public.

6 Challenges for clinical trials of NNMs in cancer therapy

Clinical trials of NNMs are costly and time-consuming. NNM innovation is more complex compared with traditional formulations (e.g., tablets, cases, and infusions) [48, 49].

6.1 Biological challenges

Conventionally, NNM advancement has been founded on a formulation-driven methodology, whereby novel delivery carriers are fabricated and characterized from a physicochemical view. It is only when adjusting NNMs to a pathological application that confinements of the clinical trial have been recognized. Understanding the connection among science and innovation, including understanding the impact of cancer pathophysiology on NNM accretion, distribution, maintenance, and adequacy, as well as the biopharmaceutical relationship between drug carrier properties and in vivo behavior in humans versus animals, is important for effective clinical trials of NNMs.

Hence, applying a disease-driven methodology by planning and fabricating NNMs that can explore pathophysiological changes in cancer biology has been proposed to upgrade clinical trials [29]. From the beginning of NNM advancement, it is fundamental to consider the relation between cancer pathophysiology and the heterogeneity of cancer in people, and the significance of physicochemical attributes of various NNMs to conquering natural barriers to empower improved focusing to malignant tissue or potentially diminished aggregation in noncancer organs. These natural barriers can be a critical hindrance for pharmaceutical industry inputs into NNMs. Moreover, animal models that reflect only a limited range of clinical diseases may provide valuable information for determining whether a medication will work in humans [29]. Interestingly, most NNM formulations and cancer clinical trials are centered on cancer targeting, involving more than 80% of NNMs research over the past 20 years alone [50].

6.2 Large-scale manufacturing of NNMs

One of the significant variables contributing to the moderate pace of clinical trials of NNMs is the morphology and physicochemical intricacy of the formulations. Platforms that require complex as well as relentless fabrication methods usually have inadequate clinical trial potential, as they can be challenging to pharmaceutically manufacturing on a commercial level [48, 49, 51, 52]. Pharmaceutical production and advancement in drug formulations are focused on quality and cost. Quality involves the production cycle and formulation stability, with NNM productivity being tested by potential issues such as:

- inadequate quality control parameters
- complex scalability
- partial removal of contaminants (e.g., organic solvents usage during manufacturing and their incomplete removal from the final formulation) [22, 53]
- surface coatings and encapsulation (ligands, multiple targeting components, or multiple drugs [48, 51, 54]

6.3 Biocompatibility and safety of NNMs

Detailed toxicology is a basic requirement in clinical trials of NNMs to determine general safety in humans [55]. Pharmaceutical regulatory bodies usually suggest that the sponsor cautiously evaluate any deviation from the standard processes in manufacturing and formulation of a drug at any stage of clinical trials, to decide whether changes in drug manufacturing directly or indirectly influence safety. CMC changes all through the IND cycle that can influence safety include [56]:

- deviation from standard synthetic pathways and formulation of reagents
- deviation from standard production strategy (e.g., chemosynthesis, fermentation, or changed source)
- deviation from the suggested method for sterilizing drug compositions
- deviated route of administration
- dosage change
- deviation from standard storing conditions (e.g., dose delivery; FDA, 2003)

7 Challenges for dietary supplements and foods in clinical trials of cancer therapy

The use of medicinal herbs for clinical applications contrasts with the failure of clinical trials of food and dietary supplements for cancer treatment (e.g., green tea, pomegranate, lycopene, soy, vitamins C, D, and E, selenium, resveratrol). In 2004, Americans spent nearly $36 billion on dietary supplements [57]. Approximately 50% of patients with tumors began taking new dietary supplements in the wake of their cancer diagnosis [58], and 58% of individuals who consume dietary supplements report they do as such for the prevention or cancer treatment [57].

Dietary supplements obtained from plants (e.g., ginger, garlic, cannabis) and animals (e.g., shark ligament, scorpion venom), just as certain fruit and vegetable products, have been advertised on television and the Internet for their implied capacity to prevent or even cure cancer.

7.1 Rigor in clinical trial design

Though cancer clinical trials of plant products frequently evaluate them as normally consumed items, they are analyzing "drug" endpoints and must satisfy strict guidelines for trial design and patient health. In the United States, design of a cancer clinical trial must be evaluated using the IND application measure from the FDA. The IND application depicts the medication's source and production process, notwithstanding aftereffects of laboratory testing that show consistency of the pharmaceutically active compound as well as identify possible contaminants [57].

7.2 Common endpoints in clinical trials

Clinical trials of food and dietary supplements may assess these products alone or in combination with authorized therapies for cancer. A few clinical trials focus on adequacy in disease alteration, and the endpoints include tumor growth anticipation, repeat free survival, general survival, or biomarkers prescient for survival, for example, prostate-explicit antigen multiplication time (PSA-DT) in prostate cancer. Other cancer trials focus on enhancements in quality of life, for example, a reduced number of side effects [59] or increased resistance to chemotherapy [60]. Dose-discovering evaluates whether a low dose has a similar impact to a high dose, and what number of pills can be taken securely and consistently [24, 57].

Pharmacokinetic (PK) endpoints can be significant, particularly in Phase I investigations, in which understanding how the body responds to the drug is basic to additional clinical phases of drug development. For instance, PK investigation in clinical trials was expected to establish that oral doses of ascorbic acid are reliably subtherapeutic and that only the intravenous route can bring about the therapeutic result [61]. However, PK studies are unrealistic for some food and dietary supplements.

7.3 Failure to receive regulatory approvals

Up to this point, the FDA has not authorized any food or dietary supplement as a drug for preventing or curing cancer. The straightforward reason for this is the lack of Phase III clinical trials demonstrating adequate activity and safety of these products for preventing or curing tumors or cancer. There have been many Phase II clinical trials with promising outcomes but few follow-ups for Phase III clinical trials. A probable reason for this is that manufacturers of dietary supplements assume a positive clinical Phase II trial outcome as enough to advertise their products without needing a high-cost Phase III trial.

7.4 Lack of financial incentives to fund Phase III clinical trials

The overall absence of patent assurance for dietary supplements and food items implies that producers usually do not profit from FDA authorization in the manner that producers of proprietary food do. Without patent assurance, producers face value pressure from competitors and are unable to acquire enough benefits from high production costs to pay for the expenses of clinical Phase III trials. Be that as it may, as a rule, producers do not need FDA authorization to sell food and dietary supplements claimed to prevent or treat malignancy. The safety of these sources is typically founded on preclinical outcomes or the results of initial-phase clinical studies that may sound as noteworthy to the overall population as cases dependent on Phase III clinical trials.

7.5 A challenge to clinical researchers from manufacturers

Researchers conducting clinical trials of food products frequently experience manufacturing issues that are not known to occur in the manufacture of pharmaceuticals at the commercial scale. For a clinical trial of grape skin (muscadine), for instance, a single producer's concentrate was chosen. Nonetheless, the researchers discovered that the substance of various bottles of the producer's concentrate may have been gathered from different sources, with various soil types and distinctive climate conditions, so they contrasted in measures of the active ingredients. To guarantee that the degree of active ingredients was reliable across the batches, researchers had to conduct extensive testing and collect a single batch of grape skin for complete clinical trials.

7.6 How supplement manufacturers elude regulatory limits

US regulatory bodies have been usually ineffective in expelling unsafe supplements and supplements with no active ingredients from the market, and even less powerful in preventing unsupported health claims.

Press releases from supplement industries have all the earmarks of being unconstrained by proof. An assessment of an official release from a dietary supplement industry official advocating supplement use based on 46 clinical investigations reported between January 1, 2005 and May 31, 2013 found that more than 90% of the examinations had revealed neither advantages nor side effects. Those official releases were referred by 148 highlights on the web-page of 6 organizations that inform producers, retailers, and shoppers about the clinical benefits of dietary supplements in cancer treatment [62]. Inference of these studies is that academic findings used to help claims are thought to be "reality" regardless of the presence of other literature with conflicting results [57].

8 Future perspectives in drug delivery for cancer therapy

To diagnose and ultimately cure cancer, new methods of drug delivery are needed along with discoveries in cancer biology and novel technologies. Sometimes solid tumors are the main hindrance in effective cancer drug delivery, and thus require passive and active approaches to eradication. The most significant approach is the development of economic, reliable, and sensitive nano-based therapies. Further exploration of these therapies will lead to the discovery of more biocompatible, degradable, and biologically active technologies for specific cancer types. Some studies analyzed nanoparticles capped with specific enzymes that can degrade the extracellular matrix found in cancer cells [63]. Other studies used inorganic nanomaterials to reduce tumor volume. The tumor microenvironment is one of the major factors affecting drug delivery, thus its degradation should be considered. Photochemotherapy along with near-infrared dye-based nanomaterials have been combined to achieve better results in cancer therapy. The primary objective of drug delivery is affected by the drug concentration which may be used in the early diagnosis and treatment of the cancer. The oncologist's main concern is to diagnose cancer at a treatable stage, thus drug delivery to tumor cells at the early stage can be helpful in early clinical detection and diagnosis. The clinically approved translatable targeted MRI is based on targeted drug delivery which can be helpful in improving the sensitivity of Magnetic Resonance Molecular Imaging in case of malignancies [64–66]. Although significant advances have been made in clinical trials for cancer therapy, there are still many challenges. Advances in genetics enrich our knowledge of cancer and cancer-causing cells. For instance, understanding RNA and its functions can help develop new precision therapies for early cancer diagnosis and detection. The development of gene regulating and editing tools like RNAi and CRISPR-Cas9 systems have been found to be effective in cancer therapy providing cancer signaling pathways and networks [67, 68]. These methods include efficient drug delivery to cancerous cells. To overcome the challenges of drug delivery in cancer several new trend-based systems need to be employed, including drug delivery systems that can accomplish controlled release of new as well as existing drugs. Clinical trials based on nanomedicines are costly and time consuming. To overcome these problems, nano-based systems can be combined with other advanced technology and formulations.

9 Conclusion

With the rising incidence of cancer cases, several clinical trials of drug delivery in cancer therapy have been conducted. In this chapter, we focused on technological advances and formulations for cancer therapy. Generally, four phases of clinical trials are adopted to formulate new drugs for cancer therapy: Phase I for determining safety, Phase II for determining effectiveness, Phase III for comparing the new drug to conventional treatments, and Phase IV in which sponsorship and marketing of the drug take place. The chapter also examined modern clinical trial designs including cluster randomized and stepped wedge methods. Further, the chapter discussed nano-based techniques that can be utilized for targeted drug delivery; however, these approaches are expensive and time consuming. Still, the effect of nanomedicines on drug delivery can be enhanced by active targeting. In addition, the chapter presented challenges and limitations of drug delivery for cancer therapy, such as barriers along the delivery path and resistance capacity of drugs. The main challenge of clinical trials is lack of resources and knowledge. As such, researchers from different fields should come together to design and develop new therapeutic drug delivery systems to fight cancer.

References

[1] Chien JY, Ho RJ. Drug delivery trends in clinical trials and translational medicine: evaluation of pharmacokinetic properties in special populations. J Pharm Sci 2011;100(1):53–8.
[2] Blackadar CB. Historical review of the causes of cancer. World J Clin Oncol 2016;7(1):54.
[3] Tran S, DeGiovanni P-J, Piel B, Rai P. Cancer nanomedicine: a review of recent success in drug delivery. Clin Transl Med 2017;6(1):44.
[4] Leo CP, Hentschel B, Szucs TD, Leo C. FDA and EMA approvals of new breast cancer drugs—a comparative regulatory analysis. Cancer 2020;12(2):437.
[5] Gehan EA. Clinical trials in cancer research. Environ Health Perspect 1979;32:31–48.
[6] McTiernan A, Schwartz RS, Potter J, Bowen D. Exercise clinical trials in cancer prevention research: a call to action. Cancer Epidemiol Prevent Biomarkers 1999;8(3):201–7.
[7] Seymour L, Ivy SP, Sargent D, Spriggs D, Baker L, Rubinstein L, et al. The design of phase II clinical trials testing cancer therapeutics: consensus recommendations from the clinical trial design task force of the national cancer institute investigational drug steering committee. Clin Cancer Res 2010;16(6):1764–9.
[8] Jenkins V, Fallowfield L. Reasons for accepting or declining to participate in randomized clinical trials for cancer therapy. Br J Cancer 2000;82(11):1783–8.
[9] Alley SC, Okeley NM, Senter PD. Antibody-drug conjugates: targeted drug delivery for cancer. Curr Opin Chem Biol 2010;14(4):529–37.
[10] Cohen MH, Williams GA, Sridhara R, Chen G, Pazdur R. FDA drug approval summary: gefitinib (ZD1839) (Iressa) tablets. Oncologist 2003;8(4):303–6.
[11] Jain K. Nanotechnology-based drug delivery for cancer. Technol Cancer Res Treat 2005;4(4):407–16.
[12] Seimetz D. Novel monoclonal antibodies for cancer treatment: the trifunctional antibody catumaxomab (Removab). J Cancer 2011;2:309.
[13] Fong L, Small EJ. Anti-cytotoxic T-lymphocyte antigen-4 antibody: the first in an emerging class of immunomodulatory antibodies for cancer treatment. J Clin Oncol 2008;26(32):5275–83.
[14] Natsume A, Niwa R, Satoh M. Improving effector functions of antibodies for cancer treatment: enhancing ADCC and CDC. Drug Des Dev Therapy 2009;3:7.
[15] Cao J, Chen D, Huang S, Deng D, Tang L, Gu Y. Multifunctional near-infrared light-triggered biodegradable micelles for chemo- and photo-thermal combination therapy. Oncotarget 2016;7(50):82170.
[16] Rejinold NS, Choi G, Choy J-H. Recent trends in nano photochemo therapy approaches and future scopes. Coord Chem Rev 2020;411:213252.
[17] Ell B, Kang Y. Transcriptional control of cancer metastasis. Trends Cell Biol 2013;23(12):603–11.
[18] Milling L, Zhang Y, Irvine DJ. Delivering safer immunotherapies for cancer. Adv Drug Deliv Rev 2017;114:79–101.
[19] Hsiao WW, Liu L. The role of traditional Chinese herbal medicines in cancer therapy—from TCM theory to mechanistic insights. Planta Med 2010;76(11):1118–31.
[20] Alexis F, Pridgen EM, Langer R, Farokhzad OC. Nanoparticle technologies for cancer therapy. Drug delivery. Springer; 2010. p. 55–86.
[21] Gardikis K, Hatziantoniou S, Bucos M, Fessas D, Signorelli M, Felekis T, et al. New drug delivery nanosystem combining liposomal and dendrimeric technology (liposomal locked-in dendrimers) for cancer therapy. J Pharm Sci 2010;99(8):3561–71.
[22] Kraft JC, Freeling JP, Wang Z, Ho RJ. Emerging research and clinical development trends of liposome and lipid nanoparticle drug delivery systems. J Pharm Sci 2014;103(1):29–52.
[23] Ellis P. Attitudes towards and participation in randomised clinical trials in oncology: a review of the literature. Ann Oncol 2000;11(8):939–46.
[24] Paller C, Ye X, Wozniak P, Gillespie B, Sieber P, Greengold R, et al. A randomized phase II study of pomegranate extract for men with rising PSA following initial therapy for localized prostate cancer. Prostate Cancer Prostatic Dis 2013;16(1):50–5.
[25] Paller CJ, Zhou XC, Heath EI, Taplin M-E, Mayer T, Stein MN, et al. Muscadine grape skin extract (MPX) in men with biochemically recurrent prostate cancer: a randomized, multicenter, placebo-controlled clinical trial. Clin Cancer Res 2018;24(2):306–15.
[26] Oncology ASoC. The state of cancer care in America, 2014: a report by the American Society of Clinical Oncology. J Oncol Pract 2014;10(2):119–42.
[27] Fisher WB, Cohen SJ, Hammond MK, Turner S, Loehrer PJ. Clinical trials in cancer therapy: efforts to improve patient enrollment by community oncologists. Med Pediatr Oncol 1991;19(3):165–8.
[28] Rizzo LY, Theek B, Storm G, Kiessling F, Lammers T. Recent progress in nanomedicine: therapeutic, diagnostic and theranostic applications. Curr Opin Biotechnol 2013;24(6):1159–66.
[29] Hare JI, Lammers T, Ashford MB, Puri S, Storm G, Barry ST. Challenges and strategies in anti-cancer nanomedicine development: an industry perspective. Adv Drug Deliv Rev 2017;108:25–38.
[30] Hashizume H, Baluk P, Morikawa S, McLean JW, Thurston G, Roberge S, et al. Openings between defective endothelial cells explain tumor vessel leakiness. Am J Pathol 2000;156(4):1363–80.
[31] Maeda H, Nakamura H, Fang J. The EPR effect for macromolecular drug delivery to solid tumors: Improvement of tumor uptake, lowering of systemic toxicity, and distinct tumor imaging in vivo. Adv Drug Deliv Rev 2013;65(1):71–9.
[32] Danhier F. To exploit the tumor microenvironment: since the EPR effect fails in the clinic, what is the future of nanomedicine? J Control Release 2016;244:108–21.
[33] Matsumura Y, Maeda H. A new concept for macromolecular therapeutics in cancer chemotherapy: mechanism of tumoritropic accumulation of proteins and the antitumor agent smancs. Cancer Res 1986;46(12 Part 1):6387–92.
[34] Hobbs SK, Monsky WL, Yuan F, Roberts WG, Griffith L, Torchilin VP, et al. Regulation of transport pathways in tumor vessels: role of tumor type and microenvironment. Proc Natl Acad Sci 1998;95(8):4607–12.
[35] van der Meel R, Vehmeijer LJ, Kok RJ, Storm G, van Gaal EV. Ligand-targeted particulate nanomedicines undergoing clinical evaluation: current status. Adv Drug Deliv Rev 2013;65(10):1284–98.
[36] Forssen E, Willis M. Ligand-targeted liposomes. Adv Drug Deliv Rev 1998;29(3):249–71.
[37] Hua S, De Matos MB, Metselaar JM, Storm G. Current trends and challenges in the clinical translation of nanoparticulate nanomedicines: pathways for translational development and commercialization. Front Pharmacol 2018;9:790.
[38] Coimbra M, Banciu M, Fens MH, de Smet L, Cabaj M, Metselaar JM, et al. Liposomal pravastatin inhibits tumor growth by targeting cancer-related inflammation. J Control Release 2010;148(3):303–10.
[39] Danhier F, Feron O, Préat V. To exploit the tumor microenvironment: passive and active tumor targeting of nanocarriers for anti-cancer drug delivery. J Control Release 2010;148(2):135–46.

[40] Kuijpers SA, Coimbra MJ, Storm G, Schiffelers RM. Liposomes targeting tumour stromal cells. Mol Membr Biol 2010;27(7):328–40.
[41] Needham D, Anyarambhatla G, Kong G, Dewhirst MW. A new temperature-sensitive liposome for use with mild hyperthermia: characterization and testing in a human tumor xenograft model. Cancer Res 2000;60(5):1197–201.
[42] Torchilin VP. Multifunctional nanocarriers. Adv Drug Deliv Rev 2006;58(14):1532–55.
[43] Wagner V, Dullaart A, Bock A-K, Zweck A. The emerging nanomedicine landscape. Nat Biotechnol 2006;24(10):1211–7.
[44] Sercombe L, Veerati T, Moheimani F, Wu SY, Sood AK, Hua S. Advances and challenges of liposome assisted drug delivery. Front Pharmacol 2015;6:286.
[45] Deamer DW. From "Banghasomes" to liposomes: a memoir of Alec Bangham, 1921–2010. FASEB J 2010;24(5):1308–10.
[46] Bertoli AM, Strusberg I, Fierro GA, Ramos M, Strusberg AM. Lack of correlation between satisfaction and knowledge in clinical trials participants: a pilot study. Contemp Clin Trials 2007;28(6):730–6.
[47] Zehir A, Benayed R, Shah RH, Syed A, Middha S, Kim HR, et al. Mutational landscape of metastatic cancer revealed from prospective clinical sequencing of 10,000 patients. Nat Med 2017;23(6):703.
[48] Tinkle S, McNeil SE, Mühlebach S, Bawa R, Borchard G, Barenholz Y, et al. Nanomedicines: addressing the scientific and regulatory gap. Ann N Y Acad Sci 2014;1313(1):35–56.
[49] Sainz V, Conniot J, Matos AI, Peres C, Zupanõiõ E, Moura L, et al. Regulatory aspects on nanomedicines. Biochem Biophys Res Commun 2015;468(3):504–10.
[50] Park K. The drug delivery field at the inflection point: time to fight its way out of the egg. J Control Release 2017;267:2–14.
[51] Kumar Teli M, Mutalik S, Rajanikant G. Nanotechnology and nanomedicine: going small means aiming big. Curr Pharm Des 2010;16(16):1882–92.
[52] Barz M, Luxenhofer R, Schillmeier M. Quo vadis nanomedicine? Future Medicine; 2015.
[53] Jaafar-Maalej C, Elaissari A, Fessi H. Lipid-based carriers: manufacturing and applications for pulmonary route. Expert Opin Drug Deliv 2012;9(9):1111–27.
[54] Svenson S. The dendrimer paradox—high medical expectations but poor clinical translation. Chem Soc Rev 2015;44(12):4131–44.
[55] Nyström AM, Fadeel B. Safety assessment of nanomaterials: implications for nanomedicine. J Control Release 2012;161(2):403–8.
[56] Hua S, Cabot PJ. Targeted nanoparticles that mimic immune cells in pain control inducing analgesic and anti-inflammatory actions: a potential novel treatment of acute and chronic pain conditions. Pain Physician 2013;16(3):E199–216.
[57] Paller CJ, Denmeade SR, Carducci MA. Challenges of conducting clinical trials of natural products to combat cancer. Clin Adv Hematol Oncol 2016;14(6):447–55.
[58] Patterson RE, Neuhouser ML, Hedderson MM, Schwartz SM, Standish LJ, Bowen DJ. Changes in diet, physical activity, and supplement use among adults diagnosed with cancer. J Am Diet Assoc 2003;103(3):323–8.
[59] Ma Y, Chapman J, Levine M, Polireddy K, Drisko J, Chen Q. High-dose parenteral ascorbate enhanced chemosensitivity of ovarian cancer and reduced toxicity of chemotherapy. Sci Transl Med 2014;6(222), 222ra18.
[60] Piao B, Wang Y, Xie G, Mansmann U, Matthes H, Beuth J, et al. Impact of complementary mistletoe extract treatment on quality of life in breast, ovarian and non-small cell lung cancer patients a prospective randomized controlled clinical trial. Anticancer Res 2004;24(1):303–10.
[61] Levine M, Padayatty SJ, Espey MG. Vitamin C: a concentration-function approach yields pharmacology and therapeutic discoveries. Adv Nutr 2011;2(2):78–88.
[62] Wang MT, Gamble G, Bolland MJ, Grey A. Press releases issued by supplements industry organisations and non-industry organisations in response to publication of clinical research findings: a case-control study. PLoS One 2014;9(7), e101533.
[63] Murty S, Gilliland T, Qiao P, Tabtieng T, Higbee E, Zaki AA, et al. Nanoparticles functionalized with collagenase exhibit improved tumor accumulation in a murine xenograft model. Part Part Syst Charact 2014;31(12):1307–12.
[64] Han Z, Li Y, Roelle S, Zhou Z, Liu Y, Sabatelle R, et al. Targeted contrast agent specific to an oncoprotein in tumor microenvironment with the potential for detection and risk stratification of prostate cancer with MRI. Bioconjug Chem 2017;28(4):1031–40.
[65] Han Z, Wu X, Roelle S, Chen C, Schiemann WP, Lu Z-R. Targeted gadofullerene for sensitive magnetic resonance imaging and risk-stratification of breast cancer. Nat Commun 2017;8(1):692.
[66] Lu Z-R. Magnetic resonance molecular imaging for non-invasive precision cancer diagnosis. Curr Opin Biomed Eng 2017;3:67–73.
[67] Platt RJ, Chen S, Zhou Y, Yim MJ, Swiech L, Kempton HR, et al. CRISPR-Cas9 knockin mice for genome editing and cancer modeling. Cell 2014;159(2):440–55.
[68] Yang J, Meng X, Pan J, Jiang N, Zhou C, Wu Z, et al. CRISPR/Cas9-mediated noncoding RNA editing in human cancers. RNA Biol 2018;15(1):35–43.

Chapter 40

Future prospects and challenges in cancer drug delivery

Deepti Malik*, Rupa Joshi, Harpinder Kaur, Ajay Prakash, and Bikash Medhi
Department of Pharmacology, PGIMER, Chandigarh, India

1 Introduction

Cancers proliferate rapidly [1]. Various stimuli like toxins, carcinogens, radiation, viruses, and prolonged inflammation coupled with weakened immune system can initiate tumor progression through continued accumulation of genetic alterations leading to increased genomic instability and mutability [1]. Loss in nuclear genome integrity, mitochondrial dysfunction, aerobic glycolysis (Warburg effect), aberrantly expressed miRNAs, loss in function of tumor suppressor genes and activation of oncogenes, chromosomal translocations, chromosomal instability, hyperdiploidy, aneuploidy, etc., enable cancer cell to acquire different manifestations like self-sufficiency in growth signals, insensitivity to growth-inhibitory signals, evasion of programmed cell death, limitless replicative potential, angiogenesis, tissue invasion, and metastasis [2]. These modifications enable more cells to turn cancerous, provide them the conditions to thrive, and allow normal cell to change into a transformed phenotype [2]. The scourge of cancer is spreading rapidly with the rate of increase in cancer expected to grow 50% year on year. It is expcted that 15 million new cases of cancer will be reported in the year 2020 [3]. The below table provides the break of new cases expected by cancer type across the world:

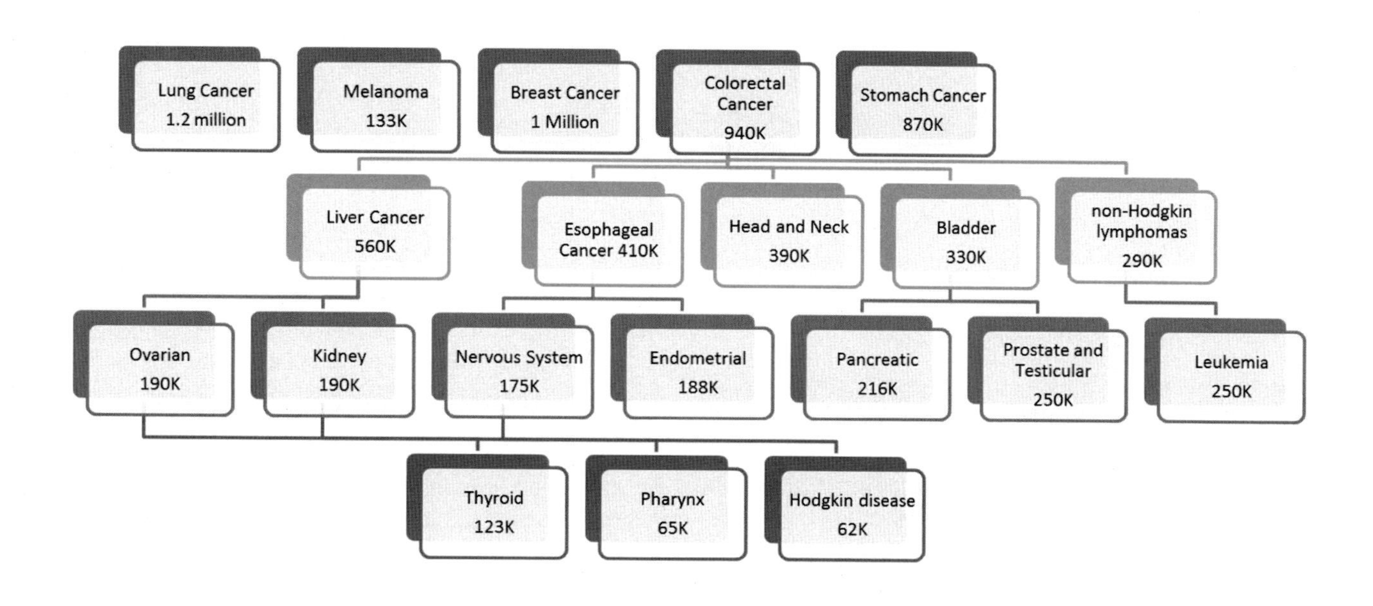

* Present Address: Department of Biochemistry, All India Institute of Medical Sciences, Bilaspur, Chhattisgarh, India

Advanced Drug Delivery Systems in the Management of Cancer. https://doi.org/10.1016/B978-0-323-85503-7.00035-3

2 Worldwide incidence of typical cancers

The story of origin of cancer is a fascinating one with no one clear reason/factor been the driving reason rather a host of factors in conjunction or independently drive the changes in the genetic markers of cells which lead to the origin and subsequent progress of cancer in multicellular organism. Here, we discuss various factors that contribute to tumor progression.

2.1 Loss in function of tumor suppressor genes contributes to the progression of cancer

Tumor suppressor genes or antioncogenes encode proteins that have a suppressive role in the formation of cancer by regulating the genes coding for cell cycle and apoptosis [4]. They inhibit the cell cycle progression when the DNA is damaged, no longer allows the cell to divide unless the DNA enzymes repair the damaged DNA in the cell, In addition, this promotes apoptosis under the conditions when the damage to DNA is excessive and has not been repaired by the DNA repair enzymes. The reduction or deletion of a gene functioning as a tumor suppressor gene leads to tumor formation [4].

Tumor suppressor genes: Function to inhibit cellular proliferation

Gene	Cancer type	Chromosomal location
P53	Involved in many cancers	17p13
RB	Retinoblastoma	13q14
DCC	Colon	8q21
APC	Colon	5q31
NF1	Neurofibromatosis	22q12
WT1	Wilms	11p13

2.2 Gain in function of oncogenes contributes to the progression of cancer

A proto-oncogene is a normal gene that encodes proteins essential for the regulation of cell growth, differentiation, and cell death [5]. Mutations within the proto-oncogene caused by various carcinogens and viruses can convert them into an effective tumor-inducing oncogene. Another prime cause for the transformation of proto-oncogene to oncogene involves chromosomal translocations. For example, translocation between chromosome 9 and 22 comprising the Philadelphia chromosome is identified in the leukemic cells of all the patients suffering with chronic myeloid leukemia [6]. Burritt's lymphoma contains the constitutively active c-Myc gene as a result of translocation of this gene from chromosome 8 to the different position near the Ig heavy chain enhancer on chromosome 14 [7]. Oncogenes are expressed at increased levels in cancers. Oncogenes can be categorized as growth factors, growth factor receptors, and transcription factors or are involved in various signaling transduction pathways. In normal cells tight regulation of growth factors, growth factor receptors, proper expression of transcription factors, and crucial scrutiny on the expression of the genes involved in various signaling pathways maintain the cellular proliferation. However, overexpression of these oncogenes observed in tumor cells leads to unbridled proliferation, invasiveness, and angiogenesis [4].

Oncogenes: Induce cellular proliferation	
src	Encode tyrosine kinases
abl	Encode tyrosine kinases
ras	Encode GTP-binding protein
myc	Encode transcription factors
jun	Encode transcription factors
fos	Encode transcription factors

2.3 Upregulation of antiapoptotic genes contributes to progression of cancer

Cell death is controlled by a phenomenon known as apoptosis that plays critical roles in multicellular organisms and has a critical role in development as well as homeostasis [8]. Cell death is regulated by phagocytes which engulf the cells based on the signals triggered by apoptosis. Cancerous cells withstand apoptosis giving them the ability to evade cytotoxic agents and avoid natural cell death leading to their survival [8].

2.4 Epigenetic abnormalities contribute to progression of cancer

The field of epigenetics has made lots of strides in recent times and one of the key finding's has been to understand that cancer cells in humans contain global epigenetic abnormalities as well as multiple genetic alternations [9]. The alterations at genetic and epigenetic levels interact among themselves across all stages of cancer development and result in growth of cancer progression [10]. Typically there are three modifications characterizing and regulating the epigenetic abnormalities which are chromatin structure and epigenetic mechanisms of gene expression, including DNA methylation, histone covalent modification, and microRNAs (miRNAs) [11]. The above three modifications are together termed as the "epigenetic code" and this code plays a key role in modulating the expression of mammalian genome in multiple cells across various stages of development of disease and in varied states including cancer [9]. The above three agents are directly responsible for regulating the expression of genes which are deregulated without altering the DNA sequences of the genes. DNA methyltransferases and histone modification enzymes, such as histone acetyl transferase (HAT) and histone deacetylase (HDAC), are well known to play a critical and central role in cancer development and progression [11]. DNA methylation, including genomic imprinting, and histone modifications play a role in altering the patterns of cancer cell distribution. The alterations caused result in permanent pattern change in gene expressions that have a key role in regulating the neoplastic phenotype [9]. Epigenetic modulations also play a major role in controlling miRNA's ability to bind to 3'-UTR sequences resulting in massive degradation of mRNA which subsequently leads to reduced gene levels [12]. Epigenetic regulation in miRNA expression in tumorigenesis causes DNA hypermethylation in different carcinomas types [13,14]. Increased levels of miR-21, miR-155, miR-23, and miR-191,miR-196b is seen in breast cancer [15, 16], miR-191,199a and miR155 are overexpressed in AML [17]. miR-2909 is suspected to play a role in immunomodulation [18], energy metabolism [19, 20] and its altered expression are suspected to be the cause of many diseases including rheumatoid arthritis [21] and pediatric B- and T-cell acute lymphoblastic leukemia [22]. Cancer cells with abnormal miRNA expression evolve the capability to sustain proliferative signaling, evade growth suppressors, resist cell death, activate invasion and metastasis, and induce angiogenesis

2.5 Current treatment and other investigational therapies in cancer treatments

As the adage saying goes prevention is better than cure, it is much better that individuals make adjustments to their life and incorporate healthy lifestyle habits to minimize the chances of getting cancer, but sedentary lifestyle and a host of related reasons like environmental pollution, poisoned food chains, and addiction to ultra-processed food are leading to record level of new cases of cancer. As a response, unprecedented level of research has been done to discover newer, safer, and effective ways of treating different kinds of cancers. Cancer treatment has made giant stripes in recent years with chemotherapy playing a central role in increasing the efficacy of treatment and becoming the first line of defense for most primary tumors along with surgery [23]. This approach paid rich dividends and was the preferred way of cancer therapy for keeping in check and treating localized tumors well into 1960s. One of the major challenges with this approach is that most cancers microscopically metastasize throughout the subjects body which leads to the reoccurrence of disease even after initial remission leading to the need of regular chemotherapy with the objective of controlling metastases [23]. With advances been made the use of cytotoxic drugs with chemotherapy has stopped with increasing use of cancer-specific agents leading to the age of "targeted therapy," sometimes also termed "molecularly targeted drugs," "molecularly targeted therapies," etc. [24].

2.6 Targeted therapy

Targeted therapy targets cancer's specific genes, proteins, and tissue environment that contribute toward bridled proliferation in cancer cells [24]. Targeted therapy works in one of the following ways:

- *Small-molecule drugs*: These drugs are microscopic in size and have the capability to infiltrate the cells and target specific sites within the cancer cells while not impacting the normal cells [25].
- *Monoclonal antibodies*: are man-made proteins that have the capability to attach themselves to specific targets within cancer cells. Monoclonal antibodies work either by marking the cancer cells so that the body's immune system can identify and destroy these cells and having no impact on normal cells/tissue. The cancer cell destruction happens by the antibodies blocking or turning off signals that tell cancer cells to grow resulting in self-destruction of these malignant cells [26].

Targeted therapies have several key advantages over chemotherapy:

- These therapies are designed in such a way as to impact only specific molecular targets within the cancer cells while chemotherapy has the same impact on normal and malignant cells.
- The primary objective of targeted therapies is to block cancerous cell proliferation (cytostatic), while chemotherapy aims to kill the cancerous cells (cytotoxic).

While targeted therapies have several key advantages over chemotherapy there are some limitations also:

- Cancer cells have the ability to mutate and become resistant to the targeted therapy rendering the ability of the targeted drug to prohibit cancer cell growth.
- Cancer cells can also find new pathways to growth that are not dependent on existing targets, rendering the therapy infective.

Keeping the limitations of targeted therapies in mind it is advisable to use targeted therapy in conjugation with multiple molecules targeting different signaling pathways or in combination with chemotherapy drugs [24].

2.7 Immunotherapy: Impacting all cancers

In targeted therapy drugs are used to impact specific cancer cells, whereas in immunotherapy drugs are used for stimulating the patient's immune system and giving it the ability to distinguish cancerous cells as invading bodies and attack these cells [27]. There are two major forms of immunotherapies:

1. CAR T-cell therapy
2. Immune checkpoint inhibitors

2.8 CAR T-cell therapy

T cells from patients are harvested and modified in the lab by creating a man-made receptor CAR (chimeric *antigen receptor*). This receptor gives the immune system the ability to identify specific cancer cell antigens. CAR T-cell therapy has aided in the fight against cancer but there can be serious side effects. As modified cells are introduced in the patient's body, they start multiplying and result in the release of significant amount of cytokines into patients' blood resulting in high fever and very low blood pressure. This phenomenon is known as cytokine release syndrome, or CRS [28].

2.9 Immune checkpoint inhibitors

Immune checkpoint signaling regulates antigen recognition of T-cell receptor (TCR). There are two kinds of immune checkpoint signal:

(1) *Costimulatory immune checkpoint*: stimulating immune progress, such as CD28, ICOS, and CD137.
(2) *Coinhibitory immune checkpoint*: inhibiting immune progress, such as PD1, CTLA-4, and VISTA.

Drugs, monoclonal antibody that blocks the interaction of programmed cell death ligand 1 (PD-L1) on tumor cells with the PD-1 (CD279) encourages activation of T-cell-mediated immune responses against tumor cells thus holds promise for treating cancer [29].

2.10 Challenges in cancer drug delivery systems

There are numerous challenges in chemotherapy and delivering an effective dosage of any cytotoxic agent to the site where the tumor is located with simultaneous reduction of unintentional harmful adverse effects is one of them. Drug delivery systems (DDS) aid in improving pharmacological characteristics of chemotherapeutics by modifying pharmacokinetics and bio-distribution of the drug. Some examples of DDS include liposomes, micelles, dendrimers, and polymeric-based systems. DDS because of their smaller size (~100nm or less) can easily extravasate from circulation via vascular gaps or defects accredited to ongoing angiogenesis characteristic of tumor sites, which is stated to be around 200nm or more. Retention of DDS within such sites is normally more because of poor lymphatic drainage seen within tumors. Further, if the size limit is lower (~20nm in diameter) it makes certain that these vehicles do not randomly penetrate the walls of normal vessels. Moreover, DDS reduces the unwanted adverse effects associated with use of conventional drugs such as peripheral neurotoxicity related to use of cisplatin and vincristine or cardiotoxicity related to use of anthracyclines (doxorubicin or daunorubicin) [30]. Various DDSs utilized for cancer drug delivery are explained in the subsequent paragraphs.

2.11 Nanoparticles in cancer therapy

After the invention of Nanotechnology, the nanoparticles were considered as boon for the DDSs in cancer therapy. It helps in improving the effectiveness of the drugs by enhancing their half-life and improves the solubility of the hydrophobic drugs. The controlled release formulations prepared by nanoparticles whose size varies from 10 to 1000nm facilitate the sustained release of drugs over time which in turn decreases the frequency of drug administration and improves patient compliance. There are almost 51 nanoparticles approved by FDA in 2016. Most common types of nanoparticles are prepared from liposomes and polymeric materials; however, micelle, metallic, and protein-based materials are also considered as more efficacious and safe [31].

2.12 Chitosan nanoparticles

The nanoparticles made of chitosan were generally considered as most affordable due to its good biocompatibility, low cost, low toxicity, and high biodegradability in the body making it an ideal pharmaceutical agent for nanoparticle formation. It is a carbohydrate polymer occurring in nature and formed from chitin deacetylation. The drugs with small molecules, proteins, and polynucleotides can be prepared by chitosan nanoparticles. It can also help in making the controlled release formulation of drugs. Moreover, the ionic cross-linking ability of chitosan due to the presence of free amine group will further help in making sustained release nanoformulations [32].

The production of chitosan nanoparticles can be done by ionotropic gelatin, emulsification solvent diffusion (PLGA), and polyelectrolyte complex (PEC) methods. Among these methods, the PEC method is used for gene therapy and the emulsification method is useful for hydrophobic solvents. In cancer therapies, the chitosan nanoparticles are very specific for tumor cells either through passive targeting by enhancing permeability and retention in the targeted tissues or by directly targeting the tumor cells through stimuli sensitive chitosan nanoparticles [33].

In the treatment of colon cancers, the delivery of 5-fluorouracil (5-FU) can be sustained up to 24h by incorporating into hyaluronic acid coupled chitosan nanoparticles (approx. 50nm). It also promises the targeted delivery of 5-FU to colon tumors within 4h of incubation at 37°C, resulting in a higher cell uptake rate by cancer cells. Tumor cells differ from healthy cells as they exhibit acidosis and hyperthermia. Thus, change in pH and temperature may serve as a signal for targeting drug delivery to the cancer or inflamed cells. For example: Nanoparticles of HTCC fabricated by the ionotropic gelatin method enhances their size from 200 nm to around 400 nm after sensing the change in pH from 7.4 to 5.0. Thus, the swelling of microgel nanoparticles will cause internalization leading to their enhanced stay to release the drug in the targeted tumor cells. For example, methotrexate disodium-loaded HTCC nanoparticles releases 93% of the drug at pH 5 after day 1 and after 5 days 30% of the drug was trapped in microgel at pH 7.4.

2.13 Silica nanoparticles

Silica nanoparticles are also known as inorganic nanoparticles. As silica is inorganic in nature and thus they are biocompatible, highly porous, and very simple to functionalize. Mesoporous silica nanoparticles (MSNs) have a larger surface area and pore volume and thus help in enhanced adsorption and drug loading. They also have an advantage of modification in surface and size, leading to enhanced efficacy and reduced toxicity. They could be employed as bioimaging probes and drug delivery systems through combination with luminescent compounds or magnetic materials. The fabrication of these MSNs can be done by different methods like solution-based method and evaporation-induced self-assembly method [34].

These MSNs act as a promising carrier for gene delivery for several cancers' therapy. MSNs being a mesoporous structure provide the matrix system for the homogenous distribution of drug particles of even different sizes. During delivery of the genes to the target cancer cells, the nuclease degradation of the genes can be avoided by deeply immersing/hiding the genes into mesopores before reaching the area of interest. Moreover, the drug loading capacity of MSNs is also high as compared to other nanoparticles like chitosan and other polymers [35]. They are also considered as promising drug delivery systems but sometimes time and cost of drug delivery system should not be ignored [36].

The binding and loading capacity of negatively charged DNA and SiRNA can be increased by coating of a cationic polymer polyethyleneimine (PEI) and thus enhances their rate of cellular reuptake. For example, SiRNA bonded to PEI-MSN can knock down GFP in HEPA-1 cells, Plasmid DNA can be delivered with MSN-PEI system, etc. On the other hand, PEI can also be sometimes toxic and can increase the cytotoxicity of MSNs delivery systems. Anticancer drug oxaliplatin in combination with tumor suppressor miRNA, i.e., with miRNA-204-5p in PEI/hyaluronic acid assembled silica nanoparticles had shown synergistic effect against the treatment of colon cancer [37,38].

Zero premature release to minimize the toxicity to surrounding tissue has been considered as prerequisite for an efficient drug delivery system in cancer treatment. The most common method is the photostimulation which is most effective and can be achieved by coumarin MSNs sensitive to UV radiations. The other methods like pH, enzymes, and magnetic fields are also used to attain zero premature release in MSNs.

2.14 Polylactide-*co*-glycolic acid (PLGA) nanoparticles

It is a copolymer of polylactic acid (PLA) and polyglycolic acid (PGA) produced via ring opening polymerization of lactic acid and glycolide. Although it is a synthetic polymer, it is biodegradable and has high purity and reproducibility as compared to various natural polymers and moreover its molecular weight can also be tailored as if required. As PLGA has already been approved by EMA and U.S. FDA for human use, it can be a most opportunistic drug delivery approach [39]. It is more stable against hydrolysis and continued to release drugs for many days to months. The in vitro potency of PLGA-chitosan nanoparticles of curcumin was found to be higher as compared to curcumin alone. Curcumin can be an effective treatment for pancreatic cancer also and its nanoparticles [40,41].

2.15 Polymeric micelles

Polymeric micelles (PM) are considered as the first polymeric self-assemblies. They are spherical in shape and composed of copolymers of amphiphilic blocks having both hydrophilic and hydrophobic ends in the aqueous medium. Thus, di-amphiphilic or tri-block copolymers are used to get self-assembled into a shell structure or a supramolecular core. A hydrophobic core and hydrophilic shell act as a reservoir for poorly water-soluble drug and aqueous milieu, respectively. Polyethylene glycol was widely used a hydrophilic block that ensures micelle solubility in an aqueous milieu. The rationale of using PM is for delivering hydrophobic anticancer drugs to targeted tissues by making them hydrophilic in nature. The aqueous solubility of the anticancer agents would increase and help in the intravenous administration of these drugs [42].

These PMs provide various physicochemical and biological features over other nano preparations. There are many formulations of PM loaded with drugs which are approved for performing clinical trials for the treatment of several cancer types. Moreover, several had also shown good results as anticancer treatment option. For example, Paclitaxel loaded PM formulation (Genexol®-PM), poly(ethylene glycol) (PEG)-poly(d, L-lactide) (PLA), etc. Both Celegene's Abraxane and Samyang's Genexol-PM had shown promising results in clinical trials and their side effect profile was reduced [43]. Other preparations in the clinical trial were NC-6004, NK105, NC-4016, NK012, NK911, SP1049C, etc. [44].

The premature drug release from the PMs caused the inadequate release of the drugs in the targeted tissues. To achieve the targeted delivery and to overcome the premature release of drugs from PMs, the stimuli responsive PMs were designed. These stimuli responsive PMs released the drugs against the particular discrete stimuli present in the tumor cells like low pH in solid tumor cells, high temperature, and high levels of glutathione in the cytoplasm. Along with these different internal triggers in tumor cells, some external stimuli like ultrasound, light, and magnetic field also played an important role in recognizing tumor cells. The examples of various stimuli responsive PMs which were being investigated for the treatment of different types of cancers include PMs responsive to redox, pH, enzyme, light, ultrasound, thermo- and magnetic field. Additionally, multiresponsive PMs were also available like dual redox/magnetic-responsive PMs, etc. [45].

2.16 Exosomes

Exosomes are defined as intracellular formed vesicles that released contents outside the cell. They basically derived from small size endosomes ranged from 40 to 100 nm. It is like a nanosphere with a bilayered membrane, consisting of many types of lipids (cholesterol, sphingolipids, ceramides, phosphoglycerides, etc.), proteins (HSP, transport proteins, tetraspanin, etc.) and nucleic acids like mRNA, miRNA, and noncoding RNA. It also served as a biomarker for various biological processes. Rose Johnstone had used the term exosomes first time in 1970 [46]. While working with maturing reticulocytes, an intracellur sac was formed filled with small membrane-enclosed structure of nearly uniform size.

Different types of biological fluids like breast milk, synovial fluid, amniotic liquid, saliva, urine, blood, and serum containing exosomes facilitate intercellular communication and triggered the biological responses. Thus, understanding of the functions of exosomes in immune system functions, intercellular communication, cell differentiation and development, neuronal cell signaling, and regeneration is very important. Being disease specific, had been used as biomarkers in various diseases like neurodegenerative diseases, viral infections, prions, cancer, etc. The endosomal-sorting complex required for transport (ESCRT) was considered as core molecular machinery for the formation of exosomes [47].

There are numerous derived exosomes like macrophage-derived exosomes, Rhabdomyosarcoma-derived exosomes, metastatic cancer cell-derived exosomes, malignant mesothelioma (MM) cell-derived exosomes, osteoclast-derived exosomes, pancreatic cancer cell (PCC)-derived exosomes, bronchial fibroblast-derived exosomes, and mesenchymal stem cell (MSC)-derived exosomes. Being secreted by number of cells, exosomes could be used for both diagnostic as well as therapeutic purposes. Angiogenesis can be stimulated by exosomes containing mRNA, miRNA, and angiogenetic proteins to microvascular endothelial cells secreted by glioblastoma cells [48,49].

They can be used to detect the number of tumors like breast, prostate, and ovarian cancers. Moreover, they are helpful in diagnosing various infections by expressing infectious RNA and proteins, detecting active and latent forms of intracellular infections. However, its ability to deliver the growth factors, miRNA, RNA, proteins, noncoding RNA, and lipids can be used as therapeutic options for tissue regeneration. They can be encapsulated for treating many types of cancers like breast, lung, pancreatic, prostate, and glioblastoma [50]. Exosomes are considered as excellent source of drug delivery systems in numerous diseases like cancers, cardiac, and neurological diseases. The exosomal miRNAs have also shown potential effect in the early diagnosis and management of different types of cancers. They can also deliver miRNA to those breast cancer cells which expressed epidermal growth factor receptors more competently as compared to other drug delivery systems [51].

Both exosomes and their mimetics can be the better option than nanoparticles, liposomes, and various polymeric drug delivery systems. The desirable features like target tissue delivery with intrinsic ability, biocompatibility, enhanced circulation half-life, minimal or low intrinsic toxicity, etc., provided by exosomes made them the better choice of drug delivery. Exosomes can be used as drug delivery systems to deliver small molecules across the blood–brain barrier, for example, delivery of curcumin, doxorubicin into tumor tissue, doxorubicin, and paclitaxel across BBB. Large molecules like proteins can also be delivered by exosomes. Catalase delivery across BBB for the treatment of Parkinson's disease is an example of delivery of large molecules through exosomes. Even nucleic acid like SiRNA into T cells, antitumor miRNA into breast cancer cells were delivered by technology of exosomes.

Exosomes, mimicking of nature's delivery systems, can circulate within human body for longer period of time by avoiding the processes of phagocytosis and macrophages degradation. Thus, enhances the transfection efficiency of siRNA by evading endosomal pathway. Despite being considered as advanced drug delivery system, it has no separate optimal purification technique for isolation of high purity exosomes and consequently making large-scale production more expensive. Alternative for this can be a hybrid type of exosome design which is further limited by its clinical efficacy and safety profile. Moreover, the issue of immunogenicity can be cumbersome in case of therapeutic cargo and functional exosomes or hybrid exosomes mimetics. For example, caspase-3 containing exosomes might inhibit cell death by apoptosis or also increase the survival of tumor cells by preventing chemotherapeutics drug accumulation. Therefore, artificial exosomes may be proved as roar to surmount the unwanted immune reactions with natural exosomes [52].

2.17 Liposomes

Liposomes were discovered in 1965 by Bangham et al. and have proved to be very resourceful tools in various disciplines including biology, biochemistry, and medicine. They are spherical shaped lipid bilayers of phospholipids and cholesterol utilized as carriers of drug and are loaded with a wide range of molecules comprising small drug molecules, nucleotides, proteins, plasmids, viruses, and various other biologically active compounds [53]. Incorporation of hydrophobic drugs is done within liposomal bilayer membranes, while that of hydrophilic drugs is done within inner aqueous core [54]. Benefits of liposomes include biocompatibility, good loading capacity, low toxicity, and controlled release kinetics. Ambisome (encapsulation of amphotericin B into conventional liposomes) is the first approved liposomal formulation [55].

Conventional liposomes had some limitations such as fusion and/or aggregation with each other which results in liposomal load's early release over time and also they undergo quick systemic clearance when taken up by cells of mononuclear phagocytic system. Surface modification approaches in which the conventional liposomes' surface is coated with inert, hydrophilic, and biocompatible polymers like polyethylene glycol (PEG) were used to solve these problems. This hydrophilic polymer provides steric stabilization to the surface of liposome by forming a protective layer around it and opsonin recognition and thus clearance of liposomes is slowed down. This has resulted in several liposomal formulations (Doxil, Caelyx, and Myocet) that have been clinically approved and utilized for cancer treatment. Although PEGylated liposomes have enhanced pharmacokinetics, but the absence of target selectivity considerably limits their therapeutic potential. Therefore, various approaches have been used to improve their performance in vivo [55]. Specialized or targeted liposomes can identify and bind to particular target molecules present on cells via surface alterations utilizing particular ligands or antibodies. Immunoliposomes are specialized liposomes having antibody as an attached moiety and attachment of an entire antibody molecule to the surface of liposomes can increase liposomal clearance from circulation [53]. Proteins, receptors, antibody Fab segments, and positive charge can also be utilized for liposomal targeting to the chosen site [53,55]. Inherent physiological conditions in the tissue that is targeted, like raised temperature or change in pH can also be exploited for producing stimuli-responsive liposomes, like thermo-sensitive and pH-sensitive liposomes [55].

The use of free doxorubicin is associated with toxicities such as gastrointestinal myelosuppression, mucositis, and alopecia [56]. In a study by Xiong et al., RGD-mimetic-modified sterically stabilized liposomes were formed and loaded with doxorubicin which facilitated uptake of the drug into the cells with melanoma by integrin-mediated endocytosis, increased cytotoxicity, and effectively retarded growth of tumor in C57BL/6 mice with melanoma B16 tumors [57]. Zhao et al. formulated pH-sensitive liposomes encapsulating doxorubicin using H7K (R2)2, a peptide responding to pH (specific to tumors), as a targeting ligand that can respond to the acidic milieu in gliomas, having cell penetrating peptide features. They witnessed a particular targeting effect elicited by an acid pH and pH-stimulated release of doxorubicin from liposomes under acidic environments in in vitro experiments. Antitumor activity of these liposomes was confirmed in mice [58]. Curcumin is extensively utilized for the treatment of various cancers but its administration has some limitations including not optimal solubility in water and less bioavailability which can be overcome by encapsulating it into liposomes [56,59]. Feng et al. have described methods of preparation of curcumin liposomes and their applications in various forms of cancer [60]. The encapsulation techniques like thin film and ethanol injection technique are intricate and utilize organic solvents. Cheng et al. described simply a scalable technique for encapsulation of curcumin which does not involve organic solvents and exploit curcumin's pH-dependent solubility properties [59].

2.18 Dendrimers

Dendrimers were discovered in 1978 by Buhleier et al. They are molecules with a core at the center and repeated branches [61]. Structure of dendrimers consists of three parts: internal core, branches that have terminal reactive groups that permit the inclusion of drugs (functionalization), and cavities amid the branches that too permit the addition of drugs and diagnostic agents [62]. Various approaches utilized for dendrimer synthesis include: divergent approach (growth is initiated from a multifunctional core molecule to periphery), convergent approach (formation of dendrimer is initiated from peripheral end and advances toward core), double stage convergent approach (formation is initiated via divergent approach and then assembly of dendrimer is via convergent approach) [56]. Molecules such as poly (propylene imine), poly (L-lysine), poly (glycerol-cosuccinic acid), melamine, etc., are normally utilized to make dendrimers. Dendrimers vary in chemical structure as well as properties which can be changed by growing branches of dendrimer or by altering groups on dendrimer surface [63]. They can be either categorized by their form (polymers, brush polymers or hyperbranched polymers) or by molecular weight (high or low molecular weight).

Structure of dendrimers offer distinctive prospects via chemical conjugation, the dendrimer–drug complexes' supramolecular structure is formed on the basis of variety of interactions like electrostatic, hydrophobic or hydrogen-bond, etc., or encapsulation, inside the central cavity and/or inside the multiple channels between the dendrons. Generally covalent linkages to attach the drug on peripheral groups of dendrimer are used to make dendrimer–drug conjugates and by this quite a few molecules of drugs can be linked to one dendrimer molecule and release of the drug molecule is in part regulated by the nature of linkages. Dendrimers have emerged as potential nanocarriers because of their definite properties and manifold linkage groups, charge, polymer size, lipid bilayer interactions, retention time in blood plasma, biodistribution, filtration, internalization, and cytotoxicity [63]. Their applications range from the delivery of drugs, where nanoparticles of dendrimers are combined with drugs and targeted to particular tissues to delivery of genes, where nanoparticles of dendrimers are combined with nucleic acids [61].

A polyethylene glycol poly-L-lysine (PEGylated PLL) dendrimer formulation consisting of docetaxel conjugated in the surface target tumors and has effectiveness against various solid tumors (ovarian, breast, lung, and prostate). It has aqueous solubility and surfactants like polysorbate 80 are not needed for dissolution which decreases the risk of anaphylactic shock and health risks. Efficacy is enhanced because of more systemic circulation time and leading the drug to the site of tumor by passing the nano-carrier through permeable tumor capillaries. The formulation decreases the toxicity of docetaxel by slowly releasing the drug. Dendrimer of PLL-PEG encapsulating cabazitaxel has also been used for solid tumors. Both formulations are in clinical trials [62]. Cationic phosphorus dendrimers can be used as nonviral vectors for optimized gene delivery for treatment of cancer [64].

2.19 Quantum dots

Quantum dots (QDs) which are semiconductor crystals in the nanometer range and constitute of a core (groups II–VI or III–V elements) and a shell (having polymer coating) protecting metal crystal core were discovered by Ekimov and Efros in the early 1980s [56]. They are potential agents which can be used as luminescent nano-probes and carriers for biological applications because they have distinctive optical properties in accordance with quantum and size effect. Dissolution, adsorption, dispersion, coupling, etc., can be used for loading of drugs into QD nano-carriers. Physical as well as chemical characteristics (crystal form, saturation, dissolution rate, solubility, particle surface hydrophobicity, and hydrophilicity), physical response, and biological properties of drugs are altered because of the role of carriers; and this in turn has effect on ADME of drugs. Therefore, QDs can increase the effectiveness and lessen the adverse effects of drug reactions to improve drug's therapeutic index and can also efficiently enhance absorption of small molecule drugs [65]. Their advantages include: lower toxicity than inorganic nanoparticles; stronger fluorescence intensity than organic fluorophores in biomedical imaging; better solubility in water after surface modification [56]. Selective targeting is a crucial aspect for the development of modified QDs. Tumor cell targeting for therapeutic as well as diagnostic applications have either focused on few ligands receptors of which are overexpressed in tumor cells (like folic acid) or on delivery of siRNA [65].

QDs can be utilized for the detection of cancer. For example, biocompatible graphene oxide QDs were developed by Shi et al. which are coated with high-luminescence magnetic nanoplatform for selectively separating as well as diagnosing Glypican-3-expressed Hep G2 liver cancer tumor CTCs from infected blood [66]. Cai et al. described pH-sensitive ZnO QDs for doxorubicin delivery. Dicarboxyl-terminated PEG was added to make them stable under physiological fluid and hyaluronic acid (targeting ligand) was conjugated for specifically binding to overexpressed glycoprotein CD44 by cancer cells. Molecules of doxorubicin were loaded through the formation of metal–doxorubicin complex (biodegraded in acidic milieu of tumors) and covalent interactions. Biodegradation of QDs lead to the controlled release of doxorubicin and the authors reported that doxorubicin combination with ZnO QDs improves apoptosis of cancer cells [67].

2.20 Carbon nanotubes

Carbon nanotubes (CNTs) are synthetic cylindrical nanomaterials having one-dimensional layer of carbon atoms (sp^2 hybridized) in a hexagonal mesh. Single-walled CNTs are designed by rolling one single graphene layer whereas multiwalled CNTs are designed by rolling multiple graphene layers [68]. Entry of single-walled CNTs into the cells is usually by direct penetration whereas multiwalled CNTs enter through endocytosis pathway [69]. Multiwalled CNTs were first made by arc-discharge evaporation method in 1991 and single walled by metal catalyst during an arc-discharge process in 1993. Properties of CNTs such as diameter, internal geometry, and physicochemical properties are determined by the way and rolling-up direction of graphene layers [68]. Unique properties of CNTs make them suitable as drug delivery carriers and include their cylindrical shape which aids in trans-membrane penetration, they can function as electrical conductors, have a very good photo-thermal feature with abilities to absorb optical intensity too, generate strong Raman signals and photoluminescence. With these benefits, CNTs can be applied in photo-thermal and photo-acoustic therapy against cancer cells. Moreover, because of their huge aspect ratio, they have very large drug loading capacity. Further, their ultrahigh surface area ratio aids in their chemical functionalization with various moieties which shows high potency of particular cells or tissues targeting, imaging, and therapy [70]. CNTs efficiently utilize the poor heat resistance property of tumor cells by converting infrared light into heat. When the temperature is >42 °C at site of tumor, cell-killing phenomenon like plasma membrane destruction, protein denaturation, irreversible destruction of tumor cells are evident whereas normal cells stay intact. The antitumor effects are significantly improved when coupled with antitumor drugs. Strategy of combining CNTs with inorganic materials can be used for diagnosis as well as treatment of cancers [71].

However, CNTs also have certain toxic effects on the body because of their distinctive physical as well as chemical properties and toxicity is associated to variation of surface, degree of aggregation in vivo, and concentration of nanoparticle. Inflammatory response, malignant mesothelioma, and biological persistence are the usually reported toxicities of CNTs [71].

Morais et al. developed multiwalled CNTs functionalized with naringenin as new carriers of drug for treatment of lung cancer. Their results showed that noncovalent interactions were involved in the functionalization of CNTs with naringenin. The release profiles revealed a pH-responsive release, exhibiting extended release in tumor pH environment. These CNTs exhibited less cytotoxicity on nonmalignant cells (hFB) as compared to free naringenin, and also better effect as anticancer on malignant lung cells (A549) as an in vitro lung cancer model [72]. Sundaram et al. synthesized single-walled CNTs by coupling with hyaluronic acid and coating walls with chlorin e6. Results revealed that the nanobiocomposite improved the capability of photodynamic therapy to be a photosensitizer carrier and induced cell death in colon cancer cells [69]. Doxorubicin can also be easily adsorbed on the surface of CNTs by π-π stacking for delivering it to cancer cells [73].

2.21 Miscellaneous

Ultrasound guided drug delivery: Ultrasound in addition to being used for diagnostic purposes can also be used for applications such as delivery of drug or gene to various tissues. Ultrasound-guided delivery of drugs permits spatially confined drug delivery to site of tumor and also reduces systemic dose as well as toxicity. Moreover, ultrasound is available easily, portable, less expensive, and can be noninvasively focused to target area with high precision and thus can be used for tumors which can be anatomically accessed by ultrasound such as liver tumors [74].

Cyclic Cell-Penetrating Peptides (CPPs): CPPs consist of short oligopeptides (5–30 amino acids) and are comparatively a new class of peptides widely explored as vehicles for intracellular delivery of therapeutics. Their limitations include poor stability, not optimum cell penetration, entrapment by endosomes, and toxicity. Various approaches are being adopted to overcome these short-comings [75].

3 Conclusion

Great advancements have been achieved in DDSs for cancer therapy; however, most of the DDSs that are being developed for cancer drug delivery are in preclinical/clinical trials and have their own limitations. Very few products are marketed and hence, there is a requirement to develop suitable candidates that can prove to be beneficial for cancer drug delivery. There is a need for more clinical data so that the advantages and limitations of these DDSs can be understood properly.

References

[1] Feitelson MA, Arzumanyan A, Kulathinal RJ, Blain SW, Holcombe RF, Mahajna J, Marino M, Martinez-Chantar ML, Nawroth R, Sanchez-Garcia I, Sharma D, Saxena NK, Singh N, Vlachostergios PJ, Guo S, Honoki K, Fujii H, Georgakilas AG, Amedei A, Nowsheen S. Sustained proliferation in cancer: mechanisms and novel therapeutic targets. Semin Cancer Biol 2015;35(Suppl):S25–54. https://doi.org/10.1016/j.semcancer.2015.02.006.

[2] Liberti MV, Locasale JW. The Warburg effect: how does it benefit Cancer cells? Trends Biochem Sci 2016;41(3):211–8. https://doi.org/10.1016/j.tibs.2015.12.001.

[3] Cancer statistics, 2020 - Siegel - 2020 - CA: A Cancer Journal for Clinicians - Wiley Online Library. Retrieved September 23, 2020, from. https://acsjournals.onlinelibrary.wiley.com/doi/full/10.3322/caac.21590.

[4] Lee EYHP, Muller WJ. Oncogenes and tumor suppressor genes. Cold Spring Harb Perspect Biol 2010;2(10). https://doi.org/10.1101/cshperspect.a003236.

[5] Shortt J, Johnstone RW. Oncogenes in cell survival and cell death. Cold Spring Harb Perspect Biol 2012;4(12). https://doi.org/10.1101/cshperspect.a009829.

[6] Shu Y, Yang W, Zhang X, Xu X. Recurrent chronic myeloid leukemia with t (9;22;16) (q34; q11; p13) treated by nilotinib. Medicine 2018;97(42). https://doi.org/10.1097/MD.0000000000012875.

[7] Spehalski E, Kovalchuk AL, Collins JT, Liang G, Dubois W, Morse HC, Ferguson DO, Casellas R, Dunnick WA. Oncogenic Myc translocations are independent of chromosomal location and orientation of the immunoglobulin heavy chain locus. Proc Natl Acad Sci U S A 2012;109(34):13728–32. https://doi.org/10.1073/pnas.1202882109.

[8] Fernald K, Kurokawa M. Evading apoptosis in cancer. Trends Cell Biol 2013;23(12):620–33. https://doi.org/10.1016/j.tcb.2013.07.006.

[9] Sharma S, Kelly TK, Jones PA. Epigenetics in cancer. Carcinogenesis 2010;31(1):27–36. https://doi.org/10.1093/carcin/bgp220.

[10] Flavahan WA, Gaskell E, Bernstein BE. Epigenetic plasticity and the hallmarks of cancer. Science (New York, NY) 2017;357(6348):2380. https://doi.org/10.1126/science.aal2380.

[11] Handy DE, Castro R, Loscalzo J. Epigenetic modifications: basic mechanisms and role in cardiovascular disease. Circulation 2011;123(19): 2145–56. https://doi.org/10.1161/CIRCULATIONAHA.110.956839.

[12] Malik D, Kaul D. KLF4 genome: a double edged sword. J Solid Tumors 2015;5:49–64.
[13] Sharma S, Kaul D, Arora M, Malik D. Oncogenic nature of a novel mutant AATF and its interactome existing within human cancer cells. Cell Biol Int 2015;39(3):326–33.
[14] Wahida F, Shehzada A, Khanb T, Kima YY. MicroRNAs: synthesis, mechanism, function, and recent clinical trials. Biochimica et Biophysica Acta - Molecular Cell Research 2010;1803(11):1231–43.
[15] Yan LX, Huang XF, Shao Q, Huang MY, Deng L, Wu QL, Zeng YX, Shao JY. MicroRNA miR-21 overexpression in human breast cancer is associated with advanced clinical stage, lymph node metastasis and patient poor prognosis. RNA 2008;14(11):2348–60.
[16] Kaul D, Malik D. MIR196B (microRNA 196b). Atlas Genet Cytogenet Oncol Haematol 2012;16(5):357–60.
[17] Ramamurthy R, Hughes M, Morris V, Bolouri H, Gerbing RB, Wang YC, Loken MR, Raimondi SC, Hirsch BA, Gamis AS, Oehler VG, Alonzo TA, Meshinchi S. miR-155 expression and correlation with clinical outcome in pediatric AML: a report from Children's oncology group. Pediatr Blood Cancer 2016;63(12):2096–103.
[18] Kaul D, Malik D, Wani S. Cellular miR-2909 RNomics governs the genes that ensure immune checkpoint regulation. Mol Cell Biochem 2019;451(1–2):37–42.
[19] Kaushik H, Malik D, Parsad D, Kaul D. Mitochondrial respiration is restricted by miR-2909 within human melanocytes. Pigment Cell Melanoma Res 2019;32(4):584–7.
[20] Malik D, Kaul D. Human cellular mitochondrial remodelling is governed by miR-2909 RNomics. PLoS ONE 2018;13(9), e0203614.
[21] Malik D, Sharma A, Raina A, Kaul D. Deregulated blood cellular miR-2909 RNomics observed in Rhematoid arthritis patients. Archives of Medicine 2015;7(1):126.
[22] Malik D, Kaul D, Chauhan N, Marwaha RK. miR-2909-mediated regulation of KLF4: a novel molecular mechanism for differentiating between B-cell and T-cell pediatric acute lymphoblastic leukemias. Molecular Cancer 2014;13(1):175.
[23] DeVita VT, Chu E. A history of Cancer chemotherapy. Cancer Res 2008;68(21):8643–53. https://doi.org/10.1158/0008-5472.CAN-07-6611.
[24] Yan L, Rosen N, Arteaga C. Targeted cancer therapies. Chin J Cancer 2011;30(1):1–4. https://doi.org/10.5732/cjc.010.10553.
[25] Pathak A, Tanwar S, Kumar V, Banarjee BD. Present and future Prospect of Small Molecule & Related Targeted Therapy against Human Cancer. Vivechan International Journal of Research 2018;9(1):36–49.
[26] Coulson A, Levy A, Gossell-Williams M. Monoclonal antibodies in Cancer therapy: mechanisms, successes and limitations. West Indian Med J 2014;63(6):650–4. https://doi.org/10.7727/wimj.2013.241.
[27] Stanculeanu D, Daniela Z, Lazescu A, Bunghez R, Anghel R. Development of new immunotherapy treatments in different cancer types. J Med Life 2016;9(3):240–8.
[28] Whilding LM, Maher J. CAR T-cell immunotherapy: the path from the by-road tothe freeway? Mol Oncol 2015;9(10):1994–2018. https://doi.org/10.1016/j.molonc.2015.10.012.
[29] Dine J, Gordon R, Shames Y, Kasler MK, Barton-Burke M. Immune checkpoint inhibitors: an innovation in immunotherapy for the treatment and Management of Patients with Cancer. Asia Pac J Oncol Nurs 2017;4(2):127–35. https://doi.org/10.4103/apjon.apjon_4_17.
[30] Cukierman E, Khan DR. The benefits and challenges associated with the use of drug delivery systems in cancer therapy. Biochem Pharmacol 2010;80(5):762–70. https://doi.org/10.1016/j.bcp.2010.04.020.
[31] Dang Y, Guan J. Nanoparticle-based drug delivery systems for cancer therapy. Smart Mater Med 2020;1:10–9.
[32] Nagpal K, Singh SK, Mishra DN. Chitosan nanoparticles: a promising system in novel drug delivery. Chem Pharm Bull (Tokyo) 2010;58(11):1423–30.
[33] Park JH, Saravanakumar G, Kim K, Kwon IC. Targeted delivery of low molecular drugs using chitosan and its derivatives. Adv Drug Deliv Rev 2010;62(1):28–41.
[34] Tang F, Li L, Chen D. Mesoporous silica nanoparticles: synthesis, biocompatibility and drug delivery. Adv Mater 2012;24(12):1504–34.
[35] Hu Y, Zhi Z, Zhao Q, Wu C, Zhao P, Jiang H, et al. 3D cubic mesoporous silica microsphere as a carrier for poorly soluble drug carvedilol. Microporous Mesoporous Mater 2012;1(147):94–101.
[36] Seljak KB, Kocbek P, Gašperlin M. Mesoporous silica nanoparticles as delivery carriers: an overview of drug loading techniques. J Drug Deliv Sci Technol 2020;59:101906.
[37] Nadrah P, Porta F, Planinšek O, Kros A, Gaberšček M. Poly(propylene imine) dendrimer caps on mesoporous silica nanoparticles for redox-responsive release: smaller is better. Phys Chem Chem Phys 2013;15(26):10740–8.
[38] Tarn D, Ashley CE, Xue M, Carnes EC, Zink JI, Brinker CJ. Mesoporous silica nanoparticle nanocarriers: biofunctionality and biocompatibility. Acc Chem Res 2013;46(3):792–801.
[39] Lai P, Daear W, Löbenberg R, Prenner EJ. Overview of the preparation of organic polymeric nanoparticles for drug delivery based on gelatine, chitosan, poly(d,l-lactide-co-glycolic acid) and polyalkylcyanoacrylate. Colloids Surf B Biointerfaces 2014;118:154–63.
[40] Arya G, Das M, Sahoo SK. Evaluation of curcumin loaded chitosan/PEG blended PLGA nanoparticles for effective treatment of pancreatic cancer. Biomed Pharmacother 2018;102:555–66.
[41] Gromisch C, Qadan M, Machado MA, Liu K, Colson Y, Grinstaff MW. Pancreatic Adenocarcinoma: unconventional approaches for an unconventional disease. Cancer Res 2020;80(16):3179–92.
[42] Cho H, Lai TC, Tomoda K, Kwon GS. Polymeric micelles for multidrug delivery in cancer. AAPS PharmSciTech 2015;16(1):10–20.
[43] Ventola CL. Progress in nanomedicine: approved and investigational nanodrugs. P T Peer-Rev. J Formul Manag 2017;42(12):742–55.
[44] Biswas S, Kumari P, Lakhani PM, Ghosh B. Recent advances in polymeric micelles for anti-cancer drug delivery. Eur J Pharm Sci 2016;83:184–202.
[45] Zhou Q, Zhang L, Yang T, Wu H. Stimuli-responsive polymeric micelles for drug delivery and cancer therapy. Int J Nanomedicine 2018;13:2921–42.
[46] Johnstone RM. Revisiting the road to the discovery of exosomes. Blood Cells Mol Dis 2005;34(3):214–9.
[47] Bunggulawa EJ, Wang W, Yin T, Wang N, Durkan C, Wang Y, et al. Recent advancements in the use of exosomes as drug delivery systems. J Nanobiotechnol 2018;16(1):81.
[48] Batrakova EV, Kim MS. Using exosomes, naturally-equipped nanocarriers, for drug delivery. J Controlled Release 2015;219:396–405.
[49] Burgio S, Noori L, Marino Gammazza A, Campanella C, Logozzi M, Fais S, et al. Extracellular vesicles-based drug delivery systems: a new challenge and the exemplum of malignant pleural mesothelioma. Int J Mol Sci 2020;21(15):5432.
[50] Pullan JE, Confeld MI, Osborn JK, Kim J, Sarkar K, Mallik S. Exosomes as drug carriers for cancer therapy. Mol Pharm 2019;16(5):1789–98.
[51] Rahbarghazi R, Jabbari N, Sani NA, Asghari R, Salimi L, Kalashani SA, et al. Tumor-derived extracellular vesicles: reliable tools for

Cancer diagnosis and clinical applications. Cell Commun Signal 2019;17(1):73.

[52] Ha D, Yang N, Nadithe V. Exosomes as therapeutic drug carriers and delivery vehicles across biological membranes: current perspectives and future challenges. Acta Pharm Sin B 2016;6(4):287–96.

[53] Sharma G, Anabousi S, Ehrhardt C, Ravi Kumar MNV. Liposomes as targeted drug delivery systems in the treatment of breast cancer. J Drug Target 2006;14(5):301–10. https://doi.org/10.1080/10611860600809112.

[54] Das SS, Bharadwaj P, Bilal M, Barani M, Rahdar A, Taboada P, Bungau S, Kyzas GZ. Stimuli-responsive polymeric Nanocarriers for drug delivery, imaging, and Theragnosis. Polymers 2020;12(6). https://doi.org/10.3390/polym12061397.

[55] Abu Lila AS, Ishida T. Liposomal delivery systems: design optimization and current applications. Biol Pharm Bull 2017;40(1):1–10. https://doi.org/10.1248/bpb.b16-00624.

[56] Montané X, Bajek A, Roszkowski K, Montornés JM, Giamberini M, Roszkowski S, et al. Encapsulation for cancer therapy. Molecules 2020;25(7):1605. https://doi.org/10.3390/molecules25071605.

[57] Xiong X-B, Huang Y, Lu W-L, Zhang X, Zhang H, Nagai T, Zhang Q. Enhanced intracellular delivery and improved antitumor efficacy of doxorubicin by sterically stabilized liposomes modified with a synthetic RGD mimetic. J Control Release 2005;107(2):262–75. https://doi.org/10.1016/j.jconrel.2005.03.030.

[58] Zhao Y, Ren W, Zhong T, Zhang S, Huang D, Guo Y, Yao X, Wang C, Zhang W-Q, Zhang X, Zhang Q. Tumor-specific pH-responsive peptide-modified pH-sensitive liposomes containing doxorubicin for enhancing glioma targeting and anti-tumor activity. J Control Release 2016;222:56–66. https://doi.org/10.1016/j.jconrel.2015.12.006.

[59] Cheng C, Peng S, Li Z, Zou L, Liu W, Liu C. Improved bioavailability of curcumin in liposomes prepared using a pH-driven, organic solvent-free, easily scalable process. RSC Adv 2017;7(42):25978–86. https://doi.org/10.1039/C7RA02861J.

[60] Feng T, Wei Y, Lee RJ, Zhao L. Liposomal curcumin and its application in cancer. Int J Nanomedicine 2017;12:6027–44. https://doi.org/10.2147/IJN.S132434.

[61] Castro RI, Forero-Doria O, Guzmán L. Perspectives of dendrimer-based nanoparticles in Cancer therapy. Anais Da Academia Brasileira De Ciencias 2018;90(2 suppl 1):2331–46. https://doi.org/10.1590/0001-3765201820170387.

[62] do Nascimento T, Todeschini AR, Santos-Oliveira R, de Bustamante Monteiro MS, de Souza VT, Ricci-Júnior E. Trends in nanomedicines for Cancer treatment. Curr Pharm Des 2020;26(29):3579–600. https://doi.org/10.2174/1381612826666200318145349.

[63] Chaturvedi VK, Singh A, Singh VK, Singh MP. Cancer nanotechnology: a new revolution for Cancer diagnosis and therapy. Curr Drug Metab 2019;20(6):416–29. https://doi.org/10.2174/1389200219666180918111528.

[64] Chen L, Li J, Fan Y, Qiu J, Cao L, Laurent R, Mignani S, Caminade A-M, Majoral J-P, Shi X. Revisiting cationic phosphorus dendrimers as a nonviral vector for optimized gene delivery toward Cancer therapy applications. Biomacromolecules 2020;21(6):2502–11. https://doi.org/10.1021/acs.biomac.0c00458.

[65] Zhao M-X, Zhu B-J. The research and applications of quantum dots as Nano-carriers for targeted drug delivery and Cancer therapy. Nanoscale Res Lett 2016;11(1):207. https://doi.org/10.1186/s11671-016-1394-9.

[66] Shi Y, Pramanik A, Tchounwou C, Pedraza F, Crouch RA, Chavva SR, Vangara A, Sinha SS, Jones S, Sardar D, Hawker C, Ray PC. Multifunctional biocompatible graphene oxide quantum dots decorated magnetic nanoplatform for efficient capture and two-photon imaging of rare tumor cells. ACS Appl Mater Interfaces 2015;7(20):10935–43. https://doi.org/10.1021/acsami.5b02199.

[67] Cai X, Luo Y, Zhang W, Du D, Lin Y. pH-sensitive ZnO quantum dots–doxorubicin nanoparticles for lung Cancer targeted drug delivery. ACS Appl Mater Interfaces 2016;8(34):22442–50. https://doi.org/10.1021/acsami.6b04933.

[68] Faraji Dizaji B, Khoshbakht S, Farboudi A, Azarbaijan MH, Irani M. Far-reaching advances in the role of carbon nanotubes in cancer therapy. Life Sci 2020;257:118059. https://doi.org/10.1016/j.lfs.2020.118059.

[69] Sundaram P, Abrahamse H. Effective photodynamic therapy for Colon Cancer cells using Chlorin e6 coated hyaluronic acid-based carbon nanotubes. Int J Mol Sci 2020;21(13). https://doi.org/10.3390/ijms21134745.

[70] Guo Q, Shen X-T, Li Y-Y, Xu S-Q. Carbon nanotubes-based drug delivery to cancer and brain. *Journal of Huazhong University of Science and Technology. Medical Sciences = Hua Zhong Ke Ji Da Xue Xue Bao. Yi Xue Ying De Wen Ban = Huazhong Keji Daxue Xuebao.* Yixue Yingdewen Ban 2017;37(5):635–41. https://doi.org/10.1007/s11596-017-1783-z.

[71] Yan H, Xue Z, Xie J, Dong Y, Ma Z, Sun X, Kebebe Borga D, Liu Z, Li J. Toxicity of carbon nanotubes as anti-tumor drug carriers. Int J Nanomedicine 2019;14:10179–94. https://doi.org/10.2147/IJN.S220087.

[72] Morais RP, Novais GB, Sangenito LS, Santos ALS, Priefer R, Morsink M, Mendonça MC, Souto EB, Severino P, Cardoso JC. Naringenin-functionalized multi-walled carbon nanotubes: a potential approach for site-specific remote-controlled anticancer delivery for the treatment of lung Cancer cells. Int J Mol Sci 2020;21(12). https://doi.org/10.3390/ijms21124557.

[73] Yaghoubi A, Ramazani A. Anticancer DOX delivery system based on CNTs: functionalization, targeting and novel technologies. J Control Release 2020;327:198–224. https://doi.org/10.1016/j.jconrel.2020.08.001.

[74] Mullick Chowdhury S, Lee T, Willmann JK. Ultrasound-guided drug delivery in cancer. Ultrasonography (Seoul, Korea) 2017;36(3):171–84. https://doi.org/10.14366/usg.17021.

[75] Park SE, Sajid MI, Parang K, Tiwari RK. Cyclic cell-penetrating peptides as efficient intracellular drug delivery tools. Mol Pharm 2019;16(9):3727–43. https://doi.org/10.1021/acs.molpharmaceut.9b00633.

Index

Note: Page numbers followed by *f* indicate figures and *t* indicate tables.

D

M

N

O

Q

R

Printed in the United States
by Baker & Taylor Publisher Services